www.devbio.com

the companion website for

Developmental Biology

Eighth Edition

devbio.com is an extensive Web companion to the textbook that is intended to supplement and enrich courses in developmental biology. It provides additional information for advanced students as well as historical, philosophical, and ethical perspectives on issues in developmental biology. Included are articles, movies, interviews, opinions (labeled as such), Web links, updates, and more. (Even a developmental biology humor page!)

devbio.com is comprised of Web topics organized by textbook chapter, and these topics are referenced throughout the textbook. You can go directly to a specific topic from the home page of the site by simply entering the number of the topic.

The following is a comprehensive list of the numbered Web topics referenced in each chapter of the book. Also included are references to the *vade mecum²* CD.

(continued on inside back cover)

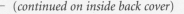

Developmental Biology
EIGHTH EDITION

Developmental Biology EIGHTH EDITION

Scott F. Gilbert
Swarthmore College

With a chapter on PLANT DEVELOPMENT *by*
Susan R. Singer *Carleton College*

Sinauer Associates, Inc., Publishers
Sunderland, Massachusetts USA

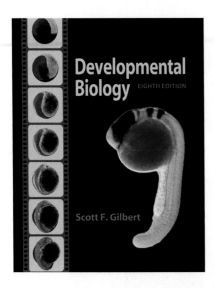

About the Cover

Zebrafish (*Danio rerio*) are excellent subjects for the study of cell movements, since the embryo is transparent and develops quickly, and its development takes place outside a uterus or eggshell. *Differentiation* of the zebrafish nervous system cells is seen in the triply stained 20-hour embryo. The micrograph localizes three gene products that help specify the regions of the brain. The *otx2* gene product (anterior dark purple region) is important in specifying the forebrain and midbrain; the *krox20* gene product (the posterior purple "stripes") is expressed in two places and is critical for specifying the hindbrain; and the products of the *engrailed* gene (gold) help form the midbrain-hindbrain border and are also seen in some muscle cells along the midline of the embryo (see Chapters 11 and 12). *Morphogenesis* (the generation of form) is shown as a "movie" in which several differently stained regions of zebrafish ectoderm migrate to form the neural tube. Numerous mutations in zebrafish can cause faulty neural tube formation, producing diseases similar to human birth defects. Stained embryo image kindly provided by Dr. G. Hauptmann and Springer-Verlag; movie images generously provided by Dr. K. Hatta.

DEVELOPMENTAL BIOLOGY, Eighth Edition

Copyright © 2006 by Sinauer Associates Inc. All rights reserved. This book may not be reproduced in whole or in part without the permission of the publisher. Sinauer Associates, Inc., 23 Plumtree Road, Sunderland, MA 01375 USA.

email: publish@sinauer.com, orders@sinauer.com
www.sinauer.com

Library of Congress Cataloging-in-Publication Data

Gilbert, Scott F., 1949-
  Developmental biology / Scott F. Gilbert ; with a chapter on plant development by Susan R. Singer— 8th ed.
    p. cm.
  Includes bibliographical references and index.
  ISBN 0-87893-250-X (hardcover)
  1. Embryology. 2. Developmental biology. I. Title.

QL955.G48 2006
571.8—dc22   2006004780

Printed in U.S.A.
Second Printing
August 2006

To Daniel, Sarah, and David

Brief Contents

Contents

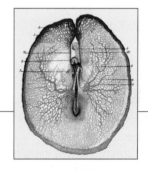

CHAPTER 6 *Cell-Cell Communication in Development* 139

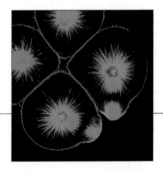

PART 2 *Early Embryonic Development*

CHAPTER 7 *Fertilization: Beginning a New Organism* 175

CHAPTER 10 *Early Development and Axis Formation in Amphibians* 291

CHAPTER 11 *The Early Development of Vertebrates: Fish, Birds, and Mammals* 325

PART 3 *Later Embryonic Development*

CHAPTER 15 *Lateral Plate Mesoderm and Endoderm 471*

CHAPTER 16 *Development of the Tetrapod Limb 505*

PART **4** *Ramifications of Developmental Biology*

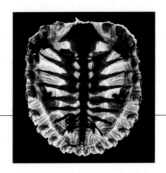

CHAPTER 20 *An Overview of Plant Development* **627**

CHAPTER 21 *Medical Implications of Developmental Biology* **655**

CHAPTER 22 *Environmental Regulation of Animal Development* 693

CHAPTER 23 *Developmental Mechanisms of Evolutionary Change* 721

Preface

" Lately it occurs to me
What a long strange trip it's been. "

Robert Hunter (1970)

This textbook is 21 years old—about the same age as most of the students taking the course. It should be taken out for a drink. The book went through a number of stages before reaching maturity, and it has had a rather tempestuous adolescence. When it first saw the light of publication in 1985, it was innocent of enhancers, signal transduction cascades, and paracrine factors. The book didn't know that Hox genes and BMPs would someday form a major part of its identity. The field of developmental biology matured enormously over the past 21 years, and the book grew as the field grew.

Hemingway said the hardest part about writing is deciding what to leave out. In writing past editions, the decisions about "what to leave out" were mainly ones of breadth: Did undergraduates really need to know about the wonderful research conducted on the development of the *Drosophila* trachea or leech segments? I knew from the start that this edition was going to be different. For the first time, I have had to make major decisions on the basis of depth. We now know more about most topics in developmental biology than undergraduates need to know. So rather than trying to "track the field," this edition takes the research as far as I think appropriate for a junior or senior undergraduate biology major.

This book is the child of three revolutions. First, the information revolution has made publishing almost instantaneous. The first edition of this book was *truly* written. By hand! Computers were just beginning to be installed into college offices, and many faculty members didn't know what to do with them. Mail took about a week to be delivered, faxes were a novelty, Federal Express was a rarely used luxury, and nobody had even heard of email. The first edition was printed on presses with an operator who would look through a monocular lens at the metallic characters to make certain none of the letters were broken. Now all the information needed to print the pages is transmitted to the press on a few computer discs.

Next, the graphics revolution changed the book's appearance. The first edition was black-and-white; only the cover image was in color. Artists drew with pens on illustrator board, and mistakes were laboriously corrected with "white-out" and mechanical erasers. Nobody talked in terms of pixels per inch, and nobody sent digital images by reply email. Instead I had an enormous file drawer full of the slides and photographs people sent me. If there had been a fire in the building, those photos were the first things I would have grabbed.

And of course, most importantly, the molecular revolution has affected developmental biology probably more than any other biological discipline. MicroRNAs, microarrays, mammalian cloning, knockout mice, transcription factors, paracrine factors, chromatin remodeling proteins, and the signaling pathways connecting them were all unknown in 1985. The subdisciplines covered in the final three chapters of the book—medical developmental biology, ecological developmental biology, and evolutionary developmental biology—did not exist then. *Arabidopsis* was an insignificant weed, and zebrafish were studied only by a few researchers in Oregon.

Reflection upon reaching 21 years old is a good thing, so I returned to assess the principles outlined in the preface to the first edition. The first principle remains as I stated it then: "This is a textbook written with juniors and seniors in mind. It does not assume a level of biology more sophisticated than that given in most good introductory biology textbooks, but a familiarity with cell biology and genetics will certainly make the going easier." This is what I still strive for.

In regard to the second principle, that of organization, the text's arrangement has changed over the course of eight editions, mostly at its 5' and 3' termini. I have added an introductory "cap" that introduces the reader to the different approaches of developmental biology, and attempts to equalize the different academic backgrounds students bring to the classroom. The introductory chapters are an attempt to bring all readers to the same level of expertise.

The 3' terminus includes the plant development chapter, the medical development chapter, and chapters on ecological and evolutionary developmental biology. These final two chapters have been the most exciting for me, since these fields didn't exist (at least not in their molecular forms) when I wrote in the first edition that "this book attempts to blur some of the lines separating developmental biology, genetics, and evolution." I'm proud to have written that in 1985. C. H. Waddington called these three areas "diachronic biology" and viewed it as a unified field. The integration of evolution, development, and genetics is now a reality, and this book could profitably be approached as a series of 21 chapters leading up to Chapters 22 and 23.

While the first principle has been maintained and the second principle modified, the third principle of the original textbook has largely been abandoned. Originally, the book covered phenomena rather than organisms. Thus there was a chapter about fertilization, followed by a chapter on cleavage, followed by a chapter on gastrulation, followed by a chapter on neurulation. Now that we know more about these phenomena (and since it is useful when laboratory work is part of the course), I decided to integrate early development in sea urchins, snails, ascidians, nematodes, *Drosophila*, and various vertebrates into a unified approach that focuses on each model organism.

The fourth principle was to respect all the approaches—molecular, cellular, and anatomical—to developmental biology. The book still tries to demonstrate the critical importance of knowing and integrating each of these three approaches, but for the past two decades, the molecular approaches have been in the forefront. To promote balance, we include the *vade mecum*[2] CD, produced by Mary Tyler and Ron Kozlowski, which emphasizes the organismal and cellular approaches to development. Similarly, the fifth principle—to emphasize science as a creative human activity, emphasizing research methods and citing the researchers who provided the data—has also been expanded by the electronic revolution. In this edition, we are able to link the literature citation to the original publications.

The final principle has been to keep the book flexible. Initially we did this by including "Sidelights & Speculations" in the original book. The book's flexibility has been further extended by the additions of the *vade mecum*[2] CD and a text-related website. The website began as a Gopher site in 1994—more than a decade ago—and was first linked to the book in its fifth edition, in 1999. The site has metamorphosed and come into its own as www.devbio.com.

Technology has also changed what teachers and students expect in a classroom presentation. For the last few editions, all the book's illustrations have been available in PowerPoint® format for instructors, who use these to enhance lectures. In this edition, the technology has advanced to enable us to include movies with this instructional material, in both QuickTime® and PowerPoint® presentations. This means students will be able to see and to appreciate aspects of motion that are so critical for understanding development. Perhaps in the next few years, the book will come with virtual reality programs so that students can sit on a cell in the dorsal blastopore lip and enjoy the vertiginous ride as they involute upside-down into the embryo and stream toward the head.

Acknowledgments

More than any previous edition, this book has been a community effort. Certainly, the students in my embryology courses and developmental genetics seminars gave me immediate feedback and inspiration. The staff and faculty of Swarthmore College have been remarkably helpful, both in answering specific questions and in sending me all sorts of weird and wonderful articles culled from the Internet. Many teachers have taken the time to email me with information about which sections of the book work well in their classes and which do not; they have provided constructive criticisms and suggested changes. I thank these people sincerely for taking their valuable time to tell me these things.

I especially want to thank all the investigators who have shared their photographs with me. Each edition of the book has been more visually stunning than the last, due in large part to the generosity of my colleagues. I thank you for myself, and on behalf of the many students who are intrigued and inspired by these beautiful images.

With so much new information appearing each week, keeping up with developmental biology has become an arduous occupation. Thus I thank Swarthmore College, the National Science Foundation, and the Howard A. Schneiderman fund at Swarthmore College for providing funds for my travel to academic meetings.

The chapter reviewers, as usual, did outstanding service despite their own demanding research and teaching responsibilities—not a single one of them had the time, but they all came through anyway. Adam Antebi, Radhika Atit, Michael Barresi, Marianne Bronner-Fraser, Blanche Capel, Judy Cebra-Thomas, Mark Cooper, Karen Crawford, Brad Davidson, Eddy De Robertis, John Fallon, Rachel Fink, Richard Harland, Margaret Hollyday, Ruth Lehmann, Kersti Linask, David McClay, Rudy Raff, Richard Schultz, Kirsi Sainio, Claudio Stern, Heather Stickney, Yoshiko Takahashi, Rocky Tuan, Andrea Ward, Gary Wessel, Kristi Wharton, Diana Wheeler, Chris Wright, and Ken Zaret all gave me important expert feedback, including steering me to new papers and offering insights on movies that could be included in our instructors' CD. Steve Small went above and beyond the call in reorganizing and updating the *Drosophila* genetics chapter; I am grateful for his very detailed input.

I have been fortunate to have the most hardworking, intelligent, and tolerant editor, Carol Wigg, who has been responsible for coordinating the production of this book since its inception. Andy Sinauer has been his usual impressive self in allowing me to change material in the book right up to the last day before it got sent to the printer. In this age of electronic media, top-flight production people become critical to a project. The book's design and layout is the result of long hours of work by Jefferson Johnson and Janice Holabird, and the result is proof of their skill.

Mary Tyler and Ron Kozlowski have been wonderful in constructing the *vade mecum*[2] CD. Susan Singer also didn't have time to write an overview of plant development, but nevertheless turned in another fantastic chapter. I could not have completed the book without the incredible help that Judy Cebra-Thomas has provided as developmental biology laboratory manager here at Swarthmore. And finally, there is no way this edition could have been contemplated, much less completed, without the help of my wife, Anne Raunio. It's been a wild three years since the last edition. Let's hope the next three go much easier.

Scott F. Gilbert

Media & Supplements to Accompany
Developmental Biology, Eighth Edition

For the Student
(See back cover for additional details)

Companion Website: www.devbio.com
Available free of charge, this website is intended to supplement and enrich courses in developmental biology. It provides more information for advanced students as well as historical, philosophical, and ethical perspectives on issues in developmental biology. Included are articles, movies, interviews, opinions, Web links, updates, and more. References to specific website topics are included throughout the textbook.

vade mecum^2: An Interactive Guide to Developmental Biology (CD-ROM)

by Mary S. Tyler and Ronald N. Kozlowski
Included in every copy of the textbook, *vade mecum^2* is a tool that helps the student understand the development of the organisms discussed in lecture and prepare them for laboratory exercises.

Developmental Biology:
A Guide for Experimental Study, Third Edition

Mary S. Tyler
(Included on the *vade mecum^2* CD)
This lab manual teaches the student to work as an independent investigator on problems in development, and provides extensive background information and instructions for each experiment. It emphasizes the study of living material, intermixing developmental anatomy in an enjoyable balance, and allows the student to make choices in their work.

For the Instructor
(Available to qualified adopters)

Instructor's Resource CD (ISBN 0-87893-939-3)
The *Developmental Biology*, Eighth Edition IRCD includes a rich collection of visual resources for use in preparing lectures and other course materials. The IRCD includes:

- All textbook art, photos, and tables in two JPEG formats, high-resolution and low-resolution
- All textbook art, photos, and tables in PowerPoint format
- A collection of videos illustrating key developmental processes
- The full set of videos from the *vade mecum^2* CD

The textbook figures have been sized and color-adjusted for optimal legibility when projected.

Also Available
The following titles are available for purchase separately or, in some cases, bundled with the textbook. Please contact Sinauer Associates for more information.

Bioethics and the New Embryology:
Springboards for Debate
Scott F. Gilbert, Anna Tyler, and Emily Zackin
2005 • Paper, 261 pages • ISBN 0-7167-7345-7

Differential Expressions^2:
Key Experiments in Developmental Biology
Mary S. Tyler, Ronald N. Kozlowski, and Scott F. Gilbert
2-DVD Set • UPC 855038001020

A Dozen Eggs:
Time-Lapse Microscopy of Normal Development
Rachel Fink
VHS • ISBN 0-87893-181-3

Fly Cycle^2
Mary S. Tyler and Ronald N. Kozlowski
DVD; ISBN 0-87893-849-4 • CD; ISBN 0-87893-848-6

Fly Cycle: The Lives of a Fly, Drosophila melanogaster
Mary S. Tyler, Jamie W. Schnetzer, and David Tartaglia
VHS • ISBN 0-87893-837-0

From Egg to Tadpole: Early Morphogenesis in Xenopus
Jeremy D. Pickett-Heaps and Julianne Pickett-Heaps
VHS • ISBN 09586081-1-3

PART 1

Principles of
Developmental
Biology

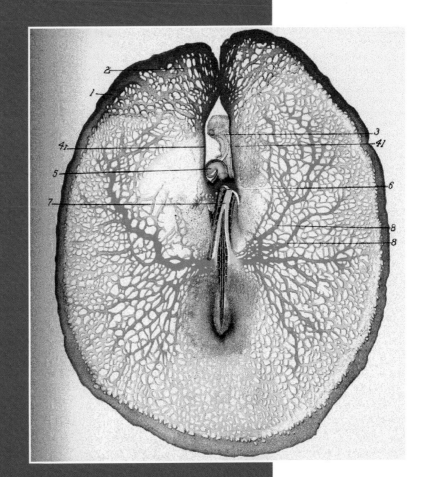

Part 1   **Principles of Developmental Biology**

PREVIOUS PAGE
A two-day chick embryo seen from the ventral surface, allowing one to view the circulation of the blood to the yolk and back. The blood leaves the embryo through the two vitelline arteries, and it returns through the vitelline veins near the head of the embryo. This figure is from F. R. Lillie's *The Embryo of the Chick* (1908).

Developmental Biology: The Anatomical Tradition

1

BETWEEN FERTILIZATION AND BIRTH, the developing organism is known as an embryo. The concept of an embryo is a staggering one, and forming an embryo is the hardest thing you will ever do. To become an embryo, you had to build yourself from a single cell. You had to respire before you had lungs, digest before you had a gut, build bones when you were pulpy, and form orderly arrays of neurons before you knew how to think. One of the critical differences between you and a machine is that a machine is never required to function until after it is built. Every animal has to function as it builds itself.

For animals, fungi, and plants, the sole way of getting from egg to adult is by developing an embryo. The embryo mediates between genotype and phenotype, between the inherited genes and the adult organism. Whereas most of biology studies adult structure and function, developmental biology finds the study of the transient stages leading up to the adult to be more interesting. Developmental biology studies the initiation and construction of organisms rather than their maintenance. It is a science of becoming, a science of process. To say that a mayfly lives but one day is profoundly inaccurate to a developmental biologist. A mayfly may be a winged adult for only a day, but it spends the other 364 days of its life as an embryo or aquatic juvenile under the waters of a pond or stream.

The questions asked by developmental biologists are often questions about becoming rather than about being. To say that XX mammals are usually females and XY mammals are usually males does not explain sex determination to a developmental biologist, who wants to know *how* the XX genotype produces a female and *how* the XY genotype produces a male. Similarly, a geneticist might ask how globin genes are transmitted from one generation to the next, and a physiologist might ask about the function of globin proteins in the body. But the developmental biologist asks how it is that the globin genes become expressed only in red blood cells and how the genes become active only at specific times in development. (We don't know the answers yet.)

Developmental biology is a great field for scientists who want to integrate different levels of biology. We can take a problem and study it on the molecular and chemical levels (e.g., How are globin genes transcribed?); on the cellular and tissue levels (Which cells are able to make globin, and how does globin mRNA leave the nucleus?); on the organ and organ-system levels (How do the capillaries form in each tissue, and how are they instructed to branch and connect?); and even at the ecological and evolutionary levels (How do differences in globin gene activation enable oxygen to flow from mother to fetus, and how do environmental factors trigger the differentiation of more red blood cells?).

Developmental biology is one of the fastest growing and most exciting fields in biology, creating a framework that integrates molecular biology, physiology,

> " Happy is the person who is able to discern the causes of things. "
>
> *VIRGIL (37 BCE)*

> " The greatest progressive minds of embryology have not looked for hypotheses; they have looked at embryos. "
>
> *JANE OPPENHEIMER (1955)*

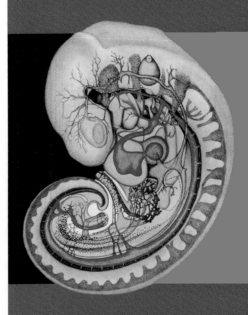

cell biology, genetics, anatomy, cancer research, neurobiology, immunology, ecology, and evolutionary biology. The study of development has become essential for understanding all other areas of biology.

The Questions of Developmental Biology

According to Aristotle, the first embryologist known to history, science begins with wonder: "It is owing to wonder that people began to philosophize, and wonder remains the beginning of knowledge" (Aristotle, *Metaphysics,* ca. 350 B.C.E.). The development of an animal from an egg has been a source of wonder throughout history. The simple procedure of cracking open a chick egg on each successive day of its 3-week incubation provides a remarkable experience as a thin band of cells is seen to give rise to an entire bird. Aristotle performed this procedure and noted the formation of the major organs. Anyone can wonder at this remarkable—yet commonplace—phenomenon, but the scientist seeks to discover how development actually occurs. And rather than dissipating wonder, new understanding increases it.

Multicellular organisms do not spring forth fully formed. Rather, they arise by a relatively slow process of progressive change that we call **development**. In nearly all cases, the development of a multicellular organism begins with a single cell—the fertilized egg, or **zygote**, which divides mitotically to produce all the cells of the body. The study of animal development has traditionally been called **embryology**, from that phase of an organism that exists between fertilization and birth. But development does not stop at birth, or even at adulthood. Most organisms never stop developing. Each day we replace more than a gram of skin cells (the older cells being sloughed off as we move), and our bone marrow sustains the development of millions of new red blood cells every minute of our lives. In addition, some animals can regenerate severed parts, and many species undergo metamorphosis (such as the transformation of a tadpole into a frog, or a caterpillar into a butterfly). Therefore, in recent years it has become customary to speak of **developmental biology** as the discipline that studies embryonic and other developmental processes.

Development accomplishes two major objectives: it generates cellular diversity and order within each generation, and it ensures the continuity of life from one generation to the next. Thus, there are two fundamental questions in developmental biology: How does the fertilized egg give rise to the adult body, and how does that adult body produce yet another body? These two huge questions have been subdivided into six general questions scrutinized by developmental biologists:

- **The question of differentiation**. A single cell, the fertilized egg, gives rise to hundreds of different cell types—muscle cells, epidermal cells, neurons, lens cells, lymphocytes, blood cells, fat cells, and so on (Figure 1.1). This generation of cellular diversity is called **differentiation**. Since each cell of the body (with very few exceptions) contains the same set of genes, how can this identical set of genetic instructions produce different types of cells? How can the fertilized egg generate so many different cell types?

- **The question of morphogenesis**. Our differentiated cells are not randomly distributed. Rather, they are organized into intricate tissues and organs. During development, cells divide, migrate, and die; tissues fold and separate. The organs so formed are arranged in a particular way: our fingers are always at the tips of our hands, never in the middle; our eyes are always in our heads, not in our

FIGURE 1.1   Some representative differentiated cell types of the vertebrate body. The progeny of the fertilized egg must diversify into hundreds of cell types. The cell types are organized according to the germ layers from which they arise. The germ cells (precursors of the sperm and egg) are set aside early in development and do not arise from the three germ layers.

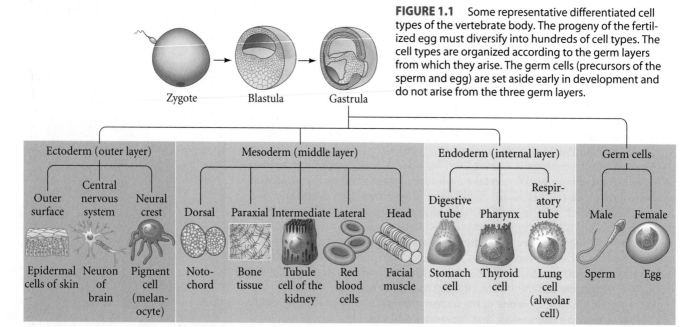

toes or gut. This creation of ordered form is called **morphogenesis**. How can the cells form such ordered structures?

- **The question of growth.** If each cell in our face were to undergo just one more cell division, we would be considered horribly malformed. If each cell in our arms underwent just one more round of cell division, we could tie our shoelaces without bending over. How do our cells know when to stop dividing? Our arms are generally the same size on both sides of the body. How is cell division so tightly regulated?
- **The question of reproduction.** The sperm and egg are very specialized cells. Only they can transmit the instructions for making an organism from one generation to the next. How are these cells set apart to form the next generation, and what are the instructions in the nucleus and cytoplasm that allow them to function this way?
- **The question of evolution.** Evolution involves inherited changes in development. When we say that today's one-toed horse had a five-toed ancestor, we are saying that changes in the development of cartilage and muscles occurred over many generations in the embryos of the horse's ancestors. How do changes in development create new body forms? Which heritable changes are possible, given the constraints imposed by the necessity of the organism to survive as it develops?
- **The question of environmental integration.** The development of many (and perhaps all) organisms is influenced by cues from the environment that surrounds the embryo or larvae. Certain butterflies, for instance, inherit the ability to produce different wing colors based on the temperature or the amount of daylight experienced by the caterpillar before it undergoes metamorphosis. Moreover, certain chemicals in the environment can disrupt normal development, causing malformations in the adult. How is the development of an organism integrated into the larger context of its habitat, and what properties enable certain chemicals to alter development?

Approaches to Developmental Biology

A field of science is defined by the questions it seeks to answer, and most of the questions in developmental biology have been bequeathed to it by its embryological heritage. There are numerous strands of embryology, each of which has predominated during a different era. Sometimes these traditions are distinct, and sometimes they blend. We can identify three major approaches to studying embryology:

- Anatomical approaches
- Experimental approaches
- Genetic approaches

While it is true that anatomical approaches gave rise to experimental approaches, and that genetic approaches built on the foundations of the earlier two approaches, all three traditions persist to this day and continue to play a major

role in developmental biology. Chapter 3 of this text discusses experimental approaches, and Chapters 4 and 5 examine the genetic approaches in greater depth. In recent years, each of these traditions has become joined with molecular genetics to produce a vigorous and multifaceted science of developmental biology. But the basis of all research in developmental biology is the changing anatomy of the organism—the subject of the remainder of this chapter.

The Anatomical Approach

What parts of the embryo form the heart? How do the cells that form the retina position themselves the proper distance from the cells that form the lens? How do the tissues that form the bird wing relate to the tissues that form the fish fin or the human hand? What organs are affected by mutations in particular genes? Are there mathematical equations that relate the growth of different organs within the body? These are the types of questions asked by developmental anatomists.

Several strands weave together to form the anatomical approaches to development. The first strand is **comparative embryology**, the study of how anatomy changes during the development of different organisms. The second strand, based on the first, is **evolutionary embryology**, the study of how changes in development may cause evolutionary change and of how an organism's ancestry may constrain the types of changes that are possible. The third strand of the anatomical approach to developmental biology is **teratology**, the study of birth defects. The fourth method used in the anatomical approach is **mathematical modeling**, which seeks to describe developmental phenomena in terms of equations.

Comparative embryology

The first known study of comparative developmental anatomy was undertaken by Aristotle in the fourth century B.C.E. In *The Generation of Animals* (ca. 350 B.C.E.), he noted the different ways that animals are born: from eggs (**oviparity**, as in birds, frogs, and most invertebrates); by live birth (**viviparity**, as in placental mammals); or by producing an egg that hatches inside the body (**ovoviviparity**, as in certain reptiles and sharks). Aristotle also identified the two major cell division patterns by which embryos are formed: the **holoblastic** pattern of cleavage (in which the entire egg is divided into smaller cells, as it is in frogs and mammals) and the **meroblastic** pattern of cleavage (as in chicks, wherein only part of the egg is destined to become the embryo, while the other portion—the yolk—serves as nutrition for the embryo). And should anyone want to know who first figured out the functions of the placenta and the umbilical cord, it was Aristotle.

There was remarkably little progress in embryology for the two thousand years following Aristotle. It was only in 1651 that William Harvey concluded that all animals—even mammals—originate from eggs. *Ex ovo omnia* ("all from the egg") was the motto on the frontispiece of Harvey's *On the Generation of Living Creatures*, and this precluded the spontaneous generation of animals from mud or excrement. This statement was not made lightly, for Harvey knew that it went against the views of Aristotle, whom Harvey still venerated. (Aristotle had thought that menstrual fluid formed the material of the embryo, while the semen gave it form and animation.) Harvey also was the first to see the blastoderm of the chick embryo (the small region of the egg containing the yolk-free cytoplasm that gives rise to the embryo), and he was the first to notice that "islands" of blood cells form before the heart does. Harvey also suggested that the amnionic fluid might function as a "shock absorber" for the embryo.

As might be expected, embryology remained little but speculation until the invention of the microscope allowed detailed observations. In 1672, Marcello Malpighi published the first microscopic account of chick development. Here, for the first time, the neural groove (precursor of the neural tube), the muscle-forming somites, and the first circulation of the arteries and veins—to and from the yolk—were identified (Figure 1.2).

Epigenesis and preformation

With Malpighi begins one of the great debates in embryology: the controversy over whether the organs of the embryo are formed *de novo* ("from scratch") at each generation, or whether the organs are already present, in miniature form, within the egg (or sperm). The first view, called **epigenesis**, was supported by Aristotle and Harvey. The second view, called **preformation**, was reinvigorated with Malpighi's support. Malpighi showed that the unincubated* chick egg already had a great deal of structure, and this observation provided him with reasons to question epigenesis. According to the preformationist view, all the organs of the adult were prefigured in miniature within the sperm or (more usually) the egg. Organisms were not seen to be "constructed," but rather "unrolled."

The preformationist hypothesis had the backing of eighteenth-century science, religion, and philosophy (Gould 1977; Roe 1981; Pinto-Correia 1997). First, because all organs were prefigured, embryonic development merely required the growth of existing structures, not the formation of new ones. No extra mysterious force was needed for embryonic development. Second, just as the adult organism was prefigured in the germ cells, another generation already existed in a prefigured state within the germ cells of the first prefigured generation. This corollary, called *emboîtment* (encapsulation), ensured that the species would remain constant. Although certain microscopists claimed

to see fully formed human miniatures within the sperm or egg, the major proponents of this hypothesis—Albrecht von Haller and Charles Bonnet—knew that organ systems develop at different rates, and that structures need not be in the same place in the embryo as they are in the newborn.

The preformationists had no cell theory to provide a lower limit to the size of their preformed organisms (the cell theory arose in the mid-1800s), nor did they view humankind's tenure on Earth as potentially infinite. Rather, said Bonnet (1764), "Nature works as small as it wishes," and the human species existed in that finite time between Creation and Resurrection. This view was in accord with the best science of its time, conforming to the French mathematician-philosopher René Descartes's principle of the infinite divisibility of a mechanical nature initiated, but not interfered with, by God. It also conformed to Enlightenment views of the Deity. The scientist-priest Nicolas Malebranche saw in preformationism the fusion of the rule-giving God of Christianity with Cartesian science (Churchill 1991; Pinto-Correia 1997).[†]

The embryological case for epigenesis was revived at the same time by Kaspar Friedrich Wolff, a German embryologist working in St. Petersburg. By carefully observing the development of chick embryos, Wolff demonstrated that the embryonic parts develop from tissues that have no counterpart in the adult organism. The heart and blood vessels (which, according to preformationism, had to be present from the beginning to ensure embryonic growth) could be seen to develop anew in each embryo. Similarly, the intestinal tube was seen to arise by the folding of an originally flat tissue. This latter observation was explicitly detailed by Wolff, who proclaimed (1767), "When the formation of the intestine in this manner has been duly weighed, almost no doubt can remain, I believe, of the truth of epigenesis." However, to explain how an organism is created anew each generation, Wolff had to postulate an unknown force, the *vis essentialis* ("essential force"), which, acting according to natural laws in the same way as gravity or magnetism, would organize embryonic development.

A reconciliation between preformationism and epigenesis was attempted by the German philosopher Immanuel Kant (1724–1804) and his colleague, biologist Johann Friedrich Blumenbach (1752–1840). Attempting to construct

---

*As was pointed out by Maître-Jan in 1722, the eggs examined by Malpighi may technically be called "unincubated," but as they were left sitting in the Bolognese sun in August, they were not unheated.

[†]Preformation was a conservative theory, emphasizing the lack of change between generations. Its principal failure was its inability to account for the variations revealed by the limited genetic evidence of the time. It was known, for instance, that matings between white and black parents produced children of intermediate skin color, an impossibility if inheritance and development were solely through either the sperm or the egg. In more controlled experiments, the German botanist Joseph Kölreuter (1766) produced hybrid tobacco plants having the characteristics of both species. Moreover, by mating the hybrid to either the male or female parent, Kölreuter was able to "revert" the hybrid back to one or the other parental type after several generations. Thus, inheritance seemed to arise from a mixture of parental components.

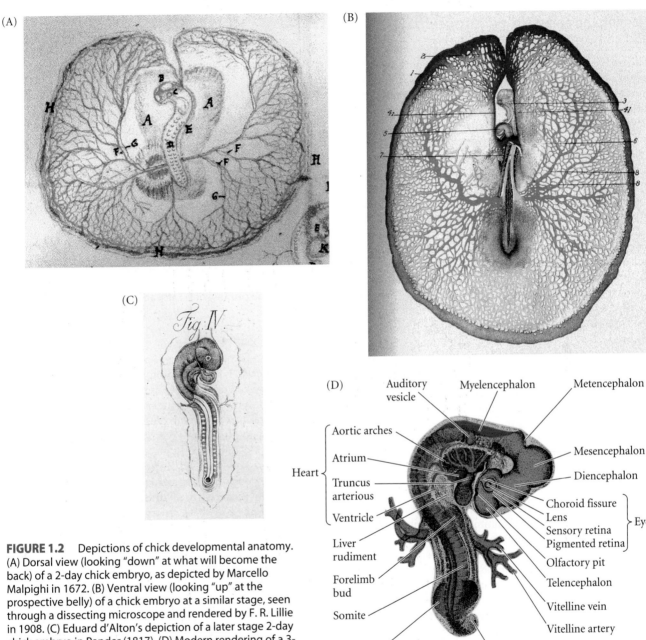

FIGURE 1.2   Depictions of chick developmental anatomy. (A) Dorsal view (looking "down" at what will become the back) of a 2-day chick embryo, as depicted by Marcello Malpighi in 1672. (B) Ventral view (looking "up" at the prospective belly) of a chick embryo at a similar stage, seen through a dissecting microscope and rendered by F. R. Lillie in 1908. (C) Eduard d'Alton's depiction of a later stage 2-day chick embryo in Pander (1817). (D) Modern rendering of a 3-day chick embryo. Details of the anatomy will be discussed in later chapters. (A from Malpighi 1672; B from Lillie 1908; C from Pander 1817, courtesy of Ernst Mayr Library of the Museum of Comparative Zoology, Harvard; D after Carlson 1981.)

a scientific theory of racial descent, Blumenbach postulated a mechanical, goal-directed force called the *Bildungstrieb* ("developmental force"). Such a force, he said, was not theoretical, but could be shown to exist by experimentation. A hydra, when cut, regenerates its amputated parts by rearranging existing elements (see Chapter 18). Some purposive organizing force could be observed in operation, and this force was a property of the organism itself. This *Bil-*

dungstrieb was thought to be inherited through the germ cells. Thus, development could proceed through a predetermined force inherent in the matter of the embryo (Cassirer 1950; Lenoir 1980). Moreover, this force was believed to be susceptible to change, as demonstrated by the left-handed variant of snail coiling (where left-coiled snails can produce right-coiled progeny). In this hypothesis, wherein epigenetic development is directed by preformed instructions, we are not far from the view held by modern biologists that most of the instructions for forming the organism are already present in the egg.

Naming the parts: The primary germ layers and early organs

The end of preformationism did not come until the 1820s, when a combination of new staining techniques, improved microscopes, and institutional reforms in German universities created a revolution in descriptive embryology. The new techniques enabled microscopists to document the epigenesis of anatomical structures, and the institutional reforms provided audiences for these reports and students to carry on the work of their teachers. Among the most talented of this new group of microscopically inclined investigators were three friends, born within a year of each other, all of whom came from the Baltic region and studied in northern Germany. The work of Christian Pander, Karl Ernst von Baer, and Heinrich Rathke transformed embryology into a specialized branch of science.

Pander studied the chick embryo for less than two years (before becoming a paleontologist), but in those 15 months, he discovered the germ layers,* three distinct regions of the embryo that give rise to the specific organ systems (see Figure 1.1).

- The **ectoderm** generates the outer layer of the embryo. It produces the surface layer (epidermis) of the skin and forms the brain and nervous system.
- The **endoderm** becomes the innermost layer of the embryo and produces the epithelium of the digestive tube and its associated organs (including the lungs).
- The **mesoderm** becomes sandwiched between the ectoderm and endoderm. It generates the blood, heart, kidney, gonads, bones, muscles, and connective tissues.

These three layers are found in the embryos of all **triploblastic** ("three-layer") animals. Some phyla, such as the cnidarians (sea anemones, hydra, jellyfish), and ctenophores (comb jellies) lack a true mesoderm and are considered **diploblastic** animals.

Pander also made observations that weighted the balance in favor of epigenesis. The germ layers, he noted, did not form their organs independently (Pander 1817). Rather, each germ layer "is not yet independent enough to indicate what it truly is; it still needs the help of its sister travelers, and therefore, although already designated for different ends, all three influence each other collectively until each has reached an appropriate level." Pander had discovered the tissue interactions that we now call **induction**. No tissue is able to construct organs by itself; it must interact with other tissues. (We will discuss the principles of induction more thoroughly in Chapter 6.) Thus, Pander felt that preformation could not be true, since the organs come into being through interactions between simpler structures.

*From the same root as *germination*, the Latin *germen* means "sprout" or "bud." The names of the three germ layers are from the Greek: ectoderm from *ektos* ("outside") plus *derma* (skin); mesoderm from *mesos* ("middle"), and endoderm from *endon* ("within").

FIGURE 1.3   Pharyngeal arches (also called branchial arches and gill arches) in the embryo of the salamander *Ambystoma mexicanum*. The surface ectoderm has been removed to permit the easy visualization of these arches (highlighted) as they form. (Photograph courtesy of P. Falck and L. Olsson.)

Interestingly, the glory of Pander's book is its engravings; the artist, Eduard d'Alton, drew details for which the vocabulary had not yet been invented. Today we can look at these drawings and see the five regions of the embryonic chick brain, even though these regions had not yet been separately defined or given names (see Figure 1.2C; Churchill 1991). The ability to make precise observations has always been among the greatest skills of embryologists, and today's developmental biologists looking at gene expression patterns are "discovering" regions of the embryo that were observed by embryologists a century ago.

Heinrich Rathke observed the development of frogs, salamanders, fish, turtles, birds, and mammals and emphasized the similarities in the development of all these vertebrate groups. During his 40 years of embryological research, he described for the first time the vertebrate **pharyngeal arches** (Figure 1.3), which become the gill apparatus of fish but in mammals become the jaws and ears (among other things, as we will see in Figure 1.14). Rathke described the formation of the vertebrate skull, and the origin of the reproductive, excretory, and respiratory systems. He also studied the development of invertebrates, especially the crayfish. He is memorialized today in the name "Rathke's pouch," the embryonic rudiment of the glandular portion of the pituitary. That he could see such a structure using the techniques available at that time is testimony to his remarkable powers of observation and his steady hand.

The four principles of Karl Ernst von Baer

Karl Ernst von Baer extended Pander's studies of the chick embryo. He discovered the **notochord**, the rod of dorsalmost mesoderm that separates the embryo into right

(A)

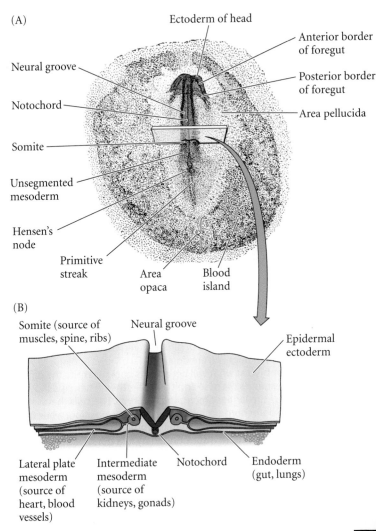

(B)

FIGURE 1.4   The notochord in the chick embryo. (A) Dorsal view of the 24-hour chick embryo. (B) A cross-section through the trunk shows the notochord and developing neural tube. By comparing Figures 1.2 and 1.4, you should see the remarkable changes between days 1, 2, and 3 of chick egg incubation. (A after Patten 1951.)

and left halves and which instructs the ectoderm above it to become the nervous system (Figure 1.4). He also discovered the mammalian egg, that long-sought cell that everyone believed existed but no one before von Baer had ever seen.*

In 1828, von Baer reported, "I have two small embryos preserved in alcohol, that I forgot to label. At present I am

*von Baer could hardly believe that he had at last found what so many others—Harvey, de Graaf, von Haller, Prevost, Dumas, and even Purkinje—had searched for and failed. "I recoiled as if struck by lightening … I had to try to relax a while before I could work up enough courage to look again, as I was afraid I had been deluded by a phantom. Is it not strange that a sight which is expected, and indeed hoped for, should be frightening when it eventually materializes?"

unable to determine the genus to which they belong. They may be lizards, small birds, or even mammals." Figure 1.5 allows us to appreciate his quandary. All vertebrate embryos (fish, reptiles, amphibians, birds, and mammals) begin with a basically similar structure. From his detailed study of chick development and his comparison of chick embryos with the embryos of other vertebrates, von Baer derived four generalizations. Now often referred to as "von Baer's laws," they are stated here with some vertebrate examples.

1. *The general features of a large group of animals appear earlier in development than do the specialized features of a smaller group.* All developing vertebrates appear very similar shortly after gastrulation. It is only later in development that the special features of class, order, and finally species emerge. All vertebrate embryos have gill arches, notochords, spinal cords, and primitive kidneys.

2. *Less general characters develop from the more general, until finally the most specialized appear.* All vertebrates initially have the same type of skin. Only later does the skin develop fish scales, reptilian scales, bird feathers, or the hair, claws, and nails of mammals. Similarly, the early development of the limb is essentially the same in all vertebrates. Only later do the differences between legs, wings, and arms become apparent.

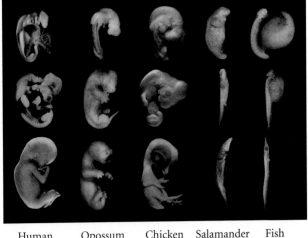

Human   Opossum   Chicken   Salamander   Fish
(axolotl)   (gar)

FIGURE 1.5   The similarities and differences among different vertebrate embryos as they proceed through development. They each begin with a basically similar structure, although they acquire this structure at different ages and sizes. As they develop, they become less like each other. (Adapted from Richardson et al. 1998; photograph courtesy of M. Richardson.)

3. *The embryo of a given species, instead of passing through the adult stages of lower animals, departs more and more from them.* * The visceral clefts of embryonic birds and mammals do not resemble the gill slits of adult fish in detail. Rather, they resemble the visceral clefts of *embryonic* fish and other *embryonic* vertebrates. Whereas fish preserve and elaborate these clefts into true gill slits, mammals convert them into structures such as the eustachian tubes (between the ear and mouth).

4. *Therefore, the early embryo of a higher animal is never like a lower animal, but only like its early embryo.* Human embryos never pass through a stage equivalent to an adult fish or bird. Rather, human embryos initially share characteristics in common with fish and avian embryos. Later, the mammalian and other embryos diverge, none of them passing through the stages of the others.

Von Baer also recognized that there is a common pattern to all vertebrate development: each of the three germ layers gives rise to the same organs whether the organism itself is a fish, a frog, or a chick.

WEBSITE 1.1 The reception of von Baer's principles. The acceptance of von Baer's principles and their interpretation over the past hundred years has varied enormously. Recent evidence suggests that one important researcher in the 1800s even fabricated data when his own theory went against these postulates.

Fate mapping the embryo

By the late 1800s, the cell had been conclusively demonstrated to be the basis for anatomy and physiology. Embryologists, too, began to base their field on the cell. One of the most important programs of descriptive embryology became the tracing of **cell lineages**: following individual cells to see what they become. In many organisms, this fine a resolution is not possible, but one can label groups of cells to see what that area of the embryo will become. By bringing such studies together, one can construct a **fate map**. These diagrams "map" the larval or adult structure onto the region of the embryo from which it arose. Fate maps are the bases for experimental embryology, since they provide researchers with information on which portions of the embryo normally become which larval or adult structures. Fate maps of some embryos at the early gastrula stage are shown in Figure 1.6. Fate maps have been generated in several ways.

OBSERVING LIVING EMBRYOS The embryos of certain invertebrates are transparent, have relatively few cells, and the daughter cells remain close to one another. In such cases,

*von Baer formulated these generalizations prior to Darwin's theory of evolution. "Lower animals" would be those appearing earlier in life's history.

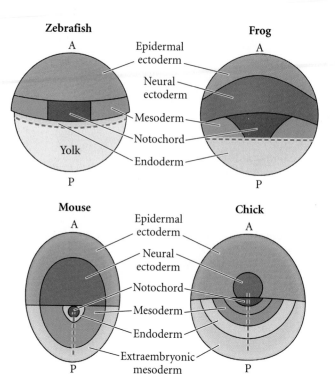

FIGURE 1.6 Fate maps of different vertebrate classes at the early gastrula stage. All are dorsal surface views (looking "down" on the embryo at what will become its back). Despite the different appearances of these adult animals, their fate maps show numerous similarities among the embryos. The cells that will form the notochord occupy a central dorsal position, while the precursors of the neural system lie immediately anterior to it. The neural ectoderm is surrounded by less dorsal ectoderm, which will form the epidermis of the skin. A indicates the anterior end of the embryo, P the posterior end. The dashed green lines indicate the site of ingression—the path cells will follow as they migrate from the exterior to the interior of the embryo.

it is actually possible to look through the microscope and trace the descendants of a particular cell into the organs they generate. This type of study was performed about a century ago by Edwin G. Conklin. In one of these studies, he took eggs of the tunicate *Styela partita*, a sea squirt that resides in the waters off the coast of Massachusetts, and patiently followed the fates of every cell in the embryo until each differentiated into particular structures (Figure 1.7; Conklin 1905). He was helped in this endeavor by a peculiarity of the *Styela* egg, wherein the different cells contain different pigments. For example, the muscle-forming cells always had a yellow color. Conklin's fate map was confirmed by cell removal experiments. Removal of the B4.1 cell (which according to the map should produce all the tail musculature), for example, resulted in a larva with no tail muscles (Reverberi and Minganti 1946).

WEBSITE 1.2 Conklin's art and science. The plates from Conklin's remarkable 1905 paper are online. Looking at them, one can see the precision of his observations and how he constructed his fate map of the tunicate embryo.

FIGURE 1.7 Fate map of the tunicate embryo. (A) The 1-cell embryo (left), shown shortly before the first cell division, with the fate of the cytoplasmic regions indicated. The 8-cell embryo on the right shows these regions after three cell divisions. (B) A linear version of the fate map, showing the fates of each cell of the embryo. Throughout this book, we will be using the color conventions of developmental anatomy: blue for ectoderm, red for mesoderm, and yellow for endoderm. (A after Nishida 1987 and Reverberi and Minganti 1946; B after Conklin 1905 and Nishida 1987.)

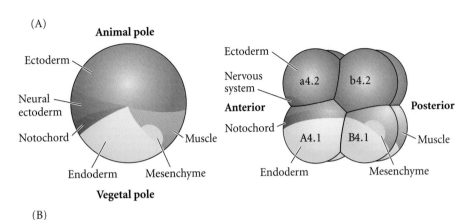

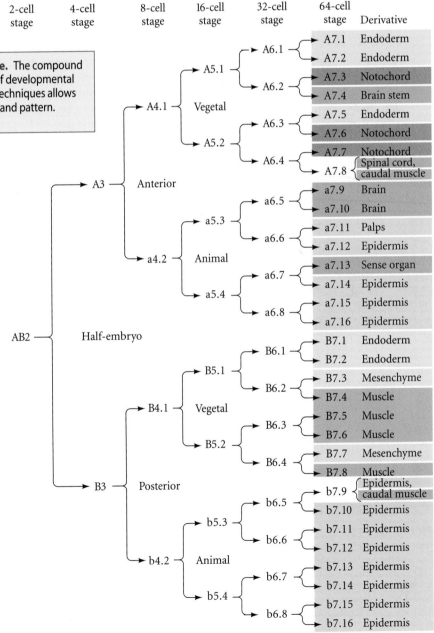

VADE MECUM² **The compound microscope.** The compound microscope has been the critical tool of developmental anatomists. Mastery of microscopic techniques allows one to enter an entire world of form and pattern. [Click on Microscope]

VITAL DYE MARKING Most embryos are not so accommodating as to have cells of different colors. Nor do all embryos have as few cells as tunicates. In the early years of the twentieth century, Vogt (1929) traced the fates of different areas of amphibian eggs by applying **vital dyes** to the region of interest. Vital dyes will stain cells but not kill them. Vogt mixed the dye with agar and spread the agar on a microscope slide to dry. The ends of the dyed agar were very thin. He cut chips from these ends and placed them onto a frog embryo. After the dye stained the cells, the agar chip was removed and cell movements within the embryo could be followed (Figure 1.8).

RADIOACTIVE LABELING AND FLUORESCENT DYES A variant of the dye-marking technique is to make one area of the embryo radioactive. To do this, a donor embryo is usually grown in a solution containing radioactive thymidine. This base becomes incorporated into the DNA of the dividing embryo. A second embryo (the host embryo) is grown under normal conditions. The region of interest is cut out from the host embryo and is replaced by a radioactive graft from the donor

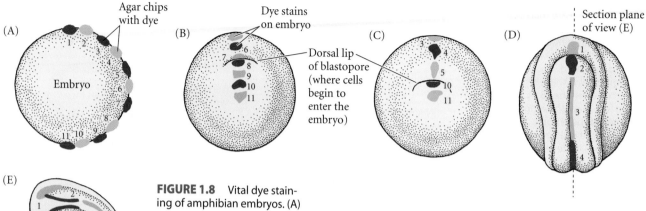

FIGURE 1.8 Vital dye staining of amphibian embryos. (A) Vogt's method for marking specific cells of the embryonic surface with vital dyes. (B–D) Dorsal surface views of stain on successively later embryos. (E) Newt embryo dissected in a medial sagittal section to show the stained cells in the interior. (After Vogt 1929.)

embryo. After some time, the host embryo is sectioned for microscopy. The cells seen to be radioactive will be the descendants of the cells of the graft, and can be distinguished using **autoradiography**. Fixed microscope slides containing the sectioned tissues are dipped into photographic emulsion. The high-energy electrons from the radioactive thymidine reduce the silver ions in the emulsion (just as light would). The result is a cluster of dark silver grains directly above the radioactive region. In this manner, the fates of different regions of the chick embryo have been determined (Rosenquist 1966).

One of the problems with both vital dyes and radioactive labels is that, as they become more diluted with each cell division, they become difficult to detect. One way around this problem is the use of **fluorescent dyes** that are so intense that once injected into individual cells, they can still be detected in the progeny of these cells many divisions later. Fluorescein-conjugated dextran, for example, can be injected into a single cell of an early embryo, and the descendants of that cell can be seen by examining the embryo under ultraviolet light (Figure 1.9). More recently, *diI*, a powerfully fluorescent molecule that becomes incorporated into lipid membranes, has also been used to follow the fates of cells and their progeny.

GENETIC MARKING One permanent way of marking cells and following their fates is to create "mosaic" embryos in which the same organism contains cells with different genetic constitutions. One of the best examples of this technique is the construction of **chimeric embryos**, consisting, for example, of a graft of quail cells inside a chick embryo. Chicks and quail develop in a very similar manner (espe-

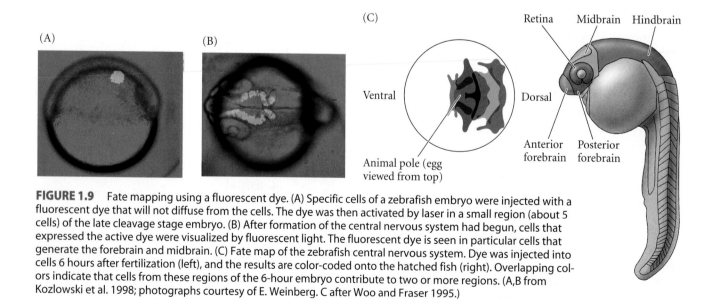

FIGURE 1.9 Fate mapping using a fluorescent dye. (A) Specific cells of a zebrafish embryo were injected with a fluorescent dye that will not diffuse from the cells. The dye was then activated by laser in a small region (about 5 cells) of the late cleavage stage embryo. (B) After formation of the central nervous system had begun, cells that expressed the active dye were visualized by fluorescent light. The fluorescent dye is seen in particular cells that generate the forebrain and midbrain. (C) Fate map of the zebrafish central nervous system. Dye was injected into cells 6 hours after fertilization (left), and the results are color-coded onto the hatched fish (right). Overlapping colors indicate that cells from these regions of the 6-hour embryo contribute to two or more regions. (A,B from Kozlowski et al. 1998; photographs courtesy of E. Weinberg. C after Woo and Fraser 1995.)

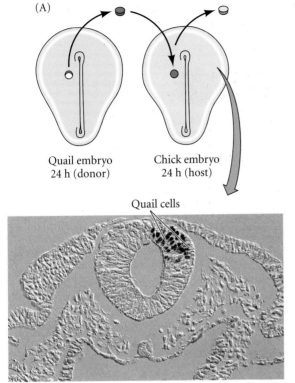

(A)

Quail embryo
24 h (donor)

Chick embryo
24 h (host)

Quail cells

Chick embryo with region of quail cells on the neural tube

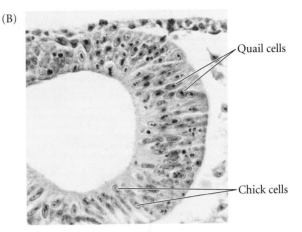

(B)

Quail cells

Chick cells

FIGURE 1.10 Genetic markers as cell lineage tracers. (A) Grafting experiment wherein the cells from a particular region of a 1-day quail embryo have been placed into a similar region of a 1-day chick embryo. After several days, the quail cells can be seen by using an antibody to quail-specific proteins. This region of the 3-day embryo produces cells that populate the neural tube. (B) Chick and quail cells can also be distinguished by the heterochromatin of their nuclei. The quail cells have a single large nucleolus (staining purple), distinguishing them from the diffuse nuclei of the chick. (After Darnell and Schoenwolf 1997; photographs courtesy of the authors.)

cially during early embryonic development), and the grafted quail cells become integrated into the chick embryo and participate in the construction of the various organs. The substitution of quail cells for chick cells can be performed on an embryo while it is still inside the egg, and the chick that hatches will have quail cells in particular sites, depending upon where the graft was placed. Quail cells differ from chick cells in two important ways. First, the quail nucleus has condensed DNA (heterochromatin) concentrated around the nucleoli, making quail nuclei easily distinguishable from chick nuclei. Second, cell-specific antigens that are quail-specific can be used to find individual quail cells, even if they are "hidden" within a large population of chick cells. In this way, fine-structure maps of the chick brain and skeletal system have been produced (Figure 1.10; Le Douarin 1969; Le Douarin and Teillet 1973).

WEBSITE **1.3** Nicole Le Douarin and chick-quail chimeras. We are fortunate to present here a movie made by Le Douarin of her chick-quail grafts. You will be able to see how these grafts are actually done.

VADE MECUM² **Histotechniques.** Most cells must be stained in order to see them; different dyes stain different types of molecules. Instructions on staining cells to observe particular structures (such as the nucleus) are given here. [Click on Histotechniques]

The Cellular Basis of Morphogenesis

One of the most important conclusions of the cell lineage studies is that cells are constantly changing during embryogenesis (Larsen and McLaughlin 1987). Cells do not remain in one place, nor do they keep the same shape. Early embryologists recognized that there were two major types of cells in the embryo: **epithelial cells**, which are tightly connected to one another in sheets or tubes; and **mesenchymal cells**, which are unconnected to one another and operate as independent units. Morphogenesis is brought about through a limited repertoire of variations in cellular processes within these two types of arrangements (Table 1.1):

• Direction and number of cell divisions
• Cell shape changes
• Cell movement
• Cell growth
• Cell death
• Changes in the composition of the cell membrane or secreted products

Cell migration

One of the most important contributions of fate maps has been their demonstration of extensive cell migration during development. Mary Rawles (1940) showed that the pig-

TABLE 1.1 Summary of major morphogenic processes regulated by mesenchymal and epithelial cells

Process	Action	Morphology	Example
MESENCHYMAL CELLS			
Condensation	Mesenchyme becomes epithelium		Cartilage mesenchyme
Cell division	Mitosis produces more cells (hyperplasia)		Limb mesenchyme
Cell death	Cells die		Interdigital mesenchyme
Migration	Cells move at particular times and places		Heart mesenchyme
Matrix secretion and degradation	Synthesis or removal of extracellular layer		Cartilage mesenchyme
Growth	Cells get larger (hypertrophy)		Fat cells
EPITHELIAL CELLS			
Dispersal	Epithelium becomes mesenchyme (entire structure)		Müllerian duct degeneration
Delamination	Epithelium becomes mesenchyme (part of structure)		Chick hypoblast
Shape change or growth	Cells remain attached as morphology is altered		Neurulation
Cell migration (intercalation)	Rows of epithelia merge to form fewer rows		Vertebrate gastrulation
Cell division	Mitosis within row or column		Vertebrate gastrulation
Matrix secretion and degradation	Synthesis or removal of extracellular matrix		Vertebrate organ formation
Migration	Formation of free edges		Chick ectoderm

ment cells (**melanocytes**) of the chick originate in the **neural crest**, a transient band of cells that joins the neural tube to the epidermis. When she transplanted small regions of neural crest-containing tissue from a pigmented strain of chickens into a similar position in an embryo from an unpigmented strain of chickens, the migrating pigment cells entered the epidermis and later entered the feathers (Figure 1.11A). Ris (1941) used similar techniques to show that while almost all of the external pigment of the chick embryo came from the migrating neural crest cells, the pig-

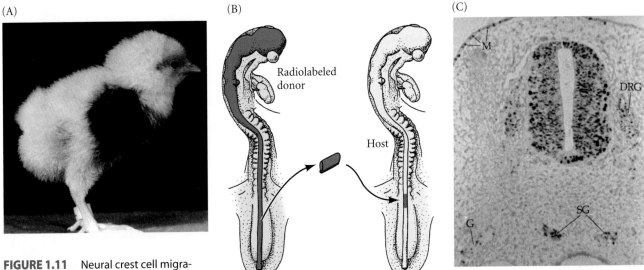

FIGURE 1.11 Neural crest cell migration. (A) Chick resulting from the transplantation of a trunk neural crest region from an embryo of a pigmented strain of chickens into the same region of an embryo of an unpigmented strain. The neural crest cells that gave rise to the pigment migrated into the wing epidermis and feathers. (B) Technique for following neural crest cells using radioactive tissue. (C) Autoradiograph showing locations of neural crest cells that have migrated from the radioactive donor cells. These cells form the pigment-forming melanocytes (M), sympathetic neural ganglia (SG), dorsal root ganglia (DRG), and glial cells (G). (A, original photograph from the archives of B. H. Willier; B after Weston 1963; C courtesy of J. Weston.)

ment of the retina formed in the retina itself and was not dependent on the migrating neural crest cells. By using radioactive marking techniques, Weston (1963) demonstrated that the migrating neural crest cells gave rise to the melanocytes, and also to the peripheral neurons and the epinephrine-secreting adrenal medulla (Figure 1.11B,C). This pattern was confirmed in chick-quail hybrids, in which the quail neural crest cells produced their own pigment and pattern in the chick feathers. More recently, fluorescent dye labeling has followed the movements of individual neural crest cells as they form their pigment, adrenal, and neuronal lineages (see Chapter 13).

In addition to the travels of pigment cells, other wide-scale migrations include those of the primordial germ cells (which migrate from the endoderm to the gonads, where they form the sperm and eggs) and the blood cell precursors (which in vertebrates undergo several migrations to colonize the liver and bone marrow).

Evolutionary Embryology

Charles Darwin's theory of evolution restructured comparative embryology and gave it a new focus. After reading Johannes Müller's summary of von Baer's laws in 1842, Darwin saw that embryonic resemblances would be a very strong argument in favor of the genetic connectedness of different animal groups. "Community of embryonic structure reveals community of descent," he would conclude in *On the Origin of Species* in 1859.

Even before Darwin, larval forms had been used for taxonomic classification. J. V. Thompson, for instance, had demonstrated that larval barnacles were almost identical to larval crabs, and he therefore counted barnacles as arthropods, not molluscs (Figure 1.12; Winsor 1969). Darwin, an expert on barnacle taxonomy, celebrated this finding: "Even the illustrious Cuvier did not perceive that a barnacle is a crustacean, but a glance at the larva shows this in an unmistakable manner." Darwin's evolutionary interpretation of von Baer's laws established a paradigm

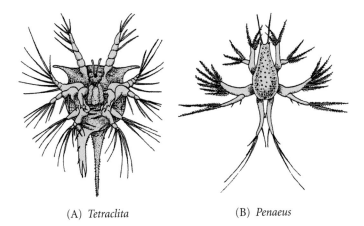

(A) *Tetraclita*   (B) *Penaeus*

FIGURE 1.12 Nauplius larvae of (A) a barnacle (*Tetraclita*, seen in ventral view) and (B) a shrimp (*Penaeus*, seen in dorsal view). The shrimp and barnacle share a similar larval stage despite their radical divergence in later development. (After Müller 1864.)

that was to be followed for many decades, namely, that relationships between groups can be discovered by finding common embryonic or larval forms. Kowalevsky (1871) would make a similar type of discovery (publicized in Darwin's *The Descent of Man*, 1874) that tunicate larvae have notochords and pharyngeal pouches, and that they form their neural tubes and other organs in a manner very similar to that of the primitive chordate *Amphioxus*. The tunicates, another enigma of classification schemes (formerly placed, along with barnacles, among the molluscs), thereby found a home with the chordates.

Darwin also noted that embryonic organisms sometimes make structures that are inappropriate for their adult form but which show their relatedness to other animals. He pointed out the existence of eyes in embryonic moles, pelvic rudiments in embryonic snakes, and teeth in embryonic baleen whales.

Darwin also argued that adaptations that depart from the "type" and allow an organism to survive in its particular environment develop late in the embryo.* He noted that the differences between species within genera become greater as development persists, as predicted by von Baer's laws. Thus, Darwin recognized two ways of looking at "descent with modification." One could emphasize the common descent by pointing out embryonic similarities between two or more groups of animals, or one could emphasize the modifications by showing how development was altered to produce structures that enabled animals to adapt to particular conditions.

Embryonic homologies

One of the most important distinctions made by the evolutionary embryologists was the difference between analogy and homology. Both terms refer to structures that appear to be similar. **Homologous** structures are those organs whose underlying similarity arises from their being derived from a common ancestral structure. For example, the wing of a bird and the forelimb of a human are homologous. Moreover, their respective parts are homologous (Figure 1.13). **Analogous** structures are those whose similarity comes from their performing a similar function, rather than their arising from a common ancestor. For example, the wing of a butterfly and the wing of a bird are analogous; the two types of wings share a common function (and therefore both are called wings), but the bird wing and insect wing did not arise from a common original ancestral structure that became modified through evolution into bird wings and butterfly wings.

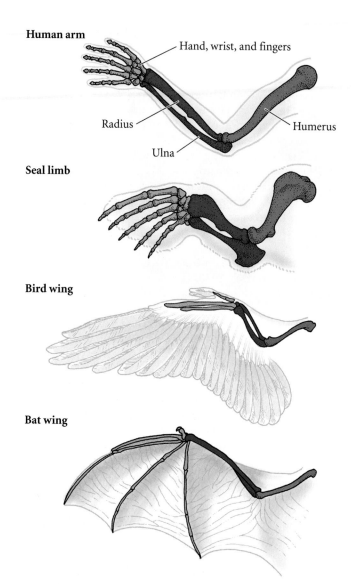

FIGURE 1.13 Homologies of structure among a human arm, a seal forelimb, a bird wing, and a bat wing; homologous supporting structures are shown in the same color. All four are homologous as forelimbs and were derived from a common tetrapod ancestor. The adaptations of bird and bat forelimbs to flight, however, evolved independently of each other, after the two lineages diverged from their common ancestor. Therefore, as wings they are not homologous, but analogous.

Homologies must be made carefully and must always refer to the level of organization being compared. For instance, the bird wing and the bat wing are homologous as forelimbs, but not as wings. In other words, they share a common underlying structure of forelimb bones because birds and mammals share a common ancestry. However, the bird wing developed independently from the bat wing. Bats descended from a long line of nonwinged mammals, and the structure of the bat wing is markedly different from that of a bird's wing.

One of the most celebrated cases of embryonic homology is that of the fish gill cartilage, the reptilian jaw, and the mammalian middle ear (reviewed in Gould 1990). In all jawed vertebrates, including fish, the first pharyngeal arch

---

*Moreover, as first noted by Weismann (1875), larvae must have their own adaptations to help them survive. The adult viceroy butterfly mimics the monarch butterfly, but the viceroy caterpillar does not resemble the beautiful larva of the monarch. Rather, the viceroy larva escapes detection by resembling bird droppings (Begon et al. 1986).

(see Figure 1.3) generates the jaw apparatus. The neural crest cells of this arch migrate to form Meckel's cartilage, the precurser of the jaw. In amphibians, reptiles, and birds, the posterior portion of this cartilage forms the quadrate bone of the upper jaw and the articular bone of the lower jaw. These bones connect to each other and are responsible for articulating the upper and lower jaws. However, in mammals, this articulation occurs at another region (the dentary and squamosal bones), thereby "freeing" these bony elements to acquire new functions.

The quadrate bone of the reptilian upper jaw evolved into the mammalian incus bone of the middle ear, and the articular bone of the reptile's lower jaw has become our malleus (Goodrich 1930; Wang et al. 2001). This latter process was first described by Reichert in 1837, when he observed in the pig embryo that the mandible (jawbone) ossifies on the side of Meckel's cartilage, while the posterior region of Meckel's cartilage ossifies, detaches from the rest of the cartilage, and enters the region of the middle ear to become the malleus.

But the story does not end here. The upper portion of the second embryonic arch supporting the gill became the hyomandibular bone of jawed fishes. This element supports the skull and links the jaw to the cranium, and may have functioned as part of the breathing apparatus (Figure 1.14A). As vertebrates moved onto land, they had a new problem: how to hear in a medium as thin as air. The hyomandibular bone happens to be near the otic (ear) capsule, and bony material is excellent for transmitting sound. Thus, while still functioning as a cranial brace, the hyomandibular bone of land-dwelling vertebrates also began functioning as a sound transducer (Figure 1.14B; Clack 1989; Brazeau and Ahlberg 2006). As the terrestrial vertebrates altered their locomotion, jaw structure, and posture, the cranium became firmly attached to the rest of the skull and did not need the hyomandibular brace. The hyomandibular bone then seems to have become specialized into the stapes bone of the middle ear (Figure 1.14C). What had been this bone's secondary function (auditory conduction) evolved to become its primary function.

Thus, the middle ear bones of the mammal are homologous to the posterior lower jaw of the reptile and to the gill arches of the fish. Chapter 23 will detail more recent information concerning the relationship of development to evolution.

Medical Embryology and Teratology

While embryologists could look at embryos to describe the evolution of life and how different animals form their organs, physicians became interested in embryos for more practical reasons. Between 2 and 5 percent of human infants are born with a readily observable anatomical abnormality (Thorogood 1997). These abnormalities may include missing limbs, missing or extra digits, cleft palate, eyes that lack certain parts, hearts that lack valves, and so forth. Physicians need know the causes of these birth defects in order to counsel parents as to the risk of having another malformed infant. In addition, the study of birth defects can tell us how the human body is normally formed. In the absence of experimental data on human embryos, we often must rely on nature's "experiments" to learn how the human body becomes organized.* Some birth defects are produced by mutant genes or chromosomes, and some are produced by environmental factors that impede development.

---

*The word "monster," used frequently in textbooks prior to the mid-twentieth century to describe malformed infants, comes from the Latin *monstrare*, "to show or point out." This is also the root of the English word "demonstrate." It was realized by Meckel (of jaw cartilage fame) that syndromes of congenital anomalies demonstrated certain principles about normal development. Parts of the body that were affected together must have some common developmental origin or mechanism that was being affected.

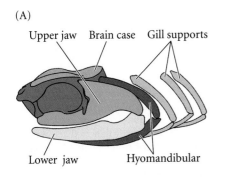

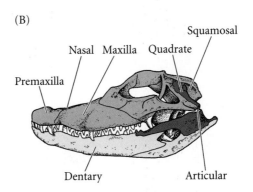

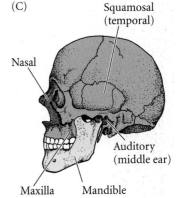

FIGURE 1.14   Jaw structure in the fish, reptile, and mammal. (A) Homologies of the jaws and gill arches as seen in the skull of the Paleozoic shark *Cobeledus aculentes*. (B) Lateral view of an alligator skull. The articular portion of the lower jaw articulates with the quadrate bone of the skull. (C) Lateral view of a human skull, showing the junction of the lower jaw with the squamosal (temporal) region of the skull. In mammals, the quadrate becomes internalized to form the incus of the middle ear. The articular bone retains its contact with the quadrate, becoming the malleus of the middle ear. (A after Zangerl and Williams 1975.)

Genetic malformations and syndromes

Abnormalities caused by genetic events (gene mutations, chromosomal aneuploidies, and translocations) are called **malformations**. Malformations often appear as **syndromes** (Greek, "running together"), in which several abnormalities occur concurrently. For instance, a human malformation called piebaldism, shown in Figure 1.15A, is due to a dominant mutation in a gene (*KIT*) on the long arm of chromosome 4 (Spritz et al. 1992). The piebald syndrome includes anemia, sterility, unpigmented regions of the skin and hair, deafness, and the absence of the nerves that cause peristalsis in the gut. The common feature underlying these conditions is that the *KIT* gene encodes a protein that is expressed in the neural crest cells and in the precursors of blood cells and germ cells. The Kit protein enables these cells to proliferate. Without this protein, the neural crest cells—which generate the pigment cells, certain ear cells, and the gut neurons—do not multiply as extensively as they should (resulting in underpigmentation, deafness, and gut malformations), nor do the precursors of the blood cells (resulting in anemia) or the germ cells (resulting in sterility).

Developmental biologists and clinical geneticists often study human syndromes (and determine their causes) by studying animals that display the same syndrome. These are called **animal models** of the disease; the mouse model for piebaldism is shown in Figure 1.15B. It has a phenotype very similar to that of the human condition, and it is caused by a mutation in the *Kit* gene of the mouse.*

Disruptions and teratogens

Abnormalities caused by exogenous agents (certain chemicals or viruses, radiation, or hyperthermia) are called **disruptions**. The agents responsible for these disruptions are called **teratogens** (Greek, "monster-formers"), and the study of how environmental agents disrupt normal development is called **teratology**. Teratogens were brought to the attention of the public in the early 1960s. In 1961, Lenz and McBride independently accumulated evidence that the drug thalidomide, prescribed as a mild sedative to many pregnant women, caused an enormous increase in a previously rare syndrome of congenital anomalies. The most noticeable of these anomalies was phocomelia, a condition in which the long bones of the limbs are deficient or absent (Figure 1.16A). Over 7000 affected infants were born to women who took thalidomide, and a woman need only have taken one tablet to produce children with all four limbs deformed (Lenz 1962, 1966; Toms 1962). Other abnormalities induced by the ingestion of this drug included heart defects, absence of the external ears, and malformed intestines.

Nowack (1965) documented the period of susceptibility during which thalidomide caused these abnormalities. The drug was found to be teratogenic only during days 34–50 after the last menstruation (20–36 days postconception). The specificity of thalidomide action is shown in Figure 1.16B. From day 34 to day 38, no limb abnormalities are seen. During this period, thalidomide can cause the absence or deficiency of ear components. Malformations of upper limbs are seen before those of the lower

(A)

(B)

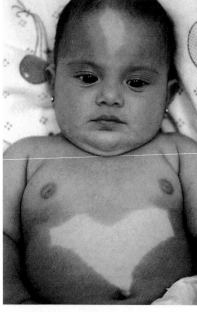 

FIGURE 1.15   Developmental anomalies caused by genetic mutation. (A) Piebaldism in a human infant. This genetically produced condition results in sterility, anemia, and underpigmented regions of the skin and hair, along with defective development of gut neurons and the ear. Piebaldism is caused by a mutation in the *KIT* gene. The Kit protein is essential for the proliferation and migration of neural crest cells, germ cell precursors, and blood cell precursors. (B) A piebald mouse with a mutation of the *Kit* gene. Mice provide important models for studying human developmental diseases. (Photographs courtesy of R. A. Fleischman.)

*The mouse *Kit* and human *KIT* genes are considered homologous by their structural similarities and their presumed common ancestry. Human genes are usually italicized and written in all capitals. Mouse genes are italicized, but only the first letter is usually capitalized. Gene products—proteins—are not italicized. If the protein has no standard biochemical or physiological name, it is usually represented with the name of the gene in Roman type, with the first letter capitalized. These rules are frequently bent, however. One is reminded of Cohen's (1982) dictum that "Academicians are more likely to share each other's toothbrush than each other's nomenclature."

(A)

(B)

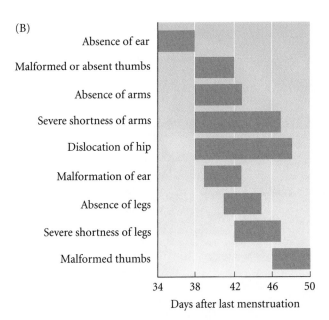

FIGURE 1.16 Developmental anomalies caused by an environmental agent. (A) Phocomelia, the lack of proper limb development, was the most visible of the birth defects that occurred in many children whose mothers took the drug thalidomide during pregnancy. (B) Thalidomide disrupts different structures at different times of human development. (Photograph © Deutsche Presse/Archive Photos; B after Nowack 1965.)

limbs, since the arms form slightly before the legs during development. The only animal models for thalidomide, however, are primates, and we still do not know for certain the mechanisms by which this drug causes human developmental disruptions (although it may work by blocking certain molecules from the developing mesoderm, thus preventing blood vessel development). Thalidomide was withdrawn from the market in November of 1961, but it is beginning to be prescribed again (although not to pregnant women), as a potential anti-tumor and anti-autoimmunity drug (Raje and Anderson 1999).

The integration of anatomical information about congenital malformations with our new knowledge of the genes responsible for development has had a revolutionary effect and is currently restructuring medicine. This integration is allowing us to discover the genes responsible for inherited malformations, and it permits us to identify the steps in development that are being disrupted by teratogens. We will see examples of this integration throughout this text, and Chapter 21 will detail some of the remarkable new discoveries in human teratology.

Mathematical Modeling of Development

Developmental biology has been described as the last domain of the mathematically incompetent scientist. This refuge, however, is not going to last. While most embryol-

ogists have been content trying to analyze specific instances of development or even formulating general principles of embryology, some researchers are now seeking quantifiable laws of development. The goal of these investigators is to base embryology on formal mathematical or physical principles (see Held 1992; Webster and Goodwin 1996; Salazar-Ciudad et al. 2000, 2001).

Furthermore, studying the formation of epithelial tubes, folding sheets, and interacting surfaces requires knowledge of the tensile strength, elasticity, shear forces, and curvature energies of the tissues involved. Anatomy is the result of "viscoelastic" soft matter, and the physical properties of embryonic structures are just beginning to be studied and appreciated in this light (see Forgacs and Newman 2005; Nanjundiah 2005). Pattern formation and growth are two areas in which mathematical modeling has given biologists insights into some underlying laws of physics as they relate to animal development.

The mathematics of organismal growth

Most animals grow by increasing their volume while retaining their proportions. Theoretically, an animal that increases its weight (volume) twofold will increase its length only 1.26 times (i.e., $1.26^3 = 2$). W. K. Brooks (1886) observed that this ratio was frequently seen in nature, and he noted that the deep-sea arthropods collected by the *Challenger* expedition increased about 1.25 times between molts. In 1904, Przibram and his colleagues performed a detailed study of mantises and found that the increase of size between molts was almost exactly 1.26 (see Przibram 1931). Even the hexagonal facets of the arthropod eye (which grow by cell expansion, not by cell division) increased by that ratio.

(A)

(B)

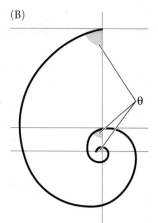

FIGURE 1.17 Equiangular spiral growth patterns. (A) A ram's horn and the shell of a chambered nautilus both show equiangular spiral growth. The nautilus shell is cut in cross section. (B) René Descartes' analysis of an equiangular spiral, showing that if the curve cuts each radius vector at a constant angle (symbolized by θ), then the curve grows continuously without ever changing its shape. (A from the author's collection; B after Thompson 1942.)

D'Arcy Thompson (1942) similarly showed that the spiral growth of shells (and fingernails) can be expressed mathematically as $r = a^\theta$, and that the ratio of the widths between two whorls of a shell can be calculated by the formula $r = e^{2\pi\cot\theta}$ (Figure 1.17 and Table 1.2). Thus, if a whorl were 1 inch in breadth at one point on a radius and the angle of the spiral were 80°, the next whorl would have a width of 3 inches on the same radius. Most gastropod (snail) and nautiloid molluscs have an angle of curvature between 80° and 85°.* Lower-angle curvatures are seen in some shells (mostly bivalves) and are common in teeth and claws.

Such growth, in which the shape is preserved because all components grow at the same rate, is called **isometric growth**. In many organisms, however, growth is not a uniform phenomenon. It is obvious that there are some periods in an organism's life during which growth is more rapid than in others. Physical growth during the first 10 years of a person's existence is much more dramatic than in the 10 years following one's graduation from college. Moreover, not all parts of the body grow at the same rate. This phenomenon of the different growth rates of parts within the same organism is called **allometric growth** (or **allometry**). Human allometry is depicted in Figure 1.18. Our arms and legs grow at a faster rate than our torso and head, such that adult proportions differ markedly from those of infants. Julian Huxley (1932) likened allometry to putting money in the bank at two different continuous interest rates.

---

*If the angle were 90°, the shell would form a circle rather than a spiral, and growth would cease. If the angle were 60°, however, the next whorl would be 4 feet on that radius, and if the angle were 17°, the next whorl would occupy a distance of some 15,000 miles!

TABLE 1.2 Constant angle of an equiangular spiral and the ratio of widths between whorls

Constant angle	Ratio of widths[a]
90°	1.0
89°8′	1.1
86°18′	1.5
83°42′	2.0
80°5′	3.0
75°38′	5.0
69°53′	10.0
64°31′	20.0
58°5′	50.0
53°46′	10^2
42°17′	10^3
34°19′	10^4
28°37′	10^5
24°28′	10^6

Source: From Thompson 1942.
[a]The ratio of widths is calculated by dividing the width of one whorl by the width of the next larger whorl.

The formula for allometric growth (or for comparing moneys invested at two different interest rates) is $y = bx^{a/c}$, where x is the initial body size, a and c are the growth rates of the two body parts, and b is the value of y when $x = 1$. If $a/c > 1$, then that part of the body represented by a is growing faster than that part of the body represented by c. In logarithmic terms (which are much easier to graph), $\log y = \log b + (a/c)\log x$.

One of the most vivid examples of allometric growth is seen in the male fiddler crab, *Uca pugnax*. In small males, the two claws are of equal weight, each constituting about 8 percent of the crab's total weight. As the crab grows larger, its chela (the large crushing claw) grows even more rapidly, eventually constituting about 38 percent of the crab's weight (Figure 1.19). When these data are plotted on double logarithmic plots (with the body mass on the x axis and

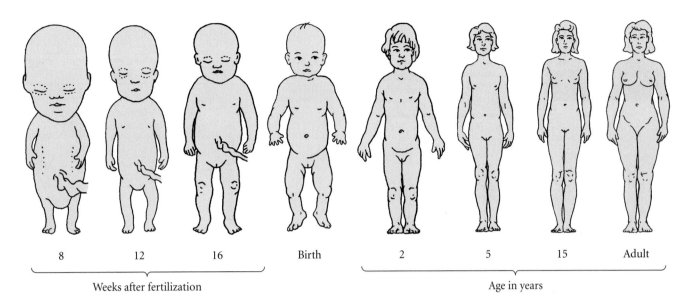

8   12   16   Birth

Weeks after fertilization

2   5   15   Adult

Age in years

FIGURE 1.18 Allometry in humans. The embryo's head is exceedingly large in proportion to the rest of the body. After the embryonic period, the head grows more slowly than the torso, hands, and legs. Human allometry has been represented in Western art only since the Renaissance. Before that, children were depicted as small adults. (After Moore 1983.)

chela mass on the *y* axis), one obtains a straight line whose slope is the *a/c* ratio. In the male *Uca pugnax*,* the *a/c* ratio is 6:1. This means that the mass of the chela increases 6 times faster than the mass of the rest of the body. In females of the species, the claw remains about 8 percent of the body weight throughout growth. It is only in the males (who use the claw for defense and display) that this allometry occurs.

Recent models of isometric and allometric growth have taken into account metabolic rates, life history changes, and cell death rates (see West et al. 2001). The relationship of growth to physical and genetic parameters and the coordination of growth rates throughout the organism remains a fascinating area that unifies development with physiology and medicine.

The mathematics of patterning

One of the most important mathematical models in developmental biology was formulated by Alan Turing (1952), one of the founders of computer science (and the mathematician who cracked the German "Enigma" code during World War II). He proposed a model wherein two homogeneously distributed solutions would interact to produce stable patterns during morphogenesis. These patterns would represent regional differences in the concentrations

of the two substances. Their interactions would produce an ordered structure out of random chaos.

Turing's **reaction-diffusion model** involves two substances. Substance P promotes the production of more substance P as well as substance S. Substance S, however, inhibits the production of substance P. Turing's mathematics show that if S diffuses more readily than P, sharp waves of concentration differences will be generated for substance

FIGURE 1.19 Male specimens of the fiddler crab, *Uca pugnax*. Allometric growth occurs only in one of the male's claws. In females (not shown), both claws retain isometric growth. (Photograph courtesy of Swarthmore College Marine Biology Laboratory.)

---

*The crab is named for the male's huge claw; *pugnax* is Latin for "fighter" or "boxer."

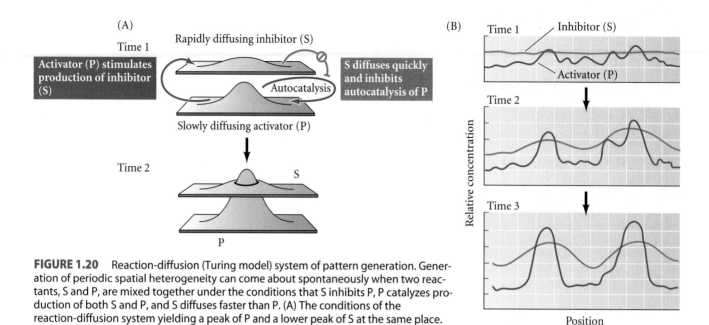

(A)

Time 1

Rapidly diffusing inhibitor (S)

Activator (P) stimulates production of inhibitor (S)

Autocatalysis

S diffuses quickly and inhibits autocatalysis of P

Slowly diffusing activator (P)

Time 2

S

P

(B)

Time 1

Inhibitor (S)

Activator (P)

Time 2

Time 3

Relative concentration

Position

FIGURE 1.20 Reaction-diffusion (Turing model) system of pattern generation. Generation of periodic spatial heterogeneity can come about spontaneously when two reactants, S and P, are mixed together under the conditions that S inhibits P, P catalyzes production of both S and P, and S diffuses faster than P. (A) The conditions of the reaction-diffusion system yielding a peak of P and a lower peak of S at the same place. (B) The distribution of the reactants is initially random, and their concentrations fluctuate over a given average. As P increases locally, it produces more S, which diffuses to inhibit more peaks of P from forming in the vicinity of its production. The result is a series of P peaks ("standing waves") at regular intervals.

P (Figure 1.20). These waves have been observed in certain chemical reactions (Prigogine and Nicolis 1967; Winfree 1974).

The reaction-diffusion model predicts alternating areas of high and low concentrations of some substance. When the concentration of such a substance is above a certain threshold level, a cell (or group of cells) may be instructed to differentiate in a certain way. An important feature of Turing's model is that particular chemical wavelengths will be amplified while all others will be suppressed. As local concentrations of P increase, the values of S form a peak centering on the P peak, but becoming broader and shallower because of S's more rapid diffusion. These S peaks inhibit other P peaks from forming. But which of the many P peaks will survive? That depends on the size and shape of the tissues in which the oscillating reaction is occurring. (This pattern is analogous to the harmonics of vibrating strings, as in a guitar. Only certain resonance vibrations are permitted, based on the boundaries of the string.)

The mathematics describing which particular wavelengths are selected consist of complex polynomial equations. Such functions have been used to model the spiral patterning of slime molds, the polar organization of the limb, and the pigment patterns of mammals, fish, and snails (Figure 1.21; Kondo and Asai 1995; Meinhardt 2003). A computer simulation based on a Turing reaction-diffusion system can successfully predict such pat-

terns, given the starting shapes and sizes of the elements involved.

Pigment patterning isn't the only property that can make use of reaction-diffusion mechanisms to produce

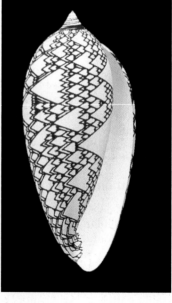

FIGURE 1.21 Photograph of the snail *Oliva porphyria* (left), and a computer model (right) in which the growth parameters of the snail's shell and its pigmentation pattern were mathematically generated. (From Meinhardt 2003; computer image courtesy of D. Fowler, P. Prusinkiewicz, and H. Meinhardt.)

(A)
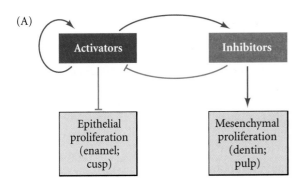

(B)
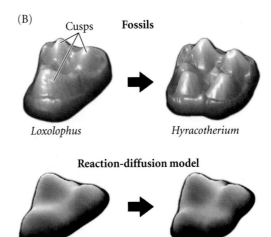

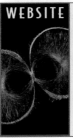

FIGURE 1.22 Mammalian tooth cusp pattern modeled by reaction-diffusion equations. (A) The reaction-diffusion mechanism serves as a motor regulating the genes responsible for slowing the growth of the enamel-forming cells and promoting the proliferation of the dentin pulp-forming cells. (B) Possible formation of the four-cusped tooth of *Hyracotherium* (a fossil horse from 55 million years ago) from the three-cusped tooth of *Loxolophus*, a mammal that may be an earlier member of the horse lineage. This transition in tooth shape can be achieved by modifying a single parameter of the reaction-diffusion equation. (Photographs courtesy of J. Jernvall).

ordered changes. The molecules of the reaction-diffusion network can be linked to those genes controlling cell proliferation (Figure 1.22A). Salazar-Ciudad and Jernvall (2002, 2003) have shown that mammalian tooth development can be analyzed using the reaction-diffusion equations, wherein the activator also acts to inhibit the proliferation of the epithelial cells (which form the tooth enamel), while the inhibitor can increase proliferation of the mesenchymal cells that form the pulp (dentin). During tooth development, the same cells make both the activator and the inhibitor. The points at which the outer cells first cease to proliferate will become cusps; it is the number and placement of cusps that characterize the teeth of the different mammalian groups. Not only can this model predict where tooth cusps form, but by changing the rates of diffusion or reaction, it demonstrates how one type of tooth might evolve into another (Figure 1.22B).

WEBSITE 1.4 The mathematical background of pattern formation. The equations modeling pattern formation are a series of partial derivatives depicting rates of synthesis, degradation, and diffusion of the activator and inhibitor molecules.

WEBSITE 1.5 How do zebras (and angelfish and mice) get their stripes? No one knows for sure, but adding the Turing equations to what's known about equine embryology allows one to model how each of the three known zebra species acquired its unique striping pattern. Similarly, changing some of the parameters of the Turing reaction enables one to predict the different pigmentation patterns of angelfish and some mice.

Coda

The anatomical approach to development is being expanded and enhanced by revolutions in microscopy, computer-aided graphical reconstructions of three-dimensional objects, and methods of applying mathematics to biology. Many of the beautiful photographs in this textbook reflect this increasingly important component of embryology.

 **Snapshot Summary** **Developmental Anatomy**

1. Organisms must function as they form their organs. They have to use one set of structures while constructing others.

2. The main question of development is, How does the egg becomes an adult? This question can be broken down into the component problems of differentiation (How do cells become different from one another and from their precursors?), morphogenesis (How is ordered form generated?), growth (How is size regulated?), reproduction (How does one generation create another generation?), evolution (How do changes in the developmental processes create new anatomical structures?), and environment (How is the developing organism affected by the physical and chemical conditions of the external environment?).

3. Epigenesis happens. New organisms are created *de novo* each generation from the relatively disordered cytoplasm of the egg.

4. Preformation is not found in the anatomical structures themselves, but in the genetic instructions that instruct their formation. The inheritance of the fertilized egg includes the genetic potentials of the organism. These preformed nuclear instructions include the ability to respond to environmental stimuli in specific ways.

5. The three germ layers give rise to specific organ systems. The ectoderm gives rise to the epidermis, nervous system, and pigment cells; the mesoderm generates the kidneys, gonads, muscle, bones, heart, and blood cells; and the endoderm forms the lining of the digestive tube and the respiratory system.

6. Karl von Baer's principles state that the general features of a large group of animals appear earlier in the embryo than do the specialized features of a smaller group. As each embryo of a given species develops, it diverges from the adult forms of other species. The early embryo of a "higher" animal species is not like the adult of a "lower" animal.

7. Labeling cells with dyes shows that some cells differentiate where they form, while others migrate from their original sites and differentiate in their new locations. Migratory cells include neural crest cells and the precursors of germ cells and blood cells.

8. "Community of embryonic structure reveals community of descent" (Charles Darwin, *On the Origin of Species*).

9. Homologous structures in different species are those organs whose similarity is due to their sharing a common ancestral structure. Analogous structures are those organs whose similarity comes from their serving a similar function (but which are not derived from a common ancestral structure).

10. Congenital anomalies can be caused by genetic factors (mutations, aneuploidies, translocations) or by environmental agents (certain chemicals, certain viruses, radiation).

11. Syndromes consist of sets of developmental abnormalities that "run together."

12. Organs that are linked in developmental syndromes share either a common origin or a common mechanism of formation.

13. If growth is isometric (i.e., all body parts grow at the same rate), a twofold change in weight will cause a 1.26-fold expansion in length.

14. Allometric growth (when different body parts grow at different rates) can create dramatic changes in the structure of organisms.

15. Complex patterns may be self-generated by reaction-diffusion events, wherein the activator of a local phenomenon stimulates the production of more of itself as well as the production of a more diffusible inhibitor.

For Further Reading

Complete bibliographical citations for all literature cited in this chapter can be found on the Vade Mecum CD that accompanies the book and at the free access website www.devbio.com

Kondo, S. and R. Asai. 1995. A reaction-diffusion wave on the skin of the marine angelfish *Pomacanthus*. *Nature* 376: 765–768.

Larsen, E. and McLaughlin, H. 1987. The morphogenetic alphabet: Lessons for simple-minded genes. *BioEssays* 7: 130–132.

Le Douarin, N. M. and M.-A. Teillet. 1973. The migration of neural crest cells to the wall of the digestive tract in the avian embryo. *J. Embryol. Exp. Morphol.* 30: 31–48.

Nishida, H. 1987. Cell lineage analysis in ascidian embryos by intracellular injection of a tracer enzyme. III. Up to the tissue-restricted stage. *Dev. Biol.* 121: 526–541.

Salazar-Ciudad, I. and J. Jernvall. 2003. How different types of pattern formation mechanisms affect the evolution of form and development. *Evol. Dev.* 6: 6–16.

Woo, K. and S. E. Fraser. 1995. Order and coherence in the fate map of the zebrafish embryo. *Development* 121: 2595–2609.

Life Cycles and the Evolution of Developmental Patterns

2

TRADITIONAL WAYS OF DEPICTING ANIMALS catalog them according to their adult structure. But, as J. T. Bonner (1965) pointed out, this is very artificial because what we consider an individual is usually just a brief slice of its life cycle. When we consider a dog, for instance, we usually picture an adult. But the dog is a "dog" from the moment of fertilization of a dog egg by a dog sperm. It remains a dog even as a senescent, dying hound. As Bonner phrased it, animals do not *have* a life cycle, they *are* a life cycle.

The life cycle has to be adapted to its abiotic (nonliving) environment, as well as being synchronized with the life cycles of other organisms. Take, for example, *Clunio marinus,* a small fly that inhabits tidal waters along the coast of western Europe. Females of this species live only 2–3 hours as adults, and within this short space of time they must mate and lay their eggs. To make matters even more precarious, they must lay their eggs on red algal mats that are exposed only during the lowest ebbing of the spring tide. Such low tides occur on four successive days shortly after the new and full moons (i.e., at about 15-day intervals). Therefore, the life cycle of these insects must coordinate with the lunar cycle as well as the daily tidal rhythms such that the flies emerge from their pupal cases during the few days of the spring tide *and* at the correct hour for its ebb (Beck 1980; Neumann and Spindler 1991).

The Circle of Life: The Stages of Animal Development

One of the major triumphs of descriptive embryology was the idea of a generalizable life cycle. Each animal, whether earthworm, eagle, or beagle, passes through similar stages of development. The life of a new individual is initiated by the fusion of genetic material from the two gametes—the sperm and the egg. This fusion, called **fertilization**, stimulates the egg to begin development. The stages of development between fertilization and hatching are collectively called **embryogenesis**. Throughout the animal kingdom, an incredible variety of embryonic types exist, but most patterns of embryogenesis are variations on five fundamental processes: cleavage, gastrulation, organogenesis, gametogenesis, and metamorphosis.

1. **Cleavage** is a series of extremely rapid mitotic divisions that immediately follows fertilization. During cleavage, the enormous volume of zygote cytoplasm is divided into numerous smaller cells called **blastomeres**. By the end of cleavage, the blastomeres have usually formed a sphere, known as a **blastula**.

2. After the rate of mitotic division slows down, the blastomeres undergo dramatic movements and change their positions relative to one another. This

series of extensive cell rearrangements is called **gastrulation**, and the embryo is said to be in the **gastrula** stage. As a result of gastrulation, the embryo contains three **germ layers**: the ectoderm, the endoderm, and the mesoderm.

3. Once the three germ layers are established, the cells interact with one another and rearrange themselves to produce tissues and organs. This process is called **organogenesis**. Many organs contain cells from more than one germ layer, and it is not unusual for the outside of an organ to be derived from one layer and the inside from another. For example, the outer layer of skin (epidermis) comes from the ectoderm, while the inner layer (the dermis) comes from the mesoderm. Also during organogenesis, certain cells undergo long migrations from their place of origin to their final location. These migrating cells include the precursors of blood cells, lymph cells, pigment cells, and the gametes. Most of the bones of our face are derived from cells that have migrated ventrally from the dorsal region of the head.

4. In many species, the organism that hatches from the egg or is born into the world is not sexually mature. Rather, the organism needs to undergo **metamorphosis** to become a sexually mature adult. In most animals, the young organism is a called a **larva**, and it may look significantly different from the adult. In many species, the larval stage is the one that lasts the longest, and is used for feeding or dispersal. In such species, the adult is a brief stage whose sole purpose is to reproduce. In silkworm moths, for instance, the adults do not have mouthparts and cannot feed; the larvae must eat enough so that the adult has the stored energy to survive and mate. Indeed, most female moths mate as soon as they eclose from their pupa, and they fly only once—to lay their eggs. Then they die.

5. In many species, a specialized portion of egg cytoplasm gives rise to cells that are the precursors of the **gametes** (sperm and egg). The gametes and their precursor cells are collectively called **germ cells**, and they are set aside for reproductive function. All the other cells of the body are called **somatic cells**. This separation of somatic cells (which give rise to the individual body) and germ cells (which contribute to the formation of a new generation) is often one of the first differentiations to occur during animal development. The germ cells eventually migrate to the gonads, where they differentiate into gametes. The development of gametes, called **gametogenesis**, is usually not completed until the organism has become physically mature. At maturity, the gametes may be released and participate in fertilization to begin a new embryo. The adult organism eventually undergoes senescence and dies.

In this chapter, we will examine some representative life cycles and discuss some of the ways that these patterns of development evolved.

FIGURE 2.1   Developmental history of the leopard frog, *Rana* ▶ *pipiens*. The stages from fertilization through hatching (birth) are known collectively as embryogenesis. The region set aside for producing germ cells is shown in bright purple. Gametogenesis, which is completed in the sexually mature adult, begins at different times during development, depending on the species. (The sizes of the varicolored wedges shown here are arbitrary and do not correspond to the proportion of the life cycle spent in each stage.)

The frog life cycle

Figure 2.1 uses the development of the leopard frog, *Rana pipiens*, to show a representative life cycle. In most frogs, gametogenesis and fertilization are seasonal events, because the frog's life depends on the plants and insects in the pond where it lives and on the temperature of the air and water. A combination of photoperiod (hours of daylight) and temperature tells the pituitary gland of the female frog that it is spring. If the female is mature, her pituitary gland secretes hormones that stimulate her ovary to make the hormone estrogen. Estrogen then instructs the liver to make and secrete yolk proteins such as vitellogenin, which are then transported through the blood into the enlarging eggs in the ovary. The yolk is transported into the bottom portion of the egg (Figure 2.2A). The bottom half of the egg usually contains more yolk than the top half and is called the **vegetal hemisphere** of the egg. Conversely, the upper half of the egg usually has less yolk and is called the **animal hemisphere**.*

Sperm formation also occurs on a seasonal basis. Male leopard frogs make their sperm in the summer, and by the time they begin hibernation in autumn they have produced all the sperm that will be available for the following spring's breeding season. In most species of frogs, fertilization is external. The male frog grabs the female's back and fertilizes the eggs as the female frog releases them (Figure 2.2B). Some species lay their eggs in pond vegetation, and the egg jelly adheres to the plants and anchors the eggs (Figure 2.2C). Other species float their eggs into the center of the pond without any support.

Fertilization accomplishes several things. First, it allows the egg to complete its second meiotic division, which provides the egg with a haploid **pronucleus**. The egg pronucleus and the sperm pronucleus meet in the egg cytoplasm to form the diploid zygote **nucleus**. Second, fertilization causes the cytoplasm of the egg to move such that different parts of the cytoplasm find themselves in new locations (Figure 2.2D). Third, fertilization activates those molecules

---

*The terms *animal* and *vegetal* reflect the movements of cells seen in some embryos (including those of frogs). The cells derived from the upper portion of the egg divide more rapidly and are actively mobile (hence, "animated"), while the yolk-filled cells of the vegetal half are seen as being immobile (hence, like plants, or vegetal).

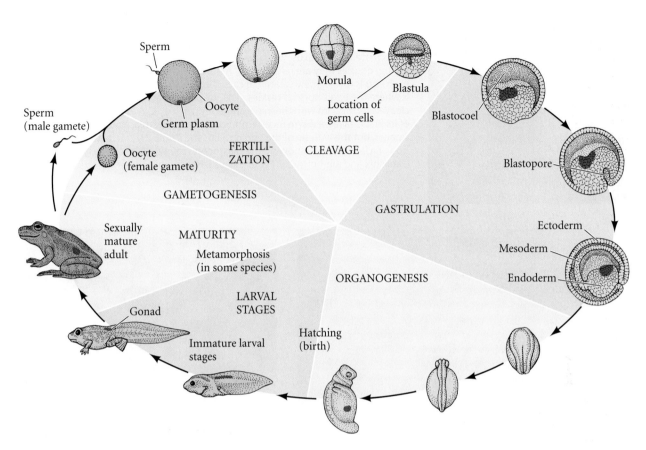

necessary to begin cell cleavage and development (Rugh 1950). The sperm and egg die quickly unless fertilization occurs.

During cleavage, the volume of the frog egg stays the same, but it is divided into tens of thousands of cells (Figure 2.2E–H). The cells in the animal hemisphere of the egg divide faster than those in the vegetal hemisphere, and the cells of the vegetal hemisphere become progressively larger the more vegetal the cytoplasm. Meanwhile, a fluid-filled cavity, the **blastocoel**, forms in the animal hemisphere (Figure 2.2I). This cavity will be important for allowing cell movements to occur during gastrulation.

Gastrulation in the frog begins at a point on the embryo surface roughly 180° opposite the point of sperm entry with the formation of a dimple, called the **blastopore**. This dimple (which will mark the future dorsal side of the embryo) expands to become a ring, and cells migrating through this blastopore ring become the mesoderm (Figure 2.3A–C). The cells remaining on the outside become the ectoderm, and this outer layer expands to enclose the entire embryo. The large, yolky cells that remain at the vegetal hemisphere (until they are encircled by the expanding ectoderm) become the endoderm. Thus, at the end of gastrulation, the ectoderm (the precursor of the epidermis and nerves) is on the outside of the embryo, the endoderm (the precursor of the gut lining) is on the inside of the embryo, and the mesoderm (the precursor of connective tissue, blood, skeleton, gonads, and kidneys) is between them.

Organogenesis begins when the notochord—a rod of mesodermal cells in the most dorsal portion of the embryo—signals the ectodermal cells above it that they are not going to become epidermis. Instead, these dorsal ectoderm cells form a tube and become the nervous system. At this stage, the embryo is called a **neurula**. The neural precursor cells elongate, stretch, and fold into the embryo, forming the **neural tube** (Figure 2.3D–F); the future epidermal cells of the back cover the neural tube.

Once the neural tube has formed, it induces changes in its neighbors, and organogenesis continues. The mesodermal tissue adjacent to the notochord becomes segmented into **somites** (Figure 2.3G,H), the precursors of the frog's back muscles, spinal vertebrae, and dermis (the inner portion of the skin). The embryo develops a mouth and an anus, and it elongates into the familiar tadpole structure (Figure 2.3I). The neurons make their connections to the muscles and to other neurons, the gills form, and the larva is ready to hatch from its egg jelly. The hatched tadpole will feed for itself as soon as the yolk supplied by its mother is exhausted.

Metamorphosis of the fully aquatic tadpole larva into an adult frog that can live on land is one of the most striking transformations in all of biology. In amphibians, metamorphosis is initiated by hormones from the tadpole's thyroid gland. (The mechanisms by which thyroid hormones accomplish these changes will be discussed in Chapter 18.) In anurans (frogs and toads), almost every organ is subject

(A)

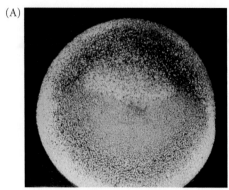

FIGURE 2.2 Early development of the frog *Xenopus laevis*. (A) As the egg matures, it accumulates yolk (here stained yellow and green) in the vegetal cytoplasm. (B) Frogs mate by amplexus, the male grasping the female around the belly and fertilizing the eggs as they are released. (C) A newly laid clutch of eggs. The brown area of each egg is the pigmented animal cap. The white spot in the middle of the pigment is where the egg's nucleus resides. (D) Cytoplasm rearrangement seen during first cleavage. Compare with the initial stage seen in (A). (E) A 2-cell embryo near the end of its first cleavage. (F) An 8-cell embryo. (G) Early blastula. Note that the cells get smaller, but the volume of the egg remains the same. (H) Late blastula. (I) Cross section of a late blastula, showing the blastocoel (cavity). (A–H courtesy of Michael Danilchik and Kimberly Ray; I courtesy of J. Heasman.)

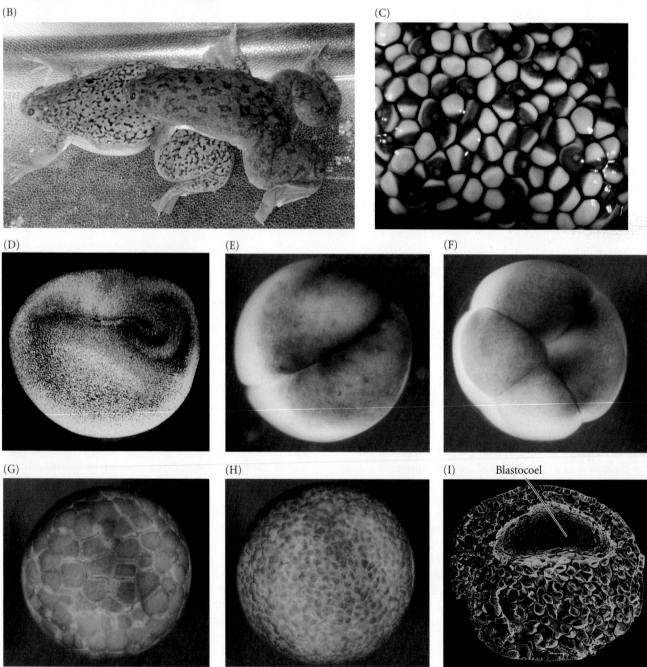

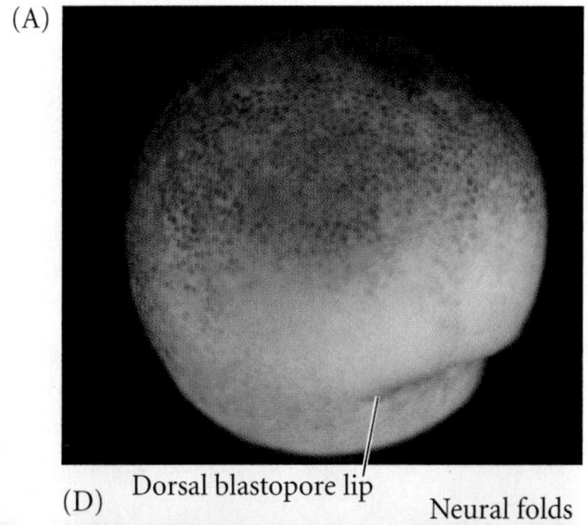

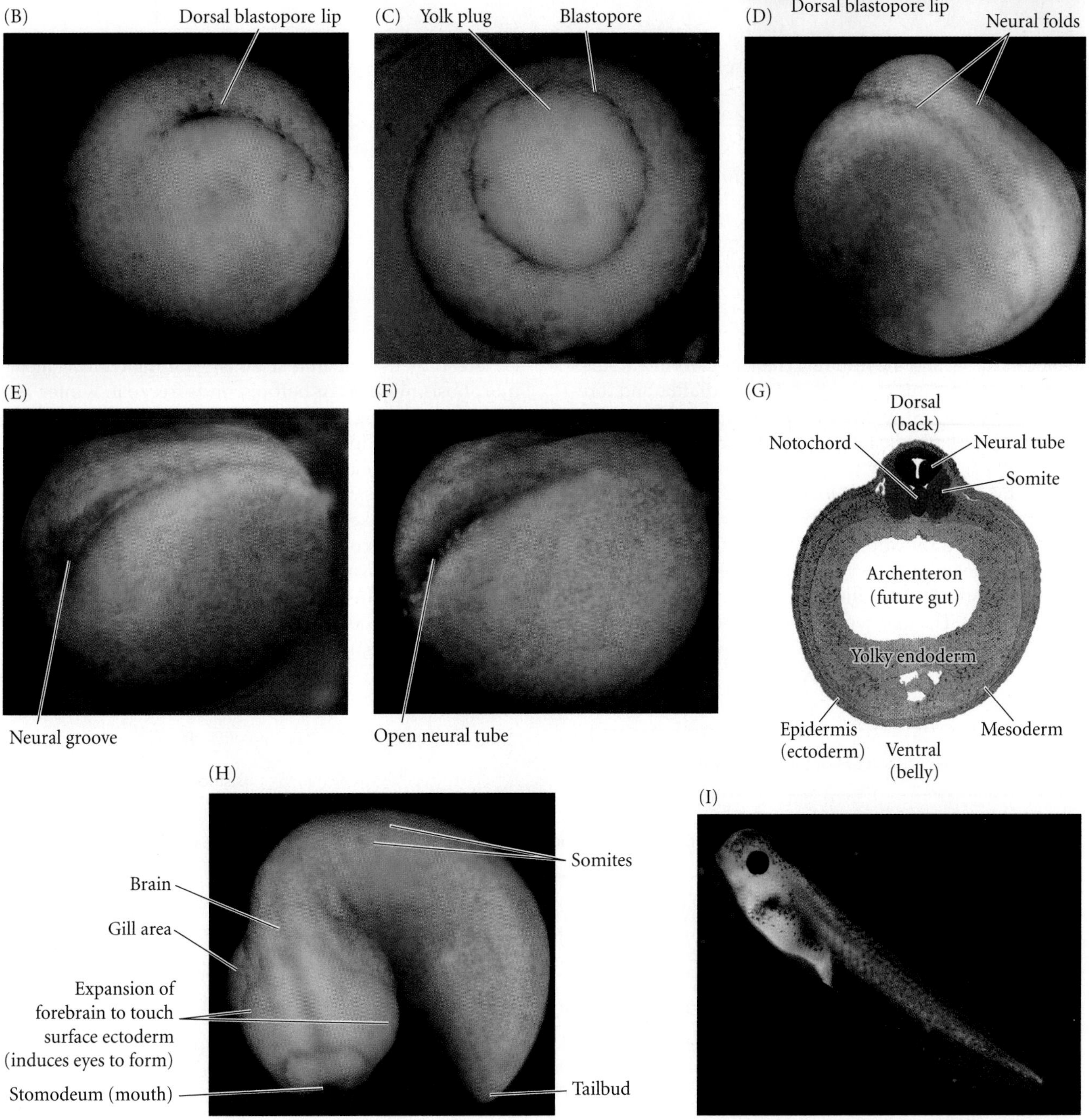

FIGURE 2.3 Continued development of *Xenopus laevis*. (A) Gastrulation begins with an invagination, or slit, in the future dorsal (top) side of the embryo. (B) This slit, the dorsal blastopore lip, as seen from the ventral surface (bottom) of the embryo. (C) The slit becomes a circle, the blastopore. Future mesoderm cells migrate into the interior of the embryo along the blastopore edges, and the ectoderm (future epidermis and nerves) migrates down the outside of the embryo. The remaining part, the yolk-filled endoderm, is eventually encircled. (D) Neural folds begin to form on the dorsal surface. (E) A groove can be seen where the bottom of the neural tube will be. (F) The neural folds come together at the dorsal midline, creating a neural tube. (G) Cross section of the *Xenopus* embryo at the neurula stage. (H) A pre-hatching tadpole, as the protrusions of the forebrain begin to induce eyes to form. (I) A mature tadpole, having swum away from the egg mass and feeding independently. (Photographs courtesy of Michael Danilchik and Kimberly Ray.)

(A)

(B)

(C)

(D)

(E)

(F)

FIGURE 2.4 Metamorphosis of the frog. (A) Huge changes are obvious when one contrasts the tadpole and the adult bullfrog. Note especially the differences in jaw structure and limbs. (B) Premetamorphic tadpole. (C) Prometamorphic tadpole, showing hindlimb growth. (D) Onset of metamorphic climax as forelimbs emerge. (E,F) Climax stages. (Photograph © Patrice Ceisel/Visuals Unlimited.)

to modification, and the resulting changes in form are striking and very obvious (Figure 2.4). The hindlimbs and forelimbs that the adult will use for locomotion differentiate as the tadpole's paddle tail recedes. The cartilaginous tadpole skull is replaced by the predominantly bony skull of the young frog. The horny teeth the tadpole uses to tear up pond plants disappear as the mouth and jaw take a new shape, and the fly-catching tongue muscle of the frog develops. Meanwhile, the large intestine characteristic of herbivores shortens to suit the more carnivorous diet of the adult frog. The gills regress and the lungs enlarge. The speed of metamorphosis is carefully keyed to environmen-

tal pressures. In temperate regions, for instance, *Rana* metamorphosis must occur before ponds freeze in winter. An adult leopard frog can burrow into the mud and survive the winter; its tadpole cannot.

As metamorphosis ends, the development of the first germ cells begins. Gametogenesis can take a long time. In *Rana pipiens*, it takes 3 years for the eggs to mature and for the female to become sexually mature. (Sperm take less time; *Rana* males are often fertile soon after metamorphosis.) In order to become sexually mature, the germ cells have to be competent to complete **meiosis**. Figure 2.5 reviews the steps of meiosis. The important things to real-

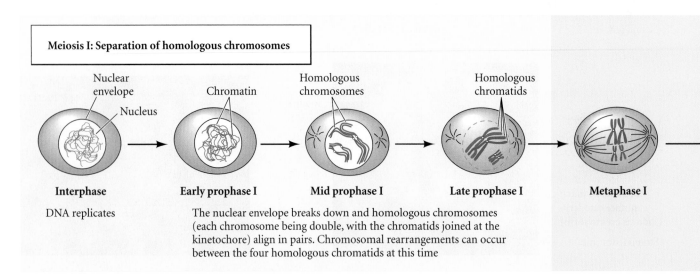

Meiosis I: Separation of homologous chromosomes

Nuclear envelope

Nucleus

Chromatin

Homologous chromosomes

Homologous chromatids

Interphase

Early prophase I

Mid prophase I

Late prophase I

Metaphase I

DNA replicates

The nuclear envelope breaks down and homologous chromosomes (each chromosome being double, with the chromatids joined at the kinetochore) align in pairs. Chromosomal rearrangements can occur between the four homologous chromatids at this time

ize about meiosis are that (1) the chromosomes replicate prior to cell division, so that each gene is represented four times; (2) the replicated chromosomes (each called a chromatid) are held together by their kinetochores (centromeres), and the four homologous chromatids pair together; (3) the first meiotic division separates the pairs from one another; (4) the second meiotic division splits the kinetochore such that each chromatid becomes a chromosome; and (5) the net result is four cells, each with a haploid nucleus. The mature sperm and egg nuclei can unite in fertilization, restoring the diploid nucleus and initiating the events that lead to development.

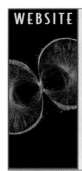

WEBSITE **2.2** **Protist differentiation.** Although containing only one cell, protists develop. These remarkable areas of protist development concern the (1) control of sex type in fission yeast, (2) the transformation of *Naegleria* amoebae into streamlined, flagellated cells, (3) cell-cell contact during *Chlamydomonas* mating, (4) nuclear and cytoplasmic control of *Acetabularia* development, (5) the cortical inheritance of the cell surface in paramecia, and (6) gene expression during *Dictyostelium* cell differentiation.

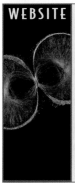

VADE MECUM[2] **The amphibian life cycle.** The life cycle is a dynamic process that is best seen as a movie. Here, frog life cycles are illustrated in labeled photographs and time-lapse videomicroscopy (as well as some great audio clips). [Click on Amphibian]

WEBSITE **2.1** **When does human life begin?** There is no consensus among developmental biologists, but different biologists have proposed that human life begins at (a) fertilization (when the new genome is formed), or (b) gastrulation (when the embryo becomes an individual and can no longer produce twins, or (c) with the acquisition of measurable brain activity (the loss of which is defined as death), or (d) the period close to or at birth (physiological independence). In addition, many scientists think the acquisition of human traits is gradual and cannot be located to one particular time.

Multicellularity: The Evolution of Differentiation and Morphogenesis

One of evolution's most important products was the multicellular organism. There appear to be several paths by which single cells evolved multicellular arrangements; we will discuss two of them here. One path involves the orderly division of the reproductive cell and the subsequent differentiation of its progeny into different cell types. This path to multicellularity can be seen among a group of multicellular organisms collectively referred to as the family Volvocaceae, or the volvocaceans (Kirk 1999, 2000).

The volvocaceans

The simpler organisms among the Volvocaceae are ordered assemblies of numerous cells, each resembling the unicel-

FIGURE 2.5 Summary of meiosis. The DNA replicates during interphase. During first meiotic prophase, the nuclear envelope breaks down and the homologous chromosomes (each chromosome is double, with its two chromatids joined at the kinetochore) align together. Chromosome rearrangements ("crossing over") can occur at this stage. After the first metaphase, the kinetochore remains unsplit, and the pairs of homologous chromosomes are sorted into different cells. During the second meiotic division, the kintochore splits, and the sister chromatids are then moved into separate cells, each with a haploid set of chromosomes.

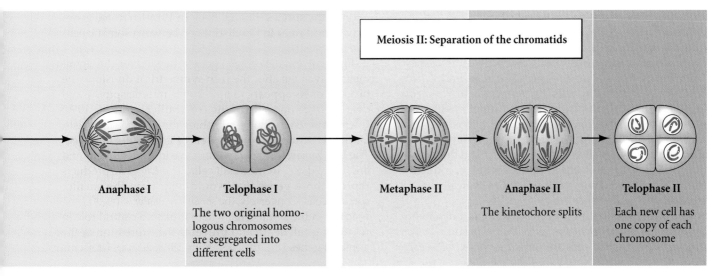

Meiosis II: Separation of the chromatids

Anaphase I — **Telophase I** — The two original homologous chromosomes are segregated into different cells

Metaphase II — **Anaphase II** — The kinetochore splits — **Telophase II** — Each new cell has one copy of each chromosome

(A)   (B)   (C)

(D)   (E)   (F)

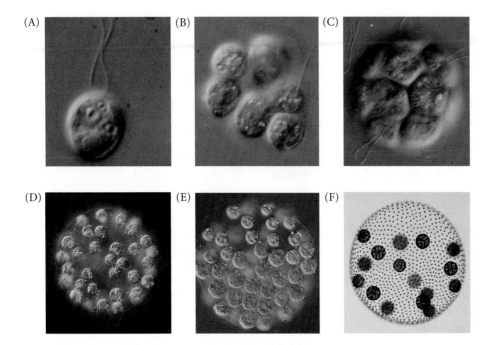

FIGURE 2.6   Representatives of the order Volvocales. All but *Chlamydomonas* are members of the family Volvocaceae. (A) The unicellular protist *Chlamydomonas reinhardtii*. (B) *Gonium pectorale*, with 8 *Chlamydomonas*-like cells in a convex disc. (C) *Pandorina morum*. (D) *Eudorina elegans*. (E) *Pleodorina californica*. Here, all 64 cells are originally similar, but the posterior ones dedifferentiate and redifferentiate as asexual reproductive cells called gonidia, while the anterior cells remain small and biflagellate, like *Chlamydomonas*. (F) *Volvox carteri*. Here, cells destined to become gonidia are set aside early in development and never have somatic characteristics. The smaller somatic cells resemble *Chlamydomonas*. Complexity increases from the single-celled *Chlamydomonas* to the multicellular *Volvox*. (Photographs courtesy of D. Kirk.)

lular protist *Chlamydomonas*, to which they are related (Figure 2.6A). A single organism of the volvocacean genus *Gonium* consists of a flat plate of 4 to 16 cells, each with its own flagellum (Figure 2.6B). In a related genus, *Pandorina*, the 16 cells form a sphere (Figure 2.6C); and in the genus *Eudorina*, the sphere contains 32 or 64 cells arranged in a regular pattern (Figure 2.6D). In these organisms, then, a very important developmental principle has been worked out: the ordered division of one cell to generate a number of cells, which are then organized in a predictable fashion. Like cleavage in most animal embryos, the cell divisions by which a single volvocacean cell produces an organism of 4 to 64 cells occur in very rapid sequence and in the absence of cell growth.

The next two genera of the volvocacean series exhibit another important principle of development: the *differentiation of cell types within an individual organism*. In these organisms, the reproductive cells become differentiated from the somatic cells. In all the genera mentioned earlier, every cell can, and normally does, produce a complete new organism by mitosis. In the genera *Pleodorina* and *Volvox*, however, relatively few cells can reproduce. In *Pleodorina californica* (Figure 2.6E), the cells in the anterior region are restricted to somatic (nonreproductive) functions, and only those cells on the posterior side can reproduce. In this species, a colony usually has 128 or 64 cells, and the ratio of the number of somatic cells to the number of reproductive cells is usually 3:5. Thus, a 128-cell colony typically has 48 somatic cells, and a 64-cell colony has 24.

In *Volvox*, almost all the cells are somatic; only a very few are able to produce new individuals. In some *Volvox* species, reproductive cells are similar to those of *Pleodori-*

na, in that they are derived from cells that look and function like somatic cells before they enlarge and divide to form new progeny. However, in other members of the genus, such as *V. carteri*, there is a complete division of labor: the reproductive cells that will create the next generation are set aside during the division of the original cell as it forms a new individual. These reproductive cells never develop functional flagella and never contribute to motility or other somatic functions; they are entirely specialized for reproduction.

Thus, although the simpler volvocaceans may be thought of as colonial organisms (because each cell is capable of independent existence and of perpetuating the species), in *V. carteri* we have a truly multicellular organism whose two distinct and interdependent cell types (somatic and reproductive) are both required for perpetuation of the species (Figure 2.6F). In *V. carteri*, three genes play critical roles in the distinction between somatic cells and germ cells.

First, the *gonidialess* (*gls*) gene is needed for the asymmetric divisions after the fifth symmetrical division. The product of *gls* appears to bind to the mitotic spindle and displace it to one side of the cell (Miller and Kirk 1999; Cheng et al. 2003), thus establishing the big and small cells. Once this is done, two other sets of genes are employed. The *Late gonidia* (*Lag*) genes are active in the large cells and are "turned off" in the small cells. The *Lag* gene products repress those genes that would help form the smaller somatic cells. Conversely, the *somatic regulator A* (*regA*) gene is active in the small cells, where it plays a central role in regulating cell death and prevents the expression of the gonidial genes (Figure 2.7A; Kirk 1988, 2001a). (It seems

(A)

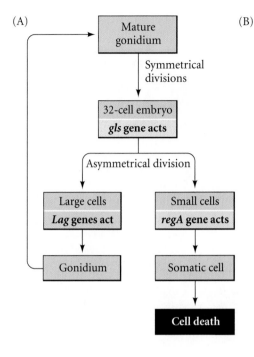

```
        ┌──────────────┐
        │    Mature    │◄──────────┐
        │   gonidium   │           │
        └──────┬───────┘           │
               │ Symmetrical       │
               │ divisions         │
        ┌──────▼───────┐           │
        │ 32-cell embryo │          │
        │ gls gene acts  │          │
        └──────┬───────┘           │
               │                   │
          Asymmetrical division    │
          ┌────┴─────┐             │
    ┌─────▼─────┐ ┌──▼──────┐      │
    │Large cells│ │Small cells│    │
    │Lag genes  │ │regA gene │     │
    │   act     │ │   acts   │     │
    └─────┬─────┘ └────┬─────┘     │
    ┌─────▼─────┐ ┌────▼─────┐     │
    │ Gonidium  │ │ Somatic  │     │
    └───────────┘ │   cell   │     │
          │        └────┬─────┘    │
          └─────────────┼──────────┘
                   ┌────▼─────┐
                   │Cell death│
                   └──────────┘
```

(B)

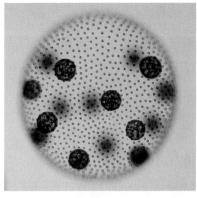 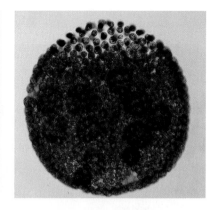

FIGURE 2.7 Germline-somatic line separation in *Volvox carteri*. (A) At the 32-cell stage, the *gonidialess* gene product causes cell division to be asymmetrical, so that large and small cells form. In the large cells, the *Lag* gene products repress the somatic cell differentiation genes, while in the small cells the *regA* gene acts to suppress the chloroplast genes, whose function is critical for cell growth and gonidial differentiation. (B) Mutation of the *regA* gene abolishes programmed cell death. A newly hatched *Volvox* carrying this mutation is indistinguishable from the wild-type spheroid (left). However, shortly before the time when the somatic cells of wild-type spheroids begin to die, the somatic cells of this mutant redifferentiate as gonidia (right). Eventually, every cell of the mutant will divide and regenerate a new spheroid in a potentially immortal developmental cycle. (A after Kirk and Kirk 2004; B photographs courtesy of D. Kirk.)

that the *regA* product suppresses cell growth by inhibiting chloroplast functioning.)

So the differentiation switch in *Volvox carteri* is controlled by negative regulators: the products of the *Lag* genes repress somatic cell genes, while the *regA* product represses the gonidial genes (Schmitt 2003; Kirk and Kirk 2004). In laboratory strains possessing mutations of the *regA* gene, somatic cells express *regA* and redifferentiate into gonidia, thus gaining the ability to reproduce asexu-

ally and become potentially immortal (Figure 2.7B). The fact that such mutants have never been found in nature indicates that cell death most likely plays an important role in the survival of *V. carteri* under natural conditions.

Although not all animals set aside the reproductive cells from the somatic cells (and plants hardly ever do), this separation of germ cells from somatic cells early in development is characteristic of many animal phyla and will be discussed in more detail in Chapter 19.

SIDELIGHTS & SPECULATIONS

Sex and Individuality in *Volvox*

Simple as it is, *Volvox* shares many of the features that characterize the life cycles and developmental histories of much more complex organisms, including ourselves. *Volvox* is among the simplest organisms to exhibit a division of labor between two completely different cell types. As a consequence, it is among the simplest organisms to include death as a regular, and genetically regulated, part of its life history.

Death and Differentiation

Unicellular organisms that reproduce by simple cell division, such as amoebae, are potentially immortal. The amoeba you see today under the microscope has no dead ancestors. When an amoeba divides, neither of the two resulting cells can be considered either ancestor or offspring; they are siblings. Death comes to an amoeba only if it is eaten or meets with a fatal accident, and

when it does, the dead cell leaves no offspring.

Death becomes an essential part of life, however, for any multicellular organism that establishes a division of labor between somatic (body) cells and germ (reproductive) cells. Consider the life history of *Volvox carteri* when it is reproducing asexually (Figure 2.8). Each asexual adult is a spheroid

(Continued on next page)

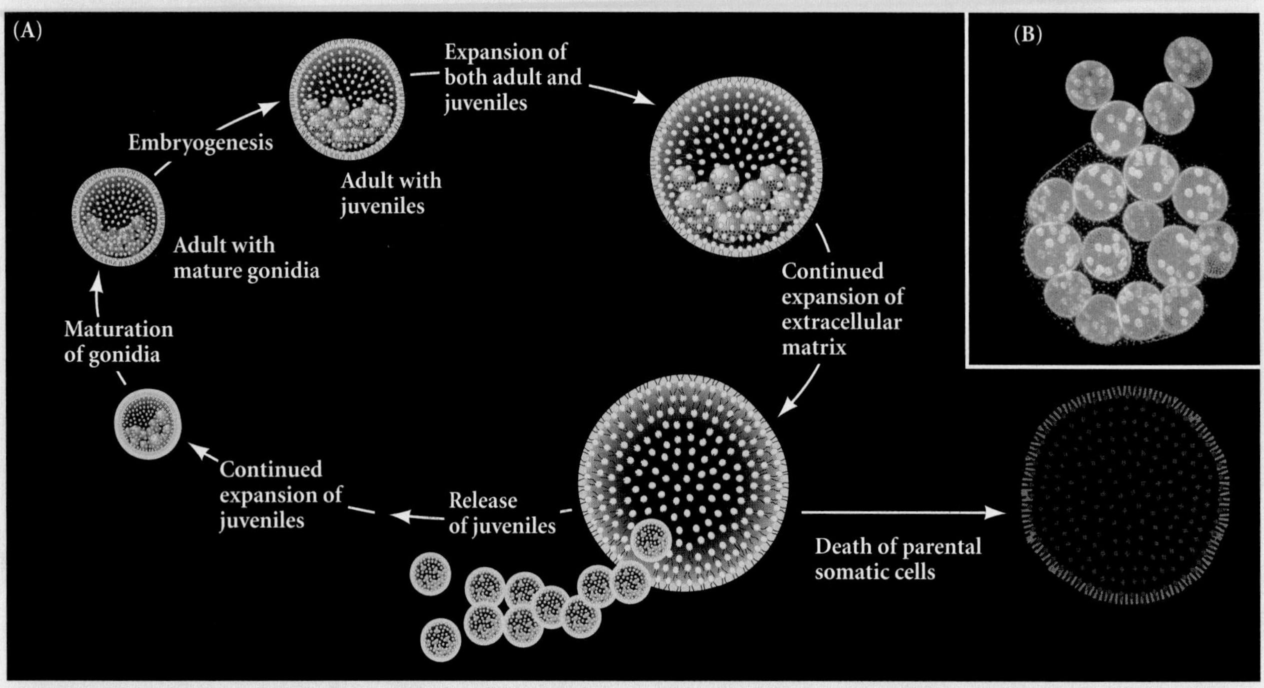

FIGURE 2.8 Asexual reproduction in *V. carteri*. (A) When reproductive cells (gonidia) are mature, they enter a cleavage-like stage of embryonic development to produce juveniles within the adult. Through a series of cell movements resembling gastrulation, the embryonic *Volvox* invert and are eventually released from the parent. The somatic cells of the parent, lacking gonidia, undergo senescence and undergo programmed cell death, while the juvenile *Volvox* mature. The entire asexual cycle takes 2 days. (B) Micrograph showing young adult spheres of *Volvox carteri* being released from parent to become free-swimming individuals. (A after Kirk 1988; B from Kirk 2001b.)

containing some 2000 small, biflagellated somatic cells along its periphery and about 16 large, asexual reproductive cells, called **gonidia**, toward one end of the interior. Each cell is haploid (like the sex cells of most animals). When mature, each gonidium divides rapidly 11 or 12 times. The first five divisions are symmetrical, resulting in a 32-cell embryo. At the sixth division, 16 of the cells divide asymmetrically producing 16 pairs of big and small sister cells. Each of these large cells will eventually produce a gonidium. At the end of cleavage, all the cells that will be present in an adult have been produced from the original gonidium.

As the cells of *V. carteri* divide, they produce an embryo that is "inside out." This embryo is a hollow sphere with its gonidia on the outside and the flagella of its somatic cells pointing toward the interior. This predicament is corrected by a process called **inversion**, in which the embryo turns itself

right side out by a set of cell movements that resemble the gastrulation movements of animal embryos. Clusters of bottle-shaped cells open a hole at one end of the embryo by producing tension on the interconnected cell sheet. At the same time, the ATPase protein kinesin accumulates near the tips of the bottle-shaped cells, providing the energy (by splitting ATP) needed for the inversion to occur (Nishii et al. 2003). The embryo everts through this hole and then closes it up. About a day after this is done, the juvenile *Volvox* are enzymatically released from the parent and swim away.

What happens to the somatic cells of the "parent" *Volvox* once its young "leave home"? Having produced offspring and being incapable of further reproduction, these somatic cells die. Actually, they "commit suicide," synthesizing proteins that cause their own death and dissolution (Pommerville and Kochert 1982). In death,

the somatic cells release for the use of others (including their own offspring) all the nutrients they stored during life. "Thus emerges," notes David Kirk, "one of the great themes of life on planet Earth: 'Some die that others may live.'"

Enter sex

Although *V. carteri* reproduces asexually much of the time, in nature it reproduces sexually once each year. When it does, one generation of individuals passes away and a new and genetically different generation is produced. The naturalist Joseph Wood Krutch (1956, pp. 28–29) put it more poetically:

The amoeba and the paramecium are potentially immortal. ... But for Volvox, death seems to be as inevitable as it is in a mouse or in a man. Volvox must die as Leeuwenhoek saw it die because it had children and is no longer needed. When its time comes it drops quietly to the bottom and joins its ancestors. As Hegner, the Johns Hopkins zoologist, once wrote, 'This is the first advent of inevitable natural death in the animal kingdom and all for the sake of sex.' And he asked: 'Is it worth it?'

SIDELIGHTS & SPECULATIONS

For *Volvox carteri*, it most assuredly is worth it. *V. carteri* lives in shallow temporary ponds that fill with spring rains but dry out in the heat of late summer. Between those times, *V. carteri* swims about, reproducing asexually. These asexual volvoxes die in minutes once the pond dries up. *V. carteri* is able to survive by turning sexual shortly before the pond disappears, producing sperm and eggs that unite to form dormant diploid zygotes that survive the heat and drought of late summer and the cold of winter. When rain fills the pond the following spring, the zygotes break their dormancy, undergo meiosis, and hatch out a new generation of haploid individuals (of both mating types) that reproduce asexually until the pond is about to dry up once more. Thus, fertilization in Volvocaceans is a survival tactic that allows the production of a dormant diploid zygote capable of surviving harsh environmental conditions.

How do these simple organisms predict the coming of adverse conditions so accurately that they can produce a sexual generation in the nick of time, year after year? The stimulus for switching from the asexual to the sexual mode of reproduction in *V. carteri* is known to be a 30-kDa sexual inducer protein. This protein is so powerful that concentrations as low as 6×10^{-17} *M* cause gonidia to undergo a modified pattern of embryonic development that results in the production

of eggs or sperm, depending on the genetic sex of the individual (Sumper et al. 1993). The sperm are released and swim to a female, where they fertilize eggs to produce dormant zygotes (Figure 2.9). The sexual inducer protein is able to work at such remarkably low concentrations by causing slight modifications of the extracellular matrix. These modifications appear to signal the transcription of a whole battery of genes that form the gametes (Sumper et al. 1993; Hallmann et al. 2001). Thus, the volvocaceans include the simplest organisms that have distingishable male and female members of the species and that have distinct developmental pathways for the production of eggs or sperm.

What is the source of this sexual inducer protein? Kirk and Kirk (1986) discovered that the sexual cycle could be initiated by heating dishes of *V. carteri* to temperatures that might be expected in a shallow pond in late summer. When this was done, the heat-

induced chemical changes in the somatic cells of the asexual volvoxes produced the sexual inducer protein (Nedelcu et al. 2003, 2004). Since the amount of sexual inducer protein secreted by one individual is sufficient to initiate sexual development in over 500 million asexual volvoxes, a single inducing *Volvox* can convert an entire pond to sexuality. This discovery explained an observation made over 90 years ago that "in the full blaze of Nebraska sunlight, *Volvox* is able to appear, multiply, and riot in sexual reproduction in pools of rainwater of scarcely a fortnight's duration" (Powers 1908). Thus, in temporary ponds formed by spring rains and dried up by summer's heat, *Volvox* has found a means of survival: it uses that heat to induce the formation of sexual individuals whose mating produces zygotes capable of surviving conditions that kill the adult organism. We see, too, that development is critically linked to the ecosystem in which the organism has adapted to survive.

FIGURE 2.9   Sexual reproduction in *V. carteri*. Males and females are indistinguishable in their asexual phase. When the sexual inducer protein is present, the gonidia of both mating types undergo a modified embryogenesis that leads to the formation of eggs in the females and sperm in the males. When the gametes are mature, sperm packets (containing 64 or 128 sperm each) are released and swim to the females. Upon reaching a female, the sperm packet breaks up into individual sperm, which can fertilize the eggs. The resulting dormant zygote has tough cell walls that can resist drying, heat, and cold. When spring rains cause the zygote to germinate, it undergoes meiosis to produce haploid males and females that reproduce asexually until heat induces the sexual cycle again.

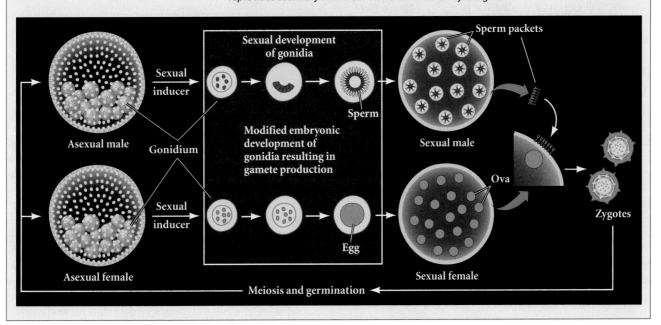

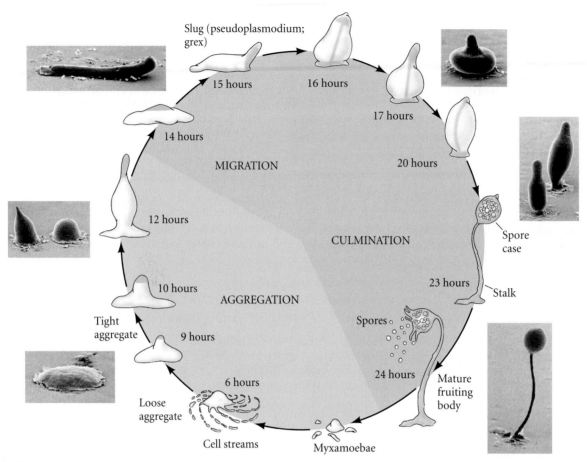

Slug (pseudoplasmodium; grex)

15 hours

16 hours

17 hours

14 hours

MIGRATION

20 hours

12 hours

CULMINATION

Spore case

10 hours

23 hours

Stalk

AGGREGATION

Tight aggregate

9 hours

Spores

6 hours

24 hours

Loose aggregate

Mature fruiting body

Cell streams

Myxamoebae

FIGURE 2.10   Life cycle of *Dictyostelium discoideum*. Haploid spores give rise to myxamoebae, which can reproduce asexually to form more haploid myxamoebae. As the food supply diminishes, aggregation occurs and a migrating slug is formed. The slug culminates in a fruiting body that releases more spores. Times refer to hours since the onset of nutrient starvation. Prestalk and stalk cells are indicated in yellow. (Photographs courtesy of R. Blanton and M. Grimson.)

WEBSITE **2.3** **Slime mold life cycle.** Check out the website to see digitized videos of the *Dictyostelium* life cycle.

VADE MECUM² **Slime mold life cycle.** The life cycle of *Dictyostelium*—the remarkable aggregation of myxamoebae, the migration of the slug, and the truly awesome culmination of the stalk and fruiting body—can best be viewed through movies. The Slime Mold segment in Vade Mecum² contains a remarkable series of videos. [Click on Slime Mold]

Differentiation and morphogenesis in *Dictyostelium*

THE LIFE CYCLE OF *DICTYOSTELIUM*   Another type of multicellular organization derived from unicellular organisms is found in *Dictyostelium discoideum*.* The life cycle of this fascinating protist is illustrated in Figure 2.10. In its asexual cycle, solitary haploid amoebae (called myxamoebae or "social amoebae" to distinguish them from amoeba species that always remain solitary) live on decaying logs, eating bacteria and reproducing by binary fission. When they have exhausted their food supply, tens of thousands of these myxamoebae join together to form moving streams of cells that converge at a central point. Here they pile atop one another to produce a conical mound called a tight aggregate. Subsequently, a tip arises at the top of this mound,

and the tight aggregate bends over to produce the migrating **slug** (with the tip at the front). The slug (often given the more dignified title of **pseudoplasmodium** or **grex**) is usually 2–4 mm long and is encased in a slimy sheath. If the environment is dark and moist, the grex begins to migrate with its anterior tip slightly raised. When it reaches an illuminated area, migration ceases and the grex differentiates into a fruiting body composed of spore cells and a stalk.

The anterior cells of the grex, representing 15–20 percent of the entire cellular population (and shown in yellow in Figure 2.10), form the tubed stalk. This process begins as some of the central anterior cells (the **prestalk cells**) begin secreting an extracellular cellulose coat and extending a tube through the grex. As the prestalk cells differentiate, they form vacuoles and enlarge, lifting up the mass

*Though colloquially called a "cellular slime mold," *Dictyostelium* is not a mold, nor is it consistently slimy. It is perhaps best to think of *Dictyostelium* as a social amoeba.

of **prespore cells** that make up the posterior four-fifths of the grex (Jermyn and Williams 1991). The stalk cells die, but the prespore cells, elevated above the stalk, become spore cells. These spore cells disperse, each one becoming a new myxamoeba.

In addition to this asexual cycle, there is a possibility of sex for *Dictyostelium*. Two myxamoebae can fuse to create a giant cell, which digests all the other cells of the aggregate. When it has eaten all its neighbors, it encysts itself in a thick wall and undergoes meiotic and mitotic divisions; eventually, new myxamoebae are liberated.

Dictyostelium has been a wonderful experimental organism for developmental biologists because initially identical cells differentiate into two alternative cell types—spore (reproductive cells) and stalk (somatic cells). It is also an organism wherein individual cells come together to form a cohesive structure composed of differentiated cell types, a process akin to tissue formation in more complex organisms. The aggregation of thousands of myxamoebae into a single organism is an incredible feat of organization that invites experimentation to answer questions about the mechanisms involved. The first of these questions is, What causes the myxamoebae to aggregate?

AGGREGATION OF *DICTYOSTELIUM* CELLS Time-lapse videomicroscopy has shown that no directed myxamoeboid movement occurs during the first 4–5 hours following nutrient starvation. During the next 5 hours, however, the cells move at speeds of about 20 mm/min, in repeating cycles of approximately 2 minutes of movement followed by 4 stationary minutes. Although the movement is directed toward a central point, it is not a simple radial movement. Rather, cells join with one another to form streams; the streams converge into larger streams, and all streams eventually merge at the center. Bonner (1947) and Shaffer (1953) showed that this movement is a result of **chemotaxis**: the cells are guided to aggregation centers by a soluble substance. This substance was later identified as **cyclic adenosine 3′,5′-monophosphate (cAMP)** (Konijn et al. 1967; Bonner et al. 1969), the chemical structure of which is shown in Figure 2.11A.

(A)

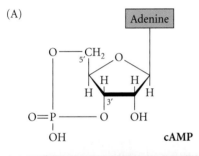

cAMP

(B)

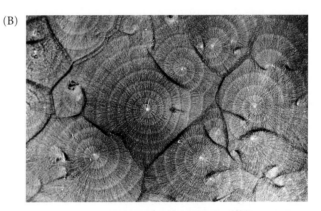

(C)

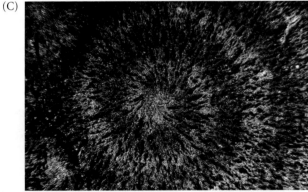

(D)

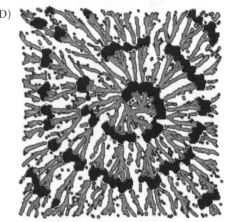

FIGURE 2.11 Chemotaxis of *Dictyostelium* myxamoebae is a result of spiral waves of cAMP. (A) Chemical structure of cAMP. (B) Visualization of several cAMP "waves." Central cells secrete cAMP at regular intervals, and each pulse diffuses outward as a concentric wave. The waves were charted by saturating filter paper with radioactive cAMP and placing it on an aggregating colony. The cAMP from the secreting cells dilutes the radioactive cAMP. When the radioactivity on the paper is recorded (by placing it over X-ray film), the regions of high cAMP concentration in the culture appear lighter than those of low cAMP concentration. (C) Spiral waves of myxamoebae moving toward the initial source of cAMP. Because moving and nonmoving cells scatter light differently, the photograph reflects cell movement. The bright bands are composed of elongated migrating cells; the dark bands are cells that have stopped moving and have rounded up. As cells form streams, the spiral of movement can still be seen moving toward the center. (D) Computer simulation of cAMP wave spreading across migrating *Dictyostelium* cells. The model takes into account the reception and release of cAMP, and changes in cell density due to the movement of the cells. The cAMP wave is plotted in dark blue. The population of amoebae goes from green (low) to red (high). Compare with the actual culture shown in (C). (B from Tomchick and Devreotes 1981; C from Siegert and Weijer 1989; D from Dallon and Othmer 1997.)

Aggregation is initiated as each of the myxamoebae begins to synthesize cAMP. There are no dominant cells that begin the secretion or control the others. Rather, the sites of aggregation are determined by the distribution of the myxamoebae (Keller and Segal 1970; Tyson and Murray 1989). Neighboring cells respond to cAMP in two ways: they initiate a movement toward the cAMP pulse for about a minute, and they release cAMP of their own (Robertson et al. 1972; Shaffer 1975). The movement of each cell is the result of a change in cytoskeletal polarity brought about by the cAMP (Parent et al. 1998; Iijima et al. 2002). Once this happens, the cell is unresponsive to further cAMP pulses for several minutes. During this time, an extracellular membrane-associated phosphodiesterase cleaves the remaining cAMP from the environment, readying the receptors to receive another pulse. In this manner a rotating spiral wave of cAMP is propagated throughout the myxamoebae population (Figure 2.11B–D). As each wave arrives, the cells take another step toward the center.

The differentiation of individual myxamoebae into either stalk (somatic) or spore (reproductive) cells is a complex matter. Raper (1940) and Bonner (1957) demonstrated that the anterior cells normally become stalk, while the remaining, posterior cells are usually destined to form spores. However, surgically removing the anterior part of a slug does not abolish its ability to form a stalk. Rather, the cells that now find themselves at the anterior end (which originally were destined to produce spores) now form the stalk (Raper 1940). Somehow a decision is made so that whichever cells are anterior become stalk cells and whichever are posterior become spores. This ability of cells to change their developmental fates according to their location within the whole organism and thereby compensate for missing parts is called **regulation**. We will see this phenomenon in many embryos, including those of mammals.

CELL ADHESION MOLECULES IN *DICTYOSTELIUM* How do individual cells stick together to form a cohesive organism? This problem is the same one that embryonic cells face, and the solution that evolved in these protists is the same one used by animal embryos: developmentally regulated cell adhesion molecules.

While growing mitotically and feeding, *Dictyostelium* cells do not adhere to one another. However, once cell division stops, the cells become increasingly adhesive, reaching a plateau of maximum adhesiveness about 8 hours after starvation. The initial cell-cell adhesion is mediated by a 24-kilodalton glycoprotein (gp24, or DdCad1; Figure 2.12A) that is absent in solitary myxamoebae but which appears shortly after mitotic division ceases (Knecht et al. 1987; Wong et al. 1996). This protein is synthesized from newly transcribed mRNA and becomes localized in the cell membranes of the myxamoebae. If myxamoebae are treated with antibodies that bind to and mask this protein, the cells will not stick to one another, and all subsequent development ceases.

Once this initial aggregation has occurred, it becomes stabilized by a second cell adhesion molecule (Figure 2.12B). This 80-kDa glycoprotein (gp80; CsaA) is also synthesized during the aggregation phase. If it is defective or absent in the cells, small slugs will form, and their fruiting bodies will be only about one-third the normal size. Thus, the second cell adhesion system seems to be needed for retaining a large enough number of cells to form large fruiting bodies (Müller and Gerisch 1978; Loomis 1988). During late aggregation, the levels of gp80 decrease, and its role is taken over by a third cell adhesion protein, a 150-kDa protein (gp150; LagC; Figure 2.12C) whose synthesis becomes apparent just prior to aggregation and which stays on the cell surface during grex migration (Wang et al. 2000). If *Dictyostelium* cells lack functional genes for gp150, development is arrested at

(A)

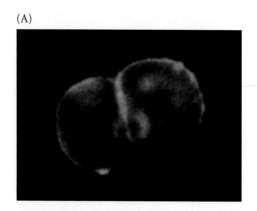

(B)

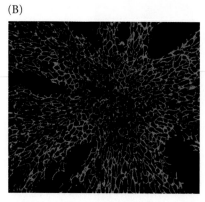

(C)

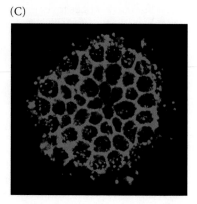

FIGURE 2.12 The three cell adhesion molecules of *Dictyostelium*. (A) *Dictyostelium* cells synthesize an adhesive 24-kDa glycoprotein (gp24) shortly after nutrient starvation. These *Dictyostelium* cells were stained with a fluorescently labeled (green) antibody that binds to gp24 and were then observed under ultraviolet light. This protein is not seen on myxamoebae that have just stopped dividing. However, as shown here—10 hours after cell division has ceased—individual myxamoebae have this protein in their cell membranes and are capable of adhering to one another. (B) The gp80 protein, stained by specific antibodies (green), is present at the cell membranes of streaming amoebae. (C) The gp150 protein (green) is present in the cells of the migrating grex (cross-sectioned). Photographs are not at the same magnification. (Photographs courtesy of W. Loomis.)

the loose aggregate stage, and the prespore and prestalk cells fail to sort out into their respective regions. Thus, *Dictyostelium* has evolved three developmentally regulated systems of cell-cell adhesion that are necessary for the morphogenesis of individual cells into a coherent organism. As we will see in subsequent chapters, metazoan cells also use cell adhesion molecules to form the tissues and organs of the embryo.

Dictyostelium is a "part-time multicellular organism" that does not form many cell types (Kay et al. 1989), and the more complex multicellular organisms do not form by the aggregation of formerly independent cells. Nevertheless, many of the principles of development demonstrated by this "simple" organism also appear in the embryos of more complex phyla (see Loomis and Insall 1999). The ability of individual cells to sense a chemical gradient (as in the myxamoeba's response to cAMP) is crucial for cell migration and morphogenesis throughout development in all animal species. The role of cell surface proteins in cell cohesion is also seen throughout the animal kingdom, and differentiation-inducing molecules are being isolated in metazoan organisms.

SIDELIGHTS & SPECULATIONS

Rules of Evidence I

Biology, like any other science, does not deal with Facts, but with evidence. Several types of evidence will be presented in this book, and they are not equivalent in strength. The first, and weakest, type of evidence is **correlative evidence**. Here, correlations are observed between two or more events, and there is an inference that one event causes the other. Correlative evidence provides a starting point for investigations, but a scientist cannot say with certainty that one event causes the other based solely on correlations. The simultaneous occurrence of the two events could even be coincidental, the events having no relationship to each other.* For example, a researcher wanting to know exactly what it is that causes *Dictyostelium* myxamoebae to coalesce might infer, based on the events described above, that the synthesis of gp24 causes the adhesion of the cells. But it is also possible that cell adhesion causes the cells to synthesize gp24, or that cell adhesion and the synthesis of gp24 are separate events initiated by the same underlying cause.

As an example, we will use the analysis of cell adhesion in *Dictyostelium*. This evidence involves the use of antibodies. Antibodies are proteins that can be made by the mammalian immune system. They are usually synthesized to attack bacteria and viruses that have entered the body, but the immune system can make antibodies to nearly any compound that is not found in an individual's body—anything that the body does not recognize as "self." So, for example, if a researcher injects *Dictyostelium* myxamoebae into rabbits, the rabbits make antibodies to these "foreign" cells. Various techniques can be used to purify antibodies so that they are directed against cells containing a single desired target (often a protein).

Once we have an antibody against a single target, many things are possible. If we wish to pinpoint those cells that contain the target, we can label the antibody with radioactive compounds or with dyes that fluoresce under ultraviolet light. If the target is present on a cell, the antibody will bind to it and the label will be detected (Figure 2.13A). (If the target is not there, the antibody has nothing to bind to, and the marker will not be detected.) When this technique was applied to *Dictyostelium*, fluorescently labeled antibodies to gp24 did not bind to dividing myxamoebae, but the antibodies did find this protein in myxamoeba cell membranes soon after the cells stop dividing and become competent to aggregate. Thus, there is a correlation between the presence of this cell membrane glycoprotein and the ability to aggregate.

But how do we get beyond mere correlation? In studying cell adhesion in *Dictyostelium*, the next step was to use the antibodies that bound to gp24 to block the adhesion of myxamoebae. Using a technique pioneered by Gerisch's laboratory (Beug et al. 1970), Knecht and co-workers (1987) isolated the antibodies' target-binding sites. When these target-binding fragments—known as Fab fragments—were added to aggregation-competent cells, the cells could not aggregate. The target-binding fragments inactivated the cells' ability to adhere to each other, presumably by binding to gp24 and blocking its function (Figure 2.13B). This is called **loss-of-function evidence**.

While stronger than correlative evidence, loss-of-function evidence still does not make other inferences impossible. For instance, perhaps the antibody fragments kill the cells outright; this would also stop the cells from adhering. Or perhaps gp24 has nothing to do with adhesion itself, but is necessary for the real adhesive molecule to function (perhaps, for example, it stabilizes membrane proteins in general). In this case, blocking the glycoprotein would similarly cause the inhibition of cell aggregation. Thus, loss-of-function evidence must be bolstered by many controls demonstrating that the agents causing the loss of function specifically knock out the particular function and nothing else.

The strongest type of evidence (especially when coupled with loss-of-function evidence) is **gain-of-function evidence**. Here, the initiation of the first event causes the second event to happen even under circumstances where neither event usually occurs. For instance, da Silva and Klein (1990) and Faix and co-workers (1990) obtained such evidence to show that the 80-kDa glycoprotein (gp80) is an adhesive

*In a tongue-in-cheek letter spoofing such correlative inferences, Sies (1988) demonstrated a remarkably good correlation between the number of storks seen in West Germany from 1965 to 1980 and the number of babies born during those same years.

(Continued on next page)

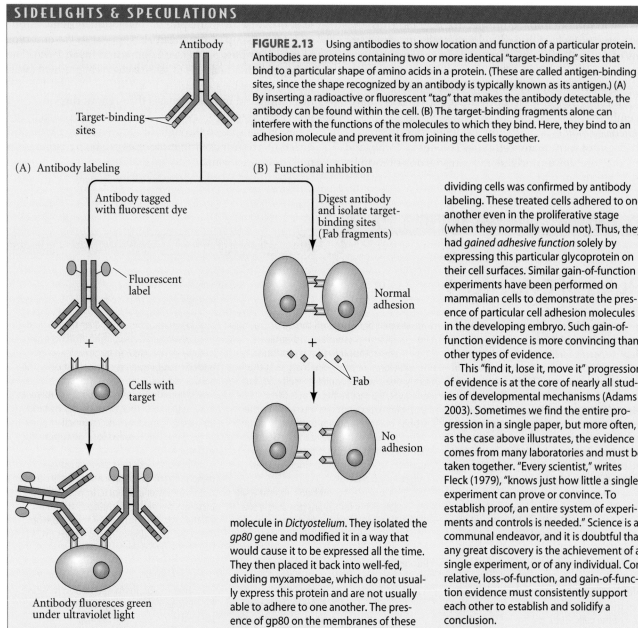

FIGURE 2.13 Using antibodies to show location and function of a particular protein. Antibodies are proteins containing two or more identical "target-binding" sites that bind to a particular shape of amino acids in a protein. (These are called antigen-binding sites, since the shape recognized by an antibody is typically known as its antigen.) (A) By inserting a radioactive or fluorescent "tag" that makes the antibody detectable, the antibody can be found within the cell. (B) The target-binding fragments alone can interfere with the functions of the molecules to which they bind. Here, they bind to an adhesion molecule and prevent it from joining the cells together.

(A) Antibody labeling

Antibody tagged with fluorescent dye

Fluorescent label

+

Cells with target

Antibody fluoresces green under ultraviolet light

(B) Functional inhibition

Digest antibody and isolate target-binding sites (Fab fragments)

Normal adhesion

+

Fab

No adhesion

molecule in *Dictyostelium*. They isolated the *gp80* gene and modified it in a way that would cause it to be expressed all the time. They then placed it back into well-fed, dividing myxamoebae, which do not usually express this protein and are not usually able to adhere to one another. The presence of gp80 on the membranes of these dividing cells was confirmed by antibody labeling. These treated cells adhered to one another even in the proliferative stage (when they normally would not). Thus, they had *gained adhesive function* solely by expressing this particular glycoprotein on their cell surfaces. Similar gain-of-function experiments have been performed on mammalian cells to demonstrate the presence of particular cell adhesion molecules in the developing embryo. Such gain-of-function evidence is more convincing than other types of evidence.

This "find it, lose it, move it" progression of evidence is at the core of nearly all studies of developmental mechanisms (Adams 2003). Sometimes we find the entire progression in a single paper, but more often, as the case above illustrates, the evidence comes from many laboratories and must be taken together. "Every scientist," writes Fleck (1979), "knows just how little a single experiment can prove or convince. To establish proof, an entire system of experiments and controls is needed." Science is a communal endeavor, and it is doubtful that any great discovery is the achievement of a single experiment, or of any individual. Correlative, loss-of-function, and gain-of-function evidence must consistently support each other to establish and solidify a conclusion.

Development in Flowering Plants

Like animals, plants are multicellular organisms. But the divergence of the plant and animal lineages occurred so early in evolutionary history that the most recent ancestor shared by both groups was almost certainly unicellular. The evolutionary divergence of plants and animals resulted in dramatically different life cycles, morphologies, biochemistry, and genetics; equally distinctive are the strategies by which members of the two kingdoms reproduce and develop.

The different groups within the plant kingdom display diverse life cycles. Among the land plants (also known as embryophytes), most of these life cycles include both diploid and haploid multicellular stages, a phenomenon referred to as **alternation of generations**. Alternation of generations results in two different multicellular body plans over the individual's life cycle. The development of angiosperm plants (flowering plants) will be detailed in Chapter 20, but some of the major elements of the process are addressed here.

Some angiosperms contain both male and female sexual organs in the same plant. These are the **monoecious** (Greek, "one house") angiosperms, such as corn and cucumbers. Many monoecious plants (such as apples,

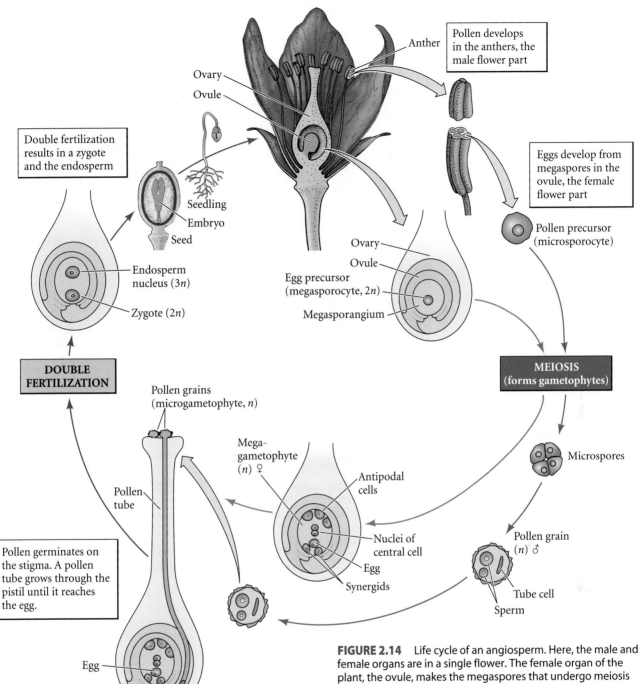

FIGURE 2.14 Life cycle of an angiosperm. Here, the male and female organs are in a single flower. The female organ of the plant, the ovule, makes the megaspores that undergo meiosis to make the egg. The male organs, the anthers, make the microspores that become pollen. Each pollen grain will eventually contain three cells: a tube cell and two sperm. The tube cell becomes a pollen tube that germinates on the ovary. The pollen tube discharges the sperm nuclei into the ovule, where one sperm nucleus fertilizes the egg to produce a diploid zygote. The second sperm nucleus fuses with the nuclei of the central cell to form the nutritive endosperm.

plums, and tomatoes) even have functional male and female parts in the same flower. Here, haploid sperm are enclosed in the **pollen** that is made by meiosis in the anther, while haploid eggs are produced by meiosis in the ovule. In **dioecious** (Greek, "two houses") plants, the male and female organs are on different individuals, such that some individuals are males (pollen producers) and others are females (fruit producers). Maples, ash, date palms, and figs are examples of dioecious plants.

Fertilization occurs when the pollen grain lands on the stigma containing the ovary. Pollen is actually multicellular, containing a tube cell and two sperm cells. The tube cell interacts with the stigma to elongate a long tube through which the sperm cells travel. When the pollen tube enters the ovule containing the egg, the two sperm cells are released and a **double fertilization** occurs (Figure 2.14). One sperm cell fuses with the egg to produce the zygote that will form the embryo. The second sperm cell fuses with a multinucleated somatic cell to produce the **triploid endosperm** that will nourish the developing embryo.

FIGURE 2.15 Phenotypic plasticity in plants. These two *Polygonum lapathifolium* plants are inbred specimens with identical genotypes. The plant on the left was grown for 8 weeks in low light conditions (20 percent of available photosynthetically active radiation), while the plant on the right was grown in high light (100 percent of available radiation). The low-light plant has large, thin leaves and few branches. The high-light plant is larger, has narrow leaves on many branches, and is more mature, as demonstrated by the senescence of the earliest leaves. (From Sultan 2000; photograph courtesy of S. E. Sultan).

The major differences between animal and plant embryos can be traced to the fundamental difference between animal and plant cells: the plant cell wall. This cellulose-based structure supports the plant cell and constrains cell expansion and mobility. It prevents cell migration in plants (plants don't undergo gastrulation, for instance) and prevents neighboring plant cells from having the types of intercellular interactions that are common in animal cells and that are so crucial in the processes of animal development. Plant embryos grow by mitosis, and in the developing plant, cell fate is determined primarily by its position. And, even though there is a general fate map, plant cell fates are not as rigidly determined as in animal embryos. Indeed, unlike most animal cells, plant cells are totipotent, and individual cells can produce an entire plant (Steward et al. 1964.)

This ability of the plant to retain totipotent cells makes possible the cyclic "death" and renewal of plants in temperate climates. The above-ground portion of many flowering plants shrivels away each fall, only to grow and bloom again in the springtime. As the plant grows, these totipotent cells form growing regions called **meristems**. The meristematic cells proliferate at the opposite tips of the plant (the root and shoot meristems) as well as along the stem (the ground meristem). This ability of a plant to retain totipotent cells throughout its life also allows one to grow an entire plant from cuttings. Thus, if you have a particularly nice dorm plant, you can often share it with a friend by cutting off a single branch and planting it in soil; an entire plant will grow from that small section. This is called **vegetative propagation**, and some plants (such as strawberries) routinely use this as a way of reproducing.

Plants also have a great deal of **developmental plasticity**. That is to say, the development of plant organs can be controlled to a large degree by the environment. Plants grown in low moisture or nutrients may decrease their root diameter compared to plants grown in high moisture and nutrient conditions. This narrowing increases the length and the surface area of the roots, allowing them to absorb more nutrients. Similarly, plants grown in the shade often change their leaf structures and the amount of branching to better harvest the small amount of sunlight (Figure 2.15; Sultan and Bazzaz 1993; Ryser and Eek 2000). This remarkable degree of plasticity may help compensate for a plant's inability to move to a different location.

Developmental Patterns among the Metazoa

Since most of the remainder of this book concerns the development of **metazoans**—multicellular animals that pass through embryonic stages of development—we will present an overview of their developmental patterns here. Figure 2.16 illustrates the major evolutionary trends of metazoan development. The most striking pattern is that life has not evolved in a straight line; rather, there are several branching evolutionary paths. We can see that metazoans belong to one of four major branches: sponges, diploblasts, protostomes, and deuterostomes.

Although most recent taxonomies will group them into the metazoans (see Degnan et al. 2005), the sponges (Porifera) develop so differently from any other animal group that some taxonomists do not consider them metazoans at all, and call them "parazoans." A sponge has three major types of somatic cells, but one of these, the **archeocyte**, can differentiate into all the other cell types in the body. Individual cells of a sponge passed through a sieve can reaggregate to form new sponges. Moreover, in some instances, such reaggregation is species-specific: if individual sponge cells from two different species are mixed together, each of the sponges that re-forms contains cells from only one species (Wilson 1907). In these cases, it is thought that the

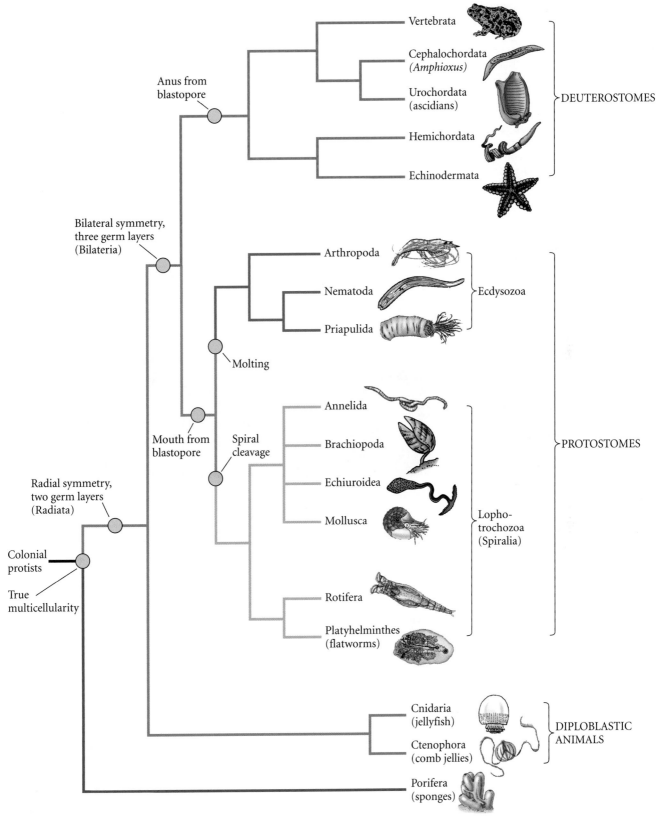

FIGURE 2.16   Major evolutionary divergences in extant animals. Other models of evolutionary relationships among the phyla are possible. This grouping of the Metazoa is based on embryonic, morphological, and molecular criteria. For alternative groupings, see Valentine 2004. (After Glenner et al. 2004.)

Totipotency among the Flatworms

The planarian flatworms are among the most primitively organized animals known. Their bilaterally symmetric bodies have three germ layers, but they lack a body cavity (the coelom). Their simple nervous system is organized into two anterior ganglia ("brains"), each connected to a ventral nerve cord. The nerve cords are interconnected by neurons that run perpendicular to the anterior-posterior body axis.

Planarians have two modes of reproduction. First, most planarians are able to reproduce asexually by fission. That is to say, the anterior portion splits from the posterior portion. The head regenerates a new tail, while the tail regenerates a new head (Figure 2.17). Some species also have sexual cycles in which there are males and females producing sperm and eggs, respectively (Hoshi et al. 2003).

The capability to regenerate missing body parts each generation is the responsibility of a group of cells called **neoblasts** (Figure 2.18; Agata and Watanabe 1999; Newmark and Alvarado 2000). Unlike *Dictyostelium*, *Volvox*, arthropods, or vertebrates, planarians do not segregate the reproductive germ cell lineage from that of the somatic lineages. The neoblasts are the only dividing cell population in the body, and they act as stem cells. That is to say, when they divide, one of the two cells pro-

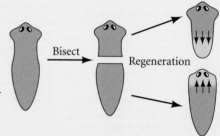

FIGURE 2.17 Regeneration in flatworms of the genus *Planaria*. If the planarian is bisected horizontally, the anterior portion of the lower segment regenerates a head, while the posterior portion of the upper segment regenerates a tail. The same tissue can generate either a head or a tail, depending on the segment in which it is located (black arrows).

duced is a differentiated cell that can form a body part, and the other is a stem cell. These neoblasts migrate to the site where the flatworm split and regenerate the missing parts.

They can form cells of all three germ layers, and in sexual species, they can form germ cells. Thus, in these organisms, the germ line is not distinct from the somatic lineages.

FIGURE 2.18 Staining for neoblasts (green) in the planarian *Girardia dorotocephala* (counterstained red). The animals were fed bromodeoxyuridine (BrdU), a substance that becomes incorporated into the DNA of dividing cells. Fluorescent antibodies (green) against BrdU were then able to locate these cells. (From Newmark and Alvarado 2000; courtesy of A. S. Alvarado.)

---

motile archeocytes collect cells from their own species and not from others (Turner 1978). Sponges contain no mesoderm, so the Porifera have no true organ systems, nor do they have a digestive tube, circulatory system, nerves, or muscles. Thus, even though they undergo gastrulation and pass through an embryonic and a larval stage, sponges are very unlike most metazoans (Fell 1997). However, sponges do share many features of development (including gene regulatory proteins and signaling cascades) with all other animal phyla, suggesting that they share a common origin (Coutinho et al. 1998; King 2004; Degnan et al. 2005).

The Diploblasts, Protostomes, and Deuterostomes

Diploblastic animals are those that have ectoderm and endoderm, but no true mesoderm. The diploblasts have

traditionally included the cnidarians (jellyfish and hydras) and the ctenophores (comb jellies). It has long been thought that these two phyla had radial symmetry and no mesoderm, while the triploblast phyla (all other animals) had bilateral symmetry and a third, mesodermal, germ layer. However, this clear-cut demarcation between them is currently being questioned. While certain cnidarians (like *Hydra*) have no true mesoderm, other cnidarians seem to have both mesoderm and bilateral symmetry (Martindale et al. 2004; Martindale 2005). The elucidation of how the gene networks used in axial organization, germ-layer formation, and cell differentiation came together and evolved is one of the most exciting areas of development and may lead to a much richer understanding of the events that led to the current diversity of multicellular life.

Most metazoans have bilateral symmetry and three germ layers and are thus considered to be **triploblastic** animals. The evolution of the mesoderm enabled greater mobility and larger bodies because it became the animal's musculature and circulatory system. The origins of mesoderm and the Bilataria will be discussed in Chapter 23. The flatworms have a fully developed mesoderm and are often considered among the most primitive of the bilateral phyla. Interestingly, the flatworms and the cnidarians do not have a strict separation of germline and somatic line (see Buss 1987). Among flatworms, for instance, are many species that have retained a population of totipotent stem cells. In other words, even as sexually mature adults, they have a population of cells that can become any cell type in the body—including gametes.

Animals of the Bilataria are further classified as either protostomes or deuterostomes. **Protostomes** (Greek, "mouth first"), which include the mollusc, arthropod, and worm phyla, are so called because the mouth is formed first, at or near the opening to the gut, which is produced during gastrulation. The anus forms later at another location. The **coelom**, or body cavity, of these animals forms from the hollowing out of a previously solid cord of mesodermal cells. There are two major branches of the protostomes. The **Ecdysozoa** includes the animals that molt their exterior skeletons.* The major constituent of this group is Arthropoda, a phylum containing the insects, arachnids, mites, crustaceans, and millipedes. The second major group of protostomes is the **Lophotrochozoa**. These animals are characterized by a common type of cleavage (spiral), a common larval form (the trochophore), and a distinctive feeding apparatus (the lophophore) found in some species. Lophotrochozoan phyla include the flatworms, bryozoans, annelids, and molluscs.

Phyla in the **deuterostome** lineage include the chordates and echinoderms. Although it may seem strange to classify humans, fish, and frogs in the same group as starfish and sea urchins, certain embryological features stress this kinship. First, in deuterostomes ("mouth second"), the oral opening is formed after the anal opening. Also, whereas protostomes generally form their body cavities by hollowing out a solid block of mesoderm (**schizocoelous** formation of the body cavity), most deuterostomes form their body cavities from mesodermal pouches extending from the gut (**enterocoelous** formation of the body cavity). It should be mentioned that there are many exceptions to these generalizations.

The evolution of organisms depends on inherited changes in their development. One of the greatest evolutionary advances—the **amniote egg**—occurred among the deuterostomes. This type of egg, exemplified by that of a chicken (Figure 2.19), is thought to have originated in the amphibian ancestors of reptiles about 255 million years ago. The amniote egg allowed vertebrates to roam on land, far from existing ponds. Whereas most amphibians must return to water to lay their eggs, the amniote egg carries its own water and food supplies. It is fertilized internally and contains yolk to nourish the developing embryo. Moreover, the amniote egg contains four sacs: the **yolk sac**, which stores nutritive proteins; the **amnion**, which contains the fluid bathing the embryo; the **allantois**, in which waste materials from embryonic metabolism collect; and the **chorion**, which interacts with the outside environment, selectively allowing materials to reach the embryo.[†] The entire structure is encased in a shell that allows the diffusion of oxygen but is hard enough to protect the embryo from environmental assaults and dehydration. A similar

[†]In mammals, the chorion is modified to form the embryonic portion of the placenta—another example of the modification of development to produce evolutionary change.

*The name Ecdysozoa is derived from the Greek *ecdysis*, "to shed" or "to get clear of." Asked to provide a more dignified job description for a "stripper," editor H. L. Mencken suggested the term "ecdysiast."

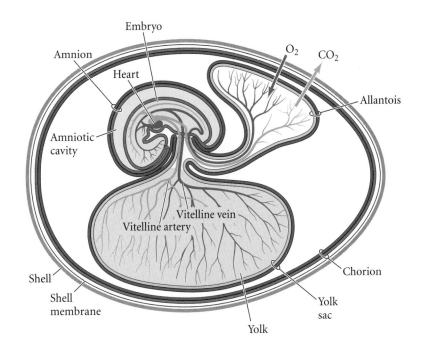

FIGURE 2.19 Diagram of the amniote egg of the chick, showing the membranes enfolding the 7-day chick embryo. The yolk is eventually surrounded by the yolk sac, which allows the entry of nutrients into the blood vessels. The chorion is derived in part from the ectoderm and extends from the embryo to the shell (where it will fuse with the blood vessel-rich allantois. This chorioallantoic membrane will exchange oxygen and carbon dioxide and absorb calcium from the shell). The amnion provides the fluid medium in which the embryo grows, and the allantois collects nitrogenous wastes that would be dangerous to the embryo. Eventually the endoderm becomes the gut tube and encircles the yolk.

VADE MECUM[2]   **The amniote egg.** The egg of the chick, as detailed in this sequence, is a beautiful and readily accessible example of the amniote egg—a remarkable adaptation to terrestrial life. [Click on Chick-early]

development of egg casings enabled arthropods to be the first terrestrial invertebrates. Thus, the final crossing of the boundary between water and land occurred with the modification of the earliest stage in development: the egg. Embryology provides an endless assortment of fascinating animals and problems to study. In this text, we will use but a small sample of them to illustrate the major principles of animal development. This sample is an incredibly small collection. We are merely observing a small tide pool within our reach, while the whole ocean of developmental phenomena lies before us.

After a brief outline of the experimental and genetic approaches to developmental biology, we will investigate the early stages of animal embryogenesis: fertilization, cleavage, gastrulation, and the establishment of the body axes. Later chapters will concentrate on the genetic and cellular mechanisms by which animal bodies are constructed. Although an attempt has been made to survey the important variations throughout the animal kingdom, a certain deuterostome chauvinism may be apparent. (For a more comprehensive survey of the diversity of animal development across the phyla, see Gilbert and Raunio 1997.)

Snapshot Summary | Life Cycles and Developmental Patterns

1. The life cycle can be considered a central unit in biology. The adult form need not be paramount. In a sense, the life cycle is the organism.

2. The basic life cycle consists of fertilization, cleavage, gastrulation, germ layer formation, organogenesis, metamorphosis, adulthood, and senescence.

3. Reproduction and sex are two separate processes that may, but do not necessarily, occur together. Some organisms, such as *Volvox* and *Dictyostelium*, exhibit both asexual reproduction and sexual reproduction.

4. Cleavage divides the zygote into numerous cells called blastomeres.

5. In animal development, gastrulation rearranges the blastomeres and forms the three germ layers.

6. Organogenesis often involves interactions between germ layers to produce distinct organs.

7. Germ cells are the precursors of the gametes. Gametogenesis forms the sperm and the eggs.

8. There are three main ways to provide nutrition to the developing embryo: (1) supply the embryo with yolk; (2) form a larval feeding stage between the embryo and the adult; or (3) create a placenta between the mother and the embryo.

9. Life cycles must be adapted to the nonliving environment and interwoven with other life cycles.

10. Don't regress your tail until you've formed your hindlimbs.

11. There are several types of evidence. Correlation between phenomenon A and phenomenon B does not imply that A causes B or that B causes A. Loss-of-function data (that is, if A is experimentally removed, B does not occur) suggests that A causes B, but other explanations are possible. Gain-of-function data (if A happens where or when it does not usually occur, then B also happens in this new time or place) is most convincing.

12. Plants and animals, though both multicellular, diverged very early in evolutionary history and their patterns of development show very little similarity.

13. The multicellular animals, or metazoans, can be divided into four major evolutionary branches: sponges, diploblasts, protostomes, and deuterostomes. The development of sponges is unique and some researchers do not consider them to be metazoans. Most diploblasts have only two germ layers, lacking true mesoderm.

14. Protostomes and deuterostomes represent two different sets of variations on development. Protostomes form the mouth first, while deuterostomes form their mouths later, usually forming the anus first.

15. The evolution of organisms depends on inherited changes in their development. One of the greatest such changes was the evolution of the amniote egg. Fertilized internally, the amniote egg contains yolk to nourish the growing embryo and allowed organisms of the deuterostome vertebrate lineage to fully exploit terrestrial environments.

For Further Reading

Complete bibliographical citations for all literature cited in this chapter can be found on the Vade Mecum CD that accompanies the book and at the free access website www.devbio.com

Adams, D. S. 2003. Teaching critical thinking in a developmental biology course at an American liberal arts college. *Int. J. Dev. Biol.* 47: 145–151.

Degnan, B. D., S. P. Leys and C. Larroux. 2005. Sponge development and the antiquity of pattern formation. *Integr. Comp. Biol.* 45: 335–341.

Glenner, H., A. J. Hansen, M. V. Sorensen, F. Ronquist, J. P. Huelsenbeck and E. Willerslev. 2004. Bayesian inference of the metazoan phylogeny: A

combined molecular and morphological approach. *Curr. Biol.* 14: 1644–1649.

Iijima, M., Y. E, Huang and P. N. Deveotes. 2002. Temporal and spatial regulation of chemotaxis. *Dev. Cell* 3: 469–478.

King, N. 2004. The unicellular ancestry of animal development. *Dev. Cell* 7: 313–325.

Kirk, D. L. 2001. Germ-soma differentiation in *Volvox*. *Dev. Biol.* 238: 213–223.

Sultan, S. E. 2000. Phenotypic plasticity for plant development, function, and life history. *Trends Plant Sci.* 5: 53–542.

Wong, E., C. Yang, J. Wang, D. Fuller, W. F. Loomis and C. H. Siu. 2002. Disruption of the gene encoding the cell adhesion molecule DdCAD-1 leads to aberrant cell sorting and cell-type proportioning during *Dictyostelium* development. *Development* 129: 3839–3850.

Principles of Experimental Embryology

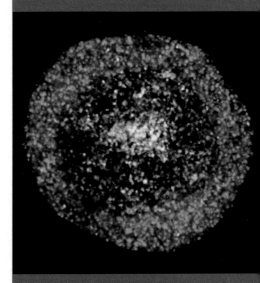

3

DESCRIPTIVE EMBRYOLOGY AND EVOLUTIONARY EMBRYOLOGY both had their roots in anatomy. At the end of the nineteenth century, however, the new biological science of physiology made inroads into embryological research. The questions of "what?" became questions of "how?" A new generation of embryologists felt that embryology should not merely be a guide to the study of anatomy and evolution, but should answer the question, "*How* does an egg become an adult?" Embryologists were urged to study the mechanisms of organ formation (morphogenesis) and differentiation. This new program was called *Entwicklungsmechanik*, often translated as "causal embryology," "physiological embryology," or "developmental mechanics." Its goals were to find the molecules and processes that caused the visible changes in embryos. Experimentation was to supplement observation in the study of embryos, and embryologists were expected to discover the properties of the embryo by seeing how the embryonic cells responded to perturbations and disruptions. Wilhelm Roux (1894), one of the founders of this branch of embryology, saw it as a grand undertaking:

> We must not hide from ourselves the fact that the causal investigation of organisms is one of the most difficult, if not the most difficult, problem which the human intellect has attempted to solve … since every new cause ascertained only gives rise to fresh questions regarding the cause of this cause.

In this chapter, we will discuss three of the major research programs in experimental embryology. The first concerns how forces outside the embryo influence its development. The second concerns how forces within the embryo cause the differentiation of its cells. The third looks at how the cells order themselves into tissues and organs.

> *It is possible, I think, by means of experimentation alone, to determine how far and in what sense we can pursue the investigation of the causes of form.*
> THOMAS HUNT MORGAN (1898)

> *The behaviour of a cell in an embryo depends on the extent to which it listens to its mother or its neighbourhood. The size and the nature of the noise, the way in which it is heard and the response are unpredictable and can only be discovered by experimentation.*
> JONATHAN BARD (1997)

 WEBSITE **3.1 Establishing experimental embryology.** The foundations of *Entwicklungsmechanik* were laid by a group of young investigators who desired a more physiological approach to embryology. These scientists disagreed with one another concerning the mechanisms of development, but they cooperated to secure places to perform and publish their research.

Environmental Developmental Biology

The developing embryo is not isolated from its environment. In numerous instances, environmental cues are a fundamental part of the organism's life cycle. The earliest tradition of experimental embryology involved altering the environmental circumstances in which the embryos developed and seeing if these perturbations changed the phenotype. In many instances, it was found that the genome did not necessarily encode a particular phenotype. Rather, the genotype demonstrated a **norm of reaction**—it encoded the information for a repertoire of possible phenotypes. The environment was then the determining factor as to which of the phenotypes within the norm of reaction would be expressed. This idea became known as **phenotypic plasticity** (Woltereck 1909; Schmalhausen 1949; Stearns et al. 1991; Schlichting and Pigliucci 1998). Phenotypic variants that result from environmental differences are often called **morphs**. A few of these experiments will be mentioned here, and many more will be detailed in Chapter 22.

Environmental sex determination

SEX DETERMINATION IN AN ECHIUROID WORM: *BONELLIA*   Although we associate sex determination in mammals with specific chromosomes (X and Y), there are numerous species for whom the environment plays the critical determining role. For instance, Baltzer (1914) showed that the sex of the echiuroid worm *Bonellia viridis* depended on where the larva settled. The female *Bonellia* worm is a marine, rock-dwelling animal, with a body about 10 cm long and a proboscis that can extend over a meter in length (Figure 3.1). The male *Bonellia*, however, is only 1–3 mm long and resides within the uterus of the female, fertilizing her eggs. Baltzer showed that if a *Bonellia* larva settles on the sea floor, it becomes a female. However, should a larva land on a female's proboscis (which apparently emits chemical signals that attract larvae), it enters the female's mouth, migrates into her uterus, and differentiates into a male. Baltzer (1914) and Leutert (1974) were able to duplicate this

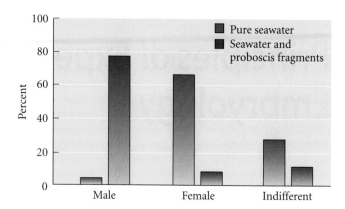

FIGURE 3.2   In vitro analysis of *Bonellia* sex determination. Larval *Bonellia* were placed either in normal seawater or in seawater containing fragments of the female proboscis. A majority of the animals cultured in the presence of the proboscis fragments became males, whereas in their absence, most became females. (After Leutert 1974.)

phenomenon in the laboratory by incubating larvae in either the absence or presence of adult females (Figure 3.2).

SEX DETERMINATION IN A VERTEBRATE: ALLIGATOR   The effects of the environment on development can have important consequences. Recent research has shown that the sex of alligators, crocodiles, and many other reptiles depends not on chromosomes, but on temperature. After studying the sex determination of the Mississippi alligator both in the laboratory and in the field, Ferguson and Joanen (1982) concluded that sex is determined by the temperature of the egg during the second and third weeks of incubation. Eggs incubated at 30°C or below during this time period produce female alligators, whereas eggs incubated at 34°C or above produce males. (At 32°C, 87 percent of the hatchlings were female.) Moreover, whereas nests built in wet marshes (close to 30°C) produce females, nests constructed on levees (close to 34°C) give rise to males. This allows alligators to escape the 1:1 sex ratio that would be imposed by chromosomal sex determination, and have a nearly 10:1 ratio of females to males in the population. These findings are obviously important to wildlife managers and farmers who wish to breed this species. They also raise questions of environmental policy, since the shade of buildings or the heat of thermal effluents can have dramatic effects on sex ratios among reptiles. We will discuss the mechanisms of temperature-dependent sex determination further in Chapter 17.

Adaptation of embryos and larvae to their environments

Another program of environmental developmental biology concerns how the embryo adapts to its particular environment. August Weismann (1875) pioneered the study of larval adaptations, and research in this area has provided

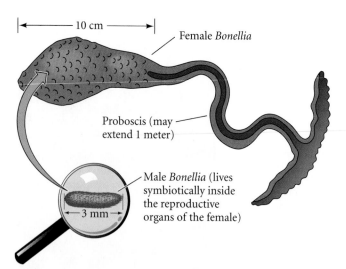

FIGURE 3.1   Sexual dimorphism in *Bonellia viridis*. The body of the mature female is about 10 cm in length, but the proboscis can extend up to a meter. The body of the symbiotic male is a minute 1–3 mm in length. While the body of the adult female is buried in the ocean sediments, her proboscis extends out of the sediments, where it can be used for feeding or attracting larvae.

FIGURE 3.3 Two morphs of *Araschnia levana*, the European map butterfly. The summer morph is represented at the top, the spring morph at the bottom. In this species, the phenotypic differences are elicited by differences in day length and temperature during the larval period. (Photographs courtesy of H. F. Nijhout.)

some fascinating insights into how an organism's development is keyed to its environment. Weismann noted that butterflies that hatched during different seasons were colored differently, and that this season-dependent coloration could be mimicked by incubating larvae at different temperatures. One example of such seasonal variation is the European map butterfly, *Araschnia levana*, which has two seasonal phenotypes so different that Linnaeus classified them as two different species (van der Weele 1995). The spring morph is bright orange with black spots, while the summer morph is mostly black with a white band (Figure 3.3). The shift from the spring to the summer morph is controlled by changes in both day length and temperature during the larval period. When researchers experimentally mimic spring conditions, summer caterpillars can give rise to "spring" butterflies (see Figure 22.11; Koch and Buchmann 1987; Nijhout 1991).

Another dramatic example of seasonal change in development occurs in the moth *Nemoria arizonaria*. This moth has a fairly typical insect life cycle. Eggs hatch in the spring, and the caterpillars feed on young oak flowers (catkins). These larvae metamorphose in the late spring, mate in the summer, and lay eggs on the oak trees, producing another brood of caterpillars. The second-brood caterpillars eat the oak leaves, metamorphose, and mate. Their

eggs overwinter to start the cycle over again the following spring. What is remarkable is that the caterpillars that hatch in the spring look nothing like their progeny that hatch in the summer. The spring caterpillars that eat oak catkins are yellow-brown, rugose, and beaded, resembling, to a great extent, an oak catkin (Figure 3.4A). They are magnificently camouflaged against predation. But what of the caterpillars that hatch in the summer, after all the catkins are gone? They, too, are well camouflaged, resembling year-old oak twigs (Figure 3.4B).

What controls this significant difference in phenotype? By doing reciprocal feeding experiments, Greene (1989) was able to convert spring caterpillars into summer morphs by feeding them oak leaves. However, the reciprocal experiment did not turn the summer morphs into catkin-like caterpillars. Thus, it appears that the catkin form is the "default state" and that something in the environment induces the twiglike morphology. That something is probably a tannin that is concentrated in oak leaves as they mature.

(A)

(B)

FIGURE 3.4 Two morphs of *Nemoria arizonaria*. (A) Caterpillars that hatch in the spring eat oak catkins and develop a cuticle that resembles these flowers. (B) Caterpillars that hatch in the summer (after the catkins are gone) eat oak leaves. These caterpillars develop a cuticle that resembles young oak twigs. (Photographs courtesy of E. Greene.)

PROTECTING THE EGG FROM UV RADIATION   Survival in their environments poses daunting challenges for embryos. Indeed, as Darwin clearly stated, most eggs and embryos fail to survive. A sea urchin may broadcast tens of thousands of eggs into the seawater, but only one or two (if any) of the resulting embryos will become adult urchins. Most eggs become food for other organisms. Moreover, if the environment changes, embryonic survival may increase or decrease dramatically. For instance, many eggs and early embryos lie in direct sunlight for long periods. If we lie in the sun for hours without sunscreen, our skin gets burned from the ultraviolet radiation; this radiation is also harmful to our DNA. So how do eggs survive all those hours of constant exposure to the sun (often on the same beaches where we sun ourselves)?

First, it seems that many eggs have evolved natural sunscreens. The eggs of many marine organisms contain high concentrations of mycosporine-like amino acid pigments, which absorb ultraviolet radiation (UV-B). Moreover, just like our skin's melanin pigment, these pigments can be induced by exposure to UV-B radiation (Jokiel and York 1982; Siebeck 1988). The eggs of tunicates are very resistant to UV-B radiation, and much of this resistance comes from extracellular coats enriched with mycosporine compounds (Mead and Epel 1995). Adams and Shick (1996, 2001) experimentally manipulated the amount of mycosporine-like amino acids in sea urchin eggs and found that embryos from eggs with more of these compounds were better protected from UV damage than embryos with less. Moreover, when mycosporine-deficient eggs were exposed to ultraviolet radiation, significant developmental anomalies were seen (Figure 3.5). Thus, these mycosporine-like amino acids appear to play an important role in protecting the developing sea urchin embryo against ultraviolet radiation.

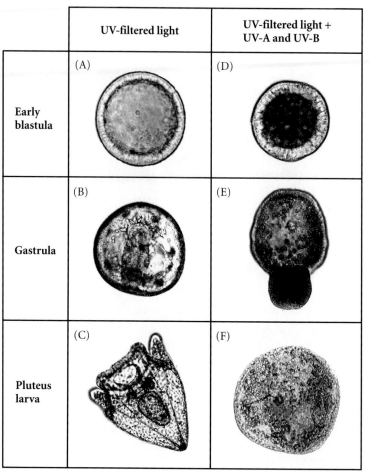

	UV-filtered light	UV-filtered light + UV-A and UV-B
Early blastula	(A)	(D)
Gastrula	(B)	(E)
Pluteus larva	(C)	(F)

FIGURE 3.5   The effect of ultraviolet (UV) radiation on embryos of the sea urchin *Strongylocentrotus droebachiensis*. Eggs were fertilized and placed in seawater lacking sources of mycosporine-like amino acids. The first column (A–C) represents embryos grown in light lacking ultraviolet radiation. The second column (D–F) shows the same stage embryos grown in the presence of such filtered light, but with ultraviolet radiation added. The red arrowheads show isolated skeletal spicules. (From Adams and Shick 2001, photographs courtesy of the authors.)

> **VADE MECUM[2]   Sea urchins and UV radiation.** This segment presents data documenting the protection of sea urchin embryos by mycosporine-like amino acids. This type of research is linking developmental biology with ecology and conservation biology. [Click on Sea Urchin-UV]

Increased UV-B exposure could be an important factor in the decline in amphibian populations seen throughout the world during the past two decades. Blaustein and his colleagues (1994; Blaustein and Beldin 2003) tested whether or not UV-B could be a factor in lowering the hatching rate of amphibian eggs. At two field sites, they divided the eggs of each of three amphibian species into three groups. The first group developed without any sun filter. The second group developed under a filter that allowed UV-B to pass through. The third group developed under a filter that blocked UV-B from reaching the eggs. For *Hyla regilla*, the filters had no effect; hatching success was excellent under all three conditions. For *Rana cascadea* and *Bufo boreas*, however, the UV-B blocking filter raised the percentage of eggs hatched from about 60 percent to close to 80 percent (Figure 3.6).

The effects of UV-B radiation in mediating amphibian population declines appears to be complex, involving climate change and fungal pathogens. Using long-term observational data and by manipulating the depth of pond water in which frogs laid their eggs, Kiesecker and colleagues (2001) showed that climate-induced reductions in pond water depth increase the exposure of eggs and embryos to UV-B radiation and, consequently, impair their immune systems, making them more vulnerable to fungal infection. The effect of UV-B on the frog immune system could not have come at a worse time, since several frog extinctions have been caused by an outbreak of a particu-

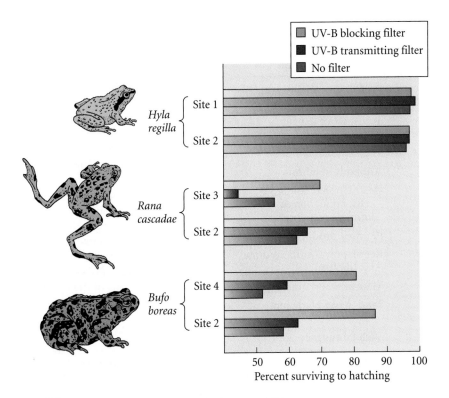

FIGURE 3.6 Hatching success rates in three amphibian species in the field. At each of two sites, eggs were placed in enclosures that were unshielded, shielded with an acetate screen that admitted UV-B radiation, or shielded with a Mylar screen that blocked UV-B radiation. Eggs of the tree frog *Hyla regilla* hatched successfully under all three conditions. Eggs of the frog *Rana cascadae* and the toad *Bufo boreas* hatched significantly better when protected from UV-B radiation. (After Blaustein et al. 1994.)

lar fungus that kills its amphibian hosts (Pounds et al. 2006). Climate warming in Central American has shifted the temperature in the mountain regions into the range that is optimum for the growth of this fungus.* The combination of UV-B exposure and global warming appears to be responsible for the devastation of many frog populations throughout the world.

The environmental programs of experimental embryology were a major part of the discipline when *Entwicklungsmechanik* was first established. However, it soon became obvious that experimental variables could be better controlled in the laboratory than in the field, and that a scientist could do many more experiments in the labora-

tory. Thus, field experimentation in embryology dwindled in the first decades of the twentieth century (see Nyhart 1995). However, with our increasing concern about the environment, this area of developmental biology has become increasingly important. Other recent work in this field will be detailed in Chapter 22.

The Developmental Dynamics of Cell Specification

The second research program of experimental embryology centered around the differentiation question. The development of specialized cell types is called **differentiation** (Table 3.1). But these changes in cell type do not happen immediately. Rather, these overt changes in cellular biochemistry and function are preceded by a process resulting in the **commitment** of the cell to a certain fate. At this point, even though the cell or tissue does not differ phenotypically from its uncommitted state, its developmental fate has become restricted.

The process of commitment can be divided into two stages (Harrison 1933; Slack 1991). The first stage is a labile phase called **specification**. The fate of a cell or a tissue is said to be specified when it is capable of differentiating autonomously when placed in a neutral environment, such as a petri dish or test tube. (The environment is neutral with respect to the developmental pathway.) At this stage, the commitment is still capable of being reversed. The second stage of commitment is **determination**. A cell or tissue is said to be determined when it is capable of differentiating autonomously even when placed into another region of the embryo. If it is able to differentiate according to its original fate even under these circumstances, it is assumed that the commitment is irreversible.[†]

Autonomous specification

Three basic modes of commitment have been described (Table 3.2). The first is called **autonomous specification**. In this case, if a particular blastomere is removed from an embryo early in its development, that isolated blastomere will produce the same types of cells that it would have made if it were still part of the embryo (Figure 3.7). More-

---

*In addition to climate change, humans were probably also responsible for spreading the *Batrachochytrium dendrobatidis* fungus throughout the world. This fungus was first seen in southern Africa, where its host was *Xenopus laevis*. The global trade in *Xenopus* for pregnancy tests in the 1950s led to the worldwide distribution of the frog, and most likely its fungal parasite as well (Blaustein and Dobson 2006).

[†]This irreversibility of determination and differentiation is only with respect to normal development. As Dolly and other cloned animals have shown, the nucleus of a differentiated cell can be reprogrammed experimentally to give rise to any cell type in the body. We will discuss this phenomenon in Chapter 4.

TABLE 3.1   Some differentiated cell types and their major products

Type of cell	Differentiated cell product	Specialized function
Keratinocyte (epidermal cell)	Keratin	Protection against abrasion, desiccation
Erythrocyte (red blood cell)	Hemoglobin	Transport of oxygen
Lens cell	Crystallins	Transmission of light
B lymphocyte	Immunoglobulins	Antibody synthesis
T lymphocyte	Cytokines	Destruction of foreign cells; regulation of immune response
Melanocyte	Melanin	Pigment production
Pancreatic islet cell	Insulin	Regulation of carbohydrate metabolism
Leydig cell (♂)	Testosterone	Male sexual characteristics
Chondrocyte (cartilage cell)	Chondroitin sulfate; type II collagen	Tendons and ligaments
Osteoblast (bone-forming cell)	Bone matrix	Skeletal support
Myocyte (muscle cell)	Muscle actin and myosin	Contraction
Hepatocyte (liver cell)	Serum albumin; numerous enzymes	Production of serum proteins and numerous enzymatic functions
Neurons	Neurotransmitters (acetylcholine, epinephrine, etc.)	Transmission of electric impulses
Tubule cell (♀) of hen oviduct	Ovalbumin	Egg white proteins for nutrition and protection of embryo
Follicle cell (♀) of insect ovary	Chorion proteins	Eggshell proteins for protection of embryo

over, the embryo from which that blastomere is taken will lack those cells (and only those cells) that would have been produced by the missing blastomere. Autonomous specification gives rise to a pattern of embryogenesis referred to as **mosaic development**, since the embryo appears to be constructed like a tile mosaic of independent, self-differentiating parts. Invertebrate embryos, especially those of molluscs, annelids, and tunicates, often use autonomous specification to determine the fates of their cells. In these embryos, **morphogenetic determinants** (certain proteins or messenger RNAs) are placed in different regions of the egg cytoplasm and are apportioned to the different cells as the embryo divides. These morphogenetic determinants specify the cell type.

Autonomous specification was first demonstrated in 1887 by a French medical student, Laurent Chabry. Chabry desired to know the causes of birth defects, and he reasoned that such malformations might be caused by the lack of certain cells. He decided to perform experiments on tunicate embryos, since they have relatively large cells and were abundant in a nearby bay. This was a fortunate choice, because tunicate embryos develop rapidly into larvae with relatively few cells and cell types (Chabry 1887; Fischer 1991).

FIGURE 3.7 Autonomous specification (mosaic development). (A–C) Differentiation of trochoblast (ciliated) cells of the mollusc *Patella*. (A) 16-cell stage seen from the side; the presumptive trochoblast cells are shaded. (B) 48-cell stage. (C) Ciliated larval stage, seen from the animal pole. (D–G) Differentiation of a *Patella* trochoblast cell isolated from the 16-cell stage and cultured in vitro. (E,F) Results of the first and second divisions in culture. (G) Ciliated products of (F). Even in isolated culture, these cells divide and become ciliated at the correct time. (After Wilson 1904.)

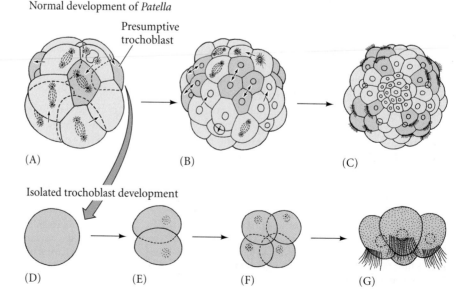

Normal development of *Patella*

Presumptive trochoblast

(A)　　(B)　　(C)

Isolated trochoblast development

(D)　　(E)　　(F)　　(G)

TABLE 3.2   Modes of cell type specification and their characteristics

I. Autonomous specification

Characteristic of most invertebrates.

Specification by differential acquisition of certain cytoplasmic molecules present in the egg.

Invariant cleavages produce the same lineages in each embryo of the species. Blastomere fates are generally invariant.

Cell type specification precedes any large-scale embryonic cell migration.

Produces "mosaic" development: cells cannot change fate if a blastomere is lost.

II. Conditional specification

Characteristic of all vertebrates and few invertebrates.

Specification by interactions between cells. Relative positions are important.

Variable cleavages produce no invariant fate assignments to cells.

Massive cell rearrangements and migrations precede or accompany specification.

Capacity for "regulative" development: allows cells to acquire different functions.

III. Syncytial specification

Characteristic of most insect classes.

Specification of body regions by interactions between cytoplasmic regions prior to cellularization of the blastoderm.

Variable cleavage produces no rigid cell fates for particular nuclei.

After cellularization, conditional specification is most often seen.

Source: After Davidson 1991.

Chabry set out to produce specific malformations by isolating or lancing specific blastomeres of the cleaving tunicate embryo. He discovered that each blastomere was responsible for producing a particular set of larval tissues (Figure 3.8). In the absence of particular blastomeres, the larvae lacked just those structures normally formed by those cells. Moreover, he observed that when particular cells were isolated from the rest of the embryo, they formed their characteristic structures apart from the context of the other cells. Thus, each of the tunicate cells appeared to develop autonomously.*

---

*This was not the answer Chabry expected, nor was it the one he had hoped to find. In nineteenth-century France, conservatives favored preformationist views, which were interpreted to support hereditary inequalities between members of a human community. What you were was determined by your lineage. Liberals, especially Socialists, favored epigenetic views, which were interpreted to indicate that everyone started off with an equal hereditary endowment, and that no one had a "right" to a higher position than any other person. Chabry, a Socialist who hated the inherited rights of the aristocrats, took pains not to extrapolate his data to anything beyond tunicate embryos (see Fischer 1991).

FIGURE 3.8   Autonomous specification in the early tunicate embryo. When the four blastomere pairs of the 8-cell embryo are dissociated, each forms the structures it would have formed had it remained in the embryo. (The fate map of the tunicate shows that the left and right sides produce identical cell lineages.) (After Reverberi and Minganti 1946.)

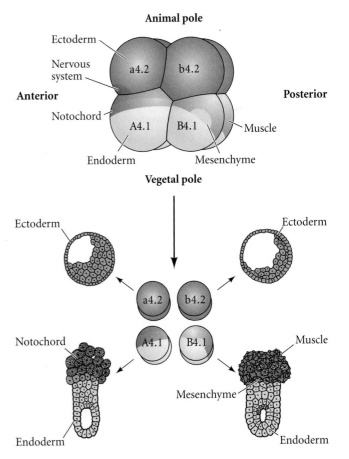

Recent studies have confirmed that when particular cells of the 8-cell tunicate embryo are removed, the embryo lacks those structures normally produced by the missing cells, and the isolated cells produce these structures away from the embryo. J. R. Whittaker provided dramatic biochemical confirmation of the cytoplasmic segregation of the morphogenetic determinants responsible for this pattern. Whittaker (1973) stained blastomeres for the presence of the enzyme acetylcholinesterase. This enzyme is found only in muscle tissue and is involved in enabling larval muscles to respond to repeated nerve impulses. From the cell lineage studies of Conklin and others (see Chapter 1), it was known that only one pair of blastomeres (the posterior vegetal pair, B4.1) in the 8-cell tunicate embryo is capable of producing tail muscle tissue. (As discussed in Chapter 1, the B4.1 blastomere pair contains the yellow crescent cytoplasm that correlates with muscle determination.) When Whittaker removed these two cells and placed them in isolation, they produced muscle tissue that stained positively for the presence of acetylcholinesterase (Figure 3.9). When he transferred some of the yellow crescent cytoplasm of the B4.1 (muscle-forming) blastomere into the b4.2 (ectoderm-forming) blastomere of an 8-cell tunicate embryo, the ectoderm-forming blastomere generated muscle cells as well as its normal ectodermal progeny (Figure 3.10; Whittaker 1982). Thus, the "find it, lose it, move it" scheme for experimentation (see "Sidelights & Speculations" in Chapter 2) revealed the autonomous nature of muscle formation in these cells. The mechanisms of autonomous specification will be detailed in Chapter 8.

Syncytial specification

Many insects use a means known as **syncytial specification** to commit cells to their fates. In early embryos of these insects, cell division is not complete. Rather, the nuclei divide within the egg cytoplasm, creating many nuclei within one large egg cell. A cytoplasm that contains many nuclei is called a **syncytium**. The egg cytoplasm, however, is not uniform. Rather, the anterior of the egg cytoplasm is markedly different from the posterior. Here, the interactions of syncytial specification occur not between cells, but between the different parts of a single cell.

Experimental embryologists have shown that each nucleus in *Drosophila* is given positional information (whether it is to become part of the anterior, posterior, or midsection of the body) by proteins called **morphogens** (Greek, "form-givers"). Morphogens are made in specific sites in the embryo, diffuse over long distances, and form **concentration gradients** where the highest concentration is at the point of synthesis and gets lower as the morphogen diffuses away from its source and degrades over time. The concentration of specific morphogens at any particular site tells the cells where they are in relation to the source of the morphogens. As we will see in Chapter 9, the anteriormost portion of the *Drosophila* embryo produces a morphogenic protein called Bicoid. The posteriormost por-

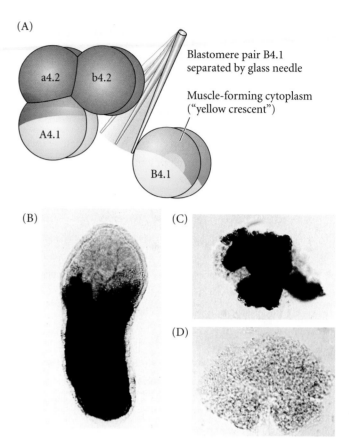

FIGURE 3.9 Acetylcholinesterase in the progeny of the muscle lineage blastomeres (B4.1) isolated from a tunicate embryo at the 8-cell stage. (A) Diagram of the isolation procedure. (B) Localization of acetylcholinesterase in the tail muscles of an intact tunicate larva. The presence of the enzyme is demonstrated by the dark staining. The same dark staining is seen in the progeny of the B4.1 blastomere pair (C), but not in the remaining 6/8 of the embryo (D) when incubated for the length of time it normally takes to form a larva. (From Whittaker 1977; photographs courtesy of J. R. Whittaker.)

tion of the egg produces a protein called Nanos. The concentration of Bicoid protein is highest in the anterior and declines toward the posterior, while that of Nanos is highest in the posterior and declines anteriorly. Thus, the long axis of the *Drosophila* egg is spanned by two opposing morphogen gradients—Bicoid coming from the anterior, and Nanos from the posterior. The Bicoid and Nanos proteins form a **coordinate system** based on their ratios, such that each region of the embryo is distinguished by a different ratio of the two proteins. As the nuclei divide and enter different regions of the syncytium's cytoplasm, they are instructed by the Bicoid:Nanos ratio as to their position along the anterior-posterior axis. Those nuclei in regions containing high amounts of Bicoid and little Nanos will be instructed to activate those genes necessary for producing the head. Nuclei in regions with slightly less Bicoid but with a small amount of Nanos are instructed to activate

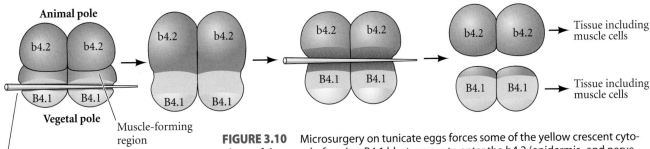

FIGURE 3.10 Microsurgery on tunicate eggs forces some of the yellow crescent cytoplasm of the muscle-forming B4.1 blastomeres to enter the b4.2 (epidermis- and nerve-producing) blastomere pair. Pressing the B4.1 blastomeres with a glass needle causes the regression of the cleavage furrow. The furrow will re-form at a more vegetal position where the cells are cut with a needle. The new furrow will thereby separate the cells in such a way that the b4.2 blastomeres receive some of the muscle-forming ("yellow crescent") B4.1 cytoplasm. These modified b4.2 cells produce muscle cells as well as their normal ectodermal progeny. (After Whittaker 1982.)

those genes that generate the thorax. Nuclei in regions that have little or no Bicoid but plenty of Nanos are instructed to form the abdominal structures (Figure 3.11; Nüsslein-Volhard et al. 1987).

That Bicoid was the head morphogen of *Drosophila* was demonstrated by the "find it, lose it, move it" experimentation scheme. Christianne Nüsslein-Volhard, Wolfgang Driever, and their colleagues (Driever and Nüsslein-Vol-hard 1988; Driever et al. 1990) showed that (1) Bicoid protein was found in a gradient, highest in the anterior (head-forming) region; (2) embryos lacking Bicoid protein could not form a head; and (3) when Bicoid was added to Bicoid-deficient embryos in different places, the place where Bicoid was injected became the head. Moreover, the areas around the site of Bicoid injection became the thorax, as expected from a concentration-dependent signal (Figure 3.12). How the Bicoid gradient works and its role in syncytial specification will be detailed in Chapter 9.

Conditional specification

THE PHENOMENA OF CONDITIONAL SPECIFICATION AND REGULATION
The third mode of commitment involves interactions among neighboring cells. In this type of specification, each cell originally has the ability to become any of many different cell types. However, interactions of the cell with other cells restrict the fate of one or more of the participants. This mode of commitment is called **conditional specification** because the fate of a cell depends upon the conditions in which the cell finds itself.

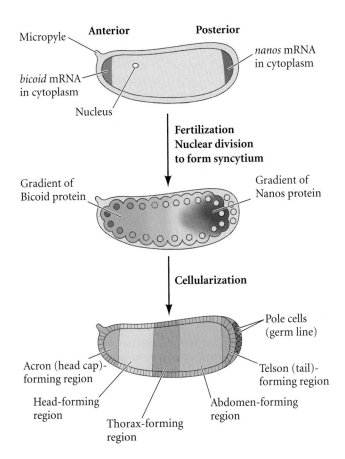

FIGURE 3.11 Syncytial specification in the fruit fly *Drosophila melanogaster*. Anterior-posterior specification originates from morphogen gradients within the egg cytoplasm. The *bicoid* mRNA is stabilized in the most anterior portion of the egg, while *nanos* mRNA is tethered to the posterior end. (The anterior can be recognized by the micropyle on the shell; this structure permits sperm to enter.) Once the egg is fertilized, these two mRNAs are translated into proteins. Bicoid protein forms a gradient that is highest at the anterior end, while Nanos forms a gradient that is highest at the posterior end. The ratio of Bicoid to Nanos forms a coordinate system that distinguishes each position along the axis from any other position. When nuclear division occurs, the amount of each protein present differentially activates transcription of the various nuclear genes that specify the segmental identities of the larva and the adult fly.

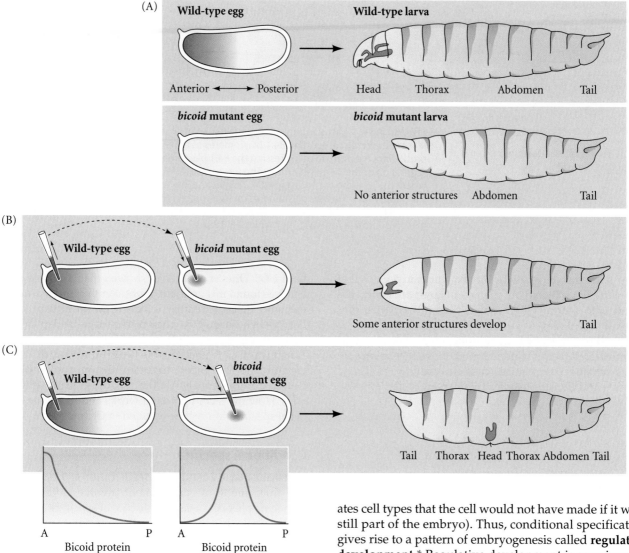

FIGURE 3.12 Bicoid protein is the head morphogen in *Drosophila* embryos. (A) "Find it": Shortly after fertilization, *bicoid* mRNA in the anterior portion of the embryo is translated into Bicoid protein, which diffuses in a gradient throughout the anterior half of the embryo. (B) "Lose it": Embryos that are genetically deficient in Bicoid protein cannot make head structures. (C) "Move it": When *bicoid* mRNA or Bicoid-containing anterior cytoplasm is placed into a Bicoid-deficient egg, head structures develop at the position where the injection is made. The areas adjacent to these anterior structures become thorax. (After Wolpert et al. 2002.)

ates cell types that the cell would not have made if it were still part of the embryo). Thus, conditional specification gives rise to a pattern of embryogenesis called **regulative development**.* Regulative development is seen in most vertebrate embryos, and it is critical in the development of identical twins. In the formation of such twins, the cleavage-stage cells of a single embryo divide into two groups, and each group of cells produces a fully developed individual (Figure 3.14).

EXPERIMENTATION AND THE ELUCIDATION OF CONDITIONAL SPECIFICATION
The research that led to the discovery of conditional specification began with the testing of a hypothesis claiming that such regulation did not exist. In the late 1800s, August Weismann proposed the first testable model of cell spe-

If a blastomere is removed from an early embryo that uses conditional specification, the remaining embryonic cells alter their fates so that the roles of the missing cells are taken over (Figure 3.13). This ability of embryonic cells to change their fates to compensate for missing parts is called **regulation**. The isolated blastomere can also give rise to a wide variety of cell types (and sometimes gener-

*Sydney Brenner (quoted in Wilkins 1993) has remarked that animal development can proceed according to either the American or the European plan. Under the European plan (autonomous specification), you are what your progenitors were. Lineage is important. Under the American plan (conditional specification), the cells start off undetermined, but with certain biases. There is a great deal of mixing, lineages are not critical, and one tends to become what one's neighbors are.

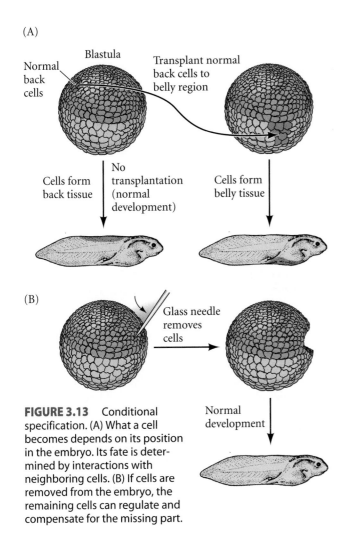

FIGURE 3.13 Conditional specification. (A) What a cell becomes depends on its position in the embryo. Its fate is determined by interactions with neighboring cells. (B) If cells are removed from the embryo, the remaining cells can regulate and compensate for the missing part.

cification, the **germ plasm theory**. Based on the scant knowledge of fertilization available at that time, Weismann boldly proposed that the sperm and egg provided equal chromosomal contributions, both quantitatively and qualitatively, to the new organism. Moreover, he postulated that the chromosomes carried the inherited potentials of this new organism.* However, not all the determinants on the chromosomes were thought to enter every cell of the embryo. Instead of dividing equally, the chromosomes were hypothesized to divide in such a way that different chromosomal determinants entered different cells. Whereas the fertilized egg would carry the full complement of determinants, certain somatic cells would retain the "blood-forming" determinants while others would retain the "muscle-forming" determinants, and so forth (Figure 3.15). Only in

the nuclei of those cells destined to become gametes (the germ cells) were all types of determinants thought to be retained. The nuclei of all other cells would have only a subset of the original determinant types.

In postulating his germ plasm model, Weismann had proposed a hypothesis of development that could be tested immediately. Based on the fate map of the frog embryo, Weismann claimed that when the first cleavage division separated the future right half of the embryo from the future left half, there would be a separation of "right" determinants from "left" determinants in the resulting blastomeres. The testing of this hypothesis pioneered the first three of these four major techniques of experimental embryology:

- The **defect experiment**, wherein one destroys a portion of the embryo and then observes the development of the impaired embryo.
- The **isolation experiment**, wherein one removes a portion of the embryo and then observes the development of both the partial embryo and the isolated part.
- The **recombination experiment**, wherein one observes the development of the embryo after replacing an original part with a part from a different region of the same embryo.
- The **transplantation experiment**, wherein one portion of the embryo is replaced by a portion from a different embryo. This fourth technique was used by some of the same scientists when they first constructed fate maps of early embryos (see Chapter 1).

One of the first scientists to test Weismann's hypothesis was Wilhelm Roux, a young German embryologist. In 1888, Roux published the results of a series of defect experiments in which he took 2- and 4-cell frog embryos and destroyed some of the cells of each embryo with a hot needle. Weis-

* Embryologists were thinking in terms of a mechanism of inheritance some 15 years before the rediscovery of Mendel's work. Weismann (1892, 1893) also speculated that these nuclear determinants of inheritance functioned by elaborating substances that became active in the cytoplasm!

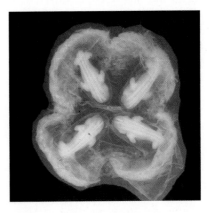

FIGURE 3.14 In the early developmental stages of many vertebrates, the separation of the embryonic cells into two parts can create twins. This phenomenon occurs sporadically in humans. However, in the nine-banded armadillo, *Dasypus novemcinctus*, the original embryo always splits into four separate groups of cells, each of which forms its own embryo. (Photograph courtesy of K. Benirschke.)

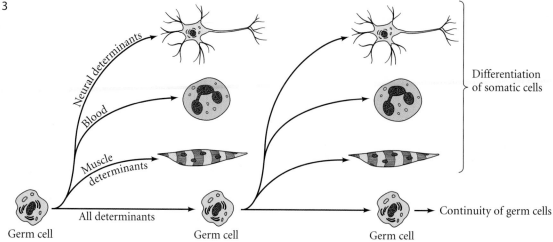

FIGURE 3.15 Weismann's germ plasm theory of inheritance. The germ cell (blue) gives rise to the differentiating somatic cells of the body, as well as to new germ cells. Weismann hypothesized that only the germ cells contained all the inherited determinants. The somatic cells were each thought to contain a subset of the determinants, and the types of determinants found in the somatic cell's nucleus would determine its differentiated type. (After Wilson 1896.)

mann's hypothesis predicted the formation of right or left half-embryos. Roux obtained half-blastulae, just as Weismann had predicted (Figure 3.16). These developed into half-neurulae having a complete right or left side, with one neural fold, one ear pit, and so on. Roux therefore concluded that the frog embryo was a "mosaic" of self-differentiating parts, and that each cell probably received a specific set of determinants and differentiated accordingly.

Nobody appreciated Roux's work and the experimental approach to embryology more than Hans Driesch. Driesch's goal was to explain development in terms of the laws of physics and mathematics. His initial investigations were similar to those of Roux. However, while Roux's studies were *defect* experiments that answered the question of how the remaining blastomeres of an embryo would develop when a subset of blastomeres was destroyed, Driesch (1892) sought to extend this research by performing *isola-*

tion experiments. He separated sea urchin blastomeres from each other by vigorous shaking (or, later, by placing them into calcium-free seawater). To Driesch's surprise, each of the blastomeres from a 2-cell embryo developed into a complete larva. Similarly, when Driesch separated the blastomeres of 4- and 8-cell embryos, some of the isolated cells produced entire pluteus larvae (Figure 3.17). Here was a result drastically different from the predictions of Weismann and Roux. Rather than self-differentiating into its future embryonic part, each isolated blastomere *regulated* its development to produce a complete organism. These experiments provided the first experimentally observable evidence of regulative development.

Driesch confirmed regulative development in sea urchin embryos by performing an intricate *recombination* experiment. In sea urchin eggs, the first two cleavage planes are normally meridional, passing through both the animal and vegetal poles, whereas the third division is equatorial, dividing the embryo into four upper and four lower cells (Figure 3.18A). Driesch (1893) changed the direction of the third cleavage by gently compressing early embryos between two glass plates, thus causing the third division to be meridional like the preceding two. After he released the pressure, the fourth division was equatorial. This procedure reshuffled the nuclei, causing a nucleus that normally would be in the region destined to form endoderm

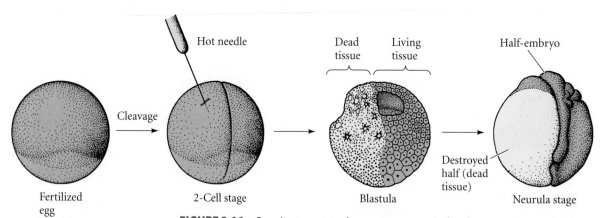

FIGURE 3.16 Roux's attempt to demonstrate mosaic development. Destroying (but not removing) one cell of a 2-cell frog embryo results in the development of only one-half of the embryo.

FIGURE 3.17 Driesch's demonstration of regulative development. (A) An intact 4-cell sea urchin embryo generates a normal pluteus larva. (B) When one removes the 4-cell embryo from its fertilization envelope and isolates each of the four cells, each cell can form a smaller, but normal, pluteus larva. (All larvae are drawn to the same scale.) Note that the four larvae derived in this way are not identical, despite their ability to generate all the necessary cell types. Such variation is also seen in adult sea urchins formed in this way (see Marcus 1979). (Photograph courtesy of G. Watchmaker.)

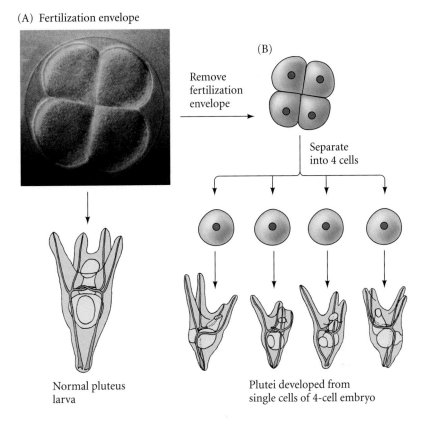

(A) Fertilization envelope

(B)

Remove fertilization envelope

Separate into 4 cells

Normal pluteus larva

Plutei developed from single cells of 4-cell embryo

to now be in the presumptive ectoderm region. Some nuclei that would normally have produced dorsal structures were now found in the ventral cells (Figure 3.18B). If segregation of nuclear determinants had occurred (as had been proposed by Weismann and Roux), the resulting embryo should have been strangely disordered. However, Driesch obtained normal larvae from these embryos. He concluded, "The relative position of a blastomere within the whole will probably in a general way determine what shall come from it."

The consequences of these experiments were momentous, both for embryology and for Driesch personally. First, Driesch had demonstrated that the prospective potency of an isolated blastomere (those cell types it was possible for it to form) is greater than its prospective fate (those cell types it would normally give rise to over the unaltered course of its development). According to Weismann and Roux, the prospective potency and the prospective fate of a blastomere should have been identical. Second, Driesch concluded that the sea urchin embryo is a

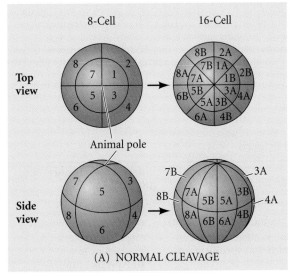

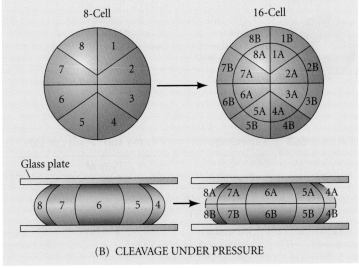

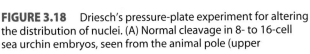

Top view

Animal pole

Side view

8-Cell   16-Cell

(A) NORMAL CLEAVAGE

8-Cell   16-Cell

Glass plate

(B) CLEAVAGE UNDER PRESSURE

FIGURE 3.18 Driesch's pressure-plate experiment for altering the distribution of nuclei. (A) Normal cleavage in 8- to 16-cell sea urchin embryos, seen from the animal pole (upper sequence) and from the side (lower sequence). (B) Abnormal cleavage planes formed under pressure, as seen from the animal pole and from the side. (After Huxley and de Beer 1934.)

"harmonious equipotential system" because all of its potentially independent parts functioned together to form a single organism. Third, he concluded that the fate of a nucleus depended solely on its location in the embryo. Driesch (1894) hypothesized a series of events wherein development proceeded by the interactions of the nucleus and cytoplasm:

> Insofar as it contains a nucleus, every cell, during development, carries the totality of all primordia; insofar as it contains a specific cytoplasmic cell body, it is specifically enabled by this to respond to specific effects only. …When nuclear material is activated, then, under its guidance, the cytoplasm of its cell that had first influenced the nucleus is in turn changed, and thus the basis is established for a new elementary process, which itself is not only the result but also a cause.

This strikingly modern concept of nuclear-cytoplasmic interaction and nuclear equivalence eventually caused Driesch to abandon science. Because the embryo could be subdivided into parts that were each capable of re-forming the entire organism, he could no longer envision it as a physical machine. In other words, Driesch came to believe that development could not be explained by physical forces. Harking back to Aristotle, he invoked a vital force, *entelechy* ("internal goal-directed force") to explain how development procedes. Essentially, he believed that the embryo was imbued with an internal psyche and the wisdom to accomplish its goals despite the obstacles embryologists placed in its path. Unable to explain his results in terms of the physics of his day, Driesch renounced the study of developmental physiology and became a philosophy professor, proclaiming vitalism (the doctrine that living things cannot be explained by physical forces alone) until his death

in 1941. However, others, especially Oscar Hertwig (1894), were able to incorporate Driesch's experiments into a more sophisticated experimental embryology.[*]

The differences between Roux's experiments and those of Driesch are summarized in Table 3.3. The difference between isolation and defect experiments and the importance of the interactions among blastomeres were highlighted in 1910, when J. F. McClendon showed that isolated frog blastomeres behave just like separated sea urchin cells. Therefore, the mosaic-like development of the first two frog blastomeres in Roux's study was an artifact of the defect experiment. Something in or on the dead blastomere still informed the live cells that it existed. Therefore, even though Weismann and Roux pioneered the study of developmental physiology, their proposition that differentiation is caused by the segregation of nuclear determinants was soon shown to be incorrect. It should be remembered that no animal so far studied was only one type of specification. *Drosophila* embryos, for instance, will use all three mechanisms to specify different groups of cells.

---

[*]Hertwig used Driesch's experiments and some of his own to strengthen, within the field of embryology, a type of materialist philosophy called **wholist organicism**. This philosophy embraces the views that (1) the properties of the whole cannot be predicted solely from the properties of the component parts, and (2) the properties of the parts are informed by their relationship to the whole. As an analogy, the meaning of a sentence depends on the meanings of its component parts (words). However, the meaning of each word depends on the entire sentence. In the sentence, "The party leaders were split on the platform," the possible meanings of the individual words are limited by the meaning of the entire sentence and by their relationships to other words within the sentence. Similarly, the phenotype of a cell in the embryo depends on its interactions within the entire embryo. The opposite materialist view is **reductionism**, which maintains that the properties of the whole can be known if all the properties of the parts are known. Embryology has traditionally espoused wholist organicism as its ontology (model of reality) while maintaining a reductionist methodology (experimental procedures) (Needham 1943; Haraway 1976; Hamburger 1988; Gilbert and Sarkar 2000).

> **VADE MECUM²   Sea urchin development.** Roux's and Driesch's experiments manipulated normal development. Normal sea urchin development is seen here in video and labeled photographs. [Click on Sea Urchin]

TABLE 3.3   Experimental procedures and results of Roux and Driesch

Investigator	Organism	Type of experiment	Conclusion	Interpretation concerning potency and fate
Roux (1888)	Frog (*Rana fusca*)	Defect	Mosaic development (autonomous specification)	Prospective potency equals prospective fate
Driesch (1892)	Sea urchin (*Echinus microtuberculatus*)	Isolation	Regulative development (conditional specification)	Prospective potency is greater than prospective fate
Driesch (1893)	Sea urchin (*Echinus* and *Paracentrotus*)	Recombination	Regulative development (conditional specification)	Prospective potency is greater than prospective fate

Morphogen gradients revisited

It seems that there is no one "right way" of specifying cell fates. There are three major modes of specification—autonomous, syncytial, and conditional—but within each of these modes there are numerous variations. In conditional specification, cells can become committed either by interactions with neighboring cells or by a gradient of morphogen. In numerous cases (many of which will be discussed in Chapter 6), specific types of proteins are secreted by adjacent cells to instruct the target cell as to its fate. In other cases, a given cell is committed to become one type of cell if it receives a certain protein in a specific concentration, but can become committed in a different direction if it receives the same protein but at a higher or lower concentration. This latter means of specification involves, as we discussed earlier in the chapter, concentration gradients of a morphogen. However, whereas our discussion of syncytial specification concerned the diffusion of morphogens *within* the cytoplasm of a single (albeit very large) cell, our present discussion focuses on morphogens secreted by one group of cells and received by other groups of nearby cells.

The concept of morphogen gradients had been used since the 1700s to model another phenomenon of regulative development: **regeneration**. It had been known since then that when hydras and planarian flatworms were cut in half, the head half would regenerate a tail from the wound site, while the tail half would regenerate a head (Pallas 1766; see "Sidelights & Speculations" in Chapter 2). Allman (1864) had called attention to the fact that this phenomenon indicated a polarity in the organization of the hydra. It was not until 1905, however, that Thomas Hunt Morgan (1905, 1906) realized that such polarity indicated an important principle in development.*

From the 1930s through the 1950s, gradient models were used to explain conditional cell specification in sea urchin and amphibian embryos (Hörstadius and Wolsky 1936; Hörstadius 1939; Toivonen and Saxén 1955). In the 1960s, these gradient models were extended to explain how cells might be told their position along an embryonic axis (Lawrence 1966; Stumpf 1966; Wolpert 1968, 1969). According to such models, a soluble substance—the morphogen—diffuses from its site of synthesis (source) to its site of degradation (sink). Wolpert (1968) illustrated such a gradient of positional information using the "French flag" analogy (Figure 3.19). Imagine a row of "flag cells," each of which is capable of differentiating into a red, white, or blue cell. Then imagine a morphogen whose source is on the left-hand edge of the blue stripe, and whose sink is at the other end of the flag, on the right-hand edge of the red stripe. A concentration gradient is thus formed, being high-

---

*Before he became a founder of genetics, T. H. Morgan was an embryologist who helped pioneer the study of flatworm regeneration. In 1900, T. H. Morgan first mentions *Drosophila*—as food for his flatworms! He was even able to "stain" the flatworms' digestive tubes by feeding them pigmented *Drosophila* eyes.

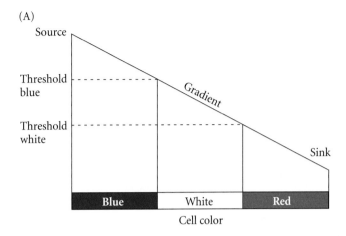

(A)

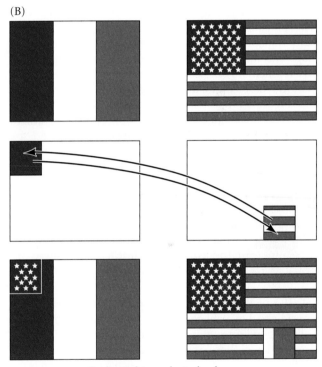

Reciprocal transplants develop
according to their final positions
in the "donor" flag

FIGURE 3.19 The "French flag" analogy for the operation of a gradient of positional information. (A) In this model, positional information is delivered by a gradient of a diffusible morphogen extending from a source to a sink. The thresholds indicated on the left are cellular properties that enable the gradient to be interpreted. For example, cells become blue at one concentration of the morphogen, but as the concentration declines below a certain threshold, cells become white. Where the concentration falls below another threshold, cells become red. The result is a pattern of three colors. (B) An important feature of this model is that a piece of tissue transplanted from one region of an embryo to another retains its identity (as to its origin), but differentiates according to its new positional instructions. This phenomenon is indicated schematically by reciprocal "grafts" between the flags of the United States and France. (After Wolpert 1978.)

est at one end of the "flag tissue" and lowest at the other. The specification of the multipotential cells in this tissue is accomplished by the concentration of the morphogen. Cells sensing a high concentration of the morphogen become blue. Then there is a threshold of morphogen concentration below which cells become white. As the declining concentration of morphogen falls below another threshold, the cells become red (see Figure 3.19A).

Different tissues may use the same gradient system, but respond to the gradient in a different way. If cells that would normally become the middle segment of a *Drosophila* leg are removed from the leg-forming area of the larva and placed in the region that will become the tip of the fly's antenna, they differentiate into claws. These cells retain their committed status as leg cells, but respond to the positional information of their environment by becoming leg tip cells—i.e., claws (Postlethwait and Schneiderman 1971). This phenomenon, said Wolpert, is analogous to reciprocally transplanting portions of the American and French flags into each other. The segments will retain their identity (French or American), but will be positionally specified (develop colors) appropriate to their new positions (see Figure 3.19B). This model is essentially the "digitalization" of an analogue signal. The concentration gradient of a morphogen is established, and somehow "read" by the cells; and the cells will become a particular cell type, depending on the level of morphogen it receives.

WEBSITE **3.2 Receptor gradients.** In addition to a gradient of morphogen, there can also be a gradient of those molecules that recognize the morphogen. The interplay of morphogen gradients and the gradients of molecules that interpret them can give rise to interesting developmental patterns.

IDENTIFYING MORPHOGENS For a diffusible molecule to be considered a morphogen, it must be demonstrated that (1) cells respond directly to that molecule, and that (2) the differentiation of those cells depends upon the concentration of that molecule. One example of such a morphogen is the zebrafish Nodal protein. As we will see in Chapter 11, *Nodal* mRNA (made by the egg before fertilization) accumulates in the blastomeres that will form the dorsal margin of the zebrafish embryo. These cells will activate the *goosecoid* gene, whose product commits these blastomeres to become the cells that instruct the anterior portion of the head to form (Gore et al. 2005). Cells slightly further away from the dorsal margin activate the *floating head* gene, which commits the cells to become notochord. And slightly further from the dorsal margin, cells activate the *no-tail* gene that is important for forming the trunk and tail muscles. Experiments carried out by Chen and Schier (2001) provide evidence that these dorsal cell fates are coordinately regulated by different amounts of Nodal protein.

Chen and Schier injected *Nodal* mRNA into a single cell of the 256-cell stage zebrafish blastulae (Figure 3.20A). This

mRNA is directly translated into Nodal protein, which is then secreted from the cell. When Chen and Schier injected 4 picograms (pg) of *Nodal* mRNA into the blastomere (Figure 3.20B), they found that the injected cell and the cells next to it expressed *goosecoid*. The cells about 5–6 cell diameters away expressed *floating head*, and the cells 7–8 cell diameters away expressed *no-tail*. This would be in agreement with the hypothesis that Nodal is a morphogen and

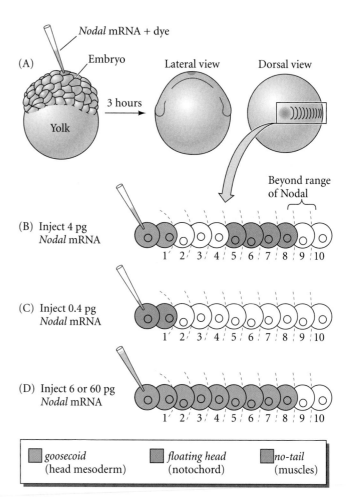

FIGURE 3.20 Nodal is a long-range morphogen in zebrafish embryos. (A) *Nodal* mRNA and a nondiffusible dye are injected into a single cell of the 256-cell zebrafish embryo. The cells are fixed and observed 3 hours later. The remainder of the figure represents one row of such cells seen from the dorsal surface. (B) When 4 picograms of *Nodal* mRNA is injected into a single cell, *goosecoid* is induced in that cell and in the cells adjacent to it. Cells about 5–6 cell diameters from the injected cell express the *floating head* gene, and cells 7–8 cells away express the *no-tail* gene. Cells farther away do not appear to respond to the Nodal protein. (C) When the concentration of injected nodal mRNA is reduced tenfold, the injected cell and those adjacent to it express *no-tail*. (D) When higher amounts of *Nodal* mRNA are injected, all the responding cells express *goosecoid*. Cells farther than 8 cells away appear to be outside the limits of Nodal diffusion.

that graded amounts of Nodal result in the commitment of cells to different fates.

Chen and Schier further tested Nodal's action by varying the amounts of mRNA injected (and thus of Nodal protein secreted). If they dropped the concentration of injected *Nodal* mRNA by tenfold, none of the cells expressed *goosecoid* or *floating head*, and only the injected cell and its closest neighbors expressed *no-tail* (Figure 3.20C). If they raised the level of injected mRNA to 6 pg, the first 8 cells all expressed *goosecoid*. No expression was seen after that (even of *floating head* or *no-tail*), suggesting that either the cells were incapable of responding to Nodal, or that Nodal did not reach them (Figure 3.20D). (Recently, Le Good and colleagues [2005] found that Nodal protein is degraded after it is secreted, so it is probable that the nonexpressive cells were simply outside its range.)

SIDELIGHTS & SPECULATIONS

Rules of Evidence II: Controls

Science is both about concluding what something *is* and about concluding what something *is not*. A piece of scientific evidence is only as good as the controls it uses to tell what something is not, and the difference between science and non-science is usually a matter of controls. Just because a particular compound causes cells in the same population to differentiate into particular cell types does not mean that it is a morphogen.

Negative controls

In *The Adventure of Black Peter*, Sherlock Holmes noted, "One should always look for a possible alternative and provide against it." What other possibilities could there be besides the morphogen hypothesis? A morphogen is a diffusible molecule that acts through a concentration gradient. However, there are at least two other possibilities besides gradients to explain the phenomenon whereby cells appear to differentiate with respect to their proximity to a source of Nodal.

First, there could be a cascade of inductions. Nodal might only induce nearby cells at high concentrations and do nothing at lower concentrations. Instead, the induced cells might put forth a second signal that induces the next set of cells, and these cells might put forth a third local signal, and so on. Such a sequence of events is sometimes called a relay cascade.

A second possibility is that the cells near the source of Nodal might respond not only by differentiating, but by migrating away to a distant position within the responding tissue (or by producing daughter cells that migrate). The stronger the signal, the farther away the induced cells or their daughters migrate.

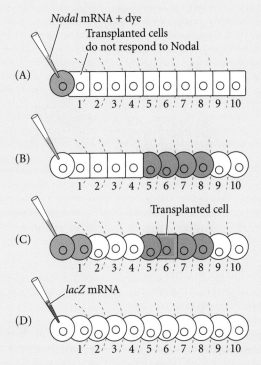

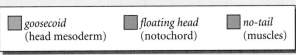

goosecoid (head mesoderm)   *floating head* (notochord)   *no-tail* (muscles)

FIGURE 3.21 Negative controls confirm that Nodal is a morphogen. (A) Transplanted cells that lack the Nodal receptor do not express the *goosecoid*, *floating head*, or *no-tail* genes when placed near the Nodal-secreting cell. (B) Such unresponsive cells do not block cells farther away from expressing the appropriate *floating head* and *no-tail* genes, indicating that Nodal diffusion is the most likely means of signal transmission. (C) A labeled responsive cell transplanted to a particular location near the Nodal-secreting cell will express the gene appropriate to the distance from the Nodal-secreting cell. (D) If a gene encoding some other substance (in this case, bacterial β-galactosidase, *lacZ*) that is not a signaling molecule, the nearby cells will not express their Nodal-responsive genes.

Chen and Schier's experiments, described in Figure 3.20 controlled for these alternative possibilities. First, to distinguish whether the effects of Nodal are due to the diffusion of Nodal or to a relay cascade, the investigators found cells that did *not* respond to Nodal. There are zebrafish with mutations in a receptor for Nodal, and blastomeres from these fish cannot respond to Nodal protein (Figure 3.21A). (These Nodal-deficient fish have numerous developmental anomalies, and the name of this mutant, *one-eyed pinhead*, suggests it has some major problems in forming its face.) The mutant blastomeres can be transplanted between the source of Nodal and the responsive blastomeres (Figure 3.21B). If

(Continued on next page)

Nodal acted through a relay cascade, the blastomeres should not express *floating head* or *no-tail*, since the cascade would have been broken by the intervening cells. However, if Nodal produced its effects through a concentration gradient, the responsive blastomeres should respond based on their distance from the source of Nodal, even if the cells between them and the source could not respond. In this experiment, the responsive blastomeres expressed *floating head* and *no-tail*, even when the cells closest to the source of Nodal could not respond to it.

Second, researchers controlled for cell movement by labeling wild-type (normal) cells with a green dye and transplanting them to a particular place in the embryo at a particular distance from the Nodal-injected cell, where they would be expected to activate the *no-tail* gene (Figure 3.21C). They expressed the *no-tail* gene at that particular location (and not at others), showing that migration of adjacent cells was not the cause of the different genes being expressed at different places.

Another negative control made certain that the morphogen effect was specific for

the mRNA being injected and not a side effect of the injection itself (or of the saline solution in which the mRNA was suspended). Thus, in another series of tests they injected mutant embryos with mRNA for the enzyme β-galactosidase and showed that this injection did not cause any cells to express the Nodal-responsive genes (Figure 3.21D).

All of the controls described above are **negative controls** because they tell the researcher what is *not* the right answer. Any scientific investigation has to have such controls; when they are not done, the investigation is suspect. For instance, a group of investigators published an article claiming that a particular wormlike structure was found in the uteruses of women with a particular disease. However, no controls were mentioned. When other scientists read the original paper, they noted that the obvious control—to look for the wormlike structure in the uteruses of women *without* the disease—was not reported. It turned out that the structure is a common entity found in most women; it is not associated with the disease (Richards et al. 1983). Similarly, when scientists inject a specific antibody into an embryo to see if it inhibits a particu-

lar phenomenon, they will inject unrelated antibodies into other embryos as negative controls. This practice shows that the inhibition of the event is *not* a result of the solvent being injected along with the antibody, the presence of nonspecific antibody protein, or a response to injection.

Positive controls

In addition to negative controls, experiments should also incorporate **positive controls** that tell scientists their techniques work, and that they are indeed measuring what they think they are measuring. For instance, the difference between the ability of normal and mutant cells to express the *no-tail, goosecoid,* or *floating head* genes told the investigators that the *Nodal* mRNA was being translated and secreted. They further tested for the efficacy of their technique by injecting a dye along with the mRNA, thereby making certain that they injected a single cell and that the injection worked. Positive controls are critical, because if you get a negative result, you need to be able to say that the null result was due to the experimental variable and not due to a fault with the technique.

MORPHOGENETIC FIELDS   One of the most interesting ideas to come from experimental embryology has been that of the **morphogenetic field**. A morphogenetic field can be described as a group of cells whose position and fate are specified with respect to the same set of boundaries (Weiss 1939; Wolpert 1977). The general fate of a morphogenetic field is determined; thus, a particular field of cells will give rise to its particular organ (forelimb, eye, heart, etc.) even when transplanted to a different part of the embryo. However, the individual cells *within* the field are not committed, and the cells of the field can regulate their fates to make up for cells missing from the field (Huxley and de Beer 1934; Opitz 1985; De Robertis et al. 1991). Moreover, as described earlier (the "French flag" analogy), if cells from one field are placed within another field, they can use the positional cues of their new location, even if they retain their original organ-specific commitment.

The morphogenetic field has been referred to as a "field of organization" (Spemann 1921) and as a "cellular ecosystem" (Weiss 1923, 1939). The ecosystem metaphor is quite appropriate in that recent studies have shown that there are webs of interactions among the cells in different regions of a morphogenetic field.

Stem cells and commitment

Another important principle derived from conditional specification is the notion of stem cells. **Stem cells** are cells that have the capacity to divide indefinitely and which can give rise to more specialized cells. When they divide, stem cells produce a more specialized type of cell *and also* generate more stem cells (Figure 3.22A).

Some single stem cells in the early embryo are capable of generating all the structures of the embryo (National Institutes of Health 2001). These cells, called **pluripotent stem cells,*** can generate ectoderm, endoderm, mesoderm, and germ cells. As well as giving rise to more pluripotent

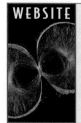 **WEBSITE 3.3 "Rediscovery" of the morphogenetic field.** The morphogenetic field was one of the most important concepts of embryology during the early twentieth century. This concept was eclipsed by research on the roles of genes in development, but it is being "rediscovered" as a consequence of those developmental genetic studies.

*There are also **totipotent** stem cells, which will be discussed in Chapters 11 and 21. Totipotent stem cells are those very early mammalian cells that can form both the entire embryo and the fetal placenta (trophoblast) around it. The pluripotent stem cells of mammals can form the embryo, but not its surrounding tissues.

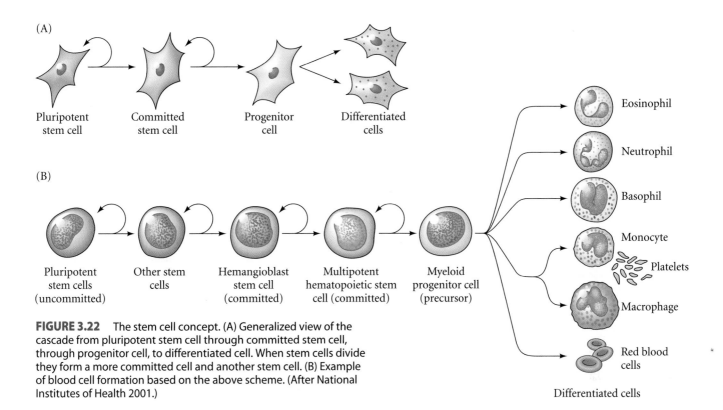

(A)

Pluripotent stem cell Committed stem cell Progenitor cell Differentiated cells

(B)

Pluripotent stem cells (uncommitted) Other stem cells Hemangioblast stem cell (committed) Multipotent hematopoietic stem cell (committed) Myeloid progenitor cell (precursor)

Eosinophil
Neutrophil
Basophil
Monocyte
Platelets
Macrophage
Red blood cells

Differentiated cells

FIGURE 3.22   The stem cell concept. (A) Generalized view of the cascade from pluripotent stem cell through committed stem cell, through progenitor cell, to differentiated cell. When stem cells divide they form a more committed cell and another stem cell. (B) Example of blood cell formation based on the above scheme. (After National Institutes of Health 2001.)

stem cells, these cells also generate **committed stem cells**. Committed stem cells can give rise to a smaller population of cells. For instance, one type of committed stem cell, the hemangioblast, gives rise to all the blood vessels, blood cells, and lymphocytes (Figure 3.22B). Another type of committed stem cell, the mesenchymal stem cell, can give rise to all the different connective tissues (cartilage, muscle, fat, etc). Committed stem cells can give rise to more specifically committed stem cells (such as the hematopoietic stem cell that generates only blood cells and lymphocytes, not blood vessels; see Figure 3.22B), or it can generate what are called **progenitor cells** or **precursor cells**. Progenitor cells are no longer stem cells, since their divisions do not create more progenitor cells. Rather, these cells divide to form one or a few related types of cells, depending on the cellular environment it is in. (For example, the myeloid progenitor cell depicted in Figure 3.22B can generate all the different types of blood cells.) A progenitor cell usually shows some evidence of differentiation, although the process is not complete until the fully differentiated cell has been formed.

The restriction on the potency of stem cells is gradual, and the potencies of these cells are determined by their surroundings. Once committed, however, they usually do not switch commitment. When placed in a new environment, they will not change the type of cells they can generate. Stem cells are critical for maintaining cell populations that last for long periods of time and must be renewed. Thus, in humans stem cells are important for the continual pro-

duction of blood, hair, epidermis, and intestinal epithelial cells.

Morphogenesis and Cell Adhesion

A body is more than a collection of randomly distributed cell types. Development involves not only the differentiation of cells, but also their organization into multicellular arrangements such as tissues and organs. When we observe the detailed anatomy of a tissue such as the neural retina of the eye, we see an intricate and precise arrangement of many types of cells. How can matter organize itself so as to create a complex structure such as a limb or an eye? The third research program for experimental embryology focused on the processes of this morphogenesis. At least five major questions confront embryologists who study morphogenesis:

1. *How are tissues formed from populations of cells?* For example, how do neural retina cells stick to other neural retina cells rather than becoming integrated into the pigmented retina or into the iris cells next to them? How are the different cell types found within the retina (the three distinct layers of photoreceptors, bipolar neurons, and ganglion cells) arranged such that the retina is functional?

2. *How are organs constructed from tissues?* The retina of the eye forms at a precise distance behind the cornea and the lens. The retina would be useless if it developed behind a bone or in the middle of the kidney. Moreover, neurons

from the retina must enter the brain to innervate the regions of the brain cortex that analyze visual information. All these connections must be precisely ordered.

3. *How do organs form in particular locations, and how do migrating cells reach their destinations?* What causes an eye to form in the head, and what stops an eye from forming in some other area of the body? Some cells—for instance, the precursors of our pigment cells, germ cells, and blood cells—must travel long distances to reach their final destinations. How are cells instructed to travel along certain routes in our embryonic bodies, and how are they told to stop once they have reached their appropriate destinations?

4. *How do organs and their cells grow, and how is their growth coordinated throughout development?* The cells of all the tissues in the eye must grow in a coordinated fashion if one is to see. Some cells, including most neurons, do not divide after birth. In contrast, the intestine is constantly shedding cells, and new intestinal cells are regenerated each day. The mitotic rate of each tissue must be carefully regulated. If the intestine generated more cells than it sloughed off, it could produce tumorous outgrowths. If it produced fewer cells than it sloughed off, it would soon become nonfunctional. What controls the rate of mitosis in the intestine?

5. *How do organs achieve polarity?* If one were to look at a cross section of the fingers, one would see a certain organized collection of tissues—bone, cartilage, muscle, fat, dermis, epidermis, blood, and neurons. Looking at a cross section of the forearm, one would find the same collection of tissues. But they are arranged very differently. How is it that the same cell types can be arranged in different ways in different parts of the same structure, and that fingers are always at the end of the arm, never in the middle?

Differential cell affinity

Many of the answers to our five questions about morphogenesis involve the properties of the cell surface. The cell surface looks pretty much the same in all cell types, and many early investigators thought that the cell surface was not even a living part of the cell. We now know that each type of cell has a different set of proteins in its cell membrane, and that some of these differences are responsible for forming the structure of the tissues and organs during development. Observations of fertilization and early embryonic development made by E. E. Just (1939) suggested that the cell membrane differed among cell types, but the experimental analysis of morphogenesis began with the experiments of Townes and Holtfreter in 1955. Taking advantage of the discovery that amphibian tissues become dissociated into single cells when placed in alkaline solutions, they prepared single-cell suspensions from each of the three germ layers of amphibian embryos soon after the neural tube had formed. Two or more of these single-cell suspensions could be combined in various ways. When the pH of the solution was normalized, the cells adhered to one another, forming aggregates on agar-coated petri dishes. By using embryos from species having cells of different sizes and colors, Townes and Holtfreter were able to follow the behavior of the recombined cells.

The results of their experiments were striking. First, they found that reaggregated cells become spatially segregated. That is, instead of the two cell types remaining mixed, each cell type sorts out into its own region. Thus, when epidermal (ectodermal) and mesodermal cells are brought together to form a mixed aggregate, the epidermal cells move to the periphery of the aggregate and the mesodermal cells move to the inside (Figure 3.23). In no case do the

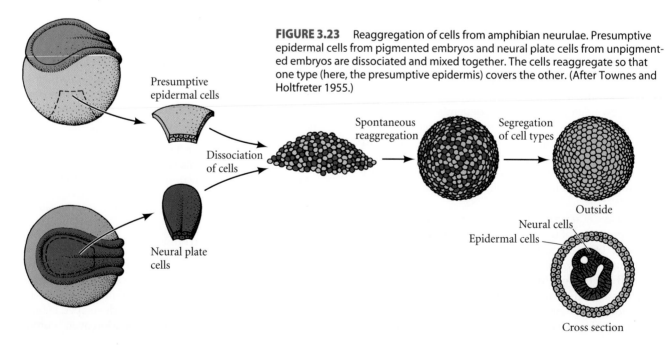

FIGURE 3.23 Reaggregation of cells from amphibian neurulae. Presumptive epidermal cells from pigmented embryos and neural plate cells from unpigmented embryos are dissociated and mixed together. The cells reaggregate so that one type (here, the presumptive epidermis) covers the other. (After Townes and Holtfreter 1955.)

Presumptive epidermal cells

Dissociation of cells

Neural plate cells

Spontaneous reaggregation

Segregation of cell types

Outside

Neural cells

Epidermal cells

Cross section

recombined cells remain randomly mixed, and in most cases, one tissue type completely envelops the other.

Second, the researchers found that the final positions of the reaggregated cells reflect their embryonic positions. The mesoderm migrates centrally with respect to the epidermis, adhering to the inner epidermal surface (Figure 3.24A). The mesoderm also migrates centrally with respect to the gut or endoderm (Figure 3.24B). However, when the three germ layers are mixed together, the endoderm separates from the ectoderm and mesoderm and is then enveloped by them (Figure 3.24C). In the final configuration, the ectoderm is on the periphery, the endoderm is internal, and the mesoderm lies in the region between them. Holtfreter interpreted this finding in terms of **selective affinity**. The inner surface of the ectoderm has a positive affinity for mesodermal cells and a negative affinity for the endoderm, while the mesoderm has positive affinities for both ectodermal and endodermal cells. Mimicry of

normal embryonic structure by cell aggregates is also seen in the recombination of epidermis and neural plate cells (Figure 3.24D; also see Figure 3.23). The presumptive epidermal cells migrate to the periphery as before; the neural plate cells migrate inward, forming a structure reminiscent of the neural tube. When axial mesoderm (notochord) cells are added to a suspension of presumptive epidermal and presumptive neural cells, cell segregation results in an external epidermal layer, a centrally located neural tissue, and a layer of mesodermal tissue between them (Figure 3.24E). Somehow, the cells are able to sort out into their proper embryonic positions.

The third conclusion of Holtfreter and his colleagues was that selective affinities change during development. Such changes should be expected, because embryonic cells do not retain a single stable relationship with other cell types. For development to occur, cells must interact differently with other cell populations at specific times. Such

FIGURE 3.24 Sorting out and reconstruction of spatial relationships in aggregates of embryonic amphibian cells. (After Townes and Holtfreter 1955.)

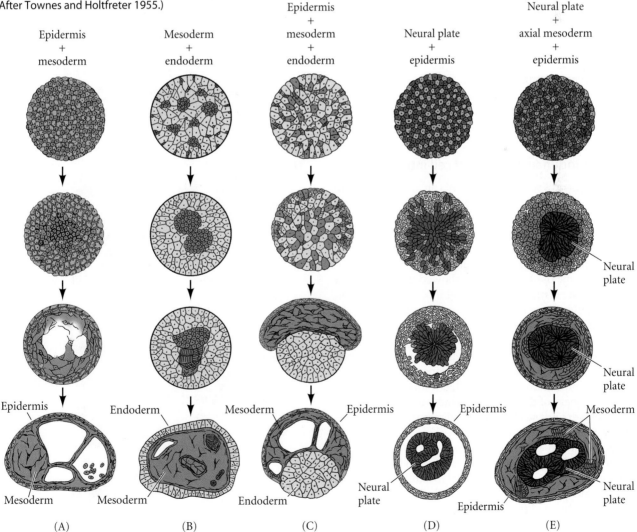

FIGURE 3.25   Reconstruction of skin from a suspension of skin cells from a 15-day embryonic mouse. (A) Section through intact embryonic skin, showing epidermis, dermis, and primary hair follicle. (B) Suspension of single skin cells from both the dermis and the epidermis. (C) Aggregates after 24 hours. (D) Section through an aggregate, showing migration of epidermal cells to the periphery. (E) Further differentiation of aggregates after 72 hours, showing reconstituted epidermis and dermis, complete with hair follicles and keratinized layer. (From Monroy and Moscona 1979, photographs courtesy of A. Moscona.)

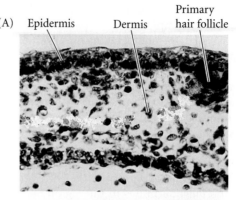

(A) Epidermis   Dermis   Primary hair follicle

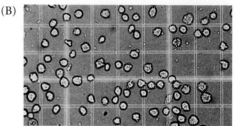

(B)

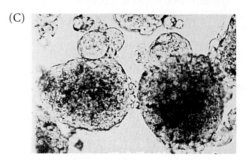

(C)

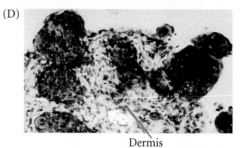

(D)

Dermis

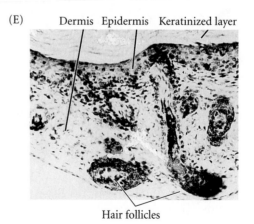

(E)   Dermis   Epidermis   Keratinized layer

Hair follicles

changes in cell affinity are extremely important in the processes of morphogenesis.

The experimental reconstruction of aggregates from cells of later mammalian embryos was accomplished by the use of the proteolytic enzyme trypsin to dissociate the cells from one another (Moscona 1952). When the resulting single cells were mixed together in a flask and swirled so that the shear force would break any nonspecific adhesions, the cells sorted themselves out according to their cell type. In so doing, researchers reconstructed the organization of the original tissue (Moscona 1961; Giudice and Just 1962). Figure 3.25 shows the "reconstruction" of skin tissue from a 15-day embryonic mouse. The dissociated epidermal cells migrate to the periphery, while the separated dermal cells migrate toward the center. Within 72 hours, the epidermis has been reconstituted, a keratin layer has formed, and interactions between these tissues form hair follicles in the dermal region. Such reconstruction of complex tissues from individual cells is called **histotypic aggregation**.

The thermodynamic model of cell interactions

Cells, then, do not sort randomly, but can actively move to create tissue organization. What forces direct cell movement during morphogenesis? In 1964, Malcolm Steinberg proposed the **differential adhesion hypothesis**, a model that sought to explain patterns of cell sorting based on thermodynamic principles. Using cells derived from trypsinized embryonic tissues, Steinberg showed that certain cell types migrate centrally when combined with some cell types, but migrate peripherally when combined with others. Figure 3.26 illustrates the interactions between pigmented retina cells and neural retina cells. When single-cell suspensions of these two cell types are mixed together, they form aggregates of randomly arranged cells. However, after several hours, no pigmented retina cells are seen on the periphery of the aggregates, and after two days, two distinct layers are seen, with the pigmented retina cells lying internal to the neural retina cells. Moreover, such interactions form a hierarchy (Steinberg 1970). If the final position of cell type A is internal to a second cell type B, and the final position of B is internal to a third cell type C, then the final position of A will always be internal to C. For example, pigmented retina cells migrate internally to neu-

(A)

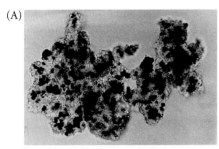

(B)

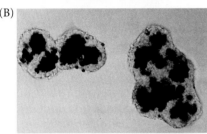

(C)

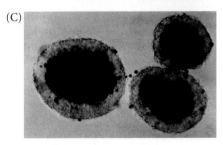

FIGURE 3.26   Aggregates formed by mixing 7-day chick embryo neural retina (unpigmented) cells with pigmented retina cells. (A) Five hours after the single-cell suspensions are mixed, aggregates of randomly distributed cells are seen. (B) At 19 hours, the pigmented retina cells are no longer seen on the periphery. (C) At 2 days, a great majority of the pigmented retina cells are located in a central internal mass, surrounded by the neural retina cells. (The scattered pigmented cells are probably dead cells.) (From Armstrong 1989, photographs courtesy of P. B. Armstrong.)

ral retina cells, and heart cells migrate internally to pigmented retina cells. Therefore, heart cells migrate internally to neural retina cells.

This observation led Steinberg to propose that cells interact so as to form an aggregate with the smallest interfacial free energy. In other words, the cells rearrange themselves into the most thermodynamically stable pattern. If cell types A and B have different strengths of adhesion, and if the strength of A-A connections is greater than the strength of A-B or B-B connections, sorting will occur, with the A cells becoming central. On the other hand, if the strength of A-A connections is less than or equal to the strength of A-B connections, then the aggregate will remain as a random mix of cells. Finally, if the strength of A-A connections is far greater than the strength of A-B connections—in other words, if A and B cells show essentially no

adhesivity toward one another—then A cells and B cells will form separate aggregates. According to this hypothesis, the early embryo can be viewed as existing in an equilibrium state until some change in gene activity changes the cell surface molecules. The movements that result seek to restore the cells to a new equilibrium configuration. All that is needed for sorting to occur is that cell types differ in the strengths of their adhesion.

In 1996, Foty and his colleagues in Steinberg's laboratory demonstrated that this was indeed the case: the cell types that had greater surface cohesion sorted within those cells that had less surface tension (Figure 3.27; Foty et al. 1996). In the simplest form of this model, all cells could have the same type of "glue" on the cell surface. The amount of this cell surface product, or the cellular architecture that allows the substance to be differentially distributed across the surface, could cause a difference in the number of stable contacts made between cell types. In a more specific version of this model, the thermodynamic differences could be caused by different types of adhesion molecules (see Moscona 1974). When Holtfreter's studies were revisited using modern techniques, Davis and colleagues (1997) found that the tissue surface tensions of the individual germ layers were precisely those required for the sorting patterns observed both in vitro and in vivo.

WEBSITE   3.4 Demonstrating the thermodynamic model. The original in vivo evidence for the thermodynamic model of cell adhesion came from studies of limb regeneration. The website goes into some of the details of these experiments and how they are interpreted.

Cadherins and cell adhesion

Recent evidence shows that boundaries between tissues can indeed be created by different cell types having both different types and different amounts of cell adhesion molecules. Several classes of molecules can mediate cell adhesion, but the major cell adhesion molecules appear to be the cadherins.

As their name suggests, **cadherins** are *ca*lcium-dependent *adhe*sion molecules. They are critical for establishing and maintaining intercellular connections, and they appear to be crucial to the spatial segregation of cell types and to the organization of animal form (Takeichi 1987). Cadherins interact with other cadherins on adjacent cells, and they are anchored into the cell by a complex of proteins called **catenins** (Figure 3.28). The cadherin-catenin complex forms the classic adherens junctions that help hold epithelial cells together. Moreover, since the catenins bind to the actin cytoskeleton of the cell, they integrate the epithelial cells into a mechanical unit. Univalent antibody fragments (Fab; see Sidelights & Speculations in Chapter 2) against cadherins will convert a three-dimensional, histotypic aggregate of cells into a single layer of cells (Takeichi et al. 1979).

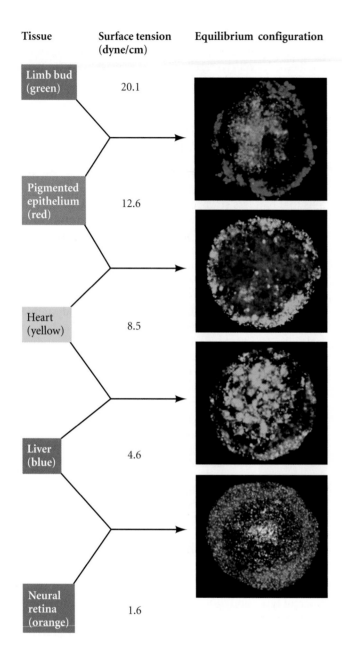

Tissue	Surface tension (dyne/cm)	Equilibrium configuration
Limb bud (green)	20.1	
Pigmented epithelium (red)	12.6	
Heart (yellow)	8.5	
Liver (blue)	4.6	
Neural retina (orange)	1.6	

FIGURE 3.27 Hierarchy of cell sorting in order of decreasing surface tensions. The equilibrium configuration reflects the strength of cell cohesion, with the cell types having the greater cell cohesion segregating inside the cells with less cohesion. The images were obtained by sectioning the aggregates and assigning colors to the cell types by computer. The black areas represent cells whose signal was edited out in the program of image optimization. (From Foty et al. 1996, photograph courtesy of M. S. Steinberg and R. A. Foty.)

ichi 1986), **R-cadherin** is critical in retina formation (Babb et al. 2005), and **B-cadherin** is expressed on many neural structures. However, cadherins are not localized to one particular set of cells. B-cadherin, for example, is found on cells of all three germ layers. In addition, cadherin expression patterns change over the course of development.

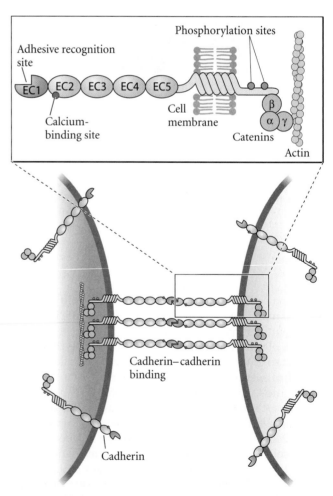

FIGURE 3.28 Cadherin-mediated cell adhesion. Cadherins are associated with three types of catenins. The catenins can become associated with the actin microfilament system within the cell. (After Takeichi 1991.)

In vertebrate embryos, several major cadherin types have been identified.* **E-cadherin** is expressed on all early mammalian embryonic cells, even at the zygote stage. Later in development, this molecule is restricted to epithelial tissues of embryos and adults. **P-cadherin** is found predominantly on the placenta, where it helps the placenta stick to the uterus (Nose and Takeichi 1986; Kadokawa et al. 1989). **N-cadherin** becomes highly expressed on the cells of the developing central nervous system (Hatta and Take-

---

*__Protocadherins__ are also calcium-dependent adhesion proteins but differ from the classic cadherins in that they lack connections to the cytoskeleton through catenins. Protocadherins are very important in separating the notochord from the other mesodermal tissues during *Xenopus* gastrulation (see Chapter 10).

Cadherins join cells together by binding cadherins on another cell. Interfering with cadherin function can prevent the formation of tissues and cause the cells to disaggregate (Figure 3.29). In most cases, it appears that different cadherins can bind to one another, and that it is the strength of cadherin-cadherin binding that mediates the formation of embryonic structures. This was first shown to be a possibility when Steinberg and Takeichi (1994) collaborated on an experiment using two cell lines that were identical except that they synthesized different amounts of P-cadherin. When these two groups of cells, each expressing a different amount of cadherin, were mixed, the cells that expressed more cadherin had a higher surface cohesion and migrated internally to the lower-expressing group of cells. This process appears to work in the embryo as well (Godt and Tepass 1998; González-Reyes and St. Johnston 1998). When cadherin genes are transfected into cells lacking cadherins (and which therefore cannot aggregate), the cells begin expressing cadherins and adhering.

Moreover, the surface tensions of these aggregates are linearly related to the amount of cadherin they are expressing on the cell surface. The cell sorting hierarchy is strictly dependent on the cadherin interactions between the cells. Moreover, the value of the cadherin-cadherin binding is remarkably strong, about 3,400 kcal/mole (about 200 times stronger than most metabolic protein-protein interactions). This free energy change associated with cadherin function could be dissipated by depolymerizing the actin skeleton. The underlying actin cyctoskeleton appears to be crucial in organizing the cadherins in a manner that allows them to form remarkably stable linkages between cells (Foty and Steinberg 2004).

Duguay and colleagues showed that both quantity and type of cadherins expressed determined which cells sorted out from one another (Duguay et al. 2003). Cells containing E-cadherin will bind to cells containing P-cadherin. Bonds between E-cadherin and E-cadherin, between P-cadherin and P-cadherin, and between E-cadherin and P-cadherin all seem to be about the same strength. Cell sorting among cells bearing E-cadherin and those bearing P-cadherin thus depends on the quantity of cadherin (of either type) on the cell surfaces. However, R-cadherin and B-cadherin do *not* bind well to each other, and in these interactions the type of cadherin expressed becomes important.

The timing of particular developmental events can also depend on cadherin expression. For instance, N-cadherin appears in the mesenchymal cells of the developing chick leg just before these cells condense and form nodules of cartilage (which are the precursors of bone tissue). N-cadherin is not seen prior to condensation, nor is it seen afterward. If the limbs are injected just prior to condensation with antibodies that block N-cadherin, the mesenchyme cells fail to condense and cartilage fails to form (Oberlander and Tuan 1994). It therefore appears that the signal to begin cartilage formation in the chick limb is the appearance of N-cadherin.

(A)

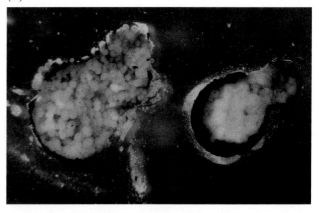

(B)

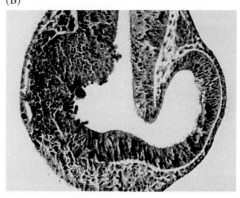

FIGURE 3.29 The importance of cadherins for maintaining cohesion between developing cells can be demonstrated by interfering with their production. (A) When an oocyte is injected with an antisense oligonucleotide against a maternally inherited cadherin mRNA (thus preventing the synthesis of the cadherin), the inner cells of the resulting embryo disperse when the animal cap is removed (left). In control embryos (right), the inner cells remain together. (B) The importance of cadherin in the separation of neural and epidermal ectoderm. At the 4-cell stage, the blastomeres that formed the left side of this *Xenopus* embryo were injected with an mRNA for N-cadherin that lacks the extracellular region of the cadherin. This mutation blocks N-cadherin function. During eye development, the cells with the mutant protein did not form a coherent retinal layer distinct from the epidermis (as can be seen on the right portion of the embryo). (A from Heasman et al. 1994, photograph courtesy of J. Heasman; B from Kintner 1993, photograph courtesy of C. Kintner.)

 WEBSITE **3.5** **Cadherins: Functional anatomy.** The cadherin molecule has several functional domains that mediate its activities, and the mechanisms of homophilic adhesion are currently being resolved.

 WEBSITE **3.6** **Other cell adhesion molecules.** There are more types of cell adhesion molecules than cadherins. The website looks at some of the other cell adhesion and substrate adhesion molecules that have been discovered.

During development, the cadherins often work with other adhesion systems. For instance, one of the most critical times in a mammal's life is when the embryo is passing through the uterus. If development is to continue, the embryo must adhere to the uterus and embed itself in the uterine wall. That is why the first differentiation event in mammalian development distinguishes the **trophoblast** cells (the outer cells that bind to the uterus) from the **inner cell mass** (those cells that will generate the embryo and eventually the mature organism). This differentiation process occurs as the embryo travels from the upper regions of the oviduct on its way to the uterus. The trophoblast cells are endowed with several adhesion molecules that anchor the embryo to the uterine wall. First, they contain both E- and P-cadherins (Kadokawa et al. 1989), and these two molecule types recognize similar cadherins on the uterine cells. Second, they have receptors (integrin proteins) for the collagen and the heparan sulfate glyco-

proteins of the uterine wall (Farach et al. 1987; Carson et al. 1988, 1993; Cross et al. 1994). Third, trophoblast cell surfaces have a modified glycosyltransferase enzyme that extends out from the cell membrane and can bind to specific carbohydrate residues on uterine glycoproteins (Dutt et al. 1987). For something as important as the implantation of the mammalian embryo, it is not surprising that several cell adhesion systems appear to work together.

Coda

Experimental embryology identified the three modes of cell specification, the cadherin mechanism of morphogenesis, and the dependency of certain characters on the environment. Subsequent chapters on cell specification, organ formation, and environmental agents of development will flesh out this outline and show the signficance of these concepts.

 **Snapshot Summary** ## Experimental Embryology

1. Norms of reaction describe an embryo's inherited ability to develop a range of phenotypes. The environment can play a role in selecting which phenotype is expressed. (Examples include temperature-dependent sex determination and seasonal phenotypic changes in caterpillars and butterflies.)

2. Developing organisms are adapted to the ecological niches in which they develop. Examples include the ability of frog eggs exposed to sunlight to repair DNA damage.

3. Before cells overtly differentiate into the many cell types of the body, they undergo a "covert" commitment to a certain fate. This commitment is first labile (the specification step) but later becomes irreversible (the determination step).

4. In some embryos, removal of a blastomere from an embryo causes the absence in the embryo of those tissues formed by that blastomere. This autonomous specification produces a mosaic pattern of development. (Examples include early snail and tunicate embryos.)

5. In autonomous specification, morphogenetic determinants in the egg cytoplasm are apportioned to different blastomeres during cleavage. (An example is the yellow crescent cytoplasm that is found in the muscle-forming cells of tunicate embryos.)

6. In syncytial specification, the fates of insect cells are determined by gradients of morphogens within the egg cytoplasm.

7. In some embryos, the removal of a blastomere can be compensated for by the other cells' changing their fates. Each cell has the potential to give rise to more cell types than it normally does. This conditional specification produces a regulative pattern of development wherein cell fates are determined relatively late.

8. In conditional specification, the fate of a cell often depends on interactions with its neighbors.

9. Cell fate in conditionally committed embryos can be specified by either qualitative or quantitative means. In some instances, particular chemicals instruct the fate of a cell. This means is responsible for much of the commitment of stem cells. In other cases, gradients of a morphogen protein specify cell types by the amount of protein they receive. This is important for the specification of the different types of mesoderm along the dorsal-ventral axis.

10. Negative controls tell investigators that a given agent does not cause an observed phenomenon. Such controls are critical in arguing that some other agent does cause the phenomenon.

11. Positive controls confirm that an investigator's protocols are working, so that if a negative result is observed, it is not merely due to faulty procedure.

12. As the embryo develops, certain cells become separated such that they can regulate within the group of cells but not outside the group. Such modules are called morphogenetic fields.

13. Stem cells are able to generate other stem cells as well as a more committed cell type. By generating more stem cells, the population continues.

14. Different cell types can sort themselves into regions by means of cell surface molecules such as cad-herins. These molecules can be critical in patterning cells into tissues and organs.

15. Differential adhesion appears to be mediated by cad-herins, and both quantity and quality of the cad-herins are involved.

For Further Reading

Complete bibliographical citations for all literature cited in this chapter can be found on the Vade Mecum CD that accompanies the book and at the free access website www.devbio.com

Chen, Y. and A. F. Schier. 2001. The zebrafish Nodal signal Squint functions as a morphogen. *Nature* 411: 607–610.

Driesch, H. 1892. The potency of the first two cleavage cells in echinoderm development: Experimental production of partial and double formations. In B. H. Willier and J. M. Oppenheimer (eds.), 1974, *Foundations of Experimental Embryology*. Hafner, New York.

Driesch, H. 1893. Zur Verlagerung der Blastomeren des Echinideneies. *Anat. Anz.* 8: 348–357.

Foty, R. A., C. M. Pfleger, G. Forgacs and M. S. Steinberg. 1996. Surface tensions of embryonic cells predict their mutual envelopment behavior. *Development* 122: 1611–1620.

Greene, E. 1989. A diet-induced developmental polymorphism in a caterpillar. *Science* 243: 643–646.

Mead, K. S. and D. Epel. 1995. Beakers versus breakers: How fertilization in the laboratory differs from fertilization in nature. *Zygote* 3: 95–99.

Roux, W. 1888. Contributions to the developmental mechanics of the embryo. On the artificial production of half-embryos by destruction of one of the first two blastomeres and the later development (postgeneration) of the missing half of the body. In B. H. Willier and J. M. Oppenheimer (eds.), 1974, *Foundations of Experimental Embryology*. Hafner, New York, pp. 2–37.

Takeichi, M. 1991. Cadherin cell adhesion receptors as a morphogenetic regulator. *Science* 251: 1451–1455.

Whittaker, J. R. 1973. Segregation during ascidian embryogenesis of egg cytoplasmic information for tissue-specific enzyme development. *Proc. Natl. Acad. Sci. USA* 70: 2096–2100.

Wolpert, L. 1978. Pattern formation in biological development. *Sci. Am.* 239(4): 154–164.

The Genetic Core of Development

4

> **The secrets that engage me—that sweep me away—are generally secrets of inheritance: how the pear seed becomes a pear tree, for instance, rather than a polar bear.**
> *CYNTHIA OZICK (1989)*

> **Here, there is one central field. Development. How the egg turns into the organism. But development ultimately includes all of biology: and it will have to be put on a molecular basis.**
> *SYDNEY BRENNER (1979)*

"BETWEEN THE CHARACTERS THAT FURNISH THE DATA for the theory, and the postulated genes, to which the characters are referred, lies the whole field of embryonic development." Here Thomas Hunt Morgan noted in 1926 that the only way to get from genotype to phenotype is through developmental processes.

In the early twentieth century, embryology and genetics were not considered separate sciences. They diverged in the 1920s, when Morgan redefined genetics as the science studying the *transmission* of inherited traits, as opposed to embryology, the science studying the *expression* of those traits. In recent years, however, the techniques of molecular biology have effected a rapprochement of embryology and genetics. In fact, the two fields have become linked to a degree that makes it necessary to discuss molecular genetics early in this text. Problems in animal development that could not be addressed a decade ago are now being solved by a set of techniques involving nucleic acid synthesis and hybridization. This chapter seeks to place these new techniques within the context of the ongoing dialogues between genetics and embryology.

The Embryological Origins of the Gene Theory

Nucleus or cytoplasm: Which controls heredity?

Mendel called them *bildungsfähigen Elemente*, "form-building elements" (Mendel 1866, p. 42); we call them genes. It is in Mendel's term, however, that we see how closely intertwined were the concepts of inheritance and development in the nineteenth century. Mendel's observations, however, did not indicate where these hereditary elements existed in the cell, or how they came to be expressed. The gene theory that was to become the cornerstone of modern genetics originated from a controversy within the field of physiological embryology (see Chapter 3). In the late 1800s, a group of scientists began to study the mechanisms by which fertilized eggs give rise to adult organisms.

Two young American embryologists, Edmund Beecher Wilson and Thomas Hunt Morgan (Figure 4.1), became part of this group of "physiological embryologists," and each became a partisan in the controversy over which of the two compartments of the fertilized egg—the nucleus or the cytoplasm—controls inheritance. Morgan led those embryologists who thought the control of development lay within the cytoplasm, while Wilson, along with Theodor Boveri, led those who felt that the nucleus contained the instructions for development (Gilbert 1978, 1987; Allen 1986). In fact, Wilson (1896, p. 262) declared that the processes of meiosis, mitosis, fertilization, and unicellular regeneration from the fragment containing the nucleus "converge to the conclusion that the chromatin

(A)

(B)

FIGURE 4.1   (A) E. B. Wilson (1856–1939; shown here around 1899), an embryologist whose work on early embryology and sex determination greatly advanced the chromosomal hypotheses of development. (Wilson was also acknowledged to be among the best amateur cellists in the United States.) (B) Thomas Hunt Morgan (1866–1945), who brought the gene theory out of embryology. This photo—taken in 1915, as the basic elements of the gene theory were coming together— shows Morgan using a hand lens to sort fruit flies. (A courtesy of W. N. Timmins; B courtesy of G. Allen.)

is the most essential element in development." He did not shrink from the consequences of this belief. Years before the rediscovery of Mendel or the gene theory, Wilson (1895, p. 4) noted,

> Now, chromatin is known to be closely similar, if not identical with, a substance known as nuclein … which analysis shows to be a tolerably definite chemical composed of a nucleic acid (a complex organic acid rich in phosphorus) and albumin. And thus we reach the remarkable conclusion that inheritance may, perhaps, be effected by the physical transmission of a particular chemical compound from parent to offspring.

Some of the major support for the chromosomal hypothesis of inheritance was coming from the embryological studies of Theodor Boveri (Figure 4.2A), a researcher at the Naples Zoological Station. Boveri fertilized sea urchin eggs with large concentrations of sperm and obtained eggs that had been fertilized by two sperm. At first cleavage, these eggs formed four mitotic poles and divided into four cells instead of two. Boveri then separated the blastomeres and demonstrated that each cell developed abnormally, and in a different way—a result of each of the cells having different types and numbers of chromosomes (see Figure 7.16). Thus, Boveri claimed that each chromosome had an individual nature and controlled different vital processes.

Adding to Boveri's evidence, in 1905, both Wilson and Nettie Stevens (Figure 4.2B) independently demonstrated a critical correlation between nuclear chromosomes and development in insects: XO or XY embryos became male; XX embryos became female. Here was a nuclear property that correlated with development. Eventually, Morgan began to obtain mutations that correlated with sex and with the X chromosome, and he began to view the genes as being physically linked to one another on the chromosomes. The embryologist Morgan had shown that nuclear chromosomes are responsible for the development of inherited characters.*

---

*Morgan's evidence for nuclear control of development went against his expectations; until 1910, he was a leading proponent of the cytoplasmic control of development. Wilson was one of Morgan's closest friends, and Morgan considered Stevens his best graduate student at that time. Both were against Morgan on this issue. Even though they disagreed, Morgan wholeheartedly supported Stevens's request for research funds, saying that her qualifications were the best possible. Wilson wrote an equally laudatory letter of support for Stevens, even though she would be a rival researcher (see Brush 1978).

(A)

(B)

FIGURE 4.2 Chromosomal uniqueness was shown by Boveri and Stevens. (A) Theodor Boveri (1862–1915), whose work, Wilson (1896) said, "accomplished the actual amalgamation between cytology, embryology, and genetics—a biological achievement which … is not second to any of our time." This photograph was taken in 1908, when Boveri's chromosomal and embryological studies were at their zenith. (B) Nettie M. Stevens (1861–1912), who trained with both Boveri and Morgan, seen here in 1904 when she was a post-doctoral student pursuing the research that correlated the number of X chromosomes with sexual development. (A from Baltzer 1967; B courtesy of the Carnegie Institute of Washington.)

WEBSITE

4.1 The embryological origin of the gene theory. The emergence of the gene theory from embryological research is a fascinating story and complements the history of genetics that begins with Mendel's experiments.

The split between embryology and genetics

Morgan's evidence provided a material basis for the concept of the gene. Originally, this type of genetics was seen as being part of embryology, but by the 1930s, genetics became its own discipline, developing its own vocabulary, journals, societies, favored research organisms, professorships, and rules of evidence. Hostility between embryologists and geneticists also emerged. Geneticists believed that embryologists were old-fashioned and that development would eventually be explained as the result of gene expression. Conversely, embryologists regarded geneticists as uninformed about how organisms actually develop and felt that genetics was irrelevant to embryological questions. Embryologists such as Frank Lillie (1927), Ross Granville Harrison (1937), Hans Spemann (1938), and Ernest E. Just (1939) (Figure 4.3) claimed that there could be no "genetic theory of development" until geneticists met at least two major challenges:

1. Geneticists had to explain how chromosomes—which were thought to be identical in every cell of the organism—produce different and changing types of cell cytoplasms.

2. Geneticists had to provide evidence that genes control the early stages of embryogenesis. Almost all the genes known at the time affected the final modeling steps in development (eye color, bristle shape, wing venation in *Drosophila*). As Just said (quoted in Harrison 1937), embryologists were interested in how a fly forms its back, not in the number of bristles on its back.

The debate became quite vehement. In rhetoric reflecting the political anxieties of the late 1930s, Harrison (1937) warned:

> Now that the necessity of relating the data of genetics to embryology is generally recognized and the Wanderlust of geneticists is beginning to urge them in our direction, it may not be inappropriate to point out a danger of this threatened invasion. The prestige of success enjoyed by the gene theory might easily become a hindrance to the understanding of development by directing our attention solely to the genom, whereas cell movements, differentiation, and in fact all of developmental processes are actually effected by cytoplasm. Already we have theories that refer the processes of development to gene action and regard the whole performance as no more than the realization of the potencies of genes. Such theories are altogether too one-sided.

Until geneticists could demonstrate the existence of inherited variants during early development, and until geneticists had a well-documented theory explaining how the same chromosomes could produce different cell types, embryologists generally felt no need to ground their science in gene action.

Early attempts at developmental genetics

Some scientists, however, felt that neither embryology nor genetics was complete without the other, and there were several attempts to synthesize the two disciplines. The first successful reintegration of genetics and embryology came in the late 1930s from Salome Gluecksohn-Schoenheimer (now S. Gluecksohn Waelsch) and Conrad Hal Waddington (Figure 4.4). Both were trained in European embryology and had learned genetics in the United States from Morgan's students. Gluecksohn-Schoenheimer and Waddington

(A)

(B)

(C)

FIGURE 4.3   Embryologists attempted to keep genetics from "taking over" their field in the 1930s. (A) Frank Lillie headed the Marine Biological Laboratory at Woods Hole and was a leader in fertilization research and reproductive endocrinology. (B) Hans Spemann (left) and Ross Harrison (right) perfected the transplantation operations used to discover when the body and limb axes are determined. They argued that geneticists had no mechanism for explaining how the same nuclear genes could create different cell types during development. (C) Ernest E. Just made critical discoveries on fertilization. He spurned genetics and emphasized the role of the cell membrane in determining the fates of cells. (A courtesy of V. Hamburger; B courtesy of T. Horder; C courtesy of the Marine Biological Laboratory, Woods Hole.)

At the same time, Waddington (1939) isolated several genes that caused wing malformations in fruit flies (*Drosophila*). He, too, analyzed these mutations in terms of how the genes might affect the developmental primordia that give rise to these structures. The *Drosophila* wing, he correctly claimed, "appears favorable for investigations on the developmental action of genes." Thus, one of the main objections of embryologists to the genetic model of development—that genes appeared to be working only on the final modeling of the embryo and not on its major outlines—was countered.

 WEBSITE   **4.2 Creating developmental genetics.** The resynthesis of embryology and genetics into developmental genetics did not come easily. These websites look at some of the ways researchers attempted to join the two fields together.

Evidence for Genomic Equivalence

The other major objection to a genetically based embryology remained: How could nuclear genes direct develop-

attempted to find mutations that affected early development and to discover the processes that these genes affected. Glucksohn-Schoenheimer (1938, 1940) showed that mutations in the *Brachyury* gene of the mouse caused the aberrant development of the posterior portion of the embryo, and she traced the effects of this mutant gene to the notochord, which would normally have helped induce the dorsal-ventral axis.

THE GENETIC CORE OF DEVELOPMENT

(A)

(B)

FIGURE 4.4 Two of the founders of developmental genetics. (A) Salome Gluecksohn-Schoenheimer (now S. Gluecksohn-Waelsch; b. 1907) received her doctorate in Spemann's laboratory. Fleeing Hitler's Germany, she brought her embryological acumen to Leslie Dunn's genetics laboratory in the United States. (B) Conrad Hal Waddington (1905–1975) did not believe in the distinction between genetics and embryology and sought to find mutations that were active during development. (A courtesy of S. Gluecksohn-Waelsch; B from Waddington 1948.)

ment when they were the same in every cell type? The existence of this **genomic equivalence** was not so much proved as assumed (because every cell is the mitotic descendant of the fertilized egg), so one of the first problems of developmental genetics was to determine whether every cell of an organism indeed had the same set of genes, or **genome**, as every other cell.

Amphibian cloning: The restriction of nuclear potency

The ultimate test of whether the nucleus of a differentiated cell has undergone any irreversible functional restriction is to have that nucleus generate every other type of differentiated cell in the body. If each cell's nucleus is identical to the zygote nucleus, then each cell's nucleus should also be **totipotent** (capable of directing the entire development of the organism) when transplanted into an activated enucleated egg. As early as 1895, the embryologist Yves Delage predicted that "If, without deterioration, the egg nucleus could be replaced by the nucleus of an ordinary embryonic cell, we should probably see this egg developing without changes" (Delage 1895, p. 738). Before such an experiment could be done, however, three techniques for transplanting nuclei into eggs had to be perfected: (1) a method for enucleating host eggs without destroying them; (2) a method for isolating intact donor nuclei; and (3) a method for transferring such nuclei into the host egg without damaging either the nucleus or the oocyte.

All three techniques were developed in the 1950s by Robert Briggs and Thomas King (see Mc Kinnell 1978; Di Berardino and Mc Kinnell 2004). First, they combined the enucleation of the host egg with its activation. When an oocyte (a developing egg cell) from the leopard frog *Rana pipiens* is pricked with a clean glass needle, the egg undergoes all the cytological and biochemical changes associated with fertilization. The internal cytoplasmic rearrangements of fertilization occur, and the completion of meiosis takes place near the animal pole of the cell. The meiotic spindle can easily be located as it pushes away the pigment granules at the animal pole, and puncturing the oocyte at this site causes the spindle and its chromosomes to flow outside the egg (Figure 4.5). The host egg is now considered both activated (the fertilization reactions necessary to initiate development have been completed) and enucleated.

The transfer of a nucleus into the egg is accomplished by disrupting a donor cell and transferring the released nucleus into the oocyte through a micropipette. Some cytoplasm accompanies the nucleus to its new home, but the ratio of donor to recipient cytoplasm is only $1:10^5$, and the donor cytoplasm does not seem to affect the outcome of the experiments. In 1952, Briggs and King, using these techniques, demonstrated that blastula cell nuclei could direct the development of complete tadpoles when transferred into the cytoplasm of an activated enucleated frog egg. This procedure is called **somatic nuclear transfer**, or more commonly, **cloning**.

What happens when nuclei from more advanced developmental stages are transferred into activated enucleated oocytes? King and Briggs (1956) found that whereas most blastula nuclei could produce entire tadpoles, there was a dramatic decrease in the ability of nuclei from later stages to direct development to the tadpole stage (Figure 4.6). When nuclei from the somatic cells of tailbud-stage tadpoles were used as donors, normal development did not occur. However, nuclei from the germ cells of tailbud-stage tadpoles (which could give rise to a complete organism after fertilization) were capable of directing normal development in 40 percent of the blastulae that developed (Smith 1956). Thus, most somatic cells appeared to lose their ability to direct development as they became determined and differentiated.

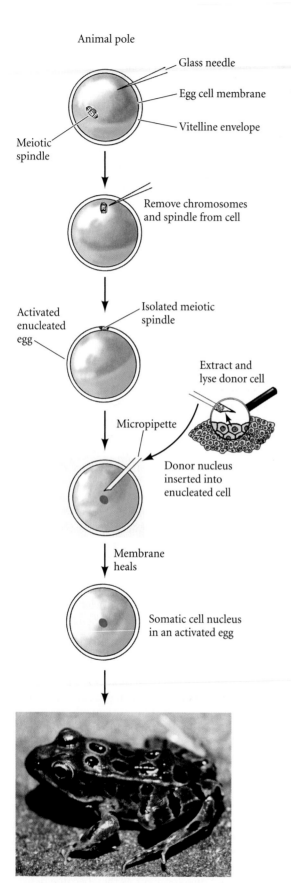

Animal pole

Glass needle

Egg cell membrane

Vitelline envelope

Meiotic spindle

Remove chromosomes and spindle from cell

Activated enucleated egg

Isolated meiotic spindle

Extract and lyse donor cell

Micropipette

Donor nucleus inserted into enucleated cell

Membrane heals

Somatic cell nucleus in an activated egg

FIGURE 4.5   Procedure for transplanting blastula nuclei into activated enucleated *Rana pipiens* eggs. The relative dimensions of the meiotic spindle have been exaggerated to show the technique. "Freddy," the handsome and mature *R. pipiens* in the photograph, was derived in this way by M. DiBerardino and N. Hoffner Orr. The vitelline envelope is the extracellular matrix surrounding the egg. (After King 1966; photograph courtesy of M. DiBerardino.)

Amphibian cloning: The totipotency of somatic cells

Is it possible that some differentiated cell nuclei differ from others in their ability to direct development? John Gurdon and his colleagues, using slightly different methods of nuclear transplantation on the frog *Xenopus*, obtained results suggesting that the nuclei of some differentiated cells can remain totipotent. Gurdon, too, found a progressive loss of potency with increasing developmental age, although *Xenopus* cells retained their potencies for a longer period than did the cells of *Rana* (Figure 4.7).

To clone amphibians from the nuclei of cells known to be differentiated, Gurdon and his colleagues cultured epithelial cells from adult frog foot webbing. These cells were shown to be differentiated: each of them contained a specific keratin, the characteristic protein of adult skin cells. When nuclei from these cells were transferred into activated enucleated *Xenopus* oocytes, none of the first-generation transfers progressed further than the formation of the

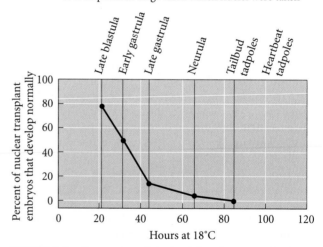

Developmental stage from which nuclei were taken

Late blastula | Early gastrula | Late gastrula | Neurula | Tailbud tadpoles | Heartbeat tadpoles

Percent of nuclear transplant embryos that develop normally

Hours at 18°C

FIGURE 4.6   Percentage of successful nuclear transplants as a function of the developmental age of the donor nucleus. The abscissa represents the developmental stage at which a donor nucleus (from *R. pipiens*) was isolated and inserted into an activated enucleated oocyte. The ordinate shows the percentage of those transplants capable of producing blastulae that could then direct development to the swimming tadpole stage. (After McKinnell 1978.)

Wild-type donor
of enucleated eggs

Albino parents
of nucleus donor

WEBSITE 4.3 Amphibian cloning: Potency and deformity. The ability of amphibian nuclei to produce normal frogs has been a controversial area. The definition of what is a differentiated cell nucleus and what constitutes a "normal" frog have both been questioned.

WEBSITE 4.4 Metaplasia. Before cloning became possible, evidence for genomic equivalence came from a phenomenon called metaplasia, the regeneration of tissue from a different source.

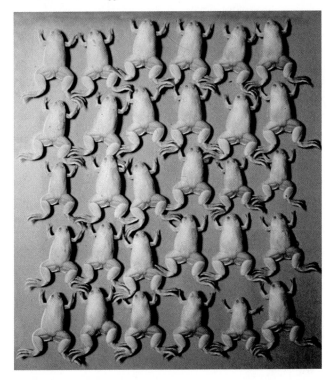

Cloning mammals

In 1997, Ian Wilmut announced that a sheep had been cloned from a somatic cell nucleus from an adult female sheep. This was the first time that an adult vertebrate had been successfully cloned from another adult.* To do this, Wilmut and his colleagues took cells from the mammary gland of an adult (6-year-old) pregnant ewe and put them into culture (Wilmut et al. 1997). The culture medium was formulated to keep the nuclei in these cells at the intact diploid stage of the cell cycle (G_1). This cell-cycle stage turned out to be critical. They then obtained oocytes from a different strain of sheep and removed their nuclei. These oocytes had to be in the second meiotic metaphase (which is the stage at which they are usually fertilized).

Fusion of the donor cell and the enucleated oocyte was accomplished by bringing the two cells together and sending electric pulses through them. The electric pulses destabilized the cell membranes, allowing the cells to fuse together. Moreover, the same pulses that fused the cells activated the egg to begin development. The resulting embryos were eventually transferred into the uteri of pregnant sheep.

Of the 434 sheep oocytes originally used in this experiment, only one survived: Dolly (Figure 4.8). DNA analysis confirmed that the nuclei of Dolly's cells were derived from the strain of sheep from which the donor nucleus was taken (Ashworth et al. 1998; Signer et al. 1998). Thus, it appears that the nuclei of vertebrate adult somatic cells can indeed be totipotent. No genes necessary for development

FIGURE 4.7 A clone of *Xenopus laevis* frogs. The nuclei for all the members of this clone came from a single individual—a female tailbud-stage tadpole whose parents (upper panel) were both marked by albino genes. The nuclei (containing these defective pigmentation genes) were transferred into activated enucleated eggs from a wild-type female (upper panel). The resulting frogs were all female and albino (lower panel). (Photographs courtesy of J. Gurdon.)

neural tube shortly after gastrulation. By serial transplantation (i.e., taking nuclei from the cloned blastulas), however, numerous tadpoles were generated (Gurdon et al. 1975). Although the tadpoles all died prior to feeding, they showed that a single differentiated cell nucleus still retained incredible potencies. A nucleus of a skin cell could produce all the cells of the young tadpole.

*The creation of Dolly was the result of a combination of scientific and social circumstances. These circumstances involved job security, people with different areas of expertise meeting each other, childrens' school holidays, international politics, and who sits near whom in a pub. The complex interconnections giving rise to Dolly are told in *The Second Creation* (Wilmut et al. 2000), a book that shows how contemporary science actually works. As Wilmut acknowledged (p. 36), "The story may seem a bit messy, but that's because life is messy, and science is a slice of life."

(A)

(B)

OOCYTE DONOR
(Scottish blackface strain)

NUCLEAR DONOR
(Finn-Dorset strain)

Eggs removed

Meiotic
spindle

Udder cells removed

Udder cells
placed in
culture,
grown in
G_1 stage.

Remove spindle

Micropipette

Enucleated
egg

Transfer cell
into enucleated egg

Egg and cell
fused with
electric current

Embryo cultured
7 days

Blastocyst
forms

Embryo transferred to
surrogate mother
(Scottish blackface)

Birth of Dolly
(Finn-Dorset lamb
genetically identical
to nuclear donor)

have been lost or mutated in a way that would make them nonfunctional.

Cloning has been confirmed in guinea pigs, rabbits, rats, mice, dogs, cats, horses, and cows. In 2003, a cloned mule became the first sterile animal to be so reproduced (Woods et al. 2003). However, certain caveats must be applied. First, although it appears that all the organs were properly formed in the cloned animals, many of the clones developed debilitating diseases as they matured (Humphreys et al. 2001; Jaenisch and Wilmut 2001; Kolata 2001). Second, the phenotype of the cloned animal is sometimes not identical to that of the animal from which the nucleus was derived. There is variability due to random chromosomal events and the effects of environment. The pigmentation of calico cats, for instance, is due to the *random inactivation* of one or the other X chromosome (a genetic mechanism that will be discussed in the next chapter) in each somatic cell of the female cat embryo. Therefore, the markings of the first cloned cat, a calico named "CC," were different from those of "Rainbow," the calico cat whose cells provided the implanted nucleus that generated the clone (Figure 4.9).

The same genotype gives rise to multiple phenotypes in cloned sheep as well. Wilmut noted that four sheep cloned from blastocyst nuclei from the same embryo "are genetically identical to each other and yet are very different in size and temperament, showing emphatically that an animal's genes do not 'determine' every detail of its physique and personality" (Wilmut et al. 2000, p. 5). Wilmut concludes that for this and other reasons, the "resurrection" of lost loved ones by cloning is not feasible.

FIGURE 4.8   Cloned mammals, whose nuclei came from adult somatic cells. (A) Dolly, the adult sheep on the left, was derived by fusing a mammary gland cell nucleus with an enucleated oocyte, which was then implanted in a surrogate mother (of a different breed of sheep) who gave birth to Dolly. Dolly has since produced a lamb (Bonnie, at right) by normal reproduction. (B) Procedure used for cloning sheep. (A, photograph by Roddy Field, © Roslin Institute; B after Wilmut et al. 2000.)

(A)

(B)

FIGURE 4.9   The kitten "CC" (A) is a clone produced using somatic nuclear transfer from "Rainbow" (B). Their markings are not identical because the pigmentation pattern in calico cats is affected by the random inactivation of the second X chromo-some (see Chapter 5). Their behaviors are also quite different. (Photographs courtesy of the College of Veterinary Medicine, Texas A&M University.)

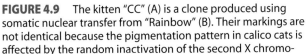

SIDELIGHTS & SPECULATIONS

Why Clone Mammals?

Given that we already knew from amphibian studies in the 1960s that nuclei were totipotent, why clone mammals? Many of the reasons are medical and commercial, and there are good reasons why these techniques were first developed by pharmaceutical companies rather than at universities. Cloning is of interest to some developmental biologists who study the relationships between nucleus and cytoplasm during fertilization, and to others who study aging (and the loss of totipotency that appears to accompany it). But cloned mammals are of special interest to the people and corporations concerned with creating **protein pharmaceuticals**.

Protein drugs such as human insulin, protease inhibitors, and clotting factors are difficult to manufacture. Because of immunological rejection problems, human proteins are usually much better tolerated by patients than proteins extracted from other species. Similarly, our bodies often reject proteins that have been synthesized by genetically engineered bacteria. The problem thus becomes how to obtain large amounts of human proteins. One of the most efficient ways to produce these proteins is to insert the human genes encoding them into the oocyte DNA of sheep, goats, or cows. Animals containing a gene from another individual (often of a different species)—a **transgene**—are called **transgenic** animals. A female sheep or cow made transgenic for the human protein gene might express this gene in her mammary tissue and secrete the protein in her milk (Figure 4.10; Prather 1991).

Producing transgenic sheep, cows, or goats is not an efficient undertaking, how-ever. Only 20 percent of the treated eggs survive the technique. Of these, only about 5 percent actually express the human gene. And of those who do express the human gene, only half are female, and only a small percentage of these females actually secrete high levels of the human protein into their milk (plus the fact that it often takes years before they first produce milk). And then, when these rare animals die after several years of milk production, their offspring are usually not as good at secreting the human protein. But if pharmaceutical companies could clone such "elite transgenic animals," the cloned females should all produce high yields of human protein in their milk. The economic incentives for such cloning are therefore enormous, with therapeutic proteins potentially becoming much cheaper

(Continued on next page)

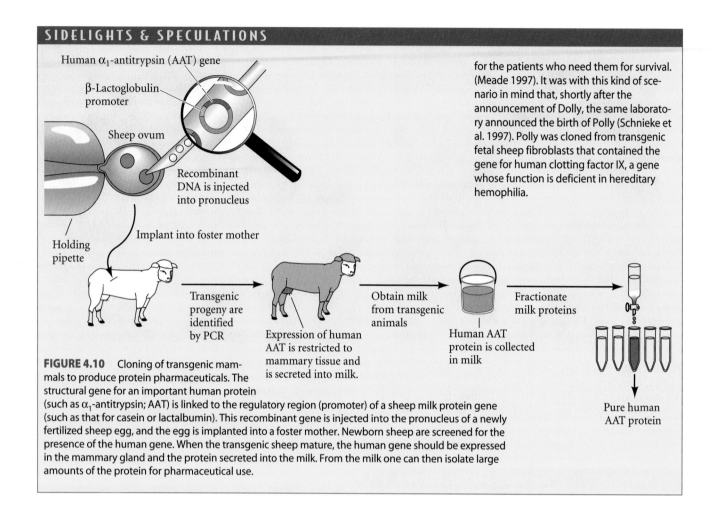

for the patients who need them for survival. (Meade 1997). It was with this kind of scenario in mind that, shortly after the announcement of Dolly, the same laboratory announced the birth of Polly (Schnieke et al. 1997). Polly was cloned from transgenic fetal sheep fibroblasts that contained the gene for human clotting factor IX, a gene whose function is deficient in hereditary hemophilia.

FIGURE 4.10 Cloning of transgenic mammals to produce protein pharmaceuticals. The structural gene for an important human protein (such as α_1-antitrypsin; AAT) is linked to the regulatory region (promoter) of a sheep milk protein gene (such as that for casein or lactalbumin). This recombinant gene is injected into the pronucleus of a newly fertilized sheep egg, and the egg is implanted into a foster mother. Newborn sheep are screened for the presence of the human gene. When the transgenic sheep mature, the human gene should be expressed in the mammary gland and the protein secreted into the milk. From the milk one can then isolate large amounts of the protein for pharmaceutical use.

Differential Gene Expression

If the genome is the same in all somatic cells within an organism (with the exception of the lymphocytes; see Website 4.5), how do the cells become different from one another? If every cell in the body contains the genes for hemoglobin and insulin proteins, how come hemoglobin proteins are made only in red blood cells, insulin proteins are made only in certain pancreas cells, and neither is made in the kidney or nervous system? Based on the embryological evidence for genomic equivalence (as well as on bacterial models of gene regulation), a consensus emerged in the 1960s that the answer lies in **differential gene expression**. The three postulates of differential gene expression are as follows:

1. Every cell nucleus contains the complete genome established in the fertilized egg. In molecular terms, the DNAs of all differentiated cells are identical.
2. The unused genes in differentiated cells are not destroyed or mutated, but retain the potential for being expressed.
3. Only a small percentage of the genome is expressed in each cell, and a portion of the RNA synthesized in each cell is specific for that cell type.

The first two postulates have already been discussed. The third postulate—that only a small portion of the genome is active in making tissue-specific products—was first tested in insect larvae. Fruit fly larvae have certain cells whose chromosomes become **polytene**. These chromosomes, beloved by *Drosophila* geneticists, undergo DNA replication in the absence of mitosis and therefore contain 512 (2^9), 1024 (2^{10}), or even more parallel DNA double helices instead of just one (Figure 4.11A). These cells do not undergo mitosis, and they grow by expanding to about 150 times their original volume. Beermann (1952) showed that the banding patterns of polytene chromosomes were identical throughout the larva, and that no loss or addition of any chromosomal region was seen when different cell types were compared. However, he and others showed that different regions of these chromosomes made different, tissue-specific mRNAs based on the differentiated cell type. In certain cell types, particular regions of the chromosomes would loosen up, "puff" out, and transcribe mRNA. In other cell types, these same regions were "silent" while different regions puffed out and synthesized mRNA.

The idea that genes were differentially expressed in different cell types was confirmed using **DNA-RNA hybrid-**

(A)

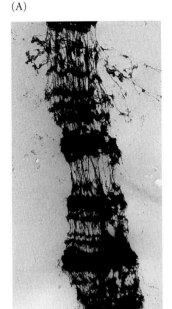

(B)

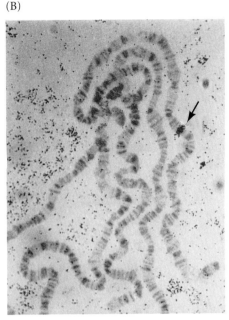

FIGURE 4.11 Polytene chromosomes. (A) Electron micrograph of a small region of a *Drosophila* polytene chromosome. The bands (dark) are highly condensed compared with the interband (lighter) regions. (B) Hybridization of a yolk protein mRNA with the polytene chromosome of a larval *Drosophila* salivary gland. The dark grains (arrow) show where the radioactive yolk protein message has bound to the chromosomes. Note that the gene for the yolk protein is present in the salivary gland chromosomes, even though yolk protein is not synthesized there. (A from Burkholder 1976, photograph courtesy of G. D. Burkholder; B from Barnett et al. 1980; photograph courtesy of P. C. Wensink.)

ization. This technique involves annealing single-stranded pieces of RNA and DNA to allow complementary strands to form double-stranded hybrids. If one of the nucleic acids is marked with a dye or radioactive tracer, it can be used to indicate the presence of its complement (Figure 4.11B). While some mRNAs from one cell type were also found in other cell types (as expected for mRNAs encoding enzymes concerned with cell metabolism), many mRNAs were found to be specific for a particular type of cell and were not expressed in other cell types, even though the genes encoding them were present (Wetmur and Davidson 1968). Thus, differential gene expression is the way a single genome, derived from the fertilized egg, can generate the hundreds of different cell types in the body.

The details of how differential gene expression occurs will be covered in Chapter 5. To understand the results that will be presented there, however, one must be familiar with some of the molecular techniques being applied to the study of development. These include techniques to determine the spatial and temporal location of specific mRNAs, as well as techniques to determine the functions of these messages.

WEBSITE **4.5 Exceptions to the rule of genomic equivalence.** While nearly all cells in our bodies have the same genome, the genomes of our lymphocytes change during development. The genes encoding our antibody proteins are different in each of our lymphocytes.

RNA Localization Techniques

Detecting specific mRNAs: RT-PCR

The **polymerase chain reaction (PCR)** is a method of in vitro gene cloning* that can generate enormous quantities of a specific DNA fragment from a small amount of starting material (Saiki et al. 1985). It can be used to clone a specific gene or to determine whether a specific gene is actively transcribing mRNA in a particular organ or cell type. Standard methods of gene cloning use living microorganisms to amplify recombinant DNA. PCR, however, can amplify a single segment of a DNA molecule several million times in a few hours, and can do it in a test tube. The techniques of PCR are extremely useful in cases where there is very little sample to study.

For instance, preimplantation mouse embryos have very tiny amounts of mRNA, and we cannot obtain millions of such embryos to study. If we wanted to know whether a single preimplantation mouse embryo contained the mRNA for a particular protein, it would be very difficult to find out using standard methods—we would have to lyse thousands of mouse embryos in order to obtain enough mRNA. However, by combining PCR techniques with the ability of the **reverse transcriptase (RT)** enzyme to make DNA out of mRNA, we can get around the problem of scarce messages. The **RT-PCR** technique allows us to convert the mRNA into DNA and to copy the specific DNA sequence of interest (Rappolee et al. 1988).

Gene cloning—making numerous copies of the same DNA sequence—should not be confused with *organism cloning*—making genetically identical copies of an organism.

FIGURE 4.12   The reverse transcriptase-polymerase chain reaction (RT-PCR) technique for determining whether a particular type of mRNA is present. First, the mRNA from a small sample is converted to double-stranded cDNA using the enzymes reverse transcriptase and RNase H. A primer is added to the cDNA and the second strand is completed using thermostable DNA polymerase from *T. aquaticus* (*Taq* polymerase). This "target" DNA is then denatured and two sets of primers are added; the primers hybridize to opposite ends of the target sequence if the sequence is present. (This will happen only if the mRNA of interest was originally present.) When *Taq* polymerase is added to the denatured DNA, each strand synthesizes its complement. These strands are then denatured, and the primers are hybridized to them, starting the cycle again. In this way, the number of new strands having the sequence of interest between the two primers increases exponentially.

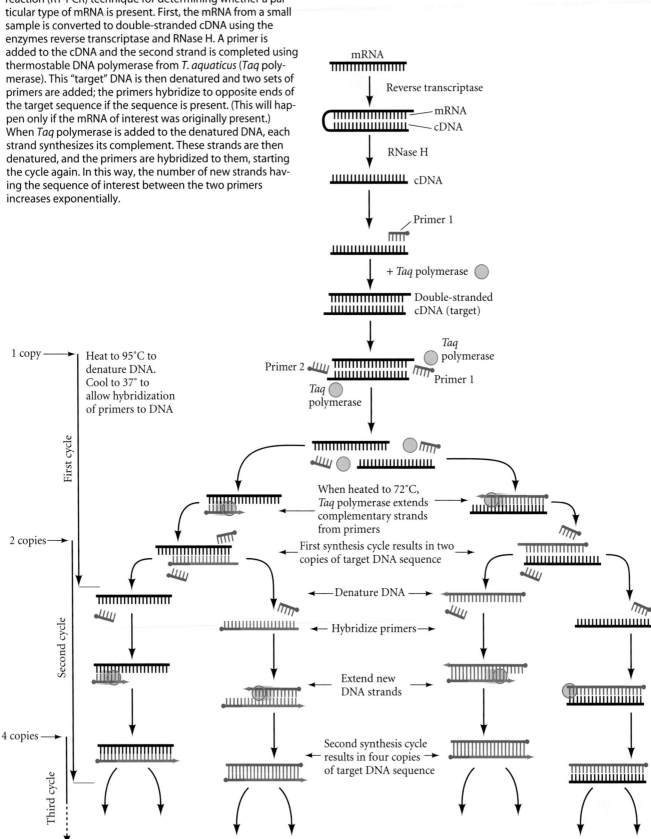

The use of RT-PCR to find rare mRNAs is illustrated in Figure 4.12. First, the mRNAs from a sample are purified and converted into complementary DNA (cDNA) using the RT enzyme. Next, a specific cDNA is targeted for amplification. Two small oligonucleotide primers that are complementary to a portion of the message being looked for are added to the population of cDNA. Oligonucleotides are relatively short stretches of DNA (about 20 bases). If the oligonucleotides bind to sequences in the cDNA, this means that the mRNA being sought was present in the original sample. The oligonucleotide primers are made so that they hybridize to opposite strands at opposite ends of the targeted sequence. (If we are trying to isolate the gene or mRNA for a specific protein of known sequence, we can synthesize oligonucleotides that are complementary to the sequences encoding the amino end and the carboxyl end of the protein.) The 3′ ends of the primers face each other, so that replication will run through the target DNA.

Once the first primer has hybridized with the cDNA, DNA polymerase can be used to synthesize a new strand. The DNA polymerase used in this process is extracted from thermophilic (heat-loving) bacteria such as *Thermus aquaticus* or *Thermococcus littoralis*.* and can withstand temperatures near boiling. Once the second strand of DNA is made, it is heat-denatured from its complement. The temperatures used would inactivate the more usual *E. coli* DNA polymerase, but the thermostable polymerases are not damaged.

The second primer is added, and now both strands can synthesize new DNA. Repeated cycles of denaturation and synthesis amplify the DNA sequence exponentially. After 20 such rounds, that specific sequence has been amplified 2^{20} (a little more than a million) times. When the DNA is subjected to electrophoresis, the presence of such an amplified fragment is easily detected. Its presence shows that there was an mRNA with the sequence of interest present in the original sample.

WEBSITE 4.6 DNA isolation techniques. The basic techniques of DNA analysis—gene cloning, sequencing, Southern blotting, and northern blotting—are discussed in most introductory biology books. This site gives a review of these procedures.

Microarrays

RT-PCR generally works on a "one gene in one experiment" basis, which means that the "whole picture" of differential gene expression between cells or tissues is difficult to obtain. In the past several years, a new technology, called **DNA microarrays**, has allowed scientists to monitor changes in the transcription of thousands of genes simultaneously (Wan et al. 1996). Microarrays combine the tech-

nology of RT-PCR with high-speed robotics. First, one takes mRNA from a tissue and, using reverse transcriptase, converts the mRNAs into their complementary DNAs. Each individual cDNA is then cloned, denatured, and amplified by the polymerase chain reaction Each of the resulting cDNA clones serves as a probe and is robotically printed onto glass slides in a particular order. The slides are subsequently hybridized to two "targets" with different fluorescent labels. These targets are pools of cDNAs that have been generated after isolating mRNA from cells or tissues in two states that one wishes to compare. For instance, if the aim is to compare cell type A with cell type B, one takes mRNAs from both cell types, converts the mRNAs into cDNAs, and labels the cDNAs from cell type A with fluorescein (green) and the cDNAs from cell type B with rhodamine (red) (Figure 4.13). The two DNA pools would then be equally mixed, and the mixture placed onto each spot of probe DNA. The resulting fluorescent intensities are produced using a laser confocal fluorescent microscope, and ratio information is obtained following image processing. By comparing the fluorescent intensities, one can tell if a particular cDNA (and hence, the mRNA of interest) is present in higher amounts in one cell type or another.

For instance, one can look at the genes active in the prospective dorsal part of a frog blastula and compare them with the genes active in the future ventral portion of the same blastula (Altmann et al. 2001). Alternatively, one can look at entire animals at different stages of their development to see which genes are active at each stage (White et al. 1999). This allows one to focus one's research on those genes whose expression differs between the two sets of cDNAs.

One can use microarrays even without knowing what gene to look for. The difference in gene expression will indicate which genes are important in the particular tissue. For instance, in Chapter 17 we will discuss the different gene expression in the cells of male (XY) and female (XX) mouse brains. Researchers did not have to guess which genes would be expressed differently—the microarrays were enough to tell them which genes were differentially active in XX and XY brains, with the results then being confirmed by RT-PCR.

WEBSITE 4.7 Microarray technology. There are many variations on the scheme described here. Microarray technology may become essential for developmental biology if we are to understand the complex changes in gene expression that occur as cell types differentiate.

Macroarrays

A less expensive modification of the microarray is the **macroarray**. The technology is similar, but the spots are larger (about 1 mm as compared to 250 μm). This means that macroarrays can be visually interpreted without a microscope, and that radioisotopes as well as fluorescent labels can be used. For instance, in Chapter 3, we saw that

---

*These bacteria normally live in hot springs (such as those in Yellowstone National Park) or in submarine thermal vents, where the temperature reaches nearly 90°C. RT-PCR takes advantage of this evolutionary adaptation.

PREPARE cDNA
"TARGETS"

PREPARE MICROARRAY
"PROBES" (by robot)

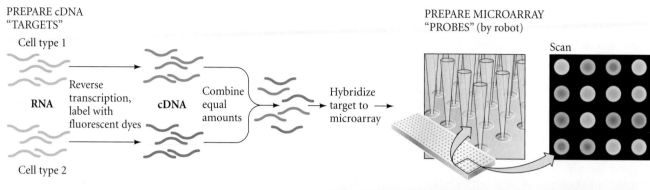

FIGURE 4.13 Microarray technique. "Target" mRNAs are prepared by isolating mRNAs from two cells types being compared (two different cell types or the same cell type at different times), making complementary DNAs from them, and adding two different types of fluorescent tags (one color for each cell type). In this case, cDNAs from cell type 1 are labeled with a green dye, while cDNAs from cell type 2 are tagged with a red dye. Microarray probes are made by taking a number of different cDNAs made from the mRNA of one cell type and adhering them to a glass microscope slide. The probes and the target are hybridized together. If the mRNA in a probe is abundant in cell type 1, the signal is green. If the mRNA is abundant in cell type 2, the signal is red. If the mRNA is present in both cell types, the signal is yellow. The bottom shows a portion of such a microarray. (After http://www.genetics.ucla.edu/microarray/instruction.html)

nodal genes are important in specifying the different regions of mesoderm in zebrafish embryos. A similar gene in *Xenopus* is called *Xenopus Nodal related-1* (*Xnr1*). To see if Xnr1 is activating particular genes, Naoto Ueno's laboratory injected *Xnr1* mRNA into some early embryos but not others. They then harvested mRNAs from the animal caps of the Xnr1-secreting and control embryos. (The animal cap would not be exposed to Xnr1 under normal development.) After converting the mRNAs into radioactively labeled cDNAs, Ueno and colleagues hybridized these probes to slides containing bound DNA. One macroarray was hybridized with radioactive cDNAs from the control animal caps; an identical macroarray was hybridized with radioactive cDNAs from the Xnr1-treated caps. One of their results is shown in Figure 4.14, in which the gene for chordin is shown to be expressed in the treated but not in the untreated animal cap.

Locating mRNAs in space and time: In situ hybridization

A truly detailed map of gene expression patterns can be obtained using **in situ hybridization**. Here, a labeled **antisense mRNA** probe (a DNA or RNA sequence that is complementary to the sequence of a specific mRNA) is hybridized with the mRNA in the organ itself. Antisense mRNA is made from a cloned gene in which the gene is reversed with respect to a promoter within the vector. (We

will see in Figure 4.21 how genes can be "reversed.") The mRNA transcribed from such a gene encodes a sequence complementary to the normal ("sense") mRNA made by that gene. Such antisense mRNA can be used as a probe, since it will recognize the "sense" mRNA in the cell.

For in situ hybridization, the antisense RNA is labeled either by being made radioactive or by being attached to a dye, allowing the probe to be visualized. Thus, one makes a sequence-specific stain that will label only those cells that have accumulated mRNAs of a particular sequence. When radioactive probes are used, embryos or organs are first fixed to preserve their structure and to prevent their mRNA from being degraded. The fixed tissues are then sectioned for microscopy and placed on a slide. When the labelled sequence is added, it binds only where the target mRNA (to which it is complementary) is present. After any unbound probe is washed off, the slide is covered with a transparent photographic emulsion for autoradiography. By using dark-field microscopy (or computer-mediated bright-field imaging), the reduced silver grains can be shown in a color that contrasts with the background stain. Thus, we can visualize those cells (or even regions within cells) that have accumulated a specific type of mRNA. Figure 4.15 shows an in situ hybridization for *Pax6* mRNA in mice. One can see that *Pax6* mRNA is found in the region where the presumptive retina meets the presumptive lens tissue. As development proceeds, it is seen in the developing retina, lens, and cornea of the eye.

In **whole-mount in situ hybridization,** the entire embryo (or a part thereof) can be stained for certain mRNAs. This technique, which uses dyes rather than radioactivity, allows researchers to look at entire embryos (or their organs) without sectioning them, thereby observing large regions of gene expression. Figure 4.16 shows an

(A)
Xnr1 mRNA or H₂O
microinjection at 2-cell stage

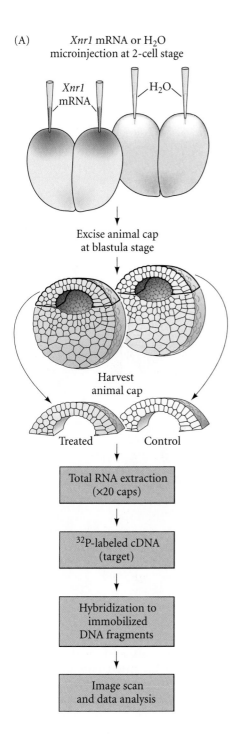

Excise animal cap
at blastula stage

Harvest
animal cap

Treated   Control

Total RNA extraction
(×20 caps)

³²P-labeled cDNA
(target)

Hybridization to
immobilized
DNA fragments

Image scan
and data analysis

(B)

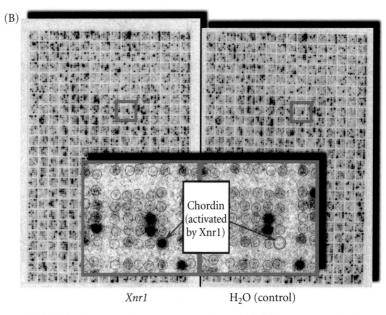

Chordin
(activated
by Xnr1)

Xnr1   H₂O (control)

FIGURE 4.14 Macroarray analysis of those genes whose expression in the early *Xenopus* embryo is caused by the activin-like protein Nodal-related 1 (Xnr1). (A) Creation of targets for macroarray analysis. (B) In the macroarrays, certain radioactive spots (representing hybridizations) were seen in samples from the Xnr1-stimulated cells, but not in the control cells. These spots represent the genes activated by Xnr1; the insert shows one of these, the *chordin* gene. Most of the thousands of genes observed were not activated. The DNA of the hybridized spots can be sequenced and identified. (Photographs courtesy of N. Ueno.)

FIGURE 4.15 In situ hybridization showing the expression of the *Pax6* gene in the developing mouse eye. Transverse microscopic sections were taken through the developing heads of 9-, 10-, and 15-day embryonic mice. At this time, the optic cup is touching the outer ectoderm and inducing it to form a lens. The radioactive *Pax6* antisense DNA probe binds only where *Pax6* mRNA is present, and can be visualized by developing the photographic emulsion. The locations where the probe has bound, depicted here as green dots (by computer imaging), show that the *Pax6* message is expressed in both the presumptive lens ectoderm and the optic stalk, which forms the retina and optic nerve. (From Grindley et al. 1995; photograph courtesy of R. E. Hill.)

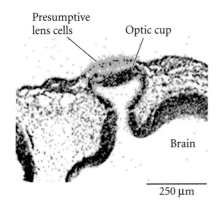

Presumptive
lens cells   Optic cup

Brain

250 μm

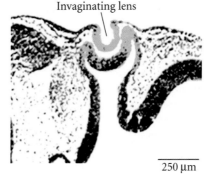

Invaginating lens

250 μm

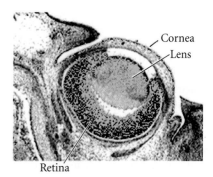

Cornea
Lens

Retina

(A)

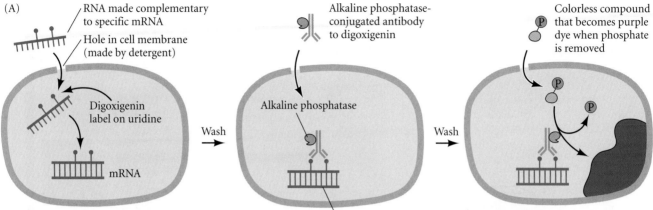

RNA made complementary to specific mRNA

Hole in cell membrane (made by detergent)

Digoxigenin label on uridine

mRNA

1. Add digoxigenin-labeled probe complementary to RNA of interest

Alkaline phosphatase-conjugated antibody to digoxigenin

Alkaline phosphatase

Wash →

2. Add alkaline phosphatase-conjugated antibody that binds to digoxigenin

Colorless compound that becomes purple dye when phosphate is removed

Wash →

3. Add chemical that becomes a dark purple dye when phosphate is removed; dye colors the cell.

FIGURE 4.16   Whole-mount in situ hybridization localizing *Pax6* mRNA in early chick embryos. (A) Schematic of the procedure. A digoxigenin-labeled antisense probe hybridizes to a specific mRNA. Alkaline phosphatase-conjugated antibodies to digoxigenin recognize the digoxigenin-labeled probe. The enzyme is able to convert a colorless compound into a dark purple precipitate. (B) *Pax6* mRNA can be seen to accumulate in the roof of the brain region that will form the eyes, as well as in the ectoderm that will form lenses. More caudal expression of this gene in the nervous system is also seen. (After Li et al. 1994; photograph courtesy of O. Sundin.)

(B)

in situ hybridization performed on a whole chick embryo that was fixed without being sectioned. (The embryo was also permeabilized by lipid and protein solvents so that the probe could get in and out of its cells.) The probe used in this experiment recognized the mRNA encoding *Pax6* in the chick embryo.

The probe shown in Figure 4.16 was labeled not with a radioactive isotope, but with a modified nucleotide. To create this probe, a region of the cloned *Pax6* gene was transcribed into mRNA, but with an important modification: in addition to the standard uridine triphosphate (UTP), the nucleotide mix also contained UTP conjugated with *digoxigenin*. Digoxigenin—a compound made by particular groups of plants and not found in animal cells—does not interfere with the coding properties of the resulting mRNA, but does make it recognizably different from any other RNA in the cell.

The digoxigenin-labeled probe was incubated with the embryo. After several hours, numerous washes removed any probe that had not bound to the embryo. Then the embryo was incubated in a solution containing an antibody against digoxigenin. The only places where digoxigenin should exist is where the probe bound (i.e., where it recognized its mRNA), so the antibody sticks in those places. This antibody, however, is not in its natural state. It has been conjugated covalently to an enzyme, such as alkaline phosphatase. After repeated washes to remove all the unbound enzyme-conjugated antibody, the embryo was incubated in another solution that was converted into a dye by the enzyme. The enzyme should be present only

where the digoxigenin is present, and the digoxigenin should be present only where the specific complementary mRNA is found.

Thus, in Figure 4.16B, the dark blue precipitate formed by the enzyme indicates the presence of the target mRNA. The figure reveals mRNA for the Pax6 protein to be present in the roof of the brain region that will form the eyes, as well as in the head ectoderm that will form the lenses. It is also expressed in a more caudal region of the neural tube as well as in the pancreas.

Determining the Function of Genes during Development

Transgenic cells and organisms

While it is important to know the sequence of a gene and its temporal-spatial pattern of expression, what's really crucial is to know the functions of that gene during development. Recently developed techniques have enabled us

THE GENETIC CORE OF DEVELOPMENT

to study gene function by moving certain genes into and out of embryonic cells.

INSERTING NEW DNA INTO A CELL   Cloned pieces of DNA can be isolated, modified (if so desired), and inserted into cells by several means. One direct technique is **microinjection**, in which a solution containing the cloned gene is injected into the nucleus of a cell (Capecchi 1980). This technique is especially useful for inserting genes into newly fertilized eggs, since the haploid nuclei of the sperm and egg are relatively large (Figure 4.17). In **transfection**, DNA fragments may be incorporated directly into cells by incubating the cells in a solution designed to make them "drink" the new DNA in. The chances of a DNA fragment being incorporated into the chromosomes in this way are relatively small, however, so the DNA of interest is usually mixed with another gene, such as a gene encoding resistance to a particular antibiotic, that enables the rare cells that *do* incorporate the DNA to "identify themselves" by surviving under culture conditions that kill all the other cells (Perucho et al. 1980; Robins et al. 1981). A similar technique is **electroporation**, in which a high-voltage pulse "pushes" the DNA into the cells.

A more "natural" way of getting genes into cells is to insert the cloned gene into a **transposable element** or **retroviral vector**. These naturally occurring mobile regions of DNA can integrate themselves into the genome of an organism. Retroviruses are RNA-containing viruses. They enter a host cell, where they make a DNA copy of themselves (using their own virally encoded reverse transcriptase); the copy then becomes double-stranded and integrates itself into a host chromosome. The integration is accomplished by two identical sequences (long terminal repeats) at the ends of the retroviral DNA. Retroviral vectors can be made by removing the viral packaging genes (needed for the exit of viruses from the cell) from the center of a mouse retrovirus. This extraction creates a vacant site where other genes can be placed. By using the appropriate restriction enzymes researchers can excise a gene of interest (such as a gene isolated by PCR) and insert it into a retroviral vector. Retroviral vectors infect mouse cells with an efficiency approaching 100 percent.

Similarly, in *Drosophila*, new genes can be carried into the fly embryo via **P elements**. These DNA sequences are naturally occurring transposable elements that can integrate like viruses into any region of the *Drosophila* genome. These elements can be isolated and cloned genes can be inserted into the center of the isolated P element. When the recombined P element is injected into a *Drosophila* oocyte, it can integrate itself into the embryo's DNA, providing the organism with the new gene (Spradling and Rubin 1982).

CHIMERIC MICE   The techniques described above have been used to transfer genes into every cell of the mouse embryo. During early mouse development, there is a stage (the blastocyst) when only two cell types are present: the outer trophoblast cells, which will form the fetal portion of the placenta, and the inner cell mass (ICM), whose cells will give rise to the embryo itself. Separation of ICM cells can lead to twins (see Chapters 3 and 11), and if an ICM blastomere from one mouse is transferred into the embryo of a second mouse, that donor ICM cell can contribute to every organ of the host embryo.

Inner cell mass blastomeres can be isolated from an embryo and cultured in vitro; such cultured cells are called **embryonic stem cells (ES cells)**. ES cells are almost totipotent, since each of them can contribute to all tissues except the trophoblast if injected into a host embryo (Gardner 1968; Moustafa and Brinster 1972). Moreover, once in culture, these cells can be treated as described in the preceding section so that they will incorporate new DNA. This added gene (the transgene) can come from any organism. A treated ES cell (the entire cell, not just the DNA) can then be injected into another early-stage embryo, and will integrate into this host. The result is a **chimeric** mouse*(Figure 4.18). Some of the chimera's cells will be derived from the host's own inner cell mass, but some portion of its cells will be derived from the treated embryonic stem cell. If the treated cells become part of the germ line of the mouse, some of its gametes will be derived from the donor cell. If such a chimeric mouse is mated with a wild-type mouse, some of its progeny will carry one copy of the inserted gene. When these heterozygous progeny are mated to one another,

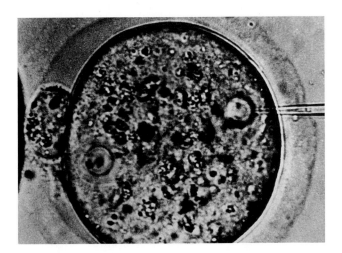

FIGURE 4.17   Insertion of new DNA into embryonic cells. Here, DNA (from cloned genes) is injected into the pronucleus of a mouse egg. (From Wagner et al. 1981; photograph courtesy of T. E. Wagner.)

---

*It is critical to note the difference between a chimera and a hybrid. A **hybrid** results from the union of two different genomes within the same cell: the offspring of an *AA* genotype parent and an *aa* genotype parent is an *Aa* hybrid. A **chimera** results when cells of different genetic constitutions appear in the same organism. The term is apt: it refers to a mythical beast with a lion's head, a goat's body, and a serpent's tail.

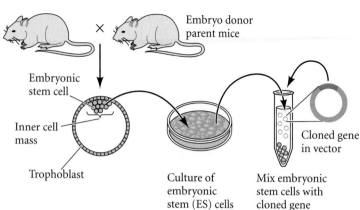

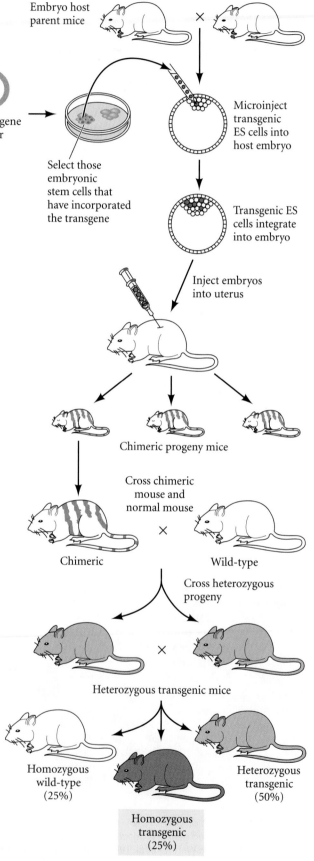

FIGURE 4.18   Production of transgenic mice. Embryonic stem cells from a mouse are cultured and their genome altered by the addition of a cloned gene. These transgenic cells are selected and then injected into the early stages of a host mouse embryo. Here, the transgenic embryonic stem cells integrate with the host's embryonic stem cells. The embryo is placed into the uterus of a pregnant mouse, where it develops into a chimeric mouse. The chimeric mouse is then crossed with a wild-type mouse. If the donor stem cells have contributed to the germ line, some of the progeny will be heterozygous for the added allele. By mating heterozygotes, a strain of transgenic mice generated that is homozygous for the added allele. The added gene (the transgene) can be from any eukaryotic source.

about 25 percent of the resulting offspring will be homozygous for (i.e., carry two copies of) the inserted gene in every cell of their bodies (Gossler et al. 1986). Thus a gene cloned from some other organism will be present in both copies of the chromosomes within these mouse genomes. Strains of such transgenic mice have been particularly useful in determining how genes are regulated during development.

GENE TARGETING ("KNOCKOUT") EXPERIMENTS   The analysis of early mammalian embryos has long been hindered by our inability to breed and select animals with mutations that affect early embryonic development. This block has been circumvented by the techniques of gene targeting (or, as it is sometimes called, **gene knockout**). These techniques are similar to those that generate transgenic mice, but instead of adding genes, gene targeting *replaces* wild-type alleles with mutant ones. As an example, we will look at a knockout of the gene for bone morphogenetic protein 7 (BMP7). Bone morphogenetic proteins are involved in numerous developmental interactions whereby one set of cells interacts with other neighboring cells to alter their properties. BMP7 has been implicated as a protein that prevents cell death and promotes cell division in several developing organs.

Dudley and his colleagues (1995) used gene targeting to find the function of BMP7 in the development of the mouse. First, they isolated the *Bmp7* gene, cut it at one site with a restriction enzyme, and inserted a bacterial gene for neomycin resistance into that site (Figure 4.19). In other

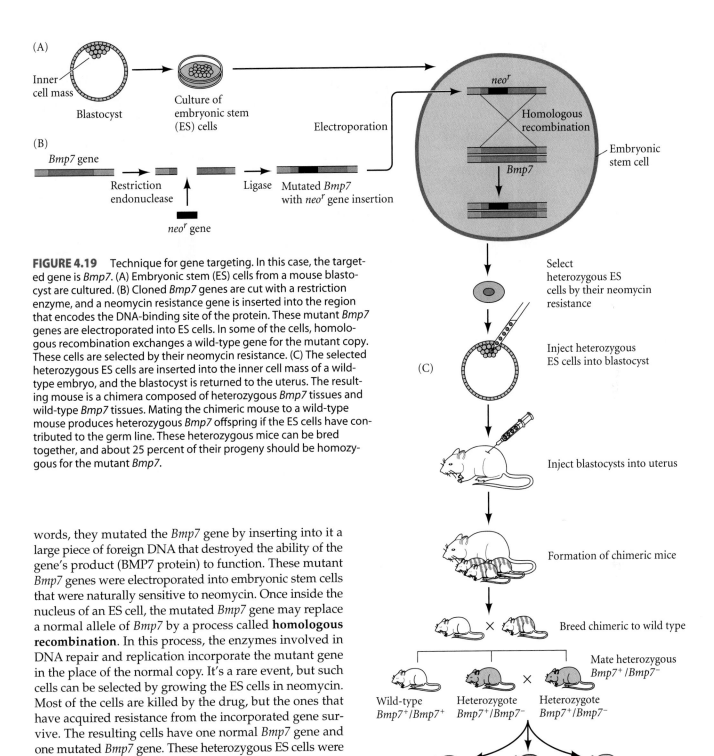

FIGURE 4.19 Technique for gene targeting. In this case, the target-ed gene is *Bmp7*. (A) Embryonic stem (ES) cells from a mouse blasto-cyst are cultured. (B) Cloned *Bmp7* genes are cut with a restriction enzyme, and a neomycin resistance gene is inserted into the region that encodes the DNA-binding site of the protein. These mutant *Bmp7* genes are electroporated into ES cells. In some of the cells, homolo-gous recombination exchanges a wild-type gene for the mutant copy. These cells are selected by their neomycin resistance. (C) The selected heterozygous ES cells are inserted into the inner cell mass of a wild-type embryo, and the blastocyst is returned to the uterus. The result-ing mouse is a chimera composed of heterozygous *Bmp7* tissues and wild-type *Bmp7* tissues. Mating the chimeric mouse to a wild-type mouse produces heterozygous *Bmp7* offspring if the ES cells have con-tributed to the germ line. These heterozygous mice can be bred together, and about 25 percent of their progeny should be homozy-gous for the mutant *Bmp7*.

words, they mutated the *Bmp7* gene by inserting into it a large piece of foreign DNA that destroyed the ability of the gene's product (BMP7 protein) to function. These mutant *Bmp7* genes were electroporated into embryonic stem cells that were naturally sensitive to neomycin. Once inside the nucleus of an ES cell, the mutated *Bmp7* gene may replace a normal allele of *Bmp7* by a process called **homologous recombination**. In this process, the enzymes involved in DNA repair and replication incorporate the mutant gene in the place of the normal copy. It's a rare event, but such cells can be selected by growing the ES cells in neomycin. Most of the cells are killed by the drug, but the ones that have acquired resistance from the incorporated gene sur-vive. The resulting cells have one normal *Bmp7* gene and one mutated *Bmp7* gene. These heterozygous ES cells were then microinjected into mouse blastocysts, where they were integrated into the cells of the embryo. The resulting chimeric mice had wild-type cells from the host embryo and heterozygous *Bmp7*-containing cells from the donor ES cells. The chimeras were mated to wild-type mice, pro-ducing progeny that were heterozygous for *Bmp7*. These heterozygous mice were then bred with each other, and about 25 percent of their progeny carried two copies of the mutated *Bmp7* gene. These homozygous mutants lacked eyes and kidneys (Figure 4.20). In the absence of function-

al BMP7 protein, it appears that many of the cells that nor-mally form these two organs stop dividing and die. In this way, gene targeting can be used to analyze the roles of par-ticular genes during mammalian development.

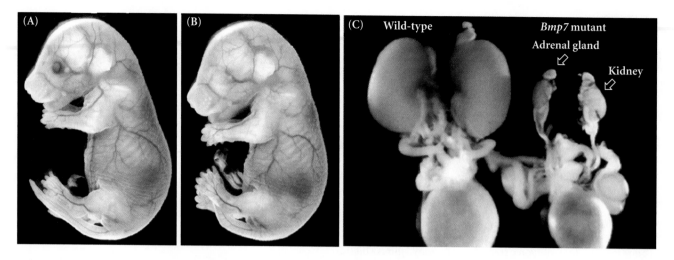

FIGURE 4.20 Morphological analysis of *Bmp7* knockout mice. (A) Wild-type and (B) homozygous *Bmp7*-deficient mouse at day 17 of their 21-day gestation. The *Bmp7*-deficient mouse lacks eyes. (C) The kidneys of these mice at day 19 of gestation. The kidney of the *Bmp7*-deficient mouse (right) is severely atrophied. Microscopic sections reveal the death of the cells that would otherwise have formed the nephrons. (From Dudley et al. 1995; photographs courtesy of E. Robertson.)

Determining the function of a message: Antisense RNA

Another method for determining the function of a gene during development is to use "antisense" copies of its message to block the function of that message. Antisense RNA allows developmental biologists to analyze the action of genes that would otherwise be inaccessible for genetic analysis.

Antisense messages can be generated by inserting cloned DNA into vectors that have promoters at both ends of the inserted gene. When the vector is incubated with nucleotide triphosphates (such as UTP) and a particular RNA polymerase, one of the promoters will initiate transcription of the message "in the wrong direction." In so doing, it synthesizes a transcript that is complementary to the natural one (Figure 4.21A). This complementary transcript is called antisense RNA because it is the complement of the original ("sense") message. When large amounts of antisense RNA are injected or transfected into cells containing the normal

FIGURE 4.21 Use of antisense RNA to examine the roles of genes in development. (A) An antisense message (in this case, to the *Krüppel* gene of *Drosophila*) is produced by placing a cloned cDNA fragment encoding the *Krüppel* message between two strong promoters (where RNA polymerases bind to initiate transcription). The two promoters are in opposite orientation with respect to the *Krüppel* cDNA. In this case, the T3 promoter is in normal orientation and the T7 promoter is reversed. These promoters are recognized by different RNA polymerases (from the T3 and T7 bacteriophages, respectively). T3 polymerase enables the transcription of "sense" mRNA, whereas T7 polymerase transcribes antisense transcripts. (B) Result of injecting the *Krüppel* antisense message into an early embryo (syncytial blastoderm stage) of *Drosophila* before the normal *Krüppel* message is produced. The central figure is a wild-type embryo just prior to hatching. Above it is a mutant lacking the *Krüppel* gene. Below it is a wild-type embryo that was injected with *Krüppel* antisense message at the syncitial blastoderm stage. Both the mutant and the antisense-treated embryos lack thoracic and anterior abdominal segments. (B after Rosenberg et al. 1985.)

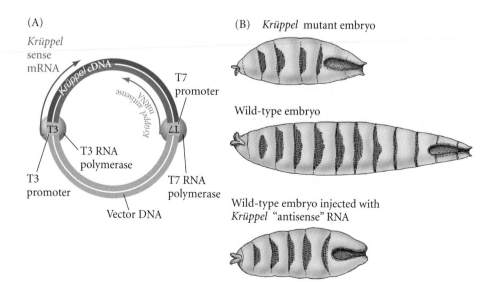

mRNA from the same gene, the antisense RNA binds to the normal message, and the resulting double-stranded nucleic acid is degraded by enzymes in the cell cytoplasm, causing a functional deletion of the message—just as if there were a deletion mutation for that gene.

The similarities between the phenotypes produced by a loss-of-function mutation and by antisense RNA treatment were demonstrated by making antisense RNA to the *Krüppel* gene of *Drosophila*. *Krüppel* is critical for forming the thorax and abdomen of the fly. If this gene is absent, fly larvae die because they lack thoracic and anterior abdominal segments (Figure 4.21B). A similar defect can be created by injecting large amounts of antisense RNA against the *Krüppel* message into early fly embryos (Rosenberg et al. 1985).

MORPHOLINO ANTISENSE OLIGOMERS   An important modification of antisense technology is the use of **morpholino antisense oligomers** (Summerton and Weller 1997). These synthetic molecules differ from traditional antisense RNA or DNA in that they contain six-member morpholine rings instead of five-member ribose or deoxyribose sugars. This structure gives them complete resistance to nucleases, enabling them to stay intact and function longer than their DNA or RNA counterparts. Moreover, they can hybridize with their target mRNAs under conditions that would prevent normal polynucleotides from binding. Their stability enables them to initiate events many cell generations after they are first injected into the cell. (Heasman et al. 2000). Morpholino antisense oligomers work by inhibiting the initiation of translation, and they are therefore made against sequences very near the translation initiation site of the message.

RNA INTERFERENCE   Another type of sequence-specific targeting of mRNA that leads to inhibition of its expression is **RNA interference (RNAi)**. Here, the introduction of a double-stranded RNA containing the same sequence as an endogenous mRNA results in phenotypes that would be expected if that endogenous message were either degraded or absent. Rather than inhibiting

translation, RNAi works by degrading the targeted message. Biologists have used RNAi to study development in nematodes, *Drosophila*, plants, fungi, and mice.

The first evidence that dsRNA could lead to gene silencing came from work on the nematode *Caenorhabditis elegans*. Guo and Kemphues (1995) were attempting to use antisense RNA to shut down the expression of the *par-1* gene in order to assess its function, and one of their controls didn't work. Or rather, it worked too well. As expected, injection of antisense RNA disrupted expression of *par-1*. However, so did the injection of the sense-strand control. The mystery was solved by Fire and colleagues (1998), who injected doublestranded RNA into *C. elegans*. Injection of dsRNA resulted in much more efficient silencing of gene expression than the injection of either the sense or the antisense strands alone. Indeed, just a few molecules of dsRNA per cell were sufficient to completely silence expression of the targeted gene. Furthermore, injecting dsRNA into the gut of the worm not only silenced the gene's expression, but also stopped its expression in the next generation. This result turned out to be a general phenomenon.

Double-stranded RNAs apparently work by activating an enzyme ("Dicer") that breaks the dsRNA strand into small pieces (Figure 4.22). These pieces (called **silencing**

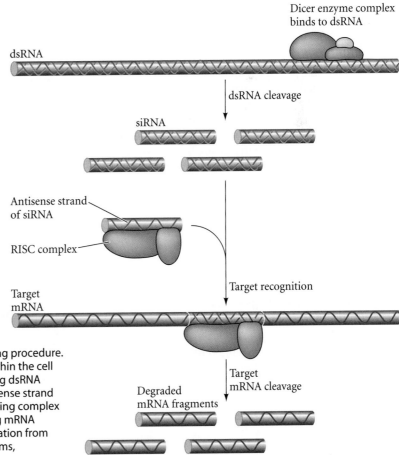

FIGURE 4.22   The dsRNA-mediated gene silencing procedure. Double-stranded RNA is recognized and bound within the cell by the Dicer protein complex. Dicer cleaves the long dsRNA strand into short siRNA oligonucleotides. The antisense strand of the siRNA is then used by an RNA-induced silencing complex (RISC) to guide mRNA cleavage, thereby promoting mRNA degradation (in invertebrates) or preventing translation from taking place (in vertebrates). (After amaxa biosystems, www.amaxa.com/138.html)

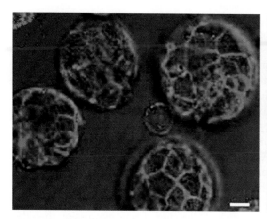
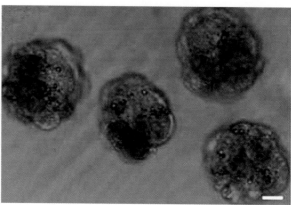

RNA, or **siRNA**) then bind to another enzyme complex ("RNA-induced silencing complex," or RISC) that destroys any RNAs bound by that small fragment (Hunter 2000; Hammond 2001). In invertebrates, adding dsRNA causes the degradation of the natural mRNA; in vertebrates, it simply blocks its translation (Figure 4.23). However, natural mRNA in vertebrate cells can be destroyed by adding synthetic siRNA.

Enzymes such as Dicer and RISC probably evolved as a means of preventing viral infections, since viruses often have double-stranded RNA intermediates. The techniques of RNAi are proving applicable in many organisms.

Coda

In the past decade, new techniques have enabled us to isolate individual genes, to localize the mRNAs of particular genes, and to delete or silence the expression of particular genes. This technology has enabled biologists to answer questions that were unanswerable only a few years ago. For the first time in human history, we are confronted with the ability to understand how the hereditary

FIGURE 4.23   Injection of dsRNA for E-cadherin into the mouse zygote blocks E-cadherin expression. (A) Staining of E-cadherin in 4-day mouse embryos. These embryos were injected with a control dsRNA that does not recognize any message in the mouse embryo. The staining was done by a fluorescent (red) antibody that binds to E-cadherin. (B) 4-day mouse embryos injected at the zygote stage with E-cadherin dsRNA. Hardly any E-cadherin is recognized by the antibody stain, and the embryos are malformed because they lack this cell adhesion protein (see Chapter 3). The same type of abnormal embryos can be produced by knocking out the E-cadherin gene. Scale bars are 20 μm. (From Wianny and Zernicka-Goetz 2000.)

potentials of the nucleus become expressed in the formation of our organs.

WEBSITE **4.8 Systems biology.** The huge amount of information concerning gene expression and regulation is the foundation of "systems biology," a new discipline that attempts to understand physiology and development by looking at the relationships between genes and the cellular environment they inhabit and help construct.

 **Snapshot Summary** **Genetic Approaches to Development**

1. Development connects genotype and phenotype.

2. A given genotype can produce a limited range of phenotypes, depending on random processes and environmental interactions with the developing organism.

3. The ability of nuclei from differentiated cells to direct the development of complete adult organisms has recently confirmed the principle of genomic equivalence.

4. Nuclear genes are not lost or mutated during development. The genome of each cell is equivalent to that of every other cell.

5. The exceptions to the rule of genomic equivalence are the lymphocytes. During differentiation, these cells rearrange their DNA to create new immunoglobulin and antigen receptor genes.

6. Only a small percentage of the genome is expressed in any particular cell.

7. Polytene chromosomes, in which the DNA has replicated without separating (as in larval *Drosophila* salivary glands), show regions where DNA is being transcribed. Different cell types show different regions of DNA being transcribed.

8. Northern blots, polymerase chain reaction techniques, and in situ hybridization can show which cells are transcribing particular genes.

9. Microarrays and macroarrays allow thousands of genes in different types of cells to be compared simultaneously.

10. The functions of a gene often can be ascertained by techniques that manipulate the gene's expression. These techniques include methods for inhibiting a gene's expression, such as antisense RNA; methods for eliminating ("knocking out") a gene; and the use of transgenes to overexpress or misexpress a gene.

11. RNA interference, or RNAi, makes use of double-stranded RNAs (dsRNA) that can specifically delete messages from a particular gene.

For Further Reading

Complete bibliographical citations for all literature cited in this chapter can be found on the Vade Mecum CD that accompanies the book and at the free access website www.devbio.com

Altmann, C. R., E. Bell, A. Sczyrba, J. Pun, S. Bekiranov, T. Gaasterland and A. H. Brivanlou. 2001. Microarray-based analysis of early development in *Xenopus laevis*. *Dev. Biol.* 236: 64–75.

Capecchi, M. R. 1980. High efficiency transformation by direct microinjection of DNA into cultured mammalian cells. *Cell* 22: 479–488.

Fire, A., S. Xu, M. K. Montgomery, S. A. Kostas, S. E. Driver and C. C. Mello. 1998. Potent and specific genetic interference by double-stranded RNA in *Caenorhabditis elegans*. *Nature* 391: 806–811.

Gurdon, J. B., R. A. Laskey and O. R. Reeves. 1975. The developmental capacity of nuclei transplanted from keratinized cells of adult frogs. *J. Embryol. Exp. Morphol.* 34: 93–112.

King, T. J. and R. Briggs. 1956. Serial transplantation of embryonic nuclei. *Cold Spring Harb. Symp. Quant. Biol.* 21: 271–289.

Moustafa, L. A. and R. L. Brinster. 1972. Induced chimaerism by transplanting embryonic cells into mouse blastocysts. *J. Exp. Zool.* 181: 193–202.

Rosenberg, U. B., A. Preiss, E. Seifert, H. Jäckle and D. C. Knipple. 1985. Production of phenocopies by *Krüppel* antisense RNA injection into *Drosophila* embryos. *Nature* 313: 703–706.

Wilmut, I., A. E. Schnieke, J. McWhir, A. J. Kind and K. H. S. Campbell. 1997. Viable offspring from fetal and adult mammalian cells. *Nature* 385: 810–814.

The Paradigm of Differential Gene Expression

5

DIFFERENT CELL TYPES MAKE DIFFERENT SETS OF PROTEINS, even when their genomes are identical. Each human being has roughly 25,000 protein-encoding genes in each nucleus (IHGSC 2004), but each cell uses only a small subset of those genes. Moreover, different cell types use different subsets of these genes. Red blood cells make globin proteins, lens cells make crystalline proteins, melanocytes make the enzymes required for synthesizing melanin pigment, and endocrine glands make their specific hormones. **Developmental genetics** is the discipline that examines how the genotype is transformed into the phenotype. The primary paradigm of developmental genetics is *differential gene expression from the same nuclear repertoire*. This chapter will discuss the mechanisms of cell differentiation: how the various genes present in the nucleus are activated or repressed at particular times and places to cause cells to produce different sets of proteins and to become different from one another.

Gene expression can be regulated at several levels such that different cell types synthesize different sets of proteins:

- **Differential gene transcription** regulates which of the nuclear genes are allowed to be transcribed into RNA
- **Selective nuclear RNA processing** regulates which of the transcribed RNAs (or which parts of such a nuclear RNA) are able to enter into the cytoplasm to become messenger RNAs
- **Selective messenger RNA translation** regulates which of the mRNAs in the cytoplasm become translated into proteins
- **Differential protein modification** regulates which proteins are allowed to remain or function in the cell

Some genes (such as those coding for the globin proteins of hemoglobin) are regulated at all of these levels.

Differential Gene Transcription

Anatomy of the gene: Active and repressed chromatin

Two fundamental differences distinguish most eukaryotic genes from prokaryotic genes. First, eukaryotic genes are contained within a complex of DNA and protein called **chromatin**. The protein component constitutes about half the weight of chromatin and is composed largely of **histones**. The **nucleosome** is the basic unit of chromatin structure. It is composed of an octamer of histone proteins (two molecules each of histones H2A, H2B, H3, and H4) wrapped with two loops containing approximately 140 base pairs of DNA (Figure 5.1A; Korn-

> " But whatever the immediate operations of the genes turn out to be, they most certainly belong to the category of developmental processes and thus belong to the province of embryology. "
>
> *C. H. WADDINGTON (1956)*

> " We have entered the cell, the mansion of our birth, and have started the inventory of our acquired wealth. "
>
> *ALBERT CLAUDE (1974)*

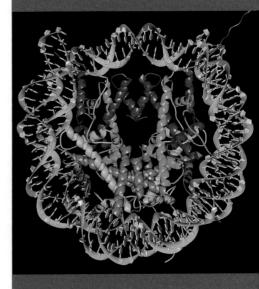

(A)

(B)

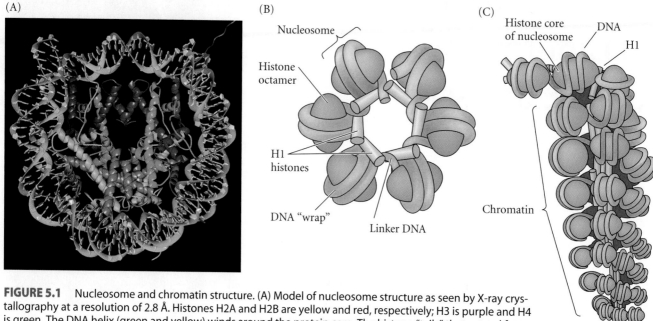

(C)

FIGURE 5.1 Nucleosome and chromatin structure. (A) Model of nucleosome structure as seen by X-ray crystallography at a resolution of 2.8 Å. Histones H2A and H2B are yellow and red, respectively; H3 is purple and H4 is green. The DNA helix (green and yellow) winds around the protein core. The histone "tails" that extend from the core are the sites of acetylation and methylation, which may disrupt or stabilize, respectively, the formation of nucleosome assemblages. (B) Histone H1 can draw nucleosomes together into compact forms. About 140 base pairs of DNA encircle each histone octamer, and about 60 base pairs of DNA link the nucleosomes together. (C) Model for the arrangement of nucleosomes in the highly compacted solenoidal chromatin structure. (A from Luger et al. 1997, photograph courtesy of the authors; B,C after Wolfe 1993.)

berg and Thomas 1974). Chromatin can be visualized as a string of nucleosome beads linked by ribbons of DNA.

Whereas classical geneticists have likened genes to "beads on a string," molecular geneticists liken genes to "string on the beads." Most of the time, the nucleosomes are themselves wound into tight "solenoids" that are stabilized by histone H1. Histone H1 is bound to the 60 or so base pairs of "linker" DNA between the nucleosomes (Figure 5.1B; Weintraub 1984). This H1-dependent conformation of nucleosomes inhibits the transcription of genes in somatic cells by packing adjacent nucleosomes together into tight arrays that prevent transcription factors and RNA polymerases from gaining access to the genes (Thoma et al. 1979; Schlissel and Brown 1984). It is generally thought, then, that the "default" condition of chromatin is a repressed state, and that tissue-specific genes become activated by local interruption of this repression (Weintraub 1985).

The histones are critical in regulating gene expression, because they are responsible for maintaining the *repression* of gene expression. This repression can be locally strengthened (so that it becomes very difficult to transcribe those genes in the nucleosomes) or relieved (so that transcribing them becomes relatively easy) by modifying the histones. In general, the addition of acetyl groups to histones loosens them and activates transcription. Enzymes (**histone acetyltransferases**) that place acetyl groups on histones (especially on lysines in H3 and H4) destabilize the nucleosomes so that they come apart easily. As might be expected, then,

enzymes that remove these acetyl groups (**histone deacetylases**) stabilize the nucleosomes and prevent transcription.

The addition of methyl groups to histones generally represses transcription even further, while removing these methyl groups brings the nucleosomes back to their baseline stability (see Strahl and Allis 2000; Cosgrove et al. 2004). The sites of these modifications, however, are very important. For instance, acetylation of the "tails" of H3 and H4 plus the methylation of the lysine at position 4 of H3 (i.e., H3K4) is usually associated with actively transcribing chromatin. In contrast, hypoacetylation of the H3 and H4 tails and the methylation of the lysine in the ninth position of H3 (H3K9) is usually associated with highly repressed chromatin (Norma et al. 2001). As we will soon see, the state of the chromatin will depend largely on transcription factor proteins that can recruit the enzymes that modify the nucleosomes in their vicinity.

 **WEBSITE** **5.1 Displacing nucleosomes.** Transcription can occur even in a region of nucleosomes. If a gene's promoter is accessible to RNA polymerase and transcription factors, the presence of nucleosomes will not inhibit the elongation of the message.

Anatomy of the gene: Exons and introns

The second difference between prokaryotic and eukaryotic genes is that eukaryotic genes are not co-linear with their peptide products. Rather, the single nucleic acid strand of eukaryotic mRNA comes from noncontiguous regions on the chromosome. Between **exons**—the regions of DNA that code for a protein*—are intervening sequences called **introns** that have nothing whatsoever to do with the amino acid sequence of the protein. The structure of a typical eukaryotic gene can be illustrated by the human β-globin gene (Figure 5.2). This gene, which encodes part of the hemoglobin protein of the red blood cells, consists of the following elements:

1. A **promoter region**, which is responsible for the binding of RNA polymerase and for the subsequent initiation of transcription. The promoter region of the human β-globin gene has three distinct units and extends from 95 to 26 base pairs before ("upstream from")† the transcription initiation site (i.e., from –95 to –26).

2. The **transcription initiation site**, which for human β-globin is ACATTTG. This site is often called the **cap sequence** because it represents the 5′ end of the RNA, which will receive a "cap" of modified nucleotides soon after it is transcribed. The specific cap sequence varies among genes.

3. The **translation initiation site**, ATG. This codon (which becomes AUG in the mRNA) is located 50 base pairs after the transcription initiation site in the human β-globin gene (although this distance differs greatly among different genes). The sequence of 50 base pairs intervening between the initiation points of transcription and translation is the **5′ untranslated region**, often called the **5′ UTR** or leader sequence. The 5′ UTR can determine the rate at which translation is initiated.

4. The first exon, which contains 90 base pairs coding for amino acids 1–30 of human β-globin.

5. An intron containing 130 base pairs with no coding sequences for the β-globin protein. The structure of this intron is important in enabling the RNA to be processed into messenger RNA and exit from the nucleus.

6. An exon containing 222 base pairs coding for amino acids 31–104.

7. A large intron—850 base pairs—having nothing to do with globin protein structure.

8. An exon containing 126 base pairs coding for amino acids 105–146.

9. A **translation termination codon**, TAA. This codon becomes UAA in the mRNA. The ribosome dissociates at this codon, and the protein is released.

10. A **3′ untranslated region** (**3′ UTR**) that, although transcribed, is not translated into protein. This region includes the sequence AATAAA, which is needed for polyadenylation: the placement of a "tail" of some 200 to 300 adenylate residues on the RNA transcript. This polyA tail (1) confers stability on the mRNA, (2) allows the mRNA to exit the nucleus, and (3) permits the mRNA to be translated into protein. The polyA tail is inserted into the RNA about 20 bases downstream of the AAUAAA sequence.

11. A **transcription termination sequence**. Transcription continues beyond the AATAAA site for about 1000 nucleotides before being terminated.

The original transcription product is called **nuclear RNA** and it contains the cap sequence, the 5′ UTR, exons, introns, and the 3′ UTR (Figure 5.3). Both ends of these transcripts are modified before these RNAs leave the nucleus. A cap consisting of methylated guanosine is placed on the 5′ end of the RNA in opposite polarity to the RNA itself. This means that there is no free 5′ phosphate group on the nuclear RNA. The 5′ cap is necessary for the binding of mRNA to the ribosome and for subsequent translation (Shatkin 1976). The 3′ terminus is usually modified in the nucleus by the addition of a polyA tail. The adenylate residues in this tail are put together enzymatically and are added to the transcript; they are not part of the gene sequence. Both the 5′ and 3′ modifications may protect the mRNA from exonucleases that would otherwise digest it (Sheiness and Darnell 1973; Gedamu and Dixon 1978). The modifications thus stabilize the message and its precursor.

 WEBSITE **5.2 What makes a DNA sequence a "gene"?** Different scientists have different definitions, and nature has given us some problematic examples of DNA sequences that may or may not be considered genes.

Anatomy of the gene: Promoters and enhancers

In addition to the protein-encoding region of the gene, there are regulatory sequences that can be located on either end of the gene (or even within it). These sequences—the promoters and enhancers—are necessary for controlling where and when a particular gene is transcribed.

Promoters are the sites where RNA polymerase binds to the DNA to initiate transcription. Promoters of genes that synthesize messenger RNAs (i.e., genes that encode proteins‡) are typically located immediately upstream from the site where the RNA polymerase initiates transcription. Most of these promoters contain the sequence TATA, to

---

*The term *exon* refers to a nucleotide sequence whose RNA "exits" the nucleus. It has taken on the functional definition of a protein-encoding nucleotide sequence. Leader sequences and 3′ UTR sequences are also derived from exons, even though they are not translated into protein.

†By convention, upstream, downstream, 5′, and 3′ directions are specified in relation to the RNA. Thus, the promoter is upstream of the gene, near its 5′ end.

---

‡There are several types of RNA that do not encode proteins. These include the ribosomal RNAs and transfer RNAs (which are used in protein synthesis) and the small nuclear RNAs (which are used in RNA processing). In addition, there are regulatory RNAs (such as *Xist* and *lin-4*, which we will discuss later in this chapter) that are involved in regulating gene expression (and which are not translated into peptides).

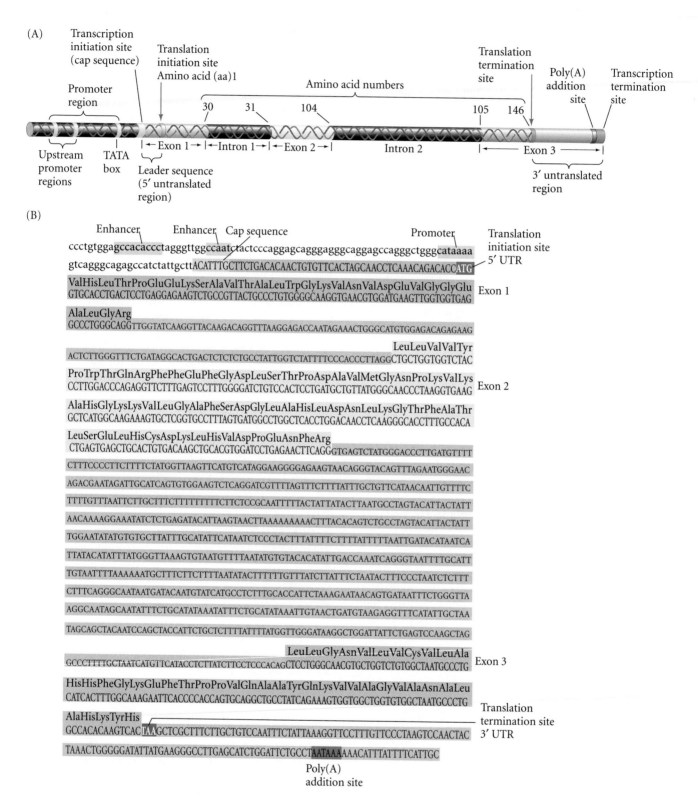

FIGURE 5.2 Nucleotide sequence of the human β-globin gene. (A) Schematic representation of the locations of the promoter region, transcription initiation site (cap sequence), 5' untranslated region (leader sequence), exons, introns, and 3' untranslated region of the human β-globin gene. Exons are shown in color; the numbers flanking them indicate the amino acid positions each exon encodes in β-globin. (B) The nucleotide sequence of the human β-globin gene, shown from the 5' end to the 3' end of the RNA. The colors correspond to their diagrammatic representation in (A). The promoter sequences are boxed, as are the translation initiation and termination codes ATG and TAA. The large capital letters boxed in color are the bases of the exons, with the amino acids for which they code abbreviated above them. Smaller capital letters indicate the intron bases. The codons after the translation termination site exist in β-globin mRNA but are not translated into proteins. Within this group is the sequence thought to be needed for polyadenylation. By convention, only the RNA-like strand of the DNA double helix is shown. (B after Lawn et al. 1980.)

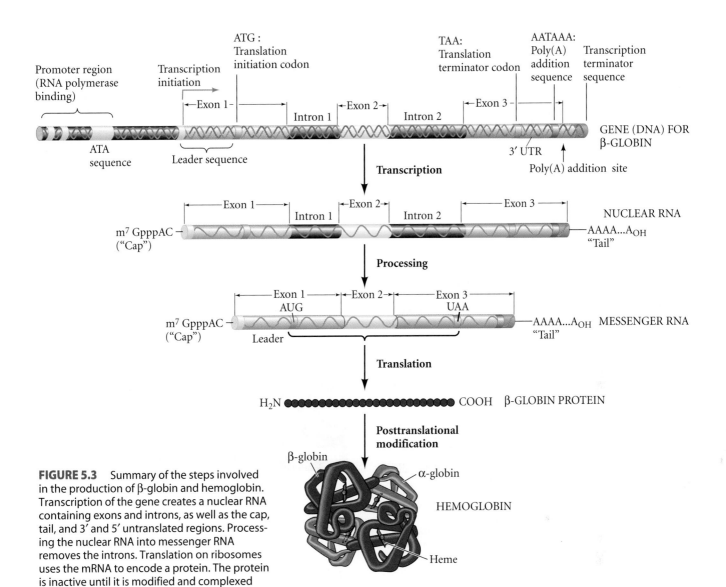

FIGURE 5.3   Summary of the steps involved in the production of β-globin and hemoglobin. Transcription of the gene creates a nuclear RNA containing exons and introns, as well as the cap, tail, and 3′ and 5′ untranslated regions. Processing the nuclear RNA into messenger RNA removes the introns. Translation on ribosomes uses the mRNA to encode a protein. The protein is inactive until it is modified and complexed with α-globin and heme to become active hemoglobin (bottom).

which RNA polymerase will be bound (Figure 5.4). This site, known as the **TATA box**, is usually about 30 base pairs upstream from the site where the first base is transcribed. Since this sequence will appear randomly in the genome at more places than just at promoter sites, other regions flanking it are also important. Many TATA box regions are flanked by **CpG islands**, regions of DNA rich in those two nucleotides (Down and Hubbard 2002). Eukaryotic RNA polymerases, however, will not bind to this naked DNA sequence. Rather, they require the presence of additional proteins, called **basal transcription factors**, to bind efficiently to the promoter. Protein-encoding genes are transcribed by RNA polymerase II, and at least six basal transcription factors have been shown to be necessary for the proper initiation of transcription by this polymerase (Buratowski et al. 1989; Sopta et al. 1989).

The first basal transcription factor, **TFIID**,* recognizes the TATA box through one of its subunits, **TATA-binding protein (TBP)**. TFIID serves as the foundation of the transcription initiation complex, and it also prevents nucleosomes from forming in this region. Once TFIID is stabilized by the basal transcription factor, **TFIIA**, it becomes able to bind **TFIIB**. Once TFIIB is in place, RNA polymerase can bind to the complex. Other transcription factors (TFIIE, F, and H) are then used to release RNA polymerase from the complex so that it can transcribe the gene, and to unwind the DNA helix so that the RNA polymerase will have a free template from which to transcribe.

The ability of the basal transcription factors to interact with RNA polymerase is regulated by two sets of proteins,

---

*TF stands for transcription factor; II indicates that the factor was first found to be needed for RNA polymerase II; and the *letter designations* refer to the active fractions from the phosphocellulose columns used to purify these proteins.

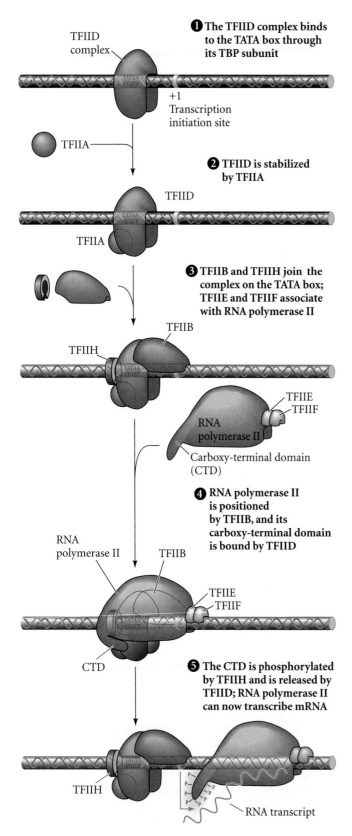

❶ The TFIID complex binds to the TATA box through its TBP subunit

TFIID complex

+1 Transcription initiation site

TFIIA

❷ TFIID is stabilized by TFIIA

TFIID

TFIIA

❸ TFIIB and TFIIH join the complex on the TATA box; TFIIE and TFIIF associate with RNA polymerase II

TFIIB

TFIIH

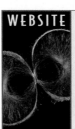

TFIIE TFIIF

RNA polymerase II

Carboxy-terminal domain (CTD)

❹ RNA polymerase II is positioned by TFIIB, and its carboxy-terminal domain is bound by TFIID

RNA polymerase II

TFIIB

TFIIE TFIIF

CTD

❺ The CTD is phosphorylated by TFIIH and is released by TFIID; RNA polymerase II can now transcribe mRNA

TFIIH

RNA transcript

FIGURE 5.4 Formation of the active eukaryotic transcription initiation complex. The diagrams represent the complex formed on the TATA box by the basal transcription factors and RNA polymerase II.

TBP-associated factors (**TAFs**) and the **mediator complex** (Myers and Kornberg 2000; Baek et al. 2002). The mediator complex contains about 25 proteins that can modulate the activity of RNA polymerase II and TFIIH (a basal transcription factor that is critical in allowing RNA polymerase II to start transcription) (Woychik and Hampsey 2002). The TAFs stabilize the TBP (Figure 5.5; Buratowski 1997; Lee and Young 1998). This function is critical for transcription, because if the TBP is not stabilized, it can fall off the small TATA sequence.

The TAFs are bound by **upstream promoter elements** on the DNA. These DNA sequences are near the TATA box (usually upstream from it). Every TAF need not be present in every cell of the body, however; **specific transcription factors** (such as the Pax6 protein mentioned in Chapter 4) can also activate the gene by stabilizing the transcription initiation complex. They can do so by binding to the TAFs, or by binding directly to other factors such as TFIIB. They can also facilitate transcription by destabilizing nucleosomes (as we will soon see).

An **enhancer** is a DNA sequence that controls the efficiency and rate of transcription of a specific promoter. In other words, enhancers tell where and when a promoter can be used, and how much of the gene product to make. Enhancers can activate only *cis*-linked promoters (i.e., promoters on the same chromosome*), but they can do so at great distances (some as great as a million bases away from the promoter). Moreover, enhancers do not need to be on the 5′ (upstream) side of the gene; they can be at the 3′ end, or even in the introns (Maniatis et al. 1987). The human β-globin gene has an enhancer in its 3′ UTR, roughly 700 base pairs downstream from the AATAAA site. This enhancer sequence is necessary for the temporal- and tissue-specific expression of the β-globin gene in adult red blood cell precursors (Trudel and Constantini 1987). Like promoters, enhancers function by binding specific regulatory proteins called transcription factors.

Cis- and *trans*-regulatory elements are so named by analogy with *E. coli* genetics and organic chemistry. There, *cis*-elements are regulatory elements that reside on the same strand of DNA (*cis*-, "on the same side as"), while *trans*-elements are those that could be supplied from another chromosome (*trans*-, "on the other side of"). The term *cis*-regulatory elements now refers to those DNA sequences that regulate a gene on the same stretch of DNA (i.e., the promoters and enhancers). *Trans*-regulatory factors are soluble molecules whose genes are located elsewhere in the genome and which bind to the *cis*-regulatory elements. They are usually transcription factors.

(A) A minimal complex of TBP and a TAF fails to activate transcription (Sp1 and NTF cannot associate with TBP)

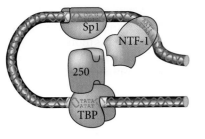

(B) Addition of the p110 TAF and the p150 TAF allows stabilization of the TBP by both NTF and Sp1

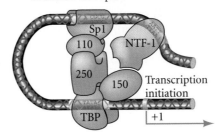

FIGURE 5.5 Model of TAF stabilization of TBP. (A) A minimal complex near the promoter, containing TBP on the TATA box of the promoter and two upstream promoter elements occupied by two transcription factors, Sp1 and NTF-1. One TAF, TAF$_{II}$250, is bound to the TBP, but this complex is not stable enough to activate transcription. (B) When other TAFs bind to these proteins, however, they form bridges that stabilize the TBP on the promoter. (After Chen et al. 1995.)

Enhancers can regulate the temporal- and tissue-specific expression of any differentially regulated gene, but different types of genes normally have different enhancers. For instance, genes encoding the *exocrine* proteins of the pancreas (the digestive proteins chymotrypsin, amylase, and trypsin) have enhancers different from that of the genes encoding the pancreatic *endocrine* proteins (such as insulin). These enhancers both lie in the 5′ flanking sequences of their genes (Walker et al. 1983).

One of the principal methods of identifying enhancer sequences is to clone DNA sequences flanking the gene of interest and fuse them to **reporter genes** whose products are both readily identifiable and not usually made in the cells of interest. Researchers can insert constructs of possible enhancers and reporter genes into embryos and then monitor the expression of the reporter gene. If the sequence contains an enhancer, the reporter gene should become active at particular times and places. For instance, the *E. coli* gene for **β-galactosidase** (the *lacZ* gene) can be used

as a reporter gene and fused to a "basal promoter" that can be activated in any cell and an enhancer that normally directs the expression of a particular mouse gene in muscles. If the resulting transgene is injected into a newly fertilized mouse egg and becomes incorporated into its DNA, β-galactosidase will be expressed in the mouse muscle cells. By staining for the presence of β-galactosidase, the expression pattern of that muscle-specific gene can be seen (Figure 5.6A). Similarly, a sequence flanking a lens crystallin protein in *Xenopus* was shown to be an enhancer. When this sequence was fused to a reporter gene for **green fluorescent protein** (**GFP**, a protein usually found only in jellyfish), GFP expression was seen solely in the lens (Figure 5.6B; Offield et al. 2000). GFP reporter genes are very useful because they can be monitored in live embryos and because the changes in gene expression can be seen in single organisms.

Enhancers are modular. As we will see, if a protein is synthesized in several tissues, there can be separate

(A)

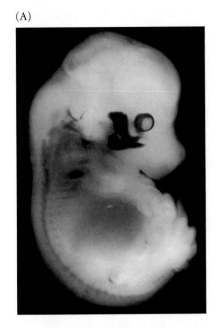

(B)

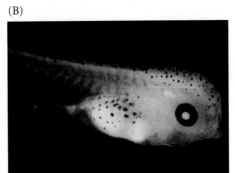

FIGURE 5.6 The genetic elements regulating tissue-specific transcription can be identified by fusing reporter genes to suspected enhancer regions of the genes expressed in particular cell types. (A) The enhancer region of the gene encoding muscle-specific protein Myf-5 is fused to a β-galactosidase reporter gene and incorporated into a mouse embryo. When stained for β-galactosidase activity (darkly staining region), the 13.5-day mouse embryo shows that the reporter gene is expressed in the eye muscles, facial muscles, forelimb muscles, neck muscles, and segmented myotomes (which give rise to the back musculature). (B) The gene for green fluorescent protein (GFP) is fused to a lens crystallin enhancer gene in *Xenopus tropicalis*. The result is the expression of GFP in the tadpole lens. (A, photograph courtesy of A. Patapoutian and B. Wold; B from Offield et al. 2000, photograph courtesy of R. Grainger.)

enhancers directing the gene to be expressed at different sites. The mouse *Pax6* gene, for instance, is expressed in the lens and retina of the eye, in the neural tube, and in the pancreas. The 5′ regulatory regions of the mouse *Pax6* gene were discovered by taking regions from its 5′ flanking sequence and introns and fusing them to a *lacZ* reporter gene. Each of these transgenes was then microinjected into newly fertilized mouse pronuclei, and the resulting embryos were stained for β-galactosidase (Figure 5.7A; Kammandel et al. 1998; Williams et al. 1998). Analysis of the results revealed that the enhancer farthest upstream from the promoter contains the regions necessary for *Pax6* expression in the pancreas, while a second enhancer activates *Pax6* expression in surface ectoderm (lens, cornea, and conjunctiva). A third enhancer resides in the leader sequence; it contains the sequences that direct *Pax6* expression in the neural tube. A fourth enhancer sequence, located in an intron shortly after the translation initiation site, determines the expression of *Pax6* in the retina (Figure 5.7B). A fifth enhancer, located in the 3′ untranslated region, was discovered based on the conservation of this sequence in mice, humans, and fish. This enhancer regulates the initial Pax6 expression in the precursors of the retina, brain, and nasal regions (Griffin et al. 2002).

Enhancers are critical in the regulation of normal development. Over the past decade, eight generalizations have emerged that emphasize their importance for differential gene expression:

1. Most genes require enhancers for their transcription.
2. Enhancers are the major determinant of differential transcription in space (cell type) and time.
3. The ability of an enhancer to function while far from the promoter means that there can be multiple signals to determine whether a given gene is transcribed. A given gene can have several enhancer sites linked to it, and each enhancer can be bound by more than one transcription factor.
4. An interaction between the transcription factors bound to the enhancer sites and the transcription initiation complex assembled at the promoter is thought to regulate transcription. The mechanism of this association is not fully known, nor do we comprehend how the promoter integrates all these signals.
5. Enhancers are combinatorial. Various DNA sequences regulate temporal and spatial gene expression, and these can be mixed and matched. For example, the enhancers for endocrine hormones such as insulin and for lens-specific proteins such as crystallins both have sites that bind Pax6 protein. But Pax6 alone doesn't tell the lens to make insulin or the pancreas to make crystallins; there are other transcription factor proteins that also must bind. It is the combination of transcription factors that causes particular genes to be transcribed.
6. Enhancers are modular. A gene can have several enhancer elements, each of which turns it on in a different set of cells.
7. Enhancers generally activate transcription by one of two means: they either remodel chromatin to expose the promoter, or they facilitate the binding of RNA polymerase to the promoter by stabilizing TAFs.
8. Enhancers can also inhibit transcription. In some cases, the same transcription factors that activate the transcription of one gene can repress the transcription of other genes. These "negative enhancers" are sometimes called **silencers**.

Transcription factors

Transcription factors are proteins that bind to enhancer or promoter regions and interact to activate or repress the transcription of a particular gene. Most transcription factors can bind to specific DNA sequences. These proteins can be grouped together in families based on similarities

(A)

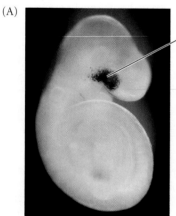

β-galactosidase (*lacZ* reporter gene)

FIGURE 5.7 Regulatory regions of the mouse *Pax6* gene. (A) A sequence in the upstream enhancer of the *Pax6* gene directs the expression of a *lacZ* reporter transgene in the surface ectoderm overlying the optic cup, as shown by the dark staining in this area. (B) Map of the enhancer sites of the *Pax6* gene, based on reporter gene studies. The gene has two major promoters (in exons 0 and 1, respectively) either of which can initiate transcription of the same RNA. The translation initiation codon (AUG) is in exon 4. (Exon 5a is an alternatively spliced exon, discussed in the text.) Regions A–D are enhancers that activate the *Pax6* gene in the pancreas, lens, neural tube, and retina, respectively. The embryo in the photograph shows the B region expression pattern. (A from Williams et al. 1998; B after Kammandel et al. 1998.)

(B)

Lens and cornea

Pancreas

Neural tube

AUG

Retina

Enhancers: A B C D

Exons: 0 1 2 3 4 5 5a 6 7

TABLE 5.1 Some major transcription factor families and subfamilies

Family	Representative transcription factors	Some functions
Homeodomain:		
Hox	Hoxa1, Hoxb2, etc.	Axis formation
POU	Pit1, Unc-86, Oct-2	Pituitary development; neural fate
LIM	Lim1, Forkhead	Head development
Pax	Pax1, 2, 3, 6, etc.	Neural specification; eye development
Basic helix-loop-helix (bHLH)	MyoD, MITF, daughterless	Muscle and nerve specification; *Drosophila* sex determination; pigmentation
Basic leucine zipper (bZip)	C/EBP, AP1	Liver differentiation; fat cell specification
Zinc finger:		
Standard	WT1, Krüppel, Engrailed	Kidney, gonad, and macrophage development; *Drosophila* segmentation
Nuclear hormone receptors	Glucocorticoid receptor, estrogen receptor, testosterone receptor, retinoic acid receptors	Secondary sex determination; craniofacial development; limb development
Sry-Sox	Sry, SoxD, Sox2	Bend DNA; mammalian primary sex determination; ectoderm differentiation

in structure (Table 5.1). The transcription factors within such a family share a common framework in their DNA-binding sites, and slight differences in the amino acids at the binding site can cause the binding site to recognize different DNA sequences.

WEBSITE 5.4 Families of transcription factors. There are several families of transcription factors grouped together by their structural similarities and their mechanisms of action. Homeodomain transcription factors are important in specifying anterior-posterior axes, and hormone receptors mediate the effects of hormones on genes.

Transcription factors have three major domains. The first is a **DNA-binding domain** that recognizes a particular DNA sequence. The second is a *trans*-**activating domain** that activates or suppresses the transcription of the gene whose promoter or enhancer it has bound. Usually, this *trans*-activating domain enables the transcription factor to interact with the proteins involved in binding RNA polymerase (such as TFIIB or TFIIE; see Sauer et al. 1995) or with enzymes that modify histones. In addition, there may be a **protein-protein interaction domain** that allows the transcription factor's activity to be modulated by TAFs or other transcription factors. We will discuss the functioning of transcription factors by analyzing two examples: MITF and Pax6.

MITF The microphthalmia (MITF) protein is a transcription factor that is active in the ear and in the pigment-forming cells of the eye and skin. Humans heterozygous for a mutation of the gene that encodes MITF are deaf, have multicolored irises, and have a white forelock in their hair

(see Figure 21.4). The MITF protein is a basic helix-loop-helix transcription factor (see Table 5.1) that has three functionally important domains (Figure 5.8). First, MITF has a protein-protein interaction domain that enables it to dimerize with another MITF protein (Ferré-D'Amaré et al. 1993). The resulting homodimer (two MITF proteins bound together) is a functional protein that can bind to DNA and activate the transcription of certain genes. The second region, the DNA-binding domain, is close to the amino-terminal end of the protein and contains numerous basic amino acids that make contact with the DNA (Hemesath et al. 1994; Steingrímsson et al. 1994). This assignment was confirmed by the discovery of various human and mouse mutations that map within the DNA-binding site for MITF and which prevent the attachment of the MITF protein to the DNA. Sequences for MITF binding have been found in the promoter regions of genes encoding three pigment-cell-specific enzymes of the tyrosinase family (Bentley et al. 1994; Yasumoto et al. 1997). Without MITF, these proteins are not synthesized properly (Figure 5.9) and melanin pigment is not made. These promoter regions all contain the same 11-base-pair sequence, including the core sequence (CATGTG) that is recognized by MITF.

The third functional region of MITF is its *trans*-activating domain. This domain includes a long stretch of amino acids in the center of the protein. When the MITF dimer is bound to its target sequence in a promoter or enhancer, the *trans*-activating region is able to bind a TAF, p300/CBP (see Figure 5.8). The p300/CPB protein is a histone acetyltransferase enzyme that can transfer acetyl groups to each histone in the nucleosomes (Ogryzko et al. 1996; Price et al. 1998). Acetylation of the nucleosomes destabilizes them and allows the genes for pigment-forming enzymes to be expressed.

Carboxyl termini

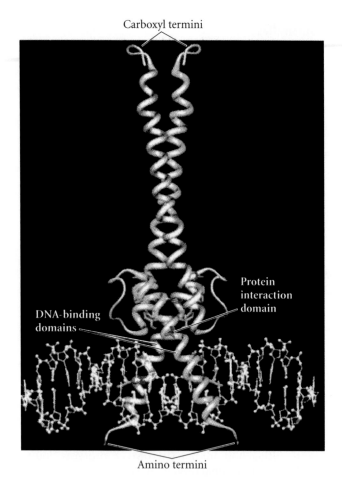

DNA-binding domains

Protein interaction domain

Amino termini

FIGURE 5.8 Three-dimensional model of the homodimeric transcription factor MITF (one protein shown in red, the other in blue) binding to a promoter element in DNA (white). The amino termini are located at the bottom of the figure and form the DNA-binding domain. The protein-protein interaction domain is located immediately above. MITF has the basic helix-loop-helix structure found in many transcription factors. The carboxyl end of the molecule is thought to be the *trans*-activating domain that binds the p300/CBP co-activator protein. (From Steingrímsson et al. 1994; photograph courtesy of N. Jenkins.)

Recent discoveries have shown that numerous transcription factors operate by recruiting histone acetyltransferases. When histone proteins are acetylated, the nucleosome disperses, allowing other transcription factors and RNA polymerase access to the site. Some TAFs, such as $TAF_{II}250$, are themselves histone acetyltransferases, so they can activate transcription both by destabilizing the nucleosome and by enabling the TBP protein to bind to the DNA to establish a site for RNA polymerase binding (Figure 5.10; Mizzen et al. 1996).

 WEBSITE **5.5 Histone acetylation.** Histone acetylation is a critical step in clearing the way for the transcription initiation complex. Derepressed chromatin is characterized by acetylated histones.

Wild type   *mitf*$^{-/-}$

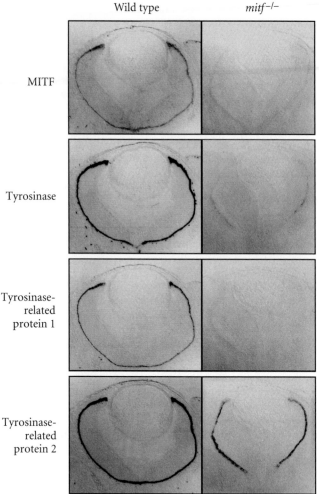

MITF

Tyrosinase

Tyrosinase-related protein 1

Tyrosinase-related protein 2

FIGURE 5.9 Serial sections of the eye in 15.5-day mouse embryos demonstrate that MITF is required for the transcription of pigmention genes. In the wild-type embryo, in situ hybridization reveals the presence of *Mitf* mRNA (dark staining) in the retinal pigment epithelial layer. In the mutant embryo (*mitf*$^{-/-}$), no MITF is present. MITF recognizes promoter sites in three tyrosinase family genes that encode enzymes involved in melanin production. In the mutant, the genes encoding these enzymes are not transcribed well compared to the wild-type mice. (From Nakayama et al. 1998; photographs courtesy of H. Arnheiter.)

PAX6 The Pax6 transcription factor, which is needed for mammalian eye, nervous system, and pancreas development, contains two potential DNA-binding domains. Its major DNA-binding domain resides at its amino-terminal end; the amino acids of this domain interact with a specific 20–26-base-pair sequence of DNA (Figure 5.11; Xu et al. 1995). Such Pax6-binding sequences have been found in the enhancers of vertebrate lens crystallin genes and in the genes expressed in the endocrine cells of the pancreas (those that release insulin, glucagon, and somatostatin) (Cvekl and Piatigorsky 1996; Andersen 1999). When Pax6 binds to a gene's enhancer or promoter, it can either activate or repress the gene. The *trans*-activating domain of Pax6 is rich in proline, threonine, and serine. Mutations in this region cause severe nervous system, pancreatic, and optic abnormalities in humans (Glaser et al. 1994).

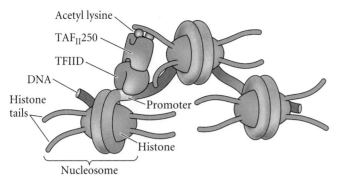

FIGURE 5.10 TAF$_{II}$250, a TAF that binds TBP, can function as a histone acetyltransferase. Thus, it can acetylate histones and thereby disrupt nucleosomes. It can also bind to acetylated lysine residues on the histones and enable the positioning of TBP on the promoter. These abilities relate to its function as proposed in Figure 5.5, where the TBP is part of the TFIID complex. (After Pennisi 2000.)

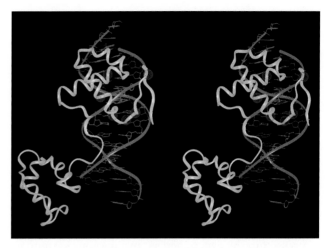

FIGURE 5.11 Stereoscopic model of Pax6 protein binding to its enhancer element in DNA. The DNA-binding region of Pax6 is shown in yellow; the DNA double helix is blue. Red dots indicate the sites of loss-of-function mutations in the *Pax6* gene that give rise to nonfunctional Pax6 proteins. It is worth trying to cross your eyes to get the central three-dimensional figure. (From Xu et al. 1995; photograph courtesy of S. O. Pääbo.)

COMBINATORIAL REGULATION OF TRANSCRIPTION The use of Pax6 by different organs demonstrates the combinatorial manner by which transcription factors work. Figure 5.12 shows two gene regulatory regions that bind Pax6. The first is that of the chick δ1 lens *crystallin* gene (Cvekl and Piatigorsky 1996; Muta et al. 2002). This gene has a promoter containing a site for TBP binding, and an upstream promoter element that binds Sp1 (a general transcriptional activator found in all cells). The gene also has an enhancer in its third intron that controls the time and place of *crystallin* gene expression. This enhancer has two Pax6-binding sites.

As mentioned in Chapter 4, Pax6 is present during early development in the central nervous system and head surface ectoderm of the chick. The *crystallin* gene will not be expressed unless Pax6 is present in the nucleus and bound to these enhancer sites (Figure 5.12A). Moreover, this enhancer has a binding site for another transcription factor, the Sox2 protein. Sox2 is not usually found in the outer ectoderm, but it appears in those outer ectodermal cells that will become lens by virtue of their being induced by the optic vesicle evaginating from the brain (Kamachi et al. 1998). Thus only those cells that contain both Sox2 and Pax6 can express the lens crystallin gene. In addition, a *third* site on the enhancer can bind either an activator (the δEF3 protein) or a repressor (the δEF1 protein) of transcription. It is thought that the repressor may be critical in pre-

venting *crystallin* expression in the nervous system, demonstrating how enhancers can function in a combinatorial manner, with several transcription factors working together to promote or inhibit transcription.

Another set of regulatory regions that use Pax6 are the enhancers regulating the transcription of the genes for insulin, glucagon, and somatostatin in the pancreas (Figure 5.12B). Here, Pax6 is essential for gene expression, and it works in cooperation with other transcription factors

FIGURE 5.12 Modular transcriptional regulatory regions using Pax6 as an activator. (A) Promoter and enhancer of the chick δ1 lens *crystallin* gene. Pax6 interacts with two other transcription factors, Sox2 and Maf, to activate this gene. (B) Promoter and enhancer of the rat somatostatin gene. Pax6 activates this gene by cooperating with the Pdx1 transcription factor. (A after Cvekl and Piatigorsky 1996; B after Andersen et al. 1999.)

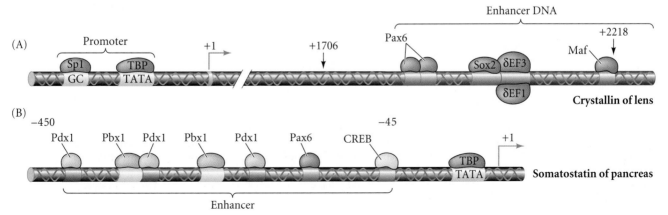

such as Pdx1 (specific for the pancreatic region of the endoderm) and Pbx1 (Andersen et al. 1999; Hussain and Habener 1999). In the absence of Pax6 (as in the homozygous *small eye* mutation in mice and rats), the endocrine cells of the pancreas do not develop properly and the production of hormones by those cells is deficient (Sander et al. 1997; Zhang et al. 2002). There are other genes that are activated by Pax6 binding, and one of them is the *Pax6* gene itself. Pax6 protein can bind to the *Pax6* gene promoter (Plaza et al. 1993). This means that once the *Pax6* gene is turned on, it will continue to be expressed, even if the signal that originally activated it is no longer given.

One can see that the genes for specific proteins use numerous transcription factors in various combinations. Thus, *enhancers are modular* (such that the *Pax6* gene is expressed in the eye, pancreas, and nervous system, as

shown in Figure 5.7); but *within these modules, transcription factors work in a combinatorial fashion* (such that in some cells Pax6 and Sox2 are both needed for the transcription of Pax6 in the lens). In this way, transcription factors can regulate the timing and place of gene expression.

Within the past decade, our knowledge of transcription factors has progressed enormously, giving us a new, dynamic view of gene expression. The gene itself is no longer seen as an independent entity controlling the synthesis of proteins. Rather, the gene both directs and is directed by protein synthesis. Natalie Angier (1992) has written, "A series of new discoveries suggests that DNA is more like a certain type of politician, surrounded by a flock of protein handlers and advisers that must vigorously massage it, twist it, and on occasion, reinvent it before the grand blueprint of the body can make any sense at all."

SIDELIGHTS & SPECULATIONS

Pioneer Transcription Factors

How can a transcription factor find its binding site, given that the enhancer might be covered by nucleosomes? Recent work has identified certain transcription factors that penetrate repressed chromatin and bind to their enhancer DNA sequences (Cirillo et al. 2002; Berkes et al. 2004). These transcription factors can bind to and displace histones 3 and 4, and they bind their DNA sequence with high affinity. They have been called "pioneer transcription factors," and they appear to be critical in establishing certain cell lineages. One of these pioneers is a homeodomain transcription factor called Pbx. Members of the Pbx family are made in every cell, and they appear to be able to find their appropriate sites even in highly compacted chromatin.

Pbx appears to be used as a "molecular beacon" for the muscle-determining transcription factor MyoD. MyoD is critical for initiating muscle development, activating hundreds of genes that are involved with establishing the muscle phenotype. However, MyoD is not able to bind to DNA without the help of Pbx proteins, which bind to DNA elements adjacent to the DNA sequence recognized by MyoD (Figure 5.13A). Here they can act as a guidepost, allowing MyoD to find and bind to its appropriate DNA recognition sequences. Indeed, Berkes and colleagues (2004) have shown that MyoD (when complexed with another transcrip-

tion factor, E12) can bind to the Pbx protein and align itself on the DNA (Figure 5.13B). Once bound there, the E12 protein can recruit histone acetyltransferases and nucleosome remodeling complexes to open up

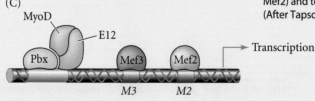

the DNA on those genes (Figure 5.13C). It is also possible that in the absence of the MyoD/E12 dimer, the Pbx protein alone recruits repressive proteins so that the gene is not inadvertently activated.

FIGURE 5.13   Model for the role of the "pioneer" transcription factor Pbx in aligning the muscle-specific transcription factor MyoD on DNA. (A) Pbx protein recognizes its DNA binding site (TGAT), even within nucleosome-rich chromatin. Pbx probably binds to transcriptional inhibitors. (B) MyoD, complexed with its E12 cofactor, is able to bind to Pbx, replacing the transcriptional inhibitors. MyoD then binds to its recognition element on the DNA. (C) The MyoD/E12 complex can then recruit the histone transacetylases and nucleosome remodeling compounds that make the chromatin accessible to other transcription factors (Mef3 and Mef2) and to RNA polymerase. (After Tapscott 2005.)

Transcription factor cascades

We now know that the tissue-specific expression of a particular gene is the result of the presence of a particular constellation of transcription factors in the cell nucleus. It is important to see that their combinatorial mode of operation means that none of these transcription factors must be tissue-specific.

But how do the transcription factors themselves come to be expressed in a tissue-specific manner? In many cases, the genes for transcription factors are activated by other transcription factors. For instance, the gene for Pax6, which is a transcription factor, is regulated (in the anterior head) by the Mbx transcription factor that is expressed at the end of gastrulation (Kawahara et al. 2002). Similarly, MyoD is expressed in the newly determined myoblast, where it initiates the muscle lineage by activating the transcription of other transcription factors, such as myogenin, which are necessary for skeletal muscle differentiation (see Chapter 14). In other cases (as we saw in the case of MITF), the transcription factor is present, but must be activated by signals provided by induction. This will be discussed more fully in the next chapter.

Wilhelm Roux (1894) described the situation eloquently in his manifesto for experimental embryology when he stated that the causal analysis of development may be the greatest problem the human intellect has attempted to solve, "since every new cause ascertained only gives rise to fresh questions concerning the cause of this cause."

SIDELIGHTS & SPECULATIONS

Studying DNA Regulatory Elements

Identifying DNA regulatory elements

How do we know that a particular DNA fragment binds a particular transcription factor? One of the simplest ways is to perform a **gel mobility shift assay**. The basis for this assay is gel electrophoresis. Recall from Chapter 4 that fragments of DNA can be placed in a depression at one end of a gel and an electric current run through the gel. The fragments will move toward the positive pole, and the distance each fragment travels in a given time will depend upon its mass and conformation. Larger fragments will run more slowly than smaller fragments. If the fragments of interest are incubated in a solution of Pax6 protein before being placed in the gel, one of two things will happen. If the Pax6 does not recognize any sequence in a DNA fragment, the DNA will not be bound and the fragment will migrate through the gel as it normally would. Alternatively, if the Pax6 protein does recognize a sequence in the DNA, it will bind to it, increasing the mass of the fragment and cause it to run more slowly through the gel. Figure 5.14A shows the use of such a gel to locate the binding site of Pax6 protein.

The results of this procedure are often confirmed by a **DNase protection assay**. If a DNA-binding protein such as Pax6 finds its target sequence in a DNA fragment, it will bind to it. If the fragment is then placed in a solution of DNase I (an enzyme that cleaves DNA randomly), the bound transcription fac-

FIGURE 5.14 Procedures for determining the DNA-binding sites of transcription factors. (A) Gel mobility shift assay. A DNA fragment containing the Pax6-binding site changes its mobility in the gel when Pax6 protein binds to it. Lanes 1 and 3 show the positions of a DNA fragment containing a Pax6-binding site. When Pax6 is added to the fragment, it moves more slowly. Lanes 2 and 4 show the positions of a similar-sized fragment that does not bind Pax6. (B) A DNase protection assay of the intron between the third and fourth exons of the chick δ1 lens crystallin gene. DNA with and without Pax6 is labeled at one end and is subjected to varying concentrations of DNase I, which randomly cleaves the DNA. The resulting fragments are run on an electrophoretic gel and autoradiographed. No cleavage is seen in the region where a bound protein (Pax6) has prevented DNase from binding. There are two sites (purple bars) where Pax6 binding is able to protect the DNA from digestion with DNase; these sites are the enhancers. (A after Beimesche et al. 1999; B from Cvekl et al. 1995.)

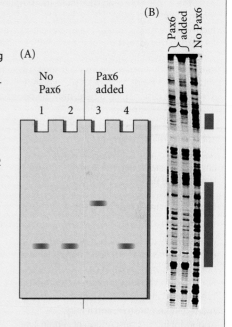

tor will protect that specific region of DNA from being cleaved. Figure 5.14B shows the results of one such assay for Pax6 binding.

Reporter genes such as the *lacZ* gene mentioned above are also used to determine the functions of various DNA fragments. If a sequence of DNA is thought to contain an enhancer, it can be fused to a reporter gene and injected into an egg by

various means. If the tested sequence is able to direct the expression of the reporter gene in the appropriate tissues, it is assumed to contain an enhancer.

One of the interesting concepts emerging from this work is that the transcription factors involved in the mechanisms of cell differentiation in a given region are often used to specify that region as a certain type of tissue. Thus, the Pax6 transcription factor is used in the specification of the eye-forming region as well as in the differentiation of the retina and lens, and the Pdx transcription factor is used both in the specification

(Continued on next page)

In most cells: No recombination

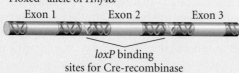

In liver cells only (expressing albumin)

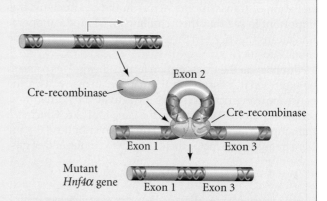

FIGURE 5.15   The Cre-lox technique for conditional mutagenesis, by which gene mutations can be generated in specific cells only. Mice are made wherein wild-type alleles (in this case, the genes encoding the Hnf4α transcription factor) have been replaced by alleles in which the second exon is flanked by *loxP* sequences. These mice are mated with mice having the gene for Cre-recombinase transferred onto a promoter that is active only in particular cells. In this case, the promoter is that of an albumin gene that functions early in liver development. In mice with both these altered alleles, Cre-recombinase is made only in the cells where that promoter was activated (i.e., in these cells synthesizing albumin). The Cre-recombinase binds to the *loxP* sequences flanking the exons and removes those exons. Thus, in the case depicted here, only the developing liver cells lack a functional *Hnf4α* gene.

of the pancreatic rudiment and in the subsequent activation of the genes for insulin and somatostatin derived from that tissue (see Hui and Perfetti 2002).

Conditional knockouts: Floxed mice

One critically important experimental use of enhancers has been the conditional elimination of gene expression in certain cell types. For example, the transcription factor Hnf4α is expressed in liver cells. It looks like a hormone-binding protein, and it might be important in the transcription of certain liver proteins. But if this gene is deleted from mouse embryos, the embryos die before they even form a liver. This is because Hnf4α is critical in forming the visceral endoderm of the yolk sac, and if this tissue fails to form properly, the animal dies very early in development. So one needs to create a mutation that will be *conditional*—that is, a mutation that will appear only in the liver and nowhere else. How can this be done?

The **Cre-Lox** technique uses homologous recombination (see Chapter 4) to place two Cre-recombinase recognition sites (*loxP* sequences) within the gene of interest, usually flanking important exons (see Kwan

2002). Such a gene is said to be "floxed" ("*loxP*-flanked"). For example, using cultured mouse ES cells, Parvis and colleagues (2002) placed two *loxP* sequences around the second exon of the mouse *Hnf4α* gene (Figure 5.15). These ES cells were then used to generate mice that had this floxed allele. A second strain of mice was generated that had a gene encoding bacteriophage Cre-recombinase (the enzyme that recognizes the *loxP* sequence) attached to the promoter of an albumin gene that is expressed very early in liver development. Thus, during mouse development, Cre-recombinase would be made only in the liver cells. When the two strains of mice were crossed, some of their offspring carried both additions. In these double-marked mice, Cre-recombinase (made only in the liver cells) bound to its recognition sites—the *loxP* sequences—flanking the second exon of the *Hnf4α* genes. It then acted as a recombinase and

deleted this second exon. The resulting DNA would encode a nonfunctional protein, since the second exon has a critical function in *HNF4α*. Thus, the *HNF4α* gene was "knocked out" only in liver cells.

Enhancer traps: The right place at the right time

The ability of an enhancer from one gene to activate other genes has been used by scientists to find new enhancers and the genes regulated by them. To do this, one makes an **enhancer trap**, consisting of a reporter gene (such as the *E. coli lacZ* gene or jellyfish *GFP* gene) fused to a relatively weak promoter. The weak promoter will not initiate the transcription of the reporter gene without the help of an enhancer. This recombinant enhancer trap is then introduced into an egg or oocyte, where it integrates randomly into the genome. If the reporter gene is expressed, it means that the reporter has come within the domain of an active enhancer (Figure 5.16). By isolating this activated region of the genome in wild-type flies or mice, the normal gene activated by the enhancer can be discovered (O'Kane and Gehring 1987).

Activating genes in all the wrong places: GAL4 activation

One of the most powerful uses of this genetic technology has been to *activate*

 WEBSITE **5.6 Other recombinase methods.** The Cre-Lox system is patented, and each researcher using it is expected to pay the owner of the patent. This has caused much legal concern over whether such procedures should be private property. It has also sparked the search for other ways of making conditional mutants.

SIDELIGHTS & SPECULATIONS

(A)

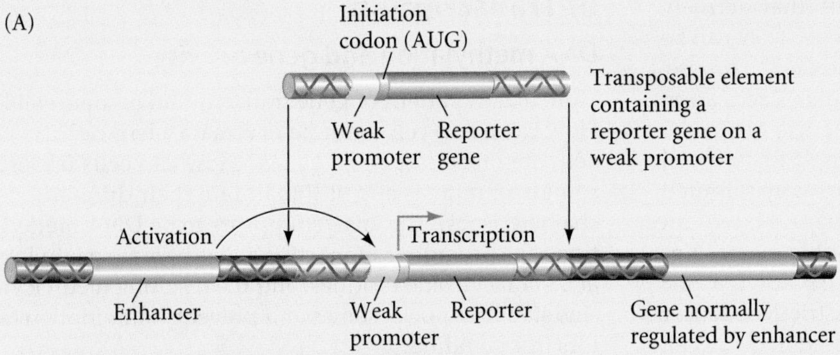

Initiation
codon (AUG)

Weak   Reporter
promoter   gene

Transposable element
containing a
reporter gene on a
weak promoter

Activation   Transcription

Enhancer   Weak   Reporter
promoter

Gene normally
regulated by enhancer

(B)

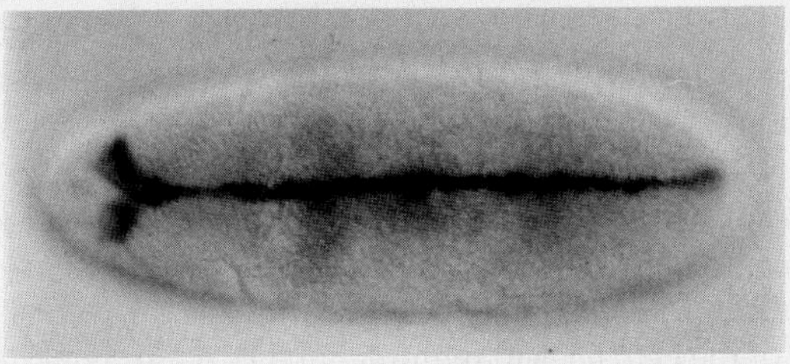

FIGURE 5.16   The enhancer trap technique. (A) A reporter gene is fused to a weak promoter that cannot direct transcription on its own. This recombinant gene is injected into the nucleus of an egg and integrates randomly into the genome. If it integrates near an enhancer, the reporter gene will be expressed when that enhancer is activated, showing the normal expression pattern of a gene normally associated with that enhancer. (B) Reporter gene expression (dark region) in a *Drosophila* embryo injected with an enhancer trap. This expression pattern demonstrated the presence of an enhancer that is active in the development of the insect nervous system and which was unrecognized before the procedure. (Photograph courtesy of Y. Hiromi.)

regulatory genes such as Pax6 in new places. Using *Drosophila* embryos, Halder and his colleagues (1995) placed a gene encoding the yeast GAL4 transcriptional activator protein downstream from an enhancer that was known to function in the labial imaginal discs (those parts of the *Drosophila* larva that become the adult mouth parts). In other words, the gene for the GAL4 transcription factor was placed next to an enhancer for genes normally expressed in the developing jaw. Therefore, GAL4 should be expressed in jaw tissue. Halder and his colleagues then constructed a second trans-

genic fly, placing the cDNA for the *Drosophila Pax6* regulatory gene downstream from a sequence composed of five GAL4-binding sites. The GAL4 protein should be made only in a particular group of cells destined to become the jaw, and when that protein was made, it should cause the transcription of *Pax6* in those particular cells (Figure 5.17A). In flies in which the *Pax6* gene was expressed in the incipient jaw cells, part of the jaw gave rise to eyes (Figure 5.17B). Pax6 in *Drosophila* and frogs (but not in mice) is able to turn several developing tissue types into eyes (Chou et al. 1999). It

appears that in *Drosophila*, Pax6 not only activates those genes that are necessary for the construction of eyes, but also represses those genes that are used to construct other organs.

FIGURE 5.17   Targeted expression of the *Pax6* gene in a *Drosophila* non-eye imaginal disc. (A) A strain of *Drosophila* was constructed wherein the gene for the yeast GAL4 transcription factor was placed downstream from an enhancer sequence that normally stimulates gene expression in the imaginal discs for mouthparts. If the embryo also contains a transgene that places GAL4-binding sites upstream of the *Pax6* gene, the *Pax6* gene will be expressed in whichever imaginal disc the GAL4 protein is made. (B) *Drosophila* ommatidia (compound eyes) emerging from the mouthparts of a fruit fly in which the *Pax6* gene was expressed in the labial (jaw) discs. (Photograph courtesy of W. Gehring and G. Halder.)

WEBSITE **5.7** **Enhancers and cancers.** Enhancer trapping may occur accidentally during development and place a new gene next to a particular regulatory element. When growth factor genes are placed next to the genes involved in making antibodies, the results are leukemias.

(A)

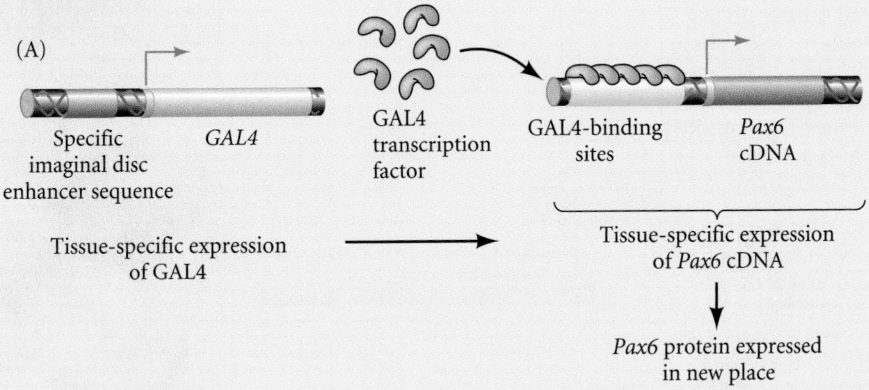

Specific   *GAL4*
imaginal disc
enhancer sequence

GAL4
transcription
factor

GAL4-binding   *Pax6*
sites   cDNA

Tissue-specific expression
of GAL4

Tissue-specific expression
of *Pax6* cDNA

Pax6 protein expressed
in new place

(B)

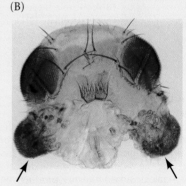

Silencers

Silencers are DNA regulatory elements that actively repress the transcription of a particular gene. They can be viewed as "negative enhacers." For instance, in the mouse, there is a sequence that prevents a promoter's activation in any tissue except neurons. This sequence, given the name **neural restrictive silencer element** (**NRSE**), has been found in several mouse genes whose expression is limited to the nervous system: those encoding synapsin I, sodium channel type II, brain-derived neurotrophic factor, Ng-CAM, and L1. The protein that binds to the NRSE is a zinc finger transcription factor called neural restrictive silencer factor (NRSF). NRSF appears to be expressed in every cell that is *not* a mature neuron (Chong et al. 1995; Schoenherr and Anderson 1995).

To test the hypothesis that the NRSE sequence is necessary for the normal repression of neural genes in non-neural cells, transgenes were made by fusing a β-galactosidase (*lacZ*) gene with part of the *L1* neural cell adhesion gene. (L1 is a protein whose function is critical for brain development, as we will see in later chapters.) In one case, the *L1* gene, from its promoter through the fourth exon, was fused to the *lacZ* sequence. A second transgene was made just like the first, except that the NRSE was deleted from the *L1* promoter. The two transgenes were separately inserted into the pronuclei of fertilized oocytes and the resulting transgenic mice were analyzed for the expression of β-galactosidase (Kallunki et al. 1995, 1997). In the embryos receiving the complete transgene (which included the NSRE), expression was seen only in the nervous system (Figure 5.18A). However, in mice whose transgene lacked the NRSE sequence, expression was seen in the heart, the limb mesenchyme and limb ectoderm, the kidney mesoderm, the ventral body wall, and the cephalic mesenchyme (Figure 5.18B).

The repressive activity of NRSF works by the binding of a histone deacetylase to the DNA-bound NRSF. Just as histone acetyltransferases release nucleosomes, histone deacetylases stabilize them, and this prevents the promoter from being recognized (Roopra et al. 2000; Jepson et al. 2000).

 WEBSITE **5.8 Further mechanisms of transcriptional regulation.** These three websites cover (1) DNase I hypersensitivity, (2) locus control regions and the mechanisms by which LCRs may regulate the temporal expression of linked genes, and (3) the association of active genes with the nuclear matrix.

FIGURE 5.18 Silencers. Analysis of β-galactosidase staining patterns in 11.5-day embryonic mice. (A) Embryo containing a transgene composed of the *L1* promoter, a portion of the *L1* gene, and a *lacZ* gene fused to the second exon (which contains the NRSE region). (B) Embryo containing a similar transgene, but lacking the NRSE sequence. The dark areas reveal the presence of β-galactosidase (the *lacZ* product). (Photographs from Kallunki et al. 1997.)

Methylation Pattern and the Control of Transcription

DNA methylation and gene activity

How does a pattern of gene transcription become stable? How can a lens cell continue to remain a lens cell and not activate muscle-specific genes? How can cells undergo rounds of mitosis and still maintain their differentiated characteristics? The answer appears to be **DNA methylation**. The promoters of inactive genes become methylated at certain cytosine residues, and the resulting methylcytosine stabilizes nucleosomes and prevents transcription factors from binding.

It is often assumed that a gene contains exactly the same nucleotides whether it is active or inactive; that is, a β-globin gene in a red blood cell precursor has the same nucleotides as a β-globin gene in a fibroblast or retinal cell of the same animal. There is, however, a subtle difference in the DNA. In 1948, R. D. Hotchkiss discovered a "fifth base" in DNA, 5-methylcytosine. In vertebrates, this base is made enzymatically after DNA is replicated. At this time, about 5 percent of the cytosines in mammalian DNA are converted to 5-methylcytosine (Figure 5.19A). This conversion can occur only when the cytosine residue is followed by a guanosine. Numerous studies have shown that the degree to which the cytosines of a gene are methylated can control the level of the gene's transcription. Cytosine methylation appears to be a major mechanism of transcriptional regulation in vertebrates; however, *Drosophila*, nematodes, and perhaps most invertebrates do not methylate their DNA.

In vertebrates, the presence of methylated cytosines in the promoter of a gene correlates with the repression of transcription from that gene. In developing human and

(A)

L1 promoter   NRSE sequence

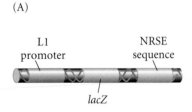

lacZ

(B)

L1 promoter   No NRSE sequence

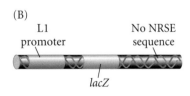

lacZ

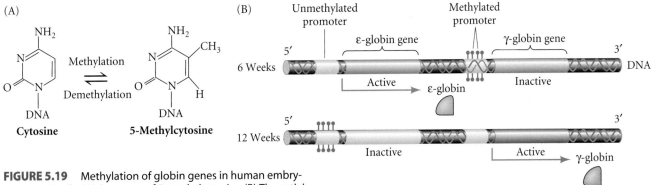

FIGURE 5.19 Methylation of globin genes in human embryonic blood cells. (A) Structure of 5-methylcytosine (B) The activity of the human β-globin genes correlates inversely with the methylation of their promoters. (After Mavilio et al. 1983.)

chick red blood cells, the DNA of the globin gene promoters is almost completely unmethylated, whereas the same promoters are highly methylated in cells that do not produce globins. Moreover, the methylation pattern changes during development (Figure 5.19B). The cells that produce hemoglobin in the human embryo have unmethylated promoters in the genes encoding the ε-globins ("embryonic globin chains") of embryonic hemoglobin. These promoters become methylated in the fetal tissue, as the genes for fetal-specific γ-globin (rather than the embryonic chains) become activated (van der Ploeg and Flavell 1980; Groudine and Weintraub 1981; Mavilio et al. 1983). Similarly, when fetal globin gives way to adult (β) globin, the γ-globin gene promoters of the fetal globin genes become methylated.

The correlation between methylated cytosines and transcriptional repression has been confirmed experimentally. By adding transgenes with different methylation patterns to cells, Busslinger and co-workers (1983) showed that methylation in the promoter or enhancer of a gene correlates extremely well with the repression of gene transcription. In vertebrate development, the absence of DNA methylation correlates well with the tissue-specific expression of many genes.

As we will see in Chapter 7, some genes are active in the early embryo only if they come from the sperm, and other genes are active in the early embryo only if they come from the egg. The differences in the activity of these genes is caused by cytosine methylation in their enhancer and promoter regions. These genes are said to be **imprinted**, and the genomic imprinting takes place differently in the sperm and in the egg (see Sidelights & Speculations).

SIDELIGHTS & SPECULATIONS

Genomic Imprinting

It is usually assumed that the genes one inherits from one's father and the genes one inherits from one's mother are equivalent. In fact, the basis for Mendelian ratios (and the Punnett square analyses used to teach them) is that it does not matter whether the genes came from the sperm or from the egg. But in mammals, there are at least 75 genes for which it *does* matter.* In these cases, only the sperm-derived or only the egg-derived allele of the gene is expressed. This means that a

*A list is maintained at http://www.mgu.har. mrc.ac.uk/research/imprinting/

severe or lethal condition arises if a mutant allele is derived from one parent, but the same mutant allele will have no deleterious effects if inherited from the other parent. It also means (as will be shown in Chapter 7) that both maternal and paternal chromosomes are required for normal mammalian development.

For instance, in mice, the gene for insulin-like growth factor II (*Igf2*) on chromosome 7 is active in early embryos only on the chromosome transmitted by the father. Conversely, the gene for the protein that binds this growth factor, *Igf2r*, is located on chromosome 17 and is active only in the

chromosome transmitted by the mother (Barlow et al. 1991; DeChiara et al. 1991; Bartolomei and Tilghman 1997). The Igf2r protein binds and degrades excess Igf2. A mouse pup that inherits a deletion of the *Igf2r* gene from its father is normal, but if the same deletion is inherited from the mother, the fetus experiences a 30 percent increase in growth and dies late in gestation. The increase in growth is caused by an excess of Igf2; the lethality is probably due to lysosomal defects, since the Igf2 receptor also targets lysosomal enzymes into that

(Continued on next page)

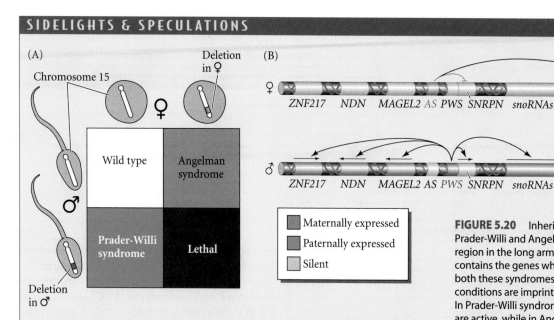

FIGURE 5.20 Inheritance patterns for Prader-Willi and Angelman syndrome. (A) A region in the long arm of chromosome 15 contains the genes whose absence causes both these syndromes. However, the two conditions are imprinted in reverse fashion. In Prader-Willi syndrome, the paternal genes are active, while in Angelman syndrome, the maternal genes are active. (B) Some of the genes and the "inactivation centers" where methylation occurs on this chromosomal region. In the maternal chromosome, the *AS* inactivation center activates *UBE3A* and suppresses *SNRPN*. Conversely, on the paternal chromosome, the *PWS* inactivation center activates *SNRPN* and several other nearby genes, as well as making antisense RNA to *UBE3A*. (B after Walker and Paulsen 2003.)

organelle. In humans, misregulation of *Igf2* methylation causes Beckwith-Wiedemann growth syndrome. Interestingly, although DNA methylation is the mechanism for imprinting this gene in both mice and humans, the mechanisms responsible for the differential *Igf2* methylation between sperm and egg appear to be very different in the two species (Ferguson-Smith et al. 2003; Walter and Paulsen 2003).

In humans, the loss of a particular segment of the long arm of chromosome 15 results in different phenotypes, depending on whether the loss is in the male- or the female-derived chromosome (Figure 5.20A). If the chromosome with the defective or missing segment comes from the father, the child is born with Prader-Willi syndrome, a disease associated with mild mental retardation, obesity, small gonads, and short stature. If the defective or missing segment comes from the mother, the child has Angelman syndrome, characterized by severe mental retardation, seizures, lack of speech, and inappropriate laughter (Knoll et al. 1989; Nicholls et al. 1998). The imprinted genes in this region are *SNRPN* and *UBE3A*.

In the egg-derived chromosome, *UBE3A* is activated and *SNRPN* is turned off, while in the sperm-derived chromosome, *SNRPN* is activated and *UBE3A* is turned off (Figure 5.20B). The expression of either maternal or paternal loci on human chromosome 15 also depends on methylation differences at specific regions in the chromosome that regulate these genes (Zeschingk et al. 1997; Ferguson-Smith and Surani 2001; Walter and Paulsen 2003).

Genomic imprinting adds information to the inherited genome that may regulate spatial and temporal patterns of gene activity. The variation of gene expression that is *not* due to changes in the DNA sequence is called **epigenetics**. Although the definition of this term has changed since Waddington first proposed the term in the 1950s (see Gilbert 2002), a modern definition of epigenetic phenomena (Wu and Morris 2001) is "a

mitotically and/or meiotically heritable change in gene function that cannot be explained by changes in DNA sequence." Differential methylation is one of the most important mechanisms of epigenetic changes. It provides a reminder that the organism cannot be explained solely by its genes. One needs knowledge of developmental parameters as well as genetic ones.

WEBSITE 5.9 Imprinting in humans and mice. The mechanisms of imprinting may be very different. In humans, methylation patterns affect the transmission of certain diseases, and unmethylating β-globin genes may be a cure for certain blood diseases.

Chromatin modification induced by DNA methylation

How are some genes methylated when others are not, and how does methylation prevent gene expression? One proposal is that there is a reciprocal interaction between the histones in chromatin and the DNA that surrounds them. There seems to be competition for modification at the lysine residue on the ninth position of histone H3. If this

site is acetylated, the histone is destabilized and can cause the nucleosome to disperse. If this site is not acetylated, it can be methylated. Methylation of this histone increases the nucleosome's stability, preventing its dissociation or movement. So the modification of the histone H3 tail might act as a "switch" between activated (dispersed nucleosomes) and inactivated (stable nucleosomes) states of a gene. The methylated histone may be able to recruit to its vicinity the enzymes that methylate the DNA (Rea et al.

2000; Tamaru and Selker 2001). Once the DNA is methylated, it may stabilize the nucleosomes even further.

There may be an intimate connection between DNA methylation and histone modification. First, a protein called MeCP2 selectively binds to methylated regions of DNA; it also binds to histone deacetylases. Thus, when MeCP2 binds to methylated DNA at one end of the protein, it can bind at another site to the histone deacetylases that can stabilize the nucleosomes. In this way, repressed chromatin becomes associated with regions where there are methylated cytosines (Keshet et al. 1986; Jones et al. 1998; Nan et al. 1998). Second, a Polycomb group protein (known to repress DNA) called EZH2 binds to DNA methyltransferases, creating an enzyme complex that might simultaneously methylate DNA and neighboring histones (Vire et al. 2005).

And finally, methylated DNA may be preferentially bound by histone H1, the histone that associates nucleosomes into higher-order folded complexes (McArthur and Thomas 1996). In this way, an inactive state can be propagated along the DNA. Transcription factors may act to prevent higher-order folding or to release the nucleosomes by demethylating and acetylating the histones, thereby allowing the DNA to become unmethylated and the promoter to become accessible.

The methyl groups that distinguish active from the inactive genes are placed on the DNA during the differentiation of the cell types. The totipotent cells of the inner cell mass, for instance, do not have methyl groups on cell type-specific genes such as those for the globins. Rather, the DNA of these early cells lacks the tissue-specific methylation patterns that characterize the differentiated cells. When mammals are cloned (see Chapter 4), we are in fact asking the egg cytoplasm to erase or ignore DNA methylation patterns that have been established in a differentiated cell's nucleus. Such reprogramming by demethylation has not been observed in cloned cow and mouse embryos derived from nuclear transfer, and most cloned mammalian and amphibian embryos show abnormal methylation patterns (Bourçhis et al. 2001; Dean et al. 2001; Byrne et al. 2002). This is probably why cloning works less than 5 percent of the time (even though the genes themselves should be normal), and why those cloned animals that do survive often have developmental anomalies (Humphreys 2001; Rideout et al. 2001).

The normal DNA methylation patterns of differentiated cells are maintained throughout cell division by enzymes called DNA (cytosine-5)-methyltransferases. During DNA replication, one strand (the template strand) retains the methylation pattern, while the newly synthesized strand does not. DNA methyltransferases have a strong preference for DNA that has one methylated strand, and when this enzyme sees a methyl-CpG on one side of the DNA, it methylates the new C on the other side (Gruenbaum et al. 1982; Bestor and Ingram 1983). DNA methyltransferases can also recruit histone deacetylases to the regions of methylated DNA (see Meehan 2003).

There are several DNA methyltransferases, and they can act in different cells and in different regions of the genome. Dnmt3a, for instance, is especially important for DNA imprinting in the mouse sperm and egg (Kaneda et al. 2004), and deleting this gene in the germ cells by the Cre-lox method resulted in abnormal methylation of the gametes. Reduction of Dnmt1 in mouse embryos (to 10 percent of the normal value) resulted in genome-wide hypomethylation, small birth size, and the formation of malignant tumors shortly after birth (Gaudet et al. 2003).

Insulators

Enchancer effects must be told where to stop. Since enhancers can work at relatively long distances, it is possible for them to activate several nearby promoters. In order to stop the spreading of an enhancer's power, there are also **insulator** sequences in the DNA (Figure 5.21A; Zhao et al. 1995; Capelson and Corces 2004). Insulators bind proteins that prevent the enhancer from activating an adjacent promoter; and they are often located between the enhancer and the promoter. For instance, in the chick, the genes of the β-globin family are flanked by insulators on both sides. On one side is an insulator that prevents the globin gene enhancers from activating odorant receptor genes (which are active in the nasal neurons), and on the other side is an insulator preventing the globin gene enhancers from activating the promoter of the folate receptor gene (Figure 5.21B).

Insulators also prevent the condensed chromatin of neighboring loci from repressing the actively transcribed globin genes. In the precursors of the red blood cells, the globin genes are active, but the adjacent folate receptor gene is in a highly condensed state. Litt and colleagues (2001) have shown that the globin genes are in an active configuration, and that the nucleosomes around the enhancer are highly acetylated. The folate receptor genes, however, are tightly packaged in condensed nucleosomes, with methyl groups on their histone H3 tails. These nucleosomes are thought to recruit methyltransferases that would extend the methylation to more nucleosomes. However, the insulator appears to recruit acetyltransferases that acetylate the histone tails and prevent the inactivation of the chromatin (Figure 5.21C).

Transcriptional regulation of an entire chromosome: Dosage compensation

In *Drosophila*, nematodes, and mammals, females are characterized as having two X chromosomes per cell, while males are characterized as having a single X chromosome per cell. Unlike the Y chromosome, the X chromosome contains thousands of genes that are essential for cell activity. Yet, despite the female's cells having double the number of X chromosomes, male and female cells contain approximately equal amounts of X chromosome-encoded gene products. This equalization phenomenon is called **dosage**

(A)

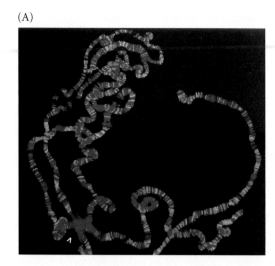

(B)

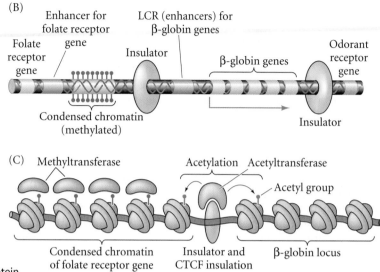

FIGURE 5.21   Insulators. (A) The BEAF32 insulator protein binds to hundreds of insulator sites in the *Drosophila* genome. The DNA is stained red with propydium iodide. The antibody to BEAF32 stains green, and the overlap of signals makes the bound BEAF32 antibody appear yellow. (B) Insulators flanking the chick β-globin genes prevent the β-globin enhancers from activating the folate receptor gene or the odorant receptor gene. (C) In developing red blood cells, the folate receptor gene is found in highly condensed chromatin, wherein the histone H3 tails are methylated. The β-globin insulator is thought to block the extension of histone methylation by recruiting histone acetyltransferases, which prevents histone methylation by the neighboring methyltransferases. (A, photograph courtesy of U. K. Laemmli; C after Litt et al. 2001.)

compensation, and it can be accomplished in three ways (Migeon 2002). In *Drosophila*, the transcription rate of the male X chromosomes is doubled so that the single male X chromosome makes the same amount of transcript as the two female X chromosomes. (Lucchesi and Manning 1987). This is accomplished by acetylation of the nucleosomes throughout the male's X chromosomes, which gives RNA polymerase more efficient access to that chromosome's promoters (Akhtar et al. 2000; Smith et al. 2001). In *C. elegans*, both X chromosomes are partially repressed (Chu et al. 2002) so that the male and female* products of the X chromosomes are equalized.

In mammals, dosage compensation occurs through the inactivation of one X chromosome in each female cell. Thus, each mammalian somatic cell, whether male or female, has only one functioning X chromosome. This phenomenon is called **X chromosome inactivation**. The chromatin of the inactive X chromosome is converted into **heterochromatin**—chromatin that remains condensed throughout most of the cell cycle and replicates later than most of the other chromatin (the **euchromatin**) of the

nucleus. The heterochromatic (inactive) X chromosome, which can often be seen on the nuclear envelope of female cells, is referred to as a **Barr body** (Figure 5.22A; Barr and Bertram 1949). By monitoring the expression of X-linked genes whose products can be detected in the early embryo, it was shown that X chromosome inactivation occurs early in development (Figure 5.22B,C). This inactivation appears to be critical. Using a mutant X chromosome that could not be inactivated, Tagaki and Abe (1990) showed that the expression of two X chromosomes per cell in mouse embryos leads to ectodermal cell death and the absence of mesoderm formation, eventually causing embryonic death at day 10 of gestation.

The early inactivation of one X chromosome per cell has important phenotypic consequences (Figure 5.23A). One of the earliest analyses of X chromosome inactivation was performed by Mary Lyon (1961), who observed coat color patterns in mice. If a mouse is heterozygous for an autosomal gene controlling hair pigmentation, then it resembles one of its two parents, or has a color intermediate between the two. In either case, the mouse is a single color. But if a female mouse is heterozygous for a pigmentation gene *on the X chromosome*, a different result is seen: patches of one parental color alternate with patches of the other parental color. Lyon proposed the following hypothesis to account for these results:

1. Very early in the development of female mammals, both X chromosomes are active.
2. As development proceeds, one X chromosome is inactivated in each cell.
3. This inactivation is random. In some cells, the paternally derived X chromosome is inactivated; in other cells, the maternally derived X chromosome is shut down.
4. This process is irreversible. Once an X chromosome has been inactivated in a cell, the same X chromosome is inactivated in all that cell's progeny. Since X inactivation happens relatively early in development, an entire

---

*As we will see in Chapter 21, the "female" is actually a hermaphrodite capable of making both sperm and eggs.

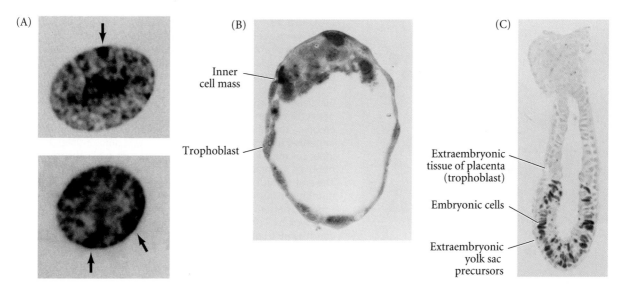

(A)   (B)   (C)

Inner cell mass

Trophoblast

Extraembryonic tissue of placenta (trophoblast)

Embryonic cells

Extraembryonic yolk sac precursors

FIGURE 5.22 Inactivation of a single X chromosome in mammalian XX cells. (A) Barr bodies in the nuclei of human oral epithelial cells stained with Cresyl violet. The top cell from a normal XX female, showing a single Barr body (arrow). In the lower cell, from a female with three X chromosomes, two Barr bodies can be seen. Only one X chromosome per cell is active. (B,C) X-chromosome inactivation in the early mouse embryo. The paternally derived X chromosome contained a *lacZ* transgene that is active in the early embryo. Those cells in which the chromosome is active make β-galactosidase and can be stained blue. The other cells are counterstained and appear pink. (B) In the early blastocyst stage (day 4), both X chromosomes are active in all cells. (C) At day 6, random inactivation of one of the chromosomes occurs. Embryonic cells in which the maternal X is active appear pink, while those where the paternal X is active stain blue. In the mouse (but not the human) trophoblast, the paternally derived X chromosome in preferentially inactivated. Thus, the trophoblast cells are uniformly pink. (A, photographs courtesy of M. L. Barr; B,C from Sugimoto et al. 2000, photographs courtesy of N. Takagi.)

region of cells derived from a single cell may all have the same X chromosome inactivated. Thus, all tissues in female mammals are mosaics of two cell types.

The Lyon hypothesis of X chromosome inactivation provides an excellent account of differential gene inactivation at the level of transcription.

X chromosome inactivation does not extend to every gene on the human X chromosome. Carrel and Willard (2005) estimate that about 15 percent of the genes on the X chromosome escape inactivation. Some interesting exceptions to the general rules further show the importance of random X-chromosome inactivation. The major exception is that X chromosome inactivation holds true only for somatic cells, not germ cells (Figure 5.23B). In female germ cells, the inactive X chromosome is reactivated shortly before the cells enter meiosis (Gartler et al. 1973; Migeon and Jelalian 1977). Thus, in early oocytes, both X chromosomes are unmethylated (and active). In each generation, X chromosome inactivation has to be established anew.

There are some additional exceptions to the rule of randomness in the inactivation pattern. For instance, the first X chromosome inactivation in the mouse is seen in the fetal portion of the placenta, where the paternally derived X chromosome is specifically inactivated (Tagaki 1974; see Figure 5.22B,C). And one exception really ends up proving the rule. There are a few male mammals with coat color patterns we would not expect to find unless the animals exhibited X chromosome inactivation. Male calico and tortoiseshell cats are among these examples. These orange and black coat patterns are normally seen in females and are thought to result from random X chromosome inactivation* (see Figure 5.23A). But rare males exhibit these coat patterns as well. How can this be? It turns out that these cats are XXY. The Y chromosome makes them male, but one X chromosome undergoes inactivation, just as in females, so there is only one active X per cell (Centerwall and Benirschke 1973). Thus, these cats undergo random X chromosome inactivation and their cells have a Barr body.

There may be an intimate connection between DNA methylation and histone modification. First, a protein called MeCP2 selectively binds to methylated regions of DNA; it also binds to histone deacetylases. Thus, when MeCP2

WEBSITE

5.10 Chromosome heterochromatization, elimination, and diminution. The inactivation or the elimination of entire chromosomes is not uncommon among invertebrates and is sometimes used as a mechanism of sex determination. Polycomb group proteins may be critical for repressing large blocks of chromatin.

*Although the terms *calico* and *tortoiseshell* are sometimes used synonymously, calico cats usually have white patches (i.e., patches with no pigment) as well.

(A)

FIGURE 5.23 X chromosome inactivation in mammals. (A) A calico cat, whose cells have one X chromosome containing an orange allele for a pigmentation gene, and the other X chromosome bearing a black allele for that pigmentation gene. The regions of different color correspond to the inactivation of one or the other X chromosome. (B) Schematic diagram illustrating X chromosome inactivation in the mouse. In the zygote, both the sperm- and egg-derived X chromosomes are active. In the blastocyst (early embryo), the paternally derived X chromosome is preferentially imprinted and inactivated. As the blastocyst differentiates, the inner cell mass loses its X-inactivation pattern, while the trophoblast maintains the inactivation of the paternal X chromosome. The embryonic cells derived from the inner cell mass then undergo random X chromosome inactivation. This establishes their transcription pattern throughout adult life (except in the germline cells, which lose their patterns of X inactivation). (A photograph courtesy of R. Loredo and G. Loredo; B after Reik and Lewis 2005.)

(B)

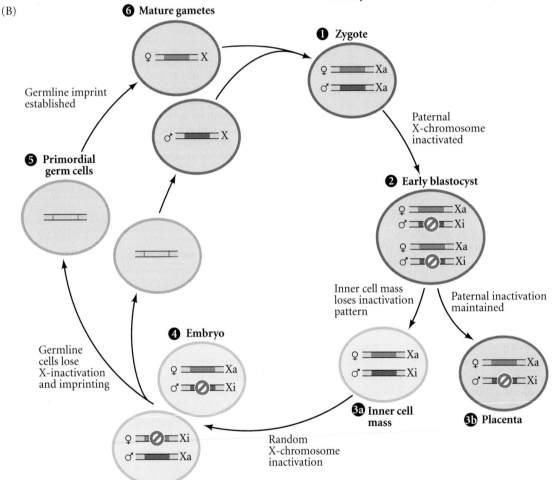

binds to methylated DNA at one end of the protein, it can bind at another site to the histone deacetylases that can stabilize the nucleosomes. In this way, repressed chromatin becomes associated with regions where there are methylated cytosines (Keshet et al. 1986; Jones et al. 1998; Nan et al. 1998). Second, a Polycomb group protein (known to repress DNA) called EZH2 binds to DNA methyltransferases, creating an enzyme complex that might simultaneously methylate DNA and neighboring histones (Vire et al. 2005).

And finally, methylated DNA may be preferentially bound by histone H1, the histone that associates nucleosomes into higher-order folded complexes (McArthur and Thomas 1996). In this way, an inactive state can be propagated along the DNA. Transcription factors may act to prevent higher-order folding or to release the nucleosomes by demethylating and acetylating the histones, thereby allowing the DNA to become unmethylated and the promoter to become accessible.

SIDELIGHTS & SPECULATIONS

The Mechanisms of X Chromosome Inactivation

The mechanisms of X chromosome inactivation are still poorly understood, but new research is giving us some indication of the factors that may be involved in initiating and maintaining a heterochromatic X chromosome.

Initiation of X chromosome inactivation: Xist RNA

In 1991, Brown and her colleagues found an RNA transcript that was made solely from the *inactive* X chromosome of humans (Brown 1991a,b). This transcript, called *XIST*, does not encode a protein. Rather, it stays within the nucleus and interacts with the inactive X chromatin, forming an *XIST*-Barr body complex (Brown et al. 1992). A similar situation exists in the mouse, in which the transcript of the *Xist* gene is seen to coat the inactive X chromosome (Figure 5.24A; Borsani et al. 1991; Brockdorrf et al. 1992).

The *Xist* transcript is an excellent candidate for the initiator of X inactivation. First, transcripts from the mouse *Xist* gene are seen in embryos prior to X chromosome inactivation, which would be expected if

this gene plays a role in initiating inactivation (Kay et al. 1993). Second, knocking out one *Xist* locus in a murine XX cell prevents X inactivation from occurring on that chromosome (Penny et al. 1996). Third, the transfer of a 450-kilobase segment containing the mouse *Xist* gene into an autosome of a male embryonic stem cell causes the random inactivation of either that autosome or the endogenous X chromosome (Lee et al. 1996). The autosome is thus "counted" as an X chromosome. Fourth, *Xist* appears to be involved in "choosing" which X chromosome is inactivated. Female mice heterozygous for a deletion of a particular region of the *Xist* gene preferentially inactivate the wild-type chromosome (Marahrens et al. 1998). *Xist* expression is needed only for the initiation of X chromosome inactivation; once inactivation occurs, *Xist* transcription is dispensable (Brown and Willard 1994).

The *Xist* RNA transcript works only in *cis*—that is, on the chromosome that made it. When cells begin to differentiate, *Xist* RNA is stabilized on one of the two X chromosomes (Figure 5.24B–D; Sheardown et al.

1997; Panning et al. 1997). However, the mechanism by which *Xist* is preferentially stabilized on one particular chromosome appears to differ between humans and mice. In the preimplantation *mouse* embryo, both X chromosomes synthesize *Xist*, but this RNA is quickly degraded on one of the X chromosomes. This differential stabilization appears to be affected by an X-linked gene called *Tsix* ("*Xist*" spelled backwards).

Tsix encodes a small, nuclear, non-protein-coding RNA that contains the *antisense* sequence to *Xist* (Lee and Lu 1999; Chao et al. 2002). *Tsix* RNA is associated with the future *active* X chromosome, where it appears to degrade *Xist*. The *human TSIX* gene, however, lacks the region that is antisense to *XIST*, and human *TSIX* does not appear to play a role in X chromosome inactivation (Migeon 2002; Migeon et al. 2002). While female mice with *Tsix* mutations die as a result of faulty X chromosome inactivation (see Lee 2000; Sado et al. 2001), human female cells (with their "naturally mutant"

(Continued on next page)

FIGURE 5.24   *Xist* RNA associates with the inactive X chromosome that made it. (A) Mouse *Xist* RNA (red) on an inactive X chromosome (blue) in metaphase. (In humans, *XIST* RNA is not seen to coat metaphase X chromosomes.) (B–D) XX murine embryonic stem cells undergoing differentiation and X chromosome inactivation. *Xist* RNA is stained light blue; mRNA from the X-linked phosphoglycerate kinase (*Pgk*) gene is stained red; the background DNA is stained blue. (B) XX cell before differentiating shows both X chromosomes transcribing *Pgk* and *Xist*. (C) As X chromosome inactivation begins, the *Xist* RNA is stabilized on one of the chromosomes. This chromosome no longer transcribes the *Pgk* gene. (D) As the cell finishes differentiating, X chromosome inactivation is complete. One X chromosome continues to make Xist protein. The other chromosome no longer makes Xist, but continues to transcribe *Pgk*. (A courtesy of R. Jaenisch; B–D from Sheardown et al. 1997, photograph courtesy of N. Brockdorff.)

(A)

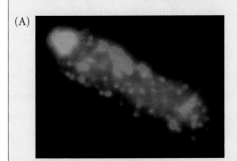

(B)

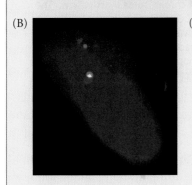

(C)

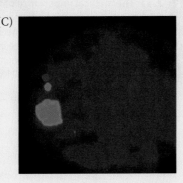

(D)

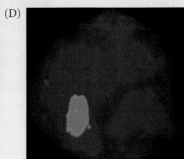

TSIX) must have some other mechanism to insure differential *XIST* stabilization.

Maintaining X chromosome inactivation: Methylation and chromatin modification

Once *Xist* initiates the inactivation of an X chromosome, the silencing of that chromosome is maintained in at least two ways. The first way involves DNA methylation. The *Xist* locus on the *active* X chromosome becomes methylated, while the active *Xist* gene (on the otherwise "inactive" X chromosome) remains unmethylated (Norris et al. 1994). Conversely, the promoter regions of numerous genes are methylated on the inactive X chromosome and unmethylated on the active X chromosome (Wolf et al. 1984; Keith et al. 1986; Migeon et al. 1991).

The second method of maintaining X-chromosome inactivation appears to involve histone modification. The mouse *Xist* gene, in a manner still unknown, recruits Polycomb group proteins that are capable of methylating lysine 27 on histone H3 (Plath et al. 2003; Kalantry et al. 2006). The methylation of lysine 27 on histone H3 occurs almost immediately after *Xist* RNA is seen to coat what will be the inactive X chromosome (see Figure 5.24A). This histone methylation occurs before transcriptional silencing is observed, and it appears to begin at the nucleosomes immediately upstream from the *Xist* gene. Soon afterward, transcription stops, and other histone modifications occur on the inactive X chromosome, including the removal of acetyl groups from histone H4 (Figure 5.25; Jeppesen and Turner 1993; Heard et al. 2001; Mermoud et al. 2002). These nucleosome changes may create the heterochromatin that is characteristic of the Barr body, and they appear to prevent the activation of the chromatin during differentiation.

We are ignorant of the mechanisms by which the *Xist* transcript regulates the state of the chromatin and by which the inactivation spreads. We still do not understand the ways in which *Xist* transcription is linked to DNA methylation. We do not yet know how the choice between the two X chromosomes is originally made, nor how the *Xist* RNA is transcribed from a region surround-

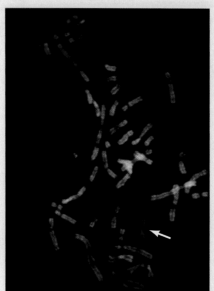

(A)

(B)

Attributes of the inactive X chromosome

XIST RNA production
XIST /Barr body complex
Histone H4 hypoacetylation
Histone H3 methylation
Histone macroH2A1 concentration
Nuclear envelope association
Late replication
Heterochromatin
Methylated promoters

FIGURE 5.25   The inactive X chromosome of human female cells contains underacetylated histone H4. (A) Chromosomes from a human female fibroblast cell stained green with fluorescent antibody to acetylated histone H4. While all the other chromosomes are stained green, the inactive X is not and thus appears red (arrow). (B) List of attributes characterizing the inactive X chromosome. (From Jeppesen and Turner 1993; photograph courtesy of the authors.)

ed by inactivated genes. There is still much to learn about dosage compensation in mammals, as well as in other animals that have chromosomal sex determination.

Nonrandom X-chromosome inactivation

One of the differences between human and mouse X-chromosome inactivation is that in mice, random X chromosome inactivation occurs in the embryo proper; but in mouse extraembryonic tissue (i.e., the placenta), it is always the paternal X chromosome that is

inactivated (see Figure 5.22C). It appears that in the mouse (but probably not in humans), the paternal X is already inactive in the zygote or very early embryo, and that this imprint is removed in the embryo but not in the extraembryonic cells (Huynh and Lee 2003; Okamoto et al. 2004; Mak et al. 2004). *Xist* RNA is seen to coat the paternal X in the extraembryonic cells, leading eventually to the chromatin changes characteristic of the inactivated X chromosome.

WEBSITE 5.11 **The medical importance of X chromosome inactivation.** The mechanisms responsible for human X chromosome inactivation may differ significantly from those that inactivate the mouse X chromosome. Moreover, random X chromosome inactivation in women heterozygous for a mutant X-linked gene (e.g., hemophilia A) may cause the predominant expression of the mutant allele, causing such women to have the disease, as would a male.

Differential RNA Processing

The regulation of gene expression is not confined to the differential transcription of DNA. Even if a particular RNA transcript is synthesized, there is no guarantee that it will create a functional protein in the cell. To become an active protein, the RNA must be (1) processed into a messenger RNA by the removal of introns, (2) translocated from the nucleus to the cytoplasm, and (3) translated by the protein-synthesizing apparatus. In some cases, the synthesized protein is not in its mature form and must be (4) posttranslationally modified to become active. Regulation during development can occur at any of these steps.

The essence of differentiation is the production of different sets of proteins in different types of cells. In bacteria, differential gene expression can be effected at the levels of transcription, translation, and protein modification. In eukaryotes, however, another possible level of regulation exists—namely, control at the level of RNA processing and transport. There are two major ways in which differential RNA processing can regulate development. The first involves "censorship"—selecting which nuclear transcripts are processed into cytoplasmic messages. Different cells select different nuclear transcripts to be processed and sent to the cytoplasm as messenger RNA. Thus, the same pool of nuclear transcripts can give rise to different populations of cytoplasmic mRNAs in different cell types (Figure 5.26A).

The second mode of differential RNA processing is the splicing of mRNA precursors into messages for different proteins by using different combinations of potential exons. If an mRNA precursor had five potential exons, one cell might use exons 1, 2, 4, and 5; a different cell might utilize exons 1, 2, and 3; and yet another cell type might use yet another combination (Figure 5.26B). Thus a single gene can produce an entire family of proteins.

Control of early development by nuclear RNA selection

In the late 1970s, numerous investigators found that mRNA was not the primary transcript from the genes. Rather, the initial transcript is **nuclear RNA (nRNA)**, sometimes called *heterogeneous nuclear RNA* (hnRNA) or *pre-messenger RNA* (pre-mRNA). This nRNA is usually many times longer than the corresponding mRNA because the nuclear RNA contains introns that get spliced out during the passage from nucleus to cytoplasm. Originally, investigators thought that whatever RNA was transcribed in the nucleus was processed into cytoplasmic mRNA. But studies of sea urchins showed that different cell types could be *transcribing* the same type of nuclear RNA, but *processing* different subsets of this population into mRNA in different types of cells (Kleene and Humphreys 1977, 1985). Wold and her colleagues (1978) showed that sequences present in sea urchin blastula *messenger* RNA, but absent in gastrula and adult tissue mRNA, were nonetheless present in the *nuclear* RNA of the gastrula and adult tissues.

More genes are transcribed in the nucleus than are allowed to become mRNAs in the cytoplasm. This "censoring" of RNA transcripts has been confirmed by probing for the introns and exons of specific genes. Gagnon and his colleagues (1992) performed such an analysis on the transcripts from the *SpecII* and *CyIIIa* genes of the sea urchin *Strongylocentrotus purpuratus*. These genes encode calcium-binding and actin proteins, respectively, which are expressed only in a particular part of the ectoderm of the sea urchin larva. Using probes that bound to an exon (which is included in the mRNA) and to an intron (which is not included in the mRNA), they found that these genes were being transcribed not only in the ectodermal cells, but also in the mesoderm and endoderm. The analysis of the *CyIIIa* gene showed that the concentration of introns was the same in both the gastrula ectoderm and the mesoderm/endoderm samples, suggesting that this gene was being transcribed at the same rate in the nuclei of all cell

(A) RNA selection   (B) Differential splicing

Cell type 1

Cell type 2

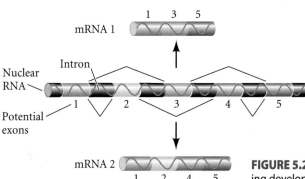

FIGURE 5.26   Roles of differential RNA processing during development. By convention, splicing paths are shown by fine V-shaped lines. (A) RNA selection, whereby the same nuclear RNA transcripts are made in two cell types, but the set that becomes cytoplasmic messenger RNA is different. (B) Differential splicing, whereby the same nuclear RNA is spliced into different mRNAs by selectively using different exons.

(A)

(C)

Ectoderm

Endoderm/
mesoderm

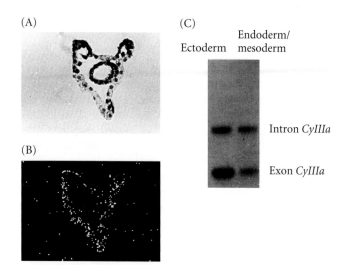

Intron *CyIIIa*

Exon *CyIIIa*

(B)

FIGURE 5.27 Regulation of ectoderm-specific gene expression by RNA processing. (A,B) *CyIIIa* mRNA is seen by autoradiography to be present only in the ectoderm. (A) Phase contrast micrograph. (B) In situ hybridization using a probe that binds to a *CyIIIa* exon. (C) The *CyIIIa* nuclear transcript, however, is found in both ectoderm and endoderm/mesoderm. The left lane of the gel represents RNA isolated from the gastrula ectodermal tissue; the right lane represents RNA isolated from endodermal and mesodermal tissues. The upper band is the RNA bound by a probe that binds to an intron sequence (which should be found only in the nucleus) of *CyIIIa*. The lower band represents the RNA bound by a probe complementary to an exon sequence. The presence of the intron indicates that the *CyIIIa* nuclear RNA is being made in both groups of cells, even if the mRNA is seen only in the ectoderm. (From Gagnon et al. 1992; photographs courtesy of R. and L. Angerer.)

types, but was made into cytoplasmic mRNA only in ectodermal cells (Figure 5.27). The unprocessed nRNA for *CyIIIa* is degraded while still in the nuclei of the endodermal and mesodermal cells.

WEBSITE **5.12 Differential nRNA censoring.** Studies of differential nRNA censoring overturned the paradigm that differential gene transcription was the ultimate means of regulating embryonic differentiation. It "freed" embryology from microbiological ways of thinking about gene expression.

WEBSITE **5.13 An inside-out gene.** Some RNAs stay in the nucleus to function. In one interesting case, the exons go outside the nucleus to be degraded, while the introns stay and help construct the nucleolus.

Creating families of proteins through differential nRNA splicing

Alternative nRNA splicing is a means of producing a wide variety of proteins from the same gene. The average ver-

tebrate nRNA consists of several relatively short exons (averaging about 140 bases) separated by introns that are usually much longer. Most mammalian nRNAs contain numerous exons. By splicing together different sets of exons, different cells can make different types of mRNAs, and hence, different proteins. Recognizing a sequence of nRNA as either an exon or an intron is a crucial step in gene regulation.

Alternative nRNA splicing is based on the determination of which sequences will be spliced out as introns. This can occur in several ways. Most genes contain "consensus sequences" at the 5′ and 3′ ends of the introns. These sequences are also called the "splice sites" of the intron. The splicing of nRNA is mediated through complexes known as **spliceosomes** that bind to these splice sites. Spliceosomes are made up of small nuclear RNAs (snRNAs) and proteins called **splicing factors** that bind to splice sites or to the areas adjacent to them. By their production of specific splicing factors, cells can differ in their ability to recognize a sequence as an intron. That is to say, a sequence that is an *exon* in one cell type may be an *intron* in another (Figure 5.28A,B). In other instances, the factors in one cell might recognize different 5′ sites (at the beginning of the intron) or different 3′ sites (at the end of the intron) (Figure 5.28C,D).

The 5′ splice site is normally recognized by small nuclear RNA U1 (U1 snRNA) and splicing factor 2 (SF2; also known as alternative splicing factor). The choice of alternative 3′ splice sites is often controlled by which splice site can best bind a protein called U2AF. The spliceosome forms when the proteins that accumulate at the 5′ splice site contact those proteins bound to the 3′ splice site. Once the 5′ and 3′ ends are brought together, the intervening intron is excised and the two exons are ligated together.

Researchers estimate that approximately 75 percent of human genes are alternatively spliced, and that such alternative splicing is a major way by which the rather limited number of genes can create a much larger array of proteins (Johnson et al. 2003). The deletion of certain potential exons in some cells but not in others enables one gene to create a family of closely related proteins. Instead of one gene-one polypeptide, one can have one gene-one family of proteins. For instance, alternative RNA splicing enables the gene for α-tropomyosin to encode brain, liver, skeletal muscle, smooth muscle, and fibroblast forms of this protein (Breitbart et al. 1987). The nuclear RNA for α-tropomyosin contains 11 potential exons, but different sets of exons are used in different cells (Figure 5.29). Such different proteins encoded by the same gene are called **splicing isoforms** of the protein.

WEBSITE **5.14 The mechanism of differential nRNA splicing.** Differential nRNA splicing depends on the assembly of the nucleosome and upon the ratio of certain proteins in the nucleus of the cell.

(A) Cassette exon: Type II procollagen

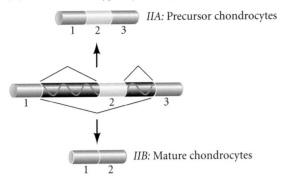

IIA: Precursor chondrocytes

IIB: Mature chondrocytes

(B) Mutually exclusive exons: FgfR2

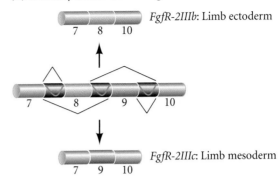

FgfR-2IIIb: Limb ectoderm

FgfR-2IIIc: Limb mesoderm

(C) Alternative 5′ splice site: *Bcl-x*

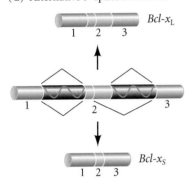

Bcl-x$_L$

Bcl-x$_S$

(D) Alternative 3′ splice site: Chordin

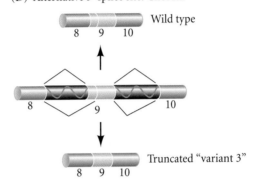

Wild type

Truncated "variant 3"

FIGURE 5.28 Some examples of alternative RNA splicing. Blue and colored portions of the bars represent exons; gray represents introns. Alternative splicing patterns are shown with V-shaped lines. (A) A "cassette" (yellow) that can be used as exon or removed as an intron distinguishes the type II collagen types of chondrocyte precursors and mature chondrocytes (cartilage cells). (B) Mutually exclusive exons distinguish fibroblast growth factor receptors found in the limb ectoderm from those found in the limb mesoderm. (C) Alternative 5′ splice site selection, such as that used to create the large and small isoforms of the protein Bcl-x. (D) Alternative 3′ splice sites are used to form the normal and truncated forms of chordin. (After McAlinden et al. 2004.)

In some instances, alternatively spliced RNAs yield proteins that play similar yet distinguishable roles in the same cell. Different isoforms of the WT1 protein perform different functions in the development of the gonads and kidneys. The isoform without the extra exon functions as a transcription factor during kidney development, while the isoform containing the extra exon appears to be involved in splicing different nRNAs and may be critical in testis development (Hastie 2001; Hammes et al. 2001).

The *Bcl-x* gene provides a good example of how alternative nRNA splicing can make a very big difference in a protein's function (Mercatante et al. 2002). If a particular DNA sequence is used as an exon, the "large Bcl-x protein" (Bcl-X$_L$) is made (see Figure 5.28C). This protein inhibits programmed cell death. However, if this sequence is seen as an intron, the "small Bcl-X protein"(Bcl-X$_S$) is made, creating a protein that induces cell death. Many tumors have a higher than normal amount of Bcl-X$_L$.

If you get the impression from this discussion that a gene with dozens of introns could create literally thousands of different, related proteins through differential

splicing, you are probably correct. The current champion at making multiple proteins from the same gene is the *Drosophila Dscam* gene. This gene encodes a membrane receptor protein involved in guiding axons to their targets during the insect's development. *Dscam* contains 115 exons. However, a dozen different adjacent DNA sequences can be selected to be exon 4. Similarly, more than 30 mutually exclusive adjacent DNA sequences can become exons 6 and 9, respectively (Figure 5.30; Schmucker et al. 2000). If all possible combinations of exons are used, this one gene can produce 38,016 different proteins, and random searches for these combinations indicate that a large fraction of them are in fact made. The nRNA of *Dscam* has been found to be alternatively spliced in different axons and may control the specificity of axon attachments (Celotto and Graveley 2001; Wojtowicz et al. 2004). The *Drosophila* genome is thought to contain only 14,000 genes, but here is a single gene that encodes three times that number of proteins!

As mentioned earlier, about 75 percent of human genes are thought to produce multiple types of mRNA. Therefore, even though the human genome may contain

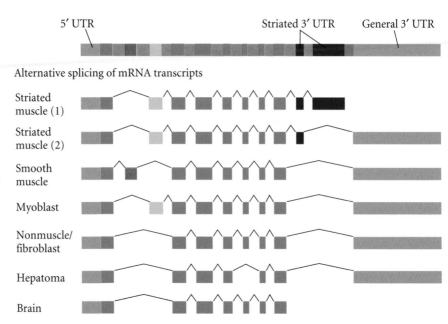

FIGURE 5.29 Alternative RNA splicing to form a family of rat α-tropomyosin proteins. The α-tropomyosin gene is represented on top. The thin lines represent the sequences that become introns and are spliced out to form the mature mRNAs. Constitutive exons (found in all tropomyosins) are shown in green. Those expressed only in smooth muscle are red; those expressed only in striated muscle are purple. Those that are variously expressed are yellow. Note that in addition to the many possible combinations of exons, two different 3′ ends ("striated" and "general") are possible. (After Breitbart et al. 1987.)

20,000–30,000 genes, its **proteome**—the number and type of proteins encoded by the genome—is far more complex.

Splicing enhancers and recognition factors

The mechanisms of differential RNA processing involve both *cis*-acting sequences on the nRNA and *trans*-acting protein factors that bind to these regions (Black 2003). The *cis*-acting sequences on nRNA are usually close to their potential 5′ or 3′ splice sites. These sequences are called "splicing enhancers," since they promote the assembly of spliceosomes at RNA cleavage sites. Conversely, these same sequences can be "splicing silencers" if they act to exclude exon from an mRNA sequence. These sequences are recognized by *trans*-acting proteins, most of which can recruit spliceosomes to that area. However, some *trans*-acting proteins, such as the polyprimidine tract-binding pro-

tein (PTP)*, repress spliceosome formation where they bind.

As might be expected, there are some splicing enhancers that appear to be specific for certain tissues. Muscle-specific *cis*-regulatory sequences have been found around those exons characterizing muscle cell messages. These are recognized by certain proteins that are found in the muscle cells early in their development (Ryan and Cooper 1996; Charlet-B et al. 2002). Their presence is able to compete with the PTP that would otherwise prevent the inclusion of the muscle-specific exon into the mature message. In this way, an entire battery of muscle-specific isoforms can be generated.

---

*PTP is involved in making the correct isoform of tropomyosin and may be especially important in determining the mRNA populations of the brain. PTP is also involved in the mutually exclusive use of exon IIIb or IIIc in the mRNA for fibroblast growth factor-2 (see Figure 5.28B) (Carstens et al. 2000; Lillevälí et al. 2001; Robinson and Smith 2006).

FIGURE 5.30 The *Dscam* gene of *Drosophila* can produce 38,016 different types of proteins by alternative nRNA splicing. The gene contains 24 exons. Exons 4, 6, 9, and 17 are encoded by sets of mutually exclusive possible sequences. Each messenger RNA will contain one of the 12 possible exon 4 sequences, one of the 48 possible exon 6 alternatives, one of the 33 possible exon 9 alternatives, and one of the 2 possible exon 17 sequences. The *Drosophila Dscam* gene is homologous to a human DNA sequence on chromosome 21 that is expressed in the human nervous system. Disturbances of this gene in humans may lead to the neurological defects of Down syndrome (Yamakawa et al. 1998; Saito 2000).

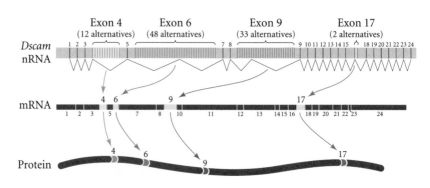

SIDELIGHTS & SPECULATIONS

Differential nRNA Processing and *Drosophila* Sex Determination

Sex determination in *Drosophila* is regulated by a cascade of RNA processing events (Baker et al. 1987; MacDougall et al. 1995). As we will see in Chapter 17, the development of the sexual phenotype in *Drosophila* is mediated by the ratio of X chromosomes to autosomes (nonsex chromosomes). The X chromosomes produce transcription factors that activate the *Sex-lethal* (*Sxl*) gene,* while the autosomes produce transcription factors that repress *Sxl*. Thus, these two sets of transcription factors—activators from the X and repressors from the autosomes—compete for the enhancer sites of *Sxl*. When the X-to-autosome ratio is 1 (i.e., when there are two X chromosomes per diploid cell), the activators dominate, the *Sxl* gene is active, and the embryo develops into a female. When the ratio is 0.5 (i.e., when the fly is XY with only one X chromosome per diploid cell), the repressors dominate, *Sxl* is not active, and the embryo develops into a male (Figure 5.31).

But what is the Sxl protein doing to determine sex? It is acting as a differential splicing factor on the nRNA transcribed from the *transformer* (*tra*) gene. Throughout the larval period, *tra* actively synthesizes a nuclear transcript that is processed into either a general mRNA (found in both females and males) or a female-specific mRNA. The female-specific message is made when Sxl protein binds to *tra* nRNA and inhibits spliceosome formation on the general 3′ splice site of the first intron (Sosnowski et al. 1989; Valcárcel et al. 1993). Instead, the spliceosome forms on another, less efficient 3′ site and allows splicing to occur there. As a result, the female form of the

*The name *Sex-lethal* comes from the deadly decoupling of the dosage compensation mechanism that arises when this gene is mutated. When this happens, the fly becomes male even if its genotype is XX. However, its two X chromosomes will be instructed to transcribe their genes at the higher (male) rate, creating regulatory defects that kill the embryo (Cline 1986).

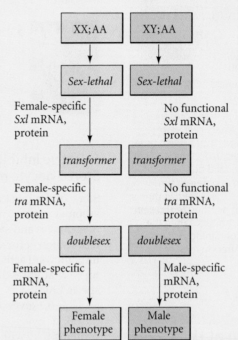

FIGURE 5.31 Sex determination in *Drosophila*. This simplified scheme shows that the X-to-autosome ratio is monitored by the *Sex-lethal* gene. If this gene is active, it processes *transformer* nRNA into a functional female-specific message. In the presence of the female-specific Transformer protein, the *doublesex* gene transcript is processed in a female-specific fashion. The female-specific Doublesex protein is a transcription factor that leads to the production of the female phenotype. If the *transformer* gene does not make a female-specific product (i.e., if the *Sex-lethal* gene is not activated), the *doublesex* transcript is spliced in the male-specific manner, leading to the formation of a male-specific Doublesex protein. This is a transcription factor that generates the male phenotype.

transformer mRNA lacks an exon found in the general form. And that difference is crucial, because this exon, which is spliced out from the female transcript, contains a translational stop codon (UGA) that causes the message to make a small, nonfunctional protein. Therefore, the general transcript makes a nonfunctional protein (Belote et al. 1989), and the female-specific mRNA is the only functional transcript for producing the Tra protein (Belote et al. 1989).

The Tra protein is itself an alternative splicing factor that regulates the splicing of the nuclear transcript of the *doublesex* (*dsx*) gene. This gene is needed for the production of either sexual phenotype, and mutations of *dsx* can reverse the expected sexual phenotype, causing XX embryos to become males or XY embryos to become females. During pupation, the *dsx* gene makes a nuclear transcript that can generate either a female-specific or a male-specific mRNA (Nagoshi et al. 1988). The first three exons of

dsx mRNA are the same in females and males (Tian and Maniatis 1992, 1993). But if Tra protein is present, it converts a weak 3′ splice site into a strong site (i.e., a more efficient binder of U2AF) and exon 4 is retained, resulting in female-specific *dsx* mRNA. If Tra protein is not present, U2AF will not bind and exon 4 will not be included, resulting in the male-specific *dsx* message.

The Doublesex proteins made by the male and female mRNAs are both transcription factors, and they recognize the same sequences of DNA. However, while female Dsx activates female-specific enhancers (such as those on the genes encoding yolk proteins), male Dsx inhibits transcription from those same enhancers (Coschigano and Wensink 1993; Jursnich and Burtis 1993). Conversely, the female protein inhibits transcription from genes that are otherwise activated by the male protein.

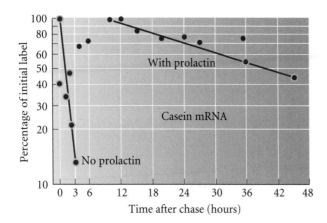

FIGURE 5.32 Degradation of casein mRNA in the presence and absence of prolactin. Cultured rat mammary cells were given radioactive RNA precursors (pulse) and, after a given time, were washed and given nonradioactive precursors (chase). This procedure labeled the casein mRNA synthesized during the pulse time. Casein mRNA was then isolated at different times following the chase and its radioactive label measured. In the absence of prolactin, the labeled (i.e., newly synthesized) casein mRNA decayed rapidly, with a half-life of 1.1 hours. When the same experiment was done in a medium containing prolactin, the half-life was extended to 28.5 hours. (After Guyette et al. 1979.)

Control of Gene Expression at the Level of Translation

The splicing of nuclear RNA is intimately connected with its export through the nuclear pores and into the cytoplasm. As the introns are being removed, specific proteins bind to the spliceosome and attach the spliceosome-RNA complex to the nuclear pores (Luo et al. 2001; Strässer and Hurt 2001). But once the RNA has reached the cytoplasm, there is still no guarantee that it will be translated. The control of gene expression at the level of translation can occur by many means; some of the most important of these are described below.

Differential mRNA longevity

The longer an mRNA persists, the more protein can be translated from it. If a message with a relatively short half-life were selectively stabilized in certain cells at certain times, it would make large amounts of its particular protein only at those times and places.

The stability of a message is often dependent on the length of its polyA tail. This, in turn, depends largely on sequences in the 3' untranslated region, certain of which allow longer polyA tails than others. If these 3' UTR regions are experimentally traded, the half-lives of the resulting mRNAs are altered: long-lived messages will decay rapidly, while normally short-lived mRNAs will remain

around longer (Shaw and Kamen 1986; Wilson and Treisman 1988; Decker and Parker 1995).

In some instances, messenger RNAs are selectively stabilized at specific times in specific cells. The mRNA for casein, the major protein of milk, has a half-life of 1.1 hours in rat mammary gland tissue. However, during periods of lactation, the presence of the hormone prolactin increases this half-life to 28.5 hours (Figure 5.32; Guyette et al. 1979).

WEBSITE

5.15 Mechanisms of mRNA translation and degradation. Translation is a complex process involving the initiation, elongation, and termination of protein synthesis. It has numerous points at which regulation can occur. Similarly, the degradation of mRNA is a tightly regulated event.

Selective inhibition of mRNA translation: Stored oocyte mRNAs

Some of the most remarkable cases of translational regulation of gene expression occur in the oocyte. The oocyte often makes and stores mRNAs that will be used only after fertilization occurs. These messages stay in a dormant state until they are activated by ion signals (discussed in Chapter 7) that spread through the egg during ovulation or sperm binding.

Table 5.2 gives a partial list of mRNAs that are stored in the oocyte cytoplasm. Some of these stored mRNAs encode proteins that will be needed during cleavage, when the embryo makes enormous amounts of chromatin, cell membranes, and cytoskeletal components. Some of them encode cyclin proteins that regulate the timing of early cell division (Rosenthal et al. 1980; Standart et al. 1986). Indeed, in many species (including sea urchins and *Drosophila*), maintenance of the normal rate and pattern of early cell divisions does not require a nucleus; rather, it requires continued protein synthesis from stored maternal mRNAs (Wagenaar and Mazia 1978; Edgar et al. 1994). Other stored messages encode proteins that determine the fates of cells. These include the *bicoid* and *nanos* messages that provide positional information in the *Drosophila* embryo, as we saw in Chapter 3, and the *glp-1* mRNA of the nematode *C. elegans*.

WEBSITE

5.16 The discovery of stored mRNAs. The existence of maternally transcribed mRNAs stored in the oocyte was one of the first discoveries of molecular embryology. Even before gene cloning became available, the identity of several of these mRNAs was known.

Most translational regulation in oocytes is negative, as the "default state" of the mRNA is to be available for translation. Therefore, there must be inhibitors preventing the translation of these mRNAs in the oocyte, and these inhibitors must somehow be removed at the appropriate times around fertilization. The 5' cap and the 3' untrans-

TABLE 5.2 Some mRNAs stored in oocyte cytoplasm and translated at or near fertilization

mRNAs encoding	Function(s)	Organism(s)
Cyclins	Cell division regulation	Sea urchin, clam, starfish, frog
Actin	Cell movement and contraction	Mouse, starfish
Tubulin	Formation of mitotic spindles, cilia, flagella	Clam, mouse
Small subunit of ribo-nucleotide reductase	DNA synthesis	Sea urchin, clam, starfish
Hypoxanthine phospho-ribosyl-transferase	Purine synthesis	Mouse
Vg1	Mesodermal determination(?)	Frog
Histones	Chromatin formation	Sea urchin, frog, clam
Cadherins	Blastomere adhesion	Frog
Metalloproteinases	Implantation in uterus	Mouse
Growth factors	Cell growth; uterine cell growth(?)	Mouse
Sex determination factor FEM-3	Sperm formation	*C. elegans*
PAR gene products	Segregate morphogenetic determinants	*C. elegans*
SKN-1 morphogen	Blastomere fate determination	*C. elegans*
Hunchback morphogen	Anterior fate determination	*Drosophila*
Caudal morphogen	Posterior fate determination	*Drosophila*
Bicoid morphogen	Anterior fate determination	*Drosophila*
Nanos morphogen	Posterior fate determination	*Drosophila*
GLP-1 morphogen	Anterior fate determination	*C. elegans*
Germ cell-less protein	Germ cell determination	*Drosophila*
Oskar protein	Germ cell localization	*Drosophila*
Ornithine transcarbamylase	Urea cycle	Frog
Elongation factor 1a	Protein synthesis	Frog
Ribosomal proteins	Protein synthesis	Frog, *Drosophila*

Compiled from numerous sources

lated region seem especially important in regulating the accessibility of mRNA to ribosomes. If the 5′ cap is not made or if the 3′ UTR lacks a polyadenylate tail, the message probably will not be translated. The oocytes of many species have "used these ends as means" to regulate the translation of their mRNAs.

It is important to realize that, unlike the usual representations of mRNA, most mRNAs probably form circles, with their 3′ end being brought to their 5′ end (Figure 5.33A). The 5′ cap is bound by **eukaryotic initiation factor-4E (eIF4E)**, a protein that is also bound to eIF4A (a helicase that unwinds double-stranded regions of RNA) and **eIF4G**, a scaffold protein that allows the mRNA to bind to the ribosome through its interaction with eIF4E (Wells et al. 1998;

Gross et al. 2003). The poly(A) binding protein, which sits on the poly(A) tail of the mRNA, also binds to the eIF4G protein. This brings the 3′ end of the message next to the 5′ end and allows the messenger RNA to be recognized by the ribosome. Thus, the 5′ cap is critical for translation, and some animal's oocytes have used this as a direct means of translational control. For instance, the oocyte of the tobacco hornworm moth makes some of its mRNAs without their methylated 5′ caps. In this state, they cannot be efficiently translated. However, at fertilization, a methyltransferase completes the formation of the caps, and these mRNAs can be translated (Kastern et al. 1982).

In amphibian oocytes, the 5′ and 3′ ends of many mRNAs are brought together by a protein called **maskin**

(A) Circularized mRNA

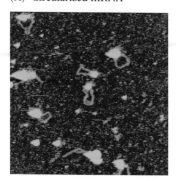

FIGURE 5.33 Translational regulation in oocytes. (A) Messenger RNAs are often found as circles, where the 5' end and the 3' end contact one another. Here, a yeast mRNA seen by atomic force microscopy is circularized by eIF4E and eIF4G (5' end) and the poly(A) binding protein (3' end). (B) In *Xenopus* oocytes, the 3' and 5' ends of the mRNA are brought together by maskin, a protein that binds to CPEB on the 3' end and translation initiation factor 4E (eIF4E) on the 5' end. Maskin blocks the initiation of translation by preventing eIF4E from binding eIF4G. (C) When stimulated by progesterone during ovulation, a kinase phosphorylates CPEB, which can then bind CPSF. CPSF can bind poly(A) polymerase and initiate growth of the poly(A) tail. Poly(A) binding protein (PABP) can bind to this tail and then bind eIF4G in a stable manner. This initiation factor can then bind eIF4E and, through its association with eIF3, position a 40S ribosomal subunit on the mRNA. (A from Wells et al. 1998; B,C after Mendez and Richter 2001.)

(B) **Translationally dormant** (C) **Translationally active**

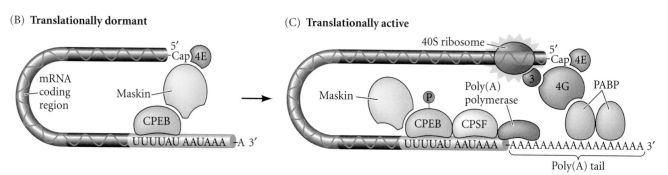

(Stebbins-Boaz et al. 1999; Mendez and Richter 2001). Maskin links the 5' and 3' ends into a circle by binding to two other proteins, each at opposite ends of the message. First, it binds to the **cytoplasmic polyadenylation-element-binding protein (CPEB)** protein attached to the UUUUAU sequence in the 3' UTR; second, maskin also binds to the eIF4E factor that is attached to the cap sequence. In this configuration, the mRNA cannot be translated (Figure 5.33B). The binding of eIF4E to maskin is thought to prevent the binding of eIF4E to eIF4G, a critically important translation initiation factor that brings the small ribosomal subunit to the mRNA.

Mendez and Richter (2001) have proposed an intricate scenario to explain how mRNAs bound together by maskin become translated about the time of fertilization. At ovulation (when the hormone progesterone stimulates the last meiotic divisions of the oocyte and the oocyte is released for fertilization), a kinase activated by progesterone phosphorylates the CPEB protein. The phosphorylated CPEB can now bind to CPSF, the cleavage and polyadenylation specificity factor (Mendez et al. 2000; Hodgman et al. 2001). The bound CPSF protein sits on a particular sequence of the 3' UTR that has been shown to be critical for polyadenylation, and it complexes with a poly(A) polymerase that elongates the poly(A) tail of the mRNA. In oocytes, a message having a short poly(A) tail is not degraded; however, such messages are not translated.

Once the poly(A) tail is extended, molecules of the poly(A) binding protein (PABP) can attach to the growing tail. PABP proteins stabilize eIF4G, allowing it to outcompete maskin for the binding site on the eIF4E protein at the 5' end of the mRNA. The eIF4G protein can then bind eIF3, which can position the small ribosomal subunit onto the mRNA. The small (40S) ribosomal subunit will then find the initiator tRNA, complex with the large ribosomal subunit, and initiate translation (Figure 5.33C).

In the *Drosophila* oocyte, Bicoid can act both as a transcription factor (activating genes such as *hunchback*) and also as a translational inhibitor (see Chapter 3). Bicoid represses the translation of *caudal* mRNA, preventing its transcription in the anterior half of the embryo. (The protein made from the *caudal* message is important in activating those genes that specify the cells to be abdomen precursors.) Bicoid inhibits *caudal* mRNA translation by binding to a "bicoid recognition element," a series of nucleotides in the 3' UTR of the caudal message (Figure 5.34). Once there, Bicoid can bind with and recruit another protein, d4EHP. The d4EHP protein can compete with eIF4E for the cap. Without eIF4E, there is no association with eIF4G and the *caudal* mRNA becomes untranslatable. As a result, *caudal* is not translated in the anterior of the embryo (where Bicoid is abundant), but is active in the posterior portion of the embryo.

STORED mRNAS IN BRAIN CELLS One of the most important areas of local translational regulation may be in the brain. The storage of long-term memory requires new protein synthesis, and the local translation of mRNAs in the dendrites of brain neurons has been proposed as a control point for increasing the strength of synaptic connections (Martin 2000; Klann et al. 2004; Wang and Tiedge 2004). The ability to increase the strength of the connections

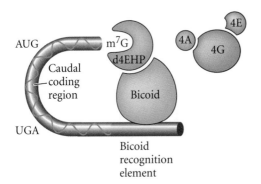

FIGURE 5.34 Protein binding in *Drosophila* oocytes. Bicoid protein binds to a recognition element in the 3′ UTR of the *caudal* message. Bicoid can bind to d4EHP, which prevents the binding of eIF4E to the cap structure. Without eIF4E, the eIF4G cannot bind and initiate translation. (After Cho et al. 2005.)

between neurons is critical in forming the original architecture of the brain and also in the ability to learn. Indeed, in recent studies of mice, Kelleher and colleagues (2004) have shown that neuronal activity-dependent memory storage depends on the activation of eIF4E and other components of protein synthesis.

Several mRNAs appear to be transported along the cytoskeleton to the dendrites of neurons (the "receiving portion" of the neuron, where synapse connections are formed with the other neurons). These messages include those mRNAs encoding receptors for neurotransmitters (needed to transmit the signals from one neuron to another); activity-regulated enzymes; and the cytoskeletal components needed to build a synapse (Figure 5.35). As we will see in later chapters, one of the proteins responsible for constructing specific synapses is **brain-derived neu-**

rotrophic factor, or **BDNF**. BDNF regulates neural activity and appears to be critical for new synapse formation. Takei and colleagues (2004) have shown that BDNF induces local translation of these neural messages in the dendrites. Thus, translational regulation in neurons might be important not only for their initial development but also for their continued ability to change due to new circumstances.

microRNAS: SPECIFIC REGULATORS OF mRNA TRANSLATION AND TRANSCRIPTION One of the most remarkable means of regulating the translation of a specific message is to make a small RNA complementary to a portion of a particular mRNA. Such a naturally occurring antisense RNA was first seen in *C. elegans* (Lee et al. 1993; Wightman et al. 1993). Here, the *lin-4* gene was found to encode a 21-nucleotide RNA that bound to multiple sites in the 3′ UTR of the *lin-14* mRNA. (Figure 5.36) The *lin-14* gene encodes a transcription factor, LIN-14, that is important during the first larval phase of *C. elegans* development. It is not needed afterward, and *C. elegans* is able to inhibit synthesis of LIN-14 from these messages by activating the *lin-4* gene. The binding of *lin-4* transcripts to the *lin-14* mRNA 3′ UTR causes degradation of the *lin-14* message (Bagga et al. 2005).

The *lin-4* RNA is now thought to be the "founding member" of a very large group of **microRNAs (miRNA)**. These RNAs of about 22 nucleotides are made from longer precursors. These precursors can be in independent transcription units (the *lin-4* gene is far apart from the *lin-14* gene), or they can reside in the introns of other genes (Aravin et al. 2003; Lagos-Quintana et al. 2003). Many of the newly discovered microRNAs have been found in the regions between genes (regions previously considered to contain "junk DNA"). The initial RNA transcript (which may contain several repeats of the miRNA sequence) form

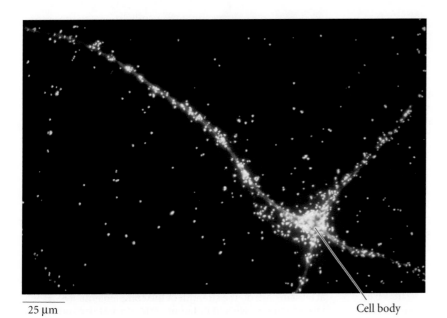

25 μm

Cell body

FIGURE 5.35 A brain-specific RNA in a cultured mammalian neuron. BC1 RNA (stained white) appears to be clustered at specific sites in the neuron (stained light blue), especially in the dendrites. (From Wang and Tiedge 2004, photograph courtesy of the authors.)

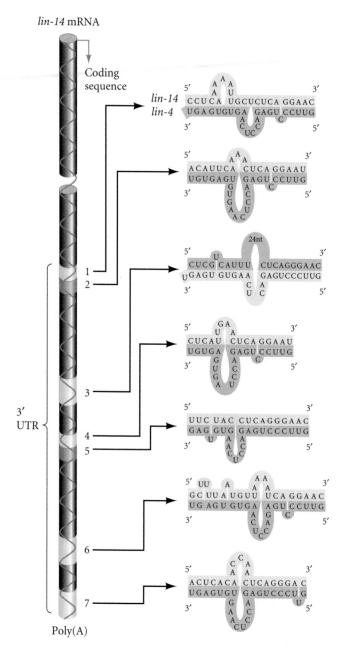

FIGURE 5.36 Hypothetical model of the regulation of *lin-14* mRNA translation by *lin-4* RNAs. The *lin-4* gene does not produce an mRNA. Rather, it produces small RNAs that are complementary to a repeated sequence in the 3′ UTR of the *lin-14* mRNA, which bind to it and prevent its translation. (After Wickens and Takayama 1995.)

hairpin loops wherein the RNA finds complementary structures within its strand. These stem-loop structures are processed by a set of RNases (Drosha and Dicer, the same enzymes used in the RNA interference technique;*) to make single-stranded microRNA (Figure 5.37). The microRNA is then packaged with a series of proteins to make an **RNA-induced silencing complex (RISC)**. Such small reg-

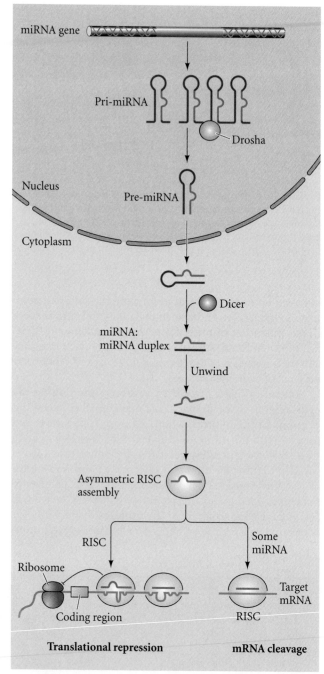

FIGURE 5.37 Current model for the formation and use of microRNAs. The miRNA gene encodes a pri-miRNA that often has several hairpin regions where the RNA finds nearby complementary bases with which to pair. The pri-miRNA is processed into individual pre-miRNA "hairpins" by the Drosha RNAase, and these are exported from the nucleus. Once in the cytoplasm, another RNAase, Dicer, eliminates the non-base-paired loop. Dicer also acts as a helicase to separate the strands of the double-stranded miRNA. One strand (probably recognized by placement of Dicer) is packaged with proteins into the RNA-induced silencing complex (RISC), which subsequently binds to the 3′ UTRs to effect translational suppression or cleavage, depending (at least in part) on the strength of the complementarity between the miRNA and its target. (After He and Hannon 2004.)

ulatory RNAs can bind to the 3′ UTR of messages and inhibit their translation. In some cases (especially when the binding of the miRNA to the 3′ UTR is tight), the site is cleaved. More usually, however, several RISCs attach to sites on the 3′UTR and prevent the message from being translated (see Bartel 2004; He and Hannon 2004).

The abundance of microRNAs and their apparent conservation among flies, nematodes, vertebrates, and even plants suggest that such RNA regulation is a previously unrecognized but potentially very important means of regulating gene expression. This hidden layer of gene regulation parallels the better known protein-level gene control mechanisms, and it may be just as important in regulating cell fate. Recent studies have shown that microRNAs are involved in mammalian heart and blood cell differentiation. During mouse heart development, the microRNA *miR1* can repress the messages encoding the Hand2 transcription factor (Zhao et al. 2005). This transcription factor is critical in the proliferation of ventricle heart muscle cells, and *miR1* may control the balance between ventricle growth and differentiation. The *miR181* miRNA is essential for committing progenitor cells to differentiate into B lymphocytes, and ectopic expression of *miR181* in mice causes a preponderance of B lymphocytes (Chen et al. 2004).

---

*A similar type of RNA, "small-interfering RNA" (siRNA) was discovered around the same time as miRNA. The siRNA, which is the basis of RNA interference technique (RNAi; see Figure 4.22) is made from double-stranded RNA and is also packaged into RISCs. Such siRNA is probably a response against viral infection.

Earlier in this chapter, we mentioned that there were mRNAs in the dendrites, and that their repression is relieved by BDNF. One of these mRNAs encodes the protein kinase Limk1, which controls dendrite length. The *Limk1* mRNA is regulated by a brain-specific microRNA, *miR-134*, that is localized in the dendritic spikes of the neuron (Schratt et al. 2006). Ruvkun and colleagues (2004) have noted, "It is now clear an extensive miRNA world was flying almost unseen by our genetic radar."

Control of RNA expression by cytoplasmic localization

Not only is the time of mRNA translation regulated, but so is the place of RNA expression. Just like the selective repression of mRNA translation, the selective localization of messages is often accomplished through their 3′ UTRs, and it is often performed in oocytes. Rebagliati and colleagues (1985) showed that there are certain mRNAs in *Xenopus* oocytes that are selectively transported to the vegetal pole (Figure 5.38). After fertilization, these messages make proteins that are found only in the vegetal blastomeres. In *Drosophila*, the *bicoid* and *nanos* messages are each localized to different ends of the oocyte. The 3′ UTR of *bicoid* mRNA allows this message to bind to the microtubules through its association with two other proteins (Swallow and Staufen). If the *bicoid* 3′ UTR is attached to some other message, that mRNA will also be bound to the anterior pole of the oocyte (Driever and Nüsslein-Volhard 1988a,b; Ferrandon et al. 1994). The 3′ UTR of the *nanos* message similarly allows it to accumulate at the posterior

---

SIDELIGHTS & SPECULATIONS

miRNA in Transcriptional Gene Regulation

In addition to its role in the translational regulation of gene expression, microRNA also appears to be able to transcriptionally silence certain genes. These genes are often located in the heterochromatin, that region of the genome where the DNA is tightly coiled and transcription inhibited by the packed nucleosomes. Volpe and colleagues (2003) discovered in yeast that if they deleted the genes encoding the appropriate RNases or RISC proteins, the heterochromatin around the centromeres became unpacked, the histones in this region lost their inhibitory methylation, and the centromeric heterochromatin started making RNA. Similar phenomena were seen when

these proteins were mutated in *Drosophila* (Pal-Bhadra et al. 2004). Indeed, in *Drosophila*, the *Suppressor-of-stellate* gene on the Y chromosome makes a microRNA that represses the transcription of the *stellate* gene on the X chromosome. (Gvozdev et al. 2003). This is important for dosage regulation of the X chromosomes in *Drosophila*.

It appears that microRNAs are able to bind to the nuclear RNA as it is being transcribed, and form a complex with the methylating and deacetylating enzymes, thus repressing the gene (Schramke et al. 2005; Kato et al. 2005). If synthetic microRNA made complementary to specific promoters is added to cultured human cells,

that microRNA is able to induce that promoter's DNA to become methylated. The lysine 9 on histone H3 also becomes methylated around the promoter, and transcription from that gene stops (Morris et al. 2004; Kawasaki and Taira 2004). Thus, microRNA directed against the 3′ end of mRNA may be able to shut down gene expression on the translational level, while microRNA directed at the promoters of genes may be able to block gene expression at the transcriptional level. The therapeutic value of these RNAs in cancer therapy is just beginning to be explored.

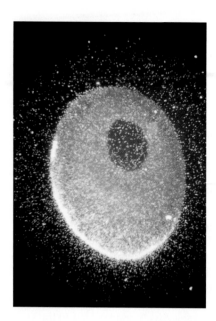

FIGURE 5.38 Localization of *Vg1* mRNA to the vegetal portion of the *Xenopus* oocyte. The white crescent at the bottom of the egg represents the tethered *Vg1* message. The black area is the haploid nucleus of the oocyte. At fertilization, the *Vg1* message is translated into an inactive protein. If that protein is processed to its active form, it can become an important signaling protein in the embryo. (Photograph courtesy of D. Melton.)

pole of the egg (Gavis and Lehmann 1994). As we saw in Chapter 3, this localization allows Bicoid protein to form a gradient with its highest amounts at the anterior pole, while Nanos forms a gradient with its peak at the posterior pole (see Figure 3.11). The ratio of these two proteins will eventually determine the anterior-posterior axis of the *Drosophila* embryo and adult. (The ability of cells to be specified by gradients of proteins is a critical phenomenon and is discussed in more detail in Chapters 3 and 9.)

WEBSITE 5.17 Other examples of translational regulation of gene expression. There are numerous other fascinating examples wherein mRNA is selectively translated under different conditions. The 1:1 ratio of α- and β-globins in adult blood comes from the differential translation of the respective globin messages. The production of heme for the hemoglobin is also regulated at the level of translation.

FIGURE 5.39 An integrated account of gene expression. Each step regulating gene expression is a subdivision of a continuous process from transcription through protein modification. The activation of transcription factors by membrane receptors will be the subject of the next chapter. (After Orphanides and Reinberg 2002.)

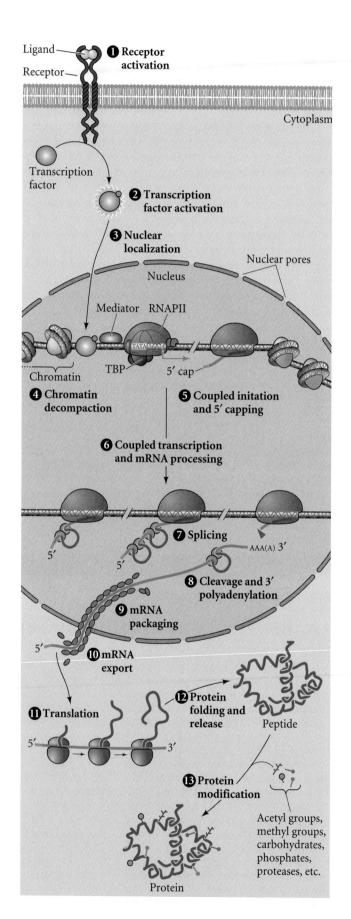

Posttranslational regulation of gene expression

When a protein is synthesized, the story is still not over (Figure 5.39). Once a protein is made, it becomes part of a larger level of organization. For instance, it may become part of the structural framework of the cell, or it may become involved in one of the myriad enzymatic pathways for the synthesis or breakdown of cellular metabolites. In any case, the individual protein is now part of a complex "ecosystem" that integrates it into a relationship with numerous other proteins. Thus, several changes can still take place that determine whether or not the protein will be active.

Some newly synthesized proteins are inactive without the cleaving away of certain inhibitory sections. This is what happens when insulin is made from its larger pro-tein precursor. Some proteins must be "addressed" to their specific intracellular destinations in order to function. Proteins are often sequestered in certain regions of the cell, such as membranes, lysosomes, nuclei, or mitochondria. Some proteins need to assemble with other proteins in order to form a functional unit. The hemoglobin protein, the microtubule, and the ribosome are all examples of numerous proteins joining together to form a functional unit. And some proteins are not active unless they bind an ion (such as Ca^{2+}), or are modified by the covalent addition of a phosphate or acetate group. This last type of protein modification will become very important in the next chapter, since many important proteins in embryonic cells are just sitting there until some signal activates them. We turn next to how the embryo develops by activating certain proteins in specific cells.

 **Snapshot Summary** # Differential Gene Expression

1. Differential gene expression from genetically identical nuclei creates different cell types. Differential gene expression can occur at the levels of gene transcription, nuclear RNA processing, mRNA translation, and protein modification. Notice that RNA processing and export occur while the RNA is still being transcribed from the gene.

2. Genes are usually repressed, and activating a gene often means inhibiting its repressor. This fact leads to thinking in double and triple negatives: Activation is often the inhibition of the inhibitor; repression is the inhibition of the inhibitor of the inhibitor.

3. Eukaryotic genes contain promoter sequences to which RNA polymerase can bind to initiate transcription. To accomplish this, the eukaryotic RNA polymerases are bound by a series of proteins called basal transcription factors.

4. Eukaryotic genes expressed in specific cell types contain enhancer sequences that regulate their transcription in time and space.

5. Specific transcription factors can recognize specific sequences of DNA in the promoter and enhancer regions. These proteins activate or repress transcription from the genes to which they have bound.

6. Enhancers work in a combinatorial fashion. The binding of several transcription factors can act to promote or inhibit transcription from a certain promoter. In some cases transcription is activated only if both factor A and factor B are present, while in other cases, transcription is activated if either factor A or factor B is present.

7. A gene encoding a transcription factor can keep itself activated if the transcription factor it encodes also activates its own promoter. Thus, a transcription factor gene can have one set of enhancer sequences to initiate its activation and a second set of enhancer sequences (which bind the encoded transcription factor) to maintain its activation.

8. Often, the same transcription factors that are used during the differentiation of a particular cell type are also used to activate the genes for that cell type's specific products.

9. Enhancers can act as silencers to suppress the transcription of a gene in inappropriate cell types.

10. Transcription factors act in different ways to regulate RNA synthesis. Some transcription factors stabilize RNA polymerase binding to the DNA, some disrupt nucleosomes, increasing the efficiency of transcription.

11. Transcription correlates with a lack of methylation on the promoter and enhancer regions of genes. Methylation differences can account for examples of genomic imprinting, wherein a gene transmitted through the sperm is expressed differently from the same gene transmitted through the egg.

12. Dosage compensation enables the X chromosome-derived products of males (which have one X chromosome per cell in fruit flies and mammals) to equal the X chromosome-derived products of females (which have two X chromosomes per cell). This compensation is accomplished at the level of transcription, either by accelerating transcription from the lone X chromosome in males (*Drosophila*), decreasing the level of transcription from each X chromosome by 50 percent (*C. elegans*) or by inactivating a large portion of one of the two X chromosomes in females (mammals).

13. X chromosome inactivation in mammals is generally random and involves the activation of the *Xist* gene on the chromosome that will be inactivated.

14. Differential nuclear RNA selection can allow certain transcripts to enter the cytoplasm and be translated while preventing other transcripts from leaving the nucleus.

15. Differential RNA splicing can create a family of related proteins by causing different regions of the nRNA to be read as exons or introns. What is an exon in one set of circumstances may be an intron in another.

16. Some messages are translated only at certain times. The oocyte, in particular, uses translational regulation to set aside certain messages that are transcribed during egg development but used only after the egg is fertilized. This activation is often accomplished either by the removal of inhibitory proteins or by the polyadenylation of the message.

17. MicroRNAs can act as translational inhibitors, binding to the 3′ UTR of the RNA.

18. Many mRNAs are localized to particular regions of the oocyte or other cells. This localization appears to be regulated by the 3′ untranslated region of the mRNA.

For Further Reading

Complete bibliographical citations for all literature cited in this chapter can be found on the Vade Mecum CD that accompanies the book and at the free access website www.devbio.com

Berkes, C. A., D. A. Bergstrom, B. H. Penn, K. J. Seaver, P. S. Knoepfler and S. J. Tapscott. 2004. Pbx marks genes for activation by MyoD indicating a role for a homeodomain protein in establishing myogenic potential. *Mol. Cell* 14: 465–477.

Celotto, A. M. and B. R. Graveley. 2001. Alternative splicing of the *Drosophila Dscam* pre-mRNA is both temporally and spatially regulated. *Genetics* 159: 599–608.

Cho, P. F. and 8 others. 2005. A new paradigm for translational control: Inhibition via 5′-3′ mRNA tethering by Bicoid and the eIF4E cognate 4EHP. *Cell* 121: 411–423.

Cosgrove, M. S., J. D. Boeke, and C. Wolberger. 2004. Regulated nucleosome mobility and the histone code. *Nature Struct. Mol. Biol.* 11: 1037–1043.

Ferguson-Smith, A., S-P. Lin, C-E. Tsai, N. Youngson, and M. Tevendale. 2003.

Genomic imprinting: Insights from studies in mice. *Seminars Cell Dev. Biol.* 14: 43–49.

Gebauer, F. and M. W. Hentze. 2004. Molecular mechanisms of translational control. *Nature Rev. Mol. Cell Biol.* 5: 827–835.

Halder, G., P. Callaerts and W. J. Gehring. 1995. Induction of ectopic eyes by targeted expression of the *eyeless* gene in *Drosophila*. *Science* 267: 1788–1792.

He, L. and G. J. Hannon. 2004. MicroRNAs: small RNAs with a big role in gene regulation. *Nature Rev. Genet.* 5: 522–531.

MacDougall, C., D. Harbison and M. Bownes. 1995. The developmental consequences of alternative splicing in sex determination and differentiation in *Drosophila*. *Dev. Biol.* 172: 353–376.

Migeon, B. R. 2002. X chromosome inactivation: Theme and variation. *Cytogen. Genome Res.* 99: 8–16.

Niessing, D., S. Blanke, and H. Jackle 2002. Bicoid associates with the 5′-cap-bound complex of caudal mRNA and represses translation. *Genes Dev.* 16: 2576–2582.

Price, E. R. and 7 others. 1998. Lineage-specific signaling in melanocytes: c-Kit stimulation recruits p300/CBP to microphthalmia. *J. Biol. Chem.* 273: 33042–33047.

Rideout, W. M. III, K. Eggan and R. Jaenisch. 2001. Nuclear cloning and epigenetic reprogramming of the genome. *Science* 293: 1093–1098.

Stebbins-Boaz, B., Q. Cao, C. H. de Moor, R. Mendez and J. D. Richter. 1999. Maskin is a CPEB-associated factor that transiently interacts with elF-4E. *Mol. Cell* 4: 1017–1027.

Tapscott, S. J. 2005. The circuitry of a master switch: MyoD and the regulation of skeletal muscle gene transcription. *Development* 132: 2685–2695.

Cell-Cell Communication in Development

6

THE FORMATION OF ORGANIZED ANIMAL BODIES has been one of the great sources of wonder for humankind. Indeed, the "miracle of life" seems just that—inanimate matter becomes organized in such a way that it lives.* While every animal starts off as a single cell, the progeny of that cell form complex structures—tissues and organs—that are themselves integrated into larger systems. Probably no one better recognizes how remarkable life actually is than the developmental biologists who get to study how all this complexity arises. In the past decade, developmental biologists have started to answer some of the most important questions of natural science: We have begun to understand how organs form.

Induction and Competence

Organs are complex structures composed of numerous types of tissues. In the vertebrate eye, for example, light is transmitted through the transparent corneal tissue and focused by the lens tissue (the diameter of which is controlled by muscle tissue), eventually impinging on the tissue of the neural retina. The precise arrangement of tissues in the eye cannot be disturbed without impairing its function. Such coordination in the construction of organs is accomplished by one group of cells changing the behavior of an adjacent set of cells, thereby causing them to change their shape, mitotic rate, or fate. This kind of interaction at close range between two or more cells or tissues of different histories and properties is called proximate interaction, or **induction**.[†] There are at least two components to every inductive interaction. The first component is the **inducer**: the tissue that produces a signal (or signals) that changes the cellular behavior of the other tissue. The second component, the tissue being induced, is the **responder**.

> " All that you touch
> You Change.
> All that you Change
> Changes you.
> The only lasting truth
> Is Change. "
> *OCTAVIA BUTLER (1998)*

> " The phenomenon of life itself negates the boundaries that customarily divide our disciplines and fields. "
> *HANS JONAS (1966)*

---

*The twelfth-century rabbi and physician Maimonides (1190) framed the question of morphogenesis beautifully when he noted that the pious men of his day believed that an angel of God had to enter the womb to form the organs of the embryo. This, the men say, is a miracle. How much more powerful a miracle would life be, he asked, if the Deity had made matter such that it could generate such remarkable order without a matter-molding angel having to intervene in every pregnancy? The idea of an angel was still part of the embryology of the Renaissance. The problem addressed today is the secular version of Maimonides' question: How can matter alone construct the organized tissues of the embryo?

[†]These inductions are often called "secondary" inductions, whereas the tissue interactions that generate the neural tube are called "primary embryonic induction." However, there is no difference in the molecular nature of "primary" and "secondary" induction. Primary embryonic induction will be detailed in Chapters 10 and 11.

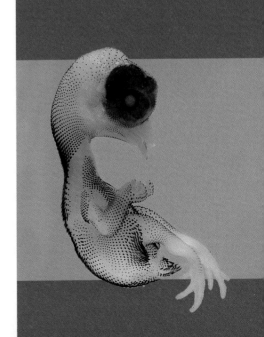

FIGURE 6.1 Ectodermal competence and the ability to respond to the optic vesicle inducer in *Xenopus*. The optic vesicle is able to induce lens formation in the anterior portion of the ectoderm (1), but not in the presumptive trunk and abdomen (2). If the optic vesicle is removed (3), the surface ectoderm forms either an abnormal lens or no lens at all. (4) Most other tissues are not able to substitute for the optic vesicle.

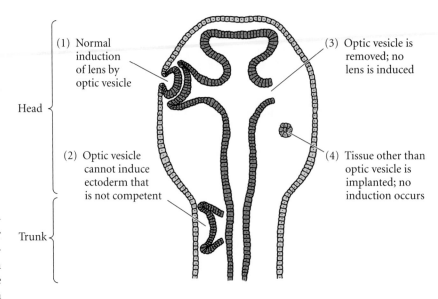

(1) Normal induction of lens by optic vesicle

(3) Optic vesicle is removed; no lens is induced

Head

(2) Optic vesicle cannot induce ectoderm that is not competent

(4) Tissue other than optic vesicle is implanted; no induction occurs

Trunk

Not all tissues can respond to the signal being produced by an inducer. For instance, if the optic vesicle (the presumptive retina) of *Xenopus laevis* is placed in an ectopic location (i.e., in a different place from where it normally forms) underneath the *head* ectoderm, it will induce that ectoderm to form lens tissue. Only the optic vesicle appears to be able to do this; therefore, it is an inducer. However, if the optic vesicle is placed beneath ectoderm in the *flank* or *abdomen* of the same organism, that ectoderm will not be able to form lens tissue. Only the head ectoderm is *competent* to respond to the signals from the optic vesicle by producing a lens* (Figure 6.1; Saha et al. 1989; Grainger 1992).

This ability to respond to a specific inductive signal is called **competence** (Waddington 1940). Competence is not a passive state but an actively acquired condition. For example, in the developing mammalian eye, the Pax6 protein appears to be important in making the ectoderm competent to respond to the inductive signal from the optic vesicle. *Pax6* gene expression is seen in the head ectoderm, which can respond to the optic vesicle by forming a lens; it is not seen in other regions of the surface ectoderm (see Figure 4.16; Li et al. 1994). The importance of Pax6 as a **competence factor** was demonstrated by recombination experiments using embryonic rat eye tissue (Fujiwara et al. 1994). The homo-

zygous *Pax6* mutant rat has a phenotype similar to the homozygous *Pax6* mutant mouse (see Chapter 4), lacking eyes and nose (Figure 6.2). It has been shown that part of this phenotype is due to the failure of lens induction. But

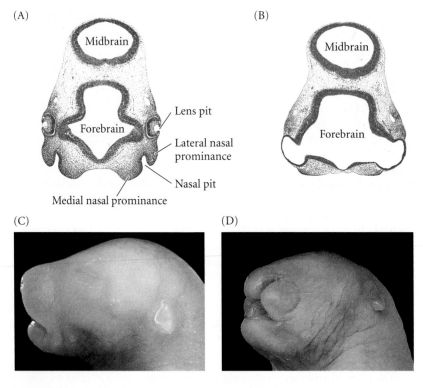

(A) Midbrain / Forebrain / Medial nasal prominance / Lens pit / Lateral nasal prominance / Nasal pit

(B) Midbrain / Forebrain

(C)

(D)

FIGURE 6.2 Induction of optic and nasal structures by Pax6 in the rat embryo. (A, B) Histology of wild-type (A) and homozygous *Pax6* mutant (B) embryos at day 12 of gestation shows induction of lenses and retinal development in the wild-type embryo, but not in the mutant. Similarly, neither the nasal pit nor the medial nasal prominence is induced in the mutant rats. (C) Newborn wild-type rats show prominent nose as well as (closed) eyes. (D) Newborn *Pax6* mutant rats show neither eyes nor nose. (From Fujiwara et al. 1994; photographs courtesy of M. Fujiwara.)

---

*When describing lens induction, one has to be careful to mention which species one is studying, because there are numerous species-specific differences. In some species, lens induction will not occur at certain temperatures. In other species, the entire ectoderm can respond to the optic vesicle by forming lenses. Such species-specific differences have made this area very difficult to study (Jacobson and Sater 1988; Saha et al. 1989; Saha 1991).

which is the defective component—the optic vesicle or the surface ectoderm?

Recombination experiments revealed that when head ectoderm from *Pax6*-mutant rat embryos was combined with a wild-type optic vesicle, no lenses were formed. However, when the head ectoderm from wild-type rat embryos was combined with a *Pax6*-mutant optic vesicle, lenses formed normally (Figure 6.3). Therefore, Pax6 is needed for the surface ectoderm to respond to the inductive signal from the optic vesicle; the inducing tissue does not need it. It is not known how Pax6 becomes expressed in the anterior ectoderm of the embryo, although it is thought that its expression is induced by the anterior regions of the neural plate. Competence to respond to the optic vesicle inducer (and *Pax6* expression) can be conferred on ectodermal tissue by incubating it next to anterior neural plate tissue (Henry and Grainger 1990; Li et al. 1994; Zygar et al. 1998).

There is no single inducer of the lens. Studies on amphibians suggest that the first inducers may be the pharyngeal endoderm and heart-forming mesoderm that underlie the lens-forming ectoderm during the early- and mid-gastrula stages (Jacobson 1963, 1966). The anterior neural plate may produce the next signals, including a signal that promotes the synthesis of Pax6 in the anterior ectoderm (Figure 6.4; Zygar et al. 1998). Thus, although the optic vesicle appears to be *the* inducer, the anterior ectoderm has already been induced by at least two other factors. (The situation is like that of the player who kicks the "winning goal" of a soccer match.) The optic vesicle appears to secrete two induction factors, one of which may be BMP4 (Furuta and Hogan 1998), a protein that induces the production of the Sox2 and Sox3 transcription factors. The other is thought to be Fgf8, a signal that induces the appearance of the L-Maf transcription factor (Ogino and Yasuda 1998; Vogel-Höpker et al. 2000). The combination of Pax6, Sox2, Sox3, and L-Maf in the ectoderm ensures the production of the lens.

Optic vesicles	Surface ectoderm	Lens induction
Wild-type	Wild-type	Yes
Pax6⁻/Pax6⁻	Wild-type	Yes
Wild-type	*Pax6⁻/Pax6⁻*	No
Pax6⁻/Pax6⁻	*Pax6⁻/Pax6⁻*	No

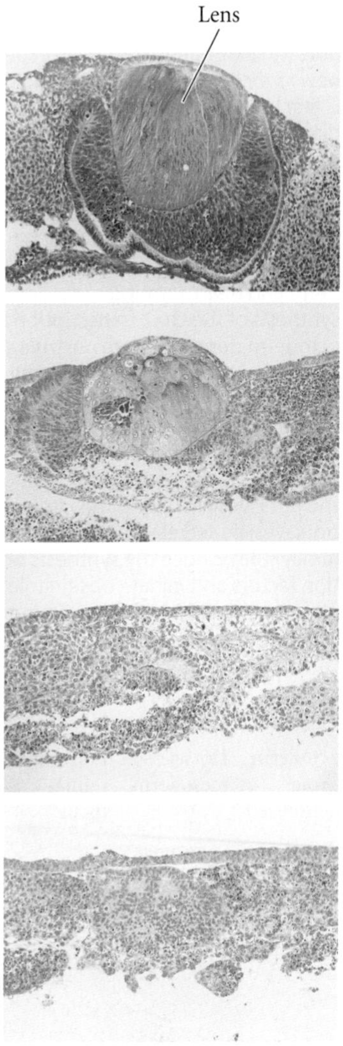

FIGURE 6.3   Recombination experiments show that the induction deficiency of *Pax6*-deficient rats is caused by the inability of the surface ectoderm to respond to the optic vesicle. (Photographs courtesy of M. Fujiwara.)

Cascades of induction: Reciprocal and sequential inductive events

Another feature of induction is the reciprocal nature of many inductive interactions. Once the lens has formed, it can then induce other tissues. One of these responding tissues is the optic vesicle itself—the inducer becomes the induced. Under the influence of factors secreted by the lens, the optic vesicle becomes the optic cup and the wall of the optic cup differentiates into two layers, the pigment-

ed retina and the neural retina (Figure 6.5; Cvekl and Piatigorsky 1996). Such interactions are called **reciprocal inductions**.

At the same time, the lens is inducing the ectoderm above it to become the cornea. Like the lens-forming ectoderm, the cornea-forming ectoderm has achieved a particular competence to respond to inductive signals, in this case the signals from the lens (Meier 1977; Thut et al. 2001). Under the influence of the lens, the corneal ectoderm cells become columnar and secrete multiple layers of collagen. Mesenchymal cells from the neural crest use this collagen matrix to enter the area and secrete a set of proteins (including the enzyme hyaluronidase) that further differentiate the cornea. A third signal, the hormone thyroxine, dehydrates the tissue and makes it transparent (Hay 1980;

FIGURE 6.4 Lens induction in amphibians. (A) The additive effects of inducers, as shown by transplantation and extirpation (removal) experiments on the newt *Taricha torosa*. The ability to produce lens tissue is first induced by pharyngeal endoderm, then by cardiac mesoderm, and finally by the optic vesicle. The optic vesicle eventually acquires the ability to induce the lens and retain its differentiation. (B) Sequence of induction postulated by similar experiments performed on embryos of the frog *Xenopus laevis*. Unidentified inducers (possibly from the pharyngeal endoderm and heart-forming mesoderm) cause the synthesis of the Otx2 transcription factor in the head ectoderm during the late gastrula stage. As the neural folds rise, inducers from the anterior neural plate (including the region that will form the retina) induce *Pax6* expression in the anterior ectoderm that can form lens tissue. Expression of Pax6 protein may constitute the competence of the surface ectoderm to respond to the optic vesicle during the late neurula stage. The optic vesicle secretes factors (probably of the BMP family) that induce the synthesis of the Sox transcription factors and initiate observable lens formation. (A after Jacobson 1966; B after Grainger 1992.)

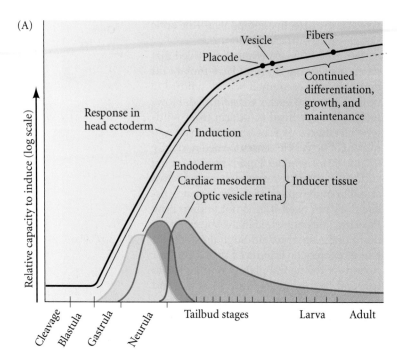

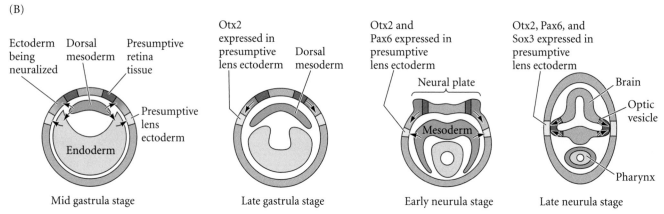

Bard 1990). Thus, there are sequential inductive events, and multiple causes for each induction.

Another principle can be seen in such reciprocal inductions: a structure does not need to be fully differentiated in order to have a function. The optic vesicle induces before it becomes the retina; the lens placode reciprocates by inducing the optic vesicle before the lens forms its characteristic fibers.

Instructive and permissive interactions

Howard Holtzer (1968) distinguished two major modes of inductive interaction. In **instructive interaction**, a signal from the inducing cell is necessary for initiating new gene expression in the responding cell. Without the inducing cell, the responding cell is not capable of differentiating in that particular way. For example, when the optic vesicle is experimentally placed under a new region of the head ectoderm and causes that region of the ectoderm to form a lens, that is an instructive interaction. Wessells (1977) has proposed three general principles characteristic of most instructive interactions:

1. In the presence of tissue A, responding tissue B develops in a certain way.
2. In the absence of tissue A, responding tissue B does not develop in that way.
3. In the absence of tissue A, but in the presence of tissue C, tissue B does not develop in that way.

The second type of inductive interaction is **permissive interaction**. Here, the responding tissue has already been specified, and needs only an environment that allows the expression of these traits. For instance, many tissues need a solid substrate containing fibronectin or laminin in order to develop. The fibronectin or laminin does not alter the type of cell that is produced, but it enables what has already been determined to be expressed.*

---

*It is easy to distinguish permissive and instructive interactions by an analogy with a more familiar situation. This textbook is made possible by both permissive and instructive interactions. A reviewer can convince me to change the material in the chapters. This is an instructive interaction, as the information expressed in the book is changed from what it would have been. However, the information in the book could not be expressed at all without permissive interactions with the publisher and printer.

Epithelial-mesenchymal interactions

Some of the best-studied cases of induction are those involving the interactions of sheets of epithelial cells with adjacent mesenchymal cells. **Epithelia** are sheets or tubes of connected cells; they can originate from any germ layer. **Mesenchyme** refers to loosely packed, unconnected cells. Mesenchymal cells are derived from the mesoderm or neural crest. All organs consist of an epithelium and an associated mesenchyme, so these **epithelial-mesenchymal interactions** are among the most important phenomena in nature. Some examples are listed in Table 6.1.

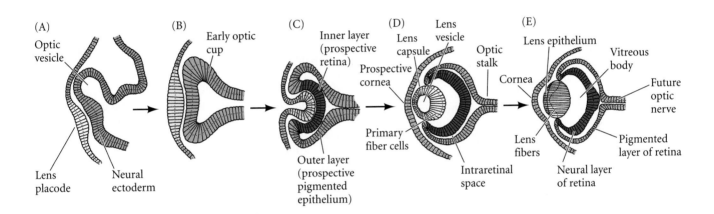

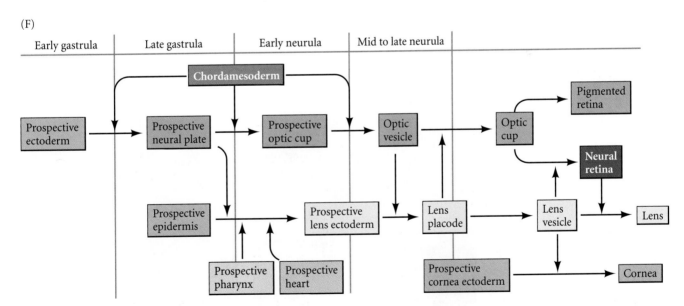

FIGURE 6.5 Schematic diagram of the induction of the mouse lens. (A) At embryonic day 9, the optic vesicle extends toward the surface ectoderm from the forebrain. The lens placode (the prospective lens) appears as a local thickening of the surface ectoderm near the optic vesicle. (B) By the middle of day 9, the lens placode has enlarged and the optic vesicle has formed an optic cup. (C) By the middle of day 10, the central portion of the lens-forming ectoderm invaginates, while the two layers of the retina become distinguished. (D) By the middle of day 11, the lens vesicle has formed. (E) By day 13, the lens consists of anterior cuboidal epithelial cells and elongating posterior fiber cells. The cornea develops in front of the lens. (F) Summary of some of the inductive interactions during eye development. (A–E after Cvekl and Piatigorsky 1996.)

TABLE 6.1 Some epithelial-mesenchymal interactions

Organ	Epithelial component	Mesenchymal component
Cutaneous structures (hair, feathers, sweat glands, mammary glands)	Epidermis (ectoderm)	Dermis (mesoderm)
Limb	Epidermis (ectoderm)	Mesenchyme (mesoderm)
Gut organs (liver, pancreas, salivary glands)	Epithelium (endoderm)	Mesenchyme (mesoderm)
Pharyngeal and respiratory associated organs (lungs, thymus, thyroid)	Epithelium (endoderm)	Mesenchyme (mesoderm)
Kidney	Ureteric bud epithelim (mesoderm)	Mesenchyme (mesoderm)
Tooth	Jaw epithelium (ectoderm)	Mesenchyme (neural crest)

REGIONAL SPECIFICITY OF INDUCTION Using the induction of cutaneous (skin) structures as our examples, we will look at the properties of epithelial-mesenchymal interactions. The first of these properties is the regional specificity of induction. Skin is composed of two main tissues: an outer epidermis (an epithelial tissue derived from ectoderm), and a dermis (a mesenchymal tissue derived from mesoderm). The chick epidermis secretes proteins that signal the underlying dermal cells to form condensations, and the condensed dermal mesenchyme responds by secreting factors that cause the epidermis to form regionally specific cutaneous structures (Figure 6.6; Nohno et al. 1995, Ting-Berreth and Chuong 1996). These structures can be the broad feathers of the wing, the narrow feathers of the thigh, or the scales and claws of the feet. As Figure 6.7 demonstrates, the dermal mesenchyme is responsible for the regional specificity of induction in the competent epidermal epithelium. Researchers can separate the embryonic epithelium and mesenchyme from each other and recombine them in different ways (Saunders et al. 1957). The same epithelium develops cutaneous structures according to the region from which the mesenchyme was taken. Here, the mesenchyme plays an instructive role, calling into play different sets of genes in the responding epithelial cells.

GENETIC SPECIFICITY OF INDUCTION The second property of epithelial-mesenchymal interactions is the genetic specificity of induction. Whereas the mesenchyme may instruct the epithelium as to what sets of genes to activate, the responding epithelium can comply with these instructions only so far as its genome permits. This property was discovered through experiments involving the transplantation of tissues from one species to another. In one of the most dramatic examples of interspecific induction, Hans Spemann and Oscar Schotté (1932) transplanted flank ectoderm from an early *frog* gastrula to the region of a *newt* gastrula destined to become parts of the mouth. Similarly, they placed presumptive flank ectodermal tissue from a *newt*

(A)

(B)

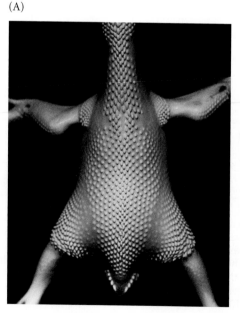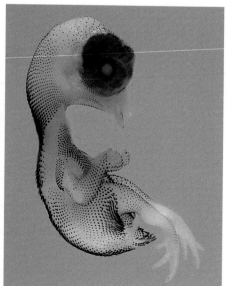

FIGURE 6.6 Feather induction in the chick. (A) Feather tracts on the dorsum of a day 9 chick embryo. Note that each feather primordium is located between the primordia of adjacent rows. (B) In situ hybridization of a day 10 chick embryo shows *Sonic hedgehog* expression (dark spots) in the ectoderm of the developing feathers and scales. (A courtesy of P. Sengal; B courtesy of W.-S. Kim and J. F. Fallon.)

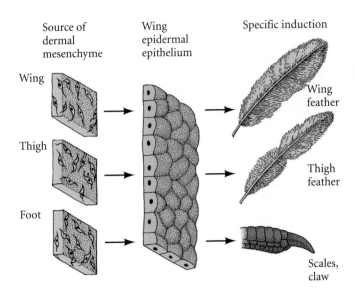

Source of dermal mesenchyme

Wing epidermal epithelium

Specific induction

Wing

Thigh

Foot

Wing feather

Thigh feather

Scales, claw

FIGURE 6.7 Regional specificity of induction in the chick. When cells from different regions of the dermis (mesenchyme) are recombined with the epidermis (epithelium), the type of cutaneous structure made by the epidermal epithelium is determined by the original source of the mesenchyme. (After Saunders 1980.)

WEBSITE **6.1** **Hen's teeth.** Some inductive events between species can bring forth lost structures. Mouse molar mesenchyme may be able to induce teeth in the bird jaw.

Paracrine Factors: The Inducer Molecules

How are the signals between inducer and responder transmitted? While studying the mechanisms of induction that produce the kidney tubules and teeth, Grobstein (1956) and others (Saxén et al. 1976; Slavkin and Bringas 1976) found that some inductive events could occur despite a filter separating the epithelial and mesenchymal cells (Figure 6.9A). Other inductions, however, were blocked by the filter. The researchers therefore concluded that some of the inductive molecules were soluble factors that could pass through the small pores of the filter, and that other inductive events required physical contact between the epithelial and mesenchymal cells. When cell membrane proteins on one cell surface interact with receptor proteins on adjacent cell surfaces, these events are called **juxtacrine interactions** (since the cell membranes are *juxtaposed*) (Figure 6.9B). When pro-

gastrula into the presumptive oral regions of *frog* embryos. The structures of the mouth region differ greatly between salamander and frog larvae. The salamander larva has club-shaped balancers beneath its mouth, whereas the frog tadpole produces mucus-secreting glands and suckers (Figure 6.8). The frog tadpole also has a horny jaw without teeth, whereas the salamander has a set of calcareous teeth in its jaw. The larvae resulting from the transplants were chimeras. The salamander larvae had froglike mouths, and the frog tadpoles had salamander teeth and balancers. In other words, the mesenchymal cells instructed the ectoderm to make a mouth, but the ectoderm responded by making the only kind of mouth it "knew" how to make, no matter how inappropriate.*

Thus the instructions sent by the mesenchymal tissue can cross species barriers. Salamanders respond to frog inducers, and chick tissue responds to mammalian inducers. The response of the epithelium, however, is species-specific. So, whereas organ-type specificity (e.g., feather or claw) is usually controlled by the mesenchyme within a species, species-specificity is usually controlled by the responding epithelium. As we will see in Chapter 23, major evolutionary changes in the phenotype can be brought about by changing the response to a particular inducer.

---

*Spemann is reported to have put it this way: "The ectoderm says to the inducer, 'you tell me to make a mouth; all right, I'll do so, but I can't make your kind of mouth; I can make my own and I'll do that'" (quoted in Harrison 1933).

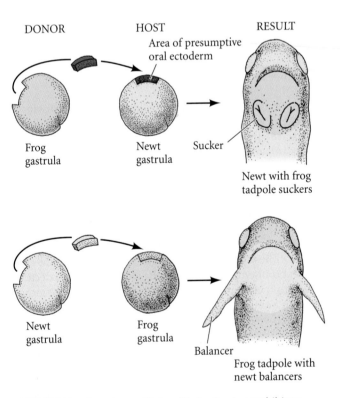

DONOR

HOST

RESULT

Area of presumptive oral ectoderm

Frog gastrula

Newt gastrula

Sucker

Newt with frog tadpole suckers

Newt gastrula

Frog gastrula

Balancer

Frog tadpole with newt balancers

FIGURE 6.8 Genetic specificity of induction in amphibians. Reciprocal transplantation between the presumptive oral ectoderm regions of salamander and frog gastrulae leads to newts with tadpole suckers and tadpoles with newt balancers. (After Hamburgh 1970.)

FIGURE 6.9   Mechanisms of inductive interaction. (A) Presumptive mouse lens ectoderm and mesenchyme were placed on a filter with presumptive retinal tissue placed beneath it. After 3 days, a lens had developed from the surface ectoderm. In the absence of a paracrine signal from the retinal tissue, the surface ectoderm would have become epidermal. (B–D) Juxtacrine and paracrine modes of signaling. (B) In juxtacrine interactions, contact is made between a signaling molecule on the surface of one cell and its receptor on another cell. (C) Paracrine modes of signaling involve the secretion of diffusible molecules from one cell and their reception by a nearby cell. (D) In some cases, the paracrine signal can come from an extracellular matrix protein secreted by a cell. (A from Muthukkarapan 1965, photograph courtesy of R. Auerbach; B–D after Grobstein 1956.)

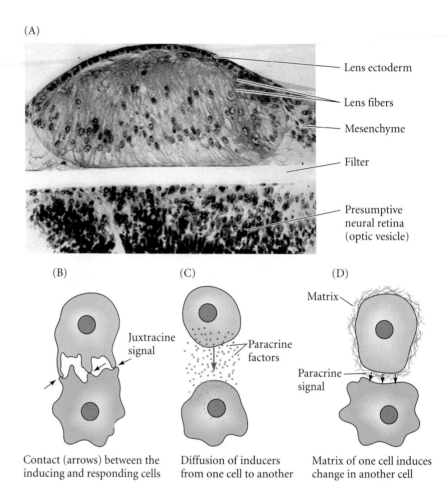

(A)

Lens ectoderm
Lens fibers
Mesenchyme
Filter
Presumptive neural retina (optic vesicle)

(B)

Juxtacrine signal

Contact (arrows) between the inducing and responding cells

(C)

Paracrine factors

Diffusion of inducers from one cell to another

(D)

Matrix

Paracrine signal

Matrix of one cell induces change in another cell

teins synthesized by one cell can diffuse over small distances to induce changes in neighboring cells, the event is called a **paracrine interaction** (Figure 6.9C,D). The diffusible proteins are called **paracrine factors** or **growth and differentiation factors** (GDFs). We will consider paracrine interactions first and return to juxtacrine interactions later in the chapter.

Whereas **endocrine factors** (hormones) travel through the blood to exert their effects, paracrine factors are secreted into the immediate spaces around the cell producing them.* These proteins are the "inducing factors" of the classic experimental embryologists. During the past decade, developmental biologists have discovered that the induction of numerous organs is actually effected by a relatively small set of paracrine factors. The embryo inherits a

rather compact "tool kit" and uses many of the same proteins to construct the heart, the kidneys, the teeth, the eyes, and other organs. Moreover, the same proteins are utilized throughout the animal kingdom; the factors active in creating the *Drosophila* eye or heart are very similar to those used in generating mammalian organs. Many of the paracrine factors can be grouped into one of four major families on the basis of their structure:

1. The fibroblast growth factor (FGF) family
2. The Hedgehog family
3. The Wingless, or Wnt, family
4. The TGF-β superfamily, encompassing the TGF-β family, the activin family, the bone morphogenetic proteins (BMPs), the Vg1 family, and several other proteins

In addition to endocrine, paracrine, and juxtacrine regulation, there is also autocrine regulation. Autocrine regulation occurs when the same cells that secrete paracrine factors also respond to them. In other words, the cell synthesizes a molecule for which it has its own receptor. Although autocrine regulation is not common, it is seen in placental cytotrophoblast cells; these cells synthesize and secrete platelet-derived growth factor, whose receptor is on the cytotrophoblast cell membrane (Goustin et al. 1985). The result is the explosive proliferation of that tissue.

---

*There is considerable debate as to the distances at which paracrine factors can operate. The proteins Nodal and activin, for instance, can diffuse over many cell diameters and can induce different sets of genes at different concentrations (Gurdon et al. 1994, 1995; see Chapter 3). The Wnt, Vg1, and BMP4 proteins, however, probably work only on their adjacent neighbors (Jones et al. 1996; Reilly and Melton 1996). These factors may induce the expression of other short-range factors from these neighbors, and a cascade of paracrine inductions can be initiated.

Signal Transduction Cascades: The Response to Inducers

Paracrine factors function by binding to a receptor that initiates a series of enzymatic reactions within the cell. These enzymatic reactions have as their end point either the regulation of transcription factors (such that different genes are expressed in the cells reacting to these paracrine factors) or the regulation of the cytoskeleton (such that the cells responding to the paracrine factors alter their shape or are permitted to migrate). These pathways of responses to the paracrine factor often have several end-points and are called **signal transduction cascades**.

The major signal transduction pathways all appear to be variations on a common and rather elegant theme, exemplified by Figure 6.10. Each receptor spans the cell membrane and has an extracellular region, a transmembrane region, and a cytoplasmic region. When a ligand (the paracrine factor) binds to its receptor in the extracellular region, that ligand induces a conformational change in the receptor's structure. This shape change is transmitted through the membrane and changes the shape of the cytoplasmic domains. The conformational change in the cytoplasmic domains gives them enzymatic activity—usually a kinase activity that can use ATP to phosphorylate specific tyrosine residues of particular proteins. Thus, this type of receptor is often called a **receptor tyrosine kinase**. The active receptor can now catalyze reactions that phosphorylate other proteins, and this phosphorylation in turn activates their latent activities. Eventually, the cascade of phosphorylation activates a dormant transcription factor or cytoskeletal protein.

Transcription factors have been divided into three catagories (Brivanlou and Darnell 2002). The first group includes those transcription factors that are constitutively present in all cells. These include the basal transcription factors such as Sp1. The second group includes those transcription factors that are active whenever a cell acquires them by cytoplasmic localization (e.g., Bicoid protein) or by induction (Pax6). The third group includes those transcription factors (such as MITF) whose functions are activated by cell signal transduction cascades.

Fibroblast growth factors and the RTK pathway

The **fibroblast growth factor** (**FGF**) gene family comprises nearly two dozen structurally related members, and these genes can generate hundreds of protein isoforms by varying their RNA splicing or initiation codons in different tissues (Lappi 1995). Fgf1 is also known as acidic FGF and appears to be important during regeneration (Yang et al. 2005); Fgf2 is sometimes called basic FGF and is very important in blood vessel formation; and Fgf7 sometimes goes by the name of keratinocyte growth factor and is critical in skin development. Although FGFs can often substitute for one another, the expression patterns of the FGFs and their receptors give them separate functions.

One member of this family, Fgf8, is especially important during limb development and lens induction. Fgf8 is usually made by the optic vesicle that contacts the outer ectoderm of the head (Figure 6.11; Vogel-Höpker et al. 2000). After contact with the outer ectoderm occurs, *fgf8* gene expression becomes concentrated in the region of the presumptive neural retina—the tissue directly apposed to the presumptive lens. Moreover, if Fgf8-containing beads are placed adjacent to head ectoderm, this ectopic Fgf8 will induce this ectoderm to produce ectopic lenses and to express the lens-associated transcription factor L-Maf (see Figure 6.11B).

FGFs often work by activating a set of receptor tyrosine kinases called the **fibroblast growth factor receptors** (FGFRs). When an FGF receptor binds an FGF (and only when it binds an FGF), the dormant kinase is activated and phosphorylates certain proteins (including other FGF receptors) within the responding cell. These proteins, once activated, can perform new functions. The **RTK pathway** was one of the first signal transduction pathways to unite various areas of developmental biology (Figure 6.12). Researchers studying *Drosophila* eyes, nematode vulvae, and human cancers found that they were all studying the same genes. The pathway begins at the cell surface, where a receptor tyrosine kinase (RTK) binds its specific ligand. Ligands that bind to RTKs include the fibroblast growth factors, epidermal growth factors, platelet-derived growth factors, and stem cell factor. Each RTK can bind only one or a small set of these ligands (the Kit RTK, for example, can bind only one ligand—stem cell factor). The RTK spans the cell membrane, and when it binds its ligand, it undergoes a conformational change that enables it to dimerize

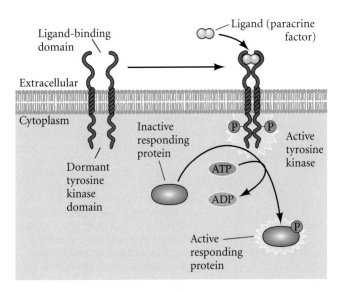

FIGURE 6.10 Structure and function of a receptor tyrosine kinase. The binding of a paracrine factor (such as Fgf8) by the extracellular portion of the receptor protein activates the dormant tyrosine kinase, whose enzyme activity phosphorylates specific tyrosine residues of certain proteins.

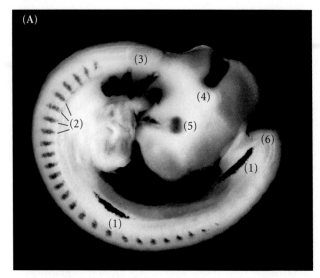

(A)

(3)

(4)

(5)

(2)

(6)

(1)

(1)

(B)

(C)

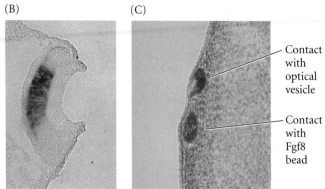

Contact with optical vesicle

Contact with Fgf8 bead

FIGURE 6.11 Fgf8 in the developing chick. (A) *Fgf8* gene expression pattern in the 3-day chick embryo, shown by in situ hybridization. Fgf8 protein (dark areas) is seen in the distalmost limb bud ectoderm (1); in the somitic mesoderm (the segmented blocks of cells along the anterior-posterior axis (2) in the branchial arches of the neck (3); at the boundary between the midbrain and hindbrain (4); in the developing eye (5); and in the tail (6). (B,C) Fgf8 function in the developing eye. (B) In situ hybridization of *fgf8* in the optic vesicle. The *fgf8* mRNA (purple) is localized to the presumptive neural retina of the optic cup and is in direct contact with the outer ectoderm cells that will become the lens. (C) Ectopic expression of L-Maf in competent ectoderm can be induced by the optic vesicle (above) and by an Fgf8-containing bead (below). (A photograph courtesy of E. Laufer, C.-Y. Yeo, and C. Tabin; B,C photographs courtesy of A. Vogel-Höpker.)

with another RTK. This conformational change activates the latent kinase activity of each RTK, and these receptors phosphorylate each other on particular tyrosine residues (see Figure 6.10). Thus, the binding of the ligand to the receptor causes the autophosphorylation of the cytoplasmic domain of the receptor.

The phosphorylated tyrosine on the receptor is then recognized by an adaptor protein. The adaptor protein serves

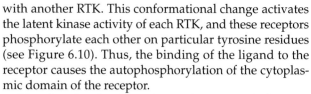

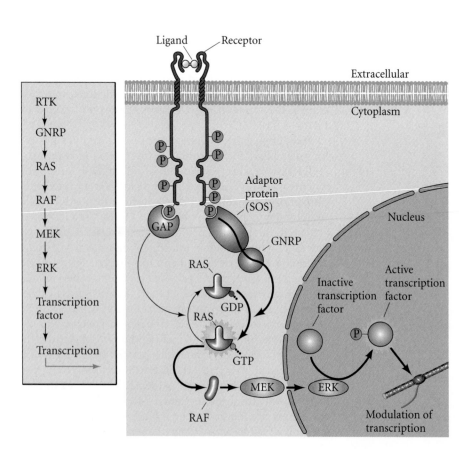

RTK
↓
GNRP
↓
RAS
↓
RAF
↓
MEK
↓
ERK
↓
Transcription factor
↓
Transcription

Ligand   Receptor

Extracellular

Cytoplasm

Adaptor protein (SOS)

GAP

GNRP

RAS

RAS

GDP

RAS

GTP

RAF

MEK

ERK

Nucleus

Inactive transcription factor

Active transcription factor

Modulation of transcription

FIGURE 6.12 The widely used RTK signal transduction pathway. The receptor tyrosine kinase is dimerized by the ligand, which causes the autophosphorylation of the receptor. The adaptor protein recognizes the phosphorylated tyrosines on the RTK and activates an intermediate protein, GNRP, which activates the Ras G protein by allowing the phosphorylation of the GDP-bound Ras. At the same time, the GAP protein stimulates the hydrolysis of this phosphate bond, returning Ras to its inactive state. The active Ras activates the Raf protein kinase C (PKC), which in turn phosphorylates a series of kinases. Eventually, the activated kinase ERK alters gene expression in the nucleus of the responding cell by phosphorylating certain transcription factors (which can then enter the nucleus to change the types of genes transcribed) and certain translation factors (which alter the level of protein synthesis). In many cases, this pathway is reinforced by the release of calcium ions. A simplified version of the pathway is depicted on the left.

as a bridge that links the phosphorylated RTK to a powerful intracellular signaling system. While binding to the phosphorylated RTK through one of its cytoplasmic domains, the adaptor protein also activates a **G protein**, such as **Ras**. Normally, the G protein is in an inactive, GDP-bound state. The activated receptor stimulates the adaptor protein to activate the **guanine nucleotide releasing factor (GNRP)**. This protein exchanges a phosphate from a GTP to transform the bound GDP into GTP. The GTP-bound G protein is an active form that transmits the signal to the next molecule. After the signal is delivered, the GTP on the G protein is hydrolyzed back into GDP. This catalysis is greatly stimulated by the complexing of the Ras protein with the **GTPase-activating protein (GAP)**. In this way, the G protein is returned to its inactive state, where it can await further signaling. Without the GAP protein,

Ras protein cannot catalyze GTP well, and so remains in its active configuration (Cales et al. 1988; McCormick 1989). Mutations in the *RAS* gene account for a large proportion of cancerous human tumors (Shih and Weinberg 1982), and the mutations of *RAS* that make it oncogenic all inhibit the binding of the GAP protein

The active Ras G protein associates with a kinase called Raf. The G protein recruits the inactive Raf protein to the cell membrane, where it becomes active (Leevers et al. 1994; Stokoe et al. 1994). The Raf protein is a kinase that activates the MEK protein by phosphorylating it. MEK is itself a kinase, which activates the ERK protein by phosphorylation. In turn, ERK is a kinase that enters the nucleus and phosphorylates certain transcription factors.

The RTK pathway is critical in numerous developmental processes. In the migrating neural crest cells of humans and mice, the pathway is important in activating the microphthalmia transcription factor (Mitf) to produce the pigment cells (Figure 6.13). Mitf, whose mechanism of action was described in Chapter 5, is transcribed in the pigment-forming melanoblast cells that migrate from the neural crest into the skin and in the melanin-forming cells of the pigmented retina. But we have not yet discussed what proteins signal this transcription factor to become active.

(A)

(B)

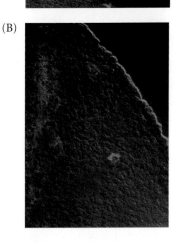

(C)

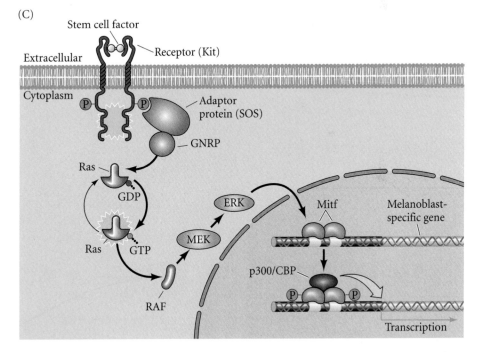

FIGURE 6.13 Activation of the Mitf transcription factor through the binding of stem cell factor by the Kit RTK protein. The information received at the cell membrane is sent to the nucleus by the RTK signal transduction pathway. (A,B) Demonstration that Kit protein and Mitf are present in the same cells. Antibodies to these proteins stain the Kit protein (red) and Mitf (green). The overlap is yellow or yellow-green. Both proteins are present in the migrating melanocyte precursor cells (melanoblasts). (A) Migrating melanoblasts can be seen in a wild-type mouse embryo at day 10.5. (B) No melanoblasts are visible in a *microphthalmia* mutant embryo of the same age. The lack of melanoblasts in the mutant is due to the relative absence of Mitf. (C) Signal transduction pathway leading from the cell membrane to the nucleus. When the receptor domain of the Kit RTK protein binds the stem cell factor, Kit dimerizes and becomes phosphorylated. This phosphorylation is used to activate the Ras G protein, which activates the chain of kinases that will phosphorylate the Mitf protein. Once phosphorylated, Mitf can bind the cofactor p300/CBP, acetylate the nucleosome histones, and initiate transcription of the genes needed for melanocyte development. (A,B from Nakayama et al. 1998, photographs courtesy of H. Arnheiter; C after Price et al. 1998.)

The clue lies in two mouse mutants whose phenotypes resemble those of mice homozygous for *microphthalmia* mutations. Like *Mitf* mutant mice, homozygous *White* mice and homozygous *Steel* mice are white because their pigment cells have failed to migrate. Could it be that all three genes (*Mitf*, *Steel*, and *White*) are on the same developmental pathway?

In 1990, several laboratories demonstrated that the *Steel* gene encodes a paracrine protein called **stem cell factor** (see Witte 1990). Stem cell factor binds to and activates the **Kit** receptor tyrosine kinase encoded by the *White* gene (Spritz et al. 1992; Wu et al. 2000). The binding of stem cell factor to the Kit RTK dimerizes the Kit protein, causing it to become phosphorylated. The phosphorylated Kit activates the pathway whereby phosphorylated ERK is able to phosphorylate the Mitf transcription factor (Hsu et al. 1997; Hemesath et al. 1998). Only the phosphorylated form of

Mitf is able to bind the p300/CBP histone acetyltransferace protein that enables it to activate transcription of the genes encoding tyrosinase and other proteins of the melanin-formation pathway (see Figure 6.13; Price et al. 1998).

 6.2 FGF binding. The binding of FGFs to their receptors is a complex acrobatic act involving an interesting cast of cell surface molecules. Glycoproteins play a major supporting role in this event.

The JAK-STAT pathway

Fibroblast growth factors can also activate the JAK-STAT cascade. This pathway is extremely important in the differentiation of blood cells, the growth of limbs, and in the activation of the casein gene during milk production (Fig-

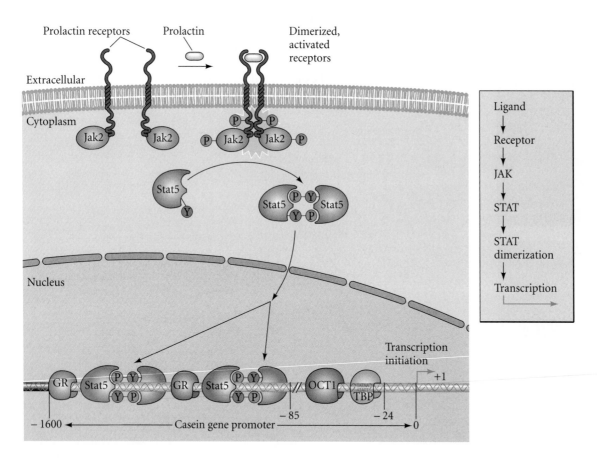

FIGURE 6.14   A STAT pathway: the casein gene activation pathway activated by prolactin. The casein gene is activated during the last (lactogenic) phase of mammary gland development, and its signal is the secretion of the hormone prolactin from the anterior pituitary gland. Prolactin causes the dimerization of prolactin receptors in the mammary duct epithelial cells. A particular JAK protein (Jak2) is "hitched" to the cytoplasmic domain of these receptors. When the receptors bind prolactin and dimerize, the JAK proteins phosphorylate each other and the dimerized receptors, activating the dormant kinase activity of the receptors. The activated receptors add a phosphate

group to a tyrosine residue (Y) of a particular STAT protein—in this case, Stat5. This allows Stat5 to dimerize, be translocated into the nucleus, and bind to particular regions of DNA. In combination with other transcription factors (which presumably have been waiting for its arrival), the Stat5 protein activates transcription of the casein gene. GR is the glucocorticoid receptor, OCT1 is a general transcription factor, and TBP is the TATA-binding protein (see Chapter 5) responsible for binding RNA polymerase. A simplified diagram is shown to the right. (For details, see Groner and Gouilleux 1995.)

ure 6.14; Briscoe et al. 1994; Groner and Gouilleux 1995). Here, the ligand is bound by either FGF receptors or other receptors that are linked to JAK (*Janus kinase*) proteins. The binding of ligand to the receptor phosphorylates the **STAT** (*signal transducers and activators of transcription*) family of transcription factors (Ihle 1996; 2001). The STAT pathway is very important in the regulation of human fetal bone growth. Mutations that prematurely activate the STAT pathway have been implicated in some severe forms of dwarfism, such as the lethal **thanatophoric dysplasia**, wherein the growth plates of the rib and limb bones fail to proliferate. The short-limbed newborn dies because its ribs cannot support breathing. The genetic lesion responsible is in the gene encoding fibroblast growth factor receptor 3 (FgfR3) (Figure 6.15; Rousseau et al. 1994; Shiang et al. 1994). This protein is expressed in the cartilage precursor cells—known as **chondrocytes**—in the growth plates of the long bones. Normally, the FgfR3 protein (a receptor tyrosine kinase) is activated by a fibroblast growth factor, and it signals the chondrocytes to stop dividing and begin differentiating into cartilage. This signal is mediated by the Stat1 protein, which is phosphorylated by activated FgfR3 and then translocated into the nucleus. Inside the nucleus, Stat1 activates the genes encoding a cell cycle inhibitor, the p21 protein (Su et al. 1997). Thus, the mutations causing thanatophoric dwarfism result from a gain-of-function phenotype, wherein the mutant FgfR3 is active constitutively—that is, without the need to be activated by an FGF (Deng et al. 1996; Webster and Donoghue 1996). The chondrocytes stop proliferating shortly after they are formed, and the bones fail to grow. Other mutations that activate FgfR3 prematurely but to a lesser degree produce **achon-**

WEBSITE **6.3 FgfR mutations.** Mutations of the human FGF receptors have been associated with several skeletal malformation syndromes, including syndromes wherein skull cartilage, rib cartilage, or limb cartilage fails to grow or differentiate.

droplasic (short-limbed) dwarfism, the most prevalent of the human dominant syndromes (Legeai-Mallet et al. 2004).

The Hedgehog family

The proteins of the Hedgehog family of paracrine factors are often used by the embryo to induce particular cell types and to create boundaries between tissues. Hedgehog proteins are processed such that only the amino-terminal two-thirds of the molecule is secreted; once this takes place, the protein must become complexed with a molecule of cholesterol in order to function. Vertebrates have at least three homologues of the *Drosophila hedgehog* gene: *sonic hedgehog* (*shh*), *desert hedgehog* (*dhh*), and *indian hedgehog* (*ihh*). The Desert hedgehog protein is expressed in the Sertoli cells of the testes, and mice homozygous for a null allele of *dhh* exhibit defective spermatogenesis. Indian hedgehog protein is expressed in the gut and cartilage and is important in postnatal bone growth (Bitgood and McMahon 1995; Bitgood et al. 1996).

Sonic hedgehog* has the greatest number of functions of the three vertebrate Hedgehog homologues (Figure 6.16). Among other important functions, this paracrine factor is

---

*Yes, it is named after the Sega Genesis character. The original *hedgehog* gene was found in *Drosophila*, in which genes are named after their mutant phenotypes. The loss-of-function *hedgehog* mutation in *Drosophila* causes the fly embryo to be covered with pointy denticles on its cuticle; hence, it looks like a hedgehog. The vertebrate *hedgehog* genes were discovered by searching vertebrate gene libraries (chick, rat, zebrafish) with probes that would find sequences similar to that of the fruit fly *hedgehog* gene. Riddle and his colleagues (1993) discovered three genes homologous to *Drosophila hedgehog*. Two were named after existing species of hedgehogs; the third was named after the animated character. In fish, two other hedgehog genes are named *echidna hedgehog* (after the spiny Australian marsupial mammal) and *Tiggywinkle hedgehog* (after Beatrix Potter's fictional hedgehog).

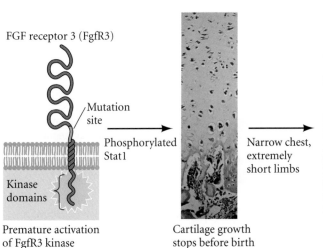

FGF receptor 3 (FgfR3)

Mutation site

Kinase domains

Premature activation of FgfR3 kinase

Phosphorylated Stat1

Cartilage growth stops before birth

Narrow chest, extremely short limbs

Thanatophoric dysplasia

FIGURE 6.15 A mutation in the gene for FgfR3 causes the premature constitutive activation of the STAT pathway and the production of phosphorylated Stat1 protein. This transcription factor activates genes that cause the premature termination of chondrocyte cell division. The result is thanatophoric dysplasia, a condition of failed bone growth that results in the death of the newborn infant because the thoracic cage cannot expand to allow breathing. (After Gilbert-Barness and Opitz 1996.)

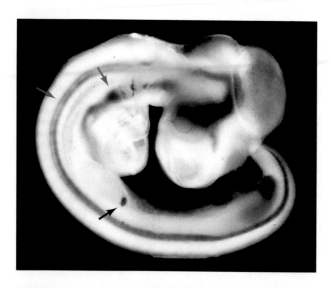

FIGURE 6.16   The *sonic hedgehog* gene is shown by in situ hybridization to be expressed in the chick nervous system (red arrow), gut (blue arrow), and limb bud (black arrow) of a 3-day chick embryo. (Photograph courtesy of C. Tabin.)

responsible for assuring that motor neurons only come from the ventral portion of the neural tube (see Chapter 12), that a portion of each somite forms the vertebrae (see Chapter 14), that the feathers of the chick form in their proper places (see Figure 6.6), and that our pinkies are always our most posterior digits (see Chapter 16). Sonic hedgehog often works with other paracrine factors, such as Wnt and FGF proteins.

THE HEDGEHOG PATHWAY   Members of the Hedgehog protein family function by binding to a receptor called Patched. The Patched protein, however, is not a signal transducer. Rather, it is bound to a signal transducer, the Smoothened protein. The Patched protein prevents Smoothened from functioning. In the absence of Hedgehog binding to Patched, Smoothened is inactive, and the Cubitus interruptus (Ci) protein is tethered to the microtubules of the responding cell. While on the microtubules, it is cleaved in such a way that a portion of it enters the nucleus and acts as a transcriptional repressor. When Hedgehog binds to Patched, the Patched protein's shape is altered such that it no longer inhibits Smoothened. Smoothened acts (probably by phosphorylation) to release the Ci protein from the microtubules and to prevent its being cleaved. The intact Ci protein can now enter the nucleus, where it acts as a transcriptional *activator* of the same genes it used to repress (Figure 6.17; Aza-Blanc et al. 1997; Lum and Beachy 2004).

The Hedgehog pathway is extremely important in vertebrate limb and neural differentiation (McMahon et al. 2003). When mice were made homozygous for a mutant allele of Sonic hedgehog, they had major limb abnormalities as well as **cyclopia***—a single eye in the center of the forehead (Chiang et al. 1996). The vertebrate homologues of the Ci protein in *Drosophila* are the Gli proteins. Severe truncations of the human *GLI3* gene produce a nonfunctional protein that gives rise to Grieg cephalopolysyndactyly, a condition involving a high forehead and extra digits. A different truncation retains the DNA-binding domain of the GLI3 protein but deletes the activator region. This mutant GLI3 protein can act only as a repressor. This protein is found in infants with Pallister-Hall syndrome, a much more severe syndrome (indeed, lethal soon after birth) involving not only extra digits, but also poor development of the pituitary gland, hypothalamus, anus, and kidneys (see Shin et al. 1999).

While mutations that inactivate the Hedgehog pathway can cause malformations, mutations that activate the pathway ectopically can cause cancers. If the Patched protein is mutated in somatic tissues such that it can no longer inhibit Smoothened, it can cause tumors of the basal cell layer of the epidermis (basal cell carcinomas). Heritable mutations of the *patched* gene cause basal cell nevus syndrome, a rare autosomal dominant condition characterized by both developmental anomalies (fused fingers, rib and facial abnormalities) and multiple malignant tumors such as basal cell carcinoma (Hahn et al. 1996; Johnson et al. 1996).

One remarkable feature of the Hedgehog signal transduction pathway is the importance of cholesterol. First, cholesterol is critical for the catalytic cleavage of Sonic hedgehog protein. Only the amino-terminal portion of the protein is functional and secreted. The cholesterol also binds to the active N-terminus of the Sonic hedgehog protein and allows this paracrine factor to diffuse over a range of a few hundred μm (about 30 cell diameters in the mouse limb). Without this cholesterol modification, diffusion is severely hampered (Lewis et al. 2001). Second, the Patched protein that binds Sonic hedgehog also needs cholesterol in order to function. Some human cyclopia syndromes are caused by mutations in genes that encode either Sonic hedgehog or the enzymes that synthesize cholesterol (Kelley et al. 1996; Roessler et al. 1996). Moreover, certain chemicals that induce cyclopia do so by interfering with the cholesterol biosynthetic enzymes (Beachy et al. 1997; Cooper et al. 1998). Two teratogens known to cause cyclopia in vertebrates are jervine and cyclopamine. Both substances are found in the plant *Veratrum californicum* and both block the synthesis of cholesterol (Figure 6.18; Keeler and Binns 1968).

VADE MECUM²   **Cyclopia induced in zebrafish.** Alcohol can act as a teratogen and can induce cyclopia in zebrafish embryos. [Click on Zebrafish]

---

*Cyclopia (having one central eye, like the Cyclops of Homer's *Odyssey*) will be discussed in more detail in Chapters 12 and 21.

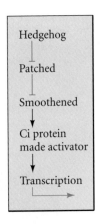

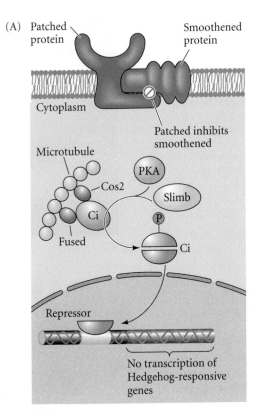

(A)

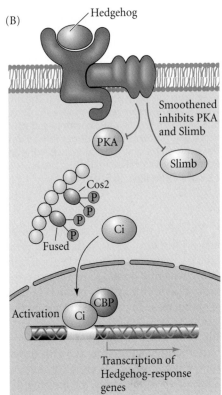

(B)

FIGURE 6.17 The Hedgehog signal transduction pathway. Patched protein in the cell membrane is an inhibitor of the Smoothened protein. (A) In the absence of Hedgehog binding to Patched, the Ci protein is tethered to the microtubules by the Cos2 and Fused proteins. This binding allows the PKA and Slimb proteins to cleave Ci into a transcriptional repressor that blocks the transcription of particular genes. (B) When Hedgehog binds to Patched, its conformation changes, releasing the inhibition of the Smoothened protein. Smoothened then releases Ci from the microtubules (probably by adding more phosphates to the Cos2 and Fused proteins) and inactivates the cleavage proteins PKA and Slimb. The Ci protein enters the nucleus, binds a CBP protein, and acts as a transcriptional activator of particular genes. (After Johnson and Scott 1998.)

The Wnt family

The Wnt's are a family of cysteine-rich glycoproteins. There are at least 15 members of this gene family in vertebrates.* Their name is a fusion of the name of the *Drosophila* segment polarity gene *wingless* with the name of one of its vertebrate homologues, *integrated*. While Sonic hedgehog is important in patterning the ventral portion of the somites (causing the cells to become cartilage), Wnt1 appears to be active in inducing the dorsal cells of the somites to become muscle and is involved in the specification of the midbrain cells (see Chapter 14; McMahon and Bradley 1990; Stern et al. 1995). Wnt proteins also are critical in establishing the polarity of insect and vertebrate limbs, promoting the proliferation of stem cells, and in several steps of urogenital system development (Figure 6.19).

*A summary of all the Wnt proteins and Wnt signaling components can be found at http://www.stanford.edu/~rnusse/ wntwindow.html

FIGURE 6.18 Head of a cyclopic lamb born of a ewe who had eaten *Veratrum californicum* early in pregnancy. The cerebral hemispheres fused, forming only one central eye and no pituitary gland. The jervine alkaloid made by this plant inhibits cholesterol synthesis, which is needed for Hedgehog production and reception. (Photograph courtesy of L. James and the USDA Poisonous Plant Laboratory.)

FIGURE 6.19 Wnt proteins play several roles in the development of the urogenital organs. Wnt4 is necessary for kidney development and for female sex determination. (A) Whole-mount in situ hybridization of *Wnt4* expression in a 14-day mouse embryonic male urogenital rudiment. Expression (dark purple-blue staining) is seen in the mesenchyme that condenses to form the kidney's nephrons. (B) The urogenital rudiment of a wild-type newborn female mouse. (C) The urogenital rudiment of a newborn female mouse with targeted knockout of the Wnt4 gene shows that the kidney fails to develop. In addition, the ovary starts synthesizing testosterone and becomes surrounded by a modified male duct system. (Photographs courtesy of J. Perasaari and S. Vainio.)

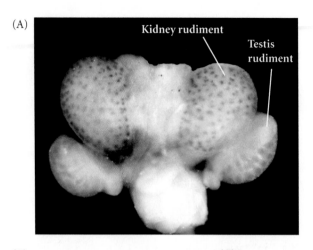

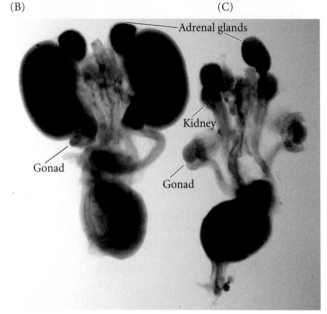

The Wnt proteins were unable to be isolated in their active form until 2003. At that time, Willert and colleagues (2003) discovered that each Wnt protein has a lipid molecule covalently bound to it. These hydrophobic molecules are critical for the activity of the Wnt proteins and probably act to increase their concentration in the cell membrane.

THE "CANONICAL" WNT PATHWAY Members of the Wnt family of paracrine factors interact with transmembrane receptors of the Frizzled family of proteins (Logan and Nusse 2004). In most instances, the binding of Wnt by a Frizzled protein causes Frizzled to activate the Disheveled protein. Once Disheveled is activated, it inhibits the activity of the glycogen synthase kinase-3 (GSK3) enzyme. GSK3, if it were active, would prevent the dissociation of the β-catenin protein from the APC protein, which targets β-catenin for degradation. However, when the Wnt signal is present and GSK3 is inhibited, β-catenin can dissociate from the APC protein and enter the nucleus. Once inside the nucleus, it can form a heterodimer with an LEF or TCF DNA-binding protein, becoming a transcription factor. This complex binds to and activates the Wnt-responsive genes (Figure 6.20A; Behrens et al. 1996; Cadigan and Nusse 1997).

This model is undoubtedly an oversimplification, because different cells use this pathway in different ways (see McEwen and Peifer 2001). Moreover, its components can have more than one function in the cell. In addition to being part of the Wnt signal transduction cascade, GSK3 is also an enzyme that regulates glycogen metabolism. The β-catenin protein was recognized as being part of the cell adhesion complex on the cell surface before it was also found to be a transcription factor. The APC protein also functions as a tumor suppressor. The transformation of normal adult colon epithelial cells into colon cancer is thought to occur when the *APC* gene is mutated and can no longer keep β-catenin out of the nucleus (Korinek et al. 1997; He et al. 1998). Once in the nucleus, β-catenin can bind with another transcription factor and activate genes for cell division.

One overriding principle is readily evident in both the Wnt pathway and the Hedgehog pathway: *activation is often accomplished by inhibiting an inhibitor.* Thus, in the Wnt pathway, the GSK3 protein is an inhibitor that is itself repressed by the Wnt signal.

THE "NONCANONICAL" WNT PATHWAYS The pathway described above is often called the "canonical" Wnt pathway because it was the first one to be discovered. However, in addition to sending signals to the nucleus, Wnt can also affect the actin and microtubular cytoskeleton. Here, Wnt activates alternative, "noncanonical," pathways. For instance, when Wnt activates Disheveled, the Disheveled protein can interact with a Rho GTPase. This GTPase can activate the kinases that phosphorylate cytoskeletal proteins and thereby alter cell shape, cell polarity (where the upper and lower portions of the cell differ), and motility (Figure 6.20B; Shulman et al. 1998; Winter et al. 2001). A third Wnt pathway diverges earlier than Disheveled. Here, the Frizzled receptor protein activates a phospholipase (PLC) that synthesizes a compound that releases calcium ions from the endoplasmic reticulum (Figure 6.20C). The released calcium can activate enzymes, transcription factors, and translation factors.

It is probable that the Frizzled proteins (of which there are many) can be used to couple different signal transduc-

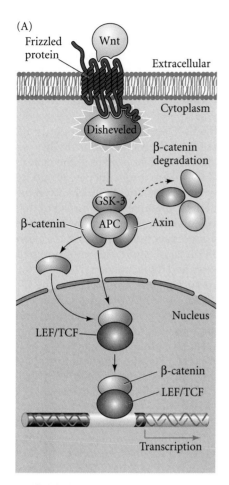

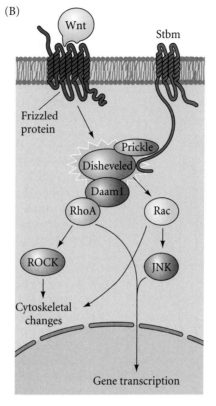

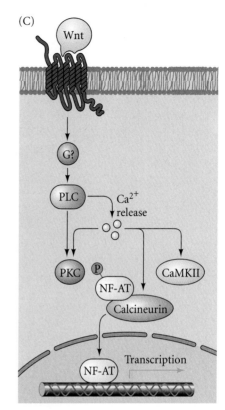

FIGURE 6.20 The Wnt signal transduction pathways. (A) The canonical Wnt pathway. The Wnt protein binds to its receptor, a member of the Frizzled family of proteins. In the case of certain Wnt proteins, the Frizzled protein then activates Disheveled, allowing it to become an inhibitor of glycogen synthase kinase 3 (GSK3). GSK3, if it were active, would prevent the dissociation of β-catenin from the APC protein. So by inhibiting GSK3, the Wnt signal frees β-catenin to associate with an LEF or TCF protein and become an active transcription factor. (B) In a pathway that regulates cell morphology, division, and movement, certain Wnt proteins activate Frizzled in a way that causes Frizzled to activate a Disheveled protein that has been tethered to the plasma membrane (through the Prickle protein). Here, it activates Rac and RhoA proteins, which coordinate the cytoskeleton and which can also regulate gene expression. (C) In a third pathway, certain Wnt proteins activate Frizzled receptors in a way that releases calcium ions and can cause Ca²⁺-dependent gene expression.

tion cascades to the Wnt signal (see Chen et al. 2005) and that different cells have evolved to use Wnt factors in different ways.

The TGF-β superfamily

There are over 30 structurally related members of the TGF-β superfamily,* and they regulate some of the most important interactions in development (Figure 6.21). The proteins encoded by TGF-β superfamily genes are processed such that the carboxy-terminal region contains the mature peptide. These peptides are dimerized into homodimers (with themselves) or heterodimers (with other TGF-β peptides) and are secreted from the cell. The TGF-β superfamily includes the TGF-β family, the activin family, the bone morphogenetic proteins (BMPs), the Vg1 family, and other proteins, including glial-derived neurotrophic factor (GDNF; necessary for kidney and enteric neuron differentiation) and Müllerian inhibitory factor (which is involved in mammalian sex determination).

---

*TGF stands for "*t*ransforming *g*rowth *f*actor." The designation "superfamily" is often given when each of the different classes of molecules constitutes a "family." The members of a superfamily all have similar structures, but are not as close as the molecules within a family are to one another.

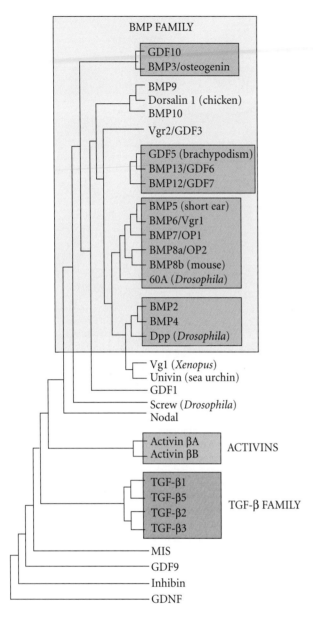

FIGURE 6.21   Relationships among members of the TGF-β superfamily. (After Hogan 1996.)

TGF-β family members TGF-β1, 2, 3, and 5 are important in regulating the formation of the extracellular matrix between cells and for regulating cell division (both positively and negatively). TGF-β1 increases the amount of extracellular matrix epithelial cells make (both by stimulating collagen and fibronectin synthesis and by inhibiting matrix degradation). TGF-β proteins may be critical in controlling where and when epithelia branch to form the ducts of kidneys, lungs, and salivary glands (Daniel 1989; Hardman et al. 1994; Ritvos et al. 1995). The effects of the indi-

vidual TGF-β family members are difficult to sort out, because members of the TGF-β family appear to function similarly and can compensate for losses of the others when expressed together. Moreover, targeted deletions of the *TGF-β1* gene in mice are difficult to interpret, since the mother can supply this factor through the placenta and milk (Letterio et al. 1994).

The members of the BMP family can be distinguished from other members of the TGF-β superfamily by having seven (rather than nine) conserved cysteines in the mature polypeptide. Because they were originally discovered by their ability to induce bone formation, they were given the name **bone morphogenetic proteins**. But it turns out that bone formation is only one of their many functions; the BMPs are extremely multifunctional.* They have been found to regulate cell division, apoptosis (programmed cell death), cell migration, and differentiation (Hogan 1996). They include proteins such as BMP4 (which in some tissues causes bone formation, in other tissues causes cell death, and in other instances specifies the epidermis) and BMP7 (which is important in neural tube polarity, kidney development, and sperm formation; see Figure 4.22). As it turns out, BMP1 is not a member of the BMP family at all; it is a protease. The *Drosophila* Decapentaplegic (Dpp) protein is homologous to vertebrate BMP4, and human BMP4 can replace Dpp and thus "rescue" *dpp*-deficient flies (Padgett et al. 1993). BMPs are thought to work by diffusion from the cells producing them. Their range is determined by the amino acids in their N-terminal region, which determine whether the specific BMP will be bound by proteoglycans, thereby restricting its diffusion (Ohkawara et al. 2002).

The Nodal and activin proteins are also members of the TGF-β superfamily. These proteins are extremely important in specifying the different regions of the mesoderm and for distinguishing the left and right sides of the vertebrate body axis.

THE SMAD PATHWAY   Members of the TGF-β superfamily of paracrine factors activate members of the Smad family of transcription factors (Heldin et al. 1997; Shi and Massague 2003). The TGF-β ligand binds to a type II TGF-β receptor, which allows that receptor to bind to a type I TGF-β receptor. Once the two receptors are in close contact, the type II

---

*One of the many reasons why humans don't seem to need an enormous genome is that the gene products—proteins—involved in our construction and development often have many functions. Many of the proteins we are familiar with in adults (such as hemoglobin, keratins, insulin, and the like) *do* have only one function, which led to the erroneous conclusion that this is the norm. Indeed, the "one-function-per-entity" concept is a longstanding one in science, having been credited to Aristotle. Philosopher John Thorp has called this *monotelism* (Greek, "one end") "Aristotle's worst idea."

Stem Cell Niches

Many organs have stem cells that undergo continual renewal. These tissues include the mammalian epidermis, hair follicles, intestinal villi, blood cells, and sperm cells, as well as *Drosophila* sperm and egg cells. These stem cells must maintain the long-term ability to divide, producing some daughter cells that are differentiated and other daughter cells that remain stem cells. The ability of a cell to become an adult stem cell is determined in large part by where it resides. The continuously proliferating stem cells are housed in compartments called **stem cell niches** (or regulatory microenvironments). There are particular places in the embryo that become stem cell niches.

Stem cell niches regulate the continuous production of stem cells and their more differentiated progeny, usually by paracrine (and sometimes juxtacrine) factors that are produced in the niche cells. These paracrine factors retain the cells in an uncommitted state. Once the cells leave the niche, the paracrine factors cannot reach them, and the cells begin differentiating.

For instance, sperm are continuously produced in the *Drosophila* testis. The stem cells for the sperm reside in a regulatory microenvironment called the **hub**. The hub consists of about a dozen somatic testes

cells and is surrounded by 5–9 germ stem cells. Those germ stem cells that remain attached to the somatic cells remain germ stem cells. However, their division is asymmetric. Those remaining attached to the hub remain the stem cell population, while those daughter cells that divide in such a way that they are not touching the hub become the gonialblast cells that will divide to become the precursors of the sperm cells.

The somatic cells of the hub are able to regulate stem cell proliferation by secreting the paracrine factor Unpaired onto the cells attached to them. Unpaired activates the JAK-STAT pathway in the adjacent germ stem cells to specify their self-renewal. Those cells that are distant from the paracrine factor cannot receive this signal, so they begin their differentiation into the sperm cell lineage (Figure 6.22; Tulina and Matunis 2001; Kiger et al. 2001). Moreover, the division of the germ stem cell always produces a cell attached to the hub and an unattached cell. The asymmetric division involves the interactions between the stem cells and the somatic cells. In the division of the stem cell, one centrosome remains attached to the cortex at the contact site between the stem cell and the somatic cells. The other centrosome moves to the opposite side, thus establishing a mitotic spindle

that will produce one daughter cell attached to the hub and one daughter cell away from it (Yamashita et al. 2003). The cell adhesion molecules linking the hub and stem cells together are probably involved in retaining one of the centrosomes in the region where the two cells touch.

FIGURE 6.22 The stem cell niche in the *Drosophila* testes. (A) The apical hub cells consist of around 12 somatic cells. Attached to them are 5–9 germ stem cells. The germ stem cells divide asymmetrically to form another germ stem cell (which remains attached to the somatic hub cells) and a gonialblast that will divide to form the sperm precursors (the spermatogonia and the spermatocyte cysts where meiosis is initiated). (B) Reporter β-galactosidase gene inserted in Unpaired, showing its transcription in the somatic hub cells. (C) Cell division pattern of the germline stem cells, wherein one of the centrosomes remains in the cortical cytoplasm near the site of hub cell adhesion, and the other centrosome migrates to the opposite pole of the germline stem cell. This results in one remaining attached to the hub while the other cell leaves it. (After Tulina and Matunis 2001, photograph courtesy of E. Matunis.)

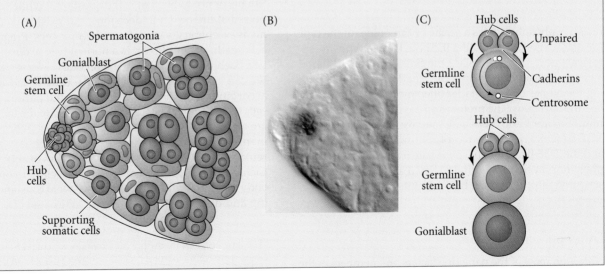

(A) Spermatogonia
Gonialblast
Germline stem cell
Hub cells
Supporting somatic cells

(B)

(C) Hub cells
Unpaired
Germline stem cell
Cadherins
Centrosome
Hub cells
Germline stem cell
Gonialblast

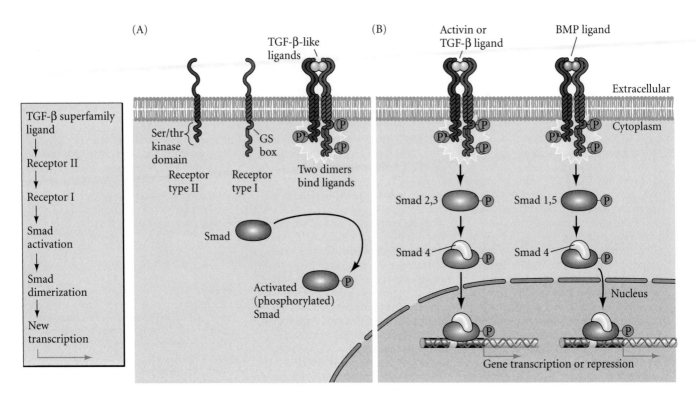

FIGURE 6.23 The Smad pathway activated by TGF-β superfamily ligands. (A) An activation complex is formed by the binding of the ligand by the type I and type II receptors. This allows the type II receptor to phosphorylate the type I receptor on particular serine or threonine residues (of the "GS box"). The phosphorylated type I receptor protein can now phosphorylate the Smad proteins. (B) Those receptors that bind TGF-β family proteins or members of the activin family phosphorylate Smads 2 and 3. Those receptors that bind to BMP family proteins phosphorylate Smads 1 and 5. These Smads can complex with Smad 4 to form active transcription factors. A simplified version of the pathway is shown at the left.

receptor phosphorylates a serine or threonine on the type I receptor, thereby activating it. The activated type I receptor can now phosphorylate the Smad proteins (Figure 6.23A). (Researchers named the Smad proteins by merging the names of the first identified members of this family—the *C. elegans* Sma protein and the *Drosophila* Mad protein.) Smads 1 and 5 are activated by the BMP family of TGF-β factors, while the receptors binding activin, Nodal, and the TGF-β family phosphorylate Smads 2 and 3. These phosphorylated Smads bind to Smad 4 and form the transcription factor complex that will enter the nucleus (Figure 6.23B).

Other paracrine factors

Although most of the paracrine factors are members of the above-described four families, some have few or no close relatives. Factors such as epidermal growth factor, hepatocyte growth factor, neurotrophins, and stem cell factor

are not included among these families, but each plays important roles during development. In addition, there are numerous factors involved almost exclusively with developing blood cells: erythropoietin, the cytokines, and the interleukins. These factors will be discussed when we detail blood cell formation in Chapter 14.

Cell Death Pathways

"To be, or not to be: that is the question." While we all are poised at life-or-death decisions, this existential dichotomy is exceptionally stark for embryonic cells. **Programmed cell death**, or apoptosis,* is a normal part of development (see Baehrecke 2002). In the nematode *C. elegans*, in which we can count the number of cells, exactly 131 cells die according to the normal developmental pattern. All the cells of this nematode are "programmed" to die unless they are actively told not to undergo apoptosis. In humans, as many as 10^{11} cells die in each adult each day and are replaced by other cells. (Indeed, the mass of cells we lose each year through normal cell death is close to our entire body weight!) Within the uterus, we were constantly making and destroying cells, and we generated about three

---

*The term *apoptosis* (both "p"s are pronounced) comes from the Greek word for the natural process of leaves falling from trees or petals falling from flowers. Apoptosis is an active process that can be subject to evolutionary selection. A second type of cell death, *necrosis*, is a pathological death caused by external factors such as inflammation or toxic injury.

times as many neurons as we eventually ended up with when we were born. Lewis Thomas (1992) has aptly noted,

> By the time I was born, more of me had died than survived. It was no wonder I cannot remember; during that time I went through brain after brain for nine months, finally contriving the one model that could be human, equipped for language.

Apoptosis is necessary not only for the proper spacing and orientation of neurons, but also for generating the middle ear space, the vaginal opening, and the spaces between our fingers and toes (Saunders and Fallon 1966; Roberts and Miller 1998; Rodriguez et al. 1997). Apoptosis prunes unneeded structures (frog tails, male mammary tissue), controls the number of cells in particular tissues (neurons in vertebrates and flies), and sculpts complex organs (palate, retina, digits, and heart).

Different tissues use different signals for apoptosis. One of the signals often used in vertebrates is bone morphogenetic protein 4 (BMP4). Some tissues, such as connective tissue, respond to BMP4 by differentiating into bone. Others, such as the frog gastrula ectoderm, respond to BMP4 by differentiating into skin. Still others, such as neural crest cells and tooth primordia, respond by degrading their DNA and dying. In the developing tooth, for instance, numerous growth and differentiation factors are secreted by the enamel knot. After the cusp has grown, the enamel knot synthesizes BMP4 and shuts itself down by apoptosis (see Chapter 13; Vaahtokari et al. 1996).

In other tissues, the cells are "programmed" to die, and will remain alive only if some growth or differentiation factor is present to "rescue" them. This happens during the development of mammalian red blood cells. The red blood cell precursors in the mouse liver need the hormone erythropoietin in order to survive. If they do not receive it, they undergo apoptosis. The erythropoietin receptor works through the JAK-STAT pathway, activating the Stat5 transcription factor. In this way, the amount of erythropoietin present can determine how many red blood cells enter the circulation.

One of the pathways for apoptosis was largely delineated through genetic studies of *C. elegans*. Indeed, the importance of this pathway was recognized by awarding a Nobel Prize to Sydney Brenner, Bob Horvitz, and Jonathan Sulston in 2002. It was found that the proteins encoded by the *ced-3* and *ced-4* genes were essential for apoptosis, and that in the cells that did not undergo apoptosis, those genes were turned off by the product of the *ced-9* gene (Figure 6.24A; Hengartner et al. 1992). The CED-4 protein is a protease-activating factor that activates CED-3, a protease that initiates the destruction of the cell. The CED-9 protein can bind to and inactivate CED-4. Mutations that inactivate the *ced-9* protein cause numerous cells that would normally survive to activate their *ced-3* and *ced-4* genes and die, leading to the death of the entire embryo. Conversely, gain-of-function mutations of *ced-9* cause CED-9 protein to be made in cells that would normally die, resulting in their

survival. Thus, the *ced-9* gene appears to be a binary switch that regulates the choice between life and death on the cellular level. It is possible that every cell in the nematode embryo is poised to die, and those cells that survive are rescued by the activation of the *ced-9* gene.

The CED-3 and CED-4 proteins form the center of the apoptosis pathway that is common to all animals studied. The trigger for apoptosis can be a developmental cue such as a particular molecule (such as BMP4 or glucocorticoids) or the loss of adhesion to a matrix. Either type of cue can activate CED-3 or CED-4 proteins or inactivate CED-9 molecules. In mammals, the homologues of the CED-9 protein are members of the **Bcl2 family** of genes. This family includes *Bcl2*, *BclX*, and similar genes (Figure 6.24B). The functional similarities are so strong that if an active human

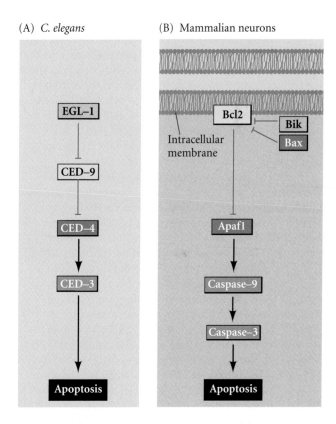

(A) *C. elegans*   (B) Mammalian neurons

FIGURE 6.24 Apoptosis pathways in nematodes and mammals. (A) In *C. elegans*, the CED-4 protein is a protease activating factor that can activate the CED-3 protease. The CED-3 protease initiates the cell destruction events. CED-9 can inhibit CED-4 (and CED-9 can be inhibited upstream by EGL-1). (B) In mammals, a similar pathway exists, and appears to function in a similar manner. In this hypothetical scheme for the regulation of apoptosis in mammalian neurons, $Bclx_L$ (a member of the Bcl2 family) binds Apaf1 and prevents it from activating the precursor of caspase-9. The signal for apoptosis allows another protein (here, Bik) to inhibit the binding of Apaf1 to $Bclx_L$. Apaf-1 is now able to bind to the caspase-9 precursor and cleave it. Caspase-9 dimerizes and activates caspase-3, which initiates apoptosis. The same colors are used to represent homologous proteins. (After Adams and Cory 1998.)

WEBSITE **6.4 The uses of apoptosis.** Apoptosis is used for numerous processes throughout development. The website explores the role of apoptosis in such phenomena as *Drosophila* germ cell development and the eyes of blind cave fish.

Bcl2 gene is placed in *C. elegans* embryos, it prevents normally occurring cell death (Vaux et al. 1992).

The mammalian homologue of CED-4 is **Apaf1** (*a*poptotic *p*rotease *a*ctivating *f*actor 1), and it participates in the cytochrome c-dependent activation of the mammalian CED-3 homologues, the proteases **caspase-9** and **caspase-3** (see Figure 6.24B; Shaham and Horvitz 1996; Cecconi et al. 1998; Yoshida et al. 1998). Activation of the caspases causes the autodigestion of the cell. Caspases are strong proteases that digest the cell from within, cleaving cellular proteins and fragmenting the DNA.

While apoptosis-deficient nematodes deficient for CED-4 are viable (despite their having 15 percent more cells than wild-type worms), mice with loss-of-function mutations for either *caspase-3* or *caspase-9* die around birth from massive cell overgrowth in the nervous system (Figure 6.25; Kuida et al. 1996, 1998; Jacobson et al. 1997). Mice homozygous for targeted deletions of *Apaf1* have similarly severe craniofacial abnormalities, brain overgrowth, and webbing between their toes.

There are instances where cell death is the normal state unless some ligand "rescues" the cells. In the chick neural tube, Patched protein (a Hedgehog receptor) will activate caspases. The binding of Sonic hedgehog (from the notochord and ventral neural tube cells) suppresses Patched, and the caspases are not activated to start apoptosis (Thibert et al. 2003). Such "dependence receptors" probably prevent neural cells from proliferating outside the proper tissue, and the loss of such receptors is associated with cancers (Porter and Dhakshinamoorty 2004). Moreover, we will soon see that certain epithelial cells must be attached to the extracellular matrix in order to function. If the cell is removed from the matrix, the apoptosis pathway is activated and the cell dies (Jan et al. 2004). This, too, is probably a mechanism that prevents cancers once cells have lost their adhesion to extracellular matrix proteins.

FIGURE 6.25 Disruption of normal brain development by blocking apoptosis. In mice in which the genes for caspase-9 have been knocked out, normal neural apoptosis fails to occur, and the overproliferation of brain neurons is obvious. (A) 16-day embryonic wild-type mouse. (B) A *caspase-9* knockout mouse of the same age. The enlarged brain protrudes above the face, and the limbs are still webbed. (C,D) This effect is confirmed by cross-sections through the forebrain at day 13.5. The knockout exhibits thickened ventricle walls and the near-obliteration of the ventricles. (From Kuida et al. 1998.)

Juxtacrine Signaling

In juxtacrine interactions, proteins from the inducing cell interact with receptor proteins of adjacent responding cells without diffusing from the cell producing it. Two of the most widely used families of juxtacrine factors are the **Notch proteins** (which bind to a family of ligands exemplified by the Delta protein) and the **eph receptors** and their **ephrin** ligands. When the ephrin on one cell binds with the eph receptor on an adjacent cell, signals are sent to each of the two cells (Davy et al. 2004; Davy and Soriano 2005). These signals are often those of either attraction or repulsion, and ephrins are often seen where cells are being told where to migrate or where boundaries are forming. We will see the ephrins and the eph receptors functioning in the formation of blood vessels, neurons, and somites. For the moment, we will look at the Notch proteins and their ligands.

The Notch pathway: Juxtaposed ligands and receptors

While most known regulators of induction are diffusible proteins, some inducing proteins remain bound to the

(A) *caspase-9*$^{+/+}$ (Normal)   (B) *caspase-9*$^{-/-}$ (Knockout)

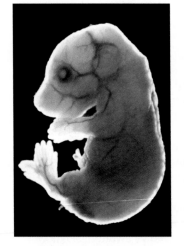 

(C) Normal   (D) Knockout

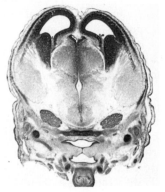 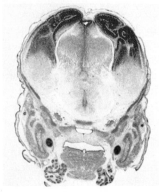

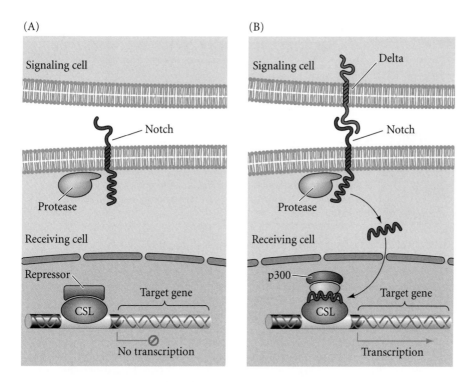

(A) Signaling cell

Notch

Protease

Receiving cell

Repressor

CSL

Target gene

No transcription

(B) Signaling cell

Delta

Notch

Protease

Receiving cell

p300

CSL

Target gene

Transcription

FIGURE 6.26 Mechanism of Notch activity. (A) Prior to Notch signaling, a CSL transcription factor (such as Suppressor of hairless or CBF1) is on the enhancer of Notch-regulated genes. The CSL binds repressors of transcription. (B) Model for the activation of Notch. A ligand (Delta, Jagged, or Serrate protein) on one cell binds to the extracellular domain of the Notch protein on an adjacent cell. This binding causes a shape change in the intracellular domain of Notch, which activates a protease. The protease cleaves Notch and allows the intracellular region of the Notch protein to enter the nucleus and bind the CSL transcription factor. This intercellular region of Notch displaces the repressor proteins and binds activators of transcription, including the histone acetyltransferase p300. The activated CSL can then transcribe its target genes (After Koziol-Dube, personal communication.)

inducing cell surface. In one such pathway, cells expressing the **Delta**, **Jagged**, or **Serrate** proteins in their cell membranes activate neighboring cells that contain the **Notch** protein in their cell membranes. Notch extends through the cell membrane, and its external surface contacts Delta, Jagged, or Serrate proteins extending out from an adjacent cell. When complexed to one of these ligands, Notch undergoes a conformational change that enables a part of its cytoplasmic domain to be cut off by the Presenilin-1 protease. The cleaved portion enters the nucleus and binds to a dormant transcription factor of the CSL family. When bound to the Notch protein, the CSL transcription factors activate their target genes (Figure 6.26; Lecourtois and Schweisguth 1998; Schroeder et al. 1998; Struhl and Adachi 1998). This activation is thought to involve the recruitment of histone acetyltransferases (Wallberg et al. 2002). Thus,

Notch can be considered as a transcription factor tethered to the cell membrane. When the attachment is broken, Notch (or a piece of it) can detach from the cell membrane and enter the nucleus (Kopan 2002).

Notch proteins are extremely important receptors in the nervous system. In both the vertebrate and *Drosophila* nervous system, the binding of Delta to Notch tells the receiving cell not to become neural (Chitnis et al. 1995; Wang et al. 1998). In the vertebrate eye, the interactions between Notch and its ligands seem to regulate which cells become optic neurons and which become glial cells (Dorsky et al. 1997; Wang et al. 1998). Notch proteins are also important in the patterning of the nematode vulva. The vulval precursor cell closest to the anchor cell (see Sidelights & Speculations) is able to inhibit its neighbors from becoming central vulval cells by signaling to them through its Notch homologue, the LIN-12 receptor. The LIN-12 signal blocks the protein kinase pathway initiated by the LIN-3 signal (Berset et al. 2001).

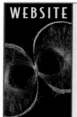 **WEBSITE** **6.5 Notch mutations.** Mutations in Notch proteins can cause nervous system abnormalities in humans. Humans have more than one Notch gene and more than one ligand. Their interactions may be critical in neural development. Moreover, the association of Notch with the presenilin protease suggests that disruption of Notch functioning might lead to Alzheimer disease.

Juxtacrine Signaling and Cell Patterning

Induction does indeed occur on the cell-to-cell level, and one of the best examples is the formation of the vulva in the nematode worm *Caenorhabditis elegans*. Remarkably, the signal transduction pathways involved turn out to be the same as those used in the formation of retinal receptors in *Drosophila*; only the targeted transcription factors are different. In both cases, an epidermal growth factor-like inducer activates the RTK pathway.

Vulval induction in *C. elegans*

Most *Caenorhabditis elegans* individuals are hermaphrodites. In their early development, they are male and the gonad produces sperm, which is stored for later use. As they grow older, they develop ovaries. The eggs "roll" through the region of sperm storage, are fertilized inside the nematode, and then

WEBSITE 6.6 Photoreceptor induction: Conserved pathways Conservation of juxtacrine signaling pathways is seen in the differentiation of photoreceptors and eyes in many species, including vertebrates. The website takes a close look at the well studied cascades of *Drosophila* photoreceptor formation, an induction pathway that is remarkably similar to the pathway used by *C.elegans* in forming vulval cells.

pass out of the body through the vulva (see Figure 8.43).

The formation of the vulva in *C. elegans* represents a case in which one inductive signal generates a variety of cell types. This organ forms during the larval stage from six cells called the **vulval precursor cells** (**VPCs**). The cell connecting the overlying gonad to the vulval precursor cells is called the **anchor cell** (see Figure 6.27). The anchor cell secretes the LIN-3 protein, a relative of epidermal growth factor (EGF) (Hill

and Sternberg 1992). If the anchor cell is destroyed (or if the *lin-3* gene is mutated), the VPCs will not form a vulva, but instead become part of the hypodermis (skin) (Kimble 1981).

The six VPCs influenced by the anchor cell form an **equivalence group**. Each member of this group is competent to become induced by the anchor cell and can assume any of three fates, depending on its proximity to the anchor cell. The cell directly beneath the anchor cell divides to form the central vulval cells. The two cells flanking that central cell divide to become the lateral vulval cells, while the three cells farther away from the anchor cell generate hypodermal cells. If the anchor cell is destroyed, all six cells of the equivalence group divide once and contribute to the hypodermal tis-

FIGURE 6.27 The *C. elegans* vulval precursor cells and their descendants. (A) Location of the gonad, anchor cell, and VPCs in the second instar larva. (B,C) Relationship of the anchor cell to the six VPCs and their subsequent lineages. 1° lineages result in the central vulval cells; 2° lineages constitute the lateral vulval cells; 3° lineages generate hypodermal cells. (C) Outline of the vulva in the fourth instar larva. The circles represent the positions of the nuclei. (D) Model for the determination of vulval cell lineages in *C. elegans*. The LIN-3 signal from the anchor cell causes the determination of the P6.p cell to generate the central vulval lineage (dark purple). Lower concentrations of LIN-3 cause the P5.p and P7.p cells to form the lateral vulval lineages. The P6.p (central lineage) cell also secretes a short-range juxtacrine signal that induces the neighboring cells to activate the LIN-12 (Notch) protein. This signal prevents the P5.p and P7.p cells from generating the primary, central vulval cell lineage. (After Katz and Sternberg 1996.)

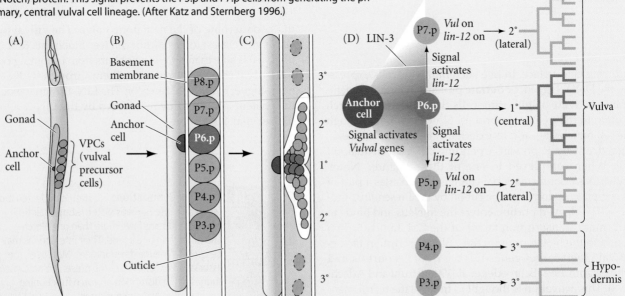

sue. If the three central VPCs are destroyed, the three outer cells, which normally form hypodermis, generate vulval cells instead.

The LIN-3 protein is received by the LET-23 receptor tyrosine kinase on the VPCs, and the signal is transferred to the nucleus through the RTK pathway. The target of the kinase cascade is the LIN-31 protein (Tan et al. 1998). When this protein is phosphorylated in the nucleus, it loses its inhibitory protein partner and is able to function as a transcription factor, promoting vulval cell fates. Two mechanisms coordinate the formation of the vulva through this induction, as shown in Figure 6.27 (Katz and Sternberg 1996):

1. The LIN-3 protein forms a concentration gradient. Here, the VPC closest to the anchor cell (i.e., the P6.p cell) receives the highest concentration of LIN-3 protein and generates the central vulval cells. The two VPCs adjacent to it (P5.p and P7.p) receive a lower amount of LIN-3 and become the lateral vulval cells. The VPCs farther away from the anchor cell do not receive enough LIN-3 to have an effect, so they become hypodermis (Katz et al. 1995).

2. In addition to forming the central vulval lineage, the VPC closest to the anchor cell also signals laterally to the two adjacent (P5.p and P7.p) cells and instructs them not to generate the central vulval lineages. The P5.p and P7.p cells receive the signal through the LIN-12 proteins on their cells membranes. The Notch signal activates a microRNA, *mir-61*, which represses the gene that would specify central vulval fate, as well as promoting those genes that are involved in forming the lateral vulval cells (Sternberg 1988; Yoo et al. 2005). The lateral cells do not instruct the peripheral VPCs to do anything, so they become hypodermis (Koga and Ohshima 1995; Simske and Kim 1995).

Cell-cell interactions and chance in the determination of cell types

The development of the vulva in *C. elegans* offers several examples of induction on the cellular level. We have already discussed the reception of the EGF-like LIN-3 signal by the cells of the equivalence group that forms the vulva. But before this induction occurs, there is an earlier interaction that forms the anchor cell. The formation of the anchor cell

is mediated by *lin-12*, the *C. elegans* homologue of the *Notch* gene. In wild-type *C. elegans* hermaphrodites, two adjacent cells, Z1.ppp and Z4.aaa, have the potential to become the anchor cell. They interact in a manner that causes one of them to become the anchor cell while the other one becomes the precursor of the uterine tissue. In loss-of-function *lin-12* mutants, both cells become anchor cells, while in gain-of-function mutations, both cells become uterine precursors (Greenwald et al. 1983). Studies using genetic mosaics and cell ablations have shown that this decision is made in the second larval stage, and that the *lin-12* gene needs to function only in that cell destined to become the uterine precursor cell. The presumptive anchor cell does not need it. Seydoux and Greenwald (1989) speculate that these two cells originally synthesize both the signal for uterine differentiation (the LAG-2 protein, homologous to Delta in *Drosophila*) and the receptor for this molecule (the LIN-12 protein, homologous to Notch; Wilkinson et al. 1994). During a particular time in larval development, the cell that, by chance, is secreting more LAG-2 causes its neighbor to cease its production of this differentiation signal and to increase its production of LIN-12 protein. The cell secreting LAG-2 becomes the gonadal anchor cell, while the cell receiving the signal through its LIN-12 protein becomes the ventral uterine precursor cell (Figure 6.28). Thus, the two cells are thought to determine each other prior to their respective differentiation events. When the LIN-12 protein is used again during vulva formation, it is activated by the primary vulval lineage to stop the lateral vulval cells from forming the central vulval phenotype (see Figure 6.27).

The anchor cell/ventral uterine precursor decision illustrates two important aspects of determination in two originally equivalent cells. First, the initial difference between the two cells is created by chance. Second, this initial difference is reinforced by feedback. Such a mechanism is also seen in the determination of which of the originally equivalent epidermal cells of the insect embryo will generate the neurons of the peripheral nervous system. Here, the choice is between becoming a skin (hypodermal) cell or a neural precursor cell (a **neuroblast**). The *Notch* gene of *Drosophila*, like its *C. elegans* homologue, *lin-12*, channels a bipotential cell into one of two alternative paths. Soon after gastrulation, a region of about 1800 ectodermal cells lies along the ventral midline of the

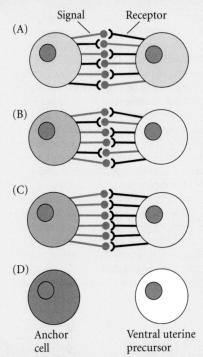

Signal   Receptor

(A)

(B)

(C)

(D)

Anchor cell   Ventral uterine precursor

FIGURE 6.28   Model for the generation of two cell types (anchor cell and ventral uterine precursor) from two equivalent cells (Z1.ppp and Z4.aaa) in *C. elegans*. (A) The cells start off as equivalent, producing fluctuating amounts of signal and receptor (inverted arrow). The *lag-2* gene is thought to encode the signal; the *lin-12* gene is thought to encode the receptor. Reception of the signal turns down LAG-2 (Delta) production and up-regulates LIN-12 (Notch). (B) A stochastic (chance) event causes one cell to produce more LAG-2 than the other cell at some particular critical time. This stimulates more LIN-12 production in the neighboring cell. (C) This difference is amplified, since the cell producing more LIN-12 produces less LAG-2. Eventually, just one cell is delivering the LAG-2 signal, and the other cell is receiving it. (D) The signaling cell becomes the anchor cell; the receiving cell becomes the ventral uterine precursor. (After Greenwald and Rubin 1992.)

Drosophila embryo. These cells, known as neurogenic ectodermal cells, all have the potential to form the ventral nerve cord of the insect. About one-fourth of these cells will become neuroblasts, while the rest will become the precursors of the hypodermis. The cells that give rise to neuroblasts are intermingled with those that become hypo-

(Continued on next page)

SIDELIGHTS & SPECULATIONS

dermal precursors. Thus, each neurogenic ectoderm cell can give rise to either hypodermal or neural precursor cells (Hartenstein and Campos-Ortega 1984). In the absence of *Notch* gene transcription in the embryo, these cells develop exclusively into neuroblasts, rather than into a mixture of hypodermal and neural precursor cells (Artavanis-Tsakonis et al. 1983; Lehmann et al. 1983). These embryos die, having a gross excess of neural cells at the expense of the ventral and head hypodermis (Poulson 1937; Hoppe and Greenspan 1986).

Heitzler and Simpson (1991) proposed that the Notch protein, like LIN-12, serves as a receptor for intercellular signals involved in the differentiation of equivalent cells. Moreover, they provided evidence that Delta is the ligand for Notch. Genetic mosaics show that, whereas Notch is need-

ed in the cells that are to become hypodermis, Delta is needed in the cells that *induce* the hypodermal phenotype.

Greenwald and Rubin (1992) have proposed a model based on the LIN-12 hypothesis to explain the spacing of neuroblasts in these proneural clusters of epidermal and neural precursors (Figure 6.29). Initially, all the neurogenic ectodermal cells have equal

potentials and produce the same signals. However, when one of the cells, by chance, produces more signal (say, Delta protein), it activates the receptors on adjacent cells and reduces their level of signaling. As their signaling levels are lowered, the neighbors of these low-signaling cells will tend to become high-level signalers. In this way, a spacing of neuroblasts is produced.

FIGURE 6.29   Model to explain the spacing pattern of neuroblasts among initially equivalent neurogenic ectodermal cells. (A) A field of equivalent cells, all of which signal and receive equally. (B) A chance event causes one of the cells (darker shading) to produce more signal. The cells surrounding it receive this higher amount of signal and reduce their own signaling level (lighter shading). (C) The rest of the pattern is now constrained. Those cells that have down-regulated their own signaling (in response to the events in B) are less likely to express more signal than their neighboring cells. The cells surrounded by down-regulated signalers are more likely to become signalers. (D,E) The fates of the cells throughout the field become specified as the amplification of the signal creates populations of signalers surrounded by populations of receivers. In the case of the neurogenic cells, the signal is thought to be the Delta protein, and the receiver is the Notch protein. (After Greenwald and Rubin 1992.)

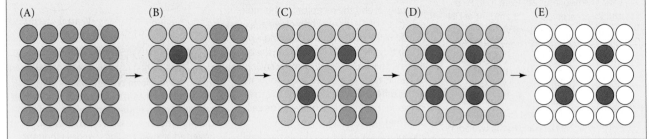

(A)   (B)   (C)   (D)   (E)

The extracellular matrix as a source of developmental signals

PROTEINS AND FUNCTIONS OF THE EXTRACELLULAR MATRIX   The **extracellular matrix** consists of macromolecules secreted by cells into their immediate environment. These macromolecules form a region of noncellular material in the interstices between the cells. The extracellular matrix is a critical region for much of animal development. Cell adhesion, cell migration, and the formation of epithelial sheets and tubes all depend on the ability of cells to form attachments to extracellular matrices. In some cases, as in the formation of epithelia, these attachments have to be extremely strong. In other instances, as when cells migrate, attachments have to be made, broken, and made again. In some cases, the extracellular matrix merely serves as a permissive substrate to which cells can adhere, or upon which they can migrate. In other cases, it provides the directions for cell movement or the signal for a developmental event.

Extracellular matrices are made up of collagen, proteoglycans, and a variety of specialized glycoprotein molecules, such as fibronectin and laminin. These large glycoproteins are responsible for organizing the matrix and the cells into an ordered structure. **Proteoglycans** play critical-

ly important roles in the delivery of the paracrine factors. These large molecules consist of core proteins (such as syndecan) with covalently attached glycosaminoglycan polysaccharide side chains. Two of the most widespread proteoglycans are heparan sulfate and chondroiton sulfate proteoglycans. Heparan sulfate proteoglycans can bind many members of the TGFβ, Wnt, and Fgf families, and they appear to be essential for presenting the paracrine factor in high concentrations to their receptors. In *Drosophila*, *C. elegans*, and mice, mutations that prevent proteoglycan protein or carbohydrate synthesis block normal cell migration, morphogenesis, and differentiation (Garcia-Garcia and Anderson 2003; Hwang et al. 2003; Kirn-Safran et al. 2004).

Fibronectin is a very large (460 kDa) glycoprotein dimer synthesized by numerous cell types. One function of fibronectin is to serve as a general adhesive molecule, linking cells to one another and to other substrates such as collagen and proteoglycans. Fibronectin has several distinct binding sites, and their interaction with the appropriate molecules results in the proper alignment of cells with their extracellular matrix (Figure 6.30). As we will see in later chapters, fibronectin also has an important role in cell

(A)

(B)

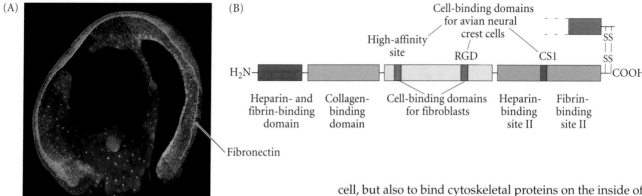

FIGURE 6.30 Fibronectin in the developing embryo. (A) Fluorescent antibodies to fibronectin show fibronectin deposition as a green band in the *Xenopus* embryo during gastrulation. The fibronectin will orient the movements of the mesoderm cells. (B) Structure and binding domains of fibronectin. The rectangles represent protease-resistant domains. The fibroblast-binding domain consists of two units, the RGD site and the high-affinity site, both of which are essential for cell binding. Another binding site for avian neural crest cells is necessary for these cells to migrate on a fibronectin substrate. Other regions of fibronectin enable it to bind to collagen, heparin, and other molecules of the extracellular matrix. (A, photograph courtesy of M. Marsden and D. W. DeSimone; B after Dufour et al. 1988.)

migration. The "roads" over which certain migrating cells travel are paved with this protein. Fibronectin paths lead germ cells to the gonads and lead heart cells to the midline of the embryo. If chick embryos are injected with antibodies to fibronectin, the heart-forming cells fail to reach the midline, and two separate hearts develop (Heasman et al. 1981; Linask and Lash 1988).

Laminin and **type IV collagen** are major components of a type of extracellular matrix called the **basal lamina**. The basal lamina is characteristic of the closely knit sheets that surround epithelial tissue (Figure 6.31). The adhesion of epithelial cells to laminin (upon which they sit) is much greater than the affinity of mesenchymal cells for fibronectin (to which they must bind and release if they are to migrate). Like fibronectin, laminin plays a role in assembling the extracellular matrix, promoting cell adhesion and growth, changing cell shape, and permitting cell migration (Hakamori et al. 1984; Morris et al. 2003).

INTEGRINS, THE RECEPTORS FOR EXTRACELLULAR MATRIX MOLECULES
The ability of a cell to bind to adhesive glycoproteins depends on its expressing membrane receptors for the cell-binding sites of these large molecules. The main fibronectin receptors were identified by using antibodies that block the attachment of cells to fibronectin (Chen et al. 1985; Knudsen et al. 1985). The fibronectin receptor complex was found not only to bind fibronectin on the outside of the

cell, but also to bind cytoskeletal proteins on the inside of the cell. Thus, the fibronectin receptor complex appears to span the cell membrane and unite two types of matrices. On the outside of the cell, it binds to the fibronectin of the extracellular matrix; on the inside of the cell, it serves as an anchorage site for the actin microfilaments that move the cell (Figure 6.32). Horwitz and co-workers (1986; Tamkun et al. 1986) have called this family of receptor proteins **integrins** because they integrate the extracellular and intracellular scaffolds, allowing them to work together. On the extracellular side, integrins bind to the sequence arginine-glycine-aspartate (RGD), found in several adhesive proteins in extracellular matrices, including fibronectin, vitronectin (found in the basal lamina of the eye), and laminin (Ruoslahti and Pierschbacher 1987). On the cytoplasmic side, integrins bind to talin and α-actinin, two proteins that connect to actin microfilaments. This dual binding enables the cell to move by contracting the actin microfilaments against the fixed extracellular matrix.

Bissell and her colleagues (1982; Martins-Green and Bissell 1995) have shown that the extracellular matrix is capable of inducing specific gene expression in developing tis-

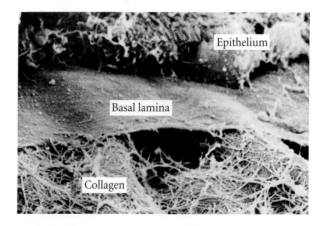

FIGURE 6.31 Location and formation of extracellular matrices in the chick embryo. This scanning electron micrograph shows the extracellular matrix at the junction of the epithelial cells (above) and mesenchymal cells (below). The epithelial cells synthesize a tight, laminin-based basal lamina, while the mesenchymal cells secrete a loose reticular lamina made primarily of collagen. (Photograph courtesy of R. L. Trelsted.)

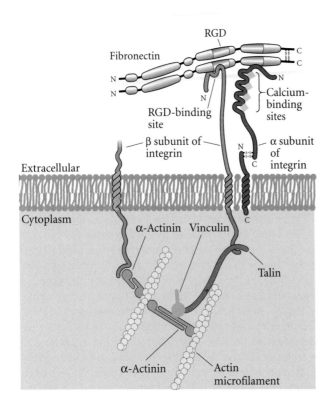

FIGURE 6.32 Speculative diagram of the fibronectin receptor complex. The integrins of the complex are membrane-spanning receptor proteins that bind fibronectin on the outside of the cell while binding cytoskeletal proteins on the inside of the cell. (After Luna and Hitt 1992.)

plastic coated with a laminin-containing basement membrane, the cells stop dividing and the differentiated genes of the mammary gland are expressed. This happens only after the integrins of the mammary gland cells bind to the laminin of the extracellular basement membrane. Then the gene for lactoferrin is expressed, as is the gene for p21, a cell division inhibitor. The c-*myc* and *cyclinD1* genes become silent. Eventually, all the genes for the developmental products of the mammary gland are expressed, and the cell division genes remain turned off. By this time, the mammary gland cells have enveloped themselves in a basal lamina, forming a secretory epithelium reminiscent of the mammary gland tissue. The binding of integrins to laminin is essential for the transcription of the casein gene, and the integrins act in concert with prolactin (see Figure 6.14) to activate that gene's expression (Roskelley et al. 1994; Muschler et al. 1999).

Several studies have shown that the binding of integrins to an extracellular matrix can stimulate the RTK pathway. When an integrin on the cell membrane of one cell binds to the fibronectin or collagen secreted by a neighboring cell, the integrin can activate the RTK cascade through an

sues, especially those of the liver, testis, and mammary gland. In these tissues, the induction of specific transcription factors depends on cell-substrate binding (Figure 6.33; Liu et al. 1991; Streuli et al. 1991; Notenboom et al. 1996). Often, the presence of bound integrin prevents the activation of genes that specify apoptosis (Montgomery et al. 1994; Frisch and Ruoslahti 1997). The chondrocytes that produce the cartilage of our vertebrae and limbs can survive and differentiate only if they are surrounded by an extracellular matrix and are joined to that matrix through their integrins (Hirsch et al. 1997). If chondrocytes from the developing chick sternum are incubated with antibodies that block the binding of integrins to the extracellular matrix, they shrivel up and die. While the mechanisms by which bound integrins inhibit apoptosis remain controversial (see Howe et al. 1998), the extracellular matrix is obviously an important source of signals that can be transduced into the nucleus to produce specific gene expression.

Some of the genes induced by matrix attachment are being identified. When plated onto tissue culture plastic, mouse mammary gland cells will divide (Figure 6.34). Indeed, the genes for cell division (c-*myc*, *cyclinD1*) are expressed, while the genes for the differentiated products of the mammary gland (casein, lactoferrin, whey acidic protein) are not expressed. If the same cells are plated onto

(A)

(B)

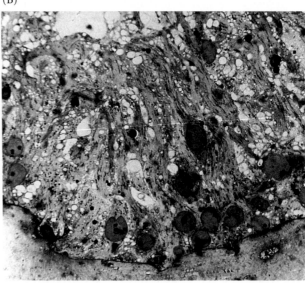

FIGURE 6.33 Role of the extracellular matrix in cell differentiation. Light micrographs of rat testis Sertoli cells grown for 2 weeks (A) on tissue culture plastic dishes and (B) on dishes coated with basal lamina. The two photographs were taken at the same magnification, 1200×. (From Hadley et al. 1985; photographs courtesy of M. Dym.)

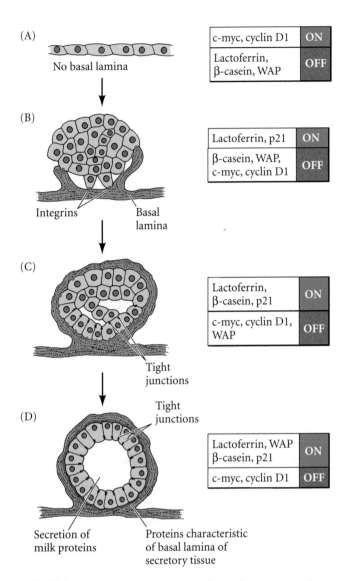

c-myc, cyclin D1	ON
Lactoferrin, β-casein, WAP	OFF

No basal lamina

Lactoferrin, p21	ON
β-casein, WAP, c-myc, cyclin D1	OFF

Integrins   Basal lamina

Lactoferrin, β-casein, p21	ON
c-myc, cyclin D1, WAP	OFF

Tight junctions

Tight junctions

Lactoferrin, WAP β-casein, p21	ON
c-myc, cyclin D1	OFF

Secretion of milk proteins

Proteins characteristic of basal lamina of secretory tissue

FIGURE 6.34 Basement membrane-directed gene expression in mammary gland tissue. (A) Mouse mammary gland tissue divides when placed on tissue culture plastic. The genes encoding cell division proteins are on, and the genes capable of synthesizing the differentiated products of the mammary gland lactoferrin, casein, whey acidic protein (WAP) are off. (B) When these cells are placed on a basement membrane that contains laminin (basal lamina), the genes for cell division proteins are turned off, while the genes encoding inhibitors of cell division (such as p21) and the gene for lactoferrin are turned on. (C,D) The mammary gland cells wrap the basal lamina around them, forming a secretory epithelium. The genes for casein and whey acidic protein are sequentially activated. (After Bissell et al. 2003.)

adaptor protein-like complex that connects the integrin to the Ras G protein (Figure 6.35A; Wary et al. 1998). Cadherins and other cell adhesion molecules (see Chapter 3) can also transmit signals by "hijacking" the FGF receptors (Williams et al. 1994b; Clark and Brugge 1995). Cadherins, for example, are able to bind to the cytoplasmic region of

FGF receptors and thereby dimerize these receptors, just as normal FGF ligands do (Figure 6.35B; Williams et al. 1994a; Doherty and Walsch 1996).

Direct transmission of signals through gap junctions

Throughout this chapter we have been discussing the transduction of signals by cell membrane receptors. In these cases, the receptor is altered in some manner by binding a ligand, so that its cytoplasmic domain transmits the signal into the cell. Another signaling mechanism transmits small, soluble signaling molecules directly through the cell membrane, for the membrane is not continuous in all places; structures called **gap junctions** serve as communication channels between adjacent cells. Cells linked by gap junctions are said to be "coupled," and small molecules (molecular weight <1500) and ions can freely pass from one cell to the other. In most embryos, at least some of the early blastomeres are connected by gap junctions. These junctions allow epithelial sheets (i.e., closely connected cells) to act together. Moreover, the ability of cells to form gap junctions with some cells and not with others creates physiological "compartments" within the developing embryo.

Gap junction channels are made of **connexin** proteins. In each cell, six identical connexins group together in the membrane to form a transmembrane channel with a central pore. The channel complex of one cell connects to the channel complex of another cell, joining the cytoplasms of both cells. Different types of connexin proteins have separate but overlapping roles in normal development. For example, connexin-43 is found in nearly every tissue of the mouse embryo. However, the mouse embryo will still develop even if the gene for connexin-43 is "knocked out" by gene targeting (see Chapter 4). It appears that most of the functions of connexin-43 can be taken over by other, related connexin proteins—most, but not all. Shortly after birth, connexin-43-deficient mice take gasping breaths, turn bluish, and die. Autopsies of these mice show that the right ventricle (the chamber that pumps blood through the pulmonary artery to the lungs) is filled with tissue that occludes the chamber and obstructs blood flow (Reaume et al. 1995; Huang et al. 1998). Although loss of the connexin-43 protein can be compensated for in many tissues, it appears to be critical for normal heart development.

 WEBSITE **6.7 Connexin mutations.** Mutations in human connexin proteins cause congenital malformations of the heart and ear. In many cases, one connexin can substitute for another, but when the connexins cannot compensate, a mutant phenotype results.

The importance of gap junctions in development has been demonstrated in amphibian and mammalian embryos (Warner et al. 1984). When antibodies to connexins

(A)

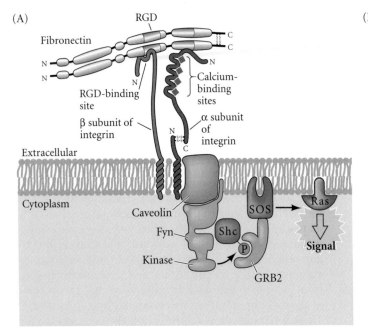

(B)

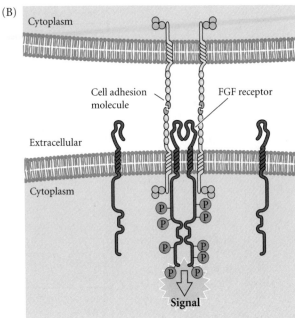

FIGURE 6.35   Two types of activation by cell adhesion molecules. (A) Cell-substrate adhesion molecules such as integrins may transmit a signal from the cytoplasmic portion of the integrin protein to the Ras G protein through a cascade involving the caveolin and Fyn proteins. (B) FGF receptors may be "hijacked" by cell adhesion molecules and dimerized. They may be brought together by the interaction of opposite cell adhesion molecules, or the "crosslinking" of FGF receptors by the apposing cell membrane may activate their kinase domains. (A after Wary et al. 1998.)

were microinjected into a single specific cell of an 8-cell *Xenopus* blastula, the progeny of that cell, which are usually coupled through gap junctions, could no longer pass ions or small molecules from cell to cell. The tadpoles that resulted from these treated blastulae showed defects specifically relating to the developmental fate of the injected cell (Figure 6.36). The progeny of the injected cell did not die, but they were unable to develop normally (Warner et al. 1984).

The first eight blastomeres of the mouse embryo are also connected to one another by gap junctions. Although these eight cells are only loosely associated, they move together to form a compacted embryo. If compaction is inhibited by antibodies against connexins, the treated blastomeres continue to divide but further development ceases (Lo and Gilula 1979; Lee et al. 1987). If antisense RNA to connexin messages is injected into one of the blastomeres of a normal mouse embryo, that cell will not form gap junctions and will not be included in the embryo (Bevilacqua et al. 1989).

Cross-Talk between Pathways

We have represented the major signal transduction pathways as if they were linear chains through which information flows in a single conduit. However, these pathways are just the major highways of information flow. Between them, avenues and streets connect one pathway with

(A)

(B)

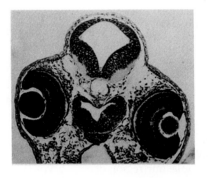
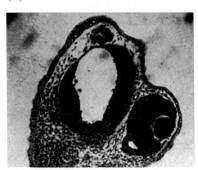

FIGURE 6.36   Developmental effects of gap junctions. Sections through *Xenopus* tadpole in which one of the blastomeres at the 8-cell stage was injected with (A) a control antibody or (B) an antibody against connexins. In the treated tadpole, the side formed by the injected blastomere lacks its eye and has abnormal brain morphology. (From Warner et al. 1984; photographs courtesy of A. E. Warner.)

another. (This may be why there are so many steps between the cell surface and the nucleus. Each step is a potential regulatory point as well as a potential intersection.) This **cross-talk** can be seen in numerous tissues, wherein two signaling pathways reinforce each other. We must remember that a cell has numerous receptors and is constantly receiving many signals simultaneously.

In some cells, gene transcription requires two different signals. This pattern is seen during lymphocyte differentiation, for which two signals are needed, each producing one of the two peptides of a transcription factor needed for the production of interleukin-2 (IL2, also known as T cell growth factor). One of these peptides, c-Fos, is induced by the binding of the T-cell receptor to an antigen (Figure 6.37). This signal activates the RTK pathway, creating a transcription factor, Elk1, that activates the *c-fos* gene to synthesize c-Fos. The second signal comes from the B7 glycoprotein on the surface of the cell presenting the antigen. This signal activates a second cascade of kinases, eventually producing c-Jun. The two peptides—c-Fos and c-Jun—join to make the AP1 protein, a transcription factor that binds to the IL2 enhancer and activates its expression (Li et al. 1996).

In other instances, the signal from one receptor can block the signal being given from a second receptor. As we will see throughout the book (for instance, in discussions of chick neural induction and limb development) FGF signals often prevent BMP signals from functioning. The MAP kinase activated by the FGF signal cascade will phosphorylate the Smad1 protein of the BMP cascade in a manner that prevents Smad1 from entering the nucleus (Pera et al. 2003).

We also have seen that one receptor can activate several different pathways. The fibroblast growth factors, for instance, can activate the RTK pathway, the STAT pathway, or even a third pathway that involves lipid turnover and increases Ca^{2+} levels in the cell. Similarly, the proteins coupled to the Frizzled receptors will determine what effects Wnt proteins will have on the cells.

Maintenance of the Differentiated State

Development obviously means more than initiating gene expression. For a cell to become committed to a particular phenotype, gene expression must be maintained. There are four major ways that nature has evolved for maintaining differentiation once it has been initiated (Figure 6.38):

1. The transcription factor whose gene is activated by a signal transduction cascade can bind to the enhancer of its own gene. In this way, once the transcription factor is made, its synthesis becomes independent of the signal that induced it originally. The MyoD transcription factor in muscle cells is produced in this manner.

2. A cell can stabilize its differentiation by synthesizing proteins that act on chromatin to keep the gene accessible. Such proteins include the Trithorax family.

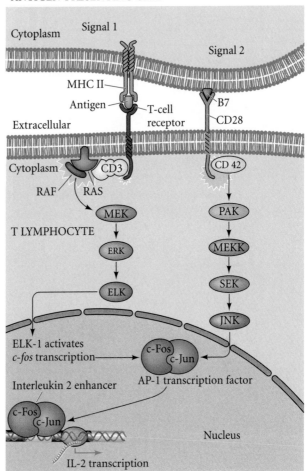

FIGURE 6.37 Two signals are needed to effect the differentiation of the T lymphocyte. The first signal comes from the receptors that bind the antigen. The second signal comes from the binding of the CD28 transmembrane protein of the T cell to the B7 protein on the surface of the antigen-presenting cell. The first signal directs the synthesis of one subunit of the AP-1 transcription factor, c-Fos. The second signal directs the synthesis of the other subunit, c-Jun. Together c-Fos and c-Jun form the AP1 transcription factor, which can activate T cell-specific enhancers such as that regulating IL2 production.

3. A cell can maintain its differentiation in an autocrine fashion. If differentiation is dependent on a particular signaling molecule, the cell can make both that signaling molecule and that molecule's receptor. The Sertoli cells of the mammalian testis may maintain their differentiation through such a self-stimulatory autocrine loop.

4. A cell may interact with its neighboring cells such that each one stimulates the differentiation of the other, and part of each neighbor's differentiated phenotype is the production of a paracrine factor that stimulates the other's phenotype. This type of I-scratch-your-back-you-scratch-mine strategy is found in the neighboring cells of the developing vertebrate limb and insect segments.

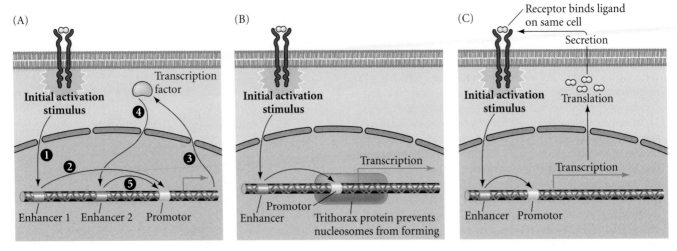

(A)

Transcription factor

Initial activation stimulus

① ② ③

⑤ ④

Enhancer 1   Enhancer 2   Promotor

(B)

Initial activation stimulus

Transcription

Promotor
Enhancer   Trithorax protein prevents nucleosomes from forming

(C)

Receptor binds ligand on same cell

Secretion

Initial activation stimulus   Translation

Transcription

Enhancer   Promotor

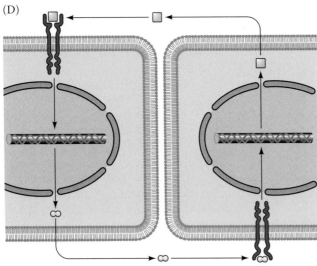

(D)

FIGURE 6.38 Four ways of maintaining differentiation after the initial signal has been given. (A) The initial stimulus (1) activates enhancer 1, which stimulates the promoter (2) to transcribe the gene. The gene product (3) is a transcription factor, and one of its targets is enhancer 2 of its own gene (4). This activated enhancer can now stimulate the promoter (5) to make more of this protein. (B) Proteins such as those of the Trithorax group prevent nucleosomes from forming so that the gene remains accessible. (C) Autocrine stimulation of the differentiated state. The cell is stimulated to activate those genes that enable it to synthesize and bind its own stimulatory proteins. (D) Paracrine loop between two cells such that a paracrine factor from one cell stimulates the differentiated state of the second cell, and that differentiated state includes the secretion of a paracrine factor that maintains the first cell's differentiated state.

COMMUNITY EFFECT: AUTOCRINE STIMULATION TO MAINTAIN DEVELOPMENT The third mechanism mentioned above leads to a phenomenon that has been called "mass effect, homotypic induction"(Grobstein 1955; Saxén and Wartiovaara 1966) and, more recently, "community effect" (Gurdon 1988). This effect is the capacity to express developmental potential only when a critical cell density of induced cells is present. In other words, once a group of cells has been induced, autocrine factors can sustain that induction and complete their differentiation. For instance, in *Xenopus* and mice, once members of the TGF-β family have instructed certain mesodermal cells to become muscles, these cells will continue to differentiate only if they are bound together in a group. They will fail to differentiate if each cell is separated from the others (Gurdon et al. 1993; Cossu et al. 1995). Similarly, presumptive neural cells from *Drosophila* will differentiate into neurons (even when transplanted to regions of epidermis) if they remain in a group; but they will not differentiate as individuals.*

In *Xenopus* muscle development, this community effect is mediated through FGF signaling. Standley and colleagues (2001) have shown (1) that FGF signaling can simulate the community effect in isolated muscle precursor cells; (2) that the muscle precursor cells have the receptors for FGFs at the critical time; and (3) that the muscle precursor cells express FGFs at this time. It thus appears that part of the developmental program for *Xenopus* muscle cells is to make an FGF protein to which they can also respond (i.e., an autocrine factor). This protein has to be in

*Community effect is also extremely important in bacterial development. Here it is called "quorum sensing," and it is critical in permitting emergent phenotypes such as light production, biofilm formation, invasiveness, and virulence. These phenotypes are expressed only in groups of bacteria and not in individuals. Each bacterium makes a small amount of a diffusible autocrine inducer that will induce the phenotype only at relatively high concentrations (see Zhu et al. 2002; Podbielski and Kreikemeyer 2004).

sufficiently high density for the continuation of the processes leading to muscle development. Autocrine FGF signaling may also be responsible for a community effect enabling adult human pre-adipocytes to continue their differentiation into fat cells (Patel et al. 2004). It has been proposed that suppressing the FGF-mediated community effect might become a mechanism for medically controlling obesity.

Coda

In 1782, the French essayist Denis Diderot posed the question of morphogenesis in the fevered dream of a noted physicist. This character could imagine that the body was formed from myriad "tiny sensitive bodies" that collected together to form an aggregate, but he could not envision how this aggregate could become an animal. A hundred years later, Charles Darwin wrote that the eye must have evolved by small useful variations, but he did not know the mechanisms by which the intricate coordination of eye development occurred.

Recent studies have shown that this coordination of cells to form organs and organisms depends on the molecules on the embryonic cell surfaces. Inducers and their competent responders interact with one another to instruct and coordinate the further development of the component parts. In subsequent chapters, we will look more closely at some of these morphogenetic interactions. As we continue our study of animal development, we will find that induction and competence provide the core of morphogenesis. Moreover, we will find the same signal transduction pathways wherever we look, both throughout the animal kingdom and within each developing embryo. We are now at the stage where we can begin our study of early embryogenesis and see the integration of the organismal, genetic, and cellular processes of animal development.

 **Cell-Cell Communication**

1. Inductive interactions involve inducing and responding tissues.

2. The ability to respond to inductive signals depends upon the competence of the responding cells.

3. Reciprocal induction occurs when the two interacting tissues are both inducers and are competent to respond to each other's signals.

4. Cascades of inductive events are responsible for organ formation.

5. Regionally specific inductions can generate different structures from the same responding tissue.

6. The specific response to an inducer is determined by the genome of the responding tissue.

7. Paracrine interactions occur when a cell or tissue secretes proteins that induce changes in neighboring cells. Juxtacrine interactions are inductive interactions that take place between the cell membranes of adjacent cells or between a cell membrane and an extracellular matrix secreted by another cell.

8. Paracrine factors are proteins secreted by inducing cells. These factors bind to cell membrane receptors in competent responding cells.

9. Competent cells respond to paracrine factors through signal transduction pathways. Competence is the ability to bind and to respond to the inducers, and it is often the result of a prior induction.

10. Signal transduction pathways begin with a paracrine or juxtacrine factor causing a conformational change in its cell membrane receptor. The new shape results in enzymatic activity in the cytoplasmic domain of the receptor protein. This activity allows the receptor to phosphorylate other cytoplasmic proteins. Eventually, a cascade of such reactions activates a transcription factor (or set of factors) that activates or represses specific gene activity.

11. Programmed cell death is one possible response to inductive stimuli. Apoptosis is a critical part of life.

12. Gap junctions allow ions and small molecules to move between cells and facilitate coordinated action of coupled cells.

13. There is cross-talk between signal transduction pathways, which allows the cell to respond to multiple inputs simultaneously.

14. The maintenance of the differentiated state can be accomplished by positive feedback loops involving transcription factors, autocrine factors, or paracrine factors.

For Further Reading

Complete bibliographical citations for all literature cited in this chapter can be found on the Vade Mecum CD that accompanies the book and at the free access website www.devbio.com

Cadigan, K. M. and R. Nusse. 1997. Wnt signaling: A common theme in animal development. *Genes Dev.* 24: 3286–3306.

Cooper, M. K., J. A. Porter, K. E. Young and P. A. Beachy. 1998. Teratogen-mediated inhibition of target tissue response to hedgehog signaling. *Science* 280: 1603–1607.

Cvekl, A. and J. Piatigorsky. 1996. Lens development and crystallin gene expression: Many roles for Pax-6. *BioEssays* 18: 621–630.

Davy, A. and P. Soriano. 2005. Ephrin signaling in vivo: Look both ways. *Dev. Dyn.* 232: 1–10.

Grainger, R. M. 1992. Embryonic lens induction: Shedding light on vertebrate tissue determination. *Trends Genet.* 8: 349–356.

Heldin, C.-H., K. Miyazono and P. ten Dijke. 1997. TGF-β signaling from cell membrane to nucleus through SMAD proteins. *Nature* 390: 465–471.

Hemesath, T. J., E. R. Price, C. Takemoto, T. Badalian and D. E. Fisher. 1998. MAP kinase links the transcription factor Microphthalmia to c-Kit signalling in melanocytes. *Nature* 391: 298–301.

Hogan, B. L. M. 1996. Bone morphogenesis proteins: Multifunctional regulators of vertebrate development. *Genes Dev.* 10: 1580–1594.

Ihle, J. N. 1996. STATs: Signal transducers and activators of transcription. *Cell* 84: 331–334.

Jacobson, A. G. 1966. Inductive processes in embryonic development. *Science* 152: 25–34.

Katz, W. S., R. J. Hill, T. R. Clandenin and P. W. Sternberg. 1995. Different levels of the *C. elegans* growth factor LIN-3 promote distinct vulval precursor fates. *Cell* 82: 297–307.

Kopan, R. 2002. Notch: a membrane-bound transcription factor. *J. Cell Sci.* 115: 1095–1097.

Lum, L. and P. A. Beachy. 2004. The hedgehog response network: sensors, switches, and routers. *Science* 304: 1755–1759.

Riddle, R. D., R. L. Johnson, E. Laufer and C. Tabin. 1993. Sonic hedgehog mediates the polarizing activity of the ZPA. *Cell* 75: 1401–1416.

Rousseau, F. and 7 others. 1994. Mutations in the gene encoding fibroblast growth factor receptor-3 in achondroplasia. *Nature* 371: 252–254.

Tulina, N. and E. Matunis. 2001. Control of stem cell self-renewal in *Drosophila* spermatogenesis by JAK-STAT signaling. *Science* 294: 2546–2549.

Vaahtokari, A., T. Aberg and I. Thesleff. 1996. Apoptosis in the developing tooth: Association with an embryonic signaling center and suppression by EGF and FGF-4. *Development* 122: 121–129.

PART 2

Early Embryonic Development

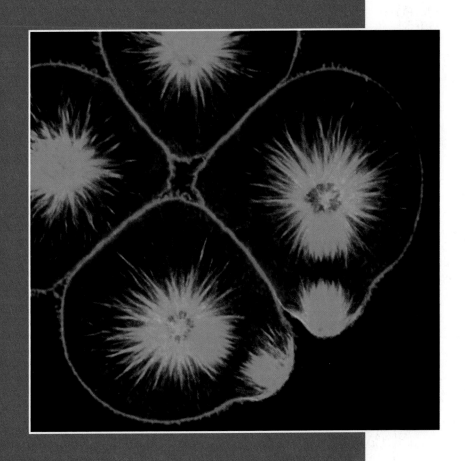

Part 2 Early Embryonic Development

PREVIOUS PAGE
This confocal micrograph shows a section through 4 cells of the 8-cell sand dollar (echinoderm) embryo, visualizing the cytoskeleton. Microtubules are stained green and microfilaments are stained red. The asymmetrical cell division seen here in the two vegetal cells (right) generates the macromeres and micromeres. Photograph courtesy of George von Dassow, Center for Cell Dynamics.

Fertilization: Beginning a New Organism

FERTILIZATION IS THE PROCESS whereby the sperm and the egg—collectively called the **gametes**—fuse together to begin the creation of a new individual whose genome is derived from both parents. Fertilization accomplishes two separate ends: sex (the combining of genes derived from two parents) and reproduction (the creation of a new organism). Thus, the first function of fertilization is to transmit genes from parent to offspring, and the second is to initiate in the egg cytoplasm those reactions that permit development to proceed.

Although the details of fertilization vary from species to species, conception generally consists of four major events:

1. Contact and recognition between sperm and egg. In most cases, this ensures that the sperm and egg are of the same species.
2. Regulation of sperm entry into the egg. Only one sperm nucleus can ultimately unite with the egg nucleus. This is usually accomplished by allowing only one sperm to enter the egg and inhibiting any others from entering.
3. Fusion of the genetic material of sperm and egg.
4. Activation of egg metabolism to start development.

Structure of the Gametes

A complex dialogue exists between egg and sperm. The egg activates the sperm metabolism that is essential for fertilization, and the sperm reciprocates by activating the egg metabolism needed for the onset of development. But before we investigate these aspects of fertilization, we need to consider the structures of the sperm and egg—the two cell types specialized for fertilization.

Sperm

It is only within the past 130 years that the sperm's role in fertilization has been known. Anton van Leeuwenhoek, the Dutch microscopist who co-discovered sperm in 1678, first believed them to be parasitic animals living within the semen (hence the term *spermatozoa*, meaning "sperm animals"). Although he originally assumed that they had nothing to do with reproducing the organism in which they were found, he later came to believe that each sperm contained a preformed embryo. Leeuwenhoek (1685) wrote that sperm were seeds (both *sperma* and *semen* mean "seed"), and that the female merely provided the nutrient soil in which the seeds were planted. In this, he was returning to a notion of procreation promulgated by Aristotle 2000 years earlier.

> " Urge and urge and urge, Always the procreant urge of the world. Out of the dimness opposite equals advance, Always substance and increase, always sex, Always a knit of identity, always distinction, Always a breed of life. "
>
> *WALT WHITMAN (1855)*

> " The final aim of all love intrigues, be they comic or tragic, is really of more importance than all other ends in human life. What it turns upon is nothing less than the composition of the next generation. "
>
> *A. SCHOPENHAUER (QUOTED BY C. DARWIN, 1871)*

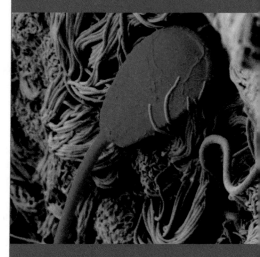

Try as he might, Leeuwenhoek was continually disappointed in his attempts to find preformed embryos within spermatozoa. Nicolas Hartsoeker, the other co-discoverer of sperm, drew a picture of what he hoped to find: a miniscule human ("homunculus") within the sperm (Figure 7.1). This belief that the sperm contained the entire embryonic organism never gained much acceptance, as it implied an enormous waste of potential life. Most investigators regarded the sperm as unimportant.*

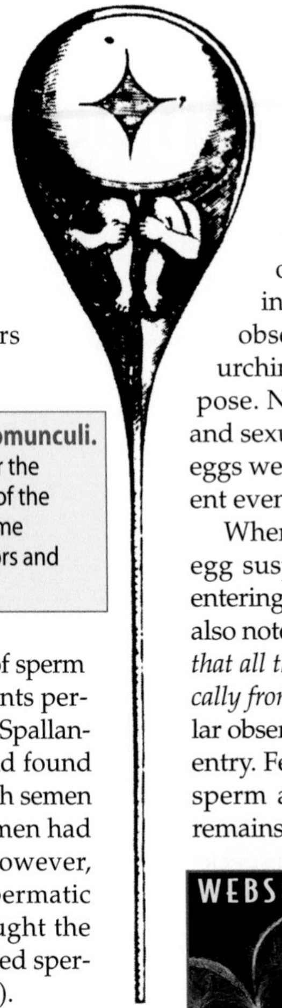

WEBSITE **7.1 Leeuwenhoek and images of homunculi.** Scholars in the 1600s thought that either the sperm or the egg carried the rudiments of the adult body. Moreover, these views became distorted by contemporary commentators and later historians.

The first evidence suggesting the importance of sperm in reproduction came from a series of experiments performed by Lazzaro Spallanzani in the late 1700s. Spallanzani put the male toads into taffeta breeches and found toad semen so filtered to be devoid of sperm; such semen did not fertilize eggs. He even showed that semen had to touch the eggs in order to be functional. However, Spallanzani, like many others, felt that the spermatic "animals" were parasites in the fluid; he thought the embryo was contained within the egg and needed spermatic fluid to activate it (see Pinto-Correia 1997).

The combination of better microscopic lenses and the elucidation of the cell theory (that all life is cellular, and all cells come from preexisting cells) led to a new appreciation of spermatic function. In 1824, J. L. Prevost and J. B. Dumas claimed that sperm were not parasites, but rather the active agents of fertilization. They noted the universal existence of sperm in sexually mature males and their absence in immature and aged individuals. These observations, coupled with the known absence of sperm in the sterile mule, convinced them that "there exists an intimate relation between their presence in the organs and the fecundating capacity of the animal." They proposed that the sperm entered the egg and contributed materially to the next generation.

These claims were largely disregarded until the 1840s, when A. von Kolliker described the formation of sperm from cells within the adult testes. He ridiculed the idea that the semen could be normal and yet support such an enormous number of parasites. Even so, von Kolliker denied there was any physical contact between sperm and egg. He believed that the sperm excited the egg to develop in much the same way as a magnet communicates its pres-

FIGURE 7.1 The human infant preformed in the sperm, as depicted by Nicolas Hartsoeker (1694).

ence to iron. It was not until 1876 that Oscar Hertwig and Herman Fol independently demonstrated sperm entry into the egg and the union of the two cells' nuclei. Hertwig had been seeking an organism suitable for detailed microscopic observations, and he found the Mediterranean sea urchin, *Toxopneustes lividus,* to be perfect for this purpose. Not only was it common throughout the region and sexually mature throughout most of the year, but its eggs were available in large numbers and were transparent even at high magnifications.

When he mixed suspensions of sperm together with egg suspensions, Hertwig repeatedly observed sperm entering the eggs and saw sperm and egg nuclei unite. He also noted that *only one sperm was seen to enter each egg, and that all the nuclei of the resulting embryo were derived mitotically from the nucleus created at fertilization.* Fol made similar observations and also detailed the mechanism of sperm entry. Fertilization was at last recognized as the union of sperm and egg, and the union of sea urchin gametes remains one of the best-studied examples of fertilization.

WEBSITE **7.2 The origins of fertilization research.** Studies by Hertwig, Fol, Boveri, and Auerbach investigated fertilization by integrating cytology with genetics. The debates over meiosis and nuclear structure were critical in these investigations of fertilization.

Each sperm cell consists of a haploid nucleus, a propulsion system to move the nucleus, and a sac of enzymes that enable the nucleus to enter the egg. In most species (the nematodes are the major exceptions), almost all of the cell's cytoplasm is eliminated during sperm maturation, leaving only certain organelles that are modified for spermatic function (Figure 7.2). During the course of sperm maturation, the haploid nucleus becomes very streamlined and its DNA becomes tightly compressed. In front of this compressed haploid nucleus lies the **acrosomal vesicle**, or **acrosome**. The acrosome is derived from the Golgi apparatus and contains enzymes that digest proteins and complex sugars; thus, the acrosome can be considered a modified secretory vesicle. The enzymes stored in the acrosome are used to lyse the outer coverings of the egg. In many species, a region of globular actin proteins lies between the sperm nucleus and the acrosomal vesicle. These proteins are used to extend a fingerlike **acrosomal process** from the sperm during the early stages of fertilization. In sea urchins and several other species, recognition between sperm and egg involves molecules on the acrosomal process. Together, the acrosome and nucleus constitute the **head** of the sperm.

The means by which sperm are propelled vary according to how the species has adapted to environmental condi-

---

*Indeed, sperm was discovered in 1676, while the events of fertilization were not elucidated until 1876. Thus, for nearly 200 years, people had no idea what the sperm actually did. See Pinto-Correia 1997 for details of this remarkable story.

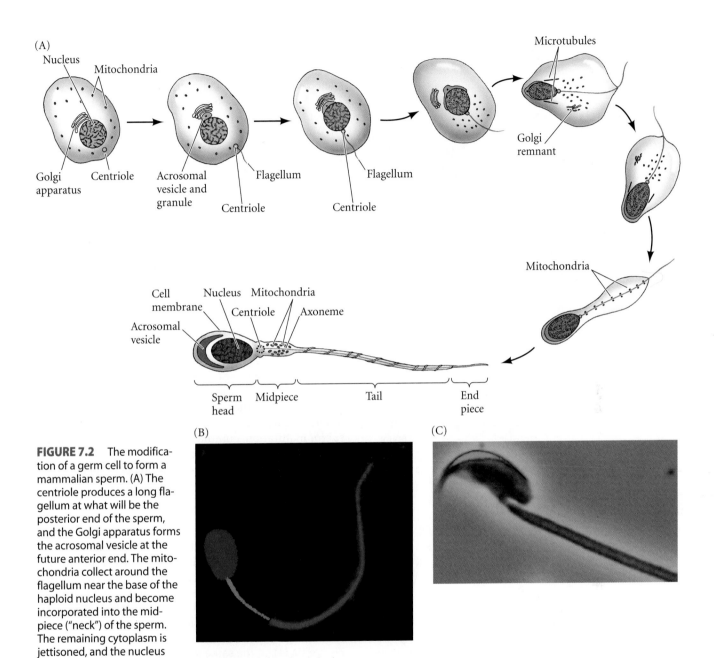

FIGURE 7.2 The modification of a germ cell to form a mammalian sperm. (A) The centriole produces a long flagellum at what will be the posterior end of the sperm, and the Golgi apparatus forms the acrosomal vesicle at the future anterior end. The mitochondria collect around the flagellum near the base of the haploid nucleus and become incorporated into the midpiece ("neck") of the sperm. The remaining cytoplasm is jettisoned, and the nucleus condenses. The size of the mature sperm has been enlarged relative to the other stages. (B) Mature bull sperm. The DNA is stained blue with DAPI; the mitochondria are stained green, and the tubulin of the flagellum is stained red. (C) Acrosome of mouse sperm, stained green by the fusion protein proacrosin-GFP. (A after Clermont and Leblond 1955; B from Sutovsky et al. 1996, photograph courtesy of G. Schatten; C, photograph courtesy of K.-S. Kim and G. L. Gerton.)

tions. In most species (the major exception once again being the nematodes, where the sperm is formed at the sites where fertilization occurs), an individual sperm is able to travel by whipping its **flagellum**. The major motor portion of the flagellum is the **axoneme**, a structure formed by microtubules emanating from the centriole at the base of the sperm nucleus (Figure 7.3A). The core of the axoneme consists of two central microtubules surrounded by a row of 9 doublet microtubules. Actually, only one microtubule of each doublet is complete, having 13 protofilaments; the other is C-shaped and has only 11 protofilaments (Figure 7.3B). The interconnected protofilaments are made exclusively of the dimeric protein **tubulin**.

Although tubulin is the basis for the structure of the flagellum, other proteins are also critical for flagellar function. The force for sperm propulsion is provided by **dynein**, a protein attached to the microtubules (see Figure 7.3B). Dynein is an ATPase, an enzyme that hydrolyzes ATP, converting the released chemical energy into mechanical energy to propel the sperm. This energy allows the active sliding of the outer doublet microtubules, causing the flagellum to bend (Ogawa et al. 1977; Shinyoji et al.

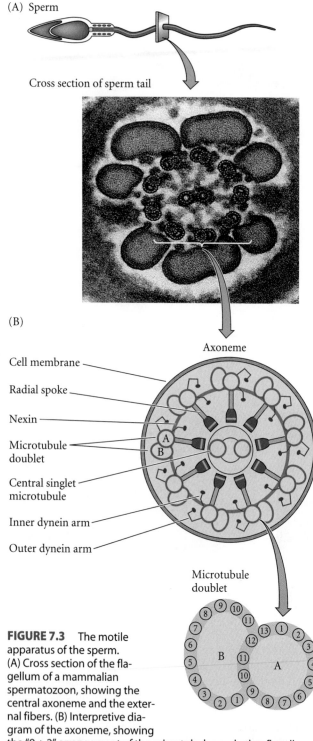

(A) Sperm

Cross section of sperm tail

(B)

Axoneme

Cell membrane

Radial spoke

Nexin

Microtubule doublet

Central singlet microtubule

Inner dynein arm

Outer dynein arm

Microtubule doublet

FIGURE 7.3   The motile apparatus of the sperm. (A) Cross section of the flagellum of a mammalian spermatozoon, showing the central axoneme and the external fibers. (B) Interpretive diagram of the axoneme, showing the "9 + 2" arrangement of the microtubules and other flagellar components. The schematic diagram through a sea urchin sperm flagellum shows the association of tubulin protofilaments into a microtubule doublet. The first ("A") portion of the doublet is a normal microtubule comprising 13 protofilaments. The second ("B") portion of the doublet contains only 11 (occasionally 10) protofilaments. The dynein arms contain the ATPases that provide the energy for flagellar movement. (A, photograph courtesy of D. M. Phillips; B after De Robertis et al. 1975 and Tilney et al. 1973.)

1998). The importance of dynein can be seen in individuals with a genetic syndrome known as the Kartagener triad. These individuals lack dynein on all their ciliated and flagellated cells, rendering these structures immotile (Afzelius 1976). Males with this disease are sterile (immotile sperm), susceptible to bronchial infections (immotile respiratory cilia), and have a 50 percent chance of having the heart on the right side of the body (immotile cilia in the center of the embryo; see Chapter 11).

The "9 + 2" microtubule arrangement with dynein arms has been conserved in axonemes throughout the eukaryotic kingdoms, suggesting that this arrangement is extremely well suited for transmitting energy for movement. The ATP needed to whip the flagellum and propel the sperm comes from rings of mitochondria located in the **midpiece** of the sperm. In many species (notably mammals), a layer of dense fibers has interposed itself between the mitochondrial sheath and the axoneme. This fiber layer stiffens the sperm tail. Because the thickness of this layer decreases toward the tip, the fibers probably prevent the sperm head from being whipped around too suddenly. Thus, the sperm cell has undergone extensive modification for the transport of its nucleus to the egg.

In mammals, the sperm released during ejaculation are able to move, but they do not yet have the capacity to bind to and fertilize an egg. The final stages of mammalian sperm maturation, cumulatively referred to as **capacitation**, do not occur until the sperm has been inside the female reproductive tract for a certain period of time.

The egg

All the material necessary for the beginning of growth and development must be stored in the mature egg, or **ovum**. Whereas the sperm eliminates most of its cytoplasm as it matures, the developing egg (called the **oocyte** before it reaches the stage of meiosis at which it is fertilized) not only conserves the material it has, but actively accumulates more. The meiotic divisions that form the oocyte conserve its cytoplasm rather than giving half of it away (see Figure 19.24); at the same time the oocyte either synthesizes or absorbs proteins such as yolk that act as food reservoirs for the developing embryo. Thus, birds' eggs are enormous single cells, swollen with accumulated yolk. Even eggs with relatively sparse yolk are large compared with sperm. The volume of a sea urchin egg is about 200 picoliters (2×10^{-4} mm^3), more than 10,000 times the volume of sea urchin sperm (Figure 7.4). So, even though sperm and egg have equal haploid *nuclear* components, the egg also accumulates a remarkable cytoplasmic storehouse during its maturation.* This cytoplasmic trove includes the following:

---

*The contents of the egg vary greatly from species to species. The synthesis and placement of these materials will be addressed in Chapter 19, when we discuss the differentiation of the germ cells.

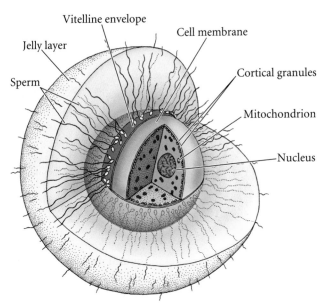

FIGURE 7.4 Structure of the sea urchin egg at fertilization. The drawing shows the relative sizes of egg and sperm. (After Epel 1977.)

tein synthesis is accomplished by ribosomes and tRNA, which exist in the egg. The developing egg has special mechanisms for synthesizing ribosomes; certain amphibian oocytes produce as many as 10^{12} ribosomes during their meiotic prophase.

- **Messenger RNA.** The oocyte not only accumulates proteins, it also accumulates mRNAs that encode proteins for the early stages of development. It is estimated that sea urchin eggs contain thousands of different types of mRNA that remain repressed until after fertilization (see Chapter 5).
- **Morphogenic factors.** Molecules that direct the differentiation of cells into certain cell types are present in the egg. These include transcription factors and paracrine factors. In many species, they are localized in different regions of the egg and become segregated into different cells during cleavage (see Chapter 8).
- **Protective chemicals.** The embryo cannot run away from predators or move to a safer environment, so it must come equipped to deal with threats. Many eggs contain ultraviolet filters and DNA repair enzymes that protect them from sunlight. Some eggs contain molecules that potential predators find distasteful, and the yolk of bird eggs even contains antibodies.

Within this enormous volume of egg cytoplasm resides a large nucleus. In a few species (such as sea urchins), the **female pronucleus** is already haploid at the time of fertilization. In other species (including many worms and most mammals), the egg nucleus is still diploid—the sperm enters before the egg's meiotic divisions are completed (Figure 7.5). In these species, the final stages of egg meiosis will take place while the sperm's nuclear material (the **male pronucleus**) is traveling toward what will be the female pronucleus.

- **Proteins.** It will be a long time before the embryo is able to feed itself or even obtain food from its mother. The early embryonic cells need a supply of energy and amino acids. In many species, this is accomplished by accumulating yolk proteins in the egg. Many of these yolk proteins are made in other organs (e.g., liver, fat bodies) and travel through the maternal blood to the oocyte.
- **Ribosomes and tRNA.** The early embryo needs to make many of its own proteins, and in some species, there is a burst of protein synthesis soon after fertilization. Pro-

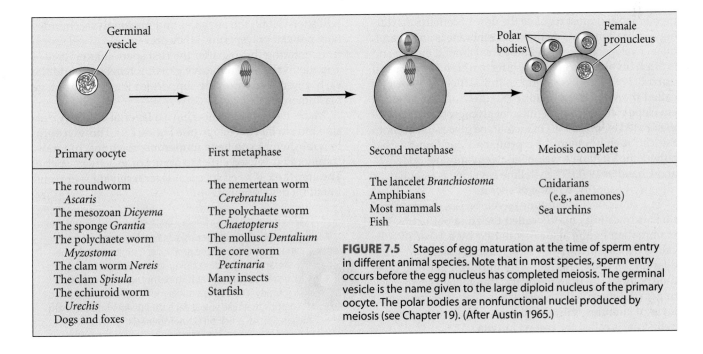

Primary oocyte	First metaphase	Second metaphase	Meiosis complete
The roundworm *Ascaris*	The nemertean worm *Cerebratulus*	The lancelet *Branchiostoma*	Cnidarians (e.g., anemones)
The mesozoan *Dicyema*	The polychaete worm *Chaetopterus*	Amphibians	Sea urchins
The sponge *Grantia*	The mollusc *Dentalium*	Most mammals	
The polychaete worm *Myzostoma*	The core worm *Pectinaria*	Fish	
The clam worm *Nereis*	Many insects		
The clam *Spisula*	Starfish		
The echiuroid worm *Urechis*			
Dogs and foxes			

FIGURE 7.5 Stages of egg maturation at the time of sperm entry in different animal species. Note that in most species, sperm entry occurs before the egg nucleus has completed meiosis. The germinal vesicle is the name given to the large diploid nucleus of the primary oocyte. The polar bodies are nonfunctional nuclei produced by meiosis (see Chapter 19). (After Austin 1965.)

(A)

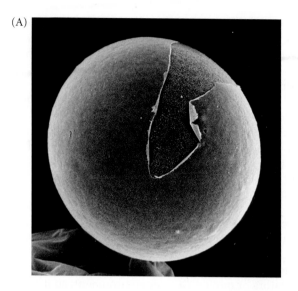

(B)

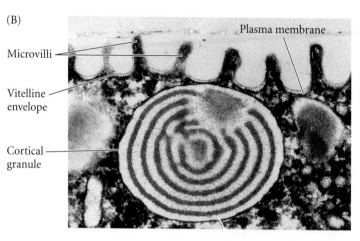

FIGURE 7.6   The sea urchin egg cell surface. (A) Scanning electron micrograph of an egg before fertilization. The plasma membrane is exposed where the vitelline envelope has been torn. (B) Transmission electron micrograph of an unfertilized egg, showing microvilli and plasma membrane, which are closely covered by the vitelline envelope. A cortical granule lies directly beneath the plasma membrane. (From Schroeder 1979; photographs courtesy of T. E. Schroeder.)

 WEBSITE   **7.3 The egg and its environment.** The laboratory is not where most eggs are found. Eggs have evolved remarkable ways to protect themselves in particular environments.

Enclosing the cytoplasm is the egg **cell membrane**. This membrane must be capable of fusing with the sperm cell membrane and must regulate the flow of certain ions during fertilization. Outside the cell membrane is an extracellular envelope that forms a fibrous mat around the egg and is often involved in sperm-egg recognition (Correia and Carroll 1997). In invertebrates, this structure is usually called the **vitelline envelope** (Figure 7.6). The vitelline envelope contains several different glycoproteins. It is supplemented by extensions of membrane glycoproteins from the cell membrane and by proteinaceous "posts" that adhere the vitelline envelope to the membrane (Mozingo and Chandler 1991). The vitelline envelope is essential for the species-specific binding of sperm.

In mammals, the vitelline envelope is a separate and thick extracellular matrix called the **zona pellucida**. The mammalian egg is also surrounded by a layer of cells called the **cumulus** (Figure 7.7), which is made up of the ovarian follicular cells that were nurturing the egg at the time of its release from the ovary. Mammalian sperm have to get past these cells to fertilize the egg. The innermost layer of cumulus cells, immediately adjacent to the zona pellucida, is called the **corona radiata**.

Lying immediately beneath the cell membrane of the egg is a thin shell (about 5 μm) of gel-like cytoplasm called the **cortex**. The cytoplasm in this region is stiffer than the internal cytoplasm and contains high concentrations of globular actin molecules. During fertilization, these actin molecules polymerize to form long cables of actin known as **microfilaments**. Microfilaments are necessary for cell division, and they are also used to extend the egg surface into small projections called **microvilli**, which may aid sperm entry into the cell (see Figure 7.6B; also see Figure 7.15).

Also within the cortex are the **cortical granules** (see Figures 7.4 and 7.6B). These membrane-bound, Golgi-derived structures contain proteolytic enzymes and are thus homologous to the acrosomal vesicle of the sperm. However, whereas a sea uchin sperm contains just one acrosomal vesicle, each sea urchin egg contains approximately 15,000 cortical granules. Moreover, in addition to digestive enzymes, the cortical granules contain mucopolysaccharides, adhesive glycoproteins, and hyalin protein. As we will soon detail, the enzymes and mucopolysaccharides help prevent polyspermy—they prevent additional sperm from entering the egg after the first sperm has entered—and the hyalin and adhesive glycoproteins surround the early embryo and provide support for the cleavage-stage blastomeres.

Many types of eggs also have a layer of **egg jelly** outside the vitelline envelope (see Figure 7.4). This glycoprotein meshwork can have numerous functions, but most commonly it is used either to attract or to activate sperm. The egg, then, is a cell specialized for receiving sperm and initiating development.

VADE MECUM²   Gametogenesis. Stained sections of testis and ovary illustrate the process of gametogenesis, the streamlining of developing sperm, and the remarkable growth of the egg as it stores nutrients for its long journey. You can see this in movies and labeled photographs that take you at each step deeper into the mammalian gonad. [Click on Gametogenesis]

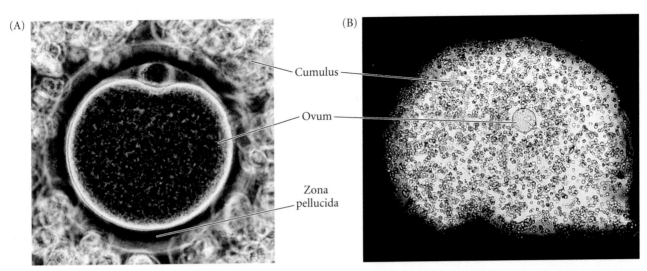

(A) (B)

Cumulus

Ovum

Zona
pellucida

FIGURE 7.7 Hamster eggs immediately before fertilization. (A) The hamster egg, or ovum, is encased in the zona pellucida. This, in turn, is surrounded by the cells of the cumulus. A polar body cell, produced during meiosis, is also visible within the zona pellucida. (B) At lower magnification, a mouse oocyte is shown surrounded by the cumulus. Colloidal carbon particles (India ink) are excluded by the hyaluronidate matrix. (Photographs courtesy of R. Yanagimachi.)

Recognition of egg and sperm

The interaction of sperm and egg generally proceeds according to five basic steps (Figure 7.8; Vacquier 1998):

1. The chemoattraction of the sperm to the egg by soluble molecules secreted by the egg
2. The exocytosis of the acrosomal vesicle to release its enzymes
3. The binding of the sperm to the extracellular envelope (vitelline layer or zona pellucida) of the egg
4. The passage of the sperm through this extracellular envelope
5. Fusion of egg and sperm cell membranes

Sometimes steps 2 and 3 are reversed (as in mammalian fertilization; see Figure 7.8B), and the sperm binds to the extracellular matrix of the egg before releasing the contents of the acrosome. After these five steps are accomplished, the haploid sperm and egg nuclei can meet and the reactions that initiate development can begin. In this chapter, we will focus on the fertilization events of sea urchins, which undergo external fertilization, and mice, which undergo internal fertilization. In subsequent chapters, the variations of fertilization will be described as we study the development of particular organisms.

External Fertilization in Sea Urchins

In many species, the meeting of sperm and egg is not a simple matter. Many marine organisms release their gametes into the environment. That environment may be as small as a tide pool or as large as an ocean (Mead and Epel 1995). Moreover, this environment is shared with other species that may shed their sex cells at the same time.

Such organisms are faced with two problems: How can sperm and eggs meet in such a dilute concentration, and how can sperm be prevented from trying to fertilize eggs of another species? Two major mechanisms have evolved to solve these problems: species-specific attraction of sperm and species-specific sperm activation. Here we describe these events as they occur in one well-studied group, the sea urchins.

Sperm attraction: Action at a distance

Species-specific sperm attraction has been documented in numerous species, including cnidarians, molluscs, echinoderms, and urochordates (Miller 1985; Yoshida et al. 1993). In many species, sperm are attracted toward eggs of their species by chemotaxis—that is, by following a gradient of a chemical secreted by the egg. In 1978, Miller demonstrated that the eggs of the cnidarian *Orthopyxis caliculata* not only secrete a chemotactic factor but also regulate the timing of its release. Developing oocytes at various stages in their maturation were fixed on microscope slides, and sperm were released at a certain distance from the eggs. Miller found that when sperm were added to oocytes that had not yet completed their second meiotic division, there was no attraction of sperm to eggs. However, after the second meiotic division was finished and the eggs were ready to be fertilized, the sperm migrated toward them. Thus, these oocytes control not only the type of sperm they attract, but also the time at which they attract them.

The mechanisms of chemotaxis differ among species (see Metz 1978; Eisenbach 2004), and the chemotactic molecules are different even in closely related species. In sea urchins, sperm motility is acquired when the sperm are spawned into seawater. As long as sperm cells are in the

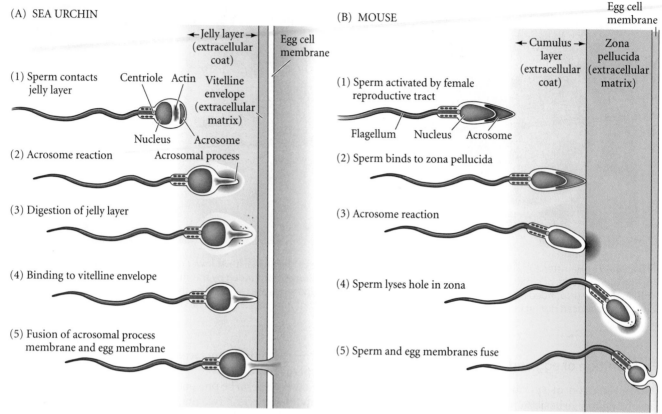

FIGURE 7.8   Summary of events leading to the fusion of egg and sperm plasma membranes in (A) the sea urchin and (B) the mouse. (A) Sea urchin fertilization is external. (1) The sperm is chemotactically attracted to and activated by the egg. (2, 3) Contact with the egg jelly triggers the acrosome reaction, allowing the acrosomal process to form and release proteolytic enzymes. (4) The sperm adheres to the vitelline envelope and lyses a hole in it. (5) The sperm adheres to the egg plasma membrane and fuses with it. The sperm pronucleus can now enter the egg cytoplasm. (B) Mammalian fertilization is internal. (1) The contents of the female reproductive tract capacitate, attract, and activate the sperm. (2) The acrosome-intact sperm binds to the zona pellucida, which is thicker than the vitelline envelope of sea urchins. (3) The acrosome reaction occurs on the zona pellucida. (4) The sperm digests a hole in the zona pellucida. (5) The sperm adheres to the egg, and their plasma membranes fuse.

testes, they cannot move because their internal pH is kept low (about pH 7.2) by the high concentrations of CO_2 in the gonad. However, once spawned into seawater, the sperm's pH is elevated to about 7.6, resulting in the activation of the dynein ATPase. The splitting of ATP provides the energy for the flagella to wave, and the sperm begin swimming vigorously (Christen et al. 1982).

But the ability to move does not provide the sperm with directions. In echinoderms, direction is provided by small chemotactic peptides such as resact. **Resact** is a 14-amino acid peptide that has been isolated from the egg jelly of the sea urchin *Arbacia punctulata* (Ward et al. 1985). Resact diffuses readily in seawater and has a profound effect at very low concentrations when added to a suspension of *Arbacia* sperm. When a drop of seawater containing *Arbacia* sperm is placed on a microscope slide, the sperm generally swim in circles about 50 µm in diameter. Within seconds after a small amount of resact is injected into the drop, sperm migrate into the region of the injection and congre-

gate there (Figure 7.9). As resact diffuses from the area of injection, more sperm are recruited into the growing cluster. Resact is specific for *A. punctulata* and does not attract sperm of other species. (An analogous compound, speract, has been isolated from the purple sea urchin, *Strongylocentrotus purpuratus*.) *A. punctulata* sperm have receptors in their cell membranes that bind resact (Ramarao and Garbers 1985; Bentley et al. 1986). When the extracellular side of the receptor binds resact, it activates latent guanylyl cyclase activity in the cytoplasmic side of the receptor (Figure 7.10). This causes the sperm cell to make more cyclic GMP, a compound that activates a calcium channel, allowing the influx of calcium ions (Ca^{2+}) from the seawater into the sperm, thus providing a directional cue (Nishigaki et al. 2000; Wood et al. 2005). Recent studies have demonstrated that the binding of a single resact molecule is able to provide direction for the sperm, which swim up a concentration gradient of this compound until they reach the egg (Kaupp et al. 2003; Kirkman-Brown et al. 2003).

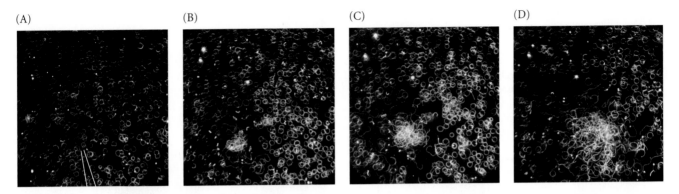

FIGURE 7.9 Sperm chemotaxis in the sea urchin *Arbacia punctulata*. One nanoliter of a 10-n*M* solution of resact is injected into a 20-μl drop of sperm suspension. (A) A 1-second photographic exposure showing sperm swimming in tight circles before the addition of resact. The position of the injection pipette is shown by the white lines. (B–D) Similar 1-second exposures showing migration of sperm to the center of the resact gradient 20, 40, and 90 seconds after injection. (From Ward et al. 1985; photographs courtesy of V. D. Vacquier.)

Resact also acts as a **sperm-activating peptide**. Sperm-activating peptides cause dramatic and immediate increases in mitochondrial respiration and sperm motility (Tombes and Shapiro 1985; Hardy et al. 1994). The increase in cyclic GMP and Ca^{2+} also activate the mitochondrial ATP-generating apparatus and the dynein ATPase that stimulates flagellar movement in the sperm (Shimomura et al. 1986; Cook and Babcock 1993). Thus, upon meeting resact, *Arbacia* sperm are told where to go and are given the motive force to get there.

The acrosome reaction

A second interaction between sperm and egg jelly is the **acrosome reaction**. In most marine invertebrates, the acrosome reaction has two components: the fusion of the acrosomal vesicle with the sperm cell membrane (an exocytosis that results in the release of the contents of the acrosomal vesicle), and the extension of the acrosomal process (Dan 1952; Colwin and Colwin 1963). The acrosome reaction in sea urchins is initiated by contact of the sperm with the egg jelly. This contact causes the exocytosis of the sperm's acrosomal vesicle and the release of proteolytic enzymes that digest a path through the jelly coat to the egg surface (Dan 1967; Franklin 1970; Levine et al. 1978). Once the sperm reaches the egg surface, the acrosomal process adheres to the vitelline envelope and tethers the sperm to the egg.

In sea urchins, the acrosome reaction is initiated by the interactions of the sperm cell membrane with a specific complex sugar in the egg jelly. These polysaccharides bind to specific receptors located on the sperm cell membrane directly above the acrosomal vesicle. The egg jelly factors

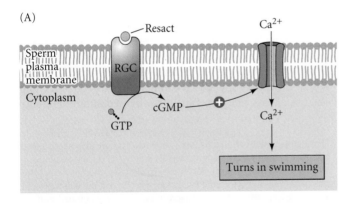

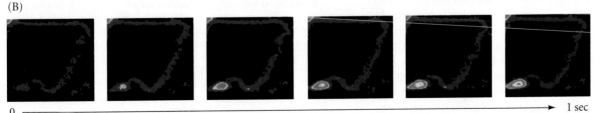

FIGURE 7.10 Model for chemotactic peptides in sea urchin sperm. (A) Resact from *Arbacia* egg jelly binds to its receptor on the sperm. This activates the receptor's guanylyl cyclase (RGC) activity, forming intracellular cGMP in the sperm. The cGMP opens calcium channels in the sperm cell membrane, allowing Ca^{2+} to enter into the sperm. The influx of Ca^{2+} activates sperm motility, and the sperm swims up the resact gradient toward the egg. (B) Ca^{2+} levels in different regions of *Strongylocentrotus purpuratus* sperm after exposure to 125 n*M* speract (this species' analog of resact). Red indicates the highest level of Ca^{2+}, blue the lowest. The sperm head reaches its peak Ca^{2+} levels within 1 second. (From Kirkman-Brown 2003.)

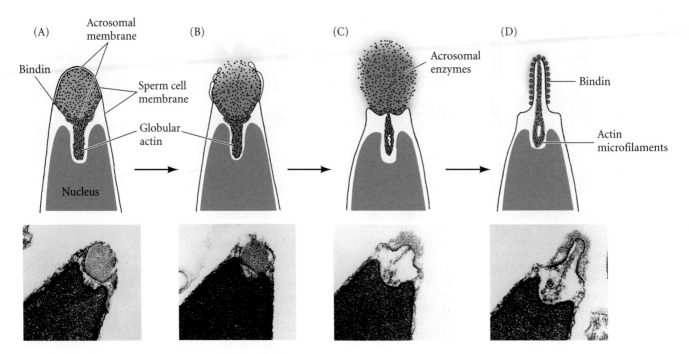

FIGURE 7.11 The acrosome reaction in sea urchin sperm. (A–C) The portion of the acrosomal membrane lying directly beneath the sperm plasma membrane fuses with the plasma membrane to release the contents of the acrosomal vesicle. (D) The actin molecules assemble to produce microfilaments, extending the acrosomal process outward. Actual photographs of the acrosome reaction in sea urchin sperm are shown below the diagrams. (After Summers and Hylander 1974; photographs courtesy of G. L. Decker and W. J. Lennarz.)

that initiate the acrosome reaction are often highly specific to each species, and egg jelly carbohydrates from one species of sea urchin fail to activate the acrosome reaction even in closely related species (Hirohashi and Vacquier 2002; Hirohashi et al. 2002; Biermann et al. 2004). Thus, the activation of the acrosome reaction constitutes a barrier to interspecies (and thus unviable) fertilizations.

In the sea urchin *Stongylocentrotus purpuratus*, the acrosome reaction is initiated by a repeating polymer of fucose sulfate. When this sulfated carbohydrate binds to its receptor on the sperm, the receptor activates three sperm membrane proteins: (1) a calcium transport channel that allows Ca^{2+} to enter the sperm head; (2) a sodium/hydrogen exchanger that pumps sodium ions (Na^+) into the sperm as it pumps hydrogen ions (H^+) out; and (3) a phospholipase enzyme that makes another second messenger, **IP$_3$** (of which we will hear much more later in the chapter). IP$_3$ is able to release Ca^{2+} from *inside* the sperm, probably from within the acrosome itself (Domino and Garbers 1988; Domino et al. 1989; Hirohashi and Vacquier 2003). The elevated calcium ion level in a relatively basic cytoplasm triggers the fusion of the acrosomal membrane with the adjacent sperm cell membrane, releasing enzymes that can lyse a path through the egg jelly to the vitelline envelope.*

The second part of the acrosome reaction involves the extension of the acrosomal process (Figure 7.11). This protrusion arises through the polymerization of globular actin molecules into actin filaments (Tilney et al. 1978). The influx of Ca^{2+} is thought to activate the protein RhoB in the acrosomal region and midpiece of the sea urchin sperm (Castellano et al. 1997). This GTP-binding protein helps organize the actin cytoskeleton in many types of cells, and it is thought to be active in polymerizing actin to make the acrosomal process.

Species-specific recognition

Although the sperm's contact with an egg's jelly coat can provide the first species-specific recognition event, another critical species-specific binding event must occur once the sea urchin sperm has penetrated the jelly and the acrosomal process of the sperm contacts the surface of the egg (Figure 7.12A). The acrosomal protein mediating this recog-

---

*Exocytotic reactions like the acrosome reaction are also seen in the release of insulin from pancreatic cells and in the release of neurotransmitters from synaptic terminals. In all cases, there is a calcium-mediated fusion between the secretory vesicle and the cell membrane. Indeed, the similarity of acrosomal vesicle exocytosis and synaptic vesicle exocytosis may actually be quite deep. Studies of acrosome reactions in sea urchins and mammals suggest that when the receptors for the sperm-activating ligands bind these molecules, they cause a depolarization of the membrane that would open voltage-dependent Ca^{2+} channels in a manner reminiscent of synaptic transmission (González-Martínez et al. 1992; Tulsani and Abou-Haila 2004). The proteins that dock the cortical granules of the egg to the cell membrane also appear to be homologous to those used in the axon terminal (Bi et al. 1995).

(A)

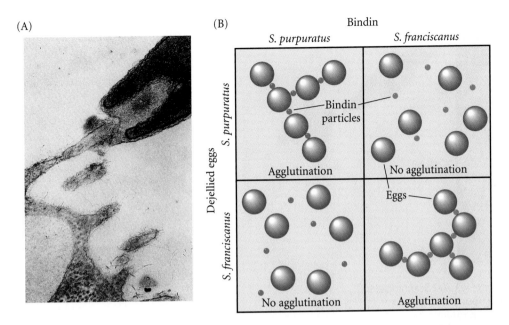

(B)

Bindin

FIGURE 7.12 Species-specific binding of acrosomal process to egg surface in sea urchins. (A) Actual contact of a sea urchin sperm acrosomal process with an egg microvillus. (B) In vitro model of species-specific binding. The agglutination of dejellied eggs by bindin was measured by adding bindin aggregates to a plastic well containing a suspension of eggs. After 2–5 minutes of gentle shaking, the wells were photographed. Each bindin bound to and agglutinated only eggs from its own species. (A from Epel 1977, photograph courtesy of F. D. Collins and D. Epel; B based on photographs in Glabe and Vacquier 1977.)

nition in sea urchins is called **bindin**. In 1977, Vacquier and co-workers isolated this nonsoluble, 30,500-Da protein from the acrosome of *Strongylocentrotus purpuratus* and found it to be capable of binding to dejellied eggs of the same species. Further, its interaction with eggs is often species-specific: bindin isolated from the acrosomes of *S. purpuratus* binds to its own dejellied eggs, but not to those of *S. franciscanus* (Figure 7.12B; Glabe and Vacquier 1977;

Glabe and Lennarz 1979). Using immunological techniques, Moy and Vacquier (1979) demonstrated that bindin is located specifically on the acrosomal process—exactly where it should be for sperm-egg recognition (Figure 7.13).

Biochemical studies have shown that the bindins of closely related sea urchin species are indeed different. This finding implies the existence of species-specific bindin *receptors* on the egg vitelline envelope. Such receptors were

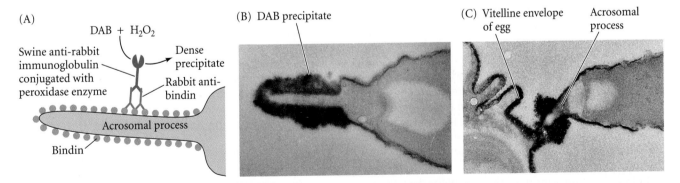

FIGURE 7.13 Localization of bindin on the acrosomal process. (A) Immunochemical technique used to localize bindin. Rabbit antibody was made to the bindin protein, and this antibody was incubated with sperm that had undergone the acrosome reaction. If bindin were present, the rabbit antibody would remain bound to the sperm. After any unbound antibody was washed off, the sperm were treated with swine antibody that had been covalently linked to peroxidase enzymes. The swine antibody bound to the rabbit antibody, placing peroxi-dase molecules wherever bindin was present. Peroxidase catalyzes the formation of a dark precipitate from diaminobenzi-dine (DAB) and hydrogen peroxide. Thus, this precipitate formed only where bindin was present. (B) Localization of bindin to the acrosomal process after the acrosome reaction (33,200×). (C) Localization of bindin to the acrosomal process at the junction of the sperm and the egg. (B and C from Moy and Vacquier 1979; photographs courtesy of V. D. Vacquier.)

(A)

(B)

(C)

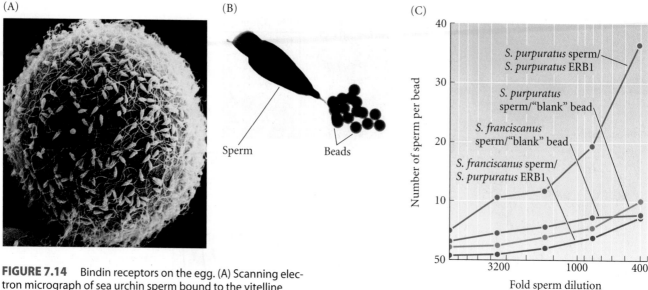

FIGURE 7.14 Bindin receptors on the egg. (A) Scanning electron micrograph of sea urchin sperm bound to the vitelline envelope of an egg. Although this egg is saturated with sperm, there appears to be room on the surface for more sperm, implying the existence of a limited number of bindin receptors. (B) *Strongylocentrotus purpuratus* sperm bind to polystyrene beads that have been coated with purified bindin receptor protein. (C) Species-specific binding of sea urchin sperm to ERB1. *S. purpuratus* sperm bound to beads coated with EBR1 bindin receptor purified from *S. purpuratus* eggs, but *S. franciscanus* sperm did not. Neither sperm bound to uncoated "blank" beads. (A photograph courtesy of C. Glabe, L. Perez, and W. J. Lennarz; B from Foltz et al. 1993; C after Kamei and Glabe 2003.)

WEBSITE 7.4 The Lillie-Loeb dispute over sperm-egg binding. In the early 1900s, fertilization research was framed by a dispute between F. R. Lillie and Jacques Loeb, who disagreed over whether the sperm recognized the egg through soluble factors or through cell-cell interactions.

also suggested by the experiments of Vacquier and Payne (1973), who saturated sea urchin eggs with sperm. As seen in Figure 7.14A, sperm binding does not occur over the entire egg surface. Even at saturating numbers of sperm (approximately 1500), there appears to be room on the ovum for more sperm heads, implying a limiting number of sperm-binding sites. **EBR1**, a 350-kDa glycoprotein that displays the properties expected of a bindin receptor, has been isolated from sea urchin eggs (Kamei and Glabe 2003). These bindin receptors are thought to be aggregated into complexes on the vitelline envelope, and hundreds of such complexes may be needed to tether the sperm to the egg (Figure 7.14B). The receptor for sperm bindin on the egg vitelline envelope appears to recognize the protein portion of bindin in a species-specific manner. Closely related species of sea urchins (different species within the same genus) have divergent bindin receptors, and eggs will adhere only to the bindin of their own species* (Figure 7.14C). Thus, species-specific recognition of sea urchin

gametes occurs at the levels of sperm attraction, sperm activation, and sperm adhesion to the egg surface.

Fusion of the egg and sperm cell membranes

Once the sperm has traveled to the egg and undergone the acrosome reaction, the fusion of the sperm cell membrane with the cell membrane of the egg can begin.

The entry of a sperm into a sea urchin egg is illustrated in Figure 7.15. Sperm-egg fusion appears to cause the polymerization of actin in the egg to form a **fertilization cone** (Summers et al. 1975). Homology between the egg and the sperm is again demonstrated, since the sperm's acrosomal process also appears to be formed by the polymerization of actin. The actin from the gametes forms a connection that widens the cytoplasmic bridge between the egg and the sperm. The sperm nucleus and tail pass through this bridge.

In the sea urchin, all regions of the egg cell membrane are capable of fusing with sperm. In several other species, certain regions of the membrane are specialized for sperm recognition and fusion (Vacquier 1979). Fusion is an active process, often mediated by specific "fusogenic" proteins. It has been suggested that sea urchin sperm bindin plays a second role as a fusogenic protein. In addition to recognizing the egg, bindin contains a long stretch of hydrophobic amino acids near its amino terminus, and this region is able to fuse phospholipid vesicles in vitro (Ulrich et al. 1999; Gage et al. 2004).

---

*Bindin and other gamete adhesion glycoproteins are among the fastest evolving proteins known (Metz and Palumbi 1996; Swanson and Vacquier 2002). Even when closely related species have near-identity of every other protein, their bindins may have diverged significantly.

(A)

(B)

(C)

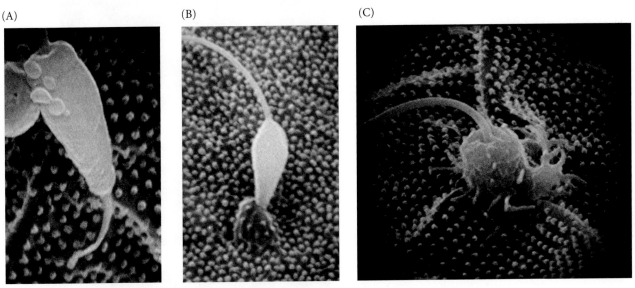

(D)

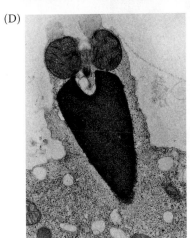

FIGURE 7.15 Scanning electron micrographs of the entry of sperm into sea urchin eggs. (A) Contact of sperm head with egg microvillus through the acrosomal process. (B) Formation of fertilization cone. (C) Internalization of sperm within the egg. (D) Transmission electron micrograph of sperm internalization through the fertilization cone. (A–C from Schatten and Mazia 1976, photographs courtesy of G. Schatten; D, photograph courtesy of F. J. Longo.)

The Prevention of Polyspermy

As soon as one sperm has entered the egg, the fusibility of the egg membrane—which was so necessary to get the sperm inside the egg—becomes a dangerous liability. In most animals, any sperm that enters the egg can provide a haploid nucleus and a centriole to the egg. In normal **monospermy**, only one sperm enters the egg, and a haploid sperm nucleus and a haploid egg nucleus combine to form the diploid nucleus of the fertilized egg (zygote), thus restoring the chromosome number appropriate for the species. The centriole provided by the sperm divides to form the two poles of the mitotic spindle during cleavage.

The entrance of multiple sperm—**polyspermy**—leads to disastrous consequences in most organisms. In the sea urchin, fertilization by two sperm results in a triploid nucleus, in which each chromosome is represented three times rather than twice. Worse, each sperm's centriole divides to form the two poles of a mitotic apparatus, so instead of a bipolar mitotic spindle separating the chromosomes into two cells, the triploid chromosomes may be divided into as many as four cells. Because there is no mechanism to ensure that each of the four cells receives the proper number and type of chromosomes, the chromosomes are apportioned unequally: some cells receive extra copies of certain chromosomes while other cells lack them (Figure 7.16). Theodor Boveri demonstrated in 1902 that such cells either die or develop abnormally.

Species have evolved ways to prevent the union of more than two haploid nuclei. The most common way is to prevent the entry of more than one sperm into the egg. The sea urchin egg has two mechanisms to avoid polyspermy: a fast reaction, accomplished by an electric change in the egg cell membrane, and a slower reaction, caused by the exocytosis of the cortical granules (Just 1919).

The fast block to polyspermy

The **fast block to polyspermy** is achieved by changing the electric potential of the egg cell membrane. This membrane provides a selective barrier between the egg cytoplasm and the outside environment, so that ion concentrations within the egg differ greatly from those of its surroundings. This concentration difference is especially significant for sodium and potassium ions. Seawater has a particularly high sodium ion (Na^+) concentration, whereas the egg cytoplasm contains relatively little Na^+. The reverse is the case with potassium ions (K^+). This condition is maintained by the cell membrane, which steadfastly inhibits the entry of Na^+ into the oocyte and prevents K^+ from leaking out into the environment. If we insert an electrode into an egg

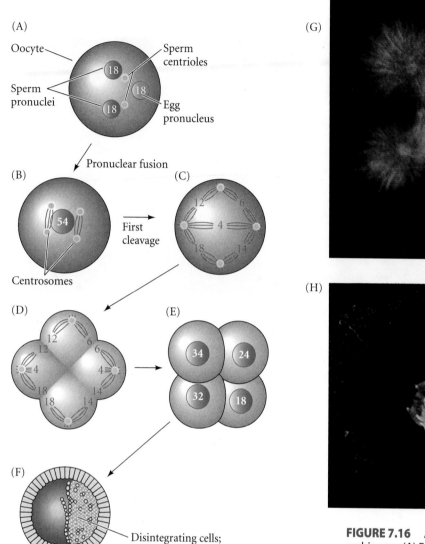

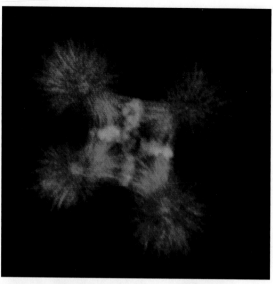

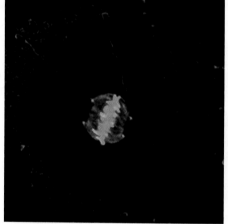

FIGURE 7.16 Aberrant development in a dispermic sea urchin egg. (A) Fusion of three haploid nuclei, each containing 18 chromosomes, and the division of the two sperm centrioles to form four centrosomes (mitotic poles). (B, C) The 54 chromosomes randomly assort on the four spindles. (D) At anaphase of the first division, the duplicated chromosomes are pulled to the four poles. (E) Four cells containing different numbers and types of chromosomes are formed, thereby causing (F) the early death of the embryo. (G) First metaphase of a dispermic sea urchin egg akin to (D). The microtubules are stained green; the DNA stain appears orange. The triploid DNA is being split into four chromosomally unbalanced cells instead of the normal two cells with equal chromosome complements. (H) Human dispermic egg at first mitosis. The four centrioles are stained yellow, while the microtubules of the spindle apparatus (and of the two sperm tails) are stained red. The three sets of chromosomes divided by these four poles are stained blue. (A–F after Boveri 1907; G photograph courtesy of J. Holy; H from Simerly et al. 1999, photograph courtesy of G. Schatten.)

and place a second electrode outside it, we can measure the constant difference in charge across the egg cell membrane. This **resting membrane potential** is generally about 70 mV, usually expressed as –70 mV because the inside of the cell is negatively charged with respect to the exterior.

Within 1–3 seconds after the binding of the first sperm, the membrane potential shifts to a positive level, about +20 mV (Longo et al. 1986). This change is caused by a small influx of Na^+ into the egg (Figure 7.17A). Although sperm can fuse with membranes having a resting potential of –70 mV, they cannot fuse with membranes having a positive resting potential, so no more sperm can fuse to the egg. It is not known whether the increased sodium permeability of the egg is due to the *binding* of the first sperm, or to the *fusion* of the first sperm with the egg (Gould and Stephano 1987, 1991; McCulloh and Chambers 1992).

The importance of Na^+ and the change in resting potential was demonstrated by Laurinda Jaffe and colleagues. They found that polyspermy can be induced if sea urchin

eggs are artificially supplied with an electric current that keeps their membrane potential negative. Conversely, fertilization can be prevented entirely by artificially keeping the membrane potential of eggs positive (Jaffe 1976). The fast block to polyspermy can also be circumvented by low-

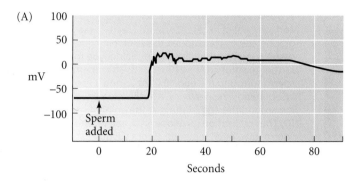

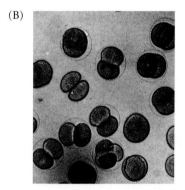

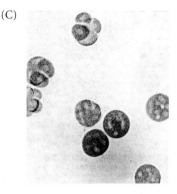

FIGURE 7.17 Membrane potential of sea urchin eggs before and after fertilization. (A) Before the addition of sperm, the potential difference across the egg plasma membrane is about –70 mV. Within 1–3 seconds after the fertilizing sperm contacts the egg, the potential shifts in a positive direction. (B, C) *Lytechinus* eggs photographed during first cleavage. (B) Control eggs developing in 490 m*M* Na⁺. (C) Polyspermy in eggs fertilized in similarly high concentrations of sperm in 120 m*M* Na⁺ (choline was substituted for sodium). (D) Table showing the rise of polyspermy with decreasing Na⁺ concentration. Salt water is about 600 m*M* NaCl. (After Jaffe 1980; photographs courtesy of L. A. Jaffe.)

(D)

Na⁺ (m*M*)	Percentage of polyspermic eggs
490	22
360	26
120	97
50	100

ering the concentration of Na⁺ in the surrounding water (Figure 7.17B–D). If the supply of sodium ions is not sufficient to cause the positive shift in membrane potential, polyspermy occurs (Gould-Somero et al. 1979; Jaffe 1980).

It is not known how the change in membrane potential acts on the sperm to block secondary fertilization. Most likely, the sperm carry a voltage-sensitive component (possibly a positively charged fusogenic protein), and the insertion of this component into the egg cell membrane could be regulated by the electric charge across the membrane (Iwao and Jaffe 1989). An electric block to polyspermy also occurs in frogs* (Cross and Elinson 1980), but probably not in most mammals (Jaffe and Cross 1983).

The slow block to polyspermy

The fast block to polyspermy is transient, since the membrane potential of the sea urchin egg remains positive for only about a minute. This brief potential shift is not sufficient to prevent polyspermy permanently, and polyspermy can still occur if the sperm bound to the vitelline envelope

---

*One might ask, as did a recent student, how amphibians could have a fast block to polyspermy, since their eggs are fertilized in pond water, which lacks high amounts of sodium ions. It turns out that the ion channels that open in frog egg membranes at fertilization are *chloride* channels instead of sodium channels as in sea urchin eggs. The concentration of Cl⁻ inside the frog egg is much higher than that of pond water. Thus, when chloride channels open at fertilization, the negatively charged chloride ions flow *out* of the cytoplasm, leaving the inside of the egg at a positive potential (see Jaffe and Schlicter 1985; Glahn and Nuccitelli 2003).

are not somehow removed (Carroll and Epel 1975). This sperm removal is accomplished by the **cortical granule reaction**, a slower, mechanical block to polyspermy that becomes active about a minute after the first successful sperm-egg fusion (Just 1919). This reaction—also known as the **slow block to polyspermy**—is found in many animal species, including most mammals.

WEBSITE 7.5 Blocks to polyspermy. Theodore Boveri's analysis of polyspermy is a classic of experimental and descriptive biology. E. E. Just's delineation of the fast and slow blocks was a critical paper in embryology. Both papers are reprinted here, along with commentaries.

VADE MECUM² E. E. Just. Blocks to polyspermy were discovered in the early 1900s by the African American embryologist Ernest Just, who became one of the few embryologists ever to be honored on a postage stamp. This segment contains videos of Just's work on sea urchin fertilization. [Click on Sea Urchin]

Directly beneath the sea urchin egg cell membrane are about 15,000 cortical granules, each about 1 μm in diameter (see Figure 7.6B). Upon sperm entry, these cortical granules fuse with the egg cell membrane and release their contents into the space between the cell membrane and the fibrous mat of vitelline envelope proteins. Several proteins are released by this cortical granule exocytosis. The first is a trypsin-like protease called **cortical granule serine protease**. This enzyme dissolves the protein posts that connect

(A)

(B)

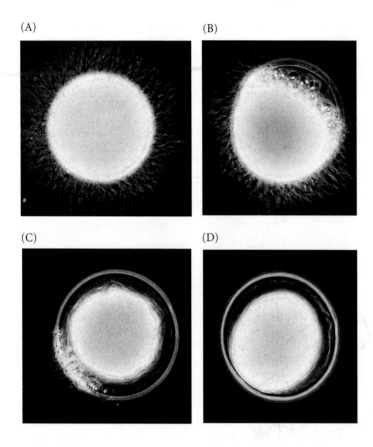

(C)

(D)

FIGURE 7.18 Formation of the fertilization envelope and removal of excess sperm. To create these photographs, sperm were added to sea urchin eggs, and the suspension was then fixed in formaldehyde to prevent further reactions. (A) At 10 seconds after sperm addition, sperm surround the egg. (B,C) At 25 and 35 seconds after insemination, respectively, a fertilization envelope is forming around the egg, starting at the point of sperm entry. (D) The fertilization envelope is complete, and excess sperm have been removed. (From Vacquier and Payne 1973; photographs courtesy of V. D. Vacquier.)

from the egg (Figures 7.18 and 7.19). A **peroxidase enzyme** released from the cortical granules hardens the fertilization envelope by crosslinking tyrosine residues on adjacent proteins. (Foerder and Shapiro 1977; Wong et al. 2004). As shown in Figure 7.18, the fertilization envelope starts to form at the site of sperm entry and continues its expansion around the egg. This process starts about 20 seconds after sperm attachment and is complete by the end of the first minute of fertilization. Finally, a fourth set of cortical granule proteins, including **hyalin**, forms a coating around the egg (Hylander and Summers 1982). The egg extends elongated microvilli whose tips attach to this **hyaline layer**. This layer provides support for the blastomeres during cleavage.

the vitelline envelope proteins to the cell membrane, and it clips off the bindin receptors and any sperm attached to them (Vacquier et al. 1973; Glabe and Vacquier 1978; Haley and Wessel 1999, 2004).

The components of the cortical granules fuse with the vitelline envelope to form a **fertilization envelope** (Wong and Wessel 2004). This fertilization envelope is elevated from the cell membrane by **mucopolysaccharides** released by the cortical granules. These sticky compounds produce an osmotic gradient that forces water to rush into the space between the cell membrane and the fertilization envelope, causing the envelope to expand and radially move away

Calcium as the initiator of the cortical granule reaction

The mechanism of the cortical granule reaction is similar to that of the acrosome reaction, and it may involve many of the same molecules. Upon fertilization, the concentration of free Ca^{2+} in the egg cytoplasm increases greatly. In this high-calcium environment, the cortical granule membranes fuse with the egg cell membrane, releasing their contents (Figure 7.19). Once the fusion of the cortical granules begins near the point of sperm entry, a wave of cortical granule exocytosis propagates around the cortex to the opposite side of the egg.

In sea urchins and mammals, the rise in Ca^{2+} concentration responsible for the cortical granule reaction is not due to an influx of calcium into the egg, but rather comes from within the egg itself. The release of calcium from intracellular storage can be monitored visually using calcium-activated luminescent dyes such as aequorin (isolated from luminescent jellyfish) or fluorescent dyes such as fura-2. These dyes emit light when they bind free Ca^{2+}. When a sea urchin egg is injected with dye and then fertilized, a striking wave of calcium release propagates across the egg (Figure 7.20). Starting at the point of sperm entry, a band of light traverses the cell (Steinhardt et al. 1977; Hafn-

WEBSITE   7.6 Building the egg's extracellular matrix. In sea urchins, the cortical granules secrete not only hyalin but a number of proteins that construct the extracellular matrix of the embryo. This highly coordinated process results in sequential layers.

VADE MECUM² Sea urchin fertilization. The remarkable reactions that prevent polyspermy in a fertilized sea urchin egg can be seen in the raising of the fertilization envelope. This segment contains movies of this event shown in real time. [Click on Sea Urchin]

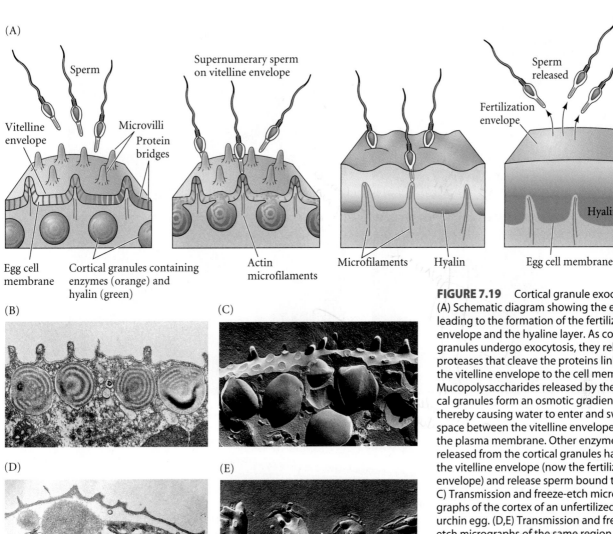

(A)

Sperm

Vitelline envelope

Microvilli
Protein bridges

Egg cell membrane

Cortical granules containing enzymes (orange) and hyalin (green)

Supernumerary sperm on vitelline envelope

Actin microfilaments

Microfilaments Hyalin

Sperm released

Fertilization envelope

Hyalin

Egg cell membrane

FIGURE 7.19 Cortical granule exocytosis. (A) Schematic diagram showing the events leading to the formation of the fertilization envelope and the hyaline layer. As cortical granules undergo exocytosis, they release proteases that cleave the proteins linking the vitelline envelope to the cell membrane. Mucopolysaccharides released by the cortical granules form an osmotic gradient, thereby causing water to enter and swell the space between the vitelline envelope and the plasma membrane. Other enzymes released from the cortical granules harden the vitelline envelope (now the fertilization envelope) and release sperm bound to it. (B, C) Transmission and freeze-etch micrographs of the cortex of an unfertilized sea urchin egg. (D,E) Transmission and freeze-etch micrographs of the same region of a recently fertilized egg, showing the raising of the fertilization envelope and the points at which the cortical granules have fused with the plasma membrane of the egg (arrows in D). (A after Austin 1965; B–E from Chandler and Heuser 1979; photographs courtesy of D. E. Chandler.)

er et al. 1988). The calcium ions do not merely diffuse across the egg from the point of sperm entry. Rather, the release of Ca^{2+} starts at one end of the cell and proceeds actively to the other end. The entire release of Ca^{2+} is complete within roughly 30 seconds in sea urchin eggs, and free Ca^{2+} is re-sequestered shortly after being released. If two sperm enter the egg cytoplasm, Ca^{2+} release can be seen starting at the two separate points of entry on the cell surface (Hafner et al. 1988).

Several experiments have demonstrated that calcium ions are directly responsible for propagating the cortical granule reaction, and that these ions are stored within the egg itself. The drug A23187 is a calcium *ionophore* (a compound that transports ions such as free Ca^{2+} across lipid

membranes, allowing them to traverse otherwise impermeable barriers). Placing unfertilized sea urchin eggs into seawater containing A23187 causes the cortical granule reaction and the elevation of the fertilization envelope. Moreover, this reaction occurs in the absence of any Ca^{2+} in the surrounding water. Therefore, the A23187 must be stimulating the release of Ca^{2+} already sequestered in organelles within the egg (Chambers et al. 1974; Steinhardt and Epel 1974).

In sea urchins and vertebrates (but not snails and worms), the calcium ions responsible for the cortical granule reaction are stored in the endoplasmic reticulum of the egg (Eisen and Reynolds 1985; Terasaki and Sardet 1991). In sea urchins and frogs, this reticulum is pronounced in the

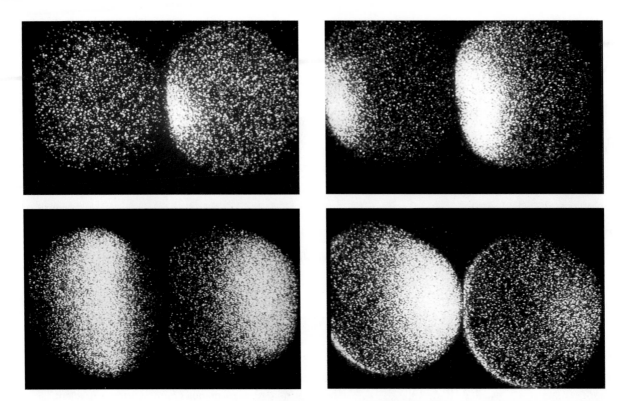

FIGURE 7.20   Wave of Ca^{2+} release across a sea urchin egg during fertilization. The egg is pre-loaded with a dye that fluoresces when it binds Ca^{2+}. When a sperm fuses with the egg, a wave of calcium release is seen, beginning at the site of sperm entry and propagating across the egg. The wave takes 30 seconds to traverse the egg. (Photograph courtesy of G. Schatten.)

cortex and surrounds the cortical granules (Figure 7.21; Gardiner and Grey 1983; Luttmer and Longo 1985). The cortical granules themselves are tethered to the cell membrane by a series of integral membrane proteins that facilitate calcium-mediated exocytosis (Conner et al. 1997; Conner and Wessel 1998). Thus, as soon as Ca^{2+} is released from the endoplasmic reticulum, the cortical granules fuse with the cell membrane above them.

In the frog *Xenopus*, the cortical endoplasmic reticulum becomes ten times more abundant during the maturation of the egg and disappears locally within a minute after the wave of cortical granule exocytosis occurs in any region of the cortex. Once initiated, the release of calcium is self-propagating. Free calcium is able to release sequestered calcium from its storage sites, thus causing a wave of Ca^{2+} release and cortical granule exocytosis.

(A)   (B)

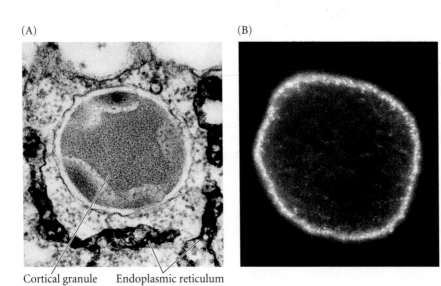

Cortical granule   Endoplasmic reticulum

FIGURE 7.21   Endoplasmic reticulum surrounding cortical granules in sea urchin eggs. (A) The endoplasmic reticulum has been stained to allow visualization by transmission electron microscopy. The cortical granule is seen to be surrounded by dark-stained endoplasmic reticulum. (B) An entire egg stained with fluorescent antibodies to calcium-dependent calcium release channels. The antibodies show these channels in the cortical endoplasmic reticulum. (A from Luttmer and Longo 1985, photograph courtesy of S. Luttmer; B from McPherson et al. 1992, photograph courtesy of F. J. Longo.)

Activation of Egg Metabolism in Sea Urchins

Although fertilization is often depicted as merely the means to merge two haploid nuclei, it has an equally important role in initiating the processes that begin development. These events happen in the cytoplasm and occur without the involvement of the parental nuclei.*

The mature sea urchin egg is a metabolically sluggish cell that is activated by the sperm. This activation is merely a stimulus that sets a preprogrammed set of metabolic events into action. The responses of the egg to the sperm can be divided into "early" responses, which occur within seconds of the cortical granule reaction, and "late" responses, which take place several minutes after fertilization begins (Table 7.1).

Early responses

As we have seen, contact or fusion between sea urchin sperm and egg activates the two major blocks to polyspermy: the fast block, initiated by sodium influx into the cell; and the slow block, initiated by the intracellular release of Ca^{2+}.

The same release of Ca^{2+} responsible for the cortical granule reaction is also responsible for the re-entry of the egg into the cell cycle and the reactivation of egg protein synthesis. Ca^{2+} levels in the egg increase from 0.1 to 1 μM, and in almost all species, this occurs as a wave or succession of waves that sweep across the egg beginning at the site of sperm-egg fusion (Jaffe 1983; Terasaki and Sardet 1991; Stricker 1999; see Figure 7.20).

Calcium release activates a whole series of metabolic reactions that initiate embryonic development (Figure 7.22). One of these is the activation of the enzyme NAD⁺ kinase, which converts NAD⁺ to NADP⁺ (Epel et al. 1981). This change may have important consequences for lipid metabolism, since NADP⁺ (but not NAD⁺) can be used as a coenzyme for lipid biosynthesis. Thus the conversion of NAD⁺ to NADP⁺ may be important in the construction of the many new cell membranes required during cleavage. NADP⁺ is also used to make NAADP (see "Sidelights & Speculations," p. 198), which appears to boost calcium release even further. Calcium release also affects oxygen consumption. A burst of oxygen reduction (to hydrogen

TABLE 7.1 Events of sea urchin fertilization

Event	Approximate time postinsemination[a]
EARLY RESPONSES	
Sperm-egg binding	0 seconds
Fertilization potential rise (fast block to polyspermy)	within 1 sec
Sperm-egg membrane fusion	within 1 sec
Calcium increase first detected	10 sec
Cortical granule exocytosis (slow block to polyspermy)	15–60 sec
LATE RESPONSES	
Activation of NAD kinase	starts at 1 min
Increase in NADP⁺ and NADPH	starts at 1 min
Increase in O_2 consumption	starts at 1 min
Sperm entry	1–2 min
Acid efflux	1–5 min
Increase in pH (remains high)	1–5 min
Sperm chromatin decondensation	2–12 min
Sperm nucleus migration to egg center	2–12 min
Egg nucleus migration to sperm nucleus	5–10 min
Activation of protein synthesis	starts at 5–10 min
Activation of amino acid transport	starts at 5–10 min
Initiation of DNA synthesis	20–40 min
Mitosis	60–80 min
First cleavage	85–95 min

Main sources: Whitaker and Steinhardt 1985; Mohri et al. 1995.
[a]Approximate times based on data from *S. purpuratus* (15–17°C), *L. pictus* (16–18°C), *A. punctulata* (18–20°C), and *L. variegatus* (22–24°C). The timing of events within the first minute is best known for *Lytechinus variegatus*, so times are listed for that species.

peroxide) is seen during fertilization, and much of this "respiratory burst" is used to crosslink the fertilization envelope. The enzyme responsible for this reduction of oxygen is also NADPH-dependent (Heinecke and Shapiro 1989; Wong et al. 2004). Lastly, NADPH helps regenerate glutathione and ovothiols, molecules that may be crucial scavengers of free radicals that could otherwise damage the DNA of the egg and early embryo (Mead and Epel 1995).

If the calcium-chelating chemical EGTA is injected into sea urchin eggs, there is no cortical granule reaction, no change in membrane resting potential, and no reinitiation of cell division upon fertilization (Runft et al. 2002). Conversely, eggs can be activated artificially in the absence of sperm by procedures that release free calcium into the oocyte. Steinhardt and Epel (1974) found that diffusion of

---

*In certain salamanders, this developmental function of fertilization has been totally divorced from the genetic function. The silver salamander (*Ambystoma platineum*) is a hybrid subspecies consisting solely of females. Each female produces an egg with an unreduced chromosome number. This egg, however, cannot develop on its own, so the silver salamander mates with a male Jefferson salamander (*A. jeffersonianum*). The sperm from the male Jefferson salamander only stimulates the egg's development; it does not contribute genetic material (Uzzell 1964). For details of this complex mechanism of procreation, see Bogart et al. 1989.

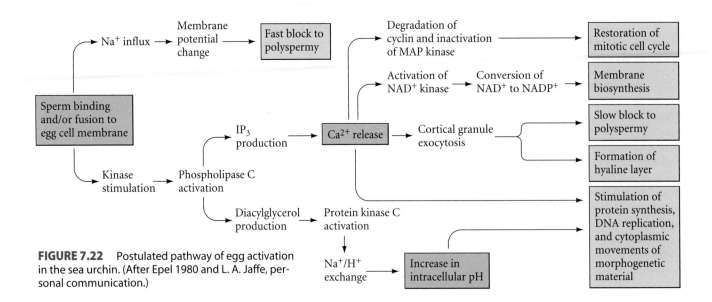

FIGURE 7.22   Postulated pathway of egg activation in the sea urchin. (After Epel 1980 and L. A. Jaffe, personal communication.)

micromolar amounts of the calcium ionophore A23187 into a sea urchin egg elicits most of the responses characteristic of a normally fertilized egg.

Late responses: Resumption of protein and DNA synthesis

The late responses of fertilization include the activation of a new burst of DNA and protein synthesis. The fusion of egg and sperm in sea urchins causes the intracellular pH to increase.* This rise in intracellular pH begins with a second influx of sodium ions, which causes a 1:1 exchange between sodium ions from the seawater and hydrogen ions from the egg. The loss of hydrogen ions causes the pH of the egg to rise (Shen and Steinhardt 1978). It is thought that pH increase and Ca^{2+} elevation act together to stimulate new DNA and protein synthesis (Winkler et al. 1980; Whitaker

---

*This is due to the production of diacylglycerol, as mentioned above. Again, variation among species may be prevalent. In the much smaller egg of the mouse, there is no elevation of pH after fertilization, and there is no dramatic increase in protein synthesis immediately following fertilization (Ben-Yosef et al. 1996).

and Steinhardt 1982; Rees et al. 1995). If one experimentally elevates the pH of an unfertilized egg to a level similar to that of a fertilized egg, DNA synthesis and nuclear envelope breakdown ensue, just as if the egg were fertilized (Miller and Epel 1999). Calcium ions are also critical to new DNA synthesis. The wave of free Ca^{2+} inactivates the enzyme MAP kinase, converting it from a phosphorylated (active) to an unphosphorylated (inactive) form, thus removing an inhibition on DNA synthesis (Carroll et al. 2000). DNA and protein synthesis can then resume.

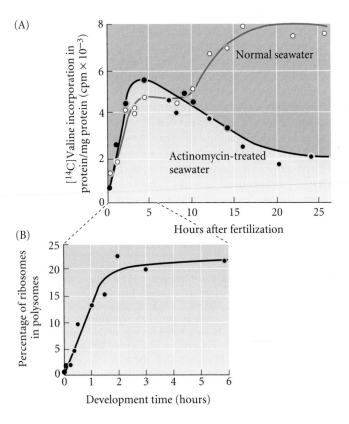

FIGURE 7.23   A burst of protein synthesis at fertilization uses mRNAs stored in the oocyte cytoplasm. (A) Protein synthesis in embryos of the sea urchin *Arbacia punctulata* fertilized in the presence or absence of actinomycin D, an inhibitor of transcription. For the first few hours, protein synthesis occurs with very little new transcription from the zygote or embryo nuclei. A second burst of protein synthesis occurs during the mid-blastula stage. This burst represents translation of newly transcribed messages, and therefore is not seen in embryos growing in actinomycin. (B) Increase in the percentage of ribosomes recruited into polysomes during the first hours of sea urchin development, especially during the first cell cycle. (A after Gross et al. 1964; B after Humphreys 1971.)

In sea urchins, a burst of protein synthesis usually occurs within several minutes after sperm entry. This protein synthesis does not depend on the synthesis of new messenger RNA; rather, it utilizes mRNAs already present in the oocyte cytoplasm (Figure 7.23; see Table 5.2). These mRNAs encode proteins such as histones, tubulins, actins, and morphogenetic factors that are used during early development. Such a burst of protein synthesis can be induced by artificially raising the pH of the cytoplasm using ammonium ions (Winkler et al. 1980).

One mechanism for this global rise in the translation of messages stored in the oocyte appears to be the release of inhibitors from the mRNA. In Chapter 5, we discussed maskin, an inhibitor of translation in the unfertilized amphibian oocyte. In sea urchins, a similar inhibitor binds translation initiation factor eIF4E at the 5′ end of several mRNAs and prevents these mRNAs from being translated. Upon fertilization, however, this inhibitor—the 4E-binding protein—becomes phosphorylated and is degraded, thus allowing translation of the stored sea urchin mRNAs (Cormier et al. 2001). One of the mRNAs "freed" by the degradation of 4E-binding protein is the message encoding cyclin B (Salaun et al. 2003, 2004). The cyclin B protein combines with Cdk1 cyclin to create **mitosis-promoting factor** (**MPF**), which is required to initiate cell division.

Thus fertilization activates pathways that target the translational inhibitory protein for degradation, and the newly accessible 5′ end of the mRNA can interact with those proteins that allow the message to be translated. One of these mRNAs encodes a protein critical for cell division. In such a manner, fertilization can initiate mitosis and the sea urchin can begin to form a multicellular organism.

SIDELIGHTS & SPECULATIONS

The Activation of Gamete Metabolism

In 1937, Yale embryologist Ross Granville Harrison lamented, "The liaison between genetics and embryology is now established, but can we say the same of embryology and physiology? Perhaps we are still under the spell of the doctrine that more than one liaison at a time is sin." Indeed, for all our knowledge of the genes that instruct the embryo, the physiological mechanisms by which the instructions are carried out have been largely ignored. This situation is changing as our knowledge of cell signaling pathways becomes integrated with our knowledge of the cytoskeleton and membrane ion channels. One of the most exciting places of integration involves the ionic mechanisms of fertilization.

IP$_3$: Releaser of calcium ions

As is obvious from the above discussion, the critical factors for gamete activation are calcium ions. It is the release of calcium ions that allows membrane fusion, the exocytosis of acrosomal and cortical vesicles, and the activation of the sperm and the egg. We still do not know all of the mechanisms whereby the union of sperm and egg initiates a massive Ca^{2+} release in the egg. However, one conclusion that has been reached is that the production of **inositol 1,4,5-trisphosphate** (**IP$_3$**) is the primary mechanism for releasing calcium ions from intracellular storage.

The IP$_3$ pathway is shown in Figure 7.24. The membrane phospholipid phosphatidylinositol 4,5-bisphosphate (PIP$_2$) is split by the enzyme **phospholipase C** (**PLC**) to yield two active compounds: **IP$_3$** and **diacylglycerol** (**DAG**). IP$_3$ is able to release Ca^{2+} into the cytoplasm by opening the Ca^{2+} channels of the endoplasmic reticulum. DAG activates protein kinase C, which in turn activates a protein that exchanges sodium ions for hydrogen ions, raising the pH of the egg (Swann and Whitaker 1986; Nishizuka 1986). This Na$^+$/H$^+$ exchange pump also requires Ca^{2+} for its activity. The result of PLC activation is therefore the liberation of Ca^{2+} and the alkalinization of the egg, and both of the compounds it creates—IP$_3$ and DAG—are involved in the initiation of development.

IP$_3$ is formed initially at the site of sperm entry in sea urchin eggs and can be detected within seconds of their being fertilized. The inhibition of IP$_3$ synthesis prevents Ca^{2+} release (Lee and Shen 1998; Carroll et al. 2000), while injected IP$_3$ can release sequestered Ca^{2+} in sea urchin eggs, leading to the cortical granule reaction (Whitaker and Irvine 1984; Busa et al. 1985). Moreover, these IP$_3$-mediated effects can be thwarted by preinjecting the egg with calcium-chelating agents (Turner et al. 1986).

IP$_3$-responsive calcium channels have been found in the egg endoplasmic reticu-lum. The IP$_3$ formed at the site of sperm entry is thought to bind to the IP$_3$ receptors of these channels, effecting a local release of calcium (Ferris et al. 1989; Furuichi et al. 1989). Once released, Ca^{2+} can diffuse directly, or they can facilitate the release of more Ca^{2+} by binding to calcium-release receptors also located in the cortical endoplasmic reticulum (McPherson et al. 1992). The binding of Ca^{2+} to these receptors releases more calcium, and this released calcium binds to more receptors, and so on. The resulting wave of calcium release is propagated throughout the cell, starting at the point of sperm entry. The cortical granules, which fuse with the cell membrane in the presence of high calcium concentrations, respond with a wave of exocytosis that follows the Ca^{2+} wave. Mohri and colleagues (1995) have shown that IP$_3$-released Ca^{2+} is both necessary and sufficient for initiating the wave of calcium release.

IP$_3$ has been similarly found to release Ca^{2+} in vertebrate eggs. As in sea urchins, waves of IP$_3$ are thought to mediate calcium release from sites within the endoplasmic reticulum (Miyazaki et al. 1992; Ducibella et al. 2002). Blocking the IP$_3$ receptor in hamster and mouse eggs prevents the release of calcium at fertilization. Xu and colleagues

Continued on next page

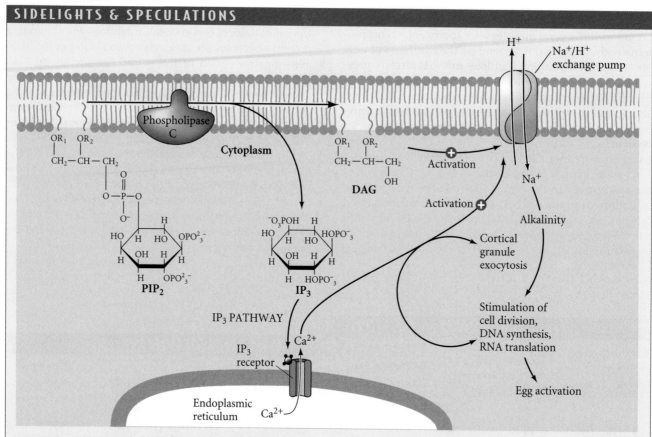

FIGURE 7.24 The roles of inositol phosphates in releasing calcium from the endoplasmic reticulum and the initiation of development. Phospholipase C splits PIP_2 into IP_3 and DAG. IP_3 releases calcium from the endoplasmic reticulum, and DAG, with assistance from the released Ca^{2+}, activates the sodium-hydrogen exchange pump in the membrane.

(1994) found that blocking IP_3-mediated calcium release blocks every aspect of sperm-induced egg activation, including cortical granule exocytosis, mRNA recruitment, and cell cycle resumption.

Phospholipase C: Generator of IP_3

The question then becomes: What initiates the production of IP_3? In other words, what activates the phospholipase C enzymes? This question has not been easy to address, since (1) there are numerous types of PLC that (2) can be activated through different pathways, and (3) different species can use different mechanisms to activate them. Results from recent studies of sea urchin eggs suggest that the active PLC in this group of animals is a member of the γ (gamma) family of PLCs (Carroll et al. 1997, 1999; Shearer et al. 1999). Inhibitors that specifically block this family of PLCs inhibit IP_3 production as well as Ca^{2+} release. More-

over, these inhibitors can be circumvented by microinjecting IP_3 into the egg. The mechanism by which IP_3 is generated in vertebrates is largely unknown, but may involve PLCs coming from the sperm nucleus (Kuretake et al. 1996; Perry et al. 2000).

Kinases: A link between sperm and PLC?

The finding that the γ class of PLC was responsible for generating IP_3 during echinoderm fertilization caused investigators to investigate exactly which proteins activated this particular class of phospholipases. Their work soon came to focus on the **Src family** of protein kinases. Src proteins are found in the cortical cytoplasm of sea urchin and starfish eggs, where they can form a complex with PLCγ. Inhibition of Src protein kinases lowered and delayed the amount of calcium released (Giusti et al. 1999, 2003; Kinsey and Shen 2000).

So what activates Src kinase activity?

One possibility is that it is activated by heterotrimeric G proteins in the cortex of the egg (Figure 7.25). Blocking these G proteins prevented calcium release (Voronina and Wessel 2003). Such G proteins are known to activate Src kinases in mammalian somatic cells, so the cortical G proteins of sea urchin eggs seem like good candidates. It is also possible that these G proteins activate PLC directly. Indeed, there may be more than one pathway and more than one way to activate the calcium release.

What is the sperm's role?

There are several additional ways that the signal from the sperm might be transmitted to activate the egg. First, the activation of the IP_3 pathway could be initiated by the contact of the sperm with the egg (Figure 7.26A). Such contact would activate a receptor, the receptor would activate the G protein, and Ca^{2+} would eventually be liberated from the endoplasmic reticulum.

A second possibility is that the activation of the IP_3 pathway is caused not by the binding of sperm and egg, but by the fusion of the sperm and egg cell membranes. McCulloh and Chambers (1992) have electrophysiological evidence that sea urchin

SIDELIGHTS & SPECULATIONS

FIGURE 7.25 G protein involvement in calcium ion entry into sea urchin eggs. (A) Mature sea urchin egg immunologically labeled for the cortical granule protein hyaline (red) and the G protein Gαq (green). (The overlap of signals produces the yellow color.) Gαq is localized to the cortex. (B) A wave of Ca^{2+} wave appears in the control egg (computer-enhanced to show relative intensities, with red being the highest), but not in the egg injected with an inhibitor of the Gαq protein. (C) Possible model for egg activation by the influx of Ca^{2+}. (After Voronina and Wessel 2003; photographs courtesy of G. M. Wessel.)

(A)
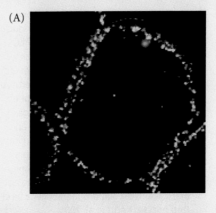

(B)

Control   Gαq inhibitor added

(C)
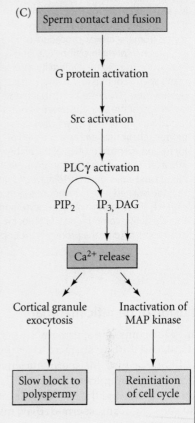

eggs are not activated until after sperm and egg cytoplasms are joined. They suggest that the egg-activating components are located either on the sperm cell membrane or in the sperm cytoplasm. It is even possible that when the fusion of gamete membranes occurs, the sperm receptor kinases or PLCs (activated by the egg jelly to initiate the acrosome reaction) activate the IP_3 cascade, resulting in Ca^{2+} release in the egg. In this scenario, shown in Figure 7.26B, bindin serves for cell-cell adhesion and membrane fusion, but not for signaling. Rather, the egg "activates itself" through the sperm.

Still another possibility is that the agent active in releasing the sequestered calcium

ACTIVATION PRIOR TO SPERM FUSION

ACTIVATION AFTER SPERM FUSION

(A)
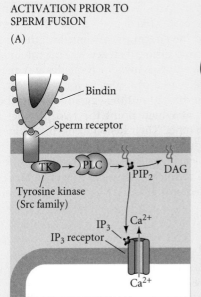

(B)
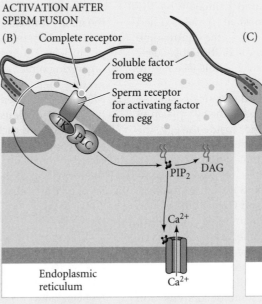

(C)
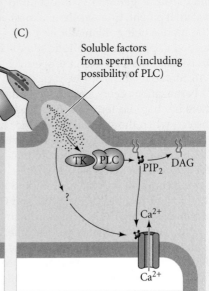

FIGURE 7.26 Possible mechanisms of egg activation. In all cases, a phospholipase C (PLC) making IP_3 is activated. (A) The bindin receptor (perhaps through a G protein) activates a Src kinase. (B) An activated G protein, Src kinase, or a PLC in the sperm plasma membrane activates the egg pathways. (C) Calcium release and egg activation by activated PLC from the sperm or by a substance from the sperm that activates egg PLC.

Continued on next page

is a small molecule from the sperm cytosol (Figure 7.26C). Support for this hypothesis comes from the clinical procedure of intracytoplasmic sperm injection (ICSI). In this procedure, used to treat infertility when a man's sperm count is low, a single intact human sperm is injected directly into the cytoplasm of the egg. The injection results in egg activation, the formation of a male pronucleus, normal embryonic development, and eventual growth of the embryo into a fertile adult (Van Steirtinghem 1994). Kimura and colleagues (1998) have shown that the isolated head of a mouse sperm is capable of activating a mouse oocyte, and that the active components of the sperm head appears to be the proteins surrounding the haploid nucleus. A set of experiments suggested that the active perinuclear factor was a sperm-specific form of PLC (Saunders et al. 2002). This protein, called PLCζ (PLC-zeta), triggered Ca^{2+} waves indistinguishable from those of normal fertilization; removal of PLCζ from mouse sperm extracts abolishes Ca^{2+} release in eggs. These observations were confirmed and extended by several experiments (Fujimoto et al. 2004; Swann et al. 2004), and sperm prevented (by RNA interference techniques) from making PLCζ fail to produce the normal calcium oscillations in the newly fertilized oocyte (Knott et al. 2005).

There is also evidence that **nicotinic acid adenine dinucleotide phosphate** (**NAADP**), a linear dinucleotide derived from NADP, serves as a sperm-borne calcium ion releaser. NAADP frees stored Ca^{2+} from membrane vesicles during muscle contraction, insulin secretion, and neurotransmitter release (Lee 2001). Upon contact with egg jelly, the concentration of NAADP increases tenfold in sea urchin sperm, reaching levels that appear to be more than sufficient to release stored Ca^{2+} within the egg (Churchill et al. 2003).

Different species have evolved different ways of obtaining Ca^{2+} during fertilization. It is possible that any particular species may use more than one means of initiating Ca^{2+} release.

Fusion of genetic material

In sea urchins, the sperm nucleus enters the egg perpendicular to the egg surface. After the sperm and egg cell membranes fuse, the sperm nucleus and its centriole separate from the mitochondria and flagellum. The sperm's mitochondria and the flagellum disintegrate inside the egg, so very few, if any, sperm-derived mitochondria are found in developing or adult organisms. Thus, although each gamete contributes a haploid genome to the zygote, the *mitochondrial* genome is transmitted primarily by the maternal parent. Conversely, in almost all animals studied (the mouse being the major exception), the centrosome needed to produce the mitotic spindle of the subsequent divisions is derived from the sperm centriole (see Figure 7.16; Sluder et al. 1989, 1993).

Fertilization occurs after the second meiotic division, so there is a haploid female pronucleus in the cytoplasm of the egg when the sperm enters. Once inside the egg, the sperm nucleus undergoes a dramatic transformation as it decondenses to form the haploid male pronucleus. First, the nuclear envelope vesiculates into small packets, exposing the compact sperm chromatin to the egg cytoplasm (Longo and Kunkle 1978; Poccia and Collas 1997). Then proteins holding the sperm chromatin in its condensed, inactive state are exchanged for other proteins derived from the egg cytoplasm. This exchange permits the decondensation of the sperm chromatin. Once decondensed, the DNA can begin transcription and replication.

In sea urchins, sperm chromosome decondensation appears to be initiated by the phosphorylation of the nuclear envelope lamin protein and the phosphorylation of two sperm-specific histones that bind tightly to the DNA. The process begins when sperm comes into contact with a certain glycoprotein in the egg jelly that elevates the level of cAMP-dependent protein kinase activity. These protein kinases phosphorylate several of the basic residues of the sperm-specific histones and thereby interfere with their binding to DNA (Garbers et al. 1980; Porter and Vacquier 1986; Stephens et al. 2002). This loosening is thought to facilitate the replacement of the sperm-specific histones with other histones that have been stored in the oocyte cytoplasm (Green and Poccia 1985).

After the sea urchin sperm enters the egg cytoplasm, the male pronucleus rotates 180 degrees so that the sperm centriole is between the sperm pronucleus and the egg pronucleus. The sperm centriole then acts as a microtubule organizing center, extending its own microtubules and integrating them with egg microtubules to form an aster.* Microtubules extend throughout the egg and contact the female pronucleus, at which point the two pronuclei migrate toward each other. Their fusion forms the diploid **zygote nucleus** (Figure 7.27). DNA synthesis can begin either in the pronuclear stage (during migration) or after the formation of the zygote nucleus, and depends on the level of Ca^{2+} released earlier in fertilization (Jaffe et al. 2001).

---

*When Oscar Hertwig observed this radial array of sperm asters forming in his newly fertilized sea urchin eggs, he called it "the sun in the egg" and thought it was the happy indication of a successful fertilization (Hertwig 1877). More recently, Simerly and co-workers (1999) found that certain types of human male infertility are due to defects in the centriole's ability to form these microtubular asters. This deficiency results in the failure of pronuclear migration and the cessation of further development.

(A)

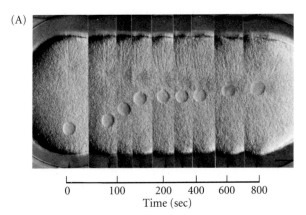

(B)

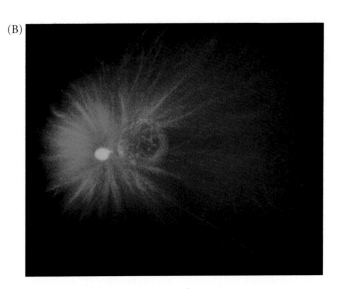

(C)

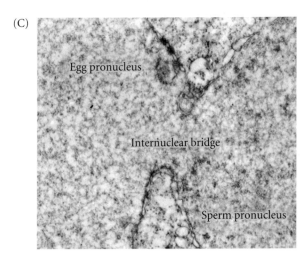

Egg pronucleus

Internuclear bridge

Sperm pronucleus

FIGURE 7.27 Nuclear events in the fertilization of the sea urchin. (A) Sequential photographs showing the migration of the egg pronucleus and the sperm pronucleus toward each other in an egg of *Clypeaster japonicus.* The sperm pronucleus is surrounded by its aster of microtubules. (B) The two pronuclei migrate toward each other on these microtubular processes. (The pronuclear DNA is stained blue by Hoechst dye.) The microtubules (stained green with fluorescent antibodies to tubulin) radiate from the centrosome associated with the (smaller) male pronucleus and reach toward the female pronucleus. (C) Fusion of pronuclei in the sea urchin egg. (A from Hamaguchi and Hiramoto 1980, courtesy of the authors; B from Holy and Schatten 1991, courtesy of J. Holy; C courtesy of F. J. Longo.)

Mammalian Fertilization

It is very difficult to study any interactions between the mammalian sperm and egg that might take place prior to these gametes making contact. One obvious reason for this is that mammalian fertilization occurs inside the oviducts of the female: while it is relatively easy to mimic the conditions surrounding sea urchin fertilization using either natural or artificial seawater, we do not yet know the components of the various natural environments that mammalian sperm encounter as they travel to the egg.

A second reason why it is difficult to study mammalian fertilization is that the sperm population ejaculated into the female is probably very heterogeneous, containing spermatozoa at different stages of maturation. Out of the 280×10^6 human sperm normally ejaculated, only about 200 reach the egg (Ralt et al. 1991). Thus, since fewer than 1 in 10,000 sperm even gets close to the egg, it is difficult to assay those molecules that might enable the sperm to swim toward the egg and become activated.

Getting the gametes into the oviduct: Translocation and capacitation

The female reproductive tract is not a passive conduit through which sperm race, but a highly specialized set of tissues that actively regulates the transport and maturity of both gametes. Both the male and female gametes use a combination of small-scale biochemical interactions and large-scale physical propulsion to get to the **ampulla,** the region of the oviduct where fertilization takes place.

A mammalian oocyte just released from the ovary is surrounded by a matrix containing cumulus cells. (Cumulus cells are the cells of the ovarian follicle to which the developing oocyte was attached; see Figure 7.7.) If this matrix is experimentally removed or significantly altered, the fimbriae of the oviduct will not "pick up" the oocyte-cumulus complex (see Figure 11.27), nor will the complex be able to adhere to or enter the oviduct (Talbot et al. 1999). Once it is picked up, a combination of ciliary beating and muscle contractions transport the oocyte-cumulus complex to the appropriate position for its fertilization in the oviduct.

The translocation of sperm from the vagina to the oviduct involves many processes that work at different times and places. Sperm motility (i.e., flagellar action) is probably a minor factor in getting the sperm into the oviduct, although sperm motility is known to be required for mouse sperm to travel through the cervical mucus, and for sperm to encounter the egg once it is in the oviduct. Sperm are found in the oviducts of mice, hamsters, guinea pigs, cows, and humans within 30 minutes of sperm deposition in the vagina—a time "too short to have been attained by even the most Olympian sperm relying on their own flagellar power" (Storey 1995). Rather, sperm appear to be transported to the oviduct by the muscular activity of the uterus.

Recent studies of sperm motility have led to several conclusions, including the following:

1. Uterine muscle contractions are critical in getting the sperm into the oviduct.
2. The region of the oviduct before the ampulla may slow down sperm and release them slowly.
3. Sperm (flagellar) motility is important once sperm arrives within the oviduct; sperm become hyperactive in the vicinity of the oocyte.
4. Sperm may receive directional cues from temperature gradients between the regions of the oviduct and from chemical cues derived from the oocyte or cumulus.
5. During this trek from the vagina to the ampullary region of the oviduct, the sperm matures such that it has the capacity to fertilize the egg when the two finally meet.

Newly ejaculated mammalian sperm are unable to undergo the acrosome reaction until they have resided for some time in the female reproductive tract (Chang 1951; Austin 1952). The set of physiological changes by which the sperm becomes competent to fertilize the egg is called **capacitation**. Sperm that are not capacitated are "held up" in the cumulus matrix and are unable to reach the egg (Austin 1960; Corselli and Talbot 1987). The requirements for capacitation vary from species to species. Capacitation can be mimicked in vitro by incubating sperm in a tissue culture medium (such media contain calcium ions, bicarbonate, and serum albumin) or in fluid taken from the oviducts.

Contrary to the opening scenes of the *Look Who's Talking* movies, "the race is not always to the swiftest." Wilcox and colleagues (1995) found that nearly all human pregnancies result from sexual intercourse during a 6-day period ending on the day of ovulation. This means that the fertilizing sperm could have taken as long as 6 days to make the journey. Although some human sperm reach the ampulla of the oviduct within a half-hour after intercourse, "speedy" sperm may have little chance of fertilizing the egg. Eisenbach (1995) has proposed a hypothesis wherein capacitation is a transient event, and sperm are given a relatively brief window of competence during which they can successfully fertilize the egg. As the sperm reach the ampulla, they acquire competence—but they lose it if they stay around too long. Sperm may also have different survival rates depending on their location within the reproductive tract, so that some late-arriving sperm may have a better chance of success than those that arrived days earlier.

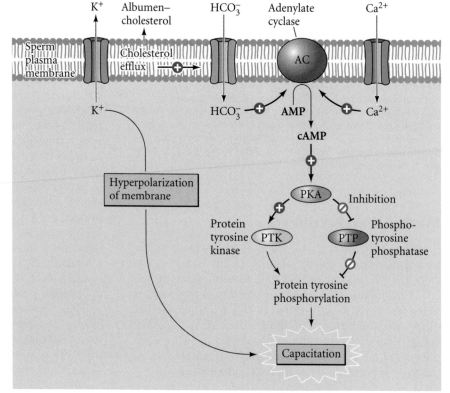

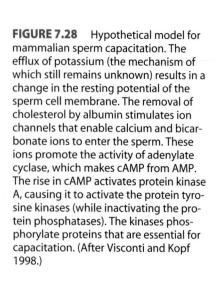

FIGURE 7.28 Hypothetical model for mammalian sperm capacitation. The efflux of potassium (the mechanism of which still remains unknown) results in a change in the resting potential of the sperm cell membrane. The removal of cholesterol by albumin stimulates ion channels that enable calcium and bicarbonate ions to enter the sperm. These ions promote the activity of adenylate cyclase, which makes cAMP from AMP. The rise in cAMP activates protein kinase A, causing it to activate the protein tyrosine kinases (while inactivating the protein phosphatases). The kinases phosphorylate proteins that are essential for capacitation. (After Visconti and Kopf 1998.)

The molecular events that take place during capacitation (Figure 7.28) have not yet been fully accounted for, but five sets of molecular changes are considered to be important:

1. The sperm cell membrane is altered by the removal of cholesterol by albumin proteins in the female reproductive tract (Cross 1998). The cholesterol efflux from the sperm cell membrane changes the location of "lipid rafts," isolated regions of the cell membrane that often contain receptor proteins. Originally located throughout the sperm cell membrane, lipid rafts now cluster over the anterior sperm head.

2. Particular proteins or carbohydrates on the sperm surface are lost during capacitation (Lopez et al. 1985; Wilson and Oliphant 1987). It is possible that these compounds block the recognition sites for the sperm proteins that bind to the zona pellucida. It has been suggested that the unmasking of these sites might be one of the effects of cholesterol depletion (Benoff 1993).

3. The membrane potential of the sperm cell membrane becomes more negative as potassium ions leave the sperm. This change in membrane potential may allow calcium channels to be opened and permit calcium to enter the sperm. Calcium and bicarbonate ions may be critical in activating cAMP production and in facilitating the membrane fusion events of the acrosome reaction (Visconti et al. 1995; Arnoult et al. 1999).

4. Protein phosphorylation occurs (Galantino-Homer et al. 1997). In particular, two chaperone (heat-shock) proteins migrate to the surface of the sperm head when they are phosphorylated, where they may play an essential role in forming the receptor that binds to the zona pellucida (Asquith et al. 2004, 2005).

5. The outer acrosomal membrane changes and comes into contact with sperm cell membrane in a way that prepares it for fusion (Tulsiani and Abou-Haila 2004).

It is uncertain whether these events are independent of one another and to what extent each of them contributes to sperm capacitation.

There may be an important connection between sperm translocation and capacitation. Smith (1998) and Suarez (1998) have documented that before entering the ampulla of the oviduct, the uncapacitated sperm bind actively to the membranes of the oviduct cells in the narrow passage (isthmus) preceding it (Figure 7.29; see also Figure 11.27). This binding is temporary and appears to be broken when the sperm become capacitated. Moreover, the life span of the sperm is significantly lengthened by this binding, and its capacitation is slowed down. This restriction of sperm entry into the ampulla, the slowing down of capacitation, and the expansion of sperm life span may have important consequences (Töpfer-Petersen et al. 2002; Gwathmey et al. 2003). The binding action may function as a block to polyspermy by preventing many sperm from reaching the egg at the same time; if the oviduct isthmus is excised in

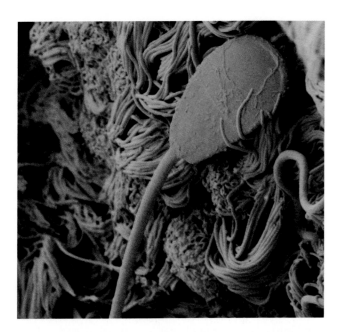

FIGURE 7.29 Scanning electron micrograph (artificially colored) showing bull sperm as it adheres to the membranes of epithelial cells in the oviduct of a cow prior to entering the ampulla. (From Lefebvre et al. 1995; photograph courtesy of S. Suarez.)

cows, a much higher rate of polyspermy results. In addition, slowing the rate of sperm capacitation and extending the active life of sperm may maximize the probability that sperm will still be available to meet the egg in the ampulla, even if ejaculation does not occur at the same time as ovulation.

In the vicinity of the oocyte: Hyperactivation, thermotaxis, and chemotaxis

Different regions of the female reproductive tract may secrete different, regionally specific molecules, and these molecules may influence sperm motility as well as capacitation. For instance, when sperm of certain mammals (especially hamsters, guinea pigs, and some strains of mice) pass from the uterus into the oviduct, they become **hyperactivated**—they swim at higher velocities and generate greater force. This hyperactivation appears to be mediated through the opening of a sperm-specific calcium channel located in the sperm tail (Quill et al. 2003). The asymmetric beating of the flagellum is changed into a rapid synchronous beat. Suarez and co-workers (1991) have shown that although this behavior is not conducive to traveling through low-viscosity fluids, it appears to be extremely well suited for linear sperm movement in the viscous fluid that sperm might encounter in the oviduct. Hyperactivation, along with a hyaluronidase enzyme on the outside of the sperm cell membrane, enable the sperm to digest a path through the extracellular matrix of the cumulus cells (Lin et al. 1994; Primakoff and Myles 2002).

An old joke claims that the reason men ejaculate so many million spermatozoa is because none of the sperm is willing to ask for directions. So what *does* provide the sperm with directions? A recent report suggests that the sperm are capable of sensing a thermal gradient of 2°C between the isthmus of the oviduct and the warmer ampullary region. Bahat and colleagues (2003) measured the temperature difference between these two regions of the oviduct and experimentally demonstrated that rabbit sperm sense the difference and preferentially swim from cooler to warmer sites (thermotaxis). Moreover, only capacitated sperm are able to sense this temperature gradient.

Once in the ampullary region, a second sensing mechanism, chemotaxis, may come into play. It appears that the oocyte and its accompanying cumulus cells secrete molecules that attract the sperm toward the egg during the last stages of sperm migration. Ralt and colleagues (1991) tested this hypothesis using follicular fluid from human follicles whose eggs were being used for in vitro fertilization. When the researchers microinjected a drop of follicular fluid into a larger drop of sperm suspension, some of the sperm changed direction and migrated toward the source of follicular fluid. Microinjection of other solutions did not have this effect. Moreover, these investigations uncovered a fascinating correlation: the fluid from only about half the follicles tested showed a chemotactic effect; in nearly every case, the egg was fertilizable if, and only if, the fluid showed chemotactic ability ($P < 0.0001$; see Eisenbach 1999; Sun et al. 2005). Further research has shown that the ability of human follicular fluid to attract human sperm only occurs if the sperm has been capacitated (Cohen-Dayag et al. 1995; Eisenbach and Tur-Kaspa 1999; Wang et al. 2001). It is possible that, like certain invertebrate eggs, the human egg secretes a chemotactic factor only when it is capable of being fertilized, and that sperm are attracted to such a compound only when they are capable of fertilizing the egg.

Recognition at the zona pellucida

Before the mammalian sperm can bind to the oocyte, it must first bind to and penetrate the egg's zona pellucida. The zona pellucida in mammals plays a role analogous to that of the vitelline envelope in invertebrates; the zona, however, is a far thicker and more dense structure than the vitelline envelope. The binding of sperm to the zona is relatively, but not absolutely, species-specific. (Species-specific gamete recognition is not a major problem when fertilization occurs internally.)

There appear to be several steps in the binding of a hyperactivated, wiggling mouse sperm to the zona pellucida. The mouse zona pellucida is made of three major glycoproteins, **ZP1**, **ZP2**, and **ZP3** (**zona proteins 1**, **2**, and **3**), along with accessory proteins that bind to the zona's integral structure. This glycoprotein matrix, which is synthesized and secreted by the growing oocyte, binds the sperm and, once the sperm is bound, initiates the acrosome reaction (Saling et al. 1979; Florman and Storey 1982; Cherr et al. 1986).

In recent years, a new model for mammalian sperm-zona binding has emerged, emphasizing sequential interactions between several sperm proteins and the components of the zona (Figure 7.30). The first step appears to be a relatively weak binding accomplished by the recognition of a sperm protein by a peripheral protein that coats the zona pellucida. This is followed by a somewhat stronger association between the zona and the sperm's SED1 protein. Last, a protein on the sperm (and possibly several other

(A)

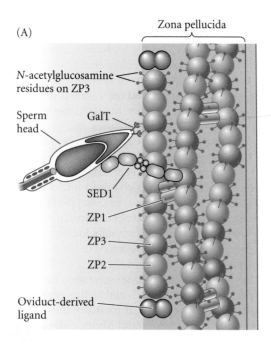

Zona pellucida

N-acetylglucosamine residues on ZP3

Sperm head

GalT

SED1

ZP1

ZP3

ZP2

Oviduct-derived ligand

(B)

FIGURE 7.30 Sperm-zona binding. (A) Possible model of proteins involved in mouse sperm-egg adhesion. First the sperm binds weakly but specifically to a ligand protein secreted by the oviduct and coating the zona pellucida. The sperm surface protein SED1 (which is localized in the correct area of the sperm head for lateral sperm adhesion) then binds to the ZP complex on the zona. Sperm galactosyltransferase (GalT) crosslinks tightly and specifically to *N*-acetylglucosamine residues on zona protein 3 (ZP3). The clustering of GalT proteins in the sperm cell membrane activates G proteins that open calcium channels and initiate the acrosome reaction. (The diagram is not drawn to scale.) (B) Electron micrograph showing sperm-zona binding in the golden hamster. (A based on data of B. Shur, courtesy of B. Shur; B photograph courtesy of R. Yanagimachi.)

(A)

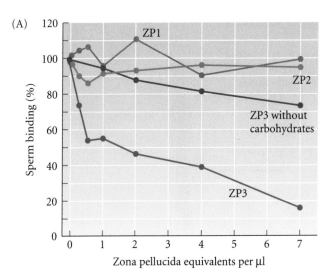

(B)

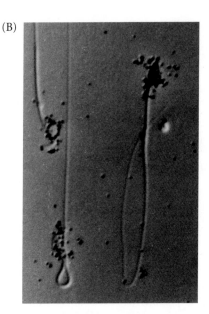

FIGURE 7.31 Mouse zona protein 3 binds sperm. (A) Inhibition assay showing a specific decrease of mouse sperm binding to zonae pellucidae. It appears from this assay that purified ZP3 (but not ZP1 or ZP2) can bind to the sperm and prevent the sperm from binding to the zona pellucida. The assay also illustrates the importance of the carbohydrate portion of ZP3 to the binding reaction. (B) Radioactively labeled ZP3 binds to capacitated mouse sperm. (A after Bleil and Wassarman 1980 and Florman and Wassarman 1985; B from Bleil and Wassarman 1986, photograph courtesy of the authors.)

factors) form strong links with the ZP3 protein of the zona. This last binding will cause the mouse sperm to undergo its acrosome reaction directly on the zona pellucida.

EARLY STAGES OF GAMETE ADHESION   Before strong and specific binding of the sperm and the egg, an initial tethering is accomplished. It appears that a 250-kDa sperm protein binds to a protein that is associated with, but not integrally part of, the zona pellucida (Rodeheffer and Shur 2004). This protein can be washed away by preparative techniques (which is probably why it wasn't discovered earlier). A second protein, the sperm cell surface adhesion protein SED1, then binds to the zona protein complex (Ensslin and Shur 2003). SED1 is found in a discrete domain of the sperm cell membrane, directly overlying the acrosome, and it only binds to the zona of unfertilized oocytes (and not to those of fertilized oocytes). Antibodies against SED1 or solubilized SED1 proteins will inhibit sperm-zona binding. Indeed, the sperm of males whose *SED1* gene has been knocked out are unable to bind to the zona pellucida.

THE FINAL STAGE OF SPERM-ZONA RECOGNITION: BINDING TO ZP3
There are several pieces of evidence demonstrating that ZP3 is the major sperm-binding glycoprotein in the mouse zona pellucida. The binding of mouse sperm to the mouse zona pellucida can be inhibited by first incubating the sperm with solubilized zona glycoproteins. Using this inhibition assay, Bleil and Wassarman (1980, 1986, 1988) found

that ZP3 was the active competitor for sperm binding sites (Figure 7.31A). This was confirmed by the finding that radiolabeled ZP3 (but not ZP1 or ZP2) bound to the heads of mouse sperm with intact acrosomes (Figure 7.31B).

The cell membrane overlying the sperm head can bind to thousands of ZP3 glycoproteins in the zona pellucida. Moreover, there appear to be several different proteins on sperm that are capable of binding ZP3 (Wassarman et al. 2001). Some of these sperm proteins bind to the serine- and threonine-linked carbohydrate chains of ZP3, and one of the zona-binding proteins, a sperm surface galactosyltransferase, recognizes *N*-acetylglucosamine residues (Miller et al. 1992; Gong et al. 1995; Lu and Shur 1997). The conclusion that the carbohydrate moieties of ZP3 are critical for sperm attachment to the zona has been confirmed by the finding that if these carbohydrate groups are removed from ZP3, it will not bind sperm as well as intact ZP3 (see Figure 7.31A; Florman and Wassarman 1985; Kopf 1998).

INDUCTION OF THE MOUSE ACROSOME REACTION BY ZP3   ZP3 is the specific glycoprotein in the mouse zona pellucida to which sperm bind. ZP3 also initiates the acrosome reaction after sperm have bound to it. The mouse zona pellucida, unlike the sea urchin vitelline envelope, is a thick structure. By undergoing the acrosome reaction on the zona pellucida, the mouse sperm can concentrate its proteolytic enzymes directly at the point of attachment and digest a hole through this extracellular layer (see Figure 7.8B). Indeed, mouse sperm that undergo the acrosome reaction before they reach the zona pellucida are unable to penetrate it (Florman et al. 1998).

The mouse acrosome reaction is induced when ZP3 crosslinks the receptors on the sperm cell membrane. One of the sperm proteins that is crosslinked is the sperm cell surface galactosyltransferase, whose active site faces out-

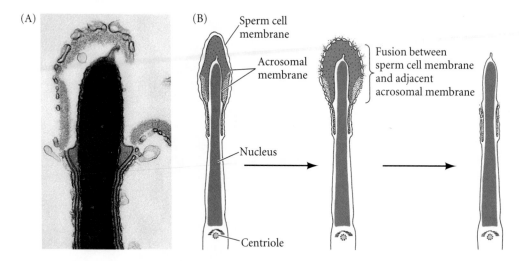

FIGURE 7.32   The acrosome reaction in hamster sperm. (A) Transmission electron micrograph of hamster sperm undergoing the acrosome reaction. The acrosomal membrane can be seen to form vesicles. (B) Interpretive diagram of electron micrographs showing the fusion of the acrosomal and cell membranes in the sperm head. (A from Meizel 1984, photograph courtesy of S. Meizel; B after Yanagimachi and Noda 1970.)

ward and binds to the carbohydrate residues of ZP3 (see Figure 7.30). This crosslinking activates specific G proteins in the sperm cell membrane, initiating a cascade that opens the membrane's calcium channels and causes the calcium-mediated exocytosis of the acrosomal vesicle (Figure 7.32; Leyton and Saling 1989; Leyton et al. 1992; Florman et al. 1998; Shi et al. 2001).

TRAVERSING THE ZONA PELLUCIDA   The exocytosis of the acrosomal vesicle releases a variety of proteases that lyse the zona pellucida. These enzymes create a hole through which the sperm can travel toward the egg. However, during the acrosome reaction, the anterior portion of the sperm cell membrane (i.e., the region containing the ZP3 binding sites) is shed from the sperm. But if sperm are going to penetrate the zona pellucida, they must somehow retain some adhesion to it. In mice, it appears that this secondary binding to the zona is accomplished by proteins in the inner acrosomal membrane that bind specifically to the ZP2 glycoprotein (Bleil et al. 1988). (Recall that in sea urchins, this inner acrosomal membrane contains the bindin proteins that are critical in sperm adhesion to the vitelline envelope). Whereas acrosome-intact sperm will not bind to ZP2, acrosome-reacted sperm will. Moreover, antibodies against the ZP2 glycoprotein will not prevent the binding of acrosome-intact sperm to the zona but will inhibit the attachment of acrosome-reacted sperm. The structure of the zona consists of repeating units of ZP3 and ZP2, occasionally crosslinked by ZP1 (see Figure 7.30). It appears that acrosome-reacted sperm transfer their binding from ZP3 to the adjacent ZP2 molecules.*

Gamete fusion and the prevention of polyspermy

In mammals, the sperm contacts the egg not at its tip (as in the case of sea urchins), but on the side of the sperm head. The acrosome reaction, in addition to expelling the enzymatic contents of the acrosome, also exposes the inner acrosomal membrane to the outside. The junction between this inner acrosomal membrane and the sperm cell membrane is called the **equatorial region**, and this is where membrane fusion between the sperm and egg begins (Figure 7.33). As in sea urchin gamete fusion, the sperm is bound to regions of the egg where actin polymerizes to extend microvilli to the sperm (Yanagimachi and Noda 1970).

The mechanism of mammalian gamete fusion is still controversial (see Primakoff and Myles 2002). Gene knockout experiments suggest that mammalian gamete fusion may depend on interaction between a sperm protein and integrin-associated CD9 protein on the egg (Le Naour et al. 2000; Miyado et al. 2000; Evans 2001). Female mice carrying gene knockouts for *CD9* are infertile because their eggs fail to fuse with sperm. This infertility can be reversed by the microinjection of mRNA encoding either mouse or human CD9 (Kaji et al. 2002). It is not known exactly how these proteins

*In guinea pigs, secondary binding to the zona is thought to be mediated by the protein PH-20. Moreover, when this inner acrosomal membrane protein was injected into adult male or female guinea pigs, 100% of them became sterile for several months (Primakoff et al. 1988). The blood sera of these sterile guinea pigs had extremely high concentrations of antibodies to PH-20. The antiserum from guinea pigs sterilized in this manner not only bound specifically to PH-20, but also blocked sperm-zona adhesion in vitro. The contraceptive effect lasted several months, after which fertility was restored. More recently, O'Rand and colleagues (2004) provided reversible immunological contraception by injecting male monkeys with eppin, a sperm-surface protein that interacts with semen components. The antibodies block these interactions, probably slowing down the sperm. These experiments show that the principle of immunological contraception is well founded.

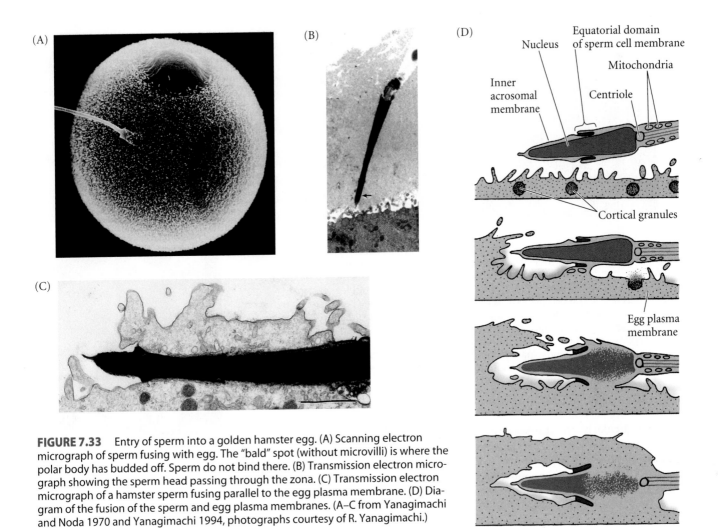

FIGURE 7.33 Entry of sperm into a golden hamster egg. (A) Scanning electron micrograph of sperm fusing with egg. The "bald" spot (without microvilli) is where the polar body has budded off. Sperm do not bind there. (B) Transmission electron micrograph showing the sperm head passing through the zona. (C) Transmission electron micrograph of a hamster sperm fusing parallel to the egg plasma membrane. (D) Diagram of the fusion of the sperm and egg plasma membranes. (A–C from Yanagimachi and Noda 1970 and Yanagimachi 1994, photographs courtesy of R. Yanagimachi.)

facilitate membrane fusion, but CD9 is also known to be critical for the fusion of myocytes (the muscle cell precursors) to form striated muscle (Tachibana and Hemler 1999).

On the sperm side of the mammalian fusion process, Inoue and colleagues (2005) have implicated the immunoglobulin-like protein Izumo (named after a Japanese shrine dedicated to marriage). Sperm from mice carrying loss-of-function mutations in the *Izumo* gene are able to bind and penetrate the zona pellucida, but are not able to fuse with the egg cell membrane. Human sperm also contain Izumo protein, and antibodies directed against Izumo prevent sperm-egg fusion in humans as well. There are other candidates for sperm fusion proteins; indeed, there may be several sperm-egg binding systems operating, and each of them may be necessary but not sufficient to insure proper gamete binding and fusion.

Polyspermy is a problem for the developing mammal as well as for the sea urchin (see Figure 7.16). In mammals, the cortical granule reaction does not create a fertilization envelope, but its ultimate effect is the same. Released enzymes modify the zona pellucida sperm receptors such that they can no longer bind sperm (Bleil and Wassarman

1980). Cortical granules of mouse eggs have been found to contain *N*-acetylglucosaminidase enzymes capable of cleaving *N*-acetylglucosamine from ZP3 carbohydrate chains. *N*-acetylglucosamine is one of the carbohydrate groups to which sperm can bind. Miller and co-workers (1992, 1993) have demonstrated that when the *N*-acetylglucosamine residues are removed at fertilization, ZP3 will no longer serve as a substrate for the binding of other sperm. ZP2 is clipped by another cortical granule protease and loses its ability to bind sperm as well (Moller and Wassarman 1989). Thus, once a sperm has entered the egg, other sperm can no longer initiate or maintain their binding to the zona pellucida and are rapidly shed.

Fusion of genetic material

As in sea urchins, the mammalian sperm that finally enters the egg carries its genetic contribution in a haploid pronucleus. In mammals, the process of pronuclear migration takes about 12 hours, compared with less than 1 hour in the sea urchin. The mammalian sperm enters almost tangentially to the surface of the egg rather than approaching

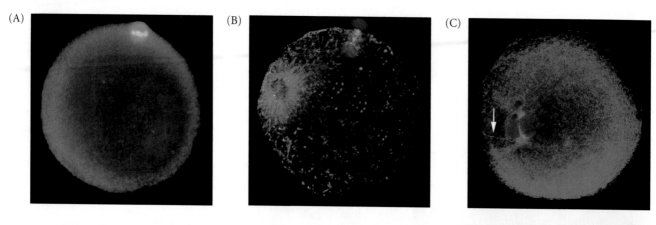

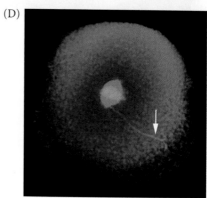

FIGURE 7.34 Pronuclear movements during human fertilization. The microtubules have been stained green, while the DNA is dyed blue. The arrows point to the sperm tail. (A) The mature unfertilized oocyte completes the first meiotic division, budding off a polar body. (B) As the sperm enters the oocyte (left side), microtubules condense around it as the oocyte completes its second meiotic division at the periphery. (C) By 15 hours after fertilization, the two pronuclei have come together, and the centrosome splits to organize a bipolar microtubule array. The sperm tail is still seen (arrow). (D) At prometaphase, chromosomes from the sperm and egg intermix on the metaphase equator and a mitotic spindle initiates the first mitotic division. The sperm tail can still be seen. (From Simerly et al. 1995, photographs courtesy of G. Schatten.)

it perpendicularly, and it fuses with numerous microvilli (see Figure 7.32). The DNA of the sperm nucleus is bound by basic proteins called protamines, which are tightly compacted through disulfide bonds. Glutathione in the egg cytoplasm reduces these disulfide bonds and allows the uncoiling of the sperm chromatin (Calvin and Bedford 1971; Kvist et al. 1980; Perreault et al. 1988).

The mammalian sperm enters the oocyte while the oocyte nucleus is "arrested" in metaphase of its second meiotic division (Figure 7.34A; see also Figure 7.5). The calcium oscillations brought about by sperm entry inactivate MAP kinase and allow DNA synthesis. But unlike the sea urchin egg, which is already in a haploid state, the mammalian oocyte still has chromosomes in the middle of meiotic metaphase. Oscillations in the level of Ca^{2+} activate another kinase that leads to the proteolysis of cyclin (thus allowing the cell cycle to continue, eventually resulting in a haploid female pronucleus) and securin (the protein holding the metaphase chromosomes together) (Watanabe et al. 1991; Johnson et al. 1998). Mammals appear to undergo several waves of Ca^{2+} release. Events that are initiated by one Ca^{2+} wave might not go to completion without additional calcium waves (Ducibella et al. 2002).

DNA synthesis occurs separately in the male and in the female pronuclei. The centrosome (new centriole) accompanying the male pronucleus produces its asters (largely from proteins stored in the oocyte). The microtubules join the two pronuclei and enable them to migrate toward one another other. Upon meeting, the two nuclear envelopes break down (Figure 7.34B). However, instead of producing a common zygote nucleus (as in sea urchins), the chromatin

condenses into chromosomes that orient themselves on a common mitotic spindle (Figure 7.34C–E). Thus, a true diploid nucleus in mammals is first seen not in the zygote, but at the two-cell stage.

Each sperm brings into the egg not only its pronucleus, but also its mitochondria, its centriole, and a small amount of cytoplasm. The sperm mitochondria and their DNA are degraded in the egg cytoplasm, so that all of the new individual's mitochondria are derived from its mother. The egg and embryo appear to get rid of the paternal mitochondria both by dilution and by actively targeting them for destruction (Cummins et al. 1998; Shitara et al. 1998; Schwartz and Vissing 2002). However, in most mammals, the sperm centriole not only survives, but it serves as the organizing agent for making the new mitotic spindle.

Several proteins and mRNAs formed by the transcription of the pronucleus in the spermatocyte are also brought into the egg. These include the mRNAs for certain transcription and paracrine factors (Krawetz 2005). The human sperm even brings in a microRNA that may downregulate an important receptor involved in cell division in the early embryo (Ostermeier et al. 2005). The significance of this small quantity of RNAs and proteins is still being debated. The one protein delivered by the sperm whose importance has been demonstrated is PLCζ. This soluble small protein surrounds the sperm head and is crucial for egg activation and Ca^{2+} release in mammals.

SIDELIGHTS & SPECULATIONS

The Nonequivalence of Mammalian Pronuclei

It is generally assumed that males and females carry equivalent haploid genomes. Indeed, one of the fundamental tenets of Mendelian genetics is that genes derived from the sperm are functionally equivalent to those derived from the egg. However, as we saw in Chapter 5, genomic imprinting can occur in mammals such that the sperm-derived genome and the egg-derived genome may be functionally different and play complementary roles during certain stages of development. This *imprinting* is thought to be caused by the different patterns of cytosine methylation on the genes.

The first evidence for nonequivalence came from studies of a human tumor called a **hydatidiform mole**, which resembles placental tissue. A majority of such moles have been shown to arise when a haploid sperm fertilizes an egg in which the female pronucleus is absent. After entering the egg, the sperm chromosomes duplicate themselves, thereby restoring the diploid chromosome number. However, the entire genome is derived from the sperm (Jacobs et al. 1980; Ohama et al. 1981). The cells survive, divide, and have a normal chromosome number, but development is abnormal. Instead of forming an embryo, the egg becomes a mass of placenta-like cells. Normal development does not occur when the entire genome comes from the male parent.

Normal development also does not occur when the genome is derived totally from the egg. The ability to develop an embryo without spermatic contribution is called **parthenogenesis** (Greek, "virgin birth"). The eggs of many invertebrates and some vertebrates are capable of developing normally in the absence of sperm (see Chapters 2 and 19). Mammals, however, do not exhibit parthenogenesis. Placing mouse oocytes in a culture medium that artificially activates the oocyte while suppressing the formation of the second polar body produces diploid mouse eggs whose genes are derived exclusively from the oocyte (Kaufman et al. 1977). These eggs divide to form embryos with spinal cords, muscles, skeletons, and organs, including beating hearts. However, development does not continue, and by day 10 or 11 (halfway through the mouse's gestation), these parthenogenetic embryos deteriorate. Neither human nor mouse development can be completed solely with egg-derived chromosomes.

The hypothesis that male and female pronuclei are both needed for normal development was also shown by pronuclear transplantation experiments (Surani and Barton 1983; Surani et al. 1986; McGrath and Solter 1984). Either male or female pronuclei can be removed from recently fertilized mouse eggs and added to other recently fertilized eggs. (The two pronuclei can be distinguished at this stage because the female pronucleus is the one beneath the polar bodies.) Thus, zygotes with two male or two female pronuclei can be constructed. Although these eggs will form diploid cells that undergo normal cleavage, eggs whose genes are derived solely from sperm nuclei or solely from oocyte nuclei do not develop to birth. Control eggs undergoing such transplantation (i.e., eggs containing one male and one female pronucleus taken from different zygotes) can develop normally (Table 7.2). Thus, for mammalian development to occur, both the sperm-derived and the egg-derived pronuclei are critical.

The importance of DNA methylation in this block to parthenogenesis was demonstrated when Kono and colleagues (2004) generated a female mouse whose genes had come exclusively from two oocytes. To accomplish this feat, however, they had to mutate the DNA methylation system in one of the mouse genomes to make it more like that of a male mouse, and then they had to perform two rounds of nuclear transfer. "Men," as one reviewer remarked, "do not need to fear becoming redundant any time soon" (Vogel 2004).

TABLE 7.2 Pronuclear transplantation experiments

Class of reconstructed zygotes	Operation	Number of successful transplants	Number of progeny surviving
Bimaternal		339	0
Bipaternal		328	0
Control		348	18

Source: McGrath and Solter 1984.

Coda

Fertilization is not a moment or an event, but a process of carefully orchestrated and coordinated events including the contact and fusion of gametes, the fusion of nuclei, and the activation of development. It is a process whereby two cells, each at the verge of death, unite to create a new organism that will have numerous cell types and organs. It is just the beginning of a series of cell-cell interactions that will characterize animal development.

 **Snapshot Summary** **Fertilization**

1. Fertilization accomplishes two separate activities: sex (the combining of genes derived from two parents), and reproduction (the creation of a new organism).

2. The events of fertilization usually include (1) contact and recognition between sperm and egg; (2) regulation of sperm entry into the egg; (3) fusion of genetic material from the two gametes; and (4) activation of egg metabolism to start development.

3. The sperm head consists of a haploid nucleus and an acrosome. The acrosome is derived from the Golgi apparatus and contains enzymes needed to digest extracellular coats surrounding the egg. The midpiece and neck of the sperm contain mitochondria and the centriole that generates the microtubules of the flagellum. Energy for flagellar motion comes from mitochondrial ATP and a dynein ATPase in the flagellum.

4. The female gamete can be an egg (with a haploid nucleus, as in sea urchins) or an oocyte (in an earlier stage of development, as in mammals). The egg (or oocyte) has a large mass of cytoplasm storing ribosomes, mRNAs, and nutritive proteins. Other mRNAs and proteins that will be used as morphogenetic factors are also stored in the egg. Many eggs also contain protective agents needed for survival in their particular environment.

5. Surrounding the egg cell membrane is an extracellular layer often used in sperm recognition. In most animals, this extracellular layer is the vitelline envelope. In mammals, it is the much thicker zona pellucida. Cortical granules lie beneath the egg's cell membrane.

6. Neither the egg nor the sperm is the "active" or "passive" partner. The sperm is activated by the egg and the egg is activated by the sperm. Both involve calcium and membrane fusions.

7. When fertilization takes place in the external environment, as in sea urchins and many other marine organisms, eggs secrete diffusible molecules that attract and activate the sperm.

8. Species-specific chemotactic molecules secreted by the egg can attract sperm that is capable of fertilizing it. In sea urchins, the chemotactic peptides resact and speract have been shown to increase sperm motility and provide a "road map" to an egg of the correct speciees.

9. The acrosome reaction releases enzymes exocytotically. These proteolytic ensymes digest the egg's protective coating, allowing the sperm to reach and fuse with the egg cell membrane. In sea urchins, this reac-

tion in the sperm is initiated by compounds in the egg jelly. Globular actin polymerizes to extend the acrosomal process. Bindin on the acrosomal process is recognized by a protein complex on the sea urchin egg surface.

10. Fusion between sperm and egg is probably mediated by protein molecules whose hydrophobic groups can merge the sperm and egg cell membranes. In sea urchins, bindin may mediate gamete fusion.

11. Polyspermy results when two or more sperm fertilize an egg. It is usually lethal, since it results in blastomeres with different numbers and types of chromosomes.

12. Many species have two blocks to polyspermy. The fast block is immediate and causes the egg membrane resting potential to rise. Sperm can no longer fuse with the egg. In sea urchins this is mediated by the influx of sodium ions. The slow block, or cortical granule reaction, is physical and is mediated by calcium ions. A wave of Ca^{2+} propagates from the point of sperm entry, causing the cortical granules to fuse with the egg cell membrane. The released contents of these granules cause the vitelline envelope to rise and harden into the fertilization envelope.

13. The fusion of sperm and egg results in the activation of crucial metabolic reactions within the egg. These reactions include reinitiation of the egg's cell cycle and subsequent mitotic division, and the resumption of DNA and protein synthesis.

14. Genetic material is carried in a male and a female pronucleus, which migrate toward each other. In sea urchins, the male and female pronuclei merge and a diploid zygote nucleus is formed. DNA replication occurs after pronuclear fusion.

15. In all species studied, free Ca^{2+}, supported by the alkalization of the egg, activates egg metabolism, protein synthesis, and DNA synthesis. Inositol 1,4,5-trisphosphate (IP_3) is responsible for releasing Ca^{2+} from storage in the endoplasmic reticulum. DAG (diacylglycerol) is thought to initiate the rise in egg pH.

16. IP_3 is generated by phospholipases. Different species may use different mechanisms to activate the phospholipases.

17. Mammalian fertilization takes place internally, within the female reproductive tract. The cells and tissues of the female reproductive tract are far from passive; they actively regulate the positioning and maturity of both the male and female gametes.

18. The translocation of sperm from the vagina to the oviduct is regulated by the muscular activity of the

uterus, by the binding of sperm in the isthmus of the oviduct, and by directional cues from the oocyte (immature egg) and/or the cumulus cells surrounding it.

19. Mammalian sperm must be capacitated in the female reproductive tract before they are capable of fertilizing the egg. Capacitation is the result of biochemical changes in the sperm cell membrane.

20. Capacitated mammalian sperm must penetrate the cumulus and bind to the zona pellucida before undergoing the acrosome reaction. In the mouse, this binding is mediated by ZP3 (zona protein 3) and several sperm proteins that recognize it.

21. ZP3 initiates the mammalian acrosome reaction on the zona pellucida, and the acrosomal enzymes are concentrated there.

22. In mammals, blocks to polyspermy include the modification of the zona proteins by the contents of the cortical granules so that sperm can no longer bind to the zona.

23. The rise in intracellular free Ca^{2+} at fertilization in amphibians and mammals causes the degradation of cyclin and the inactivation of MAP kinase, allowing the second meiotic metaphase to be completed and the formation of the haploid female pronucleus.

24. In mammals, DNA replication takes place as the pronuclei are traveling toward each other. The pronuclear membranes disintegrate as the pronuclei approach each other, and their chromosomes gather around a common metaphase plate.

25. The male and female pronuclei of mammals are not equivalent. If the zygote's genetic material is derived solely from one parent or the other, normal development will not take place. This difference in the male and female genomes is thought to be the result of different methylation patterns on the genes.

For Further Reading

Complete bibliographical citations for all literature cited in this chapter can be found on the Vade Mecum CD that accompanies the book and at the free access website www.devbio.com

Bleil, J. D. and P. M. Wassarman. 1980. Mammalian sperm and egg interaction: Identification of a glycoprotein in mouse-egg zonae pellucidae possessing receptor activity for sperm. *Cell* 20: 873–882.

Boveri, T. 1902. On multipolar mitosis as a means of analysis of the cell nucleus. [Translated by S. Gluecksohn-Waelsch.] *In* B. H. Willier and J. M. Oppenheimer (eds.), *Foundations of Experimental Embryology*. Hafner, New York 1974.

Eisenbach, M. 1999. Mammalian sperm chemotaxis and its association with capacitation. *Dev. Genet.* 25: 87–94.

Florman, H. M. and B. T. Storey. 1982. Mouse gamete interactions: The zona pellucida is the site of the acrosome reaction leading to fertilization in vitro. *Dev. Biol.* 91: 121–130.

Glabe, C. G. and V. D. Vacquier. 1978. Egg surface glycoprotein receptor for sea urchin sperm bindin. *Proc. Natl. Acad. Sci. USA* 75: 881–885.

Jaffe, L. A. 1976. Fast block to polyspermy in sea urchins is electrically mediated. *Nature* 261: 68–71.

Just, E. E. 1919. The fertilization reaction in *Echinarachinus parma. Biol. Bull.* 36: 1–10.

Vacquier, V. D. and G. W. Moy. 1977. Isolation of bindin: The protein responsible for adhesion of sperm to sea urchin eggs. *Proc. Natl. Acad. Sci. USA* 74: 2456–2460.

Early Development in Selected Invertebrates

8

REMARKABLE AS IT IS, fertilization is but the initiating step in development. The zygote, with its new genetic potential and its new arrangement of cytoplasm, now begins the production of a multicellular organism. Between the events of fertilization and the events of organ formation are two critical stages: cleavage and gastrulation. During cleavage, rapid cell divisions divide the cytoplasm of the fertilized egg into numerous cells. These cells then undergo dramatic displacements during gastrulation, a process whereby they move to different parts of the embryo and acquire new neighbors. During cleavage and gastrulation, the major axes of the embryo are determined and the embryonic cells begin to acquire their respective fates.

While cleavage always precedes gastrulation, axis formation in some species can begin as early as oocyte formation. It can be completed during cleavage (as it does in *Drosophila*) or extend all the way through gastrulation (as in *Xenopus*). Three body axes must be specified: the anterior-posterior (head-anus) axis; the dorsal-ventral (back-belly) axis; and the left-right axis. Different species specify these axes at different times, using different mechanisms.

EARLY DEVELOPMENTAL PROCESSES: AN OVERVIEW

Cleavage

Once fertilization is complete, the development of a multicellular organism proceeds by a process called **cleavage**, a series of mitotic divisions whereby the enormous volume of egg cytoplasm is divided into numerous smaller, nucleated cells. These cleavage-stage cells are called blastomeres. In most species (mammals being the chief exception), the initial rate of cell division and the placement of the blastomeres with respect to one another is under the control of the proteins and mRNAs stored in the oocyte. Only later do the rates of cell division and the placement of cells come under the control of the newly formed genome.

During the initial phase of development, when cleavage rhythms are controlled by maternal factors, the cytoplasmic volume does not increase. Rather, the enormous volume of zygote cytoplasm is divided into increasingly smaller cells. First the zygote is divided in half, then quarters, then eighths, and so forth. This division of cytoplasm without increasing its volume is accomplished by abolishing the gap periods of the cell cycle (the G_1 and G_2 phases), when growth can occur. Meanwhile, nuclear division occurs at a rapid rate never seen again (not even in tumor cells). A frog egg, for example, can divide into 37,000 cells in just 43 hours. Mitosis in cleavage-stage *Drosophila* embryos occurs every 10 minutes for over 2 hours, and some 50,000 cells form in just 12 hours.

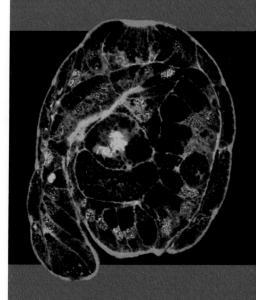

From fertilization to cleavage

The transition from fertilization to cleavage is caused by the activation of **mitosis-promoting factor** (**MPF**). MPF was first discovered as the major factor responsible for the resumption of meiotic cell divisions in the ovulated frog egg. It continues to play a role after fertilization, regulating the cell cycle of early blastomeres.

Blastomeres generally progress through a biphasic cell cycle consisting of just two steps: M (mitosis) and S (DNA synthesis) (Figure 8.1). The MPF activity of early blastomeres is highest during M and undetectable during S. The shift between the M and S phases in blastomeres is driven solely by the gain and loss of MPF activity. Cleaving cells can be experimentally trapped in S phase by incubating them in an inhibitor of protein synthesis. When MPF is microinjected into these cells, they enter M. Their nuclear envelope breaks down and their chromatin condenses into chromosomes. After an hour, MPF is degraded and the chromosomes return to S phase (Gerhart et al. 1984; Newport and Kirschner 1984).

What causes this cyclical activity of MPF? Mitosis-promoting factor contains two subunits. The large subunit, **cyclin B**, shows the cyclical behavior that is key to mitotic regulation, accumulating during S and being degraded after the cells have reached M (Evans et al. 1983; Swenson et al. 1986). Cyclin B is often encoded by mRNAs stored in the oocyte cytoplasm, and if the translation of this message is specifically inhibited, the cell will not enter mitosis (Minshull et al. 1989). Cyclin B regulates the small subunit of MPF, the **cyclin-dependent kinase**. This kinase activates mitosis by phosphorylating several target proteins, including histones, the nuclear envelope lamin proteins, and the regulatory subunit of cytoplasmic myosin. It is these actions that bring about chromatin condensation, nuclear envelope depolymerization, and the organization of the mitotic spindle, but without cyclin B, the cyclin-dependent kinase subunit of MPF will not function.

The presence of cyclin B is controlled by several proteins that ensure its periodic synthesis and degradation. In most species studied, the regulators of cyclin B (and thus of MPF) are stored in the egg cytoplasm. Therefore, the cell cycle remains independent of the nuclear genome for a number of cell divisions. These early divisions tend to be rapid and synchronous. However, as the cytoplasmic components are used up, the nucleus begins to synthesize them. In several species, the embryo now enters a **midblastula transition**, in which several new phenomena are added to the biphasic cell divisions of the embryo. First, the "gap" stages (G_1 and G_2) are added to the cell cycle (see Figure 8.1B). *Xenopus* embryos add G_1 and G_2 phases to the cell cycle shortly after the twelfth cleavage. *Drosophila* adds G_2 during cycle 14 and G_1 during cycle 17 (Newport and Kirschner 1982a; Edgar et al. 1986). Second, the synchronicity of cell division is lost, because different cells synthesize different regulators of MPF. Third, new mRNAs are transcribed. Many of these messages encode proteins that will become necessary for gastrulation. In several species, if transcription is blocked cell division will still occur at normal rates and times, but the embryo will not be able to initiate gastrulation. Many of these new messenger RNAs are also used for cell specification. As we will see in sea urchin

FIGURE 8.1   Cell cycles of somatic cells and early blastomeres. (A) The biphasic cell cycle of early amphibian blastomeres has only two states, S and M. Cyclin B synthesis allows progression to M (mitosis), while degradation of cyclin B allows cells to pass into S (synthesis) phase. (B) The complete cell cycle of a typical somatic cell. Mitosis (M) is followed by an interphase stage. Interphase is subdivided into G_1, S (synthesis), and G_2 phases. Cells that are differentiating are usually taken "out" of the cell cycle and are in an extended G_1 phase called G_0. The cyclins responsible for the progression through the cell cycle and their respective kinases are shown at their point of cell cycle regulation. (B after Nigg 1995.)

(A)

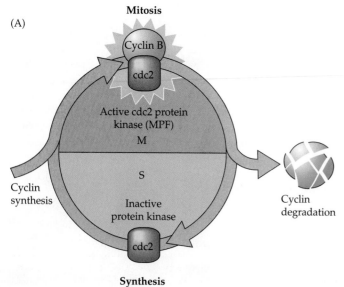

(B)

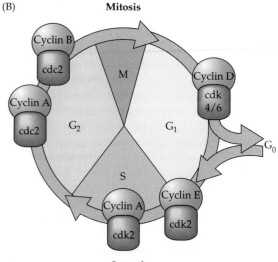

(A)

Microfilaments
(contractile ring)

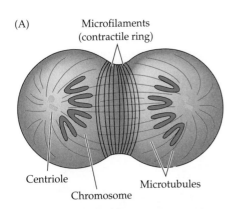

Centriole

Chromosome

Microtubules

(B)

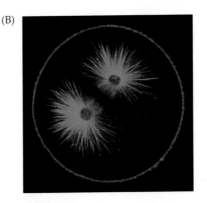

(C)

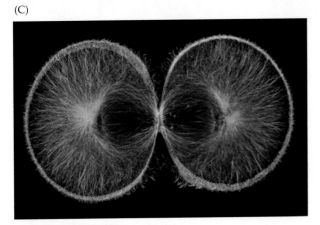

FIGURE 8.2 Role of microtubules and microfilaments in cell division. (A) Diagram of first-cleavage telophase. The chromosomes are being drawn to the centrioles by microtubules, while the cytoplasm is being pinched in by the contraction of microfilaments. (B) Confocal fluorescent image of an echinoderm embryo undergoing first cleavage (early anaphase). The microtubules are stained green, the actin microfilaments are stained red. (C) Confocal fluorescent image of sea urchin embryo at the very end of first cleavage. The microtubules are orange and the actin proteins (both unpolymerized and in microfilaments) are blue. (Photographs courtesy of G. von Dassow and the Center for Cell Dynamics.)

embryos, the new mRNA expression patterns of the midblastula transition map out territories where specific types of cells will later differentiate.

The cytoskeletal mechanisms of mitosis

Cleavage is the result of two coordinated processes. The first of these is **karyokinesis**, the mitotic division of the cell's nucleus. The mechanical agent of karyokinesis is the mitotic spindle, with its microtubules composed of tubulin (the same type of protein that makes up the sperm flagellum). The second process is **cytokinesis**: the division of the cell. The mechanical agent of cytokinesis is a contractile ring of microfilaments made of actin (the same type of protein that extends the egg microvilli and the sperm acrosomal process). Table 8.1 presents a comparison of these agents of cell division. The relationship and coordination between these two systems during cleavage is depicted in

Figure 8.2A, in which a sea urchin egg is shown undergoing first cleavage. The mitotic spindle and contractile ring are perpendicular to each other, and the spindle is internal to the contractile ring. The contractile ring creates a cleavage furrow, which eventually bisects the plane of mitosis, thereby creating two genetically equivalent blastomeres.

The actin microfilaments are found in the cortex (outer cytoplasm) of the egg rather than in the central cytoplasm. Under the electron microscope, the ring of microfilaments can be seen forming a distinct cortical band 0.1 μm wide (Figure 8.2B). This contractile ring exists only during cleavage and extends 8–10 μm into the center of the egg. It is responsible for exerting the force that splits the zygote into blastomeres; if the ring is disrupted, cytokinesis stops. Schroeder (1973) likened the contractile ring to an "intercellular purse-string," tightening about the egg as cleavage continues. This tightening of the microfilamentous ring creates the **cleavage furrow**. Microtubules are also seen near the cleavage furrow (in addition to their role in creating the mitotic spindle), since they are needed to bring membrane material to the site of membrane addition (Danilchik et al. 1998).

Although karyokinesis and cytokinesis are usually coordinated, they are sometimes separated by natural or experimental conditions. As we will learn in the next chapter, the nuclei of insect eggs undergo karyokinesis several times before cytokinesis takes place. Another way to produce this state is to treat embryos with the drug cytochalasin B. This drug inhibits the formation and organization of microfila-

TABLE 8.1 Karyokinesis and cytokinesis

Process	Mechanical agent	Major protein composition	Location	Major disruptive drug
Karyokinesis	Mitotic spindle	Tubulin microtubules	Central cytoplasm	Colchicine, nocodazole[a]
Cytokinesis	Contractile ring	Actin microfilaments	Cortical cytoplasm	Cytochalasin B

[a]Because colchicine has been found to independently inhibit several membrane functions, including osmoregulation and the transport of ions and nucleosides, nocodazole has become the major drug used to inhibit microtubule-mediated processes (see Hardin 1987).

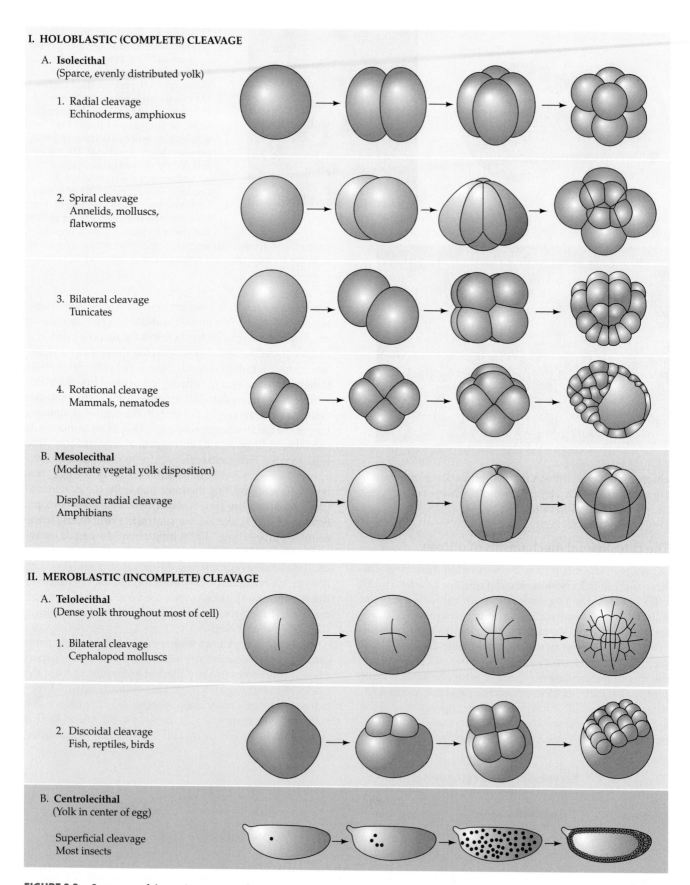

I. HOLOBLASTIC (COMPLETE) CLEAVAGE

A. **Isolecithal**
(Sparce, evenly distributed yolk)

1. Radial cleavage
   Echinoderms, amphioxus

2. Spiral cleavage
   Annelids, molluscs,
   flatworms

3. Bilateral cleavage
   Tunicates

4. Rotational cleavage
   Mammals, nematodes

B. **Mesolecithal**
(Moderate vegetal yolk disposition)

Displaced radial cleavage
Amphibians

II. MEROBLASTIC (INCOMPLETE) CLEAVAGE

A. **Telolecithal**
(Dense yolk throughout most of cell)

1. Bilateral cleavage
   Cephalopod molluscs

2. Discoidal cleavage
   Fish, reptiles, birds

B. **Centrolecithal**
(Yolk in center of egg)

Superficial cleavage
Most insects

FIGURE 8.3 Summary of the main patterns of cleavage.

ments in the contractile ring, thereby stopping cleavage without stopping karyokinesis (Schroeder 1972).

Patterns of embryonic cleavage

In 1923, embryologist E. B. Wilson reflected on how little we knew about cleavage: "To our limited intelligence, it would seem a simple task to divide a nucleus into equal parts. The cell, manifestly, entertains a very different opinion." Indeed, different organisms undergo cleavage in distinctly different ways. The pattern of embryonic cleavage peculiar to a species is determined by two major parameters: (1) the amount and distribution of yolk protein within the cytoplasm, and (2) factors in the egg cytoplasm that influence the angle of the mitotic spindle and the timing of its formation.

The amount and distribution of yolk determines where cleavage can occur and the relative size of the blastomeres. When one pole of the egg is relatively yolk-free, cellular divisions occur there at a faster rate than at the opposite pole. The yolk-rich pole is referred to as the **vegetal pole**; the yolk concentration in the **animal pole** is relatively low. The zygote nucleus is frequently displaced toward the animal pole. In general, yolk inhibits cleavage. Figure 8.3 provides a classification of cleavage types and shows the influence of yolk on cleavage symmetry and pattern.

At one extreme are the eggs of sea urchins, mammals, and snails. These eggs have sparse, equally spaced yolk and are thus **isolecithal** (Greek, "equal yolk"). In these species, cleavage is **holoblastic** (Greek *holos*, "complete"), meaning that the cleavage furrow extends through the entire egg. With little yolk, these embryos must have some other way of obtaining food. Most will generate a voracious larval form, while mammals will obtain their nutrition from the maternal placenta.

At the other extreme are the eggs of insects, fishes, reptiles, and birds. Most of their cell volumes are made up of yolk. The yolk must be sufficient to nourish these animals throughout embryonic development. Zygotes containing large accumulations of yolk undergo **meroblastic** cleavage (Greek *meros*, "part"), wherein only a portion of the cytoplasm is cleaved. The cleavage furrow does not penetrate the yolky portion of the cytoplasm because the yolk platelets impede membrane formation there. Insect eggs

have yolk in the center (i.e., they are **centrolecithal**), and the divisions of the cytoplasm occur only in the rim of cytoplasm, around the periphery of the cell (i.e., **superficial cleavage**). The eggs of birds and fishes have only one small area of the egg that is free of yolk (**telolecithal** eggs), and therefore, the cell divisions occur only in this small disc of cytoplasm, giving rise to the **discoidal** pattern of cleavage. These are general rules, however, and even closely related species have evolved different patterns of cleavage in different environments.

Yolk is just one factor influencing a species' pattern of cleavage. There are also inherited patterns of cell division superimposed upon the constraints of the yolk. This inheritance factor can readily be seen in isolecithal eggs. In the absence of a large concentration of yolk, holoblastic cleavage takes place, but four major patterns of this cleavage type can be observed: *radial*, *spiral*, *bilateral*, and *rotational* holoblastic cleavage. We will see examples of all of these cleavage patterns as this chapter takes a more detailed look at four different invertebrate groups.

Gastrulation

The blastula consists of numerous cells, the positions of which were established during cleavage. During **gastrulation**, these cells are given new positions and new neighbors, and the multilayered body plan of the organism is established. The cells that will form the endodermal and mesodermal organs are brought to the inside of the embryo, while the cells that will form the skin and nervous system are spread over its outside surface. Thus, the three germ layers—outer ectoderm, inner endoderm, and interstitial mesoderm—are first produced during gastrulation. In addition, the stage is set for the interactions of these newly positioned tissues.

Gastrulation usually involves some combination of several types of movements (Figure 8.4). These movements involve the entire embryo, and cell migrations in one part of the gastrulating embryo must be intimately coordinated with other movements that are taking place simultaneously. Although patterns of gastrulation vary enormously

FIGURE 8.4 Types of cell movements during gastrulation. The gastrulation of any particular organism is an ensemble of several of these movements.

Invagination:
Infolding of cell sheet into embryo

Example:
Sea urchin endoderm

Involution:
Inturning of cell sheet over the basal surface of an outer layer

Example:
Amphibian mesoderm

Ingression:
Migration of individual cells into the embryo

Example:
Sea urchin mesoderm, *Drosophila* neuroblasts

Delamination:
Splitting or migration of one sheet into two sheets

Example:
Mammalian and bird hypoblast formation

Epiboly:
The expansion of one cell sheet over other cells

Example: Ectoderm formation in amphibians, sea urchins, and tunicates

throughout the animal kingdom, there are only a few basic types of cell movements:

- **Invagination**. The infolding of a region of cells, much like the indenting of a soft rubber ball when it is poked.
- **Involution**. The inturning or inward movement of an expanding outer layer so that it spreads over the internal surface of the remaining external cells.
- **Ingression**. The migration of individual cells from the surface layer into the interior of the embryo. The cells become mesenchymal (i.e., they separate from one another) and migrate independently.
- **Delamination**. The splitting of one cellular sheet into two more or less parallel sheets. While on a cellular basis it resembles ingression, the result is the formation of a new sheet of cells.
- **Epiboly**. The movement of epithelial sheets (usually of ectodermal cells) that spread as a unit (rather than individually) to enclose the deeper layers of the embryo. Epiboly can occur by the cells dividing, by the cells changing their shape, or by several layers of cells intercalating into fewer layers. Often, all three mechanisms are used.

Cell Specification and Axis Formation

Cell fates can be specified by cell-cell interactions or by the asymmetric distribution of patterning molecules into particular cells. These unevenly distributed molecules are usually transcription factors that activate or repress the transcription of specific genes in those cells that acquire them. Such asymmetric distributions of patterning molecules can happen during cleavage and generally follow one of three mechanisms: (1) the molecules are bound to the egg cytoskeleton and are passively acquired by the cells that obtain this cytoplasm; (2) the molecules are actively transported along the cytoskeleton to one particular cell; or (3) the molecules become associated with a specific centrosome and follow that centrosome into one of the two mitotic sister cells (Lambert and Nagy 2002). Once asymmetry has been established, one cell can specify a neighboring cell by paracrine or juxtacrine interactions at the cell surface (see Chapter 6).

Embryos must develop three crucial axes that are the foundation of the body: the anterior-posterior axis, the dorsal-ventral axis, and the right-left axis (Figure 8.5). The **anterior-posterior** (or **anteroposterior**) **axis** is the line extending from head to tail (or mouth to anus in those organisms that lack heads and tails). The **dorsal-ventral** (**dorsoventral**) **axis** is the line extending from back (dorsum) to belly (ventrum). For instance, in vertebrates, the neural tube is a dorsal structure. In insects, the neural cord is a ventral structure. The **right-left axis** is a line between the two lateral sides of the body. Although humans (for example) may look symmetrical, recall that in most of us, the heart and liver are in the left half of the body only. Somehow, the embryo knows that some organs belong on one side and other organs go on the other.

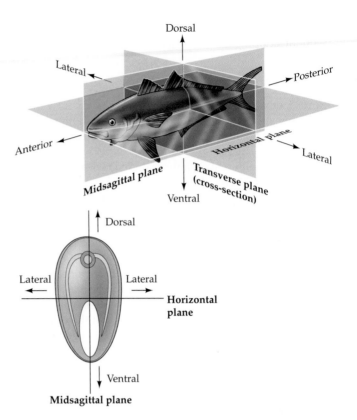

FIGURE 8.5   Axes of a bilaterally symmetrical animal. A single plane, the midsagittal plane, divides the animal into left and right halves. Cross sections are taken along the anterior-posterior axis.

In this chapter, we will look at how four invertebrates—the sea urchin (echinoderm), the ascidian (tunicate), the snail (gastropod mollusc), and *C. elegans* (a well-studied species of nematode worm)—undergo cleavage, gastrulation, axis specification, and cell fate determination. These four invertebrate groups have been important **model systems** for developmental biologists. In other words, they are easily studied in the laboratory, and they have special properties that allow their mechanisms of development to be readily observed.* They also represent a wide variety of cleavage types, patterns of gastrulation, and ways of specifying axes and cell fates.

Despite their differences, the embryos of these four invertebrate groups are all characterized by what Eric Davidson (2001) has called "Type I embryogenesis," which includes:

- the immediate activation of the zygotic genes;
- the rapid specification of the blastomeres by the products of the zygotic genes and by maternally active genes; and

---

*However, model systems—by their very ability to develop in the laboratory—sometimes preclude our asking certain questions concerning the relationship of development to its natural habitat. These questions will be addressed in Chapter 22.

• a relatively small number of cells (a few hundred or less) at gastrulation.

EARLY DEVELOPMENT IN SEA URCHINS

Sea Urchin Cleavage

Sea urchins exhibit radial holoblastic cleavage (Figures 8.6 and 8.7). The first seven cleavage divisions are "stereotypic" in that the same pattern is followed in every individual of the same species: the first and second cleavages are both meridional and are perpendicular to each other (that is to say, the cleavage furrows pass through the animal and vegetal poles). The third cleavage is equatorial, perpendicular to the first two cleavage planes, and separates the animal and vegetal hemispheres from each other (Figure 8.6, top row). The fourth cleavage, however, is very different from the first three. The four cells of the animal tier divide meridionally into eight blastomeres, each with the same volume. These eight cells are called **mesomeres**. The vegetal tier, however, undergoes an unequal equatorial cleavage to produce four large cells—the **macromeres**—and four smaller **micromeres** at the vegetal pole. As the 16-cell embryo cleaves, the eight "animal" mesomeres divide equatorially to produce two tiers, an_1 and an_2, one staggered above the other. The macromeres divide meridionally, forming a tier of eight cells below an_2. The micromeres also divide, albeit somewhat later, producing a small cluster beneath the larger tier. At the sixth division, the animal hemisphere cells divide meridionally, while the vegetal cells divide equatorially; this pattern is reversed in the seventh division (Figure 8.6, bottom row). At that time, the embryo is a 128-cell blastula, seen in the last panel of Figure 8.6. From here on, the pattern of divisions becomes less regular.

Blastula formation

The blastula stage of sea urchin development begins at the 128-cell stage. Here the cells form a hollow sphere surrounding a central cavity, or **blastocoel** (see Figure 8.7F). By this time, all the cells are the same size, the micromeres having slowed down their cell divisions. Every cell is in contact with the proteinaceous fluid of the blastocoel on the inside and with the hyaline layer on the outside. Tight junctions unite the once loosely connected blastomeres into a seamless epithelial sheet that completely encircles the blastocoel. As the cells continue to divide, the blastula remains one cell layer thick, thinning out as it expands. This is accom-

plished by the adhesion of the blastomeres to the hyaline layer and by an influx of water that expands the blastocoel (Dan 1960; Wolpert and Gustafson 1961; Ettensohn and Ingersoll 1992).

These rapid and invariant cell cleavages last through the ninth or tenth division, depending on the species. By this time, the fates of the cells have become specified (discussed in the next section) and each cell gets ciliated on the region of the cell membrane farthest from the blastocoel. This ciliated blastula begins to rotate within the fertilization envelope. Soon afterward, differences are seen in the cells. The cells at the vegetal pole of the blastula begin to thicken, forming a **vegetal plate** (see Figure 8.7F). The cells of the animal hemisphere synthesize and secrete a hatching enzyme that digests the fertilization envelope (Lepage et al. 1992). The embryo is now a free-swimming **hatched blastula**.

Fate maps and the determination of sea urchin blastomeres

The first fate maps of the sea urchin embryo followed the descendants of each of the 16-cell-stage blastomeres. More recent investigations have refined these maps by following the fates of individual cells that have been injected with fluorescent dyes such as diI. These dyes "glow" in the injected cells' progeny for many cell divisions (see Chapter 1). Such studies have shown that by the 60-cell stage, most of the embryonic cell fates are specified, but the cells are not irreversibly committed. In other words, particular blastomeres consistently produce the same cell types in each embryo, but these cells remain pluripotent and can give rise to other cell types if experimentally placed in a different part of the embryo.

FIGURE 8.6 Cleavage in the sea urchin. Planes of cleavage in the first three divisions and the formation of tiers of cells in divisions 3–6.

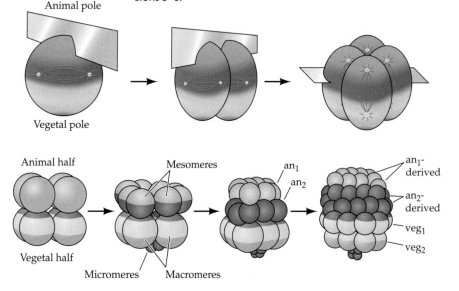

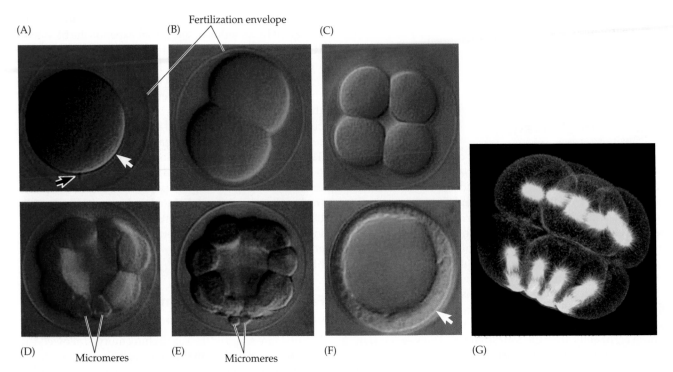

(A)

(B) Fertilization envelope

(C)

(D) Micromeres

(E) Micromeres

(F)

(G)

FIGURE 8.7 Micrographs of cleavage in live embryos of the sea urchin *Lytechinus variegatus*, seen from the side. (A) The 1-cell embryo (zygote). The site of sperm entry is marked with a black arrow; a white arrow marks the vegetal pole. The fertilization envelope surrounding the embryo is clearly visible. (B) 2-cell stage. (C) 8-cell stage. (D) 16-cell stage. Micromeres have formed at the vegetal pole. (E) 32-cell stage. (F) The blastula has hatched from the fertilization envelope. The future vegetal plate (arrow) is beginning to thicken. (G) Confocal fluorescence micrograph of the unequal cell division that initiates the 16-cell stage, highlighting the unequal equatorial cleavage of the vegetal blastomeres to produce the micromeres and macromeres. (A–F, photographs courtesy of J. Hardin; G photograph courtesy of G. van Dassow and the Center for Cell Dynamics.)

CELL FATE DETERMINATION A fate map of the 60-cell sea urchin embryo is shown in Figure 8.8. The animal half of the embryo consistently gives rise to the ectoderm—the larval skin and its neurons. The veg$_1$ layer produces cells that can enter into either the ectodermal or the endodermal organs. The veg$_2$ layer gives rise to cells that can populate three different structures—the endoderm, the coelom (internal mesodermal body wall), and secondary mesenchyme (pigment cells, immunocytes, and muscle cells). The first tier of micromeres produces the primary mesenchyme cells that form the larval skeleton, while the second tier of micromeres contributes cells to the coelom (Logan and McClay 1997, 1999; Wray 1999).

Although the early blastomeres have consistent fates in the larva, most of these fates are achieved by conditional specification. That is, a cell's fate depends on its position relative to its neighboring cells. The micromeres are able to produce a signal that tells the cells adjacent to them to become endoderm and induces them to invaginate into the embryo. Their ability to reorganize the embryonic cells is so pronounced that if the isolated micromeres are recombined with an isolated animal cap (the top two animal

tiers), the animal cap cells will generate endoderm, and a more or less normal larva will develop (Figure 8.9; Hörstadius 1939).

These skeletogenic micromeres are the only cells whose fates are determined autonomously. If these micromeres are isolated from the embryo and placed in test tubes, they will still form skeletal spicules. Moreover, if skeletogenic micromeres are transplanted into the animal region of the blastula, not only will their descendants form skeletal spicules, but the transplanted micromeres will alter the fates of nearby cells by inducing a secondary site for gastrulation. Cells that would normally have produced ectodermal skin cells will be respecified as endoderm and will produce a secondary gut (Figure 8.10; Hörstadius 1973; Ransick and Davidson 1993).

β-CATENIN AND MICROMERE SPECIFICATION The signaling molecules involved in cell specification are just now being identified. The molecule responsible for specifying the micromeres (and their ability to induce the neighboring cells) appears to be **β-catenin**. As we saw in Chapter 6, β-catenin is a transcription factor that is often activated by

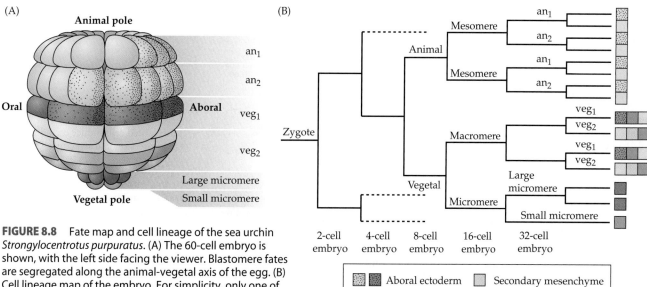

FIGURE 8.8 Fate map and cell lineage of the sea urchin *Strongylocentrotus purpuratus*. (A) The 60-cell embryo is shown, with the left side facing the viewer. Blastomere fates are segregated along the animal-vegetal axis of the egg. (B) Cell lineage map of the embryo. For simplicity, only one of the four embryonic cells is shown beyond second cleavage (solid lines). The veg_1 tier gives rise to both ectodermal and endodermal lineages, and the coelom comes from two sources: the second tier of micromeres and some veg_2 cells. (After Logan and McClay 1999; Wray 1999.)

the Wnt signaling pathway. Several pieces of evidence suggest that β-catenin specifies the micromeres. First, during normal sea urchin development, β-catenin accumulates in the nuclei of those cells fated to become endoderm and mesoderm (Figure 8.11A). This accumulation is autonomous and can occur even if the micromere precursors are separated from the rest of the embryo. Second, this nuclear accumulation appears to be responsible for specifying the vegetal half of the embryo. It is possible that the levels of nuclear β-catenin accumulation help to determine the mesodermal and endodermal fates of the vegetal cells. Treating sea urchin embryos with lithium chloride causes the accumulation of β-catenin in every cell and transforms the presumptive ectoderm into endoderm. Conversely, experimental procedures that inhibit β-catenin accumulation in the vegetal cell nuclei prevent the formation of endoderm and mesoderm (Figure 8.11B,C; Logan et al. 1998; Wikramanayake et al. 1998).

β-Catenin is critical for the specification of the micromeres and for empowering them with the ability to induce the veg_2 cells above them. The specification of the micro-

meres by β-catenin is mediated by the ***Pmar1*** gene product, a homeodomain transcription factor that acts as a transcriptional repressor. The Pmar1 protein represses an as-yet unidentified gene whose product is a general repressor of several genes that characterize primary mesenchyme

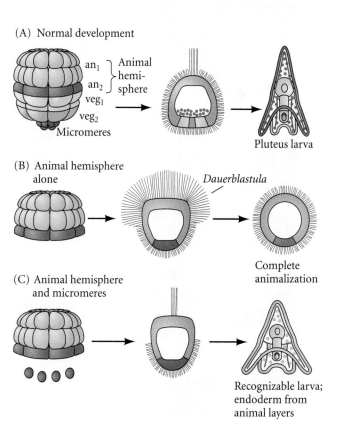

FIGURE 8.9 Ability of the micromeres to induce presumptive ectodermal cells to acquire other fates. (A) Normal development of the 64-cell sea urchin embryo, showing the fates of the different layers. (B) An isolated animal hemisphere becomes a ciliated ball of ectodermal cells. (C) When an isolated animal hemisphere is combined with isolated micromeres, a recognizable pluteus larva is formed, with all the endoderm derived from the animal hemisphere. (After Hörstadius 1939.)

FIGURE 8.10 Ability of the micromeres to induce a secondary axis in sea urchin embryos. (A) Micromeres are transplanted from the vegetal pole of a 16-cell embryo into the animal pole of a host 16-cell embryo. (B) The transplanted micromeres invaginate into the blastocoel to create a new set of primary mesenchyme cells, and they induce the animal cells next to them to become vegetal plate endoderm cells. (C) The transplanted micromeres differentiate into skeletal cables, while the induced animal cap cells form a secondary archenteron. Meanwhile, gastrulation proceeds normally from the original vegetal plate of the host. (After Ransick and Davidson 1993.)

FIGURE 8.11 The role of β-catenin in specifying the vegetal cells of the sea urchin embryo; β-catenin is stained by a fluorescently labeled antibody. (A) During normal development, β-catenin accumulates predominantly in the micromeres and somewhat less in the veg$_2$ tier cells. (B) When lithium chloride treatment permits β-catenin to accumulate in the nuclei of all blastula cells (probably by blocking the GSK-3 enzyme of the Wnt pathway), the animal cells become specified as endoderm and mesoderm. (C) When β-catenin is prevented from entering the nuclei (i.e., it remains in the cytoplasm), the vegetal cell fates are not specified and the entire embryo develops as a ciliated ectodermal ball. (After Logan et al. 1998; photographs courtesy of D. McClay.)

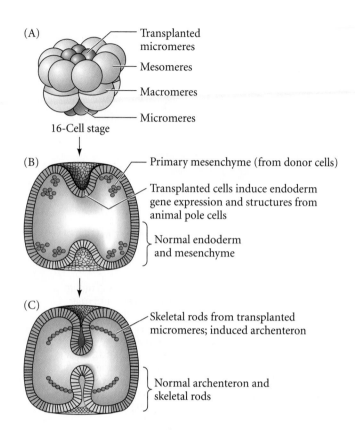

(A)
Transplanted micromeres
Mesomeres
Macromeres
Micromeres
16-Cell stage

(B)
Primary mesenchyme (from donor cells)
Transplanted cells induce endoderm gene expression and structures from animal pole cells
Normal endoderm and mesenchyme

(C)
Skeletal rods from transplanted micromeres; induced archenteron
Normal archenteron and skeletal rods

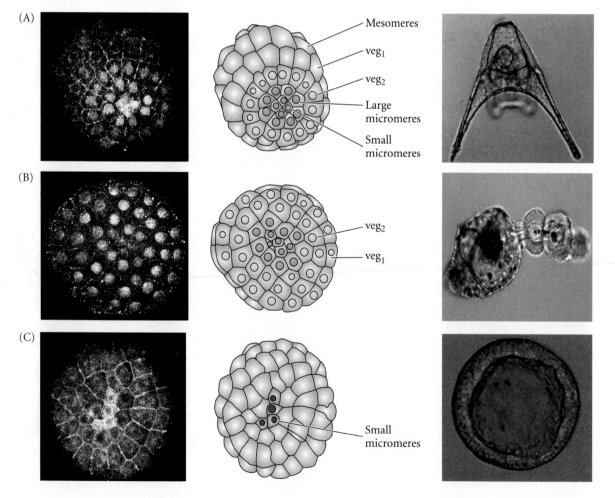

(A)
Mesomeres
veg$_1$
veg$_2$
Large micromeres
Small micromeres

(B)
veg$_2$
veg$_1$

(C)
Small micromeres

FIGURE 8.12 The micromere regulatory network proposed by Davidson and colleagues (2002, 2004).* (A) In the precursors of the primary mesenchyme cells of the 16-cell embryo, β-catenin and the ubiquitous maternal Otx transcription factor activate the *Pmar1* gene. The *Pmar1* gene product is a transcriptional repressor that specifically represses a gene encoding a "global repressor" of numerous genes that characterize the primary mesenchyme cells. Pmar1 therefore allows the de-repression of genes such as *Tbr*, *Ets*, and *Dri* (which turn on the skeleton-forming genes) and allows the de-repression of genes encoding Delta and the "early veg_2 signal" (which instruct the veg_2 layer to form endoderm and secondary mesenchyme cells). (B) In all the other cells of the embryo, *Pmar1* is not activated, since β-catenin is not expressed and maternal Otx protein is not translocated into those cells. Therefore, the repressor protein is transcribed and blocks activation of the micromere specification pathway. (After Oliveri et al. 2002.)

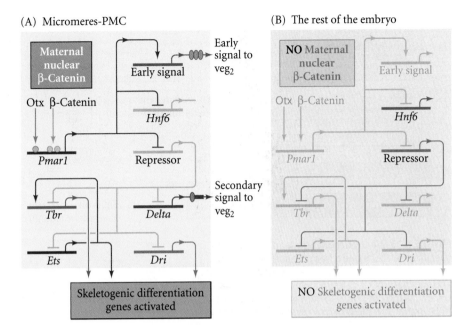

(A) Micromeres-PMC

(B) The rest of the embryo

cells (Oliveri and Davidson 2004); as mentioned earlier, one of the themes of animal development is that gene activation is often accomplished through the repression of a repressor. Some of the genes "de-repressed" by the actions of the *Pmar1* product encode transcription factors (such as Tbr, Ets, and Dri) that direct skeletogenesis (Figure 8.12). Thus, the descendants of the micromeres will form the larval skeleton. Other genes that are activated thanks to Pmar1 protein are *Delta* and *T-brain* (*Tbr*). These genes are responsible for initiating signals that tell the veg_2 layer of cells to become endoderm and secondary mesenchyme (Fuchikami et al. 2002; Revilla-i-Domingo et al. 2004).

The experiments described in Figure 8.10 demonstrated that the micromeres are able to induce a second embryonic axis when transplanted to the animal hemisphere. However, micromeres from embryos in which β-catenin was prevented from entering the nucleus were unable to induce the animal cells to form endoderm, and a second axis was not formed (Logan et al. 1998).

SPECIFICATION OF THE VEGETAL CELLS In a normal embryo, the veg_2 cells are specified by molecular signals. Before these signals are even released, the presence of moderate levels of β-catenin bias the cells to be "endomesoderm." Then an "early veg_2 signal" emanates from the micromeres immediately upon their formation at fourth cleavage and amplifies the mesendoderm specification established by β-

catenin. Next, Delta protein on the micromeres signals activation of the Notch pathway in the adjacent veg_2 cells. The Notch pathway causes these cells to become secondary mesenchyme rather than endoderm. It appears that both signals are needed, in this particular order; and if either one is not present, the veg_2 cells fail to make secondary mesenchyme or endoderm (Ransick and Davidson 1995; Sherwood and McClay 1999; Sweet et al. 1999). The identity of the "early veg_2 signal" is not yet known, but it may create competence in the veg_2 cells to respond to the Delta signal.

Finally, Wnt8 is made by the micromeres and endoderm cells (i.e., those endomesoderm cells *not* receiving the Delta signal). Wnt8 appears to act in an autocrine manner to boost the specification of both the veg_2 endoderm cells and the micromeres and to facilitate their separation into two distinct lineages. Wnt8 also acts on the endoderm precursor cells to initiate the invagination of the vegetal plate at the start of gastrulation (Wikramanayake et al. 2004).

DIFFERENTIATION: COMBINATIONS OF TRANSCRIPTION FACTORS Once transcription factors are present in their specific regions, they can activate the genes that characterize the different cell types in the embryo. One of the best-studied of these genes is that for the endodermal protein Endo16. Endo16 appears to be a secreted product of the endodermal cells and is probably an adhesion protein that allows cell rearrangement in the archenteron (Romano and Wray 2003). It serves as a convenient marker protein for gut-specific tissues. Writing about the *endo16* research, Eric Davidson notes, "What emerges is astounding: a network of logic interactions programmed into the DNA sequence that amounts essentially to a hardwired computational device" (Davidson 2001, p. 54). An outline of that computational logic is shown in Figure 8.13.

---

*The model in Figure 8.12 is continually updated. See http://sugp.caltech.edu/endomes/

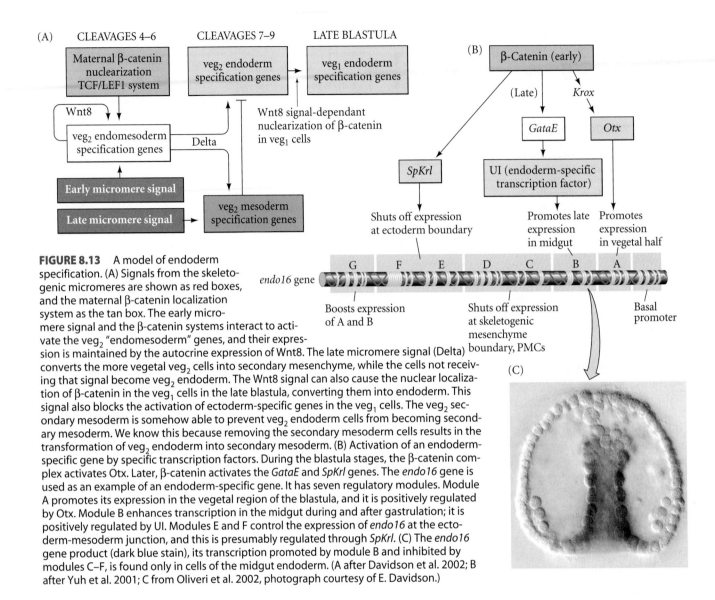

FIGURE 8.13 A model of endoderm specification. (A) Signals from the skeletogenic micromeres are shown as red boxes, and the maternal β-catenin localization system as the tan box. The early micromere signal and the β-catenin systems interact to activate the veg₂ "endomesoderm" genes, and their expression is maintained by the autocrine expression of Wnt8. The late micromere signal (Delta) converts the more vegetal veg₂ cells into secondary mesenchyme, while the cells not receiving that signal become veg₂ endoderm. The Wnt8 signal can also cause the nuclear localization of β-catenin in the veg₁ cells in the late blastula, converting them into endoderm. This signal also blocks the activation of ectoderm-specific genes in the veg₁ cells. The veg₂ secondary mesoderm is somehow able to prevent veg₂ endoderm cells from becoming secondary mesoderm. We know this because removing the secondary mesoderm cells results in the transformation of veg₂ endoderm into secondary mesoderm. (B) Activation of an endoderm-specific gene by specific transcription factors. During the blastula stages, the β-catenin complex activates Otx. Later, β-catenin activates the *GataE* and *SpKrl* genes. The *endo16* gene is used as an example of an endoderm-specific gene. It has seven regulatory modules. Module A promotes its expression in the vegetal region of the blastula, and it is positively regulated by Otx. Module B enhances transcription in the midgut during and after gastrulation; it is positively regulated by UI. Modules E and F control the expression of *endo16* at the ectoderm-mesoderm junction, and this is presumably regulated through *SpKrl*. (C) The *endo16* gene product (dark blue stain), its transcription promoted by module B and inhibited by modules C–F, is found only in cells of the midgut endoderm. (A after Davidson et al. 2002; B after Yuh et al. 2001; C from Oliveri et al. 2002, photograph courtesy of E. Davidson.)

Figure 8.13A outlines the specification of the endoderm, while Figure 8.13B shows a portion of the regulatory network that activates the *endo16* gene. The *endo16* upstream regulatory region contains 7 modular elements to which at least 13 different transcription factors bind. Some of these modular regions activate *endo16* transcription, while other modules act to inhibit *endo16* transcription. In Figure 8.13B, modules A and B activate *endo16* expression in the vegetal cells (prior to gastrulation) and in the midgut (during and after gastrulation). The inhibitory modules (C–F) are activated in those cells other than endoderm. Thus, the different regions of the *endo16* regulatory region act synergistically to make certain that the *endo16* gene is expressed only in the midgut endoderm (Figure 8.13C) and in no other cell type (Yuh et al. 2001; Davidson et al. 2002).

AXIS SPECIFICATION In the sea urchin blastula, the cell fates line up along the animal-vegetal axis established in the egg cytoplasm prior to fertilization. The animal-vegetal axis also appears to structure the future anterior-posterior axis, with the vegetal region sequestering those maternal components necessary for posterior development (Boveri 1901; Maruyama et al. 1985).

In most sea urchins, the dorsal-ventral and left-right axes are specified after fertilization, but the manner of their specification is not well understood. The oral-aboral axis (approximating a ventral-dorsal axis) of the sea urchin embryo usually is delineated by the first cleavage plane. Lineage tracer dye injected into one blastomere at the 2-cell stage demonstrated that, in nearly all cases, the oral pole of the future oral-aboral axis lay 45 degrees clockwise from the first cleavage plane as viewed from the animal pole (Cameron et al. 1989). This oral-aboral axis appears to form through a mechanism similar to that used by vertebrates to establish their right-left axes (Duboc et al. 2004; Flowers et al. 2004). The oral ectoderm is specified by the expression of the *Nodal* gene, a member of the TGF-β family.

The role of *Nodal* was discovered through the classic "find it/lose it/move it" mode of experimentation mentioned in Chapter 2. First, researchers cloned a sea urchin *Nodal* gene and, using in situ hybridization, demonstrated that *Nodal* was first expressed in the presumptive oral ectoderm at about the 60-cell stage. Nodal protein then becomes prominent on one side of the blastula and on the presumptive oral side of the gastrula. When the researchers prevented translation of the *Nodal* message, development was normal until the mesenchyme blastula stage—but the larvae never obtained bilateral symmetry, the archenteron did not bend to one side to form the mouth, and the primary mesenchyme did not separate into the two sets of spicule-forming skeleton cells. Moreover, the genes usually expressed in the oral ectoderm were not expressed. Conversely, when researchers induced ectopic expression of the *Nodal* gene throughout the ectoderm, all of the ectoderm appeared to become oral ectoderm. Thus, *Nodal* appears to be crucial in establishing the oral ectoderm. Nodal protein appears to act against a BMP. Angerer and colleagues (2001) showed that the sea urchin gene corresponding to *BMP2* and *BMP4* promoted *aboral* fates when misexpressed in oral ectoderm cells. There appears to be a gradient whereby oral fates are promoted by Nodal and aboral fates are promoted by BMP2/4.

As we will see later in this chapter and in Chapter 11, the left-right axis of deuterostomes is usually associated with an asymmetric expression of a *Nodal* gene. In sea urchins, the left-right asymmetric expression of *Nodal* is observed in the larva long after the oral-aboral axis has been established. Interestingly, sea urchin *Nodal* expression is on the *right* side of the embryo, whereas in other deuterostomes (including tunicates, *Amphioxus*, and all vertebrates studied), the asymmetric *Nodal* expression has been on the *left* side (Duboc et al. 2005). This asymmetry in sea urchins is important because only the left side of the larva develops a coelomic sac and the imaginal rudiment that will generate the adult (see Figure 8.22). *Nodal* expression on the right side of the larva activates a pathway that prevents this sac from forming, thereby ensuring that only one coelemic sac forms, and that it forms on the left portion of the larva. If *Nodal* expression is blocked, both sides will form imaginal rudiments.

WEBSITE 8.1 Sea urchin cell specification. The specification of sea urchin cells was one of the first major research projects in experimental embryology and remains a fascinating area of research. It appears that the initial signaling parses the blastula into domains characterized by the expression of specific transcription factors.

Sea Urchin Gastrulation

The late sea urchin blastula consists of a single layer of about a thousand cells that form a hollow ball, somewhat flattened at the vegetal end. The blastomeres are derived from different regions of the zygote and have different sizes and properties. Figure 8.14 and Figure 8.15 illustrate development of the blastula through gastrulation to the pluteus larva stage characteristic of sea urchins. The drawings show the fate of each cell layer during gastrulation.

Ingression of the primary mesenchyme

FUNCTION OF PRIMARY MESENCHYME CELLS Shortly after the blastula hatches from its fertilization envelope, a group of cells derived from the micromeres undergoes an epithelial-to-mesenchymal transformation. These cells change their cytoskeleton, become bottle-shaped, lose their adhesions to the cells lateral to them, and then break away from the apical layer to enter the blastocoel (Figure 8.15, 9–10

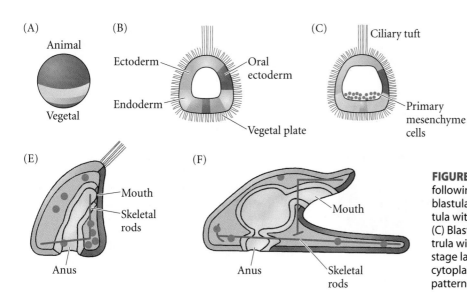

FIGURE 8.14 Normal sea urchin development, following the fate of the cellular layers of the blastula. (A) Fate map of the zygote. (B) Late blastula with ciliary tuft and flattened vegetal plate. (C) Blastula with primary mesenchyme. (D) Gastrula with secondary mesenchyme. (E) Prism-stage larva. (F) Pluteus larva. Fates of the zygote cytoplasm can be followed through the color pattern (see Figure 8.9). (Courtesy of D. McClay.)

hours). These micromere-derived cells are called the **primary mesenchyme**. Since they will form the larval skeleton, they are sometimes called the **skeletogenic mesenchyme**. These mesenchyme cells begin extending and contracting long, thin (250 nm in diameter and 25 μm long) processes called **filopodia**. At first the cells appear to move randomly along the inner blastocoel surface, actively making and breaking filopodial connections to the wall of the blastocoel. Eventually, however, they become localized within the prospective ventrolateral region of the blastocoel. Here they fuse into syncytial cables, which will form the axis of the calcium carbonate spicules of the larval skeletal rods (Figure 8.14D–F).

IMPORTANCE OF EXTRACELLULAR LAMINA INSIDE THE BLASTOCOEL

The ingression of the micromere descendants into the blastocoel is a result of these primary mesenchyme cells losing their affinity for their neighbors and for the hyaline membrane; instead they acquire a strong affinity for a group of proteins that line the blastocoel. This model of mesenchymal migration was first proposed by Gustafson and Wolpert (1967) and was confirmed in 1985, when Rachel Fink and David McClay measured the strengths of sea urchin blastomere adhesion to the hyaline layer, to the basal lamina lining the blastocoel, and to other blastomeres.

Originally, all the cells of the blastula are connected on their outer surface to the hyaline layer, and on their inner surface to a basal lamina secreted by the cells. On their lateral surfaces, each cell has another cell for a neighbor. Fink and McClay found that the prospective ectoderm and endoderm cells (descendants of the mesomeres and macromeres, respectively) bind tightly to one another and to the hyaline layer, but adhere only loosely to the basal lamina (Table 8.2). The micromeres originally display a similar pattern of binding. However, the micromere pattern changes at gastrulation. Whereas the other cells retain their tight binding to the hyaline layer and to their neighbors, the primary mesenchyme precursors lose their affinity for these structures (which drops to about 2 percent of its original value), while their affinity for components of the basal lamina and extracellular matrix (such as fibronectin) increases a hundredfold. This change in affinity causes the micromeres to release their attachments to the external hyaline layer and to their neighboring cells and, drawn in by the basal lamina, to migrate up into the blastocoel (Figure 8.16A). These changes in affinity have been correlated with changes in cell surface molecules that occur during this time (Wessel and McClay 1985), and proteins such as fibronectin, integrin, laminin, L1, and cadherins have been shown to be involved in cellular ingression.

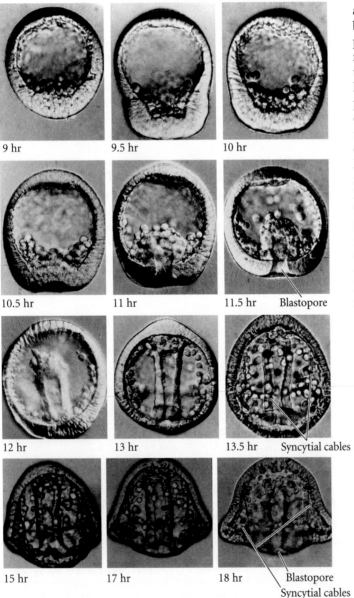

9 hr   9.5 hr   10 hr

10.5 hr   11 hr   11.5 hr   Blastopore

12 hr   13 hr   13.5 hr   Syncytial cables

15 hr   17 hr   18 hr   Blastopore
Syncytial cables

FIGURE 8.15 Entire sequence of gastrulation in *Lytechinus variegatus*. The times show the length of development at 25°C. (Photographs courtesy of J. Morrill; pluteus larva courtesy of G. Watchmaker.)

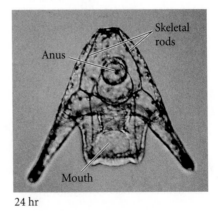

Skeletal rods

Anus

Mouth

24 hr

(A)

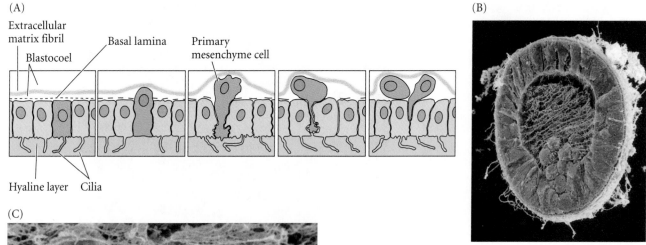

Extracellular
matrix fibril
Blastocoel
Basal lamina
Primary
mesenchyme cell
Hyaline layer Cilia

(B)

(C)

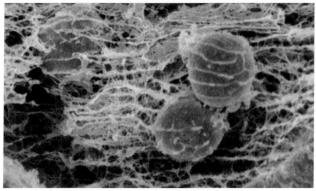

As shown in Figures 8.16B and C, there is a heavy concentration of extracellular material around the ingressing primary mesenchyme cells (Galileo and Morrill 1985; Cherr et al. 1992). Once inside the blastocoel, the primary mesenchyme cells appear to migrate along the extracellular matrix of the blastocoel wall, extending their filopodia in front of them (Galileo and Morrill 1985; Karp and Solursh 1985). Several proteins (including fibronectin and a particular sulfated glycoprotein) are necessary to initiate and maintain this migration (Wessel et al. 1984; Sugiyama 1972; Lane and Solursh 1991; Berg et al. 1996). But these guidance cues cannot be sufficient, since the migrating cells "know" when to stop their movement and form spicules near the equator of the blastocoel. The primary mesenchyme cells arrange themselves in a ring at a specific position along the animal-vegetal axis.

At two sites near the future ventral side of the larva, many of these primary mesenchyme cells cluster together, fuse with one another, and initiate spicule formation (Figure 8.17; Hodor and Ettensohn 1998). If a labeled micromere from another embryo is injected into the blastocoel of a gastrulating sea urchin embryo, it migrates to the correct location and contributes to the formation of the embry-

FIGURE 8.16 Ingression of primary mesenchyme cells. (A) Interpretative depiction of changes in the adhesive affinities of the presumptive primary mesenchyme cells (pink). These cells lose their affinities for hyalin and for their neighboring blastomeres while gaining an affinity for the proteins of the basal lamina. Nonmesenchymal blastomeres retain their original high affinities for the hyaline layer and neighboring cells. (B) SEM of primary mesenchyme cells enmeshed in the extracellular matrix of an early *Strongylocentrotus* gastrula. (C) Gastrula-stage mesenchyme cell migration. The extracellular matrix fibrils of the blastocoel lie parallel to the animal-vegetal axis and are intimately associated with the primary mesenchyme cells. (B,C from Cherr et al. 1992; photographs courtesy of the authors.)

TABLE 8.2 Affinities of mesenchymal and nonmesenchymal cells to cellular and extracellular components[a]

| Cell type | Dislodgment force (in dynes) | | |
	Hyaline	Gastrula cell monolayers	Basal lamina
16-cell-stage micromeres	5.8×10^{-5}	6.8×10^{-5}	4.8×10^{-7}
Migratory-stage mesenchyme cells	1.2×10^{-7}	1.2×10^{-7}	1.5×10^{-5}
Gastrula ectoderm and endoderm	5.0×10^{-5}	5.0×10^{-5}	5.0×10^{-7}

Source: After Fink and McClay 1985.

[a]Tested cells were allowed to adhere to plates containing hyaline, extracellular basal lamina, or cell monolayers. The plates were inverted and centrifuged at various strengths to dislodge the cells. The dislodgement force is calculated from the centrifugal force needed to remove the test cells from the substrate.

(A)
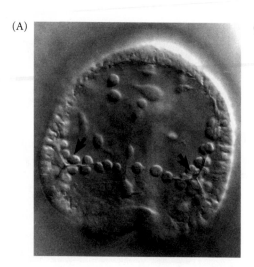

(B)
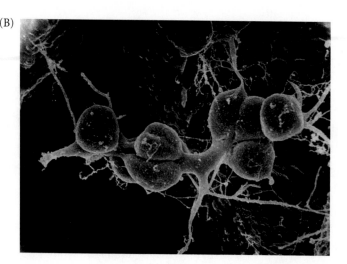

FIGURE 8.17 Formation of syncytial cables by primary mesenchyme cells of the sea urchin. (A) Primary mesenchyme cells in the early gastrula align and fuse to lay down the matrix of the calcium carbonate spicule (arrows). (B) Scanning electron micrograph of spicules formed by the fusing of primary mesenchyme cells into syncytial cables. (A from Ettensohn 1990; B from Morrill and Santos 1985.)

onic spicules (Ettensohn 1990; Peterson and McClay 2003). It is thought that the necessary positional information is provided by the prospective ectodermal cells and their basal laminae (Figure 8.18A; Harkey and Whiteley 1980; Armstrong et al. 1993; Malinda and Ettensohn 1994). Only the primary mesenchyme cells (and not other cell types or latex beads) are capable of responding to these patterning cues (Ettensohn and McClay 1986). Miller and colleagues (1995) have reported the existence of extremely fine (0.3 μm diameter) filopodia on the skeleton-forming mesenchyme cells. These filopodia are not thought to function in locomotion; rather, they appear to explore and sense the blastocoel wall and may be responsible for picking up dorsal-ventral and animal-vegetal patterning cues from the ectoderm (Figure 8.18B; Malinda et al. 1995).

First stage of archenteron invagination

As the primary mesenchyme cells leave the vegetal region of the spherical embryo, important changes are occurring

in the cells that remain there. These cells thicken and flatten to form a vegetal plate, changing the shape of the blastula (Figure 8.15, 9 hours). These vegetal plate cells remain bound to one another and to the hyaline layer of the egg, and they move to fill the gaps caused by the ingression of

(A)
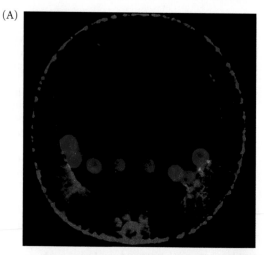

(B)
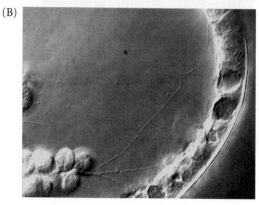

FIGURE 8.18 Localization of the primary mesenchyme cells. (A) The localization of the micromeres to form the calcium carbonate skeleton is determined by the ectodermal cells. The primary mesenchyme cells are stained green, while β-catenin is stained red. The primary mesenchyme cells appear to accumulate in those regions characterized by high β-catenin concentrations. (B) Nomarski videomicrograph showing a long, thin filopodium extending from a primary mesenchyme cell to the ectodermal wall of the gastrula, as well as a shorter filopodium extending inward from the ectoderm. The mesenchymal filopodia extend through the extracellular matrix and directly contact the cell membrane of the ectodermal cells. (B from Miller et al. 1995, photographs courtesy of J. R. Miller and D. McClay.)

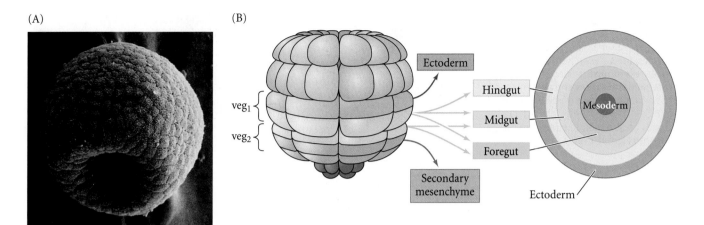

FIGURE 8.19   Invagination of the vegetal plate. (A) Vegetal plate invagination in *Lytechinus variegatus*, seen by scanning electron microscopy of the external surface of the early gastrula. The blastopore is clearly visible. (B) Fate map of the vegetal plate of the sea urchin embryo, looking "upward" at the vegetal surface. The central portion becomes the secondary mesenchyme cells, while the concentric layers around it become the foregut, midgut, and hindgut, respectively. The boundary where the endoderm meets the ectoderm marks the anus. The secondary mesenchyme and foregut come from the veg_2 layer, the midgut comes from veg_1 and veg_2 cells, and the hindgut (and the ectoderm in contact with it) comes from the veg_1 layer. (A from Morrill and Santos 1985, courtesy of J. B. Morrill; B after Logan and McClay 1999.)

the primary mesenchyme. Moreover, the vegetal plate bends inward and invaginates about one-fourth to one-half the way into the blastocoel (Figure 8.19A; see also Figure 8.15, 10.5–11.5 hours). Then invagination suddenly ceases. The invaginated region is called the **archenteron** (primitive gut), and the opening of the archenteron at the vegetal pole is called the **blastopore**.

Invagination appears to be caused by shape changes in the vegetal plate cells and in the extracellular matrix underlying them (see Kominami and Takata 2004 for a review). Kimberly and Hardin (1998) have shown that a group of vegetal plate cells surrounding the 2–8 cells at the vegetal pole become bottle-shaped, constricting their apical ends. This change causes the cells to pucker inward. Destroying these cells with lasers retards gastrulation. In addition, the hyaline layer at the vegetal plate buckles inward due to changes in its composition (Lane et al. 1993). The hyaline layer is actually made up of two layers, an outer lamina made primarily of hyalin protein and an inner lamina composed of fibropellin proteins (Hall and Vacquier 1982; Bisgrove et al. 1991). Fibropellins are stored in secretory granules within the oocyte and are secreted from those granules after cortical granule exocytosis releases the hyalin protein. By the blastula stage, the fibropellins have formed a mesh-like network over the embryo surface. At the time of invagination, the vegetal plate cells (and only those cells) secrete a chondroitin sulfate proteoglycan into the inner

lamina of the hyaline layer directly beneath them. This hygroscopic (water-absorbing) molecule swells the inner lamina, but not the outer lamina, which causes the vegetal region of the hyaline layer to buckle. Slightly later, a second force, arising from the movements of epithelial cells adjacent to the vegetal plate, may facilitate invagination by drawing the buckled layer inward (Burke et al. 1991).

At the stage when the skeletogenic mesenchyme cells begin ingressing into the blastocoel, the fates of the vegetal plate cells have already been specified (Ruffins and Ettensohn 1996). The secondary mesenchyme is the first group of cells invaginating, and it forms the tip of the archenteron, leading the way into the blastocoel. It will form the pigment cells, the musculature around the gut, and the coelomic pouches. The endodermal cells adjacent to the micromere-derived mesenchyme become foregut, migrating the farthest distance into the blastocoel. The next layer of endodermal cells becomes midgut, and the last circumferential row to invaginate forms the hindgut and anus (Figure 8.19B).

Second and third stages of archenteron invagination

The invagination of the vegetal cells occurs in discrete stages. After a brief pause following the initial invagination, the second phase of archenteron formation begins. During this stage, the archenteron extends dramatically, sometimes tripling its length. In this process of extension, the wide, short gut rudiment is transformed into a long, thin tube (Figure 8.20; see also Figure 8.15, 12 hours). To accomplish this extension, the cells of the archenteron rearrange themselves by migrating over one another and by flattening themselves (Ettensohn 1985; Hardin and Cheng 1986). This phenomenon, where cells intercalate to narrow the tissue and at the same time move it forward, is called **convergent extension** (Martins et al. 1998).

In all species of sea urchins observed, a third stage of archenteron elongation occurs. This final phase is initiated by the tension provided by secondary mesenchyme cells, which form at the tip of the archenteron and remain there. These cells extend filopodia through the blastocoel

FIGURE 8.20 Cell rearrangement during the extension of the archenteron in sea urchin embryos. In this species, the early archenteron has 20 to 30 cells around its circumference. Later in gastrulation, the archenteron has a circumference made by only 6 to 8 cells. Fluorescently labeled clones can be seen to stretch apically. (After Hardin 1990.)

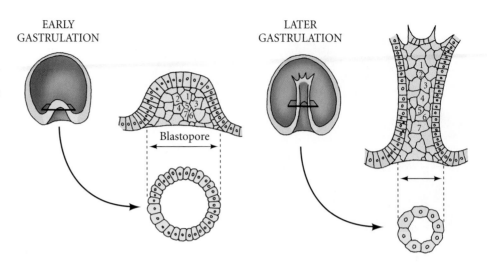

EARLY GASTRULATION

LATER GASTRULATION

Blastopore

fluid to contact the inner surface of the blastocoel wall (Dan and Okazaki 1956; Schroeder 1981). The filopodia attach to the wall at the junctions between the blastomeres and then shorten, pulling up the archenteron (Figure 8.21; see also Figure 8.15, 12 and 13 hours). Hardin (1988) ablated the secondary mesenchyme cells with a laser, with the result that the archenteron could elongate to only about two-thirds of the normal length. If a few secondary mesenchyme cells were left, elongation continued, although at a slower rate. The secondary mesenchyme cells, in this species, play an essential role in pulling the archenteron up to the blastocoel wall during the last phase of invagination.

But can the secondary mesenchyme filopodia attach to any part of the blastocoel wall, or is there a specific target in the animal hemisphere that must be present for attachment to occur? Is there a region of the blastocoel wall that is already committed to becoming the ventral side of the larva? Studies by Hardin and McClay (1990) show that there is a specific target site for the filopodia that differs from other regions of the animal hemisphere. The filopodia extend, touch the blastocoel wall at random sites, and then retract. However, when the filopodia contact a particular region of the wall, they remain attached there, flatten out against this region, and pull the archenteron toward it. When Hardin and McClay poked in the other side of the blastocoel wall so that contacts were made most readily with that region, the filopodia continued to extend and retract after touching it. Only when the filopodia found their target tissue did they cease these movements. If the gastrula was constricted so that the filopodia never reached the target area, the secondary mesenchyme cells continued

to explore until they eventually moved off the archenteron and found the target as freely migrating cells. There appears, then, to be a target region on what is to become the ventral side of the larva that is recognized by the secondary mesenchyme cells, and which positions the archenteron in the region where the mouth will form.

(A)

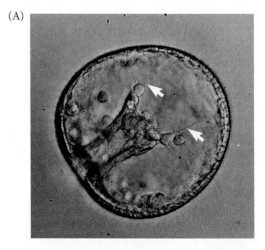

(B)

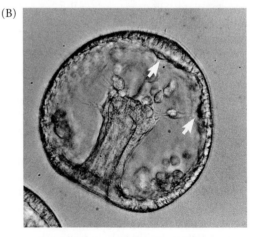

FIGURE 8.21 Mid-gastrula stage of *Lytechinus pictus*, showing filopodial extensions of secondary mesenchyme. (A) Secondary mesenchyme cells extend filopodia (arrows) from the tip of the archenteron. (B) Filopodial cables connect the blastocoel wall to the archenteron tip. The tension of the cables can be seen as they pull on the blastocoel wall at the point of attachment (arrows). (Photographs courtesy of C. Ettensohn.)

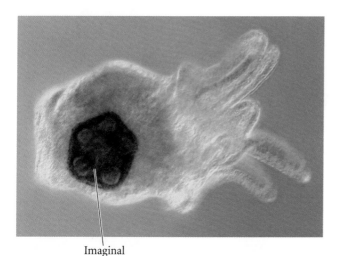

Imaginal
rudiment

FIGURE 8.22 The imaginal rudiment growing in the left side of the pluteus larva of a sea urchin. The rudiment will become the adult sea urchin, while the larval stage is jettisoned. The five-fold symmetry of the rudiment is obvious. (Photograph courtesy of G. Wray.)

As the top of the archenteron meets the blastocoel wall in the target region, the secondary mesenchyme cells disperse into the blastocoel, where they proliferate to form the mesodermal organs (see Figure 8.15, 13.5 hours). Where the archenteron contacts the wall, a mouth is eventually formed. The mouth fuses with the archenteron to create a continuous digestive tube. Thus, as is characteristic of deuterostomes, the blastopore marks the position of the anus.

As the pluteus larva elongates, the coelomic cavities form from secondary mesenchyme. Under the influence of Nodal protein, as described on page 221, the right coelomic sac remains rudimentary. However, the left coelomic sac undergoes extensive development to form many of the structures of the adult sea urchin. The left sac splits into three smaller sacs. An invagination from the ectoderm fuses with the middle sac to form the **imaginal rudiment**. This rudiment develops a fivefold symmetry (Figure 8.22), and skeletogenic mesenchyme cells enter the rudiment to synthesize the first skeletal plates of the shell. The left side of the pluteus becomes, in effect, the future oral surface of the adult sea urchin (Bury 1895; Aihara and Amemiya 2001). During metamorphosis, the imaginal rudiment separates from the larva, which then degenerates. While the imaginal rudiment (now called a juvenile) is re-forming its digestive tract and settling on the ocean floor, it is dependent on the nutrition it received from the jettisoned larva.

The echinoderm pattern of gastrulation provides the evolutionary prototype for deuterostome development. In deuterostomes (echinoderms, tunicates, cephalochordates, and vertebrates), the first opening becomes the anus while the second opening becomes the mouth (hence, *deutero stoma*, "mouth second"). Moreover, in sea urchins, we see the phenomena of convergent extension and the use of *Nodal* gene expression for the establishment of axes. We will return to the subject of deuterostome development later in this and in many of the succeeding chapters, but we will now describe development in a protostome group, the molluscs.

WEBSITE **8.2 Cloned urchins.** It seems that sea urchins and other echinoderms can clone themselves by separating off some cells of the larvae.

VADE MECUM² **Sea urchin development.** The CD-ROM provides an excellent review of sea urchin development as well as questions on the fundamentals of echinoderm cleavage and gastrulation. [Click on Sea Urchin]

THE EARLY DEVELOPMENT OF SNAILS

Cleavage in Snail Embryos

Spiral holoblastic cleavage is characteristic of several animal groups, including annelid worms, some flatworms, and most molluscs. It differs from radial cleavage in numerous ways. First, the cleavage planes are not parallel or perpendicular to the animal-vegetal axis of the egg; rather, cleavage is at oblique angles, forming a "spiral" arrangement of daughter blastomeres. Second, the cells touch one another at more places than do those of radially cleaving embryos. In fact, they assume the most thermodynamically stable packing orientation, much like adjacent soap bubbles. Third, spirally cleaving embryos usually undergo fewer divisions before they begin gastrulation, making it possible to follow the fate of each cell of the blastula. When the fates of the individual blastomeres from annelid, flatworm, and mollusc embryos were compared, many of the same cells were seen in the same places, and their general fates were identical (Wilson 1898). Blastulae produced by spiral cleavage have no blastocoel and are called **stereoblastulae**.

Figures 8.23 and 8.24 depict the cleavage pattern typical of many molluscan embryos. The first two cleavages are nearly meridional, producing four large macromeres (labeled A, B, C, and D). In many species, these four blastomeres are different sizes (D being the largest), a characteristic that allows them to be individually identified. In each successive cleavage, each macromere buds off a small micromere at its animal pole. Each successive quartet of micromeres is displaced to the right or to the left of its sis-

(A) View from animal pole

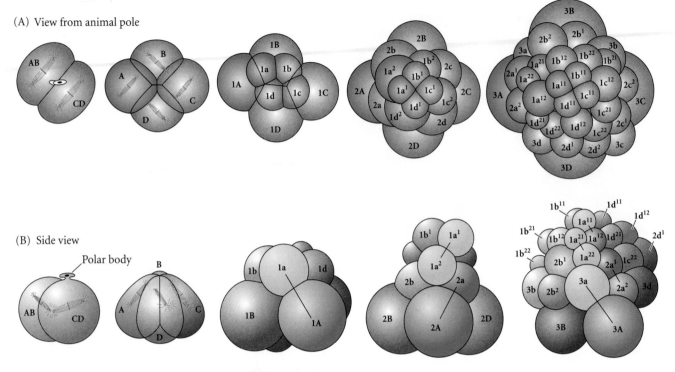

(B) Side view

Polar body

FIGURE 8.23 Spiral cleavage of the mollusc *Trochus* viewed (A) from the animal pole and (B) from one side. The cells derived from the A blastomere are shown in color. The mitotic spindles, sketched in the early stages, divide the cells unequally and at an angle to the vertical and horizontal axes. Each successive quartet of micromeres (indicated with lowercase letters) is displaced to the right or to the left of its sister macromere (uppercase letters), creating the characteristic spiral pattern.

(A)

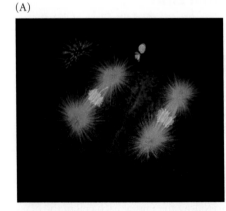

FIGURE 8.24 Spiral cleavage in molluscs. (A) The spiral nature of third cleavage can be seen in the confocal fluorescence micrograph of the 4-cell embryo of the clam *Acila castrenis*, Microtubules stain red, RNA stains green, and the DNA stains yellow. Two cells and a portion of a third cell are visible; a polar body can be seen at the top of the micrograph. (B–E). Cleavage in the mud snail *Ilyanassa obsoleta*. The D blastomere is larger than the others, allowing the identification of each cell. Cleavage is dextral. (B) 8-cell stage. PB, polar body. (C) Mid-fourth cleavage (12-cell embryo). The macromeres have already divided into large and small spirally oriented cells; 1a–d have not divided yet. (D) 32-cell embryo. (A courtesy of G. von Dassow and the Center for Cell Dynamics; B–E from Craig and Morrill 1986; photographs courtesy of the authors.)

(B)

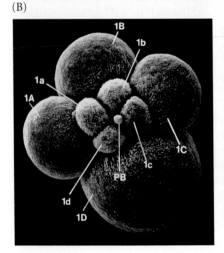

(C)

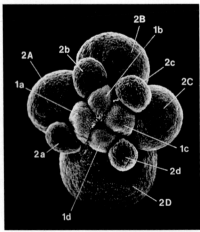

(D)

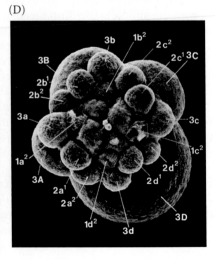

ter macromere, creating the characteristic spiral pattern. Looking down on the embryo from the animal pole, the upper ends of the mitotic spindles appear to alternate clockwise and counterclockwise (see Figure 8.24). This arrangement causes alternate micromeres to form obliquely to the left and to the right of their macromeres.

At the third cleavage, the A macromere gives rise to two daughter cells, macromere 1A and micromere 1a. The B, C, and D cells behave similarly, producing the first quartet of micromeres. In most species, these micromeres are to the *right* of their macromeres (looking down on the animal pole). At the fourth cleavage, macromere 1A divides to form macromere 2A and micromere 2a, and micromere 1a divides to form two more micromeres, $1a^1$ and $1a^2$ (see Figure 8.23). The micromeres of this second quartet are to the *left* of the macromeres. Further cleavage yields blastomeres 3A and 3a from macromere 2A, and micromere $1a^2$ divides to produce cells $1a^{21}$ and $1a^{22}$. In normal development, the first-quartet micromeres form the head structures, while the second-quartet micromeres form the statocyst (balance organ) and shell. These fates are specified both by cytoplasmic localization and by induction (Clement 1967; Cather 1967; Render 1991; Sweet 1998).

The orientation of the cleavage plane to the left or to the right is controlled by cytoplasmic factors within the oocyte. This was discovered by analyzing mutations of snail coiling. Some snails have their coils opening to the right of their shells (**dextral coiling**), whereas the coils of other snails open to the left (**sinistral coiling**). Usually the direction of coiling is the same for all members of a given species, but occasional mutants are found (i.e., in a population of right-coiling snails, a few individuals will be found with coils that open on the left). Crampton (1894) analyzed the embryos of such aberrant snails and found that their early cleavage differed from the norm. The orientation of the cells after the second cleavage was different in the sinistrally coiling snails as a result of a different orientation of the mitotic apparatus (Figure 8.25). In some species (such as the pond snail *Physa*, an entirely sinistral species), the sinistrally coiling cleavage patterns are mirror-images of the dextrally coiling pattern of the right-handed species. In other instances (such as *Lymnaea*, where about 2 percent of the snails are lefties), sinistrality is the result of a two-step process: at each division, the initial cleavage is radial; however, as the cleavage furrow forms, the blastomeres shift to the left-hand spiral position (Shibazaki et al. 2004). In Figure 8.25, one can see that the position of the 4d blastomere (which is extremely important, as its progeny will form the mesodermal organs) is different in the two types of spiraling embryos.

In snails such as *Lymnaea*, the direction of snail shell coiling is controlled by a single pair of genes (Sturtevant 1923; Boycott et al. 1930). In *Lymnaea peregra*, rare mutants exhibiting sinistral coiling were found and mated with wild-type, dextrally coiling snails. These matings showed that the right-coiling allele, *D*, is dominant to the left-coiling allele, *d*. However, the direction of cleavage is determined not by the genotype of the developing snail, but by the genotype of the snail's mother. A *dd* female snail can produce only sinistrally coiling offspring, even if the offspring's genotype is *Dd*. A *Dd* individual will coil either left or right, depending on the genotype of its mother. Such matings produce a chart like this:

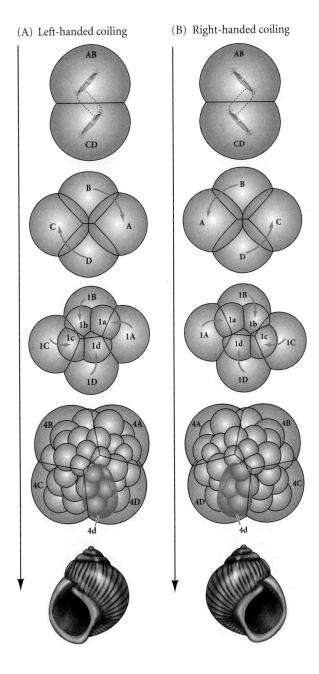

(A) Left-handed coiling (B) Right-handed coiling

FIGURE 8.25 Looking down on the animal pole of (A) left-coiling and (B) right-coiling snails. The origin of sinistral and dextral coiling can be traced to the orientation of the mitotic spindle at the second cleavage. Left- and right-coiling snails develop as mirror images of each other. (After Morgan 1927.)

	Genotype	Phenotype
$DD\,♀ \times dd\,♂ \rightarrow$	Dd	All right-coiling
$DD\,♂ \times dd\,♀ \rightarrow$	Dd	All left-coiling
$Dd \times Dd \rightarrow$	$1DD{:}2Dd{:}1dd$	All right-coiling

The genetic factors involved in snail coiling are brought to the embryo by the oocyte cytoplasm. It is the genotype of the ovary in which the oocyte develops that determines which orientation cleavage will take. When Freeman and Lundelius (1982) injected a small amount of cytoplasm from dextrally coiling snails into the eggs of *dd* mothers, the resulting embryos coiled to the right. Cytoplasm from sinistrally coiling snails did not affect right-coiling embryos. These findings confirmed that the wild-type mothers were placing a factor into their eggs that was absent or defective in the *dd* mothers.

A fate map of *Ilyanassa obsoleta*

 WEBSITE **8.3 Alfred Sturtevant and the genetics of snail coiling.** By a masterful thought experiment, Sturtevant demonstrated the power of applying genetics to embryology. To do this, he brought Mendelian genetics into the study of snail coiling.

Joanne Render (1997) constructed a detailed fate map of the snail *Ilyanassa obsoleta* by injecting specific micromeres with large polymers conjugated to the fluorescent dye Lucifer Yellow. The fluorescence is maintained over the period of embryogenesis and can be seen in the larval tissue derived from the injected cells. The results of Render's map, given in Figure 8.26, showed that the second-quartet micromeres (2a–d) generally contribute to the shell-forming mantle, the velum, the mouth, and the heart. The third-quartet micromeres (3a–d) generate large regions of the foot, velum, esophagus, and heart. The 4d cell—the mesentoblast—contributes to the larval kidney, heart, retractor muscles, and intestine.

The polar lobe: Cell determination and axis formation

Molluscs provide some of the most impressive examples of both mosaic development—in which the blastomeres are specified autonomously—and of cytoplasmic localization, wherein morphogenetic determinants are placed in a specific region of the oocyte (see Chapter 3). Mosaic development is widespread throughout the animal kingdom, especially among protostomes such as annelids, nematodes, and molluscs, all of which initiate gastrulation at the future anterior end after only a few cell divisions.

In molluscs, the mRNAs for some transcription factors and paracrine factors are placed in particular cells by associating with certain centrosomes (Figure 8.29; Lambert and

FIGURE 8.26 Fate map of *Ilyanassa obsoleta*. Beads containing Lucifer Yellow were injected into individual blastomeres at the 32-cell stage. When the embryos developed into larvae, their descendants could be identified by their fluorescence. (After Render 1997.)

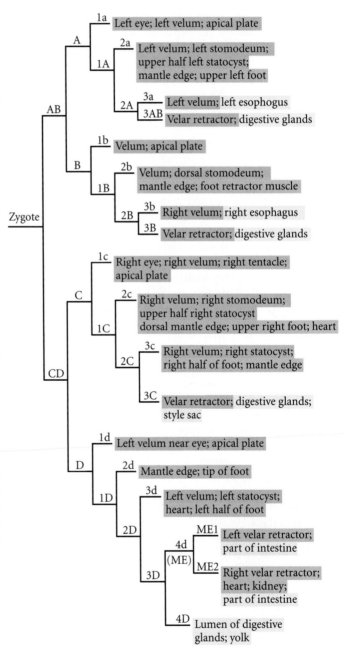

SIDELIGHTS & SPECULATIONS

Adaptation by Modifying Embryonic Cleavage

Evolution is caused by the hereditary alteration of embryonic development. Sometimes we are able to identify a specific modification of embryogenesis that has enabled the organism to survive in an otherwise inhospitable environment. One such modification, discovered by Frank Lillie in 1898, is brought about by an alteration of the typical pattern of molluscan spiral cleavage in the unionid family of clams.

Unlike most clams, *Unio* and its relatives live in swift-flowing streams. Streams create a problem for the dispersal of larvae: because the adults are sedentary, free-swimming larvae would always be carried downstream by the current. *Unio* clams have adapted to this environment via two modifications of their development. The first is an alteration in embryonic cleavage. In typical molluscan cleavage, either all the macromeres are equal in size or the 2D blastomere is the largest cell at that embryonic stage. However, cell division in *Unio* is such that the *2d* blastomere gets the largest amount of cytoplasm (Figure 8.27). This cell divides to produce most of the larval structures, including a gland capable of producing a large shell. The resulting larva is called

a **glochidium** and resembles a tiny bear trap. Glochidia have sensitive hairs that cause the valves of the shell to snap shut when they are touched by the gills or fins of a wandering fish. The larvae can thus attach themselves to the fish and "hitchhike" until they are ready to drop off and metamorphose into adult clams. In this manner, they can spread upstream as well as downstream.

In some unionid species, glochidia are released from the female's brood pouch and then wait passively for a fish to swim by. Some other species, such as *Lampsilis ventricosa*, have increased the chances of their larvae finding a fish by yet another developmental modification. Many clams develop a thin mantle that flaps around the shell and surrounds the brood pouch. In some unionids, the shape of the brood pouch (marsupium) and the undulations of the mantle mimic the shape and swimming behavior of a minnow (Welsh 1969). To make the deception even better, they develop a black "eye-spot" on one end and a flaring "tail" on the other. The "fish" in Figure 8.28 is not a fish at all, but the brood pouch and mantle of the female clam beneath it. When a preda-

FIGURE 8.28 Phony fish atop the unionid clam *Lampsilis ventricosa*. The "fish" is actually the brood pouch and mantle of the clam. (Photograph courtesy of J. H. Welsh.)

tory fish is lured within range of this "prey," the clam discharges the glochidia from the brood pouch. Thus, the modification of existing developmental patterns has permitted unionid clams to survive in challenging environments.

FIGURE 8.27 Formation of a glochidium larva by the modification of spiral cleavage. After the 8-cell embryo is formed (A), the placement of the mitotic spindle causes most of the D cytoplasm to enter the 2d blastomere (B). This large 2d blastomere divides (C), eventually giving rise to the large "bear-trap" shell of the larva (D). (After Raff and Kaufman 1983.)

 WEBSITE 8.4 Modifications of cell fate in spiralian eggs. Within the gastropods, differences in the timing of cell fate result in significantly different body plans. Furthermore, in the leeches and nemerteans, the spiralian cleavage pattern has been modified to produce new types of body plans.

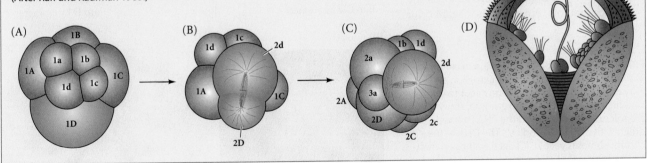

(A)　　　　　　　　　　(B)　　　　　　　　　　(C)

FIGURE 8.29 Association of decapentaplegic (dpp) mRNA with specific centrosomes of *Ilyanassa*. (A) In situ hybridization of the mRNA for the BMP-like paracrine factor Dpp in the 4-cell snail embryo shows no Dpp accumulation. (B) At prophase of the 4- to 8-cell stage, dpp mRNA (black) accumulates at one centrosome of the pair forming the mitotic spindle. (C) As mitosis continues, dpp mRNA is seen to attend the centrosome in the macromere rather than the centrosome in the micromere of each cell. The dpp message encodes a BMP-like paracrine factor critical to molluscan development. (From Lambert and Nagy 2002; photographs courtesy of L. Nagy.)

Nagy 2002). In other cases, the patterning molecules appear to be bound to a certain region of the egg that will form the **polar lobe**.

E. B. Wilson and his student H. E. Crampton observed that certain spirally cleaving embryos (mostly in the mollusc and annelid phyla) extrude a bulb of cytoplasm immediately before first cleavage. This protrusion is the polar lobe. In some species of snails, the region uniting the polar lobe to the rest of the egg becomes a fine tube. The first cleavage splits the zygote asymmetrically, so that the polar lobe is connected only to the CD blastomere (Figure 8.30A). In several species, nearly one-third of the total cytoplasmic volume is contained in this anucleate lobe, giving it the appearance of another cell (Figure 8.30B). The resulting three-lobed structure is often referred to as the **trefoil-stage** embryo (Figure 8.30C). The CD blastomere absorbs the polar lobe material, but extrudes it again prior

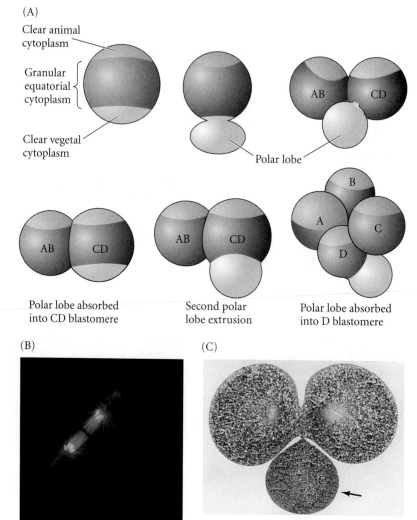

(A)

Clear animal cytoplasm

Granular equatorial cytoplasm

Clear vegetal cytoplasm

Polar lobe

AB　　CD

AB　　CD

Polar lobe absorbed into CD blastomere

AB　　CD

Second polar lobe extrusion

B
A　　C
D

Polar lobe absorbed into D blastomere

(B)　　　　　　　　(C)

FIGURE 8.30 Polar lobe formation in certain mollusc embryos. (A) Cleavage. Extrusion and reincorporation of the polar lobe occur twice. (B) Late first division of a scallop embryo, showing the microtubules (red) and the RNA (stained green with propidium iodide). The DNA of the chromosomes is yellow. (C) Section through first-cleavage, or trefoil-stage, embryo of *Dentalium*. The arrow points to the large polar lobe. (A after Wilson 1904; B courtesy of G. von Dassow and the Center for Cell Dynamics; C courtesy of M. R. Dohmen.)

FIGURE 8.31 Importance of the polar lobe in the development of *Ilyanassa*. (A) Normal trochophore larva. (B) Abnormal larva, typical of those produced when the polar lobe of the D blastomere is removed. (E, eye; F, foot; S, shell; ST, statocyst, a balancing organ; V, velum; VC, velar cilia; Y, residual yolk; ES, everted stomodeum; DV, disorganized velum.) (From Newrock and Raff 1975; photographs courtesy of K. Newrock.)

(A)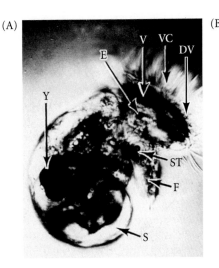

(B)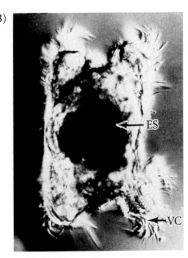

to second cleavage. After this division, the polar lobe is attached only to the D blastomere, which absorbs its material. From this point on, no polar lobe is formed.

Crampton (1896) showed that if one removes the polar lobe at the trefoil stage, the remaining cells divide normally. However, instead of a normal trochophore larva,* the result is an incomplete larva, wholly lacking its endoderm (intestine), mesodermal organs (such as the heart and retractor muscles), as well as some ectodermal organs (such as eyes) (Figure 8.31). Moreover, Crampton demonstrated that the same type of abnormal larva can be produced by removing the D blastomere from the 4-cell embryo. Crampton concluded that the polar lobe cytoplasm contains the endodermal and mesodermal determinants, and that these determinants give the D blastomere its endomesoderm-forming capacity. Crampton also showed that the localization of the mesodermal determinants is established shortly after fertilization, thereby demonstrating that a specific cytoplasmic region of the egg, destined for inclusion in the D blastomere, contains whatever factors are necessary for the special cleavage rhythms of the D blastomere and for the differentiation of the mesoderm.

Centrifugation studies demonstrated that the morphogenetic determinants sequestered within the polar lobe are probably located in the cytoskeleton or cortex, not in the lobe's diffusible cytoplasm (Clement 1968). Van den Biggelaar (1977) obtained similar results when he removed the cytoplasm from the polar lobe with a micropipette. Cytoplasm from other regions of the cell flowed into the polar lobe, replacing the portion that he had removed. The subsequent development of these embryos was normal. In addition, when he added the diffusible polar lobe cytoplasm to the B blastomere, no duplicated structures were seen (Verdonk and Cather 1983). Therefore, the diffusible

part of the polar lobe cytoplasm does not contain the morphogenetic determinants; they probably reside in the nonfluid cortical cytoplasm or on the cytoskeleton.

Clement (1962) also analyzed the further development of the D blastomere in order to observe the further appropriation of these determinants. The development of the D blastomere can be traced in Figure 8.24B–D. This macromere, having received the contents of the polar lobe, is larger than the other three. When one removes the D blastomere or its first or second macromere derivatives (1D or 2D), one obtains an incomplete larva, lacking heart, intestine, velum (the ciliated border of the larva), shell gland, eyes, and foot. This is essentially the same phenotype one gets when one removes the polar lobe. Since the D blastomeres do not directly contribute cells to many of these structures, it appears that the D-quadrant macromeres are involved in inducing other cells to have these fates.

When one removes the 3D blastomere shortly after the division of the 2D cell to form the 3D and 3d blastomeres, the larva produced looks similar to those formed by the removal of the D, 1D, or 2D macromeres. However, ablation of the 3D blastomere at a later time produces an almost normal larva, with eyes, foot, velum, and some shell gland, but no heart or intestine (see Figure 8.31). After the 4d cell is given off (by the division of the 3D blastomere), removal of the D derivative (the 4D cell) produces no qualitative difference in development. In fact, all the essential determinants for heart and intestine formation are now in the 4d blastomere, and removal of that cell results in a heartless and gutless larva (Clement 1986). The 4d blastomere is responsible for forming (at its next division) the two **mesentoblasts**, the cells that give rise to both the mesodermal (heart) and endodermal (intestine) organs.

The mesodermal and endodermal determinants of the 3D macromere, then, are transferred to the 4d blastomere, while the inductive ability of the 3D blastomere (to induce eyes and shell gland, for instance) is needed during the time the 3D cell is formed but is not required afterward. The 3D cell appears to activate the MAP kinase signaling

---

*The **trochophore** (Greek, *trochos*, "wheel") is a planktonic (free-swimming) larval form found among the molluscs and several other protostome phyla with spiral cleavage, most notably the marine annelid worms.

pathway in the micromeres above it (Lambert and Nagy 2001; see also Chapter 6). If cells are stained for activated MAP kinase, the stain is seen in those cells that require the signal from the 3D macromere for their normal differentiation (Figure 8.32). Removal of 3D prevents MAP kinase signaling, and if the MAP kinase signaling is blocked by specific inhibitors, the resulting larvae look precisely like those formed by the deletion of the D blastomeres (see Figure 8.31). Thus, the 3D macromere appears to activate the MAP kinase cascade in the ectodermal (eye- and shell gland-forming) micromeres above it.*

In addition to its role in cell differentiation, the material in the polar lobe is also responsible for specifying the dorsal-ventral (back-belly) polarity of the embryo. When polar lobe material is forced to pass into the AB blastomere as well as into the CD blastomere, twin larvae form that are joined at their ventral surfaces (Guerrier et al. 1978; Henry and Martindale 1987).

---

*The MAP kinase cascade is also seen in the 3D blastomere of equally cleaving spiralian embryos, and thus may represent an evolutionarily ancient mechanism for specifying the dorsal-ventral axis among all spirally cleaving taxa (Henry 2002; Lambert and Nagy 2003).

To summarize, experiments have demonstrated that the nondiffusible polar lobe cytoplasm is extremely important in normal molluscan development for a number of reasons:

- It contains the determinants for the proper cleavage rhythm and the cleavage orientation of the D blastomere.
- It contains certain determinants (those entering the 4d blastomere and hence leading to the mesentoblasts) for autonomous mesodermal and intestinal differentiation.
- It is responsible for permitting the inductive interactions (through the material entering the 3D blastomere) leading to the formation of the shell gland and eye.
- It contains determinants needed for specifying the dorsal-ventral axis of the embryo.

Although the polar lobe is clearly important in normal snail development, we still do not know the mechanisms for most of its effects. One possible clue has been provided by Atkinson (1987), who observed differentiated cells of the velum, digestive system, and shell gland within lobeless embryos. But even though lobeless embryos can produce these cells, they appear unable to organize them into functional tissues and organs. Tissues of the digestive tract can be found, but are not connected; individual muscle cells are scattered around the lobeless larva, but are not

FIGURE 8.32 The 3D blastomere activates MAP kinase activity in adjacent micromeres. (A) Activated MAP kinase (blue stain) can be seen in the 3D macromere and in the micromeres above it (1a–d^1, 1d^2, 2d^1, 2d^2, 3d). The nuclei are counterstained green, and the cell boundaries have been superimposed on the photographic image. Staining was done 30 minutes after the formation of the 3D macromere. (B) Control larva grown to veliger larval stage. (C) Same age larva treated with MAP kinase inhibitor 15 minutes after 3D blastomere formed. The shell, eye, statocyst, and operculum have not developed. (D) Same age larva treated with MAP kinase inhibitor at 150 minutes after 3D had formed (shortly before 4D formation). This larva had all the organs induced by the 3D macromere. (After Lambert and Nagy 2001).

(A)
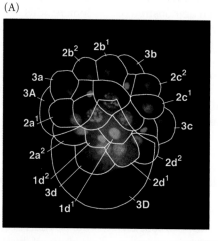

(B)
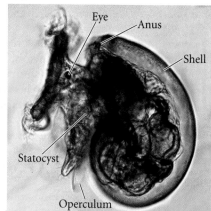

(C)
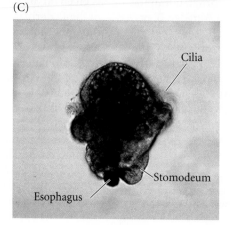

(D)
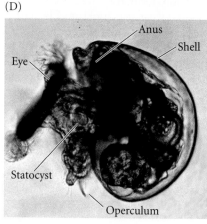

(A)

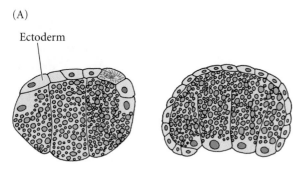

(B)

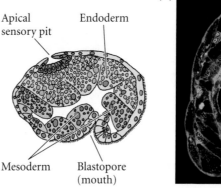

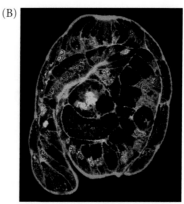

FIGURE 8.33 Gastrulation in molluscs. (A) Gastrulation in the snail *Crepidula*. The ectoderm undergoes epiboly from the animal pole and envelops the other cells of the embryo. (B) Late gastrula of the clam *Acila*, stained for actin microfilaments (orange) and nucleic acid (blue). (A after Conklin 1897; B courtesy of G. von Dassow and the Center for Cell Dynamics.)

organized into a functional muscle tissue. Thus, the developmental functions of the polar lobe are probably very complex and may be essential for axis formation.

Gastrulation in Snails

The snail stereoblastula is relatively small, and its cell fates have already been determined by the D series of macromeres. Gastrulation is accomplished primarily by epiboly, wherein the micromeres at the animal cap multiply and "overgrow" the vegetal macromeres (Collier 1997; Biggelaar and Dictus 2004). Eventually, the micromeres cover the entire embryo, leaving a small blastopore slit at the vegetal pole (Figure 8.33). Molluscs are protostomes, forming their mouth regions from the blastopore; thus this slit will become the mouth.

EARLY DEVELOPMENT IN TUNICATES

Tunicate Cleavage

Members of the tunicate subphylum are fascinating animals for several reasons, but the foremost is that they are invertebrate chordates.* Although tunicates such as ascidians lack vertebrae at all stages of their life cycles, a tunicate larva has a notochord and a dorsal nerve cord (making the tunicate a chordate). Tunicate larvae also have pharyngeal slits and an endostyle (thyroid-like gland), two other chordate characteristics. As larvae, they are free-swimming tadpoles; but when the tadpole undergoes

*The tunicate subphylum is also known as Urochordata. Its most prominent members are the ascidians ("sea squirts"), including the genera *Styela* and *Ciona*. Other, less common, tunicate groups are the thaliaceans (salps), appendicularians, and the little-known sorberaceans (Brusca and Brusca 2003).

metamorphosis, it sticks to the sea floor, its nerve cord and notochord degenerate, and it secretes a cellulose tunic (which gave the name "tunicates" to these creatures).

Ascidians are characterized by **bilateral holoblastic cleavage**, a pattern found primarily in tunicates (Figure 8.34). The most striking feature of this type of cleavage is that the first cleavage plane establishes the earliest axis of symmetry in the embryo, separating the embryo into its future right and left sides. Each successive division orients itself to this plane of symmetry, and the half-embryo formed on one side of the first cleavage plane is the mirror image of the half-embryo on the other side. The second cleavage is meridional, like the first, but unlike the first division, it does not pass through the center of the egg. Rather, it creates two large anterior cells (the A and a blastomeres) and two smaller posterior cells (blastomeres B and b). Each side now has a large and a small blastomere. During the next three divisions, differences in cell size and shape highlight the bilateral symmetry of these embryos. At the 64-cell stage, a small blastocoel is formed, and gastrulation begins from the vegetal pole.

The tunicate fate map

The fate map and cell lineages of the tunicate *Styela partita* are shown in Figure 1.7. Early tunicate embryos are specified autonomously, each cell acquiring a specific type of cytoplasm that will determine its fate. In many tunicates, the different regions of cytoplasm have distinct pigmentation, and the cell fates can easily be seen to correspond to the type of cytoplasm taken up by each cell. These cytoplasmic regions are apportioned to the egg during fertilization. In the unfertilized egg of *Styela partita*, a central gray cytoplasm is enveloped by a cortical layer containing yellow lipid inclusions (Figure 8.35A). During meiosis, the breakdown of the nucleus releases a clear substance that accumulates in the animal hemisphere of the egg. Within 5 minutes of sperm entry, the inner clear and cortical yellow cytoplasms contract into the vegetal (lower) hemisphere of the egg (Prodon et al. 2005; Sardet et al. 2005). As the male pronucleus migrates from the vegetal pole to the

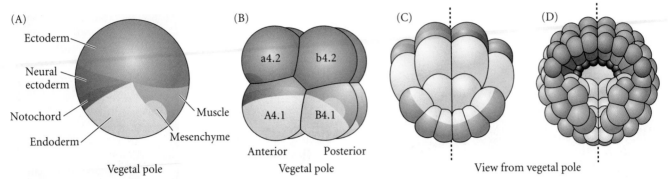

FIGURE 8.34 Bilateral symmetry in the egg of the ascidian tunicate *Styela partita*. (The cell lineages of *Styela* are shown in Figure 1.7.) (A) Uncleaved egg. The regions of cytoplasm destined to form particular organs are labeled here and coded by color throughout the diagrams. (B) 8-cell embryo, showing the blastomeres and the fates of various cells. The embryo can be viewed as two 4-cell halves; from here on, each division on the right side of the embryo has a mirror-image division on the left. (C, D) Views of later embryos from the vegetal pole. The dashed line shows the plane of bilateral symmetry. (A after Balinsky 1981.)

equator of the cell along the future posterior side of the embryo, the yellow lipid inclusions migrate with it. This migration forms the **yellow crescent**, extending from the vegetal pole to the equator (Figure 8.35B–D); this region will produce most of the tail muscles of the tunicate larva. The movement of these cytoplasmic regions depends on microtubules that are generated by the sperm centriole and on a wave of calcium ions that contracts the animal pole cytoplasm (Sawada and Schatten 1989; Speksnijder et al. 1990; Roegiers et al. 1995).

Edwin Conklin (1905) took advantage of the differing coloration of these regions of cytoplasm to follow each of the cells of the tunicate embryo to its fate in the larva (see Figure 1.7). He found that cells receiving clear cytoplasm become ectoderm; those containing yellow cytoplasm give rise to mesodermal cells; those that incorporate slate-gray inclusions become endoderm; and light gray cells become the neural tube and notochord. The cytoplasmic regions are localized bilaterally around the plane of symmetry, so they are bisected by the first cleavage furrow into the right and left halves of the embryo. The second cleavage causes the prospective mesoderm to lie in the two posterior cells, while the prospective neural ectoderm and chordamesoderm (notochord) will be formed from the two anterior cells (see Figure 8.34). The third division further partitions these cytoplasmic regions such that the mesoderm-forming cells are confined to the two vegetal posterior blastomeres, while the chordamesoderm cells are restricted to the two vegetal anterior cells.

WEBSITE **8.5 The experimental analysis of tunicate cell specification.** Researchers analyzing tunicate development are using biochemical and molecular probes to find the morphogenetic determinants that are segregated to different regions of the egg cytoplasm.

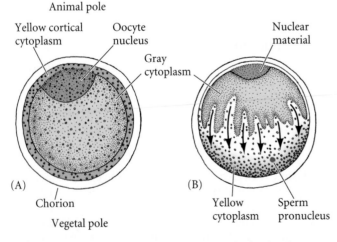

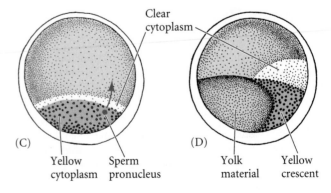

FIGURE 8.35 Cytoplasmic rearrangement in the fertilized egg of *Styela partita*. (A) Before fertilization, yellow cortical cytoplasm surrounds the gray yolky inner cytoplasm. (B) After sperm entry (in the vegetal hemisphere of the oocyte), the yellow cortical cytoplasm and the clear cytoplasm derived from the breakdown of the oocyte nucleus contract vegetally toward the sperm. (C) As the sperm pronucleus migrates animally toward the newly formed egg pronucleus, the yellow and clear cytoplasms move with it. (D) The final positions of the yellow cytoplasm marks the location where cells give rise to tail muscles. (After Conklin 1905.)

Autonomous and conditional specification of tunicate blastomeres

As mentioned in Chapter 3, the autonomous specification of tunicate blastomeres was one of the first observations in the field of experimental embryology (Chabry 1888). Cohen and Berrill (1936) confirmed Chaby's and Conklin's results, and by counting the number of notochord and muscle cells, they demonstrated that larvae derived from only one of the first two blastomeres had half the expected number of cells. Reverberi and Minganti (1946) extended this analysis in a series of isolation experiments, and they, too, observed the self-differentiation of each isolated blastomere and of the remaining embryo. The results of one of these experiments are shown in Figure 3.8. When the 8-cell embryo is separated into its four doublets (the right and left sides being equivalent), both mosaic and conditional specification are seen. The animal posterior pair of blastomeres gives rise to the ectoderm, and the vegetal posterior pair produces endoderm, mesenchyme, and muscle tissue, just as expected from the fate map. Autonomous specification is seen in the tunicate gut endoderm, muscle mesoderm, skin ectoderm, and neural cord. Conditional specification (by induction) is seen in the formation of the brain, notochord, and mesenchyme cells.

AUTONOMOUS SPECIFICATION OF THE MYOPLASM: THE YELLOW CRESCENT   From the cell lineage studies of Conklin and others, it was known that only one pair of blastomeres (posterior vegetal; B4.1) in the 8-cell embryo is capable of producing tail muscle tissue (Whittaker 1982). These cells contain the yellow crescent cytoplasm. When yellow crescent cytoplasm is transferred from the B4.1 (muscle-forming) blastomere to the b4.2 (ectoderm-forming) blastomere of an 8-cell tunicate embryo, the ectoderm-forming blastomere generates muscle cells as well as its normal ectodermal

progeny (Figure 8.36; see also Figure 3.10). Moreover, cytoplasm from the yellow crescent area of the fertilized egg can cause the a4.2 blastomere to express muscle-specific proteins (Nishida 1992a). Conversely, when larval cell nuclei are transplanted into enucleated tunicate egg fragments, the newly formed cells show the structures typical of the egg regions providing the cytoplasm, not of those cells providing the nuclei (Tung et al. 1977). We can conclude, then, that certain determinants present in the egg cytoplasm cause the formation of certain tissues. These morphogenetic determinants appear to work by selectively activating or inactivating specific genes. The determination of the blastomeres and the activation of certain genes are controlled by the spatial localization of the morphogenetic determinants within the egg cytoplasm.

Using RNA hybridization techniques, Nishida and Sawada (2001) found particular mRNAs to be highly enriched in the vegetal hemisphere of the tunicate *Halocynthia roretzi*. One of these RNA messages encodes a zinc-finger transcription factor called **macho-1**. *Macho-1* mRNA was found to be concentrated in the vegetal hemisphere of the unfertilized egg and remains present during early fertilization. It appears to migrate with the yellow crescent cytoplasm into the posterior vegetal region of the egg during the second half of the first cell cycle. By the 8-cell stage, *macho-1* mRNA is found only in the B4.1 blastomeres. At the 16- and 32-cell stages, it is seen in those blastomeres that give rise to the muscle cells* (Figure 8.37).

When antisense oligonucleotides to deplete *macho-1* mRNA were injected into unfertilized eggs, the resulting larvae lacked all the muscles usually formed by the descen-

---

*Macho-1 mRNA is also localized in the cells that become the mesenchyme. However, FGF signals from the endoderm prevent these mesenchyme precursors from developing into muscle cells, as we will see later.

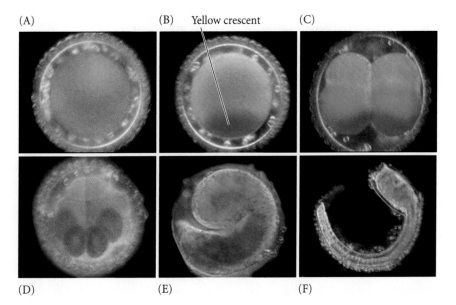

(A)   (B)   Yellow crescent   (C)

(D)   (E)   (F)

FIGURE 8.36 Cytoplasmic segregation in the egg of *Boltenia villosa*. The yellow crescent, originally seen in the vegetal pole, becomes segregated into the B4.1 blastomere pair and thence into the muscle cells. (From Swalla 2004, photographs courtesy of B. Swalla, K. Zigler, and M. Baltzley.)

FIGURE 8.37 Autonomous specification by a morphogenetic factor. The *macho-1* mRNA message is localized to the muscle-forming tunicate cytoplasm. In situ hybridization shows the *macho-1* message found first in the vegetal pole cytoplasm (A), then migrating up the presumptive posterior surface of the egg (B) and becoming localized in the B4.1 blastomere (C). (From Nishida and Sawada 2001; photographs courtesy of H. Nishida and N. Satoh.)

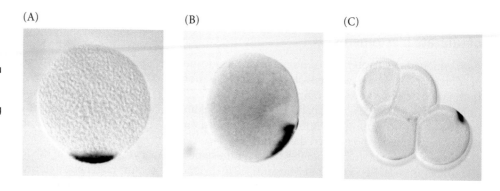

(A)   (B)   (C)

dants of the B4.1 blastomere. (They did have the secondary muscles that are generated through the interactions of A4.1 and b4.2 blastomeres.) The tails of these *macho-1*-depleted larvae were severely shortened, but the other regions of the tadpoles appeared structurally and biochemically normal. Moreover, B4.1 blastomeres isolated from *macho-1*-depleted embryos failed to produce muscle tissue. Nishida and Sawada then injected *macho-1* mRNA into cells that would *not* normally form muscle, and found that these ectoderm or endoderm precursors did generate muscle cells when given *macho-1* mRNA.

Macho-1 turns out to be a transcription factor that is required for the activation of several mesodermal genes, including *muscle actin, myosin, tbx6,* and *snail* (Sawada et al. 2005; Yagi et al. 2005a). Of these gene products, only the Tbx6 protein produced muscle differentiation (as Macho-1 did) when expressed in cells ectopically. Macho-1 thus appears to directly activate a set of *tbx6* genes, and Tbx6 proteins activate the rest of muscle development (Yagi et al. 2005b). Thus, the *macho-1* message is found at the right place and at the right time, and these experiments suggest that Macho-1 protein is both necessary and sufficient to promote muscle differentiation in certain ascidian cells.

The Macho-1 and Tbx6 proteins also appear to activate the muscle-specific gene *snail.* Snail protein is important in preventing *Brachyury* (*T*) expression in presumptive muscle cells, and is therefore needed to prevent the muscle precursors from becoming notochord cells. It appears, then, that the Macho-1 transcription factor is a critical com-

ponent of the tunicate yellow crescent, muscle-forming cytoplasm. Macho-1 activates a cascade of transcription factors that promotes muscle differentiation while at the same time inhibiting notochord specification.

AUTONOMOUS SPECIFICATION OF THE ENDODERM: β-CATENIN Presumptive endoderm originates from the vegetal A4.1 and B4.1 blastomeres. The specification of these cells coincides with the localization of β-catenin, a transcription factor discussed earlier in regard to sea urchin endoderm specification (see Figure 8.12). Inhibition of β-catenin results in the loss of endoderm and its replacement by ectoderm in the ascidian embryo (Figure 8.38; Imai et al. 2000). Conversely, increasing β-catenin synthesis causes an increase in the endoderm at the expense of the ectoderm (just as in sea urchins). The β-catenin transcription factor appears to function by activating the synthesis of the homeobox transcription factor Lhx-3. Inhibition of the *lhx-3* message prohibits the differentiation of endoderm (Satou et al. 2001).

FIGURE 8.38 Antibody staining of β-catenin protein shows its involvement with endoderm formation. (A) No β-catenin is seen in the animal pole nuclei of a 110-cell *Ciona* embryo. (B) In contrast, nuclear β-catenin is readily seen in the nuclei in the vegetal endoderm precursors at the 110-cell stage (C) When β-catenin is expressed in notochordal precursor cells, those cells will become endoderm and express endodermal markers such as alkaline phosphatase. The white arrows show normal endoderm; the black arrows show notochordal cells that are expressing endodermal enzymes. (From Imai et al. 2000; photographs courtesy of H. Nishida and N. Satoh.)

(A)   (B)   (C)

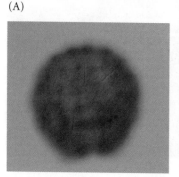
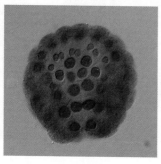
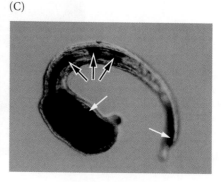

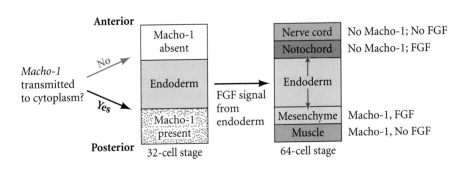

FIGURE 8.39 The two-step process for specifying the marginal cells of the tunicate embryo. The first step involves the acquisition (or nonacquisition) by the cells of the Macho-1 transcription factor. The second step involves the reception (or nonreception) of the FGF signal from the endoderm. (After Kobayashi et al. 2003.)

CONDITIONAL SPECIFICATION OF THE MESENCHYME AND NOTOCHORD BY THE ENDODERM While most of the muscles are specified autonomously from the yellow crescent cytoplasm, the most posterior muscle cells form through conditional specification by cell interactions with the descendants of the A4.1 and b4.2 blastomeres (Nishida 1987, 1992a,b). Moreover, the notochord, brain, and mesenchyme also form through inductive interactions. In fact, the notochord and mesenchyme appear to be induced by the fibroblast growth factor that is secreted by the endoderm cells (Nakatani et al. 1996; Kim et al. 2000; Imai et al. 2002).

The posterior cells that will become mesenchyme respond differently to the FGF signal due to the presence of Macho-1 in the posterior vegetal cytoplasm (Figure 8.39; Koboyashi et al. 2003). Macho-1 prevents notochord induction in the mesenchymal cell precursors by activating the *snail* gene (which will in turn suppress the activation of *Brachyury*). Thus, Macho-1 is not only a muscle-activating determinant, it is also a factor that distinguishes cell response to the FGF signal. These FGF-responding cells do not become muscle, because FGF also activates cascades that block muscle formation—another role that is conserved in vertebrates. As can be seen in Figure 8.39, the presence of Macho-1 changes the responses to endodermal FGFs, causing the anterior cells to form notochord while the posterior cells become mesenchyme.

Specification of the embryonic axes

The axes of the tunicate larva are among its earliest commitments. Indeed, all of its embryonic axes are determined by the cytoplasm of the zygote prior to first cleavage. The first axis to be determined is the dorsal-ventral axis, which is defined by the cap of cytoplasm at the vegetal pole. This vegetal cap defines the future dorsal side of the larva and the site where gastrulation is initiated (Bates and Jeffery 1988). When small regions of vegetal pole cytoplasm were removed from zygotes (between the first and second waves of zygote cytoplasmic movement), the zygotes neither gastrulated nor formed a dorsal-ventral axis.

The anterior-posterior axis is the second axis to appear and is also determined during the migration of the oocyte cytoplasm. The yellow crescent forms in the region of the egg that will become the posterior and vegetal side of the larva (see Figures 8.35 and 8.36). When roughly 10 percent

of the cytoplasm from this posterior vegetal region of the egg was removed after the second wave of cytoplasmic movement, most of the embryos failed to form an anterior-posterior axis. Rather, these embryos developed into radially symmetrical larvae with anterior fates. This posterior vegetal cytoplasm (PVC) is "dominant" to other cytoplasms in that when it was transplanted into the anterior vegetal region of zygotes that had had their own PVC removed, the anterior of the cell became the new posterior, and the axis was reversed (Nishida 1994).

The specification of the left-right axis in tunicates is poorly understood. The first cleavage divides the embryo into its future right and left sides, but we do not yet know how these sides are specified. It appears that a gene analogous to the *Nodal* gene of echinoderms (see page 227) and vertebrates becomes expressed specifically in the left-side epidermis of the tailbud-stage embryo (Morokuma et al. 2002). But how it becomes expressed there is still not known.

Gastrulation in Tunicates

Tunicates, like sea urchins and vertebrates, are deuterostomes and follow a pattern of gastrulation in which the blastopore becomes the anus. Tunicate gastrulation is characterized by the invagination of the endoderm, the involution of the mesoderm, and the epiboly of the ectoderm. About 4–5 hours after fertilization, the vegetal (endoderm) cells assume a wedge shape, expanding their apical margins and contracting near their vegetal margins (Figure 8.40). The A8.1 and B8.1 blastomere pairs appear to lead this invagination into the center of the embryo. The invagination forms a blastopore whose lips will become the mesodermal cells. The presumptive notochord cells are now on the anterior portion of the blastopore lip, while the presumptive tail muscle cells (from the yellow crescent) are on the posterior lip. The lateral lips comprise those cells that will become mesenchyme.

The second step of gastrulation involves the involution of the mesoderm. The presumptive mesoderm cells involute over the lips of the blastopore and, by migrating over the basal surfaces of the ectodermal cells, move inside the embryo. The ectodermal cells then flatten and epibolize over the mesoderm and endoderm, eventually covering the embryo. After gastrulation is complete, the embryo

(A)

Polar body

Ectoderm

Muscle

Invaginating
endoderm

Notochord

(B)

Mesenchyme

(C)

Blastopore

(D)

(E)

(F)

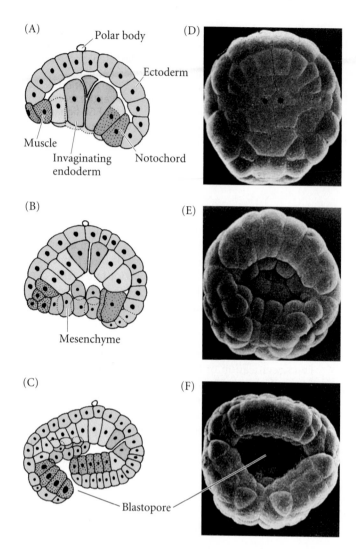

elongates along its anterior-posterior axis. The dorsal ecto-dermal cells that are the precursors of the neural tube invaginate into the embryo and are enclosed by neural folds. This process forms the neural tube, which will form a brain anteriorly and a spinal chord posteriorly. Meanwhile, the presumptive notochord cells on the right and left sides of the embryo migrate to the midline and interdigitate to form the notochord. The 40 cells of the notochord rearrange themselves from a 4-by-10 sheet of cells into a single row of 40 cells (Jiang et al. 2005). This intercalation and migration of notochord cells is called **convergent extension** (Figure 8.41; we also saw this phenomenon in our discussion of the sea urchin archenteron), and it extends the body axis along the anterior-posterior dimension. Indeed, the convergent extension of notochordal precursor cells is characteristic across all the chordate phyla.

The muscle cells of the tail differentiate on either side of the neural tube and notochord (Jeffery and Swalla 1997). This forms the tadpole-like body of the larva. At the 110-cell stage, the B7.5 blastomere pairs express the conserved heart transcription factor Mesp. The anterior daughters of these B7.5 blastomeres migrate to form two regions of cardiac mesoderm on the left and right ventral sides of the tadpole, just anterior to the tail. Like the heart precursor cells of vertebrate embryos, these two cell clusters migrate to meet at the ventral midline of the larva (Davidson and Levine 2003; Satou et al. 2004). After metamorphosis, they will form the functional heart of the adult. During this metamorphosis, the tail and brain degenerate and the tunicate no longer moves.*

*Neurobiologist Rudolfo Llinás once remarked, "What do tunicates and professors have in common? The larval forms of tunicates, equipped with a ganglion containing approximately 300 cells, go through a brief phase of free swimming. Upon finding a suitable substrate, the larva buries its head into the selected location and becomes sessile. Then it absorbs most of its small brain and nervous system and returns to a rather primitive condition—a process paralleled by some human academics upon obtaining university tenure" (Llinás 1987).

FIGURE 8.40 Gastrulation in the tunicate. Cross-sections (A–C) and scanning electron micrographs viewed from the vegetal pole (D–F) illustrate the invagination of the endoderm (A, D), the involution of the mesoderm (B, E), and the epiboly of the ectoderm (C, F). Cell fates are color-coded as in Figure 8.34. (From Satoh 1978 and Jeffery and Swalla 1997; photographs courtesy of N. Satoh.)

FIGURE 8.41 Convergent extension of the tunicate notochord. The notochord is visualized by a green fluorescent protein (GFP) probe fused to a promoter of the *Brachyury* gene, which is usually expressed in the notochord. The notochordal precursor cells converge and extend the notochord down the length of the animal's tail. (From Deschet et al. 2003, photographs courtesy of the authors.)

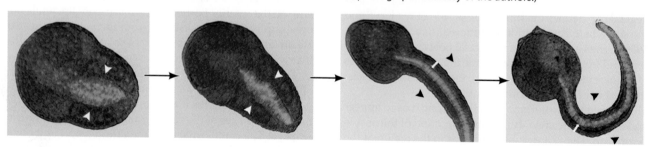

THE NEMATODE *CAENORHABDITIS ELEGANS*

Why *C. elegans*?

Our ability to analyze development requires appropriate model organisms. Sea urchins have long been a favorite of embryologists because their gametes are readily obtainable in large numbers, their eggs and embryos are transparent, and fertilization and development can occur under laboratory conditions. But sea urchins are difficult to rear in the laboratory for more than one generation, making their genetics difficult to study.

Geneticists (at least those who have worked with multicellular eukaryotes), have always favored *Drosophila*. The fruit fly's rapid life cycle, its readiness to breed, and the polytene chromosomes of the larva (which allow gene localization) make it superbly suited for hereditary analysis. But *Drosophila* development, as we will detail in Chapter 9, is complex.

A research program spearheaded by Sydney Brenner (1974) was established to identify an organism wherein it might be possible to identify each gene involved in development as well as to trace the lineage of each and every cell. Nematode roundworms seemed like a good group to start with, since embryologists such as Goldschmidt and Boveri had shown that several nematode species have a relatively small number of chromosomes and a small number of cells with invariant cell lineages.

Eventually, Brenner and his colleagues settled upon *Caenorhabditis elegans*, a small (1 mm long), free-living (i.e., non-parasitic) soil nematode (Figure 8.42A). *C. elegans* has a rapid period of embryogenesis (about 16 hours), which it can accomplish in a petri dish, and relatively few cell types. Moreover, its predominant adult form is hermaphroditic, with each individual producing both eggs and sperm. These roundworms can reproduce either by self-fertilization or by cross-fertilization with the infrequently occurring males.

The body of an adult *C. elegans* hermaphrodite contains exactly 959 somatic cells, and the entire cell lineage has been traced through its transparent cuticle (Figure 8.42B; Sulston and Horvitz 1977; Kimble and Hirsh 1979). Furthermore, unlike vertebrate cell lineages, the *C. elegans* lineage is almost entirely invariant from one individual to the next. There is little room for randomness (Sulston et al. 1983). It also has a very compact genome. Although it has about the same number of genes as human beings (*C. elegans* has between 18,000 and 20,000 genes, while *H. sapiens* has 20,000–25,000), the nematode has only about 3 percent the number of nucleotides in its genome (Hodgkin 1998, 2001).* The *C. elegans* genome has been entirely sequenced and was the first complete sequence ever obtained for a multicellular organism (*C. elegans* Sequencing Consortium 1999).

Cleavage and Axis Formation in *C. elegans*

Rotational cleavage of the *C. elegans* egg

The zygote of *Caenorhabditis* exhibits rotational holoblastic cleavage (Figure 8.42C,D). During early cleavage, each asymmetrical division produces one founder cell (denoted AB, E, MS, C, and D) that produces differentiated descendants; and one stem cell (the P1–P4 lineage). In the first cell division, the cleavage furrow is located asymmetrically along the anterior-posterior axis of the egg, closer to what will be the posterior pole. It forms an anterior founder cell (AB) and a posterior stem cell (P1). During the second division, the founder cell (AB) divides equatorially (longitudinally; 90 degrees to the anterior-posterior axis), while the P1 cell divides meridionally (transversely) to produce another founder cell (EMS) and a posterior stem cell (P2). The stem cell lineage always undergoes meridional division to produce (1) an anterior founder cell and (2) a posterior cell that will continue the stem cell lineage.

The descendants of each founder cell can be observed through the transparent cuticle and divide at specific times in ways that are nearly identical from individual to individual. In this way, the exactly 558 cells of the newly hatched larva are generated. Descendant cells are named according to their positions relative to their sister cells. For instance, ABal is the "left-hand" daughter cell of the Aba cell, and ABa is the "anterior" daughter cell of the AB cell.

Anterior-posterior axis formation

The elongated axis of the *C. elegans* egg defines the future anterior-posterior axis of the nematode's body. The decision as to which end will become the anterior and which the posterior seems to reside with the position of sperm pronucleus. When it enters the oocyte cytoplasm, the centriole associated with the sperm pronucleus initiates cytoplasmic movements that push the male pronucleus to the

---

*This similarity in gene number was rather surprising, to say the least. "What does a worm want with 20,000 genes?" wrote Jonathan Hodgkin, the curator of the *C. elegans* gene map. Humans have trillions of cells, four-chambered hearts, incredibly regionalized brains, intricate limbs, and remarkable vascular networks. *C. elegans*, on the other hand, has no hands, no chambers in its heart, and no head to speak of. Thousands of these organisms would fit under our fingernails (which *C. elegans* also lacks). Hodgkin (2001) notes that human genes and their proteins tend to be more multifunctional than their nematode counterparts. Nematode genes do not have nearly the capacity for producing alternatively spliced RNAs as do human genes. Moreover, whereas human developmental proteins often have many functions, each *C. elegans* protein appears to have just a single function. In addition, *C. elegans* may have duplicated many of its genes, thus inflating its gene number. Whatever the case, however, the mere *number* of genes does not seem to be responsible for the huge physical differences between worms and human beings.

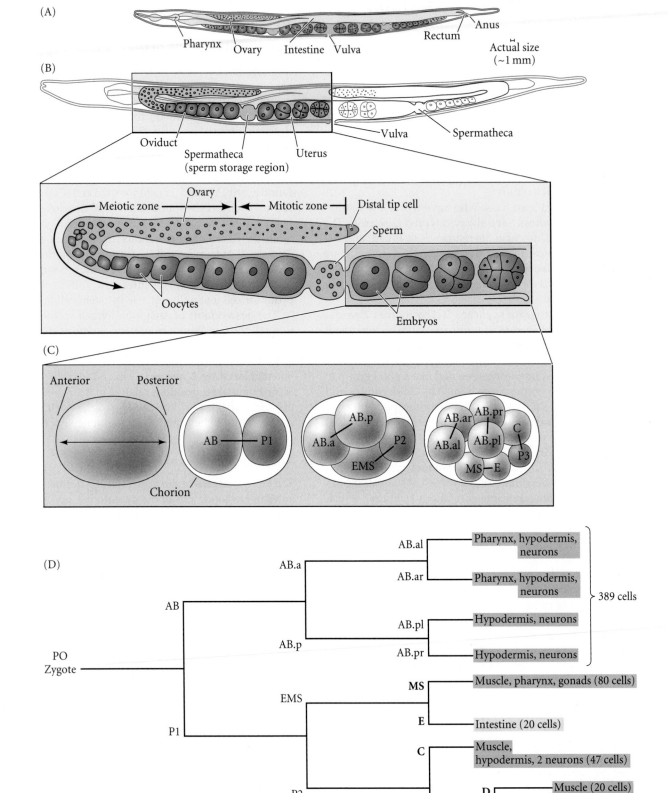

(A)

Pharynx Ovary Intestine Vulva Rectum Anus

Actual size
(~1 mm)

(B)

Oviduct Spermatheca
(sperm storage region) Uterus Vulva Spermatheca

Meiotic zone Ovary Mitotic zone Distal tip cell

Sperm

Oocytes

Embryos

(C)

Anterior Posterior

AB P1

AB.p
AB.a P2
EMS

AB.ar
AB.al AB.pr
AB.pl C
P3
MS—E

Chorion

(D)

AB.a
AB.al → Pharynx, hypodermis, neurons
AB.ar → Pharynx, hypodermis, neurons

AB

AB.p
AB.pl → Hypodermis, neurons
AB.pr → Hypodermis, neurons

389 cells

PO
Zygote

P1

EMS
MS → Muscle, pharynx, gonads (80 cells)
E → Intestine (20 cells)

P2

C → Muscle, hypodermis, 2 neurons (47 cells)

P3
D → Muscle (20 cells)
P4
Z2
Z3
Germ line

◄ **FIGURE 8.42** The nematode *Caenorhabditis elegans*. (A) Side view of adult hermaphrodite. Sperm are stored such that a mature egg must pass through the sperm on its way to the vulva. (B) The gonads. Near the distal end, the germ cells undergo mitosis. As they move farther from the distal tip, they enter meiosis. Early meioses form sperm, which are stored in the spermatheca. Later meioses form eggs, which are fertilized as they roll through the spermatheca. (C) Early development, as the egg is fertilized and moves toward the vulva. The P-lineage consists of stem cells that will eventually form the germ cells. (D) Abbreviated cell lineage chart. The germ line segregates into the posterior portion of the most posterior (P) cell. The first three cell divisions produce the AB, C, MS, and E lineages. The number of derived cells (in parentheses) refers to the 558 cells present in the newly hatched larva. Some of these continue to divide to produce the 959 somatic cells of the adult. (After Pines 1992, based on Sulston and Horvitz 1977 and Sulston et al. 1983.)

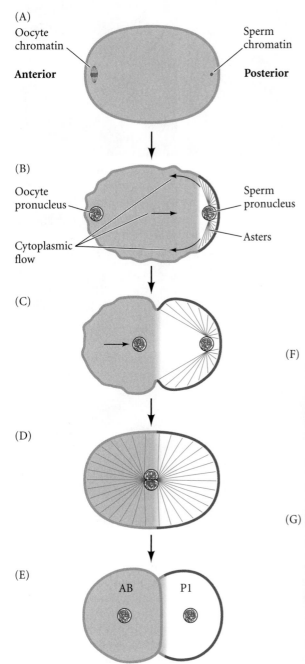

WEBSITE **8.6 P-granule migration.** Movies of P-granule migration under natural and experimental conditions were taken in the laboratory of Susan Strome. They show P-granule segregation to the P-lineage blastomeres except when perturbed by mutations or chemicals that inhibit microfilament function.

FIGURE 8.43 PAR proteins and the establishment of polarity. (A–E) PAR protein distribution. (A) When sperm enters egg, the egg nucleus is undergoing meiosis (left). The cortical cytoplasm contains PAR-3 (orange) and the internal cytoplasm contains MEX-5 (shaded gray). (B) Cytoplasm begins flowing toward the sperm pronucleus. Near the newly formed sperm asters, PAR-2 (purple) replaces PAR-3 in the cortical cytoplasm. (C) The domain of PAR-2 expression expands as the sperm nucleus migrates toward the center of the cell. (D) At the first division, about half the PAR-2, PAR-3, and MEX-5 polarity remains. (E) At the end of the first division, the AB blastomere has MEX-5 and PAR-3, while most of the P1 blastomere has very little MEX-5 and has a cortex of PAR-2 (except where the two blastomeres meet). (F) In this dividing *C. elegans* zygote, PAR-2 protein is stained green and DNA is stained blue. (G) In second division, the AB cell and the P1 cell divide 90 degrees differently. (A–E after Nance 2005; F photograph courtesy of J. Ahrenger; G photograph courtesy of J. White.)

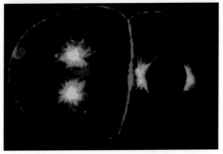

FIGURE 8.44 Segregation of the P-granules into the germ line lineage of the *C. elegans* embryo. The left column shows the cell nuclei (the DNA is stained blue by Hoescht dye); the right column shows the same embryos stained for P-granules. At each successive division, the P-granules enter the P-lineage blastomere, whose progeny will become the germ cells. (Photographs courtesy of S. Strome.)

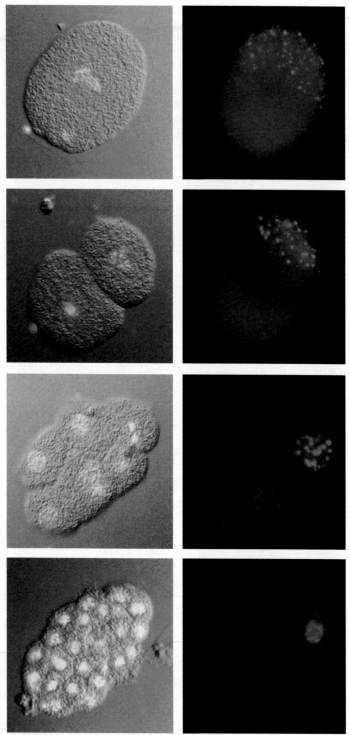

nearest end of the oblong oocyte. That end becomes the posterior pole (Goldstein and Hird 1996). The integration of cell division, cell specification, and morphogenesis is coordinated by the sperm centriole and several PAR ("partitioning") proteins. The sperm centriole is critical in initiating, but not maintaining, polarity; the centriole organizes the positioning of several maternal proteins at the future anterior and posterior ends of the egg. It is not known how the centriole accomplishes this positioning. Microtubules may play a role, but other factors must be involved as well, since inhibiting the microtubules does not inhibit the positioning of these proteins (Cowan and Hyman 2004).

Originally, PAR-3 and MEX-5 proteins are seen throughout the egg, the former in the cortex and the latter spread throughout the cytoplasm (Figure 8.43A). As the sperm centriole migrates, the protein PAR-2 becomes localized in the cortical cytoplasm at the pole closest to the sperm nucleus, displacing PAR-3 and MEX-5, as the sperm asters encompass more territory (Figure 8.43B,C). As the nuclei meet, the anterior of the cell is characterized by PAR-3 in the cortex and MEX-5 in the interior cytoplasm, while the posterior has PAR-2 in the cortex and a cytoplasm free of MEX-5 (Figure 8.43D). When the first cell division has been completed, the AB blastomere has PAR-3 and MEX-5, while all but the most anterior region of the P-cell contains cortical PAR-2 (Figure 8.43E). The PAR proteins continue to play roles in further cell divisions.

Some of the most important entities localized by the PAR proteins are the **P-granules**, ribonucleoprotein complexes that specify the germ cells. The P-granules appear to be a collection of translation regulators. The proteins of these granules include RNA helicases, polyA polymerases, and translation initiation factors (Amiri et al. 2001; Smith et al. 2002; Wang et al. 2002). Using fluorescent antibodies to a component of the P-granules, Strome and Wood (1983) discovered that shortly after fertilization, the randomly scattered P-granules move toward the posterior end of the zygote, so that they enter only the blastomere (P1) formed from the posterior cytoplasm (Figure 8.44). The P-granules of the P1 cell remain in the posterior of the P1 cell when it divides and are thereby passed to the P2 cell. During the division of P2 and P3, however, the P-granules become associated with the nucleus that enters the P3 cytoplasm. Eventually, the P-granules will reside in the P4 cell, whose progeny become the sperm and eggs of the adult. The localization of the P-granules requires microfilaments, but can occur in the absence of microtubules. Treating the zygote with cytochalasin D (a microfilament inhibitor) prevents the segregation of these granules to the posterior of the cell, whereas demecolcine (a colchicine-like microtubule inhibitor) fails to stop this movement.

FIGURE 8.45 Deficiencies of intestine and pharynx in *skn-1* mutants of *C. elegans*. Embryos derived from wild-type females (A,C) and from females homozygous for mutant *skn-1* (B,D) were tested for the presence of pharyngeal muscles (A, B) and gut-specific granules (C,D). A pharyngeal muscle-specific antibody labels the pharynx musculature of those embryos derived from wild-type females (A), but does not bind to any structure in the embryos from *skn-1* mutant females (B). Similarly, the birefringent gut granules characteristic of embryonic intestines (C) are absent from embryos derived from the *skn-1* mutant females (D). (From Bowerman et al. 1992a; photographs courtesy of B. Bowerman.)

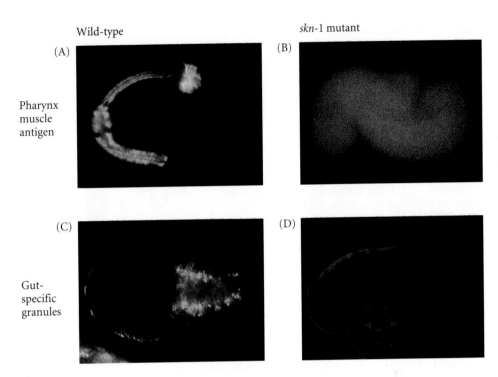

Wild-type

skn-1 mutant

(A) (B)

Pharynx muscle antigen

(C) (D)

Gut-specific granules

Formation of the dorsal-ventral and right-left axes

The dorsal-ventral axis of the nematode is seen in the division of the AB cell. As the AB cell divides, it becomes longer than the eggshell is wide. This causes the cells to slide, resulting in one AB daughter cell being anterior and one being posterior (hence their respective names, ABa and ABp; see Figure 8.42). This squeezing also causes the ABp cell to take a position above the EMS cell that results from the division of the P1 blastomere. The ABp cell defines the future dorsal side of the embryo, while the EMS cell—the precursor of the muscle and gut cells—marks the future ventral surface of the embryo. The left-right axis is specified later, at the 12-cell stage, when the MS blastomere (from the division of the EMS cell) contacts half the "granddaughters" of the ABa cell, distinguishing the right side of the body from the left side (Evans et al. 1994).

Control of blastomere identity

C. elegans demonstrates both the conditional and autonomous modes of cell specification. Both modes can be seen if the first two blastomeres are experimentally separated (Priess and Thomson 1987). The P1 cell develops autonomously without the presence of AB. It makes all the cells that it would normally make, and the result is the posterior half of an embryo. However, the AB cell, in isolation, makes only a fraction of the cell types that it would normally make. For instance, the resulting ABa blastomere fails to make the anterior pharyngeal muscles that it would have made in an intact embryo. Therefore, the specification of the AB blastomere is conditional, and it needs to interact with the descendants of the P1 cell in order to develop normally.

AUTONOMOUS SPECIFICATION The determination of the P1 lineages appears to be autonomous, with the cell fates determined by internal cytoplasmic factors rather than by interactions with neighboring cells. The P-granules are localized in a way consistent with a role as a morphogenetic determinant, and they act through translational regulation (the exact mechanism is not well understood) to specify germ cells. Meanwhile, the SKN-1, PAL-1, and PIE-1 proteins are thought to encode transcription factors that act intrinsically to determine the fates of cells derived from the four P1-derived somatic founder cells (MS, E, C, and D).

The **SKN-1** protein is a maternally expressed polypeptide that may control the fate of the EMS blastomere, which is the cell that generates the posterior pharynx. After first cleavage, only the posterior blastomere—P1—has the ability to produce pharyngeal cells when isolated. After P1 divides, only EMS is able to generate pharyngeal muscle cells in isolation (Priess and Thomson 1987). Similarly, when the EMS cell divides, only one of its progeny, MS, has the intrinsic ability to generate pharyngeal tissue. These findings suggest that pharyngeal cell fate may be determined autonomously, by maternal factors residing in the cytoplasm that are parceled out to these particular cells.

Bowerman and co-workers (1992a,b, 1993) found maternal effect mutants lacking pharyngeal cells and were able to isolate a mutation in the *skn-1* gene. Embryos from homozygous *skn-1*-deficient mothers lack both pharyngeal mesoderm and endoderm derivatives of EMS (Figure 8.45). Instead of making the normal intestinal and pharyngeal structures, these embryos seem to make extra hypodermal (skin) and body wall tissue where their intestine and pharynx should be. In other words, the EMS blastomere

FIGURE 8.46 Isolation and recombination experiments show that cell-cell interactions are required for the EMS cell to form intestinal lineage determinants. (A) When isolated shortly after its formation, the EMS blastomere cannot produce gut-specific granules. If left in place for longer periods, it can. (B) If the EMS cell is recombined with either or both derivatives of the AB blastomere, it will not form gut-specific granules. (C) If recombined with the P2 blastomere, the EMS cell gives rise to gut-specific structures. (After Goldstein 1992.)

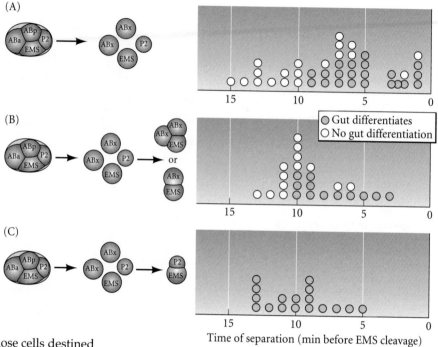

appears to be re-specified as C. Only those cells destined to form pharynx or intestine are affected by this mutation. Moreover, the protein encoded by the *skn-1* gene has a DNA-binding site motif similar to that seen in the bZip family of transcription factors (Blackwell et al. 1994).

SKN-1 is a maternal protein, and it activates the transcription of at least two genes, *med-1* and *med-2*, whose products are also transcription factors. The MED transcription factors appear to specify the entire fate of the EMS cell, since expression of the *med* genes in other cells can cause non-EMS cells to become EMS even if SKN-1 is absent (Maduro et al. 2001).

A second putative transcription factor, **PAL-1**, is also required for the differentiation of the P1 lineage. PAL-1 activity is needed for the normal development of the somatic descendants of the P2 blastomere. Embryos lacking PAL-1 have no somatic cell types derived from the C and D stem cells (Hunter and Kenyon 1996). PAL-1 is regulated by the MEX-3 protein (see Figure 8.43), an RNA-binding protein that appears to inhibit the translation of *pal-1* mRNA. Wherever *mex-3* is expressed, PAL-1 is absent. Thus, in *mex-3*-deficient mutants, PAL-1 is seen in every blastomere. SKN-1 also inhibits PAL-1 (thereby preventing it from becoming active in the EMS cell).

A third putative transcription factor, **PIE-1,** is necessary for germ line fate. PIE-1 is placed into the P blastomeres through the action of the PAR-1 protein, and it appears to inhibit both SKN-1 and PAL-1 function in the P2 and subsequent germ line cells (Hunter and Kenyon 1996). Mutations of the maternal *pie-1* gene result in germ line blastomeres adopting somatic fates, with the P2 cell behaving similarly to a wild-type EMS blastomere. The localization and the genetic properties of PIE-1 suggest that it represses the establishment of somatic cell fate and preserves the

totipotency of the germ cell lineage (Mello et al. 1996; Seydoux et al. 1996).

CONDITIONAL SPECIFICATION As mentioned earlier, the *C. elegans* embryo uses both autonomous and conditional modes of specification. Conditional specification can be seen in the development of the endoderm cell lineage. At the 4-cell stage, the EMS cell requires a signal from its neighbor (and sister), the P2 blastomere. Usually, the EMS cell divides into an MS cell (which produces mesodermal muscles) and an E cell (which produces the intestinal endoderm). If the P2 cell is removed at the early 4-cell stage, the EMS cell will divide into two MS cells, and no endoderm will be produced. If the EMS cell is recombined with the P2 blastomere, however, it will form endoderm; it will not do so, however, when combined with ABa, ABp, or both AB derivatives (Figure 8.46; Goldstein 1992).

The P2 cell produces a signal that interacts with the EMS cell and instructs the EMS daughter that is next to it to become the E cell. This message is transmitted through the Wnt signaling cascade (Figure 8.47; Rocheleau et al. 1997; Thorpe et al. 1997; Walston et al. 2004). The P2 cell produces a *C. elegans* homologue of a Wnt protein, the MOM-2 peptide. The MOM-2 peptide is received in the EMS cell by the MOM-5 protein, a *C. elegans* version of the Wnt receptor protein Frizzled. The result of this signaling cascade is to downregulate the expression of the *pop-1* gene in the EMS daughter destined to become the E cell. In *pop-1*-deficient embryos, both EMS daughter cells become E cells (Lin et al. 1995; Park et al. 2004).

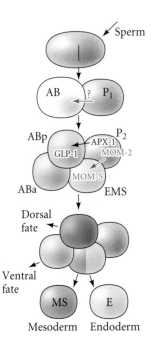

FIGURE 8.47 Cell-cell signaling in the 4-cell embryo of *C. elegans*. The P2 cell produces two signals: (1) the juxtacrine protein APX-1 (Delta), which is bound by GLP-1 (Notch) on the ABp cell, and (2) the paracrine protein MOM-2 (Wnt), which is bound by the MOM-5 (Frizzled) protein on the EMS cell. (After Han 1998.)

The P2 cell is also critical in giving the signal that distinguishes ABp from its sister, ABa (see Figure 8.47). ABa gives rise to neurons, hypodermis, and the anterior pharynx cells, while ABp makes only neurons and hypodermal cells. However, if one experimentally reverses the positions of these two cells, their fates are similarly reversed and a normal embryo forms. In other words, ABa and ABp are equivalent cells whose fate is determined by their positions within the embryo (Priess and Thomson 1987). Transplantation and genetic studies have shown that ABp becomes different from ABa through its interaction with the P2 cell. In an unperturbed embryo, both ABa and ABp contact the EMS blastomere, but only ABp contacts the P2 cell. If the P2 cell is killed at the early 4-cell stage, the ABp cell does not generate its normal complement of cells (Bowerman et al. 1992a,b). Contact between ABp and P2 is essential for the specification of ABp cell fates, and the ABa cell can be made into an ABp-type cell if it is forced into contact with P2 (Hutter and Schnabel 1994; Mello et al. 1994).

This interaction is mediated by the GLP-1 protein on the ABp cell and the APX-1 (anterior pharynx excess) protein on the P2 blastomere. In embryos whose mothers have mutant *glp-1*, ABp is transformed into an ABa cell (Hutter and Schnabel 1994; Mello et al. 1994). The GLP-1 protein is a member of a widely conserved family called the Notch proteins, which serve as cell membrane receptors in many cell-cell interactions; it is seen on both the ABa and ABp cells (Evans et al. 1994).* As mentioned in Chapter 5, one of the most important ligands for Notch proteins such as GLP-1 is another cell surface protein called Delta. In *C. elegans*, the Delta-like protein is APX-1, and it is found on the P2 cell (Mango et al. 1994a; Mello et al. 1994). This APX-1 signal breaks the symmetry between ABa and ABp, since it stimulates the GLP-1 protein solely on the AB descendant that it touches—namely, the ABp blastomere. In doing this, the P2 cell initiates the dorsal-ventral axis of *C. elegans* and confers on the ABp blastomere a fate different from that of its sister cell.

Integration of autonomous and conditional specification: Differentiation of the *C. elegans* pharynx

It should become apparent from the above discussion that the pharynx is generated by two sets of cells. One group of pharyngeal precursors comes from the EMS cell and is dependent on the maternal *skn-1* gene. The second group of pharyngeal precursors comes from the ABa blastomere and is dependent on GLP-1 signaling from the EMS cell. In both cases, the pharyngeal precursor cells (and only these cells) are instructed to activate the *pha-4* gene (Mango et al. 1994b). The *pha-4* gene encodes a transcription factor that resembles the mammalian HNF-3β protein. Microarray studies by Gaudet and Mango (2002) revealed that the PHA-4 transcription factor activates almost all of the pharynx-specific genes. It appears that the PHA-4 transcription factor may be the node that takes the maternal inputs and transforms them into a signal that transcribes the zygotic genes necessary for pharynx development.

Gastrulation in *C. elegans*

Gastrulation in *C. elegans* starts extremely early, just after the generation of the P4 cell in the 24-cell embryo (Skiba and Schierenberg 1992). At this time, the two daughters of the E cell (Ea and Ep) migrate from the ventral side into the center of the embryo. There they divide to form a gut consisting of 20 cells. There is a very small and transient blastocoel prior to the movement of the Ea and Ep cells, and their inward migration creates a tiny blastopore. The next cell to migrate through this blastopore is the P4 cell, the precursor of the germ cells. It migrates to a position beneath the gut primordium. The mesodermal cells move

---

*The GLP-1 protein is localized in the ABa and ABp blastomeres, but the maternally encoded *glp-1* mRNA is found throughout the embryo. Evans and colleagues (1994) have postulated that there might be some translational determinant in the AB blastomere that enables the *glp-1* message to be translated in its descendants. The *glp-1* gene is also active in regulating postembryonic cell-cell interactions. It is used later by the distal tip cell of the gonad to control the number of germ cells entering meiosis; hence the name GLP, for "germ line proliferation."

(A)   (B)

(C)   (D)

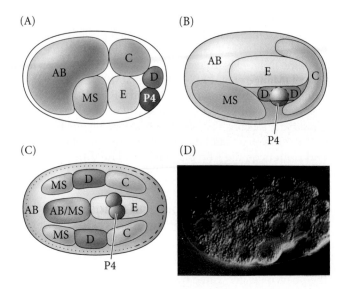

FIGURE 8.48   Gastrulation in *C. elegans*. (A) Positions of founder cells and their descendants at the 26-cell stage, just prior to gastrulation. (B) 102-cell stage, after the migration of the E, P4, and D descendants. (C) Positions of the cells near the end of gastrulation. The dotted and dashed lines represent regions of the hypodermis contributed by AB and C, respectively. (D) Early gastrulation, as the two E cells start moving inward. (After Schierenberg 1997; photograph courtesy of E. Schierenberg.)

in next: the descendants of the MS cell migrate inward from the anterior side of the blastopore, and the C- and D-derived muscle precursors enter from the posterior side. These cells flank the gut tube on the left and right sides (Figure 8.48; Schierenberg 1997). Finally, about 6 hours after fertilization, the AB-derived cells that contribute to the pharynx are brought inside, while the hypoblast (hypoder-

mal precursor) cells move ventrally by epiboly, eventually closing the blastopore. The two sides of the hypodermis are sealed by E-cadherin on the tips of the leading cells that meet at the ventral midline (Raich et al. 1999). During the next 6 hours, the cells move and develop into organs, while the ball-shaped embryo stretches out to become a worm with 558 somatic cells (see Priess and Hirsh 1986; Schierenberg 1997). An additional 115 cells will have formed but undergone apoptosis (programmed cell death; see Chapter 6). After four molts, the worm is a sexually mature, hermaphroditic adult, containing exactly 959 somatic cells as well as numerous sperm and eggs.

The *C. elegans* research program integrates genetics and embryology to provide an understanding of the networks that govern cell differentiation and morphogenesis. In addition to providing some remarkable insights into how gene expression can change during development, studies of *C. elegans* have also humbled us by demonstrating how complex these networks are. Even in an organism as "simple" as *C. elegans*, with only a few genes and cell types, the right side of the body is made in a different manner from the left! The identification of the genes mentioned above is just the beginning of our effort to understand the complex interacting systems of development.

Coda

This chapter has described early embryonic development in four invertebrate species, each of which develops in a different pattern. The largest group of animals on this planet, however, is another invertebrate group—the insects. We probably know more about the development of one particular insect, *Drosophila melanogaster*, than any other organism. The next chapter details the early development of this particularly well-studied creature.

 **Early Invertebrate Development**

1. During cleavage, most cells do not grow. Rather, the volume of the oocyte is cleaved into numerous cells. The major exceptions to this rule are mammals.

2. "Blast" vocabulary: A *blastomere* is a cell derived from cleavage in an early embryo. A *blastula* is an embryonic structure composed of blastomeres. The cavity within the blastula is the *blastocoel*. If the blastula lacks a blastocoel, it is a *stereoblastula*. (A mammalian blastula is called a *blastocyst*; see Chapter 11.) The invagination where gastrulation begins is the *blastopore*.

3. The blastomere cell cycle is governed by the synthesis and degradation of cyclin B. Cyclin B synthesis promotes the formation of mitosis-promoting factor, and MPF promotes mitosis. Degradation of cyclin B

brings the cell back to the S phase. The G phases are added at the mid-blastula transition.

4. The movements of gastrulation include invagination, involution, ingression, delamination, and epiboly.

5. Three axes form the foundations of the body: the anterior-posterior axis (head to tail or mouth to anus); the dorsal-ventral axis (back to belly); and the right-left axis (the two lateral sides of the body).

6. Body axes are established in different ways in different species. In some, such as the sea urchin and tunicate, the axes are established at fertilization through determinants in the egg cytoplasm In others, such as the nematode and snail, the axes are established by cell interactions later in development.

7. In all four invertebrates described, cleavage is holoblastic. In the sea urchin, cleavage is radial; in the snail, spiral; in the tunicate, bilateral; and in the nematode, rotational.

8. In the sea urchin and in tunicates, gastrulation occurs only after thousands of cells have formed. The blastopore becomes the anus and the mouth is formed elsewhere; this deuterostome mode of gastrulation is also characteristic of chordates (including vertebrates). In the snail and in *C. elegans*, gastrulation occurs when there are relatively few cells, and the blastopore becomes the mouth. This is the protostome mode of gastrulation.

9. Sea urchin cell fates are determined by cell-cell signaling. The micromeres constitute a major signaling center. Nuclear β-catenin is important for the inducing capacity of the micromeres.

10. Differential cell adhesion is important in regulating sea urchin gastrulation. The micromeres detach first from the vegetal plate and move into the blastocoel. They form the primary mesenchyme, which becomes the skeletal rods of the pluteus larva. The vegetal plate invaginates to form the endodermal archenteron, with a tip of secondary mesenchyme cells. The archenteron elongates by convergent extension and is guided to the future mouth region by the secondary mesenchyme.

11. Snails exhibit spiral cleavage and form stereoblastulae, with no blastocoels. The direction of spiral cleavage is regulated by a factor encoded by the mother and placed in the oocyte. Spiral cleavage can be modified by evolution, and adaptations of spiral cleavage have allowed some molluscs to survive in otherwise harsh environments.

12. The polar lobe of certain molluscs contains the morphogenetic determinants for mesoderm and endoderm. These determinants enter the D blastomere.

13. The tunicate fate map is identical on its right and left sides. The yellow cytoplasm contains muscle-forming determinants; these act autonomously. The nervous system of tunicates is formed conditionally, by interactions between blastomeres.

14. The soil nematode *Caenorhabditis elegans* was chosen as a model organism because it has a small number of cells, has a small genome, is easily bred and maintained, has a short lifespan, can be genetically manipulated, and has a cuticle through which one can see cell movements.

15. In the early divisions of the *C. elegans* zygote, one daughter cell becomes a founder cell (producing differentiated descendants) and the other becomes a stem cell (producing other founder cells and the germ line).

16. Blastomere identity in *C. elegans* is regulated by both autonomous and conditional specification.

For Further Reading

Complete bibliographical citations for all literature cited in this chapter can be found on the Vade Mecum CD that accompanies the book and at the free access website www.devbio.com

Clement, A. C. 1962. Development of *Ilyanassa* following removal of the D micromere at successive cleavage stages. *J. Exp. Zool.* 149: 193–215.

Davidson, E. H. and 24 others. 2002. A provisional regulatory gene network for specification of endomesoderm in the sea urchin embryo. *Dev. Biol.* 246: 162–190. Ongoing updates at http://sugp.caltech.edu/endomes/

Fink, R. D. and D. R. McClay. 1985. Three cell recognition changes accompany the ingression of sea urchin primary mesenchyme cells. *Dev. Biol.* 107: 66–74.

Hörstadius, S. 1939. The mechanics of sea urchin development, studied by operative methods. *Biol. Rev.* 14: 132–179.

Logan, C. Y., J. R. Miller, M. J. Ferkowicz and D. R. McClay. 1998. Nuclear β-catenin is required to specify vegetal cell fates in the sea urchin embryo. *Development* 126: 345–358.

Nishida, H. and K. Sawada. 2001. *macho-1* encodes a localized mRNA in ascidian eggs that specifies muscle fate during embryogenesis. *Nature* 409: 724–729.

Shibazaki, Y., M. Shimizu and R. Kuroda. 2004. Body handedness is directed by genetically determined cytoskeletal dynamics in the early embryo. *Curr. Biol.* 14: 1462–1467.

Sulston, J. E., J. Schierenberg, J. White and N. Thomson. 1983. The embryonic cell lineage of the nematode *Caenorhabditis elegans*. *Dev. Biol.* 100: 64–119.

The Genetics of Axis Specification in *Drosophila*

9

THANKS LARGELY TO STUDIES spearheaded by Thomas Hunt Morgan's laboratory during the first two decades of the twentieth century, we know more about the genetics of *Drosophila* than that of any other multicellular organism. The reasons have to do with both the flies themselves and with the people who first studied them. *Drosophila* is easy to breed, hardy, prolific, tolerant of diverse conditions, and the polytene chromosomes of their larvae (see Figure 4.11) are readily identified. The techniques for breeding and identifying fruit fly mutants are easy to learn. Moreover, the progress of *Drosophila* genetics was aided by the relatively free access of every scientist to the mutants and fly breeding techniques of every other researcher. Mutants were considered the property of the entire scientific community, and Morgan's laboratory established a database and exchange network whereby anyone could obtain them.

Undergraduates (starting with Calvin Bridges and Alfred Sturtevant) played important roles in *Drosophila* research, which achieved its original popularity as a source of undergraduate research projects. As historian Robert Kohler noted (1994), "Departments of biology were cash poor but rich in one resource: cheap, eager, renewable student labor." The *Drosophila* genetics program was "designed by young persons to be a young person's game," and the students set the rules for *Drosophila* research: "No trade secrets, no monopolies, no poaching, no ambushes."

But *Drosophila* was a difficult organism on which to study embryology. Although Jack Schultz (originally in Morgan's laboratory) and others following him attempted to relate the genetics of *Drosophila* to its development, the fly embryos proved too complex and intractable to study, being neither large enough to manipulate experimentally nor transparent enough to observe. It was not until the techniques of molecular biology allowed researchers to identify and manipulate the genes and RNAs of the insect that its genetics could be related to its development. And when that happened, a revolution occurred in the field of biology. The merging of our knowledge of the molecular aspects of *Drosophila* genetics with our knowledge of the fly's development built the foundations on which the current sciences of developmental genetics and evolutionary developmental biology are based.

EARLY *DROSOPHILA* DEVELOPMENT

In the last chapter, we discussed the specification of early embryonic cells by cytoplasmic determinants stored in the oocyte. The cell membranes that form during cleavage establish the region of cytoplasm incorporated into each new blastomere, and the incorporated morphogenetic determinants then direct dif-

> *Those of us who are at work on Drosophila find a particular point to the question. For the genetic material available is all that could be desired, and even embryological experiments can be done. ... It is for us to make use of these opportunities. We have a complete story to unravel, because we can work things from both ends at once.*
>
> JACK SCHULTZ (1935)

> *The chief advantage of Drosophila initially was one that historians have overlooked: it was an excellent organism for student projects.*
>
> ROBERT E. KOHLER (1994)

ferential gene expression in each cell. During *Drosophila* development, however, cellular membranes do not form until after the thirteenth nuclear division. Prior to this time, all the dividing nuclei share a common cytoplasm, and material can diffuse throughout the embryo. In these embryos, the specification of cell types along the anterior-posterior and dorsal-ventral axes is accomplished by the interactions of cytoplasmic materials *within* the single multinucleated cell. Moreover, the initiation of the anterior-posterior and dorsal-ventral differences is controlled by the position of the egg within the mother's ovary. Whereas the sperm entry site may fix the axes in ascidians and nematodes, the fly's anterior-posterior and dorsal-ventral axes are specified by interactions between the egg and its surrounding follicle cells.

Fertilization

Drosophila fertilization is not your standard sperm-meets-egg story. First, the sperm enters an egg that is already activated. Egg activation in *Drosophila* is accomplished at ovulation, a few minutes *before* fertilization begins. The oocyte nucleus has resumed its meiotic division and the cytoplasm has begun translation from stored mRNAs (Mahowald et al. 1983; Fitch and Wakimoto 1998; Heifetz et al. 2001). Second, by the time the sperm enters the egg, the egg already has begun to specify its axes; thus the sperm enters an egg that is already organizing itself as an embryo. Third, there is only one site—the **micropyle**, at the future dorsal anterior region of the embryo—where the sperm can enter the egg. The micropyle is a tunnel in the chorion (eggshell) that allows sperm to pass through it one at a time. The micropyle probably prevents polyspermy in *Drosophila*. There are no cortical granules to block polyspermy, although cortical changes are seen. Fourth, there is competition between sperm, and a sperm can be many times longer than the

adult fly. In *Drosophilia melanogaster*, the sperm tail is 1.8 mm—about as long as the adult fly, and some 300 times longer than a human sperm. This huge tail is thought to block other sperm from entering the egg. The entire sperm (huge tail and all) gets incorporated into the oocyte cytoplasm, and the sperm cell membrane does not break down until after it is fully inside the oocyte (Clark et al. 1999; Snook and Karr 1998).

WEBSITE **9.1** *Drosophila* **fertilization.** Fertilization of *Drosophila* can only occur in the region of the oocyte that will become the anterior of the embryo. Moreover, the sperm tail appears to stay in this region.

Cleavage

Most insect eggs undergo **superficial cleavage**, wherein a large mass of centrally located yolk confines cleavage to the cytoplasmic rim of the egg. One of the fascinating features of this cleavage pattern is that cells do not form until after the nuclei have divided several times. Cleavage in a *Drosophila* egg is shown in Figure 9.1. The zygote nucleus undergoes several mitotic divisions within the central portion of the egg; 256 nuclei are produced by a series of eight nuclear divisions averaging 8 minutes each. The nuclei then migrate to the periphery of the egg, where the mitoses continue, albeit at a progressively slower rate. During the ninth division cycle, about five nuclei reach the surface of the posterior pole of the embryo. These nuclei become enclosed by cell membranes and generate the **pole cells** that give rise to the gametes of the adult. Most of the other nuclei arrive at the periphery of the embryo at cycle 10 and then undergo four more divisions at progressively slower rates. During these stages of nuclear division, the embryo is called a **syncytial blastoderm**, meaning that all the cleav-

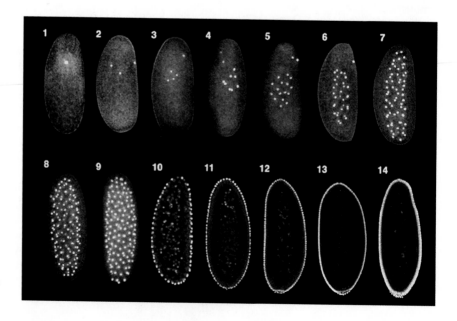

FIGURE 9.1 Laser confocal micrographs of stained chromatin showing superficial cleavage in a *Drosophila* embryo. The early nuclear divisions occur centrally. Numbers refer to the cell division cycle. At division cycle 10 (512-nucleus stage, 2 hours after fertilization), the pole cells form in the posterior, and the nuclei and their cytoplasmic islands (energids) migrate to the periphery of the cell. This creates the syncytial blastoderm. After cycle 13, the oocyte membranes ingress between the nuclei to form the cellular blastoderm. (Photographs courtesy of D. Daily and W. Sullivan.)

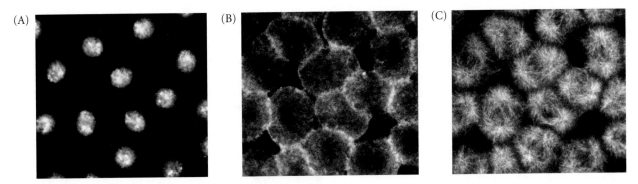

FIGURE 9.2 Localization of the cytoskeleton around nuclei in the syncytial blastoderm of *Drosophila*. A *Drosophila* embryo entering the mitotic prophase of its twelfth division was sectioned and triple-stained. (A) The nuclei were localized by a dye that binds to DNA. (B) Microfilaments were identified using a fluorescent antibody to actin. (C) Microtubules were recognized by a fluorescent antibody to tubulin. Cytoskeletal domains can be seen surrounding each nucleus. (From Karr and Alberts 1986; photographs courtesy of T. L. Karr.)

age nuclei are contained within a common cytoplasm. No cell membranes exist other than that of the egg itself.

The nuclei divide within a common cytoplasm, but this does not mean that the cytoplasm is itself uniform. Karr and Alberts (1986) have shown that each nucleus within the syncytial blastoderm is contained within its own little territory of cytoskeletal proteins. When the nuclei reach the periphery of the egg during the tenth cleavage cycle, each nucleus becomes surrounded by microtubules and microfilaments (Figure 9.2). The nuclei and their associated cytoplasmic islands are called **energids**. Following division cycle 13, the oocyte plasma membrane folds inward between the nuclei, eventually partitioning off each somatic nucleus into a single cell. This process creates the **cellular blastoderm**, in which all the cells are arranged in a single-layered jacket around the yolky core of the egg (Figure 9.3A; Turner and Mahowald 1977; Foe and Alberts 1983). Like any other cell formation, the formation of the cellular blastoderm involves a delicate interplay between microtubules and microfilaments (Figure 9.3B,C). The first phase

(A)

Egg surface

Mitotic spindle

Cleavage furrow

Aster

Nucleus

Furrow canal

Microtubules

Yolk membrane

FIGURE 9.3 Formation of the cellular blastoderm in *Drosophila*. (A) Developmental series showing progressive cellularization. (B) Confocal fluorescence photomicrographs of nuclei dividing during the cellularization of the blastoderm. While there are no cell boundaries, actin (green) can be seen forming regions within which each nucleus divides. The microtubules of the mitotic apparatus are stained red with antibodies to tubulin. (C) Cross section during cellularization. As the cells form, the domain of the actin expands into the egg. The DAH protein is in the front of the cleavage furrow and is critical for the progression of the cleavage furrow. The DNA of the nucleus is stained blue, and the tubulin at the periphery is stained green. (A after Fullilove and Jacobson 1971; B from Sullivan et al. 1993, courtesy of E. Theurkauf and W. Sullivan; C from Zhang et al. 1996, courtesy of T. Hsieh.)

(B)

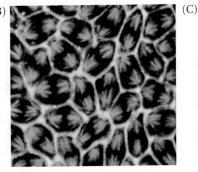

(C)

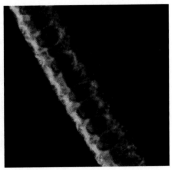

of blastoderm cellularization is characterized by the invagination of cell membranes between the nuclei to form furrow canals. This process can be inhibited by drugs that block microtubules. After the furrow canals have passed the level of the nuclei, the second phase of cellularization occurs. Here, the rate of invagination increases, and the actin-membrane complex begins to constrict at what will be the basal end of the cell (Schejter and Wieschaus 1993; Foe et al. 1993; Mazumdar and Mazumdar 2002). In *Drosophila*, the cellular blastoderm consists of approximately 6000 cells and is formed within 4 hours of fertilization.

The mid-blastula transition

After the nuclei reach the periphery, the time required to complete each of the next four divisions becomes progressively longer. While cycles 1–10 are each 8 minutes long, cycle 13, the last cycle in the syncytial blastoderm, takes 25 minutes to complete. Cycle 14, in which the *Drosophila* embryo forms cells (i.e., after 13 divisions), is asynchronous. Some groups of cells complete this cycle in 75 minutes, whereas others take 175 minutes (Figure 9.4; Foe 1989). Transcription from the nuclei (which begins around the eleventh cycle) is greatly enhanced at this stage. This slowdown of nuclear division and the concomitant increase in RNA transcription is often referred to as the **mid-blastula transition** (see Chapter 8). Such a transition is seen in the embryos of numerous vertebrate and invertebrate phyla.

The control of the mid-blastula transition (most obvious in *Xenopus* and *Drosophila* embryos) appears to be effected by the ratio of chromatin to cytoplasm (Newport and Kirschner 1982; Edgar et al. 1986a). Edgar and his colleagues compared the early development of wild-type *Drosophila* embryos with that of a haploid mutant. These haploid *Drosophila* embryos have half the wild-type quantity of chromatin at each cell division. Hence a haploid embryo

at cell division cycle 8 has the same amount of chromatin that a wild-type embryo has at cycle 7. The investigators found that, whereas wild-type embryos formed a cellular blastoderm immediately after the thirteenth division, haploid embryos underwent an extra, fourteenth, division before cellularization. Moreover, the lengths of cycles 11–14 in wild-type embryos corresponded to those of cycles 12–15 in the haploid embryos. Thus, the haploid embryos follow a pattern similar to that of the wild-type embryos—but they lag by one cell division.

WEBSITE 9.2 The regulation of *Drosophila* cleavage. The control of the cell cycle in *Drosophila* is a story of how the zygote nucleus gradually takes control from the mRNAs and proteins stored in the oocyte cytoplasm.

WEBSITE 9.3 The early development of other insects. *Drosophila* is a highly derived species. There are other insect species that develop in ways very different from the "standard" fruit fly.

Gastrulation

Gastrulation begins shortly after the mid-blastula transition. The first movements of *Drosophila* gastrulation segregate the presumptive mesoderm, endoderm, and ectoderm. The prospective mesoderm—about 1000 cells constituting the ventral midline of the embryo—folds inward to produce the **ventral furrow** (Figure 9.5A). This furrow eventually pinches off from the surface to become a ventral tube within the embryo. The prospective endoderm invaginates to form two pockets at the anterior and posterior ends of the ventral furrow. The pole cells are internalized along with the endoderm (Figure 9.5B,C). At this time, the embryo bends to form the **cephalic furrow**.

The ectodermal cells on the surface and the mesoderm undergo convergence and extension, migrating toward the ventral midline to form the **germ band**, a collection of cells along the ventral midline that includes all the cells that will form the trunk of the embryo. The germ band extends posteriorly and, perhaps because of the egg case, wraps around the top (dorsal) surface of the embryo (Figure 9.5D).

FIGURE 9.4 Differences in regional rates of cell division in *Drosophila* embryos. (A) Early gastrula embryo is stained with fluorescent antibodies to tubulin to show the microtubules of the mitotic spindles. (B) Antibodies to the cyclin A protein show that it is degraded after mitosis and is not seen in those regions undergoing cell division. (From Edgar and O'Farrell 1989; photographs courtesy of B. A. Edgar.)

(A)

(B)

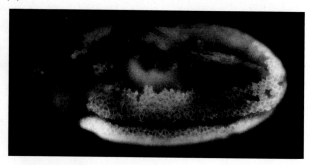

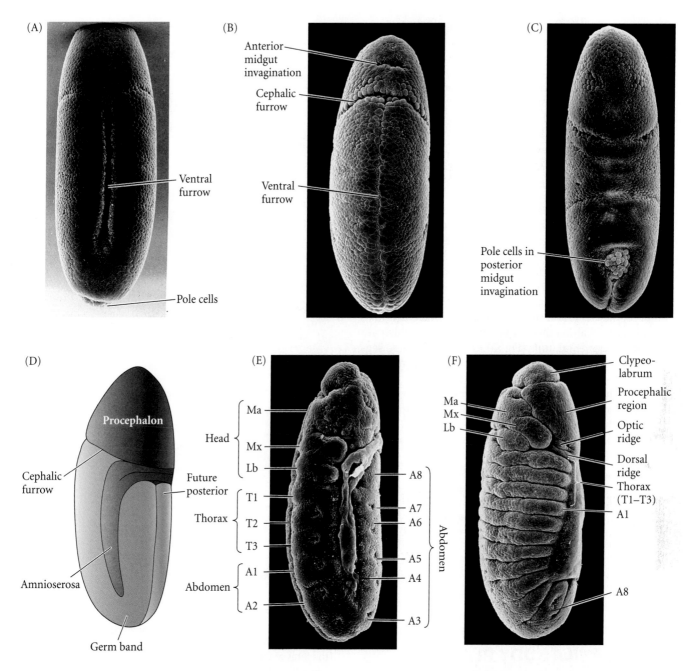

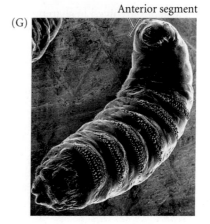

FIGURE 9.5 Gastrulation in *Drosophila*. (A) Ventral furrow beginning to form as cells flanking the ventral midline invaginate. (B) Closing of ventral furrow, with mesodermal cells placed internally and surface ectoderm flanking the ventral midline. (C) Dorsal view of a slightly older embryo, showing the pole cells and posterior endoderm sinking into the embryo. (D) Diagram of a dorsolateral view of *Drosophila* embryo at fullest germ band extension, just prior to segmentation. The cephalic furrow separates the future head region (procephalon) from the germ band, which will form the thorax and abdomen. (E) Lateral view, showing fullest extension of germ band and the beginnings of segmentation. Subtle indentations mark the incipient segments along the germ band. Ma, Mx, and Lb correspond to the mandibular, maxillary, and labial head segments; T1–T3 are the thoracic segments, and A1–A8 are the abdominal segments. (F) Germ band reversing direction. The true segments are now visible, as well as the other territories of the dorsal head, such as the clypeolabrum, procephalic region, optic ridge, and dorsal ridge. (G) Newly hatched first-instar larva. (Photographs courtesy of F. R. Turner. D after Campos-Ortega and Hartenstein 1985.)

Thus, at the end of germ band formation, the cells destined to form the most posterior larval structures are located immediately behind the future head region (Figure 9.5E). At this time, the body segments begin to appear, dividing the ectoderm and mesoderm. The germ band then retracts, placing the presumptive posterior segments at the posterior tip of the embryo (Figure 9.5F).

While the germ band is in its extended position, several key morphogenetic processes occur: organogenesis, segmentation, and the segregation of the imaginal discs* (see Figure 9.5E). In addition, the nervous system forms from two regions of ventral ectoderm. As described in Chapter 6, neuroblasts differentiate from this neurogenic ectoderm within each segment (and also from the nonsegmented region of the head ectoderm). Therefore, in insects like *Drosophila*, the nervous system is located ventrally, rather than being derived from a dorsal neural tube as in vertebrates (Figure 9.6).

The general body plan of *Drosophila* is the same in the embryo, the larva, and the adult, each of which has a distinct head end and a distinct tail end, between which are repeating segmental units (Figure 9.7). Three of these segments form the thorax, while another eight segments form the abdomen. Each segment of the adult fly has its own identity. The first thoracic segment, for example, has only legs; the second thoracic segment has legs and wings; and the third thoracic segment has legs and halteres (balancing organs). Thoracic and abdominal segments can also be distinguished from each other by differences in the cuticle of the newly hatched first instar larvae.

> **VADE MECUM²** *Drosophila* **development.** The CD-ROM contains some remarkable time-lapse sequences of *Drosophila* development, including cleavage and gastrulation. This segment also provides access to the fly life cycle. The color coding superimposed on the germ layers allows you to readily understand tissue movements. [Click on Fruit Fly]

GENES THAT PATTERN THE *DROSOPHILA* BODY PLAN

Most of the genes involved in shaping the larval and adult forms of the fly were identified in the early 1990's using a powerful "forward genetics" approach. The basic strategy was to randomly mutagenize flies and then screen for mutations that disrupted the normal formation of the body plan. Some of these mutations were quite fantastic, and included embryos and adult flies in which specific body structures were either missing or in the wrong place. These

*Imaginal discs are those cells set aside to produce the adult structures. The details of imaginal disc differentiation will be discussed in Chapter 18. For more information on *Drosophila* developmental anatomy, see Bate and Martinez-Arias 1993; Tyler and Schetzer 1996; and Schwalm 1997.

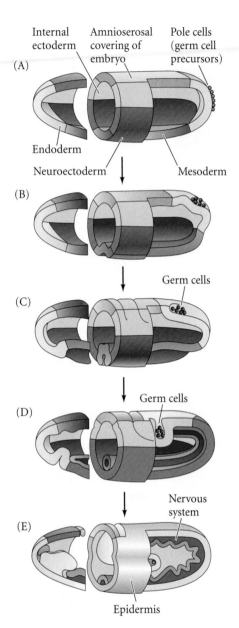

FIGURE 9.6 Schematic representation of gastrulation in *Drosophila*. (A) and (B) are surface and cutaway views showing the fates of the tissues immediately prior to gastrulation. (C) shows the beginning of gastrulation as the ventral mesoderm invaginates into the embryo. (D) corresponds to Figure 9.5A, while (E) corresponds to 9.5B and C. In (E), the neuroectoderm is largely differentiated into the nervous system and the epidermis. (After Campos-Ortega and Hartenstein 1985.)

mutant collections were distributed to many different laboratories. The genes involved in the mutant phenotypes were cloned and then characterized with respect to their expression patterns and their functions. This combined effort has led to a molecular understanding of body plan development in *Drosophila* that is unparalleled in all of biology, and in 1995 the work resulted in a Nobel Prize for

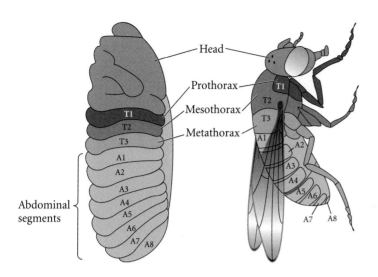

FIGURE 9.7 Comparison of larval and adult segmentation in *Drosophila*. The three thoracic segments can be distinguished by their appendages: T1 (prothorax) has legs only; T2 (mesothorax) has wings and legs; T3 (metathorax) has halteres and legs.

Edward Lewis, Christianne Nüsslein-Volhard, and Eric Wieschaus.

The rest of this chapter details the genetics of *Drosophila* development as we have come to understand it over the past two decades. First we examine how the dorsal-ventral and anterior posterior axes of the embryo are established by interactions between the developing oocyte and its surrounding follicle cells. Next we will see how dorsal-ventral patterning gradients are formed within the embryo, and how these gradients specify different tissue types. The third part of the discussion examines how segments are formed along the anterior-posterior axis, and how the different segments become specialized. Finally, we briefly show how the positioning of embryonic tissues along the two primary axes specifies these tissues to become particular organs.

Primary Axis Formation during Oogenesis

Anterior-posterior polarity in the oocyte

The processes of embryogenesis may "officially" begin at fertilization, but many of the molecular events critical for *Drosophila* embryogeneis actually occur during oogenesis. Each oocyte is descended from a single female germ cell—the **oogonium**—that is surrounded by an epithelium of follicle cells. Before oogenesis begins, the oogonium divides four times with incomplete cytokinesis, giving rise to 16 interconnected cells: 15 **nurse cells** and the single **oocyte precursor** (which is positioned at the future posterior end of the follicle; see Figure 19.4). As the oocyte precursor

develops, numerous mRNAs made in the nurse cells are transported on microtubules through the cellular interconnections into the enlarging oocyte.

The follicular epithelium surrounding the developing oocyte is initially uniform with respect to cell fate, but this uniformity is broken by two signals organized by the oocyte nucleus. Interestingly, both of these signals involve the same gene, *gurken*. The *gurken* message appears to be synthesized in the nurse cells, but becomes transported specifically to the oocyte nucleus. Here it is localized between the nucleus and the cell membrane and translated into Gurken protein (Cáceres and Nilson 2005). At this time the oocyte nucleus is very near the posterior tip of the follicle, and the Gurken signal is received by the follicle cells at that position through a receptor encoded by the *torpedo* gene* (Figure 9.8A). This signal results in the "posteriorization" of these follicle cells. As these follicle cells differentiate into dorsal follicle cells, they send a signal back into the oocyte. This signal reorganizes the cytoskeleton of the oocyte so that microtubules become oriented specifically with their minus (cap) and plus (growing) ends at the anterior and posterior ends of the oocyte, respectively (Figure 9.8B,C; Gonzalez-Reyes et al. 1995; Roth et al. 1995; Januschke et al. 2006).

This cytoskeletal rearrangement in the oocyte is accompanied by an increase in oocyte volume, owing to transfer of cytoplasmic components from the nurse cells. These components include maternal messengers such as the *bicoid* and *nanos* mRNAs. After the posterior follicle cells induce the rearrangement of the oocyte cytoskeleton, these mRNAs are carried by motor proteins along the microtubules to the anterior and posterior ends of the oocyte, respectively (Figure 9.8C,D). The protein products encoded by *bicoid* and *nanos* are critical for establishing the anterior-posterior polarity of the embryo (see Chapter 3).

Dorsal-ventral patterning in the oocyte

As the oocyte volume increases, the oocyte nucleus moves to an anterior dorsal position where a second major signaling event takes place. Here the *gurken* message becomes localized in a crescent between the oocyte nucleus and the oocyte cell membrane, and its protein product forms an anterior-posterior gradient along the dorsal surface of the oocyte (Figure 9.9; Neuman-Silberberg and Schüpbach 1993). Since it can diffuse only a short distance, Gurken protein reaches only those follicle cells closest to the oocyte nucleus, and it signals those cells to become the more columnar **dorsal follicle cells** (Montell et al. 1991; Schüpbach et al. 1991; see Figure 9.8). This establishes the dor-

---

*Molecular analysis has established that *gurken* encodes a homologue of the vertebrate epidermal growth factor (EGF), while *torpedo* encodes a homologue of the vertebrate EGF receptor (Price et al. 1989; Neuman-Silberberg and Schüpbach 1993).

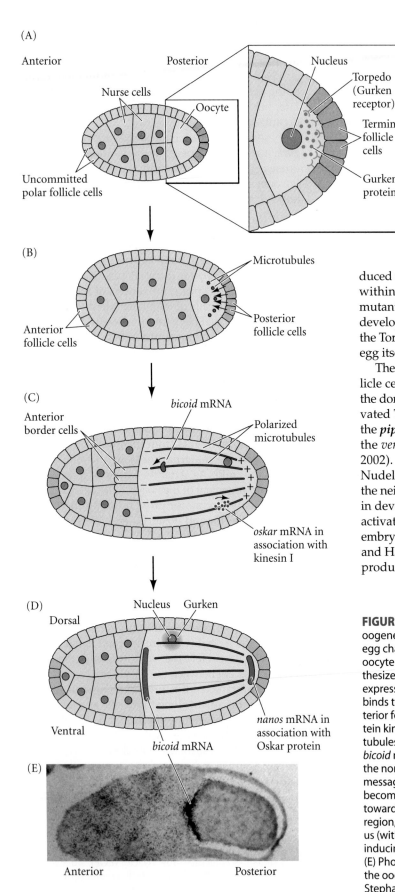

(A)

Anterior   Posterior

Nurse cells

Oocyte

Nucleus

Torpedo (Gurken receptor)

Terminal follicle cells

Gurken protein

Uncommitted polar follicle cells

(B)

Microtubules

Posterior follicle cells

Anterior follicle cells

(C)

bicoid mRNA

Anterior border cells

Polarized microtubules

oskar mRNA in association with kinesin I

(D)

Nucleus   Gurken

Dorsal

Ventral

bicoid mRNA

nanos mRNA in association with Oskar protein

(E)

Anterior   Posterior

sal-ventral polarity in the follicle cell layer that surrounds the growing oocyte.

Maternal deficiencies of either the *gurken* or the *torpedo* gene cause ventralization of the embryo. However, *gurken* is active only in the oocyte, while *torpedo* is active only in the follicle cells. This fact was revealed by experiments with germline/somatic chimeras. In one such experiment, Schüpbach (1987) transplanted germ cell precursors from wild-type embryos into embryos whose mothers carried the *torpedo* mutation. Conversely, she transplanted the germ cells from *torpedo* mutants into wild-type embryos (Figure 9.10). The wild-type eggs produced mutant, ventralized embryos when they developed within the *torpedo* mutant mothers' follicles. The *torpedo* mutant eggs were able to produce normal embryos if they developed within a wild-type ovary. Thus, unlike Gurken, the Torpedo protein is needed in the follicle cells, not in the egg itself.

The Gurken-Torpedo signal that specifies dorsalized follicle cells initiates a cascade of gene activities that creates the dorsal-ventral axis of the embryo (Figure 9.11). The activated Torpedo receptor protein inhibits the expression of the ***pipe*** gene. As a result, the Pipe protein is made only in the *ventral* follicle cells (Sen et al. 1998; Amiri and Stein 2002). Pipe (in some as yet unknown way) activates the Nudel protein, which is secreted to the cell membrane of the neighboring ventral embryonic cells. A few hours later in development, the activated Nudel protein initiates the activation of three serine proteases that are secreted by the embryo into the perivitelline fluid (see Figure 9.11C; Hong and Hashimoto 1995). These three serine proteases are the products of the *gastrulation defective* (*gd*), *snake* (*snk*), and

FIGURE 9.8 The anterior-posterior axis is specified during oogenesis. (A) The oocyte moves into the posterior region of the egg chamber, while nurse cells fill the anterior portion. The oocyte nucleus moves toward the terminal follicle cells and synthesizes Gurken protein (green). The terminal follicle cells express Torpedo, the receptor for Gurken. (B) When Gurken binds to Torpedo, the terminal follicle cells differentiate into posterior follicle cells and synthesize a molecule that activates protein kinase A in the egg. Protein kinase A orients the microtubules such that the growing end is at the posterior. (C) The *bicoid* message binds to dynein, a motor protein associated with the non-growing end of microtubules. Dynein moves the *bicoid* message to the anterior end of the egg. The *oskar* message becomes complexed with kinesin I, a motor protein that moves it toward the growing end of the microtubules at the posterior region, where Oskar can bind the *nanos* message. (D) The nucleus (with its Gurken protein) migrates along the microtubules, inducing the adjacent follicle cells to become the dorsal follicles. (E) Photomicrograph of *bicoid* mRNA (stained black) passing into the oocyte from the nurse cells during oogenesis. (E from Stephanson et al. 1988, photograph courtesy of the authors.)

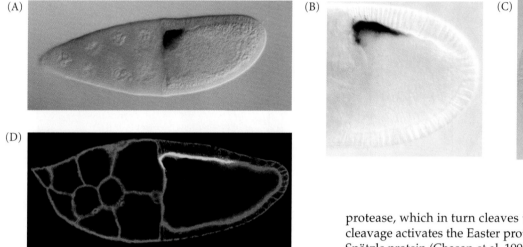

FIGURE 9.9 Expression of the *gurken* message and protein between the oocyte nucleus and the dorsal anterior cell membrane. (A) The *gurken* mRNA is localized between the oocyte nucleus and the dorsal follicle cells of the ovary. (B) The Gurken protein is similarly located (shown here is a younger stage than A). (C) Cross section of the egg through the region of Gurken protein expression. (D) A more mature oocyte, showing Gurken protein (yellow) across the dorsal region. The actin is stained red, showing cell boundaries. As the oocyte grows, follicle cells migrate across the top of the oocyte, becoming exposed to Gurken. (A from Ray and Schüpbach 1996, photograph courtesy of T. Schüpbach; B and C from Peri et al. 1999, photograph courtesy of S. Roth; D, photograph courtesy of C. van Buskirk and T. Schüpbach.)

protease, which in turn cleaves the Easter protein. This cleavage activates the Easter protease, which cleaves the Spätzle protein (Chasan et al. 1992; Hong and Hashimoto 1995; LeMosy et al. 2001).

It is obviously important that the cleavage of these proteins be limited to the most ventral portion of the embryo. This is accomplished by the secretion of a protease inhibitor from the follicle cells of the ovary. (Hashimoto et al. 2003; Ligoxygakis et al. 2003) This inhibitor of Easter and Snake is found throughout the perivitelline space surrounding the embryo. Indeed, this protein is very similar to the mammalian protease inhibitors that limit blood clotting protease cascades to the area of injury. In this way, the proteolytic cleavage of Easter and Spätzle is strictly limited to the area around the most ventral embryonic cells.

easter (*ea*) genes. Like most extracellular proteases, they are secreted in an inactive form and are activated by peptide cleavage. In a complex cascade of events, activated Nudel protein activates the Gastrulation-defective protease. This protease cleaves the Snake protein, activating the Snake

FIGURE 9.10 Germline chimeras made by interchanging pole cells (germ cell precursors) between wild-type embryos and embryos from mothers homozygous for a mutation of the *torpedo* gene. These transplants produce wild-type females whose eggs come from the mutant mothers, and *torpedo*-deficient females that lay wild-type eggs. The *torpedo*-deficient eggs produced normal embryos when they developed in the wild-type ovary, while the wild-type eggs produced ventralized embryos when they developed in the mutant mother's ovary

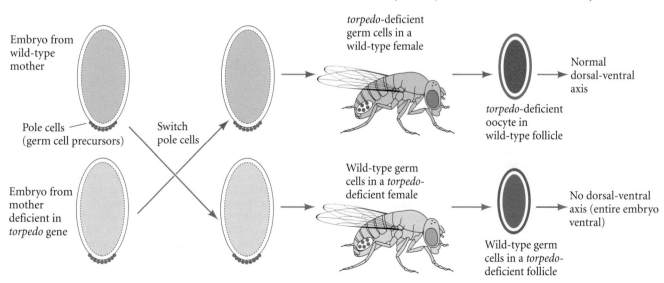

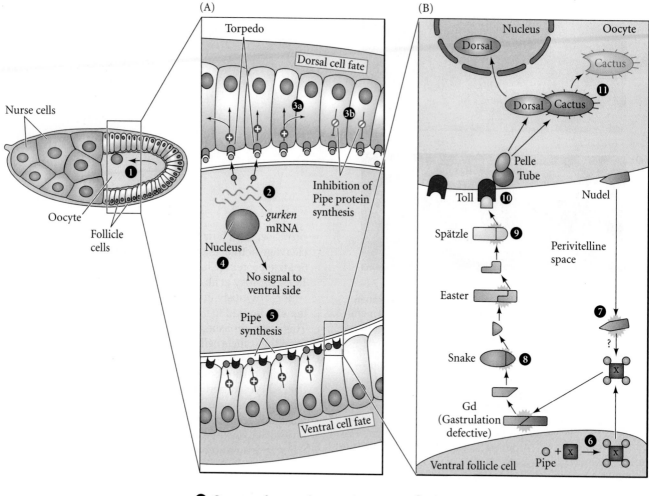

(A) (B)

❶ Oocyte nucleus travels to anterior dorsal side of oocyte where it localizes *gurken* mRNA.

❷ *gurken* messages are translated. Gurken is received by Torpedo proteins during mid-oogenesis.

❸a Torpedo signal causes follicle cells to differentiate to a dorsal morphology.

❸b Synthesis of Pipe is inhibited in dorsal follicle cells.

❹ Gurken does not diffuse to ventral side.

❺ Ventral follicle cells synthesize Pipe.

❻ In ventral follicle cells, Pipe completes the modification of an unknown factor (x).

❼ Nudel and factor (x) interact to split the Gastrulation-deficient (Gd) protein.

❽ Activated Gd splits the Snake protein, and activated Snake cleaves the Easter protein.

❾ Activated Easter splits Spätzle; activated Spätzle binds to Toll receptor protein.

❿ Toll activation activates Tube and Pelle, which phosphorylate the Cactus protein. Cactus is degraded, releasing it from Dorsal.

⓫ Dorsal protein enters the nucleus and ventralizes the cell.

FIGURE 9.11 Generating dorsal-ventral polarity in *Drosophila*. (A) The nucleus of the oocyte travels to what will become the dorsal side of the embryo. The *gurken* genes of the oocyte synthesize mRNA that becomes localized between the oocyte nucleus and the cell membrane, where it is translated into Gurken protein. The Gurken signal is received by the Torpedo receptor protein made by the follicle cells (see Figure 9.8). Given the short diffusibility of the signal, only the follicle cells closest to the oocyte nucleus (i.e., the dorsal follicle cells) receive the Torpedo signal, which causes the follicle cells to take on a characteristic dorsal follicle morphology and inhibits the synthesis of Pipe protein. Therefore, Pipe protein is made only by the *ventral* follicle cells. (B) The ventral region at a slightly later stage of develop-

ment. Pipe modifies an unknown protein (x) and allows it to be secreted from the ventral follicle cells. Nudel protein interacts with this modified factor to split the products of the gastrulation defective and snake genes to create an active enzyme that will split the inactive Easter zymogen into an active Easter protease. The Easter protease splits the Spätzle protein into a form that can bind to the Toll receptor (which is found throughout the embryonic cell membrane). This protease activity of Easter is strictly limited by the protease inhibitor found in the perivitelline space. Thus, only the ventral cells receive the Toll signal. This signal separates the Cactus protein from the Dorsal protein, allowing Dorsal to be translocated into the nuclei and ventralize the cells. (After van Eeden and St. Johnston 1999.)

The cleaved Spätzle protein is now able to bind to its receptor in the oocyte cell membrane, the product of the *toll* gene. Toll protein is a maternal product that is evenly distributed throughout the cell membrane of the egg (Hashimoto et al. 1988, 1991), but it becomes activated only by binding the Spätzle protein, which is produced only on the ventral side of the egg. Therefore, the Toll receptors on the ventral side of the egg are transducing a signal into the egg, while the Toll receptors on the dorsal side of the egg are not. This localized activation establishes the dorsal-ventral polarity of the oocyte.

Generating Dorsal-Ventral Pattern in the Embryo

Dorsal, the ventral morphogen

The protein that actually distinguishes dorsum (back) from ventrum (belly) in the embryo is the product of the *Dorsal* gene. The mRNA transcript of the mother's *Dorsal* gene is placed in the oocyte by the nurse cells. However, Dorsal protein is not synthesized from this maternal message until about 90 minutes after fertilization. When Dorsal is translated, it is found throughout the embryo, not just on the ventral or dorsal side. How can this protein act as a morphogen if it is located everywhere in the embryo?

In 1989, the surprising answer to this question was found (Roth et al. 1989; Rushlow et al. 1989; Steward 1989). While Dorsal is found throughout the syncytial blastoderm of the early *Drosophila* embryo, it is translocated into nuclei only in the ventral part of the embryo. In the nucleus, Dorsal protein acts as a transcription factor, binding to certain genes to activate or suppress their transcription. If Dorsal does not enter the nucleus, the genes responsible for specifying ventral cell types are not transcribed, the genes responsible for specifying dorsal cell types are not repressed, and all the cells of the embryo become specified as dorsal cells.

This model of dorsal-ventral axis formation in *Drosophila* is supported by analyses of maternal effect mutations that give rise to an entirely dorsalized or an entirely ventralized phenotype (Figure 9.12; Anderson and Nusslein-Volhard 1984). In those mutants in which all the cells are dorsalized (evident from their dorsal cuticle), Dorsal does not enter the nucleus in any cell. Conversely, in those mutants in which all cells have a ventral phenotype, Dorsal is found in every cell nucleus.*

Establishing a nuclear dorsal gradient

So how does the Dorsal protein enter into the nuclei only of the ventral cells? When Dorsal is first produced, it is complexed with a protein called Cactus in the cytoplasm of the

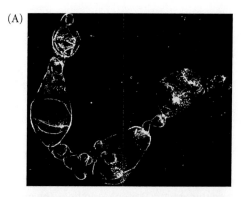

(A)

(B)

FIGURE 9.12 Effect of mutations affecting the distribution of the Dorsal protein. (A) Deformed larva consisting entirely of dorsal cells. Larvae like these developed from the eggs of a female homozygous for a mutation of the *snake* gene, one of the maternal effect genes involved in the signaling cascade that establishes a Dorsal gradient. (B) Larvae developed from *snake* mutant eggs that received injections of mRNA from wild-type eggs. These larvae have a wild-type appearance. (From Anderson and Nüsslein-Volhard 1984; photographs courtesy of C. Nüsslein-Volhard.)

syncytial blastoderm. As long as Cactus is bound to it, Dorsal remains in the cytoplasm. Dorsal enters ventral nuclei in response to a signaling pathway that frees it from Cactus (see Figure 9.11B). This separation of Dorsal from Cactus is initiated by the ventral activation of the Toll receptor. When Spätzle binds to and activates the Toll protein, Toll activates a protein kinase called Pelle. Another protein (Tube) is probably necessary for bringing Pelle to the cell membrane, where it can be activated (Galindo et al. 1995). The activated Pelle protein kinase (probably through an intermediate) can phosphorylate Cactus. Once phosphorylated, Cactus is degraded and Dorsal can enter the nucleus (Kidd 1992; Shelton and Wasserman 1993; Whalen and Steward 1993; Reach et al. 1996). Since Toll is activated by a gradient of Spätzle protein that is highest in the most ventral region, there is a corresponding gradient of Dorsal translocation in the ventral cells of the embryo, with the highest concentrations of Dorsal in the most ventral cell nuclei.[†]

---

*Remember that a gene in *Drosophila* is usually named after its mutant phenotype. Thus, the product of the *Dorsal* gene is necessary for the differentiation of ventral cells. That is, in the absence of the *Dorsal* gene, the ventral cells become dorsalized.

[†]The process described for the translocation of Dorsal protein into the nucleus is very similar to the process for the translocation of the NF-κB transcription factor into the nucleus of mammalian lymphocytes. In fact, there is substantial homology between NF-κB and Dorsal, between I-B and Cactus, between Toll and the interleukin 1 receptor, between Pelle and an IL-1-associated protein kinase, and between the DNA sequences recognized by Dorsal and by NF-κB (González-Crespo and Levine 1994; Cao et al. 1996). Thus, the biochemical pathway used to specify dorsal-ventral polarity in *Drosophila* appears to be homologous to that used to differentiate lymphocytes in mammals.

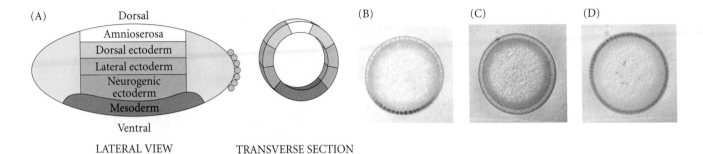

FIGURE 9.13 Translocation of Dorsal protein into ventral, but not lateral or dorsal, nuclei. (A) Fate map of a lateral cross section through the *Drosophila* embryo. The most ventral part becomes the mesoderm; the next higher portion becomes the neurogenic (ventral) ectoderm. The lateral and dorsal ectoderm can be distinguished in the cuticle, and the dorsalmost region becomes the amnioserosa, the extraembryonic layer that surrounds the embryo. (B–D) Transverse sections of embryos stained with antibody to show the presence of Dorsal protein (dark-stained area). (B) A wild-type embryo, showing Dorsal protein in the ventralmost nuclei. (C) A dorsalized mutant, showing no localization of Dorsal protein in any nucleus. (D) A ventralized mutant, in which Dorsal protein has entered the nucleus of every cell. (A after Rushlow et al. 1989; B–D from Roth et al. 1989, photographs courtesy of the authors.)

Effects of the dorsal protein gradient

What does the Dorsal protein do once it is located in the nuclei of the ventral cells? A look at the fate map of a cross section through the *Drosophila* embryo at the division cycle 14 makes it obvious that the 16 cells with the highest concentration of Dorsal are those that generate the mesoderm (Figure 9.13). The next cell up from this region generates the specialized glial and neural cells of the midline. The next two cells give rise to the ventral epidermis and ventral nerve cord, while the nine cells above them produce the dorsal epidermis. The most dorsal group of six cells generates the amnioserosal covering of the embryo (Ferguson and Anderson 1991).

This fate map is generated by the gradient of Dorsal protein in the nuclei. Large amounts of Dorsal instruct the cells to become mesoderm, while lesser amounts instruct the cells to become glial or ectodermal tissue (Jiang and Levine 1993). The first morphogenetic event of *Drosophila* gastrulation is the invagination of the 16 ventralmost cells of the embryo (Figure 9.14). All of the body muscles, fat bodies, and gonads derive from these mesodermal cells (Foe 1989). Dorsal specifies these cells to become mesoderm in two ways. First,

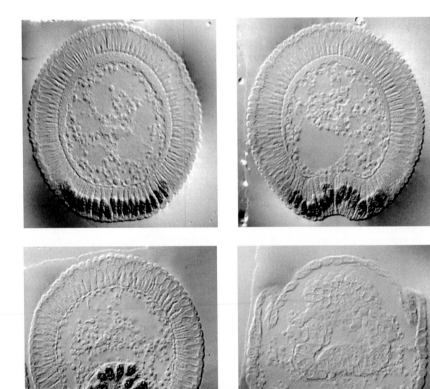

FIGURE 9.14 Gastrulation in *Drosophila*. In this cross section, the mesodermal cells at the ventral portion of the embryo buckle inward, forming a tube, which then flattens and generates the mesodermal organs. The nuclei are stained with antibody to the Twist protein. (From Leptin 1991a; photographs courtesy of M. Leptin.)

the protein activates specific genes that create the mesodermal phenotype. Five of the target genes for the Dorsal protein are *twist*, *snail*, *Fgf8* and its receptor, and *rhomboid* (Figure 9.15). These genes are transcribed only in nuclei that have received high concentrations of Dorsal, since their enhancers do not bind Dorsal with a very high affinity (Thisse et al. 1988, 1991; Jiang et al. 1991; Pan et al. 1991). Both Snail and Twist are also needed for the complete mesodermal phenotype and proper gastrulation (Leptin et al. 1991b). The Twist protein activates mesodermal genes, while the Snail protein represses particular non-mesodermal genes that might otherwise be active. The *rhomboid* and *fgf8* genes are interesting because they are activated by Dorsal but repressed by Snail. Thus, *rhomboid* and *fgf8* are not expressed in the most ventral cells (i.e., the mesodermal precursors), but are expressed in the cells adjacent to the

mesoderm. These *rhomboid*- and *fgf8*-expressing cells will become the presumptive neural ectoderm (see Figure 9.15).

The high concentration of Twist protein in the nuclei of the ventralmost cells activates the gene for the Fgf8 receptor (the product of the *heartless* gene) in the presumptive mesoderm (Jiang and Levine 1993; Gryzik and Müller 2004; Strathopoulos et al. 2004). The expression and secretion of Fgf8 by the presumptive neural ectoderm is received by its receptor on the mesoderm cells, causing these mesoderm cells to invaginate into the embryo and flatten against the ectoderm (see Figure 9.14).

Meanwhile, *intermediate* levels of nuclear Dorsal activate the transcription of the *Short gastrulation* (*Sog*) gene in two lateral stripes that flank the ventral *twist* expression domain, each 12–14 cells wide (François et al. 1994; Srinivasan et al. 2002). *Sog* encodes a protein that prevents the ectoderm in this region from becoming hypodermis and begins the processes of neural differentiation. After the mesodermal cells invaginate, these neural ectoderm cells become positioned at the bottom of the embryo (Figure 9.16).

The Dorsal protein also determines the mesoderm indirectly. In addition to activating the mesoderm-stimulating genes (*twist* and *snail*), it directly inhibits the dorsalizing genes *zerknüllt* (*zen*) and *decapentaplegic* (*dpp*). Thus, in the same cells, Dorsal can act as an activator of some genes and a repressor of others. Whether Dorsal activates or represses a given gene depends on the structure of the genes' enhancers. The *zen* enhancer has a silencer region that contains a binding site for Dorsal as well as a second binding site for two other DNA-binding proteins. These two other proteins enable Dorsal to bind a transcriptional repressor protein (Groucho) and bring it to the DNA (Valentine et al. 1998). Mutants of *Dorsal* express *dpp* and *zen* genes

(A) DORSAL PATTERNING

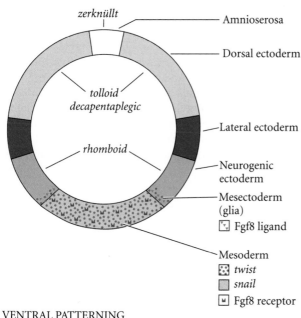

(B) VENTRAL PATTERNING

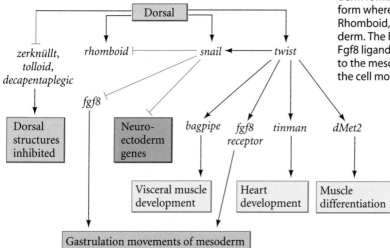

FIGURE 9.15 Subdivision of the *Drosophila* dorsal-ventral axis by the gradient of Dorsal protein in the nuclei. (A) Dorsal protein activates the zygotic genes *rhomboid*, *twist*, *fgf8*, *fgf8 receptor*, and *snail*, depending on its nuclear concentration. The mesoderm forms where Twist and Snail are present, and the glial cells form where Twist and Rhomboid interact. Those cells with Rhomboid, but no Snail or Twist, form the neurogenic ectoderm. The Fgf receptor is expressed in the mesoderm, and the Fgf8 ligands for this receptor are expressed in the glia, adjacent to the mesoderm. The binding of Fgf8 to its receptor will trigger the cell movements required for the ingression of the mesoderm. (B) Interactions in the specification of the ventral portion of the *Drosophila* embryo. Dorsal protein inhibits those genes that would give rise to dorsal structures (*tolloid*, *decapentaplegic*, and *zerknüllt*) while activating the three ventral genes. Snail protein, formed most ventrally, inhibits the transcription of *rhomboid* and prevents ectoderm formation. Twist activates *dMet2* and *bagpipe* (which activate muscle differentiation) as well as *tinman* (heart muscle development). (A after Steward and Govind 1993; B after Furlong et al. 2001; Leptin and Affolter 2004.)

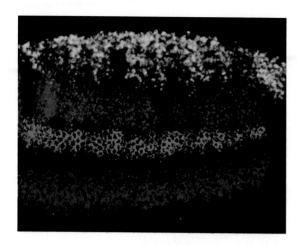

FIGURE 9.16 Dorsal-ventral patterning in *Drosophila*. The readout of the Dorsal gradient can be seen in the anterior region of a wholemount stained embryo. The expression of the most ventral gene, *ventral nervous system defective* (blue), is from the neurogenic ectoderm. The *intermediate neuroblast defective* gene (green) is expressed in lateral ectoderm. Red represents the *muscle-specific homeobox* gene, expressed in the mesoderm above the intermediate neuroblasts. The dorsalmost tissue expresses *decapentaplegic* (yellow). (From Kosman et al. 2004; courtesy of D. Kosman and E. Bier.)

throughout the embryo (Rushlow et al. 1987), and embryos deficient in *dpp* and *zen* fail to form dorsal structures (Irish and Gelbart 1987). Thus, in wild-type embryos, the mesodermal precursors express *twist* and *snail* (but not *zen* or *dpp*); precursors of the dorsal epidermis and amnioserosa express *zen* and *dpp* (but not *twist* or *snail*). Glial (mesectoderm) precursors express *twist* and *rhomboid*, while the lateral neural ectodermal precursors do not express any of these four genes (Kosman et al. 1991; Ray and Schüpbach 1996). By the cellular responses to the Dorsal protein gradient, the embryo becomes subdivided into mesoderm, neurogenic ectoderm, epidermis, and amnioserosa from the ventral to dorsal regions (see Figure 9.13A).

Segmentation and the Anterior-Posterior Body Plan

The genetic screens pioneered by Nüsslein-Volhard and Wieschaus identified a hierarchy of genes that establish anterior posterior polarity, and divide the embryo into a specific number of segments with different identities (Figure 9.17). This hierarchy is initiated by **maternal effect genes** that produce messenger RNAs that are placed in different regions of the egg. These messages encode transcriptional and translational regulatory proteins that diffuse through the syncytial blastoderm and activate or repress the expression of certain zygotic genes. The first zygotic genes to be expressed are called **gap genes** (because mutations in them cause gaps in the segmentation pattern). These genes are expressed in certain broad (about three

segments wide), partially overlapping domains. Differing combinations and concentrations of the gap gene proteins regulate the transcription of **pair-rule genes**, which divide the embryo into periodic units. The transcription of the different pair-rule genes results in a striped pattern of seven transverse bands perpendicular to the anterior-posterior axis. The pair-rule gene proteins activate the transcription of the **segment polarity genes**, whose mRNA and protein products divide the embryo into 14-segment-wide units, establishing the periodicity of the embryo. At the same time, the protein products of the gap, pair-rule, and segment polarity genes interact to regulate another class of genes, the **homeotic selector genes**, whose transcription determines the developmental fate of each segment.

Maternal gradients: Polarity regulation by oocyte cytoplasm

Classical embryological experiments demonstrated that there are at least two "organizing centers" in the insect egg, one in the anterior of the egg and one in the posterior. For instance, Klaus Sander (1975) found that if he ligated the egg early in development, separating the anterior half from the posterior half, one half developed into an anterior embryo and one half developed into a posterior embryo, but neither contained the middle segments of the embryo. The later in development the ligature was made, the fewer middle segments were missing. Thus it appeared that there were indeed morphogenetic gradients emanating from the two poles during cleavage, and that these gradients interacted to produce the positional information determining the identity of each segment.

Moreover, when the RNA in the anterior of insect eggs was destroyed (by either ultraviolet light or RNase), the resulting embryos lacked a head and thorax. Instead, these embryos developed two abdomens and telsons (tails) with mirror-image symmetry: telson-abdomen-abdomen-telson (Figure 9.18; Kalthoff and Sander 1968; Kandler-Singer and Kalthoff 1976). Sander's laboratory postulated the existence of a gradient at both ends of the egg, and hypothesized that the egg sequesters an mRNA that generates a gradient of anterior-forming material.

 WEBSITE 9.4 Evidence for gradients in insect development. The original evidence for gradients in insect development came from studies providing evidence for two "organization centers" in the egg, one located anteriorly and one located posteriorly.

The molecular model: Protein gradients in the early embryo

In the late 1980s, the gradient hypothesis was united with a genetic approach to the study of *Drosophila* embryogenesis. If there were gradients, what were the morphogens whose concentrations changed over space? What were the genes that shaped these gradients? And did these mor-

FIGURE 9.17 Generalized model of *Drosophila* anterior-posterior pattern formation. (A) The pattern is established by maternal effect genes that form gradients and regions of morphogenetic proteins. These morphogenetic determinants create a gradient of Hunchback protein that differentially activates the gap genes, which define broad territories of the embryo. The gap genes enable the expression of the pair-rule genes, each of which divides the embryo into regions about two segments wide. The segment polarity genes then divide the embryo into segment-sized units along the anterior-posterior axis. Together, the actions of these genes define the spatial domains of the homeotic genes that define the identities of each of the segments. In this way, periodicity is generated from nonperiodicity, and each segment is given a unique identity. (B) Maternal effect genes. The anterior axis is specified by the gradient of Bicoid protein (yellow through red). (C) Gap gene protein expression and overlap. The domain of Hunchback protein (orange) and the domain of Krüppel protein (green) overlap to form a region containing both transcription factors (yellow). (D) Products of the *fushi tarazu* pair-rule gene form seven bands across the embryo. (E) Products of the segment polarity gene *engrailed*, seen here at the extended germ band stage. (B courtesy of C. Nüsslein-Volhard; C courtesy of C. Rushlow and M. Levine; D courtesy of T. Karr; E courtesy of S. Carroll and S. Paddock.)

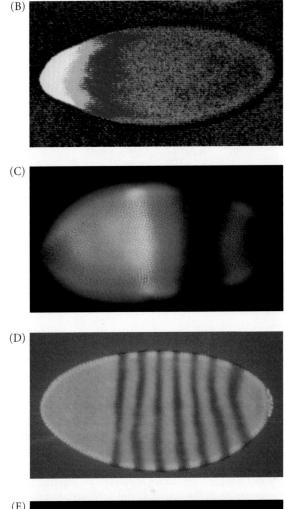

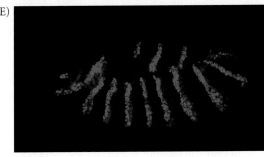

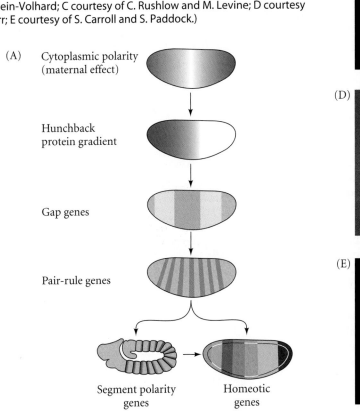

phogens act by activating or inhibiting certain genes in the areas where they were concentrated? Christiane Nüsslein-Volhard led a research program that addressed these questions. The researchers found that one set of genes encoded morphogens for the anterior part of the embryo, another set of genes encoded morphogens responsible for organizing the posterior region of the embryo, and a third set of genes encoded proteins that produced the terminal regions at both ends of the embryo (Table 9.1).

Two maternal messenger RNAs, *bicoid* and *nanos*, are most critical to the formation of the anterior-posterior axis.

The *bicoid* mRNAs are located near the anterior tip of the unfertilized egg, and *nanos* messages are located at the posterior tip (see Figure 9.8). These distributions occur as a result of the dramatic polarization of the microtubule networks in the developing oocyte. After ovulation and fertilization, the *bicoid* and *nanos* mRNAs are translated into proteins that can diffuse in the syncytial blastoderm, forming gradients that are critical for anterior posterior patterning.

BICOID mRNA LOCALIZATION AT THE ANTERIOR POLE OF THE OOCYTE.
The 3′ UTR of *bicoid* contains sequences that are critical for

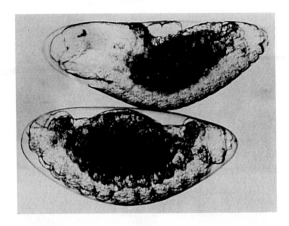

FIGURE 9.18 Normal and irradiated embryos of the midge *Smittia*. The normal embryo (top) shows a head on the left and abdominal segments on the right. The UV-irradiated embryo (bottom) has no head region, but has abdominal and tail segments at both ends. (From Kalthoff 1969; photographs courtesy of K. Kalthoff.)

its localization at the anterior pole (Ferrandon et al. 1997; Macdonald and Kerr 1998). These sequences interact with the **Exuperantia** and **Swallow** proteins while the messages are still in the nurse cells of the egg chamber (Figure 9.19A; Schnorrer et al. 2000). Experiments in which fluorescently labeled *bicoid* mRNA was microinjected into living egg chambers of wild type or mutant flies indicate that Exuperantia must be present in the nurse cells for anterior localization. But having Exuperantia in the oocyte is not sufficient to bring the *bicoid* message into the oocyte. (Cha et al 2001; Reichmann and Ephrussi 2005). The *bicoid*-protein complex is transported out of the nurse cells and into the oocyte via microtubules. The complex seems to ride on a kinesin ATPase that is critical for taking the *bicoid* message into the oocyte (Arn et al. 2003). Once inside the oocyte, the *bicoid*-mRNA complex attaches to dynein proteins that are maintained at the microtubule organizing center (the "minus end") at the anterior of the oocyte (see Figure 9.8; Cha et al. 2001).

FIGURE 9.19 Three independent genetic pathways interact to form the anterior-posterior axis of the *Drosophila* embryo. In each case, the initial asymmetry is established during oogenesis, and the pattern is organized by maternal proteins soon after fertilization. The realization of the pattern comes about when the localized maternal proteins activate or repress specific zygotic genes in different regions of the embryo. (After St. Johnston and Nüsslein-Volhard 1992.)

WEBSITE 9.5 Mechanism of *bicoid* mRNA localization. One of the most critical steps in *Drosophila* pattern formation is the binding of the *bicoid* mRNA to the anterior microtubules. Several genes are involved in this process, wherein the *bicoid* message forms a complex with several proteins.

NANOS mRNA LOCALIZATION IN THE POSTERIOR POLE OF THE OOCYTE The posterior organizing center is defined by the activities of the *nanos* gene (Lehmann and Nüsslein-Volhard 1991; Wang and Lehmann 1991; Wharton and Struhl 1991). While *bicoid* message is bound to the anchored end of the microtubules by active transport along microtubules, the

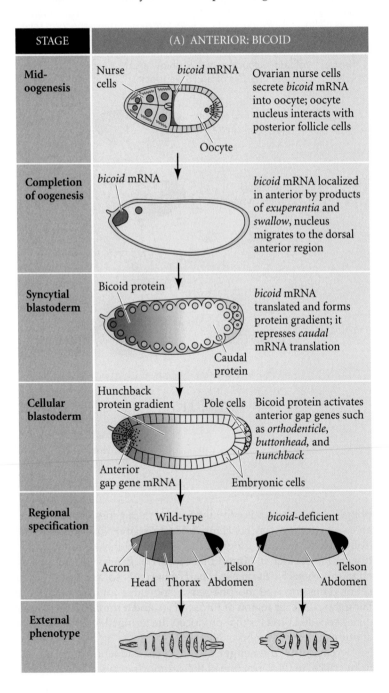

nanos message appears to get "trapped" in the posterior end of the oocyte by passive diffusion. The *nanos* message becomes bound to the cytoskeleton in the posterior region of the egg through its 3′ UTR and its association with the products of several other genes (*oskar, valois, vasa, staufen,* and *tudor*).* If *nanos* (or any other of these maternal effect genes) are absent in the mother, no abdomen forms in the embryo (Lehmann and Nüsslein-Volhard 1986; Schüpbach and Wieschaus 1986). But before the *nanos* message can get "trapped" in the posterior cortex, a *nanos* mRNA-specific trap has to be made; this trap is the Oskar protein (Ephrussi et al 1991). The *oskar* message and the Staufen protein are transported to the posterior end of the oocyte by the

motor protein kinesin I (see Figure 9.8C,D). There they become bound to the actin microfilaments of the cortex. Staufen allows the translation of the *oskar* message, and the

---

*Like the placement of the *bicoid* message, the location of the *nanos* message is determined by its 3′ untranslated region. If the *bicoid* 3′ UTR is experimentally placed on the protein-encoding region of *nanos* mRNA, the *nanos* message gets placed in the anterior of the egg. When the RNA is translated, the Nanos protein inhibits the translation of *hunchback* and *bicoid* mRNAs, and the embryo forms two abdomens—one in the anterior of the embryo and one in the posterior (Gavis and Lehmann 1992). We will see these proteins again in Chapter 19, since they are critical in forming the germ cells of *Drosophila.*

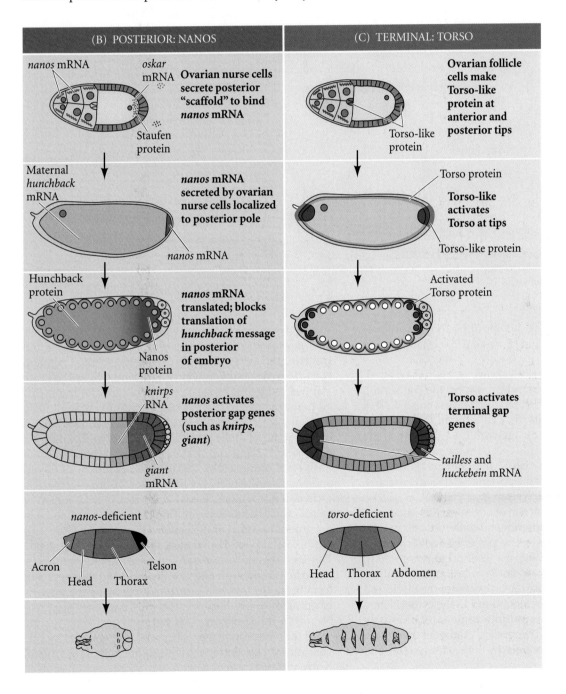

TABLE 9.1 Maternal effect genes that effect the anterior–posterior polarity of the *Drosophila* embryo

Gene	Mutant phenotype	Proposed function and structure
ANTERIOR GROUP		
bicoid (bcd)	Head and thorax deleted, replaced by inverted telson	Graded anterior morphogen; contains homeodomain; represses *caudal* mRNA
exuperantia (exu)	Anterior head structures deleted	Anchors *bicoid* mRNA
swallow (swa)	Anterior head structures deleted	Anchors *bicoid* mRNA
POSTERIOR GROUP		
nanos (nos)	No abdomen	Posterior morphogen; represses *hunchback* mRNA
tudor (tud)	No abdomen, no pole cells	Localization of Nanos protein
oskar (osk)	No abdomen, no pole cells	Localization of Nanos protein
vasa (vas)	No abdomen, no pole cells; oogenesis defective	Localization of Nanos protein
valois (val)	No abdomen, no pole cells; cellularization defective	Stabilization of the Nanos localization complex
pumilio (pum)	No abdomen	Helps Nanos protein bind *hunchback* message
caudal (cad)	No abdomen	Activates posterior terminal genes
TERMINAL GROUP		
torso (tor)	No termini	Possible morphogen for termini
trunk (trk)	No termini	Transmits Torso-like signal to Torso
fs(1)Nasrat[fs(1)N]	No termini; collapsed eggs	Transmits Torso-like signal to Torso
fs(1)polehole[fs(1)ph]	No termini; collapsed eggs	Transmits Torso-like signal to Torso

Source: After Anderson 1989.

resulting Oskar protein is capable of binding *nanos* message (Figure 9.19B; Brendza et al. 2000; Hatchet and Ephrussi 2004).

Most of the *nanos* mRNA, however, is not trapped. Rather, it is bound in the cytoplasm by the translation inhibitors Smaug and CUP. Smaug binds to the 3′UTR of *nanos* mRNA and recruits the CUP protein that prevents the association of the message with the ribosome. If the Nanos-Smaug-CUP complex reaches the posterior pole, however, Oskar can dissociate CUP from Smaug, allowing the mRNA to be bound at the posterior and ready for translation (Forrest et al. 2004; Nelson et al. 2004).

Therefore, at the completion of oogenesis, the *bicoid* message is anchored at the anterior end of the oocyte, and the *nanos* message is tethered to the posterior end (Frigerio et al. 1986; Berleth et al. 1988; Gavis and Lehmann 1992). These mRNAs are dormant until ovulation and fertilization, at which time they are translated. Since the Bicoid and Nanos *protein products* are not bound to the cytoskeleton, they diffuse toward the middle regions of the early embryo, creating the two opposing gradients that establish the anterior-posterior polarity of the embryo (see Figure 3.11).

Two other maternally provided mRNAs (*hunchback, hb*; and *caudal, cad*) are critical for patterning the anterior and posterior regions of the body plan, respectively (Lehmann et al. 1987; Wu and Lengyel 1998). These two mRNAs are synthesized by the nurse cells of the ovary and transported to the oocyte, where they are distributed ubiquitously throughout the syncytial blastoderm. But, if they are not localized, how do they mediate their localized patterning activities? It turns out that translation of the *hb* and *cad* mRNAs is repressed by the diffusion gradients of Bicoid and Nanos proteins.

In anterior regions, Bicoid binds to a specific region of *caudal*'s 3′ UTR, thereby preventing translation of Caudal in the anterior section of the embryo (Figure 9.20; Chan and Struhl 1997; Rivera-Pomar et al. 1996; Niessing et al. 2000). This suppression is necessary, because if Caudal protein is made in the embryo's anterior, the head and thorax do not form properly. Caudal is critical in specifying the posterior domains of the embryo, activating the genes responsible for the invagination of the hindgut.

At the other end of the embryo, Nanos functions by preventing *hunchback* translation (Figure 9.21; Tautz 1988). Normal translation of *hb* involves an interaction of its 3′

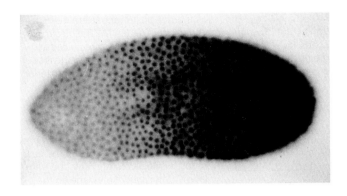

FIGURE 9.20 Gradient of Caudal protein in the syncytial blastoderm of a wild-type *Drosophila* embryo. The protein (stained darkly) enters the nuclei and helps specify posterior fates. Compare with the complementary gradient of Bicoid protein in Figure 9.23A. (From Macdonald and Struhl 1986; photograph courtesy of the authors.)

UTR with a protein called Pumilio. In the posterior of the early embryo, however, Nanos protein binds to Pumilio and deadenylates the *hunchback* mRNA, preventing its translation (Barker et al. 1992; Wreden et al. 1997).

The result of these interactions is the creation of four maternal protein gradients in the early embryo (Figure 9.22):

- An anterior-to-posterior gradient of Bicoid protein
- An anterior-to-posterior gradient of Hunchback protein
- A posterior-to-anterior gradient of Nanos protein
- A posterior-to-anterior gradient of Caudal protein

The Bicoid, Hunchback, and Caudal proteins are transcription factors whose relative concentrations can activate or repress particular zygotic genes. The stage is now set for the activation of zygotic genes in those nuclei that were

busy dividing while these four protein gradients were being established.

 WEBSITE 9.6 Christiane Nüsslein-Volhard and the molecular approach to development. The research that revolutionized developmental biology had to wait for someone to synthesize molecular biology, embryology, and *Drosophila* genetics.

The anterior organizing center: The Bicoid and Hunchback gradients

In *Drosophila*, the phenotype of the *bicoid* mutant provides valuable information about the function of morphogenetic gradients (Figure 9.23A–C). Instead of having anterior structures (acron, head, and thorax) followed by abdominal structures and a telson, the structure of the *bicoid* mutant is telson-abdomen-abdomen-telson (Figure 9.23D). It would appear that these embryos lack whatever substances are needed for the formation of anterior structures. Moreover, one could hypothesize that the substance these mutants lack is the one postulated by Sander and Kalthoff to turn on genes for the anterior structures and turn off genes for the telson structures (compare Figures 9.18 and 9.23A).

Several studies support the view that the product of the wild-type *bicoid* gene is the morphogen that controls anterior development. The first type of evidence came from experiments that altered the shape of the Bicoid protein gradient. As we have seen, the *exuperantia* and *swallow* genes are responsible for keeping the *bicoid* message at the anterior pole of the egg. In their absence, the *bicoid* message diffuses farther into the posterior of the egg, and the gradient of Bicoid protein is less steep (Driever and Nüsslein-Volhard 1988a). The phenotype produced by *exuperantia* and *swallow* mutants is similar to that of *bicoid*-deficient embryos, but less severe. These embryos lack their

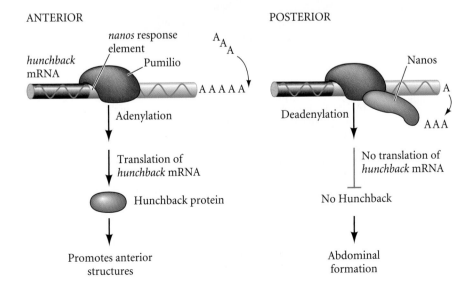

FIGURE 9.21 Control of *hunchback* mRNA translation by Nanos protein. In the anterior of the embryo, Pumilio binds to the Nanos response element (NRE) in the 3′UTR of the *hunchback* message, and the message is polyadenylated normally. The polyadenylated message can be translated into Hunchback protein. In the posterior of the embryo, the Nanos protein binds to Pumilio to cause the deadenylation of the *hunchback* message, thus preventing its translation. (After Wreden et al. 1997.)

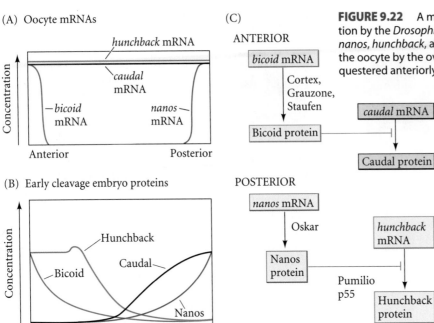

(A) Oocyte mRNAs

(B) Early cleavage embryo proteins

(C)

ANTERIOR

bicoid mRNA

↓ Cortex, Grauzone, Staufen

Bicoid protein ——⊣ Caudal protein

caudal mRNA

POSTERIOR

nanos mRNA

↓ Oskar

Nanos protein ——⊣ Hunchback protein

hunchback mRNA

Pumilio p55

FIGURE 9.22 A model of anterior-posterior pattern generation by the *Drosophila* maternal effect genes. (A) The *bicoid*, *nanos*, *hunchback*, and *caudal* messenger RNAs are placed in the oocyte by the ovarian nurse cells. The *bicoid* message is sequestered anteriorly; the *nanos* message is sent to the posterior pole. (B) Upon translation, the Bicoid protein gradient extends from anterior to posterior, while the Nanos protein gradient extends from posterior to anterior. Nanos inhibits the translation of the *hunchback* message (in the posterior), while Bicoid prevents the translation of the *caudal* message (in the anterior). This inhibition results in opposing Caudal and Hunchback gradients. The Hunchback gradient is secondarily strengthened by the transcription of the *hunchback* gene in the anterior nuclei (since Bicoid acts as a transcription factor to activate *hunchback* transcription). (C) Parallel interactions whereby translational gene regulation establishes the anterior-posterior patterning of the *Drosophila* embryo. (C after Macdonald and Smibert 1996.)

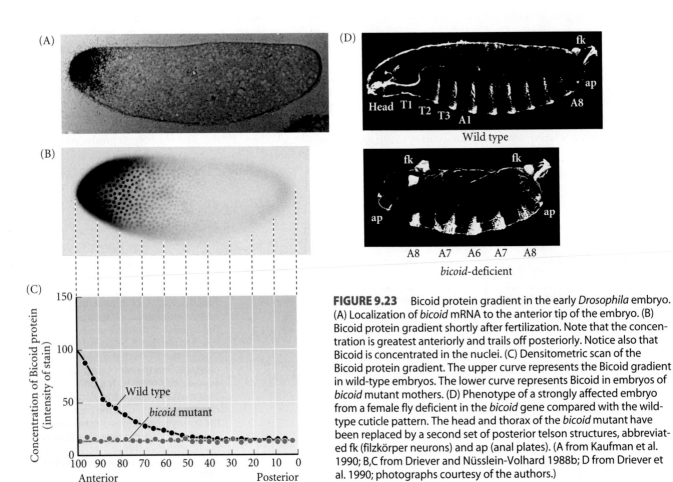

FIGURE 9.23 Bicoid protein gradient in the early *Drosophila* embryo. (A) Localization of *bicoid* mRNA to the anterior tip of the embryo. (B) Bicoid protein gradient shortly after fertilization. Note that the concentration is greatest anteriorly and trails off posteriorly. Notice also that Bicoid is concentrated in the nuclei. (C) Densitometric scan of the Bicoid protein gradient. The upper curve represents the Bicoid gradient in wild-type embryos. The lower curve represents Bicoid in embryos of *bicoid* mutant mothers. (D) Phenotype of a strongly affected embryo from a female fly deficient in the *bicoid* gene compared with the wild-type cuticle pattern. The head and thorax of the *bicoid* mutant have been replaced by a second set of posterior telson structures, abbreviated fk (filzkörper neurons) and ap (anal plates). (A from Kaufman et al. 1990; B,C from Driever and Nüsslein-Volhard 1988b; D from Driever et al. 1990; photographs courtesy of the authors.)

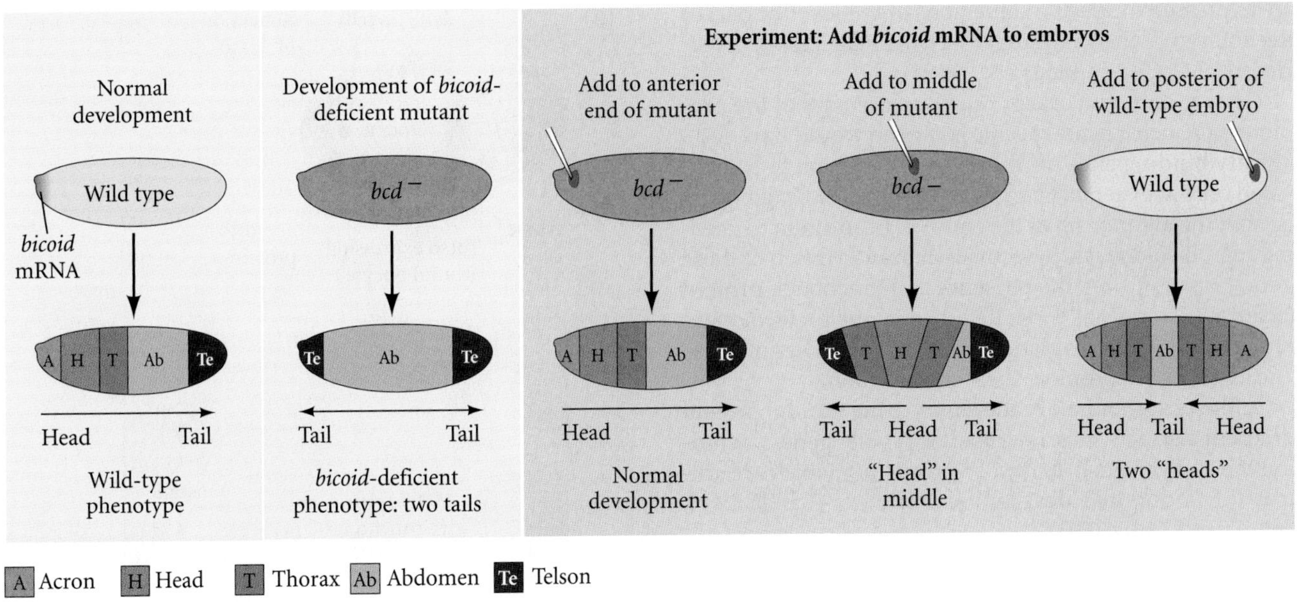

FIGURE 9.24 Schematic representation of the experiments demonstrating that the *bicoid* gene encodes the morphogen responsible for head structures in *Drosophila*. The phenotypes of *bicoid*-deficient and wild-type embryos are shown at the left. When *bicoid*-deficient embryos are injected with *bicoid* mRNA, the point of injection forms the head structures. When the posterior pole of an early-cleavage wild-type embryo is injected with *bicoid* mRNA, head structures form at both poles. (After Driever et al. 1990.)

most anterior structures and have an extended mouth and thoracic region. Also, by adding extra copies of the *bcd* gene, the Bicoid protein gradient can be extended into more posterior regions, which causes anterior structures like the cephalic furrow to be expressed in a more posterior position (Driever and Nusslein-Volhard 1988a; Struhl et al. 1989). Thus, by altering the Bicoid gradient, one correspondingly alters the fate of specific embryonic regions.

Confirmation that Bicoid is crucial for initiating head and thorax formation came from experiments in which purified *bicoid* mRNA was injected into early-cleavage embryos (Figure 9.24; see also Figure 3.12). When injected into the anterior of *bicoid*-deficient embryos (whose mothers lacked *bicoid* genes), the *bicoid* mRNA "rescued" the embryos and they developed normal anterior-posterior polarity. Moreover, any location in an embryo where the *bicoid* message was injected became the head. If *bicoid* mRNA was injected into the center of an embryo, that middle region became the head, with the regions on either side of it becoming thorax structures. If a large amount of *bicoid* mRNA was injected into the posterior end of a wild-type embryo (with its own endogenous *bicoid* message in its anterior pole), two heads emerged, one at either end (Driever et al. 1990).

How might a gradient of Bicoid protein control the determination of the anterior-posterior axis? As discussed

before, Bicoid protein acts as a translation inhibitor of *caudal*, and *caudal*'s protein product is critical for the specification of the posterior. However, Bicoid's primary function is to act as a *transcription factor* to activate the expression of target genes in the anterior part of the embryo.*

The first target gene of Bicoid activation to be discovered was *hunchback*. In the late 1980s, two laboratories independently demonstrated that Bicoid binds to and activates the *hunchback* (*hb*) gene (Driever and Nüsslein-Volhard 1989; Struhl et al. 1989). Bicoid-dependent transcription of *hb* is seen only in the anterior half of the embryo—the region where Bicoid is found. This transcription reinforces the gradient of maternal Hunchback protein produced by Nanos-dependent translational repression. Mutants deficient in maternal and zygotic *hb* genes lack mouth parts and thorax structures. Therefore, both maternal and zygotic Hunchback contribute to the anterior patterning of the embryo.

Based on two pieces of evidence, Driever and co-workers (1989) predicted that Bicoid must activate at least one other anterior gene besides *hb*. First, deletions of *hb* produced only some of the defects seen in the *bicoid* mutant phenotype. Second, the *swallow* and *exuperantia* experiments showed that only moderate levels of Bicoid protein are needed to activate thorax formation (i.e., *hunchback* gene expression), but head formation requires higher Bicoid concentrations. Since then, a large number of Bicoid target genes have been identified. These include the head gap genes *buttonhead*, *empty spiracles*, and *orthodenticle*, which

---

bicoid appears to be a relatively "new" gene that evolved in the Dipteran (fly) lineage; it has not been found in other insects. The anterior determinant in other insect groups has not yet been determined but appears to have *bicoid*-like properties (Wolff et al. 1998; Lynch and Desplan 2003).

are expressed in specific subregions of the anterior part of the embryo (Cohen and Jürgens 1990; Finkelstein and Perrimon 1990; Grossniklaus et al. 1994).

Driever and co-workers (1989) predicted that the promoters of such a head-specific gap gene would have low-affinity binding sites for Bicoid, causing them to be activated only at extremely high concentrations of Bicoid—that is, near the anterior tip of the embryo. In addition to needing high Bicoid levels for activation, transcription of these genes also requires the presence of Hunchback protein (Simpson-Brose et al. 1994; Reinitz et al. 1995). Bicoid and Hunchback act synergistically at the enhancers of these "head genes" to promote their transcription.

In the posterior half of the embryo, the Caudal protein gradient also activates a number of zygotic genes, including the gap genes *knirps* (*kni*) and *giant* (*gt*), which are critical for abdominal development (Rivera-Pomar 1995; Schulz and Tautz 1995).

The terminal gene group

In addition to the anterior and posterior morphogens, there is third set of maternal genes whose proteins generate the unsegmented extremities of the anterior-posterior axis: the **acron** (the terminal portion of the head that includes the brain) and the **telson** (tail). Mutations in these terminal genes result in the loss of the acron and the most anterior head segments as well as the telson and the most posterior abdominal segments (Degelmann et al. 1986; Klingler et al. 1988). A critical gene here appears to be *torso*, a gene encoding a receptor tyrosine kinase. The embryos of mothers with mutations of *torso* have neither acron nor telson, suggesting that the two termini of the embryo are formed through the same pathway (see Figure 9.19C). The *torso* mRNA is synthesized by the ovarian cells, deposited in the oocyte, and translated after fertilization. The transmembrane Torso protein is not spatially restricted to the ends of the egg, but is evenly distributed throughout the plasma membrane (Casanova and Struhl 1989). Indeed, a gain-of-function mutation of *torso*, which imparts constitutive activity to the receptor, converts the entire anterior half of the embryo into an acron and the entire posterior half into a telson. Thus, Torso must normally be activated only at the ends of the egg.

Stevens and her colleagues (1990) have shown that this is the case. Torso protein is activated by the follicle cells only at the two poles of the oocyte. Two pieces of evidence suggest that the activator of Torso is probably the **Torso-like** protein: first, loss-of-function mutations in the *torso-like* gene create a phenotype almost identical to that produced by *torso* mutants; and second, ectopic expression of Torso-like protein activates Torso in the new location. The *torso-like* gene is usually expressed only in the anterior and posterior follicle cells, and secreted Torso-like protein can cross the perivitelline space to activate Torso in the egg membrane (Martin et al. 1994; Furriols et al. 1998). In this

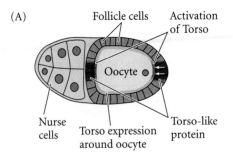

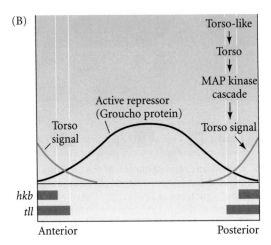

FIGURE 9.25 Formation of the unsegmented extremities by *torso* signaling. (A) Torso-like protein is expressed by the follicle cells at the poles of the oocyte. Torso protein is expressed around the entire oocyte. Torso-like activates Torso at the poles (see Casanova et al. 1995). (B) Inactivation of the transcriptional suppression of *huckebein* (*hkb*) and *tailless* (*tll*) genes. The Torso signal antagonizes the Groucho protein. Groucho acts as a repressor of *tailless* and *huckebein* expression. The gradient of Torso is thought to provide the information that allows *tailless* to be expressed farther into the embryo than *huckebein*. (A after Gabay et al. 1997; B after Paroush et al. 1997.)

manner, Torso-like activates Torso in the anterior and posterior regions of the oocyte membrane.

The end products of the RTK-kinase cascade activated by Torso diffuse into the cytoplasm at both ends of the embryo (Figure 9.25; Gabay et al. 1997; see also Chapter 6). These kinases are thought to inactivate the Groucho protein, a transcriptional inhibitor of the *tailless* and *huckebein* gap genes (Paroush et al. 1997); it is these two gap genes that specify the termini of the embryo. The distinction between the anterior and posterior termini depends on the presence of Bicoid. If *tailless* and *huckebein* act alone, the terminal region differentiates into a telson. However, if Bicoid is also present, the terminal region forms an acron (Pignoni et al. 1992).

Anterior-posterior axis specification: Summary

In summary, then, the anterior-posterior axis of the *Drosophila* embryo is specified by three sets of genes:

1. **Genes that define the anterior organizing center.** Located at the anterior end of the embryo, the anterior organizing center acts through a gradient of Bicoid protein. Bicoid functions as a *transcription factor* to activate anterior-specific gap genes and as a *translational repressor* to suppress posterior-specific gap genes (see Figure 9.19A).
2. **Genes that define the posterior organizing center.** The posterior organizing center is located at the posterior pole. This center acts *translationally* through the Nanos protein to inhibit anterior formation and *transcriptionally* through the Caudal protein to activate those genes that form the abdomen. The activation of those genes responsible for constructing the posterior is performed by Caudal, a protein whose synthesis is inhibited in the anterior portion of the embryo (see Figure 9.19B).
3. **Genes that define the terminal boundary regions.** The boundaries of the acron and telson are defined by the product of the *torso* gene, which is activated at the tips of the embryo (see Figure 9.19C).

The next step in development will be to use these gradients of transcription factors to activate specific genes along the anterior-posterior axis.

Segmentation Genes

Cell fate commitment in *Drosophila* appears to have two steps: specification and determination (Slack 1983). Early in development, the fate of a cell depends on cues provided by protein gradients. This specification of cell fate is flexible and can still be altered in response to signals from other cells. Eventually, however, the cells undergo a transition from this loose type of commitment to an irreversible determination. At this point, the fate of a cell becomes cell-intrinsic.*

The transition from specification to determination in *Drosophila* is mediated by the **segmentation genes**. These genes divide the early embryo into a repeating series of segmental primordia along the anterior-posterior axis. Segmentation genes were originally defined by zygotic mutations that disrupted the body plan, and these genes were divided into three groups based on their mutant phenotypes (Nüsslein-Volhard and Wieschaus 1980):

---

*Aficionados of information theory will recognize that the process by which the anterior-posterior information in morphogenetic gradients is transferred to discrete and different parasegments represents a transition from analog to digital specification. Specification is analog, determination digital. This process enables the transient information of the gradients in the syncytial blastoderm to be stabilized so that it can be utilized much later in development (Baumgartner and Noll 1990).

1. **Gap** mutations lacked large regions of the body (several contiguous segments; Figure 9.26A).
2. **Pair-rule** mutants lacked portions of every other segment (Figure 9.26B).
3. **Segment polarity** mutants showed defects (deletions, duplications, polarity reversals) in every segment Figure 9.26C).

(A) Gap: *Krüppel* (as an example)

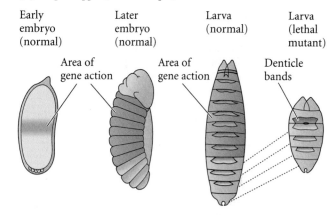

(B) Pair rule: *fushi tarazu* (as an example)

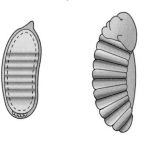
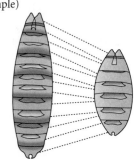

(C) Segment polarity: *engrailed* (as an example)

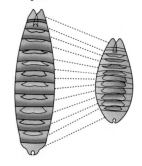

FIGURE 9.26 Three types of segmentation gene mutations. The left panel shows the early-cleavage embryo, with the region where the particular gene is normally transcribed in wild-type embryos shown in color. These areas are deleted as the mutants (right) develop.

Segments and Parasegments

Mutations in segmentation genes result in *Drosophila* embryos that lack certain segments or parts of segments. However, early researchers found a surprising aspect of these mutations: many of them did not affect actual segments. Rather, they affected the posterior compartment of one segment and the anterior compartment of the immediately posterior segment. These "transegmental" units were named **parasegments** (Figure 9.27A; Martinez-Arias and Lawrence 1985). Moreover, once the means to detect gene expression patterns were available, it was discovered that the expressions patterns in the early embryo are delineated by parasegmental boundaries—not by the boundaries of the segments. Thus, the parasegment appears to be the fundamental unit of *embryonic* gene expression.

Although parasegmental organization is also seen in the nerve cord of adult *Drosophila*, it is not seen in the adult epidermis (which is the most obvious manifestation of segmentation), nor is it found in the adult musculature. These adult structures are organized along the segmental pattern. One can think about the segmental and parasegmental organization schemes as representing different ways of organizing the **compartments** along the anterior-posterior axis of the embryo. The cells of one compartment do not mix with cells of neighboring compartments, and parasegments and segments are out of phase by one compartment.

Why should there be two modes of metamerism (sequential parts) in flies? Jean Deutsch has proposed that such a twofold way of organizing the body is needed for the coordination of movement. In every group of the Arthropoda—crustaceans, insects, myriapods, and chelicerates (spiders)—the ganglia of the ventral nerve cord are organized by parasegments, but the cuticle grooves and musculature are segmental. In *Drosophila*, the segmental grooves appear in the epidermis when the germ band is retracted, while the mesoderm becomes segmental later in development.

Viewing the segmental border as a movable hinge, this shift in frame by one com-

partment allows the muscles on both sides of any particular epidermal segment to be coordinated by the same ganglion (Figure 9.27B). This in turn allows rapid and coordinated muscle contractions for locomotion. Therefore, while parts of the body may become secondarily organized according to

segments, the parasegment is the basic unit of embryonic construction. A similar situation occurs in vertebrates, where the posterior portion of the anterior somite combines with the anterior portion of the next somite (Chapter 14).

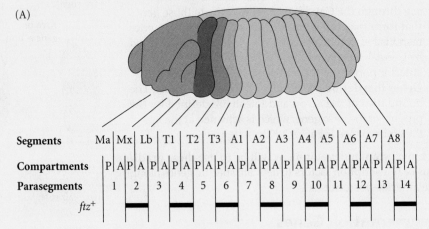

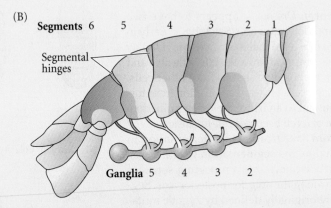

FIGURE 9.27 Overlap and integration of segments and parasegments. (A) Parasegments in the *Drosophila* embryo are shifted one compartment forward in relation to the segments. Ma, Mx, and Lb are the mandibular, maxillary, and labial head segments; T1–T3 are the thoracic segments; and A1–A9 are abdominal segments. Each segment has an anterior (A) and a posterior (P) compartment. Each parasegment (numbered 1–14) consists of the posterior compartment of one segment and the anterior compartment of the segment in the next posterior position. Black bars indicate the boundaries of gene expression observed in the *fushi tarazu* (*ftz*) mutant (see Figure 9.26B). (B) Segments and parasegments integrated in the body of an adult arthropod (the crustacean *Procambarus*). The ventral nerve cord is divided according to parasegments (color). This allows the neurons of the ganglia to regulate the ectodermal scutes and the mesodermal muscles on either side of a segmental hinge. (A after Martinez-Arias and Lawrence 1985; B after Deutsch 2004.)

The gap genes

The gap genes are activated or repressed by the maternal effect genes, and are expressed in one or two broad domains along the anterior-posterior axis. These expression patterns correlate quite well with the regions of the embryo that are missing in gap mutations. For example, the *Krüppel* gene is expressed primarily in parasegments 4–6, in the center of the *Drosophila* embryo (Figure 9.27A; see also Figure 9.17C); in the absence of the Krüppel protein, the embryo lacks segments from these and the immediately adjacent regions.

Deletions caused by mutations in three gap genes—*hunchback*, *Krüppel*, and *knirps*—span the entire segmented region of the *Drosophila* embryo. The gap gene *giant* overlaps with these three, and the gap genes *tailless* and *huckebein* are expressed in domains near the anterior and posterior ends of the embryo.

The expression patterns of the gap genes are highly dynamic. These genes usually show low levels of transcriptional activity across the entire embryo that become consolidated into discrete regions of high activity as cleavage continues (Jäckle et al. 1986). The Hunchback gradient is particularly important in establishing the initial gap gene expression patterns. By the end of nuclear division cycle 12, Hunchback is found at high levels across the anterior part of the embryo, and then forms a steep gradient through about 15 nuclei near the middle of the embryo (see Figure 9.17A and 9.22). The last third of the embryo has undetectable Hunchback levels at this time. The transcription patterns of the anterior gap genes are initiated by the different concentrations of the Hunchback and Bicoid proteins. High levels of Bicoid and Hunchback induce the expression of *giant*, while the *Krüppel* transcript appears over the region where Hunchback begins to decline. High levels of Hunchback also prevent the transcription of the posterior gap genes (such as *knirps* and *giant*) in the anterior part of the embryo (Struhl et al. 1992). It is thought that a gradient of the Caudal protein, highest at the posterior pole, is responsible for activating the abdominal gap genes *knirps* and *giant* in the posterior part of the embryo. The *giant* gene thus has two methods for its activation, one for its anterior expression band and one for its posterior expression band (Rivera-Pomar 1995; Schulz and Tautz 1995).

After the initial gap gene expression patterns have been established by the maternal effect gradients and Hunchback, they are stabilized and maintained by repressive interactions between the different gap gene products themselves.* These boundary-forming inhibitions are thought to be directly mediated by the gap gene products because all four major gap genes (*hunchback*, *giant*, *Krüppel*, and

---

*The interactions between these genes and gene products are facilitated by the fact that these reactions occur within a syncytium, in which the cell membranes have not yet formed.

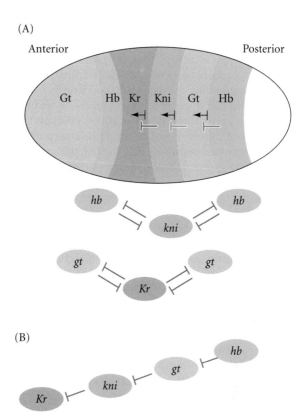

FIGURE 9.28 Expression and regulatory interactions among gap genes products. (A) Schematic expression of the gap genes during the late fourteenth cell cycle. Bars between the domains represent repression of the more anterior domain by the protein posterior to it. Arrows represent the direction in which the domains shift during the cell cycle. (For clarity, overlaps are not shown). Strong mutual repression (diagrammed below) establishes the basic pattern of gene expression. (B) Asymmetrical repression of gap genes by their posterior neighbors causes an anterior shift in the domains of expression. (After Monk 2004.)

knirps) encode DNA-binding proteins (Knipple et al. 1985; Gaul and Jäckle 1990; Capovilla et al. 1992). The major mechanism involved in this stabilization seems to be strong mutual repression between pairs of *nonadjacent* gap genes (Figure 9.28A). Gene misexpression experiments show that Giant and Kruppel are strong mutual repressors, as are Hunchback and Knirps (Kraut and Levine 1991; Clyde et al. 2003). For example, if *hunchback* activity is lacking, the posterior domain of *knirps* expands toward the anterior. Conversely, if *hunchback* is misexpressed in nuclei that normally express *knirps*, strong repression is detected. This system of strong mutual repression results in the precise placement of gap protein domains, but permits overlaps between *adjacent* gap genes.

Jaeger and colleagues (2004) used quantified gene expression data to model how stabilization of the gap gene expression patterns occurs during the thirteenth and fourteenth cleavage cycles (at around 71 minutes). Their data suggest that the patterns of gap gene expression were stabilized by three major factors. Two of these were strong mutual inhibition between Hunchback and Knirps and strong mutual inhibitions between Giant and Kruppel. The data also revealed that these inhibitory interactions are unidirectional, with each protein having a strong effect on the anterior border of the repressed genes. This latter part of the model is important because it may explain the anterior "creeping" of the gap gene transcription patterns (Figure 9.28B).

The end result of these repressive interactions is the creation of a precise system of overlapping gap mRNA expression patterns. Each domain serves as a source for diffusion of gap proteins into adjacent embryonic regions. This creates a significant overlap (at least eight nuclei, which accounts for about two segment primordia) between adjacent gap protein domains. This was demonstrated in a striking manner by Śtanojevíc and co-workers (1989). They fixed cellularizing blastoderms (see Figures 9.1 and 9.3), stained Hunchback protein with an antibody carrying a red dye, and simultaneously stained Krüppel protein with an antibody carrying a green dye. Cellularizing regions that contained both proteins bound both antibodies and stain bright yellow (see Figure 9.17C). Krüppel overlaps with Knirps in a similar manner in the posterior region of the embryo (Pankratz et al. 1990).

The pair-rule genes

The first indication of segmentation in the fly embryo comes when the pair-rule genes are expressed during cell division cycle 13. The transcription patterns of these genes divide the embryo into regions that are precursors of the segmental body plan. As can be seen in Figure 9.29 and Figure 9.17D, one vertical band of nuclei (the cells are just beginning to form) expresses a pair-rule gene, the next band of nuclei does not express it, and then the next band expresses it again. The result is a "zebra stripe" pattern along the anterior-posterior axis, dividing the embryo into 15 subunits (Hafen et al. 1984). Eight genes are currently known to be capable of dividing the early embryo in this fashion; they are listed in Table 9.2.

The primary pair-rule genes include *hairy*, *even-skipped*, and *runt*, each of which is expressed in seven stripes. All three build their striped patterns from scratch by using distinct enhancers and regulatory mechanisms for each stripe. These enhancers are often modular: control over expression in each stripe is located in a discrete region of the DNA, and these DNA regions often contain binding sites recognized by gap proteins. Thus it is thought that the different concentrations of gap proteins determine whether a pair-rule gene is transcribed or not.

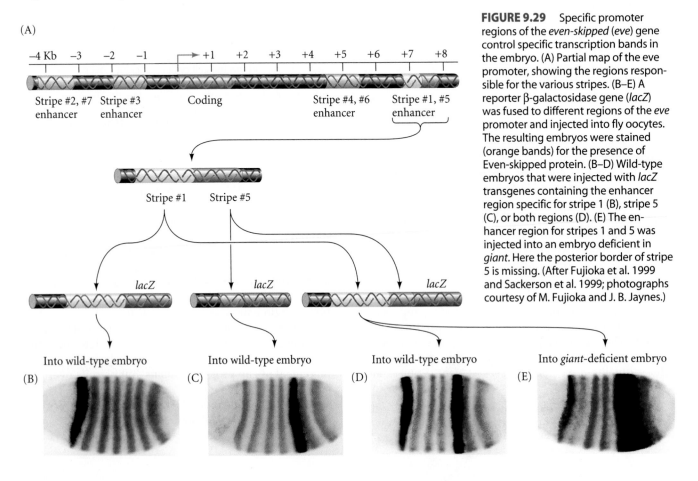

FIGURE 9.29 Specific promoter regions of the *even-skipped* (*eve*) gene control specific transcription bands in the embryo. (A) Partial map of the eve promoter, showing the regions responsible for the various stripes. (B–E) A reporter β-galactosidase gene (*lacZ*) was fused to different regions of the *eve* promoter and injected into fly oocytes. The resulting embryos were stained (orange bands) for the presence of Even-skipped protein. (B–D) Wild-type embryos that were injected with *lacZ* transgenes containing the enhancer region specific for stripe 1 (B), stripe 5 (C), or both regions (D). (E) The enhancer region for stripes 1 and 5 was injected into an embryo deficient in *giant*. Here the posterior border of stripe 5 is missing. (After Fujioka et al. 1999 and Sackerson et al. 1999; photographs courtesy of M. Fujioka and J. B. Jaynes.)

TABLE 9.2 Major genes affecting segmentation pattern in *Drosophila*

Category	Gene name
Gap genes	*Krüppel (Kr)*
	knirps (kni)
	hunchback (hb)
	giant (gt)
	tailless (tll)
	huckebein (hkb)
	buttonhead (btd)
	empty spiracles (ems)
	orthodenticle (otd)
Pair-rule genes (primary)	*hairy (h)*
	even-skipped (eve)
	runt (run)
Pair-rule genes (secondary)	*fushi tarazu (ftz)*
	odd-paired (opa)
	odd-skipped (odd)
	sloppy-paired (slp)
	paired (prd)
Segment polarity genes	*engrailed (en)*
	wingless (wg)
	cubitus interruptusD (ciD)
	hedgehog (hh)
	fused (fu)
	armadillo (arm)
	patched (ptc)
	gooseberry (gsb)
	pangolin (pan)

One of the best-studied primary pair-rule gene is *even-skipped*, illustrated in Figure 9.29. Its enhancer is composed of modular units arranged such that each unit regulates a separate stripe or a pair of stripes. For instance, *even-skipped* stripe 2 is controlled by a 500-bp region that is activated by low concentrations of Bicoid and Hunchback, and repressed by both Giant and Krüppel proteins (Figure 9.30A; Small et al. 1991, 1992; S̆tanojevíc et al. 1991). DNase I footprinting (see Chapter 5) showed that the minimal enhancer region for this stripe contains 5 binding sites for

Bicoid, 1 for Hunchback, 3 for Krüppel, and 3 for Giant. Thus, this region is thought to act as a switch that can directly sense the concentrations of these proteins and make on/off transcriptional decisions. Similarly, *even-skipped* stripe 5 is regulated negatively by Krüppel protein (on its anterior border) and by Giant protein (on its posterior border) (Fujioka et al. 1999). This suggests that these two stripes are controlled by common repressors.

Two other *even-skipped* enhancers each control the expression of two stripes. One drives expression of stripes 3+7, while the other drives expression of stripes 4+6. Clyde and colleagues (2003) showed that the borders of all four of these stripes are formed by repression mediated by Knirps and Hunchback. Knirps is expressed between stripes 4 and 6, and sets the inside borders of both stripes by differentially repressing the two enhancers. Similarly, Hunchback is expressed in anterior and posterior gradients that precisely set the outside borders of all four of these stripes. These differential repression events fix the positions of and the spacing between the even-skipped stripes (Figure 9.30B).

The importance of these enhancer elements can be shown by both genetic and biochemical means. First, a mutation in a particular enhancer can delete its particular stripe and no other. Second, if a reporter gene such as *lacZ* (encoding β-galactosidase) is fused to one of the enhancers, the *lacZ* gene is expressed only in that particular stripe (see Figure 9.29; Fujioka et al. 1999). Third, the placement of the stripes can be altered by deleting the gap genes that regulate them. Thus, the placement of the stripes of pair-rule

FIGURE 9.30 Hypothesis for the formation of the second stripe of transcription from the *even-skipped* gene. (A) Enhancer element for stripe 2 regulation, containing binding sequences for the Krüppel, Giant, Bicoid, and Hunchback proteins. Activators are shown binding on the top line, repressors are below. Note that nearly every activator site is closely linked to a repressor site, suggesting competitive interactions at these positions. (Moreover, a protein that is a repressor for stripe 2 may be an activator for stripe 5; it depends on which proteins bind next to them.) (B) The gap genes Hb and Kni mutually repress one another. These interactions fix their positions. The Hb and Kni gradients are "read" by enhancers that are differentially sensitive to repression by these two proteins. This ensures that *eve* transcription stripes 3 and 7 (shaded circles) will be symmetrically placed along these gradients, as will transcription stripes 4 and 6, and that each enhancer can form two stripes. (A after Small et al. 1992; B after Clyde et al. 2003.)

(A)

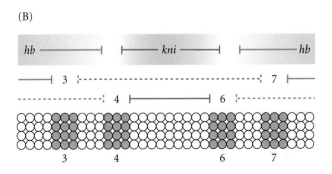

(B)

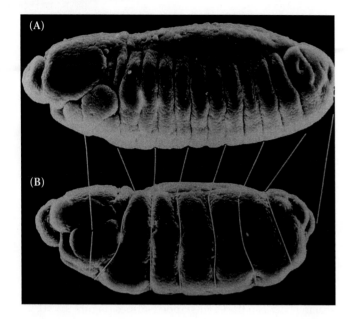

(C)

Procephalic

Maxillary

T_1 T_2 T_3 A_1 A_2 A_3 A_4 A_5 A_6 A_7 A_8

Clypolabrum   Labial

Mandibulum

FIGURE 9.31 Defects seen in the *fushi tarazu* mutant. (A) Scanning electron micrograph of a wild-type embryo, seen in lateral view. (B) Same stage of a *fushi tarazu* mutant embryo. The white lines connect the homologous portions of the segmented germ band. (C) Diagram of wild-type embryonic segmentation. The shaded areas show the parasegments of the germ band that are missing in the mutant embryo. (After Kaufman et al. 1990, photographs courtesy of T. Kaufman.)

throughout the segmented portion of the embryo. However, as the proteins from the primary pair-rule genes begin to interact with the *ftz* enhancer, the *ftz* gene is repressed in certain bands of nuclei to create interstripe regions. Meanwhile, the Ftz protein interacts with its own promoter to stimulate more transcription of *ftz* where it is already present (Figure 9.32; Edgar et al. 1986b; Karr and Kornberg 1989; Schier and Gehring 1992).

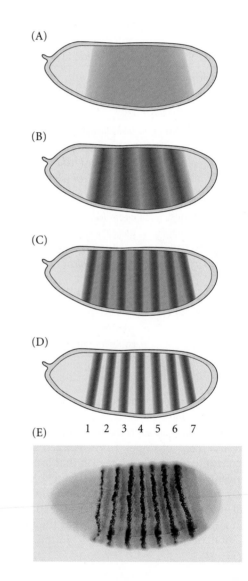

FIGURE 9.32 Transcription of the *fushi tarazu* gene in the *Drosophila* embryo. (A–D) At the beginning of division cycle 14, there is low-level transcription of *ftz* in each of the nuclei in the segmented region of the embryo. Within the next 30 minutes, the expression pattern alters as *ftz* transcription is enhanced in certain regions (which form stripes) and repressed in the interstripe regions. (E) Double labeling of the *even-skipped* (blue bands) and *fushi tarazu* (green bands) transcripts, showing that *ftz* is expressed between the *eve* bands at this stage. (A–D after Karr and Kornberg 1989; E photograph courtesy of J. B. Jaynes and M. Fujioka.)

gene expression is a result of (1) the modular *cis*-regulatory enhancer elements of the pair-rule genes, and (2) the *trans*-regulatory gap gene proteins that bind to these enhancer sites.

Once initiated by the gap gene proteins, the transcription pattern of the primary pair-rule genes becomes stabilized by interactions among their products (Levine and Harding 1989). The primary pair-rule genes also form the context that allows or inhibits the expression of the later-acting secondary pair-rule genes. One such gene is *fushi tarazu* (*ftz*), which means "too few segments" in Japanese (Figure 9.31; also see Figures 9.17D and 9.27B). Early in division cycle 14, *ftz* mRNA and its protein are seen

The eight known pair-rule genes are all expressed in striped patterns, but the patterns are not coincident with each other. Rather, each row of nuclei within a parasegment has its own array of pair-rule products that distinguishes it from any other row. These products activate the next level of segmentation genes, the segment polarity genes.

The segment polarity genes

So far our discussion has described interactions between molecules within the syncytial embryo. But once cells form, interactions take place between the cells. These interactions are mediated by the segment polarity genes, and they accomplish two important tasks. First, they reinforce the parasegmental periodicity established by the earlier transcription factors. Second, through this cell-to-cell signaling, cell fates are established within each parasegment.

The segment polarity genes encode proteins that are constituents of the Wingless and Hedgehog signal transduction pathways (see Chapter 6). Mutations in these genes lead to defects in segmentation and in gene expression pattern across each parasegment. The development of the normal pattern relies on the fact that only one row of cells in each parasegment is permitted to express the Hedgehog protein, and only one row of cells in each parasegment is permitted to express the Wingless protein. The key to this pattern is the activation of the *engrailed* gene in those cells that are going to express Hedgehog. The *engrailed* gene is activated in cells that have high levels of the Even-skipped, Fushi tarazu, or Paired transcription factors; *engrailed* is repressed in those cells with high levels of Odd-skipped, Runt, or Sloppy-paired proteins. As a result, the Engrailed protein is found in 14 stripes across the anterior-posterior axis of the embryo (see Figure 9.17E). (Indeed, in *ftz*-deficient embryos, only seven bands of *engrailed* are expressed.)

These stripes of *engrailed* transcription mark the anterior compartment of each parasegment (and the posterior compartment of each segment). The *wingless* gene is activated in those bands of cells that receive little or no Even-skipped or Fushi tarazu protein, but which do contain Sloppy-paired. This pattern causes *wingless* to be transcribed solely in the row of cells directly anterior to the cells where *engrailed* is transcribed (Figure 9.33A).

Once *wingless* and *engrailed* expression patterns are established in adjacent cells, this pattern must be maintained to retain the parasegmental periodicity of the body plan. It should be remembered that the mRNAs and proteins involved in initiating these patterns are short-lived, and that the patterns must be maintained after their initiators are no longer being synthesized. The maintenance of these patterns is regulated by reciprocal interaction between neighboring cells: cells secreting Hedgehog activate the expression of *wingless* in its neighbor; the Wingless protein is received by the cell that secreted Hedgehog and maintains *hedgehog* expression (Figure 9.33B).

In the cells transcribing the *wingless* gene, *wingless* mRNA is translocated by its 3′ UTR to the apex of the cell (Simmonds et al. 2001; Wilkie and Davis 2001). At the apex, the *wingless* message is translated and secreted from the cell. The cells expressing *engrailed* can bind this protein because they contain Frizzled, which is the *Drosophila* membrane receptor protein for Wingless (Bhanot et al. 1996). Binding of Wingless to Frizzled activates the Wnt signal transduction pathway, resulting in the continued expression of *engrailed* (Siegfried et al. 1994).

Moreover, this activation starts another portion of this reciprocal pathway. The Engrailed protein activates the transcription of the *hedgehog* gene in the *engrailed*-expressing cells. Hedgehog protein can bind to its receptor protein (Patched) on neighboring cells. When it binds to the adjacent posterior cells, it stimulates the expression of the *wingless* gene. The result is a reciprocal loop wherein the Engrailed-synthesizing cells secrete the Hedgehog protein, which maintains the expression of the *wingless* gene in the neighboring cells, while the Wingless-secreting cells maintain the expression of the *engrailed* and *hedgehog* genes in their neighbors in turn (Heemskerk et al. 1991; Ingham et al. 1991; Mohler and Vani 1992). In this way, the transcription pattern of these two types of cells is stabilized. This interaction creates a stable boundary, as well as a signaling center from which Hedgehog and Wingless proteins diffuse across the parasegment.

The diffusion of these proteins is thought to provide the gradients by which the cells of the parasegment acquire their identities. This process can be seen in the dorsal epidermis, where the rows of larval cells produce different cuticular structures depending on their position within the segment. The 1° row of cells consists of large, pigmented spikes called denticles. Posterior to these cells, the 2° row produces a smooth epidermal cuticle. The next two cell rows have a 3° fate, making small, thick hairs, and these are followed by several rows of cells that adopt the 4° fate, producing fine hairs (Figure 9.34).

The fates of the cells can be altered by experimentally increasing or decreasing the levels of Hedgehog or Wingless (Heemskerek and DiNardo 1994; Bokor and DiNardo 1996; Porter et al. 1996); thus, these two proteins appear to be necessary for elaborating the entire pattern of cell types across the parasegment. Gradients of Hedgehog and Wingless are interpreted by a second series of protein gradients within the cells. This second set of gradients provides certain cells with the receptors for Hedgehog and (often) with the receptor for Wingless (Lander et al. 2002; Casal et al. 2002). The resulting pattern of cell fates also changes the focus of patterning from parasegment to segment. There are now external markers, as the *engrailed*-expressing cells become the most posterior cells of each segment.

 WEBSITE **9.7 Asymmetrical spread of morphogens.** It is unlikely that morphogens such as Wingless spread by free diffusion. The asymmetry of Wingless diffusion suggests that neighboring cells play a crucial role in moving the protein.

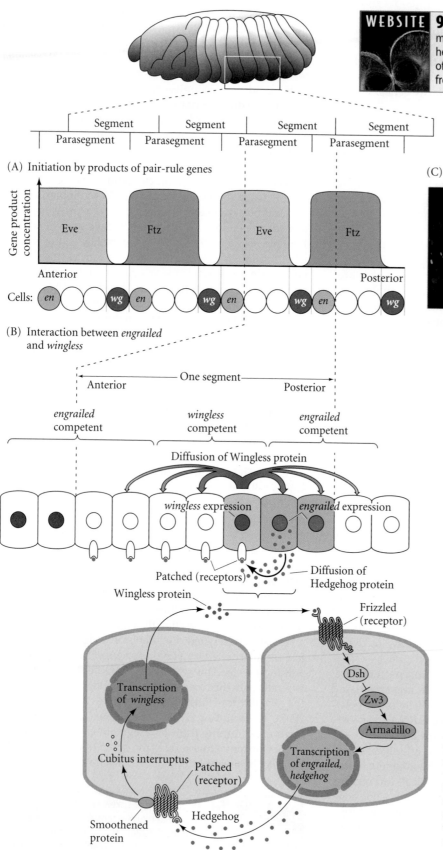

WEBSITE **9.8 Getting a head in the fly.** The segment polarity genes may act differently in the head than in the trunk. Indeed, the formation of the *Drosophila* head may differ significantly from the way the rest of the body is formed.

(A) Initiation by products of pair-rule genes

(C)

(B) Interaction between *engrailed* and *wingless*

FIGURE 9.33 Model for the transcription of the segment polarity genes *engrailed* (*en*) and *wingless* (*wg*). (A) The expression of *wg* and *en* is initiated by pair-rule genes. The *en* gene is expressed in cells that contain high concentrations of either Even-skipped or Fushi tarazu proteins. The *wg* gene is transcribed when neither *eve* or *ftz* genes are active, but a third gene (probably *sloppy-paired*) is expressed. (B) The continued expression of *wg* and *en* is maintained by interactions between the Engrailed- and Wingless-expressing cells. Wingless protein is secreted and diffuses to the surrounding cells. In those cells competent to express Engrailed (i.e., those having Eve or Ftz proteins), Wingless protein is bound by the Frizzled receptor, which enables the activation of the *en* gene via the Wnt signal transduction pathway. (Armadillo is the *Drosophila* name for β-catenin.) Engrailed protein activates the transcription of the *hedgehog* gene and also activates its own (*en*) gene transcription. Hedgehog protein diffuses from these cells and binds to the Patched receptor protein on neighboring cells. This binding prevents the Patched protein from inhibiting signaling by the Smoothened protein. The Smoothened signal enables the transcription of the *wg* gene and the subsequent secretion of the Wingless protein. (C) Localization of *wingless* transcript in the cellular blastoderm of *Drosophila*. Cell membranes have been stained blue, nuclei are stained red, and the *wingless* transcripts are stained green. (C from Simmonds et al. 2001; photographs courtesy of A. J. Simmonds.)

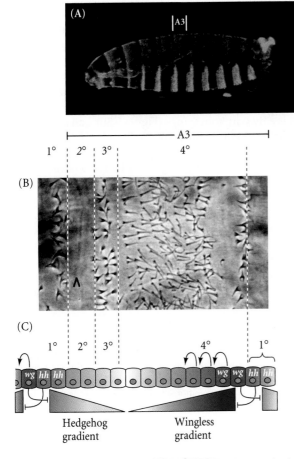

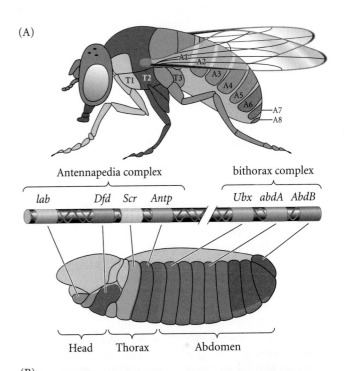

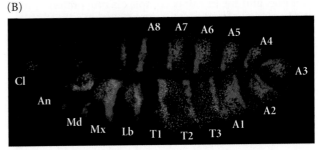

specification is accomplished by the **homeotic selector genes** (Lewis 1978). Two regions of *Drosophila* chromosome 3 contain most of these homeotic genes (Figure 9.35). One of these, the **Antennapedia complex**, contains the homeotic genes *labial (lab), Antennapedia (Antp), sex combs reduced (scr), Deformed (dfd),* and *proboscipedia (pb)*. The *labial* and *Deformed* genes specify the head segments, while *sex combs reduced* and *Antennapedia* contribute to giving the thoracic segments their identities. The *proboscipedia* gene appears

FIGURE 9.34 Cell specification by the Wingless/Hedgehog signaling center. (A) Bright-field photograph of wild-type *Drosophila* embryo, showing the position of the third abdominal segment. (B) Close-up of the dorsal area of the A3 segment, showing the different cuticular structures made by the 1°, 2°, 3°, and 4° rows of cells. The diagram below shows a model for the role of Wingless and Hedgehog. Each signal is responsible for about half the pattern. Either each signal acts in a graded manner (shown here as gradients decreasing with distance from their respective sources) to specify the fates of cells at a distance from these sources, or each signal acts locally on the neighboring cells to initiate a cascade of inductions (shown here as sequential arrows). (After Heemskerk and DiNardo 1994; photographs courtesy of the authors.)

The Homeotic Selector Genes

After the parasegmental boundaries are set, the pair-rule and gap genes interact to regulate the homeotic selector genes, which determine the identity of each segment. By the end of the cellular blastoderm stage, each segment primordium has been given an individual identity by its unique constellation of gap, pair-rule, and homeotic gene products (Levine and Harding 1989).

Patterns of homeotic gene expression

After the segmental boundaries have been established, the characteristic structures of each segment are specified. This

FIGURE 9.35 Homeotic gene expression in *Drosophila*. (A) Expression map of the homeotic genes. In the center are the genes of the Antennapedia and bithorax complexes and their functional domains. Below and above the gene map, the regions of homeotic gene expression (both mRNA and protein) in the blastoderm of the *Drosophila* embryo and the regions that form from them in the adult fly are shown. (B) In situ hybridization for four genes at a slightly later stage (the extended germ band. The engrailed (light blue) expression pattern separates the body into segments; Antennapedia (green) and Ultrabithorax (purple) separate thoracic and abdominal region; Distal-less (red) shows the placement of jaws and the beginnings of limbs. (A after Dessain et al. 1992 and Kaufman et al. 1990; B photograph courtesy of D. Kosman.)

FIGURE 9.36   A four-winged fruit fly constructed by putting together three mutations in *cis* regulators of the *Ultrabithorax* gene. These mutations effectively transform the third thoracic segment into another second thoracic segment (i.e., halteres into wings). (Photograph courtesy of E. B. Lewis.)

to act only in adults, but in its absence, the labial palps of the mouth are transformed into legs (Wakimoto et al. 1984; Kaufman et al. 1990).

The second region of homeotic genes is the **bithorax complex** (Lewis 1978). Three protein-coding genes are found in this complex: *Ultrabithorax (Ubx)*, which is required for the identity of the third thoracic segment; and the *abdominal A (abdA)* and *Abdominal B (AbdB)* genes, which are responsible for the segmental identities of the abdominal segments (Sánchez-Herrero et al. 1985). The chromosome region containing both the Antennapedia complex and the bithorax complex is often referred to as the **homeotic complex (Hom-C)**.

Because the homeotic selector genes are responsible for the specification of fly body parts, mutations in them lead to bizarre phenotypes. In 1894, William Bateson called these organisms **homeotic mutants**, and they have fascinated developmental biologists for decades.* For example, the body of the normal adult fly contains three thoracic segments, each of which produces a pair of legs. The first thoracic segment does not produce any other appendages, but the second thoracic segment produces a pair of wings

Homeo, from the Greek, means "similar." *Homeotic mutants* are mutants in which one structure is replaced by another (as where an antenna is replaced by a leg). *Homeotic genes* are those genes whose mutation can cause such transformations; thus, homeotic genes are genes that specify the identity of a particular body segment. The *homeobox* is a conserved DNA sequence of about 180 base pairs that is shared by many homeotic genes. This sequence encodes the 60-amino-acid *homeodomain*, which recognizes specific DNA sequences. The homeodomain is an important region of the transcription factors encoded by homeotic genes (see Sidelights & Speculations). However, not all genes containing homeoboxes are homeotic genes.

in addition to its legs. The third thoracic segment produces a pair of wings and a pair of balancers known as **halteres**. In homeotic mutants, these specific segmental identities can be changed. When the *Ultrabithorax* gene is deleted, the third thoracic segment (characterized by halteres) is transformed into another second thoracic segment. The result is a fly with four wings (Figure 9.36)—an embarrassing situation for a classic dipteran.†

Similarly, Antennapedia protein usually specifies the second thoracic segment of the fly. But when flies have a mutation wherein the *Antennapedia* gene is expressed in the head (as well as in the thorax), legs rather than antennae grow out of the head sockets (Figure 9.37). In the recessive mutant of *Antennapedia*, the gene fails to be expressed in the second thoracic segment, and antennae sprout in the leg positions (Struhl 1981; Frischer et al. 1986; Schneuwly et al. 1987).

The major homeotic selector genes have been cloned and their expression analyzed by in situ hybridization (Harding et al. 1985; Akam 1987). Transcripts from each

†Dipterans (two-winged insects such as flies) are thought to have evolved from four-winged insects; it is possible that this change arose via alterations in the bithorax complex. Chapter 23 includes more speculation on the relationship between the homeotic complex and evolution.

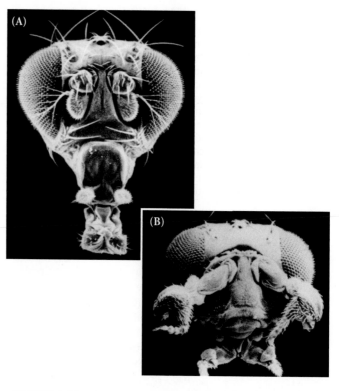

FIGURE 9.37   (A) Head of a wild-type fruit fly. (B) Head of a fly containing the *Antennapedia* mutation that converts antennae into legs. (From Kaufman et al. 1990; photographs courtesy of T. C. Kaufman.)

gene can be detected in specific regions of the embryo and are especially prominent in the central nervous system (see Figure 9.35B).

Initiating the patterns of homeotic gene expression

The initial domains of homeotic gene expression are influenced by the gap genes and pair-rule genes. For instance, expression of the *abdA* and *AbdB* genes is repressed by the gap gene proteins Hunchback and Krüppel. This inhibition prevents these abdomen-specifying genes from being expressed in the head and thorax (Casares and Sánchez-Herrero 1995). Conversely, the *Antennapedia* gene is activated by particular levels of Hunchback (needing both the maternal and the zygotically transcribed messages), so *Antennapedia* is originally transcribed in parasegment 4, specifying the mesothoracic (T2) segment (Wu et al. 2001).

Maintaining the patterns of homeotic gene expression

The expression of homeotic genes is a dynamic process. The *Antennapedia* gene, for instance, although initially expressed in presumptive parasegment 4, soon appears in parasegment 5. As the germ band expands, *Antp* expression is seen in the presumptive neural tube as far posterior as parasegment 12. During further development, the domain of *Antp* expression contracts again, and *Antp* transcripts are localized strongly to parasegments 4 and 5. Like that of other homeotic genes, *Antp* expression is negatively regulated by all the homeotic gene products expressed posterior to it (Levine and Harding 1989; González-Reyes and Morata 1990). In other words, each of the bithorax complex genes represses the expression of *Antp*. If the *Ultrabithorax* gene is deleted, *Antp* activity extends through the region that would normally have expressed *Ubx* and stops where the *Abd* region begins. (This allows the third thoracic segment to form wings like the second thoracic segment, as in Figure 9.36.) If the entire bithorax complex is deleted, *Antp* expression extends throughout the abdomen. (Such a larva does not survive, but the cuticle pattern throughout the abdomen is that of the second thoracic segment.)

As we have seen, gap gene and pair-rule gene proteins are transient, but the identities of the segments must be stabilized so that differentiation can occur. Thus, once the transcription patterns of the homeotic genes have become stabilized, they are "locked" into place by alteration of the chromatin conformation in these genes. The repression of homeotic genes appears to be maintained by the **Polycomb** family of proteins, while the active chromatin conformation appears to be maintained by the **Trithorax** proteins (Ingham and Whittle 1980; McKeon and Brock 1991; Simon et al. 1992).

Realisator genes

The homeotic genes work by activating or repressing a group of "realisator genes," those genes that are the targets of the homeotic gene proteins and which function to form the specified tissue or organ primordia. For example, the *Antennapedia* gene is expressed in the formation of the second thoracic segment. Antennapedia protein binds to and represses the enhancers of at least two genes, *homothorax* and *eyeless*, which encode transcription factors that are critical for antenna and eye formation, respectively (Casares and Mann 1998; Plaza et al. 2001). Therefore, one of Antennapedia's functions is to suppress the genes that would trigger antenna and eye development.

The Ultrabithorax protein is also able to repress the expression of the *wingless* gene in those cells that will become the halteres of the fly. One of the major differences between the dorsal imaginal disc cells of the second and the third thoracic segments is that *wingless* expression occurs in the imaginal discs cells of the second thoracic segment, but not in those of the third thoracic segment. Wingless protein acts as a growth promoter and morphogen in these tissues. Ubx protein is found in the cells of the third thoracic segment, where it prevents expression of the *wingless* gene (Figure 9.38; Weatherbee et al. 1998). Thus, one of the ways in which Ubx specifies the third thoracic segment is by preventing the expression of those genes that would generate wing tissue.

Another target of the homeotic proteins, the *distal-less* gene (itself a homeobox-containing gene; see Sidelights &

(A)

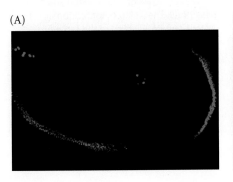

(B)

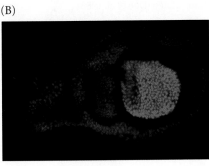

FIGURE 9.38 Antibody staining of the Ultrabithorax protein in (A) the wing disc and (B) the haltere disc of third instar *Drosophila* larvae. The cells of these discs will give rise to a wing and a haltere, respectively. In the wing disc, Ultrabithorax staining can be seen only on the cells that form the peripodial membrane, and not on those that form the wing itself. In the haltere disc, Ultrabithorax is found in those cells that will produce the major portion of the haltere. (From Weatherbee et al. 1998; photographs courtesy of S. D. Weatherbee and S. Carroll.)

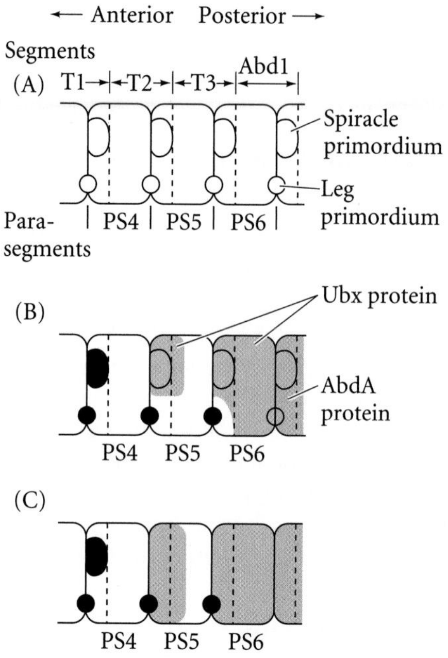

FIGURE 9.39 Schematic representation of the differences between *Ubx* expression in parasegments 5 and 6. (A) Before *Ubx* expression, each parasegment is competent to make both spiracles and legs. (B) At division cycle 10, early *Ubx* expression blocks the formation of the anterior spiracle in PS5 and PS6, and prevents limb formation in the posterior compartment of PS6. AbdA protein plays the same role in other abdominal segments. (C) At division cycle 11, the *Ubx* expression domain extends to the limb primordia of PS5 and PS6, but it is "too late" to repress *Distal-less* gene expression. (After Castelli-Gair and Akam 1995.)

Speculations), is necessary for limb development and is active solely in the thorax. *Distal-less* expression is repressed in the abdomen by a combination of Ubx and AbdA proteins and pair-rule genes, which bind to its enhancer and block its transcription (Castelli-Gair and Akam 1995; Gebelein et al. 2004). This expression pattern presents a paradox, since parasegment 5 (entirely thoracic and leg-producing) and parasegment 6 (which includes most of the legless first abdominal segment) both express Ubx. How can these two very different segments be specified by the same gene?

Castelli-Gair and Akam (1995) have shown that the mere presence of Ubx in a group of cells is not sufficient for specification. Rather, the time and place of its expression within the parasegment can be critical. Before *Ubx* gene expression, parasegments 4–6 have similar potentials. At division cycle 10, *Ubx* expression in the anterior parts of parasegments 5 and 6 prevents those parasegments from forming structures (such as the anterior spiracle) characteristic of

parasegment 4. Moreover, in the posterior compartment of parasegment 6 (but not parasegment 5), Ubx blocks the formation of the limb primordium by repressing the *Distal-less* gene. At division cycle 11, by which time Ubx has pervaded all of parasegment 6, the *Distal-less* gene has become self-regulatory and cannot be repressed by Ubx (Figure 9.39).

SIDELIGHTS & SPECULATIONS

The Homeodomain Proteins

Homeodomain proteins are a family of transcription factors characterized by a 60-amino-acid domain (the **homeodomain**) that binds to certain regions of DNA. The homeodomain was first discovered in those proteins whose absence or misregulation caused homeotic transformations of *Drosophila* segments. Homeodomain proteins are believed to activate batteries of genes that specify the particular properties of each segment. The homeodomain proteins include the products of the eight genes of the homeotic complex (Hom-C), as well as other proteins such as Fushi tarazu, Caudal, Distal-less and Bicoid. Home-odomain proteins are important in determining the anterior-posterior axes of both invertebrates and vertebrates. In *Drosophila*, the presence of certain homeodomain proteins is also necessary for the determination of specific neurons. Without these transcription factors, the fates of these neuronal cells are altered (Doe et al. 1988).

Structure of the Homeodomain

The homeodomain is encoded by a 180-base-pair DNA sequence known as the **homeobox**. The homeodomains appear to be the DNA-binding sites of these proteins, and they are critical in specifying cell fates. For instance, if a chimeric protein is constructed mostly of Antennapedia but with the carboxyl terminus (including the home-odomain) of Ultrabithorax, it can substitute for Ultrabithorax and specify the appropriate cells as parasegment 6 (Mann and Hogness 1990). The isolated homeodomain of Antennapedia will bind to the same promoters as the entire Antennapedia protein, indicating that the binding of this protein is dependent on its homeodomain (Müller et al. 1988).

The homeodomain folds into three α helices, the latter two folding into a helix-

SIDELIGHTS & SPECULATIONS

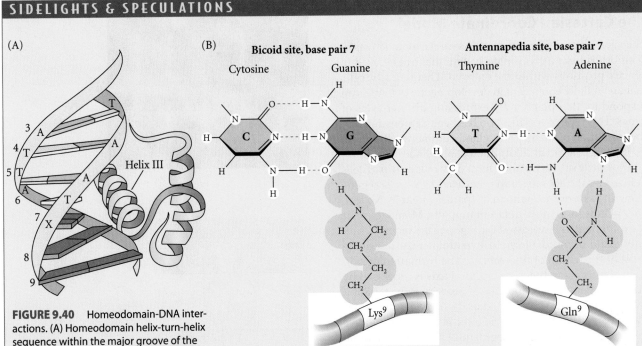

FIGURE 9.40 Homeodomain-DNA inter-actions. (A) Homeodomain helix-turn-helix sequence within the major groove of the DNA. (B) Proposed pairing between the lysine of the Bicoid homeodomain and the CG base pair of its recognition sequence, and between the glutamine of the Antenna-pedia homeodomain and the TA base pair of its recognition sequence. In both cases, the ninth amino acid of the homeodomain helix bonds with the base pair immediately following the TAAT sequence. (A after Riddi-hough 1992; B after Hanes and Brent 1991.)

turn-helix conformation that is characteristic of transcription factors that bind DNA in the major groove of the double helix (Otting et al. 1990; Percival-Smith et al. 1990). The third helix is the recognition helix (Figure 9.40A), and it is here that the amino acids make contact with the bases of the DNA. A four-base motif, TAAT, is conserved in nearly all sites recognized by homeodomains; it probably distinguishes those sites to which homeodomain proteins can bind. The 5′ terminal T appears to be critical in this recognition, as mutating it destroys all homeodomain binding. The base pairs following the TAAT motif are important in distinguishing between similar recognition sites. For instance, the next base pair is recognized by amino acid 9 of the recognition helix. Mutation studies have shown that the Bicoid and Antennapedia homeodomain proteins use lysine and glutamine, respectively, at position 9 to distinguish related recognition

sites. The lysine of the Bicoid homeodomain recognizes the G of CG pairs, while the glutamine of the Antennapedia homeodomain recognizes the A of AT pairs (Figure 9.40B,C; Hanes and Brent 1991). If the lysine in Bicoid is replaced by glutamine, the resulting protein will recognize Antennapedia-binding sites (Hanes and Brent 1989, 1991). Other homeodomain proteins show a similar pattern, in which one portion of the homeodomain recognizes the common TAAT sequence, while another portion recognizes a specific structure adjacent to it.

Cofactors for the Hom-C Genes

The genes of the *Drosophila* homeotic complex specify segmental fates, but they may need some help in doing so. The DNA-binding sites recognized by the homeodomains of the Hom-C proteins are very similar, and there is some overlap in their binding specificity. In 1990, Peifer and Wieschaus discovered that the product of the *Extradenticle* (*Exd*) gene interacts with several Hom-C proteins and may help to specify segmental identities. For instance, the Ubx protein is responsible for specifying the identity of the first abdominal segment (A1), but without Extradenticle protein, Ubx will transform this segment into A3. Moreover, Exd and

Ubx proteins are both needed for the regulation of the *decapentaplegic* (*dpp*) gene, and the structure of the *dpp* promoter suggests that the Exd protein may dimerize with the Ubx protein on the enhancer of this target gene (Raskolb and Wieschaus 1994; van Dyke and Murre 1994).

The products of the *teashirt* and *disco* genes may also be important cofactors. These zinc finger transcription factors are necessary for the functioning of the Sex combs reduced (SCR) protein. Teashirt is critical for the *specification* of anterior prothoracic (parasegment 3) identity, and it may be the gene that specifies the "ground state" condition of the homeotic complex. If the Bithorax complex and the *Antennapedia* gene are removed, all the segments become anterior prothorax. The product of the *teashirt* gene appears to work with the SCR protein to distinguish thorax from head, and also to work throughout the trunk to prevent head structures from forming. Disco protein functions anteriorly to help specify the head segments, while Teashirt protein inhibits *disco* transcription and helps SCR protein produce thorax segments. Thus, the Disco and Teashirt cofactors of SCR act to distinguish between the labial (jaw) and first thoracic segments (Roder et al. 1992; Robertson 2004).

Axes and Organ Primordia: The Cartesian Coordinate Model

The anterior-posterior and dorsal-ventral axes of *Drosophila* embryos form a coordinate system that can be used to specify positions within the embryo. Theoretically, cells that are initially equivalent in developmental potential can respond to their position by expressing different sets of genes. This type of specification has been demonstrated in the formation of the salivary gland rudiments (Panzer et al. 1992; Bradley et al. 2001; Zhou et al. 2001).

Drosophila salivary glands form only in the strip of cells defined by the activity of the *sex combs reduced* (*scr*) gene along the anterior-posterior axis (parasegment 2). No salivary glands form in *scr*-deficient mutants. Moreover, if *scr* is experimentally expressed throughout the embryo, salivary gland primordia form in a ventrolateral stripe along most of the length of the embryo. The formation of salivary glands along the dorsal-ventral axis is repressed by both Decapentaplegic and Dorsal proteins, which inhibit salivary gland formation both dorsally and ventrally. Thus, the salivary glands form at the intersection of the vertical *scr* expression band (parasegment 2) and the horizontal region in the middle of the embryo's circumference that has neither Decapentaplegic nor Dorsal (Figure 9.41). The cells that form the salivary glands are directed to do so by the intersecting gene activities along the anterior-posterior and dorsal-ventral axes.

A similar situation is seen with tissues that are found in every segment of the fly. Neuroblasts arise from 10 clusters of 4 to 6 cells each that form on each side in every segment in the strip of neural ectoderm at the midline of the embryo (Skeath and Carroll 1992). The cells in each cluster interact (via the Notch pathway discussed in Chapter 6) to generate a single neural cell from each cluster. Skeath and colleagues (1993) have shown that the pattern of neural gene transcription is imposed by a coordinate system. Their expression is repressed by the Decapentaplegic and Snail proteins along the dorsal-ventral axis, while positive enhancement by pair-rule genes along the anterior-posterior axis causes their repetition in each half-segment. It is very likely, then, that the positions of organ primordia in the fly are specified via a two-dimensional coordinate system based on the intersection of the anterior-posterior and dorsal-ventral axes.

Coda

Genetic studies on the *Drosophila* embryo have uncovered numerous genes that are responsible for the specification of the anterior-posterior and dorsal-ventral axes. We are far from a complete understanding of *Drosophila* pattern

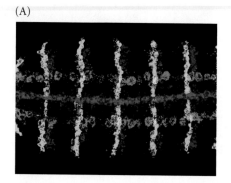

(A)

(B)

scr activates

dpp inhibits

Salivary gland

dorsal inhibits

FIGURE 9.41 Cartesian coordinate system mapped out by gene expression patterns. (A) A grid (ventral view, looking "up" at the embryo) formed by the expression of *short-gastrulation* (red), *intermediate neuroblast defective* (green), *muscle segment homeobox* (magenta), *wingless* (yellow), and *engrailed* (blue) transcripts. (B) Coordinates for the expression of genes giving rise to *Drosophila* salivary glands. These genes are activated by the protein product of the *sex combs reduced* (*scr*) homeotic gene in a narrow band along the anterior-posterior axis, and they are inhibited in the regions marked by *decapentaplegic* (*dpp*) and *dorsal* gene products along the dorsal-ventral axis. This pattern allows salivary glands to form in the midline of the embryo in the second parasegment. (A photograph courtesy of D. Kosman; B after Panzer et al. 1992.)

formation, but we are much more aware of its complexity than we were a decade ago. Mutations of *Drosophila* genes have given us our first glimpses of the multiple levels of pattern regulation in a complex organism and have enabled us to isolate these genes and their products. Most importantly, as we will see in forthcoming chapters, the *Drosophila* genes provide clues to a general mechanism of pattern formation that is used throughout the animal kingdom.

Drosophila Development and Axis Specification

1. *Drosophila* cleavage is superficial. The nuclei divide 13 times before forming cells. Before cell formation, the nuclei reside in a syncytial blastoderm. Each nucleus is surrounded by actin-filled cytoplasm.

2. When the cells form, the *Drosophila* embryo undergoes a mid-blastula transition, wherein the cleavages become asynchronous and new mRNA is made. The amount of chromatin determines the timing of this transition.

3. Gastrulation begins with the invagination of the most ventral region (the presumptive mesoderm), which causes the formation of a ventral furrow. The germ band expands such that the future posterior segments curl just behind the presumptive head.

4. The genes regulating pattern formation in *Drosophila* operate according to certain principles:

   - There are *morphogens*—such as Bicoid and Dorsal—whose gradients determine the specification of different cell types. These morphogens can be transcription factors.
   - There is a *temporal order* wherein different classes of genes are transcribed, and the products of one gene often regulate the expression of another gene.
   - *Boundaries* of gene expression can be created by the interaction between transcription factors and their gene targets. Here, the transcription factors transcribed earlier regulate the expression of the next set of genes.
   - *Translational control* is extremely important in the early embryo, and localized mRNAs are critical in patterning the embryo.
   - *Individual cell fates* are not defined immediately. Rather, there is a stepwise specification wherein a given field is divided and subdivided, eventually regulating individual cell fates.

5. Maternal effect genes are responsible for the initiation of anterior-posterior polarity. *Bicoid* mRNA is bound by its 3′ UTR to the cytoskeleton in the future anterior pole; *nanos* mRNA is sequestered by its 3′ UTR in the future posterior pole. *Hunchback* and *caudal* messages are seen throughout the embryo.

6. Dorsal-ventral polarity is regulated by the entry of Dorsal protein into the nucleus. Dorsal-ventral polarity is initiated when the nucleus moves to the dorsal-anterior of the oocyte and transcribes the *gurken* message, which is then transported to the region above the nucleus and adjacent to the follicle cells.

7. Gurken protein is secreted from the oocyte and binds to its receptor (Torpedo) on the follicle cells. This binding dorsalizes the follicle cells, preventing them from synthesizing Pipe.

8. Pipe protein in the ventral follicle cells modifies an as yet unknown factor that modifies the Nudel protein. This modification allows Nudel to activate a cascade of proteolysis in the space between the ventral follicle cells and the ventral cells of the embryo. As a result of this cascade, the Spätzle protein is activated and binds to the Toll protein on the ventral embryonic cells.

9. The activated Toll protein initiates a cascade that phosphorylates the Cactus protein, which has been bound to Dorsal. Phosphorylated Cactus is degraded, allowing Dorsal to enter the nucleus. Once in the nucleus, Dorsal activates the genes responsible for the ventral cell fates and represses those genes whose proteins would specify dorsal cell fates.

10. Dorsal protein forms a gradient as it enters the various nuclei. Those nuclei at the most ventral surface incorporate the most Dorsal protein and become mesoderm; those more lateral become neurogenic ectoderm.

11. The Bicoid and Hunchback proteins activate the genes responsible for the anterior portion of the fly; Caudal activates genes responsible for posterior development.

12. The unsegmented anterior and posterior extremities are regulated by the activation of the Torso protein at the anterior and posterior poles of the egg.

13. The gap genes respond to concentrations of the maternal effect gene proteins. Their protein products interact with each other such that each gap gene protein defines specific regions of the embryo.

14. The gap gene proteins activate and repress the pair-rule genes. The pair-rule genes have modular promoters such that they become activated in seven "stripes." Their boundaries of transcription are defined by the gap genes. The pair-rule genes form seven bands of transcription along the anterior-posterior axis, each one comprising two parasegments.

15. The pair-rule gene products activate *engrailed* and *wingless* expression in adjacent cells. The *engrailed*-expressing cells form the anterior boundary of each parasegment. These cells form a signaling center that organizes the cuticle formation and segmental structure of the embryo.

16. The homeotic selector genes are found in two complexes on chromosome 3 of *Drosophila*. Together, these regions are called Hom-C, the homeotic gene complex. The genes are arranged in the same order as their transcriptional expression. The Hom-C genes specify the individual segments, and mutations in these genes are capable of transforming one segment into another.

17. The expression of each homeotic selector gene is regulated by the gap and pair-rule genes. Their expression is refined and maintained by interactions whereby their protein products prevent the transcription of neighboring Hom-C genes.

18. The targets of the Hom-C proteins are the realisator genes. These genes include *distal-less* and *wingless* (in the thoracic segments).

19. Organs form at the intersection of dorsal-ventral and anterior-posterior regions of gene expression.

For Further Reading

Complete bibliographical citations for all literature cited in this chapter can be found on the Vade Mecum CD that accompanies the book and at the free access website www.devbio.com

Driever, W. and C. Nüsslein-Volhard. 1988a. The Bicoid protein determines position in the *Drosophila* embryo in a concentration-dependent manner. *Cell* 54: 95–104.

Fujioka, M., Y. Emi-Sarker, G. L. Yusibova, T. Goto and J. B. Jaynes. 1999. Analysis of an *even-skipped* rescue transgene reveals both composite and discrete neuronal and early blastoderm enhancers, and multi-stripe positioning by gap gene repressor gradients. *Development* 126: 2527–2538.

Lehmann, R. and C. Nüsslein-Volhard. 1991. The maternal gene *nanos* has a central role in posterior pattern formation of the *Drosophila* embryo. *Development* 112: 679–691.

Leptin, M. 1991. *twist* and *snail* as positive and negative regulators during *Drosophila* mesoderm development. *Genes Dev.* 5: 1568–1576.

Lewis, E. B. 1978. A gene complex controlling segmentation in *Drosophila*. *Nature* 276: 565–570.

Martinez-Arias, A. and P. A. Lawrence. 1985. Parasegments and compartments in the *Drosophila* embryo. *Nature* 313: 639–642.

Pankratz, M. J., E. Seifert, N. Gerwin, B. Billi, U. Nauber and H. Jäckle. 1990. Gradients of *Krüppel* and *knirps* gene products direct pair-rule gene stripe patterning in the posterior region of the *Drosophila* embryo. *Cell* 61: 309–317.

Roth, S., D. Stein and C. Nüsslein-Volhard. 1989. A gradient of nuclear localization of the dorsal protein determines dorsoventral pattern in the *Drosophila* embryo. *Cell* 59: 1189–1202.

Schüpbach, T. 1987. Germ line and soma cooperate during oogenesis to establish the dorsoventral pattern of egg shell and embryo in *Drosophila melanogaster*. *Cell* 49: 699–707.

Struhl, G. 1981. A homeotic mutation transforming leg to antenna in *Drosophila*. *Nature* 292: 635–638.

Wang, C. and R. Lehman. 1991. Nanos is the localized posterior determinate in *Drosophila*. *Cell* 66: 637–647.

Early Development and Axis Formation in Amphibians

10

AMPHIBIAN EMBRYOS ONCE DOMINATED the field of experimental embryology. With their large cells and their rapid development, salamander and frog embryos were excellently suited for transplantation experiments. However, amphibian embryos fell out of favor during the early days of developmental genetics, since frogs and salamanders undergo a long period of growth before they become fertile, and their chromosomes are often found in several copies, precluding easy mutagenesis.* However, new molecular techniques such as in situ hybridization, antisense oligonucleotides, and dominant negative proteins have allowed researchers to return to studying amphibian embryos and to integrate molecular analyses of development with earlier experimental findings. The results have been spectacular, and we are enjoying new vistas of how vertebrate bodies are patterned and structured.

Early Amphibian Development

Fertilization and cortical rotation

Fertilization can occur anywhere in the animal hemisphere of the amphibian embryo. The point of sperm entry is important because it determines the orientation of the dorsal-ventral axis of the larva (tadpole). That point will mark the ventral side of the embryo, while the site 180 degrees opposite the point of sperm entry will mark the dorsal side.[†]

The sperm centriole organizes the microtubules of the egg and causes them to arrange themselves in a parallel array in the vegetal cytoplasm, separating the cortical cytoplasm from the yolky internal cytoplasm (Figure 10.1). These microtubular tracks allow the cortical cytoplasm to rotate with respect to the inner cytoplasm. Indeed, the arrays are first seen immediately before rotation starts, and they disappear when rotation ceases (Elinson and Rowning 1988; Houliston and Elinson 1991). The cortical cytoplasm rotates 30 degrees with

> [66] We are standing and walking with parts of our body which could have been used for thinking had they developed in another part of the embryo. [99]
>
> *HANS SPEMANN (1943)*

> [66] Theories come and theories go. The frog remains. [99]
>
> *JEAN ROSTAND (1960)*

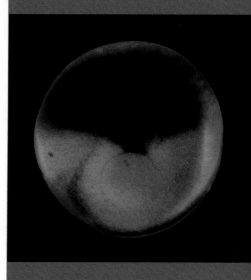

---

*In the 1960s, *Xenopus laevis* replaced *Rana* frogs and the salamanders because it could be induced to mate throughout the year. Unfortunately, *Xenopus laevis* has four copies of each chromosome rather than the more usual two, and it takes 1–2 years to reach sexual maturity. Another *Xenopus* species, *X. tropicalis*, has all the advantages of *X. laevis*, plus it is diploid and reaches sexual maturity in a mere 6 months (Gurdon and Hopwood 2000; Hirsch et al. 2002).

[†]The axis between the point of sperm entry and the dorsal side approximates, but does not exactly correspond to, the actual ventral-dorsal axis of the amphibian larva (Lane and Smith 1999; Lane and Sheets 2002). However, as the literature in the field has traditionally equated these two axes, we will use the classical terminology here.

(A)

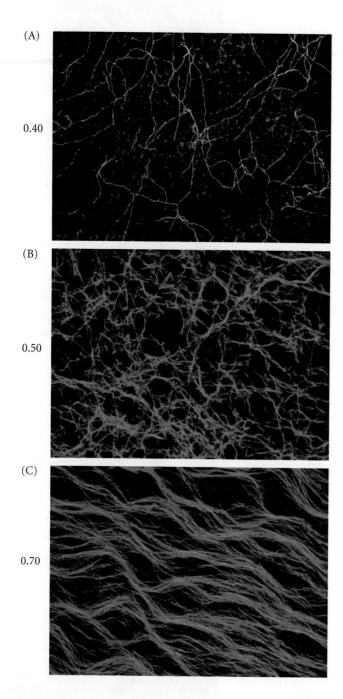

0.40

(B)

0.50

(C)

0.70

FIGURE 10.1   Formation of the parallel arrays of microtubules extend along the vegetal hemisphere along the future dorsal-ventral axis. The microtubules were visualized by using fluorescent antibodies to tubulin. (A) Vegetal view of the fertilized egg, 40 percent of the way through the first cell cycle. Microtubules appear in the shear zone between the cortex and inner cytoplasm, but they are short and disorganized. (B) With 50 percent of the first cell cycle completed, more microtubules are present, but they do not have any polarity. (C) By 70 percent of the first cell cycle, the vegetal shear zone is characterized by a parallel array of microtubules. Cortical rotation begins at this time. At the end of cytoplasmic rotation, the microtubules depolymerize. (From Cha and Gard 1999, photographs courtesy of the authors.)

Unequal radial holoblastic cleavage

Cleavage in most frog and salamander embryos is radially symmetrical and holoblastic, just like echinoderm cleavage. The amphibian egg, however, contains much more yolk. This yolk, which is concentrated in the vegetal hemisphere, is an impediment to cleavage. Thus, the first division begins at the animal pole and slowly extends down into the vegetal region (Figure 10.3; see also Figure 2.2D). In the axolotl salamander, the cleavage furrow extends through the animal hemisphere at a rate close to 1 mm per minute, slowing down to a mere 0.02–0.03 mm per minute as it approaches the vegetal pole (Hara 1977). In some species (especially salamanders and *Rana*), the first cleavage division bisects the gray crescent.

At the frog egg's first cleavage, one can see the difference in the furrow between the animal and the vegetal hemispheres (Figure 10.4A). While the cleavage furrow is still cleaving the yolky cytoplasm of the vegetal hemisphere, the second cleavage has already started near the animal pole. This cleavage is at right angles to the first one and is also meridional (Figure 10.4B). The third cleavage, as expected, is equatorial. However, because of the vegetally placed yolk, this cleavage furrow in amphibian eggs is not actually at the equator, but is displaced toward the animal pole (Valles et al. 2002). It divides the embryo into four small animal blastomeres (micromeres) and four large blastomeres (macromeres) in the vegetal region (Figure 10.4C). Despite their unequal sizes, the blastomeres continue to divide at the same rate until the twelfth cell cycle (with only a small delay of the vegetal cleavages). As cleavage progresses, the animal region becomes packed with numerous small cells, while the vegetal region contains a relatively small number of large, yolk-laden macromeres (see Figures 10.3 and 2.2E).

The cell cycles of early *Xenopus* blastomeres are regulated by the levels of mitosis-promoting factor (MPF) in the cytoplasm. As discussed in Chapter 8, there are no gap stages in the *Xenopus* cell cycle for the first 12 divisions, and there is no growth between cell divisions. (Indeed, the embryo does not increase its mass until the tadpole starts to feed.) An amphibian embryo containing 16 to 64 cells is

respect to the internal cytoplasm. In some eggs, this exposes a band of inner gray cytoplasm in the marginal region of the one-cell embryo, directly opposite the sperm entry point (Roux 1887; Ancel and Vintenberger 1948). This region, the **gray crescent**, is where gastrulation will begin (Figure 10.2). Even in *Xenopus* eggs, which do not expose a gray crescent, cortical rotation occurs and cytoplasmic movements can be seen (Manes and Elinson 1980; Vincent et al. 1986). As we will see later, the site opposite the point of sperm entry will be the place where gastrulation begins, and the microtubular array will become extremely important in initiating the dorsal-ventral and anterior-posterior axes of the larva.

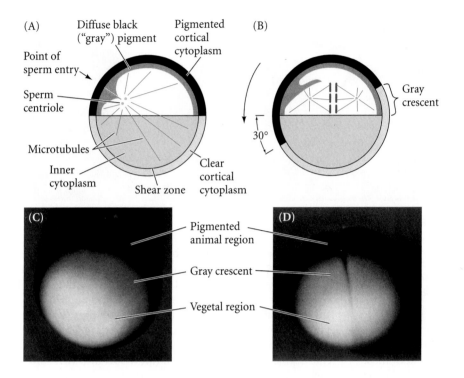

FIGURE 10.2 Reorganization of cytoplasm in the newly fertilized frog egg. (A) Schematic cross section of an egg midway through the first cleavage cycle. The egg has radial symmetry about its animal-vegetal axis. The sperm has entered at one side, and the sperm nucleus is migrating inward. The cortex represented is like that of *Rana*, with a heavily pigmented animal hemisphere and an unpigmented vegetal hemisphere. (B) About 80 percent of the way into first cleavage, the cortical cytoplasm rotates 30 degrees relative to the internal cytoplasm. Gastrulation will begin in the gray crescent, the region opposite the point of sperm entry where the greatest displacement of cytoplasm occurs. (C) Gray crescent of *R. pipiens* immediately after cortical rotation exposes the different pigmentation beneath the heavily pigmented cortical cytoplasm. (D) The first cleavage furrow bisects the gray crescent. (A,B after Gerhart et al. 1989; C,D photographs courtesy of R. P. Elinson.)

commonly called a **morula** (plural: **morulae**; Latin, "mulberry," whose shape it vaguely resembles). At the 128-cell stage, the blastocoel becomes apparent, and the embryo is considered a blastula. Actually, the formation of the blastocoel has been traced back to the very first cleavage furrow. Kalt (1971) demonstrated that in the frog *Xenopus laevis*, the first cleavage furrow widens in the animal hemisphere to create a small intercellular cavity that is sealed off from the outside by tight intercellular junctions. This cavity expands during subsequent cleavages to become the blastocoel (Figure 10.5A).

The amphibian blastocoel serves two major functions: (1) it permits cell migration during gastrulation, and (2) it prevents the cells beneath it from interacting prematurely with the cells above it. When Nieuwkoop (1973) took embryonic newt cells from the roof of the blastocoel in the animal hemisphere (a region often called the **animal cap**) and placed them next to the yolky vegetal cells from the base of the blastocoel, the cap cells differentiated into mesodermal tissue instead of ectoderm. Mesodermal tissue is normally formed from those animal cells adjacent to the vegetal endoderm precursors, because vegetal cells induce adjacent cells to differentiate into mesodermal tissues. Thus, the blastocoel prevents the contact of the vegetal cells destined to become endoderm with those cells in the ectoderm fated to give rise to the skin and nerves.

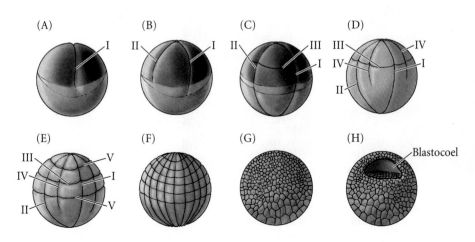

FIGURE 10.3 Cleavage of a frog egg. Cleavage furrows, designated by Roman numerals, are numbered in order of appearance. (A,B) Because the vegetal yolk impedes cleavage, the second division begins in the animal region of the egg before the first division has divided the vegetal cytoplasm. (C) The third division is displaced toward the animal pole. (D–H) The vegetal hemisphere ultimately contains larger and fewer blastomeres than the animal half. H represents a cross section through a mid-blastula stage embryo. (After Carlson 1981.)

(A)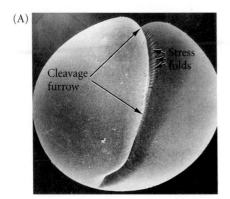

(B)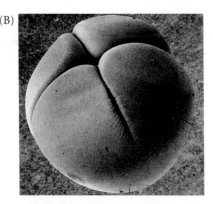

(C)

FIGURE 10.4 Scanning electron micrographs of frog egg cleavage. (A) First cleavage. (B) Second cleavage (4 cells). (C) Fourth cleavage (16 cells), showing the size discrepancy between the animal and vegetal cells after the third division. (A from Beams and Kessel 1976, photograph courtesy of the authors; B and C, photographs courtesy of L. Biedler.)

Numerous cell adhesion molecules keep the cleaving blastomeres together. One of the most important of these is EP-cadherin. The mRNA for this protein is supplied in the oocyte cytoplasm. If this message is destroyed by antisense oligonucleotides so that no EP-cadherin is made, the adhesion between blastomeres is dramatically reduced, resulting in the obliteration of the blastocoel (Figure 10.5B; see also Figure 3.28A; Heasman et al. 1994a,b).

The Mid-Blastula Transition: Preparing for Gastrulation

The first precondition for gastrulation is the activation of the genome. In *Xenopus*, only a few genes appear to be transcribed during early cleavage. For the most part, nuclear genes are not activated until late in the twelfth cell cycle (Newport and Kirschner 1982a,b; Yang et al. 2002).

At that time, different genes begin to be transcribed in different cells, and the blastomeres acquire the capacity to become motile. This dramatic change is called the **mid-blastula transition** (**MBT**; see Chapters 8 and 9). It is thought that some factor in the egg is being titrated by the newly made chromatin, because the time of this transition can be changed experimentally by altering the ratio of chromatin to cytoplasm in the cell (Newport and Kirschner 1982a,b).

One of the events that triggers the mid-blastula transition is the demethylation of certain promoters. In *Xenopus* (unlike mammals), high levels of methylated DNA are seen in both the paternally and maternally derived chromosomes. However, during the late blastula stages, there is a loss of methylation on the promoters of genes that are activated at MBT. This demethylation is not seen on promoters that are not activated at MBT, nor is it observed in the actual coding regions of MBT-activated genes. It appears, then, that demethylation of certain promoters may play a pivotal role in regulating the timing of gene expression at mid-blastula transition (Stancheva et al. 2002).

It is thought that once the chromatin at the promoters has been remodeled, various transcription factors (such as

(A)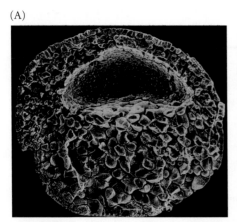

(B)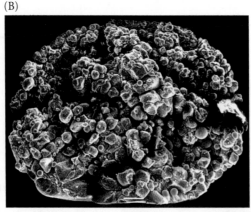

FIGURE 10.5 Depletion of EP-cadherin mRNA in the *Xenopus* oocyte results in the loss of adhesion between blastomeres and the obliteration of the blastocoel. (A) Control embryo. (B) EP-cadherin-depleted embryo. (From Heasman et al. 1994b, photographs courtesy of J. Heasman.)

the VegT protein, formed in the vegetal cytoplasm from maternal mRNA) bind to the promoters and initiate new transcription. For instance, the vegetal cells (probably under the direction of the VegT protein) become the endoderm and begin secreting the factors that induce the cells above them to become the mesoderm (Wylie et al. 1996; Agius et al. 2000).

Amphibian Gastrulation

The study of amphibian gastrulation is both one of the oldest and one of the newest areas of experimental embryology (see Beetschen 2001; Braukmann and Gilbert 2005). Even though amphibian gastrulation has been extensively studied for the past century, most of our theories concerning the mechanisms of these developmental movements have been revised over the past decade. The study of amphibian gastrulation has been complicated by the fact that there is no single way amphibians gastrulate; different species employ different means toward the same goal (Smith and Malacinski 1983; Minsuk and Keller 1996). In recent years, the most intensive investigations have focused on *Xenopus laevis*, so we will concentrate on its mode of gastrulation.

The *Xenopus* fate map

Amphibian blastulae are faced with the same tasks as the invertebrate blastulae we followed in Chapters 8 and 9—namely, to bring inside the embryo those areas destined to form the endodermal organs; to surround the embryo with cells capable of forming the ectoderm; and to place the mesodermal cells in the proper positions between them. Fate mapping by Løvtrup (1975; Landstrom and Løvtrup 1979) and by Keller (1975, 1976) has shown that cells of the *Xenopus* blastula have different fates depending on whether they are located in the deep or the superficial layers of the embryo (Figure 10.6). In *Xenopus*, the mesodermal precursors exist mostly in the deep layer of cells, while the ectoderm and endoderm arise from the superficial layer on the surface of the embryo. Most of the precursors for the notochord and other mesodermal tissues are located beneath the surface in the equatorial (marginal) region of the embryo. In urodeles (salamanders such as *Triturus* and *Ambystoma*) and in some frog species, many more of the

notochord and mesoderm precursor cells are found among the surface cells* (Purcell and Keller 1993; Shook et al. 2002).

Vegetal rotation and the invagination of the bottle cells

Gastrulation in frog embryos is initiated on the future dorsal side of the embryo, just below the equator, in the region of the gray crescent (Figure 10.7)—the region opposite the point of sperm entry. Here the cells invaginate to form a slitlike blastopore (Figure 10.8). These cells change their shape dramatically. The main body of each cell is displaced toward the inside of the embryo while maintaining contact with the outside surface by way of a slender neck. These **bottle cells** line the archenteron (primitive gut) as it forms. Thus, as in the gastrulating sea urchin, an invagination of cells initiates the formation of the archenteron. However, unlike sea urchins, gastrulation in the frog begins not in the most vegetal region, but in the **marginal zone**: the region surrounding the equator of the blastula, where the animal and vegetal hemispheres meet (see Figure 10.8). Here the endodermal cells are not as large or as yolky as the most vegetal blastomeres.

In salamanders, bottle cells appear to have an active role in the early movements of gastrulation. Johannes Holtfreter (1943, 1944) found that bottle cells from early salamander gastrulae could attach to glass coverslips and lead the movement of those cells attached to them. Even more convincing were Holtfreter's recombination experiments.

---

*In urodeles, the superficial mesoderm cells move into the deeper layers by forming a primitive streak similar to that of amniote embryos (see the next chapter). As in chick and mouse embryos, the cells leave the epithelial sheet to become mesenchymal cells (Shook et al. 2002).

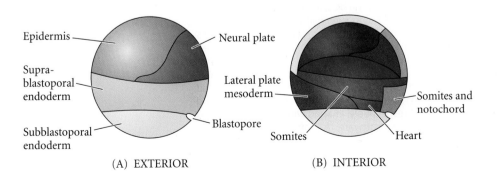

(A) EXTERIOR

Epidermis
Supra-blastoporal endoderm
Subblastoporal endoderm
Neural plate
Blastopore

(B) INTERIOR

Lateral plate mesoderm
Somites
Somites and notochord
Heart

FIGURE 10.6 Fate maps of the blastula of the frog *Xenopus laevis*. (A) Exterior. (B) Interior. Most of the mesodermal derivatives are formed from the interior cells. The placement of ventral mesoderm remains controversial (see Kumano and Smith 2002; Lane and Sheets 2002). (After Lane and Smith 1999; Newman and Krieg 1999.)

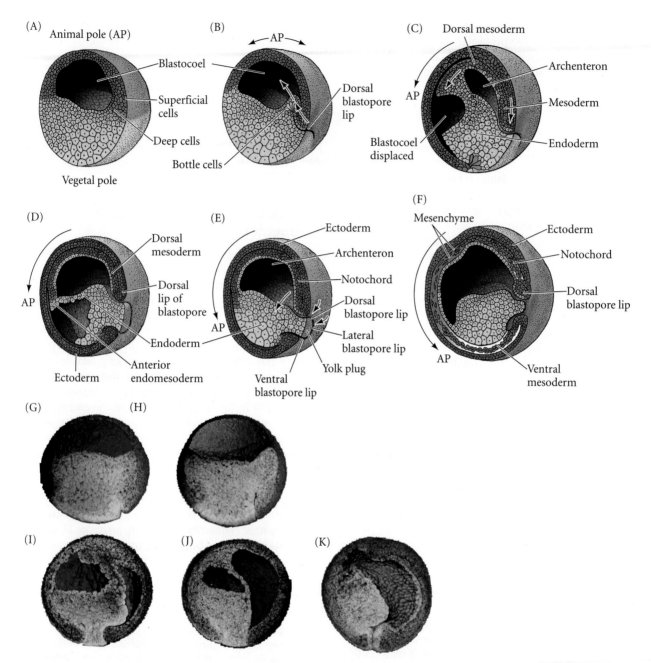

FIGURE 10.7 Cell movements during frog gastrulation. The meridional sections are cut through the middle of the embryo and positioned so that the vegetal pole is tilted toward the observer and slightly to the left. The major cell movements are indicated by arrows, and the superficial animal hemisphere cells are colored so that their movements can be followed. (A,B) Early gastrulation. The bottle cells of the margin move inward to form the dorsal lip of the blastopore, and the mesodermal precursors involute under the roof of the blastocoel. AP marks the position of the animal pole, which will change as gastrulation continues. (C,D) Mid-gastrulation. The archenteron forms and displaces the blastocoel, and cells migrate from the lateral and ventral lips of the blastopore into the embryo. The cells of the animal hemisphere migrate down toward the vegetal region, moving the blastopore to the region near the vegetal pole. (E,F) Toward the end of gastrulation, the blastocoel is obliterated, the embryo becomes surrounded by ectoderm, the endoderm has been internalized, and the mesodermal cells have been positioned between the ectoderm and endoderm. (G–K) Modern visualizations of the views shown in (A–D) and (F), respectively, imaged with a surface imaging microscope (Ewald et al. 2002) and rendered using graphics software. (A–F after Keller 1986; G–K courtesy of Andrew Ewald and Scott Fraser.)

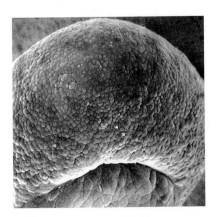

FIGURE 10.8 Surface view of an early dorsal blastopore lip of *Xenopus*. The size difference between the animal and vegetal blastomeres is readily apparent. (Micrograph courtesy of C. Phillips.)

When dorsal marginal zone cells (which would normally give rise to the dorsal lip of the blastopore) were excised and placed on inner prospective endoderm tissue, they formed bottle cells and sank below the surface of the inner endoderm (Figure 10.9). Moreover, as they sank, they created a depression or "groove" reminiscent of the early blastopore. Thus, Holtfreter claimed that the ability to invaginate and ingress into the deep endoderm is an intrinsic property of the dorsal marginal zone cells.

The situation in the frog embryo is somewhat different. Working with *Xenopus*, R. E. Keller and his students showed that the peculiar shape change of the bottle cells is needed to *initiate* gastrulation (it is the constriction of these cells that forms the slitlike blastopore); however, once involution movements are underway, bottle cells are no longer essential for the frog's gastrulation. When bottle cells are removed after their formation, involution and blastopore formation and closure continue (Keller 1981; Hardin and Keller 1988). Thus, in *Xenopus*, the major factor in the movement of cells into the embryo appears to be the involution of the subsurface cells rather than the superficial marginal cells.

Internalization of the endoderm and mesoderm in *Xenopus* is initiated by a movement called **vegetal rotation**. At least 2 hours before the bottle cells form, internal cell rearrangements propel the cells of the dorsal floor of the blastocoel toward the animal cap. This rotation places the prospective pharyngeal endoderm adjacent to the blastocoel and immediately above the involuting mesoderm. These cells then migrate along the basal surface of the blastocoel roof (Figure 10.10A–D; Nieuwkoop and Florschütz 1950; Winklbauer and Schürfeld 1999; Ibrahim and Winklbauer 2001). The superficial layer of marginal cells is pulled inward to form the endodermal lining of the archenteron, merely because it is attached to the actively migrating deep cells. While experimental removal of the bottle cells does not affect the involution of the deep or superficial marginal zone cells into the embryo, the removal of the deep involuting marginal zone (IMZ) cells stops archenteron formation.

INVOLUTION AT THE BLASTOPORE LIP The next phase of gastrulation involves the involution of the marginal zone cells while the animal cells undergo epiboly and converge at the blastopore (see Figure 10.7C,D). When the migrating marginal cells reach (and become) the **dorsal lip of the blastopore**, they turn inward and travel along the inner surface of the outer animal hemisphere cells. Thus, the cells constituting the lip of the blastopore are constantly changing. The order of the march into the embryo was determined by the vegetal rotation that abutted the prospective pharyngeal endoderm against the inside of the animal cap tissue. The first cells to compose the dorsal blastopore lip and enter into the embryo are the prospective pharyngeal endoderm of the foregut (including the bottle cells). As these first cells pass into the interior of the embryo, the dorsal blastopore lip becomes composed of cells that involute into the embryo to become the **prechordal plate** (the precursor of the head mesoderm). The next cells involuting into the embryo through the dorsal blastopore lip are the **chordamesoderm** cells. These cells will form the **notochord**, a transient mesodermal rod that plays an important role in inducing and patterning the nervous system. Thus the cells constituting the dorsal blastopore lip are constantly changing as the original cells migrate into the embryo and are replaced by cells migrating down, inward, and upward.

As the new cells enter the embryo, the blastocoel is displaced to the side opposite the dorsal lip of the blastopore. Meanwhile, the lip expands laterally and ventrally as the processes of bottle cell formation and involution continue around the blastopore. The widening blastopore "crescent"

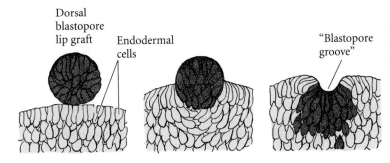

FIGURE 10.9 A graft of cells from the dorsal marginal zone of a salamander embryo sinks into a layer of endodermal cells and forms a blastopore-like groove. (After Holtfreter 1944.)

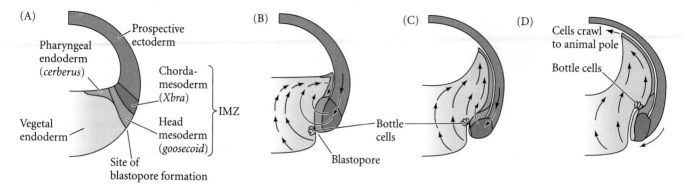

FIGURE 10.10 Early movements of *Xenopus* gastrulation. Orange represents the prospective pharyngeal endoderm (as described by *cerberus* expression). Dark orange represents the prospective head mesoderm (*goosecoid* expression), and the chordamesoderm (*Xbra* expression) is red. (A) At the beginning of gastrulation, the involuting marginal zone (IMZ) forms. (B) Vegetal rotation (arrows) pushes the prospective pharyngeal endoderm to the side of the blastocoel. (C,D) The vegetal endoderm movements push the pharyngeal endoderm forward, driving the mesoderm passively into the embryo and toward the animal pole. The ectoderm (blue) begins epiboly. (After Winklbauer and Schürfeld 1999.)

develops lateral lips and finally a ventral lip over which additional mesodermal and endodermal precursor cells pass. With the formation of the ventral lip, the blastopore has formed a ring around the large endodermal cells that remain exposed on the vegetal surface (Figure 10.11). This remaining patch of endoderm is called the **yolk plug**; it,

too, is eventually internalized. At that point, all the endodermal precursors have been brought into the interior of the embryo, the ectoderm has encircled the surface, and the mesoderm has been brought between them.

CONVERGENT EXTENSION OF THE DORSAL MESODERM Involution begins dorsally, led by the pharyngeal endoderm and the head mesoderm. These tissues will migrate most anteriorly beneath the surface ectoderm. The next tissues to enter the dorsal blastopore lip contain notochord and somite precursors.* Meanwhile, as the lip of the blastopore expands to have dorsolateral, lateral, and ventral sides, the prospec-

---

*The pharyngeal endoderm and head mesoderm cannot be separated experimentally at this stage, so they are sometimes referred to collectively as the head **endomesoderm**. The notochord is the basic unit of the dorsal mesoderm, but it is thought that the dorsal portion of the somites may have similar properties to the notochord.

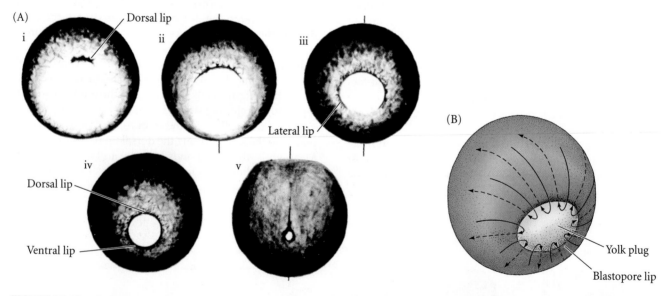

FIGURE 10.11 Epiboly of the ectoderm. (A) Changes in the region around the blastopore as the dorsal, lateral, and ventral lips are formed in succession. When the ventral lip completes the circle, the endoderm becomes progressively internalized. Numbers ii–v correspond to Figures 10.7 B–E, respectively. (B)

Summary of epiboly of the ectoderm and involution of the mesodermal cells migrating into the blastopore and then under the surface. The endoderm beneath the blastopore lip (the yolk plug) is not mobile and is enclosed by these movements. (A from Balinsky 1975, photographs courtesy of B. I. Balinsky.)

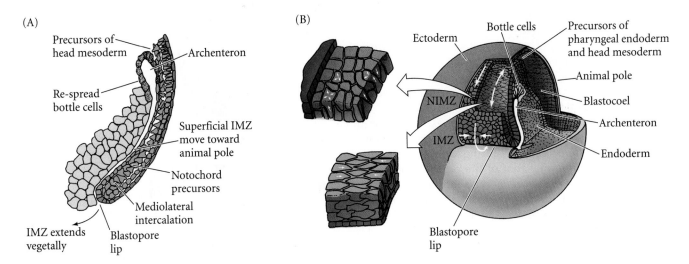

FIGURE 10.12 *Xenopus* gastrulation continues. (A) The deep marginal cells flatten and the formerly superficial cells form the wall of the archenteron. (B) Radial intercalation, looking down at the dorsal blastopore lip from the dorsal surface. In the noninvoluting marginal zone (NIMZ) and the upper portion of the IMZ, deep (mesodermal) cells are intercalating radially to make a thin band of flattened cells. This thinning of several layers into a few causes convergent extension (white arrows) toward the blastopore lip. Just above the lip, mediolateral intercalation of the cells produces stresses that pull the IMZ over the lip. After involuting over the lip, mediolateral intercalation continues, elongating and narrowing the axial mesoderm. (After Wilson and Keller 1991; Winklbauer and Schürfeld 1999.)

tive heart, kidney, and ventral mesoderms enter into the embryo.

Figure 10.12 depicts the behavior of the involuting marginal zone cells at successive stages of *Xenopus* gastrulation (Keller and Schoenwolf 1977; Hardin and Keller 1988). The IMZ is originally several layers thick. Shortly before their involution through the blastopore lip, the several layers of deep IMZ cells intercalate radially to form one thin, broad layer. This intercalation further extends the IMZ vegetally (Figure 10.12A). At the same time, the superficial cells spread out by dividing and flattening. When the deep cells reach the blastopore lip, they involute into the embryo and initiate a second type of intercalation. This intercalation causes a **convergent extension** along the mediolateral axis that integrates several mesodermal streams to form a long, narrow band (Figure 10.12B). This movement is reminiscent of traffic on a highway when several lanes must merge to form a single lane (or for that matter, the cell movements of the sea urchin archenteron). The anterior part of this band migrates toward the animal cap. Thus, the mesodermal stream continues to migrate toward the animal pole, and the overlying layer of superficial cells (including the bottle cells) is passively pulled toward the animal pole, thereby forming the endodermal roof of the archenteron (see Figures 10.7 and 10.12A). The radial and mediolateral intercalations of the deep layer of cells appear to be

responsible for the continued movement of mesoderm into the embryo.

Two major forces appear to drive convergent extension. The first force is differential cell cohesion. During gastrulation, the genes encoding adhesion proteins **paraxial protocadherin** and **axial protocadherin**, become expressed specifically in the paraxial (somite-forming) mesoderm and the notochord, respectively (Figure 10.13). An experimental dominant negative form of axial protocadherin prevents the presumptive notochord cells from sorting out from the paraxial mesoderm and blocks normal axis formation. A dominant negative paraxial protocadherin (which is secreted instead of being bound to the cell membrane) prevents convergent extension* (Kim et al. 1998; Kuroda et al. 2002). Moreover, the expression domain of paraxial protocadherin separates the trunk mesodermal cells, which undergo convergent extension, from the head mesodermal cells, which do not.

The second factor regulating convergent extension is calcium flux. Wallingford and colleagues (2001) found that dramatic waves of calcium ions surge across the dorsal tissues undergoing convergent extension, causing waves of contraction within the tissue. The calcium ions are released from intracellular stores and are required for convergent extension. If the release of calcium ions is blocked, normal cell specification still occurs, but the dorsal mesoderm neither converges nor extends. This calcium is thought to regulate the contraction of the actin microfilaments. This may help explain why the head region does not undergo convergent extension. The *Otx2* gene, the vertebrate homologue of the *Drosophila orthodenticle* gene, encodes a transcription factor expressed in the most anterior region of the embryo. Otx2 protein is critical in head formation, acti-

---

*Dominant negative proteins are mutated forms of the wild-type protein that interfere with the normal functioning of the wild-type protein. Thus, a dominant negative protein will have an effect similar to a loss-of-function mutation in the gene encoding the particular protein.

vating those genes involved in forebrain formation (see Figure 10.40). In addition to specifying the anterior tissues of the embryo, Otx2 prevents the cells expressing it from undergoing convergence and extension. One of the genes Otx2 activates is the *Xenopus calponin* gene (Morgan et al. 1999). Calponin is a protein that binds to actin and myosin and prevents actin microfilaments from contracting.

As mesodermal movement progresses, convergent extension continues to narrow and lengthen the involuting marginal zone. The involuting cells contain the prospective endodermal roof of the archenteron in its superficial layer and the prospective mesodermal cells, including those of the notochord, in its deep region. Toward the end of gastrulation, the centrally located notochord separates from the somitic mesoderm on either side of it, and the notochord elongates separately as its cells continue to intercalate (Wilson and Keller 1991). This may in part be a consequence of the different adhesion molecules in the axial and paraxial mesoderms (Figure 10.13; Kim et al. 1998). This convergent extension of the mesoderm appears to be autonomous, because the movements of these cells occur even if this region of the embryo is experimentally isolated from the rest of the embryo (Keller 1986).

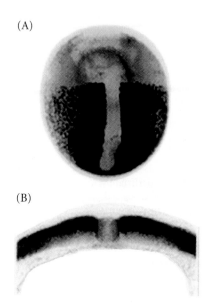

(A)

(B)

FIGURE 10.13 The expression of paraxial protocadherin. (A) Expression of paraxial protocadherin during late gastrulation (dark areas) shows the distinct downregulation in the notochord and the absence of expression in the head region. (B) Double-stained cross section through a late *Xenopus* gastrula shows the separation of notochord (brown staining for chordin) and the paraxial mesoderm (blue staining for paraxial protocadherin). (Photographs courtesy of E. M. De Robertis.)

SIDELIGHTS & SPECULATIONS

Fibronectin and the Pathways for Mesodermal Migration

How are the involuting cells informed where to go once they enter the inside of the embryo? In many amphibians, it appears that the involuting mesodermal precursors migrate toward the animal pole on a fibronectin lattice secreted by the cells of the blastocoel roof (Figure 10.14A,B). Shortly before gastrulation, the presumptive ectoderm of the blastocoel roof secretes an extracellular matrix that contains fibrils of fibronectin (Boucaut et al. 1984; Nakatsuji et al. 1985). The involuting mesoderm appears to travel along these fibronectin fibrils. Confirmation of this hypothesis was obtained by chemically synthesizing a "phony" fibronectin that can compete with the genuine fibronectin of the extracellular matrix. Cells bind to a region of the fibronectin protein that contains a three-amino acid sequence (Arg-Gly-Asp). Boucaut and co-workers injected large amounts of a small peptide containing this sequence into the blastocoels of salamander embryos shortly before gastrulation began. If fibronectin were essential for cell

migration, then cells binding this soluble peptide fragment instead of the real extracellular fibronectin should stop migrating. Unable to find their "road," the mesodermal cells should cease involution. That is precisely what happened (Figure 10.14C–F). No migrating cells were seen along the underside of the ectoderm in the experimental embryos. Instead, the mesodermal precursors remained outside the embryos, forming a convoluted cell mass. Other small synthetic peptides (including other fragments of the fibronectin molecule) did not impede migration. Thus the fibronectin-containing extracellular matrix appears to provide both a substrate for adhesion as well as cues for the direction of cell migration. Shi and colleagues (1989) showed that salamander IMZ cells would migrate in the wrong direction if extra fibronectin lattices were placed in their path.

In *Xenopus*, fibronectin is similarly secreted by the cells lining the blastocoel roof. The result is a band of fibronectin lining the roof, including Brachet's cleft, the part of the

blastocoel roof extending vegetally on the dorsal side (see Figure 10.14A,B). The vegetal rotation places the pharyngeal endoderm and involuting mesoderm into contact with these fibronectin fibrils (Winklbauer and Schürfeld 1999). Convergent extension pushes the migrating cells upward toward the animal pole. The fibronectin fibrils are necessary for the head mesodermal cells to flatten and to extend broad (lamelliform) processes in the direction of migration (Winklbauer et al. 1991; Winklbauer and Keller 1996). Studies using inhibitors of fibronectin formation have shown that fibronectin fibrils are necessary for the direction of mesoderm migration, the maintenance of intercalation of animal cap cells, and the initiation of radial intercalation in the marginal zone (Marsden and DeSimeone 2001).

The mesodermal cells are thought to adhere to fibronectin through the $\alpha_5\beta_1$ integrin protein (Alfandari et al. 1995). Mesodermal migration can also be arrested by the microinjection of antibodies against either

SIDELIGHTS & SPECULATIONS

fibronectin or the α_5 subunit of integrin, which serves as part of the fibronectin receptor (D'Arribère et al. 1988, 1990). Alfandari and colleagues (1995) have shown that the α_5 subunit of integrin appears on the mesodermal cells just prior to gastrulation, persists on their surfaces throughout gastrulation, and disappears when gastrulation ends. The integrin coordinates the interaction of the blastocel roof with actin microfilaments in the migrating mesendodermal cells. This interaction allows for increased traction and speed of migration (Davidson et al. 2002). It seems, then, that the coordinated synthesis of fibronectin and its receptor may signal the times for the mesoderm to begin, continue, and stop migration.

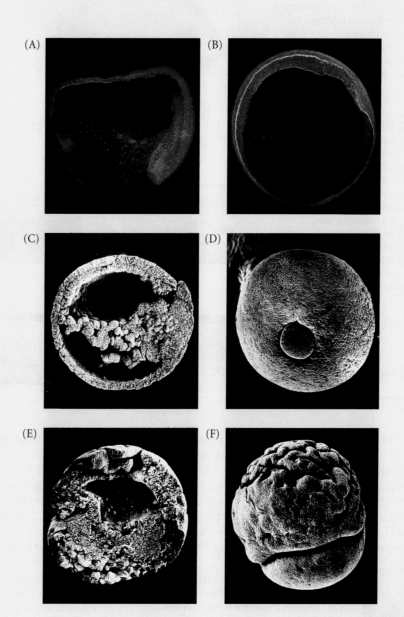

FIGURE 10.14 Fibronectin and amphibian gastrulation. (A,B) Sagittal section of *Xenopus* embryos at (A) early and (B) late gastrulation. The fibronectin lattice on the blastocoel roof is identified by fluorescent antibody labeling (yellow) while the embryonic cells are counterstained red. (C–F) Scanning electron micrographs of (C,D) a normal salamander embryo injected with a control solution at the blastula stage; and (E,F) an embryo of the same stage injected with the cell-binding fragment of fibronectin. (C) Section during mid-gastrulation. (D) The yolk plug toward the end of gastrulation. (E,F) The finishing stages of the arrested gastrulation, wherein the mesodermal precursors, having bound the synthetic fibronectin, cannot recognize the normal fibronectin-lined migration route. The archenteron fails to form, and the noninvoluted mesodermal precursors remain on the surface. (A,B from Marsden and DeSimeone 2001, photographs courtesy of the authors; C–F from Boucaut et al. 1984, photographs courtesy of J.-C. Boucaut and J.-P. Thiery.)

Epiboly of the prospective ectoderm

During gastrulation, the animal cap and **noninvoluting marginal zone** (**NIMZ**) cells expand by epiboly to cover the entire embryo (see Figure 10.12B). These cells will form the surface ectoderm. The major mechanism of epiboly in *Xenopus* gastrulation appears to be an increase in cell number (through division) coupled with a concurrent integration of several deep layers into one (Figure 10.15; Keller and Schoenwolf 1977; Keller and Danilchik 1988; Saka and Smith 2001). As the ectoderm spreads over the entire embryo, it internalizes all the endoderm within it. At this point, the ectoderm covers the embryo, the endoderm is located within the embryo, and the mesoderm is positioned between them.

The dorsal portion of the noninvoluting marginal zone extends more rapidly toward the blastopore than the ventral portion, causing the blastopore lips to move toward the ventral side. While those mesodermal cells entering through the dorsal lip of the blastopore give rise to the central dorsal mesoderm (notochord and somites), the remainder of the body mesoderm (which forms the heart, kid-

FIGURE 10.15 Epiboly of the ectoderm is accomplished by cell division and intercalation. (A,B) Cell division in the presumptive ectoderm. Cell division is shown by staining for phosphorylated histone 3, a marker of mitosis. Stained nuclei appear black. In early gastrulae (A; stage 10.5), most cell division occurs in the animal hemisphere presumptive ectoderm. In late gastrulae (B; stage 12), cell division can be seen throughout the ectodermal layer. (Interestingly, the dorsal mesoderm shows no cell division). (C) Scanning electron micrographs of the *Xenopus* blastocoel roof, showing the changes in cell shape and arrangement. Stages 8 and 9 are blastulae; stages 10–11.5 represent progressively later gastrulae. (A,B after Saka and Smith 2001, photographs courtesy of the authors; C from Keller 1980, photographs courtesy of R. E. Keller.)

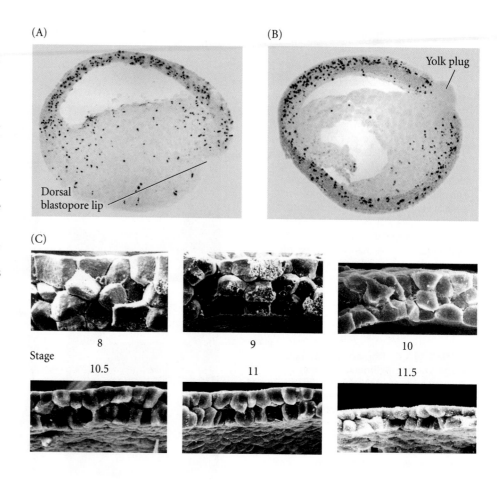

neys, bones, and parts of several other organs) enters through the ventral and lateral blastopore lips to create the **mesodermal mantle**. The endoderm is derived from the superficial cells of the involuting marginal zone that form the lining of the archenteron roof and from the subblastoporal vegetal cells that become the archenteron floor (Keller 1986). The remnant of the blastopore—where the endoderm meets the ectoderm—now becomes the anus.*

 WEBSITE **10.1** **Migration of the mesodermal mantle.** Different growth rates coupled with the intercalation of cell layers allows the mesoderm to expand in a tightly coordinated fashion.

Progressive Determination of the Amphibian Axes

As we have seen, the unfertilized egg has a polarity along the animal-vegetal axis. Thus, the germ layers can be mapped onto the oocyte even before fertilization. The animal hemisphere blastomeres will become the cells of the ectoderm (skin and nerves), the vegetal hemisphere cells will form the cells of the gut and associated organs (endoderm), and the mesodermal cells will form from the internal cytoplasm around the equator. This general fate map is thought to be imposed upon the egg by the transcription factor **VegT** and the TGF-β family paracrine factor **Vg1**. The mRNAs for these proteins are located in the cortex of the vegetal hemisphere of *Xenopus* oocytes, and they are apportioned to the vegetal cells during cleavage (see Figure 5.38). When the VegT transcripts are destroyed using antisense oligonucleotides, the entire embryo develops as an epidermis, with no mesodermal or endodermal components (Zhang et al. 1998; Taverner et al. 2005). Joseph and Melton (1998) demonstrated that embryos lacking functional Vg1 lacked endoderm and dorsal mesoderm. Thus, the allocation of cells to the three germ layers depends on pre-localized cytoplasmic determinants in the egg.

These findings tell us nothing, however, about which part of the egg will form the belly and which the back. The anterior-posterior, dorsal-ventral, and left-right axes are specified by events triggered at fertilization and realized during gastrulation.

Amphibian axis formation is an example of regulative development. In Chapter 3, we discussed the concept of regulative development, wherein (1) an isolated blastomere has a potency greater than its normal embryonic fate, and

---

*As gastrulation expert Ray Keller famously remarked, "Gastrulation is the time when a vertebrate takes its head out of its anus. "

(2) a cell's fate is determined (induced) by interactions between neighboring cells (see Chapter 6). The requirement for inductive interactions in amphibian axis determination was demonstrated in Hans Spemann's laboratory at the University of Freiburg (see De Robertis and Aréchaga 2001; Sander and Fässler 2001). Experiments by Spemann and his students framed the questions that experimental embryologists asked for most of the twentieth century and resulted in a Nobel Prize for Spemann in 1935. In recent times, the discoveries of the molecules associated with these inductive processes have provided some of the most exciting moments in contemporary science.

The experiment that began this research program was performed in 1903, when Spemann demonstrated that early newt blastomeres have identical nuclei, each capable of producing an entire larva. His procedure was ingenious: Shortly after fertilizing a newt egg, Spemann used a baby's hair (taken from his daughter) to lasso the zygote in the plane of the first cleavage. He then partially constricted the egg, causing all the nuclear divisions to remain on one side of the constriction. Eventually, often as late as the 16-cell stage, a nucleus would escape across the constriction into the non-nucleated side. Cleavage then began on this side, too, whereupon Spemann tightened the lasso until the two halves were completely separated. Twin larvae developed, one slightly more advanced than the other (Figure 10.16). Spemann concluded from this experiment that early amphibian nuclei were genetically identical and that each cell was capable of giving rise to an entire organism.

However, when Spemann performed a similar experiment with the constriction still longitudinal, but perpendicular to the plane of the first cleavage (separating the future dorsal and ventral regions rather than the right and left sides), he obtained a different result altogether. The nuclei continued to divide on both sides of the constriction, but only one side—the future dorsal side of the embryo—gave rise to a normal larva. The other side produced an unorganized tissue mass of ventral cells, which Spemann called the *Bauchstück*—the belly piece. This tissue mass was a ball of epidermal cells (ectoderm) containing blood and mesenchyme (mesoderm) and gut cells (endoderm), but no dorsal structures such as nervous system, notochord, or somites (Figure 10.17).

Why should these two experiments give different results? One possibility was that when the egg was divided perpendicular to the first cleavage plane, some *cytoplasmic* substance was not equally distributed into the two halves. Fortunately, the salamander egg was a good place to test that hypothesis. As we have seen, there are dramatic movements in the cytoplasm following the fertilization of amphibian eggs, and in some amphibians these movements expose a gray, crescent-shaped area of cytoplasm in the region directly opposite the point of sperm entry. Moreover, the first cleavage plane normally splits the gray crescent equally into the two blastomeres. If these cells are then separated, two complete larvae develop. However, should this cleavage plane be aberrant (either in the rare natural event or in an experiment), the gray crescent material passes into only one of the two blastomeres. Spemann found that when these two blastomeres are separated (such that only one of the two cells contains the crescent), only the blastomere containing the gray crescent develops normally.

WEBSITE

10.2 Embryology and individuality. One egg usually makes only one adult; however, there are exceptions to this rule. Spemann was drawn into embryology through the paradoxes of creating more than one individual from a single egg.

It appeared, then, that something in the gray crescent region was essential for proper embryonic development. But how did it function? What role did it play in normal development? The most important clue came from the fate map of this area of the egg, for it showed that the gray crescent region gives rise to the cells that initiate gastrulation. These are the cells that will form the dorsal lip of the blastopore. The cells of the dorsal lip are committed to invaginate into the blastula, thus initiating gastrulation and the

FIGURE 10.16 Spemann's demonstration of nuclear equivalence in newt cleavage. (A) When the fertilized egg of the newt *Triturus taeniatus* was constricted by a ligature, the nucleus was restricted to one half of the embryo. The cleavage on that side of the embryo reached the 8-cell stage, while the other side remained undivided. (B) At the 16-cell stage, a single nucleus entered the as yet undivided half, and the ligature was further constricted to complete the separation of the two halves. (C) After 14 days, each side had developed into a normal embryo. (After Spemann 1938.)

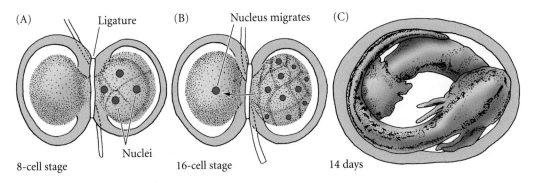

(A) Ligature   (B) Nucleus migrates   (C)

Nuclei

8-cell stage   16-cell stage   14 days

(A)   (B)

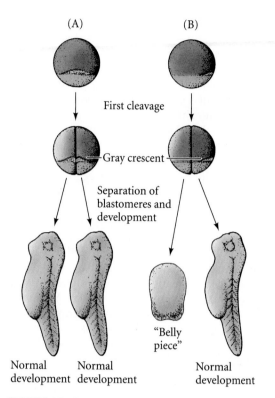

First cleavage

Gray crescent

Separation of
blastomeres and
development

"Belly
piece"

Normal
development

Normal
development

Normal
development

FIGURE 10.17 Asymmetry in the amphibian egg. (A) When the egg is divided along the plane of first cleavage into two blastomeres, each of which gets one-half of the gray crescent, each experimentally separated cell develops into a normal embryo. (B) When only one of the two blastomeres receives the entire gray crescent, it alone forms a normal embryo. The other blastomere produces a mass of unorganized tissue lacking dorsal structures. (After Spemann 1938.)

formation of the notochord. Because all future amphibian development depends on the interaction of cells rearranged during gastrulation, Spemann speculated that the importance of the gray crescent material lies in its ability to initiate gastrulation, and that crucial changes in cell potency occur during gastrulation. In 1918, he performed experiments that demonstrated this to be true. He found that the cells of the early gastrula were uncommitted, but that the fates of late gastrula cells were determined.

Spemann's demonstration involved exchanging tissues between the gastrulae of two species of newts whose embryos were differently pigmented (Figure 10.18). When a region of prospective epidermal cells from an *early* gastrula was transplanted into an area in another early gastrula where the neural tissue normally formed, the transplanted cells gave rise to neural tissue. When prospective neural tissue from early gastrulae was transplanted to the region fated to become belly skin, the neural tissue became epidermal (Table 10.1). Thus, these early newt gastrula cells were not yet committed to a specific fate. Such cells exhibit **regulative** (i.e., **conditional** or **dependent**) development

because their ultimate fates depend on their location in the embryo.

However, when the same interspecies transplantation experiments were performed on *late* gastrulae, Spemann obtained completely different results. Rather than differentiating in accordance with their new location, the transplanted cells exhibited **autonomous** (or **independent**, or **mosaic**) **development**. Their prospective fate was said to be **determined**, and the cells developed independently of their new embryonic location. Specifically, prospective neural cells now developed into brain tissue even when placed in the region of prospective epidermis, and prospective epidermis formed skin even in the region of the prospective neural tube. Within the time separating early and late gastrulation, the potencies of these groups of cells had become restricted to their eventual paths of differentiation. Something was causing them to become committed to epidermal and neural fates. What was happening?

(A) TRANSPLANTATION IN EARLY GASTRULA

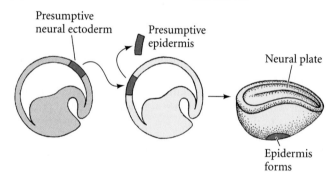

Presumptive
neural ectoderm

Presumptive
epidermis

Neural plate

Epidermis
forms

(B) TRANSPLANTATION IN LATE GASTRULA

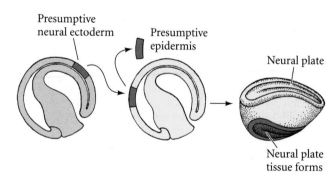

Presumptive
neural ectoderm

Presumptive
epidermis

Neural plate

Neural plate
tissue forms

FIGURE 10.18 Determination of ectoderm during newt gastrulation. Presumptive neural ectoderm from one newt embryo is transplanted into a region in another embryo that normally becomes epidermis. (A) When the tissues are transferred between early gastrulae, the presumptive neural tissue develops into epidermis, and only one neural plate is seen. (B) When the same experiment is performed using late-gastrula tissues, the presumptive neural cells form neural tissue, thereby causing two neural plates to form on the host. (After Saxén and Toivonen 1962.)

TABLE 10.1 Results of tissue transplantation during early- and late-gastrula stages in the newt

Donor region	Host region	Differentiation of donor tissue	Conclusion
EARLY GASTRULA			
Prospective neurons	Prospective epidermis	Epidermis	Dependent (conditional) development
Prospective epidermis	Prospective neurons	Neurons	Dependent (conditional) development
LATE GASTRULA			
Prospective neurons	Prospective epidermis	Neurons	Independent (autonomous) development (determined)
Prospective epidermis	Prospective neurons	Epidermis	Independent (autonomous) development (determined)

Hans Spemann and Hilde Mangold: Primary Embryonic Induction

The most spectacular transplantation experiments were published by Hans Spemann and Hilde Mangold in 1924.* They showed that, of all the tissues in the early gastrula, only one has its fate determined. This self-differentiating tissue is the dorsal lip of the blastopore, the tissue derived from the gray crescent cytoplasm. When this dorsal lip tissue was transplanted into the presumptive belly skin region of another gastrula, it not only continued to be blastopore lip, but also initiated gastrulation and embryogenesis in the surrounding tissue (Figure 10.19). Two conjoined embryos were formed instead of one!

In these experiments, Spemann and Mangold used differently pigmented embryos from two newt species—the darkly pigmented *Triturus taeniatus* and the nonpigmented *Triturus cristatus*—so that they were able to identify host and donor tissues on the basis of color. When the dorsal lip of an early *T. taeniatus* gastrula was removed and implanted into the region of an early *T. cristatus* gastrula fated to become ventral epidermis (belly skin), the dorsal lip tissue invaginated just as it would normally have done (showing self-determination) and disappeared beneath the vegetal cells. The pigmented donor tissue then continued to self-differentiate into the chordamesoderm (notochord) and other mesodermal structures that normally form from the dorsal lip. As the new donor-derived mesodermal cells moved forward, host cells began to participate in the production of a new embryo, becoming organs that normally

they never would have formed. In this secondary embryo, a somite could be seen containing both pigmented (donor) and unpigmented (host) tissue. Even more spectacularly, the dorsal lip cells were able to interact with the host tissues to form a complete neural plate from host ectoderm. Eventually, a secondary embryo formed, face to face with its host. The results of these technically difficult experiments have been confirmed many times (Capuron 1968; Smith and Slack 1983; Recanzone and Harris 1985).

 WEBSITE 10.3 Spemann, Mangold, and the organizer. Spemann did not see the importance of this work the first time he and Mangold did it. This website provides a more detailed account of why Spemann and Mangold did this experiment.

Spemann (1938) referred to the dorsal lip cells and their derivatives (notochord and head endomesoderm) as the **organizer** because (1) they induced the host's ventral tissues to change their fates to form a neural tube and dorsal mesodermal tissue (such as somites), and (2) they organized host and donor tissues into a secondary embryo with clear anterior-posterior and dorsal-ventral axes. He proposed that during normal development, these cells "organize" the dorsal ectoderm into a neural tube and transform the flanking mesoderm into the anterior-posterior body axis. It is now known (thanks largely to Spemann and his students) that the interaction of the chordamesoderm and ectoderm is not sufficient to organize the entire embryo. Rather, it initiates a series of sequential inductive events. Because there are numerous inductions during embryonic development, this key induction—in which the progeny of dorsal lip cells induce the dorsal axis and the neural tube—is traditionally called the **primary embryonic induction**.[†]

---

*Hilde Proescholdt Mangold died in a tragic accident in 1924, when her kitchen's gasoline heater exploded. At the time she was 26 years old, and her paper was just being published. Hers is one of the very few doctoral theses in biology that have directly resulted in the awarding of a Nobel Prize. For more information about Hilde Mangold, her times, and the experiments that identified the organizer, see Hamburger 1984, 1988, and Fässler and Sander 1996.

[†]This classical term has been a source of confusion because the induction of the neural tube by the notochord is no longer considered the first inductive process in the embryo. We will soon discuss inductive events that precede this "primary" induction.

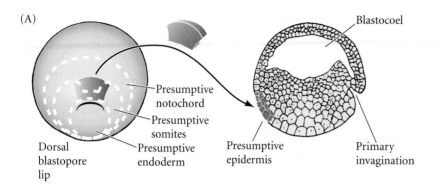

(A)

Blastocoel

Presumptive
notochord

Presumptive
somites

Dorsal
blastopore
lip

Presumptive
endoderm

Presumptive
epidermis

Primary
invagination

FIGURE 10.19 Organization of a secondary axis by dorsal blastopore lip tissue. (A) Dorsal lip tissue from an early gastrula is transplanted into another early gastrula in the region that normally becomes ventral epidermis. (B) The donor tissue invaginates and forms a second archenteron, and then a second embryonic axis. Both donor and host tissues are seen in the new neural tube, notochord, and somites. (C) Eventually, a second embryo forms that is joined to the host. (After Hamburger 1988.)

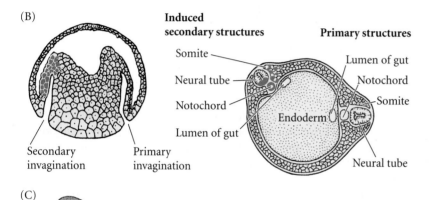

(B)

Induced secondary structures

Somite

Neural tube

Notochord

Lumen of gut

Primary structures

Lumen of gut

Notochord

Somite

Endoderm

Neural tube

Secondary
invagination

Primary
invagination

(C)

Mechanisms of Axis Determination in Amphibians

The experiments of Spemann and Mangold showed that the dorsal lip of the blastopore, and the dorsal mesoderm and pharyngeal endoderm that form from it, constituted an "organizer" that could instruct the formation of new embryonic axes. But the mechanisms by which the organizer was constructed and through which it operated remained a mystery. Indeed, it is said that Spemann and Mangold's paper posed more questions than it answered. Among these questions were:

- How did the organizer get its properties? What caused the dorsal blastopore lip to differ from any other region of the embryo?
- What factors were being secreted from the organizer to cause the formation of the neural tube and to create the anterior-posterior, dorsal-ventral, and left-right axes?

- How did the different parts of the neural tube become established, with the most anterior becoming the sensory organs and forebrain, and the most posterior becoming spinal cord?

We will take up each of these questions in turn.

How does the organizer form?

Why are the dozen or so cells of the initial organizer positioned opposite the point of sperm entry and what determines their fate so early? Recent evidence provides an unexpected answer: these cells are in the right place at the right time, at the point where two signals converge. The first signal tells the cells that they are dorsal. The second signal says that they are mesoderm.

THE DORSAL SIGNAL: β-CATENIN It turns out that one of the reasons that the organizer cells are special is that these mesodermal cells reside above a special group of vegetal cells. One of the major clues in determining how the dorsal blastopore lip obtained its properties came from the experiments of Pieter Nieuwkoop (1969, 1973, 1977) and Osamu Nakamura. Nakamura and Takasaki (1970) showed that the mesoderm arises from the marginal (equatorial) cells at the border between the animal and vegetal poles. The Nakamura and Nieuwkoop laboratories demonstrated that the properties of this newly formed mesoderm were induced by the vegetal (presumptive endoderm) cells underlying them. Nieuwkoop removed the equatorial cells (i.e., presumptive mesoderm) from a blastula and showed that neither the animal cap (presumptive ectoderm) nor the vegetal cap (presumptive endoderm) produced any mesodermal tissue. However, when the two caps were recombined, the animal cap cells were induced to form mesodermal structures such as notochord, muscles, kidney cells, and blood cells (Figure 10.20). The polarity of this induction (whether the animal cells formed dorsal mesoderm or ventral mesoderm) depended on whether the endodermal (vegetal) fragment was taken from the dorsal or ventral side. While the ventral and lateral vegetal cells

(A) Dissected blastula fragments give rise to different tissue in culture:

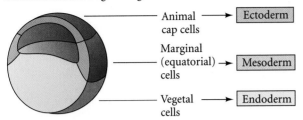

Animal cap cells → Ectoderm

Marginal (equatorial) cells → Mesoderm

Vegetal cells → Endoderm

(B) Animal and vegetal fragments give rise to mesoderm

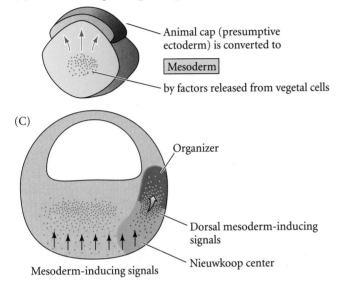

Animal cap (presumptive ectoderm) is converted to Mesoderm by factors released from vegetal cells

(C)

Organizer

Dorsal mesoderm-inducing signals

Nieuwkoop center

Mesoderm-inducing signals

FIGURE 10.20 Summary of experiments by Nieuwkoop and by Nakamura and Takasaki, showing mesodermal induction by vegetal endoderm. (A) Isolated animal cap cells become a mass of ciliated epidermis, isolated vegetal cells generate gutlike tissue, and isolated equatorial (marginal zone) cells become mesoderm. (B) If animal cap cells are combined with vegetal cap cells, many of the animal cells generate mesodermal tissue. (C) Simplified model for mesoderm induction in *Xenopus*. A ventral signal (probably a complex set of signals from activin-like TGF-β factors and FGFs) is released throughout the vegetal region of the embryo. This signal induces the marginal cells to become mesoderm. On the dorsal side (away from the point of sperm entry), a signal is released by the vegetal cells of the Nieuwkoop center. This dorsal signal induces the formation of the Spemann organizer in the overlying marginal zone cells. The possible identity of this signal will be discussed later in this chapter. (C after De Robertis et al. 1992.)

(those closer to the side of sperm entry) induced ventral (mesenchyme, blood) and intermediate (muscle, kidney) mesoderm, the dorsalmost vegetal cells specified dorsal mesoderm components (somites, notochord), including those having the properties of the organizer. These dorsalmost vegetal cells of the blastula, which are capable of inducing the organizer, have been called the **Nieuwkoop center** (Gerhart et al. 1989).

The Nieuwkoop center was demonstrated in the *Xenopus* embryo by transplantation and recombination experiments (Figure 10.21). First, Gimlich and Gerhart (Gimlich and Gerhart 1984; Gimlich 1985, 1986) performed an experiment analogous to the Spemann and Mangold studies, except that they used early blastulae rather than gastrulae. When they transplanted the dorsalmost vegetal blastomere from one blastula into the ventral vegetal side of another blastula, two embryonic axes were formed (see Figure 10.21B). Second, Dale and Slack (1987) recombined single vegetal blastomeres from a 32-cell *Xenopus* embryo with the uppermost animal tier of a fluorescently labeled embryo of the same stage. The dorsalmost vegetal cell, as expected, induced the animal pole cells to become dorsal mesoderm. The remaining vegetal cells usually induced the animal cells to produce either intermediate or ventral mesodermal tissues (Figure 10.22). Holowacz and Elinson (1993) found that cortical cytoplasm from the dorsal vegetal cells of the 16-cell *Xenopus* embryo was able to induce

the formation of secondary axes when injected into ventral vegetal cells. Thus, dorsal vegetal cells can induce animal cells to become dorsal mesodermal tissue.

So one important question became: What gives the dorsalmost vegetal cells their special properties? The major candidate for the factor that forms the Nieuwkoop center in these dorsalmost vegetal cells is **β-catenin**. β-catenin is a multifunctional protein that can act as an anchor for cell membrane cadherins (see Chapter 3) or as a nuclear transcription factor (see Chapter 8). In the sea urchin embryo, β-catenin is responsible for specifying the micromeres. β-Catenin is a key player in the formation of the dorsal tissues. Experimental depletion of β-catenin results in the lack of dorsal structures (Heasman et al. 1994a). Moreover, the injection of exogenous β-catenin into the ventral side of an embryo produces a secondary axis (Funayama et al. 1995; Guger and Gumbiner 1995).

In *Xenopus* embryos, β-catenin is initially synthesized throughout the embryo from maternal mRNA (Yost et al. 1996; Larabell et al. 1997). β-catenin begins to accumulate in the dorsal region of the egg during the cytoplasmic movements of fertilization and continues to accumulate preferentially at the dorsal side throughout early cleavage. This accumulation is seen in the nuclei of the dorsal cells and appears to cover both the Nieuwkoop center and organizer regions (Figure 10.23; Schneider et al. 1996; Larabell et al. 1997).

Since β-catenin is originally found throughout the embryo, how does it become localized specifically to the side opposite sperm entry? The answer appears to reside in the translocation of Disheveled protein from the vegetal pole to the dorsal side of the egg during fertilization. From research done on the Wnt pathway, we have learned that β-catenin is targeted for destruction by glycogen synthase kinase 3 (GSK3; see Chapter 6). Activated GSK3 blocks axis formation when added to the egg (Pierce and Kimelman 1995; He et al. 1995; Yost et al. 1996). If endogenous GSK3 is knocked out by a dominant negative protein in the ventral cells of the early embryo, a second axis forms (see Figure 10.24F).

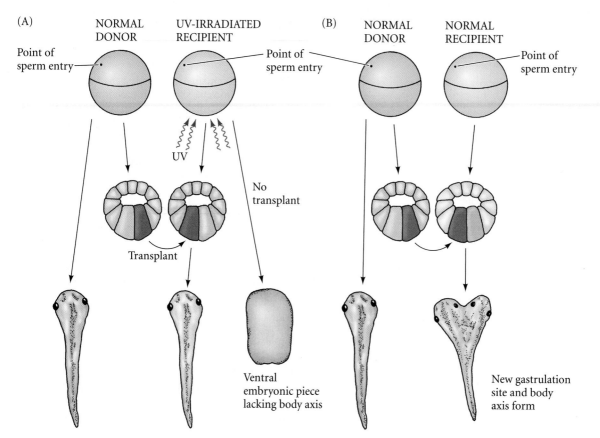

FIGURE 10.21 Transplantation experiments on 64-cell amphibian embryos demonstrated that the vegetal cells underlying the prospective dorsal blastopore lip region are responsible for causing the initiation of gastrulation. (A) Rescue of irradiated embryos by transplanting the dorsal vegetal blastomeres of a normal embryo into a cavity made by the removal of a simi-lar number of vegetal cells. An irradiated zygote without this transplant fails to undergo normal gastrulation. (B) Formation of a new gastrulation site and body axis by the transplantation of the most dorsal vegetal cells of one embryo into the ventral-most vegetal region of another embryo. (After Gimlich and Gerhart 1984.)

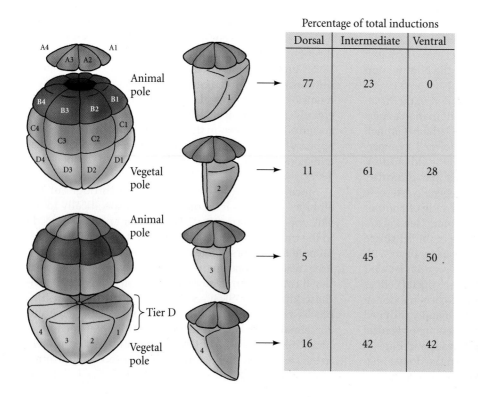

Percentage of total inductions		
Dorsal	Intermediate	Ventral
77	23	0
11	61	28
5	45	50
16	42	42

FIGURE 10.22 The regional specificity of mesoderm induction can be demonstrated by recombining blastomeres of 32-cell *Xenopus* embryos. Animal pole cells were labeled with fluorescent polymers so that their descendants could be identified, then combined with individual vegetal blastomeres. The inductions resulting from these recombinations are summarized at the right. D1, the dorsalmost vegetal blastomere, was the most likely to induce the animal pole cells to form dorsal mesoderm. (After Dale and Slack 1987.)

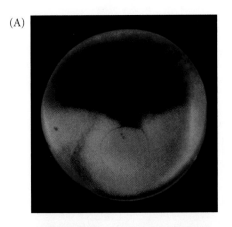

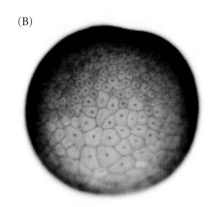

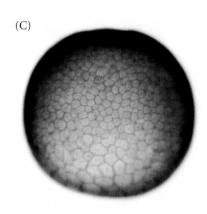

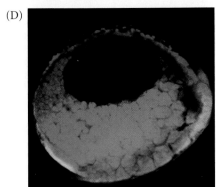

FIGURE 10.23 The role of Wnt pathway proteins in dorsal-ventral axis specification. (A–D) Differential translocation of β-catenin into *Xenopus* blastomere nuclei. (A) Early 2-cell stage of *Xenopus*, showing β-catenin (orange) predominantly at the dorsal surface. (B) Presumptive dorsal side of a *Xenopus* blastula stained for β-catenin shows nuclear localization. (C) Such nuclear localization is not seen on the ventral side of the same embryo. (D) β-Catenin dorsal localization persists through the gastrula stage. (A,D courtesy of R. T. Moon; B and C from Schneider et al. 1996, photographs courtesy of P. Hausen.)

But GSK3 can be inactivated by the GSK3-binding protein (GBP) and Disheveled (Dsh). These two proteins release GSK3 from the degradation complex and prevent it from binding β-catenin. During the first cell cycle, the microtubules form those parallel tracts in the vegetal portion of the egg. Cytoplasmic rotation occurs on these tracks, and so does the active transport of Dsh and GBP. GBP binds to kinesin, an ATPase (like dynein) that travels along microtubules. Kinesin always migrates towards the growing end of the microtubules, and in this case, that means moving to the point opposite sperm entry, i.e., the future dorsal side (Figure 10.24A–C). Disheveled, which is originally found in the vegetal pole cortex, grabs on the GPB, and it too becomes translocated along the microtubular monorail (Miller et al. 1999; Weaver et al. 2003). The cortical rotation is probably important in orienting and straightening the microtubular array and in maintaining the direction of transport when the kinesin complexes occasionally jump the track (Weaver and Kimelman 2004). Once at the site opposite the point of sperm entry, GBP and Dsh are released from the microtubules. Here, on the future dorsal side of the embryo, they inactivate GSK3, allow β-catenin to accumulate on the dorsal side while ventral β-catenin is degraded (Figure 10.24D,E).

The β-catenin transcription factor can associate with other transcription factors to give them new properties. It is known that *Xenopus* β-catenin can combine with a ubiquitous transcription factor known as **Tcf3**, and that expres-sion of a mutant form of Tcf3, lacking the β-catenin binding domain, results in embryos without dorsal structures (Molenaar et al. 1996). The β-catenin/Tcf3 complex appears to bind to the promoters of several genes whose activity is critical for axis formation. Two of these genes are *twin* and *siamois*, which encode homeodomain transcription factors and are expressed in the Nieuwkoop center immediately following the mid-blastula transition. If these genes are ectopically expressed in the ventral vegetal cells, a secondary axis emerges on the former ventral side of the embryo, and if cortical microtubular polymerization is prevented, *siamois* expression is eliminated (Lemaire et al. 1995; Brannon and Kimelman 1996). The Tcf3 protein is thought to inhibit *siamois* and *twin* transcription when it binds to those genes' promoters in the absence of β-catenin. However, when the β-catenin binds to Tcf3, the repressor is converted into an activator, and the *twin* and *siamois* genes are activated (Brannon et al. 1997). Siamois and Twin proteins are critical for the expression of organizer-specific genes (Figure 10.25; Brannon et al. 1997).

THE VEGETAL TGF-β-LIKE SIGNAL Siamois and Twin bind to the enhancers of several genes involved in organizer function (Fan and Sokol 1997; Kessler 1997). These include genes encoding the transcription factors Goosecoid and Xlim1 and the paracrine factors Cerberus and Frzb (Laurent et al. 1997; Engleka and Kessler 2001). Goosecoid appears to be essential for specifying the dorsal mesoderm. Thus one could expect that if the dorsal side of the embryo contained β-catenin, then β-catenin would allow this region to express *twin* and *siamois*, and their expression would initiate the formation of the organizer. However, these homeodomain transcription factors alone are not sufficient for

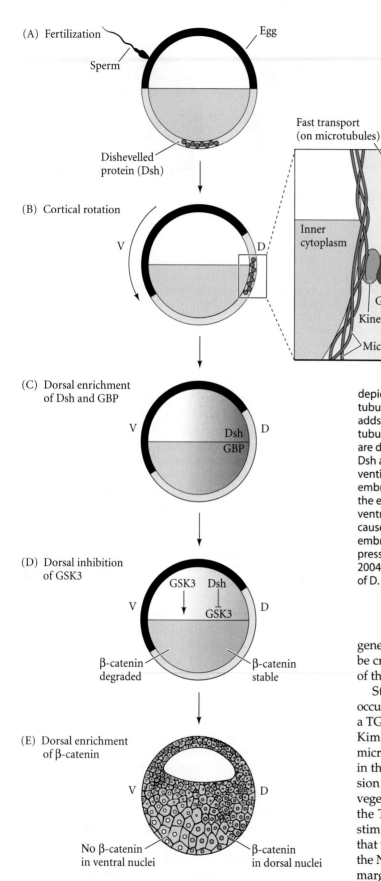

(A) Fertilization

Sperm

Egg

Dishevelled protein (Dsh)

(B) Cortical rotation

V   D

Fast transport (on microtubules)   Slow transport (cortical rotation)

Inner cytoplasm

GBP Dsh

Kinesin

Microtubules

(C) Dorsal enrichment of Dsh and GBP

V   D

Dsh

GBP

(D) Dorsal inhibition of GSK3

GSK3 Dsh

V   D

GSK3

β-catenin degraded   β-catenin stable

(E) Dorsal enrichment of β-catenin

V   D

No β-catenin in ventral nuclei   β-catenin in dorsal nuclei

(F)

FIGURE 10.24   Model of the mechanism by which the Disheveled protein stabilizes β-catenin in the dorsal portion of the amphibian egg. (A) Disheveled (Dsh) and GBP associates with kinesin at the vegetal pole of the unfertilized egg. (B) After fertilization, these protein vesicles are translocated dorsally along subcortical microtubule tracks. The insert depicts the complex of kinesin, GBP, and Dsh "riding" microtubules to the dorsal portion of the embryo. Cortical rotation adds a "slow" form of transportation to the fast-track microtubule ride. (C) Dsh and GBP are then released from kinesin and are distributed in the future dorsal third of the 1-cell embryo. (D) Dsh and GBP bind to and block the action of GSK3, thereby preventing the degradation of β-catenin on the dorsal side of the embryo. (E) The nuclei of the blastomeres in the dorsal region of the embryo receive β-catenin, while the nuclei of those in the ventral region do not. (F) Formation of a second dorsal axis caused by the injection of both blastomeres of a 2-cell *Xenopus* embryo with dominant inactive *GSK3*. Dorsal fate is actively suppressed by wild-type *GSK3*. (A–E after Weaver and Kimelman 2004; F from Pierce and Kimelman 1995, photograph courtesy of D. Kimelman.)

generating the organizer; another protein also appears to be critical in the activation of *goosecoid* and the formation of the organizer.

Studies suggest that maximum *goosecoid* expression occurs when there is synergism between these proteins and a TGF-β signal secreted by the vegetal cells (Brannon and Kimelman 1996; Engleka and Kessler 2001). While the microtubule transport system may stabilize the β-catenins in the dorsal region of the embryo and allow the expression of *siamois* and *twin* in that region, the translation of vegetally localized messages encoding paracrine factors of the TGF-β superfamily appear to induce a protein that stimulates the greatest activation of *goosecoid* in the cells that will become the organizer. A TGF-β-related protein in the Nieuwkoop center could induce the cells in the dorsal marginal zone above them to express Smad2/4 transcrip-

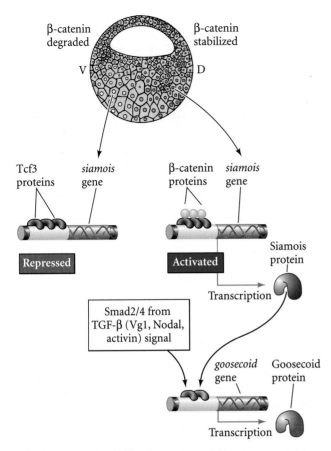

FIGURE 10.25 Summary of events hypothesized to bring about the induction of the organizer in the dorsal mesoderm. The microtubules allow the translocation of Disheveled protein to the dorsal side of the embryo. Dsh binds GSK3, thereby allowing β-catenin to accumulate in the future dorsal portion of the embryo. During cleavage, β-catenin enters the nuclei and binds with Tcf3 to form a transcription factor that activates genes encoding proteins such as Siamois and Twin. Siamois and Twin interact with transcription factors activated by the TGF-β pathway, to activate the *goosecoid* gene in the organizer. Goosecoid is a transcription factor that can activate genes whose proteins are responsible for the organizer's activities. (After Moon and Kimelman 1998.)

tion factors that would also bind to the promoter of the *goosecoid* gene and cooperate with Twin and Siamois to activate it (see Figure 10.25; Germain et al. 2000).

THE MESODERMAL SIGNAL Numerous TFG-β superfamily members in the endoderm are able to induce the cells above them to become mesodermal. Two maternal RNAs tethered to the vegetal cortex may be involved. One of these encodes Vg1, a member of the TGF-β superfamily (see Chapter 6). The other encodes VegT, a transcription factor that instructs the endoderm to synthesize and secrete activin, Derrière, and Nodal proteins (all TFG-β family members) (Latinkic et al. 1997; Smith 2001). These proteins in turn activate the

Xbra (*Brachyury*) gene encoding a transcription factor that instructs the cells to become mesoderm.

The Nodal-related protein Derrière is a particularly promising candidate for a mesoderm inducer, since it is induced by the VegT transcription factor found in the endoderm, and because it can induce animal cap cells to become mesoderm over the long-range distances predicted by Nieuwkoop's experiments (White et al. 2002). Activin is also an important candidate for a mesoderm inducer (Piepenburg et al. 2004). At moderate concentrations, activin activates the *Xbra* gene. Higher concentrations of activin induce the *goosecoid* gene to become expressed (Green and Smith 1990; Moriya et al. 1992; Shimuzu and Gurdon 1999).

Agius and colleagues (2000) have provided evidence that all of these TGF-β-like proteins may act together in a pathway, and that the Nodal-related factors are essential components in the specification of the mesoderm. When they repeated the Nieuwkoop animal-vegetal recombination experiments (see Figure 10.20) but included a specific inhibitor of Nodal-related proteins, the induction by the vegetal cells failed to occur. (The inhibitor did not block Vg1 or activin, however.)

While the TGF-β-like signal is critical for mesoderm production, the *amount* of this signal may control the *type* of mesoderm induced. During the late blastula stage, several Nodal-related proteins (including Xnr1, Xnr2, and Xnr4) are expressed in a dorsal-to-ventral gradient in the endoderm. This gradient is formed by the activation of *Xenopus* Nodal-related gene expression by the synergistic action of VegT with β-catenin. Agius and his colleagues presented a model, shown in Figure 10.26, in which the dorsally located β-catenin and the vegetally located Vg1 signals interact to create a gradient of Nodal-related proteins (Xnr1, 2, 4) across the endoderm. Those regions with little Nodal-related protein become ventral mesoderm; regions with some Nodal-related protein become lateral mesoderm; and regions with large amounts of these proteins become the organizer. Activin and Derriére gradients may behave in the same way. Thus, the initial specification of the mesoderm along the dorsal-ventral axis appears to be accomplished by Nodal-like TGF-β paracrine factors. The region with the highest concentration of these factors may provide the vegetal signal for dorsal mesoderm (organizer) specification, particularly when combined with dorsal β-catenin signal. At the moment, we do not know which (or which combination) of these paracrine factors is providing the signal for organizer formation, but we can be reasonably certain that one or more of them are doing it.

 WEBSITE 10.4 The molecular biology of organizer formation. FGFs and TGF-β family factors interact to specify both the ventral and the dorsal mesodermal components during the blastula stage. Moreover, the ectoderm protects itself from these powerful signals.

FIGURE 10.26 Model for mesoderm induction and organizer formation by the interaction of β-catenin and TGF-β proteins. (A) At late blastula stages, Vg1 and VegT are found in the vegetal hemisphere, while β-catenin is located in the dorsal region. (B) β-Catenin acts synergistically with Veg1 and VegT to activate the Nodal-related (*Xnr*) genes. This creates a gradient of Xnr proteins across the endoderm, highest in the dorsal region. (C) The mesoderm is specified by the Xnr gradient. Mesodermal regions with little or no Xnr have high levels of BMP4 and Xwnt8; they become ventral mesoderm. Those having intermediate concentrations of Xnr's become lateral mesoderm. Where there is a high concentration of Xnr's, *goosecoid* and other dorsal mesodermal genes are activated and the mesodermal tissue becomes the organizer. (These results may explain the activity concentration experiments mentioned in Chapter 3.) (After Agius et al. 2000.)

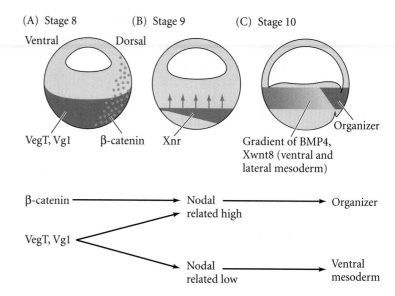

Functions of the Organizer

While the Nieuwkoop center cells remain endodermal, the cells of the organizer become the dorsal mesoderm and migrate underneath the dorsal ectoderm. There, the dorsal mesoderm induces the central nervous system to form. The properties of the organizer tissue can be divided into four major functions:

1. The ability to self-differentiate dorsal mesoderm (prechordal plate, chordamesoderm, etc.)
2. The ability to dorsalize the surrounding mesoderm into paraxial (somite-forming) mesoderm (when it would otherwise form ventral mesoderm)
3. The ability to dorsalize the ectoderm, inducing the formation of the neural tube
4. The ability to initiate the movements of gastrulation

In *Xenopus* (and in other vertebrates), the formation of the anterior-posterior axis is inextricably linked to the formation of the dorsal-ventral axis. Once the dorsal portion of the embryo is established, the movement of the involuting mesoderm establishes the anterior-posterior axis. The endomesoderm that migrates first over the dorsal blastopore lip gives rise to the anterior structures; the mesoderm migrating over the lateral and ventral lips forms the posterior structures. It is now thought that the cells of the organizer ultimately contribute to four cell types—pharyngeal endoderm, head mesoderm (prechordal plate), dorsal mesoderm (primarily the notochord), and the dorsal blastopore lip (Keller 1976; Gont et al. 1993). The pharyngeal endoderm and prechordal plate lead the migration of the organizer tissue and induce the forebrain and midbrain. The dorsal mesoderm induces the hindbrain and trunk. The dorsal blastopore lip forms the dorsal mesoderm and eventually becomes the chordaneural hinge that induces the tip of the tail.

Spemann's description of the organizer started one of the first truly international scientific research programs: the search for the organizer molecules. Researchers from Britain, Germany, France, the United States, Belgium, Finland, Japan, and the former Soviet Union all tried to find these remarkable substances (see Gilbert and Saxén 1993). R. G. Harrison referred to the amphibian gastrula as the "new Yukon to which eager miners were now rushing to dig for gold around the blastopore" (see Twitty 1966, p. 39). Unfortunately, their early picks and shovels proved too blunt to uncover the molecules involved. The proteins responsible for induction were present in concentrations too small for biochemical analyses, and the large quantity of yolk and lipids in the amphibian egg further interfered with protein purification (Grunz 1997). The analysis of organizer molecules had to wait until recombinant DNA technologies enabled investigators to make cDNA clones from blastopore lip mRNA and to see which of these clones encoded factors that could dorsalize the embryo.

 WEBSITE 10.5 Early attempts to locate the organizer molecules. While Spemann did not believe that molecules alone could organize the embryo, his students began a long quest for these factors.

The formation of the dorsal (organizer) mesoderm involves the activation of several genes. The secreted proteins of the Nieuwkoop center are thought to activate a set of transcription factors in the mesodermal cells. These transcription factors then activate the genes encoding the products secreted by the organizer. Several organizer-specific transcription factors have been found and are listed in Table 10.2.

As mentioned earlier, one of the important targets of the Nieuwkoop center appears to be the *goosecoid* gene. The area of expression of *goosecoid* mRNA correlates with the organizer domain in both normal and experimentally treated animals. When lithium chloride treatment is used to increase the organizer mesoderm throughout the margin-

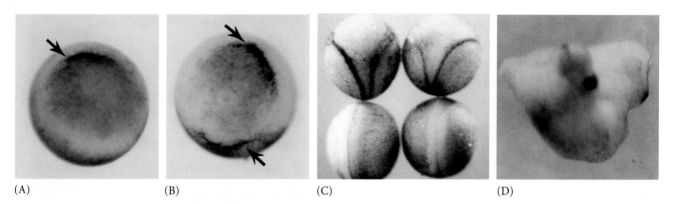

(A)   (B)   (C)   (D)

FIGURE 10.27 Ability of *goosecoid* mRNA to induce a new axis. (A) At the gastrula stage, a control embryo (either uninjected or given an injection of *goosecoid*-like mRNA but lacking the homeobox) has one dorsal blastopore lip (arrow). (B) An embryo whose ventral vegetal blastomeres were injected at the 16-cell stage with *goosecoid* message. Note the secondary dorsal lip. (C) The top two embryos, which were injected with *goosecoid* mRNA, show two dorsal axes; the bottom two control embryos do not. (D) Twinned embryo produced by *goosecoid* injection. Two complete sets of head structures have been induced. (After Cho et al. 1991a, Niehrs et al. 1993, photographs courtesy of E. De Robertis.)

al zone, *goosecoid* expression likewise is expanded. Conversely, when eggs are treated with UV light prior to first cleavage, both dorsal-anterior induction and *goosecoid* expression are significantly inhibited. Injecting the *goosecoid* message into the two ventral blastomeres of a 4-cell *Xenopus* embryo causes the progeny of those blastomeres to involute, undergo convergent extension, and form the dorsal mesoderm and head endoderm of a secondary axis (Figure 10.27). Labeling experiments have shown that such *goosecoid*-injected cells are able to recruit neighboring host cells into the dorsal axis as well (Niehrs et al. 1993). Thus, the Nieuwkoop center activates the *goosecoid* gene in the

TABLE 10.2 Proteins expressed solely or almost exclusively in the organizer (partial list)

Nuclear proteins	Secreted proteins
Xlim1	Chordin
Xnot	Dickkopf
Otx2	ADMP
XFD1	Frzb
XANF1	Noggin
Goosecoid	Follistatin
HNF3β	Sonic hedgehog
	Cerberus
	Nodal-related proteins (several)

organizer tissues, and this gene encodes a DNA-binding protein that (1) activates the migration properties (involution and convergent extension) of the dorsal blastopore lip cells; (2) autonomously determines the dorsal mesodermal fates of those cells expressing it; (3) enables the *goosecoid*-expressing cells to recruit neighboring cells into the dorsal axis; (4) activates *Otx2*, which is critical for brain formation, in the anterior mesoderm and presumptive brain ectoderm; and (5) represses *Wnt8*, whose product can ventralize the embryo (Blitz and Cho 1995, Yao and Kessler 2001).

Induction of neural ectoderm and dorsal mesoderm: BMP inhibitors

Goosecoid and β-catenin proteins work within the organizer cells. They must somehow activate (either directly or indirectly) those genes encoding the soluble proteins that organize the dorsal-ventral and anterior-posterior axis. Early evidence for such diffusible signals from the notochord came from several sources. First, Hans Holtfreter (1933) showed that if amphibian embryos are placed in a high-salt solution, the mesoderm will *evaginate* rather than invaginate, and will not underlie the ectoderm; and, since the ectoderm is not underlain by the notochord, it does not form neural structures.

Further evidence for soluble factors came from the transfilter studies of Finnish investigators (Saxén 1961; Toivonen et al. 1975; Toivonen and Wartiovaara 1976). Newt dorsal lip tissue was placed on one side of a filter fine enough so that no processes could fit through the pores, and competent gastrula ectoderm was placed on the other side. After several hours, neural structures were observed in the ectodermal tissue (Figure 10.28). The identities of the factors diffusing from the organizer, however, took another quarter of a century to identify.

It turned out that scientists were looking for the wrong thing. They had been searching for a molecule secreted by the organizer which was received by the ectoderm and converted it into neural tissue. However, molecular studies on induction have resulted in a remarkable and non-obvious conclusion: It is the *epidermis* that is induced to form, not the neural tissue. The ectoderm is induced to become epidermal tissue by binding bone morphogenetic proteins

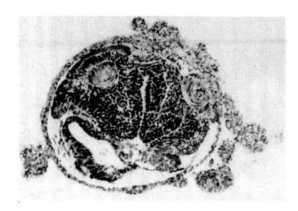

FIGURE 10.28 Neural structures induced in presumptive ectoderm by newt dorsal lip tissue, separated from the ectoderm by a nucleopore filter with an average pore diameter of 0.05 mm. Anterior neural tissues are evident, including some induced eyes. (From Toivonen 1979, photograph courtesy of L. Saxén.)

FIGURE 10.29 Rescue of dorsal structures by Noggin protein. When *Xenopus* eggs are exposed to ultraviolet radiation, cortical rotation fails to occur, and the embryos lack dorsal structures (top). If such an embryo is injected with *noggin* mRNA, it develops dorsal structures in a dosage-related fashion (top to bottom). If too much *noggin* message is injected, the embryo produces dorsal anterior tissue at the expense of ventral and posterior tissue, becoming little more than a head (bottom). (Photograph courtesy of R. M. Harland.)

(BMPs). The nervous system forms from that region of the ectoderm that is *protected* from epidermal induction (Hemmati-Brivanlou and Melton 1994, 1997). In other words, (1) the "default fate" of the ectoderm is to become neural tissue; (2) certain parts of the embryo induce the ectoderm to become epidermal tissue by secreting BMPs; and (3) the organizer tissue acts by secreting molecules that block BMPs, thereby allowing the ectoderm "protected" by these BMP inhibitors to become neural tissue. Three of the major BMP inhibitors secreted by the organizer are Noggin, chordin, and follistatin.

NOGGIN In 1992, Smith and Harland constructed a cDNA plasmid library from dorsalized (lithium chloride-treated) gastrulae. Messenger RNAs synthesized from sets of these plasmids were injected into ventralized embryos (having no neural tube) produced by irradiating early embryos with ultraviolet light. Those plasmid sets whose mRNAs rescued dorsal structures in these embryos were split into smaller sets, and so on, until single-plasmid clones were isolated whose mRNAs were able to restore the dorsal tissue in such embryos. One of these clones contained the *noggin* gene (Figure 10.29). Injection of *noggin* mRNA into 1-cell, UV-irradiated embryos completely rescued dorsal development and allowed the formation of a complete embryo.

Noggin is a secreted protein that is able to accomplish two of the major functions of the organizer: it induces dorsal ectoderm to form neural tissue, and it dorsalizes mesoderm cells that would otherwise contribute to the ventral mesoderm (Smith et al. 1993). Smith and Harland showed that newly transcribed *noggin* mRNA is first localized in the dorsal blastopore lip region and then becomes expressed in the notochord (Figure 10.30). Noggin binds to BMP4 and BMP2 and inhibits their binding to receptors (Zimmerman et al. 1996).

CHORDIN The second organizer protein found was **chordin**. It was isolated from clones of cDNA whose mRNAs were present in dorsalized, but not in ventralized, embryos (Sasai et al. 1994). These clones were tested by injecting them into ventral blastomeres and seeing whether they induced secondary axes. One of the clones capable of inducing a secondary neural tube contained the *chordin* gene. *Chordin* mRNA was found to be localized in the dorsal blastopore lip and later in the notochord (Figure 10.31). Of all organizer genes observed, chordin is the one most acutely activated by β-catenin (Wessley et al. 2004). Morpholino antisense oligomers (see Chapter 4) directed against the *chordin* message blocked the ability of an organizer graft to induce a secondary central nervous system (Oelgeschläger et al. 2003).

Like Noggin, chordin binds directly to BMP4 and BMP2 and prevents their complexing with their receptors (Piccolo et al. 1996). In zebrafish, a *chordin* loss-of-function muta-

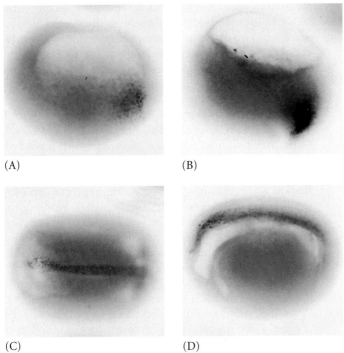

(A)

(B)

(C)

(D)

FIGURE 10.30 Localization of *noggin* mRNA in the organizer tissue, shown by in situ hybridization. (A) At gastrulation, *noggin* mRNA (dark areas) accumulates in the dorsal marginal zone. (B) When cells involute, *noggin* mRNA is seen in the dorsal blastopore lip. (C) During convergent extension, *noggin* is expressed in the precursors of the notochord, prechordal plate, and pharyngeal endoderm, which (D) extend beneath the ectoderm in the center of the embryo. (Photographs courtesy of R. M. Harland.)

tion (the *chordino* mutant) has a greatly reduced neural plate and an enlarged region of ventral mesoderm (Hammerschmidt et al. 1996).

FOLLISTATIN The mRNA for a third organizer-secreted protein, **follistatin**, is also transcribed in the dorsal blastopore lip and notochord. Follistatin was found in the organizer

through an unexpected result of an experiment that was looking for something else. Ali Hemmati-Brivanlou and Douglas Melton (1992, 1994) wanted to see whether the protein activin was required for mesoderm induction. In searching for the mesoderm inducer, they found that follistatin, an inhibitor of both activin and BMPs, would cause ectoderm to become neural tissue. They then proposed that the ectoderm would become neural unless induced to become epidermal by the BMPs. This model was supported by, and explained, some cell dissociation experiments that had also produced odd results. Three studies—by Grunz and Tacke (1989), Sato and Sargent (1989), and Godsave and Slack (1989)—had shown that when whole embryos or their animal caps were dissociated, they formed neural tissue. This result would be explainable if the "default state" of the ectoderm was not epidermal, but neural, and tissue had to be induced to have an epidermal phenotype. *The organizer, then, blocks this epidermalizing induction.*

In *Xenopus*, the epidermal inducers are **bone morphogenetic proteins: BMP4** and its close relatives BMP2, BMP7, and ADMP (anti-dorsalizing morphogenetic protein). It was known that there is an antagonistic relationship between these BMPs and the organizer. If the mRNA for BMP4 is injected into *Xenopus* eggs, all the mesoderm in the embryo becomes ventrolateral mesoderm, and no involution occurs at the blastopore lip (Dale et al. 1992; Jones et al. 1992). Conversely, overexpression of a dominant negative BMP4 receptor resulted in the formation of twinned axes (Graff et al. 1994; Suzuki et al. 1994). In 1995, Wilson and Hemmati-Brivanlou demonstrated that BMP4 induced ectodermal cells to become epidermal. By 1996, several laboratories had demonstrated

FIGURE 10.31 Localization of *chordin* mRNA. (A) Whole-mount in situ hybridization shows that just prior to gastrulation, *chordin* message (dark area) is expressed in the region that will become the dorsal blastopore lip. (B) As gastrulation begins, *chordin* is expressed at the dorsal blastopore lip. (C) In later stages of gastrulation, *chordin* message is seen in the organizer tissues. (From Sasai et al. 1994, photographs courtesy of E. De Robertis.)

(A)

(B)

(C)

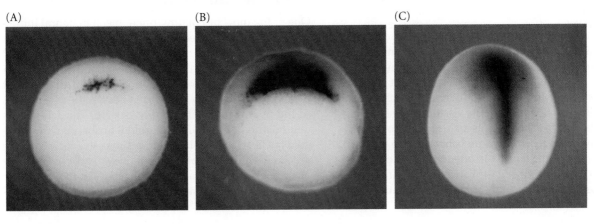

(A)

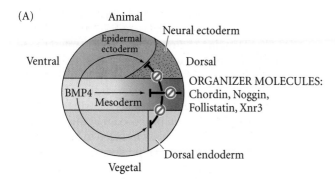

(B)

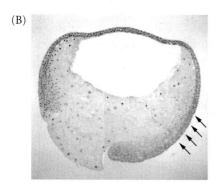

FIGURE 10.32 Model for the action of the organizer. (A) BMP4 (and certain other molecules) is a powerful ventralizing factor. Organizer proteins such as chordin, Noggin, and follistatin block the action of BMP4. The antagonistic effects of these proteins can be seen in all three germ layers. BMP4 may elicit the expression of different genes in a concentration-dependent fashion. Thus, in the regions of *noggin* and *chordin* expression, BMP4 is totally prevented from binding, and these tissues become notochord (organizer) tissue. Slightly farther away from the organizer, *myf5*, a marker for the dorsolateral muscles, is activated. As more and more BMP4 molecules are allowed to bind to the cells, the *Xvent2* (ventrolateral) and *Xvent1* (ventral) genes become expressed. (B) BMP pathway activation visualized in an early gastrulating *Xenopus* embryo. Antibodies specific to the phosphorylated form of Smad1 (i.e., Smad1 that has been activated by BMP4) can be seen binding to the ectodermal cells in the ventral portion of the embryo. Close to the dorsal lip there is a gradient of BMP4 inactivation (arrows). (After Dosch et al. 1997, De Robertis et al. 2000; B from Kurata et al. 2000, photograph courtesy of N. Ueno.)

that Noggin, chordin, and follistatin are all secreted by the organizer, and that each of them prevents BMP from binding to the ectoderm and mesoderm near the organizer (Piccolo et al. 1996; Zimmerman et al. 1996; Iemura et al. 1998).

BMP4 is initially expressed throughout the ectodermal and mesodermal regions of the late blastula. However, during gastrulation, *bmp4* transcripts are restricted to the ventrolateral marginal zone. This is probably because the Xiro1 protein is made in the dorsal mesoderm (organizer) region starting at the beginning of gastrulation, and this transcription factor represses *bmp4* transcription (Hemmati-Brivan-

lou and Thomsen 1995; Northrop et al. 1995; Glavic et al. 2001). In the ectoderm, BMPs repress the genes (such as *neurogenin*) involved in forming neural tissue, while activating other genes involved in epidermal specification (Lee et al. 1995). In the mesoderm, it appears that graded levels of BMP4 activate different sets of mesodermal genes, thereby specifying the dorsal, intermediate, and lateral mesodermal tissues (Figure 10.32; Hemmati-Brivanlou and Thomsen 1995; Gawantka et al. 1995; Dosch et al. 1997).

In 2005, two important sets of experiments provided comfirmation of the default model and of the importance of blocking BMPs to specify the nervous system. First, Khokha and colleagues (2005) used antisense morpholinos to eliminate the three BMP antagonists (i.e., Noggin, chordin, and follistatin) in *Xenopus*. The resulting embryos had catastrophic failure of dorsal development and lacked neural plates and dorsal mesoderm (Figure 10.33A,B). Second, Reversade and colleagues blocked the activity of BMPs 4, 2, and 7, and ADMP with antisense morpholinos (Reversade et al. 2005; Reversade and De Robertis 2006). When they simultaneously blocked the formation of BMPs 2, 4, and 7, the neural tube became greatly expanded, taking over a much larger region of the ectoderm (Figure 10.33C). When they did a quadruple inactivation of the three BMPs *and* ADMP, the entire ectoderm became neural, and no dorsoventral polarity was apparent (Figure 10.33D). Thus, the epidermis is instructed by BMP signaling, and the organizer works by blocking that BMP signal from reaching the ectoderm above it.

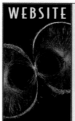 **WEBSITE 10.6 The specification of the endoderm.** Although the mesoderm is induced, and the differences between neural and epidermal ectoderm are induced, the endoderm appears to be specified autonomously. Recent studies show that TGF-β family signals and the maternally derived VegT transcription factor in the vegetal cells initiate a cascade of events leading to endoderm formation.

The Regional Specificity of Induction

The determination of regional differences

One of the most fascinating phenomena in neural induction is the regional specificity of the neural structures that are produced. Forebrain, hindbrain, and spinocaudal regions of the neural tube must all be properly organized in an anterior-to-posterior direction. The organizer tissue not only induces the neural tube, but also specifies the regions of the neural tube. This region-specific induction was demonstrated by Hilde Mangold's husband, Otto Mangold (1933). He transplanted four successive regions of the archenteron roof of late-gastrula newt embryos into the blastocoels of early-gastrula embryos. The most anterior portion of the archenteron roof (containing head meso-

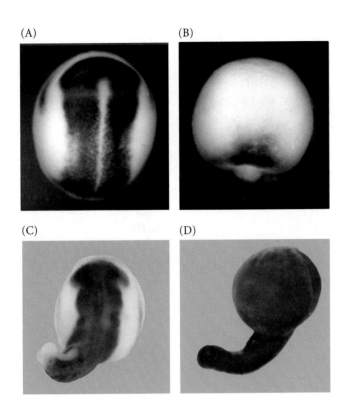

(A)

(B)

(C)

(D)

FIGURE 10.33 Control of neural specification by the levels of BMPs. (A,B) Lack of dorsal structures in *Xenopus* embryos whose BMP-inhibitor genes (*chordin*, *noggin*, and *follistatin*) were eliminated by antisense morpholino oligonucleotides. (A) Control embryo with neural folds stained for the expression of the neural gene *Sox2*. (B) Lack of neural tube and *Sox2* expression in an embryo treated with the morpholinos against all three BMP inhibitors. (C,D) Expanded neural development. (C) The neural tube, visualized by Sox2 staining, is greatly enlarged in embryos treated with antisense mopholinos that destroy BMPs 2, 4, and 7. (D) Complete transformation of the entire ectoderm into neural ectoderm (and loss of the dorsal-ventral axis) by inactivation of ADMP as well as BMPs 2, 4, and 7. (A,B from Khokha et al. 2005, photographs courtesy of R. Harland; C,D from Reversade and De Robertis 2005.)

BMP4 and Geoffroy's Lobster

The hypothesis that the organizer secretes proteins that block BMPs received further credence from an unexpected source—the emerging field of evolutionary developmental biology (see Chapter 23). Researchers have discovered that the same chordin-BMP4 interaction that instructs the formation of the neural tube in vertebrates also forms neural tissue in fruit flies (Holley et al. 1995; Schmidt et al. 1995; De Robertis and Sasai 1996). The dorsal neural tube of the vertebrate and the ventral neural cord of the fly appear to be generated by the same set of instructions.

The *Drosophila* homologue of the *bmp4* gene is *decapentaplegic* (*dpp*). As discussed in Chapter 9, Dpp protein is responsible for patterning the dorsal-ventral axis; It is present in the dorsal portion of the fly embryo and diffuses ventrally. Dpp is opposed by a protein called Short-gastrulation (Sog), which is the *Drosophila* homologue of

chordin. These insect homologues not only appear to be similar to their vertebrate counterparts, they can actually substitute for each other. When *sog* mRNA is injected into ventral regions of *Xenopus* embryos, it induces the amphibian notochord and neural tube. Injecting *chordin* mRNA into *Drosophila* embryos produces ventral nervous tissue.

Although chordin usually dorsalizes the *Xenopus* embryo, it ventralizes *Drosophila*. In *Drosophila*, Dpp is made dorsally; in *Xenopus*, BMP4 is made ventrally. In both cases, Sog/chordin makes neural tissue by blocking the effects of Dpp/BMP4. In *Drosophila*, Sog interacts with Tolloid and several other proteins to create a gradient of Sog proteins. In *Xenopus*, the homologue of the same proteins act to create a gradient of chordin (see Figure 23.7; Hawley et al. 1995; Holley et al. 1995; De Robertis et al. 2000).

In 1822, the French anatomist Étienne Geoffroy Saint-Hilaire provoked one of the most heated and critical confrontations in biology when he proposed that the lobster was but a vertebrate upside down. He claimed that the ventral side of the lobster (with its nerve cord) was homologous to the dorsal side of the vertebrate (Appel 1987). It seems that he was correct on the molecular level, if not on the anatomical level. The instructions for producing a nervous system in fact may have evolved only once, and the myriad animal lineages may all have used this same set of instructions—just in different places. The BMP4(Dpp)/chordin(Sog) interaction is an example of "homologous processes," suggesting a unity of developmental principles among all animals (Gilbert and Bolker 2001).

FIGURE 10.34 Regional specificity of induction can be demonstrated by implanting different regions (color) of the archenteron roof into early *Triturus* gastrulae. The resulting embryos develop secondary dorsal structures. (A) Head with balancers. (B) Head with balancers, eyes, and forebrain. (C) Posterior part of head, diencephalon, and otic vesicles. (D) Trunk-tail segment. (After Mangold 1933.)

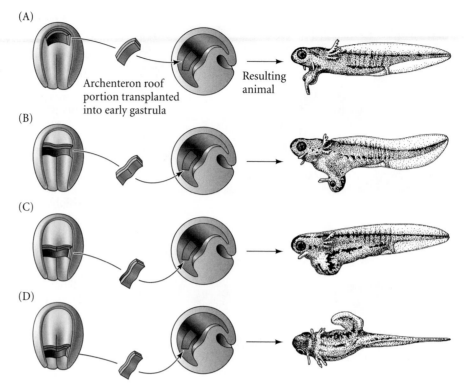

(A)

(B)

(C)

(D)

Archenteron roof portion transplanted into early gastrula

Resulting animal

derm) induced balancers and portions of the oral apparatus; the next most anterior section induced the formation of various head structures, including nose, eyes, balancers, and otic vesicles; the third section (including the notochord) induced the hindbrain structures; and the most posterior section induced the formation of dorsal trunk and tail mesoderm* (Figure 10.34). Moreover, when dorsal blastopore lips from early salamander gastrulae were transplanted into other early salamander gastrulae, they formed secondary heads. When dorsal lips from later gastrulas were transplanted into early salamander gastrulae, however, they induced the formation of secondary tails (Figure 10.35; Mangold 1933). These results show that the first cells of the organizer to enter the embryo induce the formation of brains and heads, while those cells that form the dorsal lip of later-stage embryos induce the cells above them to become spinal cords and tails.

The question then became: What molecules are being secreted by the organizer in a regional fashion such that the first cells involuting through the blastopore lip (the endomesoderm) induce head structures, while the next portion of involuting mesoderm (notochord) produces trunk and tail structures? Figure 10.36 shows a possible

model for these inductions, the elements of which we'll now describe in detail.

The head inducer: Wnt inhibitors

The most anterior regions of the head and brain are underlain not by notochord, but by pharyngeal endoderm and head (prechordal) mesoderm (see Figures 10.7C,D and 10.36A). This "endomesoderm" constitutes the leading edge of the dorsal blastopore lip. Recent studies have shown that these cells not only induce the most anterior head structures, but that they do it by blocking the Wnt pathway as well as by blocking BMP4.

CERBERUS In 1996, Bouwmeester and colleagues showed that the induction of the most anterior head structures could be accomplished by a secreted protein called **Cerberus**.[†] Unlike the other proteins secreted by the organizer, Cerberus promotes the formation of the cement gland (the most anterior region of tadpole ectoderm), eyes, and olfactory (nasal) placodes. When *cerberus* mRNA was injected into a vegetal ventral *Xenopus* blastomere at the 32-cell stage, ectopic head structures were formed (Figure 10.37). These head structures arose from the injected cell as well as from neighboring cells.

The *cerberus* gene is expressed in the pharyngeal endomesoderm cells that arise from the deep cells of the

---

*The induction of dorsal mesoderm—rather than the dorsal ectoderm of the nervous system—by the posterior end of the notochord was confirmed by Bïjtel (1931) and Spofford (1945), who showed that the posterior fifth of the neural plate gives rise to tail somites and the posterior portions of the pronephric kidney duct.

[†]"Cerberus" is named after the three-headed dog that guarded the entrance to Hades in Greek mythology.

(A) Transplantation of young gastrula dorsal lip

(B) Transplantation of advanced gastrula dorsal lip

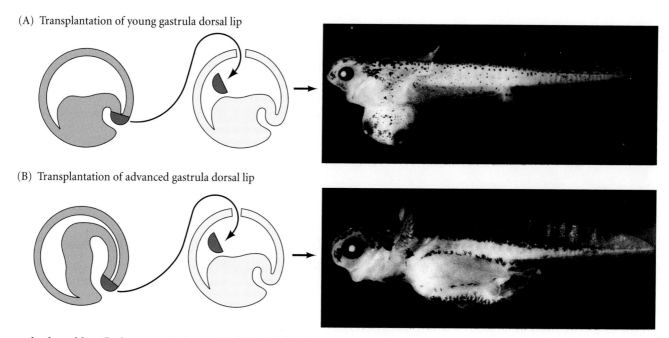

FIGURE 10.35 Regionally specific inducing action of the dorsal blastopore lip. (A) Young dorsal lips (which will form the anterior portion of the organizer) induce anterior dorsal structures when transplanted into early newt gastrulae. (B) Older dorsal lips transplanted into early newt gastrulae produce more posterior dorsal structures. (From Saxén and Toivonen 1962, photographs courtesy of L. Saxén.)

early dorsal lip. Cerberus protein can bind BMPs, Nodal-related proteins, and Xwnt8 (see Figure 10.36; Piccolo et al. 1999. When Cerberus synthesis is blocked , the levels of BMP, Nodal-related proteins, and Wnts all rise in the anterior of the embryo, and the ability of the anterior endomesoderm to induce a head is severely diminished. (Silva et al. 2003).

FRZB AND DICKKOPF Shortly after the attributes of Cerberus were demonstrated, two other proteins, Frzb and Dickkopf, were discovered to be synthesized in the involuting

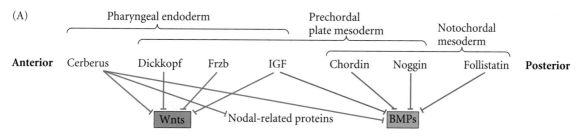

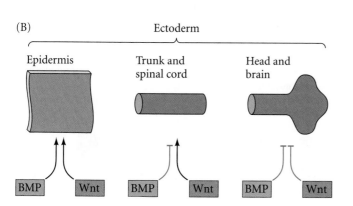

FIGURE 10.36 Paracrine factor antagonists from the organizer are able to block specific paracrine factors to distinguish head from tail. (A) The pharyngeal endoderm that underlies the head secretes Dickkopf, Frzb, and Cerberus. Dickkopf and Frzb block Wnt proteins; Cerberus blocks Wnts, Nodal-related proteins, and BMPs. The prechordal plate secretes the Wnt blockers Dickkopf and Frzb, as well as BMP-blockers chordin and Noggin. The notochord contains BMP-blockers chordin, Noggin, and follistatin, but it does not secrete Wnt-blockers. IGF from the head endomesoderm probably acts at the junction of the notochord and prechordal mesoderm. (B) Summary of paracrine antagonist function in the ectoderm. Brain formation includes inhibiting both the Wnt and BMP pathways. Spinal cord neurons are produced when Wnt functions without the presence of BMPs. Epidermis is formed by having both the Wnt and BMP pathways operating.

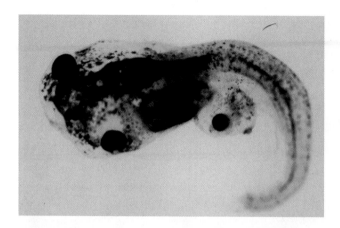

FIGURE 10.37 *Cerberus* mRNA injected into a single D4 (ventral vegetal) blastomere of a 32-cell *Xenopus* embryo induces head structures as well as a duplicated heart and liver. The secondary eye (a single cyclopic eye) and olfactory placode can be readily seen. (From Bouwmeester et al. 1996, photograph courtesy of E. M. De Robertis.)

endomesoderm. **Frzb** (pronounced "frisbee") is a small, soluble form of Frizzled (the Wnt receptor) and it is capable of binding Wnt proteins in solution (Figure 10.38; Leyns et al. 1997; Wang et al. 1997). Frzb is synthesized predominantly in the endomesoderm cells beneath the head (Figure 10.38B,C). If embryos are made to synthesize excess Frzb, Wnt signaling fails to occur throughout the embryo; such embryos lack ventral posterior structures and become "all head." The **Dickkopf** protein (German, "thick head," "stubborn") also appears to interact directly with the Wnt receptors, preventing Wnt signaling (Mao et al. 2001, 2002). Injection of antibodies against Dickkopf causes the resulting embryos to have small, deformed heads with no forebrain (Glinka et al. 1998). Therefore, the induction of trunk structures may be caused by the blockade of BMP signaling from the notochord, while Wnt signals are allowed to proceed. However, to produce a head, both the BMP signal and the Wnt signal must be blocked. This blockade comes from the endomesoderm, the most anterior portion of the organizer (Glinka et al. 1997).

INSULIN-LIKE GROWTH FACTORS In addition to those proteins that block BMP and Wnt signaling by physically binding to these paracrine factors, themselves, the head region contains yet another set of proteins that prevents BMP and

Wnt signals from reaching the nucleus. Pera and colleagues (2001) showed that insulin-like growth factors (IGFs) are required for the formation of the anterior neural tube with its brain and sensory placodes. IGFs accumulate in the dorsal midline and are especially prominent in the anterior neural tube (Figure 10.39A). When injected into ventral mesodermal blastomeres, mRNA from IGFs caused the formation of ectopic heads, while blocking the IGF receptors resulted in the lack of head formation (Figure 10.39B).

Insulin-like growth factors appear to work by initiating an FGF-like signal transduction cascade that interferes with the signal transduction pathways of both BMPs and Wnts (Richard-Parpaillon et al. 2002; Pera et al. 2003).

Trunk induction: Wnt signals and retinoic acid

Toivonen and Saxén provided evidence for a gradient of a posteriorizing factor that would act to specify the trunk

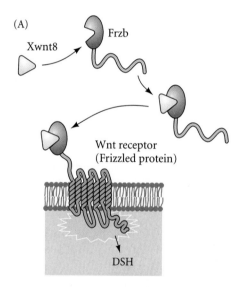

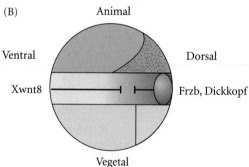

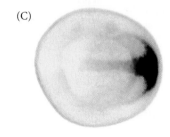

FIGURE 10.38 Xwnt8 is capable of ventralizing the mesoderm and preventing anterior head formation in the ectoderm. (A) Frzb protein is secreted by the anterior region of the organizer. It must bind to Xwnt8 before that inducer can bind to its receptor. Frzb resembles the Wnt-binding domain of the Wnt receptor (Frizzled protein), but Frzb is a soluble molecule. (B) Xwnt8 is made throughout the marginal zone. (C) Double in situ hybridization localizing Frzb (dark blue) and chordin (brown) messages. The *frzb* mRNA is seen to be transcribed in the head endomesoderm of the organizer, but not in the notochord (where chordin is expressed). (From Leyns et al. 1997, photograph courtesy of E. M. De Robertis.)

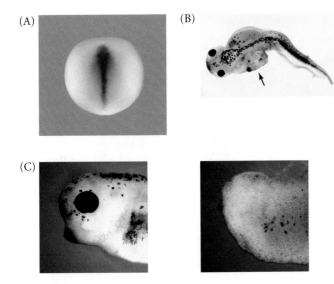

(A)

(B)

(C)

FIGURE 10.39 Insulin-like growth factors enhance anterior neural development. (A) Expression pattern of *Igf3*, showing protein accumulation in the anterior neural tube. (B) An ectopic headlike structure (complete with eyes and cement gland) formed when *Igf2* mRNA was injected into ventral marginal zone blastomeres. (C) Anterior of 3-day control tadpole (left) compared with a tadpole whose 4-cell embryonic blastomeres were injected with an IGF inhibitor. The cement gland and eyes are absent. (From Pera et al. 2001, photographs courtesy of E. M. De Robertis.)

and tail tissues of the amphibian embryo*(Toivonen and Saxén 1955, 1968; reviewed in Saxén 2001).This factor's activity would be highest in the posterior of the embryo and weakened anteriorly. Recent studies have extended this model and have proposed candidates for posteriorizing molecules. The primary protein involved in posteriorizing the neural tube is thought to be a member of the Wnt family of paracrine factors, most likely Xwnt8 (Domingos et al. 2001; Kiecker and Niehrs 2001).

It appears that a gradient of Wnt proteins is necessary for specifying the posterior region of the neural plate (the trunk and tail). (Hoppler et al. 1996; Niehrs 2004). In *Xenopus*, an endogenous gradient of Wnt signaling and β-catenin is highest in the posterior and absent in the anterior (Figure 10.40A). Moreover, if Xwnt8 is added to developing embryos, spinal cord-like neurons are seen more anteriorly in the embryo, and the most anterior markers of the forebrain are absent. Conversely, suppressing Wnt signaling (by adding Frzb or Dickkopf to the developing embryo) leads to the expression of the anteriormost

---

*The tail inducer was initially thought to be part of the trunk inducer, since transplantation of the late dorsal blastopore lip into the blastocoel often produced larvae with extra tails. However, it appears that tails are normally formed by interactions between the neural plate and the posterior mesoderm during the neurula stage (and thus are generated outside the organizer). Here, Wnt, BMPs, and Nodal signaling all seem to be required (Tucker and Slack 1995; Niehrs 2004).

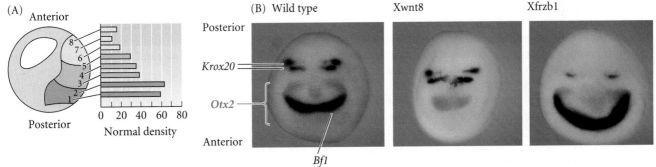

(A) Anterior

8 7 6 5 4 3 2 1

0 20 40 60 80
Normal density

Posterior

(B) Wild type Xwnt8 Xfrzb1

Posterior

Krox20

Otx2

Anterior

Bf1

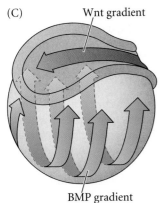

(C) Wnt gradient

BMP gradient

FIGURE 10.40 The Wnt signaling pathway and posteriorization of the neural tube. (A) The gradient of β-catenin in the presumptive neural plate during gastrulation. Gastrulating embryos were stained for β-catenin and the density of the stain compared between regions of the ectodermal cells. (B) Changing the level of Wnt signaling changes the expression of regionally specific markers. In the control neurula (left), Bf1 marks the neural cells of the forebrain and is the most anterior marker. The *Otx2* gene is expressed in the forebrain and midbrain (red), and the *Krox20* gene is expressed in two regions of the hindbrain. Microinjecting plasmids that produce more Xwnt8 (center) causes posteriorization of the neural plate, demonstrated by the disappearance of the Bf1 expression band and the reduction of *Otx2* expression in the midbrain. Microinjecting plasmids making Xfrzb1 (*Xenopus* Frzb, a Wnt antagonist; right) anteriorizes the neural plate, causing the expansion of the regions of *Bf1* and *Otx2* expression at the expense of the more posterior regions. (C) Double-gradient model whereby a gradient in BMP expression specifies the frog dorsal-ventral axis while a gradient of Wnt proteins specifies the anterior-posterior axis. (After Kiecker and Niehrs 2001; Niehrs 2004.)

markers in more posterior neural cells (Figure 10.40B). Therefore, there appear to be two major gradients in the amphibian gastrula—a BMP gradient that specifies the dorsal-ventral axis and a Wnt gradient specifying the anterior-posterior axis (Figure 10.40C). It must be remembered, too, that both these axes are established by the initial axes of Nodal-like TGF-β factors and β-catenin across the vegetal cells. The basic model of neural induction, then, looks like the diagram in Figure 10.41.

While the Wnt proteins probably play a major role in specifying the anterior-posterior axis, they are probably not the only agent involved. Fibroblast growth factors appear to be critical in allowing the cells to respond to the Wnt signal (Holowacz and Sokol 1999; Domingos et al. 2001). Retinoic acid also has been found in a gradient highest at the posterior end of the neural plate, and RA can also posteriorize the neural tube in a concentration-dependent manner (Cho and De Robertis 1990; Sive and Cheng 1991; Chen et al. 1994). Retinoic acid signaling appears to be especially important in patterning the hindbrain (Blumberg et al. 1997; Kolm et al. 1997; Dupé and Lumsden 2001).

WEBSITE **10.7 Regional specification.** The research into regional specification has been a fascinating endeavor involving scientists from all over the world. Before molecular biology gave us the tools to uncover morphogenetic proteins, embryologists developed ingenious ways of finding out what those proteins were doing.

Specifying the Left-Right Axis

Although the developing tadpole looks symmetrical from the outside, several internal organs, such as the heart and the gut tube, are not evenly balanced on the right and left sides. In other words, in addition to its dorsal-ventral and anterior-posterior axes, the embryo has a left-right axis. In all vertebrates studied so far, the crucial event in left-right axis formation is the expression of a *nodal* gene in the lateral plate mesoderm on the *left* side of the embryo. In *Xenopus*, this gene is *Xnr1* (*Xenopus nodal-related 1*). If the expression of this gene is permitted to occur on the right-hand side, the position of the heart (normally found on the left side) and the coiling of the gut are randomized.

But what limits *Xnr1* expression solely to the left-hand side? In *Xenopus*, it is possible that the first cue is given at fertilization. The microtubules involved in cytoplasmic rotation appear to be crucial, since if their formation is inhibited, no left-right axis appears (Yost 1998). One possible explanation is that the Vg1 protein, which appears to be expressed throughout the vegetal hemisphere, seems to be processed into its active form predominantly on the left-hand side of the embryo. Injecting active Vg1 protein into the left vegetal blastomeres has no effect, but adding it to the right vegetal blastomeres leads to *Xnr1* expression in

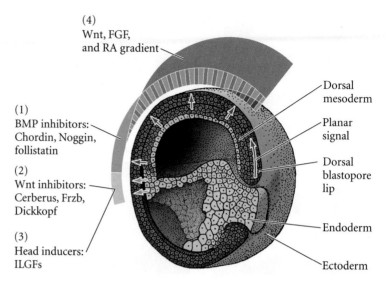

(4) Wnt, FGF, and RA gradient

(1) BMP inhibitors: Chordin, Noggin, follistatin

(2) Wnt inhibitors: Cerberus, Frzb, Dickkopf

(3) Head inducers: ILGFs

Dorsal mesoderm
Planar signal
Dorsal blastopore lip
Endoderm
Ectoderm

FIGURE 10.41 Model of organizer function and axis specification in the *Xenopus* gastrula. (1) BMP inhibitors from organizer tissue (dorsal mesoderm and pharyngeal mesendoderm) block the formation of epidermis, ventrolateral mesoderm, and ventrolateral endoderm. (2) Wnt inhibitors in the anterior of the organizer (pharyngeal endomesoderm) allow the induction of head structures. (3) Head structures are induced through insulin-like growth factor signaling. (4) A gradient of caudalizing factors (Wnts, retinoic acid) causes the regional expression of Hox genes, specifying the regions of the neural tube.

both the right and left lateral plates, and to the randomization of heart and gut positions (Hyatt et al. 1996; Kramer and Yost 2002). It is not yet known how the events at fertilization lead to the expression of *Xnr1* in the left lateral plate mesoderm during gastrulation.

The pathway by which Xnr1 instructs the heart and gut to fold properly is unknown, but one of the key genes activated by Xnr1 appears to be *pitx2*. Since it is activated by Xnr1, *pitx2* is normally expressed only on the left side of the embryo. Pitx2 protein persists on the embryo's left side as the heart and gut develop, controlling their respective positions; if Pitx2 is injected into the right side of an embryo, heart placement and gut coiling are randomized (Figure 10.42; Ryan et al. 1998). As we will see, the pathway through which Nodal protein establishes left-right polarity by activating *Pitx2* on the left side of the embryo is conserved throughout all vertebrate lineages.

Coda

We are finally putting names to the "agents" and "soluble factors" postulated by the early experimental embryologists. We are also delineating the intercellular pathways of paracrine factors and transcription factors that constitute the first steps in organogenesis. The international research program initiated by Spemann's laboratory in the 1920s is reaching fruition, and this research has revealed layers of complexity that reach beyond anything Spemann could

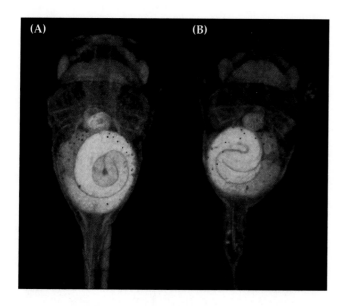

FIGURE 10.42 Pitx2 determines the direction of heart looping and gut coiling. (A) Wild-type *Xenopus* tadpole viewed from the ventral side, showing rightward heart looping and counter-clockwise gut coiling. (B) If an embryo is injected with Pitx2 so that this protein is present in the mesoderm of both the right and left sides (instead of just the left side), heart looping and gut coiling are random with respect to each other. Sometimes this treatment results in complete reversals, as in this embryo, in which the heart loops to the left and the gut coils in a clockwise manner. (From Ryan et al. 1998, photograph courtesy of J. C. Izpisúa-Belmonte.)

have conceived. Just as Spemann's experiments told us how much we didn't know, so today we are faced with a whole new set of questions generated by our answers to older ones. Surveying the field in 1927, Spemann remarked:

> We still stand in the presence of riddles, but not without hope of solving them. And riddles with the hope of solution—what more can a scientist desire?

The challenge still remains.

SIDELIGHTS & SPECULATIONS

Competence, Bias, and Neurulation

In addition to the signals coming from the underlying chordal plate and dorsal mesoderm, there may also be a bias in the cells of the dorsal part of the embryo toward becoming neural. Phillips and colleagues (London et al. 1988; Savage and Phillips 1989) have shown that the dorsal and ventral ectoderm have biases in different directions. It appears that before gastrulation, signals are given that predispose the dorsal ectoderm in a neural fate and bias the ventral and lateral ectoderm to become epidermis.

Competence to become neural ectoderm is given by β-catenin signaling, which inhibits BMP expression and activates the *chordin*, *nodal-related-3*, and *noggin* genes, even in the ectoderm (Baker et al. 1999; Kuroda et al. 2004). Thus, the ectoderm immediately above the dorsal blastopore lip becomes biased to produce the anterior-most neural ectoderm (Wessely et al. 2001; Osada et al. 2003; Kuroda et al. 2004). Moreover, this region of the ectoderm is probably exposed to FGF signals immediately before neural induction, and FGF signaling

also blocks BMP. Without FGF, the ectoderm does not appear to be able to form the neural plate (Schohl and Fagotto 2002; Delaune et al. 2005). Indeed, the ability for the dissociated ectoderm cells to become neural (rather than epidermal) is due not only to the dilution of the BMP signal, but also to the activation of the FGF signaling cascade in the dissociated cells. This MAP kinase signal phosphorylates the Smad1 protein in a manner that prevents it from functioning as a transcription factor (Kuroda et al. 2005).

 **Snapshot Summary** **Early Amphibian Development**

1. Amphibian cleavage is holoblastic, but it is unequal due to the presence of yolk in the vegetal hemisphere.

2. Amphibian gastrulation begins with the invagination of the bottle cells, followed by the coordinated involution of the mesoderm and the epiboly of the ectoderm. Vegetal rotation plays a significant role in directing the involution.

3. The driving forces for ectodermal epiboly and the convergent extension of the mesoderm are the intercalation events in which several tissue layers merge. Fibronectin plays a critical role in enabling the mesodermal cells to migrate into the embryo.

4. The dorsal lip of the blastopore forms the organizer tissue of the amphibian gastrula. This tissue dorsal-

izes the ectoderm, transforming it into neural tissue, and it transforms ventral mesoderm into lateral and dorsal mesoderm.

5. The organizer consists of pharyngeal endoderm, head mesoderm, notochord, and dorsal blastopore lip tissues. The organizer functions by secreting proteins (Noggin, chordin, and follistatin) that block the BMP signal that would otherwise ventralize the mesoderm and activate the epidermal genes in the ectoderm.

6. The organizer is itself induced by the Nieuwkoop center, located in the dorsalmost vegetal cells. This center is formed by the translocation of the Disheveled protein to the dorsal side of the egg.

7. Disheveled protein stabilizes β-catenin in the dorsal cells of the embryo. Thus, the Nieuwkoop center is formed by the accumulation of β-catenin, which can complex with Tcf3 to form a transcription factor complex that can activate the transcription of the siamois and twin genes.

8. The Siamois and Twin proteins in collaboration with transcription factors generated by a TGF-β signal can activate the *goosecoid* gene in the organizer. β-catenin, in collaboration with other transcription factors activates *chordin* and several other genes that encode the paracrine factors and paracrine factor inhibitors that mediate functions of the organizer.

9. In the head region, an additional set of proteins (Cerberus, Frzb, Dickkopf) block the Wnt signal from the ventral and lateral mesoderm.

10. Wnt signaling causes a gradient of β-catenin along the anterior-posterior axis of the neural plate. This graded signaling appears to specify the regionalization of the neural tube.

11. Insulin-like growth factors appear to transform the neural tube into anterior (forebrain) tissue.

12. The left-right axis appears to be initiated at fertilization through the Vg1 protein. In a still unknown fashion, this protein activates a Nodal protein solely on the left side of the body. In *Xenopus*, as in other vertebrates, Nodal protein activates expression of *Pitx2*, which is critical in distinguishing left-sidedness from right-sidedness in the heart and gut tubes.

For Further Reading

Complete bibliographical citations for all literature cited in this chapter can be found on the Vade Mecum CD that accompanies the book and at the free access website www.devbio.com

Cho, K. W. Y., B. Blumberg, H. Steinbeisser and E. De Robertis. 1991. Molecular nature of Spemann's organizer: The role of the *Xenopus* homeobox gene *goosecoid*. *Cell* 67: 1111–1120.

Khokha, M. K., J. Yeh, T. C. Grammer and R. M. Harland. 2005. Depletion of three BMP antagonists from Spemann's organizer leads to catastrophic loss of dorsal structures. *Dev. Cell* 8: 401–411.

Larabell, C. A. and 7 others. 1997. Establishment of the dorsal-ventral axis in *Xenopus* embryos is presaged by early asymmetries in β-catenin which are modulated by the Wnt signaling pathway. *J. Cell Biol.* 136: 1123–1136.

Niehrs, C. 2004. Regionally specific induction by the Spemann-Mangold organizer. *Nature Rev. Genet.* 5: 425–434.

Piccolo, S., E. Agius, L. Leyns, S. Bhattacharyya, H. Grunz, T. Bouwmeester and E. M. DeRobertis. 1999. The head inducer Cerberus is a multifunctional antagonist of Nodal, BMP, and Wnt signals. *Nature* 397: 707–710.

Reversade, B., H. Kuroda, H. Lee, A. Mays, and E. M. De Robertis. 2005. Deletion of BMP2, BMP4, and BMP7 and Spemann organizer signals induces massive brain formation in *Xenopus* embryos. *Development* 132: 3381–3392.

Spemann, H. and H. Mangold. 1924. Induction of embryonic primordia by implantation of organizers from a different species. (Trans. V. Hamburger). *In* B. H. Willier and J. M. Oppenheimer (eds.), *Foundations of Experimental Embryology*. Hafner, New York, pp. 144–184. Reprinted in *Int. J. Dev. Biol.* 45: 13–38.

Weaver, C. and D. Kimelman. 2004. Move it or lose it: Axis specification in *Xenopus*. *Development* 131: 3491–3499.

The Early Development of Vertebrates: Fish, Birds, and Mammals

11

THIS FINAL CHAPTER ON THE PROCESSES of early development will extend our survey of vertebrate development to include fish, birds, and mammals. The amphibian embryos described in Chapter 10 divide by means of radial holoblastic cleavage. Cleavage in bird, reptile, and teleost (bony) fish eggs is meroblastic, with only a small portion of the egg cytoplasm being used to make cells. Mammals modify their holoblastic cleavage to make a placenta, which enables the embryo to develop inside another organism. Although methods of both cleavage and gastrulation differ among the vertebrate classes, certain underlying principles are common to all vertebrates.

EARLY DEVELOPMENT IN FISH

In recent years, the teleost fish *Danio rerio*, commonly known as the zebrafish, has become a favorite model organism of those who study vertebrate development. Zebrafish have large broods, breed all year, are easily maintained, have transparent embryos that develop outside the mother (an important feature for microscopy), and can be raised so that mutants can be readily screened and propagated. In addition, they develop rapidly, so that at 24 hours after fertilization, the embryo has formed most of its organ primordia and displays the characteristic tadpole-like form (Figure 11.1; see Granato and Nüsslein-Volhard 1996; Langeland and Kimmel 1997). The ability to microinject fluorescent dyes into individual cells has also allowed scientists to follow individual cells as an organ develops. Therefore, much of the description of fish development below is based on studies of this species.

The zebrafish is the first vertebrate for which intensive mutagenesis has been attempted. By treating parents with mutagens and selectively breeding their progeny, scientists have found thousands of mutations whose normally functioning genes are critical for zebrafish development. The traditional method of genetic screening (modeled after large-scale screens in *Drosophila*) begins when the male parental fish are treated with a chemical mutagen that will cause random mutations in their germ cells (Figure 11.2). Each mutagenized male is then mated with a wild-type female fish to generate F_1 lines. Individuals in the F_1 generation carry the mutations inherited from their father. If the mutation is dominant, it will be expressed in the F_1 generation. If these mutations are recessive, the F_1 fish will not show a mutant phenotype, since the wild-type dominant allele will mask the mutation. The F_1 fish are then mated with wild-type fish to produce an F_2 generation that includes both males and females that have the mutant allele. When two F_2 parents carry the same recessive mutation, there is a 25 percent chance that their offspring will show the mutant phenotype (see Figure 11.2). Since zebrafish development occurs in the open (as opposed to within an opaque

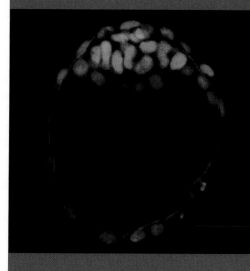

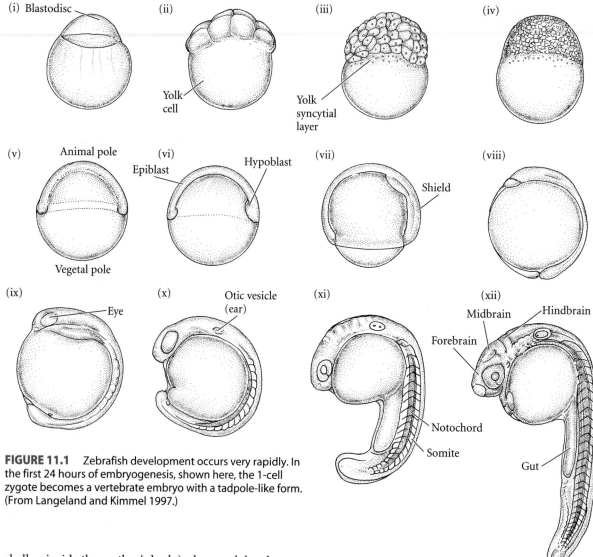

FIGURE 11.1 Zebrafish development occurs very rapidly. In the first 24 hours of embryogenesis, shown here, the 1-cell zygote becomes a vertebrate embryo with a tadpole-like form. (From Langeland and Kimmel 1997.)

shell or inside the mother's body), abnormal developmental stages can be readily observed, and the defects in development can often be traced to changes in a particular group of cells (Driever et al. 1996; Haffter et al. 1996).

Zebrafish genes are extremely amenable to study. Zebrafish embryos are susceptible to morpholino antisense molecules (Zhong et al. 2001; see Chapter 4), and researchers can use this method to test whether a particular gene is required for a particular function. Furthermore, the green fluorescent protein (GFP) reporter gene can be fused with specific zebrafish promoters and enhancers and inserted into the fish embryos. The resulting transgenic fish express GFP at the same times and places as they express the proteins controlled by these regulatory sequences. The amazing thing is that one can observe the reporter protein in living transparent embryos (Figure 11.3).

Zebrafish embryos are also permeable to small molecules placed in the water—a property that allows us to test drugs that may be deleterious to vertebrate development. For instance, zebrafish development can be altered by the addition of ethanol or retinoic acid, both of which produce malformations in the fish that resemble human developmental syndromes known to be caused by these molecules (Blader and Strähle 1998).

The similarity of developmental programs among all vertebrates and the above-mentioned ability of this fish to be genetically manipulated has given the zebrafish an important role in investigating the genes that operate during human development. The genetic screening method described above has enabled us to identify thousands of genes that may also be active in human development. For instance, the zebrafish *mariner* gene encodes the myosin VIIA protein found in the otic vesicle. Mutations of this gene in zebrafish impair hearing by preventing the sensory cells of the otic vesicle from forming properly. Humans with a disruption of the gene that encodes myosin VIIA have a syndrome of congenital deafness (Ernest et al. 2000). When developmental biologists screened zebrafish mutants for cystic kidney disease, they found 12 different

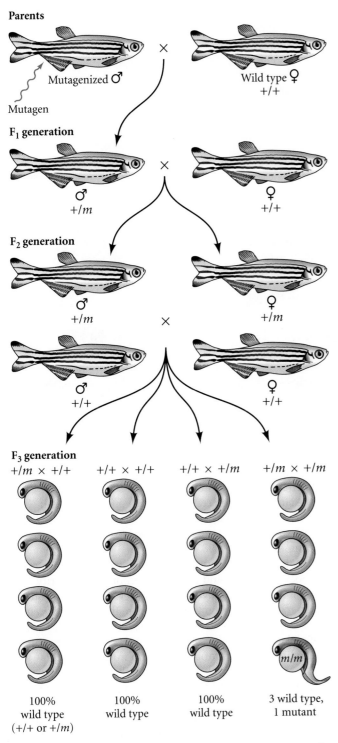

Parents

Mutagenized ♂

Mutagen

Wild type ♀
+/+

F₁ generation

♂
+/m

♀
+/+

F₂ generation

♂
+/m

♀
+/m

♂
+/+

♀
+/+

F₃ generation

+/m × +/+   +/+ × +/+   +/+ × +/m   +/m × +/m

m/m

100%
wild type
(+/+ or +/m)

100%
wild type

100%
wild type

3 wild type,
1 mutant

FIGURE 11.2 Screening protocol for identifying mutations of zebrafish development. The male parent is mutagenized and mated with a wild-type (+/+) female. If some of the male's sperm carry a recessive mutant allele (*m*), then some of the F₁ progeny of the mating will inherit that allele. F₁ individuals (here shown as a male carrying the mutant allele *m*) are then mated with wild-type partners. This creates an F₂ generation wherein some males and some females carry the recessive mutant allele. When the F₂ fish are mated, some of their progeny will show the mutant phenotype. (After Haffter et al. 1996.)

(A)   (B)

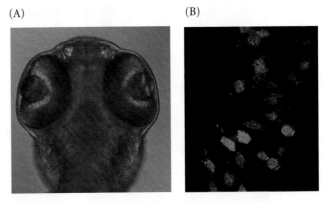

FIGURE 11.3 A reporter gene at work in living zebrafish embryos. The gene for green fluorescent protein (GFP) was fused to the regulatory region of a zebrafish *sonic hedgehog* gene. As a result, GFP was synthesized wherever the Hedgehog protein is normally expressed in the fish embryo. (A) In the head of a zebrafish embryo, GFP is seen in the developing retina and nasal placodes. (B) Because GFP is expressed by individual cells, scientists can see precisely which cells make GFP, and thus which cells normally transcribe the gene of interest (in this case, *sonic hedgehog* in the retina). (Photographs courtesy of U. Strahle and C. Neumann.)

genes. Two of these genes were known to cause human cystic kidney disease, and the other 10 were as-yet unknown genes that were found to interact with the first two in a common pathway. Moreover, that pathway, which involves the synthesis of cilia, was not what had been expected. Thus, the zebrafish studies disclosed an important and previously unknown pathway (Sun et al. 2004). As one zebrafish researcher joked, "Fish really are just little people with fins" (Bradbury 2004).

WEBSITE 11.1 GFP zebrafish movies and photographs. The ability to photograph and film living embryos expressing the GFP reporter gene driven by promoters of specific genes has opened a new dimension in developmental biology and has allowed us to link gene structure to developmental anatomy.

VADE MECUM² Zebrafish development. A full account of zebrafish development includes time-lapse movies of the beautiful and rapid development of this organism. [Click on Zebrafish]

Cleavage in Fish Eggs

The eggs of most bony fish are *telolecithal*, meaning that most of the egg cell is occupied by yolk. Cleavage can take place only in the blastodisc, a thin region of yolk-free cytoplasm at the animal pole of the egg. The cell divisions do

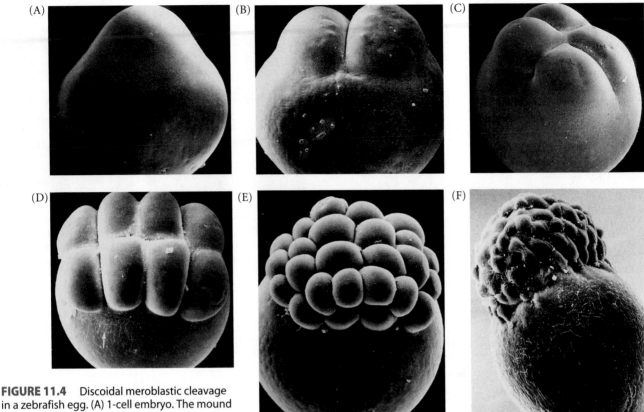

FIGURE 11.4 Discoidal meroblastic cleavage in a zebrafish egg. (A) 1-cell embryo. The mound atop the cytoplasm is the blastodisc. (B) 2-cell embryo. (C) 4-cell embryo. (D) 8-cell embryo, wherein two rows of four cells are formed. (E) 32-cell embryo. (F) 64-cell embryo, wherein the blastodisc can be seen atop the yolk cell. (From Beams and Kessel 1976; photographs courtesy of the authors.)

not completely divide the egg, so this type of cleavage is called **meroblastic** (Greek *meros*, "part"). Since only the blastodisc becomes the embryo, this type of meroblastic cleavage is called **discoidal**.

Scanning electron micrographs show beautifully the incomplete nature of discoidal meroblastic cleavage in fish eggs (Figure 11.4). The calcium waves initiated at fertilization stimulate the contraction of the actin cytoskeleton to squeeze non-yolky cytoplasm into the animal pole of the egg. This process converts the spherical egg into a pear-shaped structure with an apical blastodisc (Leung et al. 1998, 2000). In fish, there are many waves of calcium release, and they orchestrate the processes of cell division. The calcium ions coordinate the mitotic apparatus with the actin cytoskeleton; they help propagate cell division across the cell surface; they are needed to deepen the cleavage furrow; and they heal the membrane after the separation of the blastomeres (Lee et al. 2003).

The first cell divisions follow a highly reproducible pattern of meridional and equatorial cleavages. These divisions are rapid, taking about 15 minutes each. The first 12 divisions occur synchronously, forming a mound of cells

that sits at the animal pole of a large **yolk cell**. This mound of cells constitutes the **blastoderm**. Initially, all the cells maintain some open connection with one another and with the underlying yolk cell, so that moderately sized (17-kDa) molecules can pass freely from one blastomere to the next (Kimmel and Law 1985).

Maternal effect mutations have shown the importance of oocyte proteins and mRNAs in embryonic polarity, cell division, and cell cleavage (Drosch et al. 2004). Interestingly, in maternal effect mutations where the embryonic actin microfilaments fail to form (and hence cytokinesis fails to take place), the nuclei still end up in their normal positions. Moreover, the migration of certain mRNAs (such as the *vasa* message, which migrates to the cleavage furrow regions) remain unperturbed. This is not the case when the mutations are in microtubular organizing genes, suggesting that the microtubular cytoskeleton is the critical component in regulating the positioning of the cleavage furrow and of mRNAs in the early embryo (Kishimoto et al. 2004). In addition to these maternal effect genes, Wagner and colleagues (2004) have discovered several paternal effect genes, wherein the sperm proteins are critical to embryonic development.

Beginning at about the tenth cell division, the onset of the **mid-blastula transition** can be detected: zygotic gene transcription begins, cell divisions slow, and cell movement becomes evident (Kane and Kimmel 1993). At this

(A)

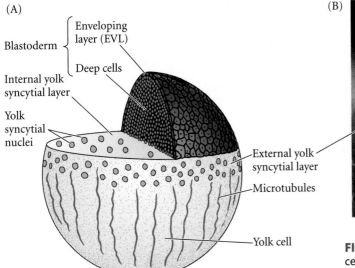

(C)

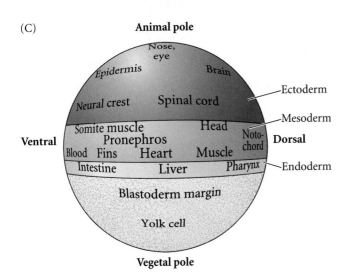

(B)

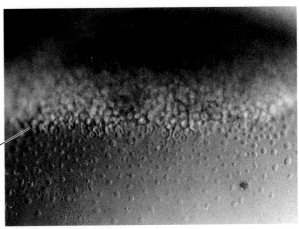

FIGURE 11.5   Fish blastula. (A) Prior to gastrulation, the deep cells are surrounded by the enveloping layer (EVL). The animal surface of the yolk cell is flat and contains the nuclei of the yolk syncytial layer (YSL). Microtubules extend through the yolky cytoplasm and through the external region of the YSL. (B) Late blastula-stage embryo of the minnow *Fundulus*, showing the external YSL. The nuclei of these cells were derived from cells at the margin of the blastoderm, which released their nuclei into the yolky cytoplasm. (C) Fate map of the deep cells after cell mixing has stopped. This is a lateral view; for the sake of clarity, not all organ fates are labeled. (A,C after Langeland and Kimmel 1997; B from Trinkaus 1993, photograph courtesy of J. P. Trinkaus.)

YSL are the **deep cells**. These are the cells that give rise to the embryo proper.

The fates of the early blastoderm cells are not determined, and cell lineage studies (in which a nondiffusible fluorescent dye is injected into a cell so that its descendants can be followed) show that there is much cell mixing during cleavage. Moreover, any one of these early blastomeres can give rise to an unpredictable variety of tissue descendants (Kimmel and Warga 1987; Helde et al. 1994). A fate map of the blastoderm cells can be made shortly before gastrulation begins. At this time, cells in specific regions of the embryo give rise to certain tissues in a highly predictable manner, although they remain plastic, and cell fates can change if tissue is grafted to a new site (Figure 11.5C; see also Figure 1.6; Kimmel et al. 1990).

Gastrulation in Fish Embryos

The first cell movement of fish gastrulation is the epiboly of the blastoderm cells over the yolk. In the initial phase of this movement, the deep cells of the blastoderm move outward to intercalate with the more superficial cells (Warga and Kimmel 1990). Later, this concatenation of cells moves vegetally over the surface of the yolk cell and envelops it completely (Figure 11.6). This downward movement toward the vegetal pole is a result of radial intercalation of the deep cells into the superficial layer. The EVL is tightly joined to the YSL and is dragged along with it. That the

time, three distinct cell populations can be distinguished (Figure 11.5A). The first of these is the **yolk syncytial layer**, or **YSL**. The YSL is formed at the ninth or tenth cell cycle, when the cells at the vegetal edge of the blastoderm fuse with the underlying yolk cell. This fusion produces a ring of nuclei within the part of the yolk cell cytoplasm that sits just beneath the blastoderm. Later, as the blastoderm expands vegetally to surround the yolk cell, some of the yolk syncytial nuclei will move under the blastoderm to form the **internal YSL**, and others will move vegetally, staying ahead of the blastoderm margin, to form the **external YSL** (Figure 11.5B). The YSL will be important for directing some of the cell movements of gastrulation.

The second cell population distinguished at the midblastula transition is the **enveloping layer (EVL)**. It is made up of the most superficial cells from the blastoderm, which form an epithelial sheet a single cell layer thick. The EVL is an extraembryonic protective covering that is sloughed off during later development. Between the EVL and the

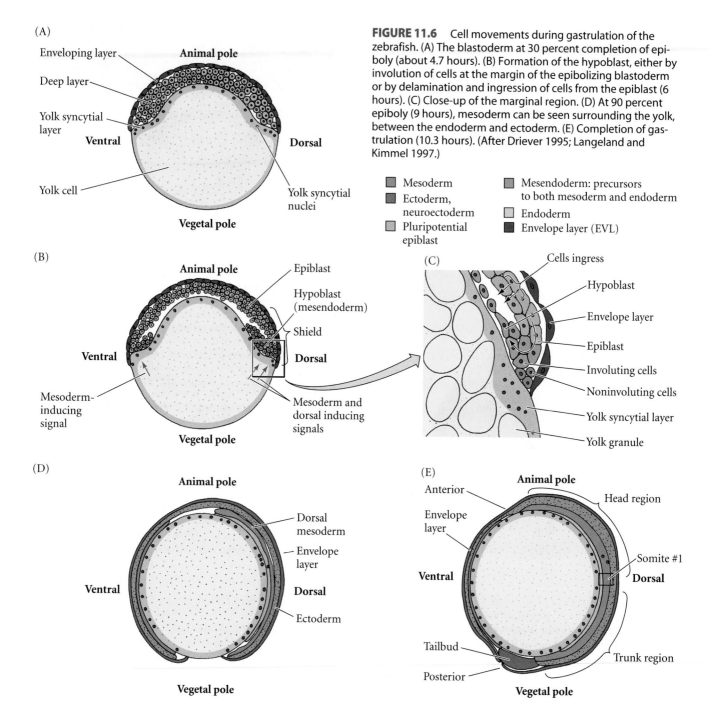

(A)

Enveloping layer
Deep layer
Yolk syncytial layer
Animal pole
Ventral
Dorsal
Yolk cell
Yolk syncytial nuclei
Vegetal pole

FIGURE 11.6 Cell movements during gastrulation of the zebrafish. (A) The blastoderm at 30 percent completion of epiboly (about 4.7 hours). (B) Formation of the hypoblast, either by involution of cells at the margin of the epibolizing blastoderm or by delamination and ingression of cells from the epiblast (6 hours). (C) Close-up of the marginal region. (D) At 90 percent epiboly (9 hours), mesoderm can be seen surrounding the yolk, between the endoderm and ectoderm. (E) Completion of gastrulation (10.3 hours). (After Driever 1995; Langeland and Kimmel 1997.)

■ Mesoderm
■ Ectoderm, neuroectoderm
■ Pluripotential epiblast
■ Mesendoderm: precursors to both mesoderm and endoderm
■ Endoderm
■ Envelope layer (EVL)

(B)

Animal pole
Epiblast
Hypoblast (mesendoderm)
Shield
Ventral
Dorsal
Mesoderm-inducing signal
Mesoderm and dorsal inducing signals
Vegetal pole

(C)

Cells ingress
Hypoblast
Envelope layer
Epiblast
Involuting cells
Noninvoluting cells
Yolk syncytial layer
Yolk granule

(D)

Animal pole
Dorsal mesoderm
Envelope layer
Ventral
Dorsal
Ectoderm
Vegetal pole

(E)

Animal pole
Anterior
Head region
Envelope layer
Somite #1
Ventral
Dorsal
Tailbud
Trunk region
Posterior
Vegetal pole

vegetal migration of the blastoderm margin is dependent on the epiboly of the YSL can be demonstrated by severing the attachments between the YSL and the EVL. When this is done, the EVL and the deep cells spring back to the top of the yolk, while the YSL continues its expansion around the yolk cell (Trinkaus 1984, 1992). E-cadherin is critical for these morphogenetic movements and for the adhesion of the EVL to the deep cells (Shimizu et al. 2005).

The expansion of the YSL depends partially on a network of microtubules within it, and radiation or drugs that block the polymerization of tubulin slow epiboly (Strahle

and Jesuthasan 1993; Solnica-Krezel and Driever 1994). During epiboly, one side of the blastoderm becomes noticeably thicker than the other. Cell labeling experiments indicate that the thicker side marks the site of the future dorsal surface of the embryo (Schmitz and Campos-Ortega 1994).

The formation of germ layers

After the blastoderm cells have covered about half the zebrafish yolk cell (this occurs earlier in fish eggs with larg-

FIGURE 11.7 Convergence and extension in the zebrafish gastrula. (A) Dorsal view of convergence and extension movements during zebrafish gastrulation. Epiboly spreads the blastoderm over the yolk; involution or ingression generates the hypoblast; convergence and extension bring the hypoblast and epiblast cells to the dorsal side to form the embryonic shield. Within the shield, intercalation extends the chordamesoderm toward the animal pole. (B) Convergent extension of the chordamesoderm is shown by those cells expressing the gene *no tail*, a gene that is expressed by notochord cells. (C) Convergent extension of paraxial mesodermal cells to flank the notochord. These cells are marked by their expression of the *snail* gene (dark areas). (From Langeland and Kimmel 1997; photographs courtesy of the authors.)

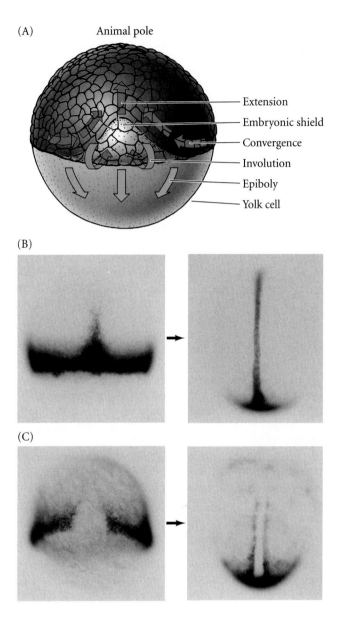

er yolks), a thickening occurs throughout the margin of the epibolizing blastoderm. This thickening, called the *germ ring*, is composed of a superficial layer, the epiblast; and an inner layer, the hypoblast. Some research groups believe that the hypoblast is formed by the involution of superficial cells from the blastoderm under the margin followed by their migration toward the animal pole (see Figure 11.6C); in this scenario, involution begins at the future dorsal portion of the embryo, but occurs all around the margin. Other laboratories claim that superficial cells ingress to form the hypoblast (see Trinkaus 1996). We do not fully understand how the hypoblast is made, and it is possible that both mechanisms are at work, with different modes of hypoblast formation predominating in different species.

Once the hypoblast has formed, cells of the epiblast and hypoblast intercalate on the future dorsal side of the embryo to form a localized thickening, the *embryonic shield* (Figure 11.7A). As we will see, the embryonic shield is functionally equivalent to the dorsal blastopore lip of amphibians, since it can organize a secondary embryonic axis when transplanted to a host embryo (Oppenheimer 1936; Ho 1992). Thus, as the cells of the blastoderm undergo epiboly around the yolk, they are also involuting or ingressing at the blastoderm margin and then converging anteriorly and dorsally toward the embryonic shield (Trinkaus 1992). The hypoblast cells of the embryonic shield itself converge and extend anteriorly, eventually narrowing along the dorsal midline of the hypoblast. This movement forms the chordamesoderm, the precursor of the notochord (Figure 11.7B). The cells adjacent to the chordamesoderm—the paraxial mesoderm cells—are the precursors of the mesodermal somites (Figure 11.7C). Concomitant convergence and extension in the epiblast brings presumptive neural cells from all over the epiblast into the dorsal midline, where they form the neural keel. Those cells remaining in the epiblast become the ectoderm.

The zebrafish fate map, then, is not much different from that of the frog seen in Chapter 10, or of the other vertebrates we will soon discuss. If you can visualize opening a *Xenopus* blastula at the vegetal pole and then stretching that opening into a marginal ring, the resulting fate map closely resembles that of the zebrafish embryo at the stage

when half of the yolk has been covered by the blastoderm (see Figure 1.6; Langeland and Kimmel 1997).

Meanwhile, the endoderm arises from the most marginal blastomeres of the late blastula-stage embryo. These blastomeres involute early in gastrulation and occupy the deep layers of the hypoblast, directly above the yolk syncytial layer.

Axis Formation in Fish Embryos

Dorsal-ventral axis formation: The embryonic shield and Nieuwkoop center

The embryonic shield is critical in establishing the dorsal-ventral axis in fish. Shield tissue can convert lateral and ventral mesoderm (blood and connective tissue precur-

FIGURE 11.8   The embryonic shield as organizer in the fish embryo. (A) A donor embryonic shield (about 100 cells from a stained embryo) is transplanted into a host embryo at the same early gastrula stage. The result is two embryonic axes joined to the host's yolk cell. In the photograph, both axes have been stained for *sonic hedgehog* mRNA, which is expressed in the ventral midline. (The embryo to the right is the secondary axis.) (B) The same effect can be achieved by activating nuclear catenin in embryos at sites opposite where the embryonic shield will form. (A after Shinya et al. 1999; photograph courtesy of the authors; B courtesy of J. C. Izpisua Belmonte.)

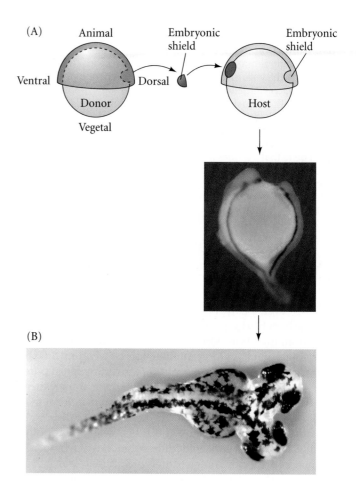

sors) into dorsal mesoderm (notochord and somites), and it can cause the ectoderm to become neural rather than epidermal. This transformative capacity was shown by transplantation experiments in which the embryonic shield of an early-gastrula embryo was transplanted to the ventral side of another (Figure 11.8; Oppenheimer 1936; Koshida et al. 1998). Two axes formed, sharing a common yolk cell. Although the prechordal plate and notochord were derived from the donor embryonic shield, the other organs of the secondary axis came from host tissues that would normally form ventral structures. The new axis had been induced by the donor cells. Such experiments are similar to those performed on amphibian gastrulae by Spemann and Mangold (see Chapter 10) and demonstrate that the fish embryonic shield is the homologue of the amphibian dorsal blastopore lip (the organizer).

Like the amphibian blastopore lip, the embryonic shield forms the prechordal plate and the notochord of the developing embryo. The precursors of these two regions are responsible for inducing ectoderm to become neural ectoderm. Moreover, the presumptive notochord and prechordal plate appear to do this in a manner very much like the homologous structures in amphibians.* In both fish and amphibians, bone morphogenetic proteins (BMPs) and certain Wnt proteins made in the ventral and lateral regions of the embryo would normally induce the ectoderm to become epidermis. The notochords of both fish and amphibians secrete factors that block this induction, thereby allowing the ectoderm to become neural. In fish, BMP2B induces embryonic cells to acquire ventral and lateral fates, and Wnt8 ventralizes, lateralizes, and posteriorizes the embryonic tissues (see Schier 2001). The protein secreted by the chordamesoderm that binds with and inactivates BMP2B is a chordin-like paracrine factor called Chordino (Figure 11.9; Kishimoto et al. 1997; Schulte-Merker et al. 1997). If the *chordino* gene is mutated, the neural tube fails to form; if the

bmp2b gene is mutated, dorsal structures (such as the notochord) expand at the expense of ventral structures.

Soluble BMP antagonists are not the only important factor for specifying the neural plate in zebrafish; the FGFs also play important roles in patterning the ectoderm into epidermal and neural domains. In the animal (prospective anterior) ectoderm, neural fate is specified by BMP antagonists coming from the organizer (just like in *Xenopus*). Also as in *Xenopus*, insulin-like growth factors (IGFs) play a role in the production of the anterior neural plate. Zebrafish IGFs appear to upregulate *chordino* and *goosecoid* while restricting the expression of *bmp2b*. Although IGFs appear to be made throughout the embryo, during gastrulation the IGF receptors are found predominantly in the anterior portion of the embryo (Eivers et al. 2004). In the caudal region of the embryo, the ectoderm is specified to become neural by FGF signaling from the germ ring (Kudoh et al. 2004). This FGF signaling may block Smad1 from getting into the nucleus (as mentioned in the previous chapter) and be another way to block BMPs (see Furthauer et al. 2004; Tsang et al. 2004).

Zebrafish have meroblastic cleavage and a huge yolk cell with syncytial nuclei. Amphibians such as *Xenopus* have holoblastic cleavage and a blastocoel. Yet both types of embryos use similar molecular tools to undergo cleav-

---

*Another similarity between the amphibian and fish organizers is that they can be duplicated by rotating the egg and changing the orientation of the microtubules (Fluck et al. 1998). One difference in the axial development of these groups is that in amphibians, the prechordal plate is necessary for inducing the anterior brain to form (see Chapter 10). In zebrafish, although the prechordal plate appears to be necessary for forming ventral neural structures, the anterior regions of the brain can form in its absence (Schier et al. 1997; Schier and Talbot 1998).

FIGURE 11.9 Axis formation in the zebrafish embryo. (A) Prior to gastrulation, the zebrafish blastoderm is arranged with the presumptive ectoderm near the animal pole, the presumptive mesoderm beneath it, and the presumptive endoderm sitting atop the yolk cell. The yolk syncytial layer (and possibly the endoderm) sends two signals to the presumptive mesoderm. One signal (lighter arrows) induces the mesoderm, while a second signal (heavy arrow) specifically induces an area of mesoderm to become the dorsal mesoderm (embryonic shield).
(B) Formation of the dorsal-ventral axis. During gastrulation, the ventral mesoderm secretes BMP2B (arrows) to induce the ventral and lateral mesodermal and epidermal differentiation. The dorsal mesoderm secretes factors (such as Chordino) that block BMP2B and dorsalize the mesoderm and ectoderm (converting the latter into neural tissue). (After Schier and Talbot 1998.)

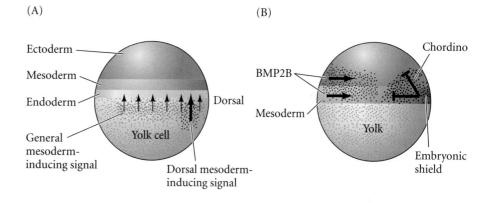

age, gastrulation, and axis specification. Both use β-catenin and Nodal-related proteins to form the dorsal mesoderm and enable this mesoderm to express the organizer genes. Both use BMPs and Wnts to lateralize and vegetalize the embryo, and in both groups the organizer genes encode proteins such as chordin, Noggin, and Dickkopf that antagonize the BMPs and Wnts. Furthermore, later in development both zebrafish and *Xenopus* use a particular Wnt protein to posteriorize the ectoderm, forming the trunk neural tube. In some instances, fish and amphibians use these proteins in different ways, but the result is a structure that is definitely recognizable as a vertebrate embryo.

The fish Nieuwkoop center

The embryonic shield appears to acquire its organizing ability in much the same way as its amphibian counterparts. In amphibians, the endoderm cells beneath the dorsal blastopore lip (i.e., the Nieuwkoop center) accumulate β-catenin synthesized from maternal messages. This protein is critical in enabling the amphibian endoderm to induce the cells above it to become the dorsal blastopore lip. In zebrafish, the nuclei in that part of the yolk syncytial layer that lies beneath the cells that will become the embryonic shield similarly accumulate β-catenin. Indeed, the presence of this protein distinguishes the dorsal YSL from the lateral and ventral YSL regions* (Figure 11.10A; Schneider et al. 1996). Inducing β-catenin accumulation on the ventral side of the egg causes dorsalization and a second embryonic axis (Kelly et al. 1995).

The β-catenin of the embryonic shield combines with the zebrafish homologue of Tcf3 to become an active transcription factor. It activates the genes encoding two mesoderm-patterning proteins, **Squint** and **Bozozok**, that are very similar to the proteins that pattern the amphibian mesoderm. Squint is a Nodal-like paracrine factor, while Bozozok[†] is a homeodomain protein similar to the amphibian Nieuwkoop center protein Siamois.

Zebrafish Bozokok protein works in several ways. First, acting alone, it can repress BMP and WNT genes that would promote ventral functions (Solnica-Krezel and Driever 2001). Second, Bozozok suppresses a transcriptional inhibitor (the *vega1* gene), allowing the organizer genes to function (Kawahara et al. 2000). Third, Bozozok and Squint act individually to activate the *chordino* gene, and they act synergistically to activate other organizer genes such as *goosecoid*, *noggin*, and *dickkopf* (Figure 11.10B; Sampath et al. 1998; Gritsman et al. 2000; Schier and Talbot 2001.) These genes encode the proteins that block BMPs and Wnts and allow the specification of the dorsal mesoderm and neural ectoderm. Thus, the embryonic shield is considered equivalent to the amphibian organizer, and the dorsal part of the yolk cell, together with the dorsal marginal blastomeres (the precursors of the Kupffer cells; see below), can be thought of as the Nieuwkoop center of the teleost fish embryo.

Maternal β-catenin also initiates a third pathway, one that coordinates the cell movements of the mesoderm and endoderm during gastrulation. Dorsal mesoderm cells derived from the embryonic shield generate the notochord posteriorly and the prechordal plate mesoderm anteriorly. The anterior cells are highly motile and lead the mesoderm into the embryo. The cells that form the notochord are not motile, but they do undergo convergent extension

---

*Some of the endodermal cells that accumulate β-catenin will become the precursors of the ciliated cells of Kupffer's vesicle (Cooper and D'Amico 1996). As we will discuss later, these cells are critical in determining the left-right axis of the embryo.

[†]*Bozozok* is Japanese slang for an arrogant youth on a motorcycle. The gene's name was derived from the severe loss-of-function phenotype wherein a single-eyed embryo curves ventrally over the yolk cell (i.e., resembling a rider on a fast motorcycle). However, this gene is also known as *Dharma* (after a famous Buddhist priest) because embryos with gain-of-function—*too much* of this protein, the result of experimentally injecting its mRNA into the embryo—develop huge eyes and head, but no trunk or tail; they thus resemble Japanese Dharma dolls (Yamanaka et al. 1998; Fekany et al. 1999).

(A)

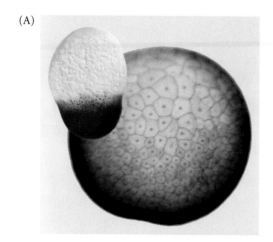

(B)

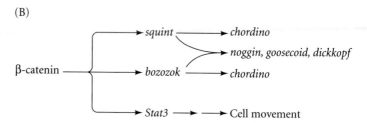

FIGURE 11.10 β-Catenin activates organizer genes in the zebrafish. (A) Nuclear localization of β-catenin marks the dorsal side of the *Xenopus* blastula (larger image) and helps form its Nieuwkoop center beneath the organizer. In the zebrafish late blastula (smaller image), nuclear localization of β-catenin is seen in the yolk syncytial layer nuclei beneath the future embryonic shield. (B) β-Catenin activates *squint* and *bozozok*, whose proteins activate organizer-specific genes as well as the *Stat3* gene whose product is necessary for gastrulation movements. (A, photograph courtesy of S. Schneider; B after Schier and Talbot 2001.)

similar to the processes seen in *Xenopus*. Both the activity of the anterior cells and the convergent extension of the notochordal cells are regulated by **Stat3**, a transcription factor subunit that is itself regulated positively by β-catenin (see Figure 11.10B; Yamashita et al. 2002). Embryos injected with antisense morpholinos to Stat3 had mispositioned heads and shortened anterior-posterior axes. Without Stat3, the anterior movements of the dorsal mesoderm and the dorsal convergence of the nonaxial mesoderm do not occur.

The Stat3 transcription factor appears to regulate the expression of small GTPases such as RhoA. These are critical components of what has been called the **planar cell polarity pathway**, whereby cells become differentiated along their medial-lateral axes. This polarization of the cells is a requirement for intercalation, and also causes the cell divisions to align themselves along the animal-vegetal axis

of the egg (Gong et al. 2004). Together, radial intercalation and oriented cell division produce the convergent extension of the mesoderm during gastrulation. In zebrafish, Wnt signals (Wnt11 in the anterior, Wnt5a in the posterior) control this cell polarization, and mutations of these Wnt genes lead to the failure of convergent extension in those specific areas (Heisenberg and Tada 2002; Miyagi et al. 2004). The signal from these Wnts, however, does not activate the canonical Wnt pathway that results in the nuclear localization of β-catenin. Rather, it is mediated through RhoA, a GTPase that organizes the actin cytoskeleton. Thus, the Stat3 transcription factor becomes a precondition for the planar cell polarity pathway through which convergent extension can occur.

Nodal signaling is also critical for dorsal axis formation. The maternal mRNA for the zebrafish Nodal protein

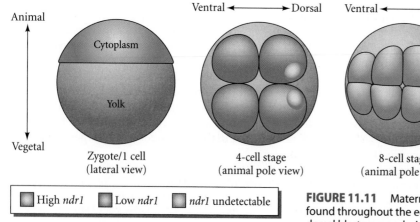

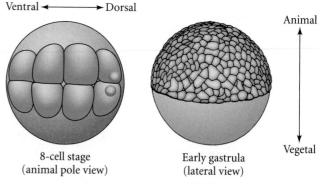

FIGURE 11.11 Maternal mRNA for Nodal-related protein is initially found throughout the egg cytoplasm. It begins to accumulate in the dorsal blastomeres during the 4-cell stage, and by the 8-cell stage, this protein is localized only in the two dorsalmost blastomeres. In the early gastrula, the mRNA for Nodal-related protein is found almost exclusively in those cells critical for specifying the dorsal region of the embryo. (After Driever 2005.)

(Nodal-related 1, or Ndr1) is transported along microtubules into the two dorsalmost blastomeres of the 8-cell embryo (Figure 11.11; Gore et al. 2005). The β-catenin and Nodal transcription factors probably interact during the early gastrula stage to specify the dorsal components of the embryo.

In addition to its function in forming the fish organizer, Nodal-like signaling is also critical for endoderm formation. Here, Squint and Cyclops (another Nodal-related protein) induce the Bon and GATA5 transcription factors. These two transcription factors are co-expressed in the marginal domain of the late blastula and regulate the expression of the downstream genes that form the endoderm (Reiter et al. 2001).

Anterior-posterior axis formation

The patterning of the neural ectoderm along the anterior-posterior axis in the zebrafish appears to be the result of the interplay of an FGF, a Wnt, and retinoic acid (RA). There are two separate processes: first, the Wnt signal represses the expression of anterior genes; and then, Wnt, retinoic acid, and FGF are required to activate the posterior genes.

This regulation of anterior-posterior identity appears to be coordinated by **retinoic acid-4-hydroxylase**, an enzyme that degrades retinoic acid (Kudoh et al. 2002; Dobbs-McAuliffe et al. 2004). The gene encoding this enzyme, *cyp26*, is expressed specifically in the region of the embryo destined to become the anterior end. Indeed, this gene's expression is first seen during the late blastula stage, and by gastrulation, it defines the presumptive anterior neural plate. Retinoic acid-4-hydroxylase prevents the accumula-

tion of retinoic acid at the embryo's anterior end, blocking the expression of the posterior genes there. This inhibition is reciprocated, since FGFs and Wnts inhibit the expression of the retinoic acid-4-hydroxylase gene, as well as inhibiting the expression of the head-specifying gene *Otx2*. This mutual inhibition creates a border between the zone of posterior gene expression and the zone of anterior gene expression. As epiboly continues, more and more of the body axis is specified to become posterior (Figure 11.12).

Left-right axis formation

In all vertebrates studied, the right and left sides differ both anatomically and developmentally. In fish, the heart is on the left side and there are different structures in the left and right regions of the brain (Figure 11.13). Moreover, as in other vertebrates, the cells on the left side of the body are given that information by Notch and Nodal signaling and by the Pitx2 transcription factor, while the cells on the right side of the body are exposed to FGF signaling. The ways the different vertebrate classes accomplish this asymmetry differ, but recent evidence suggests that the currents produced by cilia in the node may be responsible for left-right axis formation in all the vertebrate classes (Okada et al. 2005); when the gene for a dynein subunit of cilia was cloned, it was found to be expressed in the ventral portion of the node (or organizer) in mouse, chick, *Xenopus*, and

FIGURE 11.12 Model for the specification of zebrafish neural ectoderm by Fgf, Wnt, and retinoic acid (RA) signaling. The *Cyp26* gene encodes retinoic acid-4-hydroxylase, a cytochrome family enzyme that degrades retinoic acid. (A) Pathway through which a boundary can form between anterior (*Cyp26-*, *Otx-*expressing) and posterior (*hoxb1-*, *meis3-*expressing) neural ectoderm. In the posterior region, Fgf and/or Wnt suppress anterior genes such as *Otx2*. The Fgf /Wnt signal also suppresses *Cyp26* expression posteriorly, so retinoic acid accumulates (due to the absence of retinoic acid-4-hydroylase) and activates the posterior genes. (B) Temporal sequence of the posteriorization process. In late blastula stage, the *Cyp26* gene is confined to the anterior region by Fgf and Wnts from the margin. After the start of gastrulation, convergent extension takes the margin farther from the anterior. Retinoic acid accumulates in this region and activates genes associated with the posterior neural ectoderm. (After Kudoh et al. 2002.)

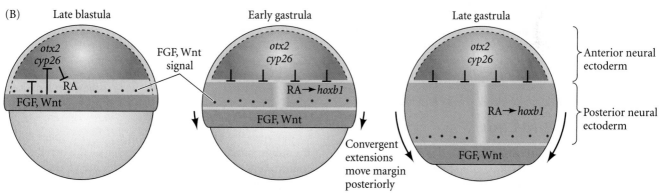

(A)

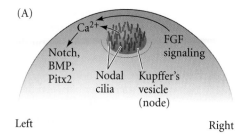

Left   Right

(B)   Left habenular nucleus   Right habenular nucleus

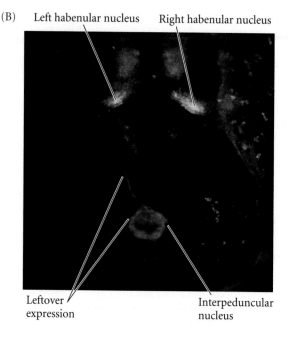

Leftover   Interpeduncular
expression   nucleus

FIGURE 11.13   Left-right asymmetry in the zebrafish embryo. (A) Model for asymmetric gene expression. Nodal cilia in Kupffer's vesicle create a current that causes the release of Ca^{2+} on the embryo's left side. Calcium ions stimulate Notch and BMP4 pathways on the left side and activate the Pitx2 transcription factor in the left-hand mesoderm (blue). FGF expression is seen predominantly on the right-hand side (red). (B) Brain asymmetry in zebrafish. Antibody staining of the Leftover (red) and Right-on (green) proteins in neurons of the habenular nucleus (a behavior-controlling region of the zebrafish forebrain) and the axonal projections to their midbrain target (the interpeduncular nucleus) reveals marked asymmetry. Most Leftover-positive axons emerge from the left habenula to innervate the target. (A after Okada et al. 2005; B from Gamse et al. 2005, photograph courtesy of M. Halpern.)

zebrafish embryos (Essner et al. 2002, 2005). Using extremely rapid (500 frames/sec) photography, Okada and colleagues (2005) showed that the clockwise rotational motion of cilia and the leftward flow of particles were conserved in all these vertebrate groups, despite the different types and shapes of their nodal structures.

In zebrafish, the nodal structure housing the cilia that control left-right asymmetry is a transient fluid-filled organ called **Kupffer's vesicle**. As mentioned earlier, Kupffer's vesicle arises from a group of dorsal cells near the embryonic shield shortly after gastrulation. Essner and colleagues (2002, 2005) were able to inject small beads into Kupffer's vesicles and see their translocation from one side of the vesicle to the other. Blocking ciliary function by preventing the synthesis of dynein or by ablating the precursors of the ciliated cells prevented normal left-right axis formation.

Although the exact mechanism for this leftward flow of the nodal current as yet to be determined, it appears to be due to the clockwise rotation of the cilia combined with the location of the cilia in the node cells (Okada et al. 2005). As we will see, in the mouse and chick, these cilia reside in Hensen's node, a cavity homologous to Kupffer's vesicle in fish.

EARLY DEVELOPMENT IN BIRDS

Cleavage in Bird Eggs

Ever since Aristotle first followed its 3-week development, the domestic chicken (*Gallus gallus*) has been a favorite organism for embryological studies. It is accessible year-round and is easily raised. Moreover, at any particular temperature, its developmental stage can be accurately predicted. Thus, large numbers of embryos can be obtained at the same stage. The chick embryo can be surgically manipulated and, since chick organ formation is accomplished by genes and cell movements similar to those of mammalian organ formation, the chick embryo has often served as an inexpensive surrogate for human embryos. In the first years of the twenty-first century, gene targeting, embryonic stem cells, and the complete sequencing of the chick genome has made the chick embryo one of the few developmental systems in which one can do both experimental and genetic manipulations (Stern 2005).

Fertilization of the chick egg occurs in the oviduct, before the albumen and shell are secreted to cover it. The egg is telolecithal (like that of the fish), with a small disc of cytoplasm sitting atop a large yolk. Like fish eggs, the yolky eggs of birds undergo discoidal meroblastic cleavage. Cleavage occurs only in the blastodisc, a small disc of cytoplasm 2–3 mm in diameter at the animal pole of the egg cell. The first cleavage furrow appears centrally in the blastodisc, and other cleavages follow to create a single-layered blastoderm (Figure 11.14). As in the fish embryo, the cleavages do not extend into the yolky cytoplasm, so the early-cleavage cells are continuous with one another and with the yolk at their bases (see Figure 11.14E). Thereafter, equatorial and vertical cleavages divide the blastoderm into a tissue five to six cell layers thick. The cells become linked together by tight junctions (Bellairs et al. 1975; Eyal-Giladi 1991).

Between the blastoderm and the yolk is a space called the **subgerminal cavity**. This space is created when the blastoderm cells absorb water from the albumen ("egg white") and secrete the fluid between themselves and the yolk

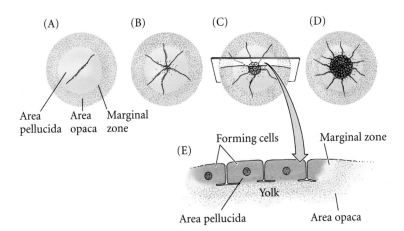

FIGURE 11.14 Discoidal meroblastic cleavage in a chick egg. (A–D) Four stages viewed from the animal pole (the future dorsal side of the embryo). (E) An early-cleavage embryo viewed from the side. (After Bellairs et al. 1978.)

(New 1956). At this stage, the deep cells in the center of the blastoderm are shed and die, leaving behind a one-cell-thick **area pellucida**; this part of the blastoderm forms most of the actual embryo. The peripheral ring of blastoderm cells that have not shed their deep cells constitutes the **area opaca**. Between the area pellucida and the area opaca is a thin layer of cells called the **marginal zone** (or marginal belt*) (Eyal-Giladi 1997; Arendt and Nübler-Jung 1999). Some of the marginal zone cells become very important in determining cell fate during early chick development.

Gastrulation of the Avian Embryo

The hypoblast

By the time a hen has laid an egg, the blastoderm contains some 20,000 cells. At this time, most of the cells of the area pellucida remain at the surface, forming the epiblast, while other area pellucida cells have delaminated and migrated individually into the subgerminal cavity to form the poly-invagination islands (primary hypoblast), an archipelago of disconnected clusters containing 5–20 cells each (Figure 11.15A). Shortly thereafter, a sheet of cells from the posterior margin of the blastoderm (distinguished from the other regions of the margin by Koller's sickle, a local thickening) migrates anteriorly and pushes the primary hypoblast cells anteriorly, thereby forming the secondary hypoblast, also known as the endoblast (11.15B–E; Eyal-Giladi et al. 1992; Bertocchini and Stern 2002). The resulting two-layered blastoderm (epiblast and hypoblast) is joined together at the marginal zone of the area opaca, and the space between the layers forms a blastocoel. Thus, although the shape and formation of the avian blastodisc differ from those of the amphibian, fish, or echinoderm blastula, the overall spatial relationships are retained.

The avian embryo comes entirely from the epiblast; the hypoblast does not contribute any cells to the developing embryo (Rosenquist 1966, 1972). Rather, the hypoblast cells form portions of the external membranes, especially the yolk sac and the stalk linking the yolk mass to the endodermal digestive tube. Hypoblast cells also provide chemical signals that specify the migration of epiblast cells. However, the three germ layers of the embryo proper (plus a considerable amount of extraembryonic membrane) are formed solely from the epiblast (Schoenwolf 1991).

The primitive streak

The major structural characteristic of avian, reptilian, and mammalian gastrulation is the **primitive streak**.[†] Dye-marking experiments and scanning electron micrographs indicate that the primitive streak cells arise within the posterior marginal region and that cells from other areas of the embryo are not involved in its formation (Lawson and Schoenwolf 2001a,b). The streak is first visible as cells accumulate in the middle layer, followed by a thickening of the epiblast at the posterior marginal zone, just anterior to Koller's sickle (Figure 11.16A). This thickening is initiated by an increase in the height (thickness) of the cells forming the center of the primitive streak. The presumptive streak cells around them become globular and motile, and they appear to digest away the extracellular matrix underlying them. This process allows their intercalation (mediolaterally) and convergent extension. Convergent extension is responsible for the progression of the streak—a doubling in streak length is accompanied by a concomitant halving of its width. Those cells that initiated streak formation (i.e., the cells that were in the midline of the epiblast, overlying Koller's sickle; see Figure 1.15D,E) appear to migrate anteriorly and may constitute an unchanging cell population that directs the movement of epiblast cells into the streak.

---

*Arendt and Nübler-Jung (1999) have argued that the region should be called the marginal *belt* to distinguish it from the marginal *zone* of amphibians. Here we will continue to use the earlier nomenclature.

[†] But as we saw in the previous chapter, a structure resembling the primitive streak has also been found in certain salamander embryos (Shook et al. 2002).

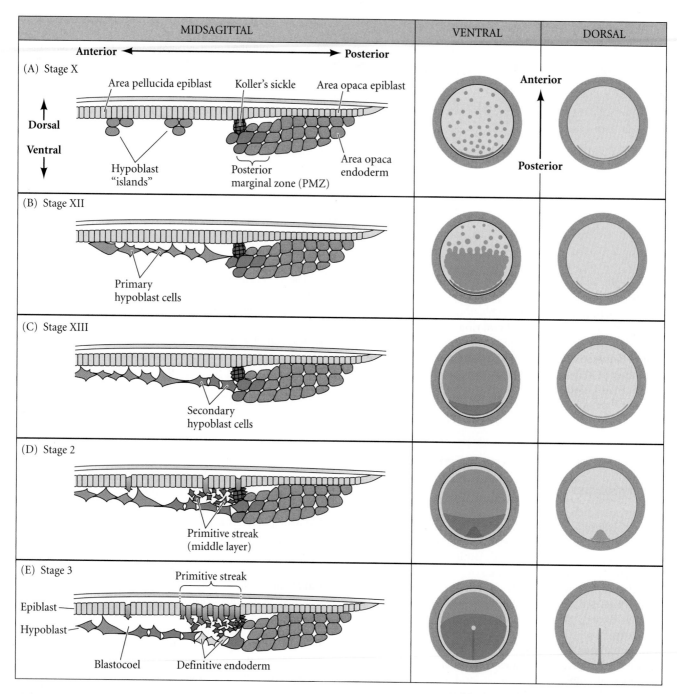

FIGURE 11.15 Formation of the three-layered blastoderm of the chick embryo. The left column depicts a diagrammatic mid-sagittal section through part of the blastoderm. The middle column depicts the entire embryo viewed from the ventral side. This shows the migration of the primary and secondary hypoblast (endoblast) cells. The right column shows the entire embryo seen from the dorsal side. (A) Stage X embryo, where islands of hypoblast cells can be seen, as well as a congregation of hypoblast cells around Koller's sickle. (B) By stage XII, the hypoblast island cells have coalesced to form the primary hypoblast layer, which meets endoblast cells and primitive streak cells at Koller's sickle. (C) By stage XIII, the secondary hypoblast cells migrate anteriorly. (D) By stage 2, the primitive streak cells form a layer between the hypoblast and epiblast cells. (E) By stage 3, the primitive streak has become a definitive region of the epiblast, with cells migrating through it to become the mesoderm and endoderm. (After Stern 2004.)

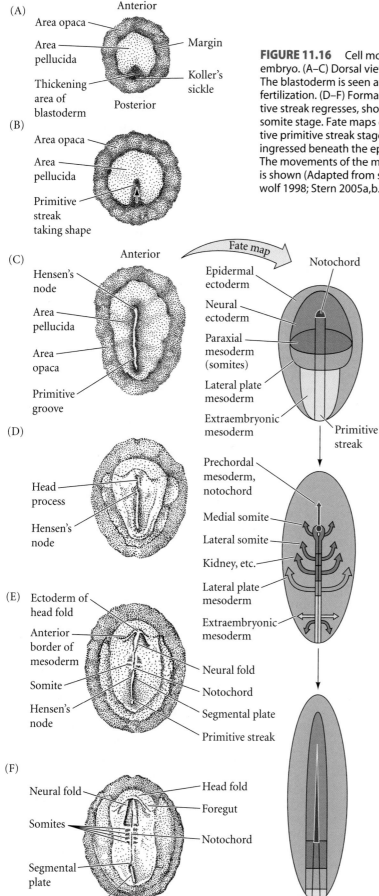

FIGURE 11.16 Cell movements of the primitive streak and fate map of the chick embryo. (A–C) Dorsal view of the formation and elongation of the primitive streak. The blastoderm is seen at (A) 3–4 hours, (B) 7–8 hours, and (C) 15–16 hours after fertilization. (D–F) Formation of notochord and mesodermal somites as the primitive streak regresses, shown at (D) 19–22 hours, (E) 23–24 hours, and (F) the four-somite stage. Fate maps of the chick epiblast are shown for two stages, the definitive primitive streak stage (C) and neurulation (F). In (F), the endoderm has already ingressed beneath the epiblast, and convergent extension is seen in the midline. The movements of the mesodermal precursors through the primitive streak at (C) is shown (Adapted from several sources, especially Spratt 1946; Smith and Schoenwolf 1998; Stern 2005a,b.)

As cells converge to form the primitive streak, a depression forms within the streak. This depression is called the **primitive groove**, and it serves as an opening through which migrating cells pass into the blastocoel. Thus, the primitive groove is homologous to the amphibian blastopore. At the anterior end of the primitive streak is a regional thickening of cells called the **primitive knot**, or **Hensen's node**. The center of this node contains a funnel-shaped depression (sometimes called the **primitive pit**) through which cells can pass into the blastocoel. Hensen's node is the functional equivalent of the dorsal lip of the amphibian blastopore (i.e., the organizer)* and the fish embryonic shield (Boettger et al. 2001).

The first cells that ingress through the primitive streak and into the blastocoel are endodermal precursors from the epiblast (Figures 11.16B; Vakaet 1984; Bellairs 1986; Eyal-Giladi et al. 1992). These cells undergo an epithelial-to-mesenchymal transformation and the basal lamina beneath them breaks down. As these cells enter the primitive streak, the streak elongates toward the future head region. Cell division adds to the length produced by convergent extension, and some of the cells from the anterior portion of the epiblast contribute to the formation of Hensen's node (Streit et al. 2000; Lawson and Schoenwolf 2001b).

At the same time, the secondary hypoblast (endoblast) cells continue to migrate anteriorly from the posterior margin of the blastoderm (see Figure 11.15E). The elongation of the primitive streak appears to be coextensive with the anterior migration of these secondary hypoblast cells, and the hypoblast directs the movement of the primitive streak (Waddington 1933; Foley et al. 2000). The streak eventually extends to 60–75 percent of the length of the area pellucida.

---

*Frank M. Balfour proposed the homology of the amphibian blastopore and the chick primitive streak in 1873, while he was still an undergraduate (Hall 2003). August Rauber (1876) soon provided further evidence for the homology between them.

The primitive streak defines the axes of the avian embryo. It extends from posterior to anterior; migrating cells enter through its dorsal side and move to its ventral side; and it separates the left portion of the embryo from the right. The axis of the streak is equivalent to the dorsal-ventral axis of amphibians. The anterior end of the streak (Hensen's node) gives rise to the prechordal mesoderm, notochords, and anterior somites. Cells that ingress through the middle of the streak give rise to the somites, heart, and kidneys. Cells becoming the posterior portion of the streak make the lateral plate and extraembryonic mesoderm (Psychoyos and Stern 1996). After the ingression of the mesoderm cells, those cells close to the streak will form the medial (central) structures, while those farther from it will be the lateral structures (Figure 11.16C–E).

As soon as the primitive streak has formed, epiblast cells begin to migrate through it and into the blastocoel (Figure 11.17). The streak thus has a continually changing cell population. Cells migrating through the anterior end pass down into the blastocoel and migrate anteriorly, forming the endoderm, head mesoderm, and notochord; cells passing through the more posterior portions of the primitive streak give rise to the majority of mesodermal tissues (Schoenwolf et al. 1992). Unlike the *Xenopus* mesoderm, which migrates as sheets of cells into the blastocoel, cells entering the inside of the avian embryo ingress as individuals after undergoing an epithelial-to-mesenchymal transformation (Stern et al. 1990; DeLuca et al. 1999).

MIGRATION THROUGH THE PRIMITIVE STREAK: FORMATION OF ENDODERM AND MESODERM   The first cells to migrate through Hensen's node are those destined to become the pharyngeal endoderm of the foregut. Once inside the blastocoel, these endodermal cells migrate anteriorly and eventually displace the hypoblast cells, causing the hypoblast cells to be confined to a region in the anterior portion of the area pellucida. This region, the **germinal crescent**, does not form any embryonic structures, but it does contain the precursors of the germ cells, which later migrate through the blood vessels to the gonads (see Chapter 19). The next cells entering the blastocoel through Hensen's node also move anteriorly, but they do not move as far ventrally as the presumptive foregut endodermal cells. Rather, they remain between the endoderm and the epiblast to form the **head mesenchyme** and the **prechordal plate mesoderm** (see Psychoyos and Stern 1996). Thus, the head of the avian embryo forms anterior (rostral) to Hensen's node. The next cells migrating through Hensen's node become the **chordamesoderm**. The chordamesoderm has two components, the head process and the notochord. The most anterior part, the **head process**, is formed by the central mesoderm cells migrating anteriorly behind the prechordal plate mesoderm towards the rostral tip of the embryo (see Figures 11.16 and 11.17). The head process will underlie those cells that form the forebrain and midbrain. As the primitive streak regresses, the cells migrating through Hensen's node become the **notochord**. The notochord starts at the level where the ears and hindbrain form, and extends caudally, where it will interact with the cells overlying it to form the spinal cord.

The migration of the mesoderm through the anterior primitive streak and its condensation to form the chordamesoderm appears to be controlled by FGF signaling. Fgf8 is expressed in the primitive streak and repels migrating cells away from the streak. Conversely, these same migrating cells are attracted to sources of Fgf4, and this protein is made by the incipient chordamesoderm. Using green fluorescent protein, Yang and colleagues (2002) were able to follow the trajectories of these cells as they migrated through Hensen's node (see Figure 11.17). They were also able to deflect these normal trajectories by using beads that secreted Fgf4 or Fgf8.

Meanwhile, cells continue migrating inwardly through the lateral portions of the primitive streak. As they enter the blastocoel, these cells separate into two layers. The deep layer joins the hypoblast along its midline and displaces the hypoblast cells to the sides. These deep-moving cells give rise to all the endodermal organs of the embryo, as well as to most of the extraembryonic membranes (the hypoblast forms the rest). The second migrating layer spreads between this endoderm and the epiblast, forming a loose layer of cells. This middle layer of cells generates the mesodermal portions of the embryo and mesoderm lining the extraembryonic membranes. The movement away from the streak appears to be motivated by Fgf8-mediated chemorepulsion, but the attractive cues for the posteriorly migrating cells have not yet been identified (Yang et al. 2002). By 22 hours of incubation, most of the presumptive endodermal cells are in the interior of the embryo, although presumptive mesodermal cells continue to migrate inward for a longer time.

REGRESSION OF THE PRIMITIVE STREAK   Now a new phase of gastrulation begins. While mesodermal ingression continues, the primitive streak starts to regress, moving Hensen's node from near the center of the area pellucida to a more posterior position (Figure 11.18). The regressing streak leaves in its wake the posterior dorsal axis of the embryo, including the notochord. As Hensen's node moves caudally, the posterior region of notochord is laid down. Finally, Hensen's node regresses to its posterior position, forming the anal region. By this time, all the presumptive endodermal and mesodermal cells have entered the embryo, and the epiblast is composed entirely of presumptive ectodermal cells.

As a consequence of the sequence in which the head mesoderm and notochord are established, avian (and mammalian, reptilian, and teleost fish) embryos exhibit a distinct anterior-to-posterior gradient of developmental maturity. While cells of the posterior portions of the embryo are undergoing gastrulation, cells at the anterior end are already starting to form organs (see Darnell et al. 1999). For the next several days, the anterior end of the embryo is more advanced in its development (having had a "head start," if you will) than the posterior end.

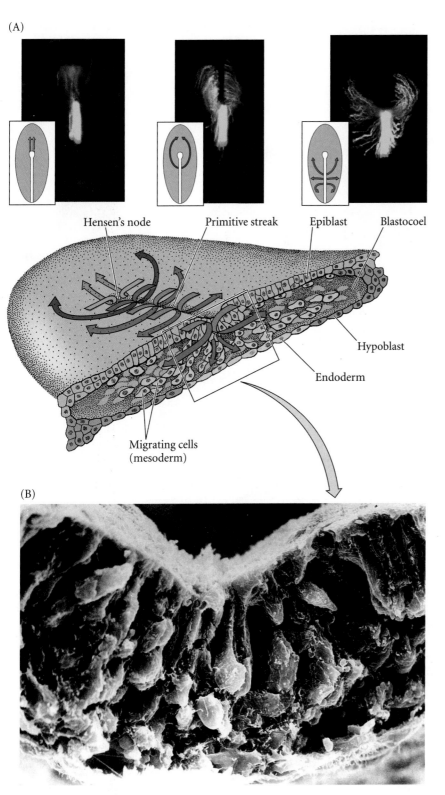

(A)

Hensen's node   Primitive streak   Epiblast   Blastocoel

Hypoblast

Endoderm

Migrating cells
(mesoderm)

(B)

FIGURE 11.17 Migration of endodermal and mesodermal cells through the primitive streak. (A) Stereogram of a gastrulating chick embryo, showing the relationship of the primitive streak, the migrating cells, and the hypoblast and epiblast of the blastoderm. The lower layer becomes a mosaic of hypoblast and endodermal cells; the hypoblast cells eventually sort out to form a layer beneath the endoderm and contribute to the yolk sac. Above each region of the stereogram are micrographs showing the tracks of GFP-labeled cells at that position in the primitive streak. Cells migrating through Hensen's node travel anteriorly to form the prechordal plate and notochord; those migrating through the next anterior region of the streak travel laterally, but converge near the midline to make notochord and somites; those from the middle of the streak form intermediate mesoderm and lateral plate mesoderm (see the fate map in Figure 11.16). Farther posterior, the cells migrating through the primitive streak make the extraembryonic mesoderm (not shown). (B) This scanning electron micrograph shows epiblast cells passing into the blastocoel and extending their apical ends to become bottle cells. (A after Balinsky 1975, with photographs from Yang et al. 2002; B from Solursh and Revel 1978; photographs courtesy of M. Solursh and C. J. Weijer.)

Epiboly of the ectoderm

While the presumptive mesodermal and endodermal cells are moving inward, the ectodermal precursors proliferate and migrate to surround the yolk by epiboly. The enclo-

sure of the yolk by the ectoderm (again reminiscent of the epiboly of the amphibian ectoderm) is a Herculean task that takes the greater part of four days to complete. It involves the continuous production of new cellular material and the migration of the presumptive ectodermal cells along the underside of the vitelline envelope (New 1959; Spratt 1963). Interestingly, only the cells of the outer margin of the area opaca attach firmly to the vitelline envelope. These cells are inherently different from the other blastoderm cells, as they can extend enormous (500 μm) cytoplasmic processes onto the vitelline envelope. These elongated filopodia are believed to be the locomotor apparatus of the marginal cells, by which the marginal cells pull other ectodermal cells around the yolk (Schlesinger 1958). The filopodia bind to fibronectin, a laminar protein that is a component of the chick vitelline envelope (see Chapter 6). If the contact between the marginal cells and the fibronectin is experimentally broken by adding a soluble polypeptide similar

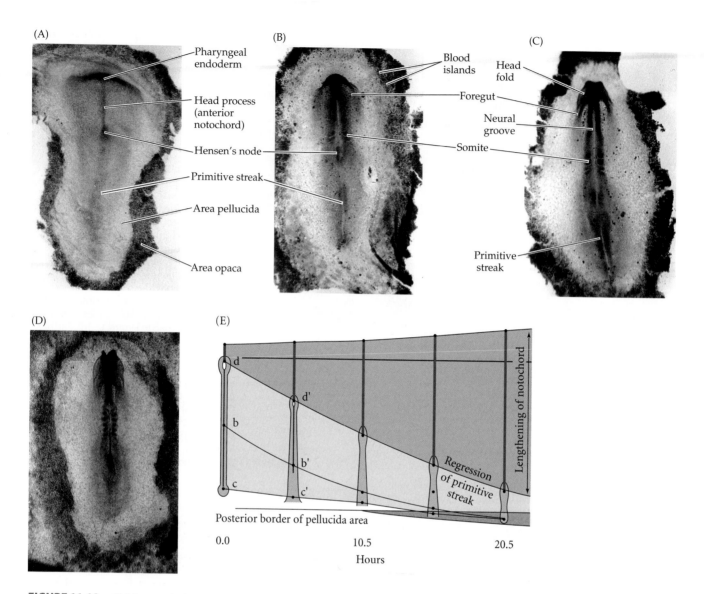

(A)
- Pharyngeal endoderm
- Head process (anterior notochord)
- Hensen's node
- Primitive streak
- Area pellucida
- Area opaca

(B)
- Blood islands
- Somite

(C)
- Head fold
- Foregut
- Neural groove
- Somite
- Primitive streak

(D)

(E)
- d
- d'
- b
- b'
- c
- c'
- Lengthening of notochord
- Regression of primitive streak
- Posterior border of pellucida area
- 0.0
- 10.5
- 20.5
- Hours

FIGURE 11.18 Chick gastrulation 24–28 hours after fertilization. (A) The primitive streak at full extension (24 hours). The head process (anterior notochord) can be seen extending from Hensen's node. (B) Two-somite stage (25 hours). Pharyngeal endoderm is seen anteriorly, while the anterior notochord pushes up the head process beneath it. The primitive streak is regressing. (C) Four-somite stage (27 hours). (D) At 28 hours, the primitive streak has regressed to the caudal portion of the embryo. (E) Regression of the primitive streak, leaving the notochord in its wake. Various points of the streak (represented by letters) were followed after it achieved its maximum length. The *x* axis (time) represents hours after achieving maximum length (the reference line is about 18 hours of incubation). (A–D, photographs courtesy of K. Linask; E after Spratt 1947.)

the hypoblast, and the mesoderm has positioned itself between these two regions. Although we have identified many of the processes involved in avian gastrulation, we are now just beginning to understand the mechanisms by which some of these processes are carried out.

WEBSITE **11.2 Epiblast cell heterogeneity.** Although the early epiblast appears uniform, different cells have different molecules on their cell surfaces. This variability allows some of them to remain in the epiblast while others migrate into the embryo.

VADE MECUM² **Chick development.** Viewing these movies of 3-D models of chick cleavage and gastrulation will help you understand these phenomena. [Click on Chick-Early]

to fibronectin, the filopodia retract and ectodermal migration ceases (Lash et al. 1990).

Thus, as avian gastrulation draws to a close, the ectoderm has surrounded the yolk, the endoderm has replaced

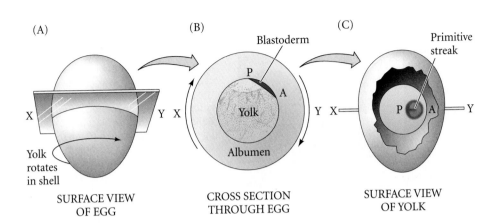

Axis Formation in the Chick Embryo

Although the formation of the chick body axes is accomplished during gastrulation, axis *specification* occurs earlier, during the cleavage stage.

The role of gravity in forming the anterior-posterior axis

The conversion of the radially symmetrical blastoderm into a bilaterally symmetrical structure is determined by gravity. As the ovum passes through the hen's reproductive tract, it is rotated for about 20 hours in the shell gland. This spinning, at a rate of 10–12 revolutions per hour, shifts the yolk such that its lighter components (probably containing stored maternal determinants for development) lie beneath one side of the blastoderm. This imbalance tips up one end of the blastoderm, and that end becomes the posterior portion of the embryo—the part where primitive streak formation begins (Figure 11.19; Kochav and Eyal-Giladi 1971; Callebaut et al. 2004).

It is not known what interactions cause this specific portion of the blastoderm to become the posterior marginal zone (PMZ) and initiate gastrulation. Early on, the ability to initiate a primitive streak is found throughout the marginal zone; if the blastoderm is separated into parts, each with its own marginal zone, each part will form its own primitive streak (Spratt and Haas 1960). However, once the PMZ has formed, it controls the other regions of the margin. Not only do the cells of the PMZ initiate gastrulation, but they also prevent other regions of the margin from forming their own primitive streaks (Khaner and Eyal-Giladi 1989; Eyal-Giladi et al. 1992). Some recent studies suggest that Nodal expression is needed to initiate the primitive streak, and that the secretion of Cerberus (an antagonist of Nodal protein) by the primary hypoblast cells prevents further primitive streak formation throughout the margin (Bertocchini and Stern 2002; Bertocchini et al. 2004). As the primary hypoblast cells move away from the posterior marginal zone, the absence of Cerberus allows Nodal protein to be expressed in the PMZ, stimulating primitive streak formation there. Once formed, the streak secretes

Lefty, another Nodal antagonist, thereby preventing any further primitive streaks from forming. Eventually, the Cerberus-secreting hypoblast cells are pushed to the future anterior of the embryo, where they will help induce head formation.

It now seems apparent that the posterior marginal zone contains cells that act as the equivalent of the amphibian Nieuwkoop center. When placed in the anterior region of the marginal zone, a graft of posterior marginal zone tissue (posterior to and not including Koller's sickle) is able to induce a primitive streak and Hensen's node without contributing cells to either structure (Bachvarova et al. 1998; Khaner 1998). Like the amphibian Nieuwkoop center, this region expresses Vg1 (Mitrani et al. 1990; Hume and Dodd 1993; Seleiro et al. 1996).

The chick "organizer"

The "organizer" of the chick embryo forms just anteriorly to the PMZ. The epiblast and middle layer cells in the anterior portion of Koller's sickle become Hensen's node (Bachvarova et al. 1998). The posterior portions of Koller's sickle contribute to the posterior portion of the primitive streak (Figure 11.20). Hensen's node has long been known to be the avian equivalent of the amphibian dorsal blastopore lip, since it is (1) the site where gastrulation begins, (2) the region whose cells become the chordamesoderm, and (3) the region whose cells can organize a second embryonic axis when transplanted into other locations of the gastrula (Figure 11.21; Waddington 1933; 1934; Dias and Schoenwolf 1990).

Gene expression in the chick organizer can be categorized into two sets of genes (Lawson et al. 2001). The first set contains those genes that are first expressed in the posterior portion of Koller's sickle and which probably help form the Nieuwkoop center-like portion of the PMZ cells. These genes, which include *Vg1* and *Nodal*, then appear throughout the entire length of the primitive streak (Figure 11.22A,B). Recent studies have shown that Vg1 plays a crucial role in forming the primitive streak, and that if Vg1 is ectopically expressed in the *anterior* marginal zone, Nodal will also be expressed and a secondary axis will

(A)

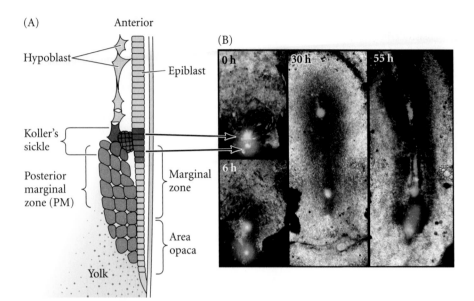

FIGURE 11.20 Formation of Hensen's node from Koller's sickle. (A) Diagram of the posterior end of an early (pre-streak) embryo, showing the cells labeled with fluorescent dyes in the photographs. (B) Just before gastrulation, cells in the anterior end of Koller's sickle (the epiblast and middle layer) were labeled with green dye. Cells of the posterior portion of Koller's sickle were labeled with red dye. As the cells migrated, the anterior cells formed Hensen's node and its notochord derivatives. The posterior cells formed the posterior region of the primitive streak. The time after dye injection is labeled on each photograph. (After Bachvarova et al. 1998; photographs courtesy of R. F. Bachvarova.)

form there (Skromne and Stern 2002). The second set of genes comprises those whose expression is confined to the anterior portion of the primitive streak, and finally to Hensen's node (Figure 11.22C). These genes include *chordin* and *sonic hedgehog*.

As is the case in all vertebrates, the dorsal mesoderm is able to induce the formation of the central nervous system in the ectoderm overlying it. The cells of Hensen's node and its derivatives act like the amphibian organizer, and they secrete BMP antagonist proteins such as chordin, Noggin, and Nodal. These proteins repress BMP signaling and dorsalize the ectoderm and mesoderm (Figure 11.23). However, repression of the BMP signal by these antagonists does not

appear to be sufficient for neural induction (see Stern 2005). Fibroblast growth factors (FGFs) synthesized in Hensen's node precursor cells just prior to gastrulation appear to be critical for preparing the epiblast to generate neuronal phenotypes, and these FGFs block BMP signaling (Alvarez et al. 1998; Storey et al. 1998; Streit et al. 2000).

FGFS: INDUCERS OF THE MESODERM AND NEURULATION Fibroblast growth factors play three fundamental roles in cell specification during gastrulation. First, as in all vertebrates, they are responsible for specifying the mesoderm. FGFs from the hypoblast (in collaboration with Nodal from the posterior marginal zone) accomplish this by activating the

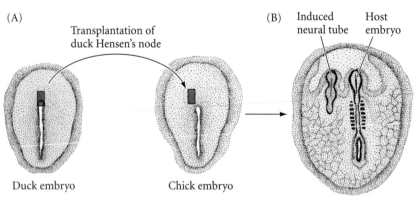

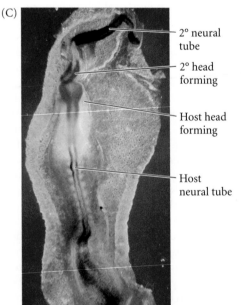

FIGURE 11.21 Induction of a new embryo by transplantation of Hensen's node. (A) A Hensen's node from a duck embryo is transplanted into the epiblast of a chick embryo. (B) A secondary embryo is induced (as is evident by the neural tube) from host tissues at the graft site. (C) Graft of Hensen's node from one embryo into the periphery of a host embryo. After further incubation, the host embryo has a neural tube whose regionalization can be seen by in situ hybridization. Probes to *otx2* (red) recognize the head region, while probes to *hoxb1* (blue) recognize the trunk neural tube. The donor node has induced the formation of a secondary axis, complete with head and trunk regions. (A,B after Waddington 1933; C from Boettger et al. 2001.)

(A)

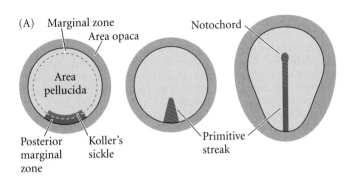

Marginal zone
Area opaca
Area pellucida
Posterior marginal zone
Koller's sickle
Notochord
Primitive streak

(B) Vg1

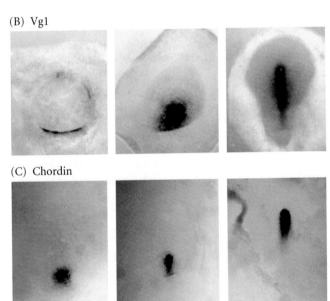

(C) Chordin

FIGURE 11.22 Gene expression in the primitive streak. (A) Schemata of the two general gene expression patterns. The early chick epiblast (left panel) shows the area opaca, area pellucida, marginal zone, Koller's sickle (red), and posterior marginal zone (blue). At a slightly later stage (middle panel), cells expressing both the Nieuwkoop center genes and organizer genes extend into the primitive streak. At later stages (right panel), the Nieuwkoop center genes are expressed throughout the streak, while the organizer genes are expressed in the most anterior region. (B) Expression of Vg1 protein as the primitive streak forms. (C) Expression of chordin as the primitive streak forms. (A after Boettger et al. 2001; B,C after Lawson et al. 2001, photographs courtesy of G. Schoenwolf.)

they are also slowly inducing activation of *Churchill* in the ectoderm. The Churchill protein (so named because the protein's two zinc fingers extend like the British prime minister's famous "V for Victory" symbol) can activate the Smad-interacting protein SIP1. SIP1 probably blocks the transmission of the Smad1 signal and prevents further ingression of cells through the primitive streak. Thus, it allows prospective neural plate cells to remain in the epiblast. Third, FGFs help bring about neurulation in the central ectodermal cells. SIP1 appears to sensitize the prospective neural plate cells to the BMP antagonists, making them more resistant to the BMP signal. Moreover, FGFs induce *ERNI* and *Sox3*, two pre-neural genes that, in the absence of BMP signals, initiate the cascade leading to the production of neural tissue, Thus, FGFs appear to be a critically important regulator of cell fate in the early chick embryo.

Brachyury and *Tbx6* genes in the cells going through the primitive streak (Figure 11.24; Sheng et al. 2003). Second, FGFs separate mesoderm formation from neurulation. The mechanism by which FGFs help end mesoderm ingression and stabilize the epiblast appears to be due to a gene called **Churchill**. While FGFs are rapidly inducing the mesoderm,

FIGURE 11.23 Possible contribution to chick neural induction by the inhibition of BMP signaling. (A) In a neurulating embryo, Noggin protein (purple) is expressed in the notochord and the pharyngeal endoderm. (B) *Bmp7* expression (dark purple), which had encompassed the entire epiblast, becomes restricted to the non-neural regions of the ectoderm. (C) Similarly, the product of BMP signaling, the phosphorylated form of Smad1 (recognized by antibodies to the phosphorylated form of the protein; dark brown) is not seen in the neural plate. (After Faure et al. 2002; photographs courtesy of the authors.)

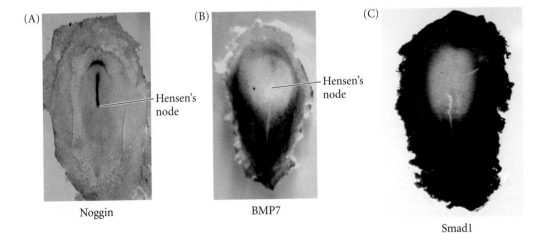

(A) Noggin

(B) BMP7 — Hensen's node

(C) Smad1

(A)

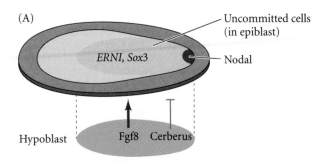

Uncommitted cells (in epiblast)

*ERNI, Sox*3

Nodal

Hypoblast

Fgf8   Cerberus

(B)

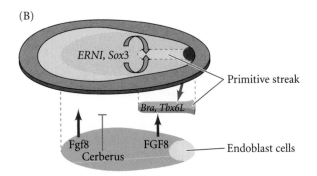

*ERNI, Sox*3

Primitive streak

Bra, Tbx6L

Fgf8   FGF8
Cerberus

Endoblast cells

(C)

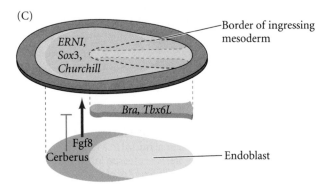

Border of ingressing mesoderm

*ERNI,
Sox*3,
Churchill

Bra, Tbx6L

Fgf8
Cerberus

Endoblast

(D)

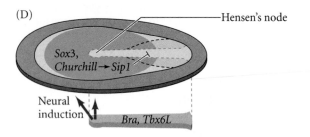

Hensen's node

*Sox*3,
Churchill → *Sip*1

Neural
induction   *Bra, Tbx6L*

FIGURE 11.24 Model by which FGFs regulate mesoderm formation and neurulation. (A) Stage XI, where the hypoblast (green) secretes Fgf8, which induces pre-neural genes *ERNI* and *Sox2* (blue) in the epiblast. The cells in this domain, however, remain uncommitted. Nodal, expressed in the posterior epiblast, cannot function; it is inhibited by the Cerberus protein secreted by the hypoblast. (B) At around stage 1, the hypoblast is displaced from the posterior edge by the endoblast (secondary hypoblast) (gold), allowing Nodal to function. Nodal plus Fgf8 induces *Brachyury* and *Tbx6* expression to specify the mesoderm and initiate the ingression of mesoderm cells through the primitive streak (red). (C) At stage 4, continued Fgf8 expression activates *Churchill* in the epiblast. (D) By end of stage 4, Churchill protein induces SIP1, which blocks *Brachyury* and *Tbx6*, preventing further ingression of epiblast cells through the streak. The remaining epiblast cells can now become sensitized to neural inducers from Hensen's node (purple). (After Sheng 2003.)

endoderm appears critical in specifying the head to form. The caudalization of the neural plate appears to be accomplished through retinoic acid that is made and secreted from the posterior mesoderm. As in fish and frog embryos, the presence of retinoic acid is a balance between the syn-

(A)   (B)

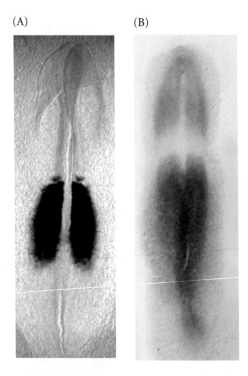

FIGURE 11.25 Anterior-posterior patterning in the chick embryo. (A) At stage 7, retinaldehyde dehydroxylase-2, the major retinoic acid-synthesizing gene, is expressed in the mesoderm surrounding the neural tube (dark stain). (B) At the same stage, the retinoic acid degradation enzymes (stained in light blue) are in the anterior. *Hoxb1* expression (which forms the anterior boundary of the posterior cells) is stained in purple. Thus, retinoic acid is allowed to accumulate only in the posterior cells. (From Blentic et al. 2003.)

ANTERIOR-POSTERIOR PATTERNING The lower layer (hypoblast and definitive endoderm) is required for establishing the anterior identity of the chick embryo. This specification appears to be separate from neural induction. A combination of experimental manipulations and gene expression markers show that while the node acts to induce neuralization, the lower layer secretes Cerberus and other head-inducers at its anterior end (Withington et al. 2001; Chapman et al. 2003). Thus, as in *Xenopus*, the anterior

thesis and degradation of this compound at various sites (Figure 11.25; Blentic et al. 2003; Molotkova et al. 2005).

Left-Right Axis Formation

As we have seen, the vertebrate body has distinct right and left sides. The heart and spleen, for instance, are generally on the left side of the body, while the liver is usually on the right. The distinction between the sides is regulated by two major proteins: the paracrine factor **Nodal** and the transcription factor **Pitx2**. However, the mechanism by which *Nodal* gene expression is activated in the left side of the body differs among vertebrate classes. The ease with which chick embryos can be manipulated has allowed scientists to elucidate the pathways of left-right determination in birds more readily than in other vertebrates.

As the primitive streak reaches its maximum length, transcription of the *sonic hedgehog (shh)* gene ceases on the right side of the embryo, inhibited by the expression of activin and its receptor (Figure 11.26A). Activin signaling, through BMP4, blocks the expression of *shh* and also activates *fgf8* on the right side of the embryo. The expressed

Fgf8 activates *snail (cSnR)*, a transcription factor that can block the expression of *Pitx2*.

Meanwhile, on the left side of the body, Sonic hedgehog activates *Cerberus* (Figure 11.26B). Cerberus (sometimes called Caronte among chick embryologists) is a paracrine factor that prevents BMPs from repressing the *Nodal* and *lefty-2* genes (Rodriguez-Esteban et al. 1999; Yokouchi et al. 1999). Nodal protein activates *Pitx2* and represses *snail*. In addition, Lefty-1 in the ventral midline prevents the Cerberus signal from passing to the right side of the embryo (Figure 11.26C,D) As in *Xenopus*, Pitx2 is crucial in directing the asymmetry of the embryonic structures. Experimentally induced expression of either *Nodal* or *Pitx2* on the right side of the chick is able to reverse the asymmetry or cause randomization of the asymmetry on the right or left sides* (Levin et al. 1995; Logan et al. 1998; Ryan et al. 1998).

---

*In humans, homozygous loss of *PITX2* causes Rieger's syndrome, a condition characterized by asymmetry anomalies. A similar condition is caused by knocking out this gene in mice (Fu et al. 1999; Lin et al. 1999).

(A)

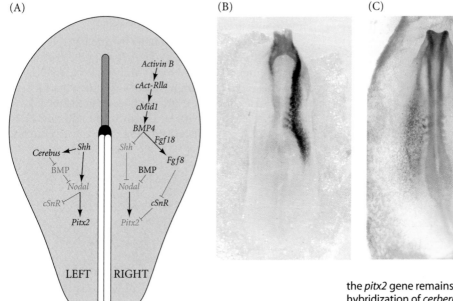

LEFT   RIGHT

(B)   (C)   (D)

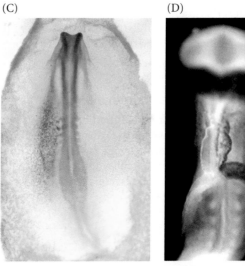

FIGURE 11.26 Pathway for left-right asymmetry in the chick embryo. (A) On the left side of Hensen's node, *sonic hedgehog (shh)* activates Cerberus, a BMP antagonist. Cerberus protein blocks expression of bone morphogenetic proteins (BMPs) on the left side, which would otherwise block the expression of *nodal*. In the presence of Nodal, the *pitx2* gene is activated and the *snail* gene (*cSnR*) is repressed. Pitx2 is active in the various organ primordia and specifies the side to be left. On the right side of the embryo, activin is expressed, along with activin receptor IIa. This activates Fgf8, a protein that blocks the expression of *cerberus*. In the absence of Cerberus protein, BMP represses the activation of *nodal*. This allows the *snail* gene to be active while

the *pitx2* gene remains repressed. (B) Whole-mount in situ hybridization of *cerberus* mRNA. This view is from the ventral surface, "from below," so the expression seems to be on the right. Dorsally, the expression pattern would be on the left. (C) Whole-mount in situ hybridization using probes for the chick *nodal* message (stained purple) shows its expression in the lateral plate mesoderm only on the left side of the embryo. This view is from the dorsal side (looking "down" at the embryo). (D) A similar in situ hybridization, using the probe for *pitx2* at a later stage of development. The embryo is seen from its ventral surface. At this stage, the heart is forming, and *pitx2* expression can be seen on the left side of the heart tube (as well as symmetrically in more anterior tissues (A after Raya and Izpisua-Belmonte 2004; B from Rodriguez-Esteban et al. 1999, photograph courtesy of J. Izpisúa-Belmonte; C, photograph courtesy of C. Stern; D from Logan et al. 1998, photograph courtesy of C. Tabin.)

The real mystery, then, is what limits the expression of activin protein to the right side of the embryo? There are several speculations as to how symmetry is broken (see Tabin and Vogan 2003; Raya 2004), but the earliest steps in the pathway leading to left-right discrimination remain to be deciphered (see p. 366).

EARLY MAMMALIAN DEVELOPMENT

Cleavage in Mammals

It is not surprising that mammalian cleavage has been the most difficult to study. Mammalian eggs are among the smallest in the animal kingdom, making them hard to manipulate experimentally. The human zygote, for instance, is only 100 µm in diameter—barely visible to the eye and less than one-thousandth the volume of a *Xenopus* egg. Also, mammalian zygotes are not produced in numbers comparable to sea urchin or frog zygotes; a female mammal usually ovulates fewer than 10 eggs at a given time, so it is difficult to obtain enough material for biochemical studies. As a final hurdle, the development of mammalian embryos is accomplished inside another organism rather than in the external environment. Only recently has it been possible to duplicate some of these internal conditions and observe mammalian development in vitro.

The unique nature of mammalian cleavage

It took time to surmount the difficulties, but our knowledge of mammalian cleavage has turned out to be worth waiting for. Mammalian cleavage is strikingly different from most other patterns of embryonic cell division.

Prior to its fertilization, the mammalian oocyte, wrapped in cumulus cells, is released from the ovary and swept by the fimbriae into the oviduct (Figure 11.27). Fertilization occurs in the **ampulla** of the oviduct, a region close to the ovary. Meiosis is completed after sperm entry, and the first cleavage begins about a day later (see Figure 7.35). Cleavages in mammalian eggs are among the slowest in the animal kingdom, taking place some 12–24 hours apart. Meanwhile, the cilia in the oviduct push the embryo toward the uterus; the first cleavages occur along this journey.

In addition to the slowness of cell division, several other features distinguish mammalian cleavage, including the unique orientation of mammalian blastomeres with relation to one another. The first cleavage is a normal meridional division; however, in the second cleavage, one of the two blastomeres divides meridionally and the other divides equatorially (Figure 11.28). This is called **rotational cleavage** (Gulyas 1975).

Another major difference between mammalian cleavage and that of most other embryos is the marked asynchrony of early cell division. Mammalian blastomeres do not all divide at the same time. Thus, mammalian embryos do not increase exponentially from 2- to 4- to 8-cell stages, but frequently contain odd numbers of cells. And, unlike almost all other animal genomes, the mammalian genome is activated during early cleavage, and it is the newly formed nuclei (rather than the oocyte cytoplasm) that produce the proteins necessary for cleavage and development. In the mouse and goat, the switch from maternal to zygotic control occurs at the 2-cell stage. In humans, the zygotic genes are first activated between the 4- and 8-cell stages (Piko and Clegg 1982; Braude et al. 1988; Prather 1989).

Most research on mammalian development has focused on the mouse, since mice are relatively easy to breed, have large litters, and can be housed easily in laboratories. Thus, most of the studies discussed here will concern murine development.

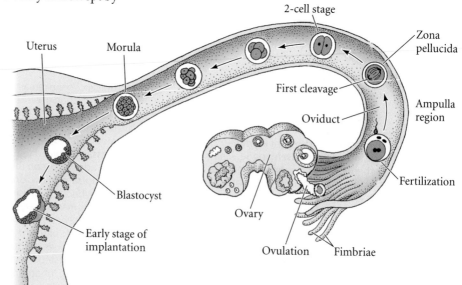

FIGURE 11.27 Development of a human embryo from fertilization to implantation. Compaction of the human embryo occurs on day 4, when it is at the 10-cell stage. The embryo "hatches" from the zona pellucida upon reaching the uterus. During its migration to the uterus, the zona prevents the embryo from prematurely adhering to the oviduct rather than traveling to the uterus. (After Tuchmann-Duplessis et al. 1972.)

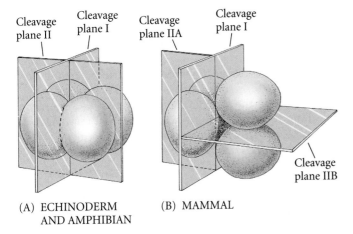

(A) ECHINODERM
AND AMPHIBIAN

(B) MAMMAL

FIGURE 11.28 Comparison of early cleavage in (A) echinoderms and amphibians (radial cleavage) and (B) mammals (rotational cleavage). Nematodes also have a rotational form of cleavage, but they do not form the blastocyst structure characteristic of mammals. (After Gulyas 1975.)

Compaction

One of the most crucial differences between mammalian cleavage and all other types involves the phenomenon of **compaction**. Mouse blastomeres through the 8-cell stage form a loose arrangement with plenty of space between them (Figure 11.29A,B). Following the third cleavage, however, the blastomeres undergo a spectacular change in their behavior. Cell adhesion proteins such as E cadherin are expressed, and the blastomeres suddenly huddle together and form a compact ball of cells (Figure 11.29C,D; Peyrieras et al. 1983; Fleming et al. 2001). This tightly packed arrangement is stabilized by tight junctions that form between the outside cells of the ball, sealing off the inside of the sphere. The cells within the sphere form gap junc-

tions, thereby enabling small molecules and ions to pass between them.

The cells of the compacted 8-cell embryo divide to produce a 16-cell **morula** (Figure 11.29E). The morula consists of a small group of internal cells surrounded by a larger group of external cells (Barlow et al. 1972). Most of the descendants of the external cells become the **trophoblast** (trophectoderm) cells. This group of cells produces no embryonic structures. Rather, it forms the tissue of the chorion, the embryonic portion of the placenta. The chorion enables the fetus to get oxygen and nourishment from the mother. It also secretes hormones that cause the mother's uterus to retain the fetus and produces regulators of the immune response so that the mother will not reject the embryo (as she would an organ graft). It is important to remember that the crucial outcome of these first divisions is to generate cells that will stick to the uterus. Thus, formation of the trophectoderm is the first differentiation event in mammalian development. These external cells first adhere to the uterine lining, then digest a path that allows the embryo to lodge itself in the uterine wall.

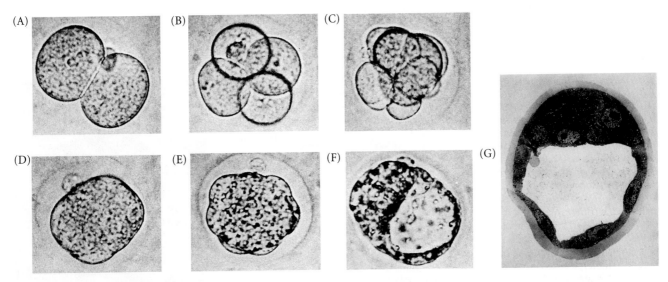

FIGURE 11.29 Cleavage of a single mouse embryo in vitro. (A) 2-cell stage. (B) 4-cell stage. (C) Early 8-cell stage. (D) Compacted 8-cell stage. (E) Morula. (F) Blastocyst. (G) Electron micrograph through the center of a mouse blastocyst (A–F from Mulnard 1967, photographs courtesy of J. G. Mulnard; G from Ducibella et al. 1975, courtesy of T. Ducibella.)

The mouse embryo proper is derived from the descendants of the inner cells of the 16-cell stage, supplemented by cells dividing from the outer cells of the morula during the transition to the 32-cell stage (Pedersen et al. 1986; Fleming 1987). These cells generate the **inner cell mass** (**ICM**), which will give rise to the embryo and its associated yolk sac, allantois, and amnion. By the 64-cell stage, the inner cell mass (approximately 13 cells) and the trophoblast cells have become separate cell layers, with neither contributing cells to the other group (Dyce et al. 1987; Fleming 1987). The inner cell mass actively supports the trophoblast, secreting proteins such as Fgf4 that cause the trophoblast cells to divide (Tanaka et al. 1998).

The earliest blastomeres (such as each blastomere of an 8-cell embryo) can form both trophoblast cells and the embryonic precursors. These very early cells are said to be **totipotent** (Latin, "capable of forming everything"). The inner cell mass is said to be **pluripotent** (Latin, "capable of many things") That is, each cell of the ICM can generate any cell type in the body, but because the distinction between ICM and trophoblast has been established, it is thought that ICM cells are not able to form the trophoblast.

Once the decision to become either trophoblast or inner cell mass is made, the cells of these two regions express different genes (Figure 11.30). The trophoblast cells synthesize the T-box transcription factor **eomesodermin** and the homeodomain-containing, caudal-like transcription factor **Cdx2**. Eomesodermin activates those proteins characteristic of the trophoblast layer (Russ et al. 2000; Hanna et al. 2002). Cdx2 is responsible for downregulating Oct4 and Nanog, two transcription factors that, along with Stat3, characterize the inner cell mass (Strumpf et al. 2005). At the 8-cell stage, Cdx2, eomesodermin, and Oct4 are each transcribed in all cells. But in the blastocyst, Cdx2 and eomesodermin are maintained in the trophoblast, whereas Oct4 is maintained in the inner cell mass (Niwa et al. 2005). The signal mediating this distinction is still unknown.

The expression of the three transcription factors characteristic of the inner cell mass—Oct4, Stat3, and Nanog—is critical for the formation of the embryo and for maintaining the pluripotency of the inner cell mass. Oct4 is expressed first, and it is expressed in the morula as well as in the inner cell mass and early epiblast. Oct4 blocks cells from taking on the trophoblastic fate. Later, Nanog* prevents the ICM blastomeres from becoming hypoblast cells, and stimulates blastomere self-renewal in the epiblast. The activated (phosphorylated) form of Stat3 also stimulates self-renewal of ICM blastomeres (see Figure 11.30; Pesce and Scholer et al. 2001; Chambers et al. 2003; Mitsui et al. 2003). If the inner cell mass blastomeres are removed in a manner that lets them retain their expression of Nanog, Oct4, and phosphorylated Stat3 proteins, these cells divide and become **embryonic stem cells**. The pluripotency of these stem cells (and the medical uses for them, as detailed in Chapter 21) is dependent on their retaining the expression of these three transcription factors.

Initially, the morula does not have an internal cavity. However, during a process called **cavitation**, the trophoblast cells secrete fluid into the morula to create a blastocoel. The membranes of trophoblast cells contains sodium pumps (an Na^+/K^+-ATPase and an Na^+/H^+ exchanger) that pump Na^+ into the central cavity. The subsequent accumulation of Na^+ draws in water osmotically, thus creating and enlarging the blastocoel (Borland 1977; Ekkert et al. 2004; Kawagishi et al. 2004). Interestingly, this sodium pumping activity appears to be stimulated by the oviduct cells on which the embryo is traveling toward the uterus (Xu et al. 2004).

As the blastocoel expands, the inner cell mass becomes positioned on one side of the ring of trophoblast cells (see

---

*The research leading to the discovery of Nanog was partially motivated by desire to convert normal human somatic cells into stem cell lines. The gene's name derives from the mythical Celtic land of perpetual youth.

FIGURE 11.30 Proposed functions of Nanog, Oct4, and Stat3 in retaining the uncommitted pluripotent fate of embryonic mouse cells. Oct4 stimulates the morula cells retaining it to become inner cell mass and not trophoblast. Nanog works at the next differentiation event, preventing the ICM cells from becoming hypoblast and promoting their becoming the pluripotent embryonic epiblast. Stat3 is probably involved in the self-renewal of these pluripotent cells. Cdx2 in the trophoblast prevents *Oct4* and *Nanog* expression, thereby stabilizing the trophoblast lineage. (B) Mouse blastocyst in which the Oct4 protein in the ICM is stained orange. (After Mitsui et al. 2003; Strumpf et al. 2005; photograph courtesy of J. Rossant.)

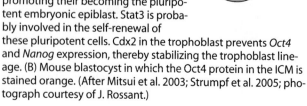

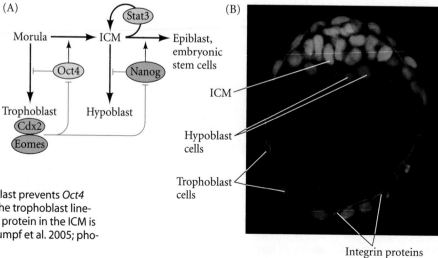

Figure 11.29F); the resulting type of blastula, called a **blastocyst**, is another hallmark of mammalian cleavage.*

Escape from the zona pellucida

While the embryo is moving through the oviduct en route to the uterus, the blastocyst expands within the zona pellucida (the extracellular matrix of the egg that was essential for sperm binding during fertilization; see Chapter 7). During this time, the zona pellucida prevents the blastocyst from adhering to the oviduct walls. (If this happens—as it sometimes does in humans—it is called an ectopic or "tubal" pregnancy, a dangerous condition because an embryo implanted in the oviduct can cause a life-threatening hemorrhage when it begins to grow.) When the embryo reaches the uterus, it must "hatch" from the zona so that it can adhere to the uterine wall.

The mouse blastocyst hatches from the zona pellucida by digesting a small hole in it and squeezing through that hole as the blastocyst expands (Figure 11.31A). A trypsin-like protease secreted by the trophoblast seems responsible for hatching the blastocyst from the zona (Perona and Wassarman 1986; O'Sullivan et al. 2001). Once out of the zona, the blastocyst can make direct contact with the uterus (Figure 11.31B,C). The uterine epithelium (endometrium) "catches" the blastocyst on an extracellular matrix containing complex sugars, collagen, laminin, fibronectin, hyaluronic acid, and heparan sulfate receptors. As in so many

intercellular adhesions during development, there appears to be a period of labile attachment, followed by a period of more stable attachment. The first attachment seems to be mediated by L-selectin on the trophoblast cells adhering to sulfated polysaccharides on the uterine cells (Genbacev et al. 2003). These sulfated polysaccharides are synthesized in response to estrogen and progesterone secreted by the corpus luteum (the remnant of the ruptured ovarian follicle).

After the initial binding, several other adhesion systems appear to coordinate their efforts to keep the blastocyst tightly bound to the uterine lining. The trophoblast cells synthesize integrins that bind to the uterine collagen, fibronectin, and laminin, and they synthesize heparan sulfate proteoglycan precisely prior to implantation (see Carson et al. 1993). Once in contact with the endometrium, the trophoblast secretes another set of proteases, including collagenase, stromelysin, and plasminogen activator. These protein-digesting enzymes digest the extracellular matrix of the uterine tissue, enabling the blastocyst to bury itself within the uterine wall (Strickland et al. 1976; Brenner et al. 1989).

Gastrulation in Mammals

Birds and mammals are both descendants of reptilian species. Therefore, it is not surprising that mammalian development parallels that of reptiles and birds. What *is* surprising is that the gastrulation movements of reptilian and avian embryos, which evolved as an adaptation to yolky eggs, are retained even in the absence of large amounts of yolk in the mammalian embryo. The mammalian inner cell mass can be envisioned as sitting atop an imaginary ball of yolk, following instructions that seem more appropriate to its reptilian ancestors.

---

*The interplay of myth and biology certainly comes to the fore when describing mammalian development. Although discovered by Rauber in 1881, the first public display of the mammalian blastocyst was probably in Gustav Klimt's 1908 painting *Danae*, in which blastocyst-like patterns are featured on the heroine's robe as she becomes impregnated by Zeus.

(A)

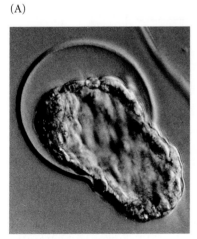

(B)

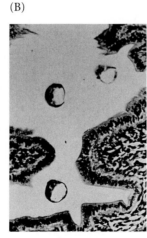

(C)

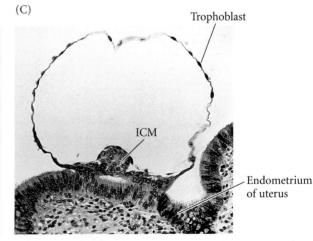

FIGURE 11.31 Hatching from the zona and implantation of the mammalian blastocyst in the uterus. (A) Mouse blastocyst hatching from the zona pellucida. (B) Mouse blastocysts entering the uterus. (C) Initial implantation of the blastocyst in a rhesus monkey. (A from Mark et al. 1985, photograph courtesy of E. Lacy; B from Rugh 1967; C, photograph courtesy of the Carnegie Institution of Washington, Chester Reather, photographer.)

Modifications for development within another organism

The mammalian embryo obtains nutrients directly from its mother and does not rely on stored yolk. This adaptation has entailed a dramatic restructuring of the maternal anatomy (such as expansion of the oviduct to form the uterus) as well as the development of a fetal organ capable of absorbing maternal nutrients. This fetal organ—the **chorion**—is derived primarily from embryonic trophoblast cells, supplemented with mesodermal cells derived from the inner cell mass. The chorion forms the fetal portion of the placenta. It also induces the uterine cells to form the maternal portion of the placenta, the **decidua**. The decidua becomes rich in the blood vessels that will provide oxygen and nutrients to the embryo.

The origins of early mammalian tissues are summarized in Figure 11.32. The first segregation of cells within the inner cell mass forms two layers. The lower layer is the **hypoblast** (sometimes called the **primitive endoderm** or **visceral endoderm**), and the remaining inner cell mass tissue above it is the **epiblast** (Figure 11.33A). The epiblast and hypoblast form a structure called the **bilaminar germ disc**. The hypoblast cells delaminate from the inner cell mass to line the blastocoel cavity, where they give rise to the **extraembryonic endoderm**, which forms the yolk sac. As in avian embryos, these cells do not produce any part of the newborn organism. The epiblast cell layer is split by small clefts that eventually coalesce to separate the **embry-**

onic epiblast from the other epiblast cells that line the **amnionic cavity** (Figure 11.33B,C). Once the lining of the amnion is completed, the amniontic cavity fills with a secretion called **amnionic fluid**, which serves as a shock absorber for the developing embryo while preventing it from drying out. The embryonic epiblast is thought to contain all the cells that will generate the actual embryo, and it is similar in many ways to the avian epiblast.

By labeling individual cells of the epiblast with horseradish peroxidase, Kirstie Lawson and her colleagues (1991) were able to construct a detailed fate map of the mouse epiblast (see Figure 1.6). Gastrulation begins at the posterior end of the embryo, and this is where the **node*** forms (Figure 11.34). Like the chick epiblast cells, the mammalian mesoderm and endoderm migrate through a primitive streak; also like their avian counterparts, the migrating cells of the mammalian epiblast lose E-cadherin, detach from their neighbors, and migrate through the streak as individual cells (Burdsal et al. 1993). Those cells migrating through the node give rise to the notochord. However, in contrast to notochord formation in the chick, the cells that form the mouse notochord are thought to become integrated into the endoderm of the primitive gut (Jurand 1974; Sulik et al. 1994). These cells can be seen as a band of small, ciliated

---

*In mammalian development, Hensen's node is usually just called "the node," despite the fact that Hensen discovered this structure in rabbit and guinea pig embryos.

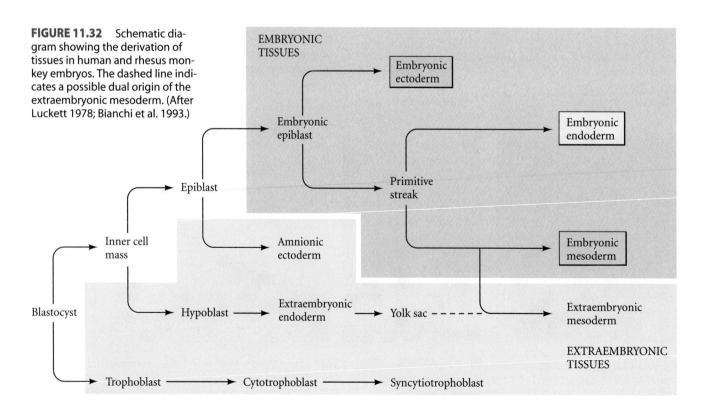

FIGURE 11.32 Schematic diagram showing the derivation of tissues in human and rhesus monkey embryos. The dashed line indicates a possible dual origin of the extraembryonic mesoderm. (After Luckett 1978; Bianchi et al. 1993.)

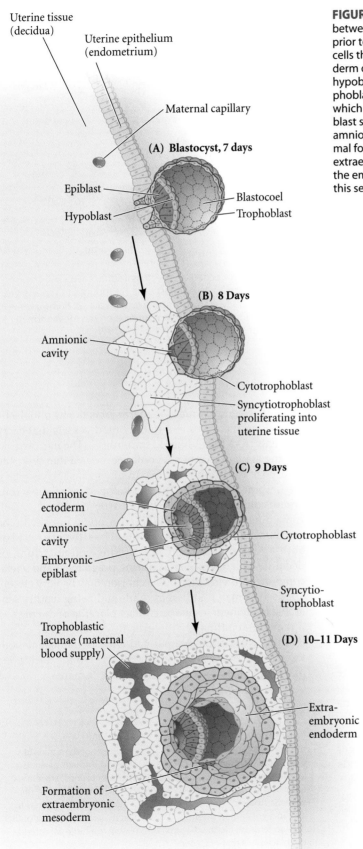

FIGURE 11.33 Tissue formation in the human embryo between days 7 and 11. (A,B) Human blastocyst immediately prior to gastrulation. The inner cell mass delaminates hypoblast cells that line the blastocoel, forming the extraembryonic endoderm of the primitive yolk sac and a two-layered (epiblast and hypoblast) blastodisc. The trophoblast divides into the cytotrophoblast, which will form the villi, and the syncytiotrophoblast, which will ingress into the uterine tissue. (C) Meanwhile, the epiblast splits into the amnionic ectoderm (which encircles the amnionic cavity) and the embryonic epiblast. The adult mammal forms from the cells of the embryonic epiblast. (D) The extraembryonic endoderm forms the yolk sac. The actual size of the embryo at this stage is about that of the period at the end of this sentence.

Uterine tissue (decidua)

Uterine epithelium (endometrium)

Maternal capillary

(A) Blastocyst, 7 days

Epiblast

Hypoblast

Blastocoel

Trophoblast

(B) 8 Days

Amnionic cavity

Cytotrophoblast

Syncytiotrophoblast proliferating into uterine tissue

(C) 9 Days

Amnionic ectoderm

Amnionic cavity

Embryonic epiblast

Cytotrophoblast

Syncytio-trophoblast

Trophoblastic lacunae (maternal blood supply)

(D) 10–11 Days

Extra-embryonic endoderm

Formation of extraembryonic mesoderm

cells extending rostrally from the node. They form the notochord by converging medially and "budding" off in a dorsal direction from the roof of the gut (see Figure 11.39C).

Cell migration and specification appear to be coordinated by fibroblast growth factors. The cells of the primitive streak appear to be capable of both synthesizing and responding to FGFs (Sun et al. 1999; Ciruna and Rossant 2001). In embryos that are homozygous for the loss of the *fgf8* gene, cells fail to migrate through the primitive streak, and neither mesoderm nor endoderm are formed. Fgf8 (and perhaps other FGFs) probably control cell movement into the primitive streak by downregulating the E-cadherin that holds the epiblast cells together. Fgf8 may also control cell specification by regulating *snail*, *Brachyury* (*T*), and *Tbx6*, three genes that are essential (as they are in the chick embryo) for mesodermal specification and patterning.

The ectodermal precursors are located anterior to the fully extended primitive streak, as in the chick epiblast; in some instances, however, a single cell gives rise to descendants in more than one germ layer, or to both embryonic and extraembryonic derivatives. Thus, at the epiblast stage, these lineages have not become separate from one another. As in avian embryos, the cells migrating into the space between the hypoblast and epiblast layers become coated with hyaluronic acid, which they synthesize as they leave the primitive streak. This substance keeps them separate while they migrate (Solursh and Morriss 1977). It is thought that the replacement of human hypoblast cells by endoderm precursors occurs on days 14–15 of gestation, while the migration of cells forming the mesoderm does not start until day 16 (see Figure 11.34C; Larsen 1993).

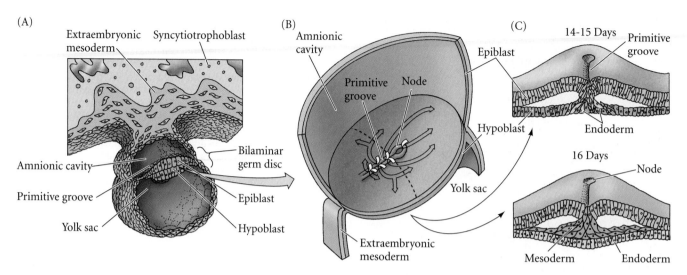

(A)

Extraembryonic mesoderm

Syncytiotrophoblast

Amnionic cavity

Primitive groove

Yolk sac

Bilaminar germ disc

Epiblast

Hypoblast

(B) Amnionic cavity

Primitive groove

Node

Epiblast

Hypoblast

Yolk sac

Extraembryonic mesoderm

(C) 14-15 Days

Primitive groove

Endoderm

16 Days

Node

Mesoderm

Endoderm

FIGURE 11.34 Amnion structure and cell movements during human gastrulation. (A,B) Human embryo and uterine connections at day 15 of gestation. (A) Sagittal section through the midline. (B) View looking down on the dorsal surface of the embryo. (C) The movements of the epiblast cells through the primitive streak and Hensen's node and underneath the epi-blast are superimposed on the dorsal surface view. At days 14 and 15, the ingressing epiblast cells are thought to replace the hypoblast cells (which contribute to the yolk sac lining), while at day 16, the ingressing cells fan out to form the mesodermal layer. (After Larsen 1993.)

Formation of extraembryonic membranes

While the embryonic epiblast is undergoing cell movements reminiscent of those seen in reptilian or avian gastrulation, the extraembryonic cells are making the distinctly mammalian tissues that enable the fetus to survive within the maternal uterus. Although the initial trophoblast cells of mice and humans divide like most other cells of the body, they give rise to a population of cells in which nuclear division occurs in the absence of cytokinesis. The original trophoblast cells constitute a layer called the **cytotrophoblast**, whereas the multinucleated cell type forms the **syncytiotrophoblast**. The cytotrophoblast initially adheres to the endometrium through a series of adhesion molecules, as we saw above. Moreover, cytotrophoblasts contain proteolytic enzymes that enable them to enter the uterine wall and remodel the uterine blood vessels so that the maternal blood bathes fetal blood vessels. The syncytiotrophoblast tissue is thought to further the progression of the embryo into the uterine wall by digesting uterine tissue. The cytotrophoblast secretes paracrine factors that attract maternal blood vessels and gradually displace their vascular tissue such that the vessels become lined with trophoblast cells (Fisher et al. 1989; Hemberger et al. 2003). Shortly thereafter, mesodermal tissue extends outward from the gastrulating embryo (see Figure 11.33D). Studies of human and rhesus monkey embryos have suggested that the yolk sac (and hence the hypoblast) as well as primitive streak-derived cells contribute this extraembryonic mesoderm (Bianchi et al. 1993).

The extraembryonic mesoderm joins the trophoblastic extensions and gives rise to the blood vessels that carry nutrients from the mother to the embryo. The narrow connecting stalk of extraembryonic mesoderm that links the embryo to the trophoblast eventually forms the vessels of the umbilical cord. The fully developed extraembryonic organ, consisting of trophoblast tissue and blood vessel-containing mesoderm, is the chorion, and it fuses with the uterine wall to create the placenta. Thus, the placenta has both a maternal portion (the uterine endometrium, or decidua, which is modified during pregnancy) and a fetal component (the chorion). The chorion may be very closely apposed to maternal tissues while still being readily separable from them (as in the contact placenta of the pig), or it may be so intimately integrated with maternal tissues that the two cannot be separated without damage to both the mother and the developing fetus (as in the deciduous placenta of most mammals, including humans).*

---

*There are numerous types of placentas, and the extraembryonic membranes form differently in different orders of mammals (see Cruz and Pedersen 1991). Although mice and humans gastrulate and implant in a similar fashion, their extraembryonic structures are distinctive. It is very risky to extrapolate developmental phenomena from one group of mammals to another. Even Leonardo da Vinci got caught (Renfree 1982). His remarkable drawing of the human fetus inside the placenta is stunning art, but poor science: the placenta is that of a cow.

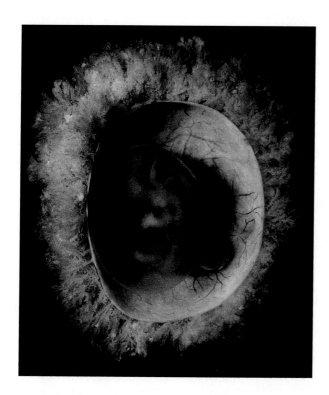

FIGURE 11.35 Human embryo and placenta after 50 days of gestation. The embryo lies within the amnion, and its blood vessels can be seen extending into the chorionic villi. The small sphere to the right of the embryo is the yolk sac. (The Carnegie Institution of Washington, courtesy of C. F. Reather.)

the villi that project from the outer surface of the chorion. These villi contain the blood vessels and allow the chorion to have a large area exposed to the maternal blood. Although fetal and maternal circulatory systems normally never merge, diffusion of soluble substances can occur through the villi (Figure 11.36). In this manner, the mother provides the fetus with nutrients and oxygen, and the fetus sends its waste products (mainly carbon dioxide and urea) into the maternal circulation. The maternal and fetal blood cells usually do not mix, although a small number of fetal red blood cells are seen in the maternal blood circulation (see Chapter 21).

Figure 11.35 shows the relationships between the embryonic and extraembryonic tissues of a 6.5-week human embryo. The embryo is seen encased in the amnion and is further shielded by the chorion. The blood vessels extending to and from the chorion are readily observable, as are

WEBSITE 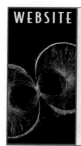 **11.5 Placental functions.** Placentas are nutritional, endocrine, and immunological organs. They provide hormones that enable the uterus to retain the pregnancy and also accelerate mammary gland development. Placentas also block the potential immune response of the mother against the developing fetus. Recent studies suggest that the placenta uses several mechanisms to block the mother's immune response.

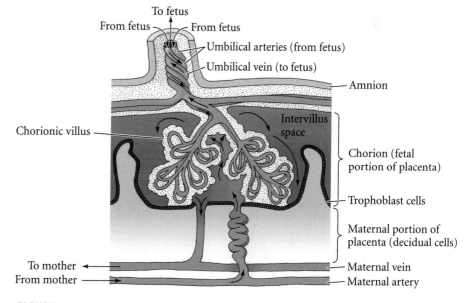

FIGURE 11.36 Relationship of the chorionic villi to the maternal blood supply in the primate uterus. In the umbilicus, there are two arteries and a single vein.

Twins and Embryonic Stem Cells

The early cells of the embryo can replace each other and compensate for a missing cell. This was first demonstrated in 1952, when Seidel destroyed one cell of a 2-cell rabbit embryo and the remaining cell produced an entire embryo. Once the inner cell mass (ICM) has become separate from the trophoblast, the ICM cells constitute an equivalence group. In other words, each ICM cell has the same potency (in this case, each cell can give rise to all the cell types of the embryo, but not to the trophoblast), and their fates will be determined by interactions among their descendants. Gardner and Rossant (1976) also showed that if cells of the ICM (but not trophoblast cells) are injected into blastocysts, they contribute to the new embryo. Since the ICM blastomeres can generate any cell type in the body, the cells of the blastocyst are referred to as pluripotent.

This regulative capacity of the ICM blastomeres is also seen in humans. Human twins are classified into two major groups: monozygotic (one-egg, or identical) twins and dizygotic (two-egg, or fraternal) twins. Fraternal twins are the result of two separate fertilization events, whereas identical twins are formed from a single embryo whose cells somehow become dissociated from one another. Identical twins may be produced by the separation of early blastomeres, or even by the separation of the inner cell mass into two regions within the same blastocyst.

Identical twins occur in roughly 0.25% of human births. About 33% of identical twins have two complete and separate chorions, indicating that separation occurred before the formation of the trophoblast tissue at day 5 (Figure 11.37A). The remaining identical twins share a common chorion, sug-

gesting that the split occurred within the inner cell mass after the trophoblast formed. By day 9, the human embryo has completed the construction of another extraembryonic layer, the lining of the amnion. This tissue forms the amnionic sac, which surrounds the embryo with amnionic fluid and pro-

FIGURE 11.37 Diagram showing the timing of human monozygotic twinning with relation to extraembryonic membranes. (A) Splitting occurs before the formation of the trophoblast, so each twin has its own chorion and amnion. (B) Splitting occurs after trophoblast formation but before amnion formation, resulting in twins having individual amnionic sacs but sharing one chorion. (C) Splitting after amnion formation leads to twins in one amnionic sac and a single chorion. (After Langman 1981).

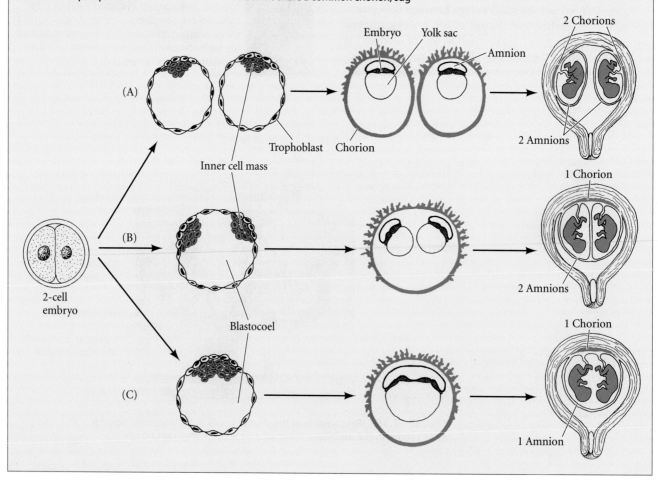

tects it from desiccation and abrupt movement. If the separation of the embryo were
to come after the formation of the chorion
on day 5 but before the formation of the
amnion on day 9, then the resulting
embryos should have one chorion and two
amnions (Figure 11.37B). This happens in
about two-thirds of human identical twins.
A small percentage of identical twins are
born within a single chorion and amnion
(Figure 11.37C). This means that the division of the embryo came after day 9. Such
newborns are at risk of being conjoined
("Siamese") twins.

(A)

Pronase   Zona
pellucida

Blastomeres

Blastocyst

Blastocysts implanted
into foster mother

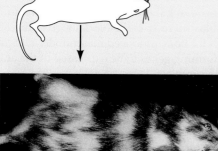

The ability to produce an entire embryo
from cells that normally would have contributed to only a portion of the embryo is
called **regulation**, and was discussed in
Chapter 3. Regulation is also seen in the
ability of two or more early embryos to form
one chimeric individual rather than twins,
triplets, or a multiheaded individual.
Chimeric mice can be produced by artificially aggregating two or more early-cleavage
(usually 4- or 8-cell) embryos to form a composite embryo. As shown in Figure 11.38A,
the zonae pellucidae of two genetically different embryos can be artificially removed
and the embryos brought together to form
a common blastocyst. These blastocysts are
then implanted into the uterus of a foster
mother. When they are born, the chimeric
offspring have some cells from each
embryo. This is readily seen when the
aggregated blastomeres come from mouse
strains that differ in their coat colors. When

blastomeres from white and black strains
are aggregated, the result is commonly a
mouse with black and white bands. Markert
and Petters (1978) have shown that three
early 8-cell embryos can unite to form a
common compacted morula and that the
resulting mouse can have the coat colors of
the three different strains. Moreover, they
showed that each of the three embryos
gave rise to precursors of the gametes.
When a chimeric (black/brown/white)
female mouse was mated to a white-furred
(recessive) male, offspring of each of the
three colors were produced (Figure 11.38B).

There is even evidence that human
embryos can form chimeras (de la Chap-
pelle et al. 1974; Mayr et al. 1979). Some
individuals have two genetically different
cell types (XX and XY) within the same body,
each with its own set of genetically defined
characteristics. The simplest explanation for
such a phenomenon is that these individu-

FIGURE 11.38 Production of chimeric mice. (A) The experimental procedures used to produce chimeric mice. Early 8-cell embryos of genetically distinct mice (here, with coat color differences) are isolated from mouse oviducts and brought together after their zonae are removed by proteolytic enzymes. The cells form a composite blastocyst, which is implanted into the uterus of a foster mother. The photograph shows one of the actual chimeric mice produced in this manner. (B) An adult female chimeric mouse (bottom) produced from the fusion of three 4-cell embryos: one from two white-furred parents, one from two black-furred parents, and one from two brown-furred parents. The resulting mouse has coat colors from all three embryos. Moreover, each embryo contributed germ line cells, as is evidenced by the three colors of offspring (above) produced when this chimeric female was mated with recessive (white-furred) males. (A photograph courtesy of B. Mintz; B from Markert and Petters 1978, photograph courtesy of C. Markert.)

(B)

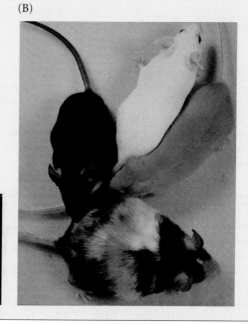

(Continued on next page)

Mammalian Anterior-Posterior Axis Formation

Two signaling centers

The formation of the mammalian anterior-posterior axis has been studied most extensively in mice. The structure of the mouse epiblast, however, differs from that of humans in that it is cup-shaped rather than disc-shaped. The dorsal surface of the epiblast (the embryonic ectoderm) contacts the amnionic cavity, while the ventral surface of the epiblast contacts the newly formed mesoderm. In this cuplike arrangement, the endoderm covers the surface of the embryo on the "outside" of the cup (Figure 11.39A).

The mammalian embryo appears to have two signaling centers: one in the node (equivalent to Hensen's node and the trunk portion of the amphibian organizer) and one in the anterior visceral endoderm (AVE; equivalent to the chick hypoblast and similar to the head portion of the amphibian organizer) (Figure 11.39B; Beddington and Robertson 1999; Foley et al. 2000). The node (at the "bottom of the cup" in the mouse) appears to be responsible for the creation of all of the body, and the two signaling centers work together to form the anterior region of the embryo (Bachiller et al. 2000). The notochord forms by the dorsal infolding of the small, ciliated cells of the node (Figure 11.39C).

The AVE originates from the visceral endoderm (hypoblast) that migrates forward. As this region migrates, it secretes two antagonists of the Nodal protein, Lefty1 and Cerberus (Brennan et al. 2001; Perea-Gomez et al. 2001; Yamamoto et al. 2004). (Lefty1 binds to the Nodal's receptors and blocks Nodal binding, and Cerberus binds to Nodal itself.) While the Nodal proteins within the epiblast

activate the expression of posterior genes that are required for mesoderm formation, the AVE creates an anterior region where Nodal cannot act. The AVE also begins expressing the anterior markers Otx2 and Wnt inhibitor Dickkopf. Studies of mutant mice indicate that the AVE promotes anterior specification by suppressing posterior patterning by Nodal and Wnt proteins. However, the AVE alone cannot induce neural tissue, as the node can (Tam and Steiner 1999).

Formation of the node is dependent on the trophoblast (Episkopou et al. 2001). As in the chick embryo, the placement of the node and the primitive streak appears to be due to the blocking of Nodal signaling by the Cerberus and Lefty-1 from the AVE (Perea-Gomez et al. 2002). Once formed, the node will secrete Chordin; the head process and notochord will later add Noggin. These two BMP antagonists are not expressed in the AVE. Dickkopf is expressed in both the AVE and in the node, but only the Dickkopf from the node is critical for head development (Mukhopadhyay et al. 2001). While knockouts of either the *chordin* or the *noggin* gene do not affect development, mice missing both genes lack a forebrain, nose, and other facial structures (Figure 11.40). It is probable that the AVE functions in the epiblast to restrict the Nodal signal, thereby cooperating with the node-produced mesendoderm to promote the head-forming genes to be expressed in the anterior portion of the epiblast.

Patterning the anterior-posterior axis: FGFs, retinoic acid, and the Hox code hypothesis

The head region of the embryo is devoid of Nodal signaling, and BMP, FGFs, and Wnts are also inhibited. The posterior region is characterized by Nodal, BMPs, Wnts, FGFs, and retinoic acid (RA). There appears to be a gradient of

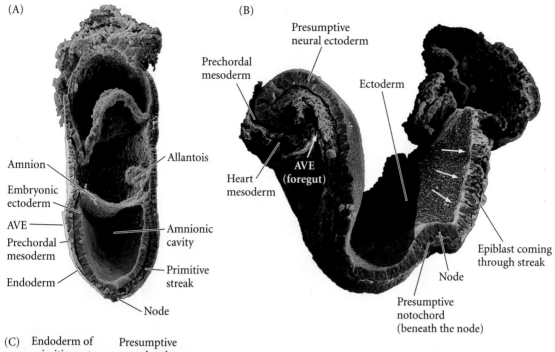

(A)

Amnion

Embryonic ectoderm

AVE

Prechordal mesoderm

Endoderm

Allantois

Amnionic cavity

Primitive streak

Node

(B)

Presumptive neural ectoderm

Prechordal mesoderm

Ectoderm

Heart mesoderm

AVE (foregut)

Epiblast coming through streak

Node

Presumptive notochord (beneath the node)

(C) Endoderm of primitive gut Presumptive notochord

Node

FIGURE 11.39 Axis and notochord formation in the mouse. (A) In the 7-day mouse embryo, the dorsal surface of the epiblast (embryonic ectoderm) is in contact with the amnionic cavity. The ventral surface of the epiblast contacts the newly formed mesoderm. In this cuplike arrangement, the endoderm covers the surface of the embryo. The node is at the bottom of the cup, and it has generated chordamesoderm. The two signaling centers, the node and the anterior visceral endoderm, are located on opposite sides of the cup. Eventually, the notochord will link them. The caudal side of the embryo is marked by the presence of the allantois. (B) By embryonic day 8, the anterior visceral endoderm lines the foregut, and the prechordal mesoderm is now in contact with the forebrain ectoderm. The node is now farther caudal, due largely to the rapid growth of the anterior portion of the embryo. The cells in the midline of the epiblast migrate through the primitive streak (white arrows). (C) Ventral surface of a 7.5-day mouse embryo. The presumptive notochord cells extend from the node into the endoderm of the primitive gut, converging medially to begin formation of the notochord. (A,B photographs courtesy of K. Sulik; C courtesy of K. Sulik and G. Schoenwolf.)

Wnt, BMP, and FGF proteins that is highest in the posterior and which drops off strongly near the anterior region. Moreover, in the anterior half of the embryo, starting at the node, there is a high concentration of antagonists that prevent BMPs and Wnts from acting (Figure 11.41A). The Fgf8 gradient is created by the decay of mRNA; *Fgf8* is expressed at the growing posterior tip of the embryo. However, the *Fgf8* message is slowly degraded in the newly formed tissues, so that there is a gradient of *Fgf8* mRNA across the posterior of the embryo (Figure 11.41B; Dubrulle and Pourquié 2004). The mRNA gradient is converted into a protein gradient of Fgf8. In addition to FGFs, the late gastrula also has a gradient of RA that is high in the posterior regions and low in the anterior portions of the embryo. This gradient (like that of chick, frog, and fish embryos)

appears to be controlled by the expression of RA-synthesizing enzymes in the embryo's posterior and RA-degrading enzymes in the anterior parts of the embryo (Sakai et al. 2001; Oosterveen et al. 2004).

The FGF gradient patterns the posterior portion of the embryo by working through the Cdx family of caudal-related genes (Figure 11.41C; Lohnes 2003). The Cdx genes, in turn, integrate the various posteriorization signals and activate particular Hox genes.

Anterior-posterior polarity in all vertebrates becomes specified by the expression of Hox genes. Hox genes are homologous to the homeotic selector genes (Hom-C genes) of the fruit fly (see Chapter 9). The *Drosophila* homeotic gene complex on chromosome 3 contains the *Antennapedia* and *bithorax* clusters of homeotic genes (see Figure 9.35),

FIGURE 11.40 Expression of BMP antagonists in the mammalian node. (A) Expression of chordin during mouse gastrulation is seen in the anterior primitive streak, the node, and axial mesoderm. It is not expressed in the anterior visceral endoderm. (B–D) Phenotypes of 12.5-day embryos. (B) Wild-type embryo. (C) Embryo with the *chordin* gene knocked out has a defective ear but an otherwise normal head. (D) Phenotype of a mouse deficient in both *chordin* and *noggin*. There is no jaw and a single centrally located eye, over which protrudes a large proboscis (nose). (After Bachiller et al. 2000; photographs courtesy of E. M. De Robertis.)

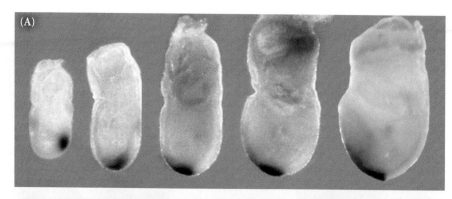

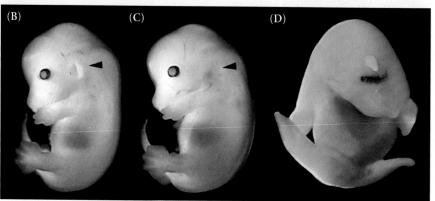

and can be seen as a single functional unit. (Indeed, in other insects, such as the flour beetle *Tribolium*, it is a single physical unit.) The mammalian genomes contain four copies of the Hox complex per haploid set, located on four different chromosomes (*Hoxa* through *Hoxd* in the mouse, *HOXA* through *HOXD* in humans; see Boncinelli et al. 1988; McGinnis and Krumlauf 1992; Scott 1992). The order of these genes on their respective chromosomes is remark-

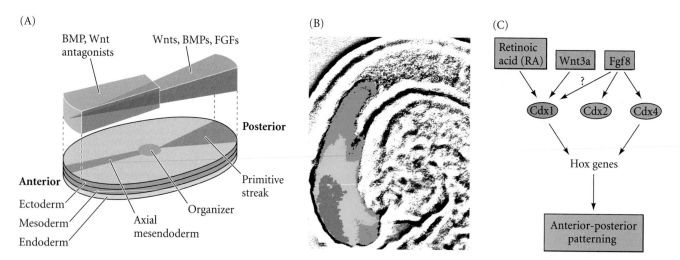

FIGURE 11.41 Anterior-posterior patterning in the mouse embryo. (A) Concentration gradients of BMPs, Wnts, and FGFs in the late gastrula mouse embryo (depicted as a flattened disc). The primitive streak and other posterior tissues are the sources of Wnt and BMP proteins, whereas the organizer and its derivatives (such as the notochord) produce antagonists. Fgf8 is expressed in the posterior tip of the gastrula and continues to be made in the tail bud. Its mRNA decays, creating a gradient across the posterior portion of the embryo. (B) Fgf8 gradient in the tailbud region of a 9-day mouse embryo. The highest amount of Fgf8 (red) is found near the tip. The gradient was determined by in situ hybridization of an Fgf8 probe and staining for increasing amounts of time. (C) Retinoic acid, Wnt3a, and Fgf8 each contribute to posterior patterning, but they are integrated by the Cdx family of proteins that regulates the activity of the Hox genes. (A after Robb and Tam 2004; B after Dubrulle and Pourquié 2004, photograph courtesy of O. Pourquié; C after Lohnes 2003.)

ably similar between insects and humans, as is the expression pattern of these genes. Those mammalian genes homologous to the *Drosophila labial, proboscipedia,* and *deformed* genes are expressed anteriorly and early, while those genes homologous to the *Drosophila Abd-B* gene are expressed posteriorly and later. As in *Drosophila,* a separate set of genes encodes the transcription factors that regulate head formation. In *Drosophila,* these are the *orthodenticle* and *empty spiracles* genes. In mice, the midbrain and forebrain are made through the expression of genes homologous to these—*Otx2* and *Emx* (see Kurokawa et al. 2004; Simeone 2004).

While Hox genes appear to specify the anterior-posterior axis throughout the vertebrates, we shall discuss mammals here (since the experimental evidence is particularly strong for this class). The mammalian Hox/HOX genes are numbered from 1 to 13, starting from that end of each complex that is expressed most anteriorly. Figure 11.42 shows the relationships between the *Drosophila* and mouse homeotic gene sets. The equivalent genes in each mouse complex (such as *Hoxa1, Hoxb1,* and *Hoxd1*) are called **paralogues**. It is thought that the four mammalian Hox complexes were formed by chromosome duplications. Because the correspondence between the *Drosophila* Hom-C genes

and mouse Hox genes is not one-to-one, it is likely that independent gene duplications and deletions have occurred since these two animal groups diverged (Hunt and Krumlauf 1992; see Chapter 23). Indeed, the most posterior mouse Hox gene (equivalent to *Drosophila AbdB*) underwent its own set of duplications in some mammalian chromosomes (see Figure 11.42).

Expression of Hox genes along the dorsal axis

Hox gene expression can be seen along the dorsal axis (in the neural tube, neural crest, paraxial mesoderm, and surface ectoderm) from the anterior boundary of the hindbrain through the tail. The regions of expression are not in register, but the 3' Hox genes (homologous to *labial, proboscopedia,* and *Deformed* of the fly) are expressed more anteriorly than the 5' Hox genes (homologous to *Ubx, AbdA,* and *AbdB*). Thus, one generally finds the genes of paralogous group 4 expressed anteriorly to those of paralogous group 5, and so forth (see Figure 11.42; Wilkinson et al. 1989; Keynes and Lumsden 1990). Mutations in the Hox genes suggest that the level of the body along the anterior-posterior axis is primarily determined by the most posterior Hox gene expressed in that region.

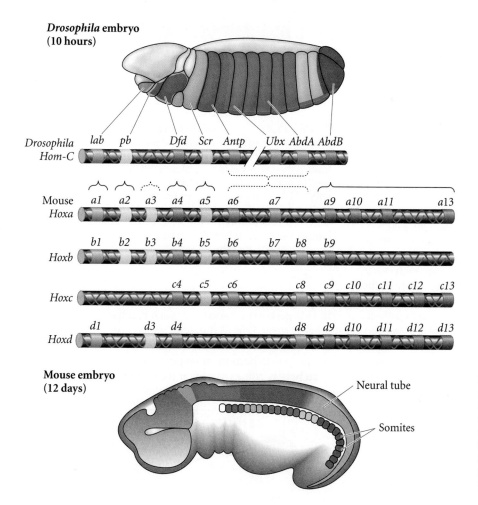

FIGURE 11.42 Evolutionary conservation of homeotic gene organization and transcriptional expression in fruit flies and mice is seen in the similarity between the *Hom-C* cluster on *Drosophila* chromosome 3 and the four Hox gene clusters in the mouse genome. The mouse genes of the higher numbered paralogous groups are those that are expressed later and more posteriorly. Genes having similar structures occupy the same relative positions on each of the four chromosomes, and paralogous gene groups display similar expression patterns. The comparison of the transcription patterns of the *Hom-C* and *Hoxb* genes of *Drosophila* and mice are shown above and below the chromosomes, respectively. (After Carroll 1995.)

Experimental analysis of the Hox code

The expression patterns of mouse Hox genes suggest a code whereby certain combinations of Hox genes specify a particular region of the anterior-posterior axis (Hunt and Krumlauf 1991). Particular sets of paralogous genes provide segmental identity along the anterior-posterior axis of the body. Evidence for such a code comes from three sources:

- Gene targeting or "knockout" experiments (see Chapter 4) in which mice are constructed that lack both copies of one or more Hox genes
- Retinoic acid teratogenesis, in which mouse embryos exposed to RA show an atypical pattern of Hox gene expression along the anterior-posterior axis and abnormal differentiation of their axial structures
- Comparative anatomy, in which the types of vertebrae in different vertebrate species are correlated with the constellation of Hox gene expression

GENE TARGETING There is a specific pattern to the vertebrae in mice. There are seven cervical (neck) vertebrae, 13 thoracic (ribbed) vertebrae, 6 lumbar (abdominal) vertebrae, 4 sacral (hip) vertebrae, and numerous (and variable) caudal (tail) vertebrae (Figure 11.43A). When all six copies of the *Hox10* paralogous group (i.e., *Hoxa10*, *c10*, and *d10* in Figure 11.42) were knocked out, no lumbar vertebrae were found. Instead, the presumptive lumbar vertebrae formed ribs and other characteristics similar to those of thoracic vertebrae (Figure 11.43B). This was a homeotic transformation comparable to those in insects; however, the redundancy of genes in the mouse made it much more difficult to produce, because the existence of even one copy of these genes prevented the transformation (Wellik and Capecchi 2003). Similarly, when all six copies of the *Hox11* group were knocked out, the thoracic and lumbar vertebrae were normal, but the sacral vertebrae failed to form and were replaced by copies of the lumbar vertebrae (Figure 11.43C).

RETINOIC ACID TERATOGENESIS Homeotic changes are also seen when mouse embryos are exposed to teratogenic doses of retinoic acid, a derivative of vitamin A. As we have seen, by day 7 of development, a gradient of RA has been established that is high in the posterior regions and low in the anterior portions of the embryo (Sakai et al. 2001). This gradient appears to be controlled by the differential synthesis or degradation of RA in the different parts of the embryo. Hox genes are responsive to retinoic acid either by virtue of having RA receptor sites in their enhancers or by being responsive to *Cdx* (the mammalian homologue of the *Drosophila Caudal* gene), which is activated by RA (Conlon and Rossant 1992; Kessel 1992; Lohnes 2003). Exogenous retinoic acid can mimic the RA concentrations normally encountered only by the posterior cells, and high doses of RA can activate Hox genes in more anterior locations along the anterior-posterior axis (Figure 11.44; Kessel and Gruss 1991; Allan et al. 2001). Thus, when excess retinoic acid is

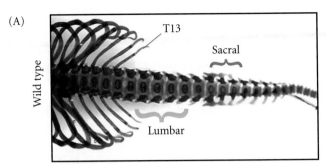

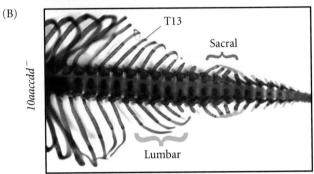

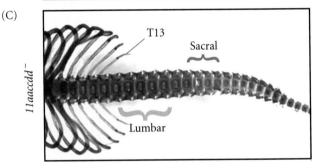

FIGURE 11.43 Axial skeletons of mice in gene knockout experiments. Each photograph is of an 18.5-day embryo, looking upward at the ventral region from the middle of the thorax toward the tail. (A) Wild-type mouse. (B) Complete knockout of *Hox10* paralogues (*Hox10aaccdd*) converts lumbar vertebrae (after the thirteenth thoracic vertebra) into more ribbed thoracic vertebrae. (C) Complete knockout of *Hox11* paralogues (*Hox11aaccdd*) transforms the sacral vertebrae into copies of lumbar vertebrae. (After Wellik and Capecchi 2003, photographs courtesy of M. Capecchi.)

administered to mouse embryos on day 8 of gestation, shifts in Hox gene expression occur such that the last cervical vertebra is turned into a thoracic (ribbed) vertebra. Conversely, impairment of retinoic acid function causes Hox gene expression to become more posterior, and the first thoracic vertebrae becomes a copy of the cervical vertebrae.

VADE MECUM² **Retinoic acid as a teratogen.** See the teratogenic effects of retinoic acid on zebrafish development. [Click on Zebrafish]

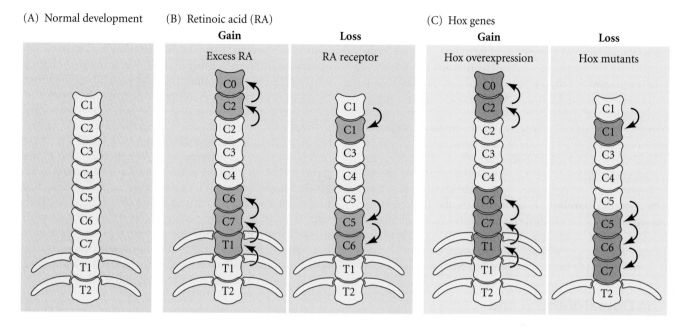

(A) Normal development   (B) Retinoic acid (RA)   (C) Hox genes

FIGURE 11.44 The effect of retinoic acid on mouse embryos. Addition of exogenous retinoic acid leads to differences in Hox gene expression and the transformation of vertebral form. Simi-larly, the loss of retinoic acid function (through mutations in the RA receptors) can lead to the opposite type of transformations. (After Houle et al. 2003.)

COMPARATIVE ANATOMY   A new type of comparative embryology is emerging based on the comparison of gene expression patterns among species. Gaunt (1994) and Burke and her collaborators (1995) have compared the vertebrae of the mouse and the chick (Figure 11.45A). Although the mouse and the chick have a similar number of vertebrae, they apportion them differently. Mice (like all mammals, be they giraffes or whales) have only 7 cervical vertebrae. These are followed by 13 thoracic vertebrae, 6 lumbar vertebrae, 4 sacral vertebrae, and a variable (20+) number of caudal vertebrae. The chick, on the other hand, has 14 cervical vertebrae, 7 thoracic vertebrae, 12 or 13 (depending

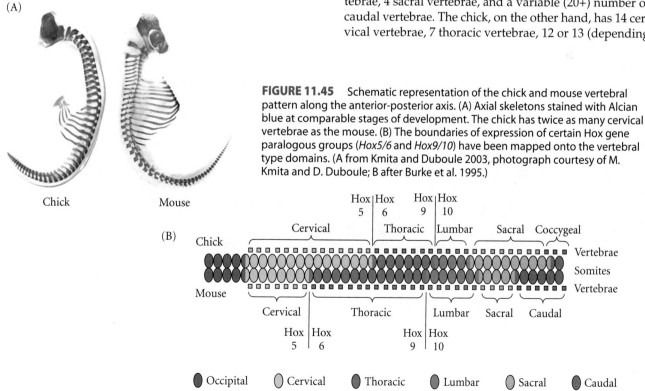

FIGURE 11.45 Schematic representation of the chick and mouse vertebral pattern along the anterior-posterior axis. (A) Axial skeletons stained with Alcian blue at comparable stages of development. The chick has twice as many cervical vertebrae as the mouse. (B) The boundaries of expression of certain Hox gene paralogous groups (*Hox5/6* and *Hox9/10*) have been mapped onto the vertebral type domains. (A from Kmita and Duboule 2003, photograph courtesy of M. Kmita and D. Duboule; B after Burke et al. 1995.)

on the strain) lumbosacral vertebrae, and 5 coccygeal (fused tail) vertebrae. The researchers asked, Does the constellation of Hox gene expression correlate with the type of vertebra formed (e.g., cervical or thoracic) or with the relative position of the vertebrae (e.g., number 8 or 9)?

The answer is that the constellation of Hox gene expression predicts the type of vertebra formed. In the mouse, the transition between cervical and thoracic vertebrae is between vertebrae 7 and 8; in the chick, it is between vertebrae 14 and 15 (Figure 11.45B). In both cases, the *Hox5* paralogues are expressed in the last cervical vertebra, while the anterior boundary of the *Hox6* paralogues extends to the first thoracic vertebra. Similarly, in both animals, the thoracic-lumbar transition is seen at the boundary between the *Hox9* and *Hox10* paralogous groups. It appears there is a code of differing Hox gene expression along the anterior-posterior axis, and that code determines the type of vertebra formed.

The Dorsal-Ventral and Right-Left Axes in Mice

The dorsal-ventral axis

Very little is known about the mechanisms of dorsal-ventral axis formation in mammals. In mice and humans, the hypoblast forms on the side of the inner cell mass that is exposed to the blastocyst fluid, while the dorsal axis forms from those ICM cells that are in contact with the trophoblast and amnionic cavity. Thus, the dorsal-ventral axis of the embryo is defined, in part, by the embryonic-abembryonic axis of the blastocyst. The embryonic region contains the ICM, while the abembryonic region is that part of the blastocyst opposite the ICM.

There is a big debate over the relationship of the first cleavage plane and the allocation of cells to the embryonic and abembryonic regions of the trophoblast. Most developmental biologists think that it is a matter of chance which cells become trophoblast and which become inner cell mass. Indeed, time-lapse photomicroscopy appears to show that the plane of the first division is totally irrelevant to the allocation of blastomeres along the embryonic/abembryonic axis, and that the descendants of both of the first two blastomeres contribute to the trophoblast and to the inner cell mass (Hiiragi and Solter 2004; Motosugi et al. 2005). These investigators conclude that the embryo has no predetermined polarity, and that the embryo reacts to variations in the environment to coordinate its axis with that of the uterus. The first dorsal-ventral polarity is seen at the blastocyst stage, and as development proceeds, the notochord maintains dorsal-ventral polarity by inducing specific patterns of gene expression in the overlying ectoderm (Goulding et al. 1993).

However, some other scientists claim that the first cleavage plane specifies the embryonic/abembryonic axis, and that one cell of the first two blastomeres seems biased to become ICM, while the other cell appears to be biased to form the cells of the trophoblast (Figure 11.46; Gardner 2001; Piotrowska and Zernicka-Goetz 2001; Plusa et al. 2002, 2005).

The left-right axis

The mammalian body is not symmetrical. Although the human heart begins its formation at the midline of the embryo, it moves to the left side of the chest cavity and loops to the right. The spleen is found solely on the left side

(A)

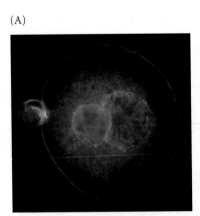

(B)

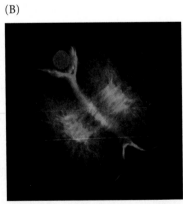

(C)

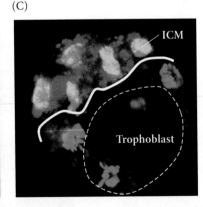

FIGURE 11.46 First cleavage and axis formation in the developing mouse. (A) Triple-stained mouse egg showing nuclear apposition. The DNA of the nuclei is stained blue, the microtubules are green, and the actin microfilaments are red. The second polar body is seen to the left of the zygote. (B) Telophase of the first cell division, stained as in (A). (C) Blastocyst of a mouse in which one of the two daughter cells was labeled with red dye while the other daughter cell was labeled with blue dye. In this case, one cell gave rise to most of the ICM cells while the other cell gave rise to most of the trophoblast. (The artifactual white line would represent the cells formed by the first cleavage division.) Some laboratories say that this is the normal condition, while other laboratories say that the orientation of the first two blastomeres is random and that each of the first cleavage blastomeres gives rise to both ICM and trophoblast cells. (A,B from Hiiragi and Solter 2004, courtesy of D. Solter; C from Plusa et al. 2005, courtesy of M. Zernicka-Goetz.)

of the abdomen, the major lobe of the liver forms on the right side of the abdomen, the large intestine loops right to left as it traverses the abdominal cavity, and the right lung has one more lobe than the left lung (Figure 11.47).

Mutations in mice have shown that there are two levels of regulation of the left-right axis: a global level, and an organ-specific level. Mutation of the gene *situs inversus viscerum* (*iv*) randomizes the left-right axis for each asymmetrical organ independently (Hummel and Chapman 1959; Layton 1976). This means that the heart may loop to the left in one homozygous animal, but to the right in another. Moreover, the direction of heart looping is not coordinated with the placement of the spleen or the stomach. This lack of coordination can cause serious problems, even death. A second gene, *inversion of embryonic turning* (*inv*), causes a more global phenotype. Mice homozygous for an

insertion mutation at this locus had all their asymmetrical organs on the wrong side of the body (Yokoyama et al. 1993).* Since all the organs were reversed, this asymmetry did not have dire consequences for the mice.

Several additional asymmetrically expressed genes have recently been discovered, and their influence on one another has enabled scientists to arrange them into a possible pathway. The end of this pathway—the activation of Nodal proteins and the Pitx2 transcription factor on the left side of

*This gene was discovered accidentally when Yokoyama and colleagues (1993) randomly inserted the transgene for the tyrosinase enzyme into the genomes of mice. In one instance, the transgene inserted itself into a region of chromosome 4, knocking out the existing *inv* gene. The resulting homozygous mice had laterality defects.

FIGURE 11.47 Left-right asymmetry in the developing human. (A) Abdominal cross sections show that the originally symmetrical organ rudiments acquire asymmetric positions by week 11. The liver moves to the right and the spleen moves to the left. (B) Not only does the heart move to the left side of the body, but the originally symmetrical veins of the heart regress differentially to form the superior and inferior venae cavae, which connect only to the right side of the heart. (C) The right lung branches into three lobes, while the left lung (near the heart) forms only two lobes. In human males, the scrotum also forms asymmetrically. (After Kosaki and Casey 1998.)

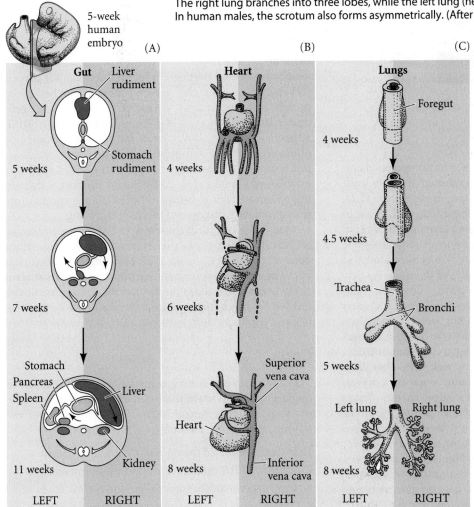

(A)

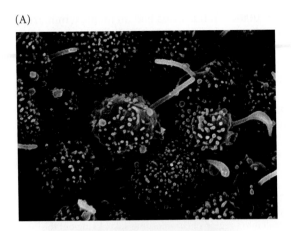

(B)

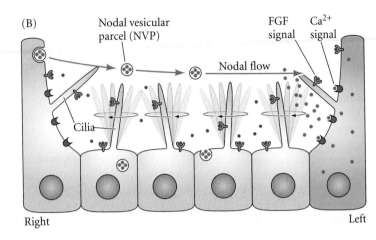

Nodal vesicular parcel (NVP)

FGF signal

Ca²⁺ signal

Nodal flow

Cilia

Right

Left

(C)

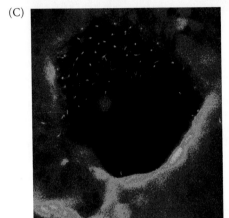

Right

Left

FIGURE 11.48 Situs formation in mammals. (A) Ciliated cells of the mammalian node. (B) Schematic drawing showing the FGF-induced secretion of nodal vesicular parcels from the cells of the node, the movement of the NVPs to the left side, motivated by the ciliary currents, and the rise in Ca²⁺ concentration on the left side of the node. (C) Calcium ions (red, green) concentrated on the left side of the node in mice. (A photograph courtesy of K. Sulik and G. C. Schoenwolf; B after Tanaka et al. 2005; C courtesy of M. Bueckner.)

the lateral plate mesoderm—appears to be the same in mammals as in frog and chick embryos, although the path leading to this point differs among the species (see Figure 11.26; Collignon et al. 1996; Lowe et al. 1996; Meno et al. 1996). In frogs, the pathway begins with the placement of Vg1; in chicks it begins with the suppression of *sonic hedgehog* expression. In mammals, the distinction between left and right sides begins in the ciliary cells of the node (Figure 11.48A). The cilia cause fluid in the node to flow from right to left. When Nonaka and colleagues (1998) knocked out a mouse gene encoding the ciliary motor protein dynein (see Chapter 7), the nodal cilia did not move, and the situs (lateral position) of each asymmetrical organ was randomized.

This finding correlated extremely well with other data. First, it had long been known that humans having a dynein deficiency had immotile cilia and a random chance of having their hearts on the left or right side of the body (Afzelius 1976). Second, when the *iv* gene described above was cloned, it was found to encode the ciliary dynein protein (Supp et al. 1997). Third, when Nonaka and colleagues (2002) cultured early mouse embryos under an artificial flow of medium from left to right, they obtained a reversal of the left-right axis. Moreover, the flow was able to

direct the polarity of the left-right axis in *iv* mutant mice, whose cilia are otherwise immotile.

Why should fluid flow be at all important to left-right asymmetry? The reason may reside in small (around 1 μm) membrane-bound particles called **nodal vesicular parcels** (**NVPs**). These "parcels," which contain Sonic hedgehog protein and retinoic acid, are secreted from the node cells under the influence of FGF signals (Figure 11.48B). It appears that ciliary flow carries the NVPs to the left side of the body; if FGF signaling is inhibited, the parcels are not secreted and left-right asymmetry fails to become established (Tanaka et al. 2005). Such a method of delivering paracrine factors from one set of cells to another represents a newly discovered mode of signaling.

One of the results of the transport of the nodal vesicular parcels is the rise of calcium ions on the left side of the node (Figure 11.48C; Levin 2003; McGrath et al. 2003). It is yet to be discovered how the expression of genes such as *Nodal* become placed under the control of these ion fluxes; but we are just beginning to understand the differences between right and left.

Coda

The different vertebrate groups have evolved variations on particular important themes during development (Figure 11.49; Solnicka-Krezel 2005). The major themes of vertebrate gastrulation include:

1. Internalization of the endoderm and mesoderm
2. Epiboly of the ectoderm around the entire embryo
3. Convergence of the internal cells to the midline
4. Extension of the body along the anterior-posterior axis

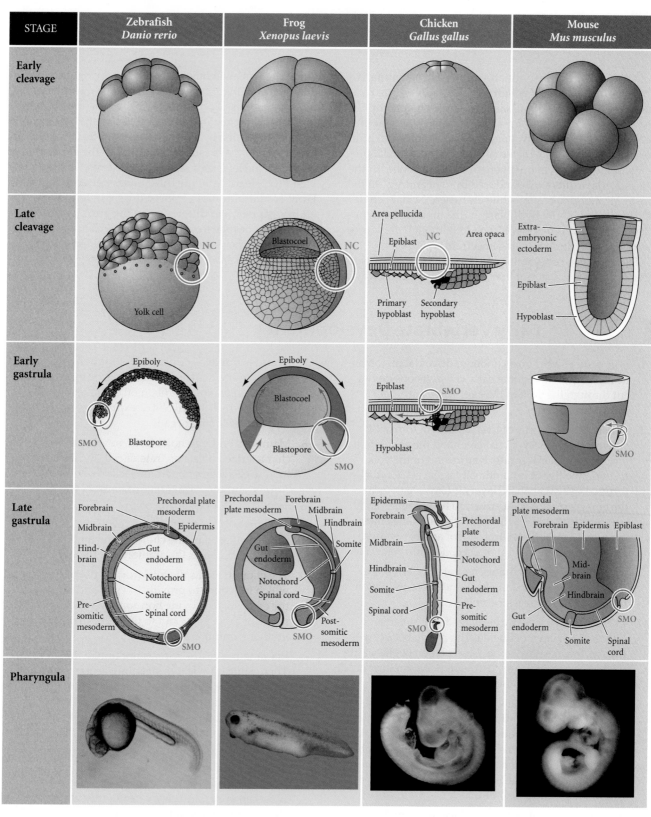

FIGURE 11.49 Overview of the early development of four vertebrate classes. Cleavage differs greatly between the four groups. Zebrafish and chicks have meroblastic discoidal cleavage; frogs have unequal holoblastic cleavage; and mammals have equal holoblastic cleavage. These cleavage patterns form different structures, but each has a Nieuwkoop center (NC; green circles). As gastrulation begins, each of the groups has cells that are equivalent to the Spemann-Mangold organizer (SMO; red circles). The SMO marks the beginning of the blasto-pore region, and the remainder of the blastopore is indicated by the red arrows extending from the organizer. By the late gastrula stage, the endoderm (yellow) is inside the embryo, the ectoderm (blue, purple) surrounds the embryo, and the mesoderm (red) is between the endoderm and ectoderm. The regionalization of the mesoderm has also begun. The photographs in the bottom row show the pharyngula stage that immediately follows gastrulation in each vertebrate group. (After Solnicka-Krezel 2005.)

Although fish, amphibian, avian, and mammalian embryos have different patterns of cleavage and gastrulation, they use many of the same molecules to accomplish the same goals. Each group uses gradients of Nodal proteins to establish polarity along the dorsal-ventral axis. In *Xenopus* and zebrafish, maternal factors induce Nodal proteins in the vegetal hemisphere or marginal zone. In the chick, Nodal expression is induced by Wnt and Vg1 in the posterior marginal zone, while elsewhere Nodal activity is suppressed by the hypoblast. In the mouse, the hypoblast similarly restricts Nodal activity, although the source of its ability to do so remains uncertain.

Each of these vertebrate groups uses BMP inhibitors to specify the dorsal axis, but they use them in different ways.

Similarly, Wnt antagonists and Otx2 expression are important in specifying the anterior regions of the embryo, but different groups of cells may be expressing these proteins. In all cases, the region of the body from the hindbrain to the tail is specified by Hox genes. Finally, the left-right axis of these embryos is established through the expression of Nodal on the left-hand side of the embryo. This will activate Pitx2, leading to the differences between the sides. How Nodal becomes expressed on the left side of the embryo appears to differ among the various vertebrate groups. But despite their initial differences in cleavage and gastrulation, the vertebrates have maintained very similar ways of establishing the three bodily axes.

Snapshot Summary **Early Vertebrate Development**

1. Fish, reptiles, and birds undergo discoidal meroblastic cleavage, wherein the early cell divisions do not cut through the yolk of the egg. These early cells form a blastoderm.

2. In fish, the deep cells of the blastoderm form between the yolk syncytial layer and the enveloping layer. These cells migrate over the top of the yolk, forming the hypoblast and epiblast layers. On the future dorsal side, these layers intercalate to form the embryonic shield, a structure homologous to the amphibian organizer. Transplantation of the embryonic shield into the ventral side of another embryo will cause the formation of a second embryonic axis.

3. In each class of vertebrates, neural ectoderm is permitted to form where the BMP-mediated induction of epidermal tissue is prevented

4. In chick embryos, early cleavage forms an area opaca and an area pellucida. The region between them is the marginal zone. Gastrulation begins at the posterior marginal zone, as the hypoblast and primitive streak both start there.

5. The primitive streak is derived from anterior epiblast cells and the central cells of the posterior marginal zone. As the primitive streak extends rostrally, Hensen's node is formed. Cells migrating through Hensen's node become prechordal mesendoderm and is followed by the head process and notochord cells.

6. The prechordal plate helps induce the formation of the forebrain; the chordamesoderm induces the formation of the midbrain, hindbrain, and spinal cord. The first cells migrating laterally through the primitive streak become endoderm, displacing the hypoblast. The mesoderm cells then migrate through the primitive streak. Meanwhile, the surface ectoderm undergoes epiboly around the entire yolk.

7. In birds, gravity is critical in determining the anterior-posterior axis. The left-right axis is formed by the expression of Nodal protein on the left side of the embryo, which signals Pitx2 expression on the left side of developing organs.

8. Mammals undergo holoblastic rotational cleavage, characterized by a slow rate of cell division, a unique cleavage orientation, lack of divisional synchrony, and the formation of a blastocyst.

9. The blastocyst forms after the blastomeres undergo compaction. It contains outer cells—the trophoblast cells—that become the chorion, and an inner cell mass that becomes the amnion and the embryo.

10. The chorion forms the fetal portion of the placenta, which functions to provide oxygen and nutrition to the embryo, to provide hormones for the maintenance of pregnancy, and to provide barriers to the mother's immune system.

11. Mammalian gastrulation is not unlike that of birds. There appear to be two signaling centers, one in the node and one in the anterior visceral endoderm. The latter center is critical for generating the forebrain, while the former is critical in inducing the axial structures caudally from the midbrain.

12. Hox genes pattern the anterior-posterior axis and help to specify positions along that axis. If Hox genes are knocked out, segment-specific malformations can arise. Similarly, causing the ectopic expression of Hox genes can alter the body axis.

13. The homology of gene structure and the similarity of expression patterns between *Drosophila* and mammalian Hox genes suggests that this patterning mechanism is extremely ancient.

14. The mammalian left-right axis is specified similarly to that of the chick, but with some significant differences in the roles of certain genes.

For Further Reading

Complete bibliographical citations for all literature cited in this chapter can be found on the Vade Mecum CD that accompanies the book and at the free access website www.devbio.com

Beddington, R. S. P. and E. J. Robertson. 1999. Axis development and early asymmetry in mammals. *Cell* 96: 195–209.

Bertocchini, F. and C. D. Stern. 2002. The hypoblast of the chick embryo positions the primitive streak by antagonizing Nodal signaling. *Dev. Cell* 3: 735–744.

Burke, A. C., A. C. Nelson, B. A. Morgan and C. Tabin. 1995. Hox genes and the evolution of vertebrate axial morphology. *Development* 121: 333–346.

Essner, J. J., J. D. Amack, M. K. Nyholm, E. B. Harris and H. J. Yost. 2005. Kupffer's vesicle is a ciliated organ of asymmetry in the zebrafish embryo that initiates left-right development of the brain, heart and gut. *Development* 132: 1247–1260.

Haffter, P. and 16 others. 1996. The identification of genes with unique and essential functions in the development of the zebrafish, *Danio rerio. Development* 123: 1–36.

Langeland, J. and C. B. Kimmel. 1997. The embryology of fish. *In* S. F. Gilbert and A. M. Raunio (eds), *Embryology: Constructing the Organism*. Sinauer Associates, Sunderland, MA, pp. 383–407.

Markert, C. L. and R. M. Petters. 1978. Manufactured hexaparental mice show that adults are derived from three embryonic cells. *Science* 202: 56–58.

Sheng, G., M. dos Reis and C. D. Stern. 2003. Churchill, a zinc finger transcriptional activator, regulates the transition between gastrulation and neurulation. *Cell* 115: 603–613.

Strumpf, D. C.-A. Mao, Y. Yamanaka, A. Ralston, K. Chawengsaksophak, F. Beck and J. Rossant. 2005. Cdx2 is required for correct cell fate specification and differentiation of trophectoderm in the mouse blastocyst. *Development* 132: 2093–2102.

PART 3

Later Embryonic Development

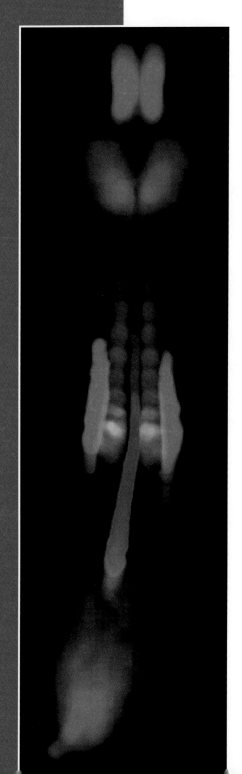

Part 3   **Later Embryonic Development**

PREVIOUS PAGE
Gene expression patterns in a 12-somite chick embryo (about 33 hours). In situ hybridization was performed with probes binding to *chordin* mRNA (blue) in the notochord, *paraxis* mRNA (green) in the somites, and *Pax2* mRNA (red) in the intermediate mesoderm, as well as in the ectoderm of the ears and the brain. From Denkers et al., *Developmental Dynamics* 229 (2004): 651–657. Photograph courtesy of T. J. Mauch.

The Emergence of the Ectoderm: Central Nervous System and Epidermis

12

"**WHAT IS PERHAPS THE MOST INTRIGUING** question of all is whether the brain is powerful enough to solve the problem of its own creation." So Gregor Eichele ended a review of research on mammalian brain development (Eichele 1992). The construction of an organ that perceives, thinks, loves, hates, remembers, changes, fools itself, and coordinates our conscious and unconscious bodily processes is undoubtedly the most challenging of all developmental enigmas. A combination of genetic, cellular, and organismal approaches is giving us a preliminary understanding of how the basic anatomy of the brain becomes ordered.

The fates of the vertebrate ectoderm are shown in Figure 12.1. In the past two chapters, we have seen how the ectoderm is instructed to form the vertebrate nervous system and epidermis. A portion of the dorsal ectoderm is specified to become neural ectoderm, and its cells become distinguishable by their columnar appearance. This region of the embryo is called the **neural plate**. The process by which this tissue forms a **neural tube**—the rudiment of the central nervous system—is called **neurulation**, and an embryo undergoing such changes is called a **neurula** (Figure 12.2). The neural tube forms the brain anteriorly and the spinal cord posteriorly. This chapter will look at the processes by which the neural tube and the epidermis arise and acquire their distinctive patterns.

Establishing the Neural Cells

In previous chapters we have looked at the competency and specification of the ectoderm. This chapter follows the story of the committed neural and epidermal precursor cells. Neural cells become specified through their interactions with other cells. There are at least four stages through which the pluripotent cells of the epiblast or blastula become neural precursor cells, or **neuroblasts** (Wilson and Edlund 2001):

- **Competence**, wherein cells can become neuroblasts if they are exposed to the appropriate combination of signals.
- **Specification**, wherein cells have received the appropriate signals to become neuroblasts, but progression along the neural differentiation pathway can still be repressed by other signals.
- **Commitment** (**determination**), wherein neuroblasts have entered the neural differentiation pathway and will become neurons even in the presence of inhibitory signals.
- **Differentiation**, wherein the neuroblasts leave the mitotic cycle and express those genes characteristic of neurons.

> 66 For the real amazement, if you wish to be amazed, is this process. You start out as a single cell derived from the coupling of a sperm and an egg; this divides in two, then four, then eight, and so on, and at a certain stage there emerges a single cell which has as all its progeny the human brain. The mere existence of such a cell should be one of the great astonishments of the earth. People ought to be walking around all day, all through their waking hours calling to each other in endless wonderment, talking of nothing except that cell. 99
>
> *LEWIS THOMAS (1979)*

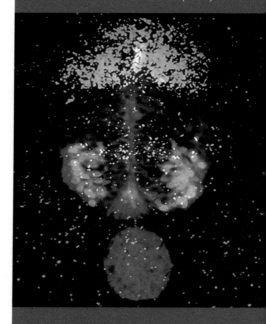

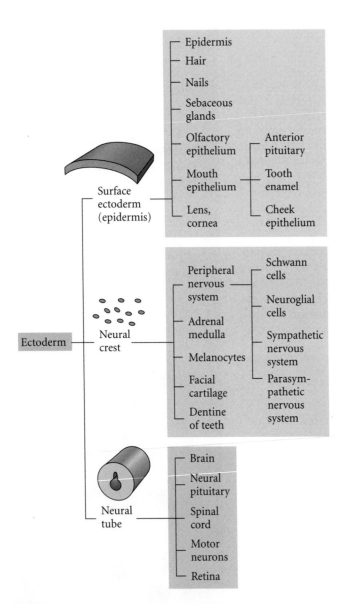

FIGURE 12.1 Major derivatives of the ectoderm germ layer. The ectoderm is divided into three major domains: the surface ectoderm (primarily epidermis), the neural crest (peripheral neurons, pigment, facial cartilage), and the neural tube (brain and spinal cord).

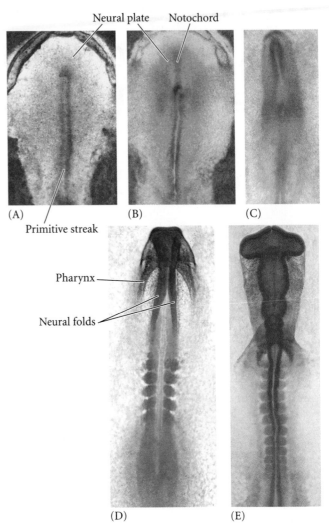

FIGURE 12.2 Neurulation in a chick embryo (dorsal view). (A) Flat neural plate. (B) Flat neural plate with underlying notochord (head process). (C) Neural groove. (D) Neural folds begin closing at the dorsalmost region, forming the incipient neural tube. (E) Neural tube, showing the three brain regions and the spinal cord. The neural tube remains open at the anterior end, and the optic bulges (which become the retinas) have extended to the lateral margins of the head. (F) 24-hour chick embryo, as in (D). The cephalic (head) region has undergone neurulation, while the caudal (tail) regions are still undergoing gastrulation. (A–E, photographs courtesy of G. C. Schoenwolf; F after Patten 1971.)

Formation of the Neural Tube

There are two major ways of converting the neural plate into a neural tube. In **primary neurulation**, the cells surrounding the neural plate direct the neural plate cells to proliferate, invaginate, and pinch off from the surface to form a hollow tube. In **secondary neurulation**, the neural tube arises from the coalescence of mesenchyme cells into a solid cord that subsequently forms cavities that coalesce to create a hollow tube. In general, the anterior portion of the neural tube is made by primary neurulation, while the

(F)

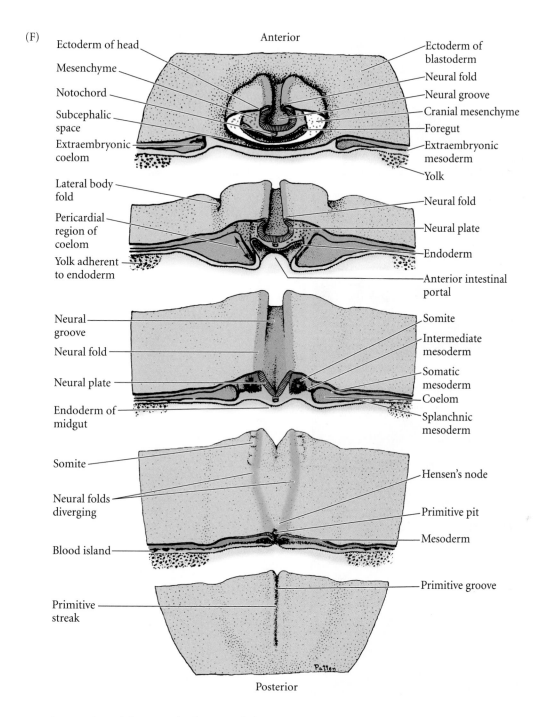

Anterior

Ectoderm of head
Mesenchyme
Notochord
Subcephalic space
Extraembryonic coelom

Ectoderm of blastoderm
Neural fold
Neural groove
Cranial mesenchyme
Foregut
Extraembryonic mesoderm
Yolk

Lateral body fold
Pericardial region of coelom
Yolk adherent to endoderm

Neural fold
Neural plate
Endoderm
Anterior intestinal portal

Neural groove
Neural fold
Neural plate
Endoderm of midgut

Somite
Intermediate mesoderm
Somatic mesoderm
Coelom
Splanchnic mesoderm

Somite
Neural folds diverging
Blood island

Hensen's node
Primitive pit
Mesoderm

Primitive streak

Primitive groove

Posterior

posterior portion of the neural tube is made by secondary neurulation. The complete neural tube forms by joining these two separately formed tubes together.

In birds, all the neural tube anterior to the twenty-eighth somite pair (i.e., everything anterior to the hindlimbs) is made by primary neurulation (Pasteels 1937; Catala et al. 1996). In mammals, secondary neurulation begins at the level of the sacral vertebrae of the tail (Schoenwolf 1984; Nievelstein et al. 1993). In amphibians such as *Xenopus*, only the tail neural tube is derived from secondary neurulation (Gont et al. 1993); the same pattern occurs in fish (whose neural tubes were formerly thought to be formed

solely by secondary neurulation) (Lowery and Sive 2004; Hatten, personal communication). The convergence of zebrafish ectoderm at the midline and the formation of its neural tube is depicted in the series of photographs on the cover of this book.

Primary neurulation

The events of primary neurulation in the chick are illustrated in Figure 12.3. This process divides the original ectoderm into three sets of cells: (1) the internally positioned neural tube, which will form the brain and spinal cord, (2)

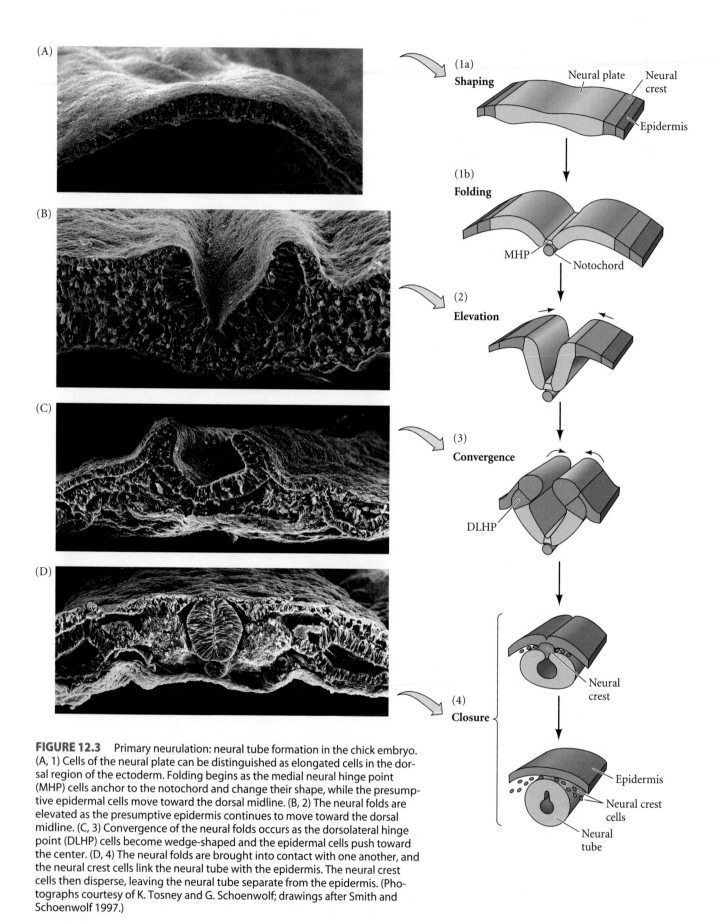

FIGURE 12.3   Primary neurulation: neural tube formation in the chick embryo. (A, 1) Cells of the neural plate can be distinguished as elongated cells in the dorsal region of the ectoderm. Folding begins as the medial neural hinge point (MHP) cells anchor to the notochord and change their shape, while the presumptive epidermal cells move toward the dorsal midline. (B, 2) The neural folds are elevated as the presumptive epidermis continues to move toward the dorsal midline. (C, 3) Convergence of the neural folds occurs as the dorsolateral hinge point (DLHP) cells become wedge-shaped and the epidermal cells push toward the center. (D, 4) The neural folds are brought into contact with one another, and the neural crest cells link the neural tube with the epidermis. The neural crest cells then disperse, leaving the neural tube separate from the epidermis. (Photographs courtesy of K. Tosney and G. Schoenwolf; drawings after Smith and Schoenwolf 1997.)

the externally positioned epidermis of the skin, and (3) the neural crest cells (see Figure 12.1). The neural crest cells form in the region that connects the neural tube and epidermis; but they migrate to new locations where they will generate the peripheral neurons and glia, the pigment cells of the skin, and several other cell types (see Chapter 13).

The process of primary neurulation appears to be similar in all vertebrates; Figure 12.4 illustrates the process in amphibians (Gallera 1971). Shortly after the neural plate has formed, its edges thicken and move upward to form the **neural folds**, while a U-shaped **neural groove** appears in the center of the plate, dividing the future right and left sides of the embryo. The neural folds migrate toward the midline of the embryo, eventually fusing to form the neural tube beneath the overlying ectoderm. The cells at the dorsalmost portion of the neural tube become the neural crest cells.

Primary neurulation can be divided into four distinct but spatially and temporally overlapping stages: (1) *forma-*

tion of the neural plate; (2) *shaping* of the neural plate; (3) *bending* of the neural plate to form the neural groove; and (4) *closure* of the neural groove to form the neural tube (see Figure 12.3; Smith and Schoenwolf 1997; Colas and Schoenwolf 2001).

> **VADE MECUM² Chick neurulation.** By 33 hours of incubation, neurulation in the chick embryo is well underway. Both whole mounts and a complete set of serial cross sections through a 33-hour chick embryo are included in this segment so that you can see this amazing event. The serial sections can be displayed either as a continuum in movie format or individually, along with labels and color-coding that designates germ layers. [Click on Chick-Mid]

FORMATION AND SHAPING OF THE NEURAL PLATE   The process of neurulation begins when the underlying dorsal mesoderm (and the pharyngeal endoderm in the head region) signals the ectodermal cells above it to elongate into columnar

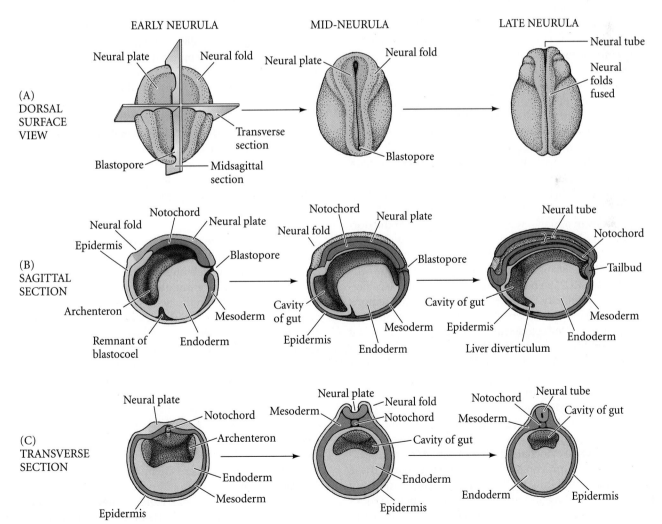

FIGURE 12.4   Three views of neurulation in an amphibian embryo, showing early (left), middle (center), and late (right) neurulae in each case. (A) Looking down on the dorsal surface of the whole embryo. (B) Sagittal section through the medial plane of the embryo. (C) Transverse section through the center of the embryo. (After Balinsky 1975.)

neural plate cells (Smith and Schoenwolf 1989; Keller et al. 1992). Their elongated shape distinguishes the cells of the prospective neural plate from the flatter pre-epidermal cells surrounding them. As much as 50 percent of the ectoderm is included in the neural plate.

The neural plate is shaped by the movements of the epidermal and neural plate regions and lengthens along the anterior-posterior axis. The neural plate lengthens and narrows by convergent extension, intercalating several layers of cells into a few layers. These convergence and extension movements are critical for the shaping of the neural plate, and mutations disturbing these movements can block neural tube closure (Ueno and Greene 2003).

In addition, divisions of the neural plate cells are preferentially in the **rostral-caudal** (beak-tail; anterior-posterior) direction (Jacobson and Sater 1988; Schoenwolf and Alvarez 1989; Sausedo et al. 1997). These events will occur even if the tissues involved are separated. If the neural plate is isolated, its cells converge and extend to make a thinner plate, but fail to roll up into a neural tube. However, if the "border region" containing both presumptive epidermis and neural plate tissue is isolated, it will form small neural folds in culture (Jacobson and Moury 1995; Moury and Schoenwolf 1995).

BENDING OF THE NEURAL PLATE The bending of the neural plate involves the formation of hinge regions where the neural plate contacts surrounding tissues. In birds and mammals, the cells at the midline of the neural plate are called the **medial hinge point (MHP) cells**. They are derived from the portion of the neural plate just anterior to Hensen's node and from the anterior midline of Hensen's node (Schoenwolf 1991a,b; Catala et al. 1996). The MHP cells become anchored to the notochord beneath them and form a hinge, which forms a furrow at the dorsal midline. The notochord induces the MHP cells to decrease their height and to become wedge-shaped (van Straaten et al. 1988; Smith and Schoenwolf 1989). The cells lateral to the MHP do not undergo such a change (see Figure 12.3B,C). Shortly thereafter, two other hinge regions form furrows near the connection of the neural plate with the remainder of the ectoderm. These regions are called the **dorsolateral hinge points (DLHPs)**, and they are anchored to the surface ectoderm. The DLHP cells, too, increase their height and become wedge-shaped.

Cell wedging is intimately linked to changes in cell shape. In the DLHPs, microtubules and microfilaments are both involved in these changes. Colchicine, an inhibitor of microtubule polymerization, inhibits the elongation of these cells, while cytochalasin B, an inhibitor of microfilament formation, prevents the apical constriction of these cells, thereby inhibiting wedge formation (Burnside 1973; Karfunkel 1972; Nagele and Lee 1987). As the neuroepithelial cells enlarge (by means of the microtubules), the microfilaments and their associated regulatory proteins accumulate at the apical ends of the cells (Zolessi and Arruti 2001). Constriction of these microfilaments allows the change of

cell shape to occur. After the initial furrowing of the neural plate, the plate bends around these hinge regions. Each hinge acts as a pivot that directs the rotation of the cells around it (Smith and Schoenwolf 1991). In *Xenopus*, the actin-binding protein Shroom is critical in initiating this apical constriction to bend the neural plate (Haigo et al. 2003).

Meanwhile, extrinsic forces are also at work. The surface ectoderm of the chick embryo pushes toward the midline of the embryo, providing another motive force for the bending of the neural plate (see Figure 12.3C; Alvarez and Schoenwolf 1992; Lawson et al. 2001). This movement of the presumptive epidermis and the anchoring of the neural plate to the underlying mesoderm may also be important for ensuring that the neural tube invaginates *inward*, or into the embryo and not outward. If small pieces of neural plate are isolated from the rest of the embryo (including the mesoderm), they tend to roll inside out (Schoenwolf 1991a). The pushing of the presumptive epidermis toward the center and the furrowing of the neural tube creates the neural folds.

CLOSURE OF THE NEURAL TUBE The neural tube closes as the paired neural folds are brought together at the dorsal midline. The folds adhere to each other, and the cells from the two folds merge. In some species, the cells at this junction form the neural crest cells. In birds, the neural crest cells do not migrate from the dorsal region until after the neural tube has closed at that site. In mammals, however, the cranial neural crest cells (which form facial and neck structures; see Chapter 13) migrate while the neural folds are still being elevated (i.e., prior to neural tube closure), whereas in the spinal cord region, the neural crest cells do not migrate until closure has occurred (Nichols 1981; Erickson and Weston 1983).

The closure of the neural tube does not occur simultaneously throughout the ectoderm. This phenomenon is best seen in those vertebrates (such as birds and mammals) whose body axis is elongated prior to neurulation. In amniotes, induction in the head starts before induction in the trunk, so in the 24-hour chick embryo, neurulation in the **cephalic** (head) region is well advanced, while the **caudal** (tail) region of the embryo is still undergoing gastrulation (see Figure 12.2E,F). The two open ends of the neural tube are called the **anterior neuropore** and the **posterior neuropore**.

Unlike neurulation in chicks, in which neural tube closure is initiated at the level of the future midbrain and "zips up" in both directions, neural tube closure in mammals is initiated at several places along the anterior-posterior axis. In humans, there are probably three sites of neural tube closure (Figure 12.5A–C; Nakatsu et al. 2000; O'Rahilly and Muller 2002). Different **neural tube defects** are caused when various parts of the neural tube fail to close. Failure to close the **posterior neuropore** around day 27 results in a condition called **spina bifida**, the severity of which depends on how much of the spinal cord remains

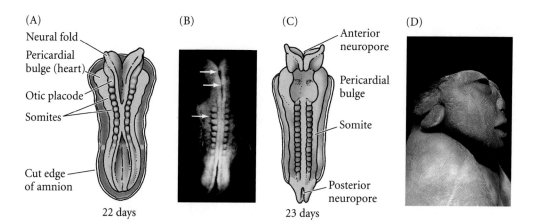

(A)

Neural fold
Pericardial bulge (heart)
Otic placode
Somites

Cut edge of amnion

22 days

(B)

(C)

Anterior neuropore
Pericardial bulge
Somite
Posterior neuropore

23 days

(D)

FIGURE 12.5   Neurulation in the human embryo. (A) Dorsal view of a 22-day (8-somite) human embryo initiating neurulation. Both anterior and posterior neuropores are open to the amniotic fluid. (B) 10-somite human embryo showing the three major sites of neural tube closure (arrows). (C) Dorsal view of a neurulating human embryo with only its neuropores open. (D) Photograph of a stillborn infant with anencephaly. (B from Nakatsu et al. 2000; D courtesy of National March of Dimes.)

exposed. Failure to close sites 2 and 3 in the rostral neural tube keeps the *anterior* neuropore open. This failure results in a lethal condition, **anencephaly**, in which the forebrain remains in contact with the amnionic fluid and subsequently degenerates (Figure 12.5D). The fetal forebrain development ceases, and the vault of the skull fails to form. The failure of the entire neural tube to close over the entire body axis is called **craniorachischisis**. Collectively, neural tube defects are not rare in humans, as they are seen in about 1 in every 1000 live births. Neural tube closure defects can often be detected during pregnancy by various physical and chemical tests.

The neural tube eventually forms a closed cylinder that separates from the surface ectoderm. This separation appears to be mediated by the expression of different cell adhesion molecules. Although the cells that will become the neural tube originally express E-cadherin, they stop producing this protein as the neural tube forms, and instead synthesize N-cadherin and N-CAM (Figure 12.6A). As a result, the surface ectoderm and neural tube tissues no longer adhere to each other. If the surface ectoderm is experimentally made to express N-cadherin (by injecting N-cadherin mRNA into one cell of a 2-cell *Xenopus* embryo), the separation of the neural tube from the presumptive epidermis is dramatically impeded (Figure 12.6B; Detrick et al. 1990; Fujimori et al. 1990).

Human neural tube closure requires a complex interplay between genetic and environmental factors (Cabrera et al. 2005). Certain genes, such as *Pax3*, *Sonic hedgehog*, and *openbrain*, are essential for the formation of the mammalian neural tube; but dietary factors, such as cholesterol and folate (also known as folic acid or vitamin B_{12}), also appear to be critical. It has been estimated that over 50 percent of human neural tube defects could be prevented by a pregnant woman's taking supplemental folate, and the U.S. Public Health Service recommends that all women of childbearing age take 0.4 milligrams of folate daily to reduce their risk of having a child with neural tube defects (Milun-

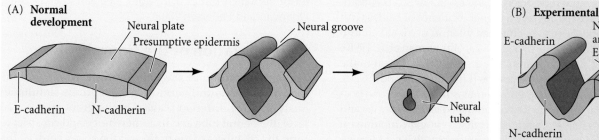

(A) **Normal development**

Neural plate
Presumptive epidermis
Neural groove

E-cadherin   N-cadherin

Neural tube

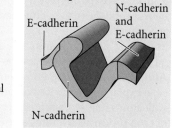

(B) **Experimental**

E-cadherin
N-cadherin and E-cadherin
N-cadherin

FIGURE 12.6   Expression of N- and E-cadherin adhesion proteins during neurulation in *Xenopus*. (A) Normal development. In the neural plate stage, N-cadherin is seen in the neural plate, while E-cadherin is seen on the presumptive epidermis. Eventually, the N-cadherin-bearing neural cells separate from the E-cadherin-containing epidermal cells. (The neural crest cells express neither N- nor E-cadherin, and they disperse.) (B) No separation of the neural tube occurs when one side of the frog embryo is injected with N-cadherin mRNA, so that N-cadherin is expressed in the epidermal cells as well as in the presumptive neural tube.

(A)   (B)   (C)

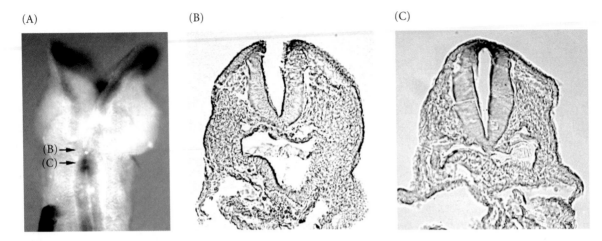

FIGURE 12.7 Folate-binding protein in the neural folds as neural tube closure occurs. (A) 10-somite mouse embryo stained for folate-binding protein mRNA. (B,C) Sections through embryo (A) at the two arrows, showing the folate-binding protein (dark blue) at the points of neural tube closure. (From Saitsu et al. 2003; photographs courtesy of K. Shiota.)

sky et al. 1989; Czeizel and Dudas 1992; Centers for Disease Control 1992). While the mechanism by which folate facilitates neural tube closure is not understood, recent studies have demonstrated that there is a folate receptor protein on the dorsalmost regions of the mouse neural tube immediately prior to fusion (Figure 12.7; Saitsu et al. 2003). Rothenberg and colleagues (2004) showed that most women who delivered babies with neural tube defects had antibodies against this protein, whereas such antibodies were seldom found in women whose babies did not have open neural tubes. Mice with mutations in the gene encoding this folate receptor were born with a high incidence of neural tube abnormalities; the incidence of these neural tube defects decreased if their mothers were given folate supplements while pregnant (Piedrahita et al. 1999; Spiegelstein et al. 2004).

However, folate deficiency appears to be only one risk factor for neural tube defects. Mothers in low socioeconomic groups appear to have a higher incidence of babies with neural tube defects, even when vitamin use is taken into account (Little and Elwood 1992a; Wasserman et al. 1998). Moreover, there appears to be a seasonal variation in the incidence of neural tube defects (Little and Elwood 1992b). Although this phenomenon remains mysterious, one possible explanation may be contaminated crops. Marasas and colleagues (2004) documented that a fungal contaminant of corn produces a teratogen (fumonisin) that causes neural tube failure by perturbing the function of several lipids and proteins—including the folate receptor protein. This fungus has been found in regions where neural tube defects are more prevalent than normal. Its teratogenic effects in mice can be reduced when pregnant mice are fed folate supplements.

Secondary neurulation

Secondary neurulation involves the segregation of the cells of the prospective **medullary cord** from cells of the prospective epidermis and prospective gut tissue (Figure 12.8A); the condensation of these mesenchymal cells forms the medullary cord beneath the surface ectoderm (Figure 12.8B); the cavitation of the central portion of this cord into several hollow spaces (Figure 12.8C); and the coalescence of all these cavities into a single central cavity (Figure 12.8D; Schoenwolf and Delongo 1980). In human and chick embryos, there appears to be a transitional region at the junction of the anterior ("primary") and posterior ("secondary") neural tubes. In this transition region in human embryos, coalescing of cavities is seen, but the neural tube also forms by the bending of neural plate cells. Knowledge of the mechanisms of secondary neurulation may be important in medicine, given the prevalence of human posterior spinal cord malformations (see Saitsu et al. 2004; Donovan and Pedersen 2005).

Differentiation of the Neural Tube

The differentiation of the neural tube into the various regions of the central nervous system occurs simultaneously in three different ways. On the gross anatomical level, the neural tube and its lumen bulge and constrict to form the chambers of the brain and the spinal cord. At the tissue level, the cell populations within the wall of the neural tube rearrange themselves to form the different functional regions of the brain and the spinal cord. Finally, on the cellular level, the neuroepithelial cells themselves differentiate into the numerous types of nerve cells (**neurons**) and supportive cells (**glia**) present in the body.

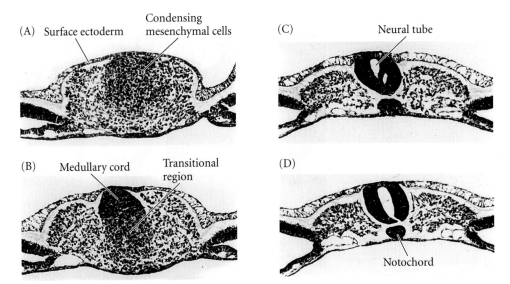

(A) Surface ectoderm / Condensing mesenchymal cells

(B) Medullary cord / Transitional region

(C) Neural tube

(D) Notochord

FIGURE 12.8 Secondary neurulation in the caudal region of a 25-somite chick embryo. (A) Mesenchymal cells condense to form the medullary cord at the most caudal end of the chick tail-bud. (B) The medullary cord at a slightly more anterior position in the tailbud. (C) The neural tube is cavitating and the noto- chord forming; note the presence of separate lumens. (D) The lumens coalesce to form the central canal of the neural tube. (From Catala et al. 1995; photographs courtesy of N. M. Le Douarin.)

The early development of most vertebrate brains is similar, but because the human brain may be the most organized piece of matter in the solar system and is arguably the most interesting organ in the animal kingdom, we will concentrate on the development that is supposed to make *Homo* sapient.

The anterior-posterior axis

The early mammalian neural tube is a straight structure. However, even before the posterior portion of the tube has formed, the most anterior portion of the tube is undergoing drastic changes. In this region, the neural tube balloons into the three primary vesicles: the forebrain (**prosencephalon**), midbrain (**mesencephalon**), and hindbrain (**rhombencephalon**) (Figure 12.9A). By the time the posterior end of the neural tube closes, secondary bulges—the

FIGURE 12.9 Early human brain development. The three primary brain vesicles are subdivided as development continues. At the right is a list of the adult derivatives formed by the walls and cavities of the brain. (After Moore and Persaud 1993.)

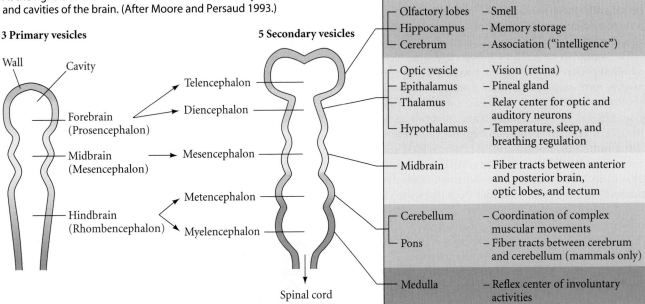

optic vesicles—have extended laterally from each side of the developing forebrain.

The prosencephalon becomes subdivided into the anterior **telencephalon** and the more caudal **diencephalon** (Figure 12.9B). The telencephalon will eventually form the **cerebral hemispheres**, and the diencephalon will form the **optic vesicles** that become the retina, as well as the **thalamic** and **hypothalamic** brain regions, which receive neural input from the retina. Indeed, the retina itself is a derivative of the diencephalon. The mesencephalon does not become subdivided, and its lumen eventually becomes the **cerebral aqueduct**. The rhombencephalon becomes subdivided into a posterior **myelencephalon** and a more anterior **metencephalon**. The myelencephalon eventually becomes the **medulla oblongata**, whose neurons generate the nerve centers ("nuclei") responsible for pain relay to the head and neck, auditory connections, tongue movement, and balance control, as well as respiratory, gastrointestinal, and cardiovascular movements. The metencephalon gives rise to the **cerebellum**, the part of the brain responsible for coordinating movements, posture, and balance.

The rhombencephalon develops a segmental pattern that specifies the places where certain nerves originate. Periodic swellings called **rhombomeres** divide the rhombencephalon into smaller compartments. The rhombomeres represent separate "territories" in that the cells within each rhombomere mix freely within it, but not with cells from adjacent rhombomeres (Guthrie and Lumsden 1991; Lumsden 2004). Moreover, each rhombomere has a different fate. The neural crest cells above each rhombomere will form **ganglia**—clusters of neuronal cell bodies whose axons form a nerve. The generation of the cranial nerves from the rhombomeres has been most extensively studied in the chick, in which the first neurons appear in the even-numbered rhombomeres, r2, r4, and r6 (Figure 12.10; Lumsden and Keynes 1989). Neurons originating from r2 ganglia form the fifth (trigeminal) cranial nerve; those from r4 form the seventh (facial) and eighth (vestibuloacoustic) cranial nerves; and the ninth (glossopharyngeal) cranial nerve exits from r6.

The zebrafish neural tube follows the same basic differentiation pattern as the mammalian neural tube (although the eyes form earlier). When red fluorescent dye was injected into the hindbrain as that vesicle was forming, the dye soon diffused into the midbrain and forebrain as well (Figure 12.11). This diffusion was dependent on the inflation of the lumen by the Na^+/K^+ ATPase, which sets up an osmotic gradient causing water to fill the ventricles (Lowery and Sive 2005). (A similar mechanism is used to fill the mammalian blastocoel.) Further shaping of the ventricles depends on the secretion of the cerebrospinal fluid and by regional cell proliferation and cell adhesion. Both the midbrain and hindbrain ventricles are shaped by lateral hinge regions where the epithelium bends sharply.

The ballooning of the early embryonic brain is remarkable in its rate, its extent, and in the fact that it is primarily the result of an increase in cavity size, not tissue growth. In

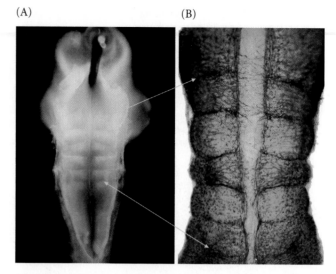

(A)   (B)

FIGURE 12.10 Rhombomeres of the chick hindbrain. (A) Hindbrain of a 3-day chick embryo. The roof plate has been removed so that the segmented morphology of the neural epithelium can be seen. The r1/r2 boundary is at the upper arrow, and the r6/r7 boundary is at the lower arrow. (B) A chick hindbrain at a similar stage stained with antibody to a neurofilament subunit. The rhombomere boundaries are emphasized because they serve as channels for neurons crossing from one side of the brain to the other. (After Lumsden 2004, photographs courtesy of A. Lumsden.)

the chick embryo, brain volume expands 30-fold between days 3 and 5 of development. This rapid expansion is thought to be caused by positive cerebrospinal fluid pressure exerted against the walls of the neural tube. It might be expected that this fluid pressure would be dissipated by the spinal cord, but this does not appear to happen. Rather, as the neural folds close in the region between the presumptive brain and the presumptive spinal cord, the surrounding dorsal tissues push inward to constrict the neural tube at the base of the brain (Figure 12.12; Schoenwolf and Desmond 1984; Desmond and Schoenwolf 1986; Desmond and Levitan 2002). This occlusion (which also occurs in the human embryo) separates the presumptive brain region from the future spinal cord (Desmond 1982). If the fluid pressure in the anterior portion of an occluded neural tube is experimentally removed, the chick brain enlarges at a much slower rate and contains many fewer cells than normal. The occluded region of the neural tube reopens after the initial rapid enlargement of the brain ventricles.

The anterior-posterior patterning of the nervous system is controlled by a series of genes that include the Hox gene complexes. These genes were discussed more fully in Chapter 11 and will be revisited in Chapter 23.

WEBSITE 12.2 Specifying the brain boundaries. The Pax transcription factors and the paracrine factor FGF8 are critical in establishing the boundaries of the forebrain, midbrain, and hindbrain.

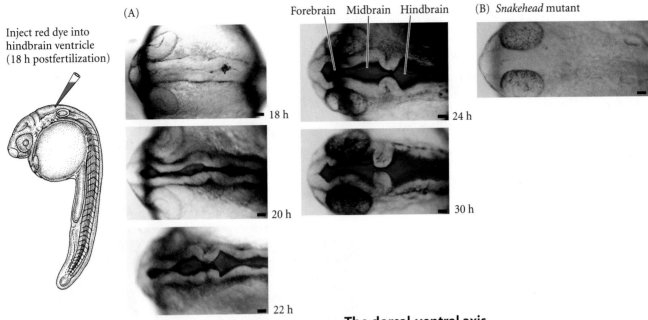

FIGURE 12.11 Brain ventricle formation in zebrafish. (A) When a fluorescent red dye is injected into the incipient hind-brain ventricle of 18-hour zebrafish embryos, the dye diffuses to the midbrain and forebrain regions. As development proceeds, the brain ventricles are shaped by regional differences in cell proliferation and adhesion. (B) In the *snakehead* mutant, Na^+/K^+ ATPase is deficient. The brain ventricles fail to form and dye injected into the presumptive hindbrain does not diffuse. (After Lowery and Sive 2005; photographs courtesy of H. Sive.)

The dorsal-ventral axis

The neural tube is polarized along its dorsal-ventral axis. In the spinal cord, for instance, the *dorsal* region is the place where the spinal neurons receive input from sensory neurons, while the *ventral* region is where the motor neurons reside. In the middle are numerous interneurons that relay information between the sensory and motor neurons.

The dorsal-ventral polarity of the neural tube is induced by signals coming from its immediate environ-

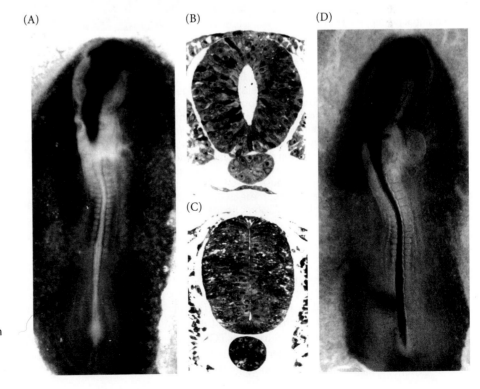

FIGURE 12.12 Occlusion of the neural tube allows expansion of the future brain region. (A) Dye injected into the anterior portion of a 3-day chick neural tube fills the brain region, but does not pass into the spinal region. (B,C) Sections of the chick neural tube at the base of the brain (B) before occlusion and (C) during occlusion. (D) Reopening of the occlusion after initial brain enlargement allows dye to pass from the brain region into the spinal cord region. (Photographs courtesy of M. Desmond.)

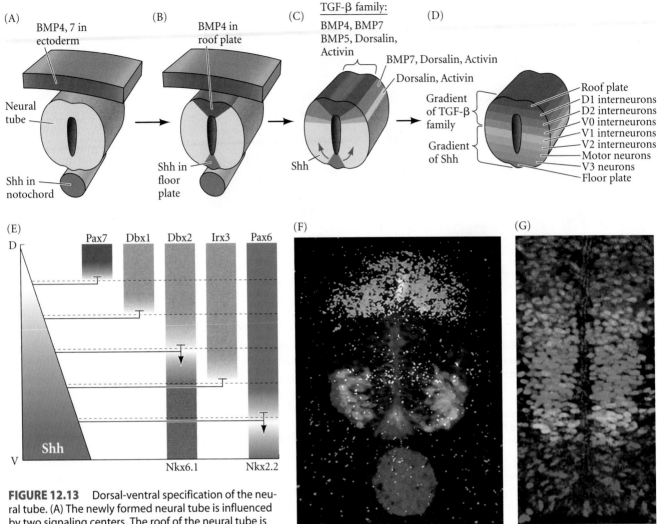

FIGURE 12.13 Dorsal-ventral specification of the neural tube. (A) The newly formed neural tube is influenced by two signaling centers. The roof of the neural tube is exposed to BMP4 and BMP7 from the epidermis, and the floor of the neural tube is exposed to Sonic hedgehog protein from the notochord. (B) Secondary signaling centers are established within the neural tube. BMP4 is expressed and secreted from the roof plate cells; Sonic hedgehog is expressed and secreted from the floor plate cells. (C) BMP4 establishes a nested cascade of TGF-β factors, spreading ventrally into the neural tube from the roof plate. Sonic hedgehog diffuses dorsally as a gradient from the floor plate cells. (D) The neurons of the spinal cord are given their identities by their exposure to these gradients of paracrine factors. The amounts and types of paracrine factors present cause different transcription factors to be activated in the nuclei of these cells, depending on their position in the neural tube. (E) Sonic hedgehog is confined to the ventral region of the neural tube by the TGF-β factors, and the gradient of Sonic hedgehog specifies the ventral neural tube by activating and inhibiting the synthesis of particular transcription factors. (F) Chick neural tube, showing areas of Sonic hedgehog (green) and the expression domain of the TGF-β-family protein dorsalin (blue). Motor neurons induced by a particular concentration of Sonic hedgehog are stained orange/yellow. (G) In situ hybridization for three transcription factors: Pax7 (blue, characteristic of the dorsal neural tube cells), Pax6 (green), and Nkx6.1 (red). Where Nkx6.1 and Pax6 overlap (yellow), the motor neurons become specified. (E courtesy of T. M. Jessell; F after Jessell 2000; G courtesy of J. Briscoe.)

ment. The ventral pattern is imposed by the notochord, while the dorsal pattern is induced by the epidermis. Specification of the axis is initiated by two major paracrine factors. The first is Sonic hedgehog protein, originating from the notochord. The second set of factors are TGF-β proteins, originating in the dorsal ectoderm (Figure 12.13). In both cases, these factors induce a second signaling center within the neural tube itself. Sonic hedgehog is secreted from the notochord and induces the medial hinge cells to become the **floor plate** of the neural tube. These floor plate cells also secrete Sonic hedgehog, and this paracrine factor from the floor plate cells forms a gradient that is highest at the most ventral portion of the neural tube (Roelink et al. 1995; Briscoe et al. 1999).

The dorsal fates of the neural tube are established by proteins of the TGF-β superfamily, especially bone morphogenetic proteins 4 and 7, dorsalin, and activin (Liem et al. 1995, 1997, 2000). Initially, BMP4 and BMP7 are found in the epidermis. Just as the notochord establishes a secondary signaling center—the floor plate cells—on the ventral side of the neural tube, the epidermis establishes a secondary signaling center by inducing BMP4 expression in the **roof plate** cells of the neural tube. The BMP4 protein from the roof plate induces a cascade of TGF-β proteins in adjacent cells (see Figure 12.13C). Different sets of cells are thus exposed to different concentrations of TGF-β proteins at different times, with the most dorsal cells being exposed to more factors at higher concentrations and at earlier times.

The paracrine factors interact to instruct the synthesis of different transcription factors along the dorsal-ventral axis of the neural tube. For instance, those cells adjacent to the floor plate that receive high concentrations of Sonic hedgehog (and hardly any TGF-β signal) synthesize the Nkx6.1 and Nkx2.2 transcription factors and become the ventral (V3) neurons. The cells dorsal to these, exposed to slightly less Sonic hedgehog and slightly more TGF-β factors, produce Nkx6.1 and Pax6 transcription factors. These cells become the motor neurons. The next two groups of cells, receiving progressively less Sonic hedgehog, become the V2 and V1 interneurons (Figure 12.14A; Lee and Pfaff 2001; Muhr et al. 2001).

The importance of Sonic hedgehog in inducing and patterning the ventral portion of the neural tube can be shown experimentally. If notochord fragments are taken from one embryo and transplanted to the lateral side of a host neural tube, the host neural tube will form another set of floor plate cells at its sides (Figure 12.14B,C). These floor plate cells, once induced, induce the formation of motor neurons on either side of them. The same results can be obtained if the notochord fragments are replaced by pellets of cultured cells secreting Sonic hedgehog (Echelard et al. 1993). Moreover, if a piece of notochord is removed from an embryo, the neural tube adjacent to the deleted region will have no floor plate cells (Placzek et al. 1990). The importance of the TGF-β superfamily factors in patterning the dorsal portion of the neural tube was demonstrated by zebrafish mutants. Those mutants deficient in certain BMPs lacked dorsal and intermediate types of neurons (Nguyen et al. 2000).

By combining information about the anterior-posterior axis with information establishing the dorsal-ventral axis, neurons can become uniquely specified for certain regions. For instance, one would expect the motor neurons innervating the forelimb to recognize different targets than the motor neurons innervating hindlimb targets. The Hox genes (and especially the HoxC paralogue group) play important roles in determining motor neuron targets (Dasen et al. 2003, 2005).

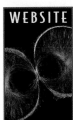

WEBSITE **12.3 Constructing the pituitary gland.** The developmental genetics of the pituitary have shown that (as in the trunk neural tube) paracrine signals elicit the expression of overlapping sets of transcription factors. These factors regulate the different cell types of the gland responsible for producing growth hormone, prolactin, gonadotropins, and other hormones.

Tissue Architecture of the Central Nervous System

The neurons of the brain are organized into layers (**cortices**) and clusters (**nuclei**), each having different functions and connections. The original neural tube is composed of a **ger-**

FIGURE 12.14 Cascade of inductions initiated by the notochord in the ventral neural tube. (A) Relationship between Sonic hedgehog concentrations, the generation of particular neuronal types in vitro, and distance from the notochord. (B) Two cell types in the newly formed neural tube. Those closest to the notochord become the floor plate neurons; motor neurons emerge on the ventrolateral sides. (C) If a second notochord, floor plate, or any other Sonic hedgehog-secreting cell is placed adjacent to the neural tube, it induces a second set of floor plate neurons, as well as two other sets of motor neurons. (A after Briscoe et al. 1999; B,C after Placzek et al. 1990.)

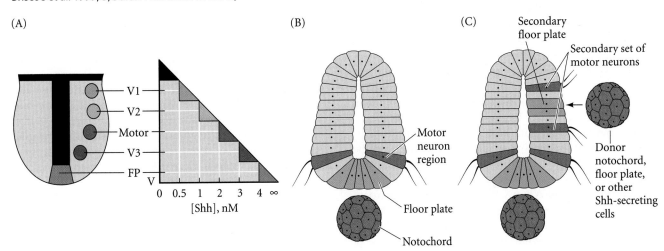

(A)

V1
V2
Motor
V3
FP
V
0 0.5 1 2 3 4 ∞
[Shh], nM

(B)

Motor neuron region

Floor plate

Notochord

(C)

Secondary floor plate

Secondary set of motor neurons

Donor notochord, floor plate, or other Shh-secreting cells

(A)

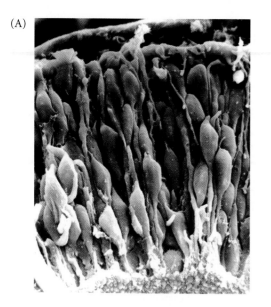

(B)

Stage of cell cycle

G_1 S G_2 M G_1

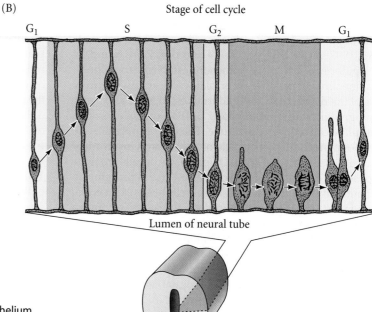

Lumen of neural tube

FIGURE 12.15 Neural stem cells in the germinal epithelium. (A) Scanning electron micrograph of a newly formed chick neural tube, showing cells at different stages of their cell cycles. (B) Schematic section of a chick embryo neural tube, showing the position of the nucleus in a neuroepithelial cell as a function of the cell cycle. Mitotic cells are found near the inner surface of the neural tube, adjacent to the lumen. (A courtesy of K. Tosney; B after Sauer 1935.)

minal **neuroepithelium**—a layer of rapidly dividing neural stem cells one cell layer thick. Sauer and others (1985) showed that the cells of the germinal epithelium are continuous from the luminal surface of the neural tube to the outside surface, but that the *nuclei* of these cells are at different heights, giving the superficial impression that the neural tube has numerous cell layers (Figure 12.15A). The nuclei move within their cells as they go through the cell cycle. DNA synthesis (S phase) occurs while the nucleus is at the outside edge of the neural tube, and the nucleus migrates luminally as the cell cycle proceeds (Figure 12.15B).

Mitosis occurs on the luminal side of the cell layer. If mammalian neural tube cells are labeled with radioactive thymidine during early development, 100 percent of them will incorporate this base into their DNA (Fujita 1964). Shortly thereafter certain cells stop incorporating these DNA precursors, indicating that they are no longer participating in DNA synthesis and mitosis. These cells then migrate and differentiate into neuronal and glial cells outside the neural tube (Fujita 1966; Jacobson 1968).

If the labeled progeny of dividing cells are found in the outer cortex in the adult brain, then those neurons must have migrated to their cortical positions from the germinal neuroepithelium. When a cell of the germinal neuroepithelium is ready to generate neurons (instead of more neural stem cells), the plane of cell division shifts. Instead of having both cells attached to the luminal surface, one of the two daughter cells remains in the epithelium while the other becomes detached. The cell connected to the luminal surface usually remains a stem cell, while the other cell migrates and differentiates* (Chenn and McConnell 1995; Hollyday 2001). The time of this vertical division is the last time the latter cell will divide, and is called that neuron's birthday. Different types of neurons and glial cells have birthdays at different times. Labeling cells at different times during development shows that the cells with the earliest birthdays migrate the shortest distances; those with later birthdays migrate to form the more superficial regions of the brain cortex. Subsequent differentiation depends on the positions the neurons occupy once outside the germinal neuroepithelium (Letourneau 1977; Jacobson 1991).

Spinal cord and medulla organization

As the cells adjacent to the lumen continue to divide, the migrating cells form a second layer around the original neural tube. This layer becomes progressively thicker as more cells are added to it from the germinal neuroepithelium. This new layer is called the **mantle** (or **intermediate**) **zone**, and the germinal epithelium is now called the **ventricular zone** (and, later, the **ependyma**) (Figure 12.16). The mantle zone cells differentiate into both neurons and glia. The neurons make connections among themselves and send forth axons away from the lumen, thereby creating a

*Compare this sequence with the discussion of *Drosophila* sperm stem cells in Chapter 6 ("Sidelights & Speculations: Stem Cell Niches"). A similar sequence of oriented cell division has been postulated for the epidermis (see p. 404).

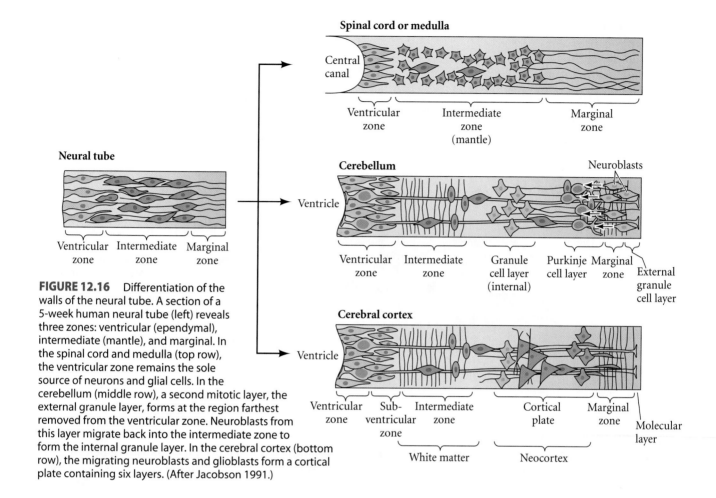

FIGURE 12.16 Differentiation of the walls of the neural tube. A section of a 5-week human neural tube (left) reveals three zones: ventricular (ependymal), intermediate (mantle), and marginal. In the spinal cord and medulla (top row), the ventricular zone remains the sole source of neurons and glial cells. In the cerebellum (middle row), a second mitotic layer, the external granule layer, forms at the region farthest removed from the ventricular zone. Neuroblasts from this layer migrate back into the intermediate zone to form the internal granule layer. In the cerebral cortex (bottom row), the migrating neuroblasts and glioblasts form a cortical plate containing six layers. (After Jacobson 1991.)

cell-poor **marginal zone**. Eventually glial cells cover many of the axons in the marginal zone in myelin sheaths, giving them a whitish appearance. Hence, the axonal marginal layer is often called the **white matter**, while the mantle zone, containing the neuronal cell bodies, is referred to as the **gray matter**.

In the spinal cord and medulla, this basic three-zone pattern of ventricular (ependymal), mantle, and marginal layers is retained throughout development. When viewed in cross-section, the gray matter (mantle) gradually becomes a butterfly-shaped structure surrounded by white matter; and both become encased in connective tissue. As the neural tube matures, a longitudinal groove—the **sulcus limitans**—divides it into dorsal and ventral halves. The dorsal portion receives input from sensory neurons, whereas the ventral portion is involved in effecting various motor functions (Figure 12.17).

Cerebellar organization

In the brain, cell migration, differential neuronal proliferation, and selective cell death produce modifications of the three-zone pattern seen in Figure 12.16. In the cerebellum, some neuronal precursors enter the marginal zone to form

clusters of neurons called **nuclei** (note that this is a completely distinct structure from the "cell nucleus"). Each nucleus works as a functional unit, serving as a relay station between the outer layers of the cerebellum and other parts of the brain. Other neuronal precursors migrate away from the germinal epithelium. These cerebellar neuroblasts migrate to the outer surface of the developing cerebellum and form a new germinal zone, the **external granule layer**, near the outer boundary of the neural tube.

At the outer boundary of the external granule layer (which is 1–2 cells thick), neuroblasts proliferate and come into contact with cells that secrete BMP factors. The BMPs specify the postmitotic products of these neuroblasts to become the **granule neurons** (Alder et al. 1999). Granule neurons migrate back toward the ventricular (ependymal) zone, where they produce a region called the **internal granule layer**. Meanwhile, the original ventricular zone of the cerebellum generates a wide variety of neurons and glial cells, including the distinctive and large **Purkinje neurons**, the major cell type of the cerebellum. Purkinje neurons secrete Sonic hedgehog, which sustains the division of granule neuron precursors in the external granule layer (Wallace 1999). Each Purkinje neuron has an enormous **dendritic arbor** that spreads like a tree above a bulblike

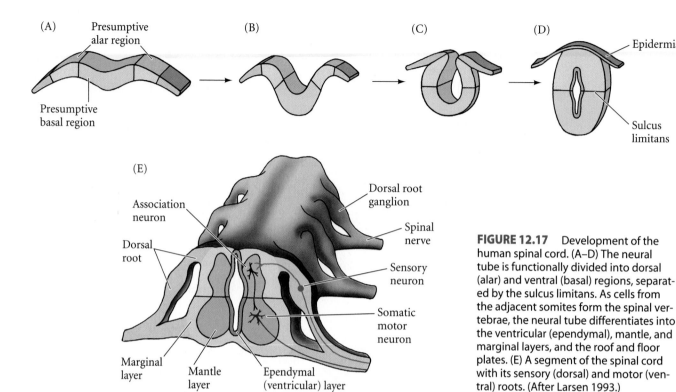

FIGURE 12.17 Development of the human spinal cord. (A–D) The neural tube is functionally divided into dorsal (alar) and ventral (basal) regions, separated by the sulcus limitans. As cells from the adjacent somites form the spinal vertebrae, the neural tube differentiates into the ventricular (ependymal), mantle, and marginal layers, and the roof and floor plates. (E) A segment of the spinal cord with its sensory (dorsal) and motor (ventral) roots. (After Larsen 1993.)

cell body (Figure 12.18). A typical Purkinje neuron may form as many as 100,000 connections (**synapses**) with other neurons—more connections than any other type of neuron studied. Each Purkinje neuron also emits a slender axon, which connects to neurons in the deep cerebellar nuclei.

Purkinje neurons are critical in the electrical pathway of the cerebellum. All electric impulses eventually regulate the activity of these neurons, which are the only output neurons of the cerebellar cortex. For this to happen, the proper cells must differentiate at the appropriate place and time. How is this accomplished?

One mechanism thought to be important for positioning young neurons within the developing mammalian brain is **glial guidance** (Rakic 1972; Hatten 1990). Throughout the cortex, neurons are seen to ride a "glial monorail" to their respective destinations. In the cerebellum, the granule cell precursors travel on the long processes of the **Bergmann glia** (Figure 12.19; Rakic and Sidman 1973; Rakic 1975). This neuronal-glial interaction is a complex and fascinating series of events, involving reciprocal recognition between glia and neuroblasts (Hatten 1990; Komuro and Rakic 1992). The neuron maintains its adhesion to the glial cell through a number of proteins, one of them an adhesion protein called **astrotactin**. If the astrotactin on a neuron is masked by antibodies to that protein, the neuron will fail to adhere to the glial processes (Edmondson et al. 1988; Fishell and Hatten 1991). Mice deficient in astrotactin have slow neuronal migration rates, abnormal Purkinje cell development, and problems coordinating their balance (Adams et al. 2002).

Much insight into the mechanisms of spatial ordering in the brain has come from the analysis of neurological mutations in mice. Over 30 mutations are known to affect the arrangement of cerebellar neurons. Many of these cerebellar mutants have been found because the phenotype of such mutants—namely, the inability to keep balance while walking—can be easily recognized. For obvious reasons, these mutations are given names such as *weaver, reeler, staggerer,* and *waltzer.*

WEBSITE **12.4 Cerebellar mutations of the mouse.** The mouse mutations affecting cerebellar function have given us remarkable insights into the ways in which the cerebellum is constructed. The *reeler* mutation, in particular, has been extremely important in our knowledge of how cerebellar neurons migrate.

Cerebral organization

The three-zone arrangement of the neural tube is also modified in the cerebrum. The cerebrum is organized in two distinct ways. First, like the cerebellum, it is organized vertically into layers that interact with one another. Certain neuroblasts from the mantle zone migrate on glial processes through the white matter to generate a second zone of neurons at the outer surface of the brain. This new layer of gray matter will become the **neocortex**. The neocortex eventually stratifies into six layers of neuronal cell bodies;

FIGURE 12.18 Cerebellar organization. These sagittal sections of fluorescently labeled rat cerebellum were photographed using dual-photon confocal microscopy. The lower view is a vast enlargement of one area of the top image. Purkinje neurons are light blue with bright green processes, Bergmann glial cells are red, and granular cells show up in dark blue. The close-up illustrates the highly structured organization of neurons and glial cells. (Photographs courtesy of T. Deerinck and M. Ellisman, University of California/San Diego.)

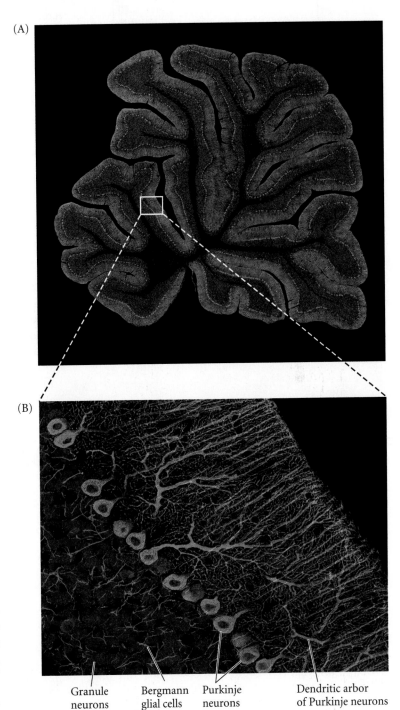

(A)

(B)

| Granule neurons | Bergmann glial cells | Purkinje neurons | Dendritic arbor of Purkinje neurons |

the adult forms of these layers are not completed until the middle of childhood. Each layer of the neocortex differs from the others in its functional properties, the types of neurons found there, and the sets of connections that they make. For instance, neurons in layer 4 receive their major input from the thalamus (a region that forms from the diencephalon), whereas neurons in layer 6 send their major output back to the thalamus.

In addition to the six vertical layers, the cerebral cortex is organized *horizontally* into over 40 regions that regulate anatomically and functionally distinct processes. For instance, neurons in layer 6 of the *visual cortex* project axons to the *lateral* geniculate nucleus of the thalamus, which is involved in vision (see Chapter 13), while layer 6 neurons of the *auditory cortex* (located more anteriorly than the visual cortex) project axons to the *medial* geniculate nucleus of the thalamus, which functions in hearing. One of the major questions in developmental neurobiology is whether the different functional regions of the cerebral cortex are already specified in the ventricular region, or if the specification is accomplished much later by the synaptic connections between the regions. Evidence that the specification is early (and that there might be some "proto-map" of the cerebral cortex) is suggested by certain human mutations that destroy the layering and functional abilities in only one part of the cortex, leaving the other regions intact (Piao et al. 2004).

Neither the vertical nor the horizontal organization of the cerebral cortex is clonally specified—that is, none of the functional units forms from the progeny of a single cell. Rather, the developing cortex forms from the mixing of cells derived from numerous stem cells. After their final mitoses, most of the neuroblasts generated in the ventricular zone migrate outward along glial processes to form the **cortical plate** at the outer surface of the brain. As in the rest of the brain, those neurons with the earliest birthdays form the layer closest to the ventricle. Subsequent neurons travel greater distances to form the more superficial layers of the cortex. This process forms an "inside-out" gradient of

development (Rakic 1974). A single stem cell in the ventricular layer can give rise to neurons (and glial cells) in any of the cortical layers (Walsh and Cepko 1988). But how do the cells know which layer to enter?

McConnell and Kaznowski (1991) have shown that the determination of laminar identity (i.e., which layer a cell migrates to) is made during the final cell division. Newly generated neuronal precursors transplanted after this last division from young brains (where they would form layer

FIGURE 12.19 Neuron-glia inter-action. (A) Diagram of a cortical neuron migrating on a glial cell process. (B) Sequential photographs of a neuron migrating on a cerebellar glial process. The leading process has several filopodial extensions. The neuron can reach speeds around 40 μm per hour as it travels. (A after Rakic 1975; B from Hatten 1990, photograph courtesy of M. Hatten.)

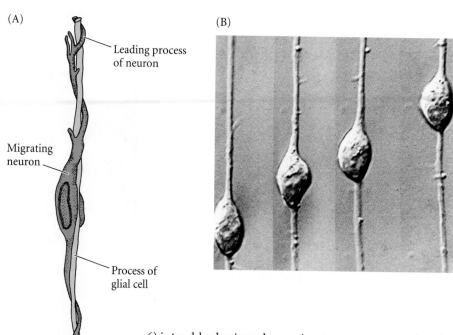

(A)

Leading process of neuron

Migrating neuron

Process of glial cell

(B)

6) into older brains whose migratory neurons are forming layer 2 are committed to their fate, and migrate only to layer 6. However, if these cells are transplanted prior to their final division (during mid-S phase), they are uncommitted, and can migrate to layer 2 (Figure 12.20). The fates of neuronal precursors from older brains are more fixed. While the neuronal precursor cells formed early in development have the potential to become any neuron (at lay-

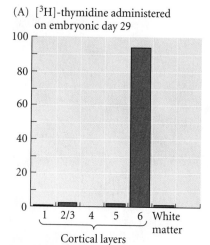

(A) [³H]-thymidine administered on embryonic day 29

Cortical layers: 1, 2/3, 4, 5, 6, White matter

FIGURE 12.20 Determination of cortical laminar identity in the ferret cerebrum. (A) "Early" neuronal precursors (birthdays on embryonic day 29) migrate to layer 6. (B) "Late" neuronal precursors (birthdays on postnatal day 1) migrate farther, into layers 2 and 3. (C) When early neuronal precursors (dark blue) are transplanted into older ventricular zones after their last mitotic S phase, the neurons they form migrate to layer 6. (D) If these precursors are transplanted before or during their last S phase, however, they migrate (with the host neurons) to layer 2. (After McConnell and Kaznowski 1991.)

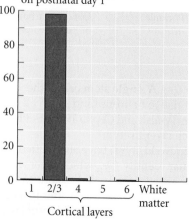

(B) [³H]-thymidine administered on postnatal day 1

Cortical layers: 1, 2/3, 4, 5, 6, White matter

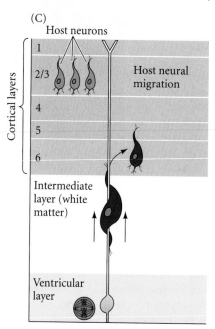

(C)

Host neurons

Cortical layers: 1, 2/3, 4, 5, 6

Host neural migration

Intermediate layer (white matter)

Ventricular layer

Cell-autonomous fate when transplanted after last S phase

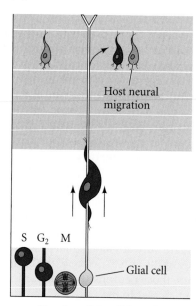

(D)

Host neural migration

S   G₂   M

Glial cell

Host (conditional) fate when transplanted in S phase

ers 2 or 6, for instance), later precursor cells give rise only to upper-level (layer 2) neurons (Frantz and McConnell 1996). Once the cells arrive at their final destination, it is thought that they produce particular adhesion molecules that organize them together as brain nuclei (Matsunami and Takeichi 1995).

We still do not know the nature of the information given to the cell as it becomes committed. However, Hanashina and her colleagues (2004) have shown that there are genetic switches that get "thrown" at these division times. One of these switches is the gene encoding the transcription factor Foxg1. When the mouse neuronal progenitor cells divide to form the first layer of cortical neurons, *Foxg1* is not expressed in the progenitor cells or in the first-formed neu-

rons. However, later, when the progenitor cells generate those neurons destined for layers 4 and 5, they express this gene. If the *Foxg1* gene is conditionally knocked out of this lineage, the neural precursor cells continually give rise to layer 1 neurons. Therefore, it seems that the Foxg1 transcription factor is required to suppress the "layer 1" neural fate.

WEBSITE **12.5** **Constructing the cerebral cortex.** Three genes have recently been shown to be necessary for the proper lamination of the mammalian brain. They appear to be important for cortical neural migration, and when mutated in humans can produce profound mental retardation.

SIDELIGHTS & SPECULATIONS

The Unique Development of the Human Brain

The development of the human neocortex is strikingly plastic and an almost constant "work in progress." Whatever distinguishes humans from other primates must reside in the unique features of human development, especially in the development of the brain. At least five phenomena have been identified that distinguish the development of the human brain from that of other species, including other primates:

1. The retention of the fetal neuronal growth rate after birth.
2. The migration of cells from the prosencephalon to the diencephalon.
3. The activity of transcription.
4. The specific form of the *FOXP2* gene.
5. The continuation of brain maturation into adulthood.

Fetal neuronal growth rate after birth

If there is one important developmental trait that distinguishes humans from the rest of the animal kingdom, it is our retention of the fetal neural growth rate. Both human and ape brains have a high growth rate before birth. After birth, however, this

rate slows greatly in the apes, whereas humans brain growth continues at a rapid rate for about 2 years (Figure 12.21A; Martin 1990). During early postnatal development, we add approximately 250,000 neurons per minute (Purves and Lichtman 1985). The ratio of brain weight to body weight at birth is also similar for great apes and humans, but by adulthood the ratio for humans is literally "off the chart" (Figure 12.21B; Bogin 1997).

At the cellular level, we find that no fewer than 30,000 synapses per second per square centimeter of cortex are formed during the first few years of human life. It is thought that this increase in neuron numbers may (1) generate new modules (addressable sites) that can acquire new functions, (2) store new memories for use in thinking and forecasting, and (3) enable learning by intercon-

necting among themselves and with prenatally generated neurons (Rose 1998; Baringa 2003). It is in this early postnatal stage

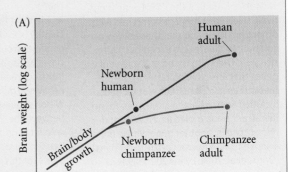

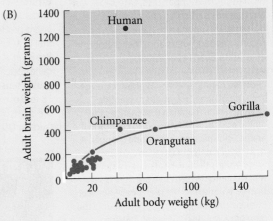

FIGURE 12.21 Retention of fetal neural growth rate in humans. (A) Whereas other primates, such as chimpanzees, complete neuronal proliferation around the time of birth, the neurons of human newborns continue to proliferate at the same rate as the fetal brain neurons. (B) The human brain/body weight ratio (encephalization index) is about 3.5 times higher than that of apes. (After Bogin 1997.)

Continued on next page

SIDELIGHTS & SPECULATIONS

that intervention is thought to be able to raise IQ (reviewed in Wickelgren 1999).

The prodigious rate of continued neuron production has important consequences. Portmann (1941), Gould (1977), and Montagu (1962) have each made the claim that we are essentially "extrauterine fetuses" for the first year of life. If one follows the charts of ape maturity, actual human gestation is 21 months. In other words, humans are at 21 months what other apes are at birth. Our "premature" birth is an evolutionary compromise between maternal pelvic width and the infant's head circumference and lung maturity. The mechanism for retaining the fetal neural growth rate has been called *hypermorphosis*—the phyletic extension of development beyond its ancestral state (Vrba 1996).

The *ASPM* gene is known to be important in human brain growth, since mutations in this gene are the most common cause of microcephaly (small brains). The *Aspm* gene in mice is expressed specifically in sites of active neurogenesis, especially in the ventricular zone of the cerebral cortex (Bond et al. 2002). Three studies (Zhang 2003; Kouprina et al. 2004; Evans et al. 2004a) indicate that the *ASPM* gene is positively selected in humans compared with other primates,* and that these differences correlate to differences in cerebral cortex size. *ASPM* was also found in a screen of genes that are evolving faster in the primates than in rodents. Other such genes included *Sonic hedgehog*, the gene for neural cell adhesion molecule (NCAM), and *microcephalin*, another gene whose mutations can cause microcephaly in humans (Dorus et al. 2004; Evans 2004b). *Microcephalin* and *ASPM* seem to have evolved particularly rapidly since the split of humans from chimpanzees. One *APSM* variant that has particularly high frequencies in humans, and appears to have arisen as recently as 5800 years ago (Mekel-Bobrov et al. 2005; Evans et al. 2005).

---

*Positive natural selection is often tested by the ratio of nonsynonymous nucleotide changes (i.e., a mutation that changes the amino acid) to synonymous (no change in the amino acid specified) nucleotide changes. A ratio significantly greater than 1.0 indicates that natural selection has favored amino acid replacements. No selective pressure is seen on the *ASPM* gene until the divergence between humans and chimpanzees, at which time the ratio grew to 1.44 (Evans et al. 2004a).

Cell migration into the dorsal thalamus

Most neurons remain within the region of the brain in which they were formed. However, in the human brain (but not in the brains of mice or monkeys), certain neurons from the telencephalon migrate into the diencephalon. Here they enter the thalamus, an area involved in memory and problem-solving that is especially large in the human brain (Letinic and Rakic 2001). Neurons from the telencephalon appear to contribute to those thalamic regions that distribute information to the cerebral cortex of the frontal lobes. The developing human thalamus chemotactically attracts the telencephalon neurons, but this chemotaxis is not seen in mice. The growth of the thalamus as a "relay station" that coordinates cortical functions and the ability of this region to attract newly grown cortical neurons may help explain the unique status of the human brain.

High transcriptional activity

In 1975, Mary-Claire King and Allan Wilson showed that the proteins of the human and chimp brain were, within experimental limits, identical. But while the nearly 99 percent identity of human, chimpanzee, and bonobo genomes would suggest that all three be classified as species within the same genus (see Goodman 1999), the morphological and behavioral differences are enormous. A. C. Wilson suggested that the difference between the species might reside in the *amount* of proteins made from their genes (see Gibbons 1998), and now there is evidence for this hypothesis. Using microarrays, several recent investigations found that while the quantity and types of genes expressed in humans and chimp livers and blood were indeed extremely similar, human *brains* produced over five times as much mRNA as chimp brains (Enard et al. 2002a; Preuss et al. 2004). Transcription of some genes (such as *SPTLC1*, a gene whose defect causes sensory nerve damage) was elevated 18-fold over the same gene's expression in the chimpanzee cortex, while other genes (such as *DDX17*, whose product is involved

in RNA processing) are expressed 10 times less in human than in chimp cortices.

Speech, language, and the *FOXP2* gene

Spoken language is a uniquely human trait and is presumed to be the prerequisite for the evolution of cultures. Speech entails the fine-scale control of the larynx (voice box) and the mouth. Individuals who are heterozygous for mutations at the *FOXP2* locus have severe problems with language articulation and with forming sentences (Vargha-Khadem et al. 1995; Lai et al. 2001). This observation has provided genetic anthropologists with an interesting gene to study. Enard and colleagues (2002b) have shown that, although the *FOXP2* gene is extremely conserved throughout most of mammalian evolution, it has a unique form in humans, having accumulated at least two amino acid-changing mutations just since our divergence from the common ancestor of humans and chimpanzees.

Foxp2 in the mouse is expressed in the developing brain, but its major site of expression is the lung (Shu et al. 2001). The human gene is also highly expressed in the brain regions assigned for speech coordination (the caudate nucleus and inferior olive nuclei). These sites are abnormal in patients with *FOXP2* deficiency (Lai et al. 2003). It is not known whether the amino acid changes in this putative transcription factor gave it a new function in the nervous system or if these changes were involved in the formation of language and culture. What is known is this protein is presently critical for forming the orofacial movements and grammar characteristic of human speech and language.

Teenage brains: Wired and unchained

Until recently, most scientists thought that after the initial growth of neurons during fetal development and early childhood, there were no more periods of rapid neural proliferation. However, magnetic resonance imaging (MRI) studies have shown that the brain keeps developing until around puberty, and that not all areas of the brain mature

 WEBSITE 12.6 Neuronal growth and the invention of childhood. An interesting hypothesis claims that the caloric requirements of this brain growth necessitated a new stage of the human life cycle—childhood—during which the child is actively fed by adults.

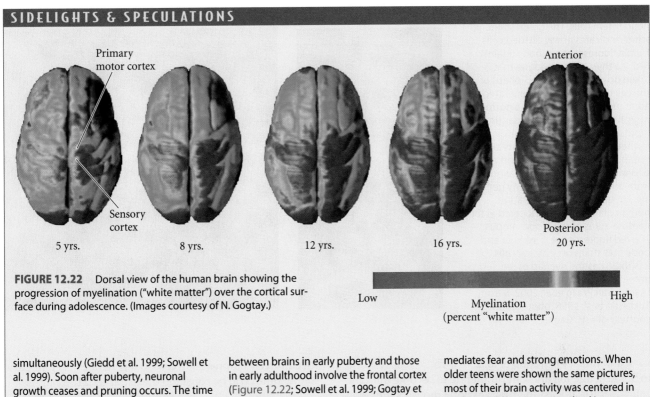

FIGURE 12.22 Dorsal view of the human brain showing the progression of myelination ("white matter") over the cortical surface during adolescence. (Images courtesy of N. Gogtay.)

simultaneously (Giedd et al. 1999; Sowell et al. 1999). Soon after puberty, neuronal growth ceases and pruning occurs. The time of this "pruning" correlates to the time when language acquisition becomes difficult (which may be why children learn language more readily than adults). There is also a wave of myelin production ("white matter" from the glial cells that ensheathes neuronal axons) at this time. Myelination is critical for the proper functioning of the neural areas, and the greatest differences

between brains in early puberty and those in early adulthood involve the frontal cortex (Figure 12.22; Sowell et al. 1999; Gogtay et al. 2004).

These differences in brain development may explain the extreme responses teens have to certain stimuli, as well as their ability to learn certain tasks. In tests that used functional MRI to scan subjects' brains while emotional pictures were shown on a computer screen, the brains of young teens showed activity in the amygdala, which

mediates fear and strong emotions. When older teens were shown the same pictures, most of their brain activity was centered in the frontal lobe, an area involved in more reasoned perceptions (Baird et al. 1999; Luna et al. 2001). The teenage brain is a complicated and dynamic entity, and (as any parent knows) one that is not easily understood. But if one survives these years, the resulting adult brain is usually capable of making reasoned decisions, even in the onslaught of emotional situations.

Adult neural stem cells

Until recently, it was generally believed that once the nervous system was mature, no new neurons were "born"—the neurons we formed *in utero* and during the first few years of life were all we could ever expect to have. However, the good news from recent studies is that the adult mammalian brain is capable of producing new neurons, and environmental stimulation can increase the number of these new neurons. In these experiments, researchers injected adult mice, rats, or marmosets with bromodeoxyuridine (BrdU), a nucleoside that resembles thymidine. BrdU will be incorporated into a cell's DNA only if the cell is undergoing DNA replication; thus, any cell labeled with BrdU must have been undergoing DNA synthesis during the time it was exposed to BrdU. This labeling technique revealed that *thousands* of new neurons were being made each day in these adult mammals. Moreover, these new brain cells integrated with other cells of the

brain, had normal neuronal morphology, and exhibited action potentials (Figure 12.23A; van Praag et al. 2002).

Injecting humans with BrdU is usually unethical, since large doses of BrdU are often lethal. However, in certain cancer patients, the progress of chemotherapy is monitored by transfusing the patient with a small amount of BrdU. Gage and colleagues (Erikkson et al. 1998) took postmortem samples from the brains of five such patients who died between 16 and 781 days after the BrdU infusion. In all five subjects, they saw new neurons in the granular cell layer of the hippocampal dentate gyrus (a part of the brain where memories may be formed). The BrdU-labeled cells also stained for neuron-specific markers (Figure 12.23B). Thus, although the rate of new neuron formation in adulthood may be relatively small, the human brain is not an anatomical *fait accompli* at birth, or even after childhood.

The existence of neural stem cells in adults is now well established for the olfactory epithelium and the hippocampus (Kempermann et al. 1997a,b; van Praag et al. 1999; Kor-

FIGURE 12.23 Evidence of adult neural stem cells. The green staining, which indicates newly divided cells, is from a fluorescent antibody against BrdU (a thymidine analogue that is taken up only during the S phase of the cell cycle). (A) Newly generated adult mouse neurons have a normal morphology and receive synaptic inputs. The green cells are newly formed neurons. The red spots are synaptophysin, a protein found on the dendrites at the synapses of axons from other neurons. (B) Newly generated neuron (arrow) in the adult human brain. This cell is located in the dentate gyrus of the hippocampus. The red fluorescence is from an antibody that stains only neural cells. Yellow indicates the overlap of red and green. Glial cells are stained purple. (A from van Praag et al. 2002; B from Erikksson et al. 1998, photograph courtesy of F. H. Gage.)

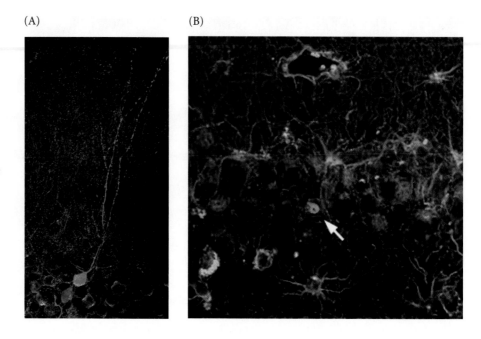

(A)   (B)

nack and Rakic 1999; Kato et al. 2001). These cells respond to Sonic hedgehog and can proliferate to become multiple cell types for at least the first year of a mouse's life (Ahn and Joyner 2005). It appears that the stem cells producing these neurons are located in the ependyma (the former ventricular zone in which the embryonic neural stem cells once resided) or in the subventricular zone adjacent to it (Doetsch et al. 1999; Johansson et al. 1999; Cassidy and Frisén 2001). These adult neural stem cells represent only about 0.3 percent of the ventricle wall cell population, but they can be distinguished from the other cells by their particular cell surface proteins (Rietze et al. 2001).*

The existence of adult neural stem cells in the cortex is more controversial. Some investigators (Gould et al. 1999a,b; Magavi et al. 2000) claim to have found them; other scientists (see Rakic 2002) question the existence of these cortical neural stem cells.

The mechanisms by which neural stem cells are kept in a state of ready quiescence well into adulthood is just beginning to be explored. Before they become neurons, neural stem cells are characterized by the NRSE translational inhibitor that prevents neuronal differentiation by binding to a silencer region of DNA (see Chapter 5). However, when neural stem cells begin to differentiate, they synthesize a small, double-stranded RNA that has the same sequence as the silencer and which might bind NRSE and thereby permit neuronal differentiation (Kuwabara et al. 2004). The use

of cultured neuronal stem cells to regenerate or repair parts of the brain will be considered in Chapter 21.

Differentiation of Neurons

The human brain consists of more than 10^{11} neurons associated with over 10^{12} glial cells. The neuroepithelial cells of the neural tube give rise to three main types of cells. First, they become the *ventricular (ependymal) cells* that remain integral components of the neural tube lining and which secrete the cerebrospinal fluid. Second, they generate the *precursors of the neurons* that conduct electric potentials and coordinate our bodily functions, our thoughts, and our sensations of the world. Third, they give rise to the *precursors of the glial cells* that aid in the construction of the nervous system, provide insulation around the neurons, and which may be important in memory storage. As we have seen, the differentiation of these precursor cells is believed to be largely determined by the environment that they enter (Rakic and Goldman 1982); and, at least in some cases, a given neuroepithelial cell can presumably give rise to both neurons and glia (Turner and Cepko 1987).

The brain contains a wide variety of neuronal and glial types (as is evident from a comparison of the relatively small granule cell with the enormous Purkinje neuron). The fine, branching extensions of the neuron that are used to pick up electric impulses from other cells are called **dendrites** (Figure 12.24). Some neurons develop only a few dendrites, whereas other cells (such as the Purkinje neurons) develop extensive dendritic trees. Very few dendrites are found on cortical neurons at birth, and one of the amazing events of the first year of human life is the increase in the number of these receptive processes. During this year, each cortical neuron develops enough dendritic surface to

---

*These neural stem cells may have particular physiological roles as well. During pregnancy, prolactin stimulates the production of neuronal progenitor cells in the subventricular zone of the adult mouse forebrain. These progenitor cells migrate to produce olfactory neurons that may be important for maternal behavior of rearing offspring (Shingo et al. 2003).

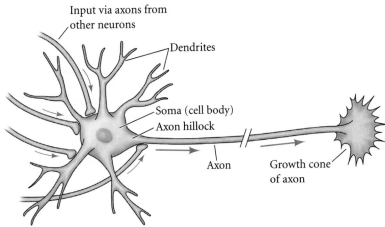

Input via axons from other neurons

Dendrites

Soma (cell body)

Axon hillock

Axon

Growth cone of axon

RECEPTOR

FIGURE 12.24 Diagram of a motor neuron. Electric impulses (red arrows) are received by the dendrites, and the stimulated neuron transmits impulses through its axon to its target tissue. The axon (which may be 2–3 feet long) is a cellular process through which the neuron sends its signals. The growth cone of the axon is both a loco-motor and a sensory apparatus that actively explores the environment and picks up directional cues telling it where to go. Eventually the growth cone will form a connection, or synapse, with the axon's target tissue.

accommodate as many as 100,000 synapses with other neurons. The average cortical neuron connects with 10,000 other neural cells. This pattern of synapses enables the human cortex to function as the center for learning, reasoning, and memory; to develop the capacity for symbolic expression; and to produce voluntary responses to interpreted stimuli.

Another important feature of a developing neuron is its **axon** (sometimes called a **neurite**). Whereas dendrites are often numerous and do not extend far from the neuronal cell body, or **soma**, axons may extend for several feet (see Figure 12.24). The pain receptors on your big toe, for example, must transmit their messages all the way to your spinal cord. One of the fundamental concepts of neurobiology is that the axon is a continuous extension of the nerve cell body. At the turn of the twentieth century, there were many competing theories of axon formation. Theodor Schwann, one of the founders of the cell theory, believed that numerous neural cells linked themselves together in a chain to form an axon. Viktor Hensen, the discoverer of the embry-

onic node, thought that the axon formed around preexisting cytoplasmic threads between the cells. Wilhelm His (1886) and Santiago Ramón y Cajal (1890) postulated that the axon was an outgrowth (albeit an extremely large one) of the neuronal soma.

In 1907, Ross Harrison demonstrated the validity of the outgrowth theory in an elegant experiment that founded both the science of developmental neurobiology and the technique of tissue culture. Harrison isolated a portion of the neural tube from a 3-mm frog tadpole. (At this stage, shortly after the closure of the neural tube, there is no visible differentiation of axons.) He placed this neuroblast-containing tissue in a drop of frog lymph on a coverslip and inverted the coverslip over a depression slide so he could watch what was happening within this "hanging drop." What Harrison saw was the emergence of axons as outgrowths from the neuroblasts, elongating at about 56 μm per hour.

Nerve outgrowth is led by the tip of the axon, called the **growth cone** (Figure 12.25A). The growth cone does not proceed in a straight line, but rather "feels" its way along the substrate. The growth cone moves by the elongation and contraction of pointed filopodia called **microspikes** (Figure 12.25B). These microspikes contain microfilaments,

(A)

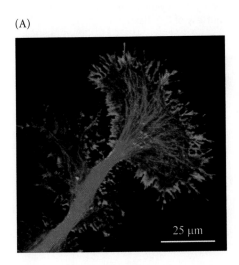

25 μm

(B)

Microspikes

Growth cone

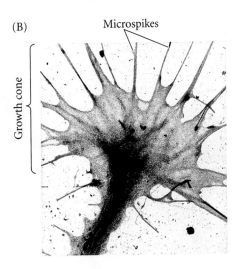

FIGURE 12.25 Axon growth cones. (A) Growth cone of the hawkmoth *Manduca sexta* during axon extension and pathfinding. The actin in the filopodia is stained green with fluorescent phalloidin, while the microtubules are stained red with a fluorescent antibody to tubulin. (B) Actin microspikes in an axon growth cone, seen by transmission electron microscopy. (A courtesy of R. B. Levin and R. Luedemanan; B from Letourneau 1979.)

which are oriented parallel to the long axis of the axon. (This mechanism is similar to that seen in the filopodial microfilaments of secondary mesenchyme cells in echinoderms; see Chapter 8.) Treating neurons with cytochalasin B destroys the actin microspikes, inhibiting their further advance (Yamada et al. 1971; Forscher and Smith 1988). Within the axon itself, structural support is provided by microtubules, and the axon will retract if the neuron is placed in a solution of colchicine. Thus, the developing neuron retains the same mechanisms we noted in the dorsolateral hinge points of the neural tube—namely, elongation by microtubules and apical shape changes by microfilaments.

As in most migrating cells, the exploratory microspikes of the growth cone attach to the substrate and exert a force that pulls the rest of the cell forward. Axons will not grow if the growth cone fails to advance (Lamoureux et al. 1989). In addition to their structural role in axonal migration, the microspikes also have a sensory function. Fanning out in front of the growth cone, each microspike samples the microenvironment and sends signals back to the soma (Davenport et al. 1993). As we will see in Chapter 13, microspikes are the fundamental organelles involved in neuronal pathfinding.

Neurons transmit electric impulses from one region of the body to another. These impulses usually go from the dendrites into the soma, where they are focused into the axon. To prevent dispersal of the electric signal and to facilitate its conduction to the target cell, that part of the axon is insulated at intervals by glial cells. Within the central nervous system, axons are insulated at intervals by processes that originate from a type of glial cell called an **oligoden-**

drocyte. The oligodendrocyte wraps itself around the developing axon, then produces a specialized cell membrane called a **myelin sheath**. In the peripheral nervous system, myelination is accomplished by a glial cell type called the **Schwann cell** (Figure 12.26). Transplantation experiments have shown that the axon, and not the glial cell, controls the thickness of the myelin sheath. Mikhailov and colleagues (2004) have demonstrated that sheath diameter is regulated by the amount of neuregulin-1 secreted by the axon.

The myelin sheath is essential for proper neural function, and demyelination of nerve fibers is associated with convulsions, paralysis, and certain debilitating afflictions such as multiple sclerosis. There are also mouse mutants where subsets of neurons are poorly myelinated. In the *trembler* mutant, the Schwann cells are unable to produce a particular protein component such that myelination is deficient in the peripheral nervous system, but normal in the central nervous system. Conversely, in the mouse mutant called *jimpy*, the central nervous system is deficient in myelin, while the peripheral nerves are unaffected (Sidman et al. 1964; Henry and Sidman 1988).

Axons are also specialized for secreting specific chemical **neurotransmitters** across the small gap (the **synaptic cleft**) that separates the axon of a neuron from the surface of its target cell. Some neurons develop the ability to synthesize and secrete acetylcholine, while others develop the enzymatic pathways for making and secreting epinephrine, norepinephrine, octopamine, serotonin, γ-aminobutyric acid (GABA), or dopamine, among other neurotransmitters. Each neuron must activate those genes

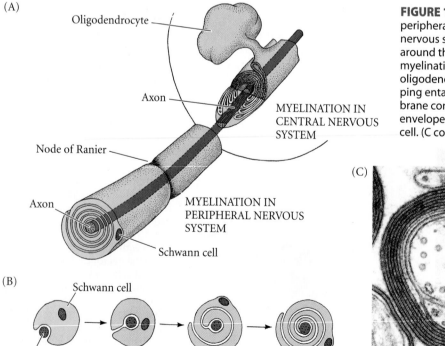

FIGURE 12.26 Myelination in the central and peripheral nervous systems. (A) In the peripheral nervous system, Schwann cells wrap themselves around the axon; in the central nervous system, myelination is accomplished by the processes of oligodendrocytes. (B) The mechanism of this wrapping entails the production of an enormous membrane complex. (C) Micrograph of an axon enveloped by the myelin membrane of a Schwann cell. (C courtesy of C. S. Raine.)

(A)

Oligodendrocyte

Axon

MYELINATION IN CENTRAL NERVOUS SYSTEM

Node of Ranier

Axon

MYELINATION IN PERIPHERAL NERVOUS SYSTEM

Schwann cell

(B)

Schwann cell

Axon

(C)

(A) 4-mm embryo

(B) 4.5-mm embryo

Lens placode

(C) 5-mm embryo

Lens vesicle

(D) 7-mm embryo

Retina   Lens

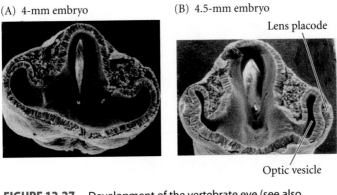

Optic vesicle

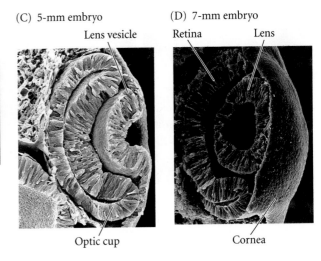

Optic cup

Cornea

FIGURE 12.27   Development of the vertebrate eye (see also Figure 6.5). (A) The optic vesicle evaginates from the brain and contacts the overlying ectoderm, inducing a lens placode. (B,C) The overlying ectoderm differentiates into lens cells as the optic vesicle folds in on itself, and the lens placode becomes the lens vesicle. (C) The optic vesicle becomes the neural and pigmented retina as the lens is internalized. (D) The lens vesicle induces the overlying ectoderm to become the cornea (A–C from Hilfer and Yang 1980, photographs courtesy of S. R. Hilfer; D, photograph courtesy of K. Tosney.)

responsible for making the enzymes that can synthesize its neurotransmitter. Thus, neuronal development involves both structural and molecular differentiation.*

Development of the Vertebrate Eye

An individual gains knowledge of its environment through its sensory organs. The major sensory organs of the head develop from interactions of the neural tube with a series of epidermal thickenings called the **cranial ectodermal placodes** (discussed in more detail in the next chapter). Most of these placodes form neurons and sensory epithelia. The two **olfactory placodes** form the nasal epithelium (smell receptors) as well as the ganglia for the olfactory nerves; similarly, the two **otic placodes** invaginate to form the inner ear labyrinth, whose neurons form the acoustic ganglia that enable us to hear. In this section, we will focus on the development of the eye from the **lens placode**.

The lens placode does not form neurons; rather, it forms the transparent lens that allow light to impinge on the retina. The interactions between the cells of the lens placode and the presumptive retina will structure the eye via a cascade of reciprocal changes that enable the construction of an intricately complex organ.

The dynamics of optic development

The induction of the eye was discussed in Chapter 6, and will only be summarized here. At gastrulation, the invo-

*The regeneration of neurons and their axons will be discussed in Chapter 18. The glial cells are probably very important in permitting or preventing axon regeneration.

luting endoderm and mesoderm interact with the adjacent prospective head ectoderm to give the head ectoderm a lens-forming bias (Saha et al. 1989). But not all parts of the head ectoderm eventually form lenses, and the lens must have a precise spatial relationship with the retina. The activation of the head ectoderm's latent lens-forming ability and the positioning of the lens in relation to the retina is accomplished by the **optic vesicle** (Figure 12.27). The optic vesicle extends from the diencephalon, and where it meets the head ectoderm, it induces the formation of a **lens placode**, which then invaginates to form the lens. The optic vesicle itself becomes the **optic cup**, whose two layers differentiate in different ways. The cells of the outer layer produce melanin pigment (being one of the few tissues other than the neural crest cells that can form this pigment) and ultimately become the **pigmented retina**. The cells of the inner layer proliferate rapidly and generate a variety of glia, ganglion cells, interneurons, and light-sensitive photoreceptor neurons. Collectively, these cells constitute the **neural retina**. The retinal ganglion cells are neurons whose axons send electric impulses to the brain. Their axons meet at the base of the eye and travel down the optic stalk, which is then called the **optic nerve**.

How is a specific region of neural ectoderm informed that it will become the optic vesicle? It appears that a group of transcription factors—Six3, Pax6, and Rx1—are expressed together in the most anterior tip of the neural plate. This single domain will later split into the bilateral regions that form the optic vesicles (Adelmann 1929; Mathers et al. 1995; Chiang et al. 1996). Again, we see the similarities between the *Drosophila* and the vertebrate nervous systems, for these three proteins are also necessary for the formation of the *Drosophila* eye.

As discussed in Chapters 5 and 6, the Pax6 protein appears to be especially important in the development of the lens and retina. Indeed, it appears to be a common denominator for photoreceptive cells in all phyla. If the mouse *Pax6* gene is inserted into the *Drosophila* genome

(A)   (B)   (C)

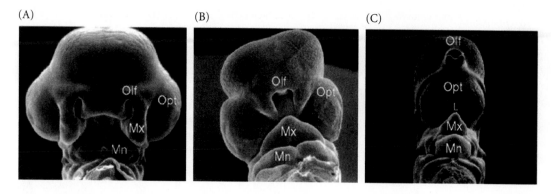

(D)

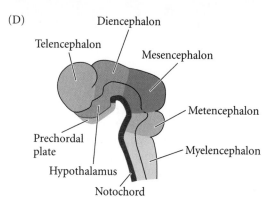

FIGURE 12.28   Sonic hedgehog separates the eye field into two bilateral fields. Jervine, an alkaloid found in certain plants, inhibits endogenous *Shh* signaling. (A) Scanning electron micrograph showing the external facial features of a normal mouse embryo. (B) Mouse embryos exposed to 10 μ*M* jervine had variable loss of midline tissue and resulting fusion of the paired, lateral olfactory processes (Olf), optic vesicles (Opt), and maxillary (Mx) and mandibular (Mn) processes of the jaw. (C) Complete fusion of the optic vesicles and lenses (L) resulted in cyclopia. (D) Drawing showing the location of the prechordal plate (the source of Shh) in the 12-day mouse embryo. (A–C from Cooper et al. 1998, photographs courtesy of P. A. Beachy.)

and activated randomly, *Drosophila* eyes form in those cells where mouse *Pax6* is expressed (see Chapter 23; Halder et al. 1995). While *Pax6* is also expressed in the murine forebrain, hindbrain, and nasal placodes, the eyes seem to be most sensitive to its absence. In humans and mice, *Pax6* heterozygotes have small eyes, while homozygotic mice and humans (and *Drosophila*) lack eyes altogether (Jordan et al. 1992; Glaser et al. 1994; Quiring et al. 1994).

The separation of the single eye field into two bilateral fields depends on the secretion of Sonic hedgehog. If the mouse *Shh* gene is mutated, or if the processing of this protein is inhibited, the single median eye field will not split. The result is **cyclopia**—a single eye in the center of the face, usually below the nose (Figure 12.28; see also Figure 6.18; Chiang et al. 1996; Kelley et al. 1996; Roessler et al. 1996). Shh from the prechordal plate suppresses *Pax6* expression in the center of the embryo, dividing the field in two. The phenomenon of *human* cyclopia (discussed in Chapter 21) also involves loss-of-function mutations that prevent Sonic hedgehog from functioning.

Conversely, if *too much* hedgehog protein is synthesized by the prechordal plate, *Pax6* is suppressed in too large an area, and the eyes fail to form at all. This is why the cavefish are blind. Yamamoto and colleagues (2004) have demonstrated that the difference between the surface population of the Mexican tetra fish *Astyanax mexicanus* and its eyeless cave-dwelling populations is that sonic hedgehog proteins from the prechordal plate downregulate *Pax6* and thereby arrest eye development (Figure 12.29). The

optic cup fails to form properly and the lens undergoes apoptosis. This can be verified experimentally: injecting *Shh* mRNA into one side of surface fish embryos blocks eye development on that side only.

Neural retina differentiation

As in other regions of the brain, the dorsal-ventral polarity of the eye depends on the gradients of BMP signals from the dorsal region and hedgehog proteins from the ventral domains. The most ventral cells express Pax2 protein and become the optic stalk; the most dorsal region responds to BMPs by expressing the MITF transcription factor and generates the pigmented epithelium. The central region of the bulge expresses the **retinal homeobox**, or *Rx* gene, and becomes the optic cup. The *Rx* gene is critical for retinal specification and the development of the eye (Figure 12.30). It is upregulated by Otx2 protein, which specifies the anterior head region, and it encodes a transcription factor that activates numerous genes, including *Pax6* and *Six3*, that are active in making the retinal neurons (Bailey et al. 2004; Voronina et al. 2004). While initial expression of *Pax6* and *Six3* is not dependent on *Rx* expression, their continued expression in the retinal progenitor cells requires Rx protein. Pax6 protein will be critical in the specification of the retinal ganglial cells (which transmit the visual information to the brain), and Six3 is critical for coordinating the number of cell divisions taken by the retinal precursor cells before they differentiate (Bene et al. 2004).

FIGURE 12.29 Surface-dwelling (A) and cave-dwelling (B) Mexican tetras (*Astayanax mexicanum*). The eye fails to form in the population that has lived in caves over 10,000 years (top). Two genes that respond to Shh proteins, *Ptc2* and *Pax2*, are expressed in broader domains in the cavefish embryos than in those of surface dwellers (center). The optic vesicles (bottom) of surface-dwelling fish embryos are normal size and have small domains of *Pax2* expression (specifying the optic stalk). The optic vesicles of the cave-dwelling embryos (where *Pax6* is usually expressed) are much smaller, and the *Pax2*-expressing region has grown at the expense of the *Pax6* region. (After Yamamoto et al. 2004; photographs courtesy of W. Jeffery.)

(A) Surface-dwelling populations

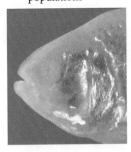

(B) Cave-dwelling populations

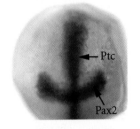

← Ptc

Pax2

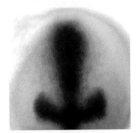

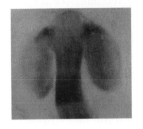

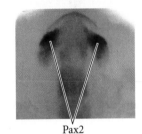

Pax2

Like the cerebral and cerebellar cortices, the neural retina develops into a layered array of different neuronal types. These layers include the light- and color-sensitive photoreceptor cells (**rods** and **cones**); the cell bodies of the ganglion cells; and **bipolar interneurons** that transmit electric stimuli from the rods and cones to the ganglion cells (Figure 12.31). In addition, the retina contains numerous **Müller glial cells** that maintain its integrity, as well as **amacrine neurons** (which lack large axons) and **horizontal neurons** that transmit electric impulses in the plane of the retina.

The neuroblasts of the retina are competent to generate all seven retinal cell types. For instance, if one injects an individual retinal neuroblast with a genetic marker, that marker will be seen in a strip that can include all the different cell types of the retina (Turner and Cepko 1987; Yang et al. 2004). But unlike the neurons in the cerebral cortices,

(A)

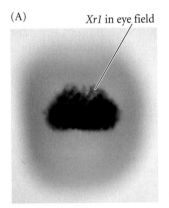

Xr1 in eye field

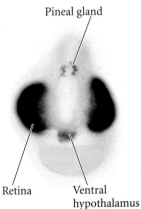

Pineal gland

Retina

Ventral hypothalamus

(B)

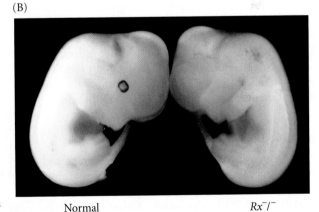

Normal   *Rx*⁻/⁻

(C)

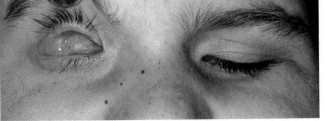

FIGURE 12.30 Expression of the *Rx* genes in vertebrate retina development. (A) Expression pattern of the *Xenopus Xr1* gene in the single eye field of the early neurula, the two developing retinas (as well as in the pineal, an organ that has a presumptive retina-like set of photoreceptors) of a newly hatched tadpole. (B) Lack of eyes in a mouse embryo whose *Rx* gene has been knocked out. (C) Absence of ocular tissue in a human patient whose blindness is due to mutant *Rx* genes. (From Bailey et al. 2004, photographs courtesy of M. Jamrich.)

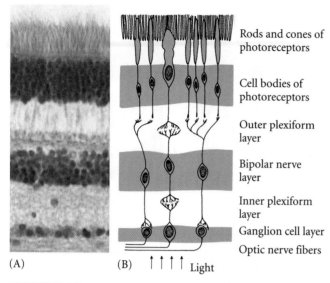

Rods and cones of photoreceptors

Cell bodies of photoreceptors

Outer plexiform layer

Bipolar nerve layer

Inner plexiform layer

Ganglion cell layer

Optic nerve fibers

(A)   (B) ↑ ↑ ↑ ↑ Light

FIGURE 12.31 Retinal neurons sort out into functional layers during development. (A) The three layers of neurons in the adult retina and the synapses between them. (B) A functional depiction of the major neuronal pathway in the retina. Light traverses the layers until it is received by the photoreceptors. The axons of the photoreceptors synapse with bipolar neurons, which transmit electric signals to the ganglion cells. The axons of the ganglion cells join to form the optic nerve, which enters the brain. (A photograph courtesy of G. Grunwald.)

birthday does not play a major role in the specification. Rather, paracrine factors act on postmitotic neurons to specify their fates.

Not all the tissue of the optic cup becomes neural. The tips of the optic cup on either side of the lens, develop into a pigmented ring of muscular tissue called the **iris**. The iris muscles control the size of the pupil (and give an individual his or her characteristic eye color). At the junction between the neural retina and the iris, the optic cup forms the **ciliary body**. This tissue secretes the **aqueous humor**, a fluid needed for the nutrition of the lens and for forming the pressure needed to stabilize the curvature of the eye and the constant distance between the lens and the cornea.

Lens and cornea differentiation

We have been focusing on the retina; but the *eye* can't focus on the retina unless it has a lens and a cornea. The differentiation of the lens tissue into a transparent membrane capable of directing light onto the retina involves changes in cell structure and shape as well as the synthesis of transparent, lens-specific proteins called crystallins.

Shortly after the lens vesicle has detached from the surface ectoderm, mesenchyme cells from the neural crest migrate into the space between the lens and the surface epithelium. These cells condense to form several flat layers of cells, eventually becoming the corneal precursor cells (Figure 12.32A,B; Cvekl and Tamm 2004). As these cells mature, they dehydrate and form tight junctions

among the cells, forming the cornea. Intraocular fluid pressure (from the aqueous humor) is necessary for the correct curvature of the cornea so that light can be focused on the retina (Coulombre 1956, 1965). Intraocular pressure is sustained by a ring of scleral bones (derived from the neural crest), which acts as an inelastic restraint. As invagination proceeds, the cells at the inner portion of the lens vesicle elongate and, under the influence of the neural retina, become the lens fibers (Piatigorsky 1981). As these fibers continue to grow, they synthesize crystallins, which eventually fill up the cell and cause the extrusion of the nucleus.

The crystallin-synthesizing fibers eventually fill the space between the two layers of the lens vesicle (Figure 12.32C). The anterior cells of the lens vesicle constitute a germinal epithelium, which keeps dividing. These dividing cells move toward the equator of the vesicle, and as they pass through the equatorial region, they, too, begin to elongate (Figure 12.32D). Thus, the lens contains three regions: an anterior zone of dividing epithelial cells, an equatorial zone of cellular elongation, and a posterior and central zone of crystallin-containing fiber cells. This arrangement persists throughout the lifetime of the animal as fibers are continuously being laid down. In the adult chicken, the process of differentiation from an epithelial cell to a lens fiber takes two years (Papaconstantinou 1967).

As mentioned in Chapter 5 and 6, the regulation of the crystallin genes is under the control of Pax6, Sox2, and L-Maf (Figure 12.32E). Pax6 appears in the head ectoderm before the lens is formed, and Sox2 (formerly called δEF2) is induced in the lens placode by BMP4 secreted from the optic vesicle. Once Pax6 and Sox2 are together in the same cells, lens differentiation begins, and the crystallin genes are activated. L-Maf also appears to be induced by the optic vesicle, and it appears later than Sox2. It may be needed for the maintenance of crystallin gene expression (Reza and Yasuda 2004; Kondoh et al. 2004).

WEBSITE **12.7 Why babies don't see well.** The retinal photoreceptors are not fully developed at birth. As the child grows older, the density of photoreceptors increases, allowing far better discrimination and nearly 350 times the light-absorbing capacity that is present at birth.

The Epidermis and the Origin of Cutaneous Structures

The origin of epidermal cells

The skin is the largest organ in our bodies. A tough, elastic, water-impermeable membrane, skin protects our body against dehydration, injury, and infection. Moreover, it is constantly renewable. The skin is composed of a lower der-

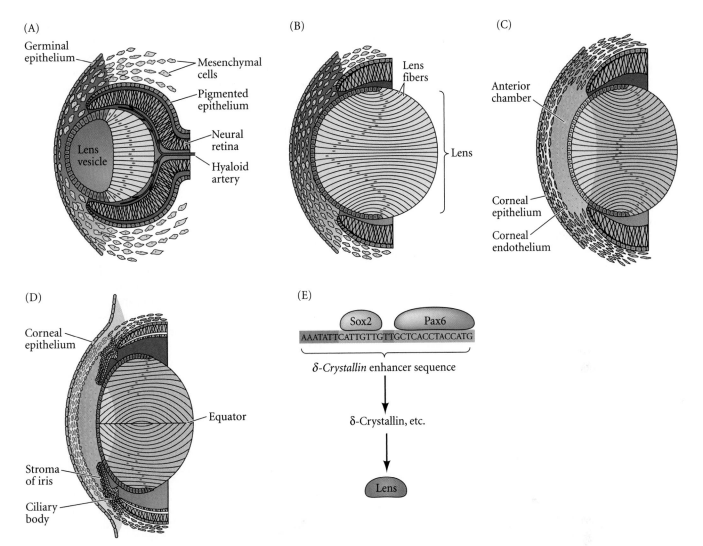

FIGURE 12.32 Differentiation of the lens and the anterior portion of the mouse eye. (A) At embryonic day 13, the lens vesicle detaches from the surface ectoderm and has invaginated into the optic cup. Mesenchyme cells from the neural crest migrate into this space. The elongation of the interior lens cells begins, producing primary lens fibers. (B) At day 14, the lens is filled with crystallin-synthesizing fibers. The mesenchyme cells between the lens and surface condense to form several layers. (C) At day 15, the lens detaches from the corneal layers, generating an anterior cavity. (D) At day 15.5, corneal layers differenti-ate and begin to become transparent. The anterior edge of the optic cup enlarges to form a non-neural region containing the iris muscles and the ciliary body. New lens cells are derived from the anterior lens epithelium. As the lens grows, the nuclei of the primary lens cells degenerate and new lens fibers grow from the epithelium on the lateral sides. (E) Close binding of the Sox2 and Pax6 transcription factors on a small region of the *δ-crystallin* enhancer. (A–D after Cvekl and Tamm 2004; E after Kondoh et al. 2004.)

mis and an outer epidermis. The epidermis originates from the ectodermal* cells covering the embryo after neurula-tion. Originally, this tissue is one cell layer thick, but in most vertebrates it soon becomes a two-layered structure.

---

*Some confusing vocabulary: *Epidermis* is the outer layer of skin. The *ectoderm* is the germ layer that forms the epidermis, neural tube, and neural crest. *Epithelial* refers to a sheet of cells that are held together tightly (as opposed to the loosely connected *mesenchy-mal* cells). Epithelia can be produced by any germ layer. The epidermis and the neural tube both happen to be ectodermal epithelia; the gut is an *endodermal* epithelium.

The outer layer gives rise to the **periderm**, a temporary covering that is shed once the inner layer differentiates to form a true epidermis. The inner layer, called the **basal layer** or **stratum germinativum**, contains proliferating cells attached to a basement membrane (Figure 12.33). The basal layer contains the **epidermal stem cells** that divide to con-tinuously replace the upper layer of epidermis that gets sloughed off. Lechler and Fuchs (2005) have shown that, like the neural stem cells of the ependymal layer, the epi-dermal stem cells divide asymmetrically: the daughter cell that remains attached to the basal lamina remains a stem cell, while the cell that leaves the basal layer migrates outward and starts differentiating, making the keratins characteristic of skin and joining them into dense intermediate filaments. These differentiated epidermal cells, the **keratinocytes**, are bound tightly together and produce a water-impermeable seal of lipid and protein.

As they age, cell division from the basal layer produces younger cells and pushes the older cells to the border of the skin. After the synthesis of the differentiated products, the cells cease transcriptional and metabolic activities. By now, the cells are dead, flattened sacs of keratin protein, and their nuclei are pushed to one edge of the cell. These cells constitute the **cornified layer** (**stratum corneum**). Throughout life, the dead keratinocytes of the cornified layer are shed—humans lose about 1.5 grams of these cells every day*—and are replaced by new cells. In mice, the journey from the basal layer to the sloughed cell takes about 2 weeks. Human epidermis turns over a bit more slowly. The proliferative ability of the basal layer is remarkable in that it can supply the cellular material to continuously replace 1–2 square meters of skin for several decades.

---

*Most of this skin becomes "house dust" on furniture and floors. If you doubt this, burn some dust; it will smell like singed skin.

Several growth factors stimulate the development of the epidermis. One of these is **transforming growth factor-α** (**TGF-α**). TGF-α is made by the basal cells and stimulates these cells' own division. When a growth factor is made by the same cell that receives it, that factor is called an **autocrine growth factor**. Another growth factor needed for epidermal development is **keratinocyte growth factor** (**KGF**), also known as **fibroblast growth factor 7** (**Fgf7**), a paracrine factor produced by the fibroblasts of the underlying (mesodermally derived) dermis. KGF is received by the basal cells of the epidermis and is thought to regulate their migration and differentiation (Guo et al. 1993; Karvinen et al. 2003).

Cutaneous appendages

The epidermis and dermis interact at specific sites to create the sweat glands and the **cutaneous appendages**: hairs, scales, or feathers (depending on the species). The formation of these appendages requires a series of reciprocal inductive interactions between the dermal mesenchyme and the ectodermal epithelium, resulting in the formation of **placodes** that are the base precursors of hair follicles. Epidermal cells in the regions capable of forming these placodes secrete Wnt protein amongst themselves, and Wnt signaling is critical for the initiation of follicle development (see Reddy et al. 2001; Andl et al. 2002).

WEBSITE **12.8 Developmental genetics of hair formation.** A complex cascade of proteins in the Wnt and FGF signaling pathways regulate the epithelial-mesenchymal interactions of hair formation. Several mutations in humans and mice show how sensitive that formation is to paracrine factors.

In mammals, the first indication that a hair follicle placode (sometimes called a hair germ or hair peg) will form at a particular place is an aggregation of cells in the basal layer of the epidermis (Figure 12.34A,B). This aggregation is directed by the underlying dermal fibroblast cells and occurs at different times and different places in the embryo. It is probable that the dermal signals cause the stabilization of β-catenin in the ectoderm (Gat et al. 1998). The basal epidermal cells elongate, divide, and sink into the dermis. The dermal fibroblasts respond to this ingression of epidermal cells by forming a small node (the **dermal papilla**) beneath the hair germ (Figure

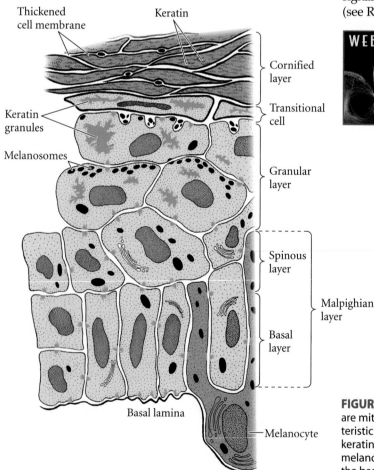

FIGURE 12.33 Layers of the human epidermis. The basal cells are mitotically active, whereas the fully keratinized cells characteristic of external skin are dead and are continually shed. The keratinocytes obtain their pigment through the transfer of melanosomes from the processes of melanocytes that reside in the basal layer. (After Montagna and Parakkal 1974.)

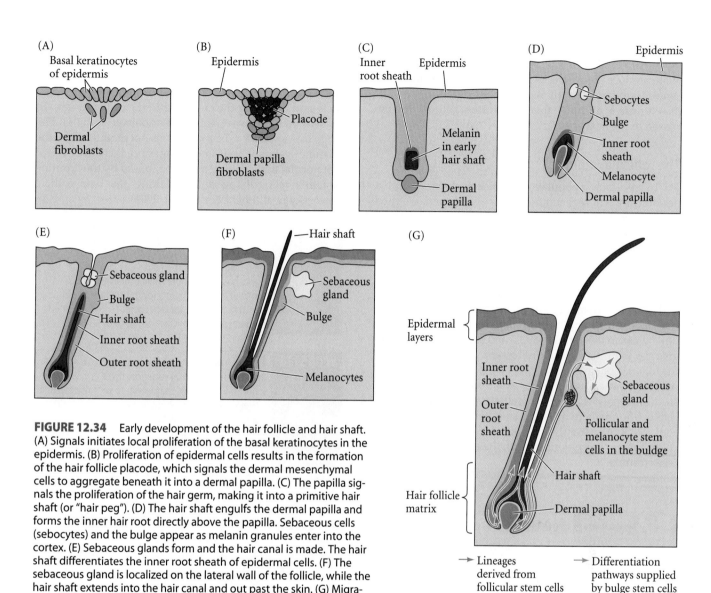

FIGURE 12.34 Early development of the hair follicle and hair shaft. (A) Signals initiates local proliferation of the basal keratinocytes in the epidermis. (B) Proliferation of epidermal cells results in the formation of the hair follicle placode, which signals the dermal mesenchymal cells to aggregate beneath it into a dermal papilla. (C) The papilla signals the proliferation of the hair germ, making it into a primitive hair shaft (or "hair peg"). (D) The hair shaft engulfs the dermal papilla and forms the inner hair root directly above the papilla. Sebaceous cells (sebocytes) and the bulge appear as melanin granules enter into the cortex. (E) Sebaceous glands form and the hair canal is made. The hair shaft differentiates the inner root sheath of epidermal cells. (F) The sebaceous gland is localized on the lateral wall of the follicle, while the hair shaft extends into the hair canal and out past the skin. (G) Migration and lineages of the epidermal stem cells (A–F after Philpott and Paus 1998; G after Alonso and Fuchs 2003.)

12.34C). The dermal papilla then pushes up on the basal stem cells and stimulates them to divide more rapidly. The basal cells respond by producing postmitotic cells that will differentiate into the keratinized hair shaft (Figure 12.34D,E; see Hardy 1992, Miller et al. 1993).

As this differentiation is taking place, two epithelial swellings begin to grow on the side of the hair germ. The cells of the upper swelling form the **sebaceous glands** (Figure 12.34F), which produce an oily secretion, **sebum**. In many mammals (including humans), sebum mixes with shed peridermal cells to form the whitish vernix caseosa, which covers the fetus at birth. The first hairs in the human embryo are of a thin, closely spaced type called **lanugo**. This type of hair is usually shed before birth and is replaced (at least partially by new follicles) by the short and silky **vellus**. Vellus remains on many parts of the human body usually considered hairless, such as the fore-

head and eyelids. In other areas of the body, vellus gives way to "terminal" hair. During a person's life, some of the follicles that produced vellus can later form terminal hair and still later revert to vellus production. The armpits, for instance, have follicles that produce vellus until adolescence, then begin producing terminal shafts. Conversely, in male pattern baldness, the follicles on the scalp revert to producing unpigmented and very fine vellus hair (Montagna and Parakkal 1974).

The lower of the two epithelial swellings, the **bulge** region of the hair follicle, was a great mystery for over 100 years. When Philipp Stöhr drew the histology of the human hair for his 1903 textbook, he showed this bulge ("Wulst") as the attachment site for the arrector pili muscles (which give a person "goosebumps" when they contract). However, in the 1990s, research suggested that the bulge in fact houses populations of at least two remarkable adult stem cells: the multipotent **hair follicle stem cell** and the **melanocyte stem cell**. Melanocyte stem cells continuously produce pigment cells for most of the mammal's lifetime, giving rise to all the pigment cells of the skin (Nishimura et al. 2002); this type of stem cell will be discussed in the next chapter.

The follicular stem cell

The bulge retains a population of epidermal stem cells that periodically will regenerate the hair shaft when it is shed (Pinkus and Mehregan 1981; Cotsarelis et al. 1990). The follicular stem cell gives rise to the hair shaft and the associated glands of the skin and hair (Cotsarelis et al. 1990; Morris and Potten 1999; Taylor et al. 2000). These follicular stem cells are usually slowly cycling and form the basic proliferative unit of the hair shaft.

The mesenchymal-epithelial interactions at the start of the hair cycle direct the epidermal stem cells to migrate downward to the dermal papilla and produce the cells that give rise to the hair shaft. Until recently it was thought that the follicular stem cell also gave rise to the epidermal stem cell (Morris et al. 2005; Tumbar et al. 2004). However, recent studies suggest that the hair follicle and epidermal stem cells are normally two distinct stem cell populations (Levy et al. 2005). Apparently, the follicular stem cells is able to contribute to the epidermis when the epidermis is subjected to trauma.

The bulge may turn out to be a very special place—a niche that allows cells to retain the quality of "stemness." Two paracrine factors that are produced near the bulge, Fgf18 and BMP6, may be responsible for slowing the cell cycle and allowing the cells within the bulge to remain responsive to external stimuli (Blanpain et al. 2004). The bulge also appears to be a site that is not monitored by the immune system. This is interesting, since there is evidence that normal male baldness may be caused by the immune system's recognizing and attacking the hair follicle (Jaworsky et al. 1992; Morris et al. 2004).

WEBSITE **12.9 Normal variation in human hair production.** The human hair has a complex life cycle. Moreover, some hairs (such as those of our eyelashes) grow short while other hairs (such as those of our scalp) grow long. The pattern of hair size and thickness (or lack thereof) is determined by paracrine and endocrine factors.

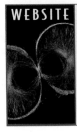

WEBSITE **12.10 Mutations of hair production.** In addition to normal variation, there are also inherited mutations that interfere with normal hair development. Some people are born without the ability to grow hair, while others develop hair over their entire bodies. These genetic conditions give us insights into the mechanisms of normal hair growth.

Snapshot Summary **Neural and Epidermal Ectoderm**

1. The neural tube forms from the shaping and folding of the neural plate. In primary neurulation, the surface ectoderm folds into a tube that separates from the surface. In secondary neurulation, the ectoderm first forms a cord, then forms a cavity within the cord.

2. Primary neurulation is regulated both by intrinsic and extrinsic forces. Intrinsic wedging occurs within cells of the hinge regions, bending the neural plate. Extrinsic forces include the migration of the surface ectoderm towards the center of the embryo.

3. Neural tube closure is also the result of a mixture of extrinsic and intrinsic forces. In humans, various diseases can result if the neural tube fails to close. Folate is important in mediating neural tube closure.

4. The neural crest cells arise at the borders of the neural tube and surface ectoderm. They become located between the neural tube and surface ectoderm, and they migrate away from this region to become peripheral neural, glial, and pigment cells.

5. There is a gradient of maturity in many embryos (especially those of amniotes) so that the anterior develops earlier than the posterior.

6. The brain forms three primary vesicles: prosencephalon (forebrain), mesencephalon (midbrain), and rhombencephalon (hindbrain). The prosencephalon and rhombencephalon will become subdivided.

7. The brain expands through fluid secretion putting positive pressure on the vesicles.

8. The dorsal-ventral patterning of the neural tube is accomplished by proteins of the TGF-β superfamily secreted from the surface ectoderm and the roof plate of the neural tube, and by Sonic hedgehog protein secreted from the notochord and floor plate cells. Gradients of these proteins trigger the synthesis of particular transcription factors that specify the neuroepithelium.

9. The neurons of the brain are organized into cortices (layers) and nuclei (clusters).

10. New neurons are formed by the division of neural stem cells in the wall of the neural tube (called the ventricular zone). The resulting neural precursors, or neuroblasts, can migrate away from the ventricular zone and form a new layer, called the mantle zone

(gray matter). Neurons forming later have to migrate through the existing layers. This process forms the cortical layers.

11. In the cerebellum, migrating neurons form a second germinal zone, called the external granule layer. Other neurons migrate out of the ventricular zone on the processes of glial cells.

12. The cerebral cortex in humans, called the neocortex, has six layers. Cell fates are often fixed as they undergo their last division. Neurons derived from the same stem cell may end up in different functional regions of the brain.

13. Human brains appear to differ from those of other primates by their retention of the fetal neuronal growth rate during early childhood, the migration of cells from the telencephalon to the diencephalon, the amount of transcriptional activity, the presence of certain *FoxP2* alleles, and by a spurt of neuronal growth and myelination that occurs during puberty.

14. Neural stem cells have been observed in the adult human brain. We now believe that humans can continue making neurons throughout life, although at nowhere near the fetal rate.

15. Dendrites receive signals from other neurons, while axons transmit signals to other neurons. The gap between cells where signals are transferred from one neuron to another (through the release of neurotransmitters) is called a synapse.

16. Axons grow from the nerve cell body, or soma. They are led by the growth cone.

17. The retina forms from an optic vesicle that extends from the brain. Pax6 plays a major role in eye formation, and the downregulation of *Pax6* by Sonic hedgehog (Shh) in the center of the brain splits the eye-forming region of the brain in half. If Shh is not expressed there, a single medial eye results.

18. The photoreceptor cells of the retina gather light and transmit an electric impulse through interneurons to the retinal ganglion cells. The axons of the retinal ganglion cells form the optic nerve.

19. The lens and cornea form from the surface ectoderm. Both must become transparent.

20. The basal layer of the surface ectoderm becomes the germinal layer of the skin. Epidermal stem cells divide to produce differentiated epidermal cells (keratinocytes) and more stem cells.

21. The follicular stem cells, which are capable of creating hair follicles, reside in the bulge of the hair follicle.

For Further Reading

Complete bibliographical citations for all literature cited in this chapter can be found on the Vade Mecum CD that accompanies the book and at the free access website www.devbio.com

Bailey, T. J., H. El-Hodiri, L. Zhang, R. Shah, P. H. Mathers and M. Jamrich. 2004. Regulation of vertebrate eye development by *Rx* genes. *Int. J. Dev. Biol.* 48: 761–770.

Catala, M., M.-A. Teillet, E. M. De Robertis and N. M. Le Douarin. 1996. A spinal cord fate map in the avian embryo: While regressing, Hensen's node lays down the notochord and floor plate thus joining the spinal cord lateral walls. *Development* 122: 2599–2610.

Colas, J.-F. and G. C. Schoenwolf. 2001. Towards a cellular and molecular understanding of neurulation. *Dev. Dynam.* 221: 117–145.

Cooper, M. K., J. A. Porter, K. E. Young and P. A. Beachy. 1998. Teratogen-mediated inhibition of target tissue response to Shh signaling. *Science* 280: 1603–1607.

Eriksson, P. S., E. Perfiliea, T. Björn-Eriksson, A.-M. Alborn, C. Nordberg, D. A. Peterson and F. H. Gage. 1998. Neurogenesis in the adult human hippocampus. *Nature Med.* 4: 1313–1317.

Gogtay, N. and 11 others. 2004. Dynamic mapping of human cortical development during childhood through early adulthood. *Proc. Nat. Acad. Sci. USA* 101: 8174–8179.

Halder, G., P. Callaerts and W. J. Gehring. 1995. Induction of ectopic eyes by targeted expression of the *eyeless* gene in *Drosophila*. *Science* 267: 1788–1792.

Hanashina, C., S. C. Li, L. Shen, E. Lai and G. Fishell. 2004. Foxg1 suppresses early cortical cell fate. *Science* 303: 56–59.

Hatten, M. E. 1990. Riding the glial monorail: A common mechanism for glial-guided neuronal migration in different regions of the mammalian brain. *Trends Neurosci.* 13: 179–184.

Jessell, T. M. 2000. Neuronal specification in the spinal cord: Inductive signals and transcriptional codes. *Nature Rev. Genet.* 1: 20–29.

Lawson, A., H. Anderson and G. C. Schoenwolf. 2001. Cellular mechanisms of neural fold formation and morphogenesis in the chick embryo. *Anat. Rec.* 262: 153–168.

Milunsky, A., H. Jick, S. S. Jick, C. L. Bruell, D. S. Maclaughlen, K. J. Rothman and W. Willett. 1989. Multivitamin folic acid supplementation in early pregnancy reduces the prevalence of neural tube defects. *J. Am. Med. Assoc.* 262: 2847–2852.

Turner, D. L. and C. L. Cepko. 1987. A common progenitor for neurons and glia persists in rat retina late in development. *Nature* 328: 131–136.

Neural Crest Cells and Axonal Specificity

13

IN THIS CHAPTER WE CONTINUE our discussion of ectodermal development, focusing here on neural crest cells and axonal guidance. Neural crest cells and axonal growth cones (the motile tips of axons) share the property of having to migrate far from their source of origin to specific places in the embryo. Both must recognize cues in order to begin this migration, and both must respond to signals that guide them along specific routes to their final destination. Recent research has revealed that neural crest cells and axonal growth cones recognize many of the same signals.

THE NEURAL CREST

Although it is derived from the ectoderm, the neural crest is so important that it has sometimes been called the "fourth germ layer." It has even been said, somewhat hyperbolically, that "the only interesting thing about vertebrates is the neural crest" (quoted in Thorogood 1989). The cells of the neural crest migrate extensively to generate a prodigious number of differentiated cell types (Table 13.1). These cell types include (1) the neurons and glial cells of the sensory, sympathetic, and parasympathetic nervous systems; (2) the epinephrine-producing (medulla) cells of the adrenal gland; (3) the pigment-containing cells of the epidermis; and (4) many of the skeletal and connective tissue components of the head. The fate of individual neural crest cells depends to a large degree on the locations to which they migrate.

Specification of the Neural Crest Cells

Neural crest cells originate at the dorsalmost region of the neural tube. The pioneering embryologist Wilhelm His (1868) first described these cells as a "band of particular material lying between the presumptive epidermis and the presumptive neural tube." In his vital dye mapping of the salamander embryo (see Chapter 1), Vogt (1925) showed that the neural crest cells formed at the boundary where the neural plate met the prospective epidermis. By transplanting tissues between pigmented and unpigmented salamander gastrulae, Moury and Jacobson (1990) showed that the salamander neural crest appears to come from both the presumptive epidermis and the presumptive neural plate cells. Similarly, transplantation experiments in which quail neural plate is grafted into chick non-neural ectoderm have shown that juxtaposing these tissues induces the formation of neural crest cells, and that both the prospective neural plate and the prospective epidermis contribute to the avian neural crest (Selleck and Bronner-Fraser 1995; see also Mancilla and Mayor 1996).

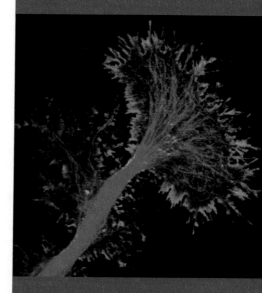

TABLE 13.1 Some derivatives of the neural crest

Derivative	Cell type or structure derived
Peripheral nervous system (PNS)	Neurons, including sensory ganglia, sympathetic and parasympathetic ganglia, and plexuses Neuroglial cells Schwann cells
Endocrine and paraendocrine derivatives	Adrenal medulla Calcitonin-secreting cells Carotid body type I cells
Pigment cells	Epidermal pigment cells
Facial cartilage and bone	Facial and anterior ventral skull cartilage and bones
Connective tissue	Corneal endothelium and stroma Tooth papillae Dermis, smooth muscle, and adipose tissue of skin, head, and neck Connective tissue of salivary, lachrymal, thymus, thyroid, and pituitary glands Connective tissue and smooth muscle in arteries of aortic arch origin

Source: After Jacobson 1991, based on multiple sources.

The specification of the neural crest at the neural plate-epidermis boundary is a multistep process (see Huang and Saint-Jeannet 2004; Meulemans and Bronner-Fraser 2004). The first step appears to be the location of this neural plate border. In amphibians, this border appears to be specified by intermediate concentrations of BMPs. Indeed, in the 1940s, Raven and Kloos (1945) showed that while the presumptive notochord could induce both the amphibian neural plate and neural crest tissue (presumably blocking nearly all BMPs), the somite mesoderm and lateral plate mesoderm could induce only the neural crest. In amniotes, the neural crest forms where moderate levels of several signaling molecules are present (Bang et al. 1999; Garcia-Castro et al. 2002; Lewis et al. 2004; Monsoro-Burq et al. 2003). These paracrine factors—BMPs, Wnts, and FGFs—induce a set of transcription factors called the **neural plate border specifiers**. These factors, including Distalless-5 and Pax3, prevent the region from becoming either neural plate or epidermis. Moreover, the border specifiers induce a second set of transcription factors, the **neural crest specifiers** (such as Slug and FoxD3) in those cells that are to become the neural crest (Figure 13.1). FoxD3 appears to be critical for the specification of ectodermal cells as neural crest. When FoxD3 is inhibited from functioning, neural crest differentiation is inhibited. Conversely, when FoxD3 is expressed ectopically (by electroporating FoxD3 into neural plate cells), those neural plate cells begin to express proteins characteristic of the neural crest. The Slug protein

appears to be required in order for neural crest cells to leave the epithelium and migrate (Nieto et al. 1994; Kos et al. 2001; Sasai et al. 2001).

 13.1 Avian neurulation. A companion to Figure 13.1, this website displays a thoughtful animation of the formation of the chick neural crest.

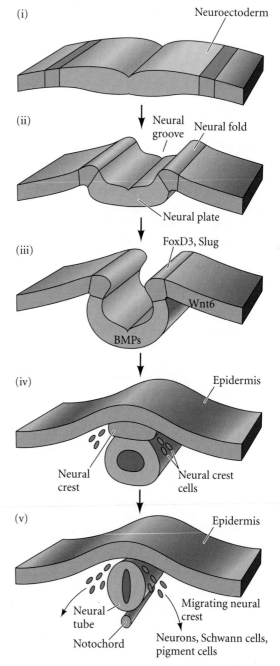

FIGURE 13.1 Schematic representation of neural crest formation in an amniote (chick) embryo, shown in cross section. Neural crest cells form at the junction between the Wnt6-expressing epidermal ectoderm and the BMPs produced by the presumptive neural ectoderm. (After Trainor and Krumlauf 2002.)

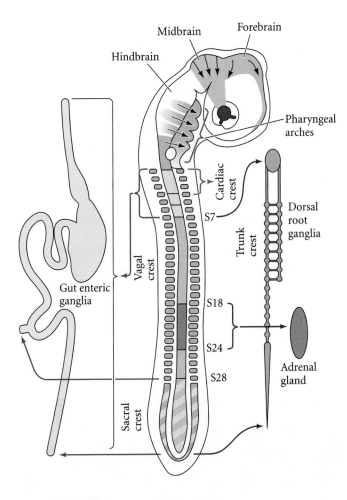

FIGURE 13.2 Regions of the chick neural crest. The cranial neural crest migrates into the pharyngeal arches and the face to form the bones and cartilage of the face and neck. It also produces the cranial nerves. The vagal neural crest (near somites 1–7) and the sacral neural crest (posterior to somite 28) form the parasympathetic nerves of the gut. The cardiac neural crest cells arise near somites 1–3; they are critical in making the division between the aorta and the pulmonary artery. Neural crest cells of the trunk (about somite 6 through the tail) make sympathetic neurons and pigment cells (melanocytes), and a subset of these (at the level of somites 18–24) form the medulla portion of the adrenal gland. (After Le Douarin 1982.)

sal root ganglia containing the sensory neurons. Those that continue more ventrally form the sympathetic ganglia, the adrenal medulla, and the nerve clusters surrounding the aorta. The later-migrating trunk neural crest cells become pigment-synthesizing **melanocytes**, migrating dorsolaterally into the ectoderm and moving through the skin toward the ventral midline of the belly.

- The **vagal** and **sacral neural crest** cells generate the **parasympathetic (enteric) ganglia** of the gut (Le Douarin and Teillet 1973; Pomeranz et al. 1991). The vagal (neck) neural crest lies opposite chick somites 1–7, while the sacral neural crest lies posterior to somite 28. Failure of neural crest cell migration from these regions to the colon results in the absence of enteric ganglia and thus to the absence of peristaltic movement in the bowels.
- The **cardiac neural crest** is a subregion of the vagal neural crest and extends from the first to the third somites (Kirby 1987; Kirby and Waldo 1990). Cardiac neural crest cells can develop into melanocytes, neurons, cartilage, or connective tissue (of the third, fourth, and sixth pharyngeal arches). In addition, this region of the neural crest produces the entire muscular-connective tissue wall of the large arteries (the "outflow tracts") as they arise from the heart, as well as contributing to the septum that separates pulmonary circulation from the aorta (Le Lièvre and Le Douarin 1975).

Regionalization of the neural crest

The neural crest is a transient structure, as its cells disperse soon after the neural tube closes. The crest can be divided into four main (but overlapping) regions, each with characteristic derivatives and functions (Figure 13.2):

- **Cranial (cephalic) neural crest** cells migrate dorsolaterally to produce the craniofacial mesenchyme, which differentiates into the cartilage, bone, cranial neurons, glia, and connective tissues of the face. These cells also enter the pharyngeal arches and pouches to give rise to thymic cells, the odontoblasts of the tooth primordia, and the bones of the middle ear and jaw.*
- **Trunk neural crest** cells take one of two major pathways. The early migratory pathway takes trunk neural crest cells ventrolaterally through the anterior half of each somitic sclerotome. (**Sclerotomes**, derived from somites, are blocks of mesodermal cells that will differentiate into the vertebral cartilage of the spine.) Those trunk neural crest cells that remain in the sclerotomes form the **dor-**

Trunk Neural Crest

Migration pathways of trunk neural crest cells

There are two major pathways taken by cells migrating from the trunk-level neural crest (Figure 13.3A). Trunk neural crest cells that leave early follow a **ventral pathway** away from the neural tube. Fate mapping experiments show that these cells become sensory (dorsal root) and sympathetic neurons, adrenomedullary cells, and Schwann cells (Weston 1963; Le Douarin and Teillet 1974). In birds and mammals (but not fish and frogs),† these cells migrate ventrally through the anterior, but not the posterior, sec-

---

*The **pharyngeal (branchial) arches** (see Figure 1.3) are outpocketings of the head and neck region into which cranial neural crest cells migrate. The **pharyngeal pouches** form between these arches and become the thyroid, parathyroid, and thymus.

---

†In the migration of fish neural crest cells, the sclerotome is not important; rather, the myotome appears to guide the migration of the crest cells ventrally (Morin-Kensicki and Eisen 1997).

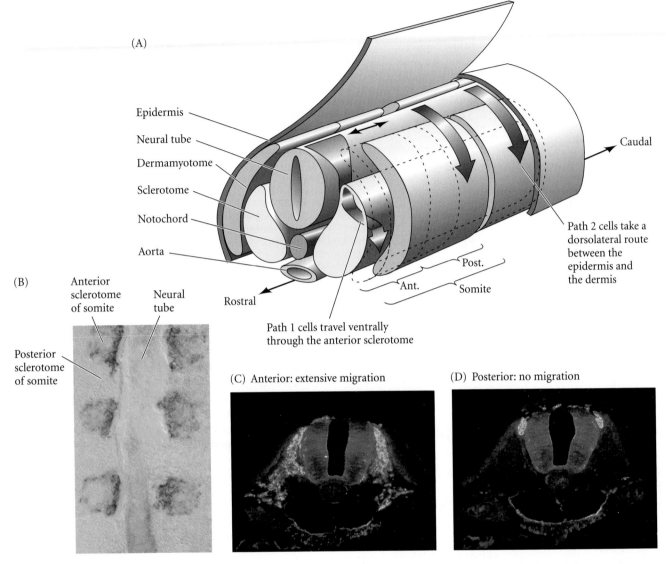

(A)

Epidermis

Neural tube

Dermamyotome

Sclerotome

Notochord

Aorta

Caudal

Path 2 cells take a dorsolateral route between the epidermis and the dermis

Post.

Ant.

Somite

Rostral

Path 1 cells travel ventrally through the anterior sclerotome

(B)

Anterior sclerotome of somite

Neural tube

Posterior sclerotome of somite

(C) Anterior: extensive migration

(D) Posterior: no migration

FIGURE 13.3 Neural crest cell migration in the trunk of the chick embryo. (A) Schematic diagram of trunk neural crest cell migration. Cells taking path 1 (the ventral pathway) travel ventrally through the anterior of the sclerotome (that portion of the somite that generates vertebral cartilage). Those cells initially opposite the posterior portion of a sclerotome migrate along the neural tube until they come to an anterior region. These cells contribute to the sympathetic and parasympathetic ganglia as well as to the adrenomedullary cells and dorsal root ganglia. Other trunk neural crest cells enter path 2 (the dorsolateral pathway) somewhat later. These cells travel along a dorsolateral route beneath the ectoderm, and become pigment-producing melanocytes. (Migration pathways are shown on only one side of the embryo.) (B) These fluorescence photomicrographs of longitudinal sections of a 2-day chick embryo are stained red with antibody to HNK-1, which selectively recognizes neural crest cells. Extensive staining is seen in the anterior, but not in the posterior, half of each sclerotome. (C,D) Cross sections through these areas, showing (C) extensive migration through the anterior portion of the sclerotome, but (D) no migration through the posterior portion. Here, the antibodies to HNK-1 are stained green. (B from Wang and Anderson 1997; C–D from Bronner-Fraser 1986; photographs courtesy of the authors.)

tion of the sclerotomes (Figure 13.3B–D; Rickmann et al. 1985; Bronner-Fraser 1986; Loring and Erickson 1987; Teillet et al. 1987).

By transplanting quail neural tubes and crests into chick embryos, Teillet and co-workers (1987) were able to mark neural crest cells both genetically and immunologically. The antibody marker recognized and labeled neural crest cells of both species; the genetic marker enabled the inves-

tigators to distinguish between quail and chick cells. These studies showed that neural crest cells initially located opposite the posterior region of a somite migrated anteriorly or posteriorly along the neural tube, and then enter the anterior region of their own or an adjacent somite. These cells join with the neural crest cells that initially were opposite the anterior portion of the somite, and they form the same structures. Thus, each dorsal root ganglion is composed of three neural crest cell populations: one from the neural crest opposite the anterior portion of the somite, and one each from the two neural crest regions opposite the posterior portions of its own and the neighboring somites.

Trunk crest cells that emigrate later travel along the **dorsolateral pathway** and become **melanocytes**, the melanin-forming pigment cells. They travel between the epidermis and the dermis, entering the ectoderm through minute holes in the basal lamina (which they themselves may create). Once in the ectoderm they colonize the skin and hair follicles (Mayer 1973; Erickson et al. 1992). The dorsolateral pathway was demonstrated in a series of classic experiments by Mary Rawles (1948), who transplanted the neural tube and neural crest from a pigmented strain of chickens into the neural tube of an albino chick embryo (see Figure 1.11).

The mechanisms of trunk neural crest migration

EMIGRATION FROM THE NEURAL TUBE Any analysis of migration (be it of birds, butterflies, or neural crest cells) has to ask these questions:

1. What signals initiate migration?
2. When does the migratory agent become competent to respond to these signals?
3. How do the migratory agents know the route to travel?
4. What signals indicate that the destination has been reached?

Neural crest cells initiate their migration from the neural folds through interactions of the neural plate with the presumptive epidermis. Originally part of an epithelium, the neural crest cells undergo an epithelial-to-mesenchymal transformation whereby they lose their attachments to other cells and migrate away from the epithelial sheet. These changes can be mimicked by culturing neural plate cells with Wnt, or bone morphogenetic proteins 4 and 7—proteins that are known to be secreted by the presumptive epidermis (see Chapter 12). In the presence of Wnt and FGF proteins (from the underlying mesoderm), BMP4 and BMP7 induce the expression of the proteins Slug and RhoB in the cells destined to become neural crest (Figure 13.4; Nieto et al. 1994; Mancilla and Mayor 1996; Liu and Jessell 1998; LaBonne and Bronner-Fraser 1998). If either Slug or RhoB is inactivated or inhibited, the crest cells fail to emigrate from the neural tube.

In order for cells to leave the neural crest, there must be both pushes and pulls. RhoB is thought to be involved in establishing the cytoskeletal conditions for migration by promoting actin polymerization into microfilaments and the attachment of these microfilaments to the cell membrane (Hall 1998). It is possible that (at least in some species) RhoB is activated by noncanonical Wnt signaling by Wnt11 (De Callisto et al. 2005). However, the crest cells cannot leave the neural tube as long as they are tightly connected to one another. One of the functions of the Slug pro-

FIGURE 13.4 All migrating neural crest cells are stained red by antibody to HNK-1. The RhoB protein (green stain) is expressed in cells as they leave the neural crest. Cells expressing both HNK-1 and RhoB appear yellow. (After Liu and Jessell 1998; photograph courtesy of T. M. Jessell.)

tein is to activate the factors that dissociate the E-cadherins binding these cells together. Originally found on the surface of the neural crest cells, these adhesion proteins are downregulated at the time of cell migration. Moreover, after migration, the migrating neural crest cells re-express cadherins as they aggregate to form the dorsal root and sympathetic ganglia (Takeichi 1988; Akitaya and Bronner-Fraser 1992; Savagne et al. 1997).

THE VENTRAL MIGRATION PATHWAY The ventral route taken by migrating trunk neural crest cells is controlled by extracellular matrices and by the chemotactic factors the cells encounter (Newgreen and Gooday 1985; Newgreen et al. 1986). The extracellular matrix is critical in patterning ventral migration through the anterior portion of the sclerotomes. One set of matrix proteins is *permissive* for migration. These include fibronectin, laminin, tenascin, various collagen molecules, and proteoglycans—all found throughout the matrix encountered by the neural crest cells. Upon migration, chick trunk neural crest cells begin to express the $\alpha 4\beta 1$ integrin protein, which binds to several of the extracellular matrix proteins. The expression of this integrin molecule is necessary for the locomotion and survival of the newly freed neural crest cells. Without it, the cells leave the neural tube, but then become disoriented and often undergo apoptosis (Testaz and Duband 2001). **Thrombospondin**, another extracellular matrix molecule, is found in the anterior, but not in the posterior, portion of the sclerotome (Tucker et al. 1999). Thrombospondin is a good substrate for neural crest cell adhesion and migration, and it may cooperate with fibronectin and laminin to promote neural crest cell migration through the anterior portion of the somite.

FIGURE 13.5 Segmental restriction of neural crest cells and motor neurons by the ephrin proteins of the sclerotome. (A) Negative correlation between regions of ephrin in the sclerotome (dark blue stain, left) and the presence of neural crest cells (green HNK-1 stain, right). (B) When neural crest cells are plated on fibronectin-containing matrices with alternating stripes of ephrin, they bind to those regions lacking ephrin. (C) Composite scheme showing the migration of spinal cord neural crest cells and motor neurons through the ephrin-deficient anterior regions of the sclerotomes. (For clarity, the neural crest cells and motor neurons are each depicted on only one side of the spinal cord.) (A and B from Krull et al. 1997; C after O'Leary and Wilkinson 1999.)

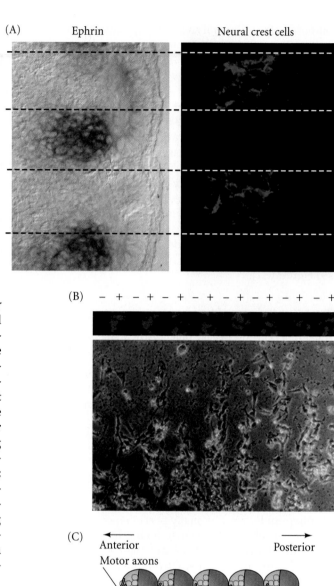

(A) Ephrin   Neural crest cells

(B) − + − + − + − + − + − + − + − +

(C) Anterior   Posterior
Motor axons
Neural crest cells   Sclerotome of somites   Spinal cord

Equally important for the patterning of neural crest cell movement are those proteins that *impede* migration. Two such proteins are the **ephrins** and **semaphorin-3F**. These proteins are expressed in the *posterior* section of each sclerotome, and wherever they are, neural crest cells do not go (Figure 13.5A). The ephrin on the posterior sclerotome is recognized by its receptor, Eph, on the neural crest cells. Similarly, semaphorin-3F on the posterior sclerotome cells is recognized by its receptor, neuropilin-2, on the migrating neural crest cells. If neural crest cells are plated on a culture dish that contains alternate stripes of immobilized cell membrane proteins with or without ephrins, the cells will leave the ephrin-containing regions and move along the stripes that lack ephrin (Figure 13.5B; Krull et al. 1997; Wang and Anderson 1997; Gammill et al. 2006). Mutant mice deficient in either semaphorin-3F or neuropilin-2 have severe neural crest migration abnormalities throughout the trunk. This patterning of neural crest cell migration generates the overall segmental character of the peripheral nervous system, reflected in the positioning of the dorsal root ganglia and other neural crest-derived structures.

The migration of particular populations of neural crest cells along the ventral pathway is often accomplished by soluble factors. The neural crest cells from the vagal and sacral regions form the enteric ganglia of the colon that control intestinal peristalsis. These crest cells are attracted to the colon by **glial-derived neurotrophic factor (GDNF)** produced by the gut mesenchyme (Young et al. 2001; Natarajan et al. 2002). If either GDNF or its receptor (RET) are deficient in mice or humans, the pup or child suffers from Hirschsprung disease, a syndrome wherein the intestine cannot properly void its solid wastes.

Neural crest cell migration is also critical for forming the adrenal gland, an organ generated from two distinct cell types. The outer portion (cortex) of the adrenal gland secretes hormones such as cortisol, while the inner portion (medulla) secretes hormones such as epinephrine. The adrenal medullary cells are derived from the trunk neural crest, while the adrenal cortex is derived from the intermediate mesoderm. The neural crest cells are directed into the medulla-forming region by chemotactic factors from those tissues of the intermediate mesoderm that form the adrenal cortex. If other regions of the intermediate mesoderm are converted into adrenal medulla (by electroporating the SF1 transcription factor into these cells), the crest cells will migrate to those ectopic places (Saito and Takahashi 2005). This migration appears to be directed by a gradient of BMP coming from the adrenal cortical precursors. If Noggin is overexpressed near the adrenal-forming region, trunk neural crest cells are not told to migrate there. Thus, both extracellular matrices and soluble chemotactic factors influence the migration of the trunk neural crest cells.

THE DORSAL MIGRATION PATHWAY   The early neural crest cells migrate along the ventral pathway described above. The neural crest cells that migrate later follow the *dorsal* path. Like the ventral pathway, the migration of neural crest cells into the dorsal pathway is controlled by the maturation of the somites adjacent to the neural tube. The first part of the somite to mature is the sclerotome (which forms the vertebral cartilage), and the first migrating neural crest cells enter the ventral pathway through the anterior portion of each sclerotome. Other neural crest cells remain above the neural tube, in what is often called the "staging area."

One of the last parts of the somite to mature is the dermotome, a region that will form the dermis of the back. Only after the dermis has formed do the neural crest cells in the staging area migrate along the dorsolateral pathway to become melanocytes (Weston and Butler 1966; Tosney 2004). The dermis originates at a place far removed from where neural crest cells reside, and it appears that the cells of the dermis secrete a chemotactic factor that attracts the crest cells. The timing of somite differentiation appears to regulate the spatial and temporal migration of the neural crest, assuring that the first migrating cells follow the ventral route and only later (i.e., after the dermis forms) will the melanocyte precursors follow the dorsal route. If dermis is transplanted into the embryo before it normally would form, it attracts neural crest cells into it—it even recruits cells that are already traveling along the ventral pathway.

Once the neural crest cells start moving, the extracellular matrix becomes important. For the melanocyte lineage, ephrin can play a *positive* role in cell migration rather than exerting the negative effect it has on ventrally moving cells. Ephrin expressed along the dorsolateral migration pathway stimulates the migration of late-migrating cells. Ephrin activates its receptor, Eph, on the neural crest cell membrane, and this Eph signaling appears to be critical for promoting this migration. Disruption of Eph signaling in late-migrating neural crest cells prevents their dorsolateral migration (Santiago and Erickson 2002). **Stem cell factor** (SCF; see Chapter 6) is also critical for this migration, allowing the continued proliferation of those neural crest cells that enter the skin; SCF may also serve as an antiapoptosis factor. If SCF is experimentally secreted from tissues, such as the cheek epithelium or the footpads, that do not usually synthesize this protein (and do not usually have melanocytes), neural crest cells will enter those regions and become melanocytes (Kunisada et al. 1998).

Neural crest cells migrating along the dorsal route become committed to forming melanocytes. As they travel through the dermis and epidermis, they eventually enter the developing hair follicles and take up residence at the base of the follicle bulge. Here, they become **melanocyte stem cells** (Mayer 1973; Nishimura et al. 2002). A portion of these cells migrate outside the bulge at the beginning of each hair development cycle to differentiate into mature melanocytes and provide pigmentation to the hair shaft (Figure 13.6A). Nishimura and colleagues (2005) have documented that the reason that the hair of mice and humans grays with age is that melanocyte stem cells become depleted from the bulge. While they are in the stem cell niche, melanocyte stem cells are inhibited from differentiating because of the strange regulation of the Mitf transcription factor (see Chapter 9). Mitf activates the genes of the melanin pathway (Figure 13.6B). The *Mitf* gene is itself activated by the combination of Sox10 and Pax3 proteins. However, on some of the genes activated by Mitf, Pax3 binds to the same place on the enhancer as the Mitf, thus competing for the site (Lang et al. 2005). So even though Pax3 stimulates melanocyte differen-

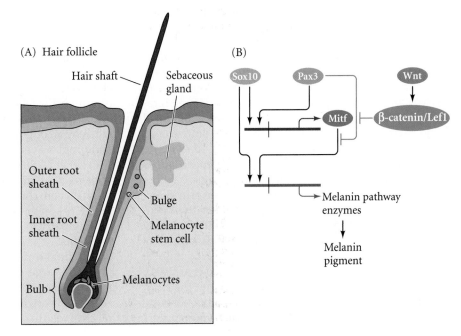

FIGURE 13.6   Hair follicle of a mouse. (A) Melanocyte stem cells are located in the lower region of the hair follicle bulge. (B) Sox10 and Pax3 proteins activate expression of *Mitf*. Mitf protein and Sox10 can activate the enzyme-encoding genes of the melanin pathway. However, Pax3 *competes* with Mitf for the enhancer site on these genes, so the genes remain unexpressed as long as Pax3 prevents Mitf binding. Once outside the niche, however, Wnt signals cause β-catenin to enter the nucleus and bind with Lef1. Lef1 is a transcription factor that can remove Pax3, allowing Mitf to bind. With Mitf and Sox10 bound, the genes encoding the melanin-promoting enzymes can be transcribed and melanin can be made. (After Lang et al. 2005; Nishimura et al. 2005.)

tiation by activating *Mitf*, it also prevents *Mitf* from functioning. Once outside the bulge, however, Wnt signaling generates β-catenin, which binds to a Lef/Tcf transcription factor and displaces the Pax3 from their sites. This allows Mitf to be expressed and activate the melanin-producing genes. In this way, Pax3 can simultaneously determine a cell's melanoblastic fate and also maintain it as a stem cell in an undifferentiated state, poised to differentiate in response to an external stimulus.

WEBSITE 13.2 The specificity of the extracellular matrix. The importance of the extracellular matrix for neural crest cell migration was first shown in a series of creative experiments using mutant salamanders.

WEBSITE 13.3 Mouse neural crest cell mutants. Some of the most important insights into neural crest cell development and migration have come from studies of mutant mice. These mice can be recognized by their altered pigmentation, resulting from abnormalities of neural crest cell proliferation, migration, or differentiation.

The Pluripotency of Neural Crest Cells

One of the most exciting features of neural crest cells is their pluripotency. A single neural crest cell can differentiate into any of several different cell types, depending on its location within the embryo. For example, the parasympathetic gut neurons formed by the vagal neural crest cells (adjacent to somites 1 and 7) produce acetylcholine as their neurotransmitter; they are therefore *cholinergic* neurons. The sympathetic neurons formed by the trunk (thoracic) neural crest cells produce norepinephrine; they are *adrenergic* neurons. But when chick vagal and thoracic neural crests are reciprocally transplanted, the former thoracic crest produces the cholinergic neurons of the parasympathetic ganglia, and the former vagal crest forms adrenergic neurons in the sympathetic ganglia (Le Douarin et al. 1975). Kahn and co-workers (1980) found that premigratory neural crest cells from both the thoracic and the vagal regions contain enzymes for synthesizing both acetylcholine and norepinephrine. Thus there is good evidence that the differentiation of a neural crest cell depends on its eventual location, and not on its place of origin.

But life isn't so simple. There are also intrinsic differences between regions of the neural crest. When *trunk* neural crest cells are transplanted into the head region, they can migrate to the sites of cartilage and cornea formation, but they make neither cartilage nor cornea (Noden 1978; Nakamura and Ayer-Le Lievre 1982; Lwigale et al. 2004). Cardiac neural crest, when transplanted into the head region, can contribute some cells to cartilage, but not nearly as many as control cranial crest cells. There thus appears

to be some restrictions on the differentiated products neural crest cells can form, especially with respect to their origin along anterior-posterior axis of the embryo. However, within general areas (such as trunk or head), there is a large degree of plasticity as to what a neural crest cell can become.

The pluripotency of some neural crest cells is such that even regions of the neural crest that never produce nerves in normal embryos can be made to do so under certain conditions. Cranial neural crest cells from the midbrain region normally migrate into the eye and interact with the pigmented retina to become scleral cartilage cells (Noden 1978). However, if this region of the neural crest is transplanted into the trunk region, it can form sensory ganglion neurons, adrenomedullary cells, glia, and Schwann cells (Schweizer et al. 1983).

The research we have just described studied the potential of populations of neural crest cells. It is still uncertain whether most of the individual cells that leave the neural crest are pluripotent or whether most are already restricted to certain fates. Bronner-Fraser and Fraser (1988, 1989) provided evidence that some, if not most, individual neural crest cells are pluripotent as they leave the crest. They injected fluorescent dextran molecules into individual avian neural crest cells while the cells were still above the neural tube, and then looked to see what types of cells their descendants became after migration. The progeny of a single neural crest cell could become sensory neurons, melanocytes, adrenomedullary cells, and glia (Figure 13.7A).

However, in some instances the neural crest cells may be determined even before migration (see Perez et al. 1999). For example, only those cells (of the vagal neural crest) expressing *Hoxb3* will colonize the gut to form enteric ganglia (Chan et al. 2005). Other studies found evidence that some populations of neural crest cells may become restricted in potential very soon after leaving the neural tube, and that transcription factors constrain which cell types they can produce. Similarly, Henion and Weston (1997) found that the initial neural crest population was a heterogeneous mixture of precursors, almost half of which generated clones containing a single cell type. Reviewing the results of numerous laboratories as well as their own studies, Nicole Le Douarin and her colleagues proposed a model whereby an original multipotential neural crest cell divides and progressively loses its developmental potentials (see Creuzet et al. 2004). These potentials are segregated into lineages that are specific to the cranial neural crest and those that are common to the cranial and trunk crests (Figure 13.7B). In this model, different cells of the neural crest differ greatly in their developmental potential and commitments.

The final differentiation of neural crest cells is determined in large part by the environment to which they migrate (Figure 13.8). For example, BMP2, a protein secreted by the heart, lungs, and dorsal aorta, influences rat neural crest cells to differentiate into autonomic neurons. Such

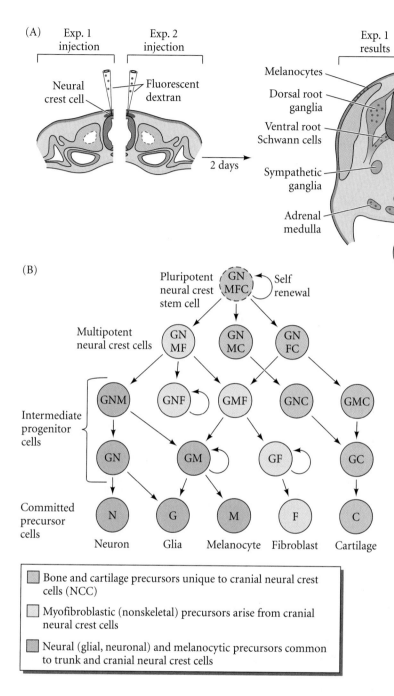

FIGURE 13.7 Pluripotency of trunk neural crest cells. (A) A single neural crest cell is injected with highly fluorescent dextran shortly before migration of the neural crest cells is initiated. The progeny of this cell each receive some of these fluorescent molecules. Two days later, neural crest-derived tissues contain dextran-labeled cells that are descended from the injected precursor. The figure summarizes data from two different experiments. (B) Model for neural crest lineage segregation and the heterogeneity of neural crest cells. The committed precursors of neurons (N), glia (G), melanocytes (M), and so forth would be derived from intermediate progenitors, some of which could act as stem cells. The multipotent neural crest stem cell (top) is hypothetical and need not exist. (A after Lumsden 1988a; B after Le Douarin 2004.)

Legend:
- Bone and cartilage precursors unique to cranial neural crest cells (NCC)
- Myofibroblastic (nonskeletal) precursors arise from cranial neural crest cells
- Neural (glial, neuronal) and melanocytic precursors common to trunk and cranial neural crest cells

neurons form the sympathetic ganglia in the region of these organs (Shah et al. 1996). While BMP2 can induce neural crest cells to become autonomic neurons, another paracrine factor, **glial growth factor** (**GGF**, also called **neuregulin**) suppresses neuronal differentiation and directs development toward glial fates (Shah et al. 1994). The neural crest cells that become the epinephrine-secreting cells of the adrenal medulla appear to be instructed to do so by the presence of the glucocorticoid hormones produced by their neighboring adrenal cortical cells (Anderson and Axel 1986; Vogel and Weston 1990). Wnt1 is able to instruct neural crest cells to become either melanocytes or sensory neu-

rons (Dorsky et al. 1998; Lee et al. 2004). Timing may be critical here; if exposed to Wnt immediately, neural crest cells appear to become sensory neurons. However, if they are exposed to Wnt while in the staging area, which has high concentrations of endothelin-3, the cells are biased to become melanocytes (Baynash et al. 1994; Lahav et al. 1996). Thus, the fate of a neural crest cell can be directed by paracrine factors secreted by the tissue environment through which it migrates.

While both trunk and cranial neural crest cells can form melanocytes, neurons, and glia, only the cells of the cranial neural crest are able to produce cartilage and bone (Table 13.2; see also Figure 13.7). Moreover, if transplanted into the trunk region, cranial crest cells participate in forming trunk cartilage that normally does not arise from neural crest components. This ability to form bone may have been a primitive property of the neural crest and may have been critical for forming the bony armor found in several extinct fish species (Smith and Hall 1993). In other words, the trunk crest has apparently lost the ability to form bone, rather than the cranial crest acquiring this ability. McGonnell and Graham (2002) have shown that bone-forming capacity may still be latent in the trunk neural crest: if cultured in certain hormones and vitamins, the trunk cells become capable of forming bone and cartilage when placed into the head region. Moreover, Abzhanov and colleagues (2003) have shown that the trunk crest cells can act like cranial crest cells (and make skeletal tissue) if the trunk cells are cultured in conditions that cause them to lose the expression of their Hox genes.

FIGURE 13.8 Paracrine factors encountered in the environment help specify the different neural crest-derived lineages in the trunk. BMP signals influence the crest cells to become autonomic neurons, while Wnt signals bias them in the direction of becoming sensory neurons. Glial growth factor specifies the neural crest cells to become Schwann cell glia, and a combinatiuon of endothelin-3 and Wnt appears to specify the melanocytes. The chromaffin cells of the adrenal medulla differentiate under the influence of glucocorticoid hormones made by their neighboring adrenal cortical tissue. (After Bronner-Fraser 2004).

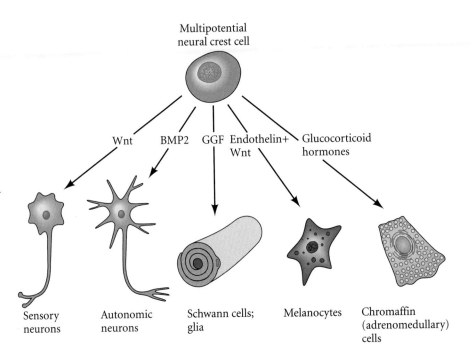

TABLE 13.2 Some derivatives of the pharyngeal arches

Pharyngeal arch	Skeletal elements (neural crest plus mesoderm)	Arches, arteries (mesoderm)	Muscles (mesoderm)	Cranial nerves (neural tube)
1	Incus and malleus (from neural crest); mandible, maxilla, and temporal bone regions (from neural crest)	Maxillary branch of the carotid artery (to the ear, nose, and jaw)	Jaw muscles; floor of mouth; muscles of the ear and soft palate	Maxillary and mandibular divisions of trigeminal nerve (V)
2	Stapes bone of the middle ear; styloid process of temporal bone; part of hyoid bone of neck (all from neural crest cartilage)	Arteries to the ear region: corticotympanic artery (adult); stapedial artery (embryo)	Muscles of facial expression; jaw and upper neck muscles	Facial nerve (VII)
3	Lower rim and greater horns of hyoid bone (from neural crest)	Common carotid artery; root of internal carotid	Stylopharyngeus (to elevate the pharynx)	Glossopharyngeal nerve (IX)
4	Laryngeal cartilages (from lateral plate mesoderm)	Arch of aorta; right subclavian artery; original spouts of pulmonary arteries	Constrictors of pharynx and vocal cords	Superior laryngeal branch of vagus nerve (X)
6*	Laryngeal cartilages (from lateral plate mesoderm)	Ductus arteriosus; roots of definitive pulmonary arteries	Intrinsic muscles of larynx	Recurrent laryngeal branch of vagus nerve (X)

Source: Based on Larsen 1993.
*The fifth arch degenerates in humans

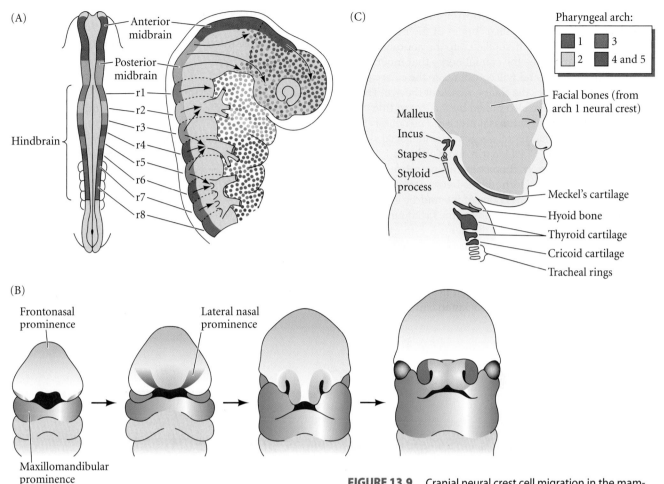

FIGURE 13.9 Cranial neural crest cell migration in the mammalian head. (A) Migrational pathways from the cranial neural crest into the pharyngeal arches and frontonasal process. (B) Continued migration of the cranial neural crest to produce the human face. The frontonasal prominence contributes to the forehead, nose, philtrum of the upper lip (the area between the lip and nose), and to the primary palate. The lateral nasal prominence generates the sides of the nose. The maxillomandibular prominences give rise to the lower jaw, much of the upper jaw, and to the sides of the middle and lower regions of the face. (C) Structures formed in the human face by the mesenchymal cells of the neural crest. The cartilaginous elements of the pharyngeal arches are indicated by colors, and the darker pink region indicates the facial skeleton produced by anterior regions of the cranial neural crest. (A after Le Douarin 2004; B after Helms et al. 2005; C after Carlson 1999.)

Cranial Neural Crest

The head, comprising the face and the skull, is the most anatomically sophisticated portion of the vertebrate body. It is the evolutionary novelty that separates the vertebrates from the other deuterostomes—echinoderms, tunicates, and lancelets (Northcutt and Gans 1983; Wilkie and Morriss-Kay 2001). Indeed, some have suggested that the vertebrates should rather be classified as "cristata"—those animals with a neural crest.

The head is largely the product of the cranial neural crest, and the evolution of jaws, teeth, and facial cartilage occurs through changes in the placement of these cells* (see Chapter 23).

---

*When Julia Platt (1893, 1897) first showed that neural crest cells formed the jaw cartilage and tooth dentine in salamanders, her work was not believed. Her claim went counter to the belief that only mesoderm could form bones and cartilage. Her hypothesis of the neural crest origin of the cranial skeleton gained acceptance only some 50 years later when confirmed by Sven Hörstadius. Unable to secure a university position, she became active in politics, and in 1931 (at the age of 74), she became mayor of Pacific Grove, California (see Zottoli and Seyfarth 1994).

As mentioned in Chapter 12, the hindbrain is segmented along the anterior-posterior axis into compartments called rhombomeres. The cranial neural crest cells migrate ventrally from those regions anterior to rhombomere 8 into the pharyngeal arches and the frontonasal process; their final destination will determine their eventual fate (Figure 13.9A; Table 13.2). The cranial crest cells follow one of three major streams:

1. Neural crest cells from the midbrain and rhombomeres 1 and 2 of the hindbrain migrate to the first pharyngeal

arch (the mandibular arch), forming the jawbones as well as the incus and malleus bones of the middle ear. These cells will also differentiate into the neurons of the trigeminal ganglion—the cranial nerve that innervates the teeth and jaw—and will contribute to the ciliary ganglion that innervates the ciliary muscle of the eye. These neural crest cells are also pulled by the expanding epidermis to form the **frontonasal process**. The neural crest cells of the frontonasal process generate the facial skeleton (Figure 13.9B,C; Le Douarin and Kalcheim 1999).

2. Neural crest cells from rhombomere 4 populate the second pharyngeal arch, forming the hyoid cartilage of the neck as well as the stapes bone of the middle ear. These cells will also form the neurons of the facial nerve.

3. Neural crest cells from rhombomeres 6–8 migrate into the third and fourth pharyngeal arches and pouches to form the hyoid cartilages, and the thymus, parathyroid, and thyroid glands (Serbedzija et al. 1992; Creuzet et al. 2005). If the neural crest is removed from those regions, these structures fail to form (Bockman and Kirby 1984).

Some of these cells migrate caudally to the clavicle (collarbone) where they settle at the sites that will be used for the attachment of certain neck muscles (McGonnell et al. 2001).

Neural crest cells from rhombomeres 3 and 5 do not migrate through the mesoderm surrounding them, but enter into the migrating streams of neural crest cells on either side of them. Cells that do not enter these streams will die (Graham et al. 1993, 1994; Sechrist et al. 1993). Observations of labeled neural crest cells from chick hindbrain, wherein individually marked cells were followed by cameras focusing through a Teflon membrane window in the egg, found that the migrating cells were "kept in line" by interactions with their environment (Kulesa and Fraser 2000). In frog embryos, there is evidence that the separate streams are kept apart by ephrins. Blocking the activity of the Eph receptors causes cells from the different streams to mix together (Smith et al. 1997; Helbling et al. 1998).

SIDELIGHTS & SPECULATIONS

Cranial Neural Crest Cell Migration and Specification

There appear to be two "waves" of cranial neural crest cell emigration. The first wave consists of a pluripotent population of neural crest-derived mesenchymal cells that migrate to form the cartilage and bones of the head and neck, as well as the stromal tissue of pharyngeal organs such as the thyroid and thymus (see Figure 13.9). Hox genes are critical for specifying these regions. The first stream of cells (down to r2) that form the jaw and face do not

express any Hox genes, while the other two streams (r3 through r8) express the first four Hox paralogue genes. If Hox gene expression is experimentally induced in the anterior neural fold, no facial skeleton is formed.

In *Xenopus*, the first paralogous Hox group (*Hoxa1, Hoxb1, Hoxd1*) is responsible for the

migration of neural crest cells into the pharyngeal arches. When antisense morpholinos knocked out the expression of all these genes, no migration of neural crest cells extended the arches, and the gill cartilages were totally absent (McNulty et al. 2005).

In the mouse, different Hox genes may play different roles in this specification. For instance, *Hox2* appears to be essential for the proper development of the second arch derivatives (Figure 13.10). When *Hoxa2* is

FIGURE 13.10 The influence of mesoderm and ectoderm on the axial identity of cranial neural crest cells and the role of *Hoxa2* in regulating second-arch morphogenesis. Neural crest cells (small black circles) from rhombomere 2 do not express Hox genes as they enter pharyngeal arch 1. *Hoxa2* is expressed in cranial crest cells migrating into pharyngeal arch 2, but not into arch 1, and it plays a major role in imposing second arch identity on second arch structures. The diagram indicates the emerging pathways by which *Hoxa2* interacts with ectodermally secreted Fgf8 and BMPs to regulate morphogenesis in the two arches. (After Trainor and Krumlauf 2001.)

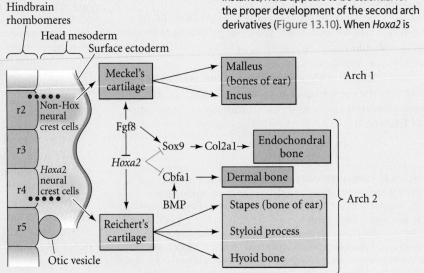

SIDELIGHTS & SPECULATIONS

knocked out from mouse embryos, the neural crest cells of the second pharyngeal arch generate Meckel's cartilage—which is the jaw cartilage characteristic of the first pharyngeal arch (Gendron-Maguire et al. 1993; Rijli et al. 1993). Conversely, if *Hoxa2* is expressed ectopically in the neural crest cells of the first arch, the crest cells of the first arch form cartilage characteristic of the second arch (Grammatopolous et al. 2000; Pasqualetti et al. 2000).

There are reciprocal interactions by which the neural crest cells and the pharyngeal arch regions mutually specify each other. First, the pharyngeal endoderm dictates the differentiation of the neural crest cells. When *individual* mouse or zebrafish cranial crest cells are transplanted from one region of the hindbrain to another, they take on the characteristic Hox gene expression pattern of their host (surrounding pharyngeal) region and lose their original gene expression pattern. However, if *groups* of cranial neural crest cells are so transplanted, they tend to keep their original identity (Trainor and Krumlauf 2000; Schilling 2001). Couly and colleagues (2002) found that removing a specific region of pharyngeal endoderm caused the loss of that particular skeletal structure formed by the neural crest. Moreover, if an extra piece of first ("mandibular") arch pharyngeal endoderm was grafted into the existing pharynx of a chick embryo, neural crest cells would enter this graft and make an extra jaw (Figure 13.11A). Reciprocally, the neural crest cells instruct gene expression in the surrounding tissues. This influences the growth patterns of the arch—e.g., whether the bird's face will have a narrow beak or a broad, ducklike bill (Figure 13.11B; Noden 1991; Schneider and Helms 2003).

Why don't birds have teeth?

One characteristic of the jaws of all birds is that they lack teeth. The dinosaurian ancestors of birds and crocodiles certainly had teeth (*T. rex* may have been a close cousin to the group of theropods that became the birds); modern crocodiles have teeth; but birds lack these structures. Instead, bird beaks have a keratinized epithelium.

Transplantation studies have shown that the oral epithelium of birds retains the ability to form teeth, since transplanting mouse neural crest-derived tooth mesenchyme will allow the oral epithelium to form toothlike structures (Kollar and Fisher 1980; Mitsiadis et al. 2003). Recently, Harris and colleagues

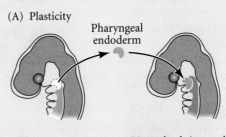

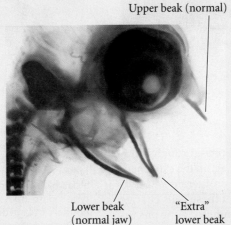

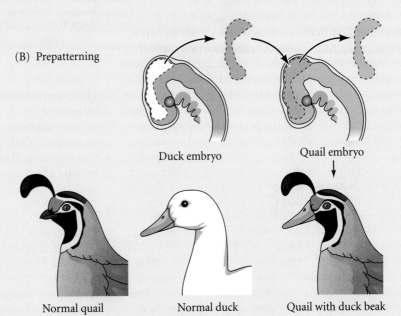

(A) Plasticity

Pharyngeal endoderm

Upper beak (normal)

Lower beak (normal jaw)

"Extra" lower beak

(B) Prepatterning

Duck embryo

Quail embryo

Normal quail

Normal duck

Quail with duck beak

FIGURE 13.11 Plasticity and prepatterning of the neural crest both play roles in beak morphology. (A) Plasticity of crest cells is shown when a piece of chick endoderm at the level of the mandibular arch is transplanted to a host chick embryo. An extra lower beak develops, as seen in the photograph. The chick has two lower beaks, one from the host and one from neural crest cells developing a jaw in the transplanted pharyngeal endoderm. (B) Prepatterning in duck and quail beaks. A quail has a narrow beak, while a duck's beak is broad and flat. When duck cranial neural crest is transplanted into a quail embryo, the quail develops a duck beak. (The reverse experiment also works, giving the duck a quail's beak.) (After Santagati and Rijli 2003; photograph courtesy of N. M. Le Douarin.)

(2006) demonstrated that the *talpid²* mutation of the chick forms teeth, and that these teeth look just like the first set of crocodile teeth. The reason teeth can form in this mutant is that the boundary between the oral and non-oral has changed. In those animals having teeth, the oral ectoderm signaling center (formed by the interaction of Fgf8, BMP4, and Sonic hedgehog) overlies the neural crest mesenchyme that is capable of forming teeth. In birds, the oral ectoderm signaling center is not positioned over the competent mesenchyme, and so the signal is not given. In the *talpid²* mutant, the oral/non-oral ectoderm boundary has again shifted, so that the oral ectoderm signaling center once more meets the competent mesenchyme. Obviously, such mutants don't occur often. In fact, they are "as rare as hen's teeth."

(A) Osteoblasts   Osteocytes

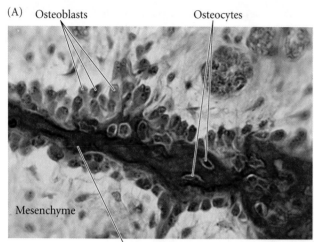

Mesenchyme

Spicule of bone

(B)

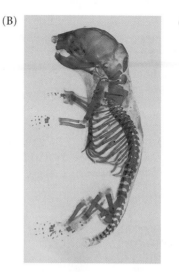

(C)

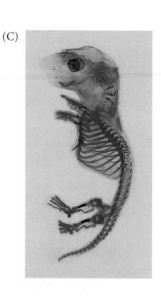

Intramembranous ossification

Some cranial neural crest cells form bones through **intra-membranous ossification** (Figure 13.12A). In the skull, neu-ral crest-derived mesenchymal cells proliferate and con-dense into compact nodules. Some of these cells change their shape to become **osteoblasts**, committed bone precur-sor cells. The osteoblasts secrete a collagen-proteoglycan **osteoid matrix** that is able to bind calcium. Osteoblasts that become embedded in the calcified matrix become **osteo-cytes** (bone cells). As calcification proceeds, bony spicules radiate out from the region where ossification began. Fur-thermore, the entire region of calcified spicules becomes surrounded by compact mesenchymal cells that form the **periosteum** (a membrane that surrounds the bone). The cells on the inner surface of the periosteum also become osteoblasts and deposit matrix parallel to the existing spicules. In this manner, many layers of bone are formed.

The mechanism of intramembranous ossification involves bone morphogenetic proteins and the activation of a transcription factor called **CBFA1** (also called Runx2). Bone morphogenetic proteins from the head epidermis are thought to instruct the neural crest-derived mesenchymal cells to become bone cells directly by causing them to express CBFA1 (Hall 1988; Ducy et al. 1997). CBFA1 appears to activate the genes for osteocalcin, osteopontin, and other bone-specific extracellular matrix proteins. This conclusion was confirmed by gene targeting experiments in which the mouse *Cbfa1* gene was knocked out (Komori et al. 1997; Otto et al. 1997). Mice homozygous for this dele-tion died shortly after birth without taking a breath, and their skeletons completely lacked bone (Figure 13.12B,C). Mice that were heterozygous for the *Cbfa1* deletion showed skeletal defects similar to those of a human syndrome called cleidocranial dysplasia (CCD). In this syndrome, the skull sutures fail to close, growth is stunted, and the clav-icle (collarbone) is often absent or deformed.* When DNA from patients with CCD was analyzed, each patient had either deletions or point mutations in the *CBFA1* gene.

FIGURE 13.12 Intramembranous ossification. (A) Mes-enchyme cells condense and change shape to produce osteoblasts, which deposit osteoid matrix. Osteoblasts then become arrayed along the calcified region of the matrix. Osteoblasts that are embedded within the calcified matrix become osteocytes. (B,C) Targeting of the *Cbfa1* gene in mice prevents bone formation. Newborn mice were stained with alcian blue (for cartilage) and alizarin red (for bone). (B) Wild-type mouse. (C) A homozygous *Cbfa1* mutant littermate shows normal cartilage development, but an absence of ossification throughout the entire body. (The mutants exhibit only the carti-laginous skeletal model for the mesodermally produced bone; this will be discussed in Chapter 14.) (From Komori et al. 1997; photographs courtesy of *Cell* and MIT Press.)

Therefore, it appears that cleidocranial dysplasia is caused by heterozygosity of the *CBFA1* gene (Mundlos et al. 1997).

The vertebrate skull, or cranium, is composed of the neu-rocranium (skull vault and base) and the viscerocranium (the jaws and other pharyngeal arch derivatives). While the neural crest origin of the viscerocranium has been well doc-umented, the contributions of cranial neural crest cells to the skull vault are more controversial. Some researchers have reported that the skull is entirely derived from the neural crest (Couly et al. 1993), whereas others have report-ed that the skull bones are derived from both the neural crest and from the head mesoderm (Le Lièvre 1978; Noden 1978, 1983). In 2002, Jiang and colleagues did much to resolve this controversy by using a nonsurgical procedure that specifically marked cranial neural crest cells. They con-structed transgenic mice that expressed β-galactosidase only

---

*CCD may have been responsible for the phenotype of Thersites, the Greek soldier described in the *Iliad* as having "both shoulders humped together, curving over his caved-in chest, and bobbing above them his skull warped to a point…" (Dickman 1997). The other type of bone formation, endochondral ossification (see Figure 13.12C), will be discussed in Chapter 14. In that type of ossification, the mesenchymal cells first become cartilage and the cartilage is later replaced by bone cells. Endochondral bone formation is seen in certain parts of the skull (such as the palatoquadrate bone of birds, which is also formed from neural crest cells).

(A)

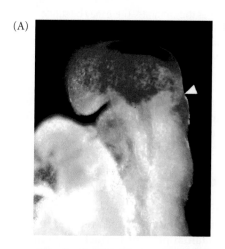

(B)

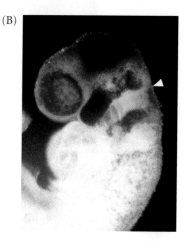

(C)

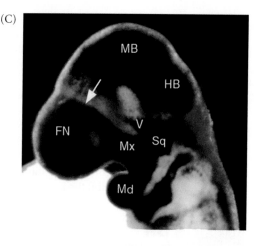

(D)

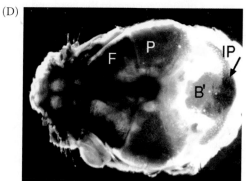

FN, frontal nasal process
MB, midbrain
HB, hindbrain
Mx, maxilla

Md, mandible
Sq, squamosal bone
F, frontal bone
P, parietal

B, mineralized bone
(non-neural crest)
IP, interparietal bone
V, trigeminal ganglion

FIGURE 13.13 Cranial neural crest cells in embryonic mice, stained for β-galactosidase expression. (A–C) Cranial neural crest migration from day 6 to day 9.5, shown by the dark blue staining neural crest cells. (D) Dorsal view of day 17.5 embryonic mice showing lack of staining in the parietal bone. (From Jiang et al. 2002; photographs courtesy of G. Morriss-Kay.)

in their cranial neural crest cells.[†] These cells turned deep blue when stained for β-galactosidase. When the embryonic mice were stained, the cells forming most of the skull—the nasal, frontal, alisphenoid, and squamosal bones—turned blue; the parietal bone did not (Figure 13.13). It appears, then, that while most of the skull is derived from the cranial neural crest, at least one major bone—the parietal—is made from head mesoderm. Although the specifics may vary among the vertebrate groups, in general the front of the head is derived from the neural crest, while the back of the skull is derived from a combination of neural crest-derived and mesodermal bones.

Given that the neural crest forms our facial skeleton, then the rates and directions of cranial crest cell division determine what we look like. Moreover, since we look more like our biological parents than our friends do (or at least, we hope this is true), then such small variations in neural crest growth must be hereditary. The regulation of our facial features is probably coordinated in large part by paracrine growth factors. Numerous paracrine factors play

critical roles in the development of the face and skull. FGFs from the pharyngeal endoderm are responsible for the attraction of the cranial neural crest cells into the arches as well as patterning the skeletal elements within the arches. Fgf8 is both a survival factor for the cranial crest cells and is critical for the proliferation cells forming the facial skeleton (Trocovic et al. 2003, 2005; Creuzet et al. 2004, 2005). The FGFs work in concert with BMPs, sometimes activating them and sometimes repressing them (Lee et al. 2001; Holleville et al. 2003). FGFs also work in concert with Sonic hedgehog (Shh). We saw in Chapter 12 that Sonic hedgehog is critical for the proper growth of the facial midline, and Shh is also crucial for shaping the neural crest derivatives of the head. The epithelium (both neural and epidermal) of the dorsal part of the frontonasal process secrete Fgf8, while the ventral epithelia of the frontonasal process secretes Shh. The crest-derived mesenchyme between the epithelia receives both signals. Where these signals meet is where a bird's beak cartilage grows out. If the region of frontnasal process containing the Fgf/Shh boundary is inverted in the chick, the beak forms in reverse direction (Hu et al. 2003; Abzhanov and Tabin 2004). The interactions among these (and other) paracrine factors to pattern the face and skull has become an important part of developmental biology.

Tooth formation

The cranial neural crest cells of the first pharyngeal arch also form the interior, dentin-secreting odontoblasts of the

---

[†]These experiments were done using the Cre-Lox technique (see Chapter 5). The mice were heterozygous for both (1) a β-galactosidase allele that could be expressed only when Cre-recombinase was activated in that cell, and (2) a Cre-recombinase allele fused to a *Wnt1* promoter. Thus, the β-galactosidase gene was activated (blue stain) only in those cells expressing Wnt1—a protein that is activated in the cranial neural crest and in certain brain cells.

teeth. The jaw epithelium becomes the outer, enamel-secreting odonotoblasts. The mouse tooth is specified by either Fgf8 (molars) or BMP4 (incisors); later, these same proteins determine where teeth will form (Mina and Kollar 1987; Lumsden 1988b; Tucker et al. 1998).

The signaling center of the tooth is the **enamel knot**, a group of cells induced in the epithelium by the neural crest-derived mesenchyme (Jernvall et al. 1994; Vaahtokari et al. 1996a,b). The enamel knot secretes a cocktail of paracrine factors, including Shh, BMPs 2, 4, and 7, and Fgf4. Shh and Fgf4 induce the proliferation of cells to form a cusp, while the BMPs inhibit the formation of new enamel knots.

WEBSITE **13.4 Tooth formation in mammals.** The formation of mammalian teeth demonstrates the interaction of numerous paracrine factors and their re-utilization for different functions during the formation of the same organ.

Cardiac Neural Crest

The heart originally forms in the neck region, directly beneath the pharyngeal arches, so it should not be surprising that it acquires cells from the neural crest. The caudal region of the cranial neural crest is sometimes called the cardiac neural crest, since its cells (and only these particular neural crest cells) generate the endothelium of the aortic arch arteries and the septum between the aorta and the pulmonary artery (Figure 13.14; Kirby 1989; Kuratani and Kirby 1991; Waldo et al. 1998).

In mice, cardiac neural crest cells are peculiar in that they express the transcription factor Pax3. Mutations of *Pax3* result in fewer cardiac neural crest cells, which in turn leads to persistent truncus arteriosus (the failure of the aorta and pulmonary artery to separate), as well as defects in the thymus, thyroid, and parathyroid glands (Conway

WEBSITE **13.5 Communication between migrating neural crest cells.** Recent research has shown that neural crest cells might cooperate with one another as they migrate. There may be subtle communication between these cells through their gap junctional complexes, and this communication may be important for heart development.

et al. 1997, 2000). Congenital heart defects in humans and mice often occur along with defects in the parathyroid, thyroid, or thymus glands. It would not be surprising to find that all these problems were linked to defects in the migration of cells from the neural crest.

Cranial Placodes

In addition to forming the cranial neural crest cells, the anterior borders between the epidermal and neural ectoderm also form the **cranial placodes**, which are local and transient thickenings of the ectoderm in the head and neck. (The cranial neural crest and the cranial placodes may have originated from the same cell population during early vertebrate evolution; see Northcutt and Gans 1983; Baker and Bronner-Fraser 1997.) These sensory placodes have neurogenic potential; they give rise to the neurons that form the distal portions of those ganglia associated with hearing, balance, taste, and smell (Figure 13.15A). The proximal neurons of these ganglia are formed from neural crest cells (see Baker and Bronner-Fraser 2001). For example, the olfactory placode gives rise to the sensory neurons involved in smell, as well as to some migratory neurons that will travel into the brain and secrete gonadotropin releasing hor-

FIGURE 13.14 The septa that separate the truncus arteriosus into the pulmonary artery and aorta form from cells of the cardiac neural crest. (A) Human cardiac neural crest cells migrate to pharyngeal arches 4 and 6 during the fifth week of gestation and enter the truncus arteriosus to generate the septa. (B) Quail cardiac crest cells were transplanted into the analogous region of a chick embryo, and the embryos were allowed to develop. The quail cardiac neural crest cells can be recognized by a quail-specific antibody, which stains them darkly. In the heart, these cells can be seen separating the truncus arteriosus into the pulmonary artery and the aorta. (A after Kirby and Waldo 1990; B from Waldo et al. 1998, photographs courtesy of K. Waldo and M. L. Kirby.)

(A)

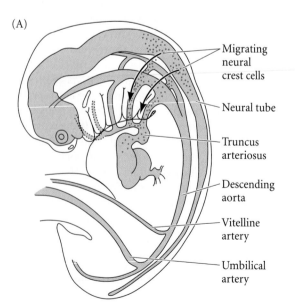

Migrating neural crest cells

Neural tube

Truncus arteriosus

Descending aorta

Vitelline artery

Umbilical artery

(B)

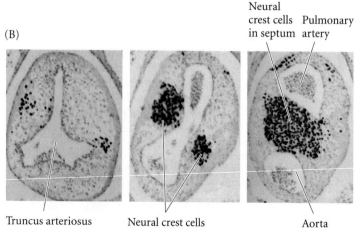

Neural crest cells in septum

Pulmonary artery

Truncus arteriosus

Neural crest cells

Aorta

(A)

Cranial placode:

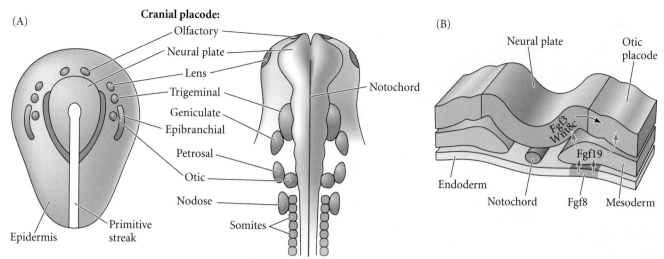

FIGURE 13.15 Cranial placodes form sensory neurons. (A) Fate map of the cranial placodes in the developing chick embryo at the neural plate (left) and 8-somite (right) stages. (B) Induction of the otic (inner ear) placode in the chick embryo. A portion of the pharyngeal endoderm secretes Fgf8, which induces the mesoderm overlying it to secrete Fgf19. Fgf19 is received by both the prospective otic placode and the adjacent neural plate. Fgf19 instructs the neural plate to secrete Wnt8c and Fgf3, two paracrine factors that work synergistically to induce *Pax2* and other genes that allow the cells to produce the otic placode and become sensory cells.

mone. The otic placode gives rise to the sensory epithelium of the ear and to neurons that help form the cochlear-vestibular ganglion. The lens placode is the only placode that does not form neurons.

The placodes are induced by the neighboring tissue, and there is evidence that the different placodes are each a small portion of what had earlier been a common "pre-placodal" territory (Streit 2004; Schlosser 2005). Histological evidence shows that the anterior neural plate is surrounded by a single thickening during the early neurula stages, and the cranial placodes may arise from a common set of inductive interactions from the pharyngeal endoderm and head mesoderm (Platt 1896; Jacobson 1966). Jacobson (1963) also showed that the pre-placodal cells adjacent to the anterior neural tube are competent to give rise to any placode. This columnar epithelium contains the transcription factors Six1, Six4, and Eya. They are maintained in all the placodes and are downregulated in the interplacodal regions (Schlosser and Ahrens 2004; Bhattacharyya et al. 2004). Later, specific interactions define the individual placodes. For instance, the chick otic placode, which develops into the sensory cells of the inner ear, is induced by a combination of FGF and Wnt signals (Ladher et al. 2000, 2005). Here, Fgf19 from the underlying cranial paraxial mesoderm is received by both the presumptive otic vesicle and the adjacent neural plate. Fgf19 induces the neural plate to secrete Wnt8c and Fgf3, which in turn act synergistically to induce formation of the otic placode. The localization of

Fgf19 to the specific region of the mesoderm is controlled by Fgf8 secreted in the endodermal region beneath it. (Figure 13.15B).

The epibranchial placodes form dorsally to where the pharyngeal pouches contact the epidermis. These structures split to form the geniculate, petrosal, and nodose placodes which give rise to the sensory neurons of the facial, glossopharyngeal, and vagal nerves, respectively. The connections made by these placodal neurons are critical, in that they enable taste and other pharyngeal sensations to be appreciated. But how do these neurons find their way into the hindbrain? The cranial neural crest cells in the second "wave" of migration do not travel ventrally to enter the pharyngeal arches; rather, they migrate dorsally to form glial cells (Weston and Butler 1966; Baker et al. 1997). These glia form tracks that guide neurons from the epibranchial placodes to the hindbrain. Indeed, if the hindbrain is removed before the neural crest cells emigrate, neurons leaving the placodes enter a crest-free environment and fail to migrate into the hindbrain (Begbie and Graham 2001). Therefore, the glial cells made by the second wave of cranial neural crest migration are critical in organizing the innervation of the hindbrain.

WEBSITE

13.6 **Human facial development syndromes.** Several human syndromes cause abnormalities of cranial neural crest cell migration. A region on chromosome 22 appears to be especially important in regulating normal facial and pharyngeal development.

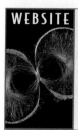

WEBSITE

13.7 **Kallmann syndrome.** Some infertile men have no sense of smell. The relationship between sense of smell and male fertility was elusive until the gene for Kallmann syndrome was identified. The gene produces a protein that is necessary for the proper migration of both olfactory axons and hormone-secreting neurons from the olfactory placode.

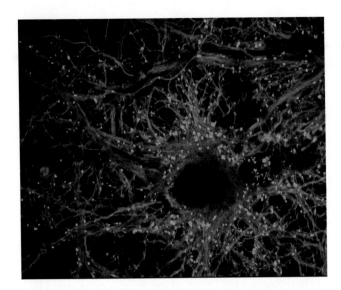

FIGURE 13.16 Connections of axons to a cultured rat hippocampal neuron. The neuron has been stained red with fluorescent antibodies to tubulin. The neuron appears to be outlined by the synaptic protein synapsin (green), which is present in the terminals of axons that contact it. (Photograph courtesy of R. Fitzsimmons and PerkinElmer Life Sciences.)

NEURONAL SPECIFICATION AND AXONAL SPECIFICITY

Not only do neuronal precursor cells and neural crest cells migrate to their place of function, but so do the axons extending from the cell bodies of neurons. Unlike most cells, whose parts all stay in the same place, the neuron can produce axons that may extend for meters. As we saw in Chapter 12, the axon has its own locomotory apparatus, which resides in the axonal growth cone. The growth cone has been called "a neural crest cell on a leash" because, like neural crest cells, it migrates and senses the environment. Moreover, it can respond to the same types of signals that migrating cells can sense. The cues for axonal migration, moreover, may be even more specific than those used to guide specific cell types to particular areas. Each of the 10^{11} neurons in the human brain has the potential to interact specifically with thousands of other neurons (Figure 13.16). A large neuron (such as a Purkinje cell or motor neuron) can receive input from more than 10^5 other neurons (Gershon et al. 1985). Understanding the generation of this stunningly ordered complexity is one of the greatest challenges to modern science.

Goodman and Doe (1993) list eight stages of neurogenesis:

1. Induction and patterning of a neuron-forming (neurogenic) region
2. Birth and migration of neurons and glia
3. Specification of cell fates

4. Guidance of axonal growth cones to specific targets
5. Formation of synaptic connections
6. Binding of trophic factors for survival and differentiation
7. Competitive rearrangement of functional synapses
8. Continued synaptic plasticity during the organism's lifetime

The first two of these processes were the topics of Chapter 12. Here, we continue our investigation of the processes of neural development.

WEBSITE 13.8 The evolution of developmental neurobiology. Santiago Ramón y Cajal, Viktor Hamburger, and Rita Levi-Montalcini helped bring order to the study of neural development by identifying some of the important questions that still occupy us today.

The Generation of Neuronal Diversity

Neurons are specified in a hierarchical manner. The first decision is whether a given ectodermal cell will become a neuron, a neural crest cell, or epidermis. If it is to become a neuron, the next decision is, What type of neuron? Will it become a motor neuron, a sensory neuron, a commissural neuron, or some other type? After this fate is determined, still another decision gives the neuron a specific target. To illustrate this process of progressive specification, we will focus on the motor neurons of vertebrates.

Vertebrates form a dorsal neural tube by blocking a BMP signal, and the specification of neural (as opposed to glial or epidermal) fate is accomplished through the Notch-Delta pathway (see Chapter 6). The specification of the *type* of neuron appears to be controlled by the position of the neuronal precursor within the neural tube and by its birthday. As described in Chapter 12, neurons at the ventrolateral margin of the vertebrate neural tube become motor neurons, while interneurons are derived from cells in the dorsal region of the tube. Since the grafting of floor plate or notochord cells (which secrete Sonic hedgehog protein, Shh) to lateral areas of the neural tube can re-specify dorsolateral cells as motor neurons, the specification of neuron type is probably a function of the cell's position relative to the floor plate. Ericson and colleagues (1996) have shown that two periods of Shh signaling are needed to specify the motor neurons: an early period wherein the cells of the ventrolateral margin are instructed to become ventral neurons, and a later period (which includes the S phase of its last cell division) that instructs a ventral neuron to become a motor neuron rather than an interneuron. The first decision is probably regulated by the secretion of Shh from the notochord, while the second is more likely regulated by Shh from the floor plate cells. Sonic hedgehog appears to specify motor neurons by inducing certain transcription factors at different concentrations (Ericson et al. 1992; Tanabe et al. 1998; see Figure 12.14).

The next decision involves target specificity. If a cell is to become a neuron and, specifically, a motor neuron, will that motor neuron be one that innervates the thigh, the forelimb, or the tongue? The anterior-posterior specification of the neural tube is regulated primarily by Hox genes from the hindbrain through the spinal cord, and by specific head genes (such as *Otx*) in the brain (Dasen et al. 2005). Within a region of the body, motor neuron specificity is regulated by the cell's age when it last divides. As discussed in Chapter 12, a neuron's birthday determines which layer of the cortex it will enter. As younger cells migrate to the periphery, they must pass through neurons that differentiated earlier in development. As younger motor neurons migrate through the region of older motor neurons in the intermediate zone, they express new transcription factors as a result of a retinoic acid (or other retinoid) signal secreted by the early-born motor neurons (Sockanathan and Jessell 1998). These transcription factors are encoded by the *Lim* genes and are structurally related to those encoded by the Hox genes.

As a result of their differing birthdays and migration patterns, motor neurons form three major groupings. The cell bodies of the motor neurons projecting to a single muscle are "pooled" in a longitudinal column of the spinal cord. This pooling is performed by different cadherins that become expressed on these different populations of cells (Landmesser 1978; Hollyday 1980; Price et al. 2002). The pools are grouped into three larger columns—the column of Terni, the lateral motor column, and the medial motor column—according to their targets. In the thorax, motor neurons in the column of Terni project ventrally into the sympathetic ganglia; motor neurons in the lateral motor column (LMC) project into the muscles (Figure 13.17A). In the limb areas, muscles are innervated by the LMC axons, with lateral neurons entering the dorsal musculature, while the motor neurons of the medial motor column innervate ventral limb musculature (Figure 13.17B; Tosney et al. 1995). This arrangement of motor neurons is consistent throughout the vertebrates.

The targets of these motor neurons are specified before their axons extend into the periphery. This was shown by Lance-Jones and Landmesser (1980), who reversed segments of the chick spinal cord so that the motor neurons were placed in new locations. The axons went to their original targets, not to the ones expected from their new positions (Figure 13.18). The molecular basis for this target specificity resides in the members of the Hox and Lim protein families that are induced during neuronal development (Tsushida et al. 1994; Sharma et al. 2000; Price and Briscoe 2004). For instance, all motor neurons express Islet1 and (slightly later) Islet2. If no other Lim protein is expressed, the neurons project to the ventral body wall muscles. However, if Lim3 protein is also synthesized, the motor neurons project dorsally to the axial muscles (see Figure 13.17). If each of the motor neuron types is made to express Lim3, they will all innervate the axial muscles (or at least attempt to; those axons that arrive late and find the sites filled will project to alternative targets). Thus, each group of neurons is characterized by a particular constellation of Lim transcription factors. These factors probably coordinate the projections of the motor neurons by specifying ephrin-A receptors that bind to the ephrin ligands on the muscle tissue (Kania and Jessell 2003).

In addition, the anterior-posterior identity of the motor neuron is established by Hox genes. By experimentally changing Hox gene expression patterns, the different neurons will innervate new (and inappropriate) muscles (Liu et al. 2001; Dasen et al. 2005).

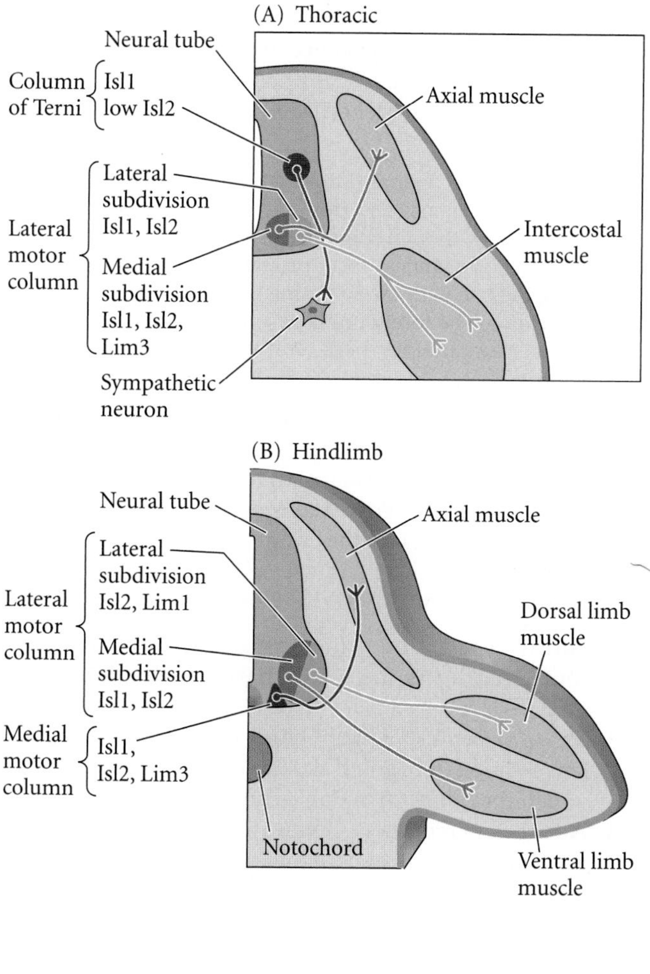

FIGURE 13.17 Motor neuron organization and Lim specification in the spinal cord. Neurons in each of three different columns express specific sets of Lim family genes (including *Isl1* and *Isl2*), and neurons within each column make similar pathfinding decisions. Motor neurons in the column of Terni (CT) project ventrally to the sympathetic ganglia. Neurons of the medial motor column project to the axial muscles, and neurons of the lateral motor column send axons to the limb musculature. Where these columns are subdivided, medial subdivisions project to ventral positions and lateral subdivisions send axons to dorsal regions of the target tissues. (After Tsushida et al. 1994; Tosney et al. 1995.)

FIGURE 13.18 Compensation for small dislocations of axonal initiation position in the chick embryo. (A) A length of spinal cord comprising segments T7–LS3 (seventh thoracic to third lumbosacral segments) is reversed in a 2.5-day embryo. (B) Normal pattern of axon projection to the muscles at 6 days. (C) Projection of axons from the reversed segment. The ectopically placed neurons eventually found their proper neural pathways and innervated the appropriate muscles. (From Lance-Jones and Landmesser 1980.)

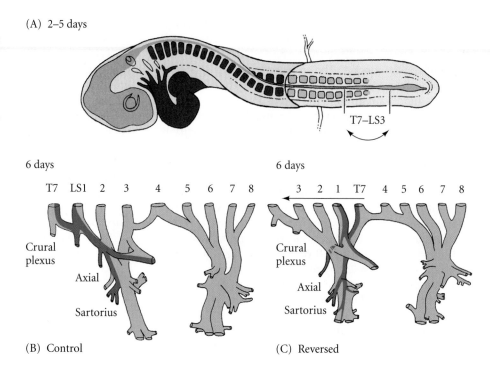

(A) 2–5 days

6 days

T7 LS1 2 3 4 5 6 7 8

Crural plexus

Axial

Sartorius

(B) Control

6 days

3 2 1 T7 4 5 6 7 8

Crural plexus

Axial

Sartorius

(C) Reversed

WEBSITE **13.9 Horseradish peroxidase staining.** Many of the fundamental discoveries of axon specificity used a technique wherein the plant enzyme horseradish peroxidase was injected into nerves.

Pattern Generation in the Nervous System

Vertebrate brain function depends not only on the differentiation and positioning of the neurons, but also on the specific connections these cells make among themselves and their peripheral targets. Nerves from a sensory organ such as the eye or nose must connect to specific neurons in the brain that can interpret stimuli from that organ, and axons from the central nervous system must cross large expanses of tissue before innervating their target tissue. How does the neuronal axon "know" how to traverse numerous potential target cells to make its specific connection?

Ross G. Harrison (see Figure 4.3) first suggested that the specificity of axonal growth is due to **pioneer nerve fibers**, axons that go ahead of other axons and serve as guides for them* (Harrison 1910). This observation simplified, but

did not solve, the problem of how neurons form appropriate patterns of interconnection. Harrison also noted, however, that axons must grow on a solid substrate, and he speculated that differences among embryonic surfaces might allow axons to travel in certain specified directions. The final connections would occur by complementary interactions on the target cell surface:

> That it must be a sort of a surface reaction between each kind of nerve fiber and the particular structure to be innervated seems clear from the fact that sensory and motor fibers, though running close together in the same bundle, nevertheless form proper peripheral connections, the one with the epidermis and the other with the muscle. … The foregoing facts suggest that there may be a certain analogy here with the union of egg and sperm cell.

Research on the specificity of neuronal connections has focused on two major systems: motor neurons, whose axons travel from the spinal cord to a specific muscle; and the optic system, wherein axons originating in the retina find their way back into the brain. In both cases, the specificity of axonal connections is seen to unfold in three steps (Goodman and Shatz 1993):

1. **Pathway selection**, wherein the axons travel along a route that leads them to a particular region of the embryo
2. **Target selection**, wherein the axons, once they reach the correct area, recognize and bind to a set of cells with which they may form stable connections
3. **Address selection**, wherein the initial patterns are refined such that each axon binds to a small subset (sometimes only one) of its possible targets

---

*The growth cones of pioneer neurons migrate to their target tissue while embryonic distances are still short and the intervening embryonic tissue is still relatively uncomplicated. Later in development, other neurons bind to pioneer neurons and thereby enter the target tissue. Klose and Bentley (1989) have shown that in some cases, pioneer neurons die after the "follow-up" neurons reach their destination. Yet if the pioneer neurons are prevented from differentiating, the other axons do not reach their target tissue.

The first two processes are independent of neuronal activity. The third process involves interactions between several active neurons and converts the overlapping projections into a fine-tuned pattern of connections.

It has been known since the 1930s that the axons of motor neurons can find their appropriate muscles even if the neural activity of the axons is blocked. Twitty (who was Harrison's student) and his colleagues found that embryos of the newt *Taricha torosa* secreted a toxin, tetrodotoxin, that blocked neural transmission in other species. By grafting pieces of *T. torosa* embryos onto embryos of other salamander species, they were able to paralyze the host embryos for days while development occurred. Normal neuronal connections were made, even though no neuronal activity could occur. At about the time the tadpoles were ready to feed, the toxin wore off, and the young salamanders swam and fed normally (Twitty and Johnson 1934; Twitty 1937). More recent experiments using zebrafish mutants with nonfunctional neurotransmitter receptors similarly demonstrated that motor neurons can establish their normal patterns of innervation in the absence of neuronal activity (Westerfield et al. 1990).

But the question remains: How are the axons instructed where to go?

Cell adhesion and contact guidance by attractive and permissive molecules

The initial pathway an axonal growth cone follows is determined by the environment the growth cone experiences. The extracellular environment can provide substrates upon which to migrate, and these substrates can provide navigational information to the growth cone. Some of the substrates the growth cone encounters will permit it to adhere to them, and thus promote axon migration. Other substrates will cause the growth cone to retract, and will not allow its axon to grow in that direction. Some of these substrates give extremely specific cues, recognized by only a small set of neuronal growth cones, while other cues are recognized by large sets of neurons. The substrate's migrational cues can come from anatomical structures, the extracellular matrix, or from adjacent cell surfaces.

Extracellular matrix cues often offer preferred adhesion substrates for migration. Growth cones prefer to migrate on surfaces that are more adhesive than their surroundings, and a track of adhesive molecules can direct them to their targets. The presence of adhesive extracellular matrix molecules delineates microscopic roads through the embryo on which axons travel (Akers et al. 1981; Gundersen 1987). Many of these roads appear to be paved with the glycoprotein **laminin** (Figure 13.19). Letourneau and co-workers (1988) have shown that the axons of certain spinal neurons travel through the neuroepithelium over a transient laminin-coated surface that precisely marks the axonal path. Similarly, there is a very good correlation between the elongation of retinal axons and the presence of laminin on the neuroepithelial and glial cells in the

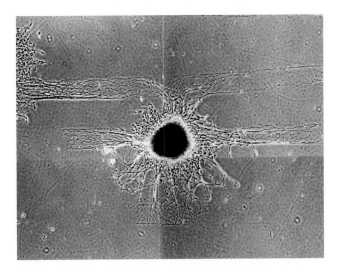

FIGURE 13.19 Outgrowth of sensory neurons placed on a patterned substrate consisting of parallel stripes of laminin applied to a background of type IV collagen. (From Gundersen 1987; photograph courtesy of R. W. Gundersen.)

embryonic mouse brain (Cohen et al. 1987; Liesi and Silver 1988). Punctate laminin deposits are seen on the glial cell surfaces along the pathway leading from the retinal ganglion cells to the optic tectum in the brain, whereas adjacent areas (where the retinal axons fail to grow) lack these laminin deposits. After the retinal axons have reached the tectum, the glial cells differentiate and lose their laminin. At this point, the retinal ganglion neurons that have formed the optic nerve lose their integrin receptor for laminin.

Guidance by specific growth cone repulsion

In addition to general extracellular matrix cues, there are protein guidance cues that are specific to certain groups of neurons. Axons in the developing nervous system respond to attractive and repulsive signals of the ephrin, semaphorin, netrin, and Split protein families. We have already seen that neural crest cells are patterned by their recognition of ephrin, and that what is an attractive cue to one set of cells (such as the presumptive melanocytes going through the dermis) can be a repulsive signal to another set of cells (such as the presumptive sympathetic ganglia). We will see that whether a guidance signal is attractive or repulsive can depend on the type of cell receiving that signal and on the time when that cell receives it.

Two of the membrane protein families involved in neural patterning are the **ephrins** and the **semaphorins**. Just as neural crest cells are inhibited from migrating across the posterior portion of a sclerotome, the axons from the dorsal root ganglia and motor neurons also pass only through the anterior portion of each sclerotome and avoid migrating through the posterior portion (Figure 13.20A; also see

FIGURE 13.20 Repulsion of dorsal root ganglion growth cones. (A) Motor axons migrating through the rostral (anterior), but not the caudal (posterior), compartments of each sclerotome. (B) In vitro assay, wherein ephrin stripes were placed on a background surface of laminin. Motor axons grew only where the ephrin was absent. (C) Inhibition of growth cones by ephrin after 10 minutes of incubation. The left-hand photograph shows a control axon subjected to a similar (but not inhibitory) compound; the right axon was exposed to an ephrin found in the posterior sclerotome. (From Wang and Anderson 1997; photographs courtesy of the authors.)

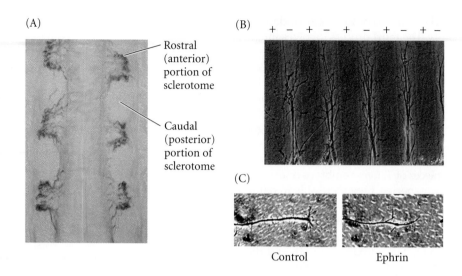

(A)

Rostral (anterior) portion of sclerotome

Caudal (posterior) portion of sclerotome

(B)

(C)

Control   Ephrin

Figure 13.5). Davies and colleagues (1990) showed that membranes isolated from the posterior portion of a somite cause the growth cones of these neurons to collapse (Figure 13.20B,C). These growth cones contain Eph receptors and neuropilin receptors that are responsive to ephrins and semaphorins on the posterior sclerotome cells (Wang and Anderson 1997; Krull et al. 1999; Kuan et al. 2004). Thus the same signals that pattern neural crest cell migration also pattern the spinal neuronal outgrowths.

Found throughout the animal kingdom, the semaphorins often guide growth cones by selective repulsion. They are especially important in forcing "turns" when an axon must change direction. Semaphorin 1 is a transmembrane protein that is expressed in a band of epithelial cells in the developing insect limb. This protein appears to inhibit the growth cones of the Ti1 sensory neurons from moving forward, thus causing them to turn (Figure 13.21; Kolodkin et al. 1992, 1993). In *Drosophila*, semaphorin 2 is secreted by a single large thoracic muscle. In this way, the thoracic muscle prevents itself from being innervated by inappropriate axons (Matthes et al. 1995).

Semaphorin 3, found in mammals and birds, is also known as collapsin. This secreted protein was found to collapse the growth cones of axons originating in the dorsal root ganglia (Luo et al. 1993). There are several types of neurons in the dorsal root ganglia whose axons enter the dorsal spinal cord. Most of these axons are prevented from

traveling farther and entering the ventral spinal cord; however, a subset of them does travel ventrally through the other neural cells (Figure 13.22). These particular axons are not inhibited by semaphorin 3, while those of the other neurons are (Messersmith et al. 1995). This finding suggests that semaphorin 3 patterns sensory projections from the dorsal root ganglia by selectively repelling certain axons so that they terminate dorsally. A similar scheme is used in the brain, wherein semaphorin made in one part of the brain is used to prevent the entry of neurons coming from another region of the brain (Marín et al. 2001).

As any psychologist knows, the line between attraction and repulsion is often thin. At the base of both phenomena is some sort of recognition event. This is also the case with neurons. When the receptor protein ephrin A7 recognizes ephrin A5 in the mouse neural tube, the result is attraction rather than repulsion. The interaction between these two proteins is critical for the closure of the neural tube. The EphA7 and EphA5 proteins cause *adhesion* of the neural plate cells, and deletion of either gene results in a condition resembling anencephaly. The change from repulsion to adhesion is caused by alternative RNA processing. By

FIGURE 13.21 The action of semaphorin 1 in the developing grasshopper limb. The axon of sensory neuron Ti1 projects toward the central nervous system. (The dark arrows represent sequential steps en route.) When it reaches a band of semaphorin 1-expressing epithelial cells, it reorients its growth cone and extends ventrally along the distal boundary of the semaphorin 1-expressing cells. When its filopodia connect to the Cx1 pair of cells, it crosses the boundary and projects into the CNS. When semaphorin 1 is blocked by antibodies, the growth cone searches randomly for the Cx1 cells. (After Kolodkin et al. 1993.)

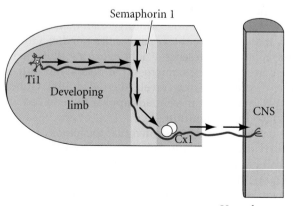

Semaphorin 1

Ti1

Developing limb

CNS

Cx1

Ventral nerve cord

(A)

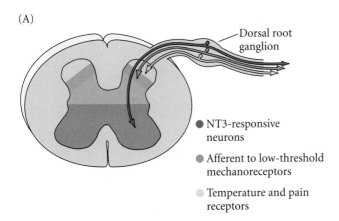

- NT3-responsive neurons
- Afferent to low-threshold mechanoreceptors
- Temperature and pain receptors

FIGURE 13.22 Semaphorin 3 as a selective inhibitor of axonal projections into the ventral spinal cord. (A) Trajectory of axons in relation to semaphorin 3 expression in the spinal cord of a 14-day embryonic rat. Neurotrophin 3-responsive neurons can travel to the ventral region of the spinal cord, but the afferent axons for the mechanoreceptors and for temperature and pain receptor neurons terminate dorsally. (B) Transgenic chick fibroblast cells that secrete semaphorin 3 inhibit the outgrowth of mechanoreceptor axons. These axons are growing in medium treated with NGF, which stimulates their growth, but are still inhibited from growing toward the source of semaphorin 3. (C) Neurons that are responsive to NT3 for growth are not inhibited from extending toward the source of semaphorin 3 when grown with NT3. (A after Marx 1995; B and C from Messersmith et al. 1995, photographs courtesy of A. Kolodkin.)

(B)

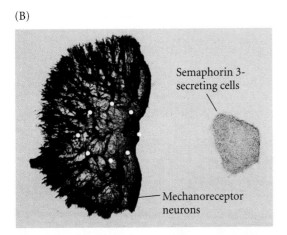

Semaphorin 3-secreting cells

Mechanoreceptor neurons

(C)

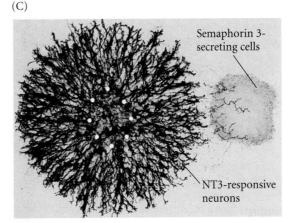

Semaphorin 3-secreting cells

NT3-responsive neurons

using a different splice site, the mouse neural plate produces EphA7 lacking the tyrosine kinase domain that transmits the repulsive signal (Holmberg et al. 2000). The result is that the cells recognize each other through these proteins, and no repulsion occurs.

In a different way, semaphorins can also be attractants. Semaphorin 3A is a classic chemorepellant for the axons coming from pyramidal neurons in the mammalian cortex. However, it is a chemoattractant for the *dendrites* of the same cells. In this way, a target can "reach out" to the dendrites of these cells without attracting their axons as well (Polleux et al. 2000).

WEBSITE

13.10 The pathways of motor neurons. To innervate the limb musculature, a motor axon extends over hundreds of cells in a complex and changing environment. Recent research has discovered several paths and several barriers that help guide these axons to their appropriate destinations.

Guidance by diffusible molecules

NETRINS AND THEIR RECEPTORS The idea that chemotactic cues guide axons in the developing nervous system was first proposed by Santiago Ramón y Cajal (1892). He suggested that the commissural neurons of the spinal cord might be told by diffusible factors to send axons from their dorsal positions to the ventral floor plate. Commissural neurons are interneurons that cross the ventral midline to coordinate right and left motor activities. Thus, they somehow must migrate to (and through) the ventral midline. The axons of these neurons begin growing ventrally down the side of the neural tube. However, about two-thirds of the way down, their direction changes, and they project through the ventrolateral (motor) neuron area of the neural tube toward the floor plate cells (Figure 13.23A).

There appear to be two systems involved in attracting the commissural neurons to the ventral midline. The first is the Sonic hedgehog (Shh) protein that is made in the floor plate (see Figure 12.14; Charron et al. 2003). If Shh is inhibited by cyclopamine or conditionally knocked out of the floor plate cells, the commissural axons have difficulty getting to the ventral midline. However, a gradient of Shh does not provide a full explanation of the migration. Some other factor is also involved. In 1994, Serafini and colleagues developed an assay that allowed them to screen for the presence of a presumptive diffusible molecule that might be guiding the commissural neurons. When dorsal spinal cord explants from chick embryos were plated onto

FIGURE 13.23 Trajectory of the commissural axons in the rat spinal cord. (A) Schematic drawing of a model wherein commissural neurons first experience a gradient of Sonic hedgehog and netrin-2, and then a steeper gradient of netrin-1. The commissural axons are chemotactically guided ventrally down the lateral margin of the spinal cord toward the floor plate. Upon reaching the floor plate, contact guidance from the floor plate cells causes the axons to change direction. (B) Autoradiographic localization of netrin-1 mRNA by in situ hybridization of antisense RNA to the hindbrain of a young rat embryo. Netrin-1 mRNA (dark area) is concentrated in the floor plate neurons. (B from Kennedy et al. 1994; photograph courtesy of M. Tessier-Lavigne.)

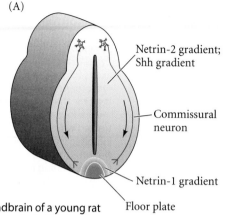

(A)

Netrin-2 gradient; Shh gradient

Commissural neuron

Netrin-1 gradient

Floor plate

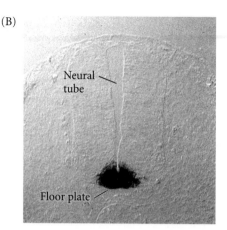

(B)

Neural tube

Floor plate

collagen gels, the presence of floor plate cells near them promoted the outgrowth of commissural axons. Serafini and his co-workers took fractions of embryonic chick brain homogenate and tested them to see if any of the proteins therein mimicked explant activity. This research resulted in the identification of two proteins, **netrin-1** and **netrin-2**. Netrin-1 is made by and secreted from the floor plate cells, whereas netrin-2 is synthesized in the lower region of the spinal cord, but not in the floor plate (Figure 13.23B). The netrins are recognized by a receptor, DCC, found in the axon growth cones.

The chemotactic effect of the netrins was demonstrated by transforming chick fibroblast cells (which usually do not make these proteins) into netrin-producing cells using a vector containing an active netrin gene (Kennedy et al. 1994). Aggregates of these netrin-secreting fibroblast cells elicited commissural axon outgrowth from rat dorsal spinal cord explants, while control cells that were given the vector without the active netrin gene did not elicit such activity (Figure 13.24). It is possible that the commissural neurons first encounter a gradient of netrin-2 and Shh, which brings them into the domain of the steeper netrin-1 gradient (see Figure 13.23A). In these experiments, both netrins

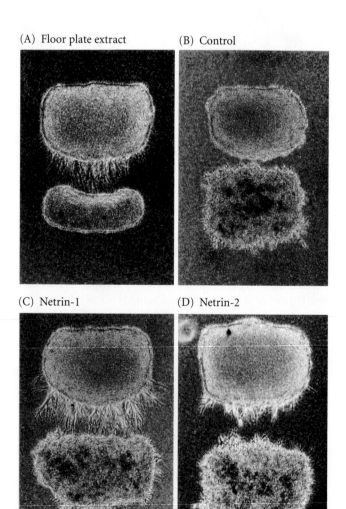

(A) Floor plate extract   (B) Control

(C) Netrin-1   (D) Netrin-2

FIGURE 13.24 Transformed chick fibroblast (COS) cells secreting netrins elicit axon outgrowth of commissural neurons from 11-day embryonic rat dorsal spinal cord explants. (A) Outgrowth of commissural axons is seen when a dorsal spinal cord explant (upper tissue) encounters a floor plate explant. (B) No outgrowth is seen when a dorsal spinal cord explant is exposed to COS cells that have been transfected with the cloning vector alone (no netrin gene). (C,D) Commissural axon outgrowth in response to COS cells that express the gene for (C) netrin-1 and (D) netrin-2. The identity of the outgrowths was confirmed by immunohistology, which showed commissural-specific antigens on the axons. (From Kennedy et al. 1994; photographs courtesy of M. Tessier-Lavigne.)

became associated with the extracellular matrix. Such an association can play important roles.*

Netrin-1 may serve as both an attractive and a repulsive signal in vertebrates. The growth cones of *Xenopus* retinal neurons, for example, are attracted to netrin-1 and are guided to the head of the optic nerve by this diffusible factor. Once there, however, the combination of netrin-1 and laminin *prevents* the axons from leaving the optic nerve. It appears that the laminin of the extracellular matrix surrounding the optic nerve converts the netrin from being an attractive molecule to being a repulsive one (Höpker et al. 1999).

---

*The binding of a soluble factor to the extracellular matrix makes for an interesting ambiguity between *chemotaxis* and migrating on preferred substrates (*haptotaxis*). Nature doesn't necessarily conform to the categories we create. There is also some confusion between the terms *neurotropic* and *neurotrophic*. Neuro*tropic* (Latin, *tropicus*, a turning movement) means that something attracts the neuron. Neuro*trophic* (Greek, *trophikos*, nursing) refers to a factor's ability to keep the neuron alive, usually by supplying growth factors. Since many agents have both properties, they are alternatively called "neurotropins" and "neurotrophins." In the recent literature, "neurotrophin" appears to be more widely used.

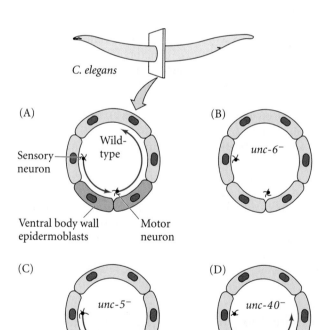

FIGURE 13.25 UNC expression and function in axonal guidance. (A) In the body of the wild-type *C. elegans* embryo, sensory neurons project ventrally and motor neurons project dorsally. The ventral body wall epidermoblasts expressing UNC-6 are darkly shaded. (B) In the *unc-6* mutant embryo, neither of these migrations occurs. (C) The *unc-5* loss-of-function mutation affects only the dorsal movements of the motor neurons. (D) The *unc-40* loss-of-function mutation affects only the ventral migration of the sensory growth cones. (After Goodman 1994.)

The netrins have numerous regions of homology with UNC-6, a protein implicated in directing the circumferential migration of axons around the body wall of *Caenorhabditis elegans*. In the wild-type nematode, UNC-6 induces axons from certain centrally located sensory neurons to move ventrally while inducing ventrally placed motor neurons to extend axons dorsally. In *unc-6* loss-of-function mutations, neither of these axonal movements occurs (Hedgecock et al. 1990; Ishii et al. 1992; Hamelin et al. 1993). Mutations of the *unc-40* gene disrupt ventral (but not the dorsal) axonal migration, while mutations of the *unc-5* gene prevent only the dorsal migration (Figure 13.25). Genetic and biochemical evidence suggests that UNC-5 and UNC-40 are portions of the UNC-6 receptor complex, and that UNC-5 can convert a UNC-40-mediated attraction into a repulsion (Leonardo et al. 1997; Hong et al. 1999; Chang 2004).

If UNC-6 is attractive to some neurons and repulsive to others, one might expect that this dual role might also be ascribed to netrins. Colamarino and Tessier-Lavigne (1995) have shown this to be the case by looking at the trajectory of the trochlear (fourth cranial) nerve. On their way to innervate an eye muscle, the axons of the trochlear nerve originate near the floor plate of the brain stem and migrate dorsally away from the floor plate region. This pathway is maintained when the brainstem region is explanted into collagen gel. The dorsal outgrowth of the trochlear neurons can be prevented by placing floor plate cells or transgenic, netrin-1-secreting chick fibroblast cells within 450 μm of the dorsal portion of the explant. This dorsal outgrowth is not prevented by dorsal explants of the neural tube or by chick fibroblast cells that do not contain the active netrin-1 gene (Figure 13.26). Therefore, netrins and UNC-6 appear to be chemotactic to some neurons and chemorepulsive to others.

There is reciprocity in science, and just as research on vertebrate netrin genes led to the discovery of their *C. elegans* homologues, research on the nematode *unc-5* gene led to the discovery of the gene encoding the human netrin receptor. This turns out to be a gene whose mutation in mice causes a disease called rostral cerebellar malformation (Ackerman et al. 1997; Leonardo et al. 1997).

WEBSITE **13.11 Genetic control of neuroblast migration in *C. elegans*.** The homeotic gene *mab-5* controls the direction in which certain neurons migrate in the nematode. The expression of this gene can alter which way a neuron travels.

SLIT AND ROBO Diffusible proteins can also provide guidance by repulsion. One important chemorepulsive molecule is the **Slit** protein. In *Drosophila*, Slit is secreted by the neural cells in the midline, and it acts to prevent most neurons from crossing the midline from either side. The growth cones of *Drosophila* neurons express a **Roundabout (Robo)** protein, which is the receptor for Slit. In this way, most *Drosophila* neurons are prevented from migrating across the

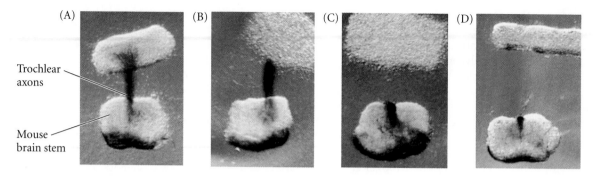

Trochlear axons

Mouse brain stem

FIGURE 13.26 Netrins inhibit the outgrowth of trochlear axons from explants of the mouse brain stem (the lower tissues). Trochlear axons emerge dorsally and are not inhibited either by (A) a dorsal spinal cord explant or (B) unaltered chick fibroblast cells. They *are* inhibited by (C) transgenic fibroblast cells secreting netrin-1 and (D) an explant from the netrin-secreting ventral floor plate of the spinal cord. (After Colamarino and Tessier-Lavigne 1995; photographs courtesy of M. Tessier-Lavigne.)

midline. However, the commissural neurons that traverse the embryo from side to side avoid this repulsion by downregulating Robo protein as they approach the midline. Once the growth cone is across the middle of the embryo, the neuron re-expresses Robo and becomes sensitive again to the midline inhibitory actions of Slit (Figure 13.27A; Brose et al. 1999; Kidd et al. 1999; Orgogozo et al. 2004).

In vertebrates, the Slit/Robo system cooperates with the Netrin/DCC system to permit the commissural neurons to cross the midline; there are several vertebrate Robo and Slit proteins, and they do different tasks (Mambetisaeva et al. 2005). As the axon extends toward its target in the developing brain, the neuron is kept from crossing the midline by Slit, which binds to *Robo1* (Figure 13.27B). Expressed Robo1 protein prevents DCC from binding to the netrin proteins. When the axon gets near the midline, however, Robo3 is expressed and blocks *Robo1*. Robo1 is no longer able to bind Slit or to block *DCC*. This enables netrin to bind *DCC* and turns the axon growth cone towards the midline. The axon grows through the midline, but after crossing it, *Robo3* is downregulated while *Robo1* is upregulated. This allows Slit to act as a chemorepellent, forcing the growth cone away from the midline. *DCC* is once again blocked, preventing the axon from going back (Woods 2004). Mutations in the human *ROBO3* gene disrupt the normal crossing of axons from one side of the brain's medulla to the other (Jen et al. 2004). Among other prob-

lems, people with this mutation are unable to coordinate their eye movements.

Target selection

Once a neuron reaches a group of cells wherein lie its potential targets, it is responsive to various proteins produced by the target cells. In addition to the proteins already mentioned, some target cells produce a set of chemotactic factors collectively called **neurotrophins**. These proteins include **nerve growth factor** (**NGF**), **brain-derived neurotrophic factor** (**BDNF**), and **neurotrophins 3 and 4/5** (**NT3, NT4/5**). These proteins are released from potential target tissues and work at short ranges as either chemotactic factors or chemorepulsive factors (Paves and Saarma 1997). Each can promote and attract the growth of some axons to its source while inhibiting other axons. For instance, sensory neurons from the rat dorsal root ganglia are attracted to sources of NT3 (Figure 13.28), but are inhibited by BDNF.

Although genetic and biochemical techniques enable us to look at the effect of one type of molecule at a time, we must remember that any growth cone is sensing a wide range of chemotactic and chemorepulsive molecules, both in solution and on the substrate upon which it migrates. Growth cones do not rely on a single type of molecule to recognize their target, but integrate the simultaneously presented attractive and repulsive cues, selecting their targets based on the combined input of these multiple cues (Winberg et al. 1998).

Forming the synapse: Activity-dependent development

When an axon contacts its target (usually either a muscle or another neuron), it forms a specialized junction called

WEBSITE **13.12** **The early evidence for chemotaxis.** Before molecular techniques, investigators used transplantation experiments and ingenuity to reveal evidence that chemotactic molecules were being released by target tissues.

WEBSITE **13.13** **The neurotrophin receptors.** Neurotrophins can bind to high-affinity receptors or to low-affinity receptors, and the pattern of binding can determine whether the signal is stimulatory or inhibitory.

(A) Slit protein   (B) Robo protein

(C) Wild type   (D) *Slit*$^{-/-}$

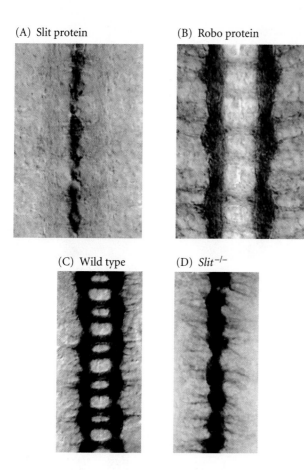

(E)

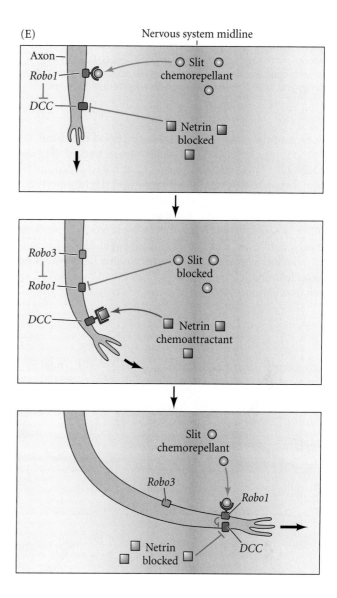

FIGURE 13.27 Robo/Slit regulation of midline crossing by neurons. (A–D) Robo and Slit in the *Drosophila* central nervous system. (A) Antibody staining reveals Slit protein in the midline neurons. (B) Robo protein appears along the neurons of the longitudinal tracts of the CNS axon scaffold. (C) Wild-type CNS axon scaffold shows the ladderlike arrangement of neurons crossing the midline. (D) Staining of the CNS axon scaffold with antibodies to all CNS neurons in a Slit loss-of-function mutant shows axons entering but failing to leave the midline (instead of running alongside it). (E) Regulation by Slit and Robo in vertebrates. Neurons are prevented from crossing the midline by Slit, which activates the *Robo1* gene. Robo1 then blocks *DCC* from binding to the netrin proteins. When the axon gets near the midline, *Robo3* is expressed and blocks these functions of Robo1, thus enabling netrin to bind to *DCC* and turning the axon growth cone towards the midline. After the cone crosses the midline, *Robo1* is upregulated and *Robo3* is downregulated. This allows Slit to act as a chemorepellent, forcing the growth cone away from the midline. (A–D from Kidd et al. 1999, photographs courtesy of C. S. Goodman; E after Woods 2004.)

a **synapse**. Neurotransmitters from the axon terminal are released at these synapses to depolarize or hyperpolarize the membrane of the target cell across the synaptic cleft.

The construction of the synapse involves several steps (Burden 1998). When motor neurons in the spinal cord extend axons to muscles, growth cones that contact newly formed muscle cells migrate over their surfaces. When a growth cone first adheres to the cell membrane of a muscle fiber, no specializations can be seen in either membrane. However, the axon terminal soon begins to accumulate neurotransmitter-containing synaptic vesicles, the membranes of both cells thicken at the region of contact, and the synaptic cleft between the cells fills with an extracellular matrix that includes a specific form of laminin (Figure 13.29A–C). This muscle-derived laminin specifically binds the growth cones of motor neurons and may act as a "stop signal" for axonal growth (Martin et al. 1995; Noakes et al. 1995). In at least some neuron-to-neuron synapses, the synapse is stabilized by N-cadherin. The activity of the synapse releases N-cadherin from storage vesicles in the growth cone (Tanaka et al. 2000).

In muscles, after the first contact is made, growth cones from other axons converge at the site to form additional synapses. During development, all mammalian muscles studied are innervated by at least two axons. However, this

FIGURE 13.28 Embryonic axon from a rat dorsal root ganglion turning in response to a source of NT3. The photographs document the turning over a 10-minute period. The same growth cone was insensitive to other neurotrophins. (After Paves and Saarma 1997, photographs courtesy of M. Saarma.)

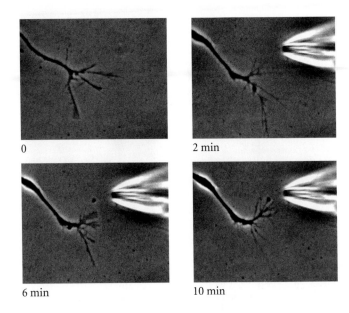

0   2 min

6 min   10 min

polyneuronal innervation is transient. Soon after birth, all but one of these axon branches are retracted (Figure 13.29D,E). This "address selection" is based on competition between the axons (Purves and Lichtman 1980; Thompson 1983; Colman et al. 1997). When one of the motor neurons is active, it suppresses the synapses of the other neurons, possibly through a nitric oxide-dependent mechanism (Dan and Poo 1992; Wang et al. 1995). Eventually, the less active synapses are eliminated. The remaining axon terminal expands and is ensheathed by a Schwann cell (see Figure 13.29E).

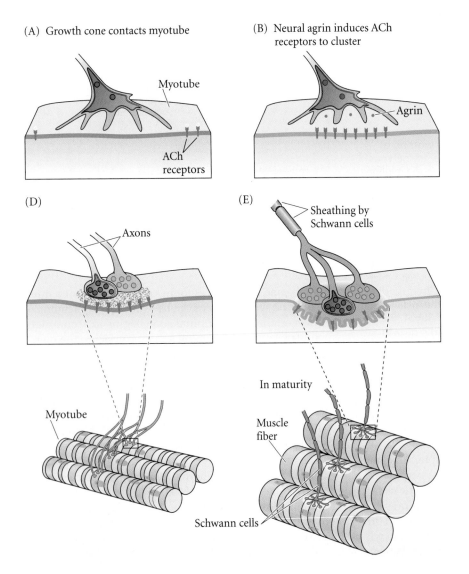

(A) Growth cone contacts myotube

Myotube

ACh receptors

(B) Neural agrin induces ACh receptors to cluster

Agrin

(C) Synaptic basal lamina forms

Extracellular matrix

Neurotransmitter vesicles

β2 laminin

(D)

Axons

Myotube

(E)

Sheathing by Schwann cells

In maturity

Muscle fiber

Schwann cells

FIGURE 13.29 Differentiation of a motor neuron synapse with a muscle. (A) A growth cone approaches a developing muscle cell. (B) The axon stops and forms an unspecialized contact on the muscle surface. Agrin, a protein released by the neuron, causes acetylcholine (ACh) receptors to cluster near the axon. (C) Neurotransmitter vesicles enter the axon terminal, and an extracellular matrix connects the axon terminal to the muscle cell as the synapse widens. This matrix contains a nerve-specific laminin. (D) Other axons converge on the same synaptic site. The wider view (below) shows muscle innervation by several axons (seen in mammals at birth). (E) All but one of the axons are eliminated. The remaining axon can branch to form a complex junction with the muscle fiber. Each axon terminal is sheathed by a Schwann cell process, and folds form in the muscle cell membrane. The overview shows muscle innervation several weeks after birth. (After Hall and Sanes 1993; Purves 1994; Hall 1995.)

Differential survival after innervation: Neurotrophic factors

One of the most puzzling phenomena in the development of the nervous system is neuronal cell death. In many parts of the vertebrate central and peripheral nervous systems, over half the neurons die during the normal course of development (see Chapter 6, especially Figure 6.25). Moreover, there do not seem to be great similarities in apoptosis patterns across species. For example, about 80 percent of a cat's retinal ganglion cells die, while in the chick retina, this figure is only 40 percent. In fish and amphibian retinas, no ganglion cells appear to die (Patterson 1992).

The apoptotic death of a neuron is not caused by any obvious defect in the neuron itself. Indeed, these neurons have differentiated and have successfully extended axons to their targets. Rather, it appears that the target tissue regulates the number of axons innervating it by limiting the supply of a neurotrophin. In addition to their roles as chemotrophic factors described in the previous section, neurotrophins regulate the survival of different subsets of neurons (Figure 13.30). NGF, for example, is necessary for the survival of sympathetic and sensory neurons. Treating mouse embryos with anti-NGF antibodies reduces the number of trigeminal sympathetic and dorsal root ganglion neurons to 20 percent of their control numbers (Levi-Montalcini and Booker 1960; Pearson et al. 1983). Furthermore, removal of these neurons' target tissues causes the death of the neurons that would have innervated them, and there is a good correlation between the amount of NGF secreted and the survival of the neurons that innervate these tissues (Korsching and Thoenen 1983; Harper and Davies 1990).

BDNF does not affect sympathetic or sensory neurons, but it can rescue fetal motor neurons in vivo from normally occurring cell death, and from induced cell death following the removal of their target tissue. The results of these in vitro studies have been corroborated by gene knockout experiments, wherein the deletion of particular neurotrophic factors causes the loss of only certain subsets of neurons (Crowley et al. 1994; Jones et al. 1994).

Neurotrophic factors are produced continuously in adults, and loss of these factors may produce debilitating diseases. BDNF is required for the survival of a particular subset of neurons in the striatum (a region of the brain involved in modulating the intensity of coordinated muscle activity such as movement, balance, and walking) and enables these neurons to differentiate and synthesize the receptor for dopamine. BDNF in this region of the brain is upregulated by Huntingtin, a protein that is mutated in **Huntington disease**. Patients with Huntington disease have decreased production of BDNF, which leads to the death of striatal neurons (Guillin et al. 2001; Zuccato et al. 2001). The result is a series of cognitive abnormalities, involuntary muscle movements, and eventual death. Another neurotrophin, glial cell line-derived neurotrophic factor (GDNF), enhances the survival of another group of neurons: the midbrain dopaminergic neurons whose destruction characterizes **Parkinson disease** (Lin et al. 1993). These neurons send axons to the striatum, whose ability to respond to their dopamine signals is dependent on BDNF. GDNF can prevent the death of these neurons in adult brains (see Lindsay 1995) and is being considered as a possible therapy for Parkinson disease (Zurn et al. 2001).

The actual survival of any given neuron in the embryo may depend on a combination of agents. Schmidt and Kater (1993) have shown that neurotrophic factors, depolarization (activation), and interactions with the substrate all combine synergistically to determine whether a neuron survives or dies. For instance, the survival of chick ciliary ganglion neurons in culture was promoted by Fgf2, laminin, and depolarization. However, Fgf2 did not promote survival when laminin was absent, and the combined effects of laminin, Fgf2, and depolarization were greater than the summed effects of each of them (Figure 13.31). The neurotrophic factors and other environmental agents appear to function by suppressing an apoptotic "suicide program" that is expressed unless repressed by these factors (Raff et al. 1993). The survival of retinal ganglion cells in culture is dependent on neurotrophic factors, but these cells can respond to these factors only if they have been

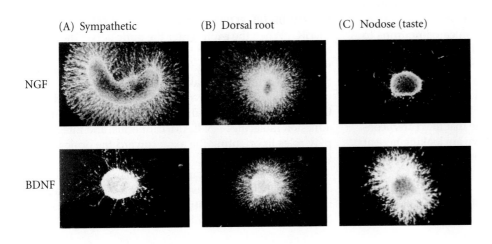

(A) Sympathetic   (B) Dorsal root   (C) Nodose (taste)

NGF

BDNF

FIGURE 13.30 Effects of NGF (top row) and BDNF (bottom) on axonal outgrowths from (A) sympathetic ganglia, (B) dorsal root ganglia, and (C) nodose (taste perception) ganglia. While both NGF and BDNF had a mild stimulatory effect on dorsal root ganglion axonal outgrowth, the sympathetic ganglia responded dramatically to NGF and hardly at all to BDNF, while the converse was true of the nodose ganglia. (From Ibáñez et al. 1991.)

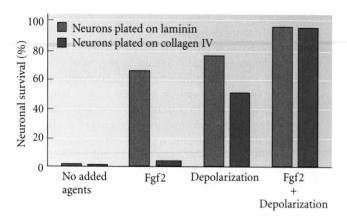

FIGURE 13.31 Interactions between the effects of substrate, depolarization, and the neurotrophin Fgf2 in the survival of ciliary ganglion neurons. Neurons were plated either on laminin (a survival-enhancing substrate) or collagen IV (a neutral substrate) and observed after 24 hours of culture in the presence or absence of depolarization or Fgf2. When cells were depolarized and grown in the presence of Fgf2, it did not matter on what substrate they grew. However, when Fgf2 was present without depolarization, the substrate made a great deal of difference. (After Schmidt and Kater 1993.)

depolarized (Meyer-Franke et al. 1995). Moreover, since neuronal activity stimulates the production of neurotrophins by active nerves, it is likely that neurons receiving a signal produce more neurotrophins (Thoenen 1995). These factors could have an effect on nearby synapses that are active (i.e., capable of responding to the neurotrophins), thereby stabilizing a set of active synapses to the exclusion of inactive ones.

Paths to glory: Migration of the retinal ganglion axons

Nearly all the mechanisms for neuronal specification and axon specificity mentioned in this chapter can be seen in the ways individual retinal neurons send axons to the vision-processing areas of the brain. Even when retinal neurons are transplanted far away from the eye, they are able to find these brain areas (Harris 1986). The ability of the brain to guide the axons of translocated neurons to their appropriate target sites implies that the guidance cues are not distributed solely along the normal pathway, but exist throughout the embryonic brain. Guiding an axon from a nerve cell body to its destination across the embryo is a complex process, and several different types of cues may be used simultaneously to ensure that the correct connections are established. Although we are looking here at non-mammalian vertebrates, the processes we describe apply to mammals as well.

GROWTH OF THE RETINAL GANGLION AXON TO THE OPTIC NERVE   The first steps in getting retinal ganglion axons to their specif-

ic regions of the optic tectum take place within the retina. As the retinal ganglion cells differentiate, their position in the inner margin of the retina is determined by cadherin molecules (N-cadherin as well as retina-specific R-cadherin) on their cell membranes (Figure 13.32A; Matsunaga et al. 1988; Inuzuka et al. 1991). Their axons grow along the inner surface of the retina toward the optic disc, the head of the optic nerve. The mature human optic nerve will contain over a million retinal ganglion axons.

The adhesion and growth of the retinal ganglion axons along the inner surface of the retina may be governed by its laminin-containing basal lamina. However, simple adhesion to laminin cannot explain the directionality of this growth. N-CAM appears to be especially important here, since the directional migration of the retinal ganglion growth cones depends on the N-CAM-expressing glial endfeet at the inner retinal surface (Stier and Schlosshauer 1995). The secretion of netrin-1 by the cells of the optic disc (where the axons are assembled to form the optic nerve) plays a role in this migration as well (Figure 13.32B). Mice lacking netrin-1 genes (or the genes for the netrin receptor found in the retinal ganglion axons) have poorly formed optic nerves, as many of the axons fail to leave the eye and grow randomly around the disc (Deiner et al. 1997). The role of netrin may change in different parts of the eye. At the entrance to the optic nerve, netrin-1 is co-expressed with laminin on the surface of the retina. Laminin converts netrin from having an attractive signal to having a repulsive signal. This repulsion might "push" the growth cone away from the retinal surface and into the head of the optic nerve, where netrin is expressed without laminin (Mann et al. 2004). Condroitin sulfate proteoglycan, a repulsive factor for retinal neurons, may also provide push toward the head of the optic nerve (Hynes and Lander 1992).

Upon their arrival at the optic nerve, the migrating axons fasciculate with axons already present there. N-CAM is critical to this fasciculation, and antibodies against N-CAM (or removal of its polysialic acid component) cause the axons to enter the optic nerve in a disorderly fashion, which in turn causes them to emerge at the wrong positions in the tectum (Figure 13.32C; Thanos et al. 1984; Yin et al. 1995).

GROWTH OF THE RETINAL GANGLION AXON THROUGH THE OPTIC CHIASM   When the axons enter the optic nerve, they grow on glial cells toward the midbrain. In non-mammalian vertebrates, the axons will go to a portion of the brain called the optic tectum. (Mammalian axons go to the lateral geniculate nuclei; this pathway will be discussed further in Chapter 22.) In vitro studies suggest that numerous cell adhesion molecules—N-CAM, cadherins, and integrins—play roles in orienting the axon toward the optic tectum (Neugebauer et al. 1988).

Upon entering the brain, mammalian retinal ganglion axons reach the optic chiasm, where they have to "decide" if they are to continue straight or if they are to turn 90 degrees and enter the other side of the brain (Figure

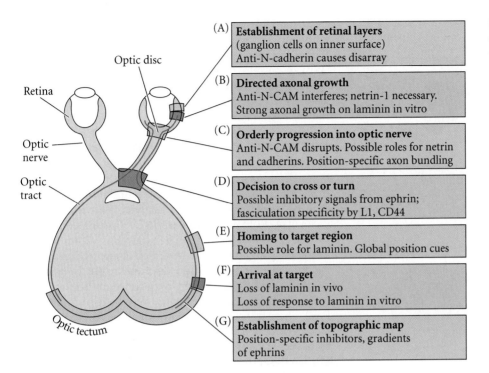

Optic disc

Retina

Optic nerve

Optic tract

Optic tectum

(A) **Establishment of retinal layers**
(ganglion cells on inner surface)
Anti-N-cadherin causes disarray

(B) **Directed axonal growth**
Anti-N-CAM interferes; netrin-1 necessary.
Strong axonal growth on laminin in vitro

(C) **Orderly progression into optic nerve**
Anti-N-CAM disrupts. Possible roles for netrin
and cadherins. Position-specific axon bundling

(D) **Decision to cross or turn**
Possible inhibitory signals from ephrin;
fasciculation specificity by L1, CD44

(E) **Homing to target region**
Possible role for laminin. Global position cues

(F) **Arrival at target**
Loss of laminin in vivo
Loss of response to laminin in vitro

(G) **Establishment of topographic map**
Position-specific inhibitors, gradients
of ephrins

FIGURE 13.32 Multiple guidance cues direct the movement of retinal ganglion axons to the optic tectum. (After Hynes and Lander 1992.)

13.32D). It appears that those axons not destined to cross to the opposite side of the brain are repulsed from doing so when they enter the optic chiasm (Godement et al. 1990). The basis of this repulsion appears to be the synthesis of ephrin on the neurons in the chiasm and ephrin receptors on the retinal axons (Cheng et al. 1995; Marcus et al. 2000).

In the mouse eye, the EphB1 receptor is expressed on those temporal axons that are repelled by the optic chiasm's EphB2, and those axons project to the side of the tectum on the same side as their eye (the ipsilateral projections); EphB1 is nearly absent on axons that are allowed to cross over. Mice lacking the *EphB1* gene show hardly any ipsilateral projections. This pattern of *EphB1* expression appears to be regulated by the Zic2 transcription factor found on those retinal axons that do form ipsilateral projections (Williams et al. 2003; Herrera et al. 2003; Pak et al. 2004).

Ephrin appears to play a similar role in the retinotectal mapping in the frog. In the developing frog, the ventral regions express the EphB receptor, while the dorsal axons do not. Before metamorphosis, both axons cross the optic chiasm. However, when the frog nervous system is being remodeled during metamorphosis, the chiasm expresses ephrin-B, which causes a subpopulation of ventral cells to be repulsed and project to the same side rather than cross the chiasm (Mann et al. 2002). (This allows the frog to have binocular vision, which is very good if one is trying to catch flies with one's tongue.)

Two guidance molecules—the L1 adhesion molecule and laminin—appear to promote crossing of the optic chiasm. On their way to the optic tectum, the axons of nonmammalian vertebrates travel on a pathway (the optic tract) over glial cells whose surfaces are coated with laminin (Figure 13.32E). Very few areas of the brain contain laminin, and the laminin in this pathway exists only when the optic nerve fibers are growing on it (Cohen et al. 1987).

TARGET SELECTION When the axons come to the end of the laminin-lined optic tract, they spread out and find their specific targets in the optic tectum (Figure 13.32F,G). Studies on frogs and fish (in which retinal neurons from each eye project to the opposite side of the brain) have indicated that each retinal ganglion axon sends its impulse to one specific site (a cell or small group of cells) within the optic tectum (Sperry 1951). There are two optic tecta in the frog brain. The axons from the right eye form synapses with the left optic tectum, while those from the left eye form synapses in the right optic tectum. The growth of axons in the *Xenopus* optic tract appears to be mediated by fibroblast growth factors secreted by the cells lining the tract. The retinal ganglion axons express FGF receptors in their growth cones. However, as the axons reach the tectum, the amount of FGF rapidly diminishes, perhaps slowing the axons down and allowing them to find their targets (McFarlane et al. 1995).

The map of retinal connections to the frog optic tectum (the **retinotectal projection**) was detailed by Marcus Jacobson (1967). Jacobson created this map by shining a narrow beam of light on a small, limited region of the retina and noting, by means of a recording electrode in the tectum, which tectal cells were being stimulated. The retinotectal projection of *Xenopus laevis* is shown in Figure 13.33. Light illuminating the ventral part of the retina stimulates cells

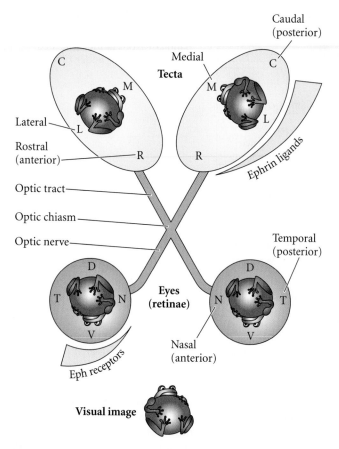

FIGURE 13.33 Map of the normal retinotectal projection in adult *Xenopus*. The right eye innervates the left tectum, and the left eye innervates the right tectum. The dorsal portion of the retina (D) innervates the lateral (L) regions of the tectum. The nasal (anterior) region of the retina projects to the caudal (C) region of the tectum. (After Holt 2002; courtesy of C. Holt.)

Current theories do not propose a point-for-point specificity between each axon and the neuron that it contacts. Rather, evidence now demonstrates that gradients of adhesivity (especially those involving repulsion) play a role in defining the territories that the axons enter, and that activity-driven competition between these neurons determines the final connection of each axon.*

ADHESIVE SPECIFICITIES IN DIFFERENT REGIONS OF THE OPTIC TECTUM
There is good evidence that retinal ganglion cells can distinguish between regions of the optic tectum. Cells taken from the ventral half of the chick neural retina preferentially adhere to dorsal (medial) halves of the tectum, and vice versa (Roth and Marchase 1976; Gottlieb et al. 1976; Halfter et al. 1981).

Retinal ganglion cells are specified along the dorsal-ventral axis by a gradient of transcription factors. Dorsal retinal cells are characterized by high concentrations of Tbx5 transcription factor, while ventral cells have high levels of Pax2. These transcription factors are induced by paracrine factors (BMP4 and retinoic acid, respectively) from nearby tissues (Koshiba-Takeuchi et al. 2000). Misexpression of Tbx5 in the early chick retina results in marked abnormalities of the retinotectal projection. Therefore, the retinal ganglial cells are specified according to their location.

One gradient that has been identified functionally is a gradient of repulsion that is highest in the posterior tectum and weakest in the anterior tectum. Bonhoeffer and colleagues (Walter et al. 1987; Baier and Bonhoeffer 1992) prepared a "carpet" of tectal membranes with alternating "stripes" of membrane derived from the posterior and the anterior tecta. They then let cells from the nasal (anterior) or temporal (posterior) regions of the retina extend axons into the carpet. The nasal ganglion cells extended axons equally well on both the anterior and posterior tectal membranes. The neurons from the temporal side of the retina, however, extended axons only on the anterior tectal membranes (Figure 13.34). When the growth cone of a temporal retinal ganglion axon contacted a posterior tectal cell membrane, the growth cone's filopodia withdrew, and the cone collapsed and retracted (Cox et al. 1990).

The basis for this specificity appears to be two sets of gradients along the tectum and retina. The first gradient set consists of ephrin proteins and their receptors. In the optic tectum, ephrin proteins (especially ephrins A2 and A5) are found in gradients that are highest in the posterior (caudal) tectum and decline anteriorly (rostrally) (Figure 13.35A). Moreover, cloned ephrin proteins have the ability to repulse axons, and ectopically expressed ephrin will prohibit axons from the temporal (but not from the nasal)

on the lateral surface of the tectum. Similarly, light focused on the temporal (posterior) part of the retina stimulates cells in the caudal portion of the tectum. These studies demonstrated a point-for-point correspondence between the cells of the retina and the cells of the tectum. When a group of retinal cells is activated, a very small and specific group of tectal cells is stimulated. Furthermore, the points form a continuum; in other words, adjacent points on the retina project onto adjacent points on the tectum. This arrangement enables the frog to see an unbroken image. This intricate specificity caused Sperry (1965) to put forward the chemoaffinity hypothesis:

> The complicated nerve fiber circuits of the brain grow, assemble, and organize themselves through the use of intricate chemical codes under genetic control. Early in development, the nerve cells, numbering in the millions, acquire and retain thereafter, individual identification tags, chemical in nature, by which they can be distinguished and recognized from one another.

*In recent years, researchers have discovered over 30 mutations in zebrafish that affect either the migration of the retinal ganglion axons to the optic tectum or the specificity of the retinotectal connections (Karlstrom et al. 1997). These mutants are still being analyzed, but they promise to provide major insights into the mechanisms by which sensory input enters the brain.

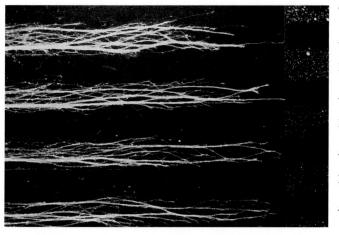

Tectal membranes

Anterior

Posterior

Anterior

Posterior

Anterior

Posterior

Anterior

FIGURE 13.34 Differential repulsion of temporal retinal ganglion axons on tectal membranes. Alternating stripes of anterior and posterior tectal membranes were absorbed onto filter paper. When axons from temporal (posterior) retinal ganglion cells were grown on such alternating carpets, they preferentially extended axons on the anterior tectal membranes. (From Walter et al. 1987.)

regions of the retina from projecting to where it is expressed (Drescher et al. 1995; Nakamoto et al. 1996). The complementary Eph receptors have been found on chick retinal ganglion cells, and they are expressed in a temporal-to-nasal gradient along the retinal ganglion axons

(Cheng et al. 1995). This gradient appears to be due to a spatially and temporally regulated expression of retinoic acid (Sen et al. 2005).

Ephrin appears to be a remarkably pliable molecule and new studies show that concentration differences in ephrinA in the tectum can account for both the smooth topographic map (wherein the position of neurons in the retina maps continuously onto the targets). Hansen and colleagues (2004) have shown that ephrinA can be attractive as well as repulsive signals for the retinal axons. Moreover, their quantitative assay for axon growth showed that an ephrin concentration-dependent transition from attractive signal to repulsive signal that was specific for the origin of the retinal axon. Axon growth is promoted by low ephrinA concentrations that are anterior to the proper target and inhibited by higher concentrations posterior to the correct target (Figure 13.35B). Each axon is thus led to the appropriate place and then told to go no further. At that eqilibrium point, there would be no growth and no inhibition, and the synapses with the target tectal neurons could be made.

The second set of gradients parallels the ephrins and Ephs. The

(A)

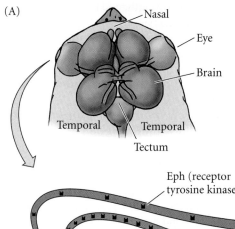

Nasal

Eye

Brain

Temporal

Temporal

Tectum

Eph (receptor tyrosine kinase)

Ephrins (ligands)

Retina

Nasal

Temporal

Anterior

Posterior

Tectum

(B)

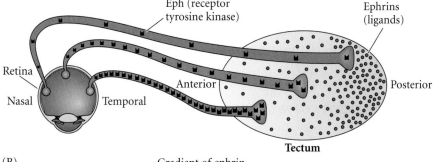

Gradient of ephrin in tectal membranes

Temporal retina

Nasal retina

Attraction

Repulsion

Attraction

Repulsion

Equilibrium (target)

Equilibrium (target)

FIGURE 13.35 Differential retinotectal adhesion is guided by gradients of Eph receptors and their ligands. (A) Representation of the dual gradients of Eph receptor tyrosine kinase in the retina and its ligands (ephrin A2 and ephrin A5) in the tectum. (B) Experiment showing that temporal, but not nasal, retinal ganglion axons respond to a gradient of ephrin ligand in tectal membranes by turning away or slowing down. An equilibrium of attractive and repulsive forces inherent in the gradient may lead specific axons to their targets. (After Barinaga 1995; Hansen et al. 2004.)

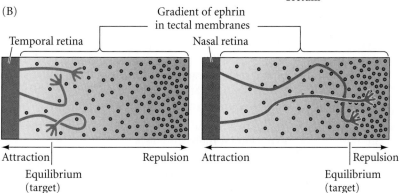

tectum has a gradient of Wnt3 that is highest at the medial region and lowest laterally (like the ephrin gradient). In the retina, a gradient of Wnt receptor is highest ventrally (like the Eph proteins). The two sets of gradients are both required to specify the position of the axon in the tectum (Schmitt et al. 2006).

The Development of Behaviors: Constancy and Plasticity

One of the most fascinating aspects of developmental neurobiology is the correlation of certain neuronal connections with certain behaviors. There are two remarkable aspects of this phenomenon. First, there are cases in which complex behavioral patterns are inherently present in the "circuitry" of the brain at birth. The heartbeat of a 19-day chick embryo quickens when it hears a distress call, and no other call will evoke this response (Gottlieb 1965). Furthermore, a newly hatched chick will immediately seek shelter if presented with the shadow of a hawk. An actual hawk is not needed; the shadow cast by a paper silhouette will suffice, and the shadow of no other bird will cause this response (Tinbergen 1951). There appear, then, to be certain neuronal connections that lead to "hard-wired" behaviors in vertebrates.

Activity-dependent synapse formation also appears to be involved in the final stages of retinal projection to the brain. In frog, bird, and rodent embryos treated with tetrodotoxin, axons will grow normally to their respective territories and will make synapses with the tectal neurons. However, the retinotectal map is a coarse one, lacking fine resolution. Just as in the final specification of motor neuron synapses, neural activity is needed for the point-for-point retinal projection onto the tectal neurons (Harris 1984; Fawcett and O'Leary 1985; Kobayashi et al. 1990). The fine-tuning of the retinotectal map involves the NMDA receptor, a protein on the tectal neurons. When the NMDA receptor is inhibited, the fine-scale resolution is not obtained (Debski et al. 1990). It appears that NMDA may

be coordinating the interaction between nitric oxide (NO) and BDNF (Wu et al. 1994; Ernst et al. 1999; Cogen and Cohen-Cory 2000). Nitric oxide is involved in the elimination of mistargeted retinal axons, while BDNF may stabilize retinal axon connections. It appears that NO induces growth cone collapse and retraction of developing retinal axons, whereas BDNF protects growth cones and axons from the effects of NO (Ernst et al. 2000). Exposure to both BDNF and NO, but not to either factor alone, stabilized growth cones and axons. Activity-dependent synapse formation is extremely important during the development of the mammalian visual system—a process that will be detailed in Chapter 22.

Equally remarkable are those instances in which the nervous system is so plastic that new experiences can modify the original set of neuronal connections, causing the creation of new neurons or the formation of new synapses between existing neurons. We will discuss neural plasticity at greater length in Chapter 22; here it suffices to say that the brain does not stop developing at birth. The Nobel Prize-winning research of Hubel and Wiesel (1962, 1963) demonstrated that there is competition between the retinal neurons of each eye for targets in the cortex, and that their connections must be strengthened by experience. As mentioned in the preceding chapter, new experiences lead to the generation of new neurons in adult birds and mammals. Thus, the nervous system continues to develop in adult life, and the pattern of neuronal connections is a product of both inherited patterning and patterning produced by experience.

As one investigator (Purves 1994) concluded his analysis of brain development:

> Although a vast majority of this construction must arise from developmental programs laid down during the evolution of each species, neural activity can modulate and instruct this process, thus storing the wealth of idiosyncratic information that each of us acquires through individual experience and practice.

 ## Neural Crest Cells and Axonal Specificity

1. The neural crest is a transitory structure. Its cells migrate to become numerous different cell types.

2. Trunk neural crest cells can migrate dorsolaterally into the ectoderm, where they become melanocytes. They can also migrate ventrally, to become sympathetic and parasympathetic neurons and adrenomedullary cells.

3. Cranial neural crest cells enter the pharyngeal arches to become the cartilage of the jaw and the bones of the middle ear. They also form the bones of the frontonasal process, the papillae of the teeth, and the cranial nerves.

4. Cardiac neural crest enters the heart and forms the septum (separating wall) between the pulmonary artery and aorta.

5. The formation of the neural crest depends on interactions between the prospective epidermis and the neural plate. Paracrine factors from these regions induce the formation of transcription factors that enable neural crest cells to emigrate.

6. The path a neural crest cell takes depends on the extracellular matrix it meets.

7. Trunk neural crest cells will migrate through the anterior portion of each sclerotome, but not through the posterior portion of a sclerotome. Semaphorin

and ephrin proteins expressed in the posterior portion of each sclerotome can prevent neural crest cell migration.

8. Some neural crest cells appear to be capable of forming a large repertoire of cell types. Other neural crest cells may be restricted even before they migrate. The final destination of the neural crest cell can sometimes change its specification.

9. The fates of the cranial neural crest cells are influenced by Hox genes. They can acquire their Hox gene expression pattern through interaction with neighboring cells.

10. Motor neurons are specified according to their position in the neural tube. The Lim family of transcription factors plays an important role in this specification.

11. The targets of motor neurons are specified before their axons extend into the periphery.

12. The growth cone is the locomotor organelle of the neuron, and it senses environmental cues. Axons can find their targets without neuronal activity.

13. Some proteins are generally permissive to neuron adhesion and provide substrates on which axons can migrate. Other substances prohibit migration.

14. Some growth cones recognize molecules that are present in very specific areas and are guided by these molecules to their respective targets.

15. Some neurons are "kept in line" by repulsive molecules. If they wander off the path to their target, these molecules send them back. Some molecules, such as the semaphorins, are selectively repulsive to particular sets of neurons.

16. Some neurons sense gradients of a protein and are brought to their target by following these gradients. The netrins may work in this fashion.

17. Target selection can be brought about by neurotrophins, proteins that are made by the target tissue and that stimulate the particular set of axons able to innervate it. In some cases, the target makes only enough of these factors to support a single axon.

18. Address selection is activity-dependent. An active neuron can suppress synapse formation by other neurons on the same target.

19. Retinal ganglion cells in frogs and chicks send axons that bind to specific regions of the optic tectum. This process is mediated by numerous interactions, and target selection appears to be mediated through ephrins.

20. Some behaviors appear to be innate ("hard-wired"), while others are learned. Experience can strengthen certain neural connections.

For Further Reading

Complete bibliographical citations for all literature cited in this chapter can be found on the Vade Mecum CD that accompanies the book and at the free access website www.devbio.com

Bronner-Fraser, M. and S. E. Fraser. 1988. Cell lineage analysis reveals multipotency of some avian neural crest cells. *Nature* 335: 161–164.

Hall, B. K. 2000. The neural crest as a fourth germ layer and vertebrates as quadroblastic not triploblastic. *Evol. Dev.* 2: 3–5.

Jiang, X., S. Iseki, R. E. Maxson, H. M. Sucov and G. Morriss-Kay. 2002. Tissue origins and interactions in the mammalian skull vault. *Dev. Biol.* 241: 106–116.

Le Douarin, N. M. 2004. The avian embryo as a model to study the development of the neural crest: A long and still ongoing study. *Mech. Dev.* 121: 1089–1102.

Mambetisaeva, E., T. W. Andrews, L. Camurri, A. Annan and V. Sundaresan.

2005. Robo family of proteins exhibit differential expression in mouse spinal cord and Robo-Slit interaction is required for midline crossing in vertebrate spinal cord. *Dev. Dynam.* 233: 41-51.

Meulemans, D. and M. Bronner-Fraser. 2004. Gene-regulatory interactions in neural crest evolution and development. *Dev. Cell* 7: 291–299.

Schlosser, G. 2005. Evolutionary origins of vertebrate placodes: Insights from developmental studies and from comparisons with other deuterostomes. *J. Exp. Zool.* 304B: 347–399.

Teillet, M.-A., C. Kalcheim and N. M. Le Douarin. 1987. Formation of the dorsal root ganglia in the avian embryo: Segmental origin and migratory behavior of neural crest progenitor cells. *Dev. Biol.* 120: 329–347.

Tosney, K. W. 2004. Long-distance cue from emerging dermis stimulates neural crest melanoblast migration. *Dev. Dynam.* 229: 99–108.

Vaahtokari, A., T. Aberg, J. Jernvall, S. Keränen and I. Thesleff. 1996. The enamel knot as a signalling center in the developing mouse tooth. *Mech. Dev.* 54: 39–43.

Waldo, K., S. Miyagawa-Tomita, D. Kumiski and M. L. Kirby. 1998. Cardiac neural crest cells provide new insight into septation of the cardiac outflow tract: Aortic sac to ventricular septal closure. *Dev. Biol.* 196: 129–144.

Walter, J., S. Henke-Fahle and F. Bonhoeffer. 1987. Avoidance of posterior tectal membranes by temporal retinal axons. *Development* 101: 909–913.

Paraxial and Intermediate Mesoderm

14

IN CHAPTERS 12 AND 13 WE FOLLOWED the various tissues formed by the verte-brate ectoderm. In this chapter and the next, we will follow the development of the mesodermal and endodermal germ layers. We will see that the endoderm forms the lining of the digestive and respiratory tubes, with their associated organs. The mesoderm generates all the organs between the ectodermal wall and the endodermal tissues.

The trunk mesoderm of a neurula-stage embryo can be subdivided into four regions (Figure 14.1A):

1. The central region of trunk mesoderm is the **chordamesoderm**. This tissue forms the notochord, a transient organ whose major functions include inducing the formation of the neural tube and establishing the anterior-posterior body axis. The formation of the notochord on the future dorsal side of the embryo was discussed in Chapters 10 and 11.
2. Flanking the notochord on both sides is the **paraxial mesoderm**, or **somitic mesoderm**. The tissues developing from this region will be located in the back of the embryo, along the spine. The cells in this region will form somites—blocks of mesodermal cells on either side of the neural tube—which will produce many of the connective tissues of the back (bone, muscle, cartilage, and dermis).
3. The **intermediate mesoderm** forms the urogenital system, consisting of the kidneys, the gonads, and their associated ducts. The outer (cortical) portion of the adrenal gland also derives from this region.
4. Farthest away from the notochord, the **lateral plate mesoderm** gives rise to the heart, blood vessels, and blood cells of the circulatory system, as well as to the lining of the body cavities and to all the mesodermal components of the limbs except the muscles. It also helps form a series of extraembryonic membranes that are important for transporting nutrients to the embryo.

These four subdivisions are thought to be specified along the mediolateral (cen-ter-to-side) axis by increasing amounts of BMPs (Pourquié et al. 1996; Tonegawa et al. 1997). The more lateral mesoderm of the chick embryo expresses higher levels of BMP4 than the midline areas, and one can change the identity of the mesodermal tissue by altering BMP expression. While it is not known how this patterning is accomplished, it is thought that the different BMP concentrations may cause differential expression of the Forkhead (Fox) family of transcription factors. *Foxf1* is transcribed in those regions that will become the lateral plate and extraembryonic mesoderm, whereas *Foxc1* and *Foxc2* are expressed in the paraxial mesoderm that will form the somites (Wilm et al. 2004). If *Foxc1* and *Foxc2* are both deleted from the mouse genome, the paraxial mesoderm is re-

> " Of physiology from top to toe
>   I sing,
> Not physiognomy alone or brain alone
>   is worthy for the Muse,
> I say the form complete is
>   worthier far,
> The Female equally with the
>   Male I sing. "
>   — *WALT WHITMAN (1867)*

> " Built of 206 bones, the skeleton
> is a living cathedral of ivory vaults,
> ribs, and buttresses—a structure at
> once light and strong. "
>   — *NATALIE ANGIER (1994)*

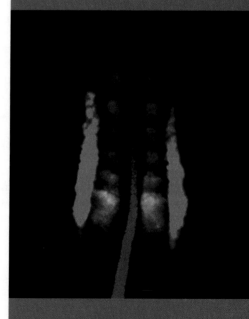

(A)

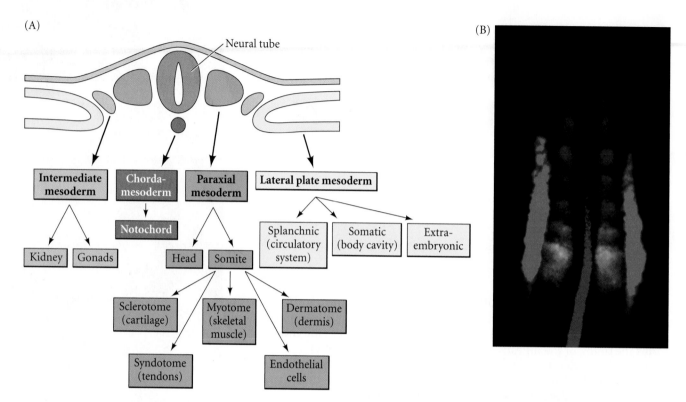

(B)

FIGURE 14.1 The major lineages of the amniote mesoderm. (A) Schematic of the mesodermal compartments of the amniote embryo. (B) Staining for the medial mesodermal compartments in the trunk of a 12-somite chick embryo (about 33 hours). In situ hybridization was performed with probes binding to *chordin* mRNA (blue) in the notochord, *paraxis* mRNA (green) in the somites, and *Pax2* mRNA (red) in the intermediate mesoderm. (B from Denkers et al. 2004, photograph courtesy of T. J. Mauch.)

specified as intermediate mesoderm and initiates the expression of *Pax2*, which encodes a major transcription factor of the intermediate mesoderm (Figure 14.1B).

Anterior to the trunk mesoderm is a fifth mesodermal region, the prechordal plate mesoderm. This region provides the head mesenchyme that forms much of the connective tissues and musculature of the face.

PARAXIAL MESODERM: THE SOMITES AND THEIR DERIVATIVES

One of the major tasks of gastrulation is to create a mesodermal layer between the endoderm and the ectoderm. As seen in Figure 14.2, the formation of mesodermal and endodermal tissues is not subsequent to neural tube formation, but occurs synchronously. The notochord extends beneath the neural tube, from the base of the head into the tail. On either side of the neural tube lie thick bands of mesodermal cells. These bands of paraxial mesoderm are referred

to either as the **segmental plate** (in chick embryos) or the **unsegmented mesoderm** (in other vertebrate embryos). As the primitive streak regresses and the neural folds begin to gather at the center of the embryo, the paraxial mesoderm separates into blocks of cells called **somites**. The paraxial mesoderm appears to be specified by the antagonism of BMP signaling by the Noggin protein. Noggin is usually synthesized by the early segmental plate mesoderm, and if Noggin-expressing cells are placed into the presumptive lateral plate mesoderm, the lateral plate tissue will be re-specified into somite-forming paraxial mesoderm (Figure 14.3; Tonegawa and Takahashi 1998).

The mature somites contain three major compartments: the **sclerotome** which forms the vertebrae and rib cartilage; the **myotome**, which forms the musculature of the back, ribs, and limbs; and the **dermatome**, which generates the dermis of the back. In addition, two smaller compartments are formed from these. The **syndetome** arising within the sclerotome generates the tendons, and an as-yet-unnamed group of cells in the posterior somite generates vascular cells of the dorsal aorta and intervertebral blood vessels (Pardanaud et al. 1996; Sato et al. 2005). In addition, it is the somites that determine the migration paths of neural crest cells and spinal nerve axons.

VADE MECUM² **Mesoderm in the vertebrate embryo.**
The organization of the mesoderm in the neurula stage is similar for all vertebrates. You can see this organization by viewing serial sections of the chick embryo. [Click on Chick-Mid]

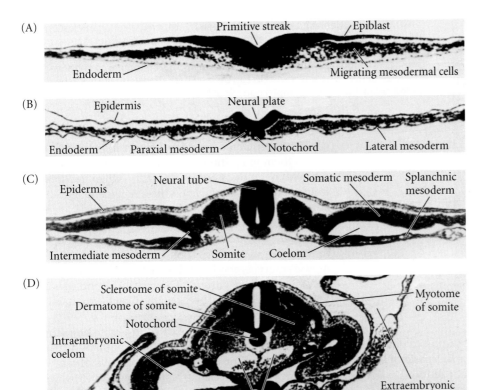

(A) Primitive streak — Epiblast — Endoderm — Migrating mesodermal cells

(B) Epidermis — Neural plate — Endoderm — Paraxial mesoderm — Notochord — Lateral mesoderm

(C) Epidermis — Neural tube — Somatic mesoderm — Splanchnic mesoderm — Intermediate mesoderm — Somite — Coelom

(D) Sclerotome of somite — Dermatome of somite — Notochord — Intraembryonic coelom — Myotome of somite — Dorsal aortae — Extraembryonic coelom

FIGURE 14.2 Gastrulation and neurulation in the chick embryo, focusing on the mesodermal component. (A) Primitive streak region, showing migrating mesodermal and endodermal precursors. (B) Formation of the notochord and paraxial mesoderm. (C,D) Differentiation of the somites, coelom, and the two aortae (which will eventually fuse). A–C, 24-hour embryos; D, 48-hour embryo.

The Formation of Somites

The periodicity of somite formation

Somite formation depends on a "clock and wave" mechanism, in which an oscillating signal (the "clock") is provided by the Notch and Wnt pathways, and a rostral-to-caudal gradient provides a moving "wave" of an FGF that sets the somite boundaries. In the chick embryo, a new somite is formed about every 90 minutes. In mouse embryos, this time frame is more variable (Tam 1981). However, somites appear at exactly the same time on both sides of the embryo, and the clock for somite formation is set when the cells first enter the presomitic mesoderm. If the presomitic mesoderm is inverted such that the caudal end is rostral and the rostral end faces the tail, somite formation will start from the caudal end and proceed rostrally. Even if isolated from the rest of the body, the presomitic mesoderm will segment at the appropriate time and in the right direction (Palmeirim et al. 1998).

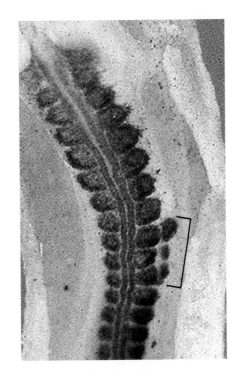

FIGURE 14.3 Specification of somites. Placing Noggin-secreting cells into a prospective region of chick lateral plate mesoderm will respecify that mesoderm into somite-forming paraxial mesoderm. Induced somites (bracketed) were detected by in situ hybridization with Pax3. (After Tonegawa and Takahashi 1998; photograph courtesy of Y. Takahashi.)

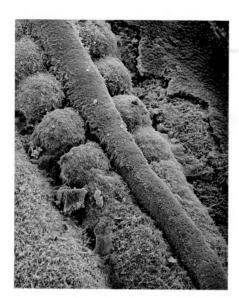

FIGURE 14.4   Neural tube and somites seen by scanning electron microscopy. When the surface ectoderm is peeled away, well-formed somites are revealed, as well as paraxial mesoderm (bottom right) that has not yet separated into distinct somites. A rounding of the paraxial mesoderm into a somitomere can be seen at the lower left, and neural crest cells can be seen migrating ventrally from the roof of the neural tube. (Photograph courtesy of K. W. Tosney.)

Even the number of somites is set at the initial stages of presomitic mesoderm formation. When *Xenopus* or mouse embryos are experimentally or genetically reduced in size, the number of somites remains the same (Tam 1981). The total number of somites formed is characteristic of a species (50 in chicks, 65 in mice, and as many as 500 in some snakes).

The important components of **somitogenesis** (somite formation) are (1) periodicity, (2) fissure formation (to separate the somites), (3) epithelialization, (4) specification, and (5) differentiation. The first somites appear in the anterior portion of the trunk, and new somites "bud off" from the rostral end of the presomitic mesoderm at regular intervals (Figure 14.4). Somite formation begins as paraxial mesoderm cells become organized into whorls of cells, sometimes called called somitomeres. The somitomeres become compacted and split apart as fissures separate them into discrete, immature somites. The mesenchymal cells making up the immature somite now change, with the outer cells joining into an epithelium while the inner cells remain mesenchymal. Because individual embryos in any species can develop at slightly different rates (as when chick embryos are incubated at slightly different temperatures), the number of somites present is usually the best indicator of how far development has proceeded.

Where somites form: The Notch pathway

Although we do not completely understand the mechanisms controlling the temporal periodicity of somite formation, one of the key agents in determining where somites form is the Notch signaling pathway. When a small group of cells from a region constituting the posterior border at the presumptive somite boundary is transplanted into a region of unsegmented mesoderm that would not ordinarily be part of the boundary area, a new boundary is created. The transplanted boundary cells instruct the cells anterior to them to epithelialize and separate. Nonboundary cells will not induce border formation when transplanted to a nonborder area. However, these nonboundary cells can acquire boundary-forming ability if an activated Notch protein is electroporated into those cells (Figure 14.5A–C; Sato et al. 2002). Morimoto and colleagues (2005) have been able to visualize the endogenous levels of Notch activity in mouse embryos, and have shown that it oscillates in a segmentally defined pattern. The somite boundaries were formed at the interface between the Notch-expressing and Notch-nonexpressing areas.

Notch has also been implicated in somite fissioning by mutations. Segmentation defects have been found in mice that are mutant for important components of the Notch pathway. These include the Notch protein itself, as well as its ligands Delta-like1 and Delta-like3 (Dll1 and Dll3). Mutations affecting Notch signaling have been shown to be responsible for aberrant vertebral formation in mice and humans. In humans, individuals with spondylocostal dysplasia have numerous vertebral and rib defects that have been linked to mutations of the *Delta-like3* gene. Mice with knockouts of *Dll3* have a phenotype similar to that of the human syndrome (Figure 14.5D,E; Bulman et al. 2000; Dunwoodie et al. 2002).

Moreover, Notch signaling follows a remarkable wave-like pattern wherein the *Notch* gene becomes highly expressed in the posterior region of the forming somite, just anterior to the "cut." *Notch* genes are transcribed in a cyclic fashion and function as an autonomous segmentation "clock" (Palmeirim et al. 1997; Jouve et al. 2000, 2002). If Notch signaling determines the placement of somite formation, then Notch protein must control a cascade of gene expression that ultimately separates the tissues.

One of the genes regulated by Notch is *hairy1*, a homologue of the *Drosophila* segmentation gene. It is expressed in the presomitic segmental plate in a cyclic, wavelike manner, cresting every 90 minutes in the chick embryo (Figure 14.6). The caudal domain of the *hairy1* expression pattern rises anteriorly, and then recedes like a wave, leaving a band of expression at what will become the posterior half of the somite. The caudal boundary of this domain is exactly where the transplantation experiments showed Notch expression to be important. Another gene that is controlled in a cyclic fashion, and that may be regulated by Notch, is *Mesp2*. Mesp2 is a transcription factor whose expression is at the next-forming fission site, and it ultimately induces EphA4, one of the compounds whose repulsive interaction separate the somites (Saga et al. 1997; Watanabe et al. 2005).

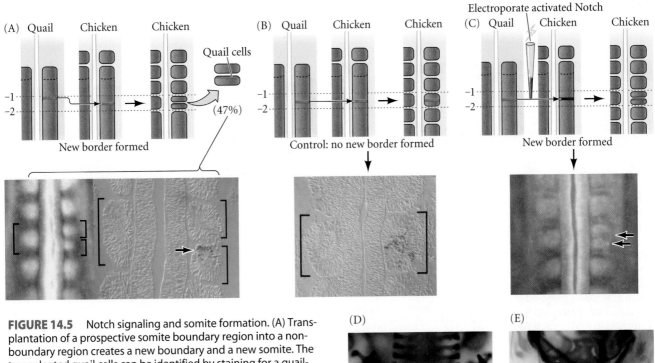

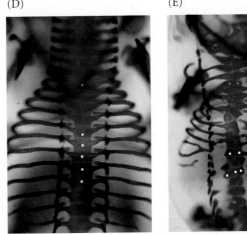

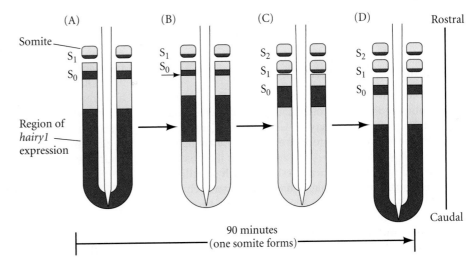

FIGURE 14.5 Notch signaling and somite formation. (A) Transplantation of a prospective somite boundary region into a nonboundary region creates a new boundary and a new somite. The transplanted quail cells can be identified by staining for a quail-specific protein. (B) Transplantation of nonboundary cells into a nonboundary region does not create a new boundary or a new somite. (C) Transplantation of a nonboundary region that has had Notch activated will cause a new somite boundary to occur. (D,E) Dorsal views of (D) control mouse and (E) littermate with a mutation in their *Delta-like3* gene (the gene encoding a Notch ligand). The mutant had several ossification centers (white dots) in a row instead of in a column, and its ribs were malformed. (A–C after Sato et al. 2002, photographs courtesy of Y. Takahashi; D,E from Dunwoodie et al. 2002, photographs courtesy of S. Dunwoodie.)

FIGURE 14.6 Somite formation correlates with the wavelike expression of the *hairy1* gene in the chick. (A) In the posterior portion of a chick embryo somite, S_1 has just budded off the presomitic mesoderm. Expression of the *hairy1* gene (purple) is seen in the caudal half of this somite, as well as in the posterior portion of the presomitic mesoderm and in a thin band that will form the caudal half of the next somite (S_0). (B) A caudal fissure (small arrow) begins to separate the new somite from the presomitic mesoderm. The posterior region of *hairy1* expression extends anteriorly. (C) The newly formed somite is now referred to as S_1; it retains the expression of *hairy1* in its caudal half, as the posterior domain of *hairy1* expression moves farther anteriorly and shortens. The former S_1 somite, now called S_2, undergoes differentiation. (D) The formation of somite S_1 is complete, and the anterior region of what had been the posterior *hairy1* expression pattern is now the anterior expression pattern. It will become the caudal domain of the next somite. The entire process takes 90 minutes.

SIDELIGHTS & SPECULATIONS

Coordinating Waves and Clocks in Somite Formation

Although the mechanisms of somite formation are being studied intensely, a consensus as to the mechanisms by which somites are formed is only vaguely visible. The basic theory of periodic somite formation starts with the concept of a negative feedback loop. Dale et al. (2003) proposed that one of the proteins activated by the Notch protein is also able to inhibit *Notch*, which would establish such a negative feedback loop. When the inhibitor is degraded, *Notch* would become active again. Such a cycle would create a "clock" whereby the *Notch* gene would be turned on and off by a protein it itself induces. These off-and-on oscillations could provide the molecular basis for the periodicity of somite segmentation. A recent report indicates that the Mesp2 protein may be such a regulator of *Notch* (Morimoto et al. 2005).

But what regulates the Notch clock and coordinates it with the FGF wave? It is possible that the Notch pathway has to interact with the Wnt pathway. The first clue was that Axin2, a negative regulator of Wnt signaling, also shows wavelike expression patterns in the presomitic mesoderm. This was interesting, since Wnt3a is involved in forming the presomitic mesoderm and is needed for the elongation of the body axis. Since Axin2 not only inhibits Wnt signaling but is positively regulated by Wnt, Aulehla and colleagues (2003) proposed that a negative feedback loop was working:

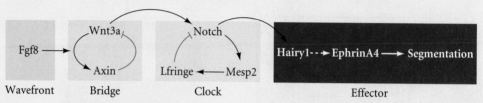

FIGURE 14.7 Possible scheme for the regulation of the clock through which an Fgf8 gradient regulates a Wnt oscillating clock, which in turn controls a Notch clock that can inhibit its own activity in a negative feedback loop. This pathway is hypothetical, and different species may use different molecules.

1. Wnt3a stimulated Axin2 production.
2. Axin2 inhibited Wnt signaling.
3. Axin2, an unstable protein, decays.
4. With the decay of Axin2, Wnt3a can signal again.

This feedback loop appears to regulate Notch activity in mice (Aulehla and Herrmann 2004; NaKaya et al. 2005).

So what regulates Wnt3a expression? Recall that in Chapter 13, we discussed the gradient of *Fgf8* expression in the chick embryo. Similar gradients have been seen in other vertebrates. Somitogenesis has long been observed to occur simultaneously with the regression of the primitive streak, and recent studies on the FGF gradient suggest that this is not mere coincidence (Dubrulle et al. 2001; Sawada et al. 2001). Hensen's node secretes Fgf8, and this paracrine factor can prevent cells from expressing *Lunatic fringe*—a possible activator of Notch signaling. Thus, according to this model, Fgf8 protein would stimulate production of Wnt3a protein; Wnt3a would

then cycle due to its production of Axin2. This Wnt cycling would control the expression of *Lunatic fringe* in the posterior, oscillating, region of the presomitic mesoderm. Lunatic fringe would control Notch expression. Thus, as the result of a series of events initiated by Fgf8 produced in Hensen's node, sharp boundaries of Notch expression appear where the somite will be cut from the unsegmented paraxial mesoderm (Figure 14.7).

But some clocks can be faster or slower than other clocks. What keeps the segmentation clocks on both sides of the body synchronized? As you might expect from previous chapters, the answer was found to be retinoic acid. If retinoic acid synthesis is inhibited in mouse embryos, the oscillation activity on the right side of the embryo becomes delayed with respect to the left side (Vermot et al. 2005). Thus, without endogenous RA synthesis, the left side of the body could mature faster than the right side.

The separation of somites from the unsegmented mesoderm

Two proteins whose roles appear to be critical for somite separation are the Eph tyrosine kinases and their ligands, the ephrin proteins. We saw in Chapter 13 that the Eph tyrosine kinase receptor proteins and their ephrin ligands are able to cause cell-cell repulsion between the posterior somite and migrating neural crest cells. Eph signaling is believed to mediate cell shape changes, and such changes could be responsible for the separation of the presomitic mesoderm at the ephrin-B2/EphA4 border. In the zebrafish, the boundary between the most recently separated somite and the presomitic mesoderm forms between

ephrin-B2 in the posterior of the somite and EphA4 in the most anterior portion of the presomitic mesoderm (Figure 14.8; Durbin et al. 1998). EphA4 is restricted to the boundary area in chick embryos as well. As somites form, this pattern of gene expression is reiterated caudally. Interfering with this signaling (by injecting embryos with RNA encoding dominant negative Ephs) leads to abnormal somite boundary formation.*

---

*A recently discovered signal from the ventral region of the somites may initiate the separation of the newly forming somite from the presomitic mesoderm (Sato and Takahashi 2005). This ventral signal moves dorsally and apparently puts the mesenchymal cells in a line so that the cut can be made.

(A)

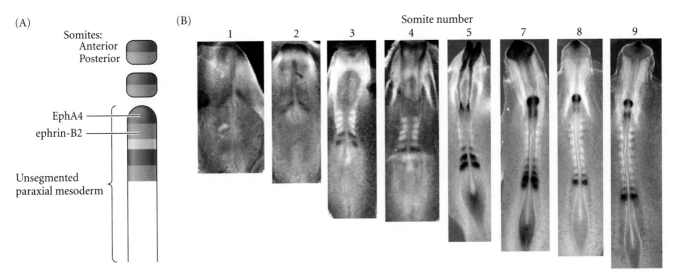

(B)

FIGURE 14.8 Ephrin and its receptor constitute a possible cut site for somite formation. (A) Expression pattern of the receptor tyrosine kinase EphA4 (blue) and its ligand, ephrin-B2 (red) as somites develop. The somite boundary forms at the junction between the region of ephrin expression on the posterior of the last somite formed and the region of EphA4 expression on the anterior of the next somite to form. In the presomitic mesoderm, the pattern is created anew as each somite buds off. The posteriormost region of the next somite to form does not express ephrin until that somite is ready to separate. (B) In situ hybridization showing EphA4 (dark blue) expression as new somites form in the chick embryo. (A after Durbin et al. 1998; B photograph courtesy of J. Kastner.)

Epithelialization of the somites

Several studies in the chick have shown that epithelialization occurs immediately after somitic fission appears. As seen in Figure 14.4, the cells of the newly formed somite are randomly organized as a mesenchymal mass. The synthesis of two cell proteins—the extracellular matrix organizing protein fibronectin, and the N-cadherin adhesion protein—rearranges the outer cells of each somite into an epithelium (Lash and Yamada 1986; Hatta et al. 1987; Saga et al. 1997). These extracellular matrix proteins, in turn, may be regulated by Paraxis and Mesp2. These transcription factors are also expressed at the rostral (anterior) end of the unsegmented mesoderm of mouse embryos, which is precisely the region that will form the somite. In mice lacking either the *Paraxis* or the *Mesp2* gene, somites segregate from the segmental plate but fail to epithelialize (Burgess et al. 1995; Barnes et al. 1997; Takahashi et al. 2005).

The *Mesp2* gene is controlled by the Notch signaling pathway. The *Paraxis* gene is regulated by processes involving the cytoskeleton. Indeed, the cytoskeletal proteins are essential for any cell shape change such as this mesenchymal-epithelial transition. In particular, the small GTPases, Rac1, and Cdc42 are able to organize the actin cytoskeleton, and all three of these can be activated by integrin when integrin binds to fibronectin (Kjoller and Hall 1999). Such cytoskeletal changes are seen in the somite cells as they make the transition from the mesenchymal to the epithelial state (Figure 14.9). Cdc42 is critical for converting the presomitic mesenchymal cells into an epithelium, and the

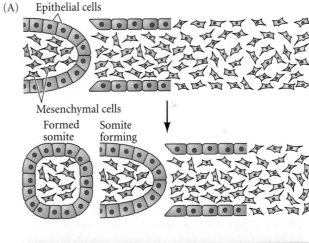

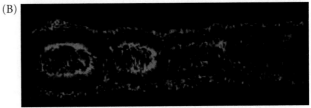

FIGURE 14.9 Epithelialization and de-epithelialization in somites of a chick embryo. (A) Changes in cell shape from mesenchymal (pink) to epithelial (gray) cells when a somite forms from presomitic mesenchyme. A formed somite is surrounded by epithelial cells, with mesenchymal cells remaining inside. In chickens, epithelialization occurs first at the posterior edge of the somite, with the anterior edge becoming epithelial later. (B) Changes in cell polarity as somites form are revealed by staining that visualizes F-actin accumulation (red). (After Nakaya et al. 2004; photograph courtesy of Y. Takahashi.)

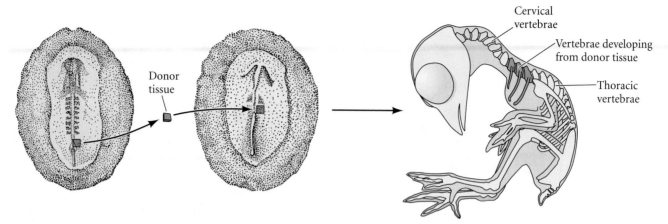

Cervical vertebrae

Vertebrae developing from donor tissue

Thoracic vertebrae

Donor tissue

proper levels of Rac1 are needed to activate *Paraxis* (Nakaya et al. 2004).

Somite specification along the anterior-posterior axis

Although all somites look identical, they will form different structures. For instance, the somites that form the cervical vertebrae of the neck and lumbar vertebrae of the abdomen are not capable of forming ribs; ribs are generated only by the somites forming the thoracic vertebrae. Moreover, the specification of the thoracic vertebrae occurs very early in development. The segmental plate mesoderm is determined by its position along the anterior-posterior axis before somitogenesis. If one isolates the region of chick segmental plate that will give rise to a thoracic somite and transplants this mesoderm into the cervical (neck) region of a younger embryo, the host embryo will develop ribs in its neck—but only on the side where the thoracic mesoderm has been transplanted (Figure 14.10; Kieny et al. 1972; Nowicki and Burke 2000).

The somites are specified according to the Hox genes they express. These Hox genes are active in the segmental plate mesoderm before it becomes organized into somites (Carapço et al. 2005). Mice that are homozygous for a loss-of-function mutation of *Hoxc8* convert a lumbar vertebra into an extra thoracic vertebra, complete with ribs (see Figure 11.43). The Hox genes are activated concomitantly with somite formation, and the embryo appears to "count somites" in setting the expression boundaries of the Hox genes. If Fgf8 levels are manipulated to create extra (albeit smaller) somites, the appropriate Hox gene expression will be activated in the appropriately numbered somite, even if it is in a different position along the anterior-posterior axis. Moreover, when mutations affect the autonomous segmentation clock, they also affect the activation of the appropriate Hox genes (Dubrulle et al. 2001; Zakany et al. 2001). Once established, each somite retains its pattern of Hox gene expression, even if that somite is transplanted into another region of the embryo (Nowicki and Burke 2000). The regulation of the Hox genes by the segmentation clock should allow coordination between the formation and the specification of the new segments.

FIGURE 14.10 When segmental plate mesoderm that would ordinarily form thoracic somites is transplanted into a region in a younger embryo (caudal to the first somite) that would ordinarily give rise to cervical (neck) somites, the grafted mesoderm differentiates according to its original position and forms ribs in the neck. (After Kieny et al. 1972.)

Derivatives of the somites

In contrast to the early commitment of the presomitic segmental plate mesoderm along the anterior-posterior axis, the commitment of the cells *within* a somite occurs relatively late, after the somite has already formed. When the somite is first separated from the presomitic mesoderm, any of its cells can become any of the somite-derived structures. These structures include:

- The cartilage of the vertebrae and ribs
- The muscles of the rib cage, limbs, abdominal wall, back, and tongue
- The tendons that connect the muscles to the bones
- The dermis of the dorsal skin
- Vascular cells that contribute to the formation of the aorta and the intervertebral blood vessels

As the somite matures, its various regions become committed to forming only certain cell types. The ventral-medial cells of the somite (those cells located farthest from the back but closest to the neural tube) undergo mitosis, lose their round epithelial characteristics, and become mesenchymal cells again. The portion of the somite that gives rise to these cells is called the **sclerotome**, and these mesenchymal cells ultimately become the cartilage cells (chondrocytes) of the vertebrae and part (if not all) of each rib (Figure 14.11A,B; see also Figure 14.2).

Fate mapping with chick-quail chimeras (Ordahl and Le Douarin 1992; Brand-Saberi et al. 1996; Kato and Aoyama 1998) has revealed that the remaining epithelial portion of the somite is arranged into three regions (Figure 14.11C,D). The central region of the dorsal layer of the dermamyotome has classically been called the **dermatome**, and it generates the mesenchymal connective tissue of the back skin: the dermis. The cells in the two lateral portions of the epitheli-

(A) 2-day embryo

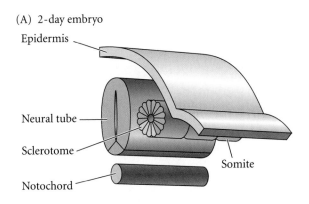

Epidermis

Neural tube

Sclerotome

Somite

Notochord

(B) 3-day embryo

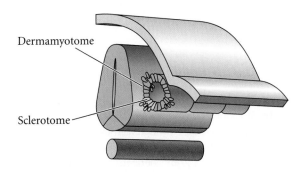

Dermamyotome

Sclerotome

(C) 4-day embryo

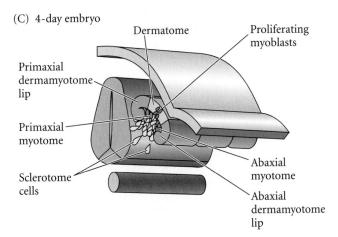

Dermatome

Proliferating myoblasts

Primaxial dermamyotome lip

Primaxial myotome

Sclerotome cells

Abaxial myotome

Abaxial dermamyotome lip

(D) Late 4-day embryo

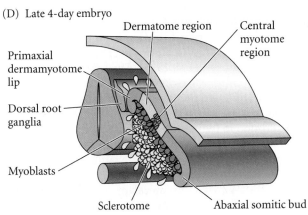

Dermatome region

Central myotome region

Primaxial dermamyotome lip

Dorsal root ganglia

Myoblasts

Sclerotome

Abaxial somitic bud

FIGURE 14.11 Diagram of a transverse section through the trunk of a chick embryo on days 2–4. (A) In the 2-day somite, the sclerotome cells can be distinguished from the rest of the somite. (B) On day 3, the sclerotome cells lose their adhesion to one another and migrate toward the neural tube. (C) On day 4, the remaining cells divide. The medial cells form a primaxial myotome beneath the dermamyotome, while the lateral cells form an abaxial myotome. (D) A layer of muscle cell precursors (the myotome) forms beneath the epithelial dermamyotome. (A,B after Langman 1981; C,D after Ordahl 1993.)

um (those regions closest to and farthest from the neural tube) constitute a muscle-forming region, the **primary myotome**. The cells of the myotome produce a lower layer of muscle precursor cells, the myoblasts. The resulting double-layered structure is called the **dermamyotome**.

Those myoblasts formed from the region closest to the neural tube form the **primaxial muscles**.* These muscles include the intercostal musculature between the ribs as well as the deep muscles of the back. The primaxial muscles are centrally located and display segmentation. Those myoblasts formed in the region farthest from the neural tube produce the **abaxial muscles** of the body wall, limbs, and tongue (see Figure 14.12). These myoblasts migrate away from the center and eventually differentiate within the lateral plate-derived dermis. The boundary between the primaxial and abaxial muscles and between the somite-derived and lateral plate-derived dermis is called the **lateral somitic frontier** (Christ and Ordahl 1995; Burke and Nowicki 2003; Nowicki et al. 2003.) Various transcription factors distinguish these two groups of muscles.

After the myoblasts from the lateral lips of the dermamyotome have formed the primary myotome beneath the dermatome, the centralmost region of the dermamyotome gives rise to a third population of muscle cells (Gros et al. 2005; Relaix et al. 2005). These cells delaminate from the dermamyotome and join the primary myotome cells to make a secondary myotome (usually referred to simply as the myotome). Unlike the marginal myoblast cells, these central cells are undifferentiated, and they proliferate rapidly to account for most of the myoblast cells. Moreover, while most of these cells differentiate to form muscles, some remain undifferentiated and surround the mature muscle cells. These undifferentiated cells become the satellite cells responsible for postnatal muscle growth and muscle repair.

---

*The terms "primaxial" and "abaxial" are used here to designate the muscles from the medial and lateral portions of the somite, respectively. The terms "epaxial" and "hypaxial" are commonly used, but these terms are derived from secondary modifications of the adult anatomy (the hypaxial muscles being innervated by the ventral regions of the spinal cord) rather than the somitic myotome lineages.

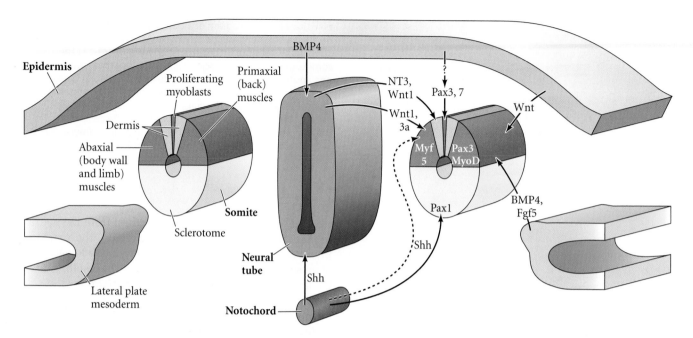

FIGURE 14.12 Model of major postulated interactions in the patterning of the somite. A combination of Wnts (probably Wnt1 and Wnt3a) is induced by BMP4 in the dorsal neural tube. These Wnt proteins, in combination with low concentrations of Shh from the notochord and floor plate, induce the primaxial myotome, which synthesizes the myogenic transcription factor Myf5. High concentrations of Shh induce Pax1 expression in those cells fated to become the sclerotome. Certain concentrations of neurotrophin-3 (NT3) from the dorsal neural tube appear to specify the dermatome, while Wnt proteins from the epidermis, in conjunction with BMP4 and Fgf5 from the lateral plate mesoderm, are thought to induce the primaxial myotome. (After Cossu et al. 1996b.)

Determination of the sclerotome and dermatome

Like the proverbial piece of real estate, the destiny of a somitic region depends on three things: location, location, and location.

The specification of the somite is accomplished by the interaction of several tissues. The ventral-medial portion of the somite is induced to become the sclerotome by paracrine factors, especially Sonic hedgehog (Shh), secreted from the notochord and the neural tube floor plate (Fan and Tessier-Lavigne 1994; Johnson et al. 1994). If portions of the notochord (or another source of Shh) are transplanted next to other regions of the somite, those regions, too, will become sclerotome cells. Sclerotome cells express another transcription factor, Pax1, that is required for their differentiation into cartilage and whose presence is necessary for formation of the vertebrae (Figure 14.12; Smith and Tuan 1996). Sclerotome cells also express I-mf, an inhibitor of the myogenic (muscle-forming) bHLH family of transcription factors (Chen et al. 1996).

The dermatome differentiates in response to two factors secreted by the neural tube: neurotrophin-3 (NT3), and Wnt1. Antibodies that block the activities of NT3 prevent the conversion of epithelial dermatome into the loose dermal mesenchyme that migrates beneath the epidermis (Brill et al. 1995). In birds, this somite-derived dermis is responsible for feather induction (see Figure 6.7), and the dorsal region of the neural tube is critical for its specification. Removing or rotating the neural tube prevents this dermis from forming (Takahashi et al. 1992; Olivera-Martinez 2002).

Determination of the myotome

In similar ways, the myotome is induced by at least two distinct signals. Studies using transplantation and knockout mice indicate that the primaxial myoblasts coming from the medial portion of the somite are induced by factors from the neural tube (probably Wnt1 and Wnt3a from the dorsal region, and low levels of Shh from the ventral region) (Münsterberg et al. 1995; Stern et al. 1995; Ikeya and Takada 1998). The abaxial myoblasts coming from the lateral edge of the somite are probably specified by a combination of Wnt proteins from the epidermis and bone morphogenetic protein-4 (BMP4) from the lateral plate mesoderm (see Figure 14.12; Cossu et al. 1996a; Pourquié et al. 1996; Dietrich et al. 1998). These factors cause the myoblasts to migrate away from the dorsal region and delay their differentiation until they are in a more ventral position.

In addition to these positive signals, there are inhibitory signals preventing the positive signals from affecting an inappropriate group of cells. For example, Shh not only activates sclerotome and myotome development, but also inhibits BMP4 from the lateral plate mesoderm from extending medially and ventrally (thus preventing the conversion of sclerotome into muscle) (Watanabe et al. 1998). Similarly, Noggin produced by the most medial portion of the dermamyotome prevents BMP4 from giving these cells the migratory characteristics of abaxial muscle (Marcelle et al. 1997).

And what happens to the notochord, that central mesodermal structure? After it has provided the axial integrity of the early embryo and has induced the formation of the dorsal neural tube, most of it degenerates by apoptosis. This apoptosis is probably signaled by mechanical forces (Aszódi et al. 1998). Wherever the sclerotome cells have formed a vertebral body, the notochordal cells die. However, in between the vertebrae, the notochordal cells form part of the intervertebral discs, the nuclei pulposi. These are the spinal discs that "slip" in certain back injuries.

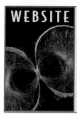

WEBSITE 14.1 Calling the competence of the somite into question. When the *tbx6* gene was knocked out from mice, the resulting embryos had three neural tubes in the posterior of their bodies. Without the *tbx6* gene, the somitic tissue responded to the notochord and epidermal signals as if it were neural ectoderm.

WEBSITE 14.2 Cranial paraxial mesoderm. Most of the head musculature does not come from somites. Rather, it comes from the cranial paraxial (prechordal plate) mesoderm. These cells originate adjacent to the sides of the brain, and they migrate to their respective destinations.

Myogenesis: The Generation of Muscle

Specification and differentiation by myogenic bHLH proteins

As we have seen, muscle cells come from two cell lineages in the somite, the primaxial and the abaxial. In both lineages, paracrine factors instruct cells to become muscles by inducing them to synthesize the MyoD protein (see Figure 14.12; Maroto et al. 1997; Tajbakhsh et al. 1997; Pownall et al. 2002). The way this happens differs slightly between the primaxial and abaxial lineages and between different vertebrate classes. In the lateral portion of the mouse dermamyotome, which forms the hypaxial muscles, factors from the surrounding environment induce the Pax3 transcription factor. In the absence of other inhibitory transcription factors (such as those found in the sclerotome cells), Pax3 activates the *myoD* gene. In the medial region of the dermamyotome, which forms the epaxial muscles, MyoD is induced by the Myf5 protein. MyoD and Myf5 belong to a family of transcription factors called the **myogenic bHLH** (basic helix-loop-helix) proteins (sometimes also referred to as the myogenic regulatory factors, or MRFs). The proteins of this family all bind to similar sites on the DNA and activate muscle-specific genes.

In the formation of skeletal muscles, MyoD establishes a temporal cascade of gene activation. First, it can bind directly to certain regulatory regions to activate gene expression. For instance, the MyoD protein appears to directly activate the muscle-specific creatine phosphokinase gene by binding to the DNA immediately upstream from it (Lassar et al. 1989). There are also two MyoD-binding sites on the DNA adjacent to the genes encoding a subunit of the chicken muscle acetylcholine receptor (Piette et al. 1990). MyoD also directly activates its own gene. Therefore, once the *myoD* gene is activated, its protein product binds to the DNA immediately upstream of *myoD* and keeps this gene active.

Second, MyoD can activate other genes whose products act as cofactors for MyoDs binding to a later group of enhancers. For instance, MyoD activates the *p38* gene (which is not muscle-specific) and the *Mef2* gene. The p38 protein facilitates the binding of MyoD and Mef2 to a new set of enhancers, activating a second set of muscle-specific genes (Penn et al. 2004).

While Pax3 is found in several other cell types, the myogenic bHLH proteins are specific to muscle cells. Any cell making a myogenic bHLH transcription factor such as MyoD or Myf5 is committed to becoming a muscle cell. Transfection of genes encoding any of these myogenic proteins into a wide range of cultured cells converts those cells into muscles (Thayer et al. 1989; Weintraub et al. 1989).

WEBSITE 14.3 Myogenic bHLH proteins and their regulators. Because the MyoD protein and its relatives are so powerful they can turn nearly any cell into a muscle cell, the synthesis of this protein has to be inhibited at numerous steps. Numerous inhibitors of MyoD family gene expression and protein function have been found.

Muscle cell fusion

The myotome cells producing the myogenic bHLH proteins are the myoblasts—committed muscle cell precursors. Experiments with chimeric mice and cultured myoblasts showed that these cells align and fuse to form the multinucleated myotubes characteristic of muscle tissue. Thus, the multinucleated myotube cells are the product of several myoblasts joining together and dissolving the cell membranes between them (Konigsberg 1963; Mintz and Baker 1967).

Muscle cell fusion begins when the myoblasts leave the cell cycle. As long as particular growth factors (particularly fibroblast growth factors) are present, the myoblasts will proliferate without differentiating. When these factors are depleted, the myoblasts stop dividing, secrete fibronectin onto their extracellular matrix, and bind to it through α5β1 integrin, their major fibronectin receptor (Menko and Boettiger 1987; Boettiger et al. 1995). If this adhesion is experimentally blocked, no further muscle development ensues,

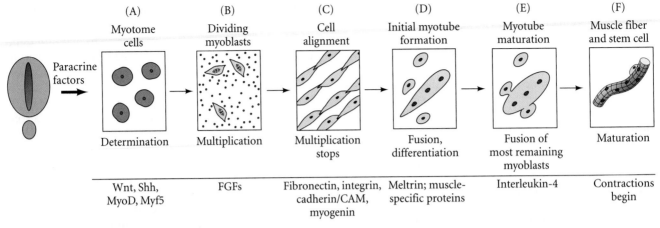

(A) Myotome cells	(B) Dividing myoblasts	(C) Cell alignment	(D) Initial myotube formation	(E) Myotube maturation	(F) Muscle fiber and stem cell
Determination	Multiplication	Multiplication stops	Fusion, differentiation	Fusion of most remaining myoblasts	Maturation
Wnt, Shh, MyoD, Myf5	FGFs	Fibronectin, integrin, cadherin/CAM, myogenin	Meltrin; muscle-specific proteins	Interleukin-4	Contractions begin

Paracrine factors

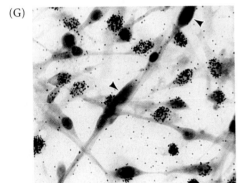

FIGURE 14.13 Conversion of myoblasts into muscles in culture. (A) Determination of myotome cells by paracrine factors. (B) Committed myoblasts divide in the presence of growth factors (primarily FGFs), but show no obvious muscle-specific proteins. (C–E) When the growth factors are used up, the myoblasts cease dividing, align, and fuse into myotubes. (F) The myotubes become organized into muscle fibers that spontaneously contract. (G) Autoradiograph showing DNA synthesis in myoblasts and the exit of fusing cells from the cell cycle. Phospholipase C can "freeze" myoblasts after they have aligned with other myoblasts, but before their membranes fuse. Cultured myoblasts were treated with phospholipase C and then exposed to radioactive thymidine. Unattached myoblasts continued to divide and thus incorporated the radioactive thymidine into their DNA. Aligned (but not yet fused) cells (arrowheads) did not incorporate the label. (A–E after Wolpert 1998; F from Nameroff and Munar 1976, photograph courtesy of M. Nameroff.)

so it appears that the signal from the integrin-fibronectin attachment is critical for instructing the myoblasts to differentiate into muscle cells (Figure 14.13).

The second step is the alignment of the myoblasts into chains. This step is mediated by cell membrane glycoproteins, including several cadherins and CAMs (Knudsen 1985; Knudsen et al. 1990). Recognition and alignment between cells takes place only if the cells are myoblasts. Fusion can occur even between chick and rat myoblasts in culture (Yaffe and Feldman 1965); the identity of the species is not critical.

The third step is the cell fusion event itself. As in most membrane fusions, calcium ions are critical, and fusion can be activated by calcium ionophores, such as A23187, that carry Ca^{2+} across cell membranes (Shainberg et al. 1969; David et al. 1981). Fusion appears to be mediated by a set of metalloproteinases called **meltrins**. These proteins were discovered during a search for myoblast proteins that would be homologous to fertilin, a protein implicated in sperm-egg membrane fusion. Yagami-Hiromasa and colleagues (1995) found that one of these meltrins (meltrin-α) is expressed in myoblasts at about the same time that fusion begins, and that antisense RNA to the meltrin-α message inhibited fusion when added to myoblasts. As the myoblasts become capable of fusing, another myogenic bHLH protein, myogenin, becomes active. While MyoD and Myf5 are active in the lineage specification of muscle

cells, myogenin appears to mediate their differentiation (Bergstrom and Tapscott 2001). Myogenin binds to the regulatory region of several muscle-specific genes and activates their expression.

After the original fusion of myoblasts to become a myotube, the myotube secretes interleukin-4 (IL4). IL4 is a paracrine factor that was originally identified as being an important signaling molecule in the adult immune system, and until 2003, it was not known to play a role in the embryo. However, Horsely and colleagues (2003) found that IL4 secreted by the new myotubes recruits other myoblasts to fuse with the tube, thereby forming the mature myotube (see Figure 14.13).

MUSCLE PROGENITOR CELLS As any athlete or sports fan knows, adult muscles can regenerate following injury. The new myofibers come from sets of stem cells or progenitor cells that reside alongside the adult muscle fibers. There may be more than one type of muscle stem cell or progenitor, and their functions may overlap (see Polesskaya et al. 2003). Lineage tracing using chick-quail chimeras indicate that these muscle progenitor cells are somitically derived myoblasts (see Figure 14.13E) that have not fused and remain potentially available throughout adult life (Armand et al. 1983). One type of putative stem cell, the **satellite cell**, is found within the basal lamina of mature myofibers. Satellite cells respond to injury or exercise by proliferating into myogenic cells that fuse and form new muscle fibers; these cells may be stem cells with the capacity to generate daughter cells for renewal or differentiation. Olguin and Olwin (2004) provide evidence that while quiescent, satellite cells express Pax7 (which specifies myoblasts in the somite), and this transcription factor inhibits MyoD expression and muscle differentiation in these cells.

Another type of muscle stem cell (which may be derived from those somitic cells that also migrate to form the dorsal aorta) is activated by Wnt signaling from the injured muscle tissue. The Wnt signal appears to activate *Pax7*, and in *these* cells, the Pax7 protein activates (rather than represses) the MyoD-family genes, thus promoting muscle differentiation. Recent studies have indicated that at least some satellite cells come from the central portion of the dermamyotome (Gros et al. 2005; Relaix et al. 2005). Most muscle precursor cells express both the *Pax7* (satellite-specific) gene and *Pax3* (which is needed for muscle determination). Muscle stem cells are a controversial field, and the number of cell types responsible for muscle regeneration and the types of mechanisms that allow these cells to stay quiescent and then differentiate when needed are still being explored.

 WEBSITE **14.4 Muscle formation.** Research on chimeric mice has shown that skeletal muscle becomes multinucleate by the fusion of cells, while heart muscle becomes multinucleate by nuclear divisions within a cell.

Osteogenesis: The Development of Bones

Three distinct lineages generate the skeleton. The somites generate the axial (vertebral) skeleton, the lateral plate mesoderm generates the limb skeleton, and the cranial neural crest gives rise to the branchial arch and craniofacial bones and cartilage.* There are two major modes of bone formation, or **osteogenesis**, and both involve the transformation of preexisting mesenchymal tissue into bone tissue. The direct conversion of mesenchymal tissue into bone is called **intramembranous** (or **dermal**) **ossification** and was discussed in Chapter 13. In other cases, the mesenchymal cells differentiate into cartilage, which is later replaced by bone. The process by which a cartilage intermediate is formed and then replaced by bone cells is called **endochondral ossification**. Endochondral ossification is seen predominantly in the vertebral column, ribs, pelvis, and limbs.

Endochondral ossification

Endochondral ossification involves the formation of cartilage tissue from aggregated mesenchymal cells, and the

---

*Craniofacial cartilage development was discussed in Chapter 13 and will be revisited in Chapter 22; the development of the limbs will be detailed in Chapter 16.

subsequent replacement of cartilage tissue by bone (Horton 1990). This is the type of bone formation characteristic of the vertebrae, ribs, and limbs. The vertebrae and ribs form from the somites, while the limb bones (discussed in Chapter 16) form from the lateral plate mesoderm.

The process of endochondral ossification can be divided into five stages. First, the mesenchymal cells commit to becoming cartilage cells. This commitment is caused by Sonic hedgehog, which induces the nearby sclerotome cells to express the Pax1 transcription factor (Cserjesi et al. 1995; Sosic et al. 1997). Pax1 initiates a cascade that is dependent on external paracrine factors and internal transcription factors (Figure 14.14A).

During the second phase of endochondral ossification, the committed mesenchyme cells condense into compact nodules and differentiate into chondrocytes, the cartilage cells (Figure 14.14B). BMPs appear to be critical in this stage. They are responsible for inducing the expression of the adhesion molecules N-cadherin and N-CAM and the transcription factor Sox9. N-cadherin appears to be important in the initiation of these condensations, and N-CAM seems to be critical for maintaining them (Oberlender and Tuan 1994; Hall and Miyake 1995). Sox9 activates other transcription factors as well as the genes encoding collagen 2 and agrican, which are critical in cartilage function. In humans, mutations of the *SOX9* gene cause camptomelic dysplasia, a rare disorder of skeletal development that results in deformities of most of the bones of the body. Most affected babies die from respiratory failure due to poorly formed tracheal and rib cartilage (Wright et al. 1995).

During the third phase of endochondral ossification (Figure 14.14C), the chondrocytes proliferate rapidly to form the cartilage model for the bone. As they divide, the chondrocytes secrete a cartilage-specific extracellular matrix. In the fourth phase, the chondrocytes stop dividing and increase their volume dramatically, becoming **hypertrophic chondrocytes** (Figure 14.14D). This step appears to be mediated by the transcription factor Runx2 (also called Cbfa1), which is necessary for the development of both intramembranous and endochondral bone (see Figure 13.12). Runx2 is itself regulated by histone deacetylase-4 (HDAC4), a form of chromatin restructuring enzyme that is expressed solely in the prehypertrophic cartilage. If HDAC4 is overexpressed in the cartilaginous ribs or limbs, ossification is seriously delayed; if HDAC4 is knocked out of the mouse genome, the limbs and ribs ossify prematurely (Vega et al. 2004).

These large chondrocytes alter the matrix they produce (by adding collagen X and more fibronectin) to enable it to become mineralized (calcified) by calcium phosphate. They also secrete the angiogenesis factor, VEGF, which can transform mesodermal mesenchyme cells into blood vessels (Gerber et al. 1999; Haigh et al. 2000; see Chapter 15). A number of events lead to the hypertrophy and mineralization (calcification) of the chondrocytes, including an initial switch from aerobic to anaerobic respiration, which alters

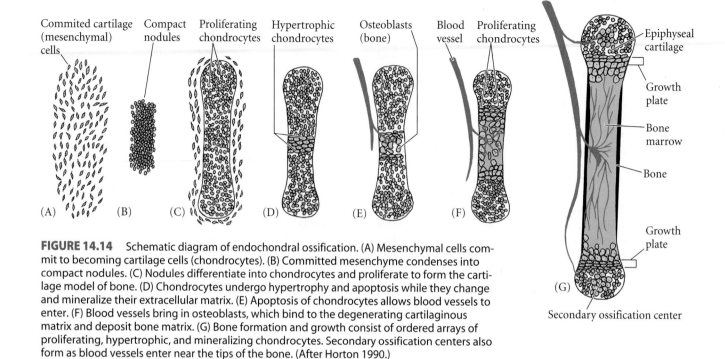

FIGURE 14.14 Schematic diagram of endochondral ossification. (A) Mesenchymal cells commit to becoming cartilage cells (chondrocytes). (B) Committed mesenchyme condenses into compact nodules. (C) Nodules differentiate into chondrocytes and proliferate to form the cartilage model of bone. (D) Chondrocytes undergo hypertrophy and apoptosis while they change and mineralize their extracellular matrix. (E) Apoptosis of chondrocytes allows blood vessels to enter. (F) Blood vessels bring in osteoblasts, which bind to the degenerating cartilaginous matrix and deposit bone matrix. (G) Bone formation and growth consist of ordered arrays of proliferating, hypertrophic, and mineralizing chondrocytes. Secondary ossification centers also form as blood vessels enter near the tips of the bone. (After Horton 1990.)

their cell metabolism and mitochondrial energy potential (Shapiro et al. 1982). Hypertrophic chondrocytes secrete numerous small membrane-bound vesicles into the extracellular matrix. These vesicles contain enzymes that are active in the generation of calcium and phosphate ions and initiate the mineralization process within the cartilaginous matrix (Wu et al. 1997). The hypertrophic chondrocytes, their metabolism and mitochondrial membranes altered, then die by apoptosis (Hatori et al. 1995; Rajpurohit et al. 1999).

In the fifth phase, the blood vessels induced by VEGF invade the cartilage model (Figure 14.14E–G). As the hypertrophic chondrocytes die, the cells that surround the cartilage model differentiate into osteoblasts. The replacement of chondrocytes by bone cells is dependent on the mineralization of the extracellular matrix. This remodeling releases VEGF, and more blood vessels are made around the dying cartilage. The blood vessels bring in both osteoblasts and chondroclasts (which eat the debris of the apoptotic chondrocytes). If the blood vessels are inhibited from forming, bone development is significantly delayed (Yin et al. 2002; see Karsenty and Wagner 2002).

The osteoblasts begin forming bone matrix on the partially degraded matrix and construct a bone collar around the dying cartilage cells (Bruder and Caplan 1989; Hatori et al. 1995; St. Jacques et al. 1999). It is thought that the osteoblasts have the same sclerotomal precursors as the chondrocytes (Figure 14.15) The osteoblasts form when Indian hedgehog (secreted by the prehypertrophic chondrocytes) causes a relatively immature cell (probably pre-

chondrocyte) to produce the transcription factor Runx2. Runx2 allows the cell to make bone matrix, but keeps the cell from becoming fully differentiated. Moreover, this osteoblast becomes responsive to Wnt signals that upregulate the transcription factor Osterix (Nakashima et al. 2002; Hu et al. 2005). Osterix instructs the cells to become bone. New bone material is added peripherally from the internal surface of the periosteum, a fibrous sheath containing connective tissue and capillaries that covers the developing bone. At the same time, there is a hollowing out of the internal region of the bone to form the bone marrow cavity. This destruction of bone tissue is carried out by osteoclasts, multinucleated cells that enter the bone through the blood vessels (Kahn and Simmons 1975; Manolagas and Jilka 1995). Osteoclasts are not derived from the somite; rather they are derived from a blood cell lineage (in the lateral plate mesoderm) and come from the same precursors as macrophage blood cells (Ash et al. 1980; Blair et al. 1986).

The importance of the mineralized extracellular matrix for bone differentiation is clearly illustrated in the developing skeleton of the chick embryo, which uses the calcium carbonate of the egg's shell as its calcium source. During development, the circulatory system of the chick embryo translocates about 120 mg of calcium from the shell to the skeleton (Tuan 1987). When chick embryos are removed from their shells at day 3 and grown in plastic wrap for the duration of their development, much of the cartilaginous skeleton fails to mature into bony tissue (Figure 14.16; Tuan and Lynch 1983).

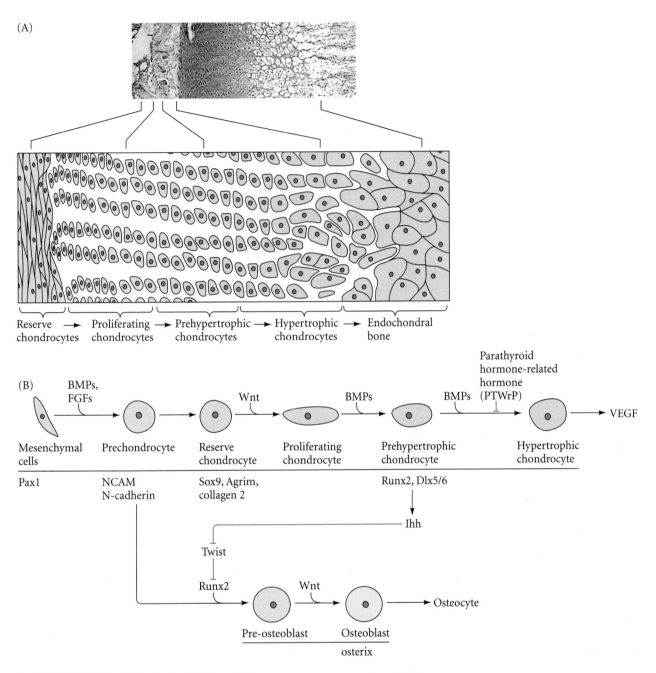

(A)

Reserve → Proliferating → Prehypertrophic → Hypertrophic → Endochondral
chondrocytes   chondrocytes   chondrocytes   chondrocytes   bone

(B)

BMPs, FGFs

Wnt

BMPs

BMPs

Parathyroid hormone-related hormone (PTWrP)

VEGF

Mesenchymal cells → Prechondrocyte → Reserve chondrocyte → Proliferating chondrocyte → Prehypertrophic chondrocyte → Hypertrophic chondrocyte

Pax1

NCAM
N-cadherin

Sox9, Agrim,
collagen 2

Runx2, Dlx5/6

Ihh

Twist

Runx2 → Pre-osteoblast → Wnt → Osteoblast → Osteocyte

osterix

FIGURE 14.15 Endochondral ossification. (A) Long bone undergoing endochondral ossification. The cartilage is stained with Alcian blue and the bone is stained with Alizarin red. Below is a diagram of the transition zone wherein the chondrocytes (cartilage cells) divide, enlarge, die, and are replaced by osteocytes (bone cells). (B) Paracrine factors and transcription factors active in the transition of cartilage to bone. The sclerotome cell can become a chondrocyte (characterized by the Sox9 transcription factor) or an osteocyte (characterized by the osterix transcription factor) depending on the types of paracrine factors it experiences. The paracrine factor Ihh (secreted by the growing chondrocytes) appears to repress Twist, an inhibitor of Runx2. Runx2 is critical for directing cell fate into the bone pathway, and activates osterix, a transcription factor that activates the bone-specific proteins.

WEBSITE **14.5 Paracrine factors, their receptors, and human bone growth.** Mutations in the genes encoding paracrine factors and their receptors cause numerous skeletal anomalies in humans and mice. The FGF and Hedgehog pathways are especially important.

Vertebrae formation

The notochord appears to induce its surrounding mesenchyme cells to secrete epimorphin, and this epimorphin attracts sclerotome cells to the region around the notochord and neural tube, where they begin to condense and differ-

(A)   (B)

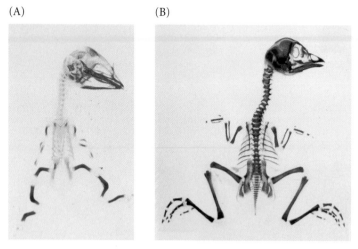

FIGURE 14.16   Skeletal mineralization in 19-day chick embryos that developed (A) in shell-less culture and (B) inside an egg during normal incubation. The embryos were fixed and stained with Alizarin red to show the calcified bone matrix. (From Tuan and Lynch 1983; photographs courtesy of R. Tuan.)

entiate into cartilage (Oka et al. 2006). However, before the sclerotome cells form a vertebra, they must split into a rostral and a caudal segment (Figure 14.17). As the motor neurons from the neural tube grow laterally to innervate the newly forming muscles, the rostral segment of each sclerotome recombines with the caudal segment of the next anterior sclerotome to form the vertebral rudiment (Remak 1850; Aoyama and Asamoto 2000; Morin-Kensicki et al.

FIGURE 14.17   Respecification of the sclerotome to form each vertebra. Each sclerotome splits into a rostral and caudal segment. As the spinal neurons grow outward to innervate the muscles from the myotome, the rostral segment of each sclerotome combines with the caudal segment of the next anterior sclerotome to form a vertebral rudiment. (After Larson 1998.)

2002). As we will see in our discussion of tendons, this **resegmentation** enables the muscles to coordinate the movement of the skeleton, permitting the body to move laterally. Resegmentation to allow coordinated movement is reminiscent of the strategy used by insects when constructing segments out of parasegments (see Chapter 9). The bending and twisting movements of the spine are permitted by the intervertebral (synovial) joints that form from specific (dorsal) regions of the sclerotome. Removal of these dorsal sclerotome cells leads to the failure of synovial joints to form and to the fusion of adjacent vertebrae (Mittapalli et al. 2005).

Tendon formation: The syndetome

In addition to the three "canonical" regions of the sclerotome, dermatome, and myotome, two further regions of the somite have recently been found. The tendons arise from a fourth compartment of the somite, the **syndetome** (Greek *syn*, "connected"). Since the tendons connect muscles to bones, it is not surprising that the syndetome is derived from the most dorsal portion of the sclerotome—that is, from sclerotome cells adjacent to the muscle-forming myotome. The tendon-forming cells of the syndetome can be visualized by their expression of the *scleraxis* gene (Figure 14.18; Schweitzer et al. 2001; Brent et al. 2003). Because there is no obvious morphological distinction between the sclerotome and syndetome cells (they are both mesenchymal), our knowledge of this somitic compartment had to wait until we had molecular markers (*Pax1* for the sclerotome; *scleraxis* for the syndetome) that could distinguish them and allow one to follow their cells' fates.

The syndetome is made from the myotome's secretion of Fgf8 onto the immediately subjacent row of sclerotome cells (Figure 14.19A; Brent et al. 2003; Brent and Tabin 2004). Other transcription factors limit the expression of *scleraxis* to the anterior and posterior portions of the syndetome, causing two stripes of *scleraxis* expression. Meanwhile, the

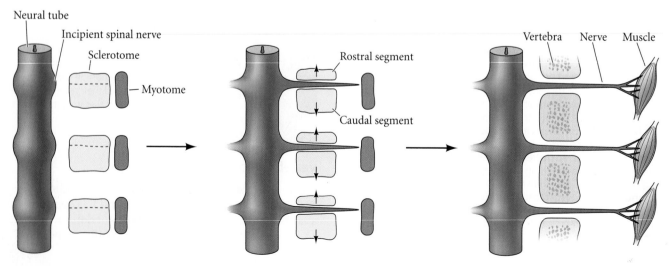

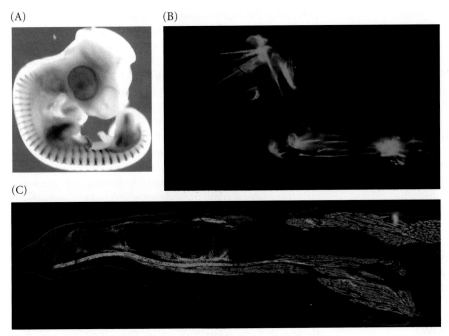

(A)

(B)

(C)

FIGURE 14.18 Scleraxis is expressed in the progenitors of the tendons. (A) In situ hybridization showing scleraxis pattern in the developing chick embryo. (B) Areas of scleraxis expression seen in the tendons of a newborn mouse forelimb (ventral view). The gene for green fluorescent protein had been fused on the scleraxis enhancer. (C) Wrist and finger of a newborn mouse showing the scleraxis in the tendons (profundus tendon and sublimis tendon) connecting those muscles to the digit and wrist. (A from Schweitzer et al. 2001, all photographs courtesy of R. Schweitzer.)

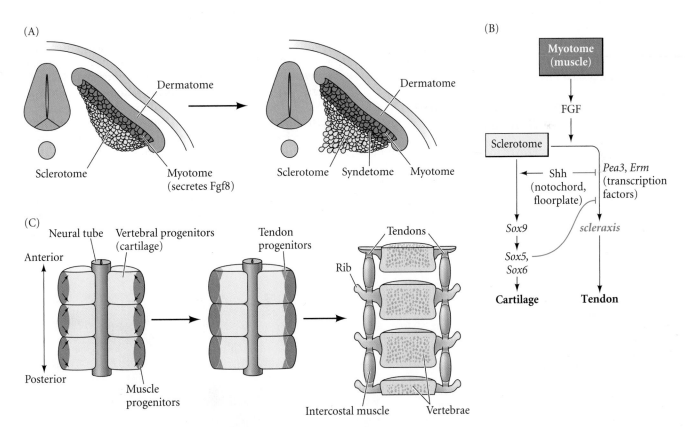

FIGURE 14.19 Induction of *scleraxis* in the chick sclerotome by Fgf8 from the myotome. (A) The dermatome, myotome, and sclerotome are established before the tendon precursors are specified. Tendon precursors (syndetome) are specified in the dorsalmost tier of sclerotome cells by Fgf8 received from the myotome. (B) Pathway by which Fgf8 signals from the muscle precursor cells induce the subjacent sclerotome cells to become tendons. (C) Syndetome cells migrate (arrows) along the developing vertebrae. They differentiate into tendons that connect the ribs to the intercostal muscles beloved by "spare-rib" devotees. (A,C after Brent et al. 2003.)

developing cartilage cells, under the influence of Sonic hedgehog from the notochord and floorplate, synthesize Sox5 and Sox6—transcription factors that block *scleraxis* transcription (Figure 14.19B). In this way, the cartilage protects itself from any spread of the Fgf8 signal. The tendons then associate with the muscles directly above them and with the skeleton (including the ribs) on either side of them (Figure 14.19C; Brent et al. 2005).

The fifth somitic compartment contains somite cells that will form the vascular walls of the aorta and the intervertebral blood vessels. A relatively small number of cells in the trunk somites of chick embryos are seen to migrate ventrally. These cells become part of the endothelial (blood vessel) lining of the dorsal aorta and the blood vessels forming between the somites. These cells are distinguished from other somite cells by having activated Notch proteins. Furthermore, somitic cells whose Notch proteins have been experimentally activated will also migrate and form aortic vascular cells (Pardanaud et al. 1996; Sato et al. 2005).

INTERMEDIATE MESODERM: THE UROGENITAL SYSTEM

The intermediate mesoderm generates the urogenital system—the kidneys, the gonads, and their respective duct systems. Saving the gonads for our discussion of sex determination in Chapter 17, we will concentrate here on the development of the mammalian kidney.

Specification of the Intermediate Mesoderm: Pax2/8 and Lim1

The intermediate mesoderm of the chick embryo acquires its ability to form kidneys through its interactions with the paraxial mesoderm. While its bias to become intermediate mesoderm is probably established through the BMP gradient mentioned earlier, specification appears to become stabilized through signals from the paraxial mesoderm. Mauch and her colleagues (2000) showed that signals from the paraxial mesoderm induced primitive kidney formation in the intermediate mesoderm of the chick embryo. They cut developing embryos such that the intermediate mesoderm could not contact the paraxial mesoderm on one side of the body. That side of the body (where contact with the paraxial mesoderm was abolished) did not form kidneys, but the undisturbed side was able to form kidneys (Figure 14.20A,B). The paraxial mesoderm appears to be both necessary and sufficient for inducing kidney-forming ability in the intermediate mesoderm, since co-culturing lateral plate mesoderm with paraxial mesoderm causes pronephric tubules to form in the lateral plate mesoderm, and no other cell type can accomplish this.

These interactions induce the expression of a set of homeodomain transcription factors, including Lim1, Pax2, and Pax8, that cause the intermediate mesoderm to form the kidney (Figure 14.20C; Karavanov et al. 1998; Kobayashi et al. 2005). In *Xenopus*, Pax8 and Lim1 have overlapping boundaries, and kidney development originates from those cells expressing both genes. Ectopic co-expression of *Pax8* and *Lim1* will produce kidneys in other tissues as well (Carroll and Vize 1999).

In the chick embryo, *Pax2* and *Lim1* are expressed in the intermediate mesoderm, starting at the level of the sixth somite (i.e., only in the trunk, not in the head); if *Pax2* is experimentally expressed in the presomitic mesoderm, it converts that paraxial mesoderm into intermediate mesoderm, expresses *Lim1*, and forms kidneys (Mauch et al. 2000; Suetsugu et al. 2005). Similarly, in mouse embryos with knockouts of both the *Pax2* and *Pax8* genes, the mesenchyme-to-epithelium transition necessary to form the kidney duct fails, the cells undergo apoptosis, and no nephric structures form (Bouchard et al. 2002). Moreover, in the mouse, *Lim1* and *Pax2* appear to induce one another.

Lim1 protein plays several roles in the formation of the mouse kidney. First, it is needed for converting the inter-

FIGURE 14.20 Signals from the paraxial mesoderm induce pronephros formation in the intermediate mesoderm of the chick embryo. (A) The paraxial mesoderm was surgically separated from the intermediate mesoderm on the right side of the body. (B) As a result, a pronephric kidney (Pax2-staining duct) developed only on the left side. (C) *Lim1* expression in an 8-day mouse embryo, showing the prospective intermediate mesoderm (A,B after Mauch et al. 2000; photograph courtesy of T. J. Mauch and G. C. Schoenwolf; C courtesy of K. Sainio and M. Hytönen)

(A)

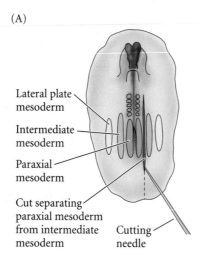

Lateral plate mesoderm

Intermediate mesoderm

Paraxial mesoderm

Cut separating paraxial mesoderm from intermediate mesoderm

Cutting needle

(B)

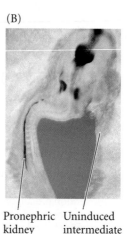

Pronephric kidney

Uninduced intermediate mesoderm

(C)

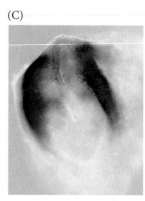

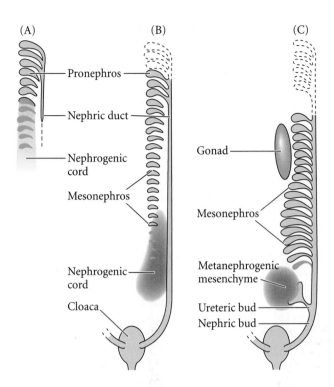

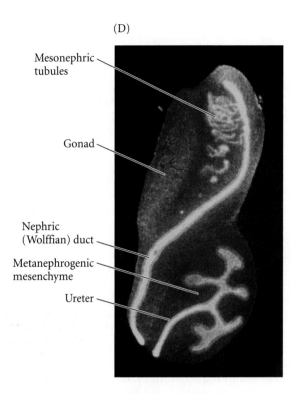

FIGURE 14.21 General scheme of development in the vertebrate kidney. (A) The original tubules, constituting the pronephros, are induced from the nephrogenic mesenchyme by the pronephric duct as it migrates caudally. (B) As the pronephros degenerates, the mesonephric tubules form. (C) The final mammalian kidney, the metanephros, is induced by the ureteric bud, which branches from the nephric duct. (D) The intermediate mesoderm of a 13-day mouse embryo showing the initiation of the metanephric kidney (bottom) while the mesonephros is still apparent. The duct tissue is stained with a fluorescent antibody to a cytokeratin found in the pronephric duct and its derivatives. (A–C after Saxén 1987; D, photograph courtesy of S. Vainio.)

mediate mesenchyme into the kidney duct (Tsang et al. 2000) and later it is required for the formation of the ureteric bud and the tubular structure both in mesonephric and metanephric mesenchyme (Karavanov et al. 1998, Shawlot and Behringer 1995, Kobayashi et al. 2005).

The anterior border of these *Lim1*- and *Pax2*-expressing cells appears to be established by the cells above a certain region losing their competence to respond to the paraxial signal. Barak and colleagues (2005) have demonstrated that chick epiblast cells fated to become intermediate mesoderm in the anterior region are competent to respond to these signals. However, they lose this competence as they migrate from the primitive streak to their final location in the anterior intermediate mesoderm.

Progression of Kidney Types

Homer Smith noted in 1953 that "our kidneys constitute the major foundation of our philosophical freedom. Only

because they work the way they do has it become possible for us to have bone, muscles, glands, and brains." While this statement may smack of hyperbole, the human kidney is an incredibly intricate organ whose importance cannot be overestimated. Its functional unit, the **nephron**, contains over 10,000 cells and at least 12 different cell types, each cell type having a specific function and thus located in a particular place in relation to the others along the length of the nephron.

The development of the mammalian kidney progresses through three major stages. The first two stages are transient; only the third and last persists as a functional kidney. Early in development (day 22 in humans; day 8 in mice), the **pronephric duct** arises in the intermediate mesoderm just ventral to the anterior somites. The cells of this duct migrate caudally, and the anterior region of the duct induces the adjacent mesenchyme to form the tubules of the initial kidney, the **pronephros** (Figure 14.21A). While the pronephric tubules form functioning kidneys in fish and in amphibian larvae, they are not thought to be active in amniotes. In mammals, the pronephric tubules and the anterior portion of the pronephric duct degenerate, but the more caudal portions of the pronephric duct persist and serve as the central component of the excretory system throughout its development (Toivonen 1945; Saxén 1987). This remaining duct is often referred to as the nephric or **Wolffian duct**.

As the pronephric tubules degenerate, the middle portion of the nephric duct induces a new set of kidney tubules in the adjacent mesenchyme. This set of tubules constitutes the **mesonephros**, or mesonephric kidney (Figure 14.21B; Sainio and Raatikainen-Ahokas 1999). In some mammalian species, the mesonephros functions briefly in urine filtration, but in mice and rats, it does not function as a working kidney. In humans, about 30 mesonephric tubules form, beginning around day 25. As more tubules are induced caudally, the anterior mesonephric tubules begin to regress through apoptosis (although in mice, the anterior tubules remain while the posterior ones regress; Figure 14.21C,D). While it remains unknown whether the human mesonephros actually filters blood and makes urine, it does provide important developmental functions during its brief existence. First, as we will see in Chapter 15, it is one of the main sources of the hematopoietic stem cells necessary for blood cell development (Medvinsky and Dzierzak 1996; Wintour et al. 1996). Second, in male mammals, some of the mesonephric tubules persist to become the sperm-carrying tubes (the vas deferens and efferent ducts) of the testes (see Chapter 17).

The permanent kidney of amniotes, the **metanephros**, is generated by some of the same components as the earlier, transient kidney types (see Figure 14.21C). It is thought to originate through a complex set of interactions between epithelial and mesenchymal components of the intermediate mesoderm. In the first steps, the metanephrogenic mesenchyme is committed and forms in the posterior regions of the intermediate mesoderm, where it induces the formation of a branch from each of the paired nephric

ducts. These epithelial branches are called the **ureteric buds**. These buds eventually separate from the nephric duct to become the collecting ducts and ureters that take the urine to the bladder. When the ureteric buds emerge from the nephric duct, they enter the metanephrogenic mesenchyme. The ureteric buds induce this mesenchymal tissue to condense around them and differentiate into the nephrons of the mammalian kidney. As this mesenchyme differentiates, it tells the uretereric bud to branch and grow.

Reciprocal Interactions of Developing Kidney Tissues

The two intermediate mesodermal tissues—the ureteric bud and the metanephrogenic mesenchyme—interact and reciprocally induce each other to form the kidney (Figure 14.22). The metanephrogenic mesenchyme causes the ureteric bud to elongate and branch. The tips of these branches induce the loose mesenchyme cells to form epithelial aggregates. Each aggregated nodule of about 20 cells proliferates and differentiates into the intricate structure of a renal nephron. Each nodule first elongates into a "comma" shape, then forms a characteristic S-shaped tube. Soon afterward, the cells of this epithelial structure begin to differentiate into regionally specific cell types, including the capsule cells, the podocytes, and the proximal and distal tubule cells. While this transformation is happening,

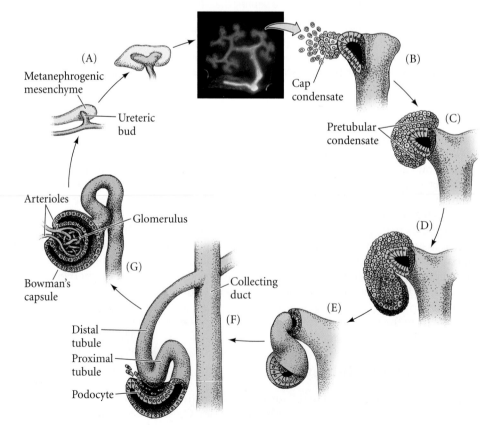

FIGURE 14.22 Reciprocal induction in the development of the mammalian kidney. (A) As the ureteric bud enters the metanephrogenic mesenchyme, the mesenchyme induces the bud to branch. (B–G) At the tips of the branches, the epithelium induces the mesenchyme to aggregate and cavitate to form the renal tubules and glomeruli (where the blood from the arteriole is filtered). When the mesenchyme has condensed into an epithelium, it digests the basal lamina of the ureteric bud cells that induced it and connects to the ureteric bud epithelium. A portion of the aggregated mesenchyme (the pretubular condensate) becomes the nephron (renal tubules and Bowman's capsule), while the ureteric bud becomes the collecting duct for the urine. (After Saxén 1987 and Sariola 2002.)

0.5 mm

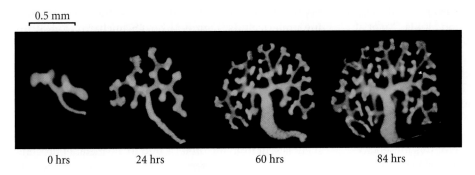

0 hrs   24 hrs   60 hrs   84 hrs

FIGURE 14.23 Kidney induction observed in vitro. (A) A kidney rudiment from an 11.5-day mouse embryo was placed into culture. This transgenic mouse had a *GFP* gene fused to a *Hoxb-7* promoter, so it expressed green fluorescent protein in the Wolffian (nephric) duct and in the ureteric buds. Since GFP can be photographed in living tissues, the kidney could be followed as it developed. (After Srinivas et al. 1999; photographs courtesy of F. Costantini.)

the epithelializing nodules break down the basal lamina of the ureteric bud ducts and fuse with them.* This fusion creates a connection between the ureteric bud and the newly formed tubule, allowing material to pass from one into the other (Bard et al. 2001). These tubules derived from the mesenchyme form the secretory nephrons of the functioning kidney, and the branched ureteric bud gives rise to the renal collecting ducts and to the ureter, which drains the urine from the kidney.

Clifford Grobstein (1955, 1956) documented this reciprocal induction in vitro. He separated the ureteric bud from the metanephrogenic mesenchyme and cultured them either individually or together. In the absence of mesenchyme, the ureteric bud does not branch. In the absence of the ureteric bud, the mesenchyme soon dies. When they are placed together, however, the ureteric bud grows and branches, and nephrons form throughout the mesenchyme (Figure 14.23).

The mechanisms of reciprocal induction

The induction of the metanephros can be viewed as a dialogue between the ureteric bud and the metanephrogenic mesenchyme. As the dialogue continues, both tissues are altered. We will eavesdrop on this dialogue more intently than we have done for other organs. This is because the kidney has become a model for organogenesis. There are many reasons for this. First, early kidney development has only two major components. Second, the identity and roles of many of the paracrine and transcription factors produced during this dialogue have been discovered from studies of knockout mice. Third, the absence of many of these transcription factors is associated with serious pathologies characterized by absent or rudimentary kidneys. While there are several simultaneous dialogues

between different groups of kidney cells, there appear to be at least eight critical sets of signals operating in the reciprocal induction of the metanephros.

STEP 1: FORMATION OF THE METANEPHROGENIC MESENCHYME: FOXES, HOXES, AND WT1 Only the metanephrogenic mesenchyme has the competence to respond to the ureteric bud and form kidney tubules; indeed this mesenchyme cannot become any tissue other than nephrons. If induced by nonnephric tissues (such as embryonic salivary gland or neural tube tissue), metanephric mesenchyme responds by forming kidney tubules and no other structures (Saxén 1970; Sariola et al. 1982).

The positional specification of the metanephrogenic mesenchyme is negatively regulated by Forkhead-winged-helix transcription factors Foxc1 and Foxc2. *Foxc1/2* double mutant mice have an expanded metanephric area that induces extra ureters and kidneys (Kume et al. 2000). Next, the permanent kidney-forming metanephrogenic mesenchyme is specified by the genes of the *Hox11* paralogue group. When *Hox11* genes are knocked out of mouse embryos, the differentiation of the mesenchyme is arrested, and the mesenchyme cannot induce the ureteric bud to form (Patterson et al. 2001; Wellik et al. 2002). The competence to respond to ureteric bud inducers is regulated by a tumor suppressor gene, *WT1*. Without *WT1*, the metanephric mesenchyme cells remain uninduced and die (Kreidberg et al. 1993). In situ hybridization shows that *WT1* is normally first expressed in the intermediate mesoderm prior to kidney formation and is then expressed in the developing kidney, gonad, and mesothelium (Pritchard-Jones et al. 1990; van Heyningen et al. 1990; Armstrong et al. 1992). Although the metanephrogenic mesenchyme appears homogeneous, it may contain both mesodermally derived tissue and some cells of neural crest origin (Le Douarin and Tiellet 1974; Sariola et al. 1989; Sainio et al. 1994).

STEP 2: THE METANEPHROGENIC MESENCHYME SECRETES GDNF TO INDUCE AND DIRECT THE URETERIC BUD. So what are all these factors doing in the metanephrogenic mesenchyme? It seems that they are setting the stage for the secretion of a paracrine factor that can induce the ureteric buds to emerge. This second signal in kidney development is **glial cell line-derived**

*The intricate coordination of nephron development with the blood capillaries that the nephrons filter is accomplished by the secretion of VEGF from the podocytes. VEGF, as we will see in the next chapter, is a powerful inducer of blood vessels, and it causes endothelial cells from the dorsal aorta to form the capillary loops of the glomerular filtration apparatus (Aitkenhead et al. 1998; Klanke et al. 2002).

neurotrophic factor (**GDNF**).* GDNF is expressed through a complex network initiated by Pax2 and Hox11 (Xu et al. 1999; Wellik et al. 2002). Pax2 and Hox11 (in concert with other transcription factors that permit this interaction) activate GDNF expression in the metanephrogenic mesenchyme (see Brodbeck and Englert 2004).

If GDNF were secreted throughout the metanephrogenic mesenchyme, numerous epithelial buds would sprout from the nephric duct. So the expression of GDNF must be limited to the posterior region of this mesenchymal tissue. This restriction of GDNF expression to the posterior portion of the metanephrogenic mesenchyme is accomplished by the Sprouty1 and Robo2 proteins. Robo2 is the same protein that helps deflect axon growth cones. If the Robo2 protein is mutated, both the anterior and the posterior regions of the metanephrogenic mesenchyme express GDNF, and the nephric duct sends out an anterior and a posterior ureter bud (Grieshammer et al. 2004). Sprouty is an inhibitor of FGF signaling. In mice whose *Sprouty1* gene has been knocked out, GDNF is produced throughout the metanephrogenic mesenchyme, initiating numerous ureter buds (Basson et al. 2005).

The *receptors* for GDNF (the Ret tyrosine kinase receptor and the GFRα1 co-receptor) are synthesized in the

---

*This is the same compound that we saw in Chapter 13 as critical for the induction of dopaminergic neurons in the mammalian brain. We will meet GDNF again when we discuss sperm cell production in Chapter 19. GDNF is one busy protein.

nephric ducts and later become concentrated in the growing ureteric buds (Figure 14.24A; Schuchardt et al. 1996). Mice whose *gdnf* or GDNF receptor genes are knocked out die soon after birth from renal agenesis—lack of kidneys (Figure 4.24B–D) (Moore et al. 1996; Pichel et al. 1996; Sánchez et al. 1996).

STEP 3: THE URETERIC BUD SECRETES FGF2 AND BMP7 TO PREVENT MESENCHYMAL APOPTOSIS   The third signal in kidney development is sent from the ureteric bud to the metanephrogenic mesenchyme, and it alters the fate of the mesenchyme cells. If left uninduced by the ureteric bud, the mesenchyme cells undergo apoptosis (Grobstein 1955; Koseki et al. 1992). However, if induced by the ureteric bud, the mesenchyme cells are rescued from the precipice of death and are converted into proliferating stem cells (Bard and Ross 1991; Bard et al. 1996). The factors secreted from the ureteric bud include Fgf2 and BMP7. Fgf2 has three modes of action in that it inhibits apoptosis, promotes the condensation of mesenchyme cells, and maintains the synthesis of WT1 (Perantoni et al. 1995). BMP7 has similar effects (see Figure 4.25; Dudley et al. 1995; Luo et al. 1995).

STEP 4: WNT9B AND WNT6 FROM THE URETERIC BUD INDUCE MESENCHYME CELLS TO AGGREGATE   The ureteric bud causes dramatic changes in the behavior of the metanephrogenic mesenchyme cells, converting them into an epithelium. The newly induced mesenchyme synthesizes E-cadherin, which

FIGURE 14.24   Ureteric bud growth is dependent on GDNF and its receptors. (A) The receptors for GDNF are concentrated in the posterior portion of the nephric duct. GDNF secreted by the metanephrogenic mesenchyme stimulates the growth of the ureteric bud from this duct. At later stages, the GDNF receptor is found exclusively at the tips of the ureteric buds. (B) The ureteric bud from a 11.5-day wild-type mouse embryonic kidney cultured for 72 hours has a characteristic branching pattern. (C) In embryonic mice heterozygous for a mutation of the gene encoding GDNF, the size of the ureteric bud and the number and length of its branches are reduced. (D) In mouse embryos missing both copies of the *gdnf* gene, the ureteric bud does not form. (A after Schuchardt et al. 1996; B–D from Pichel et al. 1996, photographs courtesy of J. G. Pichel and H. Sariola.)

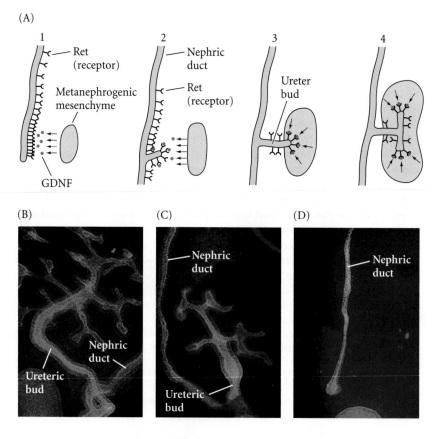

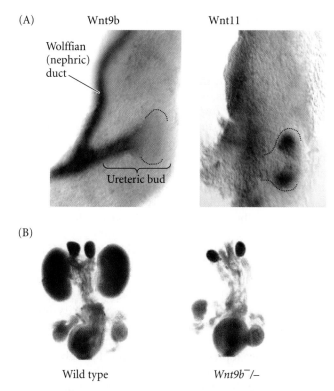

FIGURE 14.25 Wnts are critical for kidney development. (A) In the 11-day mouse kidney, Wnt9b is found on the stalk of the ureteric bud, while Wnt11 is found at the tips. Wnt9b induces the metanephrogenic mesenchyme to condense; Wnt11 will partition the metanephrogenic mesoderm to induce branching of the ureteric bud. Borders of the bud are indicated by a dashed line. (B) Wild-type 18.5-day male mice have normal kidneys, adrenal glands, and ureters. In mice deficient for *Wnt9b*, the kidneys are absent. (From Carroll et al. 2005.)

causes the mesenchyme cells to clump together. These aggregated nodes of mesenchyme now synthesize an epithelial basal lamina containing type IV collagen and laminin. At the same time, the mesenchyme cells synthesize receptors for laminin, allowing the aggregated cells to become an epithelium (Ekblom et al. 1994; Müller et al. 1997). The cytoskeleton also changes from one characteristic of mesenchyme cells to one typical of epithelial cells (Ekblom et al. 1983; Lehtonen et al. 1985).

The transition from mesenchymal to epithelial organization may be mediated by several molecules, including the expression of Pax2 in the newly induced mesenchyme cells. When antisense RNA to *Pax2* prevents the translation of the *Pax2* mRNA that is transcribed as a response to induction, the mesenchyme cells of cultured kidney rudiments fail to condense (Rothenpieler and Dressler 1993). Thus, Pax2 may play several roles during kidney formation.

In the mouse, Wnt6 and Wnt9 from the lateral sides of the ureter bud (but not from the tip) are critical for transforming the metanephrogenic mesenchyme cells into tubu-

lar epithelium. The mesenchyme has receptors for these Wnts. Wnt6 appears to promote the condensation of mesenchyme in an FGF-independent way (Itäranta et al. 2002), and Wnt9b induces Wnt4 in the mesenchyme. As we will see, Wnt4 is very important for the formation of the nephron, and mice deficient in Wnt9b lack kidneys* (Figure 14.25; Carroll et al. 2005).

STEP 5: WNT4 CONVERTS AGGREGATED MESENCHYME CELLS INTO A NEPHRON. Once induced, and after it has started to condense, the mesenchyme begins to secrete Wnt4, which acts in an autocrine fashion to complete the transition from mesenchymal mass to epithelium (Stark et al. 1994; Kispert et al. 1998). Wnt4 expression is found in the condensing mesenchyme cells, in the resulting S-shaped tubules, and in the region where the newly epithelialized cells fuse with the ureteric bud tips. In mice lacking the Wnt4 gene, the mesenchyme becomes condensed but does not form epithelia. Therefore, the ureteric bud induces the changes in the metanephrogenic mesenchyme by secreting FGFs, Wnt 9 and Wnt6; but these changes are mediated by the effects of the mesenchyme's secretion of Wnt4 on itself.

One molecule that may be involved in the transition from aggregated mesenchyme to nephrons is the Lim-1 homeodomain transcription factor (Karavanov et al. 1998; Kobayashi et al. 2005). This protein is found in the mesenchyme cells after they have condensed around the ureteric bud, and its expression persists in the developing nephron (Figure 14.26). Two other proteins that may be critical for the conversion of the aggregated cells into a nephron are polycystins 1 and 2. These proteins are the products of the genes whose loss-of-function alleles give rise to human polycystic kidney disease. Mice deficient in these genes have abnormal, swollen nephrons (Ward et al. 1996; van Adelsberg et al. 1997).

STEP 6: SIGNALS FROM THE MESENCHYME INDUCE THE BRANCHING OF THE URETERIC BUD Recent evidence has implicated several paracrine factors in the branching of the ureteric bud, and these factors probably work as pushes and pulls. Some factors may preserve the extracellular matrix surrounding the epithelium, thereby preventing branching from taking place (the "push"), while other factors may cause the digestion of this extracellular matrix, permitting branching to occur (the "pull").

The first protein regulating ureteric bud branching is GDNF (Sainio et al. 1997). GDNF from the mesenchyme not only induces the initial ureteric bud from the nephric duct

---

*Wnt9b appears to be critical for inducing the mesenchyme-to-epithelium transitions throughout the intermediate mesoderm. This includes the formation of the Wolffian and Müllerian ducts, as well as the kidney tubules. In addition to having no kidneys, Wnt9b-deficient mice have no uterus (if they are female) and no vas deferens (if male). In the rat, Leukemia inhibitory factor (LIF) appears to substitute for the Wnts 6 and 9 (Barasch et al. 1999; Carroll et al. 2005).

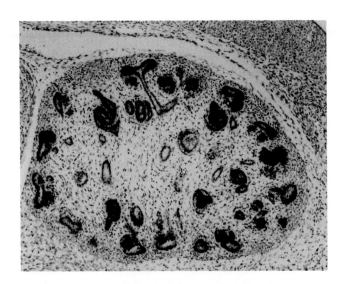

FIGURE 14.26 *Lim1* expression (dark stain) in a 19-day embryonic mouse kidney. In situ hybridization shows high levels of expression in the newly epithelialized comma-shaped and S-shaped bodies that will become nephrons. Compare with earlier *Lim1* expression shown in Figure 14.20C. (From Karavanov et al. 1998; photograph courtesy of A. A. Karavanov.)

TGFβ1 is added to cultured kidneys, it prevents the epithelium from branching (Figure 14.28A,B; Ritvos et al. 1995). TGFβ1 is known to promote the synthesis of extracellular matrix proteins and to inhibit the metalloproteinases that can digest these matrices (Penttinen et al. 1988; Nakamura et al. 1990). Thus, it is possible that TGFβ1 stabilizes branches once they form.

A third molecule that may be important in epithelial branching is BMP4 (Miyazaki et al. 2000). BMP4 is found in the mesenchymal cells surrounding the nephric duct, and BMP4 receptors are found in the epithelial tissue of the duct. BMPs act to antagonize branching signals, and thus BMP4 restricts the branching of the duct to the appropriate sites. When the BMP4 signaling cascade is activated ectopically in embryonic mouse kidney rudiments, it severely distorts the normal branching pattern (Figure 14.28C; Ritvos et al. 1995).

The fourth molecule involved in branching is collagen XVIII, which is part of the extracellular matrix induced by the mesenchyme. It may provide the specificity for the branching pattern (Lin et al. 2001). Collagen XVIII is found on the branches of the kidney epithelium, but not at the tips; in the developing lung, the reciprocal pattern is seen. This pattern is generated in part by GDNF, which downregulates collagen XVIII expression in the tips of the ureteric bud branches. When ureteric duct epithelium is incubated in lung mesenchyme, the collagen XVIII expression pattern seen is typical for that of the lung, and the branching pattern resembles that of the lung epithelium (Figure 14.28D).

STEPS 7 AND 8: DIFFERENTIATION OF THE NEPHRON AND GROWTH OF THE URETERIC BUD

The interactions we have described to this point create a cap condensate of metanephrogenic mesenchyme cells that covers the tips of the ureteric bud branches. A few cells (perhaps around five) of this aggregate, situated at the lateral edges of the cap, begin to proliferate rapidly and form the pretubular condensate. This pretubular condensate will give rise to the secretory nephron of the kidney. The other cells of the cap condensate regulate the subsequent branching of the ureteric bud so that more branches can be formed (Bard et al. 2001; Sariola 2002).

but it can also induce secondary buds from the ureteric bud once the bud has entered the mesenchyme (Figure 14.27). GDNF from the mesenchyme promotes cell division in the Ret-expressing cells at the tip of the ureter bud (Shakya et al. 2005). GDNF also appears to induce Wnt11 synthesis in these responsive cells at the tip of the bud (see Figure 14.25A), and Wnt 11 reciprocates by regulating GDNF levels. Wnt11 signals to the metanephrogenic mesenchyme cause the ureter bud to branch. The cooperation between the GDNF/Ret pathway and the Wnt pathway appears to coordinate the balance between branching and metanephrogenic cell division such that continued metanephric development is ensured (Majumdar et al. 2003).

The second candidate branch-regulating molecule is transforming growth factor β1 (TGFβ1). When exogenous

FIGURE 14.27 The effect of GDNF on the branching of the ureteric epithelium. The ureteric bud and its branches are stained orange (with antibodies to cytokeratin 18), while the nephrons are stained green (with antibodies to nephron brush border antigens). (A) 13-day embryonic mouse kidney cultured 2 days with a control bead (circle) has a normal branching pattern. (B) A similar kidney cultured 2 days with a GDNF-soaked bead shows a distorted pattern, as new branches are induced in the vicinity of the bead. (From Sainio et al. 1997; photographs courtesy of K. Sainio.)

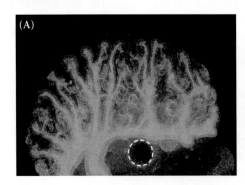
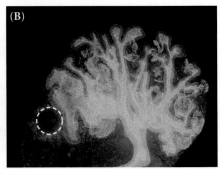

(A)

(B)

(D)

(C)

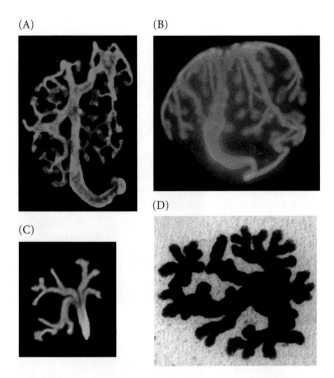

FIGURE 14.28 Signaling molecules and branching of the ureteric epithelium. (A) An 11-day embryonic mouse kidney cultured for 4 days in control medium has a normal branching pattern. (B) An 11-day mouse kidney cultured in TGFβ1 shows no branching until reaching the periphery of the mesenchyme, and the branches formed are elongated. (C) An 11-day mouse kidney cultured in activin (which activates the same receptor as BMP4) shows a marked distortion of branching. (D) Epithelial branching of kidney cells grown in lung mesenchyme takes on an appearance similar to that of lung epithelium. (A–C from Ritvos et al. 1995; D from Lin et al. 2001. Photographs courtesy of Y. Lin and S. Vainio.)

The transcription factor Foxb2 is synthesized in the cap cells. When Foxb2 is knocked out in mouse embryos, the resulting kidney lacks a branched ureteric tree (it branches only three or four times instead of the normal seven or eight, resulting in an 8- to 16-fold reduction in the number of branches), and the aggregates do not differentiate into nephrons (Hatini et al. 1996). The cap cells are also able to convert vitamin A to retinoic acid and use this RA to retain the expression of Ret (one of the GDNF receptors) in the ureteric bud (Batourina et al. 2002). The cap cells also secrete Fgf7, a growth factor whose receptor is found on the ureteric bud. Fgf7 is critical for maintaining ureteric epithelial growth and ensuring an appropriate number of nephrons in the kidney (Qiao et al. 1999; Chi et al. 2004).

STEP 9: INSERTING THE URETER INTO THE BLADDER The branching epithelium becomes the collecting system of the kidney. This epithelium collects the filtered urine from the nephron and secretes anti-diuretic hormone for the resorption of water (a process that, not so incidentally, makes life on land possible). The rest of the ureteric bud becomes the ureter, the tube that carries urine into the bladder. The junction between the ureter and bladder is extremely important, and hydronephrosis, a birth defect involving renal filtration, occurs when this junction is so tightly formed that urine cannot enter the bladder. The ureter is made into a watertight connecting duct by the condensation of mesenchymal cells around it (but not around the collecting ducts). These mesenchymal cells become smooth muscle cells whose peristalsis allows the urine to flow into the bladder; they also secrete BMP4 (Cebrian et al. 2004). BMP4 upregulates genes for uroplakin, a protein that causes differentiation of this region of the ureteric bud into the ureter. (The domain of the ureteric bud that branches to form the collecting ducts is protected by BMP inhibitors.)

The bladder develops from a portion of the cloaca (Figure 14.29A,B). The cloaca is an endodermally lined chamber at the caudal end of the embryo that will become the waste receptacle for both the intestine and the kidney. (The term "cloaca" is Latin for sewer—a bad joke by the early European anatomists.) Amphibians, reptiles, and birds still have this organ and use it to void both liquid and solid wastes. In mammals, the cloaca becomes divided by a septum into the urogenital sinus and the rectum. Part of the urogenital sinus becomes the bladder, while another part becomes the urethra (which will carry the urine out of the body). The ureter bud originally empties into the bladder via the nephric (Wolffian) duct (which grows towards the bladder by an as yet unknown mechanism). Once at the bladder, the urogenital sinus cells of the bladder wrap themselves around both the ureter and the Wolffian duct. Then, the Wolffian ducts migrate ventrally, opening into the urethra rather than into the bladder (Figure 14.29C–F; Batourina et al. 2002). In females, the Wolffian duct degenerates while the Müllerian duct opens into the vagina (see Chapter 17). In males, however, the Wolffian duct also forms the sperm outflow track, so that males expel sperm and urine through the same opening.

The Hox13 paralogue group appears to be important in specifying the distal ureter. In *HOXA13* defects, there are abnormalities of the cloaca, the male and female reproductive tracts, and the urethra (Pinsky 1974; Mortlock and Innis 1997). Since *HOXA13* is also involved in digit specification (see Chapter 16), the fingers and toes are also malformed, and the syndrome has been called the "hand-foot-genital" syndrome.

FURTHER STEPS The next stages of kidney development now occur in the separate groups. The glomerulus, for instance, forms from the epithelialized descendents of the metanephrogenic mesenchyme located distally, while the more proximal descendents, closer to the collecting ducts, become the distal and proximal convoluted tubules, and so forth. In the developing kidney, we see an epitome of the reciprocal interactions needed to form an organ. We also see that we have only begun to understand how organs form.

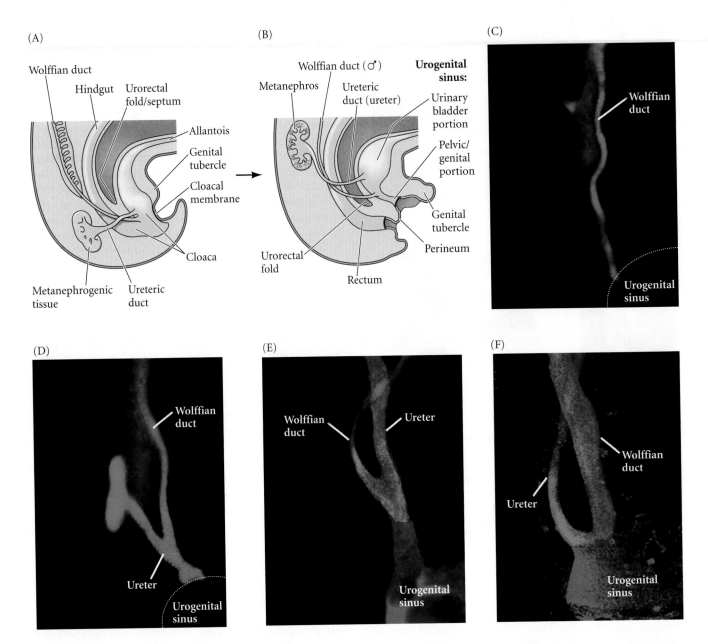

FIGURE 14.29 Development of the bladder and its connection to the kidney via the ureter. (A) The cloaca originates as an endodermal collecting area that opens into the allantois. (B) The urogenital septum divides the cloaca into the future rectum and the urogenital sinus. The bladder forms from the anterior portion of that sinus, and the urethra develops from the posterior region of the sinus. The space between the rectal opening and the urinary opening is the perineum. (C–F) Insertion of the ureter into the embryonic mouse bladder. (C) Day 10 mouse urogenital tract. The Wolffian (nephric) duct is stained with GFP attached to a Hoxb7 enhancer. (D) Urogenital tract from a day 11 embryo, after ureteric bud outgrowth. (E) Whole-mount urogenital tract from a day 12 embryo. The ducts are stained green and the urogenital sinus red. (F) The ureter separates from the nephric duct and forms a separate opening into the bladder. (A,B after Cochard 2002; C–F from Batourina et al. 2002, photographs courtesy of C. Mendelsohn.)

 **Snapshot Summary** **Paraxial and Intermediate Mesoderm**

1. The paraxial mesoderm forms blocks of tissue called somites. Somites give rise to three major divisions: the sclerotome, the myotome, and the dermatome.

2. Somites are formed from the segmental plate (unsegmented mesoderm) by the interactions of several proteins. The Notch pathway is extremely important in this process, and Eph receptor systems may be involved in the separation of the somites from the unsegmented paraxial mesoderm. N-cadherin, fibronectin, and Rac1 appear to be important in causing these cells to become epithelial.

3. The dermatome of the somite forms the back dermis. The sclerotome of the somite forms the vertebral cartilage. In thoracic vertebrae, the sclerotome cells also form the proximal portions of the ribs.

4. The primaxial myotome forms the back musculature. The abaxial myotome forms the muscles of the body wall, limb, diaphragm, and tongue.

5. The somite regions are specified by paracrine factors secreted by neighboring tissues. The sclerotome is specified to a large degree by Sonic hedgehog, which is secreted by the notochord and floor plate cells. The dermatome is specified by neurotrophin-3, secreted by the roof plate cells of the neural tube.

6. The two myotome regions are specified by different factors. The primaxial myotome is specified by Wnt proteins from the dorsal neural tube. The abaxial myotome is specified by BMP4 (and perhaps other proteins) secreted by the lateral plate mesoderm. In both instances, myogenic bHLH transcription factors are induced in the abaxial and paraxial myoblasts—the cells that will become muscles.

7. To form muscles, the myoblasts stop dividing, align themselves into myotubes, and fuse.

8. The major lineages that form the skeleton are the somites (axial skeleton), lateral plate mesoderm (appendages), and neural crest and head mesoderm (skull and face).

9. There are two major types of ossification. In intramembranous ossification, which occurs primarily in the skull and facial bones, mesenchyme is con-verted directly into bone. In endochondral ossification, mesenchyme cells become cartilage. These cartilagenous models are later replaced by bone cells.

10. The replacement of cartilage by bone during endochondral ossification depends upon the mineralization of the cartilage matrix.

11. Osteoclasts continually remodel bone throughout a person's lifetime. The hollowing out of bone for the bone marrow is accomplished by osteoclasts.

12. Tendons are formed through the conversion of the dorsalmost layer of sclerotome cells into syndetome cells by FGFs secreted by the myotome.

13. The intermediate mesoderm is specified through interactions with the paraxial mesoderm. It generates the kidneys and gonads. This specification requires Pax2, Pax6, and Lim1.

14. The metanephric kidney of mammals is formed by the reciprocal interactions of the metanephrogenic mesenchyme and a branch of the nephric duct called the ureteric bud.

15. The metanephrogenic mesenchyme becomes competent to form nephrons by expressing WT1, and it starts to secrete GDNF. GDNF is secreted by the mesoderm and induces the formation of the ureteric bud.

16. The ureteric bud secretes Fgf2 and BMP7 to prevent apoptosis in the metanephrogenic mesenchyme. Without these factors, this kidney-forming mesenchyme dies.

17. The ureteric bud secretes Wnt9 and Wnt6, which induce the competent metanephrogenic mesenchyme to form epithelial tubules. As they form these tubules, the cells secrete Wnt4, which promotes and maintains their epithelialization.

18. The condensing mesenchyme secretes paracrine factors that mediate the branching of the ureteric bud. These factors include GDNF, BMP4 and TGFβ1. The branching also depends upon the extracellular matrix of the epithelium.

For Further Reading

Complete bibliographical citations for all literature cited in this chapter can be found on the Vade Mecum CD that accompanies the book and at the free access website www.devbio.com

Brent, A. E., R. Schweitzer, and C. J. Tabin. 2003. A somitic compartment of tendon precursors. *Cell* 113: 235–248.

Gerber, H. P., T. H. Vu,. A. M. Ryan, J. Kowalski and Z. Werb. 1999. VEGF couples hypertrophic cartilage remodeling, ossification, and angiogenesis during endochondral bone formation. *Nature Med.* 5: 623–628.

Mauch, T. J., G. Yang, M. Wright, D. Smith and G. C. Schoenwolf. 2000. Signals from trunk paraxial mesoderm induce pronephros formation in chick

intermediate mesoderm. *Dev. Biol.* 220: 62–75.

Nowicki, J. L., R. Takimoto, and A. C. Burke. 2003. The lateral somitic frontier: Dorso-ventral aspects of anterior-posterior regionalization in the embryo. *Mech. Dev.* 120: 227–240.

Ordahl, C. P. and N. Le Douarin. 1992. Two myogenic lineages within the developing somite. *Development* 114: 339–353.

Palmeirim, I., D. Henrique, D. Ish-Horowicz and O. Pourquié. 1997. Avian hairy gene expression identifies a

molecular clock linked to vertebrate segmentation and somitogenesis. *Cell* 91: 639–648.

Pichel, J. G. and 11 others. 1996. Defects in enteric innervation and kidney development in mice lacking GDNF. *Nature* 382: 73–76.

Sato, Y., K. Yasuda and Y. Takahashi. 2002. Morphological boundary forms by a novel inductive event mediated by Lunatic-Fringe and Notch during somitic segmentation. *Development* 129: 3633–3644.

Lateral Plate Mesoderm and Endoderm

15

IN THE CHAOS OF THE ENGLISH CIVIL WARS, William Harvey, physician to the King and discoverer of the blastoderm, was comforted by viewing the heart as the undisputed ruler of the body, through whose divinely ordained powers the lawful growth of the organism was assured. Later embryologists looked at the heart as more of a servant than a ruler, the chamberlain of the household who assured that nutrients reached the apically located brain and the peripherally located muscles. In either metaphor, the heart, circulation, and digestive system were seen as being absolutely critical for development. As Harvey persuasively argued in 1651, the chick embryo must form its own blood without any help from the hen, and this blood is crucial in embryonic growth. How this happened was a mystery to him. "What artificer," he wrote, could create blood "when there is yet no liver in being?" The nutrition provided by the egg was also paramount to Harvey. His conclusion about the nutritive value of the yolk and albumen was "The egge is, as it were, an exposed womb; wherein there is a substance concluded as the Representative and Substitute, or Vicar of the breasts."

This chapter outlines the mechanisms by which the circulatory system, the respiratory system, and the digestive system emerge in the amniote embryo. As we will see, the lateral plate mesoderm and the endoderm interact to create both the circulatory and the digestive organs.*

LATERAL PLATE MESODERM

On the peripheral sides of the two bands of intermediate mesoderm reside the **lateral plate mesoderm** (see Figures 14.1 and 14.2). Each plate splits horizontally into two layers. The dorsal layer is the **somatic (parietal) mesoderm**, which underlies the ectoderm and, together with the ectoderm, forms the **somatopleure**. The ventral layer is the **splanchnic (visceral) mesoderm**, which overlies the endoderm and, together with the endoderm, forms the **splanchnopleure** (Figure 15.1A). The space between these two layers becomes the body cavity—the **coelom**—which stretches from the future neck region to the posterior of the body. During later development, the right-side and left-side coeloms fuse, and folds of tissue extend from the somatic mesoderm, dividing the coelom into separate

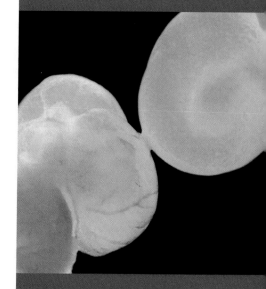

---

*Some scientists think that the mesoderm and endoderm were originally a single germ layer, the "mesendoderm," that accomplished all these functions. Recall from Chapter 8 that what in vertebrates is a broad territory of embryonic cells is actually the progeny of a single "mesentoblast" in many invertebrates. The signals that regulate the mesentoblast and the entire mesodermal and endodermal territory in vertebrates may be very similar (Maduro et al. 2001; Rodaway and Patient 2001).

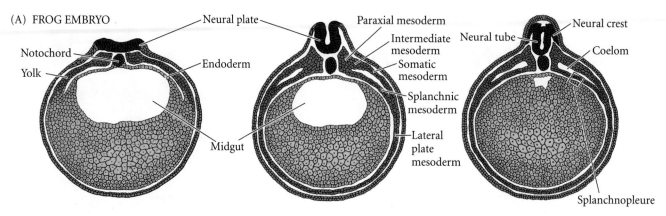

(A) FROG EMBRYO

Neural plate · Paraxial mesoderm · Intermediate mesoderm · Neural tube · Neural crest · Somatic mesoderm · Coelom · Notochord · Splanchnic mesoderm · Yolk · Endoderm · Lateral plate mesoderm · Midgut · Splanchnopleure

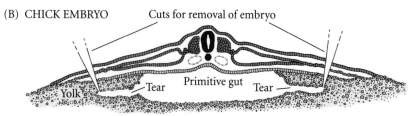

(B) CHICK EMBRYO   Cuts for removal of embryo

Yolk · Tear · Primitive gut · Tear

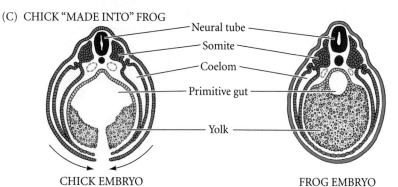

(C) CHICK "MADE INTO" FROG

Neural tube · Somite · Coelom · Primitive gut · Yolk

CHICK EMBRYO (removed from yolk, edges pulled together)   FROG EMBRYO

FIGURE 15.1 Mesodermal development in frog and chick embryos. (A) Neurula-stage frog embryos, showing progressive development of the mesoderm and coelom. (B) Transverse section of a chick embryo. (C) When the chick embryo is separated from its enormous yolk mass, it resembles the amphibian neurula at a similar stage. (A after Rugh 1951; B and C after Patten 1951.)

cavities. In mammals, the coelom is subdivided into the **pleural**, **pericardial**, and **peritoneal** cavities, enveloping the thorax, heart, and abdomen, respectively. The mechanism for creating the linings of these body cavities from the lateral plate mesoderm has changed little throughout vertebrate evolution, and the development of the chick mesoderm can be compared with similar stages of frog embryos (Figure 15.1B,C).

WEBSITE 15.1 Coelom formation. Coelom formation is readily visualized by animations. The animation presented here shows the expansion of the mesoderm during chick development.

The Heart

Consisting of a heart, blood cells, and an intricate system of blood vessels, the circulatory system provides nourish-

ment to the developing vertebrate embryo. The circulatory system is the first functional unit in the developing embryo, and the heart is the first functional organ. The vertebrate heart arises from two regions of splanchnic mesoderm—one on each side of the body—that interact with adjacent tissue to become specified for heart development.

VADE MECUM² Early heart development. The vertebrate heart begins to function early in its development. You can see this in movies of the living chick embryo at early stages when the heart is little more than a looped tube. [Click on Chick-Late]

Specification of heart tissue

In amniote vertebrates, the embryo is a flattened disc, and the lateral plate mesoderm does not completely encircle the yolk sac. The presumptive heart cells originate in the early primitive streak, just posterior to Hensen's node and extending about halfway down its length. These cells migrate through the streak and form two groups of mesodermal cells

(A)

(B)

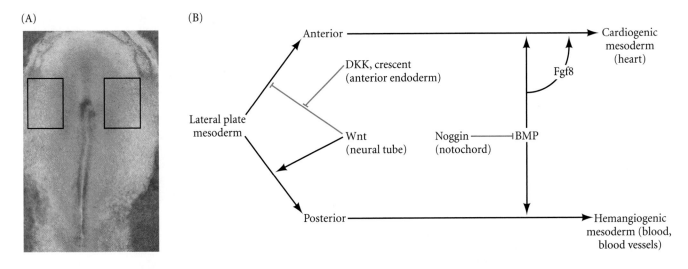

(C)

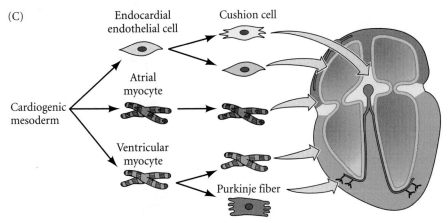

FIGURE 15.2 Heart-forming cells of the chick embryo. (A) Two symmetrical regions of cardiogenic (heart-forming) mesoderm form in the gastrulating chick embryo, determined by following individually labeled DiI-Itreated cells and by observing the expression patterns of particular genes such as smooth muscle α-actin. (B) Outline of the inductive interactions involving Wnt and BMP pathways, enabling the generation of heart and blood lineages from the lateral plate mesoderm. (C) The cardiogenic mesoderm contains the precursors of the three cell types of the endocardium and myocardium. The endocardium provides the endothelial lining of the heart as well as the cushion cells that form the valves. The atrial myocytes of the myocardium and the ventricular myocytes (some of which become the conducting Purkinje fibers) are also generated by the cardiogenic mesoderm. The neural crest cells and endothelium of the cardiac vessels are specified separately at different locations. (A after Redkar et al. 2001; C after Mikawa 1999.)

lateral to (and at the same level as) Hensen's node (Figure 15.2A; Colas et al. 2000; Redkar et al. 2001). These groups of cells are called the **cardiogenic mesoderm** (or **cardiac crescent**). The cells forming the atrial and ventricular musculature, the cushion cells of the valves, the Purkinje conducting fibers, and the endothelial lining of the heart are all generated from these two clusters (Mikawa 1999).

SPECIFICATION OF THE CARDIAC PRECURSOR CELLS   The specification of the cardiogenic mesoderm cells is induced by the endoderm adjacent to the heart through the BMP and FGF signaling pathways. The heart does not form if the anterior endoderm is removed, and the posterior endoderm cannot induce heart cells to form. Thus, isolated mesoderm from this region will form heart muscle when combined with anterior, but not posterior, endoderm (Schultheiss et al. 1995; Nascone and Mercola 1995). The endodermal signal appears to be mediated by BMPs, especially BMP2. BMPs from the endoderm promote both heart and blood development. Endodermal BMPs also induce Fgf8 synthesis in the endoderm directly beneath the cardiogenic mesoderm, and Fgf8 appears to be critical for the expression of cardiac proteins (Alsan and Schultheiss 2002).

Inhibitory signals prevent heart formation where it should not occur. The notochord secretes Noggin and chordin, blocking BMP signaling in the center of the embryo, and Wnt proteins from the neural tube, especially Wnt3a and Wnt8, *inhibit* heart formation but *promote* blood formation. The anterior endoderm, however, produces Wnt inhibitors such as Cerberus, Dickkopf, and Crescent, which prevent Wnts from binding to their receptors. In this way, cardiac precursor cells are specified in the places where BMPs (lateral mesoderm and endoderm) and Wnt antagonists (anterior endoderm) coincide (Figure 15.2B; Marvin et al. 2001; Schneider and Mercola 2001; Tzahor and Lassar 2001).

MIGRATION OF THE CARDIAC PRECURSOR CELLS   When the chick embryo is 18–20 hours old, the presumptive heart cells move anteriorly between the ectoderm and endoderm toward the middle of the embryo, remaining in close contact with the endodermal surface (Linask and Lash 1986). When these cells reach the lateral walls of the anterior gut tube, migration ceases. The directionality of this migration appears to be provided by the foregut endoderm. If the cardiac region endoderm is rotated with respect to the rest of the embryo, migration of the cardiogenic mesoderm cells is reversed. It is thought that the endodermal component responsible for this movement is an anterior-to-posterior concentration gradient of fibronectin. Antibodies against fibronectin stop the migration, while antibodies against other extracellular matrix components do not (Linask and Lash 1988a,b).

In the chick, the foregut is formed by the inward folding of the splanchnopleure (Figure 15.3). This movement brings the two cardiac tubes together. The bilateral origin of the heart can be demonstrated by surgically preventing the merger of the lateral plate mesoderm (Gräper 1907; DeHaan 1959). This manipulation results in a condition called **cardia bifida**, in which two separate hearts form, one on each side of the body (Figure 15.4A). The two endocardia lie within the common tube for a short time, but these also will fuse.

In the zebrafish, the heart precursor cells migrate actively from the lateral edges toward the midline. Several mutations affecting endoderm differentiation disrupt this process, indicating that, as in the chick, the endoderm is critical for cardiac precursor specification and migration. The *faust* gene, which encodes the GATA5 protein, is expressed in the endoderm and is required for the migration of cardiac precursor cells to the midline and also for their division and specification. It appears to be important in the pathway leading to the activation of the *Nkx2-5* gene in the cardiac precursor cells (Reiter et al. 1999). Another particularly interesting zebrafish mutation is *miles apart*. Its phenotype is limited to cardiac precursor migration to the midline and resembles the cardia bifida seen in experimentally manipulated chick embryos (Figure 15.4B,C). The *miles apart* gene encodes a receptor for a cell surface sphyngolipid molecule, and it is expressed in the endoderm on either side of the midline (Kupperman et al. 2000).

In mice, cardia bifida can also be produced by mutations of genes that are expressed in the endoderm. One of these genes, *Foxp4*, is a transcription factor expressed in the early foregut cells along the pathway the cardiac precursors travel towards the midline. In these mutants, each heart primordia develops separately, and the embryonic mouse contains two hearts, one on each side of the body (Figure15.4D,E; Li et al. 2004).

ESTABLISHMENT OF ANTERIOR AND POSTERIOR CARDIAC DOMAINS As the cardiac precursor cells migrate, the posterior region becomes exposed to increasingly higher concentrations of retinoic acid produced by the posterior mesoderm (see Figure 11.25). RA is critical in specifying the posterior cardiac

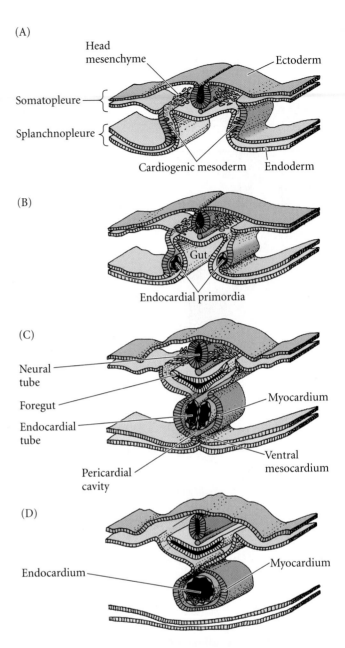

FIGURE 15.3 Formation of the chick heart from the splanchnic lateral plate mesoderm. The endocardium forms the inner lining of the heart, the myocardium forms the heart muscles, and the epicardium will eventually cover the heart. Transverse sections through the heart-forming region of the chick embryo are shown at (A) 25 hours, (B) 26 hours, (C) 28 hours, and (D) 29 hours. (After Carlson 1981.)

precursor cells into becoming the inflow, or "venous," portions of the heart—the sinus venosus and atria. Originally, these fates are not fixed, as transplantation or rotation experiments show that these precursor cells can regulate and differentiate in accordance with a new environment. But once the posterior cardiac precursors enter the realm of active RA synthesis, they express the gene for retinaldehyde dehydrogenase; they then can produce their own RA, and their posterior fate becomes committed (Figure 15.5A,B; Simões-Costa et al. 2005). This ability of retinoic

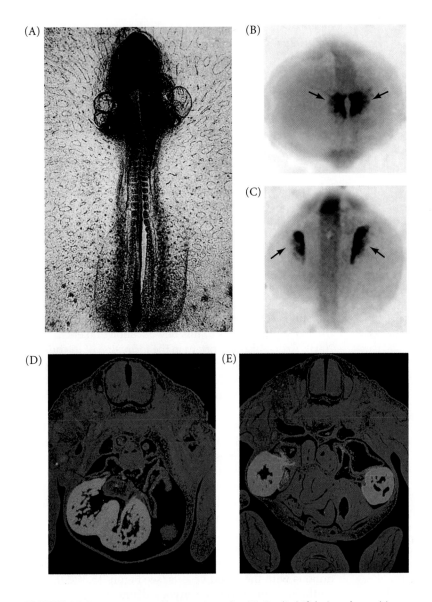

Interestingly, this relationship between heart development and retinoic acid appears to be conserved, even in embryos that form their hearts in very different ways. In tunicates, the heart is a single-layered U-shaped tube. At the 110-cell stage, the heart lineage is represented by a pair of mesodermal cells near the vegetal pole, the B7.5 blastomeres. These cells give rise to the anteriormost tail cells and the heart precursor cells (trunk ventral cells). When the B7.5 blastomeres split into the heart lineage and the anterior tail lineage, the heart precursor cells express the tunicate homologue of the *Nkx2-5* gene and migrate ventrally into the "head" of the developing tadpole (Davidson and Levine 2003; Simões-Costa et al. 2005). The anterior tail cells do not migrate, but they express retinaldehyde dehydrogenase and initiate a retinoid acid gradient (Figure 15.5C).

INITIAL CELL DIFFERENTIATION Several genes are expressed very early during heart development. *GATA4* is first seen in the precardiac cells of chicks and mice as these cells emerge from the primitive streak. *GATA4* expression is retained in all the cells making up the heart fields on both sides of the embryo. This transcription factor is necessary for the activation of numerous heart-specific genes, including the genes for atrial natriuretic factor and the cardiac-specific troponin-1 and troponin-C. It also activates expression of the N-cadherin gene that is critical for the fusion of the two heart rudiments into one tube (Zhang et al. 2003).

The BMP pathway (in the absence of the Wnt signals) is critical in inducing the synthesis of the *Nkx2-5* transcription factor in the migrating cardiogenic mesoderm; (Komuro and Izumo 1993; Lints et al. 1993; Sugi and Lough 1994; Schultheiss et al. 1995; Andrée et al. 1998). Nkx2-5 is crucial in instructing the mesoderm to become heart tissue, and it activates the synthesis of other transcription factors* (especially members of the T-box, GATA and Mef2 families).

FIGURE 15.4 Migration of heart primordia. (A) Cardia bifida (two hearts) in a chick embryo, induced by surgically cutting the ventral midline, thereby preventing the two heart primordia from fusing. (B) Wild-type zebrafish and (C) *miles apart* mutant, stained with probes for the cardiac myosin light chain. There is a lack of migration in the *miles apart* mutant. (D) Mouse heart stained with antisense RNA probe to ventricular myosin shows fusion of the heart primordia in a wild-type 12.5-day embryo. (E) Cardia bifida in a *Foxp4* deficient mouse embryo. Interestingly, each of these hearts has ventricles and atria, and they both loop and form all four chambers with normal left-right asymmetry. (A, photograph courtesy of R. L. DeHaan; B,C from Kupperman et al. 2000, photographs courtesy of Y. R. Didier; D,E from Li et al. 2004, photographs courtesy of E. E. Morrisey.)

acid to specify and commit heart precursor to become atria explains its teratogenic effects on heart development, wherein exposure of vertebrate embryos to RA can cause expansion of atrial tissues at the expense of ventricular tissues (Stainier and Fishman 1992; Hochgreb et al. 2003).

*The cells of the anterior heart field express the *Foxh1* gene, which commits these heart precursor cells to become the "arterial" portion of the heart—the ventricles and outflow tracts. In combination with the Nkx2-5 transcription factor, Foxh1 activates those genes (such as *Mef2c*) that begin the differentiation of the ventricles (von Both et al. 2004).

(A) Chick, stage 8

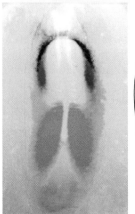
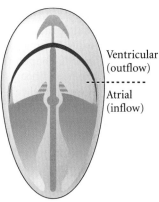

Ventricular
(outflow)

Atrial
(inflow)

(C) *Ciona*, larval stage

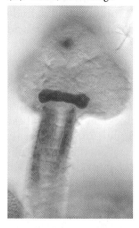

(B) Mouse, 8.0 days

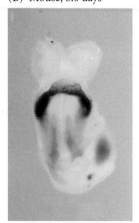

FIGURE 15.5 Double in situ hybridization for the expression of *RADH2* (encoding the retinoic acid-synthesizing enzyme retinaldehyde dehydrogenase-2 [orange] and *Tbx5*, a marker for the early heart fields [purple]). In these stages, the heart precursor cells are exposed to progressively increasing amounts of retinoic acid. (A) Chick stage 8. (B) mouse, 8.0 days. (C) *Ciona* (a tunicate) at the larval stage. (From Simões-Costa et al. 2005, photographs courtesy of J. Xavier-Neto.)

The Tbx family of genes is also critical in heart development (Plageman and Yutzey 2004). The *Tbx5* gene is expressed in the precardiac cells on both sides of the chick and mouse embryos (see Figure 15.5). In these early cells, Tbx5 acts with GATA4 and Nkx2-5 to activate numerous genes involved in heart specification.* Later, *Tbx5* expression becomes restricted to the atria and left ventricle; the ventricular septum (the wall separating the left and right ventricles) is formed at the boundary between the cells expressing *Tbx5* and those that do not. Tbx5 works antagonistically to *Tbx20*, which becomes expressed in the right ventricle. When the *Tbx5* expression domain is ectopically expanded, the location of the ventricular septum shifts to this new location (Takeuchi et al. 2003). *Tbx5* is also important in the formation of the human forelimb (see Chapter 16), and mutations in *Tbx5* cause Holt-Oram syndrome (Bruneau et al. 1999), a syndrome characterized by abnormalities of the heart and upper limbs.

Cell differentiation occurs independently in the two heart-forming primordia. As they migrate toward each other, the ventral mesoderm cells of the primordia begin to express N-cadherin on their apices, sort out from the dorsal mesoderm cells, and join together to form an epithelium. This joining will lead to the formation of the **pericardial cavity**, the sac in which the heart is formed (Linask 1992). A small population of these cells then downregulates N-cadherin and delaminates from the epithelium to form the **endocardium**, the lining of the heart that is continuous with the blood vessels (see Figure 15.3C,D).† The epithelial cells form the **myocardium** (Manasek 1968; Linask et al. 1997), which will give rise to the heart muscles that will pump for the lifetime of the organism. The endocardial cells produce many of the heart valves, secrete

GATA4 can also induce Nkx2-5, so activating one will activate the other and convert mesoderm into cardiac precursor cells (Figure 15.6). Working together, these transcription factors activate the expression of genes encoding cardiac muscle-specific proteins (such as cardiac actin, atrial natriuretic factor, and the α-myosin heavy chains) (Sepulveda et al. 1998; Kawamura et al. 2005). The Nkx2-5 transcription factor can also work as a repressor, and if *Nkx2-5* is specifically knocked out in those cells destined to become ventricles, these chambers express BMP10, causing massive overgrowth of the ventricles such that the ventricular chambers fill with muscle cells (Pashmforoush et al. 2004).

*The Nkx2-5 homeodomain transcription factor is homologous to the Tinman transcription factor active in specifying the heart tube of *Drosophila*. Moreover, neither Tinman nor Nkx2-5 alone is sufficient to complete heart development in their respective organisms. Mice lacking *Nkx2-5* start heart tube formation, but the tube fails to thicken or to loop (Lyons et al. 1995). Humans with a mutation in one of their *NKX2-5* genes have congenital heart malformations (Schott et al. 1998).

†The endocardial cell population is distinct from the myocardial cell population even before gastrulation. Cell lineage studies using retroviral markers show that clones of myocardial cells during the pre-streak stages have no endocardial cells, and that no clone of endocardial cells has any myocardial cells. Thus, the cardiogenic mesoderm already has two committed populations of cells (Cohen-Gould and Mikawa 1996).

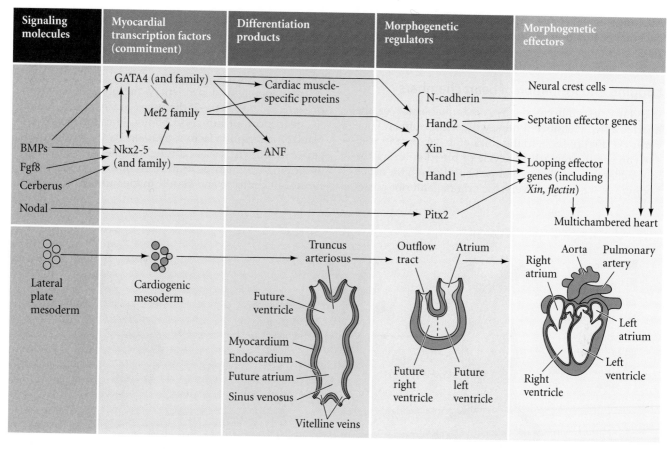

FIGURE 15.6 Cascade of heart development. A correlation can be made between the morphological stages (shown below) and the transcription factors present in the nucleus of the heart precursor cells. The cardioblasts are the committed heart precursor cells containing Nkx2-5 and GATA family proteins. These proteins convert the cardioblast into cardiomyocytes (heart muscle cells), which make cardiac muscle-specific proteins. These cardiomyocytes join together to form the cardiac tube. Under the influence of the Hand proteins, Xin, and Pitx2, the heart loops and the formation of the chambers begins.

the proteins that regulate myocardial growth, and regulate the placement of nervous tissue in the heart.

Fusion of the heart rudiments and initial heartbeats

The fusion of the two heart primordia occurs at about 29 hours in chick development and at 3 weeks in human gestation (see Figure 15.3C,D). The myocardia unite to form a single tube. The two endocardia lie within this common tube for a short while, but eventually they also fuse. The unfused posterior portions of the endocardium become the openings of the **vitelline veins** into the heart (see Figure 15.7). These veins will carry nutrients from the yolk sac into the **sinus venosus**, the posterior region where the two major veins fuse. The blood then passes through a valvelike flap into the atrial region of the heart. Contractions of the **truncus arteriosus** speed the blood into the aorta. In addition to the heart cells derived from the paired heart fields, cells from surrounding regions also appear

to form portions of the truncus. These cells appear to be derived from cells adjacent to the paired regions of cardiac mesoderm defined by *Nkx2-5* and *Tbx5* expression (de la Cruz and Markwald 1998; Abu-Issa et al. 2004; Cai et al. 2005).

Pulsations of the chick heart begin while the paired primordia are still fusing. Heart muscle cells develop an inherent ability to contract, and isolated heart cells from 7-day rat or chick embryos will continue to beat when placed in petri dishes (Harary and Farley 1963; DeHaan 1967; Imanaka-Yoshida et al. 1998). The pulsations are made possible by the appearance of the sodium-calcium exchange pump in the muscle cell membrane; inhibiting this channel's function prevents the heartbeat from starting (Wakimoto et al. 2000; Linask et al. 2001). Eventually, the rhythmicity of the heartbeat becomes coordinated by the sinus venosus. The electric impulses generated here initiate waves of muscle contraction through the tubular heart. In this way, the heart can pump blood even before its intricate system of valves has been completed. Studies of mutations of cardiac cell calcium channels have implicated these channels in the pacemaker function (Rottbauer et al. 2001; Zhang et al. 2002).

Looping and formation of heart chambers

In 3-day chick embryos and 5-week human embryos, the heart is a two-chambered tube, with one atrium and one ventricle (Figure 15.7). In the chick embryo, the unaided eye can see the remarkable cycle of blood entering the lower chamber and being pumped out through the aorta. Looping of the heart converts the original anterior-poste-rior polarity of the heart tube into the right-left polarity seen in the adult organism. When this looping is completed, the portion of the heart tube destined to become the right ventricle lies anterior to the portion that will become the left ventricle.

Heart looping is dependent on the left-right patterning proteins (Nodal, Lefty-2) discussed in Chapter 11. Within the heart primordium, Nkx2-5 regulates the Hand1 and Hand2 transcription factors. Although both Hand proteins appear to be synthesized throughout the early heart tube, as looping commences, Hand1 becomes restricted to the future left ventricle and Hand2 to the right. Without these

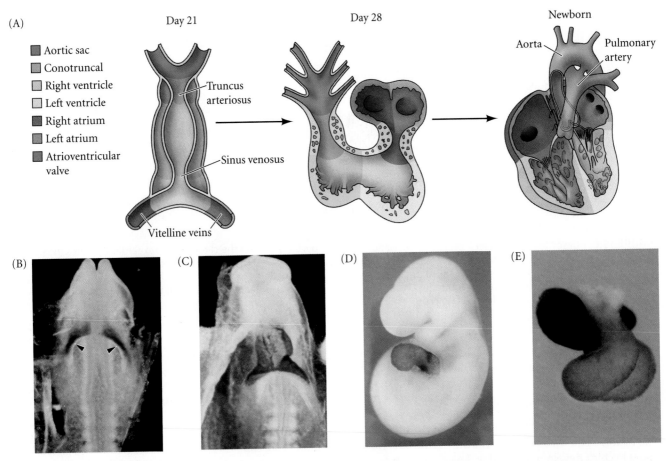

FIGURE 15.7 Cardiac looping and chamber formation. (A) Schematic diagram of cardiac morphogenesis in humans. On day 21, the heart is a single-chambered tube. The specification of the regions of the tube is shown in different colors. By day 28, cardiac looping has occurred, placing the presumptive atria anterior to the presumptive ventricles. (B,C) *Xin* expression in the fusion of left and right heart primordia. The cells fated to form the myocardium are shown by staining for the *Xin* message, whose protein product is essential for the looping of the heart tube. (B) Stage 9 chick neurula, in which expression of *Xin* (purple) is seen in the two symmetrical heart-forming fields (arrows). (C) Stage 10 chick embryo, showing the fusion of the two heart-forming regions prior to looping. (D,E) Specification of the atrium and ventricles occurs even before heart looping.

The atrium and ventricles of the mouse embryo have separate types of myosin proteins, which allows them to be differentially stained. In these photographs, the atrial myosin is stained blue, while the ventricular myosin is stained orange. (D) In the tubular heart (prior to looping), the two myosins (and their respective stains) overlap at the atrioventricular channel joining the future regions of the heart. (E) After looping, the dark blue stain is seen in the definitive atria and inflow tract, while the orange stain is seen in the ventricles. The unstained region above the ventricles is the truncus arteriosus. Derived primarily from the neural crest, the truncus arteriosus becomes separated into the aorta and pulmonary arteries. (A after Srivastava and Olson 2000; B,C from Wang et al. 1999, photographs courtesy of J. J.-C. Lin; D,E from Xavier-Neto et al. 1999, photographs courtesy of N. Rosenthal.)

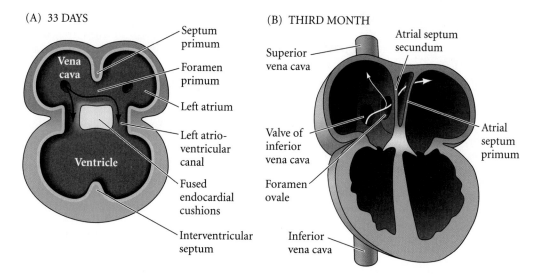

(A) 33 DAYS

- Septum primum
- Foramen primum
- Left atrium
- Left atrio-ventricular canal
- Fused endocardial cushions
- Interventricular septum

Vena cava

Ventricle

(B) THIRD MONTH

- Superior vena cava
- Atrial septum secundum
- Atrial septum primum
- Valve of inferior vena cava
- Foramen ovale
- Inferior vena cava

FIGURE 15.8 Formation of the chambers and valves of the heart. (A) Diagrammatic cross section of the human heart at 4.5 weeks. The atrial and ventricular septa are growing toward the endocardial cushion. (B) Cross section of the human heart during the third month of gestation. Blood can cross from the right side of the heart to the left side through the openings in the primary and secondary atrial septa. (After Larsen 1993.)

proteins, looping is abnormal, as the ventricles fail to form properly* (Srivastava et al. 1995; Biben and Harvey 1997).

Extracellular matrices are very important in these turning events. First, the extracellular matrix protein flectin regulates the physical tension of the heart tissues on the different sides (Tsuda et al. 1996; Linask et al. 2002). Second, transcription factors Nkx2-5 and Mef2C activate the *Xin* gene (Chinese for "heart"), whose product may mediate cytoskeletal changes essential for heart looping (Wang et al. 1999). Third, metalloproteinases, especially metalloproteinase-2 (MMP2), are critical for remodeling the cytoskeleton. If *MMP2* expression is blocked by antibodies or pharmaceutical inhibitors, the extracellular matrix fails to change, the asymmetric cell divisions (which cause the left side to grow faster than the right) fail to occur, and looping stops (Linask et al. 2005.)

Differential cell division is also important in structuring the right and left sides of the body, and the regulation of differential growth remains a major question in heart development. It is thought that many of the genes that are found in one particular area of the heart may be responsible for

differential growth. For instance, *Tbx5* is expressed in the left ventricle and atrium, while *Wnt11* is expressed in the right side of the heart. *Tbx2* is expressed throughout the heart. *Tbx2* encodes a transcriptional repressor that blocks *Tbx5* from promoting cell proliferation, and by blocking the expression of the *NMyc1* gene that activates cell division (Cai et al. 2005; Stennard et al. 2005). A series of regulators (such as *Tbx20*) probably control growth by activating or interfering with these chamber-specific factors.

The formation of the four heart valves—those leaflike flaps that must open and shut without failure once each second for the duration of our life—is just starting to be understood. In mammals, an **endocardial cushion** forms from the endocardium and divides the tube into right and left atrioventricular channels. Meanwhile, the primitive atrium is partitioned by two **septa** that grow ventrally toward the endocardial cushion (Figure 15.8A). The endocardial cushion serves as a valve during early heart development (Lamers et al. 1988), but as the heart enlarges, specialized valves develop to prevent the return of blood into the atria and to prevent the mixing of bloods from the two sides of the heart (Figure 15.8B). These valves begin to form when cells from the myocardium produce a factor that causes cells from the adjacent endocardium to detach and enter the hyaluronate-rich "cardiac jelly" extracellular matrix between the two layers (Markwald et al. 1977; Potts et al. 1991). The factors that induce this epithelial-mesenchymal transition probably differ between amniote groups. In zebrafish, the Wnt pathway appears to be critical. In mammals, some signal prevents the myocardial cells in the area of the endocardial cushion to stop synthesizing VEGF (vascular endothelial growth factor), a paracrine factor critical for differentiating the endocardium. With the VEGF signal disrupted, cells leave this region of the endocardium to develop into the valves (Hurlstone et al. 2003; Chang et al. 2004).

*Zebrafish, with only one ventricle, have only one type of Hand protein. When the gene encoding this protein is mutated, the entire ventricular portion of the heart fails to form (Srivastava and Olson 2000). Nongenetic agents are also critical in normal zebrafish heart formation. In the absence of high-pressure blood flow, heart looping, chamber formation and valve development is impaired (Hove et al. 2003).

SIDELIGHTS & SPECULATIONS

Redirecting Blood Flow in the Newborn Mammal

Although the developing mammalian fetus shares with the adult the need to get oxygen and nutrients to its tissues, the physiology of the fetus differs dramatically from that of the adult. Chief among the differences is the fetus's lack of functional lungs and intestines. All of its oxygen and nutrients must come from the placenta. This observation raises two questions. First, how does the fetus obtain oxygen from maternal blood? And second, how is blood circulation redirected to the lungs once the umbilical cord is cut and breathing becomes necessary?

Human embryonic circulation

The human embryonic circulatory system is a modification of that used in other amniotes, such as birds and reptiles. The circulatory system to and from the chick embryo and yolk sac is shown in Figure 15.9A. Blood pumped through the dorsal aorta passes over the aortic arches and down into the embryo. Some of this blood leaves the embryo through the vitelline arteries and enters the yolk sac. Nutrients and oxygen are absorbed from the yolk, and the blood returns through the vitelline veins to re-enter the heart through the sinus venosus.

Mammalian embryos obtain food and oxygen from the placenta. Thus, although the embryo has vessels analogous to the vitelline veins, the main supply of food and oxygen comes from the umbilical vein, which unites the embryo with the placenta (Figure 15.9B). This vein, which brings oxygenated, food-laden blood into the embryo, is derived from what would be the right vitelline vein in birds. The umbilical artery, carrying wastes to the placenta, is derived

from what would have become the allantoic artery of the chick. It extends from the caudal portion of the aorta and proceeds along the allantois and then out to the placenta.

Fetal hemoglobin

The solution to the fetus's problem of getting oxygen from its mother's blood involves the development of a specialized

fetal hemoglobin. The hemoglobin in fetal red blood cells differs slightly from that in adult corpuscles. Two of the four peptides—the alpha (α) chains—that make up fetal and adult hemoglobin chains are identical, but adult hemoglobin has two beta (β) chains, while the fetus has two gamma (γ) chains (Figure 15.10). The β-chains bind the natural regulator diphosphoglycerate,

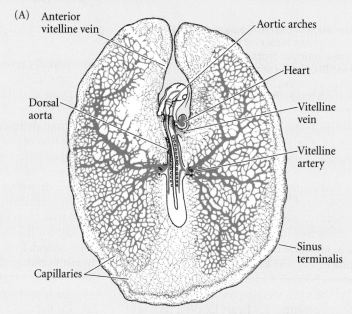

(A)

Anterior vitelline vein

Aortic arches

Heart

Dorsal aorta

Vitelline vein

Vitelline artery

Sinus terminalis

Capillaries

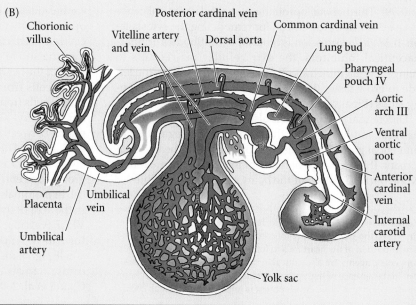

(B)

Chorionic villus

Vitelline artery and vein

Posterior cardinal vein

Dorsal aorta

Common cardinal vein

Lung bud

Pharyngeal pouch IV

Aortic arch III

Ventral aortic root

Anterior cardinal vein

Internal carotid artery

Placenta

Umbilical vein

Umbilical artery

Yolk sac

FIGURE 15.9 Embryonic circulatory systems. (A) Circulatory system of a 44-hour chick embryo. This view shows arteries in color; the veins are stippled. The sinus terminalis is the outer limit of the circulatory system and the site of blood cell generation (see also Figure 1.2). (B) Circulatory system of a 4-week human embryo. Although at this stage all the major blood vessels are paired left and right, only the right vessels are shown. Arteries are shown in red, veins in blue. (After Carlson 1981.)

which assists in the unloading of oxygen. The γ-chain isoforms do not bind diphosphoglycerate as well, and therefore have a higher affinity for oxygen. In the low-oxygen environment of the placenta, oxygen is released from adult hemoglobin. In this same environment, fetal hemoglobin does not release oxygen, but binds it. This small difference in oxygen affinity mediates the transfer of oxygen from the mother to the fetus. Within the fetus, the myoglobin of the fetal muscles has an even higher affinity for oxygen, so oxygen molecules pass from fetal hemoglobin to the fetal muscles. Fetal hemoglobin is not deleterious to the newborn, and in humans, the replacement of fetal hemoglobin-containing blood cells with adult hemoglobin-containing blood cells is not complete until about 6 months after birth. (The molecular basis for this switch in globins is discussed in Chapter 5.)

From fetal to newborn circulation

Once the fetus is no longer obtaining its oxygen from its mother, how does it restruc-

FIGURE 15.10 Adult and fetal hemoglobin molecules differ in their globin subunits. The fetal γ-chain binds diphosphoglycerate less avidly than does the adult β-chain. Consequently, fetal hemoglobin can bind oxygen more efficiently than can adult hemoglobin. In the placenta, there is a net flow (arrow) of oxygen from the mother's blood (which gives up oxygen to the tissues at lower oxygen pressures) to the fetal blood (which at the same pressure is still taking oxygen up).

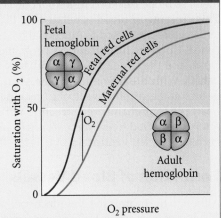

ture its circulation to get oxygen from its own lungs? During fetal development, an opening—the **ductus arteriosus**—diverts blood from the pulmonary artery into the aorta (and thus to the placenta). Because blood does not return from the pulmonary vein in the fetus, the developing mammal has to have some other way of getting blood into the left ventricle to be pumped. This is accomplished by the **foramen ovale**, an opening in the septum separating the right and left atria. Blood can enter the right atrium, pass through the foramen into the

left atrium, and then enter the left ventricle (Figure 15.11). When the first breath is drawn, blood pressure in the left side of the heart increases. This pressure closes the septa over the foramen ovale, thereby separating the pulmonary and systemic circulations. Moreover, the decrease in prostaglandins experienced by the newborn cause the muscles surrounding the ductus arteriosus to close that opening as well (Nguyen et al. 1997). Thus, when breathing begins, the respiratory circulation is shunted from the placenta to the lungs.

In some infants, the septa fail to close, and the foramen ovale is left open. Usually the opening is so small that there are no physical symptoms, and the foramen eventually closes. If it does not close completely, however, and the secondary septum fails to form, the atrial septal opening may cause enlargement of the right side of the heart, which can lead to heart failure in early adulthood.

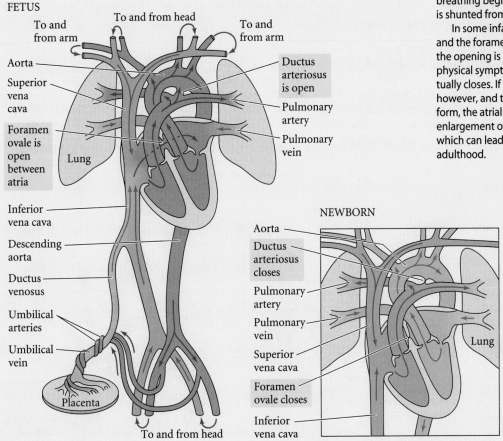

FIGURE 15.11 Redirection of human blood flow at birth. The expansion of air into the lungs causes pressure changes that redirect the flow of blood in the newborn infant. The ductus arteriosus squeezes shut, breaking off the connection between the aorta and the pulmonary artery, and the foramen ovale, a passageway between the left and right atria, also closes. In this way, pulmonary circulation is separated from systemic circulation.

The septa between the atria, however, have openings in them, so blood can still cross from one side into the other. This crossing of blood is needed for the survival of the fetus before circulation to functional lungs has been established. Upon the first breath, however, the septal openings close, and the left and right circulatory loops are established. With the formation of the septa (which usually occurs in the seventh week of human development), the heart is a four-chambered structure with the pulmonary artery connected to the right ventricle and the aorta connected to the left ventricle.

Formation of Blood Vessels

Although the heart is the first functional organ of the body, it does not even begin to pump until the vascular system of the embryo has established its first circulatory loops. Rather than sprouting from the heart, the blood vessels form independently, linking up to the heart soon afterward. Everyone's circulatory system is different, since the genome cannot encode the intricate series of connections between the arteries and veins. Indeed, chance plays a major role in establishing the microanatomy of the circulatory system. However, all circulatory systems in a given species look very much alike, because the development of the circulatory system is severely constrained by physiological, physical, and evolutionary parameters.

Constraints on the construction of blood vessels

The first constraint on vascular development is *physiological*. Unlike new machines, which do not need to function until they have left the assembly line, new organisms have to function even as they develop. The embryonic cells must obtain nourishment before there is an intestine, use oxygen before there are lungs, and excrete wastes before there are kidneys. All these functions are mediated through the

embryonic circulatory system. Therefore, the circulatory physiology of the developing embryo must differ from that of the adult organism. Food is absorbed not through the intestine, but from either the yolk or the placenta, and respiration is conducted not through the gills or lungs, but through the chorionic or allantoic membranes. The major embryonic blood vessels must be constructed to serve these extraembryonic structures.

The second constraint is *evolutionary*. The mammalian embryo extends blood vessels to the yolk sac even though there is no yolk inside (see Figure 15.9). Moreover, the blood leaving the heart via the truncus arteriosus passes through vessels that loop over the foregut to reach the dorsal aorta. Six pairs of these aortic arches loop over the pharynx (Figure 15.12). In primitive fish, these arches persist and enable the gills to oxygenate the blood. In adult birds and mammals, in which lungs oxygenate the blood, such a system makes little sense—but all six pairs of aortic arches are formed in mammalian and avian embryos before the system eventually becomes simplified into a single aortic arch. Thus, even though our physiology does not require such a structure, our embryonic condition reflects our evolutionary history.

The third set of constraints is *physical*. According to the laws of fluid movement, the most effective transport of fluids is performed by large tubes. As the radius of a blood vessel gets smaller, resistance to flow increases as r^{-4} (Poiseuille's law). A blood vessel that is half as wide as

FIGURE 15.12   The aortic arches of the human embryo. (A) Originally, the truncus arteriosus pumps blood into the aorta, which branches on either side of the foregut. The six aortic arches take blood from the truncus arteriosus and allow it to flow into the dorsal aorta. (B) As development proceeds, arches begin to disintegrate or become modified (the dotted lines indicate degenerating structures). (C) Eventually, the remaining arches are modified and the adult arterial system is formed. (After Langman 1981.)

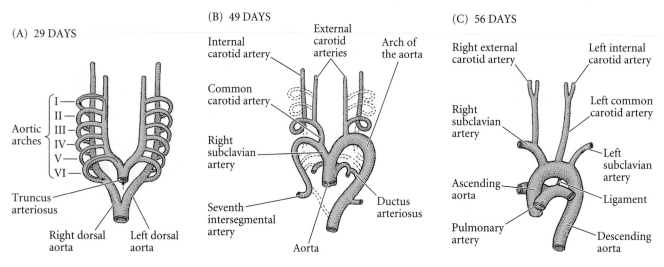

another has a resistance to flow that is 16 times greater. However, diffusion of nutrients can take place only when blood flows slowly and has access to cell membranes. So here is a paradox: the constraints of diffusion mandate that vessels be small, while the laws of hydraulics mandate that vessels be large. This paradox has been solved by the evolution of circulatory systems with a hierarchy of vessel sizes (LaBarbera 1990). In dogs, for example, blood in the large vessels (aorta and vena cava) flows over 100 times faster than it does in the capillaries. With a system of large vessels specialized for transport and small vessels specialized for diffusion (where the blood spends most of its time), nutrients and oxygen can reach the individual cells of the growing organism. This hierarchy is seen very early in development (it is already well established in the 3-day chick embryo).

But this is not the entire story. If fluid under constant pressure moves directly from a large-diameter tube into a small-diameter tube (as in a hose nozzle), the fluid velocity increases. The evolutionary solution to this problem was the emergence of many smaller vessels branching out from a larger one, making the collective cross-sectional area of all the smaller vessels greater than that of the larger vessel. Circulatory systems show a relationship (known as Murray's law) in which the cube of the radius of the parent vessel approximates the sum of the cubes of the radii of the smaller vessels. Computer models of blood vessel formation must take into account not only gene expression patterns but also the fluid dynamics of blood flow, if they are to show the branching and anastomosing of the arteries and veins (Gödde and Kurz 2001). The construction of any circulatory system negotiates among all of these physical, physiological, and evolutionary constraints.

Vasculogenesis: The initial formation of blood vessels

The development of blood vessels occurs by two temporally separate processes: **vasculogenesis** and **angiogenesis** (Figure 15.13). During vasculogenesis, a network of blood vessels is created de novo from the lateral plate mesoderm. During angiogenesis, this primary network is remodeled and pruned into a distinct capillary bed, arteries, and veins.

In the first phase of vasculogenesis, cells leaving the primitive streak in the posterior of the embryo become **hemangioblasts,**[*] the precursors of both blood cells and blood vessels. These cells, which reside in the splanchnic

mesoderm (Shalaby et al. 1997; Huber et al. 2004), condense into aggregations that are often called **blood islands.**[†] The inner cells of these blood islands become **blood progenitor cells**, while the outer cells become **angioblasts**, the progenitors of the blood vessels. In the second phase of vasculogenesis, the angioblasts multiply and differentiate into **endothelial** cells, which form the lining of the blood vessels. In the third phase, the endothelial cells form tubes and connect to form the **primary capillary plexus**, a network of capillaries.

THE SITES OF VASCULOGENESIS In amniotes, formation of the primary vascular networks occurs in two distinct and independent regions. First, **extraembryonic vasculogenesis** occurs in the blood islands of the yolk sac. These are the blood islands formed by the hemangioblasts, and they give rise to the early vasculature needed to feed the embryo and also to a primitive red blood cell that functions in the early embryo but which probably is not found in the later embryo or adult (Figure 15.14A). Second, **intraembryonic vasculogenesis** occurs within each organ.

The aggregation of hemangioblasts in the yolk sac is a critical step in amniote development, for the blood islands that line the yolk sac produce the veins that bring nutrients to the embryo and transport gases to and from the sites of respiratory exchange (Figure 15.14B). In birds, these vessels are called the **vitelline veins**; in mammals, they are the **omphalomesenteric (umbilical)** veins. In the chick, blood islands are first seen in the area opaca, when the primitive streak is at its fullest extent (Pardanaud et al. 1987). They form cords of hemangioblasts, which soon become hollow. The outer cells of the blood islands become the flat endothelial cells lining the vessels. The central cells differentiate into the embryonic blood cells. As the blood islands grow, they eventually merge to form the capillary network draining into the two vitelline veins, which bring food and blood cells to the newly formed heart.

The intraembryonic vascular networks usually arise from individual angioblast progenitor cells in the mesoderm surrounding a developing organ. These cells do not appear to be associated with blood cell formation, as is seen in the extraembryonic blood islands (Noden 1989; Risau 1995; Pardanaud et al. 1989). It is important to realize that these intraembryonic capillary networks arise within or around the organ itself and are not extensions of larger vessels. Indeed, in some cases, the developing organ produces paracrine factors that induce blood vessels to form only in its own mesenchyme (Auerbach et al. 1985; LeCouter et al. 2001).

---

[*]The prefixes *hem-* and *hemato-* refer to blood (as in hemoglobin). Similarly, the prefix *angio-* refers to blood vessels. The suffix *-blast* denotes a rapidly dividing cell, usually a stem cell. The suffixes *-poesis* and *-poietic* refer to generation or formation (*poeisis* is also the root of the word *poetry*). Thus, *hematopoietic* stem cells are those cells that generate the different types of blood cells. The Latin suffix *-genesis* (as in angiogenesis) means the same as the Greek *-poiesis*.

[†]Again, the endoderm plays a major role in lateral plate mesoderm specification. Here, the visceral endoderm of the splanchnopleure interacts with the yolk sac mesoderm to induce the blood islands. The endoderm is probably secreting Indian hedgehog, a paracrine factor that activates BMP4 expression in the mesoderm. BMP4 expression feeds back on the mesoderm itself, causing it to form hemangioblasts (Baron 2001).

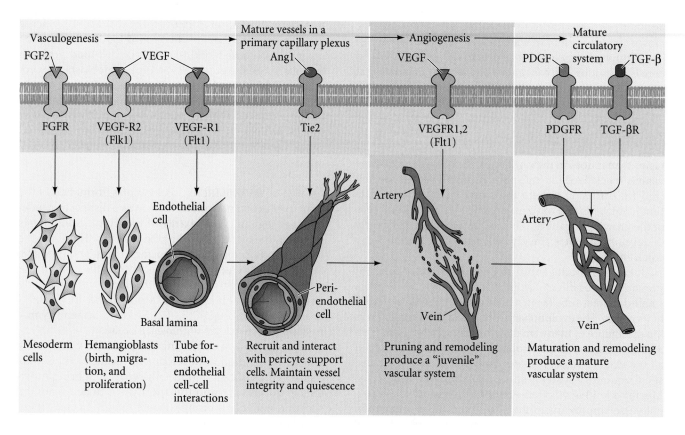

FIGURE 15.13 Vasculogenesis and angiogenesis. Vasculogenesis involves the formation of blood islands and the construction of capillary networks from them. Angiogenesis involves the formation of new blood vessels by remodeling and building on older ones. Angiogenesis finishes the circulatory connections begun by vasculogenesis. The major paracrine factors involved in each step are shown at the top of the diagram, and their receptors (on the vessel-forming cells) are shown beneath them. (After Hanahan 1997; Risau 1997.)

GROWTH FACTORS AND VASCULOGENESIS Three growth factors may be responsible for initiating vasculogenesis (see Figure 15.13). One of these, **basic fibroblast growth factor (Fgf2)**, is required for the generation of hemangioblasts from the splanchnic mesoderm. When cells from quail blastodiscs are dissociated in culture, they do not form blood islands or endothelial cells. However, when these cells are cultured in Fgf2, blood islands emerge and form endothelial cells (Flamme and Risau 1992). Fgf2 is synthesized in the chick embryonic chorioallantoic membrane and is responsible for the vascularization of this tissue (Ribatti et al. 1995).

The second protein involved in vasculogenesis is **vascular endothelial growth factor (VEGF)**. VEGF appears to enable the differentiation of the angioblasts and their multiplication to form endothelial tubes. VEGF is secreted by the mesenchymal cells near the blood islands, and hemangioblasts and angioblasts have receptors for VEGF* (Millauer et al. 1993). If mouse embryos lack the genes encoding either VEGF or the major receptor for VEGF (the Flk1 receptor tyrosine kinase), yolk sac blood islands fail to appear, and vasculogenesis fails to take place (Figure

15.15; Ferrara et al. 1996). Mice lacking genes for the VEGF receptor have blood islands and differentiated endothelial cells, but these cells are not organized into blood vessels (Fong et al. 1995; Shalaby et al. 1995). As we saw in Chapter 14, VEGF is also important in forming blood vessels to the developing bone and kidney.

---

*VEGF needs to be regulated very carefully in adults, and recent studies indicate that it can be affected by diet. The consumption of green tea has been associated with lower incidences of human cancer and the inhibition of tumor cell growth in laboratory animals. Cao and Cao (1999) have shown that green tea, and one of its components, epigallocatechin-3-gallate (EGCG), prevent angiogenesis by inhibiting VEGF. Moreover, in mice given green tea instead of water (at levels similar to humans after drinking 2–3 cups of tea), the ability of VEGF to stimulate new blood vessel formation was reduced by more than 50 percent. The drinking of moderate amounts of red wine has been correlated with reduced coronary disease. Red wine has been shown to reduce VEGF production in adults, and it appears to do so by inhibiting endothelin-1, a compound that induces VEGF and which is crucial for the formation of atherosclerotic plaques (Corder et al. 2001; Spinella et al. 2002).

(A)

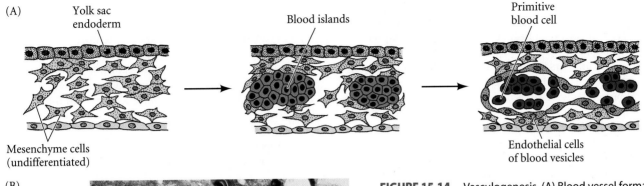

Yolk sac endoderm

Blood islands

Primitive blood cell

Mesenchyme cells (undifferentiated)

Endothelial cells of blood vesicles

(B)

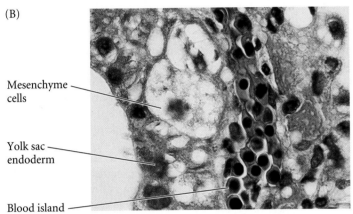

Mesenchyme cells

Yolk sac endoderm

Blood island

FIGURE 15.14 Vasculogenesis. (A) Blood vessel formation is first seen in the wall of the yolk sac, where undifferentiated mesenchyme cells cluster to form blood islands. The centers of these clusters form the blood cells, and the outsides of the clusters develop into blood vessel endothelial cells. (B) Photograph of a human blood island in the mesoderm surrounding the yolk sac. (The photograph is from a tubal pregnancy—an embryo that had to be removed because it had implanted in an oviduct rather than in the uterus.) (A after Langman 1981; B from Katayama and Kayano 1999, photograph courtesy of the authors.)

A third set of proteins, the **angiopoietins**, mediates the interaction between the endothelial cells and the **pericytes**—smooth muscle-like cells the endothelial cells recruit to cover them. Mutations of either angiopoietins or their receptor, Tie2, lead to malformed blood vessels that are deficient in the smooth muscles that usually surround them (Davis et al. 1996; Suri et al. 1996; Vikkula et al. 1996; Moyon 2001).

Angiogenesis: Sprouting of blood vessels and remodeling of vascular beds

After an initial phase of vasculogenesis, angiogenesis begins. By this process, the primary capillary networks are remodeled and veins and arteries are made (see Figure 15.13). The extracellular matrix is extremely important in regulating angiogenesis. First, VEGF acting alone on the newly formed capillaries causes a loosening of cell-cell contacts and a degradation of the extracellular matrix at certain points. The exposed endothelial cells proliferate and sprout from these regions, eventually forming a new vessel. New vessels can also be formed in the primary capillary bed by splitting an existing vessel in two. The loosening of cell-cell contacts may also allow the fusion of capillaries to form wider vessels—the arteries and veins. Eventually, the mature capillary network forms and is stabilized by TGF-β (which strengthens the extracellular matrix) and **platelet-derived growth factor (PDGF)**, which

is necessary for the recruitment of the pericyte cells that contribute to the mechanical flexibility of the capillary wall (Lindahl et al. 1997).

The developing blood vessels contain powerful inhibitors of metalloproteases (which digest extracellular matrices), and angiogenesis can only occur when these inhibitors are downregulated (Oh et al. 2001). One extreme-

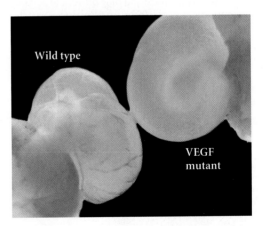

Wild type

VEGF mutant

FIGURE 15.15 Yolk sacs of a wild-type mouse and a littermate heterozygous for a loss-of-function mutation of VEGF. The mutant embryo lacks blood vessels in its yolk sac and dies. (From Ferrara and Alitalo 1999, photograph courtesy of the authors.)

ly important regulator of angiogenesis may be a particular form of collagen, **collagen XVIII**, the same collagen involved in epithelial branch formation (see Chapter 14). This collagen not only stabilizes the capillaries structurally, but also appears to defend the blood vessels chemically against external metalloproteases (Ergun et al. 2001).

When collagen XVIII is cleaved by metalloproteases, the 184 amino acids of its C-terminal form a protein called **endostatin**. Endostatin has functions very different from those of collagen. It prevents angiogenesis by inhibiting cyclin expression and by interfering with the binding of VEGF to its receptor (Zatterstrom et al. 2000; Hanai et al. 2002; Kim et al. 2002). When new blood vessels are to be made, endostatin production is downregulated (Wu et al. 2001). Moreover, endostatin may prevent tumors from growing and spreading. Tumors are "successful" only when they are able to direct blood vessels into them. Therefore, tumors secrete angiogenesis factors. The ability to inhibit such factors may have important medical applications as a way of preventing tumor growth and metastasis (Folkman et al. 1971; Fidler and Ellis 1994; see Chapter 21).

Arterial and venous differentiation

A key to our understanding of the mechanism by which veins and arteries form was the discovery that the primary capillary plexus in mice actually contains two types of endothelial cells. The precursors of the arteries contain ephrin-B2 in their cell membranes, and the precursors of the veins contain one of the receptors for this molecule, EphB4 tyrosine kinase, in their cell membranes (Wang et al. 1998). If ephrin-B2 is knocked out in mice, vasculogenesis occurs, but angiogenesis does not. It is thought that during angiogenesis EphB4 interacts with its ligand, ephrin-B2, in two ways. First, at the borders of the venous and arterial capillaries, it ensures that arterial capillaries connect only to venous ones. Second, in non-border areas, it ensures that the fusion of capillaries to make larger vessels occurs only between the same type of vessel (Figure 15.16).

In zebrafish, the separation of arterial cells and venous cells occurs very early in development. The angioblasts develop in the posterior part of the lateral mesoderm, and they migrate to the midline of the embryo, where they coalesce to form the aorta (artery) and the cardinal vein beneath it (Figure 15.17). Zhong and colleagues (2001) followed individual angioblasts and found that, contrary to expectations, all the progeny of a single angioblast formed either veins or arteries, never both. In other words, each angioblast was already specified as to whether it would form aorta or cardinal vein. This specification appears to be controlled by the Notch signaling pathway* (Lawson et al. 2001, 2002). Repression of Notch signaling resulted in the loss of ephrin-B2-expressing arteries and their replacement by veins. Conversely, activation of Notch signaling suppressed venous development, causing more arterial cells to form. Activation of the Notch proteins in the membranes of the presumptive angioblasts causes the activation of the transcription factor **Gridlock.** Gridlock in turn activates the expression of Ephrin-B2 and other arterial

---

*The coordinated use of Notch and Eph signaling pathways is also used to regulate the production of neuroblasts and somites.

FIGURE 15.16 Model of the roles of ephrin and Eph receptors during angiogenesis. (A) Primary capillary plexus produced by vasculogenesis. The arterial and venous endothelial cells have sorted themselves out by the presence of ephrin-B2 or EphB4 in their respective cell membranes. (B) A maturing vascular network wherein the ephrin-Eph interaction mediates the joining of small branches (future capillaries) and may prevent fusion laterally. (After Yancopopoulos et al. 1998.)

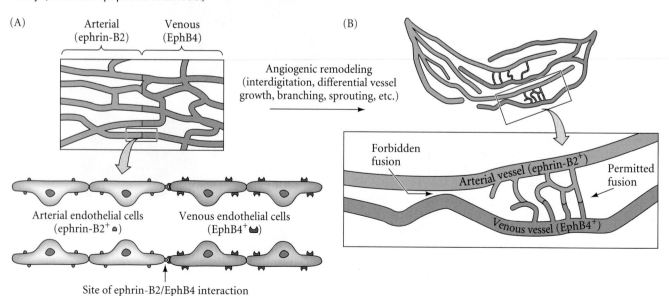

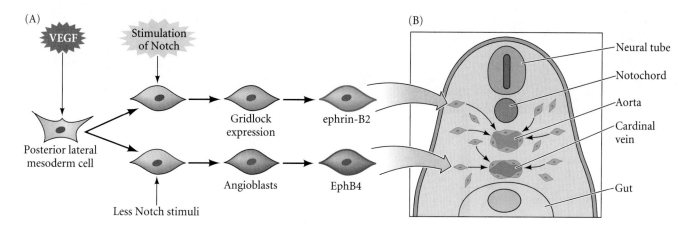

FIGURE 15.17 Blood vessel specification in the zebrafish embryo. (A) Angioblasts experiencing activation of Notch upregulate the Gridlock transcription factor. These cells express ephrin-B2 and become aorta cells. Those angioblasts experiencing significantly less Notch activation do not express Gridlock, and they become EphB4-expressing cells of the cardinal vein. (B) Once committed to forming veins or arteries, the cells migrate toward the midline of the embryo and contribute to forming the aorta or cardinal vein.

markers, while those angioblasts with low amounts of Gridlock became EphB4-expressing vein cells.

What, then, controls Notch expression? Studies in zebrafish and mice have revealed that VEGF functions earlier than originally thought, and that it is critical in inducing the differentiation of arterial vessels. Reduction of VEGF activity in zebrafish prevents Notch activation and the subsequent development of arteries, while injection of VEGF mRNA into the cells of the posterior cardinal vein induces the expression of (arterial) ephrin-B2 in these cells (Lawson et al. 2002). Weinstein and Lawson (2003) speculate that vascular beds are formed in a two-step process. First, new arteries form in response to VEGF. Second, these arteries then induce neighboring angioblasts (possibly through the ephrin/Eph interactions) to form the venous vessels that will provide the return for the arterial blood (Figure 15.18A). This speculation fits well with the detailed observations of chick vascular development done (and exquisitely drawn) by Popoff (1894) and Isida (1956), who found that the vitelline arteries appear first within the capillary network and that these capillaries appeared to induce veins on either side of them (Figure 15.18B–D).

Organ-specific angiogenesis factors

Several organs make their own angiogenesis factors. Placental cells are especially adept at creating new blood vessels, since they are responsible for bringing the fetal and maternal blood supplies into close apposition. These cells secrete several angiogenesis factors, including **leptin**, a hormone involved in appetite suppression in the adult. However, leptin can also act locally to induce angiogenesis and cause endothelial cells to organize into tubes (Antczak et al. 1997; Sierra-Honigmann et al. 1998). The kidney vasculature is mainly derived from the sprouting of endothelial cells from the dorsal aorta during the intial steps of nephrogenesis. The developing nephrons secrete VEGF, thus allowing the blood vessels to enter the developing kidney and form the capillary loops of the glomerular apparatus (Kitamoto et al. 2002).

One of the most striking examples of organ-specific angiogenesis factors is that produced by the developing peripheral nerves. Anatomists have known for decades that blood vessels follow peripheral nerves (Greenberg and Jin 2005). This allows the nerves to get oxygen and allows for hormones in the blood to regulate vasoconstriction or vasodilation. Moreover, the reason that the nerves and blood vessels are parallel is that the nerves secrete an angiogenesis factor and the blood vessels secrete a nerve growth factor. Mukouyama and colleagues (2002, 2005) have demonstrated that arteries become associated with peripheral neurons, although veins do not (Figure 15.19). Moreover, peripheral neurons induce arteries to form near them. If the peripheral neurons in the skin fail to form (due to mutations that specifically target peripheral neurons), the arteries likewise fail to form properly. Moreover, if other mutations cause the peripheral neurons to form haphazardly, the arteries will follow them. This property is due to the secretion of VEGF by the neurons and their associated Schwann cells, which is necessary for arterial formation. Thus, the peripheral nerves appear to provide a template for the organ-specific angiogenesis through their ability to secrete VEGF. This interaction is not just one-way. In some instances, the blood vessels are formed in an area first, before the neurons enter. In those cases, the vascular smooth muscle cells can secrete a compound (most likely GDNF) that allows the neuron to grow alongside it. In this way, neurons can reach their destinations by following the blood vessels (Honma et al. 2002).

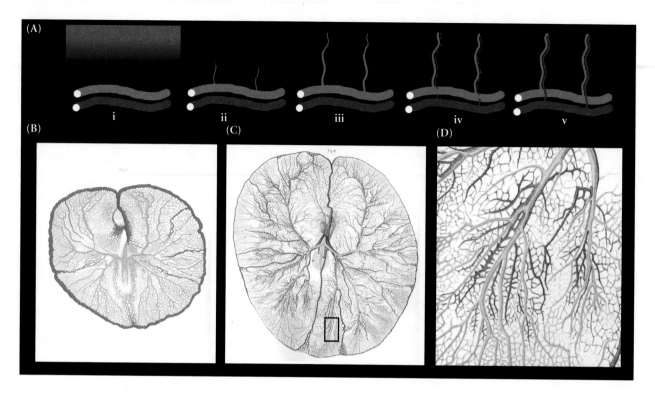

FIGURE 15.18 Model for blood vessel formation in the chick blastoderm. (A) In response to VEGF (green gradient), endothelial cells are induced to become arteries (red), and these arteries induce veins (blue) to form adjacent to them. New arterial vessels sprout from the arteries and then induce venous vessels adjacent to themselves. (B) In the chick embryo, a complex branched venous network emerges in the vascular region, with venous drainage at the periphery, via the marginal vein. (C) At later stages, collateral veins emerge adjacent to the arteries. (D) Higher magnification of the boxed region of (C). (After Weinstein and Lawson 2003; B–D modified from Popoff 1894, courtesy of N. D. Weinstein.)

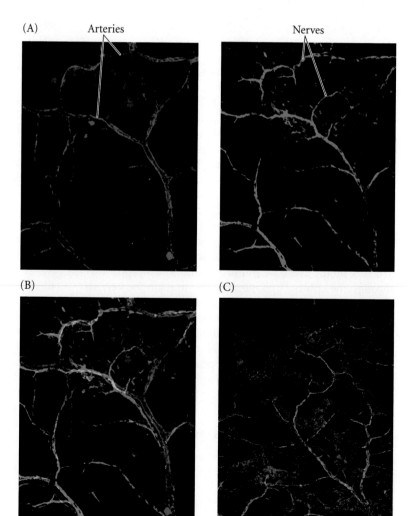

FIGURE 15.19 Arteries are specifically aligned with peripheral nerves in mouse limb skin. (A) Antibody staining of arteries (red; left) and nerves (green; right). (B) Placing the photographs together reveals that the arteries and nerves coincide. (C) Doing the same operation with stained nerves and veins reveals that the veins and nerves do not follow one another. (From Mukouyama et al. 2002, photographs courtesy of Y. Mukouyama.)

FIGURE 15.20 VEGF-C is critical for the formation of the lymphatic vessels. (A) Compared with the control, a day 15.5 mouse embryo heterozygous for a VEGF-C deficiency suffers with severe edema (bloating with excess fluid) (B) Day 16.5 mouse embryos stained for lymphatic vasculature. The lack of lymphatics in the skin of the VEGF-C mutant is obvious when compared to that of the wild-type embryo. (From Karkkainenen et al. 2003, photographs courtesy of K. Alitalo.)

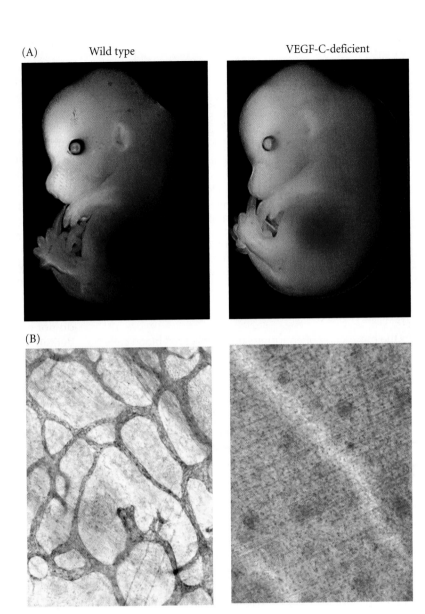

(A) Wild type VEGF-C-deficient

(B)

The lymphatic vessels

In addition to the blood vessels, there is a second circulatory system, the **lymphatic vasculature**. The lymphatic vasculature forms a separate system of vessels which is essential for draining fluid and transporting lymphocytes. The development of the lymphatic system commences when a subset of endothelial cells from the jugular vein (in the neck) sprout off to form the lymphatic sacs. After the formation of these sacs, the peripheral lymphatic vessels are generated by further sprouting (Sabin 1902; van der Putte 1975). Commitment to the lymphatic lineage appears to be mediated through the Prox1 transcription factor, which downregulates blood vessel-specific genes and upregulates genes involved in forming lymphatic vessels (Wigle et al. 1999, 2002). One of the genes upregulated in this way is *VEGFR-3*, which encodes the receptor for the paracrine factor VEGF-C. As important as VEGF is for blood vessel development, VEGF-C is equally necessary for proper lymphatic development (Figure 15.20; Karkkainen et al. 2004; Alitalo et al. 2005). VEGF-C produced in the area of the jugular vein attracts the prox1-positive endothelial cells out from the vein. It then promotes their proliferation and development into the lymphatic sacs.

The Development of Blood Cells

The stem cell concept

Many adult tissues are formed from cells that cannot be replaced. Most neurons and bones, for instance, cannot be replaced if they are damaged or lost. There are several populations of cells, however, that are constantly dying and being replaced. Each day, we lose and replace about 1.5 grams of skin cells and about 10^{11} red blood cells. The skin cells are sloughed off, and the red blood cells are killed in the spleen. Their replacements come from populations of stem cells.

As mentioned in earlier chapters, a **stem cell** is a cell that is capable of extensive proliferation, creating more stem cells (self-renewal) as well as more differentiated cellular progeny (Figure 15.21). Adult stem cells are, in effect, populations of embryonic cells within an adult organism, continuously producing both more stem cells as well as cells that can undergo further development and differentiation (Potten and Loeffler 1990). Our blood cells, intestinal crypt cells, epidermal cells, and (in males) spermatocytes are populations in steady-state equilibrium, in which cell production balances cell loss (Hay 1966). In most cases, stem cells can divide either to produce more stem cells or to replenish differentiated cells when body equilibrium is stressed by injury or by environmental factors. (This is seen

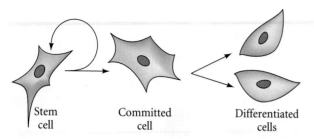

Stem   Committed   Differentiated
cell   cell   cells

FIGURE 15.21   The concept of stem cells. A stem cell divides to produce another stem cell and a committed cell. The committed cell can divide again, but its progeny have the ability to form only a restricted set of differentiated cell types. A stem cell can also produce two stem cells or two committed cells. If a significant number of stem cells divide to produce two committed cells, the longevity of the stem cell lineage becomes limited.

in the production of enormous numbers of red blood cells when the body suffers from hypoxia.)

The critical stem cell in hematopoiesis is the pluripotential hematopoietic stem cell. Often just referred to as the **hematopoietic stem cell** (**HSC**), this cell type can generate all the blood and lymph cells of the body. As we will see, the pluripotential HSC generates a series of intermediate stem cells whose potency is restricted to certain lineages.

Sites of hematopoiesis

Vertebrate blood development occurs in two phases: a transient embryonic ("primitive") phase of hematopoiesis and a subsequent definitive ("adult") phase. These phases differ in their sites of blood cell production, in the timing of hematopoiesis, in the morphology of the cells produced, and even in the type of globin genes active in the red blood cells. The embryonic phase of hematopoiesis is thought to be transitory and probably provides the embryo with its initial blood cells and capillary network to the endoderm or yolk. The definitive phase of hematopoiesis generates more cell types and provides the stem cells that will last for the lifetime of the individual.

Embryonic hematopoiesis is associated with the blood islands in the ventral mesoderm near the yolk sac. In chick embryos, the first blood cells are seen in those blood islands that form in the posterior marginal zone near the site of hypoblast initiation (Wilt 1974; Azar and Eyal-Giladi 1979). In mice, the yolk sac produces the first blood cells from the blood islands. Although the hematopoietic stem cells of these blood islands seem capable of generating all blood cell (but not lymphocyte) lineages, they usually produce red blood cells (Moore and Metcalf 1970; Rampon and Huber 2003). In *Xenopus*, the ventral mesoderm forms a large blood island that is the first site of hematopoiesis. BMPs are crucial in inducing the blood forming cells in all

vertebrates studied. Ectopic BMP2 and BMP4 can induce blood and blood vessel formation in *Xenopus*, and interference with BMP signaling prevents blood formation (Maeno et al. 1994; Hemmati-Brivanlou and Thomsen 1995). In the zebrafish, the *swirl* mutation, which prevents BMP2 signaling, also abolishes ventral mesoderm and blood cell production (Mullins et al. 1996). As mentioned above, BMP4 is critical in the formation of the blood islands in the mammalian extraembryonic mesoderm.

This embryonic hematopoietic cell population, however, is thought to be transitory. The hematopoietic stem cells that last the lifetime of the organism are derived from the mesodermal area surrounding the aorta. This was shown by a series of elegant transplantation experiments performed by Dieterlen-Lièvre, who grafted the blastoderm of chickens onto the yolk of Japanese quail (Figure 15.22A). Using these "yolk sac chimeras" (see Figure 1.10), Dieterlen-Lièvre and Martin (1981) showed that the yolk sac stem cells do not contribute cells to the adult animal. Instead, the definitive stem cells are formed within nodes of mesoderm that line the mesentery and the major blood vessels. In the 4-day chick embryo, the aortic wall appears to be the most important source of new blood cells, and it has been found to contain numerous hematopoietic stem cells (Cormier and Dieterlen-Lièvre 1988). Similar transplantation experiments in frogs have shown that their embryonic blood cells had a different source than their adult blood cells (Chen and Turpen 1995). Whether these two sources ultimately are formed from a common precursor is still a matter of contention (Walmsley et al. 2002; Lane and Sheets 2002).

Moreover, the definitive hematopoietic stem cells in fish, chicks, mammals, and frogs are each formed in the visceral (splanchnic) lateral plate mesoderm near the aorta. This domain has been called the **aorta-gonad-mesonephros** (**AGM**) region. The first blood cells in the mouse embryo appear in the mesoderm around the yolk sac, but by day 11, pluripotential hematopoietic stem cells can be found in the AGM (Kubai and Auerbach 1983; Godlin et al. 1993; Medvinsky et al. 1993). These hematopoietic stem cells later colonize the fetal liver, and around the time of birth, stem cells from the liver populate the bone marrow, which then becomes the major site of blood formation throughout adult life. A puzzling paradox, however, is that the numbers of pluripotential hematopoietic stem cells in the AGM is very small, whereas the liver contains numerous such stem cells (Kumaravelu et al. 2002).

This problem may have been solved when two laboratories independently discovered a *third* early source of blood stem cells in mammals—the placenta (Ottersbach and Dzierak 2005; Gekas et al. 2005). Pluripotential hematopoietic stem cells appear to be generated along with the endothelium of the placental blood vessels, suggesting that hemangioblasts are found in the placenta as well as in the yolk sac. Moreover, these stem cells appear in numbers large enough to account for the population of cells later found in the liver. It is possible, then, that the liver receives

(A)

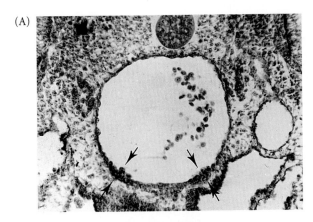

(B)

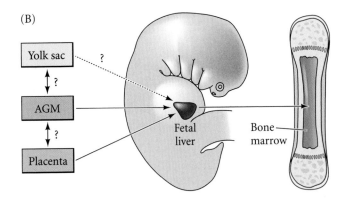

FIGURE 15.22 Sources of blood cells to the adult bone marrow. (A) Section through the aorta of a 3-day chick embryo, showing the cells (arrows) that give rise to hematopoietic stem cells in the AGM (aorta-gonad-mesonephros region) of the chick. (B) In mammals, the yolk sac, the AGM, and the placenta each probably contribute stem cells to the fetal liver. Stem cells from the fetal liver then populate the bone marrow as the hematopoietic niche in the bone marrow is constructed. There is controversy as to whether the stem cells from the yolk sac populate the AGM or the liver, or if a new set of stem cells is formed. (C) Schematic diagram of the stem cell niche by which bone marrow (endosteal) osteoblasts maintain HSCs by activating the Wnt, Notch, and receptor tyrosine kinase (RTK) pathways (red dashed arrows). (A from Dieterlen-Lièvre and Martin 1981, photograph courtesy of F. Dieterlen-Lièvre; B after Ottersbach and Dzierak 2005.)

(C)

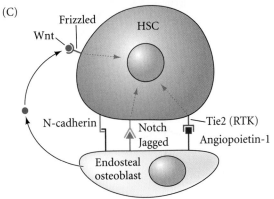

its stem cells from the placenta and the AGM, with possibly some contribution from the yolk sac hemangioblasts (Figure 15.22B).

The hematopoietic stem cells formed in the embryo are those that populate the bone marrow (and in some instances, the spleen) of adult mammals. The adult HSC "niches" in the bone and spleen make chemoattractant proteins that attract the circulating stem cells into them (Christensen et al. 2004; Gothert et al. 2005). A pluripotential HSC cell that resides in the marrow is thus a descendent of stem cells that had populated the embryonic liver, and probably the AGM or placenta.

Committed stem cells and their fates

The bone marrow HSC is a remarkable cell, in that it is the common precursor of red blood cells (erythrocytes), white blood cells (granulocytes, neutrophils, and platelets), and lymphocytes. When transplanted into inbred, irradiated mice (who are genetically identical to the donor cells and whose own stem cells have been eliminated by radiation), pluripotential HSCs can repopulate the mouse with all the blood and lymphoid cell types. It is estimated that only about 1 in every 10,000 blood cells is a pluripotential HSC (Berardi et al. 1995).

The pluripotential HSC (sometimes called the CFU-M,L) appears to be dependent on the transcription factor SCL.

Mice lacking SCL die from the absence of all blood and lymphocyte lineages. SCL is thought to specify blood cell fate in mesoderm cells, and it continues to be expressed in the HSCs (Porcher et al. 1996; Robb et al. 1996). The plurpotential HSC is also dependent on osteoblasts in the bone marrow. Osteoblasts that have finished making bone can still have a function: **endosteal osteoblasts** that line the bone marrow are responsible for providing the "niche" that attracts HSCs, prevents apoptosis, and keeps the HSCs in a state of plasticity. These osteoblasts bind HSCs (probably through N-cadherin) and provide several other signals (Calvi et al. 2003; Zhang et al. 2003). One signal is provided by the Jagged protein, which activates Notch protein on the HSC surface (Figure 15.22C). A second signal comes from angiopoietin-1 on the osteoblasts, which activates the receptor tyrosine kinase Tie2 on the surface of the HSC (Arai et al. 2004). A third signal is from the Wnt pathway, localizing β-catenin into the nucleus. Wnt proteins are made by osteoblasts, and are probably made by differentiated osteocytes as well. This Wnt pathway seems critical for the self-renewal of the HSC (Reya et al. 2003).

The HSC cells give rise to **lineage-restricted stem cells** that produce blood cells and lymphocytes. A simplified depiction of the descendants of these cells is shown in Figure 15.23. While there are several disputes about the exact lineage and time of commitment to certain cell fates (see Adolfsson et al. 2005; Hock and Orkin 2005), Figure 15.23

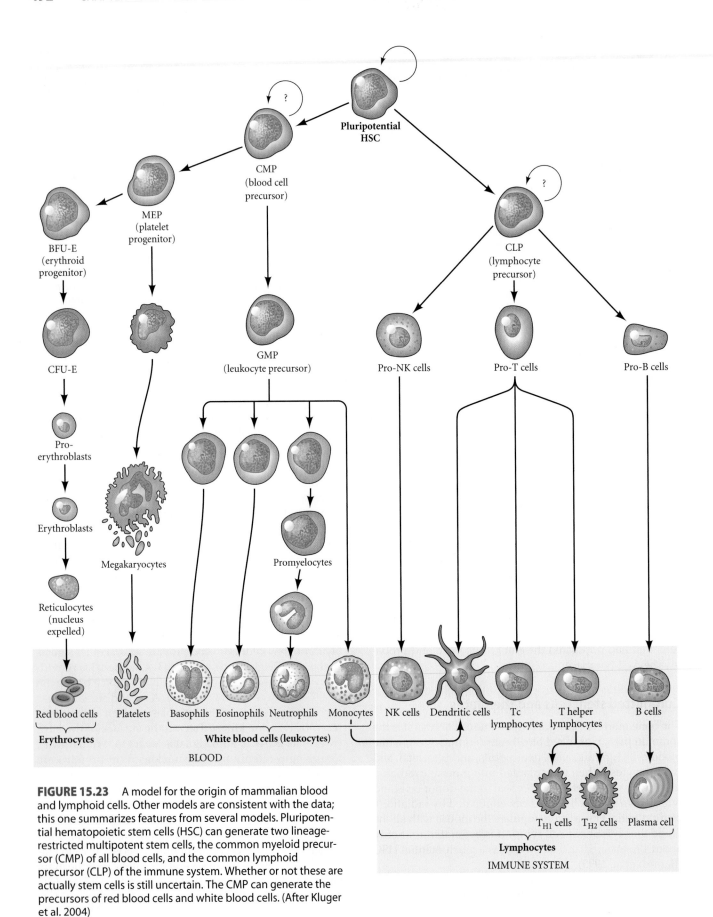

FIGURE 15.23   A model for the origin of mammalian blood and lymphoid cells. Other models are consistent with the data; this one summarizes features from several models. Pluripotential hematopoietic stem cells (HSC) can generate two lineage-restricted multipotent stem cells, the common myeloid precursor (CMP) of all blood cells, and the common lymphoid precursor (CLP) of the immune system. Whether or not these are actually stem cells is still uncertain. The CMP can generate the precursors of red blood cells and white blood cells. (After Kluger et al. 2004)

shows one plausible model. The HSC can give rise to the blood cell precursor ("common myeloid precursor cell," or CMP; sometimes called the CFU-S) or to the lymphocyte stem cell (CLP). These cells may also be stem cells, although this is not certain. The CMPs produce the megakaryocyte/erythroid precursor cell (MEP), which can generate either the red blood cell (erythrocyte) lineage or the platelet lineage. The CMP can also give rise to the granulocyte/monocyte precursor cell (GMP) which generates the basophils, eosinophils, neutrophils, and monocytes. Eventually, these cells produce **progenitor cells** that can divide but produce only *one* type of cell in addition to renewing itself. For instance, the **erythroid progenitor cell** (BFU-E) is a committed stem cell that can form only red blood cells. Its immediate progeny is capable of responding to the hormone **erythropoietin** to produce the first recognizable differentiated member of the erythrocyte lineage, the **proerythroblast**, a red blood cell precursor. This cell begins to make globin (Krantz and Goldwasser 1965). As the proerythroblast matures, it becomes an **erythroblast**, synthesizing enormous amounts of hemoglobin. Eventually, the mammalian erythroblast expels its nucleus, becoming a **reticulocyte**. Although reticulocytes, lacking a nucleus, can no longer synthesize globin mRNA, they can translate *existing* messages into globins. The final stage of differentiation is the **erythrocyte**, or mature red blood cell. In this cell, no division, RNA synthesis, or protein synthesis takes place. The DNA of the erythrocyte condenses and translates no further messages. Amphibians, fish, and birds retain the functionless nucleus; mammals extrude it from the cell.* The erythrocyte then leaves the bone marrow and enters the circulation, where it delivers oxygen to the body tissues.

Hematopoietic inductive microenvironments

Different paracrine factors are important in causing hematopoietic stem cells to differentiate along the particular pathways illustrated in Figure 15.23. The paracrine factors involved in blood cell and lymphocyte formation are called **cytokines**. Cytokines can be made by several cell types, but they are collected and concentrated by the extracellular matrix of the stromal (mesenchymal) cells at the sites of hematopoiesis (Hunt et al. 1987; Whitlock et al. 1987). For instance, granulocyte-macrophage colony-stimulating factor (GM-CSF) and the multilineage growth factor **interleukin-3 (IL3)** both bind to the heparan sulfate glycosaminoglycan of the bone marrow stroma (Gordon et al. 1987; Roberts et al. 1988). The extracellular matrix is then able to present these factors to the stem cells in concentrations high enough to bind to their receptors. At different stages of maturation, the stem cells become competent to respond to different factors. Early stem cells are able to respond to stem cell factor (SCF), while more mature stem cells lose the ability to respond to SCF, but acquire the ability to respond to IL3.

The developmental path taken by the descendant of a pluripotential HSC depends on which growth factors it meets, and is therefore determined by the stromal cells. Wolf and Trentin (1968) demonstrated that short-range interactions between stromal cells and stem cells determine the developmental fates of the stem cells' progeny. These investigators placed plugs of bone marrow in a spleen and then injected stem cells into it. Those CMPs that came to reside in the spleen formed colonies that were predominantly erythroid, whereas those that came to reside in the bone marrow formed colonies that were predominantly granulocytic. In fact, colonies that straddled the borders of the two tissue types were predominantly erythroid in the spleen and granulocytic in the marrow. Such regions of determination are referred to as **hematopoietic inductive microenvironments (HIMs)**. As expected, the HIMs induce different sets of transcription factors in these cells, and these transcription factors specify the fate of the particular cells (see Kluger et al. 2004).

Thus we see that the body never stops developing. We continue to make blood from precursors given us in our embryo. Moreover, blood cell differentiation, heart development, and vessel formation are now among the most important areas of medical science. As we will see in Chapter 21, the control of blood cell differentiation of stem cell proliferation is at the root of leukemia research, and regulating angiogenesis holds promise for preventing tumor formation. Congenital heart defects are one of the most prevalent types of birth defects, and cardiovascular disease is the most common cause of death in the industrialized nations. We ask a lot of our circulatory system. We ask a flawless flow of blood through the valves each second of our lives, we demand that the production of our blood cells be so precise that we get neither cancer nor anemia, and we demand fine-tuned coordination between our brain, heart, bone marrow, and hormones such that the cardiac muscle contractions can adapt to our physiological needs.

ENDODERM

The first of the embryonic endoderm's two major functions is to induce the formation of several mesodermal organs. As we have seen in this and earlier chapters, the endoderm is critical for instructing the formation of the notochord, the heart, the blood vessels, and even the mesodermal germ layer. The second function is to construct the linings

---

*In 1846, the young Joseph Leidy (then a struggling coroner, later the most famous biologist in America) was the first to use a microscope to solve a murder mystery. A man accused of killing a Philadelphia farmer had blood on his clothes and hatchet. The suspect claimed the blood was from chickens he had been slaughtering. Using his microscope, Leidy found no nuclei in these erythrocytes. Moreover, he found that if he let chick erythrocytes remain outside the body for hours, they did not lose their nuclei. Thus, he concluded that the blood stains could not have been chicken blood. The suspect subsequently confessed (Warren 1998).

of two tubes within the vertebrate body. The first, extending the length of the body, is the **digestive tube**. Buds from this tube form the liver, gallbladder, and pancreas. The second tube, the **respiratory tube**, forms as an outgrowth of the digestive tube and eventually bifurcates into the two lungs. The region of the digestive tube anterior to the point where the respiratory tube branches off is called the **pharynx**. Epithelial outpockets of the pharynx give rise to the tonsils, to the thyroid, thymus, and parathyroid glands, and eventually to the respiratory tube itself.

The respiratory and digestive tubes are both derived from the primitive gut (Figure 15.24). As the endoderm pinches in toward the center of the embryo, the foregut and hindgut regions are formed. At first, the oral end of the gut is blocked by a region of ectoderm called the oral plate, or stomodeum. Eventually (at about 22 days in human embryos), the stomodeum breaks, creating the oral opening of the digestive tube. The opening itself is lined by ectodermal cells. This arrangement creates an interesting situation, because the oral plate ectoderm is in contact

(A) 16 DAYS

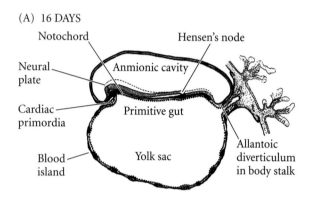

(B) 18 DAYS

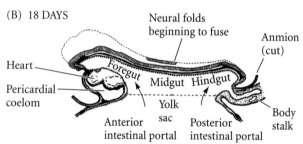

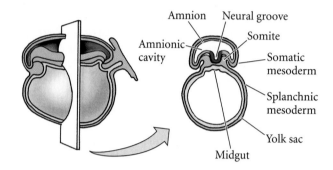

FIGURE 15.24 Formation of the human digestive system, depicted at about (A) 16 days, (B) 18 days, (C) 22 days, and (D) 28 days. (After Crelin 1961.)

(C) 22 DAYS

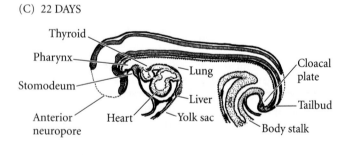

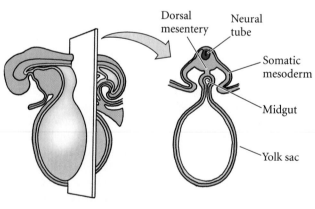

(D) 28 DAYS

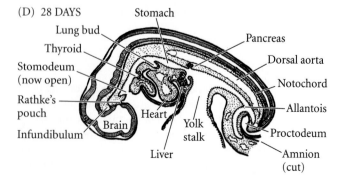

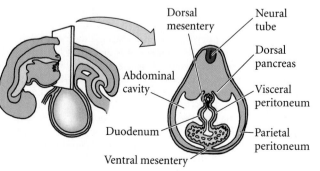

with the brain ectoderm, which has curved around toward the ventral portion of the embryo. These two ectodermal regions interact with each other, with the roof of the oral region forming Rathke's pouch and becoming the glandular portion of the pituitary gland. The neural tissue on the floor of the diencephalon gives rise to the infundibulum, which becomes the neural portion of the pituitary. Thus, the pituitary gland has a dual origin, which is reflected in its adult functions.

The Pharynx

The anterior endodermal portion of the digestive and respiratory tubes begins in the pharynx. Here, the mammalian embryo produces four pairs of **pharyngeal pouches**. Between these pouches are the **pharyngeal arches** (Figure 15.25). The first pair of pharyngeal pouches become the auditory cavities of the middle ear and the associated eustachian tubes. The second pair of pouches gives rise to the walls of the tonsils. The thymus is derived from the third pair of pharyngeal pouches; it will direct the differentiation of T lymphocytes during later stages of development. One pair of parathyroid glands is also derived from the third pair of pharyngeal pouches, while the other pair is derived from the fourth. In addition to these paired pouches, a small, central diverticulum is formed between the second pharyngeal pouches on the floor of the pharynx. This pocket of endoderm and mesenchyme will bud off from the pharynx and migrate down the neck to become the thyroid gland. The respiratory tube sprouts from the pharyngeal floor (between the fourth pair of pharyngeal pouches) to form the lungs, as we will see below.

The pharynx is where the endoderm meets the ectoderm, and the endoderm plays a critical role in determining which pouches develop. Sonic hedgehog from the endoderm appears to act as a survival factor, preventing apoptosis of the neural crest cells (Moore-Scott and Manley 2005). In zebrafish, genetic analysis combined with transplantation studies have shown that FGFs (mainly Fgf3 and Fgf8) from the ectoderm and mesoderm are important not only for the migration and survival of neural crest cells, but also for the formation of the pouches themselves. Mice with deficiencies of both *Fgf8* and *Fgf3* genes lacked all the pharyngeal pouches, even when endoderm was present. Instead of migrating laterally and ventrally to form pouches, the endoderm remained in the anterior pharynx and did not spread out (Crump et al. 2004).

The Digestive Tube and Its Derivatives

Posterior to the pharynx, the digestive tube constricts to form the esophagus, which is followed in sequence by the stomach, small intestine, and large intestine. The endodermal cells generate only the lining of the digestive tube and its glands; mesenchyme cells from the splanchnic portion of the lateral plate mesoderm will surround the tube to provide the muscles for peristalsis.

As Figure 15.26A shows, the stomach develops as a dilated region of the gut close to the pharynx. The intestines develop more caudally, and the connection between the intestine and yolk sac is eventually severed. The intestine originally ends in the endodermal cloaca; but after the cloaca separates into the bladder and rectal regions (see Chapter 14), the intestine joins with the rectum. At the cau-

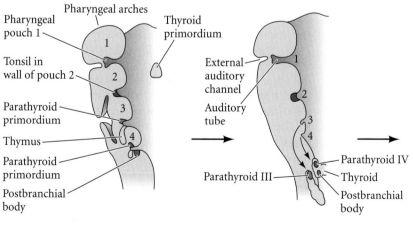

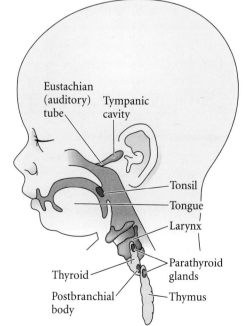

FIGURE 15.25 Formation of glandular primordia from the pharyngeal pouches. The end of each of the first pharyngeal pouches becomes the tympanic cavity of the middle ear and the Eustachian tube. The second pouches receive aggregates of lymphoid tissue and become the tonsils. The dorsal portion of the third pharyngeal pouch forms part of the parathyroid gland, while the ventral portion forms the thymus. Both migrate caudally and meet with the tissue from the fourth pharyngeal pouch to form the rest of the parathyroid and the postbranchial body. The thyroid, which had originated in the midline of the pharynx, also migrates caudally into the neck region. (After Carlson 1981.)

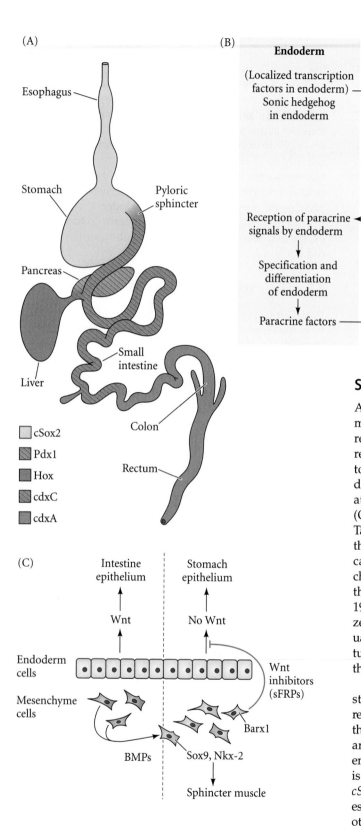

(A)

- Esophagus
- Stomach
- Pyloric sphincter
- Pancreas
- Liver
- Small intestine
- Colon
- Rectum

- cSox2
- Pdx1
- Hox
- cdxC
- cdxA

(B)

Endoderm	Mesoderm
(Localized transcription factors in endoderm) Sonic hedgehog in endoderm	→ Reception of Sonic hedgehog protein in mesoderm
	↓ Hox gene expression in mesoderm
	↓ Specification of mesoderm
Reception of paracrine signals by endoderm ←	BMPs, FGFs made by mesoderm
↓ Specification and differentiation of endoderm	
Paracrine factors →	Differentiation of mesoderm

FIGURE 15.26 Regional specification of the gut endoderm and splanchnic mesoderm through reciprocal interactions. (A) Regional transcription factors of the (mature) chick gut endoderm. These factors are seen prior to interactions with the mesoderm, but they are not stabilized. (B) Possible course of interactions between the endoderm and the splanchnic mesoderm. (C) Suggested mechanisms by which mesenchymal cells may induce gut endoderm to become either intestine or stomach. (A after Grapin-Botton et al. 2001.)

(C)

- Intestine epithelium
- Stomach epithelium
- Wnt
- No Wnt
- Endoderm cells
- Wnt inhibitors (sFRPs)
- Mesenchyme cells
- Barx1
- BMPs
- Sox9, Nkx-2
- Sphincter muscle

Specification of the gut tissue

As the endodermal tubes form in tetrapods, the endodermal epithelium is able to respond differently to different regionally specific mesodermal mesenchymes. These responses enable the digestive tube and respiratory tube to develop their very different structures. Thus, as the digestive tube meets different mesenchymes, it differentiates into esophagus, stomach, small intestine, and colon (Okada 1960; Gumpel-Pinot et al. 1978; Fukumachi and Takayama 1980; Kedinger et al. 1990). The endoderm and the splanchnic lateral plate mesoderm undergo a complicated set of interactions. However, recent research into chick and frog gut formation is leading to a consensus on the steps that regionalize the digestive tube (Ishii et al. 1997; Horb and Slack 2001; Matsushita et al. 2002). Even zebrafish gut, which is formed from assembling individual organ primordia (rather than arising from a single gut tube) shares the same molecular programs for generating the different gut tissues (Wallace and Pack 2003).

The gut appears to be regionally specified at a very early stage. Indeed, in the chick, the endoderm appears to be regionally specified even before it forms a tube. At this time, the endoderm expresses a set of transcription factors that are regionally specific. For instance, in the 1.5-day chick embryo, *CdxA*, a homologue of the *Drosophila Caudal* gene, is expressed in the region that will be the intestine, while *cSox2* is expressed in the precursors of the stomach and esophagus (Matsushita et al. 2002). Expression of these and other transcription factors will be retained throughout the development of the gut tube (Figure 15.26A).

This regional specification of the gut tube is still labile, however, and the boundaries between the regions are uncertain. The stabilization of these boundaries results from interaction with the mesoderm (Figure 15.26B). As the gut

dal end of the rectum, a depression forms where the endoderm meets the overlying ectoderm. Here, a thin **cloacal membrane** separates the two tissues. It eventually ruptures, forming the opening that will become the anus.

tube begins to form at the anterior and posterior ends, it induces the splanchnic mesoderm to become regionally specific. Roberts and colleagues (1988, 1995) have implicated Sonic hedgehog (Shh) in this specification. Early in development, *shh* expression is limited to the posterior endoderm of the chick hindgut and the pharynx. As the tubes extend toward the center of the embryo, the domains of *shh* expression increase, eventually extending throughout the gut endoderm. Shh is secreted in different concentrations at different sites, and its target appears to be the mesoderm surrounding the gut tube. The secretion of Shh by the hindgut endoderm induces a nested pattern of "posterior" Hox gene expression in the mesoderm. As in the vertebrae (see Chapter 11), the anterior borders of Hox gene expression delineate the morphological boundaries of the regions that will form the cloaca, large intestine, cecum, mid-cecum (at the midgut/hindgut border), and the posterior portion of the midgut (Roberts et al. 1995; Yokouchi et al. 1995). When Hox-expressing viruses cause the misexpression of these Hox genes in the mesoderm, the mesodermal cells alter the differentiation of the adjacent endoderm (Roberts et al. 1998). The Hox genes are thought to specify the mesoderm so that it can interact with the endodermal tube and specify its regions. Once the boundaries of the transcription factors are established, differentiation can begin. The regional differentiation of the mesoderm (into smooth muscle types) and the regional differentiation of the endoderm (into different functional units such as the stomach, duodenum, small intestine, etc.) are synchronized.

The molecular signals by which the mesenchyme influences the gut tube are just becoming known (Figure 15.26C). For instance, the mesenchyme lining the stomach-forming region of the gut tube expresses the Barx1 transcription factor, while the intestinal mesenchyme does not. Barx1 activates production of Frzb-like Wnt-blocking proteins (sFRP1 and sFRP2). Thus these Wnt antagonists block Wnt signaling in the vicinity of the stomach, but not around the intestine. Wnt signaling by the gut endoderm is critical in forming the intestine but is not used in stomach formation. (Indeed, *Barx1*-deficient mice do not develop stomachs and express intestinal markers in that tissue; Kim et al. 2005). Moreover, the intestinal mesenchyme secretes BMP4, which instructs the mesoderm anterior to it to express the Sox9 and Nkx2-5 transcription factors. These factors tell the mesoderm to become the muscles of the pyloric sphincter rather than becoming the smooth muscle that normally lines the stomach and intestine (Theodosiou and Tabin 2005).

Liver, pancreas, and gallbladder

The endoderm also forms the lining of three accessory organs that develop immediately caudal to the stomach. The **hepatic diverticulum** is a bud of endoderm that extends out from the foregut into the surrounding mesenchyme. The endoderm of this bud comes from two populations of cells—a lateral group that exclusively forms liver cells, and ventral medial endoderm cells that form several midgut regions, including the liver (Tremblay and Zaret 2005). The mesenchyme induces this endoderm to proliferate, branch, and form the glandular epithelium of the liver. A portion of the hepatic diverticulum (the region closest to the digestive tube) continues to function as the drainage duct of the liver, and a branch from this duct produces the gallbladder (Figure 15.27).

The pancreas develops from the fusion of distinct dorsal and ventral diverticula. Both of these primordia arise from the endoderm immediately caudal to the stomach, and as they grow, they come closer together and eventually fuse. In humans, only the ventral duct survives to carry digestive enzymes into the intestine. In other species (such as the dog), both the dorsal and ventral ducts empty into the intestine.

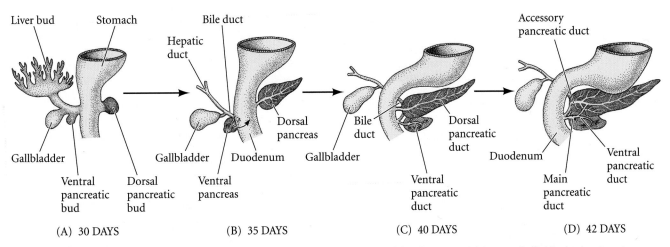

(A) 30 DAYS   (B) 35 DAYS   (C) 40 DAYS   (D) 42 DAYS

FIGURE 15.27 Pancreatic development in humans. (A) At 30 days, the ventral pancreatic bud is close to the liver primordium. (B) By 35 days, it begins migrating posteriorly, and (C) comes into contact with the dorsal pancreatic bud during the sixth week of development. (D) In most individuals, the dorsal pancreatic bud loses its duct into the duodenum; however, in about 10 percent of the population, the dual duct system persists. (After Langman 1981.)

Blood and Guts: The Specification of Liver and Pancreas

There is an intimate relationship between the splanchnic lateral plate mesoderm and the foregut endoderm. Just as the foregut endoderm is critical in specifying the cardiogenic mesoderm, the blood vessel endothelial cells induce the endodermal tube to produce the liver primordium and the pancreatic rudiments.

Liver formation

The expression of liver-specific genes (such as the genes for α-fetoprotein and albumin) can occur in any region of the gut tube that is exposed to cardiogenic mesoderm. However, this induction can occur only if the notochord is removed. If the notochord is placed by the portion of the endoderm normally induced by the cardiogenic mesoderm to become liver, the endoderm will not form liver (hepatic) tissue. Therefore, the developing heart appears to induce the liver to form, while the presence of the notochord inhibits liver formation (Figure 15.28). This induction is probably due to FGFs secreted by the developing heart cells (Le Douarin 1975; Gualdi et al. 1996; Jung et al. 1999).

However, Matsumoto and colleagues (2001) found that the heart is not the only mesodermal derivative needed to form the liver. Blood vessel endothelial cells are also critical. If endothelial cells are not present in the area around the hepatic region of the gut tube, the liver buds fail to form. This induction occurs even before the endothelial cells have formed tubes, so it does not have anything to do with getting nutrients or oxygen into the region. Thus, the heart endothelial cells have a developmental function in addition to their circulatory roles: they induce the formation of the liver bud by secreting FGFs.

However, in order to respond to the FGF signal, the endoderm has to become competent. This competence is given to the foregut endoderm by the **forkhead transcription factors**. Forkhead transcription factors Foxa1 and Foxa2 are required to open the chromatin surrounding the liver-specific genes. These proteins displace nucleosomes from the regulatory regions surrounding these genes and are required before the FGF signal is given (Lee et al. 2005). Mouse embryos lacking *Foxa1* and *Foxa2* expression in their endoderm fail to produce a liver bud or to express liver-specific enzymes.

Once the signal is given, other forkhead transcription factors, such as HNF4α, become critical. HNF4α is essential for the morphological and biochemical differentiation of the hepatic bud into liver tissue (Parviz et al. 2003). When conditional (floxed) mutants of HNF4α were made such that this factor was absent only in the developing liver, neither the tissue architecture, cellular structure, or liver-specific enzymes formed in the liver bud cells. Meanwhile, Odom and colleagues (2004) found that forkhead transcription factors were also critical for the differentiation of the endocrine islands of the pancreas. HNF4α bound to the regulatory regions of almost half the actively transcribed pancreas-specific genes in these tissues, including those involved in insulin secretion. A link between HNF4α mutations and late onset type 2 diabetes has been observed (see Kulkarni and Kahn 2004), confirming the importance of this transcription factor in pancreatic, as well as hepatic, development.

Pancreas formation

The formation of the pancreas may be the flip side of liver formation. Whereas the heart cells promote and the notochord prevents liver formation, the notochord may actively promote pancreas formation, while the heart may block the pancreas from forming. It seems that this particular region of the digestive tube has the ability to become either pancreas or liver. One set of conditions (presence of heart, absence of notochord) induces the liver, while another set of conditions (presence of notochord, absence of heart) causes the pancreas to form.

The notochord activates pancreas development by repressing *shh* expression in the endoderm (Apelqvist et al. 1997; Hebrok et al. 1998). (This was a surprising finding, since we saw in Chapter 13 that the notochord is a source of Shh protein and an inducer of further *shh* gene expression in ectodermal tissues.) Sonic hedgehog is expressed throughout the gut endoderm, *except* in the region that will form the pancreas. The notochord in this region of the embryo secretes Fgf2 and activin, which are able to downregulate *shh* expression. If *shh* is experimentally expressed in this region, the tissue reverts to being intestinal (Jonnson et al. 1994; Ahlgren et al. 1996; Offield et al. 1996).

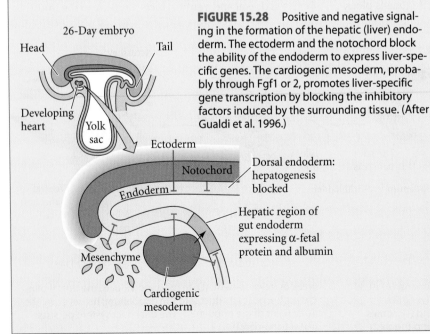

FIGURE 15.28 Positive and negative signaling in the formation of the hepatic (liver) endoderm. The ectoderm and the notochord block the ability of the endoderm to express liver-specific genes. The cardiogenic mesoderm, probably through Fgf1 or 2, promotes liver-specific gene transcription by blocking the inhibitory factors induced by the surrounding tissue. (After Gualdi et al. 1996.)

26-Day embryo

Head

Tail

Developing heart

Yolk sac

Ectoderm

Notochord

Endoderm

Dorsal endoderm: hepatogenesis blocked

Hepatic region of gut endoderm expressing α-fetal protein and albumin

Mesenchyme

Cardiogenic mesoderm

SIDELIGHTS & SPECULATIONS

The lack of Shh in this region of the gut seems to enable it to respond to signals coming from the blood vessel endothelium. Indeed, pancreatic development is initiated at precisely those three locations where the foregut endoderm contacts the endothelium of the major blood vessels. It is at these points—where the endodermal tube meets the aorta and the vitelline veins—that the transcription factor Pdx1 is expressed (Figure 15.29A–C; Lammert et al. 2001). If the blood vessels are removed from this area, the *Pdx1* expression regions fail to form, and the pancreatic endoderm fails to bud; if more blood vessels form in this area, more of the endodermal tube becomes pancreatic tissues.

The association of the pancreatic tissues with blood vessels is critical in the formation of the insulin-secreting cells of the pancreas. Pdx1 appears to act in concert with other such transcription factors to form the endocrine cells of the pancreas, the islets of Langerhans (Odom et al. 2004). The exocrine cells (which produce digestive enzymes such as chymotrypsin) and the endocrine cells appear to have the same progenitor (Fishman and Melton 2002). The islet cells secrete VEGF to attract blood vessels, and these vessels surround the developing islet (Figure 15.29D).

Pdx1 is exceptionally important in pancreatic development. This transcription factor elicits budding from the gut epithelium,

represses the expression of genes that are characteristic of other regions of the digestive tube, maintains the repression of *shh*, initiates (but does not complete) islet cell differentiation, and is necessary (but not sufficient) for insulin gene expression (Grapin-Botton et al. 2001). Using an inducible Cre-Lox system to mark the progeny of cells expressing either NGN3 or Pdx1, Gu and colleagues (2002) demonstrated that the NGN3-expressing cells are specifically islet cell progenitors, whereas the cells expressing Pdx1 give rise to all three types of pancreatic tissue (exocrine, endocrine, and duct cells) (Figure 15.29E). Moreover, Horb and colleagues (2003) have shown that Pdx1 can respecify developing liver tissue into pancreas. When *Xenopus* tadpoles were given a *pdx1* gene attached to a promoter active in liver cells, Pdx1 was made in the liver, and the liver was converted into a pancreas containing both exocrine and endocrine cells. Thus, Pdx1 appears to be the critical factor in distinguishing the liver from the pancreatic mode of development.

FIGURE 15.29 Induction of *pdx1* gene expression in the gut epithelium. (A) In the chick embryo, *pdx1* (purple) is expressed in the gut tube is induced by contact with the aorta and vitelline veins. The regions of *pdx1* expression create the dorsal and ventral anlagen of the pancreas. (B) In the mouse embryo, only the right vitelline vein survives, and it contacts the gut endothelium. Pdx expression is seen only on this side, and only one ventral pancreatic bud emerges. (C) In situ hybridization of *pdx1* mRNA in a section through the region of contact between the blood vessels and the gut tube. The regions of *pdx1* expression show as deep blue. (D) Blood vessels (stained red) direct islets (stained green with antibodies to insulin) to differentiate. The nuclei are stained deep blue. (E) Lineage of pancreatic cells. All pancreatic cells express *pdx1*. The NGN3-expressing cells form the endocrine lineages. (A–C from Lammert et al. 2001; D from Grapin-Botton et al. 2001; photographs courtesy of E. Lammert and D. A. Melton.)

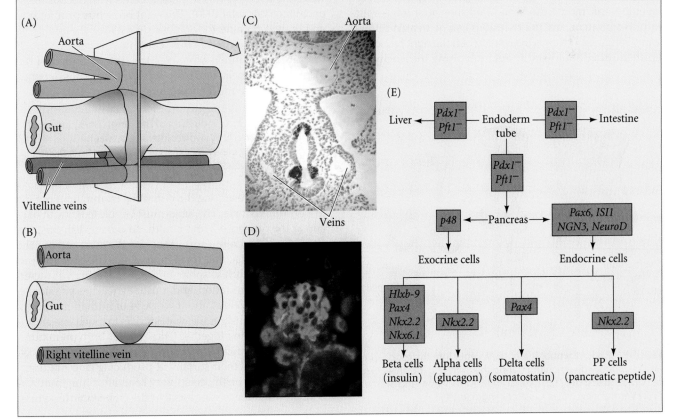

The Respiratory Tube

The lungs are a derivative of the digestive tube, even though they serve no role in digestion. In the center of the pharyngeal floor, between the fourth pair of pharyngeal pouches, the **laryngotracheal groove** extends ventrally (Figure 15.30). This groove then bifurcates into the branches that form the paired bronchi and lungs. The laryngotracheal endoderm becomes the lining of the trachea, the two bronchi, and the air sacs (alveoli) of the lungs. Sometimes this separation is not complete and a baby is born with a connection between the two tubes. This digestive and respiratory condition is called a **tracheal-esophageal fistula**, and must be surgically repaired so the baby can breathe and swallow properly.

The production of the laryngotracheal groove is correlated with the appearance of retinoic acid in the ventral mesoderm, and is probably induced by the same wave of RA that induces the posterior region of the heart. If RA is blocked, the foregut will not produce the lung bud (Desai et al. 2004). Retinoic acid probably induces the formation of Fgf10 by activating Tbx4 in the splanchnic mesoderm adjacent to the ventral foregut (Sakiyama et al. 2003).

As in the digestive tube, the regional specificity of the mesenchyme determines the differentiation of the developing respiratory tube. In the developing mammal, the respiratory epithelium responds in two distinct fashions. In the region of the neck, it grows straight, forming the trachea. After entering the thorax, it branches, forming the two bronchi and then the lungs. The respiratory epithelium of an embryonic mouse can be isolated soon after it has split into two bronchi, and the two sides can be treated differently. Figure 15.31 shows the result when the right bronchial epithelium was allowed to retain its lung mesenchyme

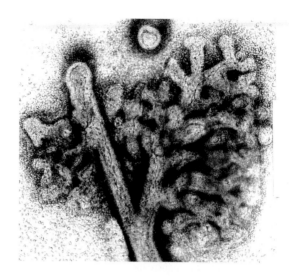

FIGURE 15.31 Ability of presumptive lung epithelium to differentiate with respect to the source of the inducing mesenchyme. After embryonic mouse lung epithelium had branched into two bronchi, the entire rudiment was excised and cultured. The right bronchus was left untouched, while the tip of the left bronchus was covered with tracheal mesenchyme. The tip of the right bronchus has formed the branches characteristic of the lung, whereas hardly any branching has occurred in the tip of the left bronchus. (From Wessells 1970; photograph courtesy of N. Wessells.)

while the left bronchus was surrounded with tracheal mesenchyme (Wessells 1970). The right bronchus proliferated and branched under the influence of the lung mesenchyme, whereas the left bronchus continued to grow in an unbranched manner. Moreover, the differentiation of the respiratory epithelia into trachea cells or lung cells depends on the mesenchyme it encounters (Shannon et al. 1998).

WEBSITE **15.2 Induction of the lung.** The induction of the lung involves interplay between FGFs and Shh. However, it appears to be different from the induction of either the pancreas or the liver.

The lungs are among the last of the mammalian organs to fully differentiate. The lungs must be able to draw in oxygen at the newborn's first breath. To accomplish this, the alveolar cells secrete a surfactant into the fluid bathing the lungs. This surfactant, consisting of specific proteins and phospholipids such as sphingomyelin and lecithin, is secreted very late in gestation, and it usually reaches physiologically useful levels at about week 34 of human gestation. The surfactant enables the alveolar cells to touch one another without sticking together. Thus, infants born prematurely often have difficulty breathing and have to be placed on respirators until their surfactant-producing cells mature.

Mammalian birth occurs very soon after lung maturation. New evidence suggests that the embryonic lung may

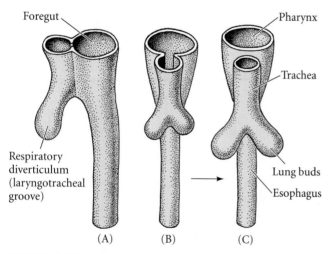

FIGURE 15.30 Partitioning of the foregut into the esophagus and respiratory diverticulum during the third and fourth weeks of human gestation. (A,B) Lateral and ventral views, end of week 3. (C) Ventral view, week 4. (After Langman 1981.)

Foregut

Respiratory diverticulum (laryngotracheal groove)

Pharynx

Trachea

Lung buds

Esophagus

(A) (B) (C)

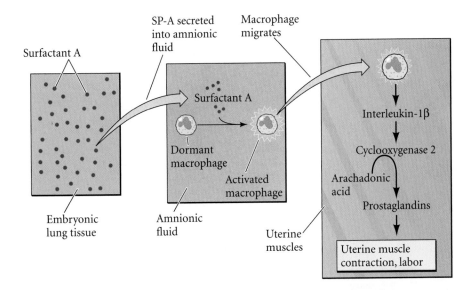

FIGURE 15.32 The immune system relays a signal from the embryonic lung. Surfactant protein-A (SP-A) activates macrophages in the amnionic fluid to migrate into the uterine muscles, where the macrophages secrete IL-1β. IL-1β stimulates production of cyclooxygenase-2, an enzyme that in turn triggers the production of the prostaglandin hormones responsible for initiating uterine muscle contractions and birth.

actually signal the mother to start delivery. Condon and colleagues (2004) have shown that surfactant-A (SP-A)—one of the final products produced by the embryonic mouse lung—activates macrophages in the amnionic fluid. These macrophages migrate from the amnion into the uterine muscle, where they produce immune system proteins such as interleukin-1β (IL-1β). IL-1β initiates the contractions of labor, both by activating cyclooxygenase-2 (which stimulates the production of the prostaglandins that contract the uterine muscle cells) and by antagonizing the progesterone receptor (Figure 15.32). Surfactant-stimulated macrophages injected into the uteri of female mice induced early labor.* Thus the signal for birth may be transmitted to the mother via her immune system.

The Extraembryonic Membranes

In reptiles, birds, and mammals, embryonic development took a new evolutionary direction—the amniote egg (see Figure 2.19). This remarkable adaptation, which allowed development to take place on dry land, evolved in the reptiles lineage and freed them to explore niches that were not close to water. This evolutionary adaptation is so significant and characteristic that reptiles, birds, and mammals are grouped together as the amniote vertebrates, or **amniotes**.

To cope with the challenges of terrestrial development, the amniote embryo produces four sets of extraembryonic membranes to mediate between it and the environment (see Chapter 11). The evolution of the placenta and internal development displaced the hard shelled egg in the

mammals, but the basic pattern of extraembryonic membranes remains the same.

In developing amniotes, there initially is no distinction between the embryonic and extraembryonic domains. However, as the body of the embryo takes shape, the epithelia at the border between the embryo and the extraembryonic domain divide unequally to create body folds that isolate the embryo from the yolk and delineate which areas are to be embryonic and which extraembryonic (Miller et al. 1994, 1999). These membranous folds are formed by the extension of ectodermal and endodermal epithelium underlain with lateral plate mesoderm. The combination of ectoderm and mesoderm, often referred to as the somatopleure (see Figure 15.3A), forms the amnion and chorion; the combination of endoderm and mesoderm—the splanchnopleure—forms the yolk sac and allantois. The endodermal or ectodermal tissue supplies functioning epithelial cells, and the mesoderm generates the essential blood supply to and from the epithelium. The formation of these folds can be followed in Figure 15.33.

The amnion and chorion

The first problem of a land-dwelling egg faces is desiccation. Embryonic cells would quickly dry out outside an aqueous environment. Such an environment is supplied by the **amnion**. The cells of this membrane secrete **amnionic fluid**; thus, embryogenesis still occurs in water.

The second problem of a terrestrial egg is gas exchange. This exchange is provided for by the **chorion**, the outermost extraembryonic membrane. In birds and reptiles, this membrane adheres to the shell, allowing the exchange of gases between the egg and the environment. In mammals, as we have seen, the chorion has developed into the **placenta**, which has evolved endocrine, immune, and nutritive functions in addition to those of respiration.

---

*IL-1β is also produced by macrophages when they attack bacterial infections, which may explain how uterine infections can cause premature labor.

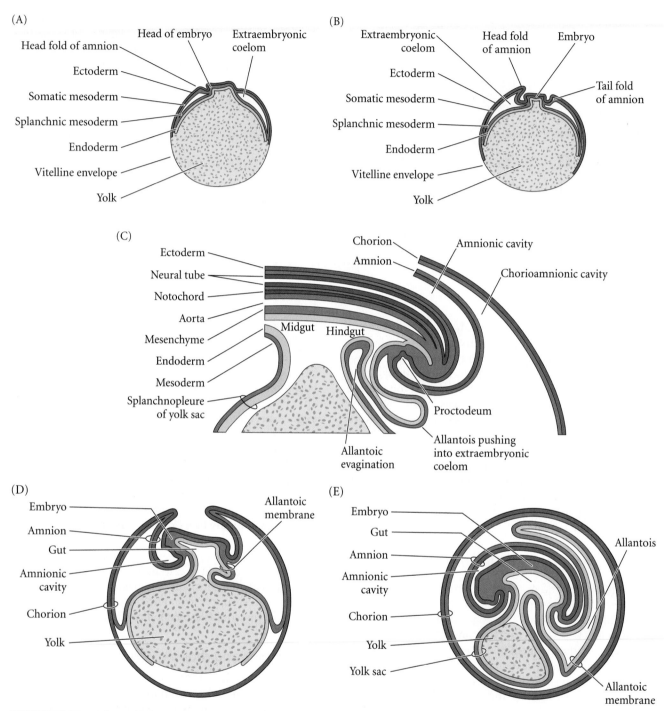

FIGURE 15.33 Schematic drawings of the extraembryonic membranes of the chick. The embryo is cut longitudinally, and the albumen and shell coatings are not shown. (A) A 2-day embryo. (B) A 3-day embryo. (C) Detailed schematic diagram of the caudal (hind) region of the chick embryo, showing the formation of the allantois. (D) A 5-day embryo. (E) A 9-day embryo. (After Carlson 1981.)

The allantois and yolk sac

The third problem for a terrestrial egg is waste disposal. The **allantois** stores urinary wastes and also helps mediate gas exchange. In reptiles and birds, the allantois becomes a large sac, as there is no other way to keep the toxic by-products of metabolism away from the developing embryo. In some amniote species, such as chickens, the mesodermal layer of the allantoic membrane reaches and fuses with the mesodermal layer of the chorion to create the **chorioallantoic membrane**. This extremely vascular envelope is crucial for chick development and is responsible for transporting calcium from the eggshell into the embryo for bone production (Tuan 1987). In mammals, the size of the allantois depends on how well nitrogenous wastes can be removed by the chorionic placenta. In humans (in which nitrogenous wastes are efficiently removed via the maternal circulation), the allantois is a vestigial sac. In pigs, however, the allantois is a large and important organ.

And finally, the land-dwelling egg must solve the problem of nutrition. The **yolk sac** is the first extraembryonic membrane to be formed, as it mediates nutrition in developing birds and reptiles. It is derived from splanchnopleur-al cells that grow over the yolk to enclose it. The yolk sac is connected to the midgut by an open tube, the **yolk duct**, so that the walls of the yolk sac and the walls of the gut are continuous. The blood vessels within the mesoderm of the splanchnopleure transport nutrients from the yolk into the body, for yolk is not taken directly into the body through the yolk duct. Rather, endodermal cells digest the protein in the yolk into soluble amino acids that can then be passed on to the blood vessels within the yolk sac. Other nutrients, including vitamins, ions, and fatty acids, are stored in the yolk sac and transported by the yolk sac blood vessels into the embryonic circulation. In these ways, the four extraembryonic membranes enable the amniote embryo to develop on land.

Snapshot Summary **Lateral Plate Mesoderm and Endoderm**

1. The lateral plate mesoderm splits into two layers. The dorsal layer is the somatic (parietal) mesoderm, which underlies the ectoderm and forms the somatopleure. The ventral layer is the splanchnic (visceral) mesoderm, which overlies the endoderm and forms the splanchnopleure.

2. The space between the two layers of lateral plate mesoderm forms the body cavity, or coelom.

3. The heart arises from splanchnic mesoderm on both sides of the body. This region of cells is called the cardiogenic mesoderm. The cardiogenic mesoderm is specified by BMPs in the absence of Wnt signals.

4. The Nkx2-5 and GATA transcription factors are important in committing the cardiogenic mesoderm to become heart cells. These cardiac precursor cells migrate from the sides to the midline of the embryo, in the neck region.

5. The cardiogenic mesoderm forms the endocardium (which is continuous with the blood vessels) and the myocardium (the muscular component of the heart).

6. The endocardial tubes form separately and then fuse. The looping of the heart transforms the original anterior-posterior polarity of the heart tube into a right-left polarity.

7. In mammals, fetal circulation differs dramatically from adult circulation. When the infant takes its first breath, changes in air pressure close the foramen ovale through which blood had passed from the right to the left atrium. At that time, the lungs, rather than the placenta, become the source of oxygen.

8. Blood vessel formation is constrained by physiological, evolutionary, and physical parameters. The subdividing of a large vessel into numerous smaller ones allows rapid transport of the blood to regions of gas and nutrient diffusion.

9. Blood vessels are constructed by two processes, vasculogenesis and angiogenesis. Vasculogenesis involves the condensing of splanchnic mesoderm cells to form blood islands. The outer cells of these islands become endothelial (blood vessel) cells. Angiogenesis involves the remodeling of existing blood vessels.

10. Numerous paracrine factors are essential in blood vessel formation. FGF2 is needed for specifying the angioblasts. VEGF is essential for the differentiation of the angioblasts. Angiopoietins allow the smooth muscle cells (and smooth muscle-like pericytes) to cover the vessels. Ephrin ligands and Eph receptor tyrosine kinases are critical for capillary bed formation.

11. The pluripotential hematopoietic stem cell generates other pluripotential stem cells, as well as lineage-restricted stem cells. It gives rise to both blood cells and lymphocytes.

12. In mammals, embryonic blood stem cells are provided by the blood islands near the yolk. The definitive adult blood stem cells come from the aorta-gonad-mesonephros region within the embryo.

13. The common myeloid precursor (CMP) is a blood stem cell that can generate the more committed stem cells for the different blood lineages. Hematopoietic inductive microenvironments determine the blood cell differentiation.

14. The endoderm constructs the digestive tube and the respiratory tube.

15. Four pairs of pharyngeal pouches become the endodermal lining of the Eustacian tubes, the tonsils, the thymus, and the parathyroid glands. The thyroid also forms in this region of endoderm.

16. The gut tissue forms by reciprocal interactions between the endoderm and the mesoderm. Sonic hedgehog from the endoderm appears to play a role in inducing a nested pattern of Hox gene expression in the mesoderm surrounding the gut. The regionalized mesoderm then instructs the endodermal tube to become the different organs of the digestive tract.

17. The endoderm helps specify the splanchnic mesoderm; the splanchnic mesoderm, especially the heart and the blood vessels, helps specify the endoderm.

18. The pancreas forms in a region of endoderm that lacks *sonic hedgehog* expression. The Pdx1 transcription factor is expressed in this region.

19. The endocrine and exocrine cells of the pancreas have a common origin. The NGN3 transcription factor probably decides endocrine fate.

20. The respiratory tube is derived as an outpocketing of the digestive tube. The regional specificity of the mesenchyme it meets determines whether the tube remains straight (as in the trachea) or branches (as in the alveoli).

21. The chorion and amnion are made by the somatopleure. In birds and reptiles, the chorion abuts the shell and allows for gas exchange. The amnion in birds, reptiles, and mammals bathes the embryo in amnionic fluid.

22. The yolk sac and allantois are derived from the splanchnopleure. The yolk sac (in birds and reptiles) allows yolk nutrients to pass into the blood. The allantois collects nitrogenous wastes.

For Further Reading

Complete bibliographical citations for all literature cited in this chapter can be found on the Vade Mecum CD that accompanies the book and at the free access website www.devbio.com.

Bruneau, B. G., M. Logan, N. Davis, T. Levi, C. J. Tabin, J. G. Seidman and C. E. Seidman. 1999. Chamber-specific cardiac expression of Tbx5 and heart defects in Holt-Oram syndrome. *Dev. Biol.* 211: 100–108.

Calvi, L. M. and 12 others. 2003. Osteoblastic cells regulate the haematopoietic stem cell niche. *Nature* 425: 841–846.

Fishman, M. P. and D. A. Melton. 2002. Pancreatic lineage analysis using a retroviral vector in embryonic mice demonstrates a common progenitor for endocrine and exocrine cells. *Int. J. Dev. Biol.* 46: 201–207.

Gothert, J. R., S. E. Gustin, M. A. Hall, B. Gottgens, D. J. Izon and C. G. Begley. 2005. In vivo fate-tracing studies using the Scl stem cell enhancer: embryonic hematopoietic stem cells significantly contribute to adult hematopoiesis. *Blood* 105: 2724–2732.

Jung, J., M. Goldfarb and K. S. Zaret. 1999. Initiation of mammalian liver development from endoderm by fibroblast growth factors. *Science* 284: 1998–2003.

Karkkainenen, M.J. and 11 others. 2004. Vascular endothelial growth factor C is required for sprouting of the first lymphatic vessels from embryonic aveins. *Nature Immunol.* 5: 74–80.

Kim, B.-M., G. Buchner, I. Miletich, P. T. Sharpe and R. A. Shivdasani. 2005. The stomach mesenchymal transcription factor Barx1 specifies gastric epithelial identity through inhibition of transient Wnt signaling. *Dev. Cell* 8: 611–622.

Odom, D. T. and 12 others. 2004. Control of pancreas and liver gene expression by HNF transcription factors. *Science* 303: 1378–1381.

Ottersbach, K. and E. Dzierak. 2005. The murine placenta contains hematopoietic stem cells within the vascular labyrinth region. *Dev. Cell* 8: 377–387.

Simões-Costa, M. S. and 8 others. 2005. The evolutionary origin of heart chambers. *Dev. Biol.* 277: 1–15.

Wang, H. U., Z.-F. Chen and D. J. Anderson. 1998. Molecular distinction and angiogenic interaction between embryonic arteries and veins revealed by ephrin-B2 and its receptor Eph-B4. *Cell* 93: 741–753.

Development of the Tetrapod Limb

16

CONSIDER YOUR LIMB. First, consider its polarity. It has fingers or toes at one end, a humerus or femur at the other. You won't find anyone with fingers in the middle of their arm. Consider also the differences between your hands and your feet. The differences are subtle, but very obvious. If your fingers were replaced by toes, you would know it. But also consider how *similar* the bones of your feet are to the bones of your hand; it's easy to see that they share a common pattern. The bones of any vertebrate limb, be it arm or leg, wing or flipper, consist of a proximal **stylopod** (humerus/femur) adjacent to the body wall; a **zeugopod** (radius-ulna/tibia-fibula) in the middle region; and a distal **autopod** (carpals-fingers/tarsals-toes) (Figure 16.1). Last, consider the growth of your limbs. Both hands are remarkably similar in size; so are both your feet. So after about 20 years of growth, each of your feet turns out, independently, to be the same length.

These commonplace phenomena present fascinating questions to the developmental biologist. How can growth be so precisely regulated? How is it that we have four limbs and not six or eight? How is it that the fingers form at one end of the limb and nowhere else? How is it that the little finger develops at one edge of the limb and the thumb at the other? How does the forelimb grow differently than the hindlimb?

All of these questions are really about pattern formation. **Pattern formation** is the set of processes by which embryonic cells form ordered spatial arrangements of differentiated tissues. Pattern formation is one of the most dramatic properties of a developing organism, and one that has provoked awe in scientists and laypeople alike. It is one thing to differentiate the chondrocytes and osteocytes that synthesize the cartilage and bone matrices, respectively; it is another thing to produce those cells in a temporal-spatial orientation that generates a functional bone. It is still another thing to make that bone a humerus and not a pelvis or a femur. The tissues of the finger—bone, cartilage, blood vessels, nerves, dermis, and epidermis—are the same in the toe and thigh; it's their arrangement that differs. The ability of limb cells to sense their relative positions and to differentiate with regard to those positions has been the subject of intense study, experimentation, and debate.

The vertebrate limb is an extremely complex organ with an asymmetrical arrangement of parts in all three dimensions. The first dimension is the **proximal** (close)-**distal** (far) **axis** (shoulder-finger or hip-toe). The bones of the limb are formed by endochondral ossification. They are initially cartilaginous, but eventually, most of the cartilage is replaced by bone. Somehow the limb cells develop differently at early stages of development (when they make the stylopod) then at later stages (when they make the autopod). The second dimension is the **anterior-posterior axis** (thumb-pinkie). Our little fingers or toes mark

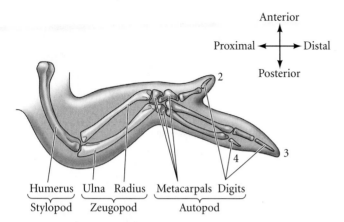

FIGURE 16.1 Skeletal pattern of the chick wing. According to convention, the digits are numbered 2, 3, and 4. The cartilage condensations forming the digits appear similar to those forming digits 2, 3, and 4 of mice and humans; however, new evidence (discussed later in the chapter) suggests that the correct designation may be 1, 2, and 3.

the *posterior* side, while our thumbs or big toes are at the *anterior* end. In humans, it is obvious that each hand develops as a mirror image of the other. One can imagine other arrangements to exist—such as the thumb developing on the left side of both hands—but these patterns do not occur. And finally, our limbs have a **dorsal-ventral axis**: the palm (ventral) is readily distinguishable from the knuckles (dorsal). In some manner, the complex three-dimensional pattern of the forelimb is routinely produced.

The fundamental problem of morphogenesis—how specific structures arise in particular places—is exemplified in limb development. Moreover, since the limbs, unlike the heart or brain, are not essential for embryonic or fetal life, one can experimentally remove or transplant parts of the developing limb, or create limb-specific mutants, without interfering with the vital processes of the organism. Such experiments have shown that the basic "morphogenetic rules" for forming a limb appear to be the same in all tetrapods. Grafted pieces of reptile or mammalian limb buds can direct the formation of chick limbs, and regions of frog limb buds can direct the patterning of salamander limbs (Fallon and Crosby 1977; Sessions et al. 1989; see Hinchliffe 1991). Moreover, as will be detailed in Chapter 18, *regenerating* salamander limbs appear to follow the same rules as developing limbs (Muneoka and Bryant 1982). But what are these morphogenetic rules?

The positional information needed to construct a limb has to function in a three-dimensional coordinate system.* During the past decade, particular proteins have been identified that play a role in the formation of each of these limb axes. The proximal-distal (shoulder-finger; hip-toe) axis

---

*Actually, it is a four-dimensional system, in which time is the fourth axis. Developmental biologists get used to seeing nature in four dimensions.

appears to be regulated by the fibroblast growth factor (FGF) family of proteins. The anterior-posterior (thumb-pinky) axis seems to be regulated by the Sonic hedgehog protein, and the dorsal-ventral (knuckle-palm) axis is regulated, at least in part, by Wnt7a. The interactions of these proteins mutually support one another and determine the differentiation of cell types.

WEBSITE 16.1 The mathematical modeling of limb development. The specification of the limb axes and the patterns in which limb outgrowth might occur were predicted mathematically before the actual molecular interactions were discovered.

Formation of the Limb Bud

Specification of the limb fields

Limbs do not form just anywhere along the body axis. Rather, there are discrete positions where limbs are generated. The mesodermal cells that give rise to a vertebrate limb can be identified by (1) removing certain groups of cells and observing that a limb does not develop in their absence (Detwiler 1918; Harrison 1918), (2) transplanting groups of cells to a new location and observing that they form a limb in this new place (Hertwig 1925), and (3) marking groups of cells with dyes or radioactive precursors and observing that their descendants partake in limb development (Rosenquist 1971).

Figure 16.2 shows the prospective forelimb area in the tail-bud stage of the salamander *Ambystoma maculatum*. The center of this disc normally gives rise to the limb itself. Adjacent to it are the cells that will form the peribrachial flank tissue and the shoulder girdle. However, if all these cells are extirpated from the embryo, a limb will still form, albeit somewhat later, from an additional ring of cells that surrounds this area (and which would not normally form a limb). If this

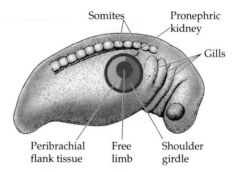

FIGURE 16.2 Prospective forelimb field of the salamander *Ambystoma maculatum*. The central area contains cells destined to form the limb *per se* (the free limb). The cells surrounding the free limb give rise to the peribrachial flank tissue and the shoulder girdle. The ring of cells outside these regions usually is not included the limb, but can form a limb if the more central tissues are extirpated. (After Stocum and Fallon 1983.)

(A)

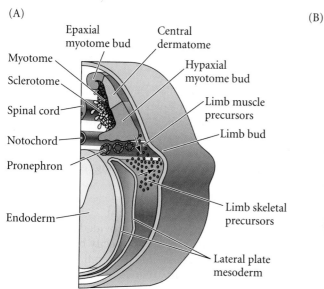

(B)

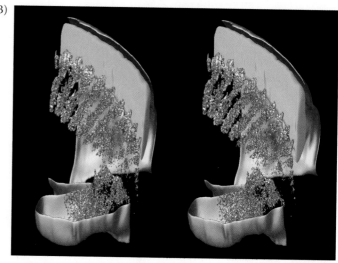

FIGURE 16.3 Proliferation of mesenchymal cells (arrows) from the somatic region of the lateral plate mesoderm causes the limb bud in the amphibian embryo to bulge outward. These cells generate the skeletal elements of the limb. Contributions of cells from the myotome provide the limb's musculature. (B) Entry of myotome cells (purple) into the limb bud. This computer stereogram was created from sections of an in situ hybridization to the myf5 mRNA found in developing muscle cells. If you can cross your eyes (or try focusing "past" the page, looking through it to your toes), the three-dimensionality of the stereogram will become apparent. (B courtesy of J. Streicher and G. Müller.)

last ring of cells is included in the extirpated tissue, no limb will develop. This larger region, representing all the cells in the area capable of forming a limb, is called the **limb field**. Limb development begins when mesenchyme cells proliferate from the somatic layer of the limb field lateral plate mesoderm (limb *skeletal* precursors) and from the somites (limb *muscle* precursors) at the same level. These mesenchymal cells accumulate under the ectodermal tissue to create a circular bulge called a **limb bud** (Figure 16.3A).

When it first forms, the limb field has the ability to regulate for lost or added parts. In the tailbud stage of *Ambystoma*, any half of the limb disc is able to generate an entire limb when grafted to a new site (Harrison 1918). This potential can also be shown by splitting the limb disc vertically into two or more segments and placing thin barriers between the segments to prevent their reunion. When this is done, each segment develops into a full limb. Thus, like an early sea urchin embryo, the limb field represents a "harmonious equipotential system" wherein a cell can be instructed to form any part of the limb.

In land vertebrates, there are only four limb buds per embryo, and they are always opposite each other with respect to the midline. Although the limbs of different vertebrates differ with respect to the somite level at which they arise, their position is constant with respect to the level of

Hox gene expression along the anterior-posterior axis (see Chapter 11). For instance, in fish (in which the pectoral and pelvic fins correspond to the anterior and posterior limbs, respectively), amphibians, birds, and mammals, the forelimb buds are found at the most anterior expression region of *Hoxc6*, the position of the first thoracic vertebra* (Oliver et al. 1988; Molven et al. 1990; Burke et al. 1995). The lateral plate mesoderm in the limb field is also special in that it induces myoblasts to migrate out from the somites and enter the limb bud to become the limb musculature (Figure 16.3B). No other region of the lateral plate mesoderm can do that (Hayashi and Ozawa 1995).

The regulative ability of the limb bud has recently been highlighted by a remarkable experiment of nature. In numerous ponds in the United States, multilegged frogs and salamanders have been found (Figure 16.4). The presence of these extra appendages has been linked to the infestation of the larval abdomen by parasitic trematode worms. The eggs of these worms apparently split the developing tadpole limb buds in several places, and the limb bud fragments develop as multiple limbs (Sessions and Ruth 1990; Sessions et al. 1999).

Induction of the early limb bud: Wnt proteins and fibroblast growth factors

Molecular studies on the earliest stages of limb formation have shown that the signal for limb bud formation comes from the lateral plate mesoderm cells that will become the prospective limb skeleton. These cells secrete the paracrine factor **Fgf10**. Fgf10 is capable of initiating the limb-form-

---

*Interestingly, Hox gene expression in at least some snakes (such as *Python*) creates a pattern in which each somite is specified to become a thoracic (ribbed) vertebra. The patterns of Hox gene expression associated with limb-forming regions are not seen (Cohn and Tickle 1999; see Chapter 23).

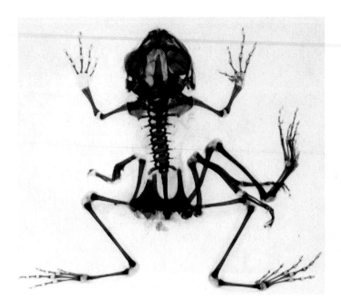

FIGURE 16.4 A multilimbed Pacific tree frog, *Hyla regilla*, is the result of infestation of the tadpole-stage developing limb buds by trematode cysts. The parasitic cysts apparently split the developing limb buds in several places, resulting in extra limbs. In this adult frog's skeleton, the cartilage is stained blue; the bones are stained red. (Photograph courtesy of S. Sessions.)

Specification of forelimb or hindlimb: Tbx4 and Tbx5

The limb buds must be specified as either forelimb or hindlimb. How are these two limb bud types distinguished? In 1996, Gibson-Brown and colleagues made a tantalizing correlation: The gene encoding the **Tbx5** transcription factor is transcribed in mouse forelimbs, while the gene encoding the closely related transcription factor **Tbx4** is expressed in hindlimbs.* Could these two transcription factors be involved in directing forelimb versus hindlimb specificity?

In 1998 and 1999, several laboratories (Ohuchi et al. 1998; Logan et al. 1998; Takeuchi et al. 1999; Rodriguez-Esteban et al. 1999, among others) provided gain-of-function evidence that Tbx4 and Tbx5 specify hindlimbs and forelimbs, respectively. First, if FGF-secreting beads were used to induce an ectopic limb between the chick hindlimb and forelimb buds, the type of limb produced was determined by the Tbx protein expressed. Those buds induced by placing FGF beads close to the hindlimb (opposite somite 25) expressed *Tbx4* and became hindlimbs. Those buds induced close to the forelimb (opposite somite 17) expressed *Tbx5* and developed as forelimbs (wings). Those buds induced in the center of the flank tissue expressed *Tbx5* in the anterior portion of the limb and *Tbx4* in the posterior portion of the limb. These limbs developed as chimeric structures, with the anterior resembling a forelimb and the posterior resembling a hindlimb (Figure 16.7). Moreover, when a chick embryo was made to express *Tbx4* throughout the flank tissue (by infecting the tissue with a virus that expressed *Tbx4*), limbs induced in the anterior region of the flank often became legs instead of wings.

---

*Tbx stands for T-box. The *T* (*Brachyury*) gene and its relatives have a sequence that encodes this specific DNA-binding domain. Humans heterozygous for the *TBX5* gene have Holt-Oram syndrome, characterized by abnormalities of the heart and upper limbs (Basson et al. 1996; Li et al. 1996).

ing interactions between the ectoderm and the mesoderm. If beads containing Fgf10 are placed ectopically beneath the flank ectoderm, extra limbs emerge (Figure 16.5; Ohuchi et al. 1997; Sekine et al. 1999). Fgf10 is originally produced throughout the lateral plate mesoderm, but immediately prior to limb formation, it becomes restricted to the regions of the lateral plate mesoderm where the limbs will form. This restriction in placement appears to be due to the actions of Wnt proteins (Wnt2b in the chick forelimb region; Wnt8c in the chick hindlimb region), which stabilize Fgf10 expression at these sites (Figure 16.6; Ohuchi et al. 1997; Kawakami et al. 2001).

(A)

(B)

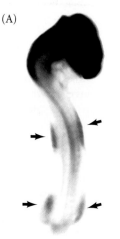
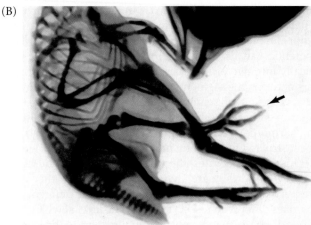

FIGURE 16.5 Fgf10 expression and action in the developing chick limb. (A) Fgf10 becomes expressed in the lateral plate mesoderm in precisely those positions (arrows) where limbs normally form. (B) When transgenic cells that secrete Fgf10 are placed in the flanks of a chick embryo, the Fgf10 can cause the formation of an ectopic limb (arrow). (From Ohuchi et al. 1997, courtesy of S. Noji.)

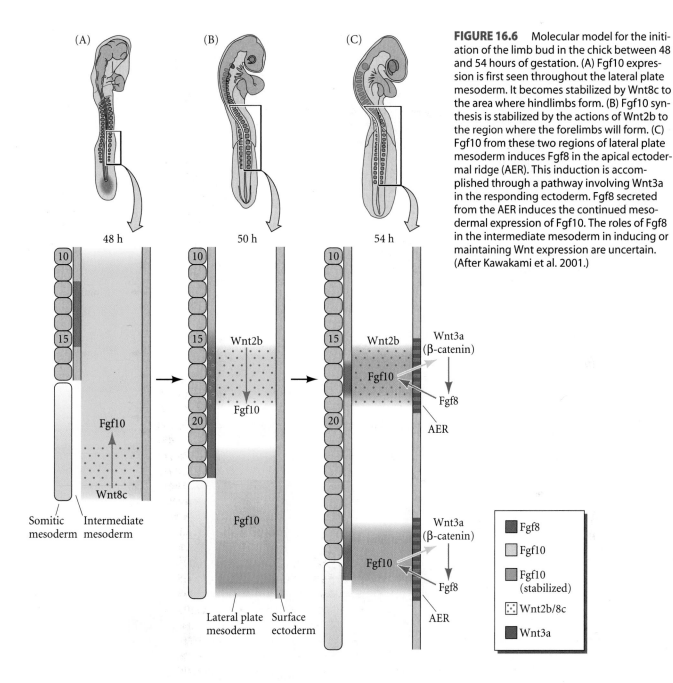

FIGURE 16.6 Molecular model for the initiation of the limb bud in the chick between 48 and 54 hours of gestation. (A) Fgf10 expression is first seen throughout the lateral plate mesoderm. It becomes stabilized by Wnt8c to the area where hindlimbs form. (B) Fgf10 synthesis is stabilized by the actions of Wnt2b to the region where the forelimbs will form. (C) Fgf10 from these two regions of lateral plate mesoderm induces Fgf8 in the apical ectodermal ridge (AER). This induction is accomplished through a pathway involving Wnt3a in the responding ectoderm. Fgf8 secreted from the AER induces the continued mesodermal expression of Fgf10. The roles of Fgf8 in the intermediate mesoderm in inducing or maintaining Wnt expression are uncertain. (After Kawakami et al. 2001.)

Thus, Tbx4 and Tbx5 appear to be critical in instructing the limbs to become hindlimbs and forelimbs, respectively.

However, the Tbx genes are not the complete story of limb specification. Transplantation experiments have shown that limb specification occurs earlier than the first appearance of Tbx4 or Tbx5 protein, and that there must be other genes that activate the *Tbx4* and *Tbx5* genes in their respective positions (Saito et al. 2002). Research is still going on to discover the mechanisms by which the forelimb becomes distinct from the hindlimb.

Generating the Proximal-Distal Axis of the Limb

The apical ectodermal ridge

When mesenchyme cells enter the limb field, they secrete Fgf10 that induces the overlying ectoderm to form a structure called the **apical ectodermal ridge** (**AER**) (Figure 16.8). The AER runs along the distal margin of the limb bud and

WEBSITE
16.2 Specifying forelimbs and hindlimbs. We know that Tbx4 and Tbx5 are central to limb type specification, we still need to know how these two transcription factors become expressed in their respective limb buds, and what they do to make the limbs different.

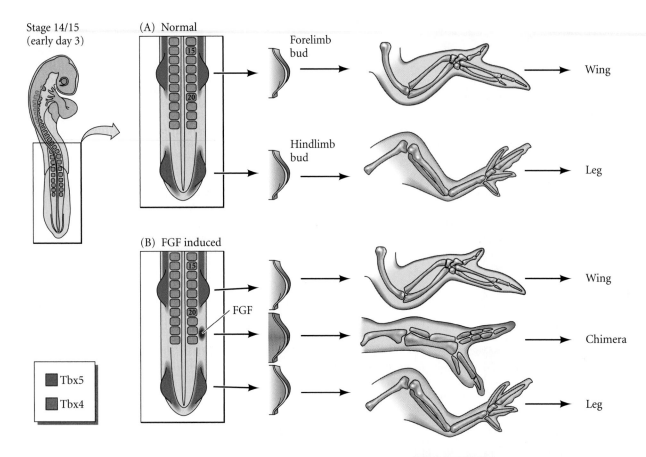

Stage 14/15
(early day 3)

(A) Normal

Forelimb
bud

Wing

Hindlimb
bud

Leg

(B) FGF induced

FGF

Wing

Chimera

Leg

■ Tbx5
■ Tbx4

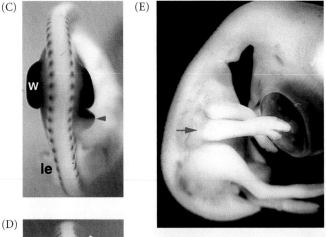

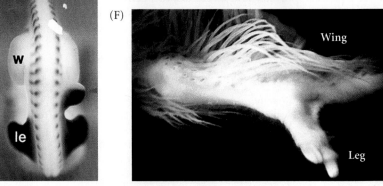

FIGURE 16.7 Specification of limb type by Tbx4 and Tbx5. (A) In situ hybridizations show that during normal chick development, Tbx5 (blue) is found in the anterior lateral plate mesoderm, while Tbx4 (red) is found in the posterior lateral plate mesoderm. Tbx5-containing limb buds produce wings, while Tbx4-containing limb buds generate legs. (B) If a new limb bud is induced with an FGF-secreting bead, the type of limb formed depends on which Tbx gene is expressed in the limb bud. If placed between the regions of *Tbx4* and *Tbx5* expression, the bead will induce the expression of *Tbx4* posteriorly and *Tbx5* anteriorly. The resulting limb bud will also express *Tbx5* anteriorly and *Tbx4* posteriorly and will generate a chimeric limb. (C) Expression of *Tbx5* in the forelimb (w, wing) buds and in the anterior portion of a limb bud induced by an FGF-secreting bead. (The somite level can be determined by staining for *Mrf4* mRNA, which is localized to the myotomes.) (D) Expression of *Tbx4* in the hindlimb (le, leg) buds and in the posterior portion of an FGF-induced limb bud. (E,F) A chimeric limb induced by an FGF bead contains anterior wing structures and posterior leg structures. (F) is at a later developmental stage, after feathers form. (A,B after Ohuchi et al. 1998, Ohuchi and Noji 1999; C–F, photographs courtesy of S. Noji.)

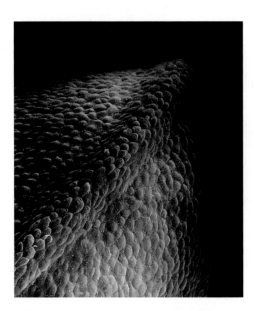

FIGURE 16.8 Scanning electron micrograph of an early chick forelimb bud, with its apical ectodermal ridge in the foreground. (Photograph courtesy of K. W. Tosney.)

1. If the AER is removed at any time during limb development, further development of distal limb skeletal elements ceases.
2. If an extra AER is grafted onto an existing limb bud, supernumerary structures are formed, usually toward the distal end of the limb.
3. If leg mesenchyme is placed directly beneath the wing AER, distal hindlimb structures (toes) develop at the end of the limb. (However, if this mesenchyme is placed farther from the AER, the hindlimb [leg] mesenchyme becomes integrated into wing structures.)
4. If limb mesenchyme is replaced by nonlimb mesenchyme beneath the AER, the AER regresses and limb development ceases.

Thus, although the mesenchyme cells induce and sustain the AER and determine the type of limb to be formed, the AER is responsible for the sustained outgrowth and development of the limb (Zwilling 1955; Saunders et al. 1957; Saunders 1972; Krabbenhoft and Fallon 1989). The AER keeps the mesenchyme cells directly beneath it in a state of mitotic proliferation and prevents them from forming cartilage. Hurle and co-workers (1989) found that if they

will become a major signaling center for the developing limb (Saunders 1948; Kieny 1960; Saunders and Reuss 1974). Its roles include (1) maintaining the mesenchyme beneath it in a plastic, proliferating state that enables the linear (proximal-distal, shoulder-finger) growth of the limb; (2) maintaining the expression of those molecules that generate the anterior-posterior (thumb-pinkie) axis; and (3) interacting with the proteins specifying the anterior-posterior and dorsal-ventral (knuckle-palm) axes so that each cell is given instructions on how to differentiate.

The proximal-distal growth and differentiation of the limb bud is made possible by a series of interactions between the AER and the limb bud mesenchyme directly (200 µm) beneath it. This distal mesenchyme is often called the **progress zone (PZ)** mesenchyme, since its proliferative activity extends the limb bud (Harrison 1918; Saunders 1948). These interactions were demonstrated by the results of several experiments on chick embryos (Figure 16.9):

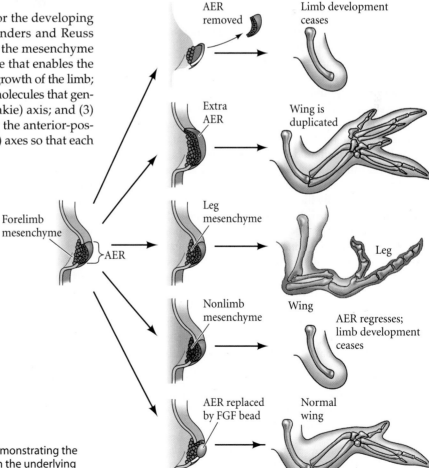

FIGURE 16.9 Summary of experiments demonstrating the effect of the apical ectodermal ridge (AER) on the underlying mesenchyme. (After Wessells 1977.)

FIGURE 16.10   Fgf8 in the apical ectodermal ridge. (A) In situ hybridization showing expression of *fgf8* message in the ectoderm as the limb bud begins to form. (B) Expression of *fgf8* RNA in the AER, the source of mitotic signals to the underlying mesoderm. (C) In the normal 3-day chick embryo, *fgf8* is expressed in the apical ectodermal ridge of both the forelimb and hindlimb buds. It is also expressed in several other places in the embryo, including the pharyngeal arches. (A,B courtesy of J. C. Izpisúa-Belmonte; C courtesy of A. López-Martínez and J. F. Fallon.)

(A)   (B)   (C)

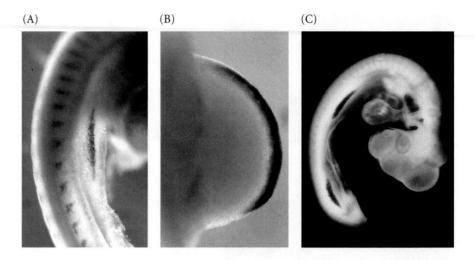

cut away a small portion of the AER in a region that would normally fall between the digits of the chick leg, an extra digit emerged at that position.*

FGFs in the induction and maintenance of the AER

FGFs are critical for the induction, maintenance, and function of the apical ectodermal ridge. The principal protein inducing the AER is Fgf10 (Xu et al. 1998; Yonei-Tamura et al. 1999). Fgf10 is capable of inducing the AER in the competent ectoderm between the dorsal and ventral sides of the embryo. The boundary where dorsal and ventral ectoderm meet is critical to the placement of the AER. In mutants in which the limb bud is dorsalized and there is no dorsal-ventral junction (as in the chick mutant *limbless*), the AER fails to form and limb development ceases (Carrington and Fallon 1988; Laufer et al. 1997; Rodriguez-Esteban et al. 1997; Tanaka et al. 1997).

Fgf10 produced by the underlying mesenchyme induces the formation of the AER, and at the same time induces the AER to synthesize and secrete Fgf8 (Figure 16.10). Fgf8 stimulates mitosis in the mesenchyme cells beneath it and causes these cells to keep expressing Fgf10. Thus, a positive feedback loop is established: Fgf10 in the mesenchyme induces Fgf8 in the AER, and Fgf8 in the AER maintains Fgf10 expression in the mesenchyme. Each FGF activates the synthesis of the other (see Figure 16.6; Mahmood et al. 1995; Crossley et al. 1996; Vogel et al. 1996; Ohuchi et al. 1997; Kawakami et al. 2001). The continued expression of

---

*When referring to the hand, one has an orderly set of names to specify each digit (*digitus pollicis, d. indicis, d. medius, d. annularis,* and *d. minimus,* respectively, from thumb to little finger). No such nomenclature exists for the pedal digits, but the plan proposed by Phillips (1991) has much merit. The pedal digits, from *hallux* to small toe, would be named *porcellus fori, p. domi, p. carnivorus, p. non voratus,* and *p. plorans domi,* respectively.

WEBSITE   **16.3 Induction of the AER.** The induction of the AER is a complex event involving the interaction between the dorsal and ventral compartments of the ectoderm. The Notch pathway may be critical in this process. Misexpression of the genes in this pathway can cause absence or duplication of limbs.

these FGFs maintains mitosis in the mesenchyme beneath the AER.

Specifying the limb mesoderm: Determining the proximal-distal polarity of the limb

In 1948, John Saunders made a simple and profound observation: if the AER is removed from an early-stage wing bud, only a humerus forms. If the AER is removed slightly later, humerus, radius, and ulna form (Figure 16.11; Saunders 1948; Iten 1982; Rowe and Fallon 1982). Explaining how this happens has not been easy. First it had to be determined whether the positional information for proximal-distal polarity resided in the AER or in the progress zone mesenchyme. Through a series of reciprocal transplantations, this specificity was found to reside in the mesenchyme. If the *AER* had provided the positional information—somehow instructing the undifferentiated mesoderm beneath it as to what structures to make—then older AERs combined with younger mesoderm should have produced limbs with deletions in the middle, while younger AERs combined with older mesenchyme should have produced duplications of structures. This was not found to be the case; rather, normal limbs form in both experiments (Rubin and Saunders 1972). But when the entire progress zone (including both the mesoderm and the AER) from an early embryo is placed on the limb bud of a later-stage embryo, new proximal structures are produced beyond those already present. Conversely, when old progress zones are added to young limb buds, distal structures develop imme-

(A)

FIGURE 16.11 The AER is necessary for wing development. (A) Skeleton of a normal chick wing (dorsal view). (B) Dorsal views of skeletal patterns after removal of the entire AER from the right wing bud of chick embryos at various stages. (From Iten 1982, photographs courtesy of L. Iten.)

(B)

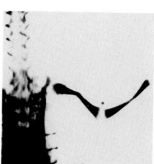

diately, so that digits are seen to emerge from the humerus without an intervening ulna and radius (Figure 16.12; Summerbell and Lewis 1975).

But how does the mesenchyme specify the proximal-distal axis? Two major models have been proposed to account for this phenomenon. One emphasizes the dimension of *time*, while the other emphasizes the dimension of *space*.

(A)

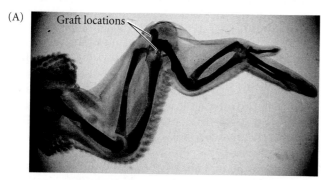

(B)

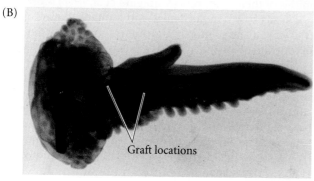

The **progress zone model** postulates that each mesoderm cell is specified by the amount of time it spends dividing in the progress zone. The longer a cell spends in the progress zone, the more mitoses it achieves, and the more distal its specification becomes. Since the progress zone remains the same size, cells must exit the PZ constantly as the limb grows; once they leave, they can form cartilage. Thus, the first cells leaving the progress zone become the stylopod, while the last cells to leave it become the autopod (Figure 16.13A; Summerbell 1974; Wolpert et al. 1979). According to this model, removing the AER means that the progress zone cells can no longer divide and be further specified, so only the proximal structures form.

Although the progress zone model has been the prevailing means of explaining limb polarity, it does not explain recent data from experiments in which the progress zone mesenchyme was prevented from dividing in mice lacking both *fgf8* and *fgf4* (both of which are expressed in the AER). In the limbs of these mice, the proximal elements were altogether missing, while the distal elements were present and often normal. This result could not be explained by a model wherein the specification of distal elements depended upon the number of mitoses in the

FIGURE 16.12 Control of proximal-distal specification by the progress zone mesenchyme. (A) Extra set of ulna and radius formed when an early wing-bud progress zone was transplanted to a late wing bud that had already formed ulna and radius. (B) Lack of intermediate structures seen when a late wing-bud progress zone was transplanted to an early wing bud. (From Summerbell and Lewis 1975, photographs courtesy of D. Summerbell.)

FIGURE 16.13 Two models for the mesodermal specification of the proximal-distal axis of the limb. (A) Progress zone model, wherein the length of time a mesodermal cell remains in the progress zone specifies its position. (B) Early allocation and progenitor expansion model, wherein the territories of the limb bud are established very early and the cells grow in each of these areas.

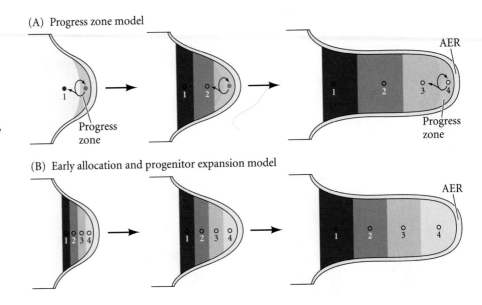

(A) Progress zone model

(B) Early allocation and progenitor expansion model

progress zone (Sun et al. 2002). Moreover, when individual early PZ cells were marked with a dye, the progeny of each cell were limited to a particular segment of the limb. The tip of the early limb bud provided cells only to the autopod, for instance (Dudley et al. 2002). This finding suggested that specification was early and that the time spent in the progress zone did not specify cell position. Moreover, cell division is seen throughout the limb bud, not solely in the progress zone.

Gail Martin and Cliff Tabin have proposed an **early allocation and progenitor expansion model** as an alternative to the progress zone model. Here, the cells of the entire early limb bud are already specified; subsequent cell divisions simply expand these cell populations (Figure 16.13B). The effects of AER removal are explained by the observation of Rowe et al. (1982) that when the AER is taken off the limb bud, there is apoptosis for approximately 200 μm. If one removes the AER of an early limb bud, before the mesenchymal cell populations have expanded, the cells specified to be zygopod and autopod die, and only the cells already specified to be stylopod remain. In later embryonic stages, the limb bud grows and the 200 μm region of apoptosis merely deletes the autopod. Research is presently underway to distinguish the contributions of these mechanisms to anterior-posterior axis specification in the limb.

The Hox specification code for the limb

The products of the Hox genes have already played a role in specifying the location where the limbs will form. Now they will play a second role in specifying whether a particular mesenchymal cell will become stylopod, zeugopod, or autopod.

The 5' (*AbdB*-like) portions (paralogues 9–13) of the *HoxA* and *HoxD* gene complexes appear to be active in the forelimb buds of mice. Based on the expression patterns of

these genes, and on naturally occurring and gene knock-out mutations, Mario Capecchi's laboratory (Davis et al. 1995) proposed a model wherein these Hox genes specify the identity of a limb region (Figure 16.14A,B). Here, *Hox9* and *Hox10* paralogues specify the stylopod, *Hox11* paralogues specify the zeugopods, and *Hox12* and *Hox13* paralogues specify the autopods. This scenario has been confirmed by numerous recent experiments. For instance, when Wellik and Capecchi (2003) knocked out all six *Hox10* paralogues (*Hox10aaccdd*) from mouse embryos, the resulting mice not only had severe axial skeletal defects (see Chapter 14), they also had no femur or patella. (These mice did have humeruses, because the *Hox9* paralogues are expressed in the *forelimb* stylopod but not in the *hindlimb* stylopod.) When all six *Hox11* paralogues were knocked out, the resulting hindlimbs had femurs but neither tibias or fibulas (and the forelimbs lacked the ulnas and radii). Thus the *Hox11* knockouts got rid of the zeugopods (Figure 16.14C). Similarly, knocking out all four *Hoxa13* and *Hoxd13* loci resulted in loss of the autopod (Fromental-Ramain et al. 1996). Humans homozygous for a *HOXD13* mutation show abnormalities of the hands and feet wherein the digits fuse (Figure 16.14D), and human patients with homozygous mutant alleles of *HOXA13* also have deformities of their autopods (Muragaki et al. 1996; Mortlock and Innis 1997). In both mice and humans, the autopod (the most distal portion of the limb) is affected by the loss of function of the most 5' Hox genes.

The mechanism by which Hox genes can specify the proximal-distal axis is not yet understood, but one clue comes from the analysis of chicken *Hoxa13*. Ectopic expression of this gene (which is usually expressed in the distal ends of developing chick limbs) appears to make the cells expressing it "stickier." This property might cause the cartilaginous nodules to condense in specific ways (Yokouchi et al. 1995; Newman 1996).

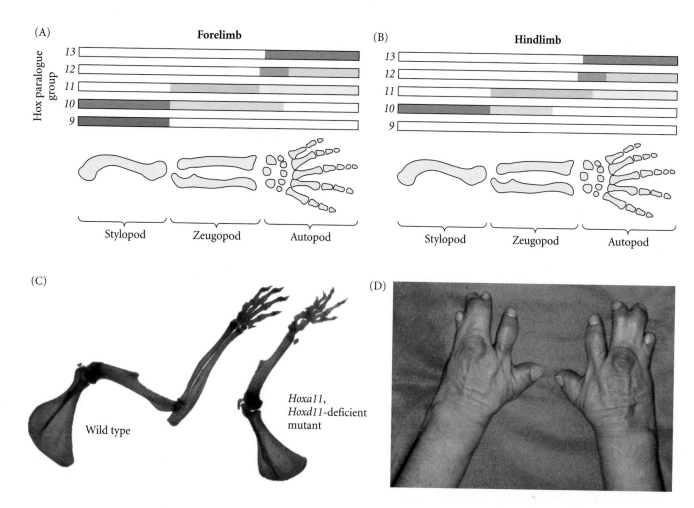

FIGURE 16.14 Deletion of limb bone elements by the deletion of paralogous Hox genes. (A) 5' Hox gene patterning of the forelimb. *Hox9* and *Hox10* specify the humerus, and *Hox10* paralogues are expressed to a lesser extent in the zeugopod. *Hox11* paralogues are chiefly responsible for patterning the zeugopod; Hox 12 paralogues function in the wrist, with a little patterning of the autopod; and the Hox13 paralogue group functions in the autopod. (B) A similar but somewhat differing pattern is seen in the hindlimb. (C) Wild-type mouse forelimb and the forelimb of mouse made doubly mutant such that it lacked functional *Hoxa11* and *Hoxd11* genes. The ulna and radius are absent in the mutant. (D) Human polysyndactyly ("many fingers joined together") syndrome resulting from homozygosity for a mutation at the *HOXD13* loci. This human syndrome also includes malformations of the urogenital system, which also expresses *HOXD13*. (A,B after Wellik and Capecchi 2003; C from Davis et al. 1995, photographs courtesy of M. Capecchi; D from Muragaki et al. 1996, photograph courtesy of B. Olsen.)

Specification of the Anterior-Posterior Limb Axis

The zone of polarizing activity

The specification of the anterior-posterior axis of the limb is the earliest change from the pluripotent condition. In the chick, this axis is specified shortly before a limb bud is recognizable. Hamburger (1938) showed that as early as the 16-somite stage, prospective wing mesoderm transplanted to the flank area develops into a limb with the anterior-posterior and dorsal-ventral polarities of the donor graft, not those of the host tissue.

Several experiments (Saunders and Gasseling 1968; Tickle et al. 1975) suggest that the anterior-posterior axis is specified by a small block of mesodermal tissue near the posterior junction of the young limb bud and the body wall. When tissue from this region is taken from a young limb bud and transplanted to a position on the anterior side of another limb bud, the number of digits of the resulting wing is doubled (Figure 16.15). Moreover, the structures of the extra set of digits are mirror images of the normally produced structures. The polarity has been maintained, but the information is now coming from both an anterior and a posterior direction. Thus this region of the mesoderm has been called the **zone of polarizing activity (ZPA)**.

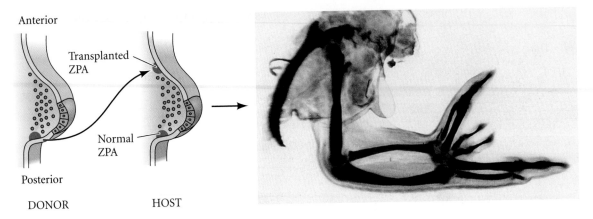

FIGURE 16.15 When a ZPA is grafted to anterior limb bud mesoderm, duplicated digits emerge as a mirror image of the normal digits. (From Honig and Summerbell 1985, photograph courtesy of D. Summerbell.)

SONIC HEDGEHOG DEFINES THE ZPA The search for the molecule(s) conferring this polarizing activity on the ZPA became one of the most intensive quests in developmental biology. In 1993, Riddle and colleagues showed by in situ hybridization that *sonic hedgehog* (*shh*), a vertebrate homologue of the *Drosophila hedgehog* gene, was expressed specifically in that region of the limb bud known to be the ZPA (Figure 16.16A).

As evidence that this association between the ZPA and sonic hedgehog was more than just a correlation, Riddle and co-workers (1993) demonstrated that the secretion of Sonic hedgehog protein is sufficient for polarizing activity. They transfected embryonic chick fibroblasts (which normally would never synthesize Shh) with a viral vector containing the *shh* gene (Figure 16.16B). The gene became expressed, translated, and secreted in these fibroblasts, which were then inserted under the anterior ectoderm of

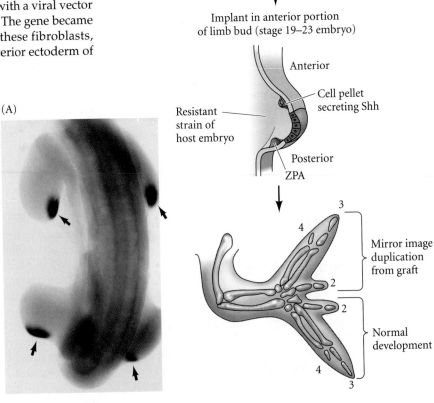

FIGURE 16.16 Sonic hedgehog protein is expressed in the ZPA. (A) In situ hybridization showing the sites of *sonic hedgehog* expression (arrows) in the posterior mesoderm of the chick limb buds. These are precisely the regions that transplantation experiments defined as the ZPA. (B) Assay for polarizing activity of Sonic hedgehog protein. The *shh* gene was inserted adjacent to an active promoter of a chicken virus, and the recombinant virus was placed into cultured chick embryo fibroblast cells. The virally transfected cells were pelleted and implanted in the anterior margin of a limb bud of a chick embryo. The resulting limbs produced mirror-image digits, showing that the secreted protein had polarizing activity. (A, photograph courtesy of R. D. Riddle; B after Riddle et al. 1993.)

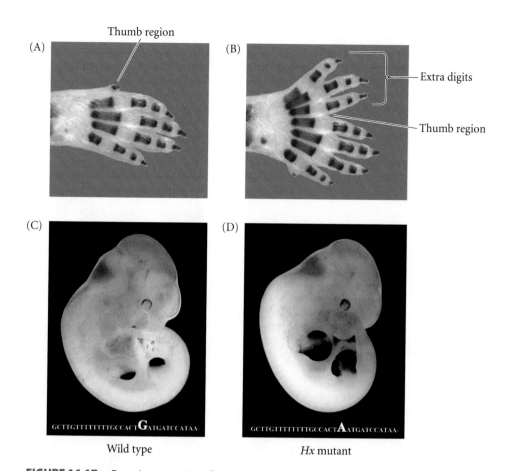

Thumb region

(A)

(B)

Extra digits

Thumb region

(C)

(D)

GCTTGTTTTTTTTGCCACT**G**ATGATCCATAA-

GCTTGTTTTTTTTGCCACT**A**ATGATCCATAA-

Wild type   *Hx* mutant

FIGURE 16.17   Ectopic expression of mouse *sonic hedgehog* in the anterior limb causes extra digit formation. (A) Wild-type mouse paw. The bones are stained with Alizarin red. (B) *Hx* (hemimelic extra-toes) mutant mouse paws, showing the extra digits associated with the anterior ("thumb") region. (The small extra nodule of posterior bone is peculiar to the *Hx* phenotype on the genetic background used and is not seen on other genetic backgrounds.) (C) Reporter constructs from wild-type *shh* limb enhancer direct transcription solely to the posterior part of each limb bud. (D) Reporter constructs from the *Hx* mutant direct transcription to both the anterior and posterior regions of each limb bud. The wild-type and mutant *shh* limb-specific enhancer region DNA sequences are shown below and highlight the single G-to-A nucleotide substitution that differentiates the two. (From Maas and Fallon 2005, photographs courtesy of B. Robert, Y. Lallemand, S. A. Maas and J. F. Fallon.)

an early chick limb bud. Mirror-image digit duplications like those induced by ZPA transplants were the result. More recently, beads containing Sonic hedgehog protein were shown to cause the same duplications (López-Martínez et al. 1995). Thus, Sonic hedgehog appears to be the active agent of the ZPA.

This fact was confirmed by a remarkable gain-of-function mutation. The *hemimelic-extra toes* (*Hx*) mutant of mice has extra digits on the thumb side of the paws (Fig-

ure 16.17A,B). This phenotype is associated with a single base-pair difference in the limb-specific *shh* enhancer, a highly conserved region located a long distance (about a million base pairs) upstream from the *shh* gene itself (Lettice et al. 2003; Sagai et al. 2004). Maas and Fallon (2005) made a reporter construct by fusing a β-galactosidase gene to this "long-range" limb enhancer region from both wild-type and *Hx* mutant genes. They injected these reporter constructs into the pronuclei of newly fertilized mouse eggs to obtain transgenic embryos. In the transgenic mouse embryos carrying the reporter gene with wild-type limb enhancer, staining for β-galactosidase activity revealed a single patch of expression in the posterior mesoderm of each limb bud (Figure 16.17C). However, the mice carrying the mutant *Hx* reporter construct showed β-galactosidase activity in *both* the anterior and posterior regions of the limb bud (Figure 16.17D). It appears that this enhancer has both positive and negative functions, and that in the anterior region of the limb bud, some factor represses the ability of this enhancer to activate *shh* transcription. The inhibitors probably cannot bind to the mutated enhancer, and thus *shh* expression is seen in both the anterior and posterior regions of the limb bud. This anterior *shh* expression, in turn, causes extra digits to develop in the mutant mice.

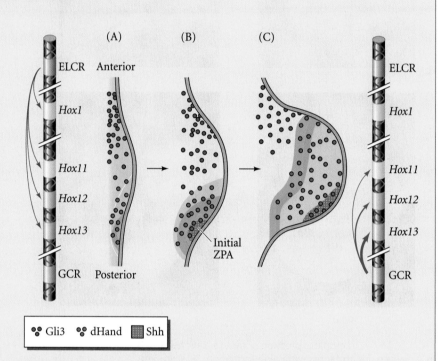

SIDELIGHTS & SPECULATIONS

Hox Gene Changes during Development and Formation of the ZPA

The limb bud has an anterior-posterior polarity as soon as it is formed. The Gli3 transcription repressor is prominent in the anterior region of the limb bud, while the dHAND transcription activator is present in the posterior mesoderm. These factors are expressed in the mesenchyme immediately before the *Sonic hedgehog* gene is expressed (Figure 16.18A). In mice lacking *dHAND*, *shh* is not expressed in the limb buds and the limbs are severely truncated. Conversely, ectopic expression of dHAND protein in the anterior region of the mouse limb bud generates a second ZPA and duplications of the digits (Charité et al. 2000).

In addition, there is a polar arrangement of Hox genes in the nascent limb bud. Within the domain established by dHAND, *Hoxd13* is expressed in the posterior region of the bud, surrounded by a layer of *Hoxd12*-expressing cells and another layer of *Hoxd11*-expressing cells (Figure 16.18B). The center and anterior regions of the limb bud (not expressing dHAND) expresses *Hoxd11*. In the forelimb, Hoxb8 activates Shh expression in the posterior mesoderm, and ectopic *Hoxb8* expression in the anterior mesoderm of the mouse forelimb will create a second ZPA (Charité et al. 1994). Probably a combination of the dHAND, Hoxd12, and Hoxb8 transcription factors (each of which is expressed in the posterior mesenchyme of the limb bud and is capable of inducing ectopic expression of *shh* when activated by Fgf8 from the AER) operate on

FIGURE 16.18 Hox gene expression changes during the formation of the tetrapod limb. (A) At the early limb bud stage (day 9 in chicks), the Gli3 repressor transcription factor is in the anterior region, while the dHand activating transcription factor is in the posterior. A nested set of HoxD transcription factors arises in the posterior (under the control of the ELCR enhancer), with Hoxd13 being the most limited, followed by Hoxd12 and Hoxd11. The remainder of the limb bud has Hoxd1 expression. (B) A short time later, Sonic hedgehog protein is seen in the region of cells expressing all of these Hox genes at the posterior of the limb bud mesenchyme. This is the initial ZPA. (C) Under the influence of Sonic hedgehog, a new enhancer (GCR) is activated, and it *reverses* the Hox expression domains. Now, *Hoxd13* is expressed throughout the distal mesenchyme of the limb bud. The zone of *Hoxd12* transcription falls a little more distally, and the domain of *Hoxd11* transcription is a bit more circumscribed still. (After Deschamps 2004.)

Specifying digit identity by Sonic hedgehog

How does Sonic hedgehog specify the identities of the digits? Surprisingly, when scientists were able to perform fine-scale fate mapping experiments on the Shh-secreting cells of the ZPA (using recombinase to express a label only in those cells expressing *shh*), they found that *shh*-expressing cells do not undergo apoptosis in the way that the AER does after it finishes its job. Rather, the Shh-secreting cells become the bone, muscle, and skin of the posterior limb

(Ahn and Joyner 2004; Harfe et al. 2004). Indeed, digits 5 and 4 (and part of digit 3) of the mouse hindlimb are formed from these Shh-secreting cells (Figure 16.20).

It seems that the specification of the digits is dependent on the amount of time Shh is expressed and only a little bit by the concentration of Shh that other cells receive. The difference between digits 4 and 5 is that the cells of the more posterior digit 5 express Shh longer and are exposed to Shh (in an autocrine manner) for a longer time. Digit 3 has some cells that secrete Shh for a shorter period of time than

FIGURE 16.19 Feedback between the AER and the ZPA in the forelimb bud. Stage 16 (about 54 hours gestation) represents the same stage as (C) in Figure 16.6. The other two stages are about 5 hours apart. At stage 17, the newly induced AER secretes Fgf8 into the underlying mesenchyme. The mesenchyme expressing *Hoxb8* and *dHAND* is induced to express *shh*, thereby forming the ZPA in the posterior region of the forelimb bud. At stage 18, Shh protein maintains FGF expression in the AER, while the FGFs from the AER maintain *shh* gene expression.

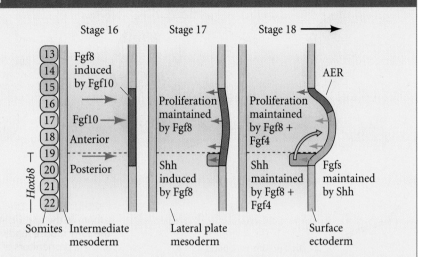

the "long-range" *shh* enhancer to activate the *shh* gene and limit its expression to the posterior mesenchyme (Lettice et al. 2003; Sagai et al. 2004; Maas and Fallon 2005).

The expression of *shh* reverses the Hox gene expression pattern. Under a new enhancer, the domain of Hox gene expression expands, and the *Hoxd13* gene is expressed throughout the distal region of the limb bud (Figure 16.18C). Closer to the source of Shh, the *Hox12* gene is also tran-

scribed, and the region immediately surrounding the zone of Shh-secreting cells expresses *Hoxd11*. Thus, *Hoxd13* is expressed alone in the anteriormost portion of the distal limb bud, while the region expressing Sonic hedgehog has at least three 5′ *Hoxd* genes expressed (Zákány et al. 2004). Moreover, transplantation of either the ZPA or other Shh-secreting cells to the anterior margin of the limb bud at this stage leads to the formation of mirror-image pat-

terns of *Hoxd* gene expression and results in mirror-image digit patterns (Izpisúa-Belmonte et al. 1991; Nohno et al. 1991; Riddle et al. 1993). In normal embryos, the *shh* gene is kept active by the FGFs of the limb bud, while FGF genes in the AER are kept active by Shh protein. Thus, once the AER and ZPA are established, they mutually support one another (Figure 16.19).

those of digit 4, and they also depend on Shh secretion by diffusion from the ZPA (since this digit is lost when Shh is modified such that it can't diffuse away from cells). Digit 2 is dependent entirely on Shh diffusion for its specification, and digit 1 is specified independently of Shh. Indeed, in a naturally occurring chick mutant that lacks *shh* expression in the limb, the only digit that forms is digit 1. Furthermore, when the genes for Shh and Gli3 are conditionally knocked out of the mouse limb, the resulting limbs have numerous digits, but the digits have obvious specificity (see Figure 16.21G; Ros et al. 2003; Litingtung et al. 2002). Vargas and Fallon (2005) propose that digit 1 is specified by Hoxd13 in the absence of Hoxd12. Forced expression of Hoxd12 in all the digits leads to the transformation of digit 1 into a more posterior digit (Knezevic et al. 1997).

Although Sonic hedgehog appears to specify the digits directly, it is also possible that it is working through BMPs. Shh initiates and sustains a gradient of proteins such as BMP2 and BMP7 across the limb bud, and there is evidence that the concentration of these BMPs also can specify the digits (Laufer et al. 1994; Kawakami et al. 1996; Drossopoulou et al. 2000). Digit identity is not specified directly in each digit primordium, however. Rather, the identity of each digit is determined by the *interdigital* mesoderm. In

other words, the identity of each digit is specified by the webbing between the digits—that region of mesenchyme that will shortly undergo apoptosis to free each digit. The interdigital tissue specifies the identity of the digit forming anteriorly to it (toward the thumb or big toe). Thus, when Dahn and Fallon (2000) removed the webbing between the cartilaginous condensations forming chick hindlimb digits 2 and 3, the second digit was changed into a copy of digit 1. Similarly, when the webbing on the other side of digit 3 was removed, the third digit formed a copy of digit 2 (Figure 16.21A–C). Moreover, the positional value of the webbing could be altered by changing the BMP levels in it. When beads containing BMP antagonists such as Noggin were placed in the webbing between digits 3 and 4, digit 3 was anteriorly transformed into digit 2 (Figure 16.21 D–F).

Generation of the Dorsal-Ventral Axis

The third axis of the limb distinguishes the dorsal half of the limb (knuckles, nails) from the ventral half (pads, soles). In 1974, MacCabe and co-workers demonstrated that the dorsal-ventral polarity of the limb bud is determined by the ectoderm encasing it. If the ectoderm is rotated 180°

(A)

(B)

(C)

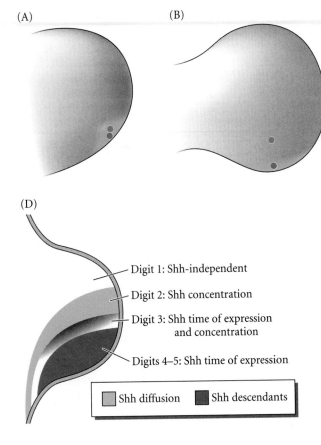

(D)

Digit 1: Shh-independent

Digit 2: Shh concentration

Digit 3: Shh time of expression
and concentration

Digits 4–5: Shh time of expression

☐ Shh diffusion ■ Shh descendants

FIGURE 16.20 The Shh-secreting cells form digits 4 and 5, and contribute to the specification of digits 2 and 3 in the mouse limb. (A) In the early mouse hindlimb bud, the progenitors of digit 4 (green dot) and the progenitors of digit 5 (red dot) are both in the ZPA and express *sonic hedgehog* (purple). (B) At later stages of the limb development, the cells forming digit 5 are still expressing *shh* in the ZPA, but the cells forming digit 4 no longer do. (C) When the digits form, the cells in digit 5 will have seen high levels of Shh protein for a longer time than the cells in digit 4. (D) Schematic by which digits 4 and 5 are specified by the amount of time they were exposed to Shh in an autocrine fashion; digit 3 is specified by the amount of time the cells were exposed to Shh both in an autocrine and paracrine fashion. Digit 2 is specified by the concentration of Shh its cells received by paracrine diffusion, and digit 1 is specified independently of Shh. (After Harfe et al. 2004.)

with respect to the limb bud mesenchyme, the dorsal-ventral axis is partially reversed; the distal elements (digits) are "upside down." This suggested that the late specification of the dorsal-ventral axis of the limb is regulated by its ectodermal component.

One molecule that appears to be particularly important in specifying dorsal-ventral polarity is **Wnt7a**. The *Wnt7a* gene is expressed in the dorsal (but not the ventral) ectoderm of chick and mouse limb buds (Dealy 1993; Parr et al. 1993). When Parr and McMahon (1995) knocked out the *Wnt7a* gene, the resulting mouse embryos had ventral footpads on both surfaces of their paws, showing that Wnt7a is needed for the dorsal patterning of the limb (Figure 16.22).

Wnt7a is the first of the dorsal-ventral axis genes expressed in limb development. It induces activation of the *Lmx1* gene in the dorsal mesenchyme. *Lmx1* encodes a transcription factor that appears to be essential for specifying dorsal cell fates in the limb. If Lmx1 protein is expressed in the ventral mesenchyme cells, those cells develop a dorsal phenotype (Riddle et al. 1995; Vogel et al. 1995; Altabef and Tickle 2002). Mutants of *Lmx1* in humans and mice also show this gene's importance for specifying dorsal limb fates. Knockouts in mice produce a syndrome in which the dorsal limb phenotype is lacking, and loss-of-function mutations in humans produce the nail-patella syndrome,

a condition in which the dorsal sides of the limbs have been ventralized (e.g., no nails on the digits, no kneecaps) (Chen et al. 1998; Dreyer et al. 1998).

Coordinating the Three Axes

The three axes of the tetrapod limb are all interrelated and coordinated. Some of the principal interactions among the mechanisms specifying the axes are shown in Figure 16.23. Indeed, the molecules that define one of these axes are often used to maintain another axis. For instance, Sonic hedgehog in the ZPA activates the expression of the *fgf4* gene in the posterior region of the AER (see Figure 16.19). Expression of Fgf4 protein is important in recruiting mesenchyme cells into the progress zone, and it is also partially responsible (along with Fgf8) in maintaining the expression of Shh in the ZPA (Li and Muneoka 1999). Shh protein in the ZPA then sustains the AER FGFs by inducing Gremlin, a BMP inhibitor that prevents BMPs in the mesoderm from inhibiting the FGFs of the AER (Zúñiga et al. 1999). Therefore, the AER and the ZPA mutually support each other through the positive loop of Sonic hedgehog and FGFs (Todt and Fallon 1987; Laufer et al. 1994; Niswander et al. 1994).

The *Wnt7a*-deficient mice described earlier lacked not only dorsal limb structures but also posterior digits, sug-

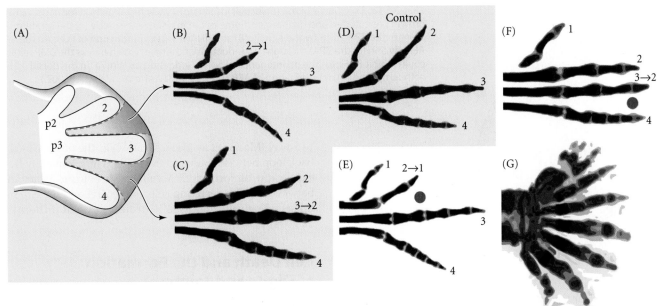

FIGURE 16.21 Regulation of digit identity by BMP concentrations in the interdigital space anterior to the digit and by Gli3. (A) Scheme for removal of interdigital (ID) regions. The results are shown in (B) and (C), respectively. (B) Removal of interdigital region 2 between digit primordia 2 (p2) and 3 (p3) causes digit 2 to change to the structure of digit 1. (C) Removing interdigital region 3 (between digit primordia 3 and 4) causes digit 3 to form the structures of digit 2. (D) Control digits and their interdigital spaces. (E,F) The same transformations as in (B) and (C) can be obtained by adding beads containing the BMP inhibitor Noggin to the interdigital regions. (E) When a Noggin-containing bead (green dot) is placed in interdigital region 2, the second digit is transformed into a copy of digit 1. (F) When the Noggin bead is placed in interdigital region 3, the third digit is transformed into a copy of digit 2. (G) Forelimb of a mouse homozygous for deletions of both *gli3* and *shh* is characterized by extra digits of no specific type. (After Dahn and Fallon 2000; Litingtung et al. 2002; photographs courtesy of R. D. Dahn and J. F. Fallon.)

gesting that Wnt7a is also needed for the anterior-posterior axis. Yang and Niswander (1995) made a similar set of observations in chick embryos. These investigators removed the dorsal ectoderm from developing limbs and found that such an operation resulted in the loss of poste-

rior skeletal elements from the limbs. The reason that these limbs lacked posterior digits was that *shh* expression was greatly reduced. Viral-induced expression of *Wnt7a* was able to replace the dorsal ectoderm and restore *shh* expression and posterior phenotypes. These findings showed that the synthesis of Sonic hedgehog is stimulated by the combination of Fgf4 and Wnt7a proteins. Conversely, overactive Wnt signaling in the dorsal ectoderm causes an overgrowth of the AER and extra digits, indicating that the proximal-distal patterning is not independent of dorsal-ventral patterning either (Adamska et al. 2004).

This coordination of activities is critical in regulating limb growth. (Imagine what would happen if the limb stopped growing along one axis but continued to grow along the others.) At the end of limb patterning, BMPs are

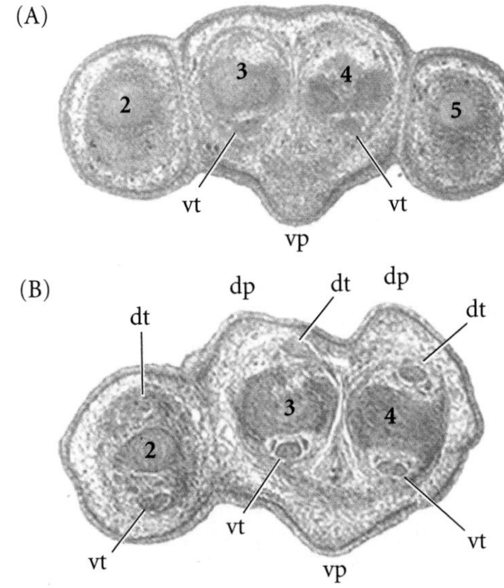

FIGURE 16.22 Dorsal-to-ventral transformations of limb regions in mice deficient for both *Wnt7a* genes. (A) Histological section (stained with hemotoxylin and eosin) of wild-type 15.5-day embryonic mouse forelimb paw. The ventral tendons and ventral footpads are readily seen. (B) Same section through a mutant embryo deficient in *Wnt7a*. Ventral tendons and footpads are duplicated on what would normally be the dorsal surface of the paw. dt, dorsal tendons; dp, dorsal footpad; vp, ventral footpad; vt, ventral tendon. Numbers indicate digit identity. (From Parr and McMahon 1995; photographs courtesy of the authors.)

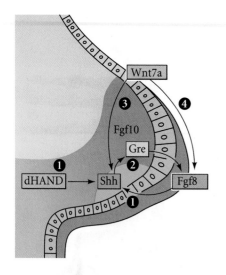

FIGURE 16.23 Some of the molecular interactions by which limb bud formation and growth are initiated and maintained. Some of the major loops include (1) the establishment of the ZPA (Shh) by the AER (Fgf8) and dHAND; (2) the induction of FGFs (in the AER) by Shh (in the ZPA) through Shh induction of Gremlin, which blocks the BMPs that could inhibit FGFs; (3) the maintenance of Sonic hedgehog by Wnt7a (in the dorsal ectoderm); and (4) determination of AER size by Wnt7a. (After te Welscher 2002.)

pathway (Merino et al. 1999). Meanwhile, the positive feedback loop between Shh and the FGFs is terminated by the expansion of former ZPA cells, which form a barrier between the Shh-secreting cells of the ZPA and the Gremlin-secreting cells that usually protect the AER (Scherz et al. 2004).

Cell Death and the Formation of Digits and Joints

Sculpting the autopod

Cell death plays a role in sculpting the tetrapod limb. Indeed, cell death is essential if our joints are to form and if our fingers are to become separate (Zaleske 1985; Zuzarte-Luis and Hurle 2005). The death (or lack of death) of specific cells in the vertebrate limb is genetically programmed and has been selected for over the course of evolution. The difference between a chicken's foot and the webbed foot of a duck is the presence or absence of cell death between the digits (Figure 16.24). Saunders and coworkers have shown that after a certain stage, chick cells between the digit cartilage are destined to die, and will do

responsible for shutting down the signaling from the AER and for simultaneously inhibiting the Wnt7a signal along the dorsal-ventral axis (Pizette et al. 2001). The BMP signal eliminates growth and patterning along all three axes. When exogenous BMP is applied to the AER, the elongated epithelium of the AER reverts to a cuboidal epithelium and ceases to produce FGFs; and when BMP is inhibited by Noggin, the AER continues to persist days after it would normally have regressed (Gañan et al. 1998; Pizette and Niswander 1999). The mechanism by which the BMP signal is regulated is not yet fully understood, but it appears to involve the BMP antagonist protein Gremlin. Gremlin works against BMP to preserve the AER and the Wnt7a

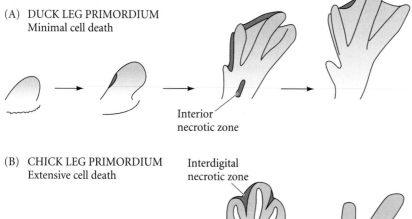
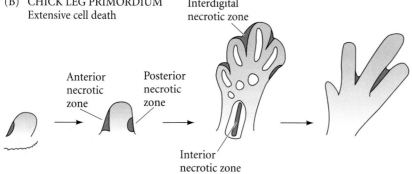

FIGURE 16.24 Patterns of cell death in leg primordia of (A) duck and (B) chick embryos. Shading indicates areas of cell death. In the duck, the regions of cell death are very small, whereas there are extensive regions of cell death in the interdigital tissue of the chick leg. (After Saunders and Fallon 1966.)

so even if transplanted to another region of the embryo or placed in culture (Saunders et al. 1962; Saunders and Fallon 1966). Before that time, however, transplantation to a duck limb will save them. Between the time when the cell's death is determined and when death actually takes place, levels of DNA, RNA, and protein synthesis in the cell decrease dramatically (Pollak and Fallon 1976).

In addition to the **interdigital necrotic zone**, three other regions of the limb are "sculpted" by cell death. The ulna and radius are separated from each other by an **interior necrotic zone**, and two other regions, the **anterior** and **posterior necrotic zones**, further shape the end of the limb (see Figure 16.24B; Saunders and Fallon 1966). Although these zones are referred to as "necrotic," this term is a holdover from the days when no distinction was made between necrotic cell death and apoptotic cell death (see Chapter 6). These cells die by apoptosis, and the death of the inter-

digital tissue is associated with the fragmentation of their DNA (Mori et al. 1995).

The signal for apoptosis in the autopod is probably provided by the BMP proteins. BMP2, BMP4, and BMP7 are each expressed in the interdigital mesenchyme, and blocking BMP signaling (by infecting progress zone cells with retroviruses carrying dominant negative BMP receptors) prevents interdigital apoptosis (Zou and Niswander 1996; Yokouchi et al. 1996). Since these BMPs are expressed throughout the progress zone mesenchyme, it is thought that cell death would be the "default" state unless there were active suppression of the BMPs. This suppression may come from the Noggin protein, which is made in the developing cartilage of the digits and in the perichondrial cells surrounding it (Capdevila and Johnson 1998; Merino et al. 1998). If Noggin is expressed throughout the limb bud, no apoptosis is seen.

SIDELIGHTS & SPECULATIONS

Limb Development and Evolution

The importance of development to the study of evolution was articulated by C. H. Waddington. He wrote that when we say that a modern horse has one toe and that it came from an ancestor with five toes, what we are saying is that during the evolution of the horse lineage, developmental changes occurred in the cartilage deposition of the embryonic horses' feet. These changes were selected over time to the point of today's *Equus* species. While we still do not know much about the details of evolutionary change in horse limbs, we have found several remarkable cases of limb evolution caused by developmental changes.*

Web-footed friends

We can start with the sculpting of the autopod. The regulation of BMPs is critical in creating the webbed feet of ducks (Merino et al. 1999). The interdigital regions of duck feet exhibit the same pattern of BMP expression as the webbing of chick feet. However, whereas the interdigital regions of

*Earlier in this chapter, it was noted that developmental biologists get used to thinking in four dimensions. Evolutionary developmental biologists have to think in *five* dimensions: the three standard dimensions of space, the dimension of developmental time (hours or days), and the dimension of evolutionary time (millions of years).

(A)

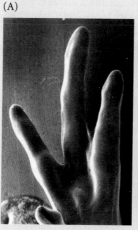

(B)

FIGURE 16.25 Inhibition of cell death by inhibiting BMPs. (A) Control chick limbs show extensive apoptosis in the space between the digits, leading to the absence of webbing. (B) However, when beads soaked with Gremlin protein are placed in the interdigital mesoderm, the webbing persists and generates a duck-like pattern. (After Merino et al. 1999; photographs courtesy of E. Hurle.)

the chick feet appear to undergo BMP-mediated apoptosis, developing duck feet synthesize the *BMP* inhibitor Gremlin and block this regional cell death (see Figure 23.8). Moreover, the webbing of chick feet can be preserved if Gremlin-soaked beads are placed in the interdigital regions (Figure 16.25). Thus the evolution of webbed-footed birds probably involved the inhibition of BMP-mediated apoptosis in the interdigital regions.

Dinosaurs and chicken fingers

Next, there is that postulated connection between birds and dinosaurs. Are birds real-

ly the descendants of dinosaurs? Although Thomas Huxley (1868, 1870) proposed in the late nineteenth century that the birds descended from dinosaurs, it was J. H. Ostrom's 1969 description of the dinosaur *Deinonychus antirrhopus* and its similarities to the fossils of the first known bird, *Archaeopteryx*, that was critical in making the dinosaur-to-bird hypothesis acceptable. Ostrom listed 22 similarities between *Deinonychus* and *Archaeopteryx*, similarities found in no other groups and linking birds and dinosaurs.

(Continued on next page)

Without its feathers, *Archaeopteryx* looks exactly like a small coelurosaur (such as Jurassic Park's *Velociraptor*). Indeed, one specimen of *Archaeopteryx* was misidentified as a coelurosaur for over 100 years until its feathers were noticed by Peter Wellnhofer (1993). Gauthier's cladistic work in the mid-1980s (see Gauthier 1986) provided systematic support for the theory that birds are the descendants of coelurosaurian dinosaurs. He listed 17 anatomical features shared between dinosaurs and birds and no other group, including a shifted pubic bone in the pelvis, clawed hands, large eye openings, a flexible wrist with a particularly shaped wrist bone, a strap-like scapula, clavicles fused to form a furcula (wishbone), and even feathers (Ji et al. 1998; Sereno 1999; Xu et al. 2003). Unlike any other reptiles, both birds and theropod dinosaurs (of which the coelurosaurs are a group) have a 3-fingered grasping hand and a four-toed foot supported by three main toes. There are also numerous transition forms which show the emergence of birds from the coelosaurian dinosaurs.* Thus Padian and Chiappe (1998a) conclude that "in fact, living birds are nothing less than small, feathered, short-tailed theropod dinosaurs."

However, whereas paleontologists were nearly unanimous in their appraisal that birds are the direct descendants of dinosaurs, some developmental biologists harbored serious doubts. Fossil evidence unambiguously identified the theropod-like birds as having wing digits 1, 2, and 3 (Padian and Chiappe 1998b), but embryological evidence suggested that the wing digits of current birds are 2, 3, and 4. For instance, Burke and Feduccia (1997) found digit primordia in the fingers of early and present-day birds correspond to the index, middle and ring fingers (2-3-4). Moreover, the arrangement of cartilaginous condensations is the one expected for the 2-3-4 pattern,

*For a detailed analysis of these transition forms, see http://www.ucmp.berkeley.edu/diapsids/avians.html

not the 1-2-3 pattern. This would mean that the similarity of dinosaur and bird digits is based on independent selection for three digits (convergent evolution) and is not based on shared ancestry. This developmental critique of the bird-dinosaur link has been made by other scientists studying chick limb development (Galis 2005; Welten et al. 2005). They point out that bird feet have reversed toes used for perching on branches (something dinosaurs never developed), and that theropods had a characteristic joint in their lower jaws for grasping prey (something never found in birds). Alan Feduccia has called the notion that birds arose directly from dinosaurs a "delusional fantasy by which one can vicariously study dinosaurs at the backyard bird feeder" (Feduccia 1997).

However, a new developmental study by Vargas and Fallon (2005a,b) suggests that embryologists have been wrong in their assessment of bird digits. Although the condensations of the digits *look* like those expected for digits 2, 3, and 4, the Hox gene expression patterns suggest that the actual digits are indeed 1, 2, and 3, just as in the theropod dinosaurs. Fallon and Vargas claim that digit 1 (thumb/hallux) is uniquely characterized (at least in the chicken hindlimb and the mouse forelimb and hindlimb) by *Hoxd13* expression in the absence of *Hoxd12* expression (see Figure 16.18). All other digit primordia express both *Hoxd12* and *Hoxd13*. Moreover, this *Hoxd13*/no *Hoxd12* is the pattern seen in the most anterior chicken wing digit. Even when an extra digit forms more anterior to the digit that is normally most anterior, its identity depends upon the Hox code. In *silkie* mutants of chicks, this extra digit is (by anatomy and also by Hox codes) an extra digit 2. Thus, Vargas and Fallon proposed that the wing digits of chickens are actually 1-2-3, and that this digit arrangement is further proof rather than a rebuttal of the idea that birds are the descendants of dinosaurs.

The Vargas and Fallon findings are still speculative, needing the support of more

species to show that digit 1 is characterized by the expression of *Hoxd13* in the absence of *Hoxd12* throughout the reptile and mammalian clades. If this developmental genetic observation is confirmed, then the "greatest challenge to the theropod-bird link" (Zhou 2004) will have been eliminated. Developmental biologist Richard Hinchliffe (1994, 1997) sees the argument in a larger context. While virtually all evolutionary biologists agree that birds and dinosaurs evolved from the same class of prehistoric creatures, he says, "the only question we are arguing about is whether [birds] derived very late in time from a specific group of theropod dinosaurs, the so-called raptors, or are they derived from a common-stem ancestor with dinosaurs."

The fin-to-limb transition

Before the appearance of the dinosaurs, a major transition in the vertebrate appendage had already taken place. This was the transition from the fish fin to the amphibian limb, and this time it's the HoxA cluster that seems to be critical. Paleontological and anatomical evidence suggest that the autopod is an evolutionarily new feature that is not present in fish. Whereas the stylopod and zeugopod have homologous parts in the fish fin, the autopod does not (Sordino and Duboule 1996; Mabee and Noordsy 2004). Looking at the Hox gene expression patterns in a primitive fish (the paddlefish *Polyodon spathula*), Metscher and colleagues (2005) discovered that while the early expression of *Hoxa13* is similar in this fish and in tetrapods, the later expression of *Hoxa13* is different. In tetrapods, *Hoxa13* is excluded from the distalmost region because it is repressed by the Hoxd11 protein (see Figure 16.18C). This repression is not seen in the primitive fish, where *Hoxd11* and *Hoxd13* are both expressed together near the tip of the limb bud. The next task will be to find out if such *Hoxa13* expression in the distal region of the limb bud brings the autopod into being.

Forming the joints

The function first ascribed to BMPs was the formation, not the destruction, of bone and cartilage tissue. In the developing limb, BMPs induce the mesenchymal cells either to undergo apoptosis or to become cartilage-producing chon-

drocytes—depending on the stage of development. The same BMPs can induce death or differentiation, depending on the age of the target cell. This "context dependency" of signal action is a critical concept in developmental biology. It is also critical for the formation of joints. Macias and colleagues (1997) have shown that during early limb

(A)

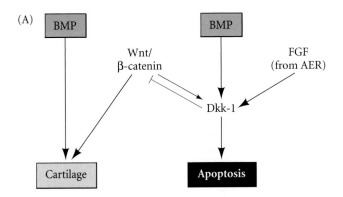

(B)

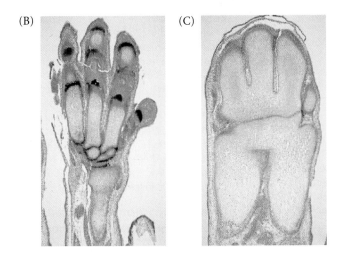

(C)

FIGURE 16.26 Possible involvement of BMPs in stabilizing cartilage and apoptosis. (A) Model for the dual role of BMP signals in limb mesodermal cells. BMP can be received in the presence of FGFs (to produce apoptosis) or Wnts (to induce bone). When FGFs from the AER are present, Dickkopf (DKK1) is activated. This protein mediates apoptosis and at the same time inhibits Wnt from aiding in skeleton formation. (B,C) The effects of Noggin. (B) 16.5-day autopod from a wild-type mouse, showing GDF5 expression (dark blue) at the joints. (C) 16.5-day *noggin*-deficient mutant mouse autopod, showing neither joints nor GDF5 expression. Presumably, in the absence of Noggin, BMP7 was able to convert nearly all the mesenchyme into cartilage. (A after Grotewold and Rüther 2002; B,C from Brunet et al. 1998, photographs courtesy of A. P. McMahon.)

bud stages (before cartilage condensation), beads secreting BMP2 or BMP7 cause apoptosis. Two days later, the same beads cause the limb bud cells to form cartilage.

In the normally developing limb, BMPs use both of these properties to form joints. BMP7 is made in the perichondrial cells surrounding the condensing chondrocytes and promotes cartilage formation (Figure 16.26A; Macias et al. 1997).

Two other BMP proteins, BMP2 and GDF5, are expressed at the regions between the bones, where joints will form (Figure 16.26B; Macias et al. 1997; Brunet et al. 1998). Mouse mutations have suggested that the function of these proteins in joint formation is critical. Mutations of *Gdf5* produce brachypodia, a condition characterized by a lack of limb joints (Storm and Kingsley 1999). In mice homozygous for loss-of-function alleles of *noggin*, no joints form, either. It appears that the BMP7 in these *noggin*-defective embryos is able to recruit nearly all the surrounding mesenchyme into the digits (Figure 16.26C; Brunet et al. 1998).

Wnt proteins and blood vessels also appear to be critical in joint formation. The conversion of mesenchyme cells into nodules of cartilage-forming tissue establishes where the bone boundaries are. The mesenchyme will not form such nodules in the presence of blood vessels, and one of the first indications of cartilage formation is the regression of blood vessels in the region wherein the nodule will form (Yin and Pacifici 2001). Wnt proteins can then initiate the changes in cell adhesion molecules (such as N-cadherin) that cause the mesenchyme cells to initiate chondrogenesis. The placement of these molecules may determine the number, shape, and size of these condensations (Hartmann and Tabin 2001; Tufan and Tuan 2001).

SIDELIGHTS & SPECULATIONS

Continued Limb Growth: Epiphyseal Plates

If all of our cartilage were turned into bone before birth, we could not grow any larger, and our bones would be only as large as the original cartilaginous model. However, as the ossification front nears the ends of the cartilage model, the chondrocytes near the ossification front proliferate prior to undergoing hypertrophy, pushing out the cartilaginous ends of the bone. In

the long bones of many mammals (including humans), endochondral ossification spreads outward in both directions from the center of the bone. These cartilaginous areas at the ends of the long bones are called **epiphyseal growth plates**. As we saw in Chapter 14, these plates contain three regions: a region of chondrocyte proliferation, a region of mature chondrocytes,

and a region of hypertrophic chondrocytes (see Figure 14.14; Chen et al. 1995). As the inner cartilage hypertrophies and the ossification front extends farther outward, the remaining cartilage in the epiphyseal growth plate proliferates. As long as the epiphyseal growth plates are able to produce chondrocytes, the bone continues to grow.

(Continued on next page)

SIDELIGHTS & SPECULATIONS

Fibroblast growth factor receptors: Dwarfism

Recent discoveries of human and murine mutations resulting in abnormal skeletal development have provided remarkable insights into how the differentiation, proliferation, and patterning of chondrocytes are regulated.

The proliferation of the epiphyseal growth plate cells and facial cartilage can be halted by the presence of fibroblast growth factors (Deng et al. 1996; Webster and Donoghue 1996). These factors appear to instruct the cartilage precursors to differentiate rather than to divide. In humans, mutations of the receptors for fibroblast growth factors can cause these receptors to become activated prematurely. Such mutations give rise to the major types of human dwarfism. Achondroplasia is a dominant condition caused by mutations in the transmembrane region of fibroblast growth factor receptor 3 (FgfR3). Roughly 95 percent of achondroplastic dwarfs have the same mutation of FgfR3, a base pair substitution that converts glycine to arginine at position 380 in the transmembrane region of the protein. In addition, mutations in the extracellular portion of the FgfR3 protein or in the tyrosine kinase intracellular domain may result in thanatophoric dysplasia, a lethal form of dwarfism that resembles homozygous achondroplasia (see Figure 6.15; Bellus et al. 1995; Tavormina et al. 1995). Mutations in FGFR1 can cause Pfeiffer syndrome, characterized by limb defects and premature fusion of the cranial sutures (craniosynostosis), which results in abnormal skull and facial shape. Different mutations in FGFR2 can give rise to various abnormalities of the limbs and face (Park et al. 1995; Wilkie et al. 1995).

Insulin-like growth factors: Pygmies

The epiphyseal growth plate cells are very responsive to hormones, and their proliferation is stimulated by growth hormone and insulin-like growth factors. Nilsson and colleagues (1986) showed that growth hormone stimulates the production of insulin-like growth factor I (IGF-I) in the epiphyseal chondrocytes, and that these chondrocytes respond to it by proliferating. When they added growth hormone to the tibial growth plates of young mice who could not manufacture their own growth hormone (because their pituitaries had been removed), it stimulated the formation of IGF-I in the chondrocytes of the proliferative zone (see Figure 14.16). The combination of growth hormone and IGF-I appears to provide an extremely strong mitotic signal. It appears that IGF-I is essential for the normal growth spurt at puberty. The pygmies of the Ituri Forest of Zaire have normal growth hormone and IGF-I levels until puberty. However, at puberty, their IGF-I levels fall to about one-third that of other adolescents (Merimee et al. 1987).

Estrogen receptors: Not just for women

The pubertal growth spurt and the subsequent cessation of growth are induced by sex hormones (Kaplan and Grumbach 1990). At the end of puberty, high levels of estrogen or testosterone cause the remaining epiphyseal plate cartilage to undergo hypertrophy. These cartilage cells grow, die, and are replaced by bone. Without any further cartilage formation, growth of these bones ceases, a process known as **growth plate closure**.

In conditions of precocious puberty, there is an initial growth spurt (making the individual taller than his or her peers), followed by the cessation of epiphyseal cell division (allowing that person's peers to catch up and surpass his or her height). In males, it was not thought that estrogen played any role in these events. However, in 1994, Smith and colleagues published the case history of a man whose growth was still linear despite his having undergone normal puberty. His epiphyseal plates had not matured, and at 28 years of age he still had proliferating chondrocytes. His "bone age"—the amount of ephiphyseal cartilage he retained—was roughly half his chronological age. This man was found to lack functional estrogen receptors. At present, at least three human males have been reported who either cannot make estrogens or who lack the estrogen receptor. All three are close to 7 feet tall and are still growing (Sharpe 1997). These cases show that estrogen plays a role in epiphyseal maturation in males as well as in females.

Parathyroid hormone-related peptide and Indian Hedgehog

Parathyroid hormone-related peptide (PTHrP) maintains cell division of chondrocytes (by activating cyclin D1 synthesis) and prevents their hypertrophy. In humans, loss-of-function mutations in the protein encoding PTHrP result in severe growth defects due to the lack of limb growth (Provot and Schipiani 2005). PTHrP is stimulated in the cartilage by Indian Hedgehog. Indian hedgehog also stimulates BMP production, coordinating the rates of cell division and matrix deposition (Vortkamp et al. 1996).

Coda

FGF proteins generate the proximal-distal axis of the vertebrate limb; Shh and BMPs generate the anterior-posterior axis; and Wnts appear to mediate formation of the dorsal ventral axis. Thus all the major paracrine factor families act in coordination to build the limb. While many of the "executives" of the limb bud formation have been identified, we still remain largely ignorant about how the orders of these paracrine factors and transcription factors are carried out. Niswander (2002) writes:

There is a very large gap in our understanding of how the activity of Shh, BMP, FGF, and Wnt genes influences, for example, where the cartilaginous condensation will

form, how the elements are sculpted, how the number of phalangeal elements are specified, and where the tendon/muscle will insert.

Limb development is an exciting meeting place for developmental biology, evolutionary biology, and medicine. Within the next decade, we can expect to know the bases for numerous congenital diseases of limb formation, and perhaps we will understand how limbs are modified into flippers, wings, hands, and legs.

 **Snapshot Summary** **The Tetrapod Limb**

1. The positions where limbs emerge from the body axis depend upon Hox gene expression.

2. The specification of a limb bud as hindlimb or forelimb is determined by Tbx4 and Tbx5 expression.

3. The proximal-distal axis of the developing limb is determined by the induction of the ectoderm at the dorsal-ventral boundary by an FGF (probably Fgf10) from the mesenchyme. This induction forms the apical ectodermal ridge (AER). The AER secretes Fgf8, which keeps the underlying mesenchyme proliferative and undifferentiated. This area of mesenchyme is called the progress zone.

4. As the limb grows outward, the stylopod forms first, then the zeugopod, and the autopod is formed last. Each phase of limb development is characterized by a specific pattern of Hox gene expression. The evolution of the autopod involved a reversal of Hox gene expression that distinguishes fish fins from tetrapod limbs.

5. The anterior-posterior axis is defined by the expression of Sonic hedgehog in the posterior mesoderm of the limb bud. This region is called the zone of polarizing activity (ZPA). If ZPA or Sonic hedgehog-secreting cells or beads are placed in the anterior margin of a limb bud, they establish a second, mirror-image pattern of Hox gene expression and a corresponding mirror-image duplication of the digits.

6. The ZPA is established by the interaction of FGFs from the AER with mesenchyme made competent to express Sonic hedgehog by its expression of dHAND

and particular Hox genes. Sonic hedgehog acts, probably in an indirect manner, probably through the Gli factors, to change the expression of the Hox genes in the limb bud.

7. The identity of each digit is specified by BMP activity in the interdigital region posterior to it, and probably by the ratio of activator and repressor forms of Gli3.

8. The dorsal-ventral axis is formed in part by the expression of Wnt7a in the dorsal portion of the limb ectoderm. Wnt7a also maintains the expression level of Sonic hedgehog in the ZPA and of Fgf4 in the posterior AER. Fgf4 and Sonic hedgehog reciprocally maintain each other's expression.

9. Cell death in the limb is necessary for the formation of digits and joints. It is mediated by BMPs. Differences between the unwebbed chicken foot and the webbed duck foot can be explained by differences in the expression of Gremlin, a protein that antagonizes BMPs.

10. The BMPs are invoved both in inducing apoptosis and in differentiating the mesenchymal cells into cartilage. The effects of BMPs can be regulated by the Noggin and Gremlin proteins.

11. The ends of the long bones of humans and other mammals contain cartilagenous regions called epiphyseal growth plates. The cartilage in these regions proliferates so that the bone grows larger. Eventually, the cartilage is replaced by bone and growth stops.

For Further Reading

Complete bibliographical citations for all literature cited in this chapter can be found on the Vade Mecum CD that accompanies the book and at the free access website www.devbio.com

Dahn, R. D. and J. F. Fallon. 2000. Interdigital regulation of digit identity and homeotic transformation by modulating BMP signaling. *Science* 289: 438–441.

Mahmood, R. and 9 others. 1995. A role for FGF-8 in the initiation and maintenance of vertebrate limb outgrowth. *Curr. Biol.* 5: 797–806.

Merino, R, J. Rodriguez-Leon, D. Macias, Y. Gañan, A. N. Economides and J. M. Hurle. 1999. The BMP antagonist Gremlin regulates outgrowth, chondrogenesis and programmed cell death in the developing limb. *Development* 126: 5515–5522.

Niswander, L., S. Jeffrey, G. R. Martin and C. Tickle. 1994. A positive feedback loop coordinates growth and patterning in the vertebrate limb. *Nature* 371: 609–612.

Ohuchi, H. and 7 others. 1998. Correlation of wing-leg identity in ectopic FGF-induced chimeric limbs with the differential expression of chick *Tbx5* and *Tbx4*. *Development* 125: 51–60.

Riddle, R. D., R. L. Johnson, E. Laufer and C. Tabin. 1993. Sonic hedgehog mediates the polarizing activity of the ZPA. *Cell* 75: 1401–1416.

Sekine, K. and 10 others. 1999. Fgf10 is essential for limb and lung formation. *Nature Genet.* 21: 138–141.

Vortkamp, A, K. Lee, B. Lanske, G. V. Segre, H. M. Kronenberg and C. J. Tabin. 1996. Regulation of rate of cartilage differentiation by Indian hedgehog and PTH-related protein. *Science* 273: 613–622.

Wellik, D. M. and M. R. Capecchi. 2003. *Hox10* and *Hox11* genes are required to globally pattern the mammalian skeleton. *Science* 301: 363–367.

Sex Determination

17

HOW AN INDIVIDUAL'S SEX IS DETERMINED has been one of the great questions of embryology since antiquity. Aristotle, who collected and dissected embryos, claimed that sex was determined by the heat of the male partner during intercourse (Aristotle, ca. 335 B.C.E.). The more heated the passion, the greater the probability of male offspring. He also counseled elderly men to conceive in the summer if they wished to have male heirs.

Aristotle promulgated a very straightforward hypothesis of sex determination: women were men whose development was arrested too early. The female was "a mutilated male" whose development had stopped because the coldness of the mother's womb overcame the heat of the father's semen. Women were therefore colder and more passive than men, and female sex organs did not mature to the point where they could provide active seeds.

Historical Views on Sex Determination

Aristotle's views on sex determination were accepted both by the Christian Church and by the Graeco-Roman physician Galen, whose anatomy texts were the standard for over a thousand years.* Around the year 190 C.E., Galen wrote:

> Just as mankind is the most perfect of all animals, so within mankind, the man is more perfect than the woman, and the reason for this perfection is his excess heat, for heat is Nature's primary instrument … the woman is less perfect than the man in respect to the generative parts. For the parts were formed within her when she was still a fetus, but could not because of the defect in heat emerge and project on the outside.

The view that women were but poorly developed men and that their genitalia were like men's, only turned inside out, remained popular for over a thousand years. As late as 1543, Andreas Vesalius, the Paduan anatomist who overturned much of Galen's anatomy (and who risked censure by the church when he argued that men and women have the same number of ribs), held this view. The illustrations from his two major works, *De Humani Corporis Fabrica* and *Tabulae Sex*, show that he saw the female genitalia as internal representations of the male genitalia. Nevertheless, Vesalius' books sparked a revolution in anatomy, and by the end of the 1500s, anatomists had dismissed Galen's representation of female anatomy. During the 1600s and 1700s, it became accepted that females produced eggs

---

*The imperial physician to Emperors Marcus Aurelius and Commodus, Galen first achieved fame as a physician to gladiators, from whose wounds and corpses he undoubtedly learned much anatomy.

> " Sexual reproduction is … the masterpiece of nature. "
> *ERASMUS DARWIN (1791)*

> " It is quaint to notice that the number of speculations connected with the nature of sex have well-nigh doubled since Drelincourt, in the eighteenth century, brought together two hundred and sixty-two "groundless hypotheses," and since Blumenbach caustically remarked that nothing was more certain than that Drelincourt's own theory formed the two hundred and sixty-third. "
> *J. A. THOMSON (1926)*

♂   ♀

that could transmit parental traits, and the physiology of the sex organs began to be studied. Still, there was no consensus about how a person's sex became determined (see Horowitz 1976; Tuana 1988; Schiebinger 1989).

Until the twentieth century, the environment—temperature and nutrition, in particular—was believed to be important in determining sex. In 1890, Geddes and Thomson summarized all available data on sex determination and came to the conclusion that the "constitution, age, nutrition, and environment of the parents must be especially considered" in any such analysis. They argued that factors favoring the storage of energy and nutrients predisposed one to have female offspring, whereas factors favoring the utilization of energy and nutrients influenced one to have male offspring.

This environmental view of sex determination remained the only major scientific theory until the rediscovery of Mendel's work in 1900 and the rediscovery of the sex chromosome by McClung in 1902. It was not until 1905, however, that the correlation in insects of the female sex with XX sex chromosomes and the male sex with XY or XO chromosomes was established (Stevens 1905; Wilson 1905). This correlation suggested strongly that a specific nuclear component was responsible for directing the development of the sexual phenotype. Evidence that sex determination occurs by nuclear inheritance rather than by environmental happenstance continued to accumulate.

Today we know that both environmental and internal mechanisms of sex determination can operate in different species. We will first discuss the chromosomal mechanisms of sex determination and then consider the ways in which the environment regulates the sexual phenotype.

CHROMOSOMAL SEX DETERMINATION

There are many ways chromosomes can determine the sex of an embryo. In mammals, the presence of either a second X chromosome or a Y chromosome determines whether the embryo is to be female (XX) or male (XY). In birds, the situation is reversed (Smith and Sinclair 2001): the male has the two similar sex chromosomes (ZZ), while the female has the unmatched pair (ZW). In flies, the Y chromosome plays no role in sex determination, but the ratio of X chromosomes to autosomes (the non-sex chromosomes) determines the sexual phenotype. In other insects (especially hymenopterans such as bees, wasps, and ants), fertilized, diploid eggs develop into females, while the unfertilized, haploid eggs become male (Beukeboom 1995).

This chapter will discuss only two of the many chromosomal modes of sex determination: sex determination in placental mammals and sex determination in *Drosophila*.

The Mammalian Pattern: Primary and Secondary Sex Determination

Primary sex determination is the determination of the gonads—the ovaries or testes. In mammals, primary sex determination is chromosomal and is not usually influenced by the environment. The formation of ovaries and of testes are both active, gene-directed processes. There is no "default state" in mammalian primary sex determination. Moreover, as we shall see, both the male and female gonads diverge from a common precursor, the **bipotential gonad**.

In most cases, the female's karyotype (chromosome complement) is XX and the male's is XY (Figure 17.1A). Every individual must carry at least one X chromosome. Since the female is XX, each of her haploid eggs has a single X chromosome. The male, being XY, generates two types of haploid sperm: half will bear an X chromosome, half a Y. If at fertilization the egg receives a second X chromosome from the sperm, the resulting individual is XX, forms ovaries, and is female; if the egg receives a Y chromosome from the sperm, the individual is XY, forms testes, and is male.

The Y chromosome is a crucial factor for determining sex in mammals. The Y chromosome carries a gene that encodes a **testis-determining factor**, which organizes the gonad into a testis rather than an ovary. A person with five X chromosomes and one Y chromosome (XXXXXY) would be male. Furthermore, an individual with a single X chromosome and no second X or Y (i.e., XO) develops as a female and begins making ovaries (although the ovarian follicles cannot be maintained; for a complete ovary, a second X chromosome is needed).

Secondary sex determination affects the phenotype outside the gonads. This includes the male or female duct systems and external genitalia. A male mammal has a penis, a scrotum (testicle sac), seminal vesicles, and prostate gland. A female mammal has a vagina, clitoris, labia, cervix, uterus, oviducts, and mammary glands. In many species, each sex has a sex-specific body size, vocal cartilage, and musculature. Secondary sex characteristics are usually determined by hormones secreted from the gonads. In the absence of gonads, however, the female phenotype is generated. When Jost (1947, 1953) removed fetal rabbit gonads before they had differentiated, the resulting rabbits had a female phenotype, regardless of whether their genotype was XX or XY.

The general scheme of mammalian sex determination is shown in Figure 17.1B. If the Y chromosome is absent, the gonadal primordia develop into ovaries. The ovaries produce **estrogen**, a hormone that enables the development of the **Müllerian duct** into the uterus, oviducts, and upper end of the vagina (Fisher et al. 1998; Couse et al.

(A)

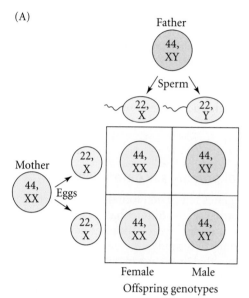

Father
44, XY

Sperm

22, X   22, Y

Mother
44, XX

Eggs

	22, X	22, X
22, X	44, XX	44, XY
22, X	44, XX	44, XY

Female   Male

Offspring genotypes

FIGURE 17.1 Sex determination in mammals. (A) Mammalian chromosomal sex determination results in approximately equal numbers of male and female offspring. (B) Postulated cascades leading to the male and female phenotypes in mammals. The conversion of the genital ridge into the bipotential gonad requires the *LHX9*, *SF1*, and *WT1* genes, since mice lacking any of these genes lack gonads. The bipotential gonad appears to be moved into the female pathway (ovary development) by the *WNT4* and *DAX1* genes, and into the male pathway (testis development) by the *SRY* gene (on the Y chromosome) in conjunction with autosomal genes such as *SOX9*. (Lower levels of Dax1 and Wnt4 are also present in the male gonad.) The ovary makes thecal cells and granulosa cells, which together are capable of synthesizing estrogen. Under the influence of estrogen (first from the mother, then from the fetal gonads), the Müllerian duct differentiates into the female reproductive tract, the internal and external genitalia develop, and the offspring develops the secondary sex characteristics of a female. The testis makes two major hormones. The first, anti-Müllerian hormone (AMH), causes the Müllerian duct to regress. The second, testosterone, causes the differentiation of the Wolffian duct into the male internal genitalia. In the urogenital region, testosterone is converted into dihydrotestosterone (DHT), and this hormone causes the morphogenesis of the penis and prostate gland. (B after Marx 1995 and Birk et al. 2000.)

(B)

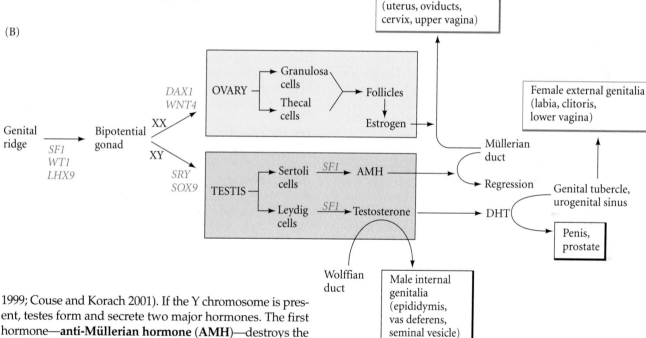

1999; Couse and Korach 2001). If the Y chromosome is present, testes form and secrete two major hormones. The first hormone—**anti-Müllerian hormone (AMH)**—destroys the Müllerian duct. The second hormone—**testosterone**—masculinizes the fetus, stimulating the formation of the penis, male duct system, scrotum, and other portions of the male anatomy, as well as inhibiting the development of the breast primordia. Thus, the body has the female phenotype unless it is changed by the two hormones secreted by the fetal testes. We will now take a more detailed look at these events.

Primary Sex Determination

The developing gonads

Mammalian gonads embody a unique embryological situation. All other organ rudiments normally can differentiate into only one type of organ. A lung rudiment can only

become a lung, and a liver rudiment can develop only into a liver. The gonadal rudiment, however, has two options: it can develop into either an ovary or a testis. The path of differentiation taken by this rudiment is dictated by the genotype and determines the future sexual development of the organism (Lillie 1917). But before this decision is made, the mammalian gonad first develops through a **bipotential** or **indifferent stage**, during which time it has neither female nor male characteristics.

In humans, the bipotential gonadal rudiments appear during week 4 and remain sexually indifferent until week 7. The gonadal rudiments are paired regions of the intermediate mesoderm; they form adjacent to the developing kidneys. The ventral portions of these rudiments comprise the genital ridge epithelium. During the indifferent stage, the genital ridge epithelium proliferates (Figure 17.2A,B). These epithelial layers will form the somatic (i.e., non-germ cell) component of the gonads. The germ cells migrate into the gonad during week 6, and are surrounded by the somatic cells.

MALE GONADAL DEVELOPMENT   If the fetus is XY, the somatic cells continue to proliferate through the eighth week, and then initiate their differentiation into **Sertoli cells**. During week 8, the developing Sertoli cells organize themselves into the **testis cords**. These cords form loops in the medullary (central) region of the developing testes and are connected to a network of thin canals, called the **rete testis**, located near the mesonephric duct (Figure 17.2C,D). Eventually, the testis sex cords become separated from the surface epithelium by a thick extracellular matrix, the **tunica albuginea**. Thus, when the germ cells enter the male gonads, they will develop within the testis cords, inside the organ.

The Sertoli cells of the fetal testis cords secrete the anti-Müllerian hormone that blocks development of the female ducts, and these cells will later support the development of sperm throughout the lifetime of the male mammal. Meanwhile, during fetal development, the interstitial mesenchyme cells of the testes differentiate into **Leydig cells**, which make testosterone.

At puberty, the testis cords hollow out to form the **seminiferous tubules**; the germ cells migrate to the periphery of these tubules, where they begin to differentiate into sperm (see Figure 19.21). In the mature seminiferous tubule, the sperm are transported from the inside of the testis through the rete testis, which joins the **efferent ducts**. These efferent ducts are the remodeled tubules of the mesonephric kidney. They link the testis to the **Wolffian duct**, which used to be the collecting tube of the mesonephric kidney* (see Chapter 14). In males, the Wolffian duct differentiates to become the **epididymis** (adjacent to the testis) and the **vas deferens**, the tube through which the sperm pass into the urethra and out of the body.

FEMALE GONADAL DEVELOPMENT   In females, the germ cells accumulate near the outer surface of the gonad, interspersed with the gonadal somatic cells. Near the time of birth, each individual germ cell is surrounded by somatic cells (Figure 17.2E,F). The germ cells will become the ova, and the surrounding cortical somatic cells will differentiate into **granulosa cells**. The mesenchyme cells of the ovary differentiate into **thecal cells**. Together, the thecal and granulosa cells form **follicles** that envelop the germ cells and secrete steroid hormones. Each follicle will contain a single germ cell—an oogonium (egg precursor)—which will enter meiosis at this time. These germ cells are required for the gonadal cells to complete their differentiation into ovarian tissue[†] (McLaren 1991). In females, the Müllerian duct remains intact and differentiates into the oviducts, uterus, cervix, and upper vagina. In the absence of adequate testosterone, the Wolffian duct degenerates. A summary of the development of mammalian reproductive systems is shown in Figure 17.3.

Mechanisms of primary sex determination

Several human genes have been found whose function is necessary for normal sexual differentiation. Since the phenotype of mutations in sex-determining genes is often sterility, clinical infertility studies have been useful in identifying those genes that are active in determining whether humans become male or female. Experimental manipulations to confirm the functions of these genes can then be done in mice.

Although the story unfolded in the following paragraphs demonstrates the remarkable progress that has been made in recent years, we still do not fully understand how all these gonad-determining genes interact. The problem of primary sex determination remains (as it has since prehistory) one of biology's great unsolved mysteries.

SRY: THE Y CHROMOSOME SEX DETERMINANT   In humans, the major gene for testis determination resides on the short arm of the Y chromosome. Individuals who are born with the short but not the long arm of the Y chromosome are male, while individuals born with the long arm of the Y chromosome but not the short arm are female. By analyzing the DNA of rare XX men and XY women, the position of the testis-determining gene was narrowed down to a 35,000-base-pair region of the Y chromosome located near the tip of the short arm. In this region, Sinclair and colleagues (1990) found a male-specific DNA sequence that could encode a peptide of 223 amino acids. This peptide is probably a transcription factor, since it contains a DNA-binding domain called the **high-mobility group**, or **HMG**, box. The HMG box is found in several transcription factors and nonhistone chromatin proteins, and it induces bending in the region of DNA to which it binds (Giese et al. 1992). This gene is called *SRY* (*sex-determining region of*

---

*As discussed in Chapter 14, the mesonephric kidney is one of the three kidney types seen during mammalian development, but it does not function as a kidney in most mammals.

[†]There is a reciprocal relationship between the germ cells and the gonadal somatic cells. The germ cells are originally bipotential and can become either sperm or eggs. Once in the male or female sex cords, however, they are instructed either to begin (and remain in) meiosis and become eggs, or to remain mitotically dormant and become spermatogonia (McLaren 1995; Brennan and Capel 2004). In XX gonads, germ cells are essential for the maintenance of ovarian follicles. Without germ cells, the follicles degenerate into cord-like structures and express male-specific markers. In XY gonads, the germ cells help support Sertoli cells differentiation, although testes cords will form without them, albeit a bit later.

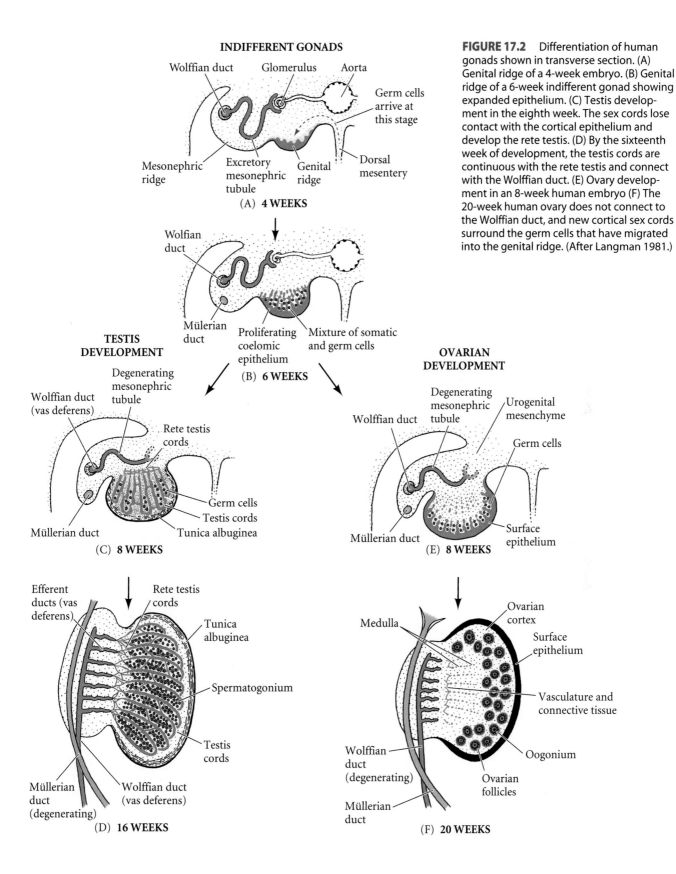

INDIFFERENT GONADS

Wolffian duct   Glomerulus   Aorta

Germ cells arrive at this stage

Mesonephric ridge   Excretory mesonephric tubule   Genital ridge   Dorsal mesentery

(A) **4 WEEKS**

Wolfian duct

Müelerian duct   Proliferating coelomic epithelium   Mixture of somatic and germ cells

(B) **6 WEEKS**

TESTIS DEVELOPMENT

Degenerating mesonephric tubule

Wolffian duct (vas deferens)

Rete testis cords

Germ cells
Testis cords
Tunica albuginea

Müllerian duct

(C) **8 WEEKS**

Efferent ducts (vas deferens)   Rete testis cords

Tunica albuginea

Spermatogonium

Testis cords

Müllerian duct (degenerating)   Wolffian duct (vas deferens)

(D) **16 WEEKS**

OVARIAN DEVELOPMENT

Degenerating mesonephric tubule   Urogenital mesenchyme

Wolffian duct

Germ cells

Müllerian duct   Surface epithelium

(E) **8 WEEKS**

Medulla   Ovarian cortex
Surface epithelium

Vasculature and connective tissue

Oogonium

Wolffian duct (degenerating)   Ovarian follicles

Müllerian duct

(F) **20 WEEKS**

FIGURE 17.2 Differentiation of human gonads shown in transverse section. (A) Genital ridge of a 4-week embryo. (B) Genital ridge of a 6-week indifferent gonad showing expanded epithelium. (C) Testis development in the eighth week. The sex cords lose contact with the cortical epithelium and develop the rete testis. (D) By the sixteenth week of development, the testis cords are continuous with the rete testis and connect with the Wolffian duct. (E) Ovary development in an 8-week human embryo (F) The 20-week human ovary does not connect to the Wolffian duct, and new cortical sex cords surround the germ cells that have migrated into the genital ridge. (After Langman 1981.)

FIGURE 17.3 The development of the gonads and their ducts in mammals. Note that both the Wolffian and Müllerian ducts are present at the indifferent gonad stage.

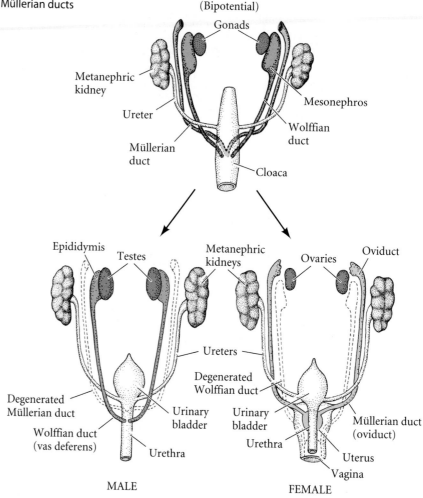

the Y chromosome), and there is extensive evidence that it is indeed the gene that encodes the human testis-determining factor.

SRY is found in normal XY males and also in the rare XX males; it is absent from normal XX females and from many XY females. Another group of XY females have the *SRY* gene, but their copies of the gene contain point or frameshift mutations that prevent the SRY protein from binding to or bending DNA (Pontiggia et al. 1994; Werner et al. 1995).

If the *SRY* gene actually does encode the major testis-determining factor, one would expect that it would act in the genital ridge immediately before or during testis differentiation. This prediction has been found to be the case in studies of the homologous gene in mice. The mouse gene (*Sry*) also correlates with the presence of testes; it is present in XX males and absent in XY females (Gubbay et al. 1990; Koopman et al. 1990). The *Sry* gene is expressed in the somatic cells of the bipotential mouse gonads of XY mice immediately before the differentiation of these cells into Sertoli cells; its expression then disappears (Hacker et al. 1995; Sekido et al. 2004).

The most impressive evidence for *Sry* being the gene for testis-determining factor comes from transgenic mice. If *Sry* induces testis formation, then inserting *Sry* DNA into the genome of a normal XX mouse zygote should cause that XX mouse to form testes. Koopman and colleagues (1991) took the 14-kilobase region of DNA that includes the *Sry* gene (and presumably its regulatory elements) and microinjected this sequence into the pronuclei of newly fertilized mouse zygotes. In several instances, XX embryos injected with this sequence developed testes, male accessory organs, and penises (Figure 17.4).* There are thus good reasons to think that *Sry/SRY* is the major gene on the Y chromosome for testis determination in mammals.

Although there must be interaction between Sry protein and some other factor(s), the cofactors and targets of the *Sry* gene product are not yet known. It had been speculated that Sry would bind to DNA, and that it would

---

*Functional sperm were not formed—but they were not expected, either. The presence of two X chromosomes prevents sperm formation in XXY mice and men, and the transgenic mice lacked the rest of the Y chromosome, which contains genes needed for spermatogenesis.

GONADS		
Gonadal type	Testis	Ovary
Germ cell location	Inside testis cords (in medulla of testis)	Inside follicles of ovarian cortex
DUCTS		
Remaining duct	Wolffian	Müllerian
Duct differentiation	Vas deferens, epididymis, seminal vesicle	Oviduct, uterus, cervix, upper portion of vagina

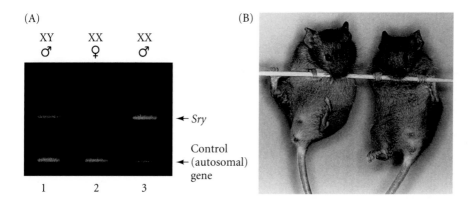

FIGURE 17.4 An XX mouse transgenic for *Sry* is male. (A) Polymerase chain reaction followed by electrophoresis shows the presence of the *Sry* gene in normal XY males and in a transgenic XX/*Sry* mouse. The gene is absent in a female XX littermate. (B) The external genitalia of the transgenic mouse are male (right) and are essentially the same as those in an XY male (left). (From Koopman et al. 1991; photographs courtesy of the authors.)

bend the DNA to bring distantly bound proteins of the transcription apparatus into close contact, enabling these proteins to interact and initiate transcription (Pontiggia et al. 1994; Werner et al. 1995). But no such proteins have been found. Recent evidence suggests that Sry might act in spliceosomes at the level of RNA processing. Ohe and colleagues (2002) showed that human SRY protein co-localizes with spliceosomes and that it functions in processing pre-mRNA. Such a speckled localization pattern (characteristic of splicing factors) has been seen in Sertoli cells (Poulat et al. 1995).

 WEBSITE **17.2 Finding the male-determining genes.** The mapping of the testis-determining factor to the *SRY* region took scientists more than 50 years to accomplish. Moreover, other testis-forming genes that act downstream of *SRY* have been found on autosomes.

***SOX9:* AN AUTOSOMAL TESTIS-DETERMINING GENE** SRY may have more than one mode of action in converting the bipotential gonads into testes. It had been assumed for the past decade that SRY worked directly in the genital ridge to convert the somatic epithelial cells of the bipotential gonad into male-specific Sertoli cells. However, no mouse genes have been found whose expression is activated by the binding of Sry to their promoters, enhancers, or to the splicing regions of their transcripts. The most promising candidate for this role is another HMG-box protein, **SOX9**. *SOX9* is an autosomal gene that is involved in several developmental processes (most notably bone formation). It can also induce testis formation. XX humans who have an extra copy of *SOX9* develop as males, even if they have no *SRY* gene, and XX mice made transgenic for *Sox9* develop testes (Figure 17.5; Huang et al. 1999; Qin and Bishop 2005). Knocking out the *Sox9* genes in the gonads of XY mice causes complete sex reversal (Barrionuevo et al. 2006). Thus, even if Sry is present, mouse gonads cannot form testes if Sox9 is absent. Therefore, it appears that *SOX9* can replace *SRY* in testis formation. This is not altogether surprising: While *Sry* is found specifically in mammals, *Sox9* is found throughout the vertebrate phyla. *Sox9* may be the

older and more central sex determination gene, and in mammals it may be activated by its relative, *Sry*. Expression of the *Sox9* gene is specifically upregulated by the transient expression of *Sry* in Sertoli cell precursors. Thus, *Sry* may act merely as a "switch" to activate *Sox9*, and the Sox9 protein may initiate the conserved evolutionary pathway to testis formation (Pask and Graves 1999; Sekido et al. 2004).

Sox9 protein may act as both a splicing factor and a transcriptional regulator. Sox9 migrates into the nucleus at the time of sex determination. Here, it binds to a promoter site on the gene for the anti-Müllerian hormone, providing a critical link in the pathway toward a male phenotype (Arango et al. 1999; de Santa Barbara et al. 2000). SOX9 also appears to be involved in RNA splicing and can replace missing splicing factors in experimental splicing assays (Ohe et al. 2002). The search is on for the genes that Sry or Sox9 might regulate (Koopman 2001).

Some major questions concerning male sex determination remain to be answered, including: (1) What activates the *Sry* gene? (2) What does the Sry protein activate? (3) How might this activation work? (4) Is the *Sox9* gene activated by Sry? and (5) What genes does Sox9 activate? As we will see, sex determination in *Drosophila* works largely via RNA processing; if Sry and Sox9 act as RNA splicing factors, then the mammalian and *Drosophila* schemes of sex determination may have more in common than originally thought.

FIBROBLAST GROWTH FACTOR 9 Another gene that may be regulated (directly or indirectly) by Sry or Sox9 is the gene for fibroblast growth factor-9 (*Fgf9*; Capel et al. 1999; Colvin et al. 2001). When the *Fgf9* gene is knocked out in mice, the homozygous mutants are almost all female. Fgf9 protein accomplishes several functions. First, it causes the proliferation of the Sertoli cell precursors and stimulates their differentiation (Schmahl et al. 2004; Willerton et al. 2004). Second, it acts to bring mesonephric cells into the gonad. Incubating XX gonads in Fgf9 allowed them to attract mesonephric cells into the gonad (Figure 17.6). These mesonephric cells are important in forming the testis cords; they may contribute directly to the cords and may

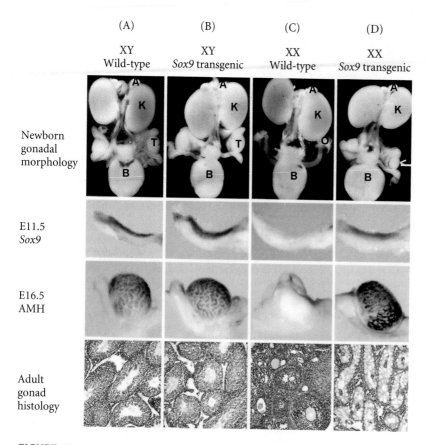

| | (A)
XY
Wild-type | (B)
XY
Sox9 transgenic | (C)
XX
Wild-type | (D)
XX
Sox9 transgenic |

Newborn gonadal morphology

E11.5 *Sox9*

E16.5 AMH

Adult gonad histology

FIGURE 17.5   Ability of *Sox9* to generate testes. (A) A wild-type XY mouse embryo expresses *Sox9* in the genital ridge at 11.5 days, anti-Müllerian duct hormone (AMH) in the embryonic gonad Sertoli cells at 16.5 days, and eventually forms descended testes (T) with seminiferous tubules. (B) An XY embryo with the *Sox9* transgene (a control for the effects of the transgene) also shows *Sox9* expression, AMH expression, and descended testes with seminiferous tubules. (C) The wild-type XX embryo shows neither *Sox9* expression nor AMH. It constructs ovaries with mature follicle cells. (D) An XX embryo with the *Sox9* transgene expresses the *Sox9* gene and has AMH in its 16.5-day Sertoli cells. It has descended testes, but the seminiferous tubules lack sperm (due to the presence of two X chromosomes in the Sertoli cells). K, kidneys; A, adrenal glands; B, bladder; T, testis; O, ovary. (From Vidal et al. 2001; photographs courtesy of A. Schedl.)

also help organize the Sertoli cell precursors to form the cords. (As mentioned earlier, this mesonephric duct is critical for the exit of the sperm cells from the testes).

SF1: THE LINK BETWEEN SRY AND THE MALE DEVELOPMENTAL PATHWAYS   Another protein that may be directly or indirectly activated by Sry is the transcription factor **Sf1** (steroidogenic factor 1). Sf1 is necessary to make the bipotential gonad, but whereas Sf1 levels decline in the genital ridge of XX mouse embryos, the *Sf1* gene stays on in the developing testis. Sf1 appears to be active in masculinizing both the Leydig cells and the Sertoli cells. In the Sertoli cells, Sf1, working in collaboration with Sox9, is needed to elevate the levels of anti-Müllerian hormone transcription (Shen et al. 1994; Arango et al. 1999). In the Leydig cells, Sf1 activates the genes encoding the enzymes that make testos-

terone. In humans, the importance of SF1 for testis development and AMH regulation is demonstrated by an XY patient who is heterozygous for *SF1*. Although the genes for SRY and SOX9 are normal, this individual has malformed fibrous gonads and retains fully developed Müllerian duct structures (Achermann et al. 1999). It is thought that SRY (directly or indirectly) maintains *SF1* expression, and that the SF1 protein is then involved in the production of testosterone in the Leydig cells, and (in collaboration with SOX9), the activation of AMH in the Sertoli cells.

***DAX1:* A POTENTIAL TESTIS-SUPPRESSING GENE ON THE X CHROMOSOME**   All the above mentioned genes may effect the formation of the testes, but what about genes that produce the ovary? At present, we know of no gene that is ovary-specific. However, there are genes that may affect the amount and

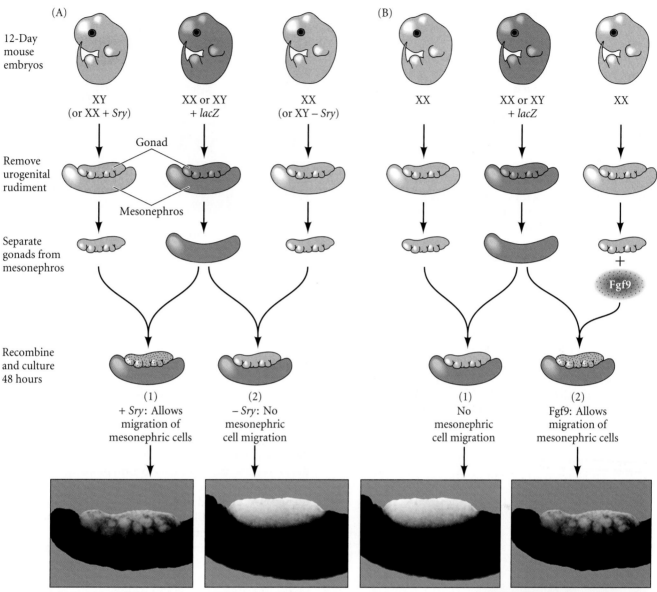

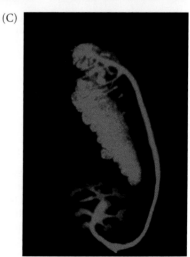

FIGURE 17.6 Migration of the mesonephric cells into *Sry*⁺ gonadal rudiments. In the experiment diagrammed, urogenital ridges (containing both the mesonephric kidneys and bipotential gonadal rudiments) were collected from 12-day embryonic mice. Some of the mice were marked with a β-galactosidase transgene (*lacZ*) that is active in every cell. Thus, every cell of these mice turned blue when stained for β-galactosidase. The gonad and mesonephros were separated and recombined, using gonadal tissue from unlabeled mice and mesonephros from labeled mice. (A) Migration of mesonephric cells into the gonad was seen (1) when the gonadal cells were XY or (2) when they were XX with an *Sry* transgene. No migration of mesonephric tissue into the gonad was seen when the gonad contained either XX cells or XY cells in which the Y chromosome had a deletion in the *Sry* gene. The sex chromosomes of the mesonephros did not affect the migration. (B) Gonadal rudiments for XX mice could induce mesonephric cell migration if these rudiments had been incubated with Fgf9. (C) Intimate relation between the mesonephric ducts and the developing gonad in the 16-day male mouse embryo. The duct tissue has been stained for cytokeratin-8. WD, Wolffian duct. (A,B after Capel et al. 1999, photographs courtesy of B. Capel; C from Sariola and Saarma 1999, photograph courtesy of H. Sariola.)

pattern of the proteins encoded by the testis-forming genes in ways that cause the development of the female gonads. One of these genes is *DAX1*.

In 1980, Bernstein and her colleagues described two sisters who were genetically XY. Their Y chromosomes were normal, but they had a duplication of a small portion of the short arm of the X chromosome. When similar cases were found, it was concluded that if there are two copies of this region on an active X chromosome, the SRY signal is inhibited (Figure 17.7). Bardoni and her colleagues (1994) proposed that this region might contain a gene encoding a protein that competes with the SRY factor and that is important in directing the development of the ovary. In XY embryos, this gene would be suppressed, but having two active copies of the gene would override this suppression. This gene, *Dax1*, has been cloned in mice and shown to encode a member of the nuclear hormone receptor family (Muscatelli et al. 1994; Zanaria 1994). *Dax1* is initially expressed in the genital ridges of both male and female mouse embryos, and it is seen in male mice shortly after *Sry* expression. Indeed, in XY mice, *Sry* and *Dax1* are expressed in the same cells, and both these genes are needed for testis formation (Meeks et al. 2003; Bauma et al. 2005). Eventually, however, *Dax1* is expressed solely in the XX gonadal rudiment (Figure 17.8). *Dax1* appears to antagonize the function of *Sry* and *Sox9*, and it downregulates *Sf1* expression (Nachtigal et al. 1998; Swain et al. 1998; Iyer and McCabe 2004).

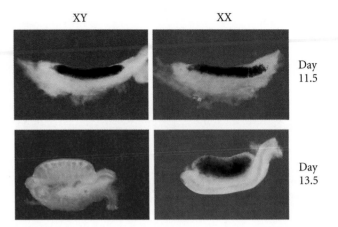

FIGURE 17.8 *Dax1* expression in the mouse genital ridge. A *lacZ* gene was fused to the regulatory region of *Dax1* and the resulting transgene inserted into XY (left) and XX (right) mouse embryos. In both instances, *Dax1* was expressed at day 11.5 in the genital ridges. By day 13.5, however, *Dax1* expression persisted in the XX gonadal rudiments, but not in those of the XY embryos. (After Swain et al. 1998; photographs courtesy of R. Lovell-Badge.)

WNT4: A POTENTIAL OVARY-DETERMINING GENE ON AN AUTOSOME Wnt4 is a paracrine factor that represses male development in the female gonad. It is expressed in the mouse genital ridge while it is still in its bipotential stage. Wnt4 expression then becomes undetectable in XY gonads (which become testes), whereas it is maintained in XX gonads as they begin to form ovaries. In transgenic XX mice that lack the *Wnt4* gene, the ovary fails to form properly, and its cells express testis-specific markers, including testosterone-producing enzymes and AMH (Vainio et al. 1999; Heikkila et al. 2005). One possible target for Wnt4 is the gene encoding TAFII 105 (Freiman et al. 2002). This subunit of the TATA-binding protein for RNA polymerase binding is seen only in ovarian follicle cells. Female mice lacking this subunit have no ovaries. In XY humans having a duplication of the *WNT4* region, DAX1 is overproduced and the gonads develop into ovaries (Jordan et al. 2001). SRY may form testes by repressing *WNT4* expression in the genital ridge, as well as by promoting *FGF9*. One possible model is shown in Figure 17.9.

The right time and the right place

Having the right genes doesn't necessarily mean you'll get the organ you expected. Timing is critical. Studies on mice have shown that the *Sry* gene of some strains of mice failed to produce testes when placed in a different strain of mouse (Eicher and Washburn 1983; Washburn and Eicher 1989; Eicher et al. 1996). The transplanted *Sry* gene is late in becoming expressed in the host mouse, delaying the onset of *Sox9* expression. By the time *Sox9* gets turned on, it is too late—the gonad is already following the path to become an ovary (Bullejos and Koopman 2005). This provides an important clue to how primary sex determination

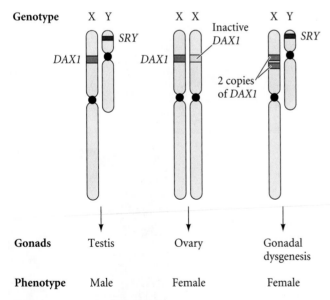

FIGURE 17.7 Phenotypic sex reversal in humans having two copies of the *DAX1* locus. *DAX1* (on the X chromosome) plus *SRY* (on the Y chromosome) produces testes. *DAX1* without *SRY* (since the other *DAX1* locus is on the inactive X chromosome) produces ovaries. Two active copies of *DAX1* (on the active X chromosome) plus *SRY* (on the Y chromosome) lead to a poorly formed gonad. Since the gonad makes neither AMH nor testosterone, the phenotype is female. (After Genetics Review Group 1995.)

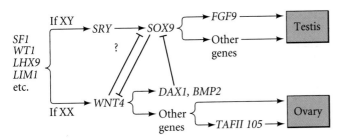

FIGURE 17.9 Possible mechanism for primary sex determination in mammals. While we do not know the specific interactions involved, this model attempts to organize the data into a coherent sequence. Other models are possible. In this model, *SRY* and *WNT4* are both activated in the gonad rudiment. If no SRY protein is present, WNT4 activates ovary-forming genes, and may be responsible for suppressing *SOX9* expression. If SRY is present, it activates *SOX9*, which in turn activates testes-forming genes such as *SF1* and *AMH*, as well as suppressing *WNT4*.

may take place. There may be a brief window of time wherein the testis-forming genes can function. If these genes are not turned on, the ovary-forming pathway is activated.

Secondary Sex Determination in Mammals: Hormonal Regulation of the Sexual Phenotype

Primary sex determination is the formation of either an ovary or a testis from the bipotential gonad. This process, however, does not give the complete sexual phenotype. Secondary sex determination in mammals is the development of the female and male phenotypes in response to hormones secreted by the ovaries and testes. Both female and male secondary sex determination have two major temporal phases. The first phase occurs within the embryo during organogenesis; the second occurs at puberty.

As mentioned earlier, if the bipotential gonads are removed from an embryonic mammal, the female phenotype is realized: the Müllerian ducts develop while the Wolffian ducts degenerate. This pattern is also seen in certain humans who are born without functional gonads. Individuals whose cells have only one X chromosome (and no Y chromosome) originally develop ovaries, but their ovaries atrophy before birth, and their germ cells die before puberty. However, under the influence of prenatal estrogen, derived first from the ovary and then from the mother and placenta, these XO infants are born with a female genital tract (Langman and Wilson 1982).

The formation of the male phenotype involves the secretion of two testicular hormones. The first of these hormones is AMH, the hormone made by the Sertoli cells that causes the degeneration of the Müllerian duct. The second is the steroid hormone testosterone, which is secreted from the fetal Leydig cells. Testosterone causes the Wolffian duct

to differentiate into the epididymis, vas deferens, and seminal vesicles, and it causes the urogenital swellings to develop into the scrotum and penis.

The existence of separate and independent AMH and testosterone pathways of masculinization is demonstrated by people with **androgen insensitivity syndrome**. These XY individuals have the *SRY* gene, and thus have testes that make testosterone and AMH. However, they *lack* the receptor protein for testosterone, and therefore cannot respond to the testosterone made by their testes (Meyer et al. 1975). However, they are able to respond to the estrogen made by their adrenal glands (which is normal for both XX and XY individuals), so they develop the female phenotype (Figure 17.10). Despite their distinctly female appearance, these individuals have testes, and even though they cannot respond to testosterone, they produce and respond to AMH. Thus, their Müllerian ducts degenerate. These people develop as normal but sterile women, lacking a uterus and oviducts and having testes in the abdomen.

Conditions in which male and female traits are seen in the same individual are called **intersex** conditions. Androgen insensitivity syndrome is one of several intersex conditions that have been labeled **pseudohermaphroditism**. In pseudohermaphrodites, there is only one type of gonad, but the secondary sex characteristics differ from what would be expected from the gonadal sex. In humans, male pseudohermaphroditism (wherein the gonadal sex is male

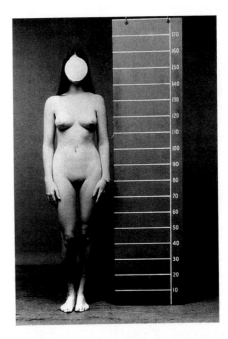

FIGURE 17.10 An XY individual with androgen insensitivity syndrome. Despite the XY karyotype and the presence of testes, such individuals develop female secondary sex characteristics. Internally, however, these women lack the Müllerian duct derivatives and have undescended testes. (Photograph courtesy of C. B. Hammond.)

and the secondary sex characteristics are female) can be caused by mutations in the androgen (testosterone) receptor or by mutations affecting testosterone synthesis (Geissler et al. 1994). Female pseudohermaphroditism (in which the gonadal sex is female but the person is outwardly male) can be caused by the overproduction of androgens in the ovary or adrenal gland. The most common cause of this latter condition is **congenital adrenal hyperplasia**, in which there is a genetic deficiency of an enzyme that metabolizes cortisol steroids in the adrenal gland. In the absence of this enzyme, testosterone-like steroids accumulate and can bind to the androgen receptor, thus masculinizing the fetus (Migeon and Wisniewski 2001; Merke et al. 2002). Thus, pseudohermaphroditism is the result of abnormalities of *secondary* sex determination.

"True" hermaphrodites, on the other hand, contain both male and female gonadal tissue. Thus, in mammals, they result from abnormalities of *primary* sex determination. True hermaphroditism can result when a Y chromosome is translocated to an X chromosome. In those tissues where the translocated Y is on the active X chromosome, the Y chromosome will be active and the *SRY* gene will be transcribed; in those cells where the Y chromosome is on the inactive X chromosome, the Y chromosome will be inactive (Berkovitz et al. 1992; Margarit et al. 2000). A gonadal mosaic for expression of *SRY* can develop into a testis, an ovary, or an ovotestis (having both tissue types) depending on the percentage of cells expressing *SRY* in the Sertoli cell precursors (see Brennan and Capel 2004).*

Testosterone and dihydrotestosterone

Although testosterone is one of the two primary masculinizing hormones, there is evidence that it might not be the active masculinizing hormone in certain tissues. Testosterone appears to be responsible for promoting the formation of the male reproductive structures (the epididymis, seminal vesicles, and vas deferens) that develop from the Wolffian duct primordium. However, it does not directly masculinize the male urethra, prostate, penis, or scrotum. These latter functions are controlled by **5α-dihydrotestosterone**, or **DHT** (Figure 17.11). Siiteri and Wilson (1974) showed that testosterone is converted to DHT in the urogenital sinus and swellings, but not in the Wolffian duct. DHT appears to be a more potent hormone than testosterone.

The importance of DHT was demonstrated by Imperato-McGinley and her colleagues (1974). They found a small community in the Dominican Republic in which several inhabitants lacked a functional gene for the enzyme 5α-ketosteroid reductase 2—the enzyme that converts testosterone to DHT (Andersson et al. 1991; Thigpen et al. 1992). Although XY children with this syndrome have function-

ing testes, they have a blind vaginal pouch and an enlarged clitoris. They appear to be girls, and are raised as such. Their internal anatomy, however, is male: they have testes, Wolffian duct development, and Müllerian duct degeneration. Thus it appears that the formation of the external genitalia is under the control of dihydrotestosterone, while Wolffian duct differentiation is controlled by testosterone itself.

Interestingly, at puberty, when the testes of children affected with this syndrome produce high levels of testosterone, their external genitalia are able to respond to the hormone and differentiate. The penis enlarges, the scrotum descends, and the person originally believed to be a girl is revealed to be a young man.

WEBSITE 17.3 Dihydrotestosterone in adult men. The drug finasteride, which inhibits the conversion of testosterone to dihydrotestosterone, is being used to treat prostate growth and male pattern baldness.

WEBSITE 17.4 Insulin-like hormone 3. In addition to testosterone, the Leydig cells secrete insulin-like hormone 3 (INSL3). This hormone is required for the descent of the gonads into the scrotum. Males lacking INSL3 are infertile because the testes do not descend. In females, lack of this hormone deregulates the menstrual cycle.

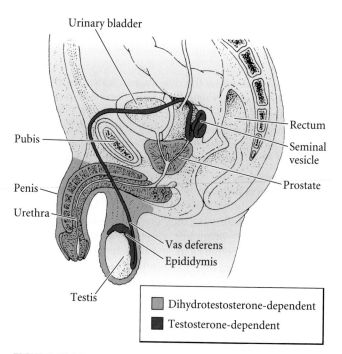

FIGURE 17.11 Testosterone- and dihydrotestosterone-dependent regions of the human male genital system. (After Imperato-McGinley et al. 1974.)

---

*For a more detailed analysis of intersexuality, see Gilbert et al. 2005 or the Intersex Society of North America website, http://isna.org

Anti-Müllerian hormone

Anti-Müllerian hormone (**AMH**; also called **Müllerian inhibiting factor**, or **MIF**) is a member of the TGF-β family of growth and differentiation factors. It is secreted from the fetal Sertoli cells and causes the degeneration of the Müllerian duct (Tran et al. 1977; Cate et al. 1986). AMH is thought to bind to the mesenchyme cells surrounding the Müllerian duct, causing these cells to secrete a paracrine factor that induces apoptosis in the duct's epithelium (Trelstad et al. 1982; Roberts et al. 1999).

 WEBSITE **17.5** **Roles of AMH.** AMH may have other roles in sex determination besides causing the breakdown of the Müllerian ducts. It may cause sex reversal in some mammals, and may become useful as an anti-tumor drug.

Estrogen

The steroid hormone **estrogen** is needed for complete development of both the Müllerian and the Wolffian ducts, and is necessary for fertility in both males and females. In females, estrogen secreted from the fetal ovaries appears sufficient to induce the differentiation of the Müllerian duct into its various components: the uterus, oviducts, and cervix. The extreme sensitivity of the Müllerian duct to estrogenic compounds is demonstrated by the teratogenic effects of **diethylstilbesterol** (**DES**), a powerful synthetic estrogen that can cause infertility by changing the patterning of the Müllerian duct (see Chapter 21).

In female mice with knockouts of the genes for estrogen receptors, the germ cells die in the adult and the granulosa cells that had enveloped them start developing like Sertoli cells (Couse et al. 1999). In male mice with knockouts of estrogen receptor genes, few sperm are made. One of the functions of the efferent duct (vas efferens) cells is to absorb most of the water from the lumen of the rete testis. This absorption of water, which is regulated by estrogen, concentrates the sperm, giving them a longer life span and providing more sperm per ejaculate. If estrogen or its receptor is absent in mice, this water is not absorbed and the mouse is sterile (Hess et al. 1997). While blood concentrations of estrogen are higher in females than in males, the concentration of estrogen in the rete testis is even higher than that in female blood.

 **WEBSITE** **17.6** **Breast development.** Breast tissue has a sexually dimorphic mode of development. Testosterone inhibits breast development, while estrogen promotes it. Most breast development is accomplished after birth, and different hormones act during puberty and pregnancy to cause breast enlargement and differentiation.

Brain sex: Secondary sex determination through another pathway?

In addition to our gonads, another organ—the brain—may also experience direct regulation by the X and Y chromosomes. While it has long been known that the brain, like other tissues, is responsive to the steroid hormones produced by the gonads (see below), new evidence suggests that sex differences in the brain can be observed even before the gonads mature (Arnold and Burgoyne 2004). The first indication that something besides testosterone and estrogens was important in forming sexually different structures in the brain came from studies on Parkinson disease, during which embryonic rat brains were dissected before the gonads matured. These studies indicated that brains from XX embryos had more epinephrine-secreting neurons than XY embryonic brains (Beyer et al. 1991). Using microarrays and PCR, Dewing and colleagues (2003) demonstrated that over 50 genes in mouse brains were expressed in sexually dimorphic patterns *before gonad differentiation had occurred*. Moreover, the mouse *Sry* gene, in addition to being expressed in the embryonic testes, is also expressed in the fetal and adult brain. The human *SRY* gene appears to be expressed in the adult brain as well (Lahr et al. 1995; Mayer et al. 1998, 2000).

Stunning demonstrations that sexual dimorphism in the brain can be caused before gonadal hormone synthesis come from natural and experimental conditions in birds. One big difference between male and female finches is that the males have large regions of their brain devoted to producing songs. Male finches sing; the females do not. While hormones are important in the formation of the song centers in finches (and, when added experimentally, can cause female birds to sing), blocking those hormones in males does not prevent normal development of the song centers or singing. Genetically male birds form these brain regions even without male hormones (Mathews and Arnold 1990).

A natural experiment presented itself in the form of a bird that was half male and half female, divided down the middle (Figure 17.12). Such animals, where some body parts are male and others female, are called **gynandromorphs** (*gynos*, female; *andros*, male; *morpho*, form). Agate and colleagues (2002) showed that the gynandromorph finch had ZZ (male) sex chromosomes on its right side and ZW chromosomes (female) on its left. Its testes made testosterone, and the bird sang like a male and copulated with females. However, although many brain structures were similar on both sides, some brain regions differed between the male and female halves. The song circuits on the right side had a more masculine phenotype than similar structures on the left, showing that both intrinsic and hormonal influences were important.

Gahr (2003) generated his own avian sexual chimeras. He surgically switched the forebrain regions (which control adult sexual behaviors) between ZZ and ZW quail embryos before their gonads had matured. If hormones

FIGURE 17.12 Gynandromorph finch with ZZ (male) cells on its right side and ZW (female) cells on its left side. Since plumage is controlled by genes on the sex chromosomes, the adult finch has male plumage on its right and female plumage on its left. Micrographs show the difference in brain regions between the right and left sides, indicate by staining of the neurons of the HVC nucleus (a neuron cluster involved in bird song production). (From Agate et al. 2002.)

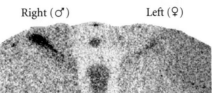

Right (♂)   Left (♀)

were all that mattered, the brains of the resulting birds would be appropriate to the gonad that developed. For the females given male forebrains, this was indeed the case: they looked and behaved like normal female quail. However, male birds given female forebrains did not act normally. They did not crow to attract mates, nor did they attempt copulation. Moreover, their testes failed to develop normally, suggesting that (in quail, at least) a genetically male brain is needed to complete development of the testes.

Thus, although brain sex is usually correlated with gonadal sex, it seems likely that this harmony is created both by intrinsic, cell-autonomous differences as well as hormonal regulation from outside the cell.

SIDELIGHTS & SPECULATIONS

Sex Determination and Behaviors

Does prenatal (or neonatal) exposure to particular steroid hormones impose permanent sex-specific changes on the central nervous system? Such sex-specific neural changes have been shown in regions of the brain that regulate involuntary sexual physiology. The cyclic secretion of luteinizing hormone by the pituitary in an adult female, for example, is dependent on the lack of testosterone during the first week of the animal's life. The luteinizing hormone secretion of female rats can be made noncyclical by giving them testosterone 4 days after birth. Conversely, the luteinizing hormone secretion of males can be made cyclical by removing their testes within a day of birth (Barraclough and Gorski 1962).

It is thought that sex hormones may act during the fetal or neonatal stage of a mammal's life to organize the nervous system in a sex-specific manner; and that during adult life, the same hormones may have transitory, activational effects. This idea is called the **organization/activation hypothesis**. Although it works to explain many of the effects of hormones on rodent develop-

ment, one of its fundamental assumptions—that α-fetoprotein strongly binds estrogens during prenatal development—does not in fact hold true for humans.

Ironically, the hormone chiefly responsible for determining the male neural pattern is **estradiol**, a type of estrogen.* Testosterone in fetal or neonatal blood can be converted into estradiol by the enzyme **P450 aromatase**. This conversion occurs in the hypothalamus and limbic system—two areas of the brain known to regulate hormone secretion and reproductive behavior (Reddy et al. 1974; McEwen et al. 1977). Thus, testosterone exerts its effects on the nervous system by being converted into estradiol in the brain. But the fetal environ-

---

*The terms *estrogen* and *estradiol* are often used interchangeably. However, estrogen refers to a class of steroid hormones responsible (among other functions) for establishing and maintaining specific female characteristics. Estradiol is one of these hormones, and in most mammals (including humans) it is the most potent of the estrogens.

---

ment is rich in estrogens from the gonads and placenta. What stops these estrogens from masculinizing the nervous system of a female fetus? In both male and female rats, fetal estrogen is bound by **α-fetoprotein**. α-Fetoprotein binds and inactivates estrogen, but not testosterone. Human fetuses, however, do not make a strong estrogen-binding protein and have a much higher level of free estrogen than do rodent embryos (see Nagel and vom Saal 2003).

Once converted into estradiol, how does testosterone work? As Ottem and colleagues (2004) have written, "if testosterone is the executive who barks out the orders ('Be a man!'), we would like to identify the underlings who scurry around to implement them." It appears that many of testosterone's orders are carried out by **prostaglandin E2** (**PGE2**). PGE2 is made from arachidonic acid by the enzyme **cyclooxygenase-2** (**COX2**); COX2 is induced by estrogen in the brain. So estrogen acts to produce more PGE2 (Figure 17.13A). Recent studies by Amateau and McCarthy (2004) have demonstrated that

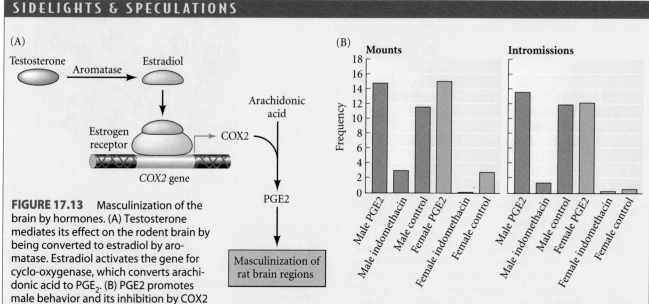

FIGURE 17.13 Masculinization of the brain by hormones. (A) Testosterone mediates its effect on the rodent brain by being converted to estradiol by aromatase. Estradiol activates the gene for cyclo-oxygenase, which converts arachidonic acid to PGE_2. (B) PGE2 promotes male behavior and its inhibition by COX2 inhibitors prevents that behavior. Males treated with indomethacin (COX2 inhibitor) did not display the mounting behavior expected of males, nor did they attempt intercourse; $P < 0.001$ between control and treated males). Conversely, females given PGE_2 acted like control males in both cases and had significantly more mounting and mating behavior than control females ($P < 0.001$). (After Amateau and McCarthy 2004.)

PGE2 is stimulated by estradiol in the newborn rat brain, and that this PGE2 induces the growth and differentiation of the regions of the brain involved in male sexual behaviors. They found that PGE2 was just as effective as estrogen in masculinizing these brain regions, and that PGE2 induces male-specific morphology and behaviors when injected into female rat brains. Moreover, when male rats were given COX-2 inhibitors, their brains became similar to those of females. Not only did their brain anatomy change, but so did their behaviors. The PGE2-treated females acted like males, attempting to mount and copulate with other females; and male mice treated with COX2 inhibitors lost their sexual drive and did not exhibit male behaviors (Figure 17.13B).

Knowledge of the mechanisms by which PGE2 causes the differentiation and growth of male-specific brain regions and promotes male-specific sexual behaviors in laboratory rats is just emerging. What is particularly important about Amateau and McCarthy's paper is the identity of the COX2 inhibitors: none other than aspirin and indomethacin, two leading pain relievers that are frequently used by pregnant women and by children. However, the lack of a strong estrogen-binding protein in human fetuses at this stage suggests that humans might not use this type of mechanism to coordinate gonadal and brain sex, if indeed the two are coordinated in humans at all.

Extrapolating from rats to humans is a very risky business, as no sex-specific behavior has yet been identified in humans, and what is "masculine" in one culture may be considered "feminine" in another (see Jacklin 1981; Bleier 1984; Fausto-Sterling 1992; Kandel et al. 1995).

Chromosomal Sex Determination in *Drosophila*

Although both mammals and fruit flies produce XX females and XY males, their chromosomes achieve these ends using very different means. In mammals, the Y chromosome plays a pivotal role in determining the male sex. Thus, XO mammals are females, with ovaries, a uterus, and oviducts (but usually very few, if any, ova).

In *Drosophila*, the Y chromosome is not involved in determining sex. A fruit fly's sex is determined by a balance of female determinants on the X chromosome and male determinants on the autosomes. Normally, flies have either one or two X chromosomes and two sets of autosomes. If there is only one X chromosome in a diploid cell (1X:2A), the fly is male. If there are two X chromosomes in a diploid cell (2X:2A), the fly is female (Bridges 1921, 1925). In flies, the Y chromosome appears to be important in sperm cell differentiation, not sex determination; it contains genes that are active in forming sperm in adults. Thus, XO *Drosophila* are sterile males.

In *Drosophila*, and in insects in general, one can observe gynandromorphs—animals in which certain regions of the body are male and other regions are female (Figure 17.14; see also Figure 17.12). Gynandromorph fruit flies result when an X chromosome is lost from one embryonic nucleus. The cells descended from that cell, instead of being XX (female), are XO (male). Because there are no sex hormones in insects that integrate the phenotype for the entire body, each cell makes its own sexual "decision." The XO cells display male characteristics, whereas the XX cells display female traits. This situation provides a beautiful

(A)

(B)

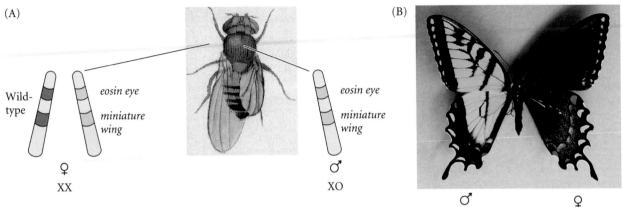

Wild-type

eosin eye

miniature wing

♀
XX

eosin eye

miniature wing

♂
XO

♂   ♀

FIGURE 17.14 Gynandromorph insects. (A) *D. melanogaster* in which the left side is female (XX) and the right side is male (XO). The male side has lost an X chromosome bearing the wild-type alleles of eye color and wing shape, thereby allowing the expression of the recessive alleles *eosin eye* and *miniature wing* on the remaining X chromosome. (B) Tiger swallowtail butterfly *Paplio glaucus*. The left half (yellow) is male while the right half (blue-black) is female. (A, drawing by Edith Wallace from Morgan and Bridges 1919; B photograph courtesy of J. Adams © 2005.)

example of the association between insect X chromosomes and sex.

Any theory of *Drosophila* sex determination must explain how the X-to-autosome (X:A) ratio is "read" and how this information is transmitted to the genes controlling the male or female phenotypes (Cline 1993). As we will see, the X chromosome produces transcription factors that activate the *Sex-lethal* gene, whereas the autosomes produce transcription factors that repress *Sex-lethal*. If there are *two* X chromosomes, however, the transcription factors encoded by those chromosomes prevail against the transcriptional repressors encoded by the autosomes. The *Sex-lethal* gene is thereby activated; its product then acts as a RNA processing factor that permits the synthesis of certain other proteins. These proteins in turn act as RNA processing factors to modify the nuclear RNA (pre-mRNA) of the *doublesex* gene. Depending on how the *doublesex* RNA is spliced, two types of Doublesex proteins can be formed (see Figure 5.31). Female Doublesex protein is formed in those flies where the two X chromosomes have activated the *Sex-lethal* gene; male Doublesex protein is formed in those flies where autosomes have prevented the activation of the *Sex-lethal* gene. The Doublesex proteins are transcription factors involved in producing the sexual phenotype of the fly.

The gene cascade of *Drosophila* sex determination

Research done over the past two decades has revolutionized our view of *Drosophila* sex determination. Several genes with roles in sex determination have been found. Loss-of-function mutations in most of these genes—*Sex-*

lethal (Sxl), transformer (tra), and *transformer-2 (tra2)*—transform XX individuals into males. Such mutations have no effect on sex determination in XY males. Homozygosity of the *intersex (ix)* gene causes XX flies to develop an intersex phenotype having portions of male and female tissue in the same organ. The *doublesex (dsx)* gene is important for the sexual differentiation of both sexes. If *dsx* is absent, both XX and XY flies turn into intersexes (Baker and Ridge 1980; Belote et al. 1985a). The positioning of these genes in a developmental pathway is based on (1) the interpretation of genetic crosses resulting in flies bearing two or more mutations of these genes, and (2) the determination of what happens when there is a complete absence of the products of one of these genes. Such studies have generated the model of the regulatory cascade shown in Figure 17.15.

The Sex-lethal gene as the pivot for sex determination

INTERPRETING THE X:A RATIO The first phase of *Drosophila* sex determination is reading the X:A ratio. What elements on the X chromosome are "counted," and how is this information used? It has been shown that high values of the X:A ratio are responsible for activating the feminizing switch gene **Sex-lethal (Sxl)**. In XY cells, *Sxl* remains inactive during the early stages of development (Cline 1983; Salz et al. 1987). However, in XX *Drosophila*, *Sxl* is activated during the first 2 hours after fertilization, and this gene transcribes a particular embryonic type of *Sxl* mRNA that is found for only about 2 hours more (Salz et al. 1989). Once activated, the *Sxl* gene remains active because its protein product is able to bind to and activate its own promoter (Bell et al. 1991).

The female-specific activation of *Sxl* is thought to be stimulated by **numerator proteins** encoded by the X chromosome. These proteins are the "X" part (i.e., the numerator) of the X:A ratio. The numerator proteins bind to the "early" promoter of the *Sxl* gene to activate its transcription shortly after fertilization. Cline (1988) has demonstrated that these proteins include Sisterless-a and Sisterless-b. The "A" part of the equation, or the **denominator proteins**,

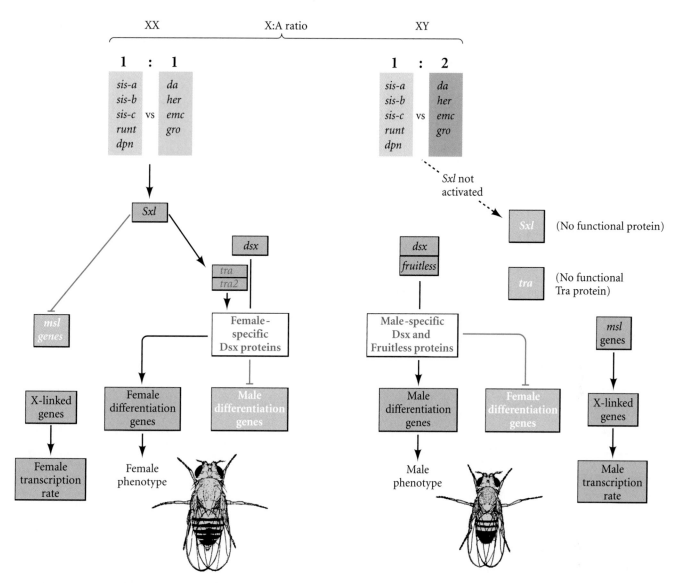

FIGURE 17.15 Proposed regulatory cascade for *Drosophila* somatic sex determination. Transcription factors from the X chromosomes and autosomes compete to activate or repress the *Sxl* gene, which becomes active in females (XX) and inactive in males (XY). The Sex lethal protein performs three main functions. First, it activates its own transcription, ensuring further Sxl production. Second, it represses the translation of *msl-2* mRNA, a factor that facilitates transcription from the X chromosome. This equalizes the amount of transcription from the two X chromosomes in females with that of the single X chromosome in the males. Third, Sxl activates the *transformer* (*tra*) genes. The Tra proteins process *doublesex* pre-mRNA in a female-specific manner that provides most of the female body with its sexual fate. It also processes the *fruitless* pre-mRNA in a female-specific manner, giving the fly female-specific behavior. In the absence of Sxl (and thus the Tra proteins), *dsx* and *fruitless* pre-mRNAs are processed in the male-specific manner. (After Baker et al. 1987.)

are autosomally encoded transcription factors such as Deadpan and Extramacrochaetae (see Figure 17.15). These proteins block the binding or activity of the numerator proteins (Van Doren et al. 1991; Younger-Shepherd et al. 1992). The denominator proteins may actually be able to form inactive heterodimers with the numerator proteins. It appears, then, that the X:A ratio is measured by competition between X-encoded activators and autosomally encoded repressors of the early promoter of the *Sxl* gene.

MAINTENANCE OF *SXL* FUNCTION Shortly after the initial *Sxl* transcription has taken place, a second, "late" promoter on the *Sex-lethal* gene is activated. The gene is now transcribed in both males and females. However, the *Sxl* mRNA of males differs from the *Sxl* mRNA of females (Bell et al. 1988). This difference is the result of differential RNA processing. Moreover, the Sxl protein appears to bind to its own mRNA precursor to splice it in the female manner. Since males do not have any available Sxl protein when

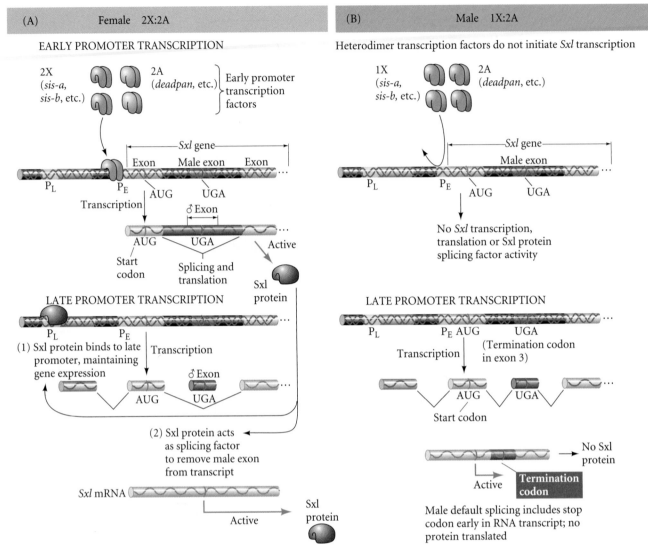

FIGURE 17.16 Differential activation of the *Sxl* gene in females and males. (A) In wild-type *Drosophila* with two X chromosomes and two sets of autosomes (2X:2A), the "numerator" proteins encoded on the X chromosomes (Sisterless-a, Sisterless-b, etc.) are not all bound by inhibitory "denominator" proteins derived from genes on the autosomes (such as *deadpan*). The numerator proteins activate the early promoter (P_E) of the *Sxl* gene. Eventually, in both males and females, constitutive transcription of *Sxl* eventually starts from the late promoter (P_L). If Sxl protein is already available (i.e., from early transcription),

the *Sxl* pre-mRNA is spliced to form the functional female-specific message. (B) In wild-type *Drosophila* with one X chromosome and two sets of autosomes (1X:2A), the numerator proteins are bound by the denominator proteins and cannot activate the early promoter. When the *Sxl* gene is transcribed from the late promoter, RNA splicing does not exclude the male-specific exon in the mRNA. The resulting message encodes a truncated and nonfunctional peptide, since the male-specific exon contains a translation termination codon. (After Keyes et al. 1992.)

the late promoter is activated, their new *Sxl* transcripts are processed in the male manner (Keyes et al. 1992).

Male *Sxl* mRNA is nonfunctional. While the female-specific *Sxl* message encodes a protein of 354 amino acids (Figure 17.16A), the male-specific *Sxl* transcript contains a translation termination codon (UGA) after amino acid 48 (Figure 17.16B). In males, the nuclear transcript is spliced in a manner that yields 8 exons, and the termination codon

is within exon 3. In females, RNA processing yields only 7 exons, and the male-specific exon 3 is spliced out as part of a large intron. Thus, the female-specific mRNA lacks the termination codon. The protein made by the female-specific *Sxl* transcript contains regions that are important for binding to RNA. There appear to be two major RNA targets to which the female-specific *Sxl* transcript binds. One of these is the pre-mRNA of *Sxl* itself; the other target is

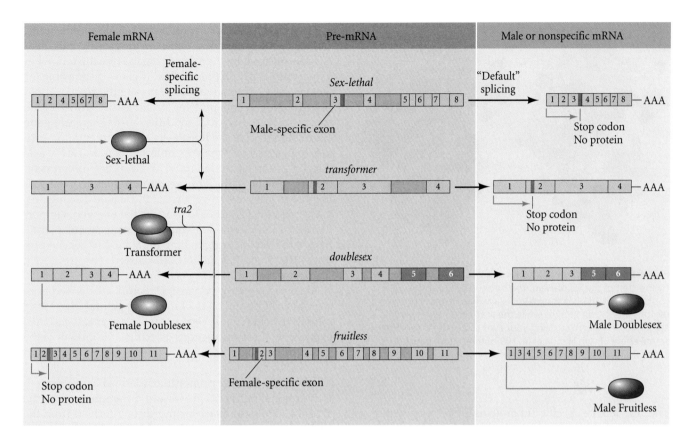

FIGURE 17.17 Sex-specific RNA splicing in four major *Drosophila* sex-determining genes. The pre-mRNAs (shown in the center of the diagram) are identical in both male and female nuclei. In each case, the female-specific transcript is shown at the left, while the default transcript (whether male or nonspecific) is shown to the right. Exons are numbered, and the positions of termination codons are marked. *Sex-lethal, transformer,* and *Doublesex* are all part of the genetic cascade of primary sex determination. The transcription pattern of *fruitless* determines the secondary characteristic of courtship behavior (see pp. 549–550). (After Baker 1989 and Baker et al. 2001.)

the pre-mRNA of *transformer*—the next gene in the cascade (Bell et al. 1988; Nagoshi et al. 1988).

WEBSITE

17.7 Other sex determination proteins in *Drosophila*. Sex-lethal does not work alone, but in concert with several other proteins whose presence is essential for its function. Many of these proteins have other roles during development.

THE *TRANSFORMER* GENES   The *Sxl* gene regulates somatic sex determination by controlling the processing of the ***transformer*** (***tra***) gene transcript. The *tra* pre-mRNA is made in both male and female cells; however, in the presence of Sxl protein, the *tra* transcript is alternatively spliced to create a female-specific mRNA, as well as a nonspecific mRNA that is found in both females and males. Like the male *Sxl* mes-

sage, the nonspecific *tra* mRNA contains a termination codon early in the message that renders the protein non-functional (Boggs et al. 1987). In *tra*, the second exon of the nonspecific mRNA contains the termination codon. This exon is not utilized in the female-specific message (Figure 17.17).

How is it that females make a different transcript than males? The female-specific Sxl protein activates a 3′ splice site in the *transformer* pre-mRNA, causing it to be processed in a way that splices out the second exon. To do this, Sxl protein blocks the binding of splicing factor U2AF to the nonspecific splice site of the *tra* message by specifically binding to the polypyrimidine tract adjacent to it (Figure 17.18; Handa et al. 1999). This causes U2AF to bind to the lower-affinity (female-specific) 3′ splice site and generate a female-specific mRNA (Valcárcel et al. 1993). The female-specific Tra protein acts in concert with the product of the *transformer-2* (*tra2*) gene to help generate the female phenotype.

Doublesex: The switch gene of sex determination

The ***doublesex*** (***dsx***) gene is active in both males and females, but its primary transcript is processed in a sex-specific manner (Baker et al. 1987). This alternative RNA processing is the result of the action of the *tra* and *tra2* gene products on the *dsx* gene (see Figures 17.17 and 5.31). If the

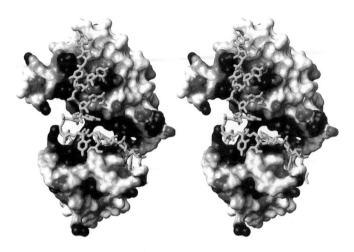

FIGURE 17.18 Stereogram showing binding of *tra* pre-mRNA by the cleft of the Sxl protein. The bound 12-nucleotide RNA (GUUGUUUUUUUU) is shown in yellow. The strongly positive regions are shown in blue, while the scattered negative regions are in red. It is worth crossing your eyes to get the three-dimensional effect. (From Handa et al. 1999; stereogram courtesy of S. Yokoyama.)

Tra2 and female-specific Tra proteins are both present, the *dsx* transcript is processed in a female-specific manner (Ryner and Baker 1991). The female splicing pattern produces a female-specific protein that activates female-specific genes (such as those of the yolk proteins) and inhibits male development. If no functional Tra is produced, a male-specific transcript of *dsx* is made. The male transcript encodes an active transcription factor that inhibits female traits and promotes male traits.

In XX flies, the female Doublesex protein (Dsx^F) combines with the product of the *intersex* gene to make a tran-

scription factor that is responsible for female-specific traits. This "Doublesex complex" enhances the growth of the genital disc to make the female sex organs (see the following paragraph). It also represses the *Fgf* genes responsible for making male accessory organs, activates the genes responsible for making yolk proteins, promotes the growth of the sperm storage duct, and modifies *bric-a-brac* (*bab*) gene expression to give the female-specific pigmentation profile. In contrast, the male Doublesex protein (Dsx^M) acts directly as a transcription factor and directs the expression of male-specific traits. It causes the male region of the genital disc to grow at the expense of the female disc regions, it activates the *Fgf* genes to produce the paragonia (a portion of the male genitalia), converts certain cuticular structures into claspers, and modifies the *bric-a-brac* gene to produce the male pigmentation pattern (Figure 17.19; Christiansen et al. 2002).

Male and female genitalia in *Drosophila* are derived from separate cell populations of the larval **genital disc**. Dsx^M inhibits *wingless*, while Dsx^F inhibits *Decapentaplegic* (*Dpp*) expression; these interactions determine the growth of the genitalia and sex ducts.

- In XY flies, the *male* primordium (derived from the ninth abdominal segment) grows and differentiates into the testes, while the *female* primordium (derived from the eighth abdominal segment), which needs the *wingless* product in order to grow, is stunted.
- In XX flies, the *female* primordium of the genital disc grows and differentiates into ovaries, while the male portion of that disc becomes (which needs *Dpp* in the ninth abdominal segment) is stunted.

If the *dsx* gene is absent (meaning neither transcript is made), both the male and the female primordia develop, and intersexual genitalia are produced.

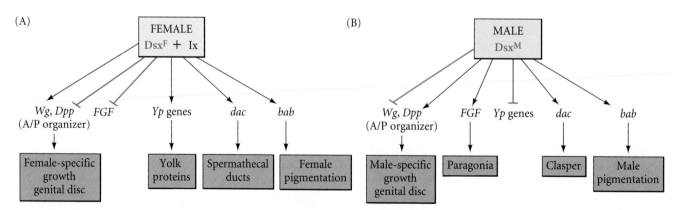

FIGURE 17.19 The roles of the Dsx^M and Dsx^F proteins in *Drosophila* sexual development. (A) Dsx^F functions with Intersex (Ix) to promote female-specific expression of those genes that control the growth of the genital disc, the synthesis of yolk proteins, the formation of spermathecal ducts (which keep sperm stored after mating), and pigment patterning. (B) Conversely, Dsx^M acts as a transcription factor to promote the male-specific

growth of the genital disc, the formation of male genitalia, the conversion of cuticle into claspers, and the male-specific pigmentation pattern. In addition, Dsx^F represses certain genes involved in specifying male-specific traits (such as the paragonia), and Dsx^M represses certain genes involved synthesizing female-specific proteins such as yolk protein. (After Christiansen et al. 2002.)

According to this model, the result of the sex determination cascade comes down to what type of mRNA is processed from the *dsx* transcript. If the X:A ratio is 1, then *Sxl* makes a female-specific splicing factor that causes the *tra* gene transcript to be spliced in a female-specific manner. This female-specific protein interacts with the *tra2* splicing factor to cause the *doublesex* pre-mRNA to be spliced in a female-specific manner. If the *doublesex* transcript is not acted on in this way, it will be processed in a "default" manner to make the male-specific message.

Brain Sex in *Drosophila*: Secondary Sex Determination through another Pathway

Our discussion of sexual dimorphism in *Drosophila* has so far been limited to nonbehavioral aspects of development. However, there appears to be a "brain-sex" pathway in *Drosophila* that provides individuals of each sex with the appropriate set of courtship behaviors. Among *Drosophila*, there are no parents to teach one "proper" mating behavior, and mating takes place soon after the flies emerge from their pupal cases. So the behaviors must be "hardwired" into the insect genome.*

The behaviors of *Drosophila* courtship and mating are quite complicated. A courting male must first confirm that

---

*This is not to say that flies don't learn; indeed, one thing they *do* learn is to avoid bad sexual encounters. A male who has been brushed off (quite literally) by a female who has recently mated hesitates before starting to court another female (Siegel and Hall 1979; MacBride et al. 1999).

the individual he is approaching is indeed a female. Once this is established, he must orient his body toward the female and follow a specific series of movements that include following the female, tapping the female, playing a species-specific courtship song by vibrating his wings, licking the female, and finally, curling his abdomen so that he is in a position to mate. Each of these sex-specific behaviors in *Drosophila* courtship appear to be regulated by the products of *fruitless*, a gene expressed in certain sets of neurons (comprising about 2 percent of all the neurons in the adult male) that are involved with male sexual behaviors (Figure 17.20). These include subsets of neurons involved in taste, hearing, smell, and touch (Lee et al. 2000; Billeter and Goodwin 2004; Stockinger et al. 2005). Fruitless also retains certain male-specific neural circuits; the neurons in these circuits die during female development (and in *fruitless* mutants; see Kimura et al. 2005).

As with *Doublesex* pre-mRNA, the Tra and Tra2 proteins splice *fruitless* pre-mRNA into a female-specific message; the default splicing pattern is male (see Figure 17.17). So the female makes Tra protein and processes the *fruitless* pre-mRNA in one way, whereas the male, lacking the Tra protein, processes the *fruitless* message in another way. The female *fruitless* mRNA includes a termination sequence in an early exon; therefore the female does not make functional Fruitless protein. The male, however, makes an mRNA that does not contain the stop codon (Heinrichs et al. 1998), and the protein it transcribes is a zinc-finger transcription factor. Using homologous recombination to force the transcription of particular splicing forms, Demir and Dickson (2005) showed that it is Fruitless, and not the flies' anatomy, that controls their sexual behavior. When female flies were induced to make the male-specific Fruitless pro-

(A)

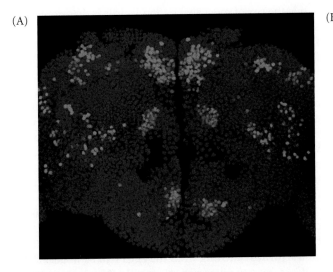

(B)

FIGURE 17.20 Subsets of neurons expressing the male-specific form of *fruitless*. (A) Central nervous system of a 48-hour pupa (the time at which sex determination of the brain is established), showing male-specific *fruitless* mRNA in particular cells (pink) of the anterior brain that are involved in courtship behaviors. (B) Some peripheral nervous system cells express the male-specific *fruitless* message. Fruitless protein is stained green in these taste neurons in the labella (the spongy part of the fly mouth). (From Stockinger et al. 2005, photographs courtesy of B. J. Dickson.)

tein, they performed the entire male courtship ritual and tried to mate with other females.

In normal females, the courtship ritual is not as involved as the male's. However, females have the ability to be receptive to the male's entreaties or to rebuff them. The product of the *retained* gene (*rtn*) is critical in this female mating behavior. Both sexes express this gene, since it is also involved in axon pathfinding. However, female flies with a loss-of-function allele of the *retained* gene resist male courtship and are thus rendered barren by their own behavior (Ditch et al. 2005).

Dosage Compensation

In animals whose sex is determined by sex chromosomes, there has to be some mechanism by which the amount of X chromosome gene expression is equalized for both males and females. This mechanism is known as **dosage compensation**. In mammals, we discussed X-chromosome inactivation, whereby one of the X chromosomes is inactivated so that the transcription product level is the same in both XX cells and XY cells (see Chapter 5). In *C. elegans*, dosage compensation occurs by lowering the transcription rates of *both* X chromosomes so that product levels are the same as those of XO individuals.

In *Drosophila*, the female X chromosomes are not suppressed; rather the male's single X chromosome is hyperactivated. This "hypertranscription" is accomplished at the level of translation, and it is mediated by the Sxl protein. Sxl protein (which you will recall is made by the female cells) binds to the 5′ leader sequence and the 3′ untranslated regions (UTRs) of the *mls-2* message. The bound Sxl inhibits the attachment of *msl-2* mRNA to the ribosome and prevents the ribosome from getting to the mRNA's coding region (Beckman et al. 2005). The result is that female cells do not produce Msl-2 protein (see Figure 17.15). But it is made in male cells, in which Sxl is not present. Mls-2 is part of a protein/mRNA complex that finds the X chromosome

WEBSITE **17.8 Conservation of sex-determining Genes.** While the pathways of sex determination appear to differ in humans and fruit flies, the discovery of a human gene similar to *doublesex* and the discovery of FGF signaling for the recruitment of mesodermal cells into the testes of both humans and flies suggest common themes in the two pathways.

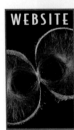

WEBSITE **17.9 Hermaphrodites.** In *C. elegans* and many other invertebrates, hermaphroditism is the general rule. These animals may be born with both ovaries and testes, or they may develop one set of gonads first and the other later (sequential hermaphroditism). In some fish, sequential hermaphroditism is seen, with an individual fish being female in some seasons and male in others.

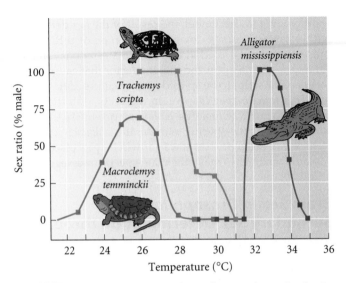

FIGURE 17.21   Temperature-dependent sex determination in three reptile species: the American alligator (*Alligator mississippiensis*), the red-eared slider turtle (*Trachemys scripta elegans*), and the alligator snapping turtle (*Macroclemys temminckii*). (After Crain and Guillette 1998.)

and loosens chromatin structure by acetylating histone 4. In this way, transcription factors gain access to the X chromosome at a much higher frequency in males than in females—hence, "hypertranscription."

ENVIRONMENTAL SEX DETERMINATION

Temperature-Dependent Sex Determination in Reptiles

While the sex of most snakes and lizards is determined by sex chromosomes at the time of fertilization, the sex of most turtles and all species of crocodilians is determined *after* fertilization, by the embryonic environment. In these reptiles, the temperature of the eggs during a certain period of development is the deciding factor in determining sex, and small changes in temperature can cause dramatic changes in the sex ratio (Bull 1980; Crews 2003). Often, eggs incubated at low temperatures (22–27°C) produce one sex, whereas eggs incubated at higher temperatures (30°C and above) produce the other. There is only a small range of temperatures that permits both males and females to hatch from the same brood of eggs.*

Figure 17.21 shows the abrupt temperature-induced change in sex ratios for the red-eared slider turtle. If a brood of eggs is incubated at a temperature below 28°C, all the turtles hatching from the eggs will be male. Above 31°C, every egg gives rise to a female. At temperatures in between, the brood will give rise to individuals of both

---

*The evolutionary advantages and disadvantages of temperature-dependent sex determination are discussed in Chapter 22.

sexes. Variations on this theme also exist. The eggs of the snapping turtle *Macroclemys*, for instance, become female at either cool (22°C or lower) or hot (28°C or above) temperatures. Between these extremes, males predominate.

One of the best-studied reptiles is the European pond turtle, *Emys obicularis*. In laboratory studies, incubating *Emys* eggs at temperatures above 30°C produces all females, while temperatures below 25°C produce all-male broods. The threshold temperature (at which the sex ratio is even) is 28.5°C (Pieau et al. 1994). The developmental "window" during which sex determination occurs can be discovered by incubating eggs at the male-producing temperature for a certain amount of time and then shifting them to an incubator at the female-producing temperature (and vice versa). In *Emys*, the last third of development appears to be the most critical for sex determination, and it is believed that the turtles cannot reverse their sex after this period.

Aromatase and cell proliferation during sex determination

The pathways to maleness and femaleness in reptiles are just being delineated. There are two major hypotheses, one of which focuses on the production of estrogen and the other that targets cell proliferation. While they are not mutually exclusive, these two approaches are often thought of as separate explanations.

CELL PROLIFERATION AND MALENESS The cell proliferation hypothesis holds that the unifying feature in testis determination is the rapid proliferation of Sertoli cell precursors (Schmahl and Capel 2003). In mice, one of the first distinguishing features of testis development is the increase in cell proliferation seen immediately after *Sry* expression. There appears to be a threshold number of Sertoli cells needed for mammalian testis development (Schmahl et al. 2000). In turtles and alligators, the same rapid division of Sertoli cell precursors is initiated during the critical stage for the forming the males (Mittwoch 1986; Schmahl et al. 2003). In this perspective, anything that will increase the division rate of Sertoli cell precursors will direct gonadogenesis in the male direction.

ESTROGEN AND AROMATASE The hormonal hypothesis is based on the observations that, unlike the situation in mammals, primary sex determination in reptiles (and birds) can be influenced by hormones, and that estrogen is essential for ovarian development. In reptiles, estrogen can override temperature and induce ovarian differentiation even at masculinizing temperatures. Similarly, injecting eggs with inhibitors of estrogen synthesis produces male offspring, even if the eggs are incubated at temperatures that usually produce females (Dorizzi et al. 1994; Rhen and Lang 1994). Moreover, the sensitive time for the effects of estrogens and their inhibitors coincides with the time when sex determination usually occurs (Bull et al. 1988; Gutzke and Chymiy 1988).

It appears that the enzyme **aromatase**, which converts testosterone into estrogen, may be critical in temperature-dependent sex determination. The estrogen synthesis inhibitors used in the experiments mentioned above worked by blocking the aromatase enzyme, showing that experimentally low aromatase levels yield male offspring.* This correlation appears to hold under natural conditions as well. The aromatase activity of *Emys* is very low at the male-promoting temperature of 25°C. At the female-promoting temperature of 30°C, aromatase activity increases dramatically during the critical period for sex determination (Desvages et al. 1993; Pieau et al. 1994). Temperature-dependent aromatase activity is also seen in diamondback terrapins, and its inhibition masculinizes their gonads (Jeyasuria et al. 1994).

When turtle gonads are taken out of the embryos and placed in culture, these isolated gonads will become testes or ovaries based on the temperature at which they are incubated (Moreno-Mendoza et al. 2001; Porter et al. 2005). This shows that sex determination is a local activity of the gonadal primordia and not a global activity directed by the pituitary or the brain. By analyzing the timing and regulation of sex-determining genes in these cultured gonads, Shoemaker and colleagues (2005) proposed that female temperatures not only activate aromatase (which would cause ovary formation), but also activate the *Wnt4* gene (which suppresses the formation of testes). Conversely, male temperatures appear to activate *Sox9* to a higher level in the male than in the female, promoting testis development (Moreno-Mendoza et al. 2001).

Sex reversal, aromatase, and conservation biology

Whether or not aromatase is critical in the initial specification of the gonad, it is important in maintaining the sex of the gonad. Studies have shown that polychlorinated biphenyl compounds (PCBs), a class of pollutants widely introduced into the environment by humans, can act like estrogens (see Bergeron et al. 1994, 1999). PCBs can reverse the sex of turtles raised at "male" temperatures. This knowledge may have important consequences in environmental conservation efforts to protect endangered species such as turtles and amphibians, in which hormones can effect changes in primary sex determination. Indeed, some reptile conservation biologists advocate using hormonal treatments to elevate the percentage of females in endangered species (www.reptileconservation.org).

In addition to deaths due to ozone depletion and fungal infections (see Chapter 3), amphibian populations may be at risk from herbicides that promote or destroy estrogens. One such case involves the development of hermaphroditic and demasculinized frogs after exposure to extreme-

---

*One remarkable finding is that the injection of an aromatase inhibitor into the eggs of an all-female, parthenogenetic species of lizards causes the formation of males (Wibbels and Crews 1994).

(A)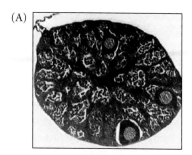

(B)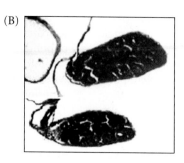

(C)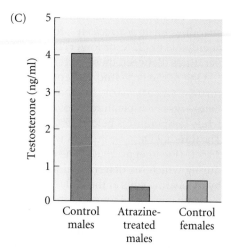

FIGURE 17.22 Demasculinization of frogs by low amounts of atrazine. (A) Testis of a frog from a natural site having 0.5 parts per billion (ppb) atrazine. The testis contains three lobules that are developing both sperm and an oocyte. (B) Two testes of a frog from a natural site containing 0.8 ppb atrazine. These organs show severe testicular dysgenesis that characterized 28 percent of the frogs found at that site. (C) Effect of a 46-day exposure to 25 ppb atrazine on plasma testosterone levels in sexually mature male *Xenopus*. Levels in control males were some tenfold higher than in control females; atrazine-treated males had plasma testosterone levels at or below those of the control females. (A,B after Hayes et al. 2003, photographs courtesy of T. Hayes; C after Hayes et al. 2002a.)

ly low doses of the weed killer atrazine. Atrazine is the most widely used herbicide in the world; the United States alone uses 60 million pounds of it annually. Atrazine is thought to induce aromatase, which as described above is capable of converting testosterone into estrogen.

Hayes and colleagues (2002a) found that exposing tadpoles to atrazine concentrations as low as 0.1 part per billion produced gonadal and other sexual anomalies in male frogs (Figure 17.22). At concentrations as low as 0.1 part per billion, leopard frogs displayed testicular dysgenesis (stunted growth) or conversion to ovaries. In many examples, oocytes were found in the testes. At concentrations of 1 part per billion, the vocal sacs of the male frogs (which the male must have in order to signal potential mates) failed to develop properly.

The testosterone levels of adult male frogs were reduced nearly 90 percent (to control female levels) by the exposure of the frogs to 25 parts per billion atrazine for 46 days. These are ecologically very relevant doses. The allowable amount of atrazine in our drinking water is 3 parts per billion, and atrazine levels can be as high as 224 parts per billion in streams in the midwestern United States (Battaglin et al. 2000; Barbash et al. 2001).

Given the amount of atrazine in the water and the sensitivity of frogs to this compound, this herbicide could be devastating to wild populations. Hayes and colleagues collected leopard frogs and water at eight sites across the central United States (Hayes et al. 2002b, 2003). They sent the water samples to two separate laboratories for the determination of atrazine, and they coded the frog specimens so that the technicians dissecting the gonads did not know from which site the animals came. The results showed that

all but one site contained atrazine, and this was the only site from which the frogs had no gonadal abnormalities.

Concern over atrazine's apparent ability to disrupt sex hormones in both wildlife and humans has resulted in bans on the use of this herbicide by France, Germany, Italy, Norway, Sweden, and Switzerland (Dalton 2002). This issue will be discussed further in Chapter 22.

Location-Dependent Sex Determination

As mentioned in Chapter 3, the sex of the echiuroid worm *Bonellia viridis* depends on where a larva settles. If a *Bonellia* larva lands on the ocean floor, it develops into a 10-cm long female. If the larva is attracted to a female's proboscis, it travels along the tube until it enters the female's body. Therein it differentiates into a minute (1–3 mm long) male that is essentially a sperm-producing symbiont of the female (see Figure 3.1).

Another species in which sex determination is affected by the location of the organism is the slipper snail *Crepidula fornicata*. In this species, individuals pile up on top of one another to form a mound. Young individuals are always male, but this phase is followed by the degeneration of the male reproductive system and a period of lability. The next phase can be either male or female, depending on the animal's position in the mound. If the snail is attached to a female, it will become male. If such a snail is removed from its attachment, it will become female. Similarly, the presence of large numbers of males will cause some of the males to become females. However, once an individual becomes female, it will not revert to being male (Coe 1936; Collin 1995; Warner et al. 1996). More examples of context-dependent sex determination will be studied in Chapter 22.

Nature has provided many variations on her masterpiece. In some species, including most mammals and insects, sex is determined by chromosomes; in other species, sex is a matter of environmental conditions. We are finally beginning to understand the mechanisms by which this masterpiece is created.

 **Snapshot Summary** ## Sex Determination

1. In mammals, primary sex determination (the determination of gonadal sex) is a function of the sex chromosomes. XX individuals are females, XY individuals are males.

2. The Y chromosome plays a key role in male sex determination. XY and XX mammals both have a bipotential gonad. In XY animals, Sertoli cells differentiate and enclose the germ cells within testis cords. The interstitial mesenchyme becomes the Leydig cells.

3. In XX individuals, the germ cells become surrounded by follicle cells in the cortex of the gonadal rudiment. The epithelium of the follicles becomes the granulosa cells; the mesenchyme becomes the thecal cells.

4. In humans, the *SRY* gene is the testis-determining factor on the Y chromosome. It synthesizes a nucleic acid-binding protein that may function as either a transcription factor or as an RNA splicing factor. It is thought to activate *SOX9*.

5. The *SOX9* gene can also initiate testes formation. It may also function as both a transcription and a splicing factor. It binds to the gene encoding anti-Müllerian duct hormone (AMH), and it may also be responsible for activating *Fgf9*.

6. *Wnt4* is involved in ovary formation. Part of its function appears to be inhibiting the testis-forming pathway. The genes generating ovarian tissues are not as well characterized as those inducing testis formation.

7. Secondary sex determination in mammals involves the hormones produced by the developing gonads. In females, the Müllerian duct differentiates into the oviducts, uterus, cervix, and upper portion of the vagina. In male mammals, the Müllerian duct is destroyed by the AMH produced by the Sertoli cells, while the testosterone produced by the Leydig cells enables the Wolffian duct to differentiate into the vas deferens and seminal vesicle. In female mammals, the Wolffian duct degenerates because of the lack of testosterone, while estrogen permits the differentiation of the Müllerian duct.

8. The conversion of testosterone to dihydrotestosterone in the genital rudiment and prostate gland precursor enables the differentiation of the penis, scrotum, and prostate gland.

9. Individuals with mutations of these hormones or their receptors may have a discordance between their primary and secondary sex characteristics.

10. In *Drosophila*, sex is determined by the ratio of X chromosomes to autosomes, and the Y chromosome does not play a role in sex determination. There are no sex hormones, so each cell makes a sex determination decision. However, paracrine factors play important roles in forming the genital structures.

11. The *Drosophila Sex-lethal* gene is activated in females (by proteins encoded on the X chromosomes) but repressed in males (by factors encoded on the autosomes). Sxl protein acts as an RNA splicing factor to splice an inhibitory exon from the *transformer* transcript. Therefore, female flies have an active Tra protein, while males do not.

12. The Tra protein also acts as an RNA splicing factor to splice exons from the *doublesex* transcript. The *dsx* gene is transcribed in both XX and XY cells, but its pre-mRNA is processed to form different mRNAs, depending on whether Tra protein is present. The proteins translated from both *dsx* messages are active, and they activate or inhibit transcription of a set of genes involved in producing the sexually dimorphic traits of the fly.

13. Sex determination of the brain may have different downstream agents than in other regions of the body. *Drosophila* Tra proteins also activate the *fruitless* gene in males (and not in females); in mammals, the Y chromosome may activate brain sexual differentiation independently from the hormonal pathways.

14. Dosage compensation is critical for the regulation of gene expression in the embryo. With the same number of autosomes, the transcription from the X chromosome must be equalized for XX females and XY males. In mammals, one X chromosome of XX females is inactivated. In *Drosophila*, the single X chromosomes of XY males is hyperactivated.

15. In turtles and alligators, sex is often determined by the temperature experienced by the embryo during the time of gonad determination. Because estrogen is necessary for ovary development, it is possible that differing levels of aromatase (an enzyme that can convert testosterone into estrogen) distinguish male from female patterns of gonadal differentiation.

16. Aromatase may be activated by environmental compounds, causing demasculinization of the male gonads in those animals where primary sex determination can be effected by hormones.

17. In some species, such as *Bonellia* and *Crepidula*, sex is determined by the position of the individual with regard to other individuals of the same species.

For Further Reading

Complete bibliographical citations for all literature cited in this chapter can be found on the Vade Mecum CD that accompanies the book and at the free access website www.devbio.com

Arnold, A. P. and P. S. Burgoyne. 2004. Are XX and XY brain cells intrinsically different? *Trends Endocr. Metab.* 15: 6–11.

Bell, L. R., J. I. Horabin, P. Schedl and T. W. Cline. 1991. Positive autoregulation of *Sex-lethal* by alternative splicing maintains the female determined state in *Drosophila. Cell* 65: 229–239.

Bullejos, M. and P. Koopman. 2005. Delayed *Sry* and *Sox9* expression in developing mouse gonads underlie B6-YDOM sex reversal. *Dev. Biol.* 278: 473–481.

Hayes, T. B., A. Collins, M. Lee, M. Mendoza, N. Noriega, A. Stuart and A. Vonk. 2002. Hermaphroditic, demasculinized frogs after exposure to the herbicide atrazine at low ecologically relevant doses. *Proc. Natl. Acad. Sci. USA* 99: 5476–5480.

Imperato-McGinley, J., L. Guerrero, T. Gautier and R. E. Peterson. 1974. Steroid 5α-reductase deficiency in man: An inherited form of male pseudohermaphroditism. *Science* 186: 1213–1215.

Koopman, P., J. Gubbay, N. Vivian, P. Goodfellow and R. Lovell-Badge. 1991. Male development of chromosomally female mice transgenic for *Sry. Nature* 351: 117–121.

Stockinger, P., D. Kvitsiani, S. Rotkopf, L. Tirián and B. J. Dickson. 2005. Neural circuitry that governs *Drosophila* male courtship behavior. *Cell* 121: 795–807.

Postembryonic Development: Metamorphosis, Regeneration, and Aging

18

DEVELOPMENT NEVER CEASES. Throughout life, we continuously generate new blood cells, lymphocytes, keratinocytes, and digestive tract epithelium from stem cells. In addition to these continuous daily changes, there are instances in which postembryonic development is obvious—sometimes even startling. One of these instances is metamorphosis, the transition from a larval stage to an adult stage. In many species that undergo metamorphosis, a large proportion of the animal's structure changes, and the larva and the adult are unrecognizable as being the same individual (see Figure 2.4). Another startling type of development is regeneration, the creation of a new organ after the original one has been removed from an adult animal. Some adult salamanders, for instance, can regrow limbs after these appendages have been amputated.

The third category of developmental change is a more controversial area. It encompasses those alterations of form associated with aging in adult organisms. Some scientists believe that the processes of age-associated degeneration are not properly part of the study of developmental biology; others point to genetically determined, species-specific patterns of aging that are an important part of the life cycle and believe that gerontology—the scientific study of aging—is rightly included in developmental biology. Whatever their relationship to embryonic development, metamorphosis, regeneration, and aging are poised to be critical topics for the biology of the twenty-first century.

METAMORPHOSIS: THE HORMONAL REACTIVATION OF DEVELOPMENT

In most animal species, embryonic development leads to a larval stage with characteristics very different from those of the adult organism. Very often, larval forms are specialized for some function, such as growth or dispersal. The pluteus larva of the sea urchin, for instance, can travel on ocean currents, whereas the adult urchin leads a sedentary existence. The division of functions between larva and adult can be remarkably distinct (Wald 1981). *Cecropia* moths, for example, hatch from eggs and develop as wingless juveniles—caterpillars—for several months. This metamorphosis enables them to spend a day or so as fully developed winged insects, and to mate (quickly) before they die. The adults never eat, and in fact have no mouthparts during this brief reproductive phase of the life cycle. As might be expected, the juvenile and adult forms often live in different environments.

During metamorphosis, developmental processes are reactivated by specific hormones, and the entire organism changes morphologically, physiologically,

66 The earth-bound early stages built enormous digestive tracts and hauled them around on caterpillar treads. Later in the life-history these assets could be liquidated and reinvested in the construction of an entirely new organism—a flying-machine devoted to sex. 99

CARROLL M. WILLIAMS (1958)

66 I'd give my right arm to know the secret of regeneration. 99

OSCAR E. SCHOTTÉ (1950)

TABLE 18.1 Summary of some metamorphic changes in anurans

System	Larva	Adult
Locomotory	Aquatic; tail fins	Terrestrial; tailless tetrapod
Respiratory	Gills, skin, lungs; larval hemoglobins	Skin, lungs; adult hemoglobins
Circulatory	Aortic arches; aorta; anterior, posterior, and common jugular veins	Carotid arch; systemic arch; cardinal veins
Nutritional	Herbivorous: long spiral gut; intestinal symbionts; small mouth, horny jaws, labial teeth	Carnivorous: short gut; proteases; large mouth with long tongue
Nervous	Lack of nictitating membrane; porphyropsin, lateral line system, Mauthner's neurons	Development of ocular muscles, nictitating membrane, rhodopsin; loss of lateral line system, degeneration of Mauthner's neurons; tympanic membrane
Excretory	Largely ammonia, some urea (ammonotelic)	Largely urea; high activity of enzymes of ornithine-urea cycle (ureotelic)
Integumental	Thin, bilayered epidermis with thin dermis; no mucous glands or granular glands	Stratified squamous epidermis with adult keratins; well-developed dermis contains mucous glands and granular glands secreting antimicrobial peptides

Source: Data from Turner and Bagnara 1976 and Reilly et al. 1994.

and behaviorally to prepare itself for its new mode of existence. These changes are not solely ones of form. In amphibian tadpoles, for example, metamorphosis involves the maturation of liver enzymes, hemoglobin, and eye pigments, as well as the remodeling of the nervous and digestive systems. Thus, metamorphosis is often a time of dramatic developmental change affecting nearly the entire organism.

Amphibian Metamorphosis

Amphibians are named for their ability to undergo metamorphosis, their appellation coming from the Greek *amphi* ("double") and *bios* ("life"). Amphibian metamorphosis is generally associated with morphological changes that prepare an aquatic organism for a primarily terrestrial existence. In **urodeles** (salamanders), these changes include the resorption of the tail fin, the destruction of the external gills, and a change in skin structure. In **anurans** (frogs and toads), the metamorphic changes are more dramatic, with almost every organ subject to modification (Table 18.1; see also Figure 2.1). The changes in amphibian metamorphosis are initiated by thyroid hormones such as thyroxine (T_4) and tri-iodothyronine (T_3) that travel through the blood to reach all the organs of the larva. When the larval organs encounter these thyroid hormones, they can respond in any of four ways: growth, death, remodeling, and respecification.

Morphological changes associated with metamorphosis

GROWTH OF NEW STRUCTURES   The hormone tri-iodothyronine induces certain adult-specific organs to form. The limbs of the adult frog emerge from specific sites on the metamorphosing tadpole. In the eye, both nictitating membranes and eyelids emerge. Moreover, T_3 induces the proliferation and differentiation of new neurons to serve these organs. As the limbs grow out from the body axis, new neurons proliferate and differentiate in the spinal cord. These neurons send axons to the newly formed limb musculature (Marsh-Armstrong et al. 2004). Blocking T_3 activity prevents these neurons from forming and causes paralysis of the limbs.

One readily observed consequence of anuran metamorphosis is the movement of the eyes to the front of the head from their originally lateral position (Figure 18.1).* The lateral eyes of the tadpole are typical of preyed-upon herbivores, whereas the frontally located eyes of the frog befit its more predatory lifestyle. To catch its prey, the frog needs to see in three dimensions. That is, it has to acquire a *binocular field of vision*, where input from both eyes converges in the brain (see Chapter 13). In the tadpole, the right eye innervates the left side of the brain, and vice versa; there are no ipsilateral (same-side) projections of the retinal neurons. During metamorphosis, however, ipsilateral pathways emerge, enabling input from both eyes to reach the same area of the brain (Currie and Cowan 1974; Hoskins and Grobstein 1985a).

---

*One of the most spectacular movements of eyes during metamorphosis occurs in flatfish such as flounder. Originally, a flounder's eyes, like the lateral eyes of other fish species, are on opposite sides of its face. However, during metamorphosis, one of the eyes migrates across the head to meet the eye on the other side (see Hashimoto et al. 2002; Bao et al. 2005). This allows the flatfish to dwell on the ocean bottom, looking upward.

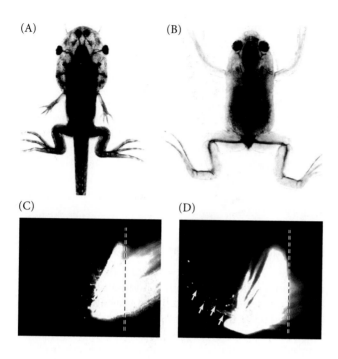

(A)   (B)

(C)   (D)

FIGURE 18.1 Eye migration and associated neuronal changes during metamorphosis of the *Xenopus laevis* tadpole. (A) The eyes of the tadpole are laterally placed, so there is relatively little binocular field of vision. (B) The eyes migrate dorsally and rostrally during metamorphosis, creating a large binocular field for the adult frog. (C,D). Retinal projections of metamorphosing tadpole. The dye DiI was placed on a cut stump of the optic nerve to label the retinal projection. (C) In early and middle stages of metamorphosis, axons project across the midline (dashed line) from one side of the brain to the other. (D) In late metamorphosis, ephrin-B is produced in the optic chiasm as certain neurons (arrows) are formed that project ipsilaterally. (A,B from Hoskins and Grobstein 1984; photographs courtesy of P. Grobstein; C,D from Nakagawa et al. 2000, photographs courtesy of C. E. Holt.)

In *Xenopus*, these new neuronal pathways result not from the remodeling of existing neurons, but from the formation of new neurons that differentiate in response to thyroid hormones (Hoskins and Grobstein 1985a,b). The ability of these axons to project ipsilaterally results from the induction of ephrin-B in the optic chiasm by the thyroid hormones (Nakagawa et al. 2000). Ephrin-B is also found in the optic chiasm of mammals (which have ipsilateral projections throughout life) but not in the chiasm of fish and birds (which have only contralateral projections). As shown in Chapter 13, ephrins can repel certain neurons, causing them to project in one direction rather than in another.

CELL DEATH DURING METAMORPHOSIS T_3 actually induces certain larval-specific structures to die. Thus, T_3 causes the degeneration of the paddle-like tail and the oxygen-procuring gills that were important for larval (but not adult) movement and respiration. While it is obvious that the tadpole's tail muscles and skin die, is this death murder or suicide? In other words, is T_3 telling the cells to kill themselves, or is T_3 telling something else to kill the cells? Recent evidence suggests that the first part of tail regression is caused by suicide, but that the last remnants of the tadpole tail must be killed off by other means. When tadpole muscle cells were injected with a dominant negative T_3 receptor (and therefore could not respond to T_3), the muscle cells survived, indicating that T_3 told them to kill themselves by apoptosis (Nakajima and Yaoita 2003; Nakajima et al. 2005). This was confirmed by the demonstration that the apoptosis-inducing enzyme caspase-9 is important in causing cell death in the tadpole muscle cells (Rowe et al. 2005). However, later in metamorphosis, the tail mus-

cles appear to be destroyed by phagocytosis, perhaps because the extracellular matrix that supported the muscle cells has been digested.

Death also comes to the tadpole red blood cell. During metamorphosis, tadpole hemoglobin is changed into an adult hemoglobin that binds oxygen more slowly and releases it more rapidly (McCutcheon 1936; Riggs 1951). The red blood cells carrying the tadpole hemoglobin have a different shape than the adult red blood cells, and these larval red blood cells are specifically digested—"eaten," if you will—by macrophages in the liver and spleen (Hasebe et al. 1999).

REMODELING DURING METAMORPHOSIS Among frogs and toads, certain larval structures are remodeled for adult needs. Thus, the larval intestine, with its numerous coils for digesting plant material, is converted into a shorter intestine for a carnivorous diet. Schrieber and his colleagues (2005) have demonstrated that the new cells of the adult intestine are derived from functioning cells of the larval intestine (instead of there being a subpopulation of stem cells that give rise to the adult intestine). The formation and differentiation of this new intestinal epithelium is probably triggered by the digestion of the old extracellular matrix by the metalloproteinase stromelysin-3, and by the new transcription of the *bmp4* and *sonic hedgehog* genes (Stolow and Shi 1995; Ishizuya-Oka et al. 2001; Fu et al. 2005). The elimination of the original extracellular matrix probably causes the apoptosis of those epithelial cells that were attached to it.* Therefore, the regional remodeling of the organs formed during metamorphosis may be generated by the reappearance of some of the same paracrine factors that modeled those organs in the embryo.

Much of the nervous system is remodeled as neurons grow and innervate new targets. The change in the optic nerve pathway was described earlier. Other larval neurons,

---

*Many epithelial cells are dependent on their attachment to the extracellular matrix to prevent apoptosis. The rapid apoptosis that occurs with the loss of extracellular matrix attachment has a special designation, **anoikis** (Frisch and Screaton 2001).

such as certain motor neurons in the tadpole jaw, switch their allegiances from larval muscle to newly formed adult muscle (Alley and Barnes 1983). Still others, such as the cells innervating the tongue muscle (a newly formed muscle not present in the larva), have lain dormant during the tadpole stage and form their first synapses during metamorphosis (Grobstein 1987). The lateral line system of the tadpole (which allows the tadpole to sense water movement and helps it to hear) degenerates, and the ears undergo further differentiation (see Fritzsch et al. 1988). The middle ear develops, as does the tympanic membrane characteristic of frog and toad outer ears. Tadpoles experience a brief period of deafness as the neurons change targets (Boatright-Horowitz and Simmons 1997). Thus, the anuran nervous system undergoes enormous restructuring as some neurons die and new ones are born, while other nerve cells change their specificity.

The shape of the anuran skull also changes significantly as practically every structural component of the head is remodeled (Trueb and Hanken 1992; Berry et al. 1998). The most obvious change is that new bone is being made. The tadpole skull is primarily neural crest-derived cartilage; the adult skull is primarily neural crest-derived bone (Gross and Hanken 2005; Figure 18.2). Another outstanding change is the formation of the lower jaw. Here, Meckel's cartilage elongates to nearly double its original length, and dermal bone forms around it. While Meckel's cartilage is growing, the gills and branchial arch cartilage (which were necessary for aquatic respiration in the tadpole) degenerate. Other cartilage, such as the ceratohyal cartilage (which will anchor the tongue), is extensively remodeled. Thus, as in the nervous system, some skeletal elements proliferate, some die, and some are remolded.

BIOCHEMICAL RESPECIFICATION   In addition to the obvious morphological changes, important biochemical transformations occur during metamorphosis as T_3 induces a new set of proteins in existing cells. One of the most dramatic bio-

chemical changes occurs in the liver. Tadpoles, like most freshwater fish, are ammonotelic—that is, they excrete ammonia. Like most terrestrial vertebrates, many adult frogs (such as the genus *Rana*, although not the more aquatic *Xenopus*) are ureotelic: they excrete urea, which requires less water than ammonia excretion. During metamorphosis, the liver begins to synthesize the enzymes necessary to create urea from carbon dioxide and ammonia (Figure 18.3). T_3 may regulate this change by inducing a transcription factor that specifically activates expression of the urea-cycle genes while suppressing the genes responsible for ammonia synthesis (Cohen 1970; Atkinson et al. 1996, 1998).

Hormonal control of amphibian metamorphosis

The control of metamorphosis by thyroid hormones was first demonstrated in 1912 by Gudernatsch, who discovered that tadpoles metamorphosed prematurely when fed powdered horse thyroid glands. In a complementary study, Allen (1916) found that when he removed or destroyed the thyroid rudiment of early tadpoles (thyroidectomy), the larvae never metamorphosed but instead grew into giant tadpoles. Subsequent studies showed that the sequential steps of anuran metamophosis are regulated by increasing

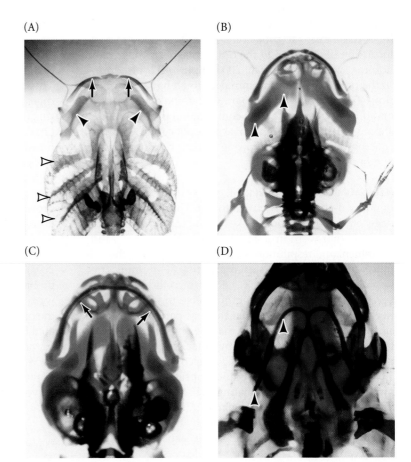

(A) (B)

(C) (D)

FIGURE 18.2   Changes in the *Xenopus* skull during metamorphosis. Wholemounts were stained with Alcian blue to stain cartilage and alizarin red to stain bone. (A) Prior to metamorphosis, the branchial arch cartilages (open arrowheads) are prominent, Meckel's cartilage (arrows) are at the tip of the head, and the ceratohyal cartilages (arrowheads) are relatively wide and anteriorly placed. (B–D) As metamorphosis ensues, the branchial arch cartilage disappears, Meckel's cartilage elongates, the mandible (lower jawbone) forms around Meckel's cartilage, and the ceratohyal cartilage narrows and becomes more posteriorly located. (From Berry et al. 1998, photographs courtesy of D. D. Brown.)

(A)

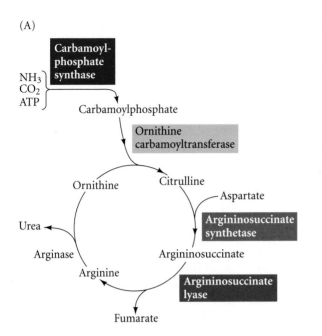

(B)

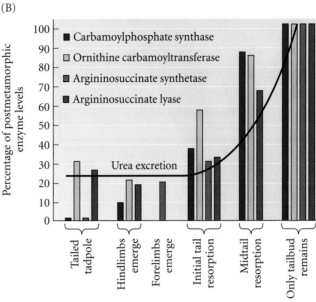

FIGURE 18.3 Development of the urea cycle during anuran metamorphosis. (A) Major features of the urea cycle, by which nitrogenous wastes are detoxified and excreted with minimal water loss. (B) The emergence of urea cycle enzyme activities correlates with metamorphic changes in the frog *Rana catesbeiana*. (After Cohen 1970.)

amounts of thyroid hormone (see Saxén et al. 1957; Kollros 1961; Hanken and Hall 1988). Some events (such as the development of limbs) occur early, when the concentration of thyroid hormones is low; other events (such as the regression of the tail and the remodeling of the intestine) occur later, after the hormones have reached higher concentrations. These observations gave rise to a **threshold model**, wherein the different events of metamorphosis are triggered by different concentrations of thyroid hormones. Although the threshold model remains useful, molecular studies have shown that the timing of the events of amphibian metamorphosis is more complex than just increasing hormone concentrations.

The metamorphic changes of frog development are brought about by (1) the secretion of the hormone **thyroxine** (T_4) into the blood by the thyroid gland; (2) the conversion of T_4 into a more active hormone, **tri-iodothyronine** (T_3) by the target tissues; and (3) the degradation of T_3 in the target tissues (Figure 18.4). T_3 binds to the nuclear **thyroid hormone receptors** (**TRs**) with much higher affinity than does T_4, and causes these receptors to become transcriptional activators of gene expression. Thus, the levels of both T_3 and TRs in the target tissues are essential for producing the metamorphic response in each tissue (Kistler et al. 1977; Robinson et al. 1977; Becker et al. 1997).

The concentration of T_3 in each tissue is regulated by the concentration of T_4 in the blood and by two critical intracellular enzymes that remove iodine atoms from T_4

and T_3. **Type II deiodinase** removes an iodine atom from the outer ring of the precursor (T_4) to convert it into the more active hormone T_3. **Type III deiodinase** removes an iodine atom from the inner ring of T_3 to convert it into an inactive compound that will eventually be metabolized to tyrosine (Becker et al. 1997). Tadpoles that are genetically modified to overexpress type III deiodinase in their target tissues never complete metamorphosis (Huang et al. 1999).

There are two types of thyroid hormone receptors. In *Xenopus*, **thyroid hormone receptor α** (**TRα**) is widely distributed throughout all tissues and is present even before the organism has a thyroid gland. **Thyroid hormone receptor β** (**TRβ**), however, is the product of a gene that is directly activated by thyroid hormones. TRβ levels are very low before the advent of metamorphosis; as the levels of thyroid hormone increase during metamorphosis, so do the intracellular levels of TRβ (Yaoita and Brown 1990; Eliceiri and Brown 1994).

The TRs do not work alone, however, but form dimers with the retinoid receptor, RXR. These dimers bind thyroid hormones and can effect transcription (Mangelsdorf and Evans 1995; Wong and Shi 1995; Wolffe and Shi 1999). The TR-RXR complex appears to be physically associated with appropriate promoters and enhancers even before it binds T_3. In its unbound state, the TR-RXR is a transcriptional repressor, recruiting histone deacetylases to the region of these genes. However, when T_3 is added to the complex, the T_3-TR-RXR complex activates those same genes by recruiting histone acetyltransferases (Sachs et al. 2001; Buchholz et al. 2003; Havis et al. 2003, Paul and Shi 2003).

Metamorphosis is often divided into stages. During the first stage, **premetamorphosis**, the thyroid gland has begun to mature and is secreting low levels of T_4 (and very low levels of T_3). The initiation of T_4 secretion may be brought about by corticotropin releasing hormone (CRH,

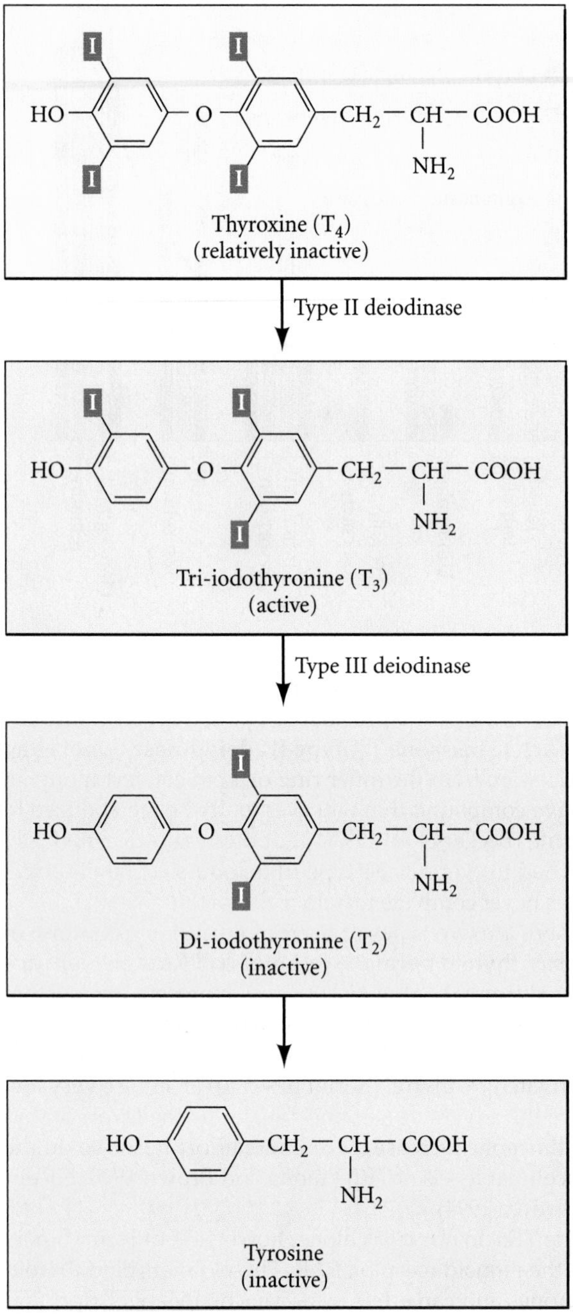

FIGURE 18.4 Metabolism of thyroxine (T_4) and tri-iodothyronine (T_3). Thyroxine serves as a prohormone. It is converted in the peripheral tissues to the active hormone T_3 by deiodinase II. T_3 can be inactivated by deiodinase III, which converts T_3 into di-iodothyronine and then to tyrosine.

T_3 and use it immediately through the TRα receptor. Thus, during the early stage of metamorphosis, the limb rudiments are able to receive thyroid hormone and use it to start leg growth (Becker et al. 1997; Huang et al. 2001; Schreiber et al. 2001).

As the thyroid matures to the stage of **prometamorphosis**, it secretes more thyroid hormones. However, many major changes (such as tail resorption, gill resorption, and intestinal remodeling) must wait until the **metamorphic climax** stage. At that time, the concentration of T_4 rises dramatically, and TRβ levels peak inside the cells. Since one of the target genes of T_3 is the TRβ gene, TRβ may be the principal receptor that mediates the metamorphic climax. In the tail, there is only a small amount of TRα during prometamorphosis, and deiodinase II is not detectable then. However, during prometamorphosis, the rising levels of thyroid hormones induce higher levels of TRβ. At metamorphic climax, deiodinase II is expressed, and the tail begins to be resorbed. In this way, the tail undergoes its absorption only *after* the legs are functional (otherwise, the poor amphibian would have no means of locomotion). The wisdom of the frog is simple: never get rid of your tail before your legs are working.

Some tissues do not seem to be responsive to thyroid hormones. For instance, thyroid hormones instruct the *ventral* retina to express ephrin-B and to generate the ipsilateral neurons shown in Figure 18.1D. The *dorsal* retina, however, is not responsive to thyroid hormones and does not generate new neurons. The dorsal retina appears to be insulated from thyroid hormones by expressing deiodinase III, which degrades the T_3 produced by deiodinase II. If deiodinase III is activated in the ventral retina, neurons will not proliferate and no ipsilateral axons will be formed (Marsh-Armstrong et al. 1999; Kawahara et al. 1999).

The frog brain also undergoes changes during metamorphosis, and one of the brain's functions is to downregulate metamorphosis once metamorphic climax has been reached. Thyroid hormones eventually produce a negative feedback loop, shutting down the pituitary cells that instruct the thyroid to secrete them (Saxén et al. 1957; Kollros 1961; White and Nicoll 1981). Huang and colleagues (2001) have recently shown that at the climax of metamorphosis, deiodinase II expression is seen in those cells of the anterior pituitary that secrete thyrotropin, the hormone that activates thyroid hormone expression. The resulting T_3 suppresses the transcription of the thyrotropin gene, thereby initiating the negative feedback loop so that less thyroid hormone is made.

which in mammals initiates the stress response). CRH may act directly on the frog pituitary, instructing it to release thyroid stimulating hormone (TSH), or it may act generally to make the body cells responsive to low amounts of T_3 (Denver 1993, 2003).

The tissues that respond earliest to the thyroid hormones are those that express high levels of deiodinase II, and can thereby convert T_4 directly into T_3 (Cai and Brown 2004). For instance, the limb rudiments, which have high levels of both deiodinase II and TRα, can convert T_4 into

(A)

Tail tip transplanted to trunk

Tail

(B)

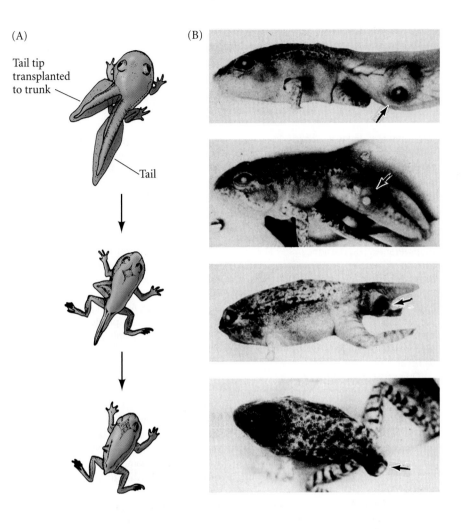

FIGURE 18.5 Regional specificity during frog metamorphosis. (A) Tail tips regress even when transplanted to the trunk. (B) Eye cups remain intact even when transplanted into the regressing tail. (After Schwind 1933.)

Regionally specific developmental programs

By regulating the amount of T_3 and TRs in their cells, the different regions of the body can respond to thyroid hormones at different times. The type of response (proliferation, apoptosis, migration) is determined by other factors already present in the different tissues. The same stimulus causes some tissues to degenerate while causing others to develop and differentiate, as exemplified by the process of tail degeneration.

The degeneration of the tadpole's tail structures is relatively rapid, since the bony skeleton does not extend to the tail (Wassersug 1989). The regression of the tail is brought about by apoptosis. After apoptosis occurs, macrophages collect in the tail region, digesting the debris with their enzymes, especially collagenases and metalloproteinases. The result is that the tail becomes a large sac of proteolytic enzymes* (Kaltenbach et al. 1979; Oofusa and Yoshizato 1991; Patterson et al. 1995). The tail epider-

mis acts differently than the head or trunk epidermis. During metamorphic climax, the larval skin is instructed to undergo apoptosis. The tadpole head and body are able to generate a new epidermis from epithelial stem cells. The tail epidermis, however, lacks these stem cells and fails to generate new skin (Suzuki et al. 2002).

Organ-specific responses to thyroid hormones have been dramatically demonstrated by transplanting a tail tip to the trunk region or by placing an eye cup in the tail (Schwind 1933; Geigy 1941). Tail tip tissue placed in the trunk is not protected from degeneration, but the eye cup retains its integrity despite the fact that it lies within the degenerating tail (Figure 18.5). Thus, the degeneration of the tail represents an organ-specific programmed cell death response, and only specific tissues die when the signal is given. Such programmed cell deaths are important in molding the body.

---

*Interestingly, the degeneration of the human tail during week 4 of gestation resembles the regression of the tadpole tail (see Fallon and Simandl 1978).

VADE MECUM² **Amphibian metamorphosis and frog calls.**
For photographs of amphibian metamorphosis (and for the sounds of the adult frogs), check out the metamorphosis and frog call sections of the CD-ROM. [Click on Amphibian]

SIDELIGHTS & SPECULATIONS

Variations on the Theme of Amphibian Metamorphosis

Heterochrony

Although most animal species develop through a larval stage, some have modified their life cycles by either greatly extending or shortening their larval period. The phenomenon wherein animals change the relative time of appearance and rate of development of characters present in their ancestors is called **heterochrony**. Here we will describe three extreme types of heterochrony:

1. *Neoteny* refers to the retention of the juvenile form as a result of retarded body development relative to the development of the germ cells and gonads (which achieve maturity at the normal time).
2. *Progenesis* also involves the retention of the juvenile form, but in this case, the gonads and germ cells develop at a faster rate than normal, becoming sexually mature while the rest of the body is still in a juvenile phase.
3. In *direct development*, the embryo abandons the stages of larval development entirely and proceeds to construct a small adult.

Neoteny

In certain salamanders, the reproductive system and germ cells mature while the rest of the body retains its juvenile form throughout life. In most such species, metamorphosis fails to occur and sexual maturity takes place in a "larval" body.

The Mexican axolotl, *Ambystoma mexicanum*, does not undergo metamorphosis in nature because its pituitary gland does not release the thyrotropin (thyroid-stimulating hormone) that would activate T_4 synthesis (Prahlad and DeLanney 1965; Norris et al. 1973; Taurog et al. 1974). The axolotl *does* synthesize functional thyroid hormone receptors, however, and when investigators administered either thyroid hormones or thyrotropin, they found that the salamander metamorphosed into an adult form not seen in nature (Figure 18.6; Huxley 1920; Safi et al. 2004).

Other species of *Ambystoma*, such as *A. tigrinum*, metamorphose only in response to cues from the environment. In parts of its range, *A. tigrinum* is neotenic: its gonads and germ cells mature and the salamander mates successfully while the rest of the body retains its aquatic larval form. However, in other regions of its range, the larval form is transitory, leading to the land-dwelling adult tiger salamander. The ability to remain aquatic is highly adaptive in locations where the terrestrial environment is too dry to sustain the adult form of this salamander (Duellman and Trueb 1986).

Some salamanders are permanently neotenic, even in the laboratory. Whereas T_3 is able to produce the long-lost adult form of *A. mexicanum*, the neotenic species of *Necturus* and *Siren* remain unresponsive to thyroid hormones (Frieden 1981). Their permanent neoteny is probably due to the lack of TRβ receptors in those tissues normally responsive to T_3 (Safi et al. 1997).

De Beer (1940) and Gould (1977) have speculated that neoteny is a major factor in the evolution of more complex taxa. By retarding the development of somatic tissues, neoteny may give natural selection a flexible substrate. According to Gould (1977, p. 283), neoteny may "provide an escape from specialization. Animals can relinquish their highly specialized adult forms, return to the lability of youth, and prepare themselves for new evolutionary directions."

Progenesis

In progenesis, gonadal maturation is accelerated while the rest of the body develops normally to a certain stage. Progenesis has enabled some salamander species to find new ecological niches. *Bolitoglossa occidentalis* is a tropical salamander that, unlike other members of its genus, lives in trees. This salamander's webbed feet and small body size suit it for arboreal existence, the webbed feet producing suction for climbing and the small body making such traction efficient. Alberch and Alberch (1981) showed that *B. occidentalis* resembles juveniles of the

(A)

(B)

FIGURE 18.6   Metamorphosis in *Ambystoma*. (A) Normal adult *Ambystoma*, with prominent gills and broad tail. (B) Metamorphosed *Ambystoma* not seen in natural populations. This individual was grown in water supplemented with thyroxine. Its gills have regressed and its skin has changed significantly. (Photographs courtesy of K. Crawford.)

(A)

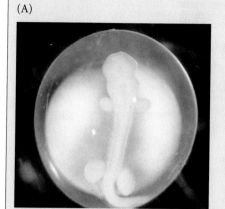

(B)

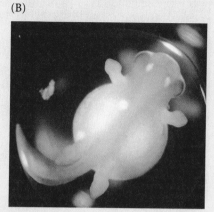

(C)

FIGURE 18.7 Direct development of the frog *Eleutherodactylus coqui*. (A) Limb buds are seen as the embryo develops on the yolk. (B) As the yolk is used up, the limb buds are easily seen. (C) Three weeks after fertilization, tiny froglets hatch. They are seen here in a petri dish and on a Canadian dime. (Photographs courtesy of R. P. Elinson.)

related species *B. subpalmata* and *B. rostrata* (whose young are small, with digits that have not yet grown past their webbing). *B. occidentalis* reaches sexual maturity at a much smaller size than its relatives, and this appears to have given it a phenotype that made tree-dwelling possible.

Direct development

While some animals have extended their larval period of life, others have "accelerated" their development by abandoning their larval forms. This latter phenomenon, called **direct development**, is typified by frog species that lack tadpoles and by sea urchins that lack pluteus larvae. Elinson and his colleagues (del Pino and Elinson 1983; Elinson 1987) have studied *Eleutherodactylus coqui*, a small frog that is one of the most abundant vertebrates on the island of Puerto Rico. Unlike the eggs of *Rana* and *Xenopus*, the eggs of *E. coqui* are fertilized while they are still within the female's body. Each egg is about 3.5 mm in diameter (roughly 20 times the volume of a *Xenopus* egg). After the eggs are laid, the male gently sits on the developing embryos, protecting them from predators and desiccation (Taigen et al. 1984).

Early *E. coqui* development is like that of most frogs. Cleavage is holoblastic, gastrulation is initiated at a subequatorial position, and the neural folds become elevated from the surface. However, shortly after the neural tube closes, limb buds appear on the surface (Figure 18.7A,B). This early emergence of limb buds is the first indication that this

animal will not pass through the usual limbless tadpole stage. Moreover, the development of *E. coqui* is modified such that the modeling of most of its features—including its limbs—does not depend on thyroid hormones. Its thyroid gland does develop, however, and thyroid hormones appear to be critical for the eventual resorption of the tail (which is used as a respiratory rather than a locomotor organ) and of the primitive kidney (Lynn and Peadon 1955). It appears that the thyroid-dependent phase has been pushed back into embryonic growth (Hanken et al. 1992; Callery et al. 2001). What emerges from the egg jelly 3 weeks after fertilization is not a tadpole, but a tiny frog (Figure 18.7C).

Direct-developing frogs do not need ponds for their larval stages and can therefore colonize habitats that are inaccessible to other frogs. Direct development also occurs in other phyla, in which it is also correlated with a large egg. It seems that if nutrition can be provided in the egg, the life cycle need not have a food-gathering larval stage.

Tadpole-rearing behaviors

Most temperate-zone frogs do not invest time or energy in providing for their tadpoles. However, among tropical frogs, there are numerous species in which adult frogs take painstaking care of their tadpoles. An example is the poison arrow frog, *Dendrobates*, found in the rain forests of Central and South America. Most of the time, these highly toxic frogs live in the leaf litter of the forest floor. After the eggs are laid in a damp

leaf, a parent (sometimes the male, sometimes the female, according to the species) stands guard over the eggs. If the ground gets too dry, the frog will urinate on the eggs to keep them moist. When the eggs mature into tadpoles, the guarding parent allows them to wriggle onto its back (Figure 18.8A). The parent then climbs into the canopy until it finds a bromeliad plant with a small pool of water in its leaf base. Here it deposits one of its tadpoles, then goes back for another, and so on until the entire brood has been placed in numerous small pools. The female returns each day to these pools and deposits a small number of unfertilized eggs into them, thus replenishing the tadpoles' food supply until they complete metamorphosis (Mitchell 1988; van Wijngaarden and Bolanos 1992; Brust 1993). It is not known how the female frog remembers—or is informed about—where the tadpoles have been deposited.

Brooding frogs carry their developing eggs in depressions in their skin. Some species brood their tadpoles in their mouths. When their tadpoles undergo metamorphosis, these frogs spit out their progeny. Even more impressive, the gastric-brooding frogs of Australia, *Rheobatrachus silus* and *R. vitellinus*, eat their eggs. The eggs develop into larvae, and the larvae undergo metamorphosis in the mother's stomach. About 8 weeks after being swallowed alive, about two dozen small frogs emerge from the female's mouth (Figure 18.8B; Corben et al. 1974; Tyler 1983). What stops the *Rheobatrachus* eggs from being digested or excreted? It appears that the eggs secrete prostaglandins that stop acid secretion and prevent peristaltic contrac-

(Continued on next page)

tions in the stomach (Tyler et al. 1983). During this time, the stomach is fundamentally a uterus, and the frog does not eat. After the oral birth, the parent's stomach morphology and function return to normal. Unfortunately, both of these remarkable frog species are now feared extinct. No member of either *Rheobatrachus* species has been seen since the mid-1980s.

These are a few examples of how frog development has been the source of adaptive changes during evolution.

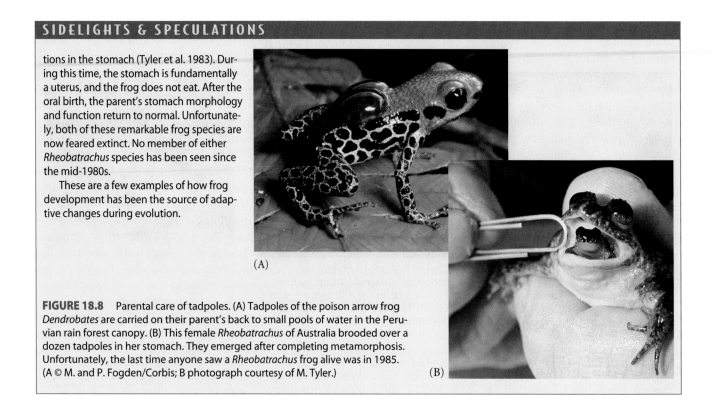

(A)

(B)

FIGURE 18.8 Parental care of tadpoles. (A) Tadpoles of the poison arrow frog *Dendrobates* are carried on their parent's back to small pools of water in the Peruvian rain forest canopy. (B) This female *Rheobatrachus* of Australia brooded over a dozen tadpoles in her stomach. They emerged after completing metamorphosis. Unfortunately, the last time anyone saw a *Rheobatrachus* frog alive was in 1985. (A © M. and P. Fogden/Corbis; B photograph courtesy of M. Tyler.)

Metamorphosis in Insects

Whereas amphibian metamorphosis is largely characterized by the remodeling of existing tissues, insect metamorphosis primarily involves the destruction of larval tissues and their replacement by an entirely different population of cells. Insects grow by molting—shedding their cuticle—and forming a new cuticle as their size increases. There are three major patterns of insect development. A few insects, such as springtails, have no larval stage and undergo direct development. These are called the **ametabolous** insects (Figure 18.9A). Immediately after they hatch, these insects have a **pronymph** stage bearing the structures that enabled it to get out of the egg. But after this transitory stage, the insect looks like a small adult; it grows larger after each molt, but is unchanged in form (Truman and Riddiford 1999).

Other insects, notably grasshoppers and bugs, undergo a gradual, **hemimetabolous** metamorphosis (Figure 18.9B). After spending a very brief period of time as a pronymph (whose cuticle is often shed as the insect hatches), the insect looks like an immature adult and is called a **nymph**. The rudiments of the wings, genital organs, and other adult structures are present and become progressively more mature with each molt. At the final molt, the emerging insect is a winged and sexually mature adult, or **imago**.

In the **holometabolous** insects such as flies, beetles, moths, and butterflies, there is no pronymph stage (Figure 18.9C). The juvenile form that hatches from the egg is called a **larva**. The larva (a caterpillar, grub, or maggot) undergoes a series of molts as it becomes larger. The stages between these larval molts are called **instars**. The number of larval molts before becoming an adult is characteristic of a species, although environmental factors can increase or decrease the number. The larval instars grow in a stepwise fashion, each instar being larger than the previous one. Finally, there is a dramatic and sudden transformation between the larval and adult stages: after the final instar, the larva undergoes a **metamorphic molt** to become a **pupa**. The pupa does not feed, and its energy must come from those foods it ingested as a larva. During pupation, adult structures form and replace the larval structures. Eventually, an **imaginal molt** enables the adult (imago) to shed its pupal case and emerge. While the larva is said to *hatch* from an egg, the imago is said to *eclose* from the pupa.

Imaginal discs

In holometabolous insects, the transformation from juvenile into adult occurs within the pupal cuticle. Most of the larval body is systematically destroyed by programmed cell death, while new adult organs develop from undifferentiated nests of **imaginal cells**. Thus, within any larva, there are two distinct populations of cells: the larval cells, which are used for the functions of the juvenile insect; and thousands of imaginal cells, which lie within the larva in clusters, awaiting the signal to differentiate.

FIGURE 18.9 Modes of insect development. Molts are represented as arrows. (A) Ametabolous (direct) development in a silverfish. After a brief pronymph stage, the insect looks like a small adult. (B) Hemimetabolous (gradual) metamorphosis in a cockroach. After a very brief pronymph phase, the insect becomes a nymph. After each molt, the next nymphal instar looks more like an adult, gradually growing wings and genital organs. (C) Holometabolous (complete) metamorphosis in a moth. After hatching as a larva, the insect undergoes successive larval molts until a metamorphic molt causes it to enter the pupal stage. Then an imaginal molt turns it into an adult.

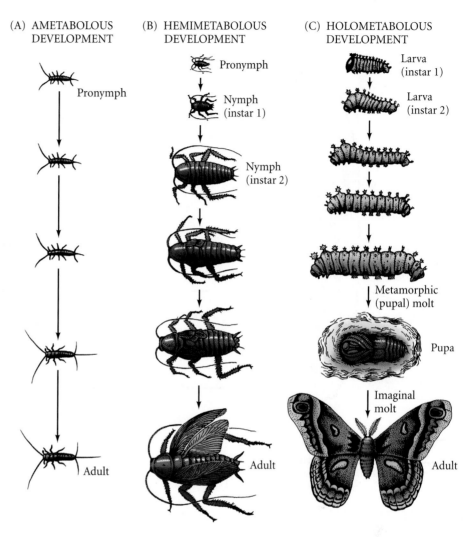

(A) AMETABOLOUS DEVELOPMENT

Pronymph

Adult

(B) HEMIMETABOLOUS DEVELOPMENT

Pronymph

Nymph (instar 1)

Nymph (instar 2)

Adult

(C) HOLOMETABOLOUS DEVELOPMENT

Larva (instar 1)

Larva (instar 2)

Metamorphic (pupal) molt

Pupa

Imaginal molt

Adult

There are three main types of imaginal cells:

1. The **imaginal discs**, whose cells will form the cuticular structures of the adult, including the wings, legs, antennae, eyes, head, thorax, and genitalia (Figure 18.10).
2. The **histoblast nests** are clusters of imaginal cells that will form the adult abdomen.
3. Clusters of imaginal cells within each organ, which will proliferate to form the adult organ as the larval organ degenerates.

The imaginal discs can be seen in the newly hatched larva as local thickenings of the epidermis. Whereas most larval cells have a very limited mitotic capacity, imaginal discs divide rapidly at specific characteristic times. As their cells proliferate, they form a tubular epithelium that folds in upon itself in a compact spiral (Figure 18.11A). At metamorphosis, these cells proliferate even further as they differentiate, and elongate (Figure 18.11B).

The fate map and elongation sequence of one of the six *Drosophila* leg disc is shown in Figure 18.12. At the end of the third instar, just before pupation, the leg disc is an epithelial sac connected by a thin stalk to the larval epidermis. On one side of the sac, the epithelium is coiled into a series of concentric folds "reminiscent of a Danish pastry" (Kalm et al. 1995). As pupation begins, the cells at the center of the disc telescope out to become the most distal portions of the leg—the claws and the tarsus. The outer cells become the proximal structures—the coxa and the adjoining epidermis (Schubiger 1968). After differentiating, the cells of the appendages and epidermis secrete a cuticle appropriate for each specific region. Although the disc is composed primarily of epidermal cells, a small number of **adepithelial cells** migrate into the disc early in development. During the pupal stage, these cells give rise to the muscles and nerves that serve the leg.

SPECIFICATION AND PROLIFERATION The specification of the general cell fates (i.e., that the disc is to be a leg disc and not a wing disc) occurs in the embryo. The more specific cell fates are specified in the larval stages, as the cells proliferate (Kalm et al. 1995). The type of leg structure generated is determined by the interactions between several

Discs for:

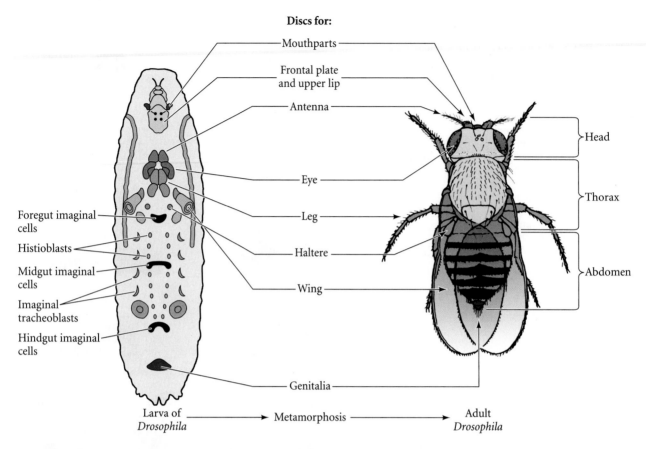

FIGURE 18.10   The locations and developmental fates of the imaginal discs and imaginal tissues in the third instar larva of *Drosophila melanogaster.* (After Kalm et al. 1995.)

genes in the imaginal disc. Figure 18.13 shows the expression of three genes involved in determining the proximal-distal axis of the fly leg. In the third-instar leg disc, the center of the disc secretes the highest concentration of two morphogens, Wingless (Wg) and Decapentaplegic (Dpp). High concentrations of these paracrine factors cause the expression of the *Distal-less* gene. Moderate concentrations cause the expression of the *dachshund* gene, and lower concentrations cause the expression of the *homothorax* gene.

Those cells expressing *Distal-less* telescope out to become the most distal structures of the leg—the claw and distal tarsal segments. Those expressing *homothorax* become the most proximal structure, the coxa. Cells expressing *dachshund* become the femur and proximal tibia. Areas where the transcription factors overlap produce the trochanter and distal tibia (Abu-Shaar and Mann 1998). These regions of gene expression are stabilized by inhibito-

ry interactions between the protein products of these genes and of the neighboring genes. In this manner, the gradient of Wg and Dpp proteins is converted into discrete domains of gene expression that specify the different regions of the *Drosophila* leg.

(A)

(B)

FIGURE 18.11   Imaginal disc elongation. Scanning electron micrograph of *Drosophila* third-instar leg disc (A) before and (B) after elongation. (From Fristrom et al. 1977; photograph courtesy of D. Fristrom.)

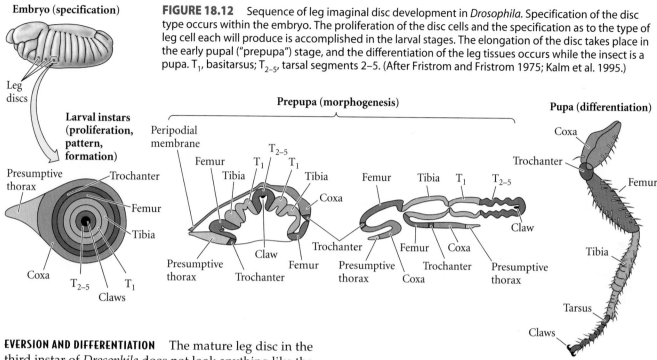

FIGURE 18.12   Sequence of leg imaginal disc development in *Drosophila*. Specification of the disc type occurs within the embryo. The proliferation of the disc cells and the specification as to the type of leg cell each will produce is accomplished in the larval stages. The elongation of the disc takes place in the early pupal ("prepupa") stage, and the differentiation of the leg tissues occurs while the insect is a pupa. T_1, basitarsus; T_{2-5}, tarsal segments 2–5. (After Fristrom and Fristrom 1975; Kalm et al. 1995.)

EVERSION AND DIFFERENTIATION   The mature leg disc in the third instar of *Drosophila* does not look anything like the adult structure. It is determined, but not yet differentiated; its differentiation requires a signal, in the form of a set of pulses of the "molting" hormone 20-hydroxyecdysone (20E; see Figure 18.17). The first pulse, occurring in the late larval stages, initiates the formation of the pupa, arrests cell division in the disc, and initiates the cell shape changes that drive the eversion of the leg. Studies by Condic and her colleagues have demonstrated that the elongation of imaginal discs occurs without cell division and is due primarily to cell shape changes within the disc epithelium (Condic et al. 1990). Using fluorescently labeled phalloidin to stain the peripheral microfilaments of leg disc cells, they showed that the cells of early third-instar discs are tightly arranged along the proximal-distal axis. When the hormonal signal to differentiate is given, the cells change their shape, and the leg is everted, the central cells of the disc becoming the

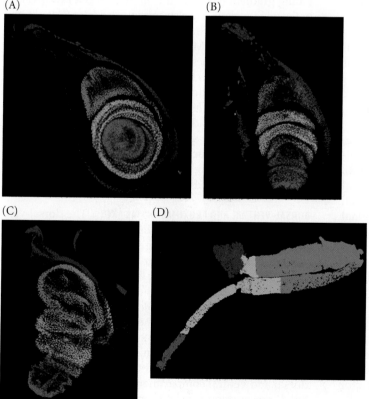

FIGURE 18.13   The fates of the imaginal disc cells are directed by transcription factors found in different regions. (A–D) Expression of transcription factor genes in the *Drosophila* leg disc. At the periphery, the *homothorax* gene (purple) establishes the boundary for the coxa. The expression of the *dachshund* gene (green) locates the femur and proximal tibia. The most distal structures, the claw and lower tarsal segments, arise from the expression domain of *Distal-less* (red) in the center of the imaginal disc. The overlap of *dachshund* and *Distal-less* appears yellow and specifies the distal tibia and upper tarsal segments. (A–C) Gene expression at successively later stages of pupal development. (D) Localization of expression domains of the genes onto a leg immediately prior to eclosion. The areas where there is overlap between expression domains are shown in yellow, aqua, and orange. (From Abu-Shaar and Mann 1998; photographs courtesy of R. S. Mann.)

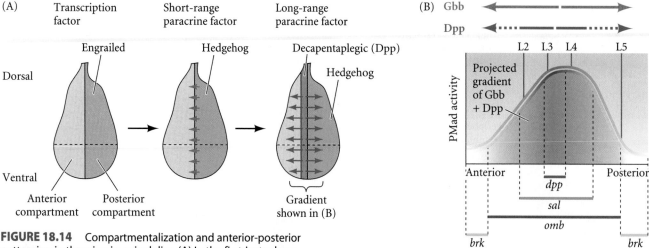

FIGURE 18.14 Compartmentalization and anterior-posterior patterning in the wing imaginal disc. (A) In the first-instar larva, the anterior-posterior axis has been formed and can be recognized by the expression of the *engrailed* gene in the posterior compartment. Engrailed, a transcription factor, activates the *hedgehog* gene. Hedgehog acts as a short-range paracrine factor to activate *decapentaplegic* in the anterior cells adjacent to the posterior compartment, where Dpp and a related protein, Glass-bottom boat (Gbb), act over a longer range. (B) Dpp and Gbb proteins create a concentration gradient of BMP-like signaling, measured by the phosphorylation of Mad (pMad). High concentrations of Dpp plus Gbb near the source activate both the *sal* and *omb* genes. Lower concentrations (near the periphery) activate *omb* but not *sal*. When Dpp plus Gbb levels drop below a certain threshold, *brinker* (*brk*) is no longer repressed. L2 – L5 mark the longitudinal wing veins, with L2 being the most anterior. (After Bangi and Wharton 2006.)

most distal (claw) cells of the limb. The leg structures will differentiate within the pupa, so that by the time the adult fly ecloses, they are fully formed and functional.

Determination of the wing imaginal discs

The largest of *Drosophila's* imaginal discs is that of the wing, containing some 60,000 cells (in contrast, the leg and haltere discs contain about 10,000 cells each; Fristrom 1972). The wing discs are distinguished from the other imaginal discs by the expression of the *vestigial* gene (Kim et al. 1996). When this gene is expressed in any other imaginal disc, wing tissue emerges.

ANTERIOR AND POSTERIOR COMPARTMENTS The axes of the wing are specified by gene expression patterns that divide the embryo into discrete but interacting compartments (Figure 18.14A; Meinhart 1980; Causo et al. 1993; Tabata et al. 1995). In the first instar, expression of the *engrailed* gene distinguishes the posterior compartment of the wing from the anterior compartment. The Engrailed transcription factor is expressed only in the posterior compartment, and in those cells, it activates the gene for the paracrine factor Hedgehog. Hedgehog functions only when cells have the

receptor (Patched) to receive it. In a complex manner, the diffusion of Hedgehog activates the gene encoding Decapentaplegic (Dpp), a BMP-like paracrine factor, in a narrow stripe of cells in the anterior region of the wing disc (Ho et al. 2005).

Dpp and a co-expressed BMP called Glass-bottom boat (Gbb) act to establish a gradient of BMP signaling activity. BMPs activate the Mad transcription factor by phosphorylating it, so this gradient can be measured by the phosphorylation of Mad. Dpp is a short-range paracrine factor, while Gbb exhibits a much longer range of diffusion to create a gradient (Figure 18.14B; Bangi and Wharton 2006). This signaling gradient regulates the amount of cell proliferation in the wing regions and also specifies cell fates (Rogulja and Irvine 2005). Several transcription factor genes respond differently to activated Mad. At high levels, the *spalt* and *optomotor blind* (*omb*) genes are activated, whereas at low levels (where Gbb provides the primary signal), only *omb* is activated. Below a particular level of phosphorylated Mad activity, the *brinker* (*brk*) gene is no longer inhibited; thus *brk* is expressed outside the signaling domain. Specific cell fates of the wing are specified in response to the action of these transcription factors. (For example, the fifth longitudinal vein of the wing is formed at the border of *optomotor-blind* and *brinker*; see Figure 8.14B).

DORSAL-VENTRAL AND PROXIMAL-DISTAL AXES The dorsal-ventral axis of the wing is formed at the second instar stage by the expression of the *apterous* gene in the prospective dorsal cells of the wing disc (Blair 1993; Diaz-Benjumea and Cohen 1993). Here, the upper layer of the wing is distinguished from the lower layer of the wing blade (Bryant 1970; Garcia-Bellido et al. 1973). The *vestigial* gene remains "on" in the ventral portion of the wing disc (Figure 18.15A). The dorsal portion of the wing synthesizes transmembrane proteins that prevent the intermixing of the dorsal and ventral cells (Milán et al. 2005). At the boundary between the dorsal and ventral compartments, the Apterous and Vestigial transcription factors interact to activate the *wingless*

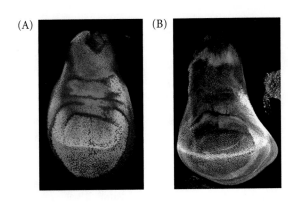

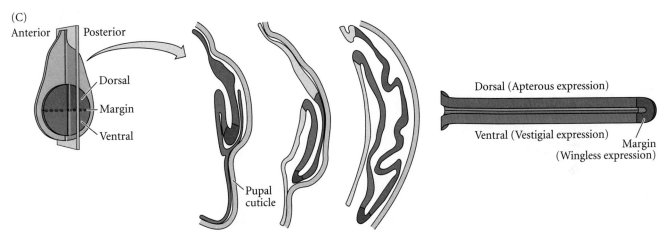

FIGURE 18.15 Determining the dorsal-ventral axis. (A) The prospective ventral surface of the wing is stained by antibodies to Vestigial protein (green), while the prospective dorsal surface is stained by antibodies to Apterous protein (red). The region of yellow illustrates where the two proteins overlap in the margin. (B) Wingless protein (purple) synthesized at the marginal juncture organizes the wing disc along the dorsal-ventral axis. The expression of Vestigial (green) is seen in cells close to those expressing Wingless. (C) The dorsal and ventral portions of the wing disc telescope out to form the two-layered wing. Gene expression patterns are indicated on the double-layered wing. (Photographs courtesy of S. Carroll and S. Paddock.)

gene (Figure 18.15B). Neumann and Cohen (1996) showed that Wingless protein acts as a growth factor to promote the cell proliferation that extends the wing.* Wingless also helps establish the proximal-distal axis of the wing: high levels of Wingless activate the *Distal-less* gene, which specifies the most distal regions of the wing (Zecca et al. 1996; Neumann and Cohen 1996, 1997). This is in the central region of the disc and "telescopes" outward as the distal margin of the wing blade.

WEBSITE **18.1** **The molecular biology of wing formation.** Formation of the *Drosophila* wing involves the interaction of more than 200 genes. This site discusses some of these gene interactions.

WEBSITE **18.2** **Homologous specification.** If a group of cells in one imaginal disc are mutated such that they give rise to a structure characteristic of another imaginal disc (for instance, cells from a leg disc giving rise to antennal structures), the regional specification of those structures will be in accordance with their position in the original disc.

Hormonal control of insect metamorphosis

Although the details of insect metamorphosis differ among species, the general pattern of hormonal action is very similar. Like amphibian metamorphosis, the metamorphosis of insects is regulated by systemic hormonal signals, which are controlled by neurohormones from the brain (for reviews, see Gilbert and Goodman 1981; Riddiford 1996). Insect molting and metamorphosis are controlled by two effector hormones: the steroid **20-hydroxyecdysone** (**20E**) and the lipid **juvenile hormone** (**JH**) (Figure 18.16A). 20-hydroxyecdysone[†] initiates and coordinates each molt and regulates the changes in gene expression that occur during metamorphosis. Juvenile hormone prevents the ecdysone-induced changes in gene expression that are necessary for metamorphosis. Thus, its presence during a molt ensures that the result of that molt is another larval instar, not a pupa or an adult.

The molting process is initiated in the brain, where neurosecretory cells release **prothoracicotropic hormone** (**PTTH**) in response to neural, hormonal, or environmental signals. PTTH is a peptide hormone with a molecular weight of approximately 40,000, and it stimulates the production of

---

*The diffusion of paracrine factors such as Wingless and Hedgehog is facilitated when these factors cluster on lipid spheres that can travel between cells without getting caught in the extracellular matrix (Glise et al. 2005; Gorfinkiel et al. 2005; Panáková et al. 2005).

[†]Since its discovery in 1954, when Butenandt and Karlson isolated 25 mg of ecdysone from 500 kg of silkworm moth pupae, 20-hydroxyecdysone has gone under several names, including ecdysterone, β-ecdysone, and crustecdysone.

(A)

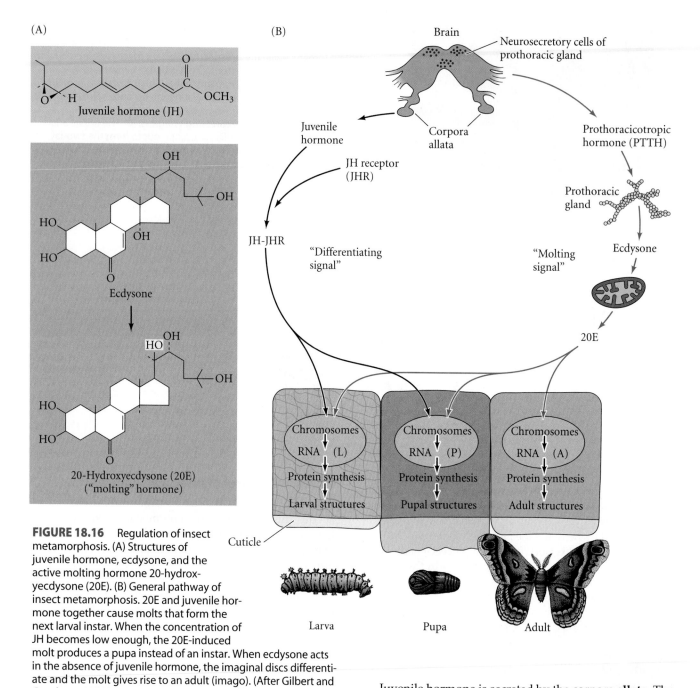

FIGURE 18.16 Regulation of insect metamorphosis. (A) Structures of juvenile hormone, ecdysone, and the active molting hormone 20-hydroxyecdysone (20E). (B) General pathway of insect metamorphosis. 20E and juvenile hormone together cause molts that form the next larval instar. When the concentration of JH becomes low enough, the 20E-induced molt produces a pupa instead of an instar. When ecdysone acts in the absence of juvenile hormone, the imaginal discs differentiate and the molt gives rise to an adult (imago). (After Gilbert and Goodman 1981.)

ecdysone by the **prothoracic gland**. Ecdysone is modified in peripheral tissues to become the active molting hormone 20E. Each molt is initiated by one or more pulses of 20E. For a larval molt, the first pulse produces a small rise in the 20E concentration in the larval hemolymph (blood) and elicits a change in cellular commitment in the epidermis. A second, larger pulse of 20E initiates the differentiation events associated with molting. These pulses of 20E commit and stimulate the epidermal cells to synthesize enzymes that digest the old cuticle and synthesize a new one.

Juvenile hormone is secreted by the **corpora allata**. The secretory cells of the corpora allata are active during larval molts but inactive during the metamorphic molt and the imaginal molt. As long as JH is present, the 20E-stimulated molts result in a new larval instar. In the last larval instar, however, the medial nerve from the brain to the corpora allata inhibits these glands from producing JH, and there is a simultaneous increase in the body's ability to degrade existing JH (Safranek and Williams 1989). Both these mechanisms cause JH levels to drop below a critical threshold value, triggering the release of PTTH from the brain (Nijhout and Williams 1974; Rountree and Bollenbacher 1986). PTTH, in turn, stimulates the prothoracic glands to secrete a small amount of ecdysone. The resulting pulse of 20E, in the absence of high levels of JH, commits the epidermal cells to pupal development. Larva-specific mRNAs are not replaced, and new mRNAs are synthesized whose protein products inhibit the transcription of the larval messages.

There are two major pulses of 20E during *Drosophila* metamorphosis. The first pulse occurs in the third instar larva and triggers the initiation of ("prepupal") morphogenesis of the leg and wing imaginal discs (as well as the death of the larval hindgut). The larva stops eating and migrates to find a site to begin pupation. The second 20E pulse occurs from 10–12 hours later and tells the "prepupa" to become a pupa. The head inverts and the salivary glands degenerate (Riddiford 1982; Nijhout 1994). It appears, then, that the first ecdysone pulse during the last larval instar triggers the processes that inactivate the larva-specific genes and initiates the morphogenesis of imaginal disc structures. The second pulse transcribes pupa-specific genes and initiates the molt (Nijhout 1994). At the imaginal molt, when 20E acts in the absence of juvenile hormone, the imaginal discs fully differentiate and the molt gives rise to an adult.

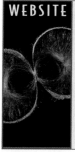

WEBSITE **18.3** **Insect metamorphosis.** Four websites discuss (1) the experiments of Wigglesworth and others who identified the hormones of metamorphosis and the glands producing them; (2) the variations that *Drosophila* and other insects play on the general theme of metamorphosis; (3) the remodeling of the insect nervous system during metamorphosis, and (4) a microarray analysis of *Drosophila* metamorphosis wherein several thousand genes were simultaneously screened.

The molecular biology of 20-hydroxyecdysone activity

ECDYSONE RECEPTORS 20-Hydroxyecdysone cannot bind to DNA by itself. Like amphibian thyroid hormones, 20E first binds to nuclear receptors. These proteins, called ecdysone receptors (EcRs), are almost identical in structure to the thyroid hormone receptors of amphibians. An EcR protein forms an active molecule by pairing with an Ultraspiracle (Usp) protein, the homologue of the amphibian RXR that helps form the active thyroid hormone receptor (Koelle et al. 1991; Yao et al. 1992; Thomas et al. 1993). In the absence of the hormone-bound EcR, the Usp protein binds to the ecdysone-responsive genes and inhibits their transcription.[*] This inhibition is converted into activation when the ecdysone receptor binds to the Usp (Schubiger and Truman 2000).

Although there is only one gene for EcR, the EcR mRNA transcript can be spliced in at least three different ways to form three distinct proteins. All three EcR proteins have the same domains for 20E and DNA binding, but they differ in their amino-terminal domains. The type of EcR present in a cell may inform that cell how to act when it receives a hormonal signal (Talbot et al. 1993; Truman et al. 1994). All cells appear to have some EcRs of each type, but the strictly larval tissues and neurons that die when exposed to 20E are characterized by their abundance of the EcR-B1 isoform of the ecdysone receptor. Imaginal discs and differentiating neurons, on the other hand, show a preponderance of the EcR-A isoform. Mutations in specific codons that are found in only some of the splicing isoforms indicate that the different forms of EcR play different roles in metamophosis and that the different receptors activate different sets of genes when they bind 20E (Davis et al. 2005).

BINDING OF 20-HYDROXYECDYSONE TO DNA During molting and metamorphosis, certain regions of the polytene chromosomes of *Drosophila* puff out in the cells of certain organs at certain times (see Figure 4.11; Clever 1966; Ashburner 1972; Ashburner and Berondes 1978). These chromosome puffs are areas where DNA is being actively transcribed. Moreover, these organ-specific patterns of chromosome puffing can be reproduced by culturing larval tissue and adding hormones to the medium, or by adding 20E to an early-stage larva. When 20E is added to larval salivary glands, certain puffs are produced and others regress (Figure 18.17). The puffing is mediated by the binding of 20E at specific places on the chromosomes; fluorescent antibodies against 20E find this hormone localized to the regions of the genome that are sensitive to it (Gronemeyer and Pongs 1980). At these sites, the ecdysone-bound receptor complex recruits a histone methyltransferase that methylates lysine-4 of histone H3, thereby loosening the nucleosomes in that area (Sedkov et al. 2003).

20E-regulated chromosome puffing occurs during the late stages of the third-instar *Drosophila* larva, as it prepares to form the pupa. The puffs can be divided into three categories: "early" puffs that 20E induces rapidly; "intermolt" puffs that 20E causes to regress; and "late" puffs that are first seen several hours after 20E stimulation. For example, in the larval salivary gland, about six puffs emerge within a few minutes of hydroxyecdysone treatment. No new protein has to be made in order for these early puffs to be induced. A much larger set of puffs is induced later in development, and these late puffs do need protein synthesis to become transcribed. Ashburner (1974, 1990) hypothesized that the "early puff" genes make a protein product that is essential for the activation of the "late puff" genes and that, moreover, this early regulatory protein itself turns off the transcription of the early genes.[†] These insights have been confirmed by molecular analyses.

---

[*]The Ultraspiracle protein may be a receptor for juvenile hormone, or JHR (see Figure 18.16), suggesting mechanisms whereby JH can block 20E at the level of transcription (Jones et al. 2001; Sasorith et al. 2002).

[†]The observation that 20E controlled the transcriptional units of chromosomes was an extremely important and exciting discovery. This was our first real glimpse of gene regulation in eukaryotic organisms. At the time when this discovery was made, the only examples of transcriptional gene regulation were in bacteria.

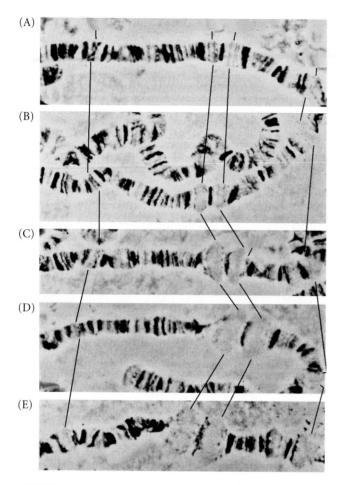

(A)

(B)

(C)

(D)

(E)

FIGURE 18.17 20E-induced puffs in cultured salivary gland cells of *D. melanogaster*. The chromosome region here is the same as that shown in Figure 4.12. (A) Uninduced control. (B–E) 20E-stimulated chromosomes at (B) 25 minutes, (C) 1 hour, (D) 2 hours, and (E) 4 hours. (Photographs courtesy of M. Ashburner.)

Figure 18.18A shows a simplified schematic for the framework of metamorphosis in *Drosophila*. 20E binds to the EcR/USP receptor complex. It activates the "early response genes" including *E74A*, *E75A* (the puffs shown in Figure 18.18) as well as *Broad* and the *EcR* gene, itself. The transcription factors encoded by these genes activate a second series of genes, such as *E75B*, *DHR4*, and *DHR3*. The products of these genes are transcription factors that work together. First, they activate *βFTZ-F1*, a gene encoding a transcription factor that enables a new set of genes to respond to the second burst of 20E. Secondly, the products of these genes shut off the early genes so that they do not interfere with the second burst of 20E. Moreover, DHR4 coordinates growth and behavior in the larva. It allows the larva to stop feeding once it reaches a certain weight and to begin searching for a place to glue itself to and form a pupa (Urness and Thummel 1995; Crossgrove et al. 1996; King-Jones et al. 2005).

The effects of these two 20E pulses can be extremely different. One example of this is the ecdysone-mediated changes in the larval salivary gland. The early pulse of 20E activates the *Broad* gene, which encodes a family of transcription factors through differential RNA splicing. The targets of the Broad complex proteins include those the genes that encode the salivary gland "glue proteins." The glue proteins allow the larva to adhere to a solid surface where it becomes a pupa (Guay and Guild 1991). So the first 20E pulse stimulates the function of the larval salivary gland. However, the second pulse of 20E calls for the destruction of this larval organ (Jiang et al. 2000; Buszczak and Segraves 2000). Here, 20E binds to the EcR-A form of the ecdysone receptor (Figure 18.18B). When complexed with USP, it activates the transcription of early response genes *E74A*, *E75B*, and *Broad*. But now a different set of targets are activated. These transcription factors activate the genes encoding the apoptosis-promoting proteins Hid and Reaper, as well as blocking the expression of the *diap2* gene (which would otherwise repress apoptosis). Thus, the first 20E pulse activates the salivary gland, and the second pulse of 20E destroys it.

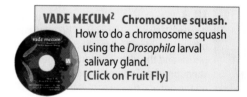

VADE MECUM² Chromosome squash. How to do a chromosome squash using the *Drosophila* larval salivary gland. [Click on Fruit Fly]

Like the ecdysone receptor gene, the *Broad* gene can generate several different transcription factor proteins through differentially initiated and spliced messages. Moreover, the variants of the ecdysone receptor may induce the synthesis of particular variants of the Broad proteins. Organs such as the larval salivary gland that are destined for death during metamorphosis express the Z1 isoform; imaginal discs destined for differentiation express the Z2 isoform; and the central nervous system (which undergoes marked remodeling during metamorphosis) expresses all isoforms, with Z3 predominating (Emery et al. 1994; Crossgrove et al. 1996). Juvenile hormone may act to prevent ecdysone-inducible gene expression by interfering with the Broad complex of proteins (Riddiford 1972; Restifo and White 1991).

WEBSITE 18.4 Precocenes and synthetic JH. Given the voracity of insect larvae, it's amazing that any plant exists. However, many plants get revenge on their predators by making compounds that alter their metamorphoses and prevent the animals from developing or reproducing.

COORDINATION OF RECEPTOR AND LIGAND Like those of amphibian metamorphosis, the stories of insect metamorphosis involve complex interactions between ligands and receptors. The "target tissues" are not mere passive recipients

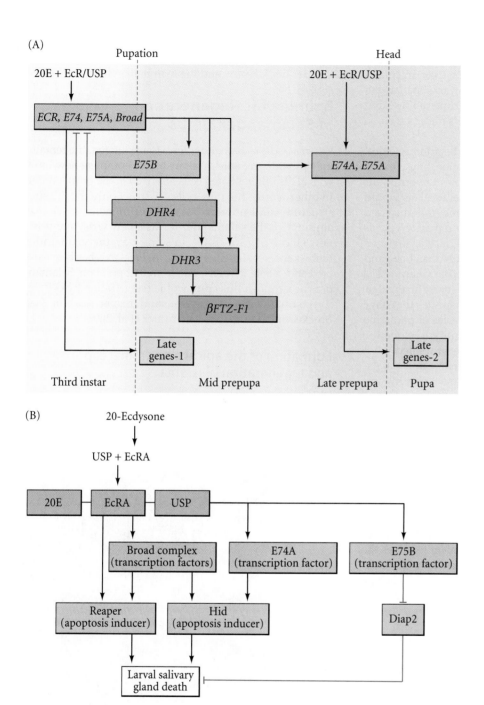

FIGURE 18.18 20-Hydroxyecdysone initiates developmental cascades. (A) Schematic of the major gene expression cascade in *Drosophila* metamorphosis. When 20E binds to the EcR/USP receptor complex, it activates the early response genes, including *E74A*, *E75A*, and *Broad*. Their products activate the "late genes." The activated EcR/USP complex will also activate a series of genes whose products are transcription factors and which activate the *βFTZ-F1* gene. The βFTZ-F1 protein modifies the chromatin so that the next 20E pulse activates a different set of late genes. The products of these genes also inhibit the early-expressed genes, including the EcR receptor. (B) Postulated cascade leading from ecdysone reception to the death of the larval salivary gland. Ecdysone binds to the EcR-A isoform of the ecdysone receptor. After complexing with USP, the activated transcription factor complex stimulates the transcription of the early response genes *E74A*, *E75B*, and the *Broad* complex. These make transcription factors that promote apoptosis in the salivary gland cells. (A after King-Jones et al. 2005; B after Buszczak and Segraves 2000.)

of hormonal signals. Rather, they become responsive to hormones only at particular times. For example, when there is a pulse of 20E at the middle of the fourth instar of the tobacco hornworm moth *Manduca*, the epidermis is able to respond because this tissue is expressing ecdysone receptors. The wing discs, however, are unaffected by ecdysone until the prepupal stage, at which time they synthesize ecdysone receptors, grow, and differentiate (Nijhout 1999). Thus, the timing of metamorphic events in insects can be controlled by the synthesis of receptors in the target tissues.

REGENERATION

Regeneration is the reactivation of development in postembryonic life to restore missing tissues. The ability to regenerate amputated body parts or nonfunctioning organs is so "unhuman" that it has been a source of fascination to humans since the beginnings of biological science. It is difficult to behold the phenomenon of limb regeneration in newts or starfish without wondering why we cannot grow back our own arms and legs. What gives salamanders this ability we so sorely lack?

Experimental biology was born in the efforts of eighteenth-century naturalists to document regeneration and to answer this question. The regeneration experiments of Tremblay (hydra), Réaumur (crustaceans), and Spallanzani (salamanders) set the standard for experimental research and for the intelligent discussion of one's data (see Dinsmore 1991).*

More than two centuries later, we are beginning to find answers to the great problem of regeneration, and we may soon be able to alter the human body so as to permit our own limbs, nerves, and organs to regenerate. This would mean that severed limbs could be restored, that diseased organs could be removed and regrown, and that nerve cells altered by age, disease, or trauma could once again function normally. (The new attempts to coax human bone and neural tissue to regenerate are discussed in Chapter 21.) To bring these treatments to humanity, we must first understand how regeneration occurs in those species that have this ability. Our new knowledge of the roles of paracrine factors in organ formation, and our ability to clone the genes that produce those factors, have propelled what Susan Bryant (1999) has called "a regeneration renaissance." Since "renaissance" literally means "rebirth," and since regeneration can be seen as a return to the embryonic state, the term is apt in many ways.

There are four major ways by which regeneration can occur. The first involves the use of stem cells to regrow organs or tissues that have been lost. We have mentioned numerous cases of this type of normal regeneration throughout the book. Examples of **stem cell-mediated regeneration** include the regrowth of hair shafts from the follicular stem cells in the hair bulge and the continual replacement of blood cells from the hematopoietic stem cells in the bone marrow. The second mechanism involves the *de*differentiation of adult structures to form an undifferentiated mass of cells that then become respecified. This type of regeneration is called **epimorphosis** and is characteristic of planarian flatworm regeneration (see Chapter 3) and also of regenerating amphibian limbs. The third mechanism is called **morphallaxis**. Here, regeneration occurs through the repatterning of existing tissues, and there is little new growth. Such regeneration is seen in *Hydra* (a cnidarian).

A fourth type of regeneration can be thought of as **compensatory regeneration**. Here, the differentiated cells divide but maintain their differentiated functions. The new cells do not come from stem cells, nor do they come from the dedifferentiation of the adult cells. Each cell produces cells similar to itself; no mass of undifferentiated tissue forms.

This type of regeneration is characteristic of the mammalian liver. Here we will concentrate on regeneration in the salamander limb, *Hydra*, and the mammalian liver.

Epimorphic Regeneration of Salamander Limbs

When an adult salamander limb is amputated, the remaining limb cells are able to reconstruct a complete new limb, with all its differentiated cells arranged in the proper order. In other words, the new cells construct only the missing structures and no more. For example, when a wrist is amputated, the salamander forms a new wrist and not a new elbow (Figure 18.19). In some way, the salamander limb "knows" where the proximal-distal axis has been severed and is able to regenerate from that point on. Salamanders accomplish this epimorphic regeneration by cell dedifferentiation, proliferation, and respecification (see Brockes and Kumar 2002; Gardiner et al. 2002).

Formation of the apical ectodermal cap and regeneration blastema

When a salamander limb is amputated, a plasma clot forms; within 6–12 hours, epidermal cells from the remain-

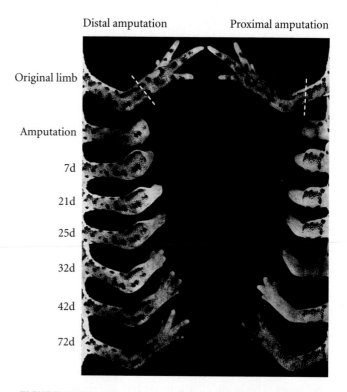

FIGURE 18.19 Regeneration of a salamander forelimb. The amputation shown on the left was made below the elbow; the amputation shown on the right cut through the humerus. In both instances, the correct positional information is respecified and a normal limb is regenerated. (From Goss 1969; photograph courtesy of R. J. Goss.)

---

*Tremblay's advice to researchers who would enter this new field is pertinent even today. He tells us to go directly to nature and to avoid the prejudices that our education has given us. Moreover, "one should not become disheartened by want of success, but should try anew whatever has failed. It is even good to repeat successful experiments a number of times. All that is possible to see is not discovered, and often cannot be discovered, the first time" (quoted in Dinsmore 1991).

ing stump migrate to cover the wound surface, forming the **wound epidermis**. This single-layered structure then proliferates to form the **apical ectodermal cap**. Thus, in contrast to wound healing in mammals, no scar forms, and the dermis does not move with the epidermis to cover the site of amputation. The nerves innervating the limb degenerate for a short distance proximal to the plane of amputation (see Chernoff and Stocum 1995).

During the next 4 days, the cells beneath the developing apical cap undergo dramatic dedifferentiation: bone cells, cartilage cells, fibroblasts, myocytes, and neural cells all lose their differentiated characteristics and become detached from one another. Genes that are expressed in differentiated tissues (such as the *mrf4* and *myf5* genes expressed in muscle cells) are downregulated, while there is a dramatic increase in the expression of genes such as *msx1* that are associated with the proliferating progress zone mesenchyme of the embryonic limb (Simon et al. 1995). Thus, the previously well-structured limb region at the cut edge of the stump forms a proliferating mass of indistinguishable, dedifferentiated cells just beneath the apical ectodermal cap. This dedifferentiated cell mass is the **regeneration blastema**. These cells will continue to proliferate, and will eventually redifferentiate to form the new structures of the limb (Figure 18.20; Butler 1935).

The creation of the regeneration blastema depends on the formation of single, mononucleated cells from the multinucleated muscle cells and epithelial cells. These cells also have to proliferate. But many of these cells are differentiated and have left the cell cycle. How do they regain the ability to divide? Microscopy and tracer dye studies show that when newt multinucleated myotubes (whose nuclei have been removed from the cell cycle: see Chapter 14) are implanted into a blastema, they give rise to labeled mononucleated cells that proliferate and can differentiate into many tissues of the regenerated limb (Hay 1959; Lo et al. 1993). It appears that myotube nuclei are forced to re-enter the cell cycle by a serum factor created by thrombin—the same protease that is involved in forming clots. Thrombin is released when the amputation occurs, and when serum is exposed to thrombin, it forms a factor capable of inducing cultured newt myotubes to enter the cell cycle. Mouse myotubes, however, do not respond to this protein, although they will dedifferentiate in response to an extract of regenerating newt blastemas. This difference in respon-

siveness may relate directly to the difference in regenerative ability between salamanders and mammals* (Tanaka et al. 1999; McGann et al. 2001). The creation of the amphibian regeneration blastema may also depend on the maintenance of ion currents driven through the stump: if this electric field is suppressed, the regeneration blastema fails to form (Altizer et al. 2002).

Proliferation of the blastema cells: The requirement for nerves

The proliferation of the regeneration blastema depends on the presence of nerves. Singer (1954) demonstrated that a minimum number of nerve fibers must be present for regeneration to take place. The neurons are believed to release factors that increase the proliferation of the blastema cells (Singer and Caston 1972; Mescher and Tassava 1975). One of these factors, **glial growth factor**, appears to be needed for maintaining a high rate of cell proliferation (Brockes and Kinter 1986; Wang et al. 2000). **Fibroblast growth factor 2** (Fgf2), another factor found in the axons innervating the amphibian limb, also appears to be critical to this process. Fgf2 may play more than one role in regeneration. First, it may serve as an angiogenesis factor, since the regenerating tissues need to develop a blood supply very shortly after amputation (Rageh et al. 2002). Fgf2 may also act to promote mitosis and patterning of the regenerating limb. Mescher and Gospodarowicz (1979) were able to restore mitosis to denervated blastemas by infusing FGFs into them, and Mullen and coworkers (1996) were able to induce regeneration in denervated blastemas by implanting Fgf2-containing beads in them. Fgf2 restored *Distal-less* gene expression to the ectoderm, probably making it responsive to some factor produced by the mesenchyme.

Proliferation of the blastema cells: The requirement for Fgf10

The mesenchymal factor that *distal-less* responds to is probably Fgf10. Fgf10 is expressed in the mesenchyme of normal embryonic limb buds and makes a critical difference in amphibian limb regeneration. In *Xenopus*, for instance, premetamorphic tadpoles are able to regenerate their hindlimbs, but later-stage tadpoles and adults are not. By

---

*One would think that being able to regenerate our limbs would be a good thing. However, nature has not seen fit for mammals to retain this ability. In addition to not being able to respond to an as-yet unidentified thrombin-produced factor, there are other differences between those vertebrates that are capable of substantial regeneration and those that are not. A major difference might involve our immune system. Frog tadpoles and larval and adult salamanders have minimal inflammatory responses, and if mice are immunosuppressed, they can regenerate damaged heart and skin tissue better than the wild-type mice (Robert and Cohen 1998; Heber-Katz et al. 2004). Mescher and Neff (2005) have speculated

that regeneration cannot occur in regions of the body that are monitored by the immune system and that our lack of limb regeneration ability is a trade-off for having a highly specific immune system.

Mammals do have a small amount of regenerational ability. The tips of rodent and human digits can regenerate if the animal is young enough, an ability that has been correlated with the expression of the homeodomain transcription factor Msx1 (Han et al. 2003; Kumar et al. 2004). Apparently, amputated human fetal digit tips express the MSX1 transcription factor in the migrating epidermis and subjacent mesenchyme, just as regenerating amphibian limbs do (Allan et al. 2005).

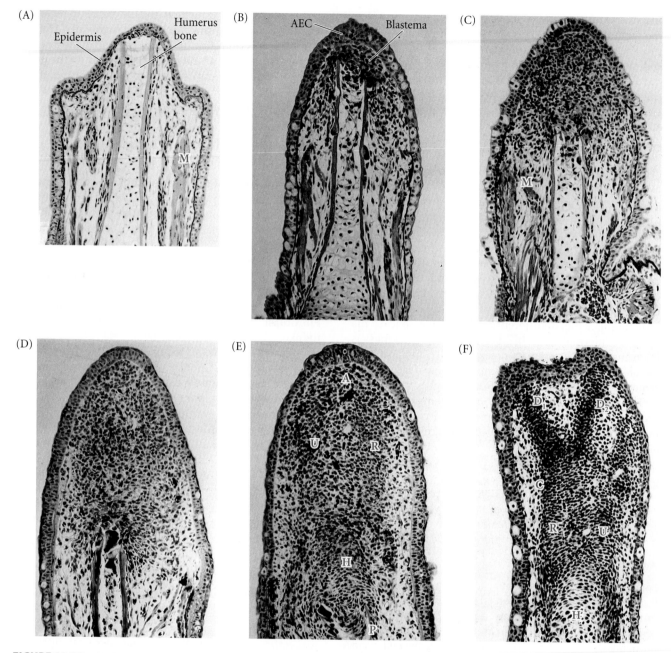

FIGURE 18.20 Regeneration in the larval forelimb of the spotted salamander *Ambystoma maculatum*. (A) Longitudinal section of the upper arm, 2 days after amputation. The skin and muscle (M) have retracted from the tip of the humerus. (B) At 5 days after amputation, a thin accumulation of blastema cells is seen beneath a thickened epidermis. (C) At 7 days, a large population of mitotically active blastema cells lies distal to the humerus. (D) At 8 days, the blastema elongates by mitotic activity; much dedifferetiation has occured. (E) At 9 days, early redifferentiation can be seen. Chondrogenesis has begun in the proximal part of the regenerating humerus, H. The letter A marks the apical mesenchyme of the blastema, and U and R are the precartilaginous condensations that will form the ulna and radius, respectively. P represents the stump where the amputation was made. (F) At 10 days after amputation, the precartilaginous condensations for the carpal bones (ankle, C), and the first two digits (D_1, D_2) can also be seen. (From Stocum 1979; photographs courtesy of D. L. Stocum.)

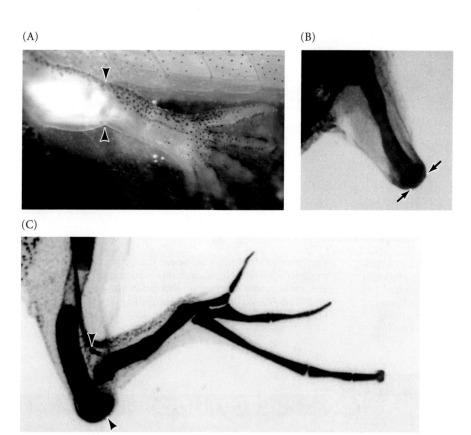

FIGURE 18.21 Restoration of regenerative ability of the *Xenopus* hindlimb by Fgf10. (A) A late tadpole limb is amputated at knee level. The arrowheads show the site of amputation. (B) If beads lacking Fgf10 are placed on the amputation site, there is no regeneration. Arrows mark the amputation site, and the limb cartilage is stained with alcian blue. (C) Beads containing Fgf10 placed on the amputation site (arrowheads) result in the regeneration of complete distal structures (some of which are clawed). Arrowheads mark the amputation site. (After Yokoyama et al. 2001; photographs courtesy of Dr. K. Tamura.)

means of reciprocal transplantation experiments, Yokoyama and colleagues (2000) found that the mesenchyme, not the ectoderm, was critical in regulating regeneration. A hindlimb would regenerate if it had the mesenchyme of a premetamorphic hindlimb and the ectoderm of an older hindlimb; it would not regenerate if it had premetamorphic ectoderm and older mesenchyme. The later hindlimb mesenchyme did not synthesize Fgf10. When Yokoyama and colleagues (2001) added beads of Fgf10 to older hindlimbs, these limbs regenerated (Figure 18.21). Moreover, the Fgf10 beads induced *Fgf8* expression in the ectoderm overlying the bead.

Thus, the regeneration of the amphibian limb may be very similar to its original development. In both cases, there is feedback between Fgf10 produced in the mesenchyme and Fgf8 produced in the overlying ectoderm (the apical ectodermal ridge in amniote limb development;

the apical ectodermal cap in amphibian regeneration). The initial condition is the expression of FGFs to make the ectoderm competent to express its own FGF. This function appears to be carried out by the lateral plate mesoderm during development and by neurons during regeneration.

A two-step model

Our knowledge of the order of the nerve-dependent and mesenchyme-dependent processes in early regeneration has been elucidated using a new assay procedure based on ectopic limb formation. Usually, if a limb is wounded, the skin heals. However, if a nerve is rerouted to the wound site along with skin fibroblasts from the other side of the limb, an ectopic limb will form (Lheureux 1977; Reynolds et al. 1983). Endo and colleagues (2004) showed that deviating the nerve to the wound site caused a bump-like blastema to form. These blastemas are composed largely of dermal fibroblasts. The neural signal caused these cells to proliferate and dedifferentiate. However, without fibroblasts from the contralateral region, these blastemas regress. It is known that fibroblasts from different regions of the limb interact with each other to produce signals needed for pattern formation (French et al. 1976). Although these bumplike blastemas are not "natural" regeneration blastemas, they do indicate that there is a temporal sequence to the formation of the regeneration blastema (Figure 18.22).

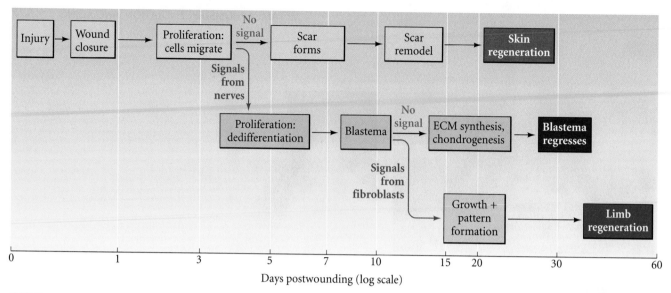

Days postwounding (log scale)

FIGURE 18.22 Two-step model for amphibian limb regeneration. After injury, dermal cells proliferate and migrate. Without signals from the nerves, these proliferating cells form a scar and allow for skin regeneration (i.e., healing). If nerve signals are present, however, the cells at the wound site proliferate and dedifferentiate to form a blastema. Then, if signals from the mesenchymal fibroblasts (probably including Fgf10) are present, the blastema is patterned and grows into a regenerated limb. If no FGF signal is present, the blastema regresses. (After Endo et al. 2004.)

SIDELIGHTS & SPECULATIONS

Pattern Formation in the Regeneration Blastema

The regeneration blastema resembles the progress zone mesenchyme of the developing limb (see Chapter 16) in many ways. The dorsal-ventral and anterior-posterior axes between the stump and the regenerating tissue are conserved, and cellular and molecular studies have confirmed that the patterning mechanisms of developing and regenerating limbs are very similar. The limb blastema appears to be specified early, and labeling cells of the blastema with fluorescent dyes shows, for instance, that cells at the distal tip of the blastema are fated to become autopod structures (Echeverri and Tanaka 2005). By transplanting regenerating limb blastemas onto developing limb buds, Muneoka and Bryant (1982) showed that the blastema cells could respond to limb bud signals and contribute to the developing limb. At the molecular level, just as Sonic hedgehog is seen in the posterior region of the developing limb progress zone mesenchyme, it is seen in the early posterior regeneration blastema (Imokawa and Yoshizato 1997; Torok et al. 1999).

The initial pattern of Hox gene expression in regenerating limbs is not the same as that in developing limbs. However, the nested pattern of *Hoxa* and *Hoxd* gene expres-

sion characteristic of limb development is established as the limb regenerates (Torok et al. 1998). Hox gene expression may be controlled by retinoic acid. If regenerating animals are treated with sufficient concentrations of retinoic acid (or other retinoids), their regenerated limbs will have duplications along the proximal-distal axis (Figure 18.23; Niazi and Saxena 1978; Maden 1982). This response is dose-dependent and at maximal dosage can result in a complete new limb regenerating (starting at the most proximal girdle), regardless of the original level of amputation. Dosages higher than this maximal result in the *inhibition* of regeneration. It appears that the retinoic acid

causes the cells to be respecified to a more proximal position (Figure 18.24; Crawford and Stocum 1988b; Pecorino et al. 1996).

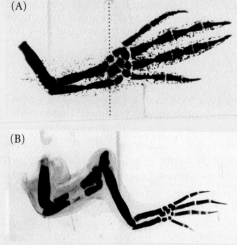

FIGURE 18.23 Effects of vitamin A (a retinoid) on regenerating salamander limbs. (A) Normal regenerated *Ambystoma mexicanum* limb (9×) with humerus, paired radius and ulna, carpals, and digits. The dotted line shows the plane of amputation through the carpal area. (B) Regeneration after amputation at the same location (5×), but after the regenerating animal had been placed in retinol palmitate (vitamin A) for 15 days. A new humerus, ulna, radius, carpal set, and digit set have emerged. (From Maden et al. 1982; photographs courtesy of M. Maden.)

Moreover, the cells of the blastema can undergo **transdifferentiation** (Okada 1991; Tsonis et al. 1995)—cells that had been differentiated muscles can become cartilage tissue.

Retinoic acid is synthesized in the wound epidermis of the regenerating limb and forms a gradient along the proximal-distal axis of the blastema (Brockes 1992; Scadding and Maden 1994; Viviano et al. 1995). This RA gradient is thought to facilitate two major processes that might inform cells of their position along that axis in the limb. First, retinoic acid can activate the *Hoxa* genes differentially across the blastema, resulting in the specification of pattern in the regenerating limb. Gardiner and colleagues (1995) have shown that the expression pattern of certain *Hoxa* genes in the distal cells of the regeneration blastema is changed by exogenous retinoic acid into an expression pattern characteristic of more proximal cells. It is probable that during normal regeneration, the wound epidermis/apical ectodermal cap secretes retinoic acid, which activates the genes needed for cell proliferation, down-regulates the genes that are specific for differentiated cells, and activates a set of Hox genes that tells the cells where they are in the limb and how much they need to grow. The mechanisms by which the Hox genes do this is not known, but changes in cell-cell adhesion have been observed (Nardi and Stocum 1983; Stocum and Crawford 1987; Bryant and Gardiner 1992). These changes are similar to those postulated to mediate the actions of Hox genes in the developing limb (see Chapter 16).

It is possible that one of the critical changes in the cell surface mediated by *Hoxa* genes is the upregulation of the cell adhesion protein CD59 in response to retinoic acid (Morais Da Silva et al. 2002; Echeverri and Tanaka 2005). Although cells at the distal tip of the blastema usually became autopod cells, when CD59 was overexpressed in the distalmost blastema cells (by electroporating plasmids encoding activated CD59 into them), these cells were found in proximal positions (i.e., in limb structures closer to the shoulder or pelvic girdle). When a large percentage of blastema cells overexpressed CD59 protein, the patterning of the regenerating limb was severely disrupted and the distal structures—fingers, ulna and radius—were frequently missing.

WEBSITE 18.5 The polar coordinate and boundary models. The phenomena of epimorphic regeneration can be seen formally as events that reestablish continuity among tissues that the amputation has severed. The polar coordinate and boundary models attempt to explain the numerous phenomena of limb regeneration.

Retinoic acid may also act through a second pathway. The *Meis1* and *Meis2* genes encode homeodomain proteins associated with the proximal (stylopod) portion of the developing limb. Originally, RA establishes a domain of *Meis* gene expression across the entire limb bud. However, FGFs secreted by the apical ectodermal ridge suppress *Meis* gene activation, limiting *Meis* gene products to the proximal region of the limbs. During regeneration of the salamander limb, the *Meis* genes are activated by retinoic acid and also appear to be associated with the proximal identities of the limb bones. Overexpression of *Meis* genes in the distal blastema cells causes these cells to relocalize in proximal locations, whereas antisense oligonucleotides that block *Meis* expression inhibit retinoic acid from proximalizing the regenerating limbs (Mercader et al. 2000, 2005). *Hoxa* and *Meis* genes appear to be the targets of RAs ability to specify proximal cell fate during regeneration. Thus, in salamander limb regeneration, adult cells can go "back to the future," returning to an "embryonic" condition to begin the formation of the limb anew.

WEBSITE 18.6 Regeneration in annelid worms. An easy laboratory exercise can discover the rules by which worms regenerate their segments. This website details some of those experiments.

(A)

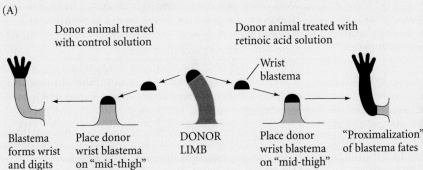

Donor animal treated with control solution

Donor animal treated with retinoic acid solution

Wrist blastema

DONOR LIMB

Blastema forms wrist and digits only

Place donor wrist blastema on "mid-thigh" region of cut host limb

Place donor wrist blastema on "mid-thigh" region of cut host limb

"Proximalization" of blastema fates

(B)

FIGURE 18.24 Proximalization of blastema respecification by retinoic acid. (A) When a wrist blastema from a recently cut axolotl forelimb is placed on a host hindlimb cut at the mid-thigh level, it will generate only the wrist. The host (whose own leg was removed) will fill the gap and regenerate the limb up to the wrist. However, if the donor animal is treated with retinoic acid, the wrist blastema will regenerate a complete limb and, when grafted, will fail to cause the host to fill the gap. (B) A wrist blastema from a darkly pigmented axolotl was treated with retinoic acid and placed on the amputated mid-thigh region of a golden axolotl. The treated blastema regenerated a complete limb. (After Crawford and Stocum 1998a,b; photograph courtesy of K. Crawford.)

(A)

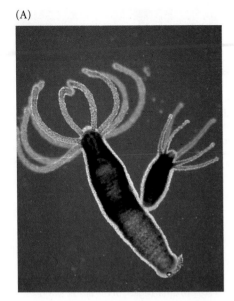

(B)

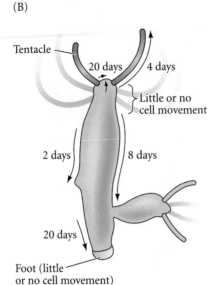

Tentacle

20 days   4 days

Little or no
cell movement

2 days   8 days

20 days

Foot (little
or no cell movement)

FIGURE 18.25 Budding in *Hydra*. (A) A new individual buds from the right side of an adult. (B) Cell movements in *Hydra* were traced by following the migration of labeled tissues. The arrows indicate the starting and leaving positions of the labeled cells. The brackets indicate regions in which no net cell movement took place. Cell division takes place throughout the body column except at the tentacles and foot (shaded). (A, photograph © Biophoto/Photo Researchers Inc.; B after Steele 2002.)

Morphallactic Regeneration in *Hydra*

Hydra* is a genus of freshwater cnidarians. Most hydras are tiny—about 0.5 cm long. A hydra has a tubular body, with a "head" at its distal end and a "foot" at its proximal end. The "foot," or **basal disc**, enables the animal to stick to rocks or the undersides of pond plants. The "head" consists of a conical **hypostome** region (containing the mouth) and a ring of tentacles (that catch food) beneath it. *Hydra*, a simple cnidarian, has only two epithelial cell layers, the ectoderm and endoderm; they lack a true mesoderm. Hydras can reproduce sexually, but do so only under adverse conditions (such as crowding or cold weather). They usually multiply asexually, by budding off a new individual (Figure 18.25A). The buds form about two-thirds of the way down the animal's body axis.

The body of the hydra is not as "stable" as are those of most organisms. In humans and flies, for instance, a skin cell in the body's trunk is not expected to migrate and eventually be sloughed off from the face or foot. But that is precisely what does happen in *Hydra*. The cells of the body column are constantly undergoing mitosis and are eventually displaced to the extremities of the column, from which they are shed (Figure 18.25B; Campbell 1967a,b). Thus, each cell gets to play several roles, depending on how old it is; and the signals specifying cell fate must be active all the time.

If a hydra's body column is cut into several pieces, each piece will regenerate a head at its original apical end and a foot at its original basal end. No cell division is required for this to happen, and the result is a small hydra. Since each cell retains its plasticity, each piece can re-form a smaller organism. This type of regeneration is known as **morphallaxis**.

The head activation gradient

Every portion of the hydra body column along the apical-basal axis is potentially able to form both a head and a foot. However, the polarity of the hydra is coordinated by a series of morphogenetic gradients that permit the head to form only at one place and the basal disc to form only at another. Evidence for such gradients was first obtained from grafting experiments begun by Ethel Browne in the early 1900s. When hypostome tissue from one hydra is transplanted into the middle of another hydra, the transplanted tissue forms a new apical-basal axis, with the hypostome extending outward (Figure 18.26A). When a basal disc is grafted to the middle of a host hydra, a new axis also forms, but with the opposite polarity, extending a basal disc (Figure 18.26B). When tissues from both ends are transplanted simultaneously into the middle of a host, no new axis is formed, or the new axis has little polarity (Figure 18.26C; Browne 1909; Newman 1974). These experiments have been interpreted to indicate the existence of a **head activation gradient** (highest at the hypostome) and a **foot activation gradient** (highest at the basal disc).

The head activation gradient can be measured by implanting rings of tissue from various levels of a donor hydra into a particular region of the host trunk (Wilby and Webster 1970; Herlands and Bode 1974; MacWilliams 1983b). The higher the level of head activator in the donor tissue, the greater the percentage of implants that will induce the formation of new heads. The head activation

---

*The Hydra is another character from Greek mythology. Whenever one of this serpent's many heads was chopped off, it regenerated two new ones. Hercules finally defeated the Hydra by cauterizing the stumps of its heads with fire. Hercules seems to have had a signficant interest in regeneration—he also finally freed the bound Prometheus, thus stopping his daily hepatectomies (see p. 582).

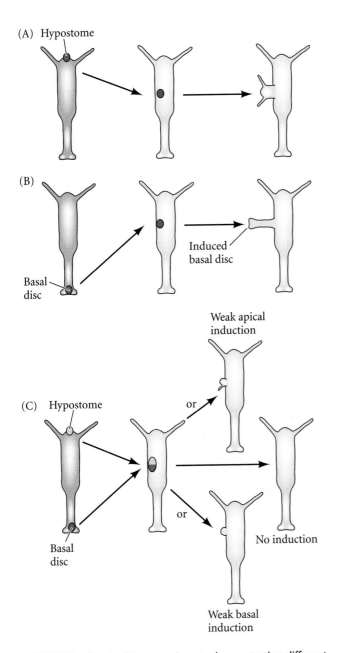

FIGURE 18.26 Grafting experiments demonstrating different morphogenetic capabilities in different regions of the *Hydra* apical-basal axis. (A) Hypostome tissue grafted onto a host trunk induces a secondary axis with an extended hypostome. (B) Basal disc tissue grafted onto a host trunk induces a secondary axis with an extended basal disc. (C) If hypostome and basal disc tissues are transplanted together, only weak (if any) inductions are seen. (After Newman 1974.)

factor is concentrated in the head and decreases linearly toward the basal disc. Three peptides have been associated with this head activation gradient. Two of them, Heady and Head Activator, are critical for head formation and the initiation of the bud. The other, Hym-301, regulates the number of tentacles formed (Takahashi et al. 2005).

The head inhibition gradient

If the tissue of the *Hydra* body column is capable of forming a head, why doesn't it do so? In 1926, Rand and colleagues showed that the normal regeneration of the hypostome is inhibited when an intact hypostome is grafted adjacent to the amputation site. Moreover, if a graft of subhypostomal tissue (from the region just below the hypostome, where there is a relatively high concentration of head activator) is placed in the same region of a host hydra, no secondary axis forms (Figure 18.27A). The host head appears to make an inhibitor that prevents the grafted tissue from forming a head and secondary axis. However, if one grafts subhypostomal tissue to a decapitated host

(A) Intact host: No secondary axis induced

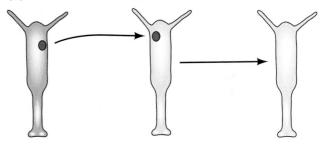

(B) Host's head removed: Secondary axis induced

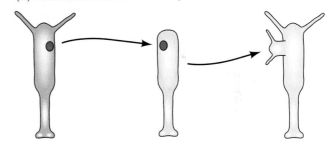

(C) Intact host: Graft away from head region induces secondary axis

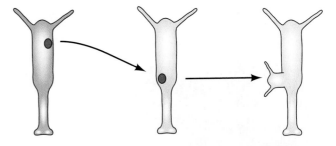

FIGURE 18.27 Grafting experiments providing evidence for a head inhibition gradient. (A) Subhypostomal tissue does not generate a new head when placed close to an existing host head. (B) Subhypostomal tissue generates a head if the existing host head is removed. A head also forms at the site where the host's head was amputated. (C) Subhypostomal tissue generates a new head when placed far away from an existing host head. (After Newman 1974.)

hydra, a second axis does form (Figure 18.27B). A gradient of this inhibitor appears to extend from the head down the body column, and can be measured by grafting subhypostomal tissue into various regions along the trunks of host hydras. This tissue will not produce a head when implanted into the apical area of an intact host hydra, but it will form a head if placed lower on the host (Figure 18.27C). Thus, there is a gradient of head inhibitor as well as head activator (Wilby and Webster 1970; MacWilliams 1983a).

The hypostome as an "organizer"

Ethel Browne (1909; Lenhoff 1991) noted that the hypostome acted as an "organizer" of the hydra. This notion has been confirmed by Broun and Bode (2002), who demonstrated that (1) when transplanted, the hypostome can induce host tissue to form a second body axis, (2) the hypostome produces both the head activation and head inhibition signals, (3) the hypostome is the only "self-differentiating" region of the hydra, and (4) the head inhibition signal is actually a signal to inhibit the formation of new organizing centers.

By inserting small pieces of hypostome tissue into a host hydra who cells were labeled with India ink (colloidal carbon), Broun and Bode found that the hypostome induced a new body axis and that almost all of the resulting head tissue came from *host* tissue, not from the differentiation of the donor tissue (Figure 18.28A). In contrast, when tissues from other regions (such as the subhypostomal region) were grafted into a host trunk, the head and apical trunk of the new hydra were made from the grafted *donor* tissue (Figure 18.28B). In other words, only the hypostome region could alter the fates of the trunk cells and cause them to become head cells. Broun and Bode also found that the signal did not have to emanate from a permanent graft. Even transient contact with the hypostome

region was sufficient to induce a new axis from a host hydra. In these cases, *all* the tissue of the new axis came from the host. The head inhibitor appears to repress the effect of the inducing signal from the donor hypostome, and it normally functions to prevent other portions of the hydra from having such organizing abilities.

At least three genes are known to be active in the hypostome organizer area, and their expression in the hypostome suggests an evolutionarily conserved set of signals that function as organizers throughout the animal kingdom. First, a *Hydra* Wnt protein is seen in the apical end of the early bud, and it defines the hypostome region as the bud elongates. This protein acts to form the head organizer, and it signals through the canonical Wnt pathway, inhibiting GSK-3 to stabilize β-catenin in the cell nucleus (Figure 18.29; Hobmayer et al. 2000; Broun et al. 2005). If GSK-3 is inhibited throughout the body axis, ectopic tentacles form at all levels, and each piece of the trunk has the ability to stimulate the outgrowth of new buds. Second, the expression of another vertebrate organizer molecule, Goosecoid, is restricted to the *Hydra* hypostome region. Moreover, when the hypostome is brought into contact with the trunk of an adult hydra, it induces the expression of *Brachyury* (even though hydras lack mesoderm), just as vertebrate organizers do (Broun et al. 1999; Broun and Bode 2002).

WEBSITE **18.7 Ethel Browne and the organizer.** Spemann and Mangold brought the notion of the Organizer into embryology, and Spemann's laboratory helped make this idea a unifying notion in embryology. However, it has been argued that this idea had its origins in the experiments of Ethel Browne on *Hydra*.

(A)

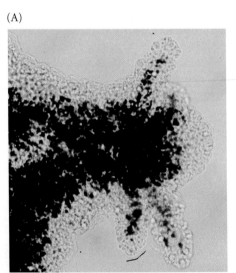

(B)

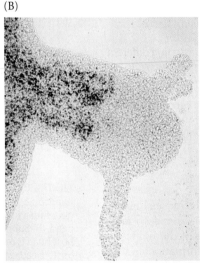

FIGURE 18.28 Formation of secondary axes following the transplantation of head regions into the trunk of a hydra. The host endoderm was stained with India ink. (A) Hypostome tissue grafted onto the trunk induces the host's own trunk tissue to become tentacles and head. (B) Subhypostomal donor tissue placed on the host trunk self-differentiates into a head and upper trunk. (From Broun and Bode 2002; photographs courtesy of H. R. Bode.)

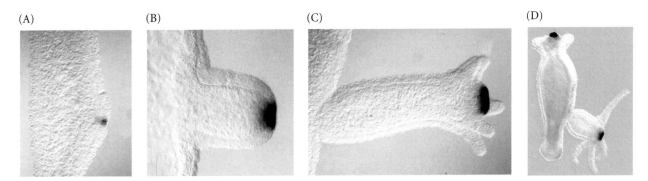

FIGURE 18.29 Expression of the *Hydra Wnt* gene during budding. (A) Early bud. (B) Mid-stage bud. (C) Bud with early tentacles. (D) Adult with late bud. (From Hobmayer et al. 2000; photographs courtesy of T. W. Holstein and B. Hobmeyer.)

The basal disc activation and inhibition gradients

Certain properties of the basal disc suggest that it is the source of both a foot inhibition and a foot activation gradient (MacWilliams et al. 1970; Hicklin and Wolpert 1973; Schmidt and Schaller 1976; Meinhardt 1993; Grens et al. 1999). The inhibition gradients for the head and the foot may be important in determining where and when a bud can form. In young adult hydras, the gradients of head and foot inhibitors appear to block bud formation. However, as the hydra grows, the sources of these labile substances grow farther apart, creating a region of tissue about two-thirds down the trunk where levels of both inhibitors are minimal. This region is where the bud forms (Figure 18.30A; Shostak 1974; Bode and Bode 1984; Schiliro et al. 1999).

Certain mutants of *Hydra* have defects in their ability to form buds, and these defects can be explained by alter-ations of the inhibition gradients. The *L4* mutant of *Hydra magnipapillata*, for instance, forms buds very slowly, and only after reaching a size about twice as long as wild-type individuals. The amount of head inhibitor in these mutants was found to be much greater than in wild-type *Hydra* (Takano and Sugiyama 1983).

Several small peptides have been found to activate foot formation, and researchers are just beginning to sort out the mechanisms by which these proteins arise and function (see Harafuji et al. 2001; Siebert et al. 2005). However, the specification of cells as they migrate from the basal region through the body column may be mediated by a gradient of tyrosine kinase. The product of the *shinguard* gene is a tyrosine kinase that extends in a gradient from the ectoderm just above the basal disc through the lower region of the trunk. Buds appear to form where this gradient fades (Figure 18.30B). The *shinguard* gene appears to be activated through the product of the *manacle* gene, a putative transcription factor that is expressed earlier in the basal disc ectoderm.

The inhibition and activation gradients also inform the hydra "which end is up" and specify positional values along the apical-basal axis. When the head is removed, the head inhibitor no longer is made, causing the head activa-

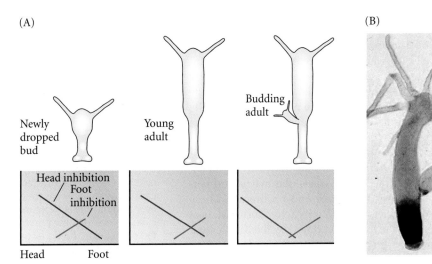

FIGURE 18.30 Bud location as a function of head and foot inhibition gradients. (A) Head inhibition (blue) and foot inhibition (red) gradients in newly dropped buds, young adults, and budding adults. (B) Expression of the Shinguard protein in a graded fashion in a budding hydra (A after Bode and Bode 1984; B after Bridge et al. 2000.)

tor to induce a new head. The region with the most head activator (i.e., those cells directly beneath the amputation site) will form the head. Once the head is made, it generates the head inhibitor, and thus equilibrium is restored.

Compensatory Regeneration in the Mammalian Liver

Compensatory regeneration—wherein differentiated cells divide to recover the structure and function of an injured organ—has been demonstrated in the mammalian liver and in the zebrafish heart (Poss et al. 2002).

According to Greek mythology, Prometheus's punishment for giving fire to humans was to be chained to a rock and to have an eagle eat a portion of his liver each day. His liver then regenerated each night, providing a continuous food supply for the eagle and eternal punishment for Prometheus. Today the standard assay for liver regeneration is a **partial hepatectomy**. In this procedure, specific lobes of the liver are removed (after administering anesthesia), leaving the other hepatic lobes intact. Although the removed lobe does not grow back, the remaining lobes enlarge to compensate for the loss of the missing tissue (Higgins and Anderson 1931). The amount of liver regenerated is equivalent to the amount of liver removed.

The human liver regenerates by the proliferation of existing tissue. Surprisingly, the regenerating liver cells do not fully dedifferentiate when they reenter the cell cycle. No regeneration blastema is formed. Rather, the five types of liver cells—hepatocytes, duct cells, fat-storing (Ito) cells, endothelial cells, and Kupffer macrophages—all begin dividing to produce more of themselves. Each cell type retains its cellular identity, and the liver retains its ability to synthesize the liver-specific enzymes necessary for glucose regulation, toxin degradation, bile synthesis, albumin production, and other hepatic functions (Michalopoulos and DeFrances 1997).

The removal or injury of the liver is sensed through the bloodstream, as some liver-specific factors are lost while others (such as gut lipopolysaccharides) increase. These lipopolysaccharides activate two of the nonhepatocyte cells to secrete paracrine factors that allow the hepatocytes to reenter the cell cycle. The Kupffer cell secretes interleukin-6 (IL-6) and tumor necrosis factor-α (which are usually involved with activating the adult immune system), while the stellate cells secrete **hepatocyte growth factor** (**HGF**, or **scatter factor**) and TGF-β. However, hepatocytes that are still connected to one another in an epithelium cannot respond to HGF.

The trauma of partial hepatectomy may activate metalloproteinases that digest the extracellular matrix and permit the hepatocytes to separate and proliferate. These enzymes also may cleave HGF to its active form (Mars et al. 1995). Together, the factors produced by the Kupffer and stellate cells allow the hepatocytes to divide by preventing apoptosis, activating cyclins D and E, and repressing cyclin inhibitors such as p27 (Figure 18.31; see Taub 2004).

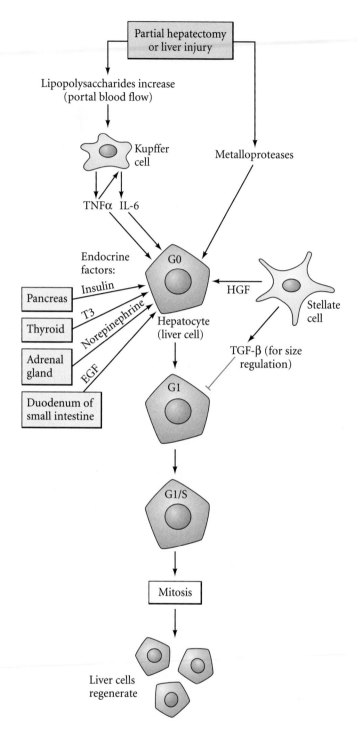

FIGURE 18.31   Liver regeneration is activated by factors produced by the Kupffer cell and the stellate cell in the liver. These cells appear to monitor the size of the liver (from factors in the blood). Partial hepatectomy or liver injury causes the Kupffer cells to release paracrine factors TNFα and IL-6. Trauma also causes the stellate cells to secrete HGF and TGF-β. Moreover, liver trauma may activate metalloproteases that allow hepatocytes to receive these paracrine factors. Together, the paracrine factors activate the signal transduction pathways leading to the activation of cyclins and the degradation of cyclin inhibitors. Endocrine factors from other organs assist in helping the hepatocyte get over the block to cell division. (After Taub 2004.)

The liver stops growing at the appropriate size; the mechanism for how this is achieved this is not yet known. One clue, however, is that the addition of extracellular factors such as follistatin (an inhibitor of TGF-β family molecules; see Chapter 10) can increase liver size during regeneration (Takabe et al. 2003).

Because human livers have the power to regenerate, a patient's diseased liver can be replaced by compatible liver tissue from a living donor (usually a relative). The donor's liver has always grown back. Human livers regenerate more slowly than those of mice, but function is restored quickly (Pascher et al. 2002; Olthoff 2003). In addition, mammalian livers possess a "second line" of regenerative ability. If the hepatocytes are unable to regenerate the liver sufficiently within a certain amount of time, the **oval cells** divide to form new hepatocytes. Oval cells are a small progenitor cell population that can produce hepatocytes and bile duct cells. They appear to be kept in reserve and used only *after* the hepatocytes have attempted to heal the liver (Fausto and Campbell 2005; Knight et al. 2005). The mechanisms by which these factors interact and by which the liver is first told to begin regenerating and then to stop regenerating after reaching the appropriate size remain to be discovered.

AGING: THE BIOLOGY OF SENESCENCE

Entropy always wins. Each multicellular organism is able to develop and maintain its identity for only so long. Then deterioration prevails over synthesis, and the organism ages. **Aging** can be defined as the time-related deterioration of the physiological functions necessary for survival and fertility. The characteristics of aging—as distinguished from diseases of aging, such as cancer and heart disease—affect all the individuals of a species. The aging process has two major facets. The first is simply how long an organism lives; the second concerns the physiological deterioration, or **senescence**, that characterizes old age. These topics are often viewed as being interrelated.

Many evolutionary biologists have denied that aging is part of the genetic repertoire of an animal (see Medawar 1952; Kirkwood 1977). Rather, they consider senescence to be a default state that occurs after an animal has fulfilled the requirements of natural selection. After its offspring are born and raised, the animal can die. Indeed, in many organisms, from moths to salmon, this is exactly what happens: as soon as its eggs are fertilized and laid, the adult organism dies. However, new data indicate that, in many organisms, genetic components regulate the rate of aging, and that altering the activity or expression of these genes can alter an individual's life span. Indeed, recent evidence (see Kenyon 2001) indicates that certain such genetic components might even be evolutionarily conserved: flies, worms, and mammals all appear to use the same set of genes to promote survival and longevity.

Maximum Life Span and Life Expectancy

The maximum life span is a characteristic of a species; it is the maximum number of years a member of a given species has been known to survive. The maximum human life span is estimated to be 121 years (Arking 1998). The life spans of tortoises and lake trout are both unknown, but are estimated to be more than 150 years. The maximum life span of a domestic dog is about 20 years, and that of a laboratory mouse is 4.5 years. If a *Drosophila* fruit fly survives to eclose (in the wild, over 90 percent die as larvae), it has a maximum life span of 3 months.

However, most people cannot expect to live 121 years, and most mice in the wild do not live to celebrate even their first birthday. **Life expectancy**—the length of time an average individual of a given species can expect to live—is not characteristic of species, but of populations. It is usually defined as the age at which half the population still survives. A baby born in England during the 1780s could expect to live to be 35 years old. In Massachusetts during that same time, the life expectancy was 28 years. This was the normal range of human life expectancy for most of the human race in most times (Arking 1998). Even today, in some areas of the world (Cambodia, Togo, Afghanistan, and several other countries) life expectancy is less than 40 years. In the United States, a male born in 1986 can expect to live close to 74 years, while females have a life expectancy of around 80 years.*

Given that, in most times and places, humans did not live much past 40, our awareness of human aging is relatively new. A 65-year-old person was rare in colonial America but is commonplace today. In 1900, 50 percent of American women were dead by age 58; in 1980, 50 percent of American women were dead by age 81 (Figure 18.32). Thus, the phenomena of senescence and the diseases of aging are much more common today than they were a century ago. In 1900, people did not have the "luxury" of dying from heart attacks or cancers, because these conditions are most likely to affect people over the age of 50. Rather, people died (as they are still dying in many parts of the world) from infections, infectious diseases, and parasites (Arking 1998). Similarly, until recently, relatively few people exhibited the more general human sensecent phenotype: gray hair, sagging and wrinkling skin, stiff joints, osteoporosis (loss of bone calcium), loss of muscle fibers and muscular strength, memory loss, eyesight deterioration, and slowed sexual responsiveness. As the melancholy

---

*You can see why the funding of Social Security is problematic in the United States. When it was created in 1935, the average working citizen died *before* age 65. Thus, he (and it usually was a he) was not expected to get back as much as he had paid into the system. Similarly, marriage "until death do us part" was an easier feat when death occurred in the third or fourth decade of life. The death rate of young women due to infections associated with childbirth was high throughout the world before antibiotics.

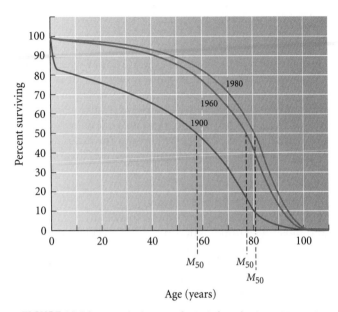

FIGURE 18.32 Survival curves for U.S. females in 1900, 1960, and 1980. M_{50} represents the age at which 50 percent of the individuals of each population survived. (After Arking 1998.)

Jacques notes in Shakespeare's *As You Like It*, those who did survive to senescence left the world "*sans* teeth, *sans* eyes, *sans* taste, *sans* everything."

Causes of aging

The general senescent phenotype is characteristic of each species. But what causes it? This question can be asked at many levels. Here we will be looking primarily at the cellular level of organization. While there is not yet a consensus on what causes aging (even at the cellular level), a theory is emerging that includes oxidative stress, hormones, and DNA damage.*

WEAR-AND-TEAR AND GENETIC INSTABILITY "Wear-and-tear" theories of aging are among the oldest hypotheses proposed to account for the human senescent phenotype (Weismann 1891; Szilard 1959). As one gets older, small

traumas to the body and its genome build up. At the molecular level, the number of point mutations increases, and the efficiency of the enzymes encoded by our genes decreases. If mutations occur in a part of the protein synthetic apparatus, the cell produces a large percentage of faulty proteins (Orgel 1963). If mutations were to arise in the DNA-synthesizing enzymes, the overall rate of mutation in the organism would be expected to increase markedly; Murray and Holliday (1981) have documented such faulty DNA polymerases in senescent cells.

Likewise, DNA repair and synthesis may be important in preventing senescence. Individuals of species whose cells have more efficient DNA repair enzymes live longer (Figure 18.33; Hart and Setlow 1974). Certain premature aging (**progeria**) syndromes in humans appear to be caused by mutations in such DNA repair enzymes (Sun et al. 1998; Shen and Loeb 2001). In humans, Hutchinson-Gilford progeria syndrome causes children to age rapidly and to die (usually of heart failure) as early as 12 years of age (Figure 18.34). This progeria is caused by a dominant mutant gene that encodes a nuclear membrane protein. Its symptoms include thin skin with age spots, resorbed bone mass, hair loss, and arteriosclerosis—all characteristics of the human senescent phenotype. Werner syndrome is another genetic premature aging disease in humans, and it is caused by a mutant helicase involved in DNA replication (Kipling et al. 2004). It is still not known what relationship these mutant genes have to the normal mechanisms of senescence.[†]

OXIDATIVE DAMAGE One major theory views metabolism as the cause of aging. According to this theory, aging is a result of metabolism and its by-products, **reactive oxygen species (ROS)**. The ROS produced by normal metabolism can oxidize and damage cell membranes, proteins, and nucleic acids. Some 2–3 percent of the oxygen atoms taken up by our mitochondria are reduced insufficiently and form ROS: superoxide ions, hydroxyl ("free") radicals, and hydrogen peroxide. Evidence that ROS molecules are critical in the aging process includes the observation that fruit flies overexpressing the enzymes that destroy ROS (catalase and superoxide dismutase) live 30–40 percent longer than do control flies (Orr and Sohal 1994; Parkes et al. 1998; Sun and Tower 1999).

*There is a popular proposal that the shortening of telomeres—repeated DNA sequences at the ends of chromosomes—is responsible for senescence. Telomere shortening has been connected to a decrease in the ability of cells to divide. However, no correlation between telomere length and the life span of an animal (humans have much shorter telomeres than mice) has been found, nor is there a correlation between human telomere length and a person's age (Cristofalo et al. 1998; Rudolph et al. 1999; Karlseder et al. 2002). Nematodes can have mutations that extend or shorten longevity, and the length of the telomere does not correlate with the age in these roundworms (Raices et al. 2005).Telomeres appear to be critical in stem cell maintenance, and the telomere-dependent inhibition of cell division might serve primarily as a defense against cancer (see Blasco 2005 and Flores et al. 2005).

[†]Another progeria is caused by loss-of-function mutations of the *Klotho* gene in mice (Kuro-o et al. 1997). *Klotho* (named after the Greek Fate who spun the thread of life; the other two Fates, Lachesis and Atropos, measured and cut life's thread, respectively) may be a very important gene, since its gain-of-function phenotype (causing its overexpression) can prolong a mouse's longevity by 30 percent (Kurosu et al. 2005). *Klotho* appears to encode a hormone that downregulates insulin signaling. As we shall soon see, the suppression of signaling by insulin and insulin-like growth factor-1 (IGF-1) is one of the ways that life span can be extended in many species.

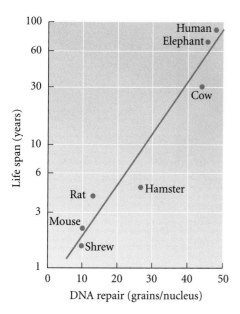

FIGURE 18.33 Correlation between life span and the ability of fibroblasts from various mammalian species to repair DNA. Repair capacity is represented in autoradiography by the number of grains from radioactive thymidine per cell nucleus. Note that the *y* axis (life span) is a logarithmic scale. (After Hart and Setlow 1974.)

Moreover, flies with mutations in the *methuselah* gene (named after the Biblical fellow said to have lived 969 years) live 35 percent longer than wild-type flies. These mutants have enhanced resistance to ROS (Lin et al. 1998). In *C. elegans*, too, individuals with mutations that result in either the degradation of ROS or the prevention of ROS formation live much longer than wild-type nematodes

(Larsen 1993; Vanfleteren and De Vreese 1996; Feng et al. 2001). These findings not only suggest that aging is under genetic control, but also provide evidence for the role of ROS in the aging process.

MITOCHONDRIAL GENOME DAMAGE The mutation rate in mitochondria is 10–20 times that of the nuclear DNA mutation rate (Johnson et al. 1999). It is thought that mutations in mitochondria could (1) lead to defects in energy production, (2) lead to the production of ROS by faulty electron transport, and/or (3) induce apoptosis. Age-dependent declines in mitochondrial function are seen in many animals, including humans (Boffoli et al. 1994). Moreover, the mutations may not only allow more ROS to be made, but also may make the mitochondrial DNA more susceptible to ROS-mediated damage.

The relationship between mitochondrial mutation and ROS in aging has become a new focus for research. Trifunovic and her colleagues (2004) induced a mutation in the mouse mitochondrial DNA polymerase such that it misread the codons more frequently than the wild-type enzyme. The increase in mitochondrial DNA mutations coincided with a premature aging syndrome (weight loss, reduced subcutaneous fat, hair loss, spinal curvature, osteoporosis, and reduced fertility) as well as premature death (Figure 18.35). The study findings are consistent with the ROS hypothesis for aging, since ROS production can result from faulty respiratory chain proteins in the mitochondria.

Conversely, Schriner and his colleagues (2005) demonstrated that the longevity of mice could be lengthened by targeting the enzyme catalase to the mitochondria. Catalase removes hydrogen peroxide and blocks the synthesis of ROS. Mice with this targeted catalase had lower ROS, better heart function, and better eyesight as they aged. They also lived nearly 20 percent longer than wild-type mice.

But the production of ROS might not be the only means by which mitochondrial mutations cause aging. When Kujoth and colleagues (2005) produced mutations in the mitochondrial DNA polymerase, the animals aged, as observed by earlier studies. However, no oxidative stress was observed. Rather, caspase-3 and DNA fragments—markers of apoptosis—were found in cells undergoing cell division. Thus, the loss of critical cells through apoptosis may be a critical part of the symptoms associated with aging.

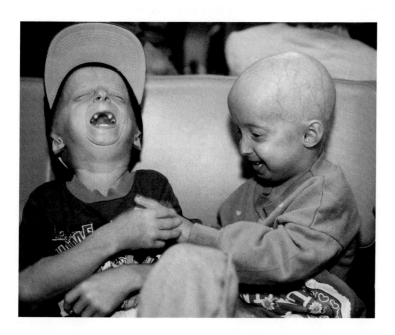

FIGURE 18.34 Children with progeria. Although not yet 8 years old, these children have a phenotype similar to that of an aged person. The hair loss, fat distribution, and transparency of the skin are characteristic of the normal human aging pattern seen in elderly adults. (Photograph © Associated Press.)

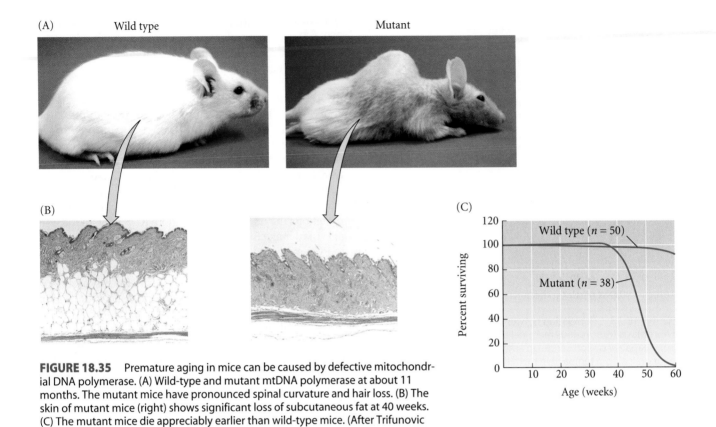

FIGURE 18.35 Premature aging in mice can be caused by defective mitochondrial DNA polymerase. (A) Wild-type and mutant mtDNA polymerase at about 11 months. The mutant mice have pronounced spinal curvature and hair loss. (B) The skin of mutant mice (right) shows significant loss of subcutaneous fat at 40 weeks. (C) The mutant mice die appreciably earlier than wild-type mice. (After Trifunovic et al. 2004; photographs courtesy of N.-G. Larsson.)

Genetically regulated aging: The insulin pathway

One of the criticisms of the idea of genetic "programs" for aging asks how evolution could have selected for them. Once the organism has passed maturity and raised its offspring, it is "an excrescence" on the tree of life (Rostand 1962); natural selection presumably cannot act on traits that affect an organism only *after* it has reproduced. But "How can evolution select for a way to degenerate?" may be the wrong question. Evolution probably can't select for such traits. The right question may be, "How can evolution select for phenotypes that can postpone reproduction or sexual maturity?" There is often a trade-off between reproduction and maintenance, and in many species reproduction and senescence are closely linked.

Recent studies of mice, *Caenorhabditis elegans*, and *Drosophila* suggest that there *is* a conserved genetic pathway that regulates aging, and that it can be selected for during evolution. This pathway involves the response to insulin or insulin-like growth factors. The insulin response pathway may be responsible for putting these organisms into a type of suspended animation during periods of adverse environmental conditions. Here, maintenance prevails over reproduction. This state is called **diapause**, and numerous organisms exhibit such a condition.

A newly hatched *C. elegans* larva proceeds through four larval stages, after which it can become an adult; or, if the nematodes are overcrowded or if there is insufficient food, the larva can enter a nonfeeding, metabolically dormant **dauer larva** stage. This dauer larva is the diapause condition for *C. elegans*. It does not feed and it is resistant to oxidative stress. The nematode can remain in this diapause state for up to 6 months, rather than becoming an adult that lives only a few weeks. In this dauer state, the nematode has increased resistance to ROS. If some of the genes in the pathway leading to dauer larva formation are mutated, adult development is allowed, but resistance to ROS is still in place. The resulting adults live two to four times as long as wild-type adults (Friedman and Johnson 1988; Kenyon et al. 1993).

The pathway that regulates both dauer larva formation and longevity has been identified as the insulin signaling pathway (Kimura et al. 1997; Guarente and Kenyon 2000; Gerisch et al. 2001; Pierce et al. 2001). Favorable environments signal the activation of the insulin receptor homologue DAF-2, and this receptor stimulates the onset of adulthood (Figure 18.36A). Poor environments fail to activate the DAF-2 receptor, and dauer formation ensues. While severe loss-of-function alleles in this pathway cause the formation of dauer larvae in any environment, weak mutations in the insulin signaling pathway enable the ani-

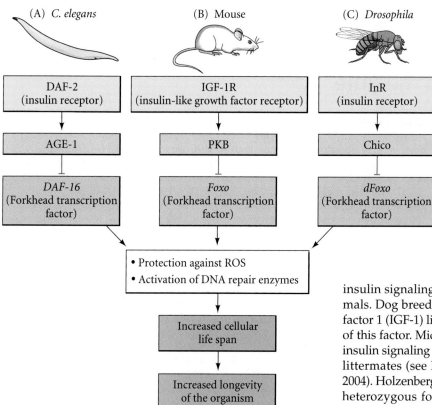

(A) *C. elegans*   (B) Mouse   (C) *Drosophila*

| DAF-2 (insulin receptor) | IGF-1R (insulin-like growth factor receptor) | InR (insulin receptor) |

| AGE-1 | PKB | Chico |

| *DAF-16* (Forkhead transcription factor) | *Foxo* (Forkhead transcription factor) | *dFoxo* (Forkhead transcription factor) |

- Protection against ROS
- Activation of DNA repair enzymes

Increased cellular life span

Increased longevity of the organism

FIGURE 18.36 A possible pathway for regulating longevity. In each case, the insulin signaling pathway inhibits the synthesis of proteins that would otherwise protect cells against oxidative damage due to ROS. These protective proteins may be particularly important in mitochondria. When insulin signaling is downregulated, forkhead transcription factors may activate DNA repair enzymes that may protect against mutations caused randomly by ROS or other agents. Such protection against ROS and mutation may increase the functional life span of the cells and the longevity of the organism.

mals to reach adulthood and live longer than wild-type animals.

Downregulation of the insulin signaling pathway also has several other functions. First, it appears to influence metabolism, decreasing mitochondrial electron transport. When the DAF-2 receptor is not active, organisms have decreased sensitivity to ROS (Feng et al. 2001; Scott et al. 2002). Second, it increases the production of enzymes that prevent oxidative damage, as well as DNA repair enzymes (Honda and Honda 1999; Tran et al. 2002). Third, it decreases fertility (Gems et al. 1998). This increase in DNA synthetic enzymes and in enzymes that protect against ROS is due to the DAF-16 transcription factor. This forkhead-type transcription factor is inhibited by the insulin receptor (DAF-2) signal. When that signal is absent, DAF-16 can function, and this factor appears to activate the genes encoding several enzymes (such as catalases, superoxide dismutases) that are involved in reducing ROS, several enzymes that increase protein and lipid turnover, and several stress proteins (Lee et al. 2003; Libina et al. 2003; Murphy et al. 2003).

It is possible that this system also operates in mammals, but the mammalian insulin and insulin-like growth factor pathways are so integrated with embryonic development and adult metabolism that mutations often have numerous and deleterious affects (such as diabetes or Donahue syndrome). However, there is some evidence that the insulin signaling pathway does affect life span in mammals. Dog breeds with low levels of insulin-like growth factor 1 (IGF-1) live longer than breeds with higher levels of this factor. Mice with loss-of-function mutations of the insulin signaling pathway live longer than their wild-type littermates (see Partridge and Gems 2002; Blüher et al. 2004). Holzenberger and colleagues (2003) found that mice heterozygous for the insulin-like growth factor type 1 receptor (IGF-1R) not only lived about 30 percent longer than their wild-type littermates, they also had greater resistance to oxidative stress. In addition, mice lacking one copy of their IGF-1 receptor gene live about 25 percent longer than wild-type mice (and have higher ROS resistance, but otherwise normal physiology and fertility). And finally, caloric restriction—one of the few known ways of increasing mammalian longevity (again, at the expense of fertility)—may extend life by reducing the levels of circulating insulin and by reducing IGF-1 (Kenyon 2001; Roth et al. 2002; Holzenberger et al. 2003). The insulin pathway in mammals also negatively regulates a Foxo4, a Forkhead transcription factor, that activates such ROS-protective enzymes (Figure 18.36B; Essers et al. 2004). Just as research is linking mitochondrial mutation and ROS, other research is linking ROS with the insulin receptor pathway.

The insulin signaling pathway also appears to regulate life span in *Drosophila* (Figure 18.36C). Flies with weak loss-of-function mutations of the insulin receptor gene or genes in the insulin signaling pathway (such as *chico*) live nearly 85 percent longer than wild-type flies (Clancy et al. 2001; Tatar et al. 2001). These long-lived mutants are sterile, and their metabolism resembles that of flies that are in diapause (Kenyon 2001). The insulin receptor in *Drosophila* is thought to regulate a Forkhead transcription factor (dFoxo) similar to the DAF-16 protein of *C. elegans*. When the *Drosophila* dFoxo is activated in the fat body, it can lengthen the fly's life span (Giannakou et al. 2004; Hwangbo et al. 2004). Although it has not been demonstrated that dFoxo regulates the enzymes that protect against ROS, independent studies have shown that when these enzymes (such as

Aging: Exceptions to the Rule

There are some species where aging seems to be optional, and these may hold some important clues to how animals can live longer and retain their health. Turtles, for instance, are a symbol of longevity in many cultures. Many turtle species not only live a long time, but they do not undergo the typical aging syndrome. In these species, older females lay as many, if not more eggs, than their younger counterparts. Miller (2001) has shown that a 60-year-old female three-toed box turtle lays as many eggs annually as she ever did. Interestingly, turtles have special adaptations against oxygen deprivation; these enzymes also protect against ROS (Congdon et al. 2003; Lutz et al. 2003).

In monarch butterflies (*Danaus plexippus*), adults that migrate to overwintering grounds in the mountains of central Mexico live several months (August–March), whereas their summer counterparts live only about 2 months (May–July). The regulation of this difference appears to be juvenile hormone (Herman and Tatar 2001). The migrating butterflies are sterile due to suppressed synthesis of JH. If migrants are given JH in the laboratory, they regain fertility but lose their longevity. Conversely, when summer monarchs have their corpora allata removed (so they no longer make juvenile hormone), their longevity increases 100 percent. Mutations in the insulin signaling pathway of *Drosophila* likewise decrease juvenile hormone synthesis (Tu et al. 2005). This decrease in JH makes them small, sterile, and long-lived, adding to whatever longevity-producing effect protection against ROS might have.

Finally, there may be organisms that have actually cheated death. The hydrozoan cnidarian *Turritopsis nutricula* may be such an immortal animal. Most hydrozoans have a complex life cycle in which a colonial (polyp) stage asexually buds off the sexually mature, solitary, adult medusa (usually called a "jellyfish"). Eggs and sperm from the medusa develop into an embryo and then a planula larva. Planula larvae then form a colonial polyp stage. Medusae, like the polyps, have a limited life span and in most hydrozoans they die shortly after releasing their gametes (Martin 1997). *Turritopsis*, however, has evolved a remarkable variation on this theme. The solitary medusa of this species can revert to its polyp stage *after* becoming sexually mature (Bavestrello et al. 1992; Piraino et al. 1996). In the laboratory, 100 percent of *Turritopsis* medusae regularly undergo this change.

How does the jellyfish accomplish this feat? Apparently, it can alter the differentiated state of a cell, transforming it into another cell type. Such a phenomenon is called **transdifferentiation**, and is usually seen only when parts of an organ regenerate. However, it appears to occur normally in the *Turritopsis* life cycle. In the transdifferentiation process, the medusa is transformed into the stolons and polyps of a hydroid colony. These polyps feed on zooplankton and soon are budding off new medusae. Thus, it is possible that organismic death does not occur in this species.

---

superoxide dismutase) are downregulated by mutation or by RNAi, the resulting flies die early, have increased oxidative stress, and display higher levels of DNA damage (Kirby et al. 2002; Woodruff et al. 2004). Conversely, over-expression of superoxide dismutase genes can lengthen the *Drosophila* life span (Parkes et al. 1998). While some evidence points to a correlation between longer life span, lower insulin signaling, and elevated ROS protection in *Drosophila* (Broughton et al. 2005), other studies suggest that some flies and other insects can obtain longer life spans without increasing the enzymes known to protect against oxidative stress (Le Bourg and Fournier 2004; Parker et al. 2004).

From an evolutionary point of view, the insulin pathway may mediate a trade-off between reproduction and survival/maintenance. Many (although not all) of the long-lived mutants have reduced fertility. Thus, it was interesting that another longevity signal originates in the gonad. When the germline cells are removed from *C. elegans*, the animals live longer. It is thought that the germline stem cells produce a substance that blocks the effects of a longevity-inducing steroid hormone (Hsin and Kenyon 1999; Gerisch et al. 2001; Arantes-Oliviera 2002). Conversely, ROS appears to promote germline development at the expense of somatic development in *C. elegans*. The oxidation of certain lipids accelerates germ cell development, while those same lipids, in their unoxidized form, prevent germ cell proliferation (Shibata et al. 2003).

Promoting longevity

Several interacting agents may promote longevity. These include caloric restriction, protection against oxidative stress, and the factors activated by a suppressed insulin pathway. It is not yet known how these factors interact—whether they are part of a single "longevity pathway," or if they act separately. Moreover, genetics and diet do not appear to be the full answer to aging. Chance, it seems, still plays a role. When clonally identical *C. elegans* are fed an identical diet, some organisms still live longer than others, and different organs deteriorate more rapidly in different individuals (Herndon et al. 2002). Mutations are randomly occurring events, and they may play a role in the aging process.

As human life expectancy increases due to advances in our ability to prevent and cure disease, we are still left with a general aging syndrome that is characteristic of our species. Unless attention is paid to this general aging syndrome, we risk ending up like Tithonios, the miserable wretch of Greek mythology to whom the gods awarded eternal life, but not eternal youth. However, our new knowledge of regeneration is being put to use by medicine, and we may soon be able to ameliorate some of the symptoms of aging. The potentially far-reaching consequences of this interaction of developmental biology and medicine will be discussed in Chapter 21.

Snapshot Summary **Metamorphosis, Regeneration, and Aging**

1. Amphibian metamorphosis includes both morphological and biochemical changes. Some structures are remodeled, some are replaced, and some new structures are formed.

2. The hormone responsible for amphibian metamorphosis is triiodothyronine (T_3). The synthesis of T_3 from thyroxine and the degradation of T_3 by deiodinases can regulate metamorphosis in different tissues. T_3 binds to thyroid hormone receptors and acts predominantly at the transcriptional level.

3. Many changes during amphibian metamorphosis are regionally specific. The tail muscles degenerate; the trunk muscles persist. An eye will persist even if transplanted into a degenerating tail.

4. Heterochrony involves changes in the relative rates of development of different parts of the animal. In neoteny, the larval form is retained while the gonads and germ cells mature at their normal rate. In progenesis, the gonads and germ cells mature rapidly while the rest of the body matures normally. In both instances, the animal can mate while retaining its larval or juvenile form.

5. In animals with direct development, the larval stage has been lost. Some frogs, for instance, form limbs while in the egg.

6. Ametabolous insects undergo direct development. Hemimetabolous insects pass through nymph stages wherein the immature organism is usually a smaller version of the adult.

7. In holometabolous insects, there is a dramatic metamorphosis from larva to pupa to sexually mature adult. In the stages between larval molts, the larva is called an instar. After the last instar, the larva undergoes a metamorphic molt to become a pupa. The pupa undergoes an imaginal molt to become an adult.

8. During the pupal stage, the imaginal discs and histoblasts grow and differentiate to produce the structures of the adult body.

9. The anterior-posterior, dorsal-ventral, and proximal-distal axes are sequentially specified by interactions between different compartments in the imaginal discs. The disc "telescopes out" during development, its central regions becoming distal.

10. Molting is caused by the hormone 20-hydroxy-ecdysone (20E). In the presence of high levels of juvenile hormone, the molt gives rise to another larval instar. In low concentrations of juvenile hormone, the molt produces a pupa; if no juvenile hormone is present, the molt is an imaginal molt.

11. The ecdysone receptor gene produces a nuclear RNA that can form at least three different proteins. The types of ecdysone receptors in a cell may influence the response of that cell to 20E. The ecdysone receptors bind to DNA to activate or repress transcription.

12. There are three major types of regeneration. In epimorphosis (such as regenerating limbs), tissues dedifferentiate into a regeneration blastema, divide, and redifferentiate into the new structure. In morphallaxis (characteristic of hydra), there is a repatterning of existing tissue with little or no growth. In compensatory regeneration (such as in the liver), cells divide but retain their differentiated state.

13. In the regenerating salamander limb, the epidermis forms an apical ectodermal cap. The cells beneath it dedifferentiate to form a blastema. The differentiated cells lose their adhesions and re-enter the cell cycle. Fgf10 and factors from neurons appear to be critical in permitting regeneration to occur.

14. Salamander limb regeneration appears to use the same pattern formation system as the developing limb.

15. In *Hydra*, there appears to be a head activation gradient, a head inhibition gradient, a foot activation gradient, and a foot inhibition gradient. Budding occurs where these gradients are minimal.

16. The hypostome region of *Hydra* appears to be an organizer region that secretes paracrine factors to alter the fates of surrounding tissue.

17. The maximum life span of a species is the longest time an individual of that species has been observed to survive. Life expectancy is the age at which approximately 50 percent of the members of a given population still survive.

18. Aging is the time-related deterioration of the physiological functions necessary for survival and reproduction. The phenotypic changes of senescence (which affect all members of a species) are not to be confused with diseases of senescence, such as cancer and heart disease (which affect some individuals but not others).

19. Reactive oxygen species (ROS) can damage cell membranes, inactivate proteins, and mutate DNA. Mutations that alter the ability to make or degrade ROS can change the life span.

20. Mitochondria may be a target for proteins that regulate aging.

21. An insulin signaling pathway, involving a receptor for insulin and insulin-like proteins, may be an important component of genetically limited life spans.

For Further Reading

Complete bibliographical citations for all literature cited in this chapter can be found on the Vade Mecum CD that accompanies the book and at the free access website www.devbio.com

Becker, K. B., K. C. Stephens, J. C. Davey, M. J. Schneider and V. A. Galton. 1997. The type 2 and type 3 iodothyronine deiodinases play important roles in coordinating development in *Rana catesbeiana* tadpoles. *Endocrinology* 138: 2989–2997.

Broun, M., L. Gee, B. Reinhardt, and H. R. Bode. 2005. Formation of the head organizer in hydra involves the canonical Wnt pathway. *Development* 132: 2907.

Brockes, J. P. and A. Kumar. 2002. Plasticity and reprogramming of differentiated cells in amphibian regeneration. *Nature Rev. Mol. Cell Biol.* 3: 566–574.

Cai, L. and D. D. Brown. 2004. Expression of type II iodothyronine deiodinase marks the time that a tissue responds to thyroid hormone-induced metamorphosis in *Xenopus laevis*. *Dev. Biol.* 266: 87–95.

Gardiner, D. M., T. Endo and S. Bryant. 2002. The molecular basis of amphibian limb regeneration: Integrating the old with the new. *Semin. Cell Dev. Biol.* 13: 345–352.

Jiang, C., A. F. Lamblin, H. Steller and C. S. Thummel. 2000. A steroid-triggered transcriptional hierarchy controls salivary gland cell death during *Drosophila* metamorphosis. *Mol. Cell* 5: 445–455.

Schubiger, M. and J. W. Truman. 2000. The RXR ortholog USP suppresses early metamorphic processes in *Drosophila* in the absence of ecdysteroids. *Development* 127: 1151–1159.

Taub, R. 2004. Liver regeneration: from myth to mechanism. *Nature Rev. Mol. Cell Biol.* 5: 836–847.

Trifunovic, A. and 12 others. 2004. Premature ageing in mice expressing defective mitochondrial DNA polymerase. *Nature* 429: 417–423.

The Saga of the Germ Line

19

WE BEGAN OUR ANALYSIS OF ANIMAL DEVELOPMENT by discussing fertilization, and we will finish our studies of individual development by investigating **gametogenesis**, the processes by which the sperm and the egg are formed. Germ cells provide the continuity of life between generations, and the mitotic ancestors of our own germ cells once resided in the gonads of reptiles, amphibians, fish, and invertebrates.

In many animals, such as insects, roundworms, and vertebrates, there is a clear and early separation of germ cells from somatic cell types. In several other animal phyla (and throughout the entire plant kingdom), this division is not as well established. In these species (which include cnidarians, flatworms, and tunicates), somatic cells can readily become germ cells even in adult organisms. The zooids, buds, and polyps of many invertebrate phyla testify to the ability of somatic cells to give rise to new individuals (Liu and Berrill 1948; Buss 1987).

In those organisms in which there is an established germ line that separates from the somatic cells early in development, the germ cells do not arise within the gonad itself. Rather, their precursors—the **primordial germ cells (PGCs)**—arise elsewhere and migrate into the developing gonads. The first step in gametogenesis, then, involves forming the PGCs and getting them into the genital ridge as the gonad is forming. Our discussion of gametogenesis will include:

1. The formation of the germ plasm and the determination of the PGCs
2. The migration of the PGCs into the developing gonads
3. The process of meiosis and the modifications of meiosis for forming sperm and eggs
4. The differentiation of the sperm and egg
5. The hormonal control of gamete maturation and ovulation

Germ Plasm and the Determination of the Primordial Germ Cells

All sexually reproducing animals arise from the fusion of gametes—sperm and eggs. All gametes arise from primordial germ cells. In many instances (including frogs, nematodes, and flies), the primordial germ cells are specified autonomously by cytoplasmic determinants in the egg that are then parceled out to specific cells during cleavage. In other instances (such as salamanders and mammals), the germ cells are specified by interactions among neighboring cells. In those species wherein the determination of the primordial germ cells is brought about by the autonomous localization of specific proteins and mRNAs, these cytoplasmic components are collectively referred to as the **germ plasm**.

"And the end of all our exploring Will be to arrive where we started And know the place for the first time."

T. S. ELIOT (1942)

"When the spermatozoon enters the egg, it enters a cell system which has already achieved a certain degree of organization."

ERNST HADORN (1955)

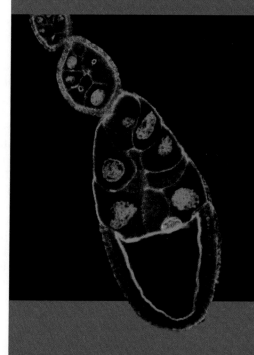

(A)

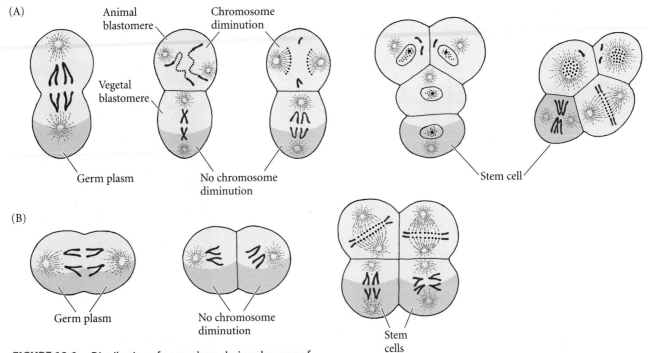

(B)

FIGURE 19.1 Distribution of germ plasm during cleavage of normal and centrifuged zygotes of *Parascaris*. (A) In normal cleavage, the germ plasm is localized in the vegetalmost blastomere, as shown by the lack of chromosomal diminution in that particular cell. Thus, at the 4-cell stage, the embryo has a single stem cell for its gametes. (B) When centrifugation is used to displace the first cleavage by 90 degrees, both of the resulting cells have vegetal germ plasm, and neither cell undergoes chromosome diminution. After the second cleavage, both of these two cells give rise to germinal stem cells. (After Waddington 1966.)

Germ cell determination in nematodes

BOVERI'S EXPERIMENTS ON *PARASCARIS* Theodor Boveri (1862–1915; see Figure 4.2) was the first person to observe an organism's chromosomes throughout its development. In so doing, he discovered a fascinating feature in the development of the roundworm *Parascaris aequorum* (formerly known as *Ascaris megalocephala*). This nematode worm has only two chromosomes per haploid cell, allowing detailed observations of its individual chromosomes. The cleavage plane of the first embryonic division is unusual in that it is equatorial, separating the animal half from the vegetal half of the zygote (Figure 19.1A). More bizarre, however, is the behavior of the chromosomes in the subsequent division of these first two blastomeres. The chromosomes in the animal blastomere fragment into dozens of pieces just before this cell divides. This phenomenon is called **chromosome diminution**, because only a portion of the original chromosome survives. Numerous genes are lost when the chromosomes fragment, and these genes are not included in the newly formed nuclei (Tobler et al. 1972; Müller et al. 1996).

Meanwhile, in the vegetal blastomere, the chromosomes remain normal. During second cleavage, the animal cell splits meridionally while the vegetal cell again divides equatorially. Both vegetally derived cells have normal chromosomes. However, the chromosomes of the more animally located of these two vegetal blastomeres fragment before the third cleavage. Thus, at the 4-cell stage, only one cell—the most vegetal—contains a full set of genes. At successive cleavages, nuclei with diminished chromosomes are given off from this vegetalmost line until the 16-cell stage, when there are only two cells with undiminished chromosomes. One of these two blastomeres gives rise to the germ cells; the other eventually undergoes chromosome diminution and forms more somatic cells. The chromosomes are kept intact only in those cells destined to form the germ line. If this were not the case, the genetic information would degenerate from one generation to the next. The cells that have undergone chromosome diminution generate the somatic cells.

Boveri has been called the last of the great observers of embryology and the first of the great experimenters. Not content with observing the retention of the full chromosome complement by the germ cell precursors, he set out to test whether a specific region of cytoplasm protects the nuclei within it from diminution. If so, any nucleus happening to reside in this region should remain undiminished. In 1910, Boveri tested this hypothesis by centrifuging *Parascaris* eggs shortly before their first cleavage. This treatment shifted the orientation of the mitotic spindle. When the spindle forms perpendicular to its normal orientation, both resulting blastomeres contain some of the vegetal cytoplasm (Figure 19.1B). Boveri found that after the first division, neither nucleus underwent chromosomal diminution. However, the next division was equatorial along the animal-vegetal axis. Here the resulting animal blastomeres both underwent diminution, whereas the two vegetal cells did not. Boveri concluded that the vegetal cytoplasm contains a factor (or factors) that protects nuclei from chromosomal diminution and determines germ cells.

C. ELEGANS   In the nematode *C. elegans*, the germline precursor cell is the P4 blastomere. The **P-granules** that enter this cell are critical for instructing it to become the germline precursor (see Figure 8.44). The P-granule protein repertoire includes several transcriptional inhibitors and RNA-binding proteins, including homologues of *Drosophila* Vasa, Piwi, and Nanos, whose functions we will discuss below (Kawasaki et al. 1998; Seydoux and Strome 1999; Subramanian and Seydoux 1999). In addition, as discussed in Chapter 8, the PIE-1 protein is critical for preventing the germ line from differentiating into somatic cells. PIE-1 blocks the transcription of nearly all *C. elegans* genes in the germ cells, probably by blocking the chromatin remodeling proteins (Unhavaithaya et al. 2002). Germ cell differentiation cannot commence until the disappearance of PIE-1 in later embryonic stages. Until that time, the germline nuclei are silenced (Figure 19.2).

WEBSITE 19.1 Mechanisms of chromosome diminution. The somatic cells do not lose DNA randomly. Rather, specific regions of DNA are lost during chromosome diminution.

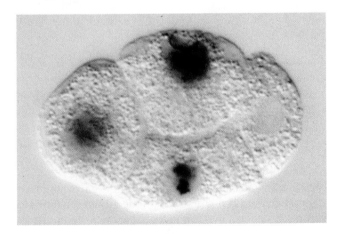

FIGURE 19.2   Inhibition of transcription in germ cell precursors of *C. elegans*. The photograph shows in situ hybridization to β-galactosidase mRNA expressed under control of the *pes-10* promoter. The *pes-10* gene is one of the earliest genes expressed in *C. elegans*. The P-blastomere that gives rise to the germ cells (far right) does not transcribe the gene. (From Seydoux and Fire 1994; photograph courtesy of G. Seydoux.)

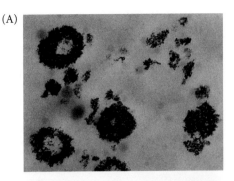

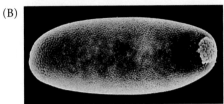

FIGURE 19.3   The pole plasm of *Drosophila*. (A) Electron micrograph of polar granules from particulate fraction of *Drosophila* pole cells. (B) Scanning electron micrograph of a *Drosophila* embryo just prior to completion of cleavage. The pole cells can be seen at the right of the photograph. (Photographs courtesy of A. P. Mahowald.)

Germ cell determination in insects

In *Drosophila*, PGCs form as a group of **pole cells** at the posterior pole of the cellularizing blastoderm. These nuclei migrate into the posterior region at the ninth nuclear division and become surrounded by the **pole plasm**, a complex collection of mitochondria, fibrils, and **polar granules** (Figure 19.3; Mahowald 1971a,b; Schubiger and Wood 1977). If the pole cell nuclei are prevented from reaching the pole plasm, no germ cells will be made (Mahowald et al. 1979). The germ cells are responsible for forming the **germline stem cells**. As mentioned in Chapter 6, each germline stem cell divides asymmetrically to produce another stem cell and a cystoblast. Cystoblasts undergo four mitotic divisions with incomplete cytokinesis to form a cluster of 16 cells interconnected by cytoplasmic bridges called **ring canals**. Only those two cells having four interconnections are capable of developing into oocytes, and of those two, only one becomes the egg (the other begins meiosis but does not complete it). Thus, only one of the 16 cystocytes becomes an ovum; the remaining 15 cells become **nurse cells** (Figure 19.4).

As it turns out, the cell destined to become the oocyte is that cell residing at the most posterior tip of the egg chamber, or **ovariole**, that encloses the 16-cell clone. However, since the nurse cells are connected to the oocyte by the ring canals, the entire complex can be seen as one egg-producing unit. The nurse cells produce numerous RNAs and proteins that ultimately are transported into the oocyte through the ring canals.

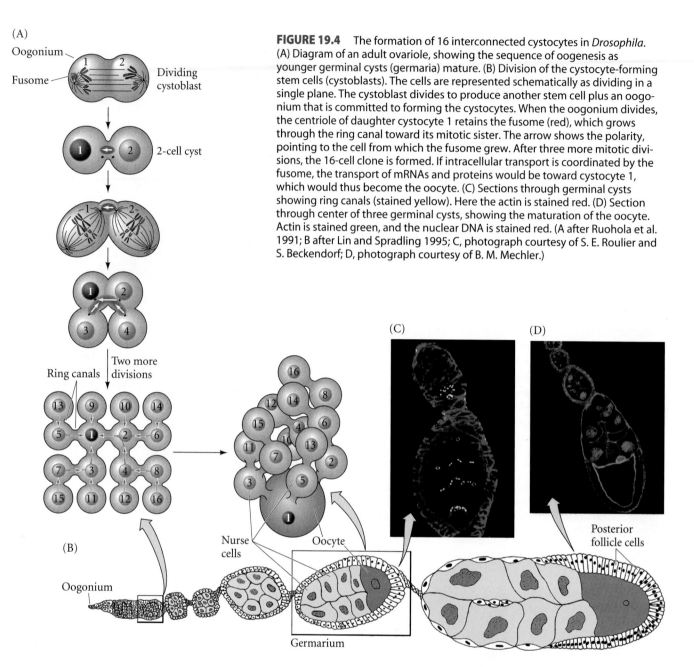

(A)

Oogonium

Fusome

Dividing cystoblast

2-cell cyst

Ring canals

Two more divisions

(B)

Oogonium

13 9 10 14
5 1 2 6
7 3 4 8
15 11 12 16

16
12 14 8
15 4 6
11 10 13
7 2
3 5 1

Nurse cells

Oocyte

Germarium

(C)

(D)

Posterior follicle cells

FIGURE 19.4 The formation of 16 interconnected cystocytes in *Drosophila*. (A) Diagram of an adult ovariole, showing the sequence of oogenesis as younger germinal cysts (germaria) mature. (B) Division of the cystocyte-forming stem cells (cystoblasts). The cells are represented schematically as dividing in a single plane. The cystoblast divides to produce another stem cell plus an oogonium that is committed to forming the cystocytes. When the oogonium divides, the centriole of daughter cystocyte 1 retains the fusome (red), which grows through the ring canal toward its mitotic sister. The arrow shows the polarity, pointing to the cell from which the fusome grew. After three more mitotic divisions, the 16-cell clone is formed. If intracellular transport is coordinated by the fusome, the transport of mRNAs and proteins would be toward cystocyte 1, which would thus become the oocyte. (C) Sections through germinal cysts showing ring canals (stained yellow). Here the actin is stained red. (D) Section through center of three germinal cysts, showing the maturation of the oocyte. Actin is stained green, and the nuclear DNA is stained red. (A after Ruohola et al. 1991; B after Lin and Spradling 1995; C, photograph courtesy of S. E. Roulier and S. Beckendorf; D, photograph courtesy of B. M. Mechler.)

The components of the germ plasm are responsible for specifying these cells to be germ cells and for inhibiting somatic gene expression in these cells. These two functions might be interrelated, since the inhibition of gene transcription appears to be essential for germ cell determination (see Santos and Lehmann 2004). One of the components of the pole plasm is the mRNA of the **germ cell-less** (**gcl**) gene. This gene was discovered by Jongens and his colleagues (1992) when they mutated *Drosophila* and screened for females who did not have "grandoffspring." They assumed that if a female did not place functional pole plasm in her eggs, she could still have offspring—but those offspring would lack germ cells and would be sterile. The wild-type

gcl gene is transcribed in the nurse cells of the fly's ovary, and its mRNA is transported into the egg. Once inside the egg, it is transported to the posteriormost portion and resides within what will become the pole plasm. This message is translated into protein during the early stages of cleavage (Figure 19.5A). The *gcl*-encoded protein appears to enter the nucleus, and it is essential for pole cell production. Flies with mutations of this gene lack germ cells (Figure 19.5B). The *gcl* gene encodes a nuclear envelope protein that prevents gene transcription and is critical for specifying the pole cells (Leatherman et al. 2002). Homologues of the *gcl* gene have been found in the germ cells of mice and humans, and the mouse *Gcl* gene represses tran-

FIGURE 19.5 Localization of *germ cell-less* gene products in the posterior of the egg and embryo. The *gcl* mRNA can be seen in the posterior pole of early-cleavage embryos produced by wild-type females (A), but not in embryos produced by *gcl*-deficient mutant females (B). The protein encoded by the *gcl* gene can be detected in the germ cells at the cellular blastoderm stage of embryos produced by wild-type females, but not in embryos from mutant females (arrows). (From Jongens et al. 1992; photographs courtesy of T. A. Jongens.)

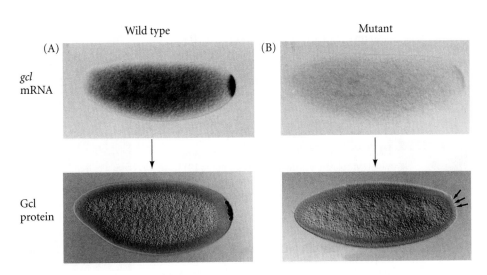

scription. Human males with mutant *GCL* genes have defective spermatozoa and are often sterile (Nili et al. 2001; Kleiman et al. 2003).

A second component of *Drosophila* pole plasm (and one that becomes localized in the polar granules) is a noncoding RNA called **polar granule component** (**Pgc**). *Pgc* also inhibits transcription, and it does so by preventing the phosphorylation of RNA polymerase II by TFIIH (Martinho et al. 2004; see Figure 5.4). Without this phosphorylation, the RNA polymerase cannot transcribe any genes. If maternal *Pgc* gene is mutated, the germ cells begin expressing the genes characteristic of their neighboring somatic cells.

A third set of pole plasm components are the **posterior group determinants**. The Oskar protein (see Chapter 9) appears to be the critical member of this group, since expression of *oskar* mRNA in ectopic sites will cause the nuclei in those areas to form germ cells. The genes that restrict Oskar to the posterior pole are also necessary for germ cell formation (Ephrussi and Lehmann 1992; Newmark et al. 1997; Riechmann et al. 2002). Moreover, Oskar appears to be the limiting step of germ cell formation, since adding more *oskar* message to the oocyte causes more germ cells to form. Oskar functions by localizing the proteins and RNAs necessary for germ cell formation (such as *germ cell-less*) to the posterior pole (Ephrussi and Lehmann 1992; Snee and Macdonald 2004).

One of the mRNAs localized by Oskar is the *Nanos* message, whose product is essential for posterior segment formation and germ cell specification. Pole cells lacking Nanos do not migrate into the gonads and fail to become gametes. While Gcl and Pgc appear to be critical in regulating transcription, Nanos appears to be essential for inhibiting the translation of certain messages. In embryos lacking Nanos, the germline cells usually die; but if inhibited from dying, these germline cells can become somatic cells (see Chapter 9). Nanos thus prevents the pole cells

from activating the pathway that would lead to the formation of somatic cells (Hayashi et al. 2004).

Another of the posterior mRNAs encodes **Vasa**, an RNA-binding protein. The mRNAs for this protein are seen in the germ plasm of many species, and Vasa is critical for initiating germ cell differentiation and meiosis (Ghabrial and Schüpbach 1999). Two other members of this group, **Piwi*** and its relative, **Aubergine**, are also found in the pole plasm. They, too, have the ability to repress transcription. Piwi will later become critical in establishing the germ cell as a stem cell in the gonad (Cox et al. 1998).

There are numerous components of the pole plasm which we know little about (see Santos and Lehmann 2004). For instance, mitochondrial ribosomes are seen transiently in the *Drosophila* pole plasm. Kobayashi and Okada (1989) demonstrated its importance by showing that injecting mitochondrial RNA into embryos formed from ultraviolet-irradiated eggs restored their ability to form pole cells. It is possible that some of the pole plasm mRNAs are being translated by these mitochondrial ribosomes. Inhibiting protein synthesis by these ribosomes impairs production of the Gcl protein. (Amikura et al. 2005).

---

*Piwi appears to be required for stem cell maintenance and proliferation throughout the eukaryotic kingdoms. In addition to being present in germ stem cells, *Piwi* genes have also been found expressed in the totipotent stem cells of planaria and regenerating annelids. Inhibiting *Piwi* in the adult flatworm blocks its regeneration (Reddien 2004). *Piwi* is also expressed in the somatic stem cells of jellyfish and is upregulated immediately before transdifferentiation. The continuous low expression of *Piwi* in differentiated cells of jellyfish may underlie their ability to remodel their bodies so profoundly (Seipel et al. 2004.) Piwi may even be responsible for stem cell maintenance across kingdoms: two *Piwi* genes in *Arabidopsis* are crucial for maintaining meristem proliferation at the root and shoot of the plant (Bohmert et al. 1998; Moussian et al. 1998).

Germ cell determination in frogs and fish

FROGS Cytoplasmic localization of germ cell determinants has also been observed in vertebrate embryos. Bounoure (1934) showed that the vegetal region of fertilized frog eggs contains material with staining properties similar to those of *Drosophila* pole plasm. He was able to trace this cortical cytoplasm into the few cells in the presumptive endoderm that would normally migrate into the genital ridge. By transplanting genetically marked cells from one embryo into another of a differently marked strain, Blackler (1962) showed that these cells are the primordial germ cell precursors.

The germ plasm of amphibians consists of germinal granules and a matrix around them. It contains many of the RNAs and proteins (including the large and small mitochondrial ribosomal RNAs) as the pole plasm of *Drosophila*, and they appear to repress transcription and translation (Kloc et al. 2002). The early movements of amphibian germ plasm have been analyzed in detail by Savage and Danilchik (1993), who labeled the germ plasm with a fluorescent dye. They found that the germ plasm of unfertilized eggs consists of tiny "islands" that appear to be tethered to the yolk mass near the vegetal cortex. These islands move with the vegetal yolk mass during the cortical rotation just after fertilization. After this rotation, the islands are released from the yolk mass and begin fusing together and migrating to the vegetal pole. Their aggregation depends on microtubules, and their movement to the vegetal pole depends on a kinesin-like protein that may act as the motor for germ plasm movement (Robb et al. 1996; Quaas and Wylie 2002). Savage and Danilchik (1993) found that UV light prevents vegetal surface contractions and inhibits the migration of germ plasm to the vegetal pole. Furthermore, the *Xenopus* homologues of *Nanos* and *Vasa* messages are specifically localized to the vegetal region (Figure 19.6; Forristall et al. 1995; Ikenishi et al. 1996; Zhou and King 1996).

ZEBRAFISH In zebrafish, the germ plasm forms a nuage-like structure characterized by polar granules, mitochondria, and concentrated mRNAs. Two of these mRNAs are *Vasa* and Nanos. These messages are maternally supplied and they appear to be associated with the cleavage furrows

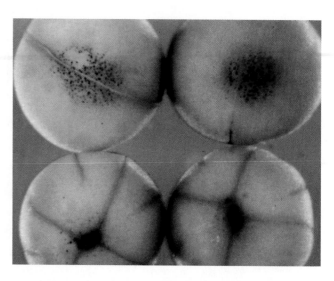

FIGURE 19.6 Germ plasm at the vegetal pole of frog embryos. In situ hybridization to the mRNA for *Xcat2* (the *Xenopus* homologue of *Nanos*) localizes the message in the vegetal cortex of first-cleavage (upper) and fourth-cleavage (lower) embryos. (After Kloc et al. 1998; photograph courtesy of L. Etkin.)

of the early dividing egg (Yoon et al. 1997). *Vasa* mRNA and other components of the germ plasm form a compact structure that is inherited by only one of the two daughter cells at each division. Thus, at late cleavage (around 1000 cells), only four cells have the germ plasm. However, after this stage, the germ plasm is distributed evenly at cell division, creating four clusters of primordial germ cells (see Figure 19.12).

Germ cell determination in mammals

In insects, frogs, nematodes, and flies, the germ cells are determined by material in the egg cytoplasm. However, in mammals, there is no obvious germ plasm. Rather, germ cells are induced in the embryo (Sutasurya and Nieuwkoop 1974; Wakahara 1996). In mice, the germ cells form at the posterior region of the epiblast, at the junction of the extraembryonic ectoderm, epiblast, primitive streak, and allantois (Figure 19.7A). This is called the posterior proximal epiblast because it is close (proximal) to the extraembryonic ectoderm, and it will be at the posterior of the embryo.

Around day 6.5 of embryonic development, BMP4 and BMP8b from the extraembryonic ectoderm give certain cells in this area the ability to produce germ cells (Figure 19.7B; Lawson et al. 1999; Ying et al. 2000). The cluster of cells capable of generating PGCs express *fragilis*, a gene encoding a particular transmembrane protein. However, these *fragilis*-expressing cells can form both PGCs and some somatic cells. In the center of this cluster of cells is a small group of about 20 cells that also expresses *blimp1* and *stella*. While Stella protein is involved in early cleavage

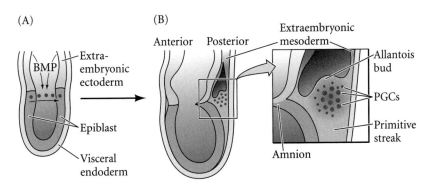

FIGURE 19.7 Specification and migration of mammalian primordial germ cells. (A) In the mouse embryo, BMP signals (blue) from the extraembryonic ectoderm induce neighboring epiblast cells (purple circles) to become precursors of PGCs and extraembryonic mesoderm. During gastrulation (arrow), these cells come to reside in the posterior epiblast. (B) On embryonic day 7, these cells emerge from the posterior primitive streak. These cells express *fragilis* (green), and in the center of this cluster are PGC precursors that also express *stella* and *blimp1* (dark green and red). (After Hogan 2002.)

events,* Blimp1 is a general transcriptional inhibitor that may be critical for specifying the PGCs. Cells that express

blimp1 are restricted to the germ cell fate (Saitou et al. 2002; Bortvin et al. 2004; Ohinata et al. 2005). The requirement for germ cell induction was shown by transplanting clumps of tissue from the distal portions of the epiblast to the proximal posterior portion of the epiblast. These cells then gave rise to PGCs (Tam and Zhou 1996). Such induction may be the more common route to germ cell formation than the segregation of an autonomously specified germ plasm (Extravour and Akam 2003).

---

*It is not known what the Stella protein does. Knocking out its gene does not affect the germ cells, so Stella probably is not involved in specifying cells to become PGCs. *Stella*-deficient mouse oocytes, however, can only develop through the first few cleavages (Bortvin et al. 2004). The molecular mechanisms by which the mammalian germ line is established remains to be discovered.

SIDELIGHTS & SPECULATIONS

Pluripotency, Germ Cells, and Embryonic Stem Cells

Primordial germ cells and embryonic stem cells are both characterized by their ability to generate any cell type in the embryo. Embryonic stem cells (ES cells) are derived from the inner cell masses of mammalian blastocysts and are believed to be the functional equivalent of the inner cell mass blastomeres (see Chapter 11). One of the best pieces of evidence for this is that when ES cells are injected into the ICMs of mouse blastocysts, they behave like mouse blastocyst cells and contribute cells to the embryo. One of the interesting species-specific differences between human and mouse ES cells is that human ES cells appear to contribute to the trophoblast, while mouse ES cells do not (Xu et al. 2002).

Although no germ plasm has been found in mammals, the retention of totipotency or pluripotency has been correlated with the expression of three nuclear transcription factors, Oct4, Stat3, and Nanog (see Chapter 11). Oct4 is a homeodomain transcription factor expressed in all early-cleavage blastomere nuclei, but its expression becomes restricted to the inner cell mass. During gastrulation, Oct4 becomes expressed solely in those posterior epiblast cells thought to give rise to the primordial germ cells. After that, Oct4 is seen only in the primordial germ cells, and later in oocytes (Figure 19.8; see also Figure 11.30). Oct4 is not seen in the developing sperm after the germ cells reach the testes and become committed to sperm production (Yeom et al. 1996; Pesce et al. 1998).

Nanog is is another homeodomain transcription factor found in the pluripotential cells of mouse blastocyst, as well as in embryonic stem cells and germline tumors. Nanog expression is highly expressed in PGCs of certain mouse embryos (Yamaguchi et al. 2005; Hatano et al. 2005). Knockout experiments indicate that Nanog is critical in maintaining the pluripotency of stem cells, and overexpression experiments demonstrate that elevated Nanog negates the need for Stat3 and by itself maintains Oct4 transcription in ES cells (Chambers et al. 2003; Mitsui et al. 2003). This pluripotency and transcription factor expression pattern is seen not only in the PGCs but in two derivatives of the PGCs—cultured embryonic germ cells, and tumorous germ cells called teratocarcinomas.

Embryonic germ (EG) cells

When PGCs are first placed into culture, they resemble ES cells. Stem cell factor increases the proliferation of migrating mouse primordial germ cells in culture, and this proliferation can be further increased by adding another growth factor, leukemia inhibition factor (LIF). However, the life span of these PGCs is short and the cells soon die. But if an additional mitotic regulator—basic fibroblast growth factor, Fgf2—is added, a remarkable change takes place. The cells continue to proliferate, producing pluripotent embryonic stem cells with characteristics resembling those of the inner cell mass (Matsui et al. 1992; Resnick et al. 1992; Rohwedel et al. 1996). These PGC-derived cells are called **embryonic germ (EG) cells**,

(Continued on next page)

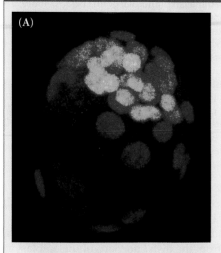

(A)

(B)

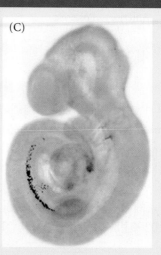

(C)

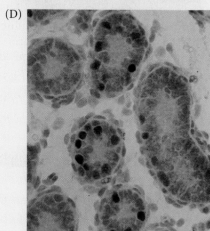

(D)

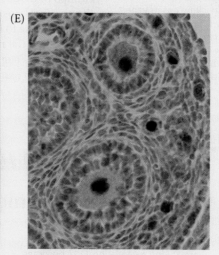

(E)

FIGURE 19.8 Expression of *Oct4* mRNA correlates with totipotency and ability to form germ cells. (A) Oct4 transcription factor is stained green with a fluorescent antibody, while all cell nuclei are stained red with propidium iodide. The overlap (indicated by the yellow color) shows that Oct4 is found only in the inner cell mass. (B, C) An *Oct4/lacZ* transgene driven by the *Oct4* promoter region shows its expression (dark color) (B) in the posterior epiblast of the 8.5-day mouse embryo and (C) in migrating PGCs in the 10.5-day embryo. (D, E) Labeled antibody (brown) staining shows Oct4 protein in the nuclei of (D) spermatogonia in postnatal testes and (E) oogonia in postnatal ovaries. (A–C after Yeom et al. 1996; D,E from Pesce et al. 1998; photographs courtesy of H. R. Schöler.)

and they have the potential to differentiate into all the cell types of the body.

In 1998, researchers in John Gerhart's laboratory cultured human EG cells (Shamblott et al. 1998). These cells were able to generate differentiated cells from all three primary germ layers, so they are presumably pluripotent. Such cells could be used medically to create neural or hematopoietic stem cells, which might be used to regenerate damaged neural or blood tissues (see Chapter 4). EG cells are often considered embryonic stem (ES) cells, and the distinction of their origin is ignored.

Embryonal carcinoma (EC) cells

What would happen if a PGC became malignant? In one type of tumor, the germ cells become embryonic stem cells, much like the Fgf2-treated PGCs in the experiment above. This type of tumor is called a **teratocarcinoma**. Whether spontaneous or exper-

imentally produced, teratocarcinomas contain an undifferentiated stem cell population that has biochemical and developmental properties remarkably similar to those of the inner cell mass (Graham 1977; see Parson 2004). Moreover, these stem cells not only divide, but can also differentiate into a wide variety of tissues, including gut and respiratory epithelia, muscle, nerve, cartilage, and bone (Figure 19.9). These undifferentiated pluripotential stem cells are called **embryonal carcinoma (EC) cells**. Once differentiated, these cells no longer divide, and are therefore no longer malignant. Such tumors can give rise to most of the tissue types in the body (Stevens and Little 1954; Kleinsmith and Pierce 1964; Kahan and Ephrussi 1970). Thus, the teratocarcinoma stem cells mimic early mammalian development, but the tumor they form is characterized by random, haphazard development.

In 1981, Stewart and Mintz formed a mouse from cells derived in part from a teratocarcinoma stem cell. Stem cells that had arisen in a teratocarcinoma of an agouti (yellow-tipped) strain of mice were cultured for several cell generations and were seen to maintain the characteristic chromosome complement of the parental mouse. Individual stem cells descended from the tumor were injected into the blastocysts of black-furred mice. The blastocysts were then transferred to the uterus of a foster mother, and live mice were born. Some of these mice had coats of two colors, indicating that the tumor cell had integrated itself into the embryo. This, in itself, is a remarkable demonstration that the tissue context is critical for the phenotype of a cell—a malignant cell was made nonmalignant.

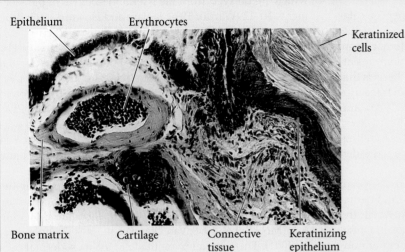

Epithelium Erythrocytes Keratinized cells

Bone matrix Cartilage Connective tissue Keratinizing epithelium

FIGURE 19.9 Photomicrograph of a section through a teratocarcinoma, showing numerous differentiated cell types. (From Gardner 1982; photograph by C. Graham, courtesy of R. L. Gardner.)

Thus, germ cell tumors can retain their pluripotency.

It is possible that ES cells, EG cells, and EC cells have a common origin in the presumptive PGC cells. Zwaka and Thomson (2005) hypothesize that the ES cells are actually the equivalent of PGCs and not of the inner cell mass. Not every inner cell mass blastomere can become an ES cell, and Zwaka and Thomson suggest that perhaps the successful stem cells are those that have been positioned next to trophoblast at the future posterior proximal region of the embryo. In other words, the blastomeres that become the ES cells might actually be the presumptive PGCs. While this idea remains hypothetical, it would relate these four pluripotent cell types.

But the story does not end here. When these chimeric mice were mated to mice carrying alleles recessive to those of the original tumor cell, the alleles of the tumor cell were expressed in many of the offspring. This means that the originally malignant tumor cell had produced many, if not all, types of normal somatic cells, and had even produced normal, functional germ cells! When such mice (being heterozygous for tumor cell genes) were mated with each other, the resulting litter contained mice that were homozygous for a large number of genes from the tumor cell (Figure 19.10).

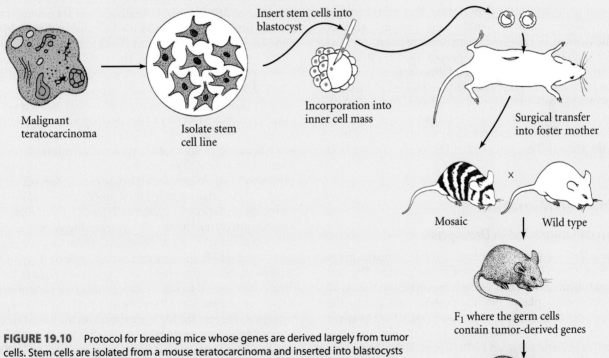

Insert stem cells into blastocyst

Malignant teratocarcinoma → Isolate stem cell line → Incorporation into inner cell mass → Surgical transfer into foster mother

Mosaic × Wild type

F_1 where the germ cells contain tumor-derived genes

"New strain" (F_2) formed when two F_1 mice are mated

FIGURE 19.10 Protocol for breeding mice whose genes are derived largely from tumor cells. Stem cells are isolated from a mouse teratocarcinoma and inserted into blastocysts from a different strain of mouse. The chimeric blastocysts are implanted in a foster mother. If the tumor cells are integrated into a blastocyst, the mouse that develops will have many of its cells derived from the tumor. If the tumor has given rise to germ cells, the chimeric mouse can be mated to normal mice to produce an F1 generation. The F1 mice should be heterozygous for all the chromosomes of the tumor cell. Matings between F1 mice should produce F2 mice homozygous for some genes derived from the tumor, and F2 mice should express many tumor cell genes. (After Stewart and Mintz 1981.)

The inert genome hypothesis

Not all the components of the germ plasm have been catalogued. Indeed, in birds and mammals, such a list has hardly even been started. Moreover, we still do not know the functions of the many of the proteins and nontranslated RNAs found in the germ plasm. One hypothesis is that the components of the germ plasm inhibit both transcription and translation, thereby preventing the cells containing it from differentiating into anything else (Nieuwkoop and Sutasurya 1981; Wylie 1999). According this hypothesis, the cells become germ cells because they are forbidden to become any other type of cell.

This suppression of transcription is seen in the germ cells of several species, including flies, frogs, and nematodes (see Figure 19.2). Many of the components in the germ plasm (such as Gcl, Pgc, Piwi, and Nanos in *Drosophila*, and PIE-1 in *C. elegans*) act by inhibiting either transcription or translation (Leatherman et al. 2002). Many of these proteins are found throughout the animal kingdom, although some appear to be relatively specific. For example, *Nanos3*, a homologue of *Drosophila Nanos*, is found in the PGCs of mice and encodes a protein that is critical for maintaining the PGCs. Nanos is also seen in the PGCs of zebrafish, *C. elegans*, *Xenopus*, and humans. Similarly, homologues of Vasa are found in all groups of animals studied as markers for the germ cells (Tsuda et al. 2003; Fabioux et al. 2004). It is interesting that when animal germ cells are separated from the somatic cells—whether in chicks, mice, or flies—the germ cells are often specified *outside* of the developing body proper. Perhaps this exile into an extraembryonic "enclave" insulates the primordial germ cells from paracrine signaling taking place within the somatic cells of the growing embryo (Dickson 1994). Once the repression of somatic gene expression is accomplished, the germ cells can return to the embryo and travel to the gonads. This germ cell migration will be our next topic.

Germ Cell Migration

Germ cell migration in *Drosophila*

During *Drosophila* embryogenesis, the primordial germ cells move from the posterior pole to the gonads. The first step in this migration is a passive one, wherein the 30–40 pole cells are displaced into the posterior midgut by the movements of gastrulation (Figure 19.11A,B). The germ cells are actively prevented from migrating during this stage (Jaglarz and Howard 1994; Li et al. 2003). In the second step, the gut endoderm triggers the germ cells to actively migrate by diapedesis (i.e., squeezing amoebically) through the blind end of the posterior midgut (Kunwar et al. 2003). The germ cells migrate from the endoderm into the visceral mesoderm. In the third step, the PGCs split into two groups, each of which will become associated with a developing gonad primordium.

In the fourth step, the germ cells migrate to the gonads, which are derived from the lateral mesoderm of parasegments 10–12 (Warrior 1994; Jaglarz and Howard 1995; Broihier et al. 1998). This step involves both attraction and repulsion. The products of the *wunen* genes appear to be responsible for directing the migration of the primordial germ cells from the endoderm into the mesoderm and their division into two streams (Figure 19.11C–E). This protein is expressed in the endoderm immediately before PGC migration and in many other tissues which the germ cells avoid, and it repels the germ cells. In loss-of-function mutants of this gene, the PGCs wander randomly (Zhang et al. 1997; Sano et al. 2005).

Two proteins appear to be critical for the attracting the *Drosophila* PGCs to the gonads. One is the HMG-CoA reductase, the product of the *columbus* gene, the other is Hedgehog (Moore et al. 1998; Van Doren et al. 1998; Deshpande et al. 2001). These proteins are made in the mesodermal cells of the gonads. In loss-of-function mutants of either gene, the PGCs wander randomly from the endoderm, and if either gene is expressed in other tissues (such as the nerve cord), those tissues will attract the PGCs. In the last step, the gonad coalesces around the germ cells, allowing the germ cells to divide and mature into gametes (Figure 19.11F). This step requires E-cadherin (Jenkins et al. 2003).

Neither the gonads nor the germ cells differentiate until metamorphosis. During the larval stages, both the PGCs and the somatic gonadal cells divide, but they remain relatively undifferentiated. At the larval-pupal transition, gonadal morphogenesis occurs (Godt and Laski 1995; King 1970). During this transition, those PGCs in the anterior region of the gonad become stem cells (Asaoka and Lin 2004), dividing asymmetrically to produce both another stem cell and a differentiated daughter cell called a cystoblast. The cystoblasts eventually develop into an egg chamber (King 1970; Zhu and Xie 2003; see Chapter 9). We are just beginning to understand how the germline stem cells retain their stem cell properties in the gonad (Gilboa and Lehmann 2004). As mentioned in Chapter 6, stem cells must be in a "niche" that supports their proliferation and inhibits their differentiation. Daughter cells that travel outside this niche begin differentiating. In ovaries, the germline stem cells are attached to the stromal cap, where they are maintained by the BMP4-like factor Decapentaplegic (Dpp). Dpp protein represses the gene encoding a transcription factor (Bag-of-marbles) that initiates oogenesis; Dpp cannot reach cells that leave the leave the stromal cap. Without Decapentaplegic to repress the *bag-of marbles gene*, the germline cell begins the developmental cascade that produces the 15 nurse cells and the single oocyte (Chen and McKearin 2003; Decotto and Spradling 2005).

The stem cells of the male germ line are connected to "hub" cells that create a stem cell microenvironment by secreting BMP signals as well as the Unpaired protein. Unpaired activates the JAK-STAT pathway in the germline stem cells (see Figure 6.14). If the JAK-STAT signaling pathway is disrupted, the germline stem cells differentiate into

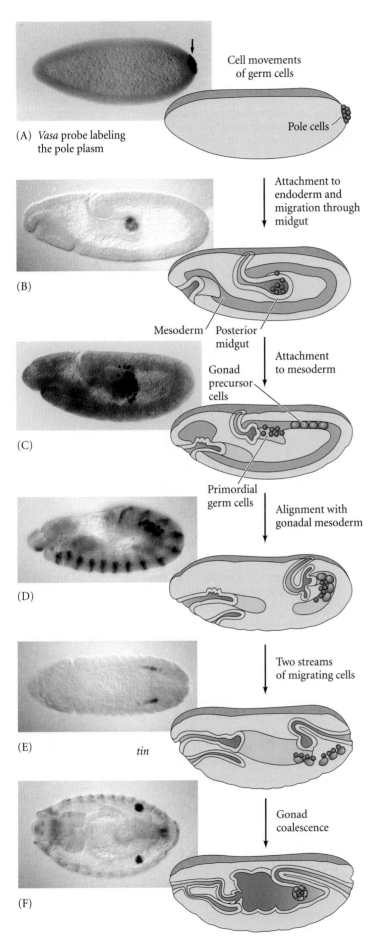

(A) *Vasa* probe labeling the pole plasm

Cell movements of germ cells

Pole cells

Attachment to endoderm and migration through midgut

(B)

Mesoderm   Posterior midgut

Attachment to mesoderm

Gonad precursor cells

(C)

Primordial germ cells

Alignment with gonadal mesoderm

(D)

Two streams of migrating cells

(E)   *tin*

Gonad coalescence

(F)

spermatogonia without any self-renewal (Kiger et al. 2001; Tulina and Matunis 2001). Thus the male and female germline stem cells are in similar niches. When these cells move out of their niche, they divide to form the gametes and (in the case of females) the nurse cells as well.

Germ cell migration in vertebrates

ZEBRAFISH   Whereas *Drosophila* PGC migration is motivated by chemorepulsion of the germ cell precursors, zebrafish PGCs arrive at the gonads via chemoattraction. Using the *Vasa* message as a marker, Weidinger and colleagues (1999) detailed the migration of the four clusters of zebrafish PGCs (Figure 19.12). These PGC clusters followed different routes, but by the end of the first day of development (at the 1-somite stage), the PGCs are found in two discrete clusters along the border of the trunk mesoderm. From there, they migrate posteriorly into the developing gonad. In zebrafish, the primordial germ cells follow a gradient of the Sdf1 protein that is secreted by the developing gonad. The receptor for this protein is the CXCR4 protein on the PGC surface (Doitsidou et al. 2002; Knaut et al. 2003). This Sdf1/CXCR4 chemotactic guidance system is known to be important in the migration of lymphocytes and hematopoietic progenitor cells. Loss of either CXCR4 from the PGCs or Sdf1 from the somatic cells results in random migration of the zebrafish primordial germ cells.

FROGS   The germ plasm of anuran amphibians (frogs and toads) collects around the vegetal pole in the zygote (see Figure 19.4). During cleavage, this material is brought upward through the yolky cytoplasm. Periodic contractions of the vegetal cell surface appear to push it along the cleavage furrows of the

FIGURE 19.11   Migration of germ cells in the *Drosophila* embryo. The left column shows the germ plasm as stained by antibodies to Vasa, a protein component of the germ plasm (D has been counterstained with antibodies to Engrailed protein to show the segmentation, and E and F are dorsal views.) The right column diagrams the movements of the germ cells. (A) The germ cells originate from the pole plasm at the posterior end of the egg. (B) Passive movements carry the PGCs into the posterior midgut. (C) The PGCs move through the endoderm and into the caudal visceral mesoderm by diapedesis. The *wunen* (*wun*) gene product expressed in the endoderm expels the PGCs, while the product of the *columbus* (*clb*) gene expressed in the caudal mesoderm attracts them. (D–F) The movements of the mesoderm bring the PGCs into the region of the tenth through twelfth segments, where the mesoderm coalesces around them to form the gonads. (Photographs from Warrior et al. 1994, courtesy of R. Warrior; diagrams after Howard 1998.)

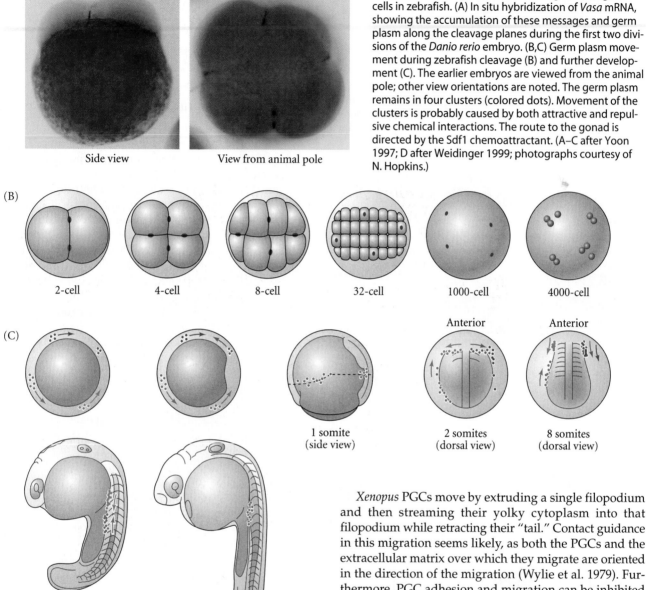

(A)

Side view   View from animal pole

FIGURE 19.12 Specification and migration of germ cells in zebrafish. (A) In situ hybridization of *Vasa* mRNA, showing the accumulation of these messages and germ plasm along the cleavage planes during the first two divisions of the *Danio rerio* embryo. (B,C) Germ plasm movement during zebrafish cleavage (B) and further development (C). The earlier embryos are viewed from the animal pole; other view orientations are noted. The germ plasm remains in four clusters (colored dots). Movement of the clusters is probably caused by both attractive and repulsive chemical interactions. The route to the gonad is directed by the Sdf1 chemoattractant. (A–C after Yoon 1997; D after Weidinger 1999; photographs courtesy of N. Hopkins.)

(B)

2-cell   4-cell   8-cell   32-cell   1000-cell   4000-cell

(C)

1 somite
(side view)

Anterior
2 somites
(dorsal view)

Anterior
8 somites
(dorsal view)

19 somites
(side view)

1 day
(side view)

Xenopus PGCs move by extruding a single filopodium and then streaming their yolky cytoplasm into that filopodium while retracting their "tail." Contact guidance in this migration seems likely, as both the PGCs and the extracellular matrix over which they migrate are oriented in the direction of the migration (Wylie et al. 1979). Furthermore, PGC adhesion and migration can be inhibited if the mesentery is treated with antibodies against *Xenopus* fibronectin (Heasman et al. 1981). Thus, the pathway for germ cell migration in these frogs appears to be composed of an oriented fibronectin-containing extracellular matrix. The fibrils over which the PGCs travel lose this polarity soon after migration has ended. As they migrate, *Xenopus* PGCs divide about three times, so that approximately 30 PGCs will colonize the gonads (Whitington and Dixon 1975; Wylie and Heasman 1993). These cells will divide to form the germ cells. The mechanism by which the *Xenopus* PGCs are directed to the gonad remains unknown, but Nishiumi and colleagues (2005) have identified a CXCR4-like molecule on the surface of amphibian PGCs after gastrulation.

MAMMALS Based on differential staining of fixed tissue, it had long been thought that the mouse germ cell precursors

newly formed blastomeres. Germ plasm eventually becomes associated with the endodermal cells lining the floor of the blastocoel (Figure 19.13; Bounoure 1934; Ressom and Dixon 1988; Kloc et al. 1993). The PGCs become concentrated in the posterior region of the larval gut, and as the abdominal cavity forms, they migrate along the dorsal side of the gut, first along the dorsal mesentery (which connects the gut to the region where the mesodermal organs are forming; see Figure 19.13E) and then along the abdominal wall and into the genital ridges. They migrate up this tissue until they reach the developing gonads.

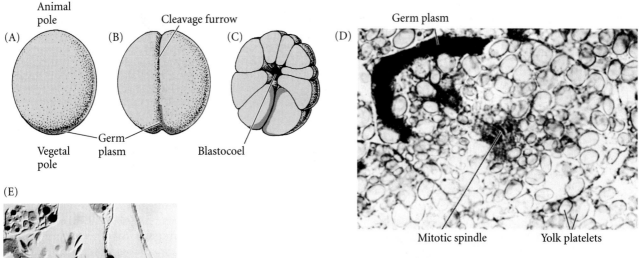

FIGURE 19.13 Migration of *Xenopus* germ plasm. (A–C) Changes in the position of the germ plasm (color) in an early frog embryo. Originally located near the vegetal pole of the uncleaved egg (A), the germ plasm advances along the cleavage furrows (B) until it becomes localized at the floor of the blastocoel (C). (D) A germ plasm-containing cell in the endodermal region of a blastula in mitotic anaphase. Note the germ plasm entering into only one of the two yolk-laden daughter cells. (E) Migration of two primordial germ cells (arrows) along the dorsal mesentery connecting the gut region to the gonadal mesoderm. (A–C after Bounoure 1934; D courtesy of A. Blackler; E from Heasman et al. 1977, courtesy of the authors.)

migrated from the epiblast into the extraembryonic mesoderm and then back again into the embryo by way of the allantois (see Chiquoine 1954; Mintz 1957). However, the ability to label mouse primordial germ cells with green fluorescent protein and to watch these living cells migrate has caused a reevaluation of the germ cell migration pathway in mammals (Anderson et al. 2000; Molyneaux et al. 2001; Tanaka et al. 2005). First, it appears that mammalian PGCs migrate directly into the endoderm from the posterior region of the primitive streak. (The cells that enter the allantois are believed to die.) These *stella*-expressing cells find themselves in the hindgut (Figure 19.14A). Although they move actively, they cannot get out of the gut until about embryonic day 9. At that time, the PGCs exit the gut but do not yet migrate toward the genital ridges. By the following day, however, PGCs are seen migrating into the genital ridges (Figure 19.14B,C). By embryonic day 11.5, the PGCs enter the developing gonads. During this trek, they have proliferated from an initial population of 10–100 cells to the 2500–5000 PGCs present in the gonads by day 12.

Like the PGCs of *Xenopus*, mammalian PGCs appear to be closely associated with the cells over which they migrate, and they move by extending filopodia over the underlying cell surfaces. Mammalian PGCs are also capable of penetrating cell monolayers and migrating through cell sheets (Stott and Wylie 1986). The mechanism by which these cells know the route of their journey is still unknown. Fibronectin is likely to be an important substrate for PGC

migration (ffrench-Constant et al. 1991), and germ cells lacking the integrin receptor for such extracellular matrix proteins cannot migrate to the gonads (Anderson et al. 1999). Directionality during the latter stages of migration may be provided by a gradient of soluble protein. In vitro evidence suggests that the genital ridges of 10.5-day mouse embryos secrete the paracrine factor Sdf1 to get from the mesentery into the gonads (Ara et al. 2003; Molyneaux et al. 2003; Stebler et al. 2004). As in zebrafish, Sdf1 binds to the receptor CXCR4 on the PGCs.

The proliferation of the PGCs as they migrate appears to be promoted by the paracrine factors Fgf7 and stem cell factor (SCF, the same growth factor needed for the proliferation of neural crest-derived melanoblasts, hematopoietic stem cells, and EG cells; see Chapter 6). Adding these paracrine factors causes PGC proliferation, whereas blocking either factor causes apoptotic death of the PGCs (Godin et al. 1991; Pesce et al. 1993; Takeuchi et al. 2005). SCF is produced by the cells lining the migration pathway and remains bound to their cell membranes. It appears that the presentation of SCF on cell membranes is important for this factor's activity. Mice homozygous for mutations in the genes for either SCF or its receptor (c-Kit) are deficient in germ cells (as well as in melanocytes and blood cells; see Dolci et al. 1991; Matsui et al. 1991).

BIRDS AND REPTILES   In birds and reptiles, the primordial germ cells are derived from epiblast cells that migrate from

(A) Migration of
PGCs to endoderm

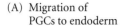
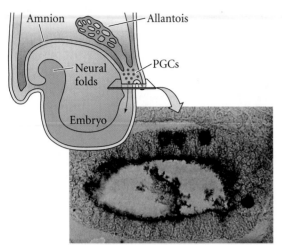

FIGURE 19.14 Primoridal germ cell migration in the mouse. (A) On day 8, the PGCs established in the posterior epiblast (see Figure 19.7) migrate into the definitive endoderm of the embryo. The photo shows four large PGCs (stained for alkaline phosphatase) in the hindgut of a mouse embryo. (B) The PGCs migrate through the gut and, dorsally, into the genital ridges. (C) Alkaline phosphatase-staining cells are seen entering the genital ridges around embryonic day 11. (A photograph from Heath 1978; C from Mintz 1957, photograph courtesy of the authors.)

(B) Migration of PGCs into gonad   (C)

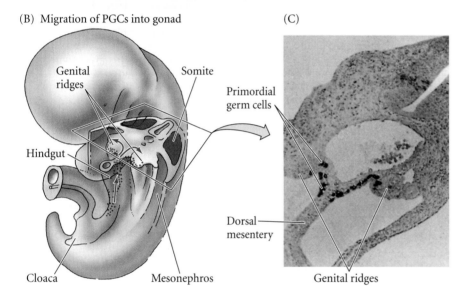

vessels. In some as-yet undiscovered way, the PGCs are instructed to exit the blood vessels and enter the gonads (Pasteels 1953; Dubois 1969). Evidence for chemotaxis comes from studies in which circulating chick PGCs were isolated from the blood and cultured between gonadal rudiments and other embryonic tissues (Kuwana et al. 1986). During a 3-hour incubation, the PGCs migrated specifically into the gonadal rudiments.

The molecules that chick PGCs use for chemotaxis may be the same Sdf1/CXCR4 chemotactic system seen in zebrafish and mammals. Like mammals, chicks only use chemotaxis during the latter stages of migration. Thus, after they leave the blood vessels, chick PGCs appear to utilize Sdf1 gradients to reach the gonads (Stebler et al. 2004). Indeed, if Sdf1-secreting cells are transplanted into late-stage chick embryos, the PGCs will be attracted to them.

the central region of the area pellucida to a crescent-shaped zone in the hypoblast at the anterior border of the area pellucida (Figure 19.15; Eyal-Giladi et al. 1981; Ginsburg and Eyal-Giladi 1987). This extraembryonic region is called the **germinal crescent**, and the PGCs multiply there.

Unlike those of amphibians and mammals, the PGCs of birds and reptiles migrate to the gonads primarily by means of the bloodstream (Figure 19.16). When blood vessels form in the germinal crescent, the PGCs enter those vessels and are carried by the circulation to the region where the hindgut is forming. Here they leave the circulation, become associated with the mesentery, and migrate into the genital ridges (Swift 1914; Mayer 1964; Kuwana 1993; Tsunekawa et al. 2000).

The PGCs of the germinal crescent appear to enter the blood vessels by diapedesis, a type of amoeboid movement common to lymphocytes and macrophages that enables cells to squeeze between the endothelial cells of small blood

Meiosis

Once in the gonad, the primordial germ cells continue to divide mitotically, producing millions of potential gamete precursors. The germ cells of both male and female gonads are then faced with the necessity of reducing their chromosomes from the diploid to the haploid condition. In the haploid condition, each chromosome is represented by only one copy, whereas diploid cells have two copies of each chromosome. To accomplish this reduction, the germ cells undergo **meiosis** (see Figure 2.5). Meiosis differs from mitosis in that (1) meiotic cells undergo two cell divisions without an intervening period of DNA replication, and (2) homologous chromosomes (each consisting of two sister chromatids joined at a kinetochore) pair together and recombine genetic material. Meiosis is therefore at the center of sexual reproduction. Villeneuve and Hillers (2001) conclude that "the very essence of sex is meiotic recombi-

(A)

Germinal
crescent

(B)

Germ
cells

FIGURE 19.15 The germinal crescent of the chick embryo. (A) Germ cells of a stage 4 (definitive primitive streak-stage, roughly 18 h) chick embryo, stained purple for the chick Vasa homologue. The stained cells are confined to the germinal crescent. (B) Higher magnification of the stage 4 germinal crescent region, showing germ cells (stained brown) in the thickened epiblast. (From Tsunekawa et al. 2000; photographs courtesy of N. Tsunekawa.)

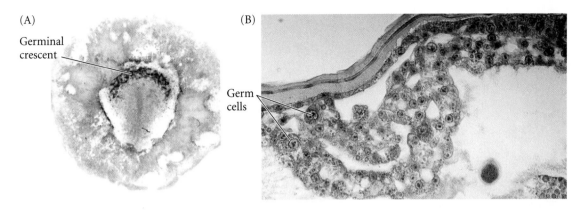

FIGURE 19.16 Migration of primordial germ cells in the chick embryo. (A) Scanning electron micrograph of a chick PGC in a capillary of a gastrulating embryo. Note the larger size of the PGC, as well as the microvilli on its surface. (B) Diagram of transverse section near the prospective gonadal region of a chick embryo. Several PGCs within a blood vessel cluster next to the gonadal epithelium. One PGC is crossing through the blood vessel endothelium, and another PGC is already located adjacent to the gonadal epithelium. (C) Having passed through the endothelium of the dorsal aorta, chick germ cells (arrowheads) migrate toward the genital ridges of the embryo. (A from Kuwana 1993, courtesy of T. Kuwana; B after Romanoff 1960; C from Tsunekawa et al. 2000, courtesy of N. Tsunekawa.)

(A)

Primordial
germ cell
(PGC)

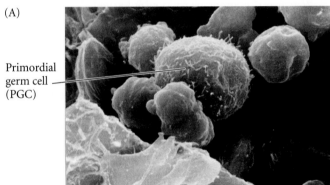

(B)

Blood vessel

Blood cells

Primordial
germ cells

Gonadal
epithelium
(genital ridge)

Mesenchyme

(C)

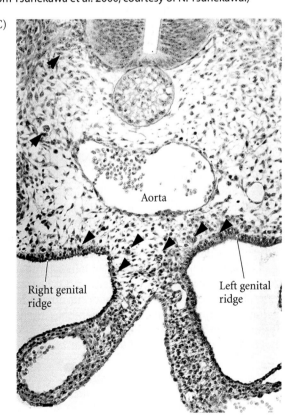

Aorta

Right genital
ridge

Left genital
ridge

(A)

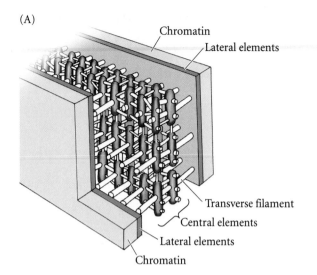

(B)

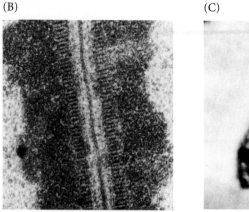

(C)

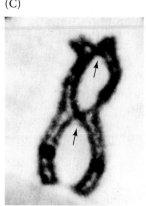

FIGURE 19.17 The synaptonemal complex and recombination. (A) Interpretive diagram of the ladderlike synaptonemal complex structure. (B) Homologous chromosomes held together in a synaptonemal complex during the zyogotene phase of the first meiotic prophase in a *Neottiella* (mushroom) oocyte. (C) Chiasmata in diplotene bivalent chromosomes of salamander oocytes. Kinetochores are visible as darkly stained circles; the arrows point to the two chiasmata. (A after Schmekel and Daneholt 1995; B from von Wettstein 1971, courtesy of D. von Wettstein; C, photograph courtesy of J. Kezer)

nation." Yet for all its centrality in genetics, development, and evolution, we know surprisingly little about meiosis.

After the germ cell's last mitotic division, a period of DNA synthesis occurs, so that the cell initiating meiosis doubles the amount of DNA in its nucleus. In this state, each chromosome consists of two sister **chromatids** attached at a common kinetochore.* (In other words, the diploid nucleus contains four copies of each chromosome.) Meiosis entails two cell divisions. In the first division, homologous chromosomes (for example, the two copies of chromosome 3 in the diploid cell) come together and are then separated into different cells. Hence, the first meiotic division *splits two homologous chromosomes* between two daughter cells such that each daughter cell has only one copy of the chromosome. But each of the chromosomes has already replicated (i.e., each has two chromatids). The second meiotic division *separates the two sister chromatids from each other.* Consequently, each of the four cells produced by meiosis has a single (haploid) copy of each chromosome.

The first meiotic division begins with a long prophase, which is subdivided into five stages. During the **leptotene** (Greek, "thin thread") stage, the chromatin of the chromatids is stretched out very thinly, and it is not possible to identify individual chromosomes. DNA replication has already occurred, however, and each chromosome consists of two parallel chromatids. At the **zygotene** (Greek, "yoked threads") stage, homologous chromosomes pair side by side. This pairing is called **synapsis**, and it is characteristic of meiosis; such pairing does not occur during mitotic divisions. Although the mechanism whereby each chromosome recognizes its homologue is not known, synapsis seems to require the presence of the nuclear membrane and

the formation of a proteinaceous ribbon called the **synaptonemal complex**. This complex is a ladder-like structure with a central element and two lateral bars (von Wettstein 1984; Schmekel and Daneholt 1995). The chromatin becomes associated with the two lateral bars, and the chromosomes are thus joined together (Figure 19.17A,B).

Examinations of meiotic cell nuclei with the electron microscope (Moses 1968; Moens 1969) suggest that paired chromosomes are bound to the nuclear membrane, and Comings (1968) has suggested that the nuclear envelope helps bring together the homologous chromosomes. The configuration formed by the four chromatids and the synaptonemal complex is referred to as a **tetrad** or a **bivalent**.

During the next stage of meiotic prophase, the chromatids thicken and shorten. This stage has therefore been called the **pachytene** (Greek, "thick thread") stage. Individual chromatids can now be distinguished under the light microscope, and crossing over may occur. **Crossing over** represents an exchange of genetic material whereby genes from one chromatid are exchanged with homologous genes from another. Crossing over may continue into the next stage, the **diplotene** (Greek, "double threads") stage. Here, the synaptonemal complex breaks down, and the two homologous chromosomes start to separate. Usually, however, they remain attached at various points called **chiasmata**, which are thought to represent regions where crossing over is occurring (Figure 19.17C). The diplotene stage is characterized by a high level of gene transcription. In some species, the chromosomes of both male and female germ

---

*Although the terms "centromere" and "kinetochore" are often used interchangeably, the *kinetochore* is the complex protein structure that assembles on a sequence of DNA known as the *centromere*.

cells take on the "lampbrush" appearance characteristic of chromosomes that are actively making RNA (see below).

During the next stage, **diakinesis** (Greek, "moving apart"), the kinetochores move away from each other, and the chromosomes remain joined only at the tips of the chromatids. This last stage of meiotic prophase ends with the breakdown of the nuclear membrane and the migration of the chromosomes to the **metaphase plate**. Anaphase of meiosis I doesn't commence until the chromosomes are properly aligned on the mitotic spindle fibers. This alignment is accomplished by proteins that prevent cyclin B from degrading until after all the chromosomes are securely fastened to microtubules. If these proteins are deficient, aneuploidies such as Down syndrome can occur (Homer et al. 2005; Steuerwald et al. 2005).

During anaphase I, the homologous chromosomes are separated from each other in an independent fashion. This stage leads to telophase I, during which two daughter cells are formed, each cell containing one partner of each homologous chromosome pair. After a brief **interkinesis**, the second division of meiosis takes place. During this division, the kinetochore of each chromosome divides during anaphase so that each of the new cells gets one of the two chromatids, the final result being the creation of four haploid cells. Note that meiosis has also reassorted the chromosomes into new groupings. First, each of the four haploid cells has a different assortment of chromosomes.

Humans have 23 different chromosome pairs; thus 2^{23} (nearly 10 million) different haploid cells can be formed from the genome of a single person. In addition, the crossing-over that occurs during the pachytene and diplotene stages of prophase I further increases genetic diversity and makes the number of potential different gametes incalculably large.

If meiosis is the core of sexual reproduction, and sexual reproduction predominates among the eukaryotic kingdoms, then meiosis must be one of the central unifying themes of life. The study of meiosis is being facilitated by the observation that nearly all the genes and proteins used in yeast meiosis also function in mammalian meiosis. This observation has allowed the identification of a "core meiotic recombination complex" used by plants, fungi, and animals. This meiotic recombination complex is built on the backbone of the **cohesin complex**, the protein glue that keeps sister chromatids *together* during *mitotic* prophase and metaphase. In *meiotic* cells, this complex recruits another set of proteins that help promote pairing between the homologous chromosomes and allow recombination to occur (Pelttari et al. 2001; Villeneuve and Hillers 2001). These recombination-inducing proteins are involved in making and repairing double-stranded DNA breaks.

Although the relationship between the synaptonemal complex and the cohesin-recruited recombination complex is not clear, it appears in mammals that the synaptonemal complex stabilizes the associations initiated by the recombination complex, giving a morphological scaffolding to the tenuous protein connections (Pelttari et al. 2001). If the synaptonemal complex fails to form, the germ cells arrest at the pachytene stage and their chromosomes fragment (Figure 19.18). If the murine synaptonemal complex forms but lacks certain proteins, chiasmata formation fails, and the germ cells are often aneuploid (having multiple copies of one or more chromosomes) (Tay and Richter 2001; Yuan et al. 2002).

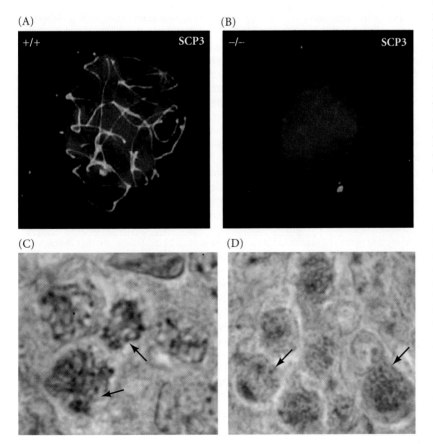

FIGURE 19.18 Importance of the synaptonemal complex. The CPEB protein is a constituent of the synaptonemal complex, and when this gene is knocked out in mice, the synaptonemal complex fails to form. (A,B) Staining for synaptonemal complex proteins in wild-type (A) and CPEB-deficient (B) mice. The synaptonemal complex is stained green, the DNA is stained blue. The synaptonemal complex is absent in the mutant nuclei, and their DNA is not organized into discrete chromosomes. (C, D) In preparations stained with hematoxylin-eosin, the DNA can be seen to be in the pachytene stage in the wild-type cells (C), and fragmented in the mutant cells (D). (After Tay and Richter 2001; photographs courtesy of J. D. Richter.)

The events of meiosis appear to be coordinated through cytoplasmic connections between the dividing cells. Whereas the daughter cells formed by mitosis routinely separate from each other, the products of the meiotic cell divisions remain coupled to each other by **cytoplasmic bridges**. These bridges are seen during the formation of sperm and eggs throughout the animal kingdom (Pepling and Spradling 1998).

SIDELIGHTS & SPECULATIONS

Big Decisions: Mitosis or Meiosis? Sperm or Egg?

In many species, the germ cells migrating into the gonad are bipotential and can differentiate into either sperm or eggs, depending on their gonadal environment. When the ovaries of salamanders are experimentally transformed into testes, the resident germ cells cease their oogenic differentiation and begin developing as sperm (Burns 1930; Humphrey 1931). Similarly, in the housefly and mouse, the gonad is able to direct the differentiation of the germ cells (McLaren 1983; Inoue and Hiroyoshi 1986). Thus, in most organisms, the sex of the gonad and that of its germ cells is the same.

But what about hermaphroditic animals, in which the change from sperm production to egg production is a naturally occurring physiological event? How is the same animal capable of producing sperm during one part of its life and oocytes during another part? Using *Caenorhabditis elegans*, Kimble and her colleagues identified two "decisions" that presumptive germ cells have to make. The first is whether to enter meiosis or to remain a mitotically dividing stem cell. The second is whether to become an egg or a sperm.

There is evidence that these decisions are intimately linked. The mitotic/meiotic decision in *C. elegans* is controlled by a single nondividing cell—the **distal tip cell**—located at the end of each gonad. The germ cell precursors near this cell divide mitotically, forming the pool of germ cells; but as these cells get farther away from the distal tip cell, they enter meiosis (Figure 19.19A). If the distal tip cell is destroyed by a focused laser beam, all the germ cells enter meiosis (Figure 19.19B); and if the distal tip cell is placed in a different location in the gonad, germline stem cells are generated near its

new position (Kimble 1981; Kimble and White 1981). The distal tip cell extends long filaments that touch the distal germ cells. The extensions contain in their cell membranes the LAG-2 protein, a *C. elegans* homologue of Delta (Henderson et al. 1994; Tax et al. 1994; Hall et al. 1999). LAG-2 maintains the germ cells in mitosis and inhibits their meiotic differentiation.

Austin and Kimble (1987) isolated a mutation that mimics the phenotype obtained when the distal tip cell is removed. It is not surprising that this mutation involves the gene encoding GLP-1, the *C. elegans* homologue of Notch—the receptor for Delta. All the germ cell precursors of nematodes homozygous for the recessive mutation of *glp-1* initiate meiosis, leaving no mitotic population (see Figure 19.18B). Instead of the 1500 germ cells usually found in the fourth larval stage of hermaphroditic development, these mutants produce only 5–8 sperm cells. When genetic chimeras are made in which wild-type germ cell precursors are found within a mutant larva, the wild-type cells are able to respond to the distal tip cells and undergo mitosis. However, when mutant germ cell precursors are found within wild-type larvae, they all enter meiosis. Thus, the *GLP-1* gene appears to be responsible for enabling the germ cells to respond to the distal tip cell's signal.*

---

*The *glp-1* gene appears to be involved in a number of inductive interactions in *C. elegans*. You may recall that GLP-1 is also needed by the AB blastomere for it to receive inductive signals from the EMS blastomere to form pharyngeal muscles (see Chapter 8).

As is usual in development, the binary decision entails both a push and a pull. The decision to enter meiosis must be amplified by a decision to end mitosis. This appears to be accomplished by the FBF proteins (which are similar to the *Drosophila* Pumilio RNA-binding proteins mentioned in Chapter 9). Notch appears to activate FBFs, which are translational repressors of the GLD (germline development) proteins. GLD-1 (in combination with a Nanos protein) suppresses the translation of mitosis-specific messages. This includes suppressing the translation of *GLP-1* mRNA (Marin and Evans 2003; Eckmann et al. 2002, 2004). FBF also represses the translation of GLD-2 and GLD-3, two proteins necessary for polyadenylating meiosis-specific mRNAs, allowing them to be translated. Thus, the Notch signal, acting through FBF, simultaneously promotes mitosis and blocks meiosis (Figure 19.19C).

After the germ cells begin their meiotic divisions, they still must become either sperm or ova. Generally, in each hermaphrodite gonad (called an ovotestis), the most proximal germ cells produce sperm, while the most distal (near the tip) become eggs (Hirsh et al. 1976). This means that the germ cells entering meiosis early become sperm, and those entering meiosis later become eggs. The genetics of this switch are currently being analyzed. The laboratories of Hodgkin (1985) and Kimble (Kimble et al. 1986) have isolated several genes needed for germ cell pathway selection, but the switch appears to involve the activity or inactivity of *FEM-3* mRNA. Figure 19.20 presents a scheme for how these genes might function.

During early development, the FEM genes, especially *FEM-3*, are critical for the

SIDELIGHTS & SPECULATIONS

(A) Intact gonad

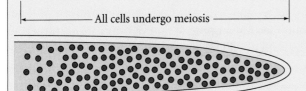

(B) Distal tip cell removed or *glp-1* mutation

All cells undergo meiosis

Distal tip cell

FIGURE 19.19 Regulation of the mitosis-or-meiosis decision by the distal tip cell of the *C. elegans* ovotestis. (A) Intact gonad early in development with regions of mitosis (lighter-colored cells) and meiosis. The membrane of the distal tip cell's extensions contains the *C. elegans* homologue of the Delta protein, while the PGCs contain the *C. elegans* homologue of Notch. (B) Gonad after laser ablation of the distal tip cell or mutation of the *glp-1* gene. All germ cells enter meiosis. (C) Hypothetical pathway whereby Notch, acting through translational inhibitor FBF, stimulates mitosis (by inhibiting an inhibitor) and simultaneously blocks meiosis (by inhibiting an activator). (C after Eckmann et al. 2004.)

(C)

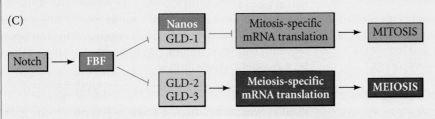

specification of sperm cells. Loss-of-function mutations of these genes convert XX *C. elegans* into females (i.e., spermless hermaphrodites). As long as the FEM proteins are made in the germ cells, sperm are produced. FEM protein is thought to activate the *FOG* genes (whose loss-of-function mutations cause the feminization of the germ line and eliminate spermatogenesis). The *FOG* gene products activate the genes involved in transforming the germ cell into sperm and also inhibit those genes that would otherwise direct the germ cells to initiate oogenesis.

Oogenesis can begin only when fem activity is suppressed. This suppression appears to act at the level of RNA translation. The 3′ untranslated region of *FEM-3* mRNA contains a sequence that binds a repressor protein during normal development. If this region is mutated such that the repressor cannot bind, the *FEM-3* mRNA remains translatable and oogenesis never takes place. The result is a hermaphrodite body that produces only sperm (Ahringer and Kimble 1991; Ahringer et al. 1992). The *trans*-acting repressor of the *FEM-3* message is a combination FBF with the Nanos and

Pumilio proteins (the same combination that represses *hunchback* message translation in *Drosophila*).

As in the meiosis/mitosis decision, there are pushes and pulls in the sex-determining pathways. The same Nanos and FBF signal that inhibits the sperm-producing fem-3 message also inhibits the oocyte-inhibiting gld-1 message. Thus, the Nanos/FBF signal simultaneously blocks sperm production (by inhibiting an activator) while promoting oocyte production (by inhibiting the inhibitor). The use of the same proteins in both the mitosis/meiosis decision and the sperm/egg decision has allowed Eckmann and colleagues (2004) to speculate that these two pathways evolved from a single original pathway whose function was to regulate the balance between cell growth and cell differentiation.

WEBSITE 19.5 Germ line sex determination in *C. elegans*. The establishment of whether a germ cell is to become a sperm or an egg involves multiple levels of inhibition. Translational regulation is seen in several of these steps.

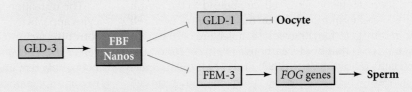

FIGURE 19.20 Model of sex determination switch in the germ line of *C. elegans* hermaphrodites. FBF and Nanos simultaneously promote oogenesis (by blocking an inhibitor) and inhibit spermatogenesis (by blocking an activator). Expression of the *FBF* and *Nanos* genes appears to be regulated by a GLD-3 protein. For as long as GLD-3 is made, sperm are produced. After GLD-3 production stops, the germ cells become oocytes. (After Eckmann et al. 2004.)

TABLE 19.1 Sexual dimorphism in mammalian meioses

Female oogenesis	Male spermatogenesis
Meiosis initiated once in a finite population of cells	Meiosis initiated continuously in a mitotically dividing stem cell population
One gamete produced per meiosis	Four gametes produced per meiosis
Completion of meiosis delayed for months or years	Meiosis completed in days or weeks
Meiosis arrested at first meiotic prophase and reinitiated in a smaller population of cells	Meiosis and differentiation proceed continuously without cell cycle arrest
Differentiation of gamete occurs while diploid, in first meiotic prophase	Differentiation of gamete occurs while haploid, after meiosis ends
All chromosomes exhibit equivalent transcription and recombination during meiotic prophase	Sex chromosomes excluded from recombination and transcription during first meiotic prophase

Source: Handel and Eppig 1998.

Spermatogenesis in Mammals

While the reductive divisions of meiosis are conserved in every eukaryotic kingdom of life, the regulation of meiosis can differ dramatically between males and females. The differences between **oogenesis**, the production of eggs, and **spermatogenesis**, the production of sperm, are outlined in Table 19.1. The processes by which the PGCs generate sperm and ova have been studied in several organisms; we will focus first on spermatogenesis in mammals.

Once mammalian PGCs arrive at the genital ridge of a male embryo, they become incorporated into the sex cords. They remain there until maturity, at which time the sex cords hollow out to form the **seminiferous tubules**, and the epithelium of the tubules differentiates into the **Sertoli cells** that will nourish and protect the developing sperm cells. The initiation of spermatogenesis during puberty is probably regulated by the synthesis of BMP8b by the spermatogenic germ cells, the **spermatogonia**. When BMP8b reaches a critical concentration, the germ cells begin to differentiate. The differentiating cells produce high levels of BMP8b, which can then further stimulate their differentiation. Mice lacking BMP8b do not initiate spermatogenesis at puberty (Zhao et al. 1996).

The spermatogenic germ cells are bound to the Sertoli cells by N-cadherin molecules on the surfaces of both cell types, and by galactosyltransferase molecules on the spermatogenic cells that bind a carbohydrate receptor on the Sertoli cells (Newton et al. 1993; Pratt et al. 1993). Spermatogenesis—the developmental pathway from germ cell to mature sperm—occurs in the recesses between the Sertoli cells. (Figure 19.21).

Forming the haploid spermatid

After reaching the testis, the PGCs divide to form **type A_1 spermatogonia**. These cells are smaller than the PGCs and are characterized by an ovoid nucleus that contains chromatin associated with the nuclear membrane. Type A_1 spermatogonia are found adjacent to the outer basement membrane of the sex cords. They are stem cells, and upon reaching maturity are thought to divide to make another type A_1 spermatogonium as well as a second, paler type of cell, the type A_2 spermatogonium. Thus, each type A_1 spermatogonium is a stem cell capable of regenerating itself as well as producing a new cell type. The A_2 spermatogonia divide to produce type A_3 spermatogonia, which then beget type A_4 spermatogonia. It is possible that each of the type A spermatogonia are stem cells, capable of self-renewal.

An A_4 spermatogonium has three options: it can form another A_4 spermatogonium (self-renewal); it can undergo cell death (apoptosis); or it can differentiate into the first committed stem cell type, the **intermediate spermatogonium**. Intermediate spermatogonia are committed to becoming spermatozoa, and they divide mitotically once to form **type B spermatogonia** (see Figure 19.21). These cells are the precursors of the spermatocytes and are the last cells of the line that undergo mitosis. They divide once to generate the **primary spermatocytes**—the cells that enter meiosis.

The transition between spermatogonia and spermatocytes appears to be mediated by glial cell line-derived neurotrophic factor (GDNF), which is secreted by the Sertoli cells. GDNF levels determine whether the dividing spermatogonia remain spermatogonia or enter the pathway to become spematocytes. Low levels of GDNF favor the differentiation of the spermatogonia, while high levels favor self-renewal of the stem cells (Meng et al. 2000). Since GDNF is upregulated by follicle-stimulating hormone (FSH), GDNF may serve as a link between the Sertoli cells and the endocrine system, and it provides a mechanism for FSH to instruct the testes to produce more sperm (Tadokoro et al. 2002). Keeping the stem cells in equilibri-

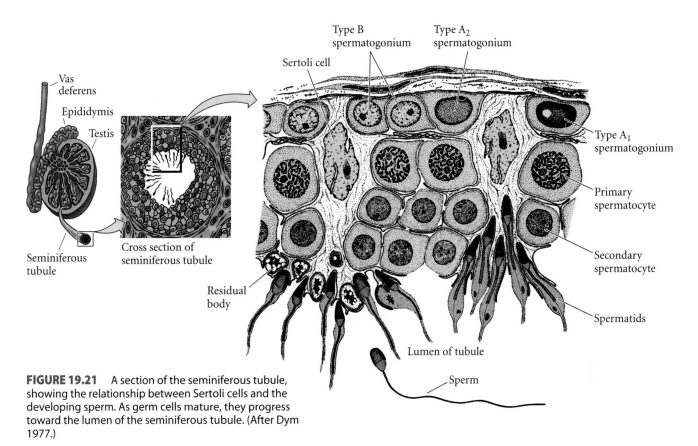

FIGURE 19.21 A section of the seminiferous tubule, showing the relationship between Sertoli cells and the developing sperm. As germ cells mature, they progress toward the lumen of the seminiferous tubule. (After Dym 1977.)

um—producing neither too many undifferentiated cells nor too many differentiated cells—is not easy. Mice with the *luxoid* mutation are sterile because they lack a transcription factor that regulates this division. All their spermatogonia become sperm at once, leaving the testes devoid of stem cells (Buass et al. 2004; Costoya et al. 2004).

Looking at Figure 19.22, we find that during the spermatogonial divisions, cytokinesis is not complete. Rather, the cells form a syncytium in which each cell communicates with the others via cytoplasmic bridges about 1 μm in diameter (Dym and Fawcett 1971). The successive divisions produce clones of interconnected cells, and because ions and molecules readily pass through these cytoplasmic bridges, each cohort matures synchronously. During this time, the spermatocyte nucleus often transcribes genes whose products will be used later to form the axoneme and acrosome.

Each primary spermatocyte undergoes the first meiotic division to yield a pair of **secondary spermatocytes**, which complete the second division of meiosis. The haploid cells thus formed are called **spermatids**, and they are still connected to one another through their cytoplasmic bridges. The spermatids that are connected in this manner have haploid nuclei, but are functionally diploid, since a gene product made in one cell can readily diffuse into the cytoplasm of its neighbors (Braun et al. 1989).

During the divisions from type A_1 spermatogonium to spermatid, the cells move farther and farther away from the basement membrane of the seminiferous tubule and closer to its lumen (see Figure 19.21; Siu and Cheng 2004). Thus, each type of cell can be found in a particular layer of the tubule. The spermatids are located at the border of the lumen, and here they lose their cytoplasmic connections and differentiate into spermatozoa. In humans, the progression from spermatogonial stem cell to mature spermatozoa takes 65 days (Dym 1994).

The processes of spermatogenesis require a very specialized network of gene expression (Sassone-Corsi 2002). Not only are histones substantially remodeled and replaced by sperm-specific variants (see below), but even the basal RNA polymerase II transcription factors are exchanged for sperm-specific variants. The TFIID complex, which contains the TATA-binding protein and 14 TAFs, functions in the recognition of RNA polymerase. One of these TAFs, TAF4b, is a sperm-specific TAF required for mouse spermatogenesis (Falender et al. 2005). Without this factor, the spermatogonial stem cells fail to make Ret (the receptor for GDNF) or the luxoid transcription factor, and spermatogenesis fails to occur.

 WEBSITE 19.6 Gonial syncytia: Bridges to the future. The products of meiotic divisions are connected by cytoplasmic bridges. The functions of these connections may differ between those cells producing sperm and those producing eggs.

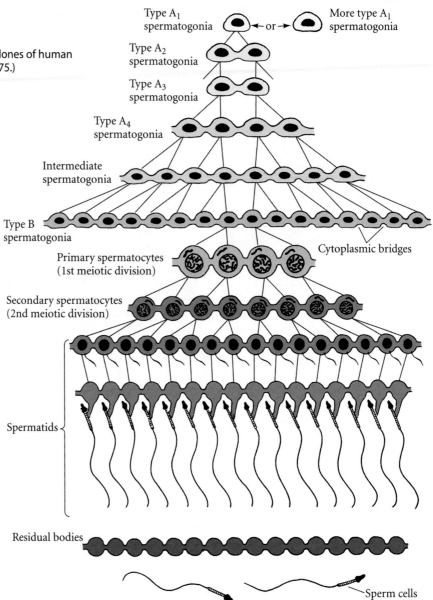

FIGURE 19.22 The formation of syncytial clones of human male germ cells. (After Bloom and Fawcett 1975.)

Type A₁ spermatogonia · or · More type A₁ spermatogonia

Type A₂ spermatogonia

Type A₃ spermatogonia

Type A₄ spermatogonia

Intermediate spermatogonia

Type B spermatogonia

Primary spermatocytes (1st meiotic division)

Cytoplasmic bridges

Secondary spermatocytes (2nd meiotic division)

Spermatids

Residual bodies

Sperm cells

Spermiogenesis: The differentiation of the sperm

The mammalian haploid spermatid is a round, unflagellated cell that looks nothing like the mature vertebrate sperm. The next step in sperm maturation, then, is **spermiogenesis** (or **spermateliosis**), the differentiation of the sperm cell. For fertilization to occur, the sperm has to meet and bind with an egg, and spermiogenesis prepares the sperm for these functions of motility and interaction. The process of mammalian sperm differentiation was shown in Figure 7.2. The first step is the construction of the acrosomal vesicle from the Golgi apparatus. The acrosome forms a cap that covers the sperm nucleus. As the acrosomal cap is formed, the nucleus rotates so that the cap will be facing the basement membrane of the seminiferous tubule. This rotation is necessary because the flagellum, which is beginning to form from the centriole on the other side of the nucleus, will extend into the lumen. During the last stage of spermiogenesis, the nucleus flattens and condenses, the remaining cytoplasm (the residual body or "cytoplasmic droplet") is jettisoned, and the mitochondria form a ring around the base of the flagellum.

During spermiogenesis, the histones of the spermatogonia are often replaced by histone variants, and widespread nucleosome dissociation takes place. This remodeling of nucleosomes might also be the point at which the PGC pattern of methylation is removed and the male genome-specific pattern of methylation is established on the sperm DNA (see Wilkins 2005). As spermiogenesis ends, the histones of the haploid nucleus are eventually replaced by protamines.* This replacement results in the complete shutdown of transcription in the nucleus and facilitates the nucleus assuming an almost crystalline structure (Govin et al. 2004). The resulting sperm then enter the lumen of the seminiferous tubule.

In the mouse, development from stem cell to spermatozoon takes 34.5 days: the spermatogonial stages last 8 days, meiosis lasts 13 days, and spermiogenesis takes another 13.5 days. Human sperm development takes nearly twice as long. Because type A₁ spermatogonia are stem cells, spermatogenesis can occur continuously. Each day, some 100 million sperm are made in each human testicle, and each ejaculation releases 200 million sperm. Unused sperm are either resorbed or passed out of the body in urine. During his lifetime, a human male can produce 10^{12} to 10^{13} sperm (Reijo et al. 1995).

---

*Protamines are relatively small proteins that are over 60 percent arginine. Transcription of the genes for protamines is seen in the early haploid spermatids, although translation is delayed for several days (Peschon et al. 1987).

WEBSITE

19.7 Gene expression during spermatogenesis. Transcription occurs both from the diploid spermatocyte nucleus and from the haploid spermatid nucleus. Posttranscriptional control is also important in regulating sperm gene expression.

Oogenesis

Oogenesis—the differentiation of the ovum—differs from spermatogenesis in several ways (see Table 19.1). Whereas the gamete formed by spermatogenesis is essentially a motile nucleus, the gamete formed by oogenesis contains all the materials needed to initiate and maintain metabolism and development. Therefore, in addition to forming a haploid nucleus, oogenesis also builds up a store of cytoplasmic enzymes, mRNAs, organelles, and metabolic substrates. While the sperm becomes differentiated for motility, the egg develops a remarkably complex cytoplasm.

The mechanisms of oogenesis vary among species more than those of spermatogenesis. This variation should not be surprising, since patterns of reproduction vary so greatly among species. In some species, such as sea urchins and frogs, the female routinely produces hundreds or thousands of eggs at a time, whereas in other species, such as humans and most other mammals, only a few eggs are produced during an indivdual's lifetime. In those species that produce thousands of ova, the germ cells, called **oogonia**, are self-renewing stem cells that endure for the lifetime of the organism. In those species that produce fewer eggs, the oogonia divide to form a limited number of egg precursor cells.

Oogenic meiosis

In the human embryo, the thousand or so oogonia divide rapidly from the second to the seventh month of gestation to form roughly 7 million germ cells (Figure 19.23). After the seventh month of embryonic development, however, the number of germ cells drops precipitously. Most oogonia die during this period, while the remaining oogonia enter the first meiotic division (Pinkerton et al. 1961). These latter cells, called **primary oocytes**, progress through the first meiotic prophase until the diplotene stage, at which point they are maintained until the female matures. With the onset of puberty, groups of oocytes periodically resume meiosis. Thus, in the human female, the first part of meiosis begins in the embryo, and the signal to resume meiosis is not given until roughly 12 years later. In fact, some oocytes are maintained in meiotic prophase for nearly 50 years. As Figure 19.23 illustrates, primary oocytes continue to die. Of the millions of primary oocytes present at her birth, only about 400 mature during a woman's lifetime.*

Oogenic meiosis differs from spermatogenic meiosis in its placement of the metaphase plate. When the primary oocyte divides, its nucleus, called the **germinal vesicle**, breaks down, and the metaphase spindle migrates to the periphery of the cell. At telophase, one of the two daughter cells contains hardly any cytoplasm, whereas the other cell retains nearly the entire volume of cellular constituents (Figure 19.24). The smaller cell is called the **first polar body**, and the larger cell is referred to as the **secondary oocyte**. During the second division of meiosis, a similar unequal cytokinesis takes place. Most of the cytoplasm is retained by the mature egg (the ovum), and a second polar body receives little more than a haploid nucleus. Thus, oogenic meiosis conserves the volume of oocyte cytoplasm in a single cell rather than splitting it equally among four progeny.

In a few species of animals, meiosis is greatly modified such that the resulting gamete is diploid and need not be fertilized to develop. Such animals are said to be **parthenogenetic** (Greek, "virgin birth"). In the fly *Drosophila mangabeirai*, one of the polar bodies acts as a sperm and "fertilizes" the oocyte after the second meiotic division. In

---

*Until very recently, it was assumed that the number of oocytes in a female mammal (including women) was established during embryogenesis and never increased. However, members of the Tilly laboratory, in attempting to ascertain whether ovarian cancer therapies reduced the number of oocytes, were surprised to find that at any given time, a normal mouse ovary appeared to be losing 1200 oocytes. Since female mice are born with only 5000 oocytes, there should be no oocytes left after a week! So researchers searched for anything resembling PGCs in the adult mouse ovary. They believed that such PGC-like cells had been located, and that these cells appeared to produce oocytes (Johnson et al. 2004). Such a finding would mean that the loss of female fertility in mammals might not be due solely to the aging of oocytes but also to the depletion of the PGC stem cells. However, these experiments have been criticized, casting doubt that any PGCs remain in the mammalian ovary (Johnson et al. 2005; Telfer et al. 2005). Still, the ovary may have some regenerative abilities that remain unexplored.

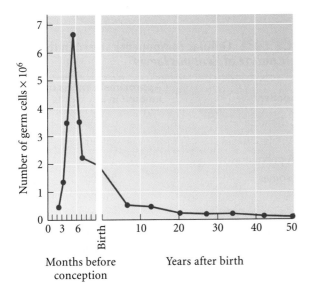

FIGURE 19.23 Changes in the number of germ cells in the human ovary over the life span. (After Baker 1970.)

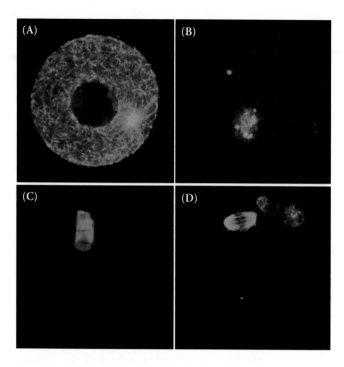

FIGURE 19.24   Meiosis in the mouse oocyte. The tubulin of the microtubules is stained green; the DNA is stained blue. (A) Mouse oocyte in meiotic prophase. The large haploid nucleus (the germinal vesicle) is still intact. (B) The nuclear envelope of the germinal vesicle breaks down as metaphase begins. (C) Meiotic anaphase I, wherein the spindle migrates to the periphery of the egg and releases a small polar body. (D) Meiotic metaphase II wherein the second polar body is given off (the first polar body has also divided). (From De Vos 2002; photographs courtesy of L. De Vos.)

some other insects and in the lizard *Cnemidophorus uniparens*, the oogonia double their chromosome number before meiosis, so that the halving of the chromosomes restores the diploid number. The germ cells of the grasshopper *Pycnoscelus surinamensis* dispense with meiosis altogether, forming diploid ova by two mitotic divisions (Swanson et al. 1981). All of these species consist entirely of females. In other species, haploid parthenogenesis is widely used not only as a means of reproduction, but also as a mechanism of sex determination. In the Hymenoptera (bees, wasps, and ants), unfertilized haploid eggs develop into males, whereas fertilized eggs, being diploid, develop into females. The haploid males are able to produce sperm by abandoning the first meiotic division, thereby forming two sperm cells through second meiosis.

But oogenesis is more than getting the nucleus haploid. All the organelles involved in fertilization have to be constructed, all the mRNAs and proteins have to positioned properly in the oocyte, and all the membrane proteins involved in coordinating the interactions with sperm have to be synthesized and in place. The accumulated material in the oocyte cytoplasm includes energy sources and ener-

gy-producing organelles (the yolk and mitochondria); the enzymes and precursors for DNA, RNA, and protein syntheses; stored messenger RNAs; structural proteins; and morphogenetic regulatory factors that control early embryogenesis. A partial catalogue of the materials stored in the oocyte cytoplasm is shown in Table 19.2, while a partial list of stored mRNAs is shown in Table 5.2.

Maturation of the oocytes in frogs

The eggs of sea urchins, fish, and amphibians are derived from an oogonial stem cell population that can generate a new cohort of oocytes each year. In the frog *Rana pipiens*, oogenesis takes 3 years. During the first 2 years, the oocyte increases its size very gradually. During the third year, however, the rapid accumulation of yolk in the oocyte causes the egg to swell to its characteristically large size (Figure 19.25). Eggs mature in yearly batches, with the first cohort maturing shortly after metamorphosis; the next group matures a year later.

Vitellogenesis—the accumulation of yolk proteins—occurs when the oocyte reaches the diplotene stage of meiotic prophase. Yolk is not a single substance, but a mixture of materials for embryonic nutrition. The major yolk component in frog eggs is a 470-kDa protein called **vitellogenin**. It is not made in the frog oocyte (as are the major yolk proteins of organisms such as annelids and crayfish), but is synthesized in the liver and carried by the bloodstream to the ovary (Flickinger and Rounds 1956; Danilchik and Gerhart 1987).

COMPLETION OF AMPHIBIAN MEIOSIS: PROGESTERONE AND FERTILIZATION   Amphibian primary oocytes can remain in the diplotene stage of meiotic prophase for years. This state resembles the G_2 phase of the mitotic cell division cycle (see Chapter 8). Resumption of meiosis in the amphibian oocyte requires progesterone. This hormone is secreted by the follicle cells in response to gonadotropic hormones secreted by the pituitary gland. Within 6 hours of proges-

TABLE 19.2   Cellular components stored in the mature oocyte of *Xenopus laevis*

Component	Approximate excess over amount in larval cells
Mitochondria	100,000
RNA polymerases	60,000–100,000
DNA polymerases	100,000
Ribosomes	200,000
tRNA	10,000
Histones	15,000
Deoxyribonucleoside triphosphates	2,500

Source: After Laskey 1979.

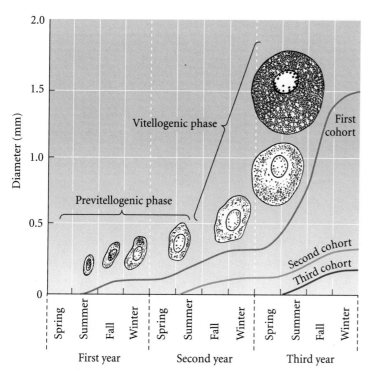

FIGURE 19.25 Growth of oocytes in the frog. During the first 3 years of life, three cohorts of oocytes are produced. The drawings follow the growth of the first-generation oocytes. (After Grant 1953.)

terone stimulation, **germinal vesicle breakdown (GVBD)** occurs, the microvilli retract, the nucleoli disintegrate, and the chromosomes contract and migrate to the animal pole to begin division. Soon afterward, the first meiotic division occurs, and the mature ovum is released from the ovary by a process called **ovulation**. The ovulated egg is in second meiotic metaphase when it is released (Figure 19.26).

How does progesterone enable the egg to break its dormancy and resume meiosis? To understand the mechanisms by which this activation is accomplished, it is necessary to briefly review the model for early blastomere division (see Chapter 8). Entry into the mitotic (M) phase of the cell cycle (in both meiosis and mitosis) is regulated by **mitosis-promoting factor**, or **MPF** (originally called "maturation-promoting factor," after its meiotic function). MPF contains two subunits, **cyclin B** and the **p34** protein. The p34 protein is a cyclin-dependent-kinase—its activity is dependent upon the presence of cyclin. Since all the components of MPF are present in the amphibian oocyte, it is generally thought that progesterone somehow converts a pre-MPF complex into active MPF.

The mediator of the progesterone signal is the **c-mos** protein. Progesterone reinitiates meiosis by causing the egg to polyadenylate the maternal

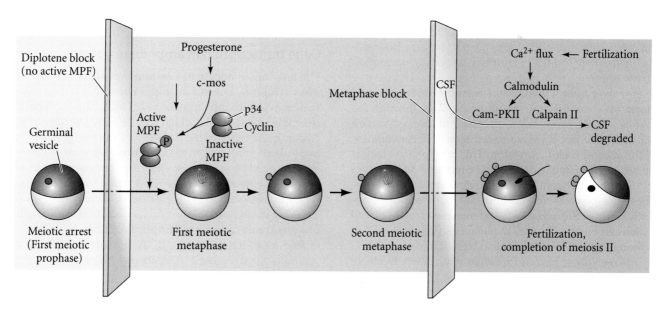

FIGURE 19.26 Schematic representation of *Xenopus* oocyte maturation, showing the regulation of meiotic cell division by progesterone and fertilization. Oocyte maturation is arrested at the diplotene stage of first meiotic prophase by the lack of active MPF. Progesterone activates the production of the c-mos protein. This protein initiates a cascade of phosphorylation that eventually phosphorylates the p34 subunit of MPF, allowing the MPF to become active. The MPF drives the cell cycle through the first meiotic division, but further division is blocked by CSF, a compound containing c-mos and cdk2. Upon fertilization, calcium ions released into the cytoplasm are bound by calmodulin and are used to activate two enzymes, calmodulin-dependent protein kinase II and calpain II, which inactivate and degrade CSF. Second meiosis is completed, and the two haploid pronuclei can fuse. At this time, cyclin B is resynthesized, allowing the first cell cycle of cleavage to begin.

FIGURE 19.27 In amphibian oocytes, lampbrush chromosomes are active in the diplotene germinal vesicle during first meiotic prophase. (A) Autoradiograph of chromosome I of the newt *Triturus cristatus* after in situ hybridization with radioactive histone mRNA. A histone gene (or set of histone genes) is being transcribed (arrow) on one of the loops of this lampbrush chromosome. (B) Lampbrush chromosome of the salamander *Notophthalmus viridescens*. Extended DNA (white) loops out and is transcribed into RNA (red). (A from Old et al. 1977, courtesy of H. G. Callan; B courtesy of M. B. Roth and J. Gall.)

(A)

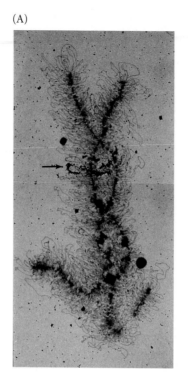

(B)

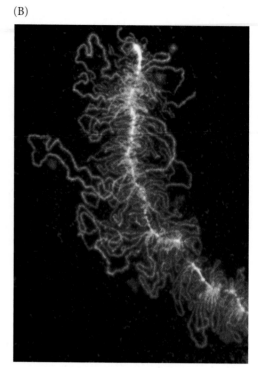

c-mos mRNA that has been stored in its cytoplasm (Sagata et al. 1988; Sheets et al. 1995; Mendez et al. 2000). This message is translated into a 39-kDa phosphoprotein. This c-mos protein is detectable only during oocyte maturation and is destroyed quickly upon fertilization. Yet during its brief lifetime, it plays a major role in releasing the egg from its dormancy. The c-mos protein activates a phosphorylation cascade that phosphorylates and activates the p34 subunit of MPF (Ferrell and Machleder 1998; Ferrell 1999). The active MPF allows the germinal vesicle to break down and the chromosomes to divide. If the translation of *c-mos* is inhibited by injecting *c-mos* antisense mRNA into the oocyte, germinal vesicle breakdown and the resumption of oocyte maturation do not occur.

However, oocyte maturation then encounters a second block. MPF can take the chromosomes only through the first meiotic division and the prophase of the second meiotic division. The oocyte is arrested again in the metaphase of the second meiotic division. This metaphase block is caused by the combined actions of c-mos and another protein, cyclin-dependent kinase 2 (cdk2; Gabrielli et al. 1993). These two proteins are subunits that together form **cytostatic factor (CSF)**, which is found in mature frog eggs, and which can block cell cycles in metaphase (Matsui 1974). It is thought that CSF prevents the degradation of cyclin.

The metaphase block is broken by fertilization. Evidence suggests that the calcium ion flux attending fertilization enables the calcium-binding protein **calmodulin** to become active. Calmodulin, in turn, can activate two enzymes that inactivate CSF: calmodulin-dependent protein kinase II, which inactivates the cdk2 kinase, and calpain II, a calcium-dependent protease that degrades c-mos (Watanabe et al. 1989; Lorca et al. 1993). This action promotes cell division in two ways. First, without CSF, cyclin can be degraded, and the meiotic division can be completed. Second, calcium-dependent protein kinase II also allows the centro-

some to duplicate, thus forming the poles of the meiotic spindle (Matsumoto and Maller 2002). The coordination of fertilization and meiosis appears to be intimately coordinated through the release and binding of calcium ions.

Gene transcription in amphibian oocytes

The amphibian oocyte has certain periods of very active RNA synthesis. During the diplotene stage, certain chromosomes stretch out large loops of DNA, causing them to resemble a lampbrush (which was a handy instrument for cleaning test tubes in the days before microfuges). In situ hybridization reveals these **lampbrush chromosomes** to be sites of RNA synthesis. Oocyte chromosomes can be incubated with a radioactive RNA probe and autoradiography used to visualize the precise locations where genes are being transcribed (Figure 19.27A). Electron micrographs of gene transcripts from lampbrush chromosomes also enable one to see chains of mRNA coming off each gene as it is transcribed (Figure 19.27B; also see Hill and MacGregor 1980).

In addition to mRNA synthesis, ribosomal RNA and transfer RNA are also transcribed during oogenesis. Figure 19.28A shows the pattern of rRNA and tRNA synthesis during *Xenopus* oogenesis. Transcription appears to begin in early (stage I, 25–40 μm) oocytes, during the diplotene stage of meiosis. At this time, all the rRNAs and tRNAs needed for protein synthesis until the mid-blastula stage are made, and all the maternal mRNAs needed for early development are transcribed. This stage lasts for months in *Xenopus*. The rate of ribosomal RNA production is prodigious. The *Xenopus* oocyte genome has over 1800 genes encoding 18S and 28S rRNA (the two large RNAs that form the ribosomes), and these genes are selectively amplified such that there are over 500,000 genes making ribosomal RNA in the oocyte (Figure 19.28B; Brown and Dawid 1968). When the mature (stage VI) oocyte reaches a certain size, its chromosomes condense, and the rRNA genes are no longer transcribed. This "mature oocyte" condition can also last for months. Upon hormonal stimulation, the oocyte completes its first meiotic division and is ovulated and fertilized. The mRNAs stored by the oocyte now join with the ribosomes to initiate protein synthesis. Within hours, the second meiotic division has begun, and the secondary oocyte has been fertilized. The embryo's genes do not begin active transcription until the mid-blastula transition (Newport and Kirschner 1982).

As we saw in Chapter 5, the oocytes of several species make two classes of mRNAs—those for immediate use in the oocyte, and those that are stored for use during early development. In frogs, the translation of stored oocyte messages (maternal mRNAs) is initiated by progesterone as the egg is about to be ovulated. One of the results of the MPF activity induced by progesterone may be the phosphorylation of proteins on the 3′ UTR of stored oocyte mRNAs. The phosphorylation of these factors is associated with the lengthening of the polyA tails of the stored messages and their subsequent translation (Paris et al. 1991).

WEBSITE 19.9 Synthesizing oocyte ribosomes. Ribosomes are almost a "differentiated product" of the oocyte, and the *Xenopus* oocyte contains 20,000 times as many ribosomes as somatic cells do. Gene repetition and gene amplification are both used to transcribe these enormous amounts of rRNA.

Meroistic oogenesis in insects

There are several types of oogenesis in insects, but most studies have focused on those insects (including *Drosophila* and moths) that undergo **meroistic oogenesis**, in which cytoplasmic connections remain between the cells produced by the oogonium.

The oocytes of meroistic insects do not pass through a transcriptionally active stage, nor do they have lampbrush chromosomes. Rather, RNA synthesis is largely confined to the nurse cells, and the RNA made by those cells is actively transported into the oocyte cytoplasm (see Figure 9.8). Oogenesis takes place in only 12 days, so the nurse cells are very metabolically active during this time. Nurse cells are aided in their transcriptional efficiency by becoming polytene—instead of having two copies of each chromosome, they replicate their chromosomes until they have produced 512 copies. The 15 nurse cells pass ribosomal and messenger RNAs as well as proteins into the oocyte cytoplasm, and entire ribosomes may be transported as well. The mRNAs do not associate with polysomes, and they are not immediately active in protein synthesis (Paglia et al. 1976; Telfer et al. 1981).

(A)

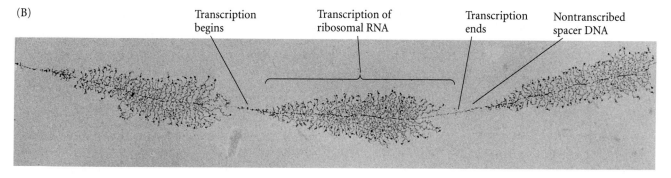

FIGURE 19.28 Ribosomal RNA production in *Xenopus* oocytes. (A) Relative rates of DNA, tRNA, and rRNA synthesis in amphibian oogenesis during the last 3 months before ovulation. (B) Transcription of the large RNA precursor of the 28S, 18S, and 5.8S ribosomal RNAs. These units are tandemly linked together, with some 450 per haploid genome. (A after Gurdon 1976; B courtesy of O. L. Miller, Jr.)

(B)

The meroistic ovary confronts us with some interesting problems. If all the cystocytes are connected so that proteins and RNAs shuttle freely among them, why should the cystocytes have different developmental fates? Why should one cell become the oocyte while the other 15 become "RNA-synthesizing factories"? Why is the flow of protein and RNA in one direction only?

As the cystocytes divide, a large, spectrin-rich structure called the **fusome** forms and spans the ring canals between the cells (see Figure 19.4A). It is constructed asymmetrically, as it always grows from the spindle pole that remains in one of the cells after the first division (Lin and Spradling 1995; de Cuevas and Spradling 1998). The cell that retains the greater part of the fusome during the first division becomes the oocyte. It is not yet known if the fusome contains oogenic determinants, or if it directs the traffic of materials into this particular cell.

Once the patterns of transport are established, the cytoskeleton becomes actively involved in transporting mRNAs from the nurse cells into the oocyte cytoplasm (Cooley and Theurkauf 1994). An array of microtubules that extends through the ring canals (see Figure 19.4C) is critical for oocyte determination. In the nurse cells, the Exuperantia protein binds *bicoid* message to the microtubules and transports it to the anterior of the oocyte (Cha et al. 2001; see Chapter 9). If the microtubular array is disrupted (either chemically or by mutations such as *bicaudal-D* or *egalitarian*), the nurse cell gene products are transmitted in all directions, and all 16 cells differentiate into nurse cells (Gutzeit 1986; Theurkauf et al. 1992, 1993; Spradling 1993).

The Bicaudal-D and Egalitarian proteins are probably core components of a dynein motor system that transports mRNAs and proteins throughout the oocyte (Bullock and Ish-Horowicz 2001). It is possible that some compounds transported from the nurse cells into the oocyte become associated with transport proteins such as dynein and kinesin, which would enable them to travel along the tracks of microtubules extending through the ring canals (Theurkauf et al. 1992; Sun and Wyman 1993). The *oskar* message, for instance, is linked to kinesin through the Barentsz protein, and kinesin can transport the *oskar* message to the posterior of the oocyte (van Eeden et al. 2001; see Figure 9.8).

Actin may become important for maintaining the polarity of transport during later stages of oogenesis. Mutations that prevent actin microfilaments from lining the ring canals prevent the transport of mRNAs from the nurse cells to the oocyte, and disruption of the actin microfilaments randomizes the distribution of mRNA (Cooley et al. 1992; Watson et al. 1993). Thus, the cytoskeleton controls the movement of organelles and RNAs between nurse cells and oocyte such that developmental cues are exchanged only in the appropriate direction.

Maturation of the mammalian oocyte

Ovulation in mammals follows one of two patterns, depending on the species. One type of ovulation is stimulated by the act of copulation. Physical stimulation of the cervix triggers the release of gonadotropins from the pituitary. These gonadotropins signal the egg to resume meiosis and initiate the events that will expel it from the ovary. This mechanism ensures that most copulations will result in fertilized ova, and animals that utilize this method of ovulation—such as rabbits and minks—have a reputation for procreative success.

Most mammals, however, have a periodic ovulation pattern, in which the female ovulates only at specific times of the year. This ovulatory period is called **estrus** (or its English equivalent, "heat"). In these animals, environmental cues (most notably the amount and type of light during the day) stimulate the hypothalamus to release **gonadotropin-releasing hormone** (**GRH**). GRH stimulates the pituitary to release the gonadotropins—**follicle-stimulating hormone** (**FSH**) and **luteinizing hormone** (**LH**)—that cause the ovarian follicle cells to proliferate and secrete estrogen. Estrogen enters certain neurons and evokes the pattern of mating behavior characteristic of the species. The gonadotropins also stimulate follicular growth and initiate ovulation. Thus, mating behavior and ovulation occur close together.

Humans have a variation on the theme of periodic ovulation. Although human females have cyclical ovulation (averaging about once every 29.5 days) and no definitive yearly estrus, most of human reproductive physiology is shared with other primates. The characteristic primate periodicity in maturing and releasing ova is called the **menstrual cycle** because it entails the periodic shedding of blood and endothelial tissue from the uterus at monthly intervals.* The menstrual cycle represents the integration of three very different cycles:

1. The **ovarian cycle**, the function of which is to mature and release an oocyte.

---

*The periodic shedding of the uterine lining is a controversial topic. Some scientists speculate that menstruation is an active process, with adaptive significance in evolution. Profet (1993) proposed that menstruation is a crucial immunological adaptation, protecting the uterus against infections contracted from semen or other environmental agents. Strassmann (1996) suggested that the cyclicity of the endometrium is an energy-saving adaptation that is important in times of poor nutrition. Vaginal bleeding would be a side effect of this adaptive process. Finn (1998) claimed that menstruation has no adaptive value and is necessitated by the immunological crises that are a consequence of bringing two genetically dissimilar organisms together in the uterus. Martin (1992) pointed out that it might even be wrong to think of there being a single function of menstruation, and that its roles might change during a woman's life cycle.

WEBSITE 19.10 *Drosophila* spermatogenesis. *Drosophila* sperm are derived from stem cells that maintain a long-term capacity to divide. The renewal of stem cells is specified by the STAT pathway, which responds to a specific support cell signal.

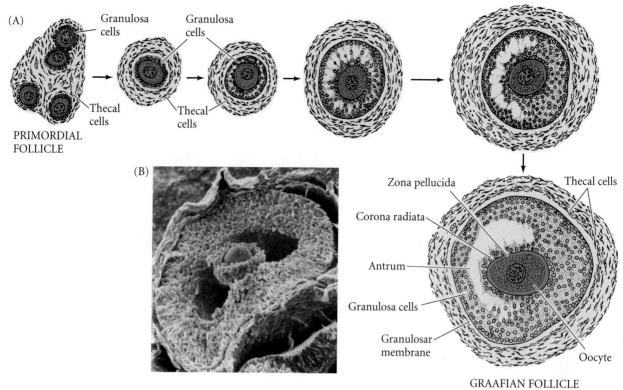

FIGURE 19.29 The ovarian follicle of mammals. (A) Maturation of the ovarian follicle. When mature, it is often called a Graafian follicle. (B) Scanning electron micrograph of a mature follicle in the rat. The oocyte (center) is surrounded by the smaller granulosa cells that will make up the cumulus. (A after Carlson 1981; B courtesy of P. Bagavandoss.)

2. The **uterine cycle**, the function of which is to provide the appropriate environment for the developing blastocyst.
3. The **cervical cycle**, the function of which is to allow sperm to enter the female reproductive tract only at the appropriate time.

These three functions are integrated through the hormones of the pituitary, hypothalamus, and ovary.

VADE MECUM² **Oogenesis in mammals.** The development of the mammalian ovum and its remarkable growth during the primary oocyte stage are the subject of photographs and QuickTime movies of histological sections through a mammalian ovary. [Click on Gametogenesis]

The majority of the oocytes within the adult human ovary are maintained in the diplotene stage of the first meiotic prophase, often referred to as the **dictyate state**. Each oocyte is enveloped by a primordial follicle consisting of a single layer of epithelial granulosa cells and a less organized layer of mesenchymal thecal cells (Figure 19.29). Peri-

odically, a group of primordial follicles enters a stage of follicular growth. During this time, the oocyte undergoes a 500-fold increase in volume (corresponding to an increase in oocyte diameter from 10 μm in a primordial follicle to 80 μm in a fully developed follicle).

Concomitant with oocyte growth is an increase in the number of **granulosa cells**, which form concentric layers around the oocyte. This proliferation of granulosa cells is mediated by a paracrine factor, GDF9, a member of the TGF-β family (Dong et al. 1996). Throughout this growth period, the oocyte remains in the dictyate stage. The fully grown follicle thus contains a large oocyte surrounded by several layers of granulosa cells. The innermost of these cells will stay with the ovulated egg, forming the **cumulus**, which surrounds the egg in the oviduct. In addition, during the growth of the follicle, an **antrum** (cavity) forms and becomes filled with a complex mixture of proteins, hormones, and other molecules.

Just as the maturing oocyte synthesizes paracrine factors that allow the follicle cells to proliferate, the follicle cells secrete growth and differentiation factors (TGF-β2, VEGF, leptin, Fgf2) that allow the oocyte to grow and bring blood vessels into the follicular region (Antczak et al. 1997). The oocytes are maintained in the dictyate stage by the ovarian follicle cells. Signals from the follicles activate a G-protein linked receptor that stimulates adenyl cyclase to elevate levels of cAMP in the oocyte (Mehlmann et al. 2004). The release from this dictyate stage and the reinitiation of meiosis is driven by lutenizing hormone (LH) from

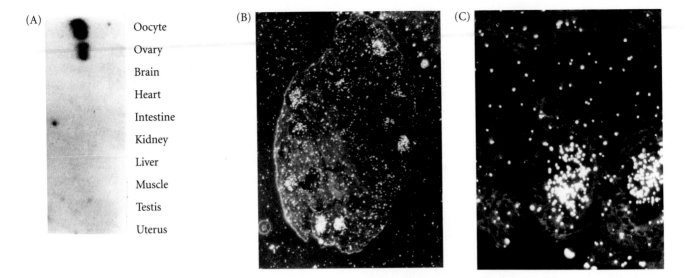

(A)

Oocyte
Ovary
Brain
Heart
Intestine
Kidney
Liver
Muscle
Testis
Uterus

FIGURE 19.30 Expression of the *ZP3* gene in the developing mouse oocyte. (A) Northern blot of *ZP3* mRNA accumulation in the tissues of a 13-day mouse embryo. A radioactive probe to the ZP3 message found it expressed only in the ovary, and specifically in the oocytes. (B) When the luciferinase reporter gene is placed onto the *ZP3* promoter and inserted into the mouse genome, luciferinase message is seen only in the developing oocytes of the ovary. (C) Higher magnification of a section of (B), showing two of the ovarian follicles containing maturing oocytes. (A from Roller et al. 1989; B and C from Lira et al. 1990; photographs courtesy of P. Wassarman.)

the pituitary. Meanwhile, the growing oocyte is actively transcribing genes whose products are necessary for cell metabolism, for oocyte-specific processes, or for early development before the zygote-derived nuclei begin to function. In mice, for instance, the growing diplotene oocyte is actively transcribing the genes for zona pellucida proteins ZP1, ZP2, and ZP3 (see Figure 7.30). Moreover, these genes are transcribed only in the oocyte and not in any other cell type, as the proteins essential for fertilization are being synthesized (Figure 19.30; Roller et al. 1989; Lira et al. 1990; Epifano et al. 1995).

Coda

We are now back where we began: the stage is set for fertilization to take place. The egg and the sperm will both die if they do not meet. As F. R. Lillie recognized in 1919, "The elements that unite are single cells, each on the point of death; but by their union a rejuvenated individual is formed, which constitutes a link in the eternal process of Life."

WEBSITE 19.11 Hormones and mammalian egg maturation. To survive, the follicle and its oocyte have to "catch the wave" of conadotropic hormone release. The hormones of the menstrual cycle synchronize egg maturation with the anatomical changes of the uterus and cervix.

WEBSITE 19.12 The reinitiation of mammalian meiosis. The hormone-mediated disruption of communication between the oocyte and its surrounding follicle cells may be critical in the resumption of meiosis in female mammals.

Snapshot Summary **The Germ Line**

1. The precursors of the gametes—the sperm and eggs—are the primordial germ cells. They form outside the gonads and migrate into the gonads during development.

2. In many species, a distinctive germ plasm exists. It often contains the Oskar, Vasa, and Nanos proteins or the mRNAs encoding them.

3. In *Drosophila*, the germ plasm becomes localized in the posterior of the embryo and forms pole cells, the precursors of the gametes. In frogs, the germ plasm originates in the vegetal portion of the oocyte.

4. In amphibians, the germ cells migrate on fibronectin matrices from the posterior larval gut to the gonads. In mammals, a similar migration is seen, and

fibronectin pathways may also be used. Stem cell factor is critical in this migration, and the germ cells proliferate as they travel.

5. In birds, the germ plasm is first seen in the germinal crescent. The germ cells migrate through the blood, then leave the blood vessels and migrate into the genital ridges.

6. Germ cell migration in *Drosophila* occurs in several steps involving transepithelial migration, repulsion from the endoderm, and attraction to the gonads.

7. Before meiosis, the DNA is replicated and the resulting sister chromatids remain bound at the kinetochore. Homologous chromosomes are connected through the synaptonemal complex.

8. The first division of meiosis separates the homologous chromosomes. The second division of meiosis splits the kinetochore and separates the chromatids.

9. The meiosis/mitosis decision in nematodes is regulated by a Delta protein homologue in the membrane of the distal tip cell. The decision for a germ cell to become either a sperm or an egg is regulated at the level of translation of the *fem-3* message.

10. Spermatogenic meiosis in mammals is characterized by the production of four gametes per meiosis and

by the absence of meiotic arrest. Oogenic meiosis is characterized by the production of one gamete per meiosis and by an arrest at first meiotic prophase to allow the egg to grow.

11. In some species, meiosis is modified such that a diploid egg is formed. Such species can produce a new generation parthenogenetically, without fertilization.

12. The egg not only synthesizes numerous compounds, but also absorbs material produced by other cells. Moreover, it localizes many proteins and messages to specific regions of the cytoplasm, often tethering them to the cytoskeleton.

13. The *Xenopus* oocyte transcribes actively from lampbrush chromosomes during the first meiotic prophase.

14. In *Drosophila* , nurse cells make mRNAs that enter the developing oocyte. Which of the cells derived from the primordial germ cell becomes the oocyte and which become nurse cells is determined by the fusome and the pattern of divisions.

15. In many if not all cases, those cells destined to become germline cells are transcriptionally and translationally silenced.

For Further Reading

Complete bibliographical citations for all literature cited in this chapter can be found on the Vade Mecum CD that accompanies the book and at the free access website www.devbio.com

Decotto, E. and A. C. Spradling. 2005. The *Drosophila* ovarian and testis stem cell niches: similar somatic stem cells and signals. *Dev. Cell* 9: 501–510.

Doitsidou, M. and 8 others. 2002. Guidance of primordial germ cell migration by the chemokine SDF-1. *Cell* 111: 647–659.

Ephrussi, A. and R. Lehmann. 1992. Induction of germ cell formation by oskar. *Nature* 358: 387–392.

Extravour, C. G. and M. Akam. 2003. Mechanisms of germ cell specification across the metazoans: epigenesis and preformation. *Development* 130: 5869–5884.

Hayashi, Y., M. Hayashi and S. Kobayashi. 2004. Nanos suppresses somatic cell fate in *Drosophila* germ line. *Proc. Natl. Acad. Sci. USA* 101: 10338–10342.

Knaut, H., C. Werz, R. Geisler, The Tübingen 2000 screen consortium, and C. Nüsslein-Volhard. 2003. A zebrafish homologue of the chemokine receptor Cxcr4 is a germ-cell guidance receptor. *Nature* 421: 279–282.

Molyneaux, K. A., J. Stallock, K. Schaible and C. Wylie. 2001. Time-lapse analysis of living mouse germ cell migration. *Dev. Biol.* 240: 488–498.

Ohinata, Y. and 11 others. 2005. Blimp1 is a critical determinant of the germ cell lineage in mice. *Nature* 436: 207–213.

Seydoux, G. and S. Strome. 1999. Launching the germline in *Caenorhabditis elegans:* Regulation of gene expression in early germ cells. *Development* 126: 3275–3283.

Stewart, T. A. and B. Mintz. 1981. Successful generations of mice produced from an established culture line of euploid teratocarcinoma cells. *Proc. Natl. Acad. Sci. USA* 78: 6314–6318.

Weidinger, G., U. Wolke, M. Koprunner, M. Klinger and E. Raz. 1999. Identification of tissues and patterning events required for distinct steps in early migration of zebrafish primordial germ cells. *Development* 126: 5295–5307.

PART 4

Ramifications of Developmental Biology

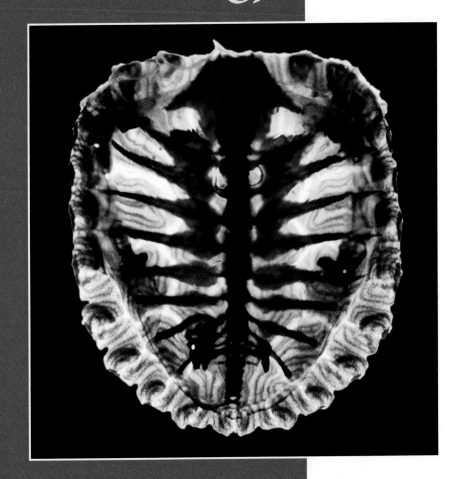

Part 4 **Ramifications of Developmental Biology**

PREVIOUS PAGE
Changes in development can result in the evolution of novel structures. The ribs of the red-eared slider turtle *Trachemys scripta* enter the dermis of its back. It is probable that these ribs are active in transforming that dermis into bone, thereby forming the carapace of the shell, a structure unique to the turtle lineage. (The embryo has been treated with Alizarin red, a dye that stains bone.) Photograph courtesy of G. Loredo, A. Brukman, and S. F. Gilbert.

An Overview of Plant Development

THE DEVELOPMENTAL STRATEGIES OF PLANTS have evolved separately from those of the animals over millions of years. The two kingdoms have many commonalities (and the land plants are sometimes referred to as "embryophytes," calling attention to the significance of the embryo in their life histories), but some of the challenges and solutions found in plants are sufficiently unique to warrant separate discussion in this chapter. What are the fundamental differences between development in animals and development in the land plants?

- **Plant cells do not migrate**. Plant cells are trapped within rigid cellulose walls that generally prevent cell and tissue migration. Plants, like most metazoan animals, develop three basic tissue systems (dermal, ground, and vascular), but do not rely on gastrulation to establish this layered system of tissues. Plant development is highly regulated by the environment, a strategy that is adaptive for a stationary organism.

- **Plants have sporic meiosis rather than gametic meiosis.** That is, meiosis in plants produces spores, not gametes. Plant gametes are produced by mitotic divisions following meiosis.

- **The life cycle of land plants (as well as many other plants) includes both diploid and haploid multicellular stages.** This type of life cycle is referred to as *alternation of generations* and results in two different multicellular body plans over the life cycle of an individual.

- **Plant germ cells are not set aside early in development.** While this is also the case in several animal phyla, it is the case for all plants.

- **Plants undergo extended morphogenesis.** Clusters of actively dividing cells called **meristems**, which are similar to stem cells in animals, persist long after maturity. Meristems allow for iterative development and the formation of new structures throughout the life of the plant.

- **Plants have tremendous developmental plasticity.** Many plant cells are highly plastic. While cloning in animals also demonstrates plasticity, plants depend far more heavily on this developmental strategy. For example, if a shoot is grazed by herbivores, meristems in the leaf often grow out to replace the lost part. (This strategy has similarities to the regeneration seen in some animals.) Whole plants can be regenerated from some single cells. In addition, a plant's form (including branching, height, and relative amounts of vegetative and reproductive structures) is greatly influenced by environmental factors such as light and temperature, and a wide range of morphologies can result from the same genotype (see Figure 2.15). The amazing level of plasticity found among the plants may help compensate for their lack of mobility.

> *The search for differences or fundamental contrasts ... has occupied many men's minds, while the search for commonality of principle or essential similarities, has been pursued by few; the contrasts are apt to loom too large, great though they may be.*
>
> *D'ARCY THOMPSON (1942)*

> *Do not quench your inspiration and your imagination; do not become the slave of your model.*
>
> *VINCENT VAN GOGH*

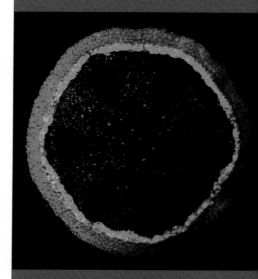

- **Developmental mechanisms evolved independently in plants and animals.** The last common ancestor of plants and animals was a single-celled eukaryote. Genome-level comparisons indicate that there is minimal homology between the genes and proteins used to establish body plans in plants and animals (Meyerowitz 2002). While both homeobox and MADS box genes were present in the last common ancestor of plants and animals, the MADS box family controls major developmental regulatory processes in plants, but not in animals.

Despite the major differences among many plants and animals, developmental genetic studies are revealing some commonalities in the logic of their pattern formation, along with evolutionarily distinct solutions to the problem of creating three-dimensional form from a single cell.

The green plants include many organisms, from algae to flowering plants (angiosperms). Recent phylogenetic studies show a common lineage for all green plants, distinct from the red and brown plants. While comparisons of developmental strategies among diverse plants is both fascinating and informative, this chapter focuses primarily on the flowering plants (angiosperms). The goal is to examine plant development within the larger context of developmental biology.

Gamete Production in Angiosperms

Plants have both multicellular haploid and multicellular diploid stages in their life cycles, and embryonic development is seen only in the diploid generation. The embryo, however, is produced by the fusion of gametes, which are formed only by the haploid generation. Understanding the relationship between the two generations is important in the study of plant development.

Gametophytes

Unlike animals, plants have multicellular haploid and multicellular diploid stages in their life cycles (see Chapter 2). Gametes develop in the multicellular haploid **gametophyte** (from the Greek *phyton*, "plant"). Fertilization gives rise to a multicellular diploid **sporophyte**, which produces haploid spores via meiosis. This type of life cycle is called a **haplodiplontic** life cycle (Figure 20.1). It differs from the **diplontic** life cycle of animals, in which only the gametes are in the haploid state.

In a haplodiplontic life cycle, gametes are not the direct result of a meiotic division. Diploid sporophyte cells undergo meiosis to produce haploid **spores**. Each spore goes through mitotic divisions to yield a multicellular, haploid gametophyte. There are two types of spores in angiosperms. **Megaspores** produce female gametophytes, while **microspores** produce male gametophytes. Male and female gametophytes have distinct morphologies. Wind or mem-

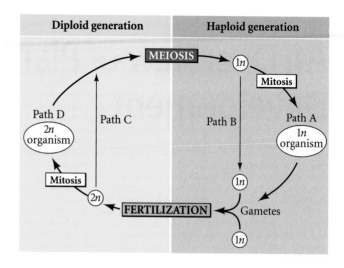

FIGURE 20.1   Plants have haplodiplontic life cycles that involve mitotic divisions (resulting in multicellularity) in both the haploid and diploid generations (paths A and D). Most animals are diplontic and undergo mitosis only in the diploid generation (paths B and D). Multicellular organisms with haplontic life cycles follow paths A and C.

WEBSITE

20.1 Plant life cycles. In plants there is an evolutionary trend from sporophytes that are nutritionally dependent on autotrophic gametophytes to the opposite—gametophytes that are dependent on autotrophic sporophytes. This trend is exemplified by comparing the life cycles of mosses and ferns to that of angiosperms.

bers of the animal kingdom deliver the male gametophyte—**pollen**—to the female gametophyte. Mitotic divisions within the gametophytes are required to produce gametes. The diploid sporophyte results from the fusion of two gametes. Among land plants, the gametophytes and sporophytes of a species have distinct morphologies, and how a single genome can be used to create two unique morphologies is an intriguing puzzle.

At first glance, angiosperms may appear to have diplontic life cycles because the gametophyte generation has been reduced to just a few cells (Figure 20.2). However, mitotic division follows meiosis in the sporophyte, resulting in a multicellular gametophyte, which produces eggs or sperm. All of this takes place in the organ that is characteristic of the angiosperms: the **flower**.

POLLEN   The pollen grain is an extremely simple multicellular structure (Figure 20.3). The outer wall of the pollen grain, the exine, is composed of resistant material provided by both the tapetum (sporophyte generation that provides nourishment for developing pollen) and the microspore (gametophyte generation). The inner wall, the

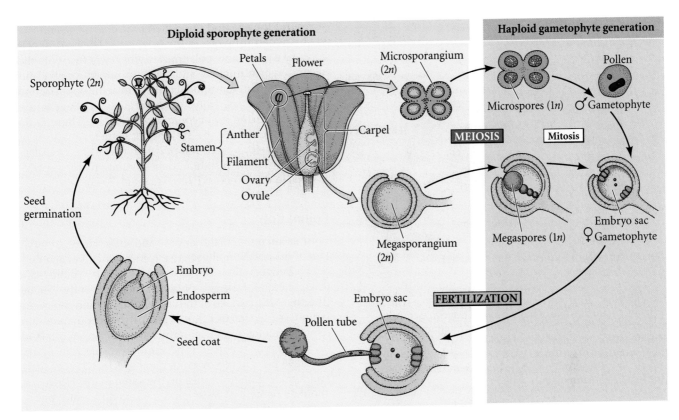

Diploid sporophyte generation

Haploid gametophyte generation

FIGURE 20.2 Life cycle of an angiosperm, represented here by a pea plant (genus *Pisum*). The sporophyte is the dominant generation, but multicellular male and female gametophytes are produced within the flowers of the sporophyte. Cells of the microsporangium within the anther undergo meiosis to produce microspores. Subsequent mitotic divisions are limited, but the end result is a multicellular pollen grain. Integuments and the ovary wall protect the megasporangium. Within the mega- sporangium, meiosis yields four megaspores—three small and one large. Only the large megaspore survives to produce the female gametophtye (the embryo sac). Fertilization occurs when the male gametophyte (pollen) germinates and the pollen tube grows toward the embryo sac. The sporophyte gen- eration may be maintained in a dormant state, protected by the seed coat.

intine, is produced by the microspore. A mature pollen grain consists of two cells: a **tube cell**, and a **generative cell** within the tube cell. The nucleus of the tube cell guides pollen germination and the growth of the pollen tube after the pollen lands on the stigma of a female gametophyte. The generative cell divides to produce two sperm. One of the two sperm will fuse with the egg cell to produce the next sporophyte generation. The second sperm will par-

(A)

(B)

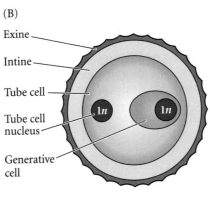

Exine

Intine

Tube cell

Tube cell nucleus

Generative cell

FIGURE 20.3 (A) Pollen grains have intricate surface patterns, as seen in this scanning electron micrograph of aster pollen. (B) A pollen grain consists of a cell within a cell. The generative cell will undergo division to produce two sperm cells. One will fertilize the egg, and the other will join with the polar nuclei, yielding the endosperm.

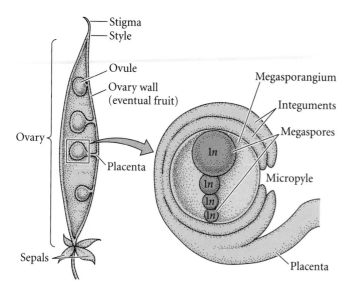

FIGURE 20.4   The carpel consists of the stigma, the style, and an ovary containing one or more ovules. Each ovule contains megasporangia protected by two layers of integument cells. The megasporangia divide meiotically to produce haploid megaspores. All of the carpel is diploid except for the megaspores, which divide mitotically to produce the embryo sac (the female gametophyte).

ticipate in the formation of the endosperm, a structure that provides nourishment for the embryo.

THE OVARY   The angiosperm ovary is part of the **carpel** of a flower, which gives rise to the female gametophyte (Figure 20.4). The carpel consists of the **stigma** (where the pollen lands), the **style**, and the **ovary**. Following fertilization, the ovary wall will develop into the **fruit**. This unique angiosperm structure provides further protection for the developing embryo and also enhances seed dispersal by frugivores (fruit-eating animals). Within the ovary are one or more **ovules** attached by a **placenta** to the ovary wall. Fully developed ovules are called **seeds**.

The ovule has one or two outer layers of cells, called the **integuments**. The integuments enclose the **megasporangium**, which contains sporophyte cells that undergo meiosis to produce **megaspores** (see Figure 20.2). There is a small opening in the integuments called the **micropyle**, through which the pollen tube will grow. The integuments develop into the seed coat, a waterproof physical barrier that protects the embryo. When the mature embryo disperses from the parent plant, diploid sporophyte tissue accompanies the embryo in the form of the seed coat and the fruit.

Within the ovule, meiosis and unequal cytokinesis yield four megaspores. The largest of these megaspores undergoes three mitotic divisions to produce the female gametophyte, a seven-celled **embryo sac** with eight nuclei (Figure 20.5). One of these cells is the egg. Two **synergid cells**

surround the egg and the pollen tube enters the embryo sac by penetrating one of the synergids. The **central cell** contains two or more polar nuclei, which will fuse with the second sperm nucleus and develop into the polyploid endosperm. Three **antipodal cells** form at the opposite end of the embryo sac from the synergids and degenerate before or during embryonic development. There is no known function for the antipodals. Genetic analyses of female gametophyte development in maize and *Arabidopsis**** are providing insight into the regulation of the specific steps in this process (Pagnussat et al. 2005).

Pollination

Pollination refers to the landing and subsequent germination of the pollen on the stigma. Hence it involves an interaction between the gametophytic generation of the male (the pollen) and the sporophytic generation of the female (the stigmatic surface of the carpel). Pollination can occur within a single **perfect flower** that contains both male and female gametophytes (self-fertilization), or pollen can land on a different flower on the same or a different plant. About 96 percent of flowering plant species produce male and female gametophytes on the same plant. However, about 25 percent of these produce two different types of flowers on the same plant, rather than perfect flowers.

Staminate flowers lack carpels, while **carpellate** flowers lack stamens. Maize plants, for example, have staminate (tassel) and carpellate (ear) flowers on the same plant. Such species, which include the majority of the angiosperms, are considered to be **monoecious** (Greek *mono,*

---

*A small weed in the mustard family, *Arabidopsis* is used as a model organism because of its very small genome.

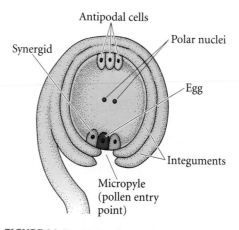

FIGURE 20.5   The embryo sac is the product of three mitotic divisions of the haploid megaspore; it comprises seven cells and eight haploid nuclei. The two polar nuclei in the central cell will fuse with the second sperm nucleus and produce the endosperm that will nourish the egg. The other six cells, including the egg, contain one haploid nucleus each.

"one"; *oecos*, "house"). The remaining 4 percent of species (e.g., willows, maples, and date palms*) produce staminate and carpellate flowers on separate plants. These species are considered to be **dioecious** ("two houses"). Only a few plant species have true sex chromosomes, yet they arose several times in flowering plant evolution (Charlesworth 2002). The terms "male" and "female" are most correctly applied only to the gametophyte generation, not to the sporophyte (Cruden and Lloyd 1995).

SELF-INCOMPATIBILITY   The arrival of a viable pollen grain on a receptive stigma does not guarantee fertilization. **Interspecific incompatibility** refers to the failure of pollen from one species to germinate and/or grow on the stigma of another species. **Intraspecific incompatibility** is incompatibility that occurs within a species. **Self-incompatibility**—incompatibility between the pollen and the stigmas of the same individual—is an example of intraspecific incompatibility (see Kao and Tsuikamoto 2004). Self-incompatibility blocks fertilization between two genetically similar gametes, increasing the probability of new gene combinations by promoting outcrossing (pollination by a different individual of the same species). Groups of closely related plants can contain a mix of self-compatible and self-incompatible species.

Several different self-incompatibility systems have evolved. Recognition of self depends on the multiallelic self-incompatibility (*S*) locus (Nasrallah 2002). *Gametophytic self-incompatibility* occurs when the *S* allele of the pollen grain matches either of the *S* alleles of the stigma (remember that the stigma is part of the diploid sporophyte generation, which has two *S* alleles, while a single pollen grain carries one *S* allele). In this case, the pollen tube begins

developing but stops before reaching the micropyle (Figure 20.6A). *Sporophytic self-incompatibility* occurs when one of the two *S* alleles of the pollen-producing sporophyte (not the gametophyte) matches one of the *S* alleles of the stigma (Figure 20.6B). Most likely, sporophyte contributions to the pollen exine are responsible for this type of self-incompatibility.

The *S* locus consists of several physically linked genes that regulate recognition and rejection of pollen. An *S* gene has been cloned that codes for an RNase (called *S* RNase) that is sufficient, in the gametophytically self-incompatible petunia pistil, to recognize and reject self-pollen (Lee et al. 1994). The pollen component of gametophytic self-incompatibility in the petunia, *SLF* (*S*-locus, *F*-box), is an F-box gene[†] within the *S* locus (Sijacic et al. 2004).

A different, more rapid gametophytic response to self-incompatibility has been investigated in poppies, a relative of the more basal flowering plants. Calcium ions accumulate in the tip of the pollen tubes, where open calcium channels are concentrated (Jaffe et al. 1975; Trewavas and Malho 1998). There is direct evidence that pollen tube growth in the poppy is regulated by a slow-moving Ca^{2+} wave controlled by the phosphoinositide signaling pathway (Figure 20.7; Franklin-Tong et al. 1996). Ca^{2+} influx occurs at both the tip of the pollen tube and on the shanks. Altered calcium influx is observed when the pollen tube is self-incompatible with the style, which leads to F-actin depolymerization, destabilization of the pollen cytoskeleton, and cessation of pollen tube growth (Franklin-Tong et al. 2002; Franklin-Tong and Franklin 2003). The incompatible pollen tube then undergoes programmed cell death (Thomas and Franklin-Tong 2004).

In sporophytic self-incompatibility, a ligand on the pollen is thought to bind to a membrane-bound kinase receptor in the stigma, starting a signaling process that

---

*The discovery that plants had sexes was important to the economy of date palms in the ancient Near East over two thousand years ago. Since only the female trees bore fruit, date farmers planted just a few male trees, then hand-pollinated the many female trees. This practice greatly increased the fruit yield per acre, and such pollination events became associated with spring fertility festivals.

[†]Members of the F-box family of genes share a common "F-box domain" for binding transcription factors. Although some F-box genes have been found in other eukaryote groups, most of these genes are unique to the plants.

(A) Gametophytic self-incompatibility

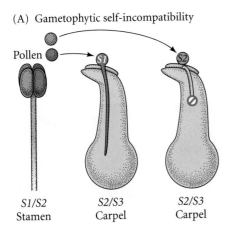

Pollen

S1/S2
Stamen

S2/S3
Carpel

S2/S3
Carpel

(B) Sporophytic self-incompatibility

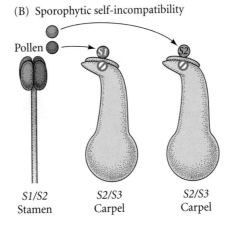

Pollen

S1/S2
Stamen

S2/S3
Carpel

S2/S3
Carpel

FIGURE 20.6   Self-incompatibility. *S1*, *S2*, and *S3* are different alleles of the self-incompatibility (*S*) locus. (A) Plants with gametophytic self-incompatibility reject pollen only when the genotype of the pollen (i.e., the gametophyte) matches either one of the carpel's two alleles. (B) In sporophytic self-incompatibility, the genotype of the pollen parent (i.e., the sporophyte), not just that of the haploid pollen grain, can trigger an incompatibility response.

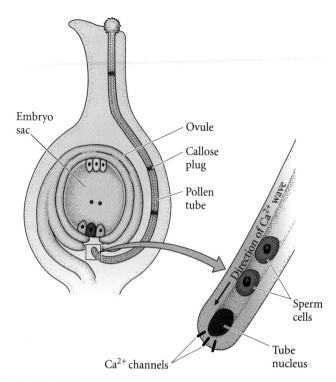

FIGURE 20.7 Calcium and pollen tube tip growth. After compatible pollen germinates, the pollen tube grows toward the micropyle. Waves of calcium ions play a key role in this growth of the tube. (After Franklin-Tong et al. 1996.)

leads to pollen rejection. In *Brassica,* one of the genes of the *S* locus encodes a transmembrane serine-threonine kinase (SRK) that functions in the epidermis of the stigma and binds a cysteine-rich peptide (SCR) from the pollen (Figure 20.8; Kachroo et al. 2001).

There are numerous examples of plant populations that have switched from self-incompatible to self-fertilizing systems. Changes in the *S* locus, specifically the *SKR* and *SCR* genes, could account for these evolutionary changes. The Nasrallahs (2002) created self-incompatible *Arabidopsis thaliana* plants (which are normally self-compatible) by introducing the *SKR* and *SCR* genes that encode self-recognizing proteins from *A. lyrata* (a self-incompatible species). This experiment demonstrates that *A. thaliana* still has all of the downstream components of the signal cascade that can lead to pollen degradation. The mechanism of pollen degradation is unclear, but appears to be highly specific.

FIGURE 20.8 Receptor-ligand self-recognition is the key to self-incompatibility in *Brassicas.* Allelic variability in both the *SRK* and *SRC* genes leads to a variety of possible combinations of ligand and receptor proteins. Unlike the common self-recognition systems of animals, including immunity and mating, self-incompatibility results from the binding of SRK and SRC proteins of self (from allelic *S* loci) rather than nonself. (After Nasrallah 2002; photographs courtesy of J. Nasrallah.)

POLLEN GERMINATION If the pollen and the stigma are compatible, the pollen takes up water (hydrates) and the pollen tube emerges. The pollen tube grows down the style of the carpel toward the micropyle (Figure 20.9). The tube nucleus and the sperm cells are kept at the growing tip by bands of callose (a complex carbohydrate). It is possible that this may be an exception to the "plant cells do not move" rule, as the generative cell(s) appear to move forward via adhesive molecules (Lord 2000). Pollen tube growth is quite slow in gymnosperms (up to a year), while in some angiosperms the tube can grow as rapidly as 1 cm per hour.

Genetic approaches have been useful in investigating how the growing pollen tube is guided toward unfertilized ovules. In *Arabidopsis,* the pollen tube appears to be guided by a long-distance signal from the ovule (Hulskamp et al. 1995; Wilhelmi and Preuss 1999). Analysis of pollen tube growth in ovule mutants of *Arabidopsis* indicates that the haploid embryo sac is particularly important in the long-range guidance of pollen tube growth. Mutants

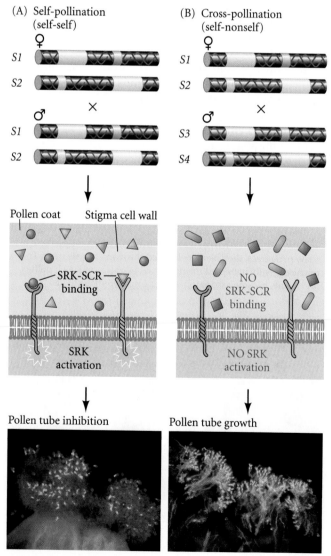

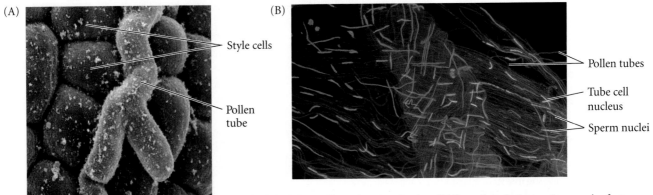

FIGURE 20.9 Pollen tube germination. (A) Scanning electron micrograph of an *Arabidopsis* pollen tube en route to the ovule for fertilization. (B) Lily pollen tubes grown in vivo and removed from the ovary. Each green strand is an individual pollen tube and contains two sperm nuclei (bright blue stain) and a fainter (lighter blue) tube cell nucleus. Note the huge number of pollen tubes, all "racing" to fertilize a single egg. (Photographs courtesy of E. Lord.)

with defective sporophyte tissue in the ovule but a normal haploid embryo sac appear to stimulate normal pollen tube development.

While the evidence points primarily to the role of the gametophyte generation in pollen tube guidance, diploid cells may make some contribution. The *Arabidopsis* gene *POP2* encodes an enzyme that degrades γ-amino butyric acid (GABA) and establishes a gradient of GABA in the style up to the micropyle (Palanivelu et al. 2003). The *pop2* mutant has misguided pollen tube growth, presumably because there is no GABA gradient. *POP2* is expressed in both the pollen and the style, which may explain why wild-type pollen tubes find their way to the micropyle when the style has a *pop2* genotype. The wild-type enzyme in the pollen tube may degrade GABA and create a sufficient gradient to guide itself to the micropyle.

As the final step in pollen guidance, the two synergid cells in the embryo sac may attract the pollen tube. In *Torenia fournieri* (wishbone flower), the embryo sac protrudes from the micropyle and can be cultured. In vitro, it can attract a pollen tube. Higashiyama and colleagues (2001) used a laser beam to destroy individual cells in the embryo sac and then tested whether or not pollen tubes were still attracted to the embryo sac. A single synergid was sufficient to guide pollen tubes; however, when both synergids were destroyed, pollen tubes were not attracted to the sac.

Fertilization

The growing pollen tube enters the embryo sac through the micropyle and grows through one of the synergids. The two sperm cells are released, and a **double fertilization** event occurs (see Southworth 1996). One sperm cell fuses with the egg, producing the zygote that will develop into the sporophyte. The second sperm cell fuses with the bi- or multinucleate central cell, giving rise to the **endosperm**, which nourishes the developing embryo. This second event is not true fertilization in the sense of male and female gametes undergoing syngamy (fusion)—that is, it does not result in a zygote, but in nutritionally supportive endoderm. When you eat popcorn, you are actually eating "popped" endosperm. The other accessory cells in the embryo sac degenerate after fertilization.

The zygote of the angiosperm produces only a single embryo.* Double fertilization, first identified a century ago, is generally restricted to the angiosperms, but it has also been found in the gymnosperm genera *Ephedra* and *Gnetum*, although no endosperm forms. Friedman (1998) has suggested that endosperm may have evolved from a second zygote "sacrificed" as a food supply in an early gymnosperm lineage with double fertilization.

Investigations of *Amborella*, the most closely related extant relative of the basal angiosperm (Figure 20.10A), is providing information on the evolutionary origin of the endosperm (Brown 1999). It is probable that the first angiosperm had a four-nucleus embryo sac (Williams and Friedman 2002, 2004). The critical cell to consider is the central cell, which is fertilized by the second sperm to create the endosperm. In eight-nuclei embryo sacs, there are seven cells. The central cell contains two nuclei and, when fertilized, produces a triploid endosperm. In *Nuphar*, a basal angiosperm, the embryo sac consists of four nuclei, and the central cell has a single nucleus that, when fertilized, develops into a 2*n* endosperm (Figure 20.10B). The 2*n* endosperm provides convincing evidence that the four-celled embryo sac in *Nuphar* does not result from the degradation of four nuclei. If other cells had degraded, a 3*n* endosperm would be predicted.

---

*The gymnosperm zygote, on the other hand, produces two or more embryos after cell division begins, by a process known as cleavage embryogenesis.

FIGURE 20.10 Ancestral angiosperms. (A) *Amborella trichopoda*. This plant is more closely related to the first angiosperm than any other extant species. (B) Ancestral angiosperms probably had 2*n* endosperms. The embryo sac of the basal angiosperm *Nuphar* (yellow water lily) has a single nucleus in its central cell, which when fertilized will produce a 2*n* endosperm. DAPI staining shows that the DNA content is 1*n*, not *n*. Because this is a section of tissue, the egg cell is hidden behind the two synergids and is shown in the insert. (A photograph courtesy of Sandra K. Floyd; B photograph courtesy of William Freidman.)

(A)

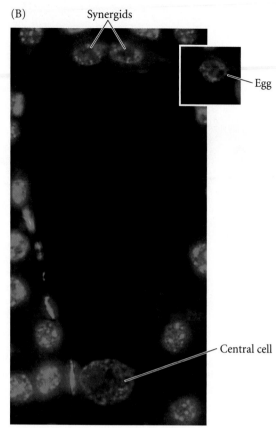

(B) Synergids

Egg

Central cell

Fertilization is not an absolute prerequisite for angiosperm embryonic development (Mogie 1992). Embryos can form within embryo sacs from haploid eggs and from cells that did not divide meiotically. This phenomenon is called **apomixis** (Greek, "without mixing"), and results in viable seeds. The viability of the resulting haploid sporophytes indicates that ploidy alone does not account for the morphological distinctions between the gametophyte and the sporophyte. Embryos can also develop from cultured sporophytic tissue. These embryos develop with no associated endosperm, and they lack a seed coat.

Embryonic Development

Embryogenesis

In plants, the term *embryogenesis* covers development from the time of fertilization until dormancy occurs. The basic body plan of the sporophyte is established during embryogenesis; however, this plan is reiterated and elaborated after dormancy is broken. The major challenges of plant embryogenesis are:

- To establish the basic body plan. **Radial patterning** produces three tissue systems (dermal, ground, and vascular), and **axial patterning** establishes the apical-basal (shoot-root) axis.
- To set aside meristematic tissue for postembryonic elaboration of the body structure (leaves, roots, flowers, etc.).
- To establish an accessible food reserve for the germinating embryo until the embryo becomes autotrophic.

Embryogenesis is similar in all angiosperms in terms of the establishment of the basic body plan. There are differences in pattern elaboration, however, including differences in the precision of cell division patterns, the extent of endosperm development, cotyledon development, and the extent of shoot meristem development (Esau 1977; Steeves and Sussex 1989; Johri et al. 1992).

MATERNAL EFFECTS IN EARLY EMBRYOGENESIS Maternal effect genes play a key role in establishing embryonic patterns in animals (see, for example, the discussion of *Drosophila* in Chapter 9). The extent of extrazygotic gene involvement in plant embryogenesis is an open question, complicated by at least three potential sources of influence: sporophytic tissue, gametophytic tissue, and the polyploid endosperm. All of these tissues are in close association with the egg/zygote (Ray 1998). Endosperm development could also be affected by maternal genes. Sporophytic and gametophytic maternal effect genes have been identified in *Arabidopsis*, and it is probable that the endosperm genome influences the zygote as well.

The first maternal effect gene identified, *SHORT INTEGUMENTS 1* (*SIN1*), must be expressed in the sporophyte for normal embryonic development to occur (Ray et al. 1996). Two transcription factors (FBP7 and FBP11) are needed in the petunia sporophyte for normal endosperm development (Columbo et al. 1997). A female gametophytic maternal effect gene, *MEDEA*,* has protein domains similar to those of a *Drosophila* maternal effect gene (Grossniklaus et al. 1998). Curiously, *MEDEA* is in the Polycomb gene group (see Chapter 9), whose products alter chromatin, directly or indirectly, and affect transcription. *MEDEA* affects an imprinted gene (see Chapter 5) that is

---

*Another name from Greek mythology, after Euripides' Medea, who killed her own children.

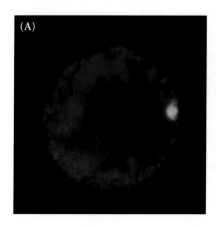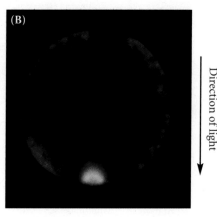

FIGURE 20.11 Axis formation in the brown alga *Pelvetia compressa*. (A) An F-actin patch (orange) is first formed at the point of sperm entry; the blue spot marks the sperm pronucleus. (B) Later, light was shone in the direction of the arrow. The sperm-induced axis was overridden, and an F-actin patch formed on the dark side, where the rhizoid will later form. (Photographs courtesy of W. Hables.)

Direction of light

expressed by the female gametophyte and by maternally inherited alleles in the zygote, but not by paternally inherited alleles (Vielle-Calzada et al. 1999). The significance of maternal effect genes in establishing the sporophyte body plan has been highlighted by Pagnussat and co-workers' (2005) screen of 130 female gametophytic mutants. Nearly half the mutations were in a maternal gene, further implicating the female gametophyte or maternal genome in embryo development.

FIRST ASYMMETRIC DIVISION: BROWN ALGAE Polarity is established in the first cell division following fertilization. Because angiosperm embryos are deeply embedded in multiple layers of tissue, the establishment of polarity is also investigated in brown algae, a model system with external fertilization (Belanger and Quatrano 2000; Brownlee 2004). The zygotes of these plants are independent of other tissues and are amenable to manipulation. The initial cell division results in one smaller cell, which will form

the rhizoid (root homologue) and anchor the rest of the plant, and one larger cell, which gives rise to the thallus (the main body of the sporophyte). The point of sperm entry fixes the position of the rhizoid end of the apical-basal axis. This axis is perpendicular to the plane of the first cell division. F-actin accumulates at the rhizoid pole (Figure 20.11A; Kropf et al. 1999). However, light or gravity can override this fixing of the axis and establish a new position for cell division (Figure 20.11B; Alessa and Kropf 1999). Once the apical-basal axis is established, secretory vesicles are targeted to the rhizoid pole of the zygote (Figure 20.12). These vesicles contain material for rhizoid outgrowth, with a cell wall of distinct macromolecular com-

FIGURE 20.12 Asymmetrical cell division in brown algae. Time course from 8 to 25 hours after fertilization, showing algal cells stained with a vital membrane dye to visualize secretory vesicles, which appear first, and the cell plate, which begins to appear about halfway through this sequence. (Photographs courtesy of K. Belanger.)

8 hours after fertilization

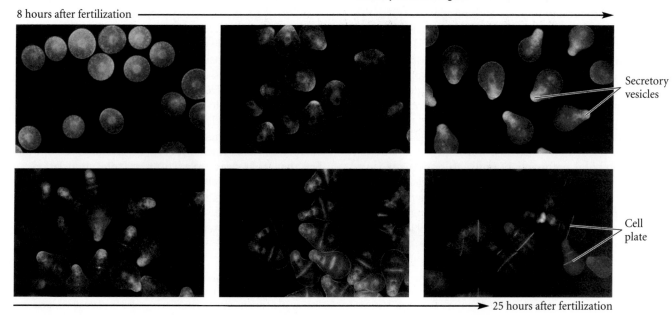

Secretory vesicles

Cell plate

25 hours after fertilization

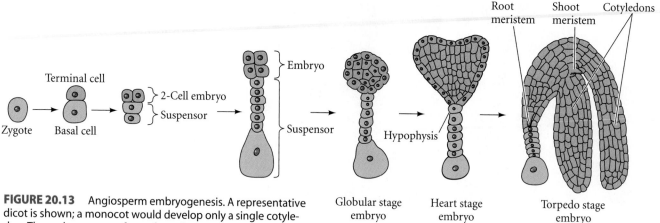

FIGURE 20.13 Angiosperm embryogenesis. A representative dicot is shown; a monocot would develop only a single cotyledon. The embryo proper forms from the terminal cell; the basal cell divides to form the suspensor, which will degenerate as development progresses. The point of interface between the suspensor and the embryo is the hypophysis. While there are basic patterns of embryogenesis in angiosperms, there is tremendous morphological variation among species.

position. Targeted secretion may also help orient the first plane of cell division. Maintenance of rhizoid versus thallus fate early in development depends on information in the cell walls (Brownlee and Berger 1995). Such cell wall information also appears to be important in angiosperms (see Scheres and Benfey 1999).

FIRST ASYMMETRIC DIVISION: ANGIOSPERMS The basic body plan of the angiosperm laid down during embryogenesis also begins with an asymmetrical* cell division, giving rise to a terminal cell and a basal cell (Figure 20.13). The terminal cell gives rise to the embryo proper. The basal cell forms closest to the micropyle and gives rise to the suspensor. The hypophysis is found at the interface between the suspensor and the embryo proper. In some species it gives rise to a portion of the root cells. (The suspensor cells divide to form a filamentous or spherical organ that degenerates later in embryogenesis.) In both gymnosperms and angiosperms, the suspensor orients the absorptive surface of the embryo toward its food source; in angiosperms, it also appears to serve as a nutrient conduit for the developing embryo.

The study of embryo mutants in maize and *Arabidopsis* has been particularly helpful in sorting out the different developmental pathways of embryos and suspensors. Investigations of suspensor mutants (*sus1, sus2,* and *raspberry1*) of *Arabidopsis* have provided genetic evidence that the suspensor has the capacity to develop embryo-like structures (Figure 20.14A,B; Schwartz et al. 1994; Yadegari

et al. 1994). In these mutants, abnormalities in the embryo proper appear prior to suspensor abnormalities.[†] Earlier experiments in which the embryo proper was removed also demonstrated that suspensors could develop like embryos (Haccius 1963). A signal from the embryo proper to the suspensor may be important in maintaining suspensor identity and blocking the development of the suspensor as an embryo. Molecular analyses of these and other genes are providing insight into the mechanisms of communication between the suspensor and the embryo proper (Figure 20.14C).

The *SUS1* gene has been renamed *DCL1* (*DICER-LIKE1*) because its predicted protein sequence is structurally like that of *Dicer* in *Drosophila melanogaster* and *DCR-1* in *Caenorhabditis elegans* (Schauer et al. 2002). These proteins may control the translation of developmentally important mRNAs. This is an exciting discovery that will lead to a better understanding of the regulation of development beyond the level of transcriptional control. Intriguingly, *DCL1* has several alleles that were originally assumed to be completely different genes regulating very different developmental processes in plants. *DCL1* alleles include *sin1* alleles. These mutants affect ovule development (discussed in the next section) and the transition from vegetative to reproductive development (discussed later in the chapter). The *carpel factory* (*caf1*) allele of *DCL1* causes indeterminancy in floral meristems leading to extra whorls of carpels. Extrapolating from *Drosophila* Dicer function, DCL1 protein may be involved in cleaving small, noncoding RNAs into even smaller, 21- to 25-nucleotide, single-stranded RNA products that could cleave to mRNAs and affect translation.

Many questions about the role of microRNAs as possible developmental signals are arising from the work being done on *DCL1* alleles and on leaf asymmetry genes, which will be discussed later (Kidner and Martienssen 2005).

---

*Asymmetrical cell division is also important in later angiosperm development, including the formation of guard cells of leaf stomata and of different cell types in the ground and vascular tissue systems.

[†]Another intriguing characteristic of these mutants is that cell differentiation occurs in the absence of morphogenesis. Thus, cell differentiation and morphogenesis can be uncoupled in plant development.

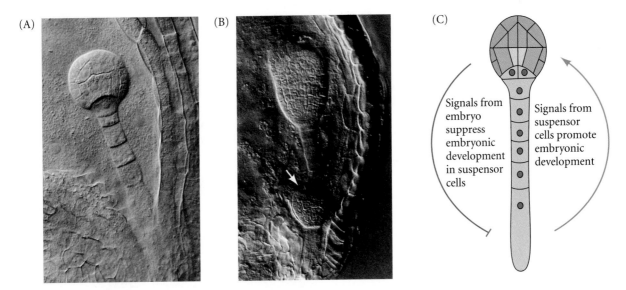

FIGURE 20.14 The *SUS* gene (a *DCL1* allele) suppresses embryonic development in the suspensor. (A) Wild-type embryo and suspensor. (B) The suspensor of a *sus* mutant develops like an embryo (arrow). (C) Model showing how the embryo proper suppresses embryonic development in the suspensor and the suspensor provides feedback information to the embryo. (Photographs courtesy of D. Meinke.)

RADIAL AND AXIAL PATTERNING Radial and axial patterns develop as cell division and differentiation continue (Figure 20.15). The cells of the embryo proper divide in transverse and longitudinal planes to form a globular stage embryo with several tiers of cells. Superficially, this stage bears some resemblance to cleavage in animals, but the nucleus/cytoplasm ratio does not necessarily increase. The emerging shape of the embryo depends on regulation of the planes of cell division and expansion, since the cells are not able to move and reshape the embryo. Cell division planes in the outer layer of cells become restricted, and this layer, called the **protoderm**, becomes distinct. Radial patterning emerges at the globular stage as the three tissue systems (dermal, ground, and vascular) of the plant are initiated (Figure 20.15A). The dermal tissue (epidermis) will form from the protoderm and contribute to the outer protective layers of the plant. Ground tissue (cortex and pith) forms from the ground meristem, which lies beneath the protoderm. The procambium, which forms at the core of the embryo, gives rise to the vascular tissue (xylem and phloem), which will function in support and transport. The differentiation of each tissue system is at least partially independent. For example, in the *keule* mutant of *Arabidopsis*, the dermal system is defective while the inner tissue systems develop normally (Mayer et al. 1991).

The globular shape of the embryo is lost as **cotyledons** ("first leaves") begin to form. Dicots have two cotyledons, which give the embryo a heart-shaped appearance as they form. In monocots, such as maize, only a single cotyledon emerges. The axial body plan is evident by this **heart stage** of development (Figure 20.15B). Hormones (specifically, auxins) may mediate the transition from radial to bilateral symmetry (Liu et al. 1993).

AUXIN AND THE APICAL-BASAL AXIS The hormone auxin plays a key role in establishing the apical-basal axis of the embryo. Very early in embryogenesis, the family of pin-formed (PIN) auxin efflux carriers are asymmetrically distributed (Figure 20.16A; Friml et al. 2003, 2004). The pinoid (PID) protein kinase aids in the asymmetric localization of PIN proteins. At the 4-cell stage, auxin moves apically, but by the globular stage, the efflux carriers direct the flow of auxin to the hypophysis, which organizes as the root meristem (Figure 20.16B). Disruption of the auxin morphogenetic gradient, either through mutation or overexpression of

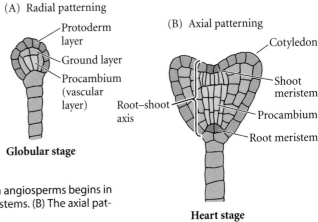

(A) Radial patterning

(B) Axial patterning

FIGURE 20.15 Radial and axial patterning. (A) Radial patterning in angiosperms begins in the globular stage and results in the establishment of three tissue systems. (B) The axial pattern (shoot-root axis) is established by the heart stage.

(A)

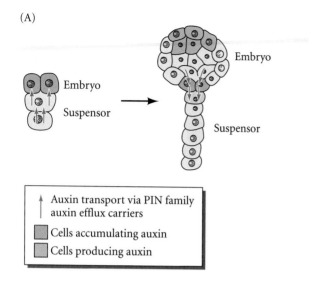

Embryo

Suspensor

Embryo

Suspensor

↑ Auxin transport via PIN family
  auxin efflux carriers

� Cells accumulating auxin

▢ Cells producing auxin

FIGURE 20.16   An auxin gradient specifies the shoot-root axis. (A) A family of PIN auxin efflux carriers are responsible for the early apical flow of auxin in the embryo and the shift to basal auxin flow as the apical end of the globular stage embryo begins to produce auxin. (B,C) The PID protein kinase localizes the PIN auxin efflux carriers. (B) In wild-type globular embryos, PID expression leads to the accumulation of PIN1 at the basal end of the embryo. When an auxin-response gene is fused with GFP and expressed, basal accumulation and normal embryonic development occur. (C) Ectopic overexpression of the *PID* gene (induced by fusion to a viral promoter) eliminates basal accumulation of PIN1. The GFP-labeled auxin response gene is then expressed throughout the embryo proper rather than in the basal portion of the embryo.

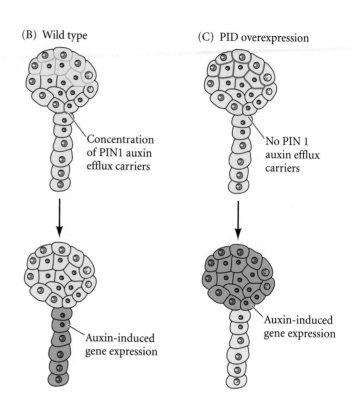

(B) Wild type

(C) PID overexpression

Concentration of PIN1 auxin efflux carriers

No PIN 1 auxin efflux carriers

Auxin-induced gene expression

Auxin-induced gene expression

PID allows auxin to accumulate more apically and root development is abnormal or inhibited (Figure 20.16C). Although auxin accumulation is necessary for apical-basal polarity, an extensive study of the PIN proteins reveals that auxin alone may not be sufficient to establish polarity in the embryo (Weijers and Jürgens 2005).

The discovery of the auxin receptor provided a link between the asymmetric distribution of auxin and the expression of auxin-induced genes (Dharmasiri et al. 2005; Kipinski and Leyser 2005). In the absence of auxin, the auxin response factors (ARF) are in a repressed state (Figure 20.17A). These genes form a heterodimer with Aux/IAA protein, which inhibits the transcription of auxin-induced genes. In the presence of auxin, the enzyme SCF binds Aux/IAA, catalyzing Aux/IAA ubiquitination (Figure 20.17B). Ubiquitination targets Aux/IAA for degradation by the 26S proteosome. Freed from the Aux/IAA repressor, ARF can act on its own to stimulate transcription or form a homo/heterodimer with another ARF to further modulate gene expression.

COTYLEDONS   In many plants, the cotyledons aid in nourishing the plant by becoming photosynthetic after germina-

tion (although those of some species never emerge from the ground). In some cases—peas, for example—the food reserve in the endosperm is used up before germination, and the cotyledons serve as the nutrient source for the germinating seedling.* Even in the presence of a persistent endosperm (as in maize), the cotyledons store food reserves such as starch, lipids, and proteins. In many monocots, the cotyledon grows into a large organ pressed against the endosperm and aids in nutrient transfer to the seedling. Upright cotyledons can give the embryo a torpedo shape. In some plants, the cotyledons grow sufficiently long that they must bend to fit within the confines of the seed coat. The embryo then looks like a walking stick. By this point, the suspensor is degenerating.

The *Arabidopsis LEAFY COTYLEDON1* (*LEC1*) gene, first identified by a mutant with leaflike cotyledons (Meinke 1994), is necessary to maintain the suspensor early in development, to specify cotyledeon identity, to initiate maturation, and to prevent early germination of the seed. *LEC1* belongs to a gene class that is unique among the embryogenesis genes in acting throughout the course of embryo development (Harada 2001; Kwong et al. 2003).

MERISTEM ESTABLISHMENT   The **shoot apical meristem** and **root apical meristem** are clusters of stem cells that will per-

---

*Mendel's famous wrinkled-seed mutant (the *rugosus* or *r* allele) has a defect in a starch branching enzyme that affects starch, lipid, and protein biosynthesis in the seed and leads to defective cotyledons (Bhattacharyya et al. 1990).

sist in the postembryonic plant and give rise to most of the sporophyte body (see Jürgens 2001 for a review of apical-basal pattern formation). The root meristem is partially derived from the hypophysis in some species (see Figure 20.15B). All other parts of the sporophyte body are derived from the embryo proper.

Genetic evidence indicates that the formation of the shoot and root meristems is regulated independently. From an evolutionary perspective, this is not surprising. One of the major adaptations to terrestrial life involved the evolution of a root system.* This independence is demonstrated by the *dek23* maize mutant and the *shootmeristemless* (*STM*) mutant of *Arabidopsis*, both of which form a root meristem but fail to initiate a shoot meristem (Clark and Sheridan 1986; Barton and Poethig 1993). The *STM* gene, which has a homeodomain, is expressed in the late globular stage, before cotyledons form. Genes have also been identified that specifically affect the development of the root axis during embryogenesis. Mutations of the *HOBBIT* gene in *Arabidopsis*, for example, affect the hypophysis derivatives and eliminate root meristem function (Willemson et al. 1998). While it is clear that root and shoot developmental programs are different, what triggers root or

---

*It should be noted, however, that the nonvascular land plants, including the mosses, did not develop root systems.

shoot development is less tractable. The *TOPLESS* gene in *Arabidopsis* may provide some clues. A single gene mutation has been identified that converts a shoot into a root, but how the wild-type gene functions is still a puzzle (Long et al. 2002).

The shoot apical meristem will initiate leaves after germination and, ultimately, the transition to reproductive development. In *Arabidopsis*, the cotyledons are produced from general embryonic tissue, not from the shoot meristem (Barton and Poethig 1993). In many angiosperms, a few leaves are initiated during embryogenesis. In the case of *Arabidopsis*, clonal analysis points to the presence of leaves in the mature embryo, even though they are not morphologically well developed (Irish and Sussex 1992). Clonal analysis has demonstrated that the cotyledons and the first two true leaves of cotton plants are derived from embryonic tissue rather than an organized meristem (Christianson 1986).

Clonal analysis experiments provide information on cell fates, but do not necessarily indicate whether or not cells are determined for a particular fate. Clonal analysis has demonstrated that cells that divide in the wrong plane and "move" to a different tissue layer often differentiate according to their new position. Position, rather than clonal origin, appears to be the critical factor in embryo pattern formation, suggesting some type of cell-cell communication (Laux and Jürgens 1994).

Dormancy

From the earliest stages of embryogenesis, there is a high level of zygotic gene expression. As the embryo reaches maturity, there is a shift from constructing the basic body plan to creating a food reserve by accumulating storage carbohydrates, proteins, and lipids. Genes coding for seed storage proteins were among the first to be characterized by plant molecular biologists because of the high levels of specific storage protein mRNAs that are present at different times in embryonic development. The high level of

(A) Auxin (Indole-3-acetic acid)

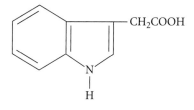

(B) No Auxin present

(C) Auxin present

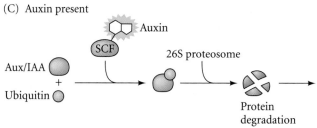

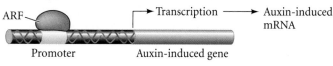

FIGURE 20.17 Mechanism of auxin action. (A) In the absence of auxin, auxin response factors (ARF) are in a repressed state. These genes form a heterodimer with Aux/IAA protein, which inhibits the transcription of auxin-induced genes. (B) In the presence of auxin, the enzyme SCF binds Aux/IAA, catalyzing Aux/IAA ubiquitination. Ubiquitination targets Aux/IAA for degradation by the 26S proteosome. Freed from the Aux/IAA repressor, ARF can act on its own to stimulate transcription or form a homo or heterodimer with another ARF to further modulate gene expression.

FIGURE 20.18 *Viviparous* maize mutant. Each kernel on this maize ear contains an embryo. *Viviparous* embryos do not go through a dormant phase, but begin germinating while still on the ear.

metabolic activity in the developing embryo is fueled by continuous input from the parent plant into the ovule. Eventually metabolism slows, and the connection of the seed to the ovary is severed by the degeneration of the adjacent supporting sporophyte cells. The seed desiccates (loses water), and the integuments harden to form a tough seed coat. The seed has entered **dormancy**, officially ending embryogenesis. The embryo can persist in a dormant state for weeks or years, a phenomenon that affords tremendous survival value. There are even cases where seeds found stored in ancient archaeological sites have germinated after thousands of years of dormancy.

Maturation leading to dormancy is the result of a precisely regulated program. The *viviparous* mutation in maize, for example, produces genetic lesions that block dormancy (Steeves and Sussex 1989). The apical meristems of *viviparous* mutants behave like those of ferns, with no pause before producing postembryonic structures. The embryo continues to develop, and seedlings emerge from the kernels on the ear attached to the parent plant (Figure 20.18). A group of plant genes have been identified that belong to the Polycomb group, which regulates early development in mammals, nematodes, and insects (Preuss 1999). These genes encode chromatin silencing factors, which may play an important role in seed formation.

Plant hormones are critical in dormancy, and linking them to genetic mechanisms is an active area of research. The hormone **abscisic acid** is important in maintaining dormancy in many species. **Gibberellins**, another class of hormones, are important in breaking dormancy.

Germination

The postembryonic phase of plant development begins with **germination**. Some dormant seeds require a period of **after-ripening**, during which low-level metabolic activities continue to prepare the embryo for germination. Highly evolved interactions between the seed and its environment increase the odds that the germinating seedling will survive to produce another generation.

Temperature, water, light, and oxygen are all key in determining the success of germination. **Stratification** is the requirement for chilling (5°C) to break dormancy in some seeds. In temperate climates, this adaptation ensures that germination takes place only after the winter months have passed. In addition, seeds have maximum germination rates at moderate temperatures of 25–30°C and often will not germinate at extreme temperatures. Seeds such as lettuce require light (specifically, the red light wavelengths) for germination; thus seeds will not germinate so far below ground that they use up their food reserves before photosynthesis is possible.

Desiccated seeds may be only 5–20 percent water. **Imbibition** is the process by which the seed rehydrates, soaking up large volumes of water and swelling to many times its original size. The **radicle** (primary embryonic root) emerges from the seed first to enhance water uptake; it is protected by a root cap produced by the root apical meristem. Water is essential for metabolic activity, but so is oxygen; a seed sitting in a glass of water will not survive. Some species have such hard protective seed coats that they must be **scarified** (scratched or etched) before water and oxygen can cross the barrier. Scarification can occur when the seed is exposed to the weather and other natural elements over time, or by its exposure to acid as the seed passes through the gut of a frugivore (fruit eater). The frugivore thus prepares the seed for germination, as well as dispersing it to a site where germination can take place.

During germination, the plant draws on the nutrient reserves in the endosperm or cotyledons. Interactions between the embryo and endosperm in monocots use gibberellin as a signal to trigger the breakdown of starch into sugar. As the shoot reaches the surface, the differentiation of chloroplasts is triggered by light. Seedlings that germinate in the dark have long, spindly stems and do not produce chlorophyll. This environmental response allows plants to use their limited resources to reach the soil surface, where photosynthesis will be productive.

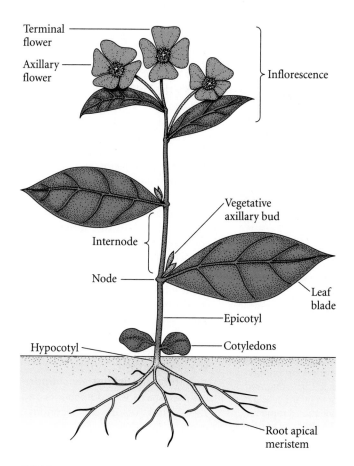

FIGURE 20.19 Morphology of a generalized angiosperm sporophyte.

Vegetative Growth

When the shoot emerges from the soil, most of the sporophyte body plan remains to be elaborated. Figure 20.19 shows the basic parts of the mature sporophyte plant, which will emerge from meristems.

Meristems

As has been mentioned, **meristems** are clusters of cells that allow the basic body pattern established during embryogenesis to be reiterated and extended after germination. Meristematic cells are similar to stem cells in animals.* They divide to give rise to one daughter cell that continues to be meristematic and another that differentiates. Meristems fall into three categories: apical, lateral, and intercalary.

Apical meristems occur at the growing shoot and root tips. Root apical meristems produce the root cap, which consists of lubricated cells that are sloughed off as the meristem is pushed through the soil by cell division and

---

*The similarities between plant meristem cells and animal stem cells may extend to the molecular level, indicating that stem cells existed before plants and animals pursued separate phylogenetic pathways. Homology has been found between genes required for plant meristems to persist and genes expressed in *Drosophila* germ line stem cells (Cox et al. 1998).

elongation in more proximal cells. The root apical meristem also gives rise to daughter cells that produce the three tissue systems of the root (Figure 20.20A). New root apical meristems are initiated from tissue within the core of the root and emerge through the ground tissue and dermal tissue. Root meristems can also be derived secondarily from the stem of the plant; in the case of maize, this is the major source of root mass.

The shoot apical meristem produces stems, leaves, and reproductive structures. In addition to the shoot apical meristem initiated during embryogenesis, axillary shoot apical meristems (axillary buds; Figure 20.20B) derived from the original one form in the axils (the angles between leaf and stem). Unlike new root meristems, these arise from the surface layers of the meristem.

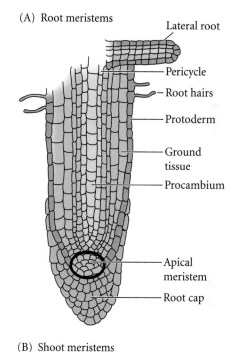

FIGURE 20.20 Both shoots and roots develop from apical meristems, with undifferentiated cells clustered at their tips. (A) The root meristem is protected by the root cap as it pushes through the soil. Lateral roots are derived from pericycle cells deep within the root. (B) The lateral organs of the shoot (leaves and axillary branches) have a superficial origin in the shoot apical meristem.

(A)

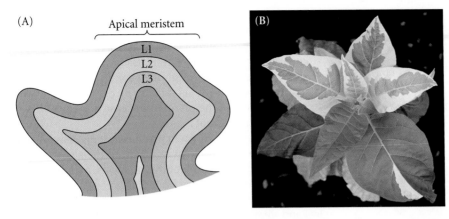

(B)

FIGURE 20.21 Organization of the shoot apical meristem. (A) Angiosperm meristems have two or three outer layers of cells that are histologically distinct (L1, L2, and L3). While cells in certain layers tend to have certain fates, they are not necessarily committed to those fates. If a cell is shifted to a new layer, it generally develops like the other cells in that layer. (B) The fates of the cell layers can be seen in a chimeric tobacco plant. One portion of the meristem contains three layers of wild-type cells, while the other portion has an L2 that lacks chlorophyll. This section of the meristem has given rise to the variegated leaves. In wild-type plants, the L1 layer always lacks chlorophyll (except in guard cells), but in this plant the L2, too, is genetically unable to produce chlorophyll; the L3 remains green. The L3 does not contribute to the outer edges of leaves, which is why they appear white in this plant. (Photograph courtesy of M. Marcotrigiano.)

Angiosperm shoot apical meristems are composed of up to three layers of cells (labeled L1, L2, and L3) on the plant surface (Figure 20.21A). One way of investigating the contributions of different layers to plant structure is by constructing chimeras (Figure 20.21B). Plant chimeras are composed of layers having distinct genotypes with discernible markers. When L2, for example, has a different genotype than L1 or L3, all pollen will have the L2 genotype, indicating that pollen is derived from L2. Chimeras have also been used to demonstrate classical induction in plants, in which (as in animal development) one layer influences the developmental pathway of an adjacent layer.

The size of the shoot apical meristem is precisely controlled by intercellular signals, most likely between layers of the meristem (see Bäurle and Laux 2003). Mutations in the *Arabidopsis CLAVATA* (*CLV*) genes, for example, lead to increased meristem size and the production of extra organs. *CLV1*, *CLV2*, and *CLV3* all limit the number of undifferentiated, stem cells in both vegetative and floral meristems (Figure 20.22). CLV1 is a serenine-threonine kinase that, along with the receptor-like transmembrane CLV2 protein, form a receptor for CLV3, which is localized in the extracellular space between cell layers of the meristem (Clark et al. 1997; Jeong et al. 1999; Rojo et al. 2002). Reddy and Meyerowitz (2005) have demonstrated that *CLV3* restricts its own domain of expression to the central zone by preventing the differentiation of surrounding cells. *POLTERGEIST* (*POL*) and *WUSCHEL* (*WUS*) are redundant in function and appear to keep genes in an undifferentiated state (Yu et al. 2000). *WUS* specifies stem cell iden-

tity in the cells positioned above, and the CLV signaling pathway is proposed to negatively feed back on *WUS* expression to control the size of the meristem. It is possible that *POL* expression is a target in CLV1 signal transduction. *STM*, like *POL* and *WUS*, plays an important role in maintaining an undifferentiated population of meristematic cells, and *STM* may positively regulate *WUS* (Clark and Schiefelbein 1997). The interactions of these gene products balance the rate of cell division (which enlarges the meristem) and the rate of cell differentiation in the periphery of the meristem (which decreases meristem size) (Meyerowitz 1997).

Lateral meristems are cylindrical meristems found in shoots and roots that result in secondary growth (an increase in stem and root girth by the production of vascular tissues). Monocots do not have lateral meristems, but often have **intercalary meristems** inserted in their stems between mature tissues. The popping sound you can hear in a cornfield on a summer night is caused by the rapid increase in stem length due to intercalary meristems.

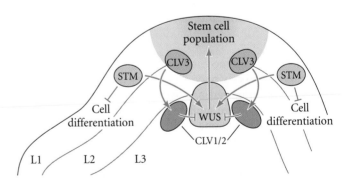

FIGURE 20.22 The WUS and STM proteins act to keep meristem cells in an undifferentiated state, while the products of the *CLAVATA* genes *CLV1*, *CLV2*, and *CLV3* all limit the number of undifferentiated meristem cells. The presence of WUS indirectly induces expression (upward arrow) of CLV3 in L1 and L2. CLV3 is a ligand that binds to the CLV1/CLV2 receptor in L3 cells. This binding triggers a negative feedback signal cascade that inhibits WUS expression and regulates the number of undifferentiated cells in the meristem, thus counterbalancing the roles of STM and WUS in keeping cells undifferentiated and dividing.

Root development

Radial and axial patterning in roots begins during embryogenesis and continues throughout development as the primary root grows and lateral roots emerge from the pericycle cells deep within the root. Both clonal analyses and laser ablation experiments that elilminate single cells have demonstrated that cells are plastic, and that position is the primary determinant of fate in early root development. Analyses of mutations in root radial organization have revealed genes with layer-specific activity (Scheres et al. 1995; Scheres and Heidstra 1999). We will illustrate these findings by looking at two *Arabidopsis* genes that regulate ground tissue fate.

In wild-type *Arabidopsis*, there are two layers of root ground tissue. The outer layer becomes the cortex while the inner layer becomes the endodermis, which forms a tube around the vascular tissue core. The *SCARECROW* (*SCR*) and *SHORT-ROOT* (*SHR*) genes have mutant phenotypes with only a single layer of root ground tissue (Benfey et al. 1993). The *SCR* gene is necessary for an asymmetrical cell division in the initial layer of cells, yielding a smaller endodermal cell and a larger cortex cell (Figure 20.23). The *scr* mutant expresses markers for both cortex and endodermal cells, indicating that differentiation progresses in the absence of cell division (Di Laurenzio et al. 1996). *SHR* is responsible for endodermal cell specification. Cells in the *shr* mutant do not develop endodermal features.

Axial patterning in roots may be morphogen-dependent, paralleling some aspects of animal development. A variety of experiments have established that the distribution of the plant hormone auxin organizes the axial pattern. A peak in auxin concentration at the root tip must be perceived for normal axial patterning (Sabatini et al. 1999; Ueda et al. 2005).

As discussed earlier, distinct genes specifying root and shoot meristem formation have been identified; however, root and shoot development may share common groups of genes that regulate cell fate and patterning (Benfey 1999). This appears to be the case for the *SCR* and *SHR* genes. In the shoot, these genes are necessary for the normal gravitropic response, which is dependent on normal endodermis formation (a defect in mutants of both genes; see Figure 20.29C). It is important to keep in mind that there are a number of steps between establishment of the basic pattern and elaboration of that pattern into anatomical and morphological structure. Uncovering the underlying control mechanisms is likely to be the most productive strategy in understanding how roots and shoots develop.

Shoot development

The unique aboveground architectures of different plant species have their origins in shoot meristems. Shoot architecture is affected by the amount of axillary bud outgrowth. Branching patterns are regulated by the shoot tip—a phenomenon called apical dominance—and plant hormones

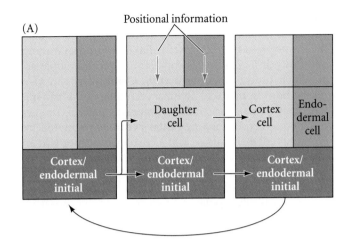

(A)

Positional information

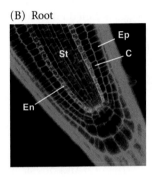
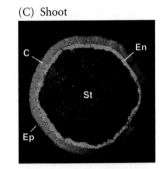

(B) Root

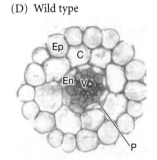

(C) Shoot

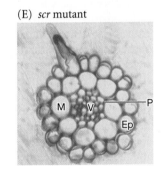

(D) Wild type

(E) *scr* mutant

(F) *shr* mutant

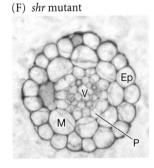

FIGURE 20.23 *SCR* and *SHR* regulate endodermal differentiation in root radial development. (A) Diagram of normal cell division yielding cortical and endodermal cells. *SCR* regulates this asymmetrical cell division. (B,C) *SCR* expression in root and shoot. The *SCR* promoter is linked to the gene for GFP (green fluorescent protein). (D–F) Cross sections of primary roots of (D) wild-type *Arabidopsis*, (E) *scr* mutant, and (F) *shr* mutant. Ep, epidermis; C, cortex; En, endodermis; M, mutant layer; P, pericycle; V, vascular tissue; St, stele. (A after Scheres and Heidstra 1999; B–F photographs courtesy of P. Benfey.)

appear to be the regulating factors. Auxin is produced by young leaves and transported toward the base of the leaf. It can suppress the outgrowth of axillary buds. Grazing and flowering often release buds from apical dominance, at which time branching occurs. **Cytokinins** can also release buds from apical dominance. Axillary buds can initiate their own axillary buds, so branching patterns can get quite complex. Branching patterns can be regulated by environmental signals so that an expansive canopy in an open area maximizes light capture. Asymmetrical tree crowns form when two trees grow very close to each other. In addition to its environmental plasticity, shoot architecture is genetically regulated. In several species, genes have now been identified that regulate branching patterns.

Leaf primordia (clusters of cells that will form leaves) are initiated at the periphery of the shoot meristem (see Figure 20.20). The union of a leaf and the stem is called a **node**, and stem tissue between nodes is called an **internode** (see Figure 20.19). In a simplistic sense, the mature sporophyte is created by stacking node/internode units together. **Phyllotaxy**, the positioning of leaves on the stem, involves communication among existing and newly forming leaf primordia. Leaves may be arranged in various patterns, including a spiral, 180° alternation of single leaves, pairs, and whorls of three or more leaves at a node (Jean and Barabé 1998). Experimentation has revealed a number of mechanisms for maintaining geometrically regular spacing of leaves on a plant, including chemical and physical interactions of new leaf primordia with the shoot apex and with existing primordia (Steeves and Sussex 1989).

It is not clear how a specific pattern of phyllotaxy gets started. Descriptive mathematical models can replicate the observed patterns, but reveal nothing about the mechanism. Biophysical models (e.g., of the effects of stress or strain on deposition of cell wall material, which affects cell division and elongation) attempt to bridge this gap. Developmental genetics approaches are promising, but few phyllotactic mutants have been identified.* Currently there is much interest in the role of local auxin maxima and minima in determining where the next leaf primordium will form on a meristem. The *PIN* gene family that plays an important role in embryo axis formation has also been implicated in phyllotaxy because of the correlation between localization of PIN auxin efflux carriers and primordia siting (Fleming 2005).

Leaf development

Leaf development includes the cells' commitment to become a leaf; establishment of the leaf axes; and morphogenesis, which gives rise to a tremendous diversity of leaf shapes. Culture experiments have assessed when leaf primordia become determined for leaf development. Research on ferns and angiosperms indicates that the youngest visible leaf primordia are not determined to make a leaf; rather, these young primordia can develop as shoots in culture (Steeves 1966; Smith 1984). The programming for leaf development occurs later. The radial symmetry of the leaf primordium becomes dorsal-ventral, or flattened, in all leaves. Two other axes, the proximal-distal and lateral axes, are also established.

The unique shapes of leaves result from regulation of cell division and cell expansion as the leaf blade develops. There are some cases in which selective cell death (apoptosis) is involved in the shaping of a leaf, but differential cell growth appears to be a more common mechanism (Gifford and Foster 1989).

DORSAL-VENTRAL PATTERNING IN LEAVES   Plant biologists refer to the surface of the leaf that is closest to the stem as the adaxial side and the more distant surface is the abaxial side. As the leaf begins to form, the *Arabidopsis* genes *PHABULOSA* (*PHB*) and *PHAVOLUTA* (*PHV*) initially have uniformly expressed RNA throughout the primordium (Figure 20.24A). The PHB and PHV proteins are postulated to be receptors for an adaxial signal, which leads to the accumulation of PHB and PHV on the adaxial leaf surface (McConnell et al. 2001; Byrne 2005). Exclusion of PHB and PHV from the abaxial side is caused by microRNA binding to the transcripts of the two genes, which leads to their degradation (Kidner and Martienssen 2005). In addition, the *PHB*- and *PHV*-specific microRNAs appear to increase methylation of their complementary *PHB* and *PHV* DNA sequences, suppressing transcription (Bao et al. 2004). Dominant gain-of-function mutants have disrupted microRNA binding sites such that these genes are expressed on both sides of the primordium, leading to a leaf with two adaxial sides. The sequences of *PHB* and *PHV* make it likely that they are activated by a lipid ligand and that this activation is followed by the development of adaxial cells.

KANADI (*KAN*) genes initiate abaxial cell differentiation in *Arabidopsis* (Kerstetter et al. 2001). In situ hybridization shows that *KAN* is transiently expressed on the abaxial side of cotyledons, leaves, and initiating floral organ primordia (Figure 20.24B). *KAN* contains a GARP domain found in transcription factors. *KAN* and *PHB/PHV* mutually suppress each other to maintain abaxial and adaxial fates, respectively.

Abaxial fate also appears to be specified by three members of the *YABBY* gene family (Siegfried et al. 1999). The exact mechanism of *YABBY* genes in leaf polarity is unknown, although promoter deletion experiments reveal that *FILAMENTOUS*, a gene in the *YABBY* family, is actively excluded from the adaxial side of the primordium. *KAN* and *YABBY* likely have redundant functions. *KAN* gene expression is restricted to the abaxial side by *PHB* and *PHV*, while *KAN* restricts *PHB* and *PHV* expression to the adaxial side (Figure 20.24C).

---

*One candidate phyllotactic mutant is the *terminal ear* mutant in maize, which has irregular phyllotaxy. The wild-type gene is expressed in a horseshoe-shaped region, with a gap where the leaf will be initiated (Veit et al. 1998). The plane of the horseshoe is perpendicular to the axis of the stem.

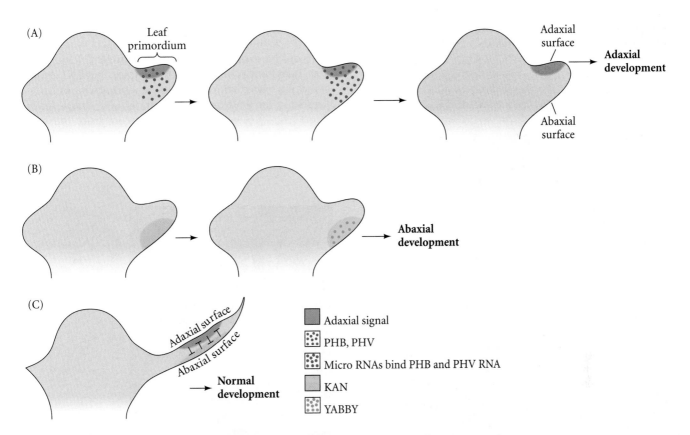

FIGURE 20.24 Patterning of adaxial and abaxial leaf surfaces of *Arabidopsis*. (A) PHB and PHV proteins are uniformly distributed throughout the leaf primordium. MicroRNAs expressed on the abaxial side specifically degrade *PHB* and *PHV*, leading to degradation by DICER and blocking translation. In addition, microRNAs increase methylation of the complementary regions in *PHB* and *PHV* DNA, suppressing transcription. (B) Transient expression of *KAN* on the abaxial side of the leaf leads to the expression of *YABBY* genes, which in turn lead to the transcription of other abaxial genes. (C) PHB and PHV block KAN adaxial activity. Expression of PHB, PHV, and KAN is transient and occurs early in leaf development.

Leaves fall into two categories, simple and compound (Figure 20.25; see review by Sinha 1999). There is much variety in simple leaf shape, from smooth-edged leaves to deeply lobed oak leaves. Compound leaves are composed of individual leaflets (and sometimes tendrils) rather than a single leaf blade. Whether simple and compound leaves develop by the same mechanism is an open question. One perspective is that compound leaves are highly lobed simple leaves. An alternative perspective is that compound leaves are modified shoots. The ancestral state for seed plants is believed to be compound, but for angiosperms it is simple. Compound leaves have arisen multiple times in the angiosperms, and it is not clear if these are reversions to the ancestral state.

Developmental genetic approaches are being applied to leaf morphogenesis. The Class I *KNOX* genes are homeobox genes that include *STM* and the *KNOTTED 1* (*KN1*) gene in maize. Gain-of-function mutations of *KN1* cause meristem-like bumps to form on maize leaves. In wild-type plants, this gene is expressed in meristems. *KNOX* genes stimulate meristem initiation and growth. Although some YABBY genes specify abaxial leaf identity (see Figure 20.24), others function by downregulating *KNOX* genes,

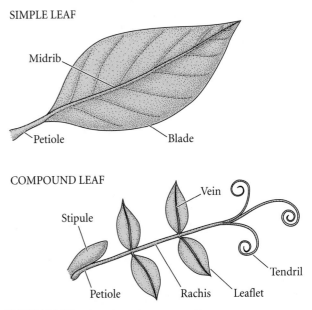

FIGURE 20.25 Simple and compound leaves.

FIGURE 20.26 Overexpression of Class 1 *KNOX* genes in tomato. The photograph shows the single leaves of (A) a wild-type plant, (B) a *mouse ears* mutant, with increased leaf complexity, and (C) a transgenic plant that uses a viral promoter to overexpress the tomato homologue (*LeT6*) of the *KN1* gene from maize. (Photographs courtesy of N. Sinha.)

development, however, is consistent with the hypothesis that compound leaves are modified shoots. Looking at patterns of *KN1* expression over a broad range of angiosperm taxa, Bharathan et al. (2002) demonstrated that *KN1* expression is associated with compound leaf primordia, but is not present in the developing leaf primordia of simple leaves (Figure 20.27). These data support the conclusion that compound leaves maintain shootlike activity, at least for some time. How *KN1* works in meristems and in leaf primordia is an area of active research. In tomato leaf, *KN1* causes a

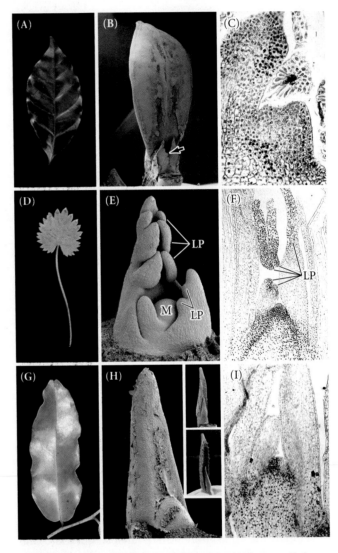

FIGURE 20.27 *KNOX* is known to occur in leaf primordia in species making complex leaves. Here we show that species with simple leaves can also express *KNOX* in leaf primordia that start out complex but undergo secondary alteration to become simple. (A) Coffee simple leaf. (B) SEM of primordia (arrow) showing early complexity. (C) Expression of *KNOX* genes in these early leaves. (D) Anise simple leaf. (E). SEM showing early complexity. (F) *KNOX* expression in these primordia. (G) *Amborella* showing a simple leaf that is simple throughout development (H) and has no *KNOX* expression (I). LP, leaf primordia; M, meristem. (From Bharathan et al. 2002; photographs courtesy of N. Sinha.)

thereby restricting where meristems form (Kumaran et al. 2002). If too much *KN1* is present, *YABBY* is insufficient to control *KN1* expression.

When *KN1*, or the tomato homologue *LeT6*, has its promoter replaced with a promoter from cauliflower mosaic virus and is inserted into the genome of tomato, the gene is expressed at high levels throughout the plant, and the leaves become "super compound" (Figure 20.26; Hareven et al. 1996; Janssen et al. 1998). Simple leaves become more lobed (but not compound) in response to overexpression of *KN1*, consistent with the hypothesis that compound leaves may be an extreme case of lobing in simple leaves (Jackson 1996). The role of *KN1* in shoot meristem and leaf

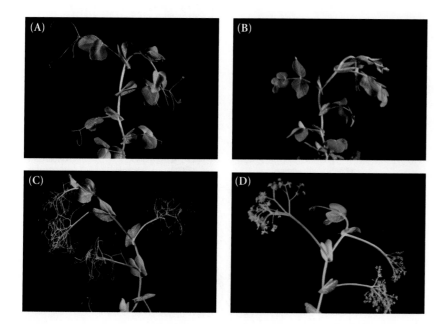

FIGURE 20.28 Leaf morphology mutants in peas. (A) Wild-type pea plant. (B) The *tl* mutant, in which tendrils are converted to leaflets. (C) The *af* mutant, in which leaflets are converted to tendrils. (D) An *af tl* double mutant, which results in a "parsley leaf" phenotype. (Photographs courtesy of S. Singer.)

decrease in the expression of a gene needed for the production of the plant hormone gibberellic acid (Hay et al. 2002).

A second gene, *LEAFY*, is essential for the transition from vegetative to reproductive development and also appears to play a role in compound leaf development. *LEAFY* was identified in *Arabidopsis* and snapdragons (in which it is called *FLORICAULA*) and has homologues in other angiosperms. The pea homologue, *UNIFOLIATA*, has a mutant phenotype in which compound leaves are reduced to simple leaves (Hofer and Ellis 1998). This finding is also indicative of a regulatory relationship between shoots and compound leaves. In an intriguing evolutionary twist, *UNI* appears to have taken on the role of *KN1* genes in leaf development in peas (Sinha 2002). It appears to have been co-opted from its role in flowering in this highly derived legume.

In some compound leaves, developmental decisions about leaf versus tendril formation are also made. Mutations of two leaf-shape genes can individually and in sum dramatically alter the morphology of the compound pea leaf. The *acacia* (*tl*) mutant converts tendrils to leaflets; *afila* (*af*) converts leaflet to tendrils (Marx 1987). The *af tl* double mutant has a complex architecture and resembles a parsley leaf (Figure 20.28).

At a more microscopic level, the patterning of stomata (openings for gas and water exchange) and trichomes (hairs) across the leaf is also being investigated. In monocots, the stomata form in parallel files, while in dicots the distribution appears more random. In both cases, the patterns appear to maximize the evenness of stomata distribution. Genetic analysis is providing insight into the mechanisms regulating this distribution. A common gene group appears to be working in both shoots and roots, affecting the distribution pattern of both trichomes and root hairs (Benfey 1999).

The Vegetative-to-Reproductive Transition

Unlike most animal systems, in which the germ line is set aside during early embryogenesis, the germ line in plants is established only after the transition from vegetative to reproductive development (flowering). The vegetative and reproductive structures of the shoot are all derived from the shoot meristem formed during embryogenesis. Clonal analysis indicates that no cells are set aside in the shoot meristem of the embryo to be used solely in the creation of reproductive structures (McDaniel and Poethig 1988). In maize, irradiating seeds causes changes in the pigmentation of some cells. These seeds give rise to plants that have visually distinguishable sectors descended from the mutant cells. Such sectors may extend from the vegetative portion of the plant into the reproductive regions (Figure 20.29), indicating that maize embryos do not have distinct reproductive compartments.

Maximal reproductive success in angiosperms depends on the timing of flowering and on balancing the number of seeds produced with the resources allocated to individual seeds. As in animals, different strategies work best for different organisms in different environments. There is a great diversity of flowering patterns among the over 300,000 angiosperm species, yet there appears to be an underlying evolutionary conservation of flowering genes and common patterns of flowering regulation.

A simplistic explanation of the flowering process is that a signal from the leaves moves to the shoot apex and induces flowering. In some species, this flowering signal is a response to environmental conditions. The developmental pathways leading to flowering are regulated at numerous control points in different plant organs (roots, cotyledons, leaves, and shoot apices) in various species,

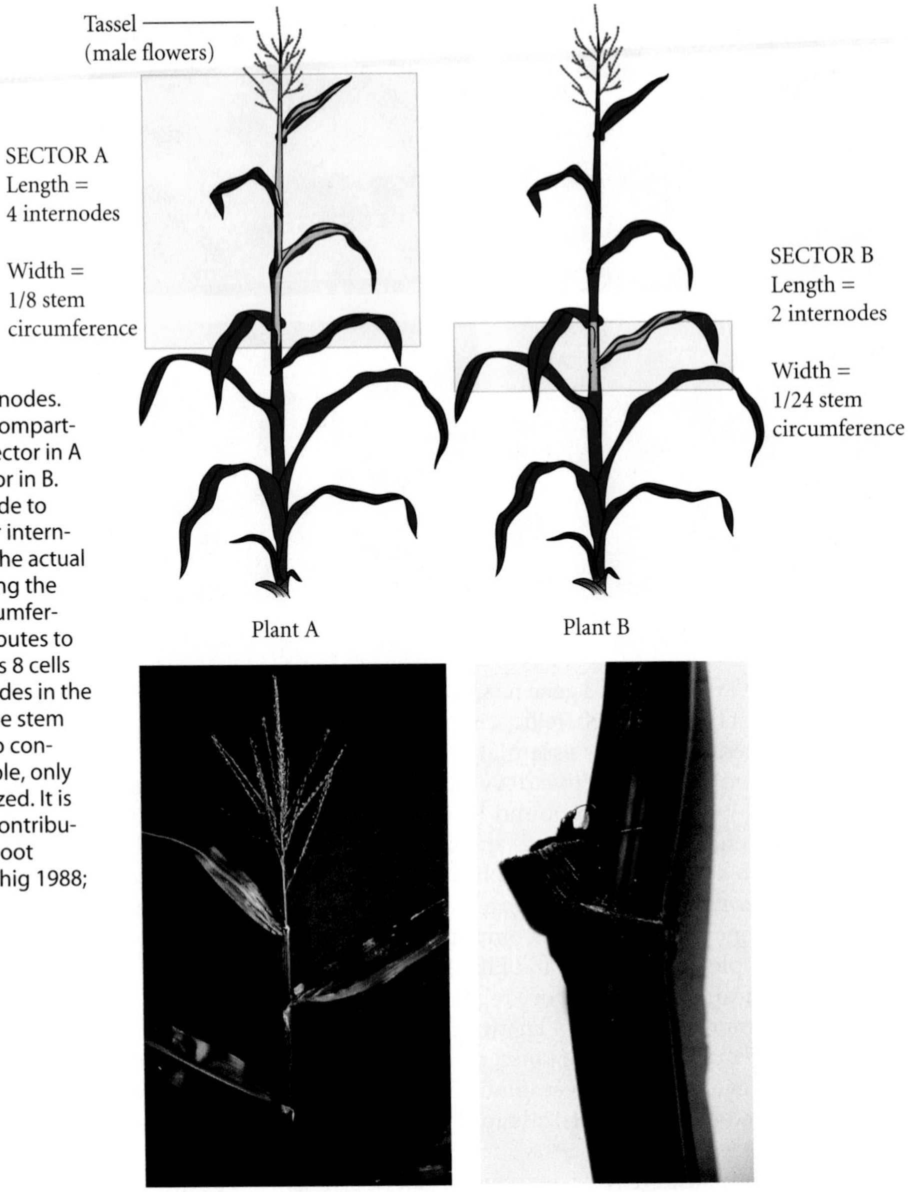

FIGURE 20.29 Clonal analysis can be used to construct a fate map of a shoot apical meristem in maize. Seeds that are heterozygous for certain pigment genes (anthocyanins) are irradiated so that the dominant allele is lost in a few cells (a chance occurrence). All cells derived from the somatic mutant will be visually distinct from the nonmutant cells. Plants A and B have mutant sectors that reveal the fate of cells in the shoot meristem of the seed. The mutant sector in A includes both vegetative and reproductive (tassel) internodes. Thus there is no distinct developmental compartment that forms the tassel. The mutant sector in A is longer and wider than the mutant sector in B. This indicates that more cells were set aside to contribute to the lower than to the upper internodes in the shoot meristem in the seed. The actual number of cells can be calculated by taking the reciprocal of the fraction of the stem circumference the sector occupies. Sector A contributes to 1/8 of the circumference of the stem; thus 8 cells were fated to contribute to these internodes in the seed meristem. Sector B is only 1/24 of the stem circumference; thus 24 cells were fated to contribute to these internodes. In this example, only cells derived from the L1 are being analyzed. It is also important to consider the possible contributions of the L2 and L3 cell layers of the shoot meristem. (Data from McDaniel and Poethig 1988; photographs courtesy of C. McDaniel.)

resulting in a diversity of flowering times and reproductive architectures. The nature of the flowering signal, however, remains unknown. There is now evidence that RNA with developmental functions can move through the phloem, and the role of very small pieces of RNA in signaling developmental mechanisms has recently been recognized.

Juvenility

Some plants, especially woody perennials, go through a **juvenile phase** during which the plant cannot produce reproductive structures even if all the appropriate environmental signals are present (Lawson and Poethig 1995). The transition from the juvenile to the adult stage may require the acquisition of competence by the leaves or meristem to respond to an internal or external signal (McDaniel et al. 1992; Singer et al. 1992; Huala and Sussex 1993). Even in herbaceous plants, there is a phase change from juvenile to adult vegetative growth. For example, the *EARLY PHASE CHANGE (EPC)* gene in maize is required to maintain the juvenile state (Vega et al. 2002). Mutant *epc* plants flower early because they have fewer juvenile leaves which are distinguished from adult leaves by the presence of wax and the absence of hairs. These *epc* mutants, however, have the same number of adult leaves as wild-type plants. Juvenile traits are not expressed in the absence of *EPC*. The mechanisms of phase change genes are just beginning to be understood. The *Arabidopsis* juvenility gene *HASTY (HST)*, is necessary for microRNA processing and export from the nucleus. Loss-of-function mutants undergo early phase change because microRNAs are trapped in the nucleus (Park et al. 2005).

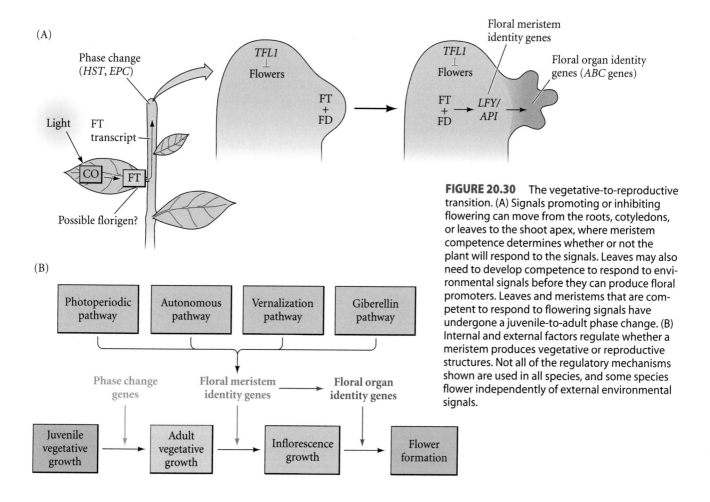

FIGURE 20.30 The vegetative-to-reproductive transition. (A) Signals promoting or inhibiting flowering can move from the roots, cotyledons, or leaves to the shoot apex, where meristem competence determines whether or not the plant will respond to the signals. Leaves may also need to develop competence to respond to environmental signals before they can produce floral promoters. Leaves and meristems that are competent to respond to flowering signals have undergone a juvenile-to-adult phase change. (B) Internal and external factors regulate whether a meristem produces vegetative or reproductive structures. Not all of the regulatory mechanisms shown are used in all species, and some species flower independently of external environmental signals.

Floral signals

Grafting and organ culture experiments, mutant analyses, and molecular analyses give us a framework for describing the reproductive transition in plants. Grafting experiments have identified the sources of signals that promote or inhibit flowering and have provided information on the developmental acquisition of meristem competence to respond to these signals (Lang et al. 1977; Singer et al. 1992; Reid et al. 1996). Analyses of mutants and molecular characterization of genes are yielding information on the mechanics of these signal-response mechanisms (Levy and Dean 1998; Hempel et al. 2000; Hecht et al. 2005).

Leaves produce a graft-transmissible substance that induces flowering. In some species, this signal is produced only under specific **photoperiods** (day lengths), while other species are day-neutral and will flower under any photoperiod (Zeevaart 1984). Not all leaves may be competent to perceive or pass on photoperiodic signals. The **phytochrome** pigments transduce these signals from the external environment. The structure of phytochrome is modified by red and far-red light, and these changes can initiate a cascade of events leading to the production of either floral promoter or floral inhibitor (Deng and Quail 1999). Leaves, cotyledons, and roots have been identified as sources of floral inhibitors in some species (McDaniel et al. 1992; Reid et al. 1996). A critical balance between inhibitor and promoter is needed for the reproductive transition. In addition, **vernalization**, a period of chilling, can enhance the competence of shoots and leaves to perceive or produce a flowering signal.

The "black box" between environmental signals and the production of a flower is vanishing rapidly, especially in the model plant *Arabidopsis* (Searle and Coupland 2004; Achard et al. 2006). The signaling pathways from light via different phytochromes to key flowering genes are being elucidated (Figure 20.30A). *CONSTANS* (*CO*) responds to day length, promoting flowering under long-day conditions. CO activates transcription of a gene called *FLOWERING LOCUS T* (*FT*). *FT* is transcribed only in the leaves, but the transcript travels through the phloem (transport tissue) from the leaf to the shoot (see Blázquez 2005). *FT* is likely translated when it arrives at the shoot meristem. The FT protein interacts with the transcription factor FD, which is expressed only in the shoot tip. Together these two proteins are responsible for the expression of regulatory genes that lead to the production of flowers.

Is FT the long-sought florigen? Possibly, but we don't know if other proteins aid in the movement of FT, or if other transcripts must also move through the phloem before flowering can occur.

Some of the genes that CONSTANS protein activates in addition to *FT* are also activated by transcription factors in other, non-light-dependent, flowering pathways. Molecular explanations are revealing redundant pathways that ensure that flowering will occur. In *Arabidopsis*, four separable pathways leading to flowering are being elucidated. Light-dependent, vernalization-dependent, gibberellin, and autonomous pathways that regulate the floral transition have been genetically dissected (Figure 20.30B). The availability of the *Arabidopsis* genome sequence now makes it possible to search on a broader scale for classes of genes that regulate critical events such as time of flowering (Ratcliffe and Riechmann 2002). Schmid and colleagues (2005) created a developmental gene expression profile of almost all the *Arabidopsis* genes using microarrays. The current challenge is to integrate the vast amount of expression data with functional analysis.

One promising approach is the search for quantitative trait loci (QTL) that control responses to environmental and hormonal factors. In *Arabidopsis*, two lines, including a wild accession with natural variation in light and hormone responses, were used to identify new genes (Borevitz et al. 2002). Once QTL are mapped, the *Arabidopsis* genome map makes it more likely that researchers will be able to associate function with a specific gene.

Inflorescence development

The ancestral angiosperm is believed to have formed a terminal flower directly from the terminal shoot apex (Stebbins 1974). In modern angiosperms, a variety of flowering patterns exist in which the terminal shoot apex is indeterminate, but axillary buds produce flowers. This observation introduces an intermediate step into the reproductive process: the transition of a vegetative meristem to an **inflorescence meristem**, which initiates axillary meristems that can produce floral organs, but does not directly produce floral parts itself. The inflorescence is the reproductive backbone (stem) that displays the flowers (see Figure 20.19).

The inflorescence meristem probably arises through the action of a gene that suppresses terminal flower formation. The *CENTRORADIALUS* (*CEN*) gene in snapdragons suppresses terminal flower formation by suppressing expression of *FLORICAULA* (*FLO*), which specifies floral meristem identity (Bradley et al. 1996). Curiously, the expression of *FLO* is necessary for *CEN* to be turned on. The *Arabidopsis* homolog of *CEN* (*TERMINAL FLOWER 1* or *TFL1*) is expressed during the vegetative phase of development as well, and has the additional function of delaying the commitment to inflorescence development (Bradley et al. 1997). Overexpression of *TFL1* in transgenic *Arabidopsis* extends the time before a terminal flower forms (Ratcliffe et al. 1998). *TFL1* must delay the reproductive transition.

Floral meristem identity

The next step in the reproductive process is the specification of floral meristems—those meristems that will actually produce flowers (Weigel 1995). In *Arabidopsis*, *LEAFY* (*LFY*), *APETALA 1* (*AP1*), and *CAULIFLOWER* (*CAL*) are **floral meristem identity genes** (Figure 20.31). *LFY* is the homologue of *FLO* in snapdragons, and its upregulation during development is key to the transition to reproductive development (Blázquez et al. 1997; Blázquez and Weigel 2000). The LEAFY transcription factor can actually move between cell layers in the meristem before activating other flowering genes (Sessions et al. 2000). Analysis of the promoter of *LFY* reveals that there are separate sites for activation by the photoperiod pathway and the autonomous pathway. The autonomous pathway works through the binding of giberellin to the *LFY* promoter.

(A) (B) (C) (D)

FIGURE 20.31 Floral meristem identity mutants. (A) Wild-type *Arabidopsis*. (B) The *leafy* mutant. (C) The *apetala1* mutant. (D) The *leafy apetala1* double mutant. (Photographs courtesy of J. Bowman.)

FIGURE 20.32 *Arabidopsis* double mutant of *ap1* and *cal*. Since *cal* alone gives a wild-type phenotype, the double mutant demonstrates the redundancy of these two genes in the flowering pathway. (Photograph courtesy of J. Bowman.)

AP1 gene family and most likely arose as a result of gene duplication and divergence. The *FUL* gene is partially redundant to *AP1* and *CAL* (Ferrandiz et al. 2000). *CAL*, however, arose only within the brassicas through an *AP1* gene duplication event, and appears to have accompanied the domestication of cauliflower and broccoli (Purugganan et al. 2000).

Floral meristem identity genes initiate a cascade of gene expression that turns on region-specifying (**cadastral**) genes, which further specify pattern by initiating transcription of floral organ identity genes (Weigel 1995). *SUPERMAN (SUP)* is an example of a cadastral gene in *Arabidopsis* that plays a role in specifying boundaries for the expression of **organ identity genes**. Three classes (A, B, and C) of organ identity genes are necessary to specify the four whorls of floral organs (Figure 20.33; Coen and Meyerowitz 1991). They are homeotic genes (but not Hox genes; rather, most are members of the MADS box gene family that had its origins before the divergence of animals and plants) and include *AP2*, *AGAMOUS (AG)*, *AP3*, and *PISTILLATA (PI)* in *Arabidopsis*. Class A genes (*AP2*) alone specify sepal development. Class A genes and class B genes (*AP3* and *PI*) together specify petals. Class B and class C (*AG*) genes are necessary for stamen formation; class C

Expression of floral meristem identity genes is necessary for the transition from an inflorescence meristem to a floral meristem. Strong *lfy* mutants tend to form leafy shoots in the axils where flowers form in wild-type plants; they are unable to make the transition to floral development. If *LFY* is overexpressed, flowering occurs early. For example, when aspen was transformed with an *LFY* gene that was expressed throughout the plant, the time to flowering was dramatically shortened from years to months (Weigel and Nilsson 1995).

AP1 and *CAL* are closely related and redundant genes. The *cal* mutant looks like the wild-type plant, but *ap1 cal* double mutants produce inflorescences that look like cauliflower heads (Figure 20.32). More recently another gene, *FRUITFUL (FUL)*, with close sequence similarity to *AP1* and *CAL*, has been characterized. The FRUITFUL gene family is closely related to the

FIGURE 20.33 The ABC model for floral organ specification. Three classes of genes—*A*, *B*, and *C*—regulate organ identity in flowers. The central diagram represents the wild-type flower; surrounding diagrams represent mutants that are missing one or more of these gene functions (indicated by the lowercase *a*, *b*, or *c*). Se, sepal; Pe, petal; St, stamen; Ca, carpel; Pe/St, a hybrid petal/stamen; Lf, leaf; Se*, a modified sepal indicating that other genes (possibly ovule genes) also regulate floral organ specification. (Model of Coen and Meyerowitz 1991.)

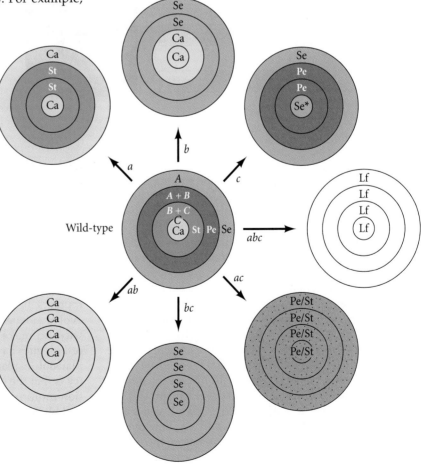

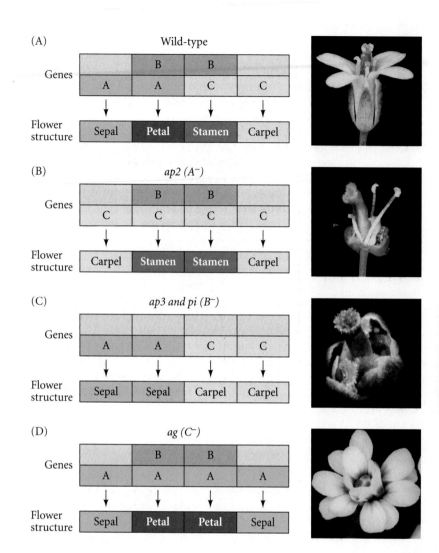

(A) Wild-type

Genes

	B	B	
A	A	C	C

Flower structure

| Sepal | Petal | Stamen | Carpel |

(B) *ap2 (A⁻)*

Genes

	B	B	
C	C	C	C

Flower structure

| Carpel | Stamen | Stamen | Carpel |

(C) *ap3 and pi (B⁻)*

Genes

| A | A | C | C |

Flower structure

| Sepal | Sepal | Carpel | Carpel |

(D) *ag (C⁻)*

Genes

	B	B	
A	A	A	A

Flower structure

| Sepal | Petal | Petal | Sepal |

FIGURE 20.34 Wild-type and mutant phenotypes of the *Arabidopsis* class A (*ap2*), class B (*ap3*, *pi*), and class C (*ag*) floral organ identity genes. (Photographs courtesy of J. Bowman.)

genes alone specify carpel formation. When all of these homeotic genes are not expressed in a developing flower, floral parts become leaflike (Figure 20.34). The ABC genes code for transcription factors that initiate a cascade of events leading to the actual production of floral parts.

While the ABC model is compelling, it is not sufficient to support the hypothesis that flowers evolved from leaves. Overexpressing the ABC genes in leaves does not produce petals or other flower parts. About a decade after the ABC model was proposed, a fourth class of floral organ identity genes, *SEPALLATA* (*SEP*), was identified (see Jack 2001). These MADS box genes can convert a leaf into a petal when ectopically expressed in the leaf. In the absence of *SEP* function, flowers become whorls of sepals (Figure 20.35). SEP transcription factors form dimers with ABC or other SEP transcription factors to initiate floral development.

In addition to the ABC and SEP genes, class D genes are now being investigated that specifically regulate ovule development. The ovule evolved long before the other angiosperm floral parts, and while its development is coor-

dinated with that of the carpel, one would expect more ancient, independent pathways to exist.

Transcription of floral organ identity genes is actually the beginning rather than the end of flower development. One of the amazing attributes of angiosperms is the tremendous diversity of flower phenotypes, many of which attract specific pollinators. Adaxial/abaxial asymmetry in flowers is one contributing factor to diverse floral morphologies that attract pollinators. Phylogenetic evidence indicates that this trait arose independently many times, as well as being lost many times. The cloning of the *CYCLOIDEA* (*CYC*) gene in snapdragons has led to extensive discussion of the molecular mechanisms involved in the evolution of asymmetry (Donoghue et al. 1998; Cubas et al. 2001). In snapdragons, *CYC* expression is observed on the adaxial side of the floral primordium early in development. In *cyc* mutants, flowers have a more radial symmetry. Did *CYC* evolve independently numerous times, or are there many ways to make an asymmetrical flower? Putative *CYC* orthologues have been identified in other species, and one plausible explanation for the multiple origins of asymmetry is that the *CYC* gene was recruited for the same function multiple times. The combination of developmental and phylogenetic approaches to the study of patterning in plants promises to provide insight into the origins of the myriad morphological novelties that are found among the angiosperms.

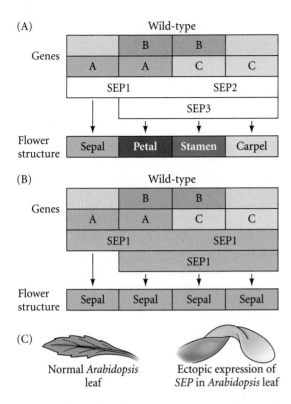

FIGURE 20.35 *SEPALLATA (SEP)* genes are a fourth class of organ identity genes. *Arabidopsis* plants with the triple mutation *sep1⁻sep 2⁻sep 3⁻* produce whorls of sepals and lack the other three floral organs. Ectopic expression of all three *SEP* genes in leaves causes them to convert to petals.

Senescence

Senescence, a developmental program leading to death, is closely linked to flowering in many angiosperms. In some species, individual flower petals senesce following pollination; orchids, which can stay fresh for long periods of time unless they are pollinated, are a good example. Fruit ripening (and ultimately overripening) is an example of organ senescence. Whole-plant senescence leads to the death of the entire sporophyte generation. **Monocarpic** plants flower once and then senesce. **Polycarpic** plants, such as the bristlecone pine, can live thousands of years (4,900 years is the current record) and flower repeatedly. In polycarpic plants, death is accidental; in monocarpic plants, it appears to be genetically programmed.

Flowers and fruits play a key role in senescence, and their removal can sometimes delay the process. In some legumes, senescence can be delayed by removing the developing seed—in other words, the embryo may trigger senescence in the parent plant. During flowering and fruit development, nutrients are reallocated from other parts of the plant to support the development of the next generation. The reproductive structures become a nutrient sink, and this can lead to whole-plant senescence.

Snapshot Summary — Plant Development

1. Plants are characterized by alternation of generations; that is, their life cycle includes both diploid and haploid multicellular generations.

2. Land plants have evolved mechanisms to protect embryos. Angiosperm embryos develop deeply embedded in parent tissue. The parent tissue provides nutrients and some patterning information. This evolutionary theme of an increasingly protected embryo is shared by both plants and animals.

3. A multicellular diploid sporophyte produces haploid spores via meiosis. These spores divide mitotically to produce a haploid gametophyte. Mitotic divisions within the gametophyte produce the gametes. The diploid sporophyte results from the fusion of two gametes.

4. In angiosperms, the male gamete, pollen, arrives at the style of the female gametophyte and effects fertilization through the pollen tube. Two sperm cells move through the pollen tube: one joins with the ovum to form the zygote, and the other is involved in the formation of the endosperm.

5. Early embryogenesis is characterized by the establishment of the shoot-root axis and by radial patterning yielding three tissue systems. Pattern emerges by regulation of planes of cell division and the directions of cell expansion, since plant cells do not move during development.

6. As the angiosperm embryo matures, a food reserve is established. Only the rudiments of the basic body plan are established by the time embryogenesis ceases and the seed enters dormancy.

7. Pattern is elaborated during postembryonic development, when meristems construct the reiterative structures of the plant.

8. Unlike most animals, the germ line is not set aside early in plant development. Coordination of signaling among leaves, roots, and shoot meristems regulates the transition to the reproductive state.

9. Floral meristem identity genes and organ identity genes enable the angiosperm flower to display a tremendous amount of morphological diversity.

10. Reproduction may be followed by genetically programmed senescence of the parent plant.

For Further Reading

Complete bibliographical citations for all literature cited in this chapter can be found on the Vade Mecum CD that accompanies the book and at the free access website www.devbio.com

Bao, N., K.-W. Lye and M. K. Barton. 2004. MicroRNA binding sites in *Arabidopsis* class III HD-ZIP mRNAs are required for methylation of the template chromosome. *Dev. Cell* 7: 653–662.

Blázquez, M. A. 2005. The right time and place for making flowers. *Science* 309: 1024–1025.

Brownlee, C. 2004. From polarity to pattern: Early development of in fucoid algae. *Ann. Plant Rev.* 12: 138–156.

Byrne, M. E. 2005. Networks in leaf development. *Curr. Opin. Plant Biol.* 8: 59–66.

Dharmasiri, N., S. Dharmasiri, and M. Estelle. 2005. The F-box protein TIR1 is an auxin receptor. *Nature* 435: 441–445.

Fleming, A. J. 2005. Formation of primordia and phyllotaxy. *Curr. Opin. Plant Biol.* 8: 53–58.

Friml, J., X. Yang, M. Michniewicz, D. Weijers, A. Quint, O. Tietz, R. Benjamins, et al. 2004. A PINOID-dependent binary switch in apical-basal PIN polar targeting directs auxin efflux. *Science* 306: 862–865.

Jack, T. 2001. Relearning our ABCs: New twists on an old model. *Trends Plant Sci.* 6: 310–316.

Medical Implications of Developmental Biology

21

OUR UNDERSTANDING OF MAMMALIAN DEVELOPMENT has expanded exponentially over the last three decades. This newly acquired knowledge has numerous medical applications, including:

- The identification of genetic defects that affect development
- The identification of factors in the maternal environment that may influence the health of the fetus
- The regulation of fertility
- The identification of teratogenic compounds that affect development
- The realization that cancers can be disruptions of developmental regulation and might be cured through developmental processes
- Attempts to detect developmental diseases (including cancers) and cure them through cloning, stem cell therapy, and genetic engineering
- Attempts to cure traumatic and degenerative disease through regeneration

When one discusses the medical aspects of any science,* one must take social issues as well as purely scientific concerns into account. Discussions concerning how genetic diseases, birth defects, cancers, or infertility might be controlled also raise issues concerning social, political, and economic motives and opportunities.

> "The amazing thing about mammalian development is not that it sometimes goes wrong, but that it ever succeeds."
> *VERONICA VAN HEYNINGEN (2000)*

> "The future is already here. It's just not evenly distributed yet."
> *WILLIAM GIBSON (1999)*

GENETIC ERRORS OF HUMAN DEVELOPMENT

If you think it is amazing that any one of us survives to be born, you are correct. It is estimated that one-half to two-thirds of all human conceptions do not develop successfully to term (Figure 21.1). Many of these embryos express their abnormality so early that they fail to implant in the uterus. Others implant, but fail to establish a successful pregnancy. Thus, most embryos are spontaneously aborted, often before the woman even knows she is pregnant (Boué et al. 1985). Using a sensitive immunological test that detects the presence of human chorionic gonadotropin (hCG) as early as 8 or 9 days after fertilization, Edmonds and co-workers (1982) monitored 112 pregnancies in normal women. Of these hCG-determined pregnancies, 67 (about 59 percent) failed to be maintained.

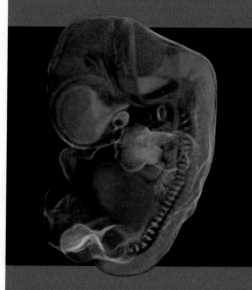

---

*While this chapter seeks to provide general information about medical topics, it is not intended to provide medical advice for specific persons or disorders. This chapter also does not propose to condense all of medical embryology into a single unit. Rather, it attempts to summarize particular principles that will be useful in studying the normal or abnormal development of any human organ. For more complete medical embryology texts, see Dudek and Fix 1998; Moore et al. 1998; Sadler and Langman 2000; Larsen et al. 2001; and Carlson 2005.

FIGURE 21.1 The fate of 20 hypothetical human eggs in the United States and western Europe. Under normal conditions, only 6.2, or fewer, of the original 20 eggs would be expected to develop successfully to term. (After Volpe 1987.)

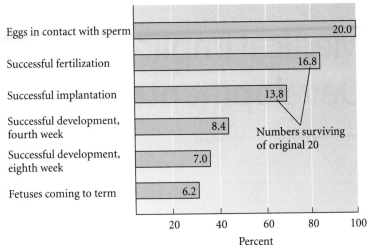

There are also defects that may not be deleterious to the fetus (which does not depend on certain organs, such as limbs and lungs, while inside the mother) but can threaten life once the baby is born. Over 5 percent of all babies born have recognizable malformations, ranging from mild to very severe (McKeown 1976).

Congenital ("present at birth") abnormalities and losses of the fetus prior to birth have both intrinsic and extrinsic causes. Those abnormalities that are caused by genetic events (mutations, aneuploidies, translocations) are called **malformations** (Opitz et al. 1987). The aneuploidy **trisomy 21**, for example, causes several malformations (most of them minor, such as facial muscle changes, but some of them major, such as heart abnormalities, gut abnormalities, and cognitive problems), collectively called **Down syndrome** (Figure 21.2). Certain genes on chromosome 21 are thought to encode transcription factors, and the extra copy of chromosome 21 probably causes an overproduction of these regulatory proteins. Such overproduction would cause the misregulation of genes necessary for heart, muscle, and nerve formation (Nishigaki et al. 2002).

Most early embryonic and fetal demise is probably due to chromosomal abnormalities that interfere with normal developmental processes. Even an extra copy of the extremely small chromosome 21 causes the misregulation of numerous developmental functions. Indeed, people

with Down syndrome are among the few individuals with autosomal trisomies to survive beyond the first weeks of infancy.*

Until recently, the molecular study of human genetics focused almost exclusively on inborn errors in metabolic and structural proteins. Thus, diseases of enzymes, collagens, and globins predominated. But these proteins are the final products of differentiated cells. Errors in the proteins involved with development—transcription factors, paracrine factors, and elements of signal transduction pathways—were little understood. The causes of diseases resulting from such errors, often categorized as "congen-

*Some infants born with trisomies 13 and 18 (Patau syndrome and Edward syndrome, respectively), can live for years with proper medical care, although they usually suffer from lung defects, intestinal defects, and heart malformations.

(A)

(B)

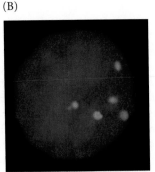

FIGURE 21.2 Down syndrome. (A) Down syndrome, caused by a third copy of chromosome 21, is characterized by a particular facial pattern and by mental retardation, the absence of a nasal bone, and often heart and gastrointestinal defects. (B) The procedure shown here tests for chromosome number using fluorescently labeled probes that bind to DNA on chromosomes 21 (pink) and 13 (blue). This person has Down syndrome (trisomy 21), but has normal two copies of chromosome 13. (A, photograph ©Laura Dwight/Corbis; B, photograph courtesy of Vysis, Inc.)

(A)

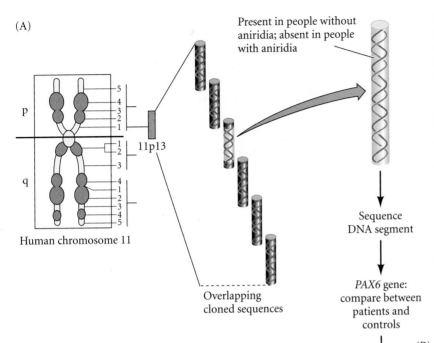

Present in people without aniridia; absent in people with aniridia

11p13

Human chromosome 11

Overlapping cloned sequences

Sequence DNA segment

PAX6 gene: compare between patients and controls

Make probe for in situ hybridization

FIGURE 21.3 Positional cloning of the human *PAX6* gene. (A) Principles of positional cloning. A gene is determined, by pedigree analysis and somatic cell genetics, to be located in a certain region of the genome—in this case, band 13 of the short arm of chromosome 11. Overlapping cloned sequences that span this region are compared between people who have the condition of interest and people who do not. In this case, certain people with aniridia lacked a particular region of DNA. This region was sequenced and found to contain the human homologue of the *PAX6* gene known in mice. Comparisons of the sequences found in controls with those found in people with aniridia indicated that the people with aniridia either lacked a copy of this gene or had mutations in one of their copies. (B) In situ hybridization of an antisense RNA probe with a human fetal eye shows *PAX6* expression (yellow) in the retina, presumptive iris, and surface ectoderm. (From Ton et al. 1991; photograph courtesy of G. F. Saunders.)

(B)

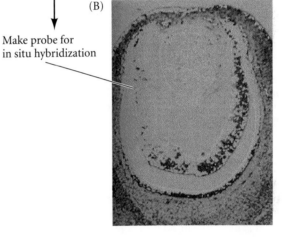

ital anomalies," remained unknown. In the past decade, however, we have made great gains in our knowledge of human development, and we now know that many of these malformations are caused by mutations of genes encoding transcription factors and signal transduction proteins (Epstein et al. 2003).

Identifying the Genes for Human Developmental Anomalies

One cannot experiment on human embryos, nor can one selectively breed humans to express a particular mutant phenotype. How, then, can we find the genes that are involved in normal human development and whose mutations cause malformations? Two techniques have revolutionized human embryology within the past decade.

The first technique is **positional gene cloning** (Collins 1992; Scambler 1997). Here, linkage (pedigree) analysis highlights a region of the genome where a particular mutant gene is thought to reside. By cloning overlapping sequences from that region to make a DNA map, researchers can hope to find a region of DNA that differs between those people who have the malformation and people who lack it. Then, by sequencing that region of the genome, the gene can be located.

As an example, we will look at the discovery of the human gene whose absence causes the defects in normal eye development known as *aniridia*. People heterozygous for this gene have little or no iris in their eyes. A similar phenotype is seen in rats and mice, and mice homozygous for this condition are stillborn and have neither eyes nor nose. Using DNA from people with aniridia, Ton and his colleagues (1991) found a region of chromosome 11 that was present in unaffected individuals but completely or

partially absent in individuals with aniridia. Moreover, this region of DNA contained a region that could be a gene (i.e., it had a promoter site as well as sequences connoting introns and exons). To determine whether this region was indeed a gene that was active in eye development, they did a Northern blot and in situ hybridizations, using this region of DNA as a probe. As noted in Chapter 4 (see Figure 4.16), mRNA complementary to the probe was found primarily in tissues from the brain and eye regions of the body. Thus this fragment of DNA contained a gene that was indeed expressed in the brain and eye. Sequencing the fragment showed that it was the human *PAX6* gene—the human homologue of the *Pax6* gene already known in mice (see Chapters 5 and 6). This finding was confirmed through in situ hybridization (Figure 21.3).

The positional gene cloning approach is currently being used to map those genes responsible for the congenital

(A)

(B)

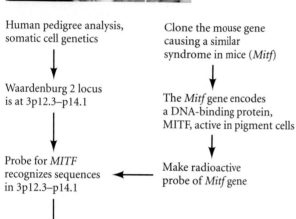

Human pedigree analysis,
somatic cell genetics

↓

Waardenburg 2 locus
is at 3p12.3–p14.1

↓

Probe for *MITF*
recognizes sequences
in 3p12.3–p14.1

↓

Mutations in *MITF*
gene found in human
subjects with Waarden-
burg syndrome 2 and
not in people without
this syndrome

Clone the mouse gene
causing a similar
syndrome in mice (*Mitf*)

↓

The *Mitf* gene encodes
a DNA-binding protein,
MITF, active in pigment cells

↓

Make radioactive
probe of *Mitf* gene

←

FIGURE 21.4 Microphthalmia syndrome in humans and mice. (A) Human patients with Waardenburg syndrome type 2. (B) Mice with *microphthalmia* mutation. The relationship between these two syndromes was ascertained by showing that patients with Waardenburg syndrome type 2 had mutations in a gene that is homologous with the mouse *Mitf* gene. (A photograph from Partington 1959; B photograph courtesy of D. Fischer.)

anomalies seen in Down syndrome. Down syndrome is the most frequent genetic disorder in humans, and as mentioned above, it is usually caused by a complete extra copy of chromosome 21. However, there are rare cases of Down syndrome in which only a small region of chromosome 21 is present in three copies. This region has been called the **Down syndrome critical region (DSCR)**, and one of the genes identified in this region is *DSCR1*. The *DSCR1* gene is predominantly expressed in the brain, heart, and skeletal muscle. Recent studies have shown that it is overexpressed in people with Down syndrome* and that it encodes a protein that inhibits calcineurin, a major calcium regulatory protein (Fuentes et al. 2000; Ermak et al. 2002).

Candidate gene mapping is a gene identification approach that is similar to positional cloning. In candidate mapping, however, one starts with a correlation between the genetic mapping of a particular syndrome and a gene associated with a similar phenotype in other species. As

an example, we will examine Waardenburg syndrome type 2—another condition that affects the eyes. This autosomal dominant condition is characterized by deafness, small eyes with heterochromatic (multicolored) irises, and a white forelock (Figure 21.4A). By analyzing the pedigrees of several families whose members had this condition, Hughes and her colleagues (1994) showed that it is caused by a mutation in a gene on the short arm of chromosome 3. A strikingly similar condition is found in mice, in which it is called microphthalmia. Mutations of the mouse *microphthalmia* gene (*Mitf*) cause a dominant syndrome involving deafness, a white patch of fur, and eye abnormalities (Figure 21.4B; see Chapter 5).

Could Waardenburg syndrome type 2 be caused by mutations in the human equivalent of the mouse *Mitf* gene? The mouse *Mitf* gene was cloned, and was found (by sequencing and in situ hybridization) to encode a DNA-binding protein that is expressed in the pigment cells of the eyes, ears, and hair follicles of embryonic mice (Hemesath et al. 1994; Hodgkinson et al. 1994; Nakayama et al. 1998). The human version of *Mitf* thus became a candidate gene for Waardenburg syndrome. The mouse *Mitf* gene was then used to make a probe to look for similar genes in the human genome. The probe found a human homologue of the mouse *microphthalmia* gene sequence, and this sequence (*MITF*) mapped to the exact same region of chromosome 3 as the mutation causing Waardenburg syndrome type 2 (Tachibana et al. 1994; Tassabehji et al. 1994). Subsequent tests on people with Waardenburg syndrome revealed that they each had mutations of the *MITF* gene. Thus, Waardenburg syndrome type 2 was correlated with mutations in a DNA-binding protein encoded by the human *MITF* locus.

Recently, candidate gene mapping has been supplemented with data from developmental biology. For

---

*Interestingly, overproduction of the DSCR1 protein in adults has also been implicated in Alzheimer disease (Ermak et al. 2001).

instance, the phenotype of knockouts of the mouse *Sonic hedgehog* (*shh*) gene resembles that of humans with holoprosencephaly (cyclopia; see Figure 6.18). This finding made *shh* a candidate gene for holoprosencephaly. By searching for *shh*, it was determined that several (but not all) families with holoprosencephaly have mutations in this gene (Nanni et al. 1999; Odent et al. 1999).

WEBSITE **21.1 Human embryology and genetics.** This website links you to other websites that connect you to tutorials in human development, as well as to the Online Mendelian Inheritance in Man (OMIM), which details all human genetic conditions.

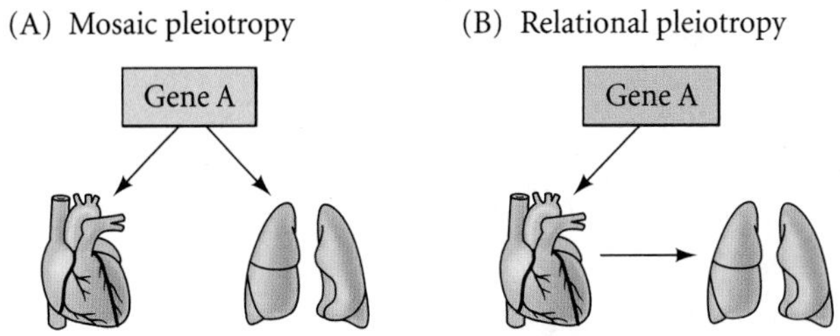

The Nature of Human Syndromes

Human malformations, which range from life-threatening to relatively benign, are often linked into **syndromes**, as in the Down and Waardenburg syndromes discussed above. As mentioned in Chapter 1, the word "syndrome" comes from the Greek *syndromos*, "working together." Such genetic events show pleiotropy.

Pleiotropy

The production of several effects by one gene or pair of genes is called **pleiotropy** (see Grüneberg 1938; Hadorn 1955). For instance, Waardenburg syndrome type 2 involves iris defects, pigmentation abnormalities, deafness, and inability to produce the normal number of mast cells (a type of blood cell). The skin pigment, the iris of the eye, the inner ear tissue, and the mast cells of the blood are not related to one another in such a way that the absence of one would produce the absence of the others. Rather, these four body parts *independently* use the MITF protein as a transcription factor. This phenomenon has been called **mosaic pleiotropy** because the affected organ systems are separately affected by the abnormal gene function (Figure 21.5A).

While the eye pigment, body pigment, and mast cell manifestations of Waardenburg syndrome type 2 are separate events, other aspects of the syndrome are not. The failure of *MITF* expression in the pigmented retina prevents this structure from fully differentiating. This, in turn, causes a malformation of the choroid fissure of the eye, resulting in the drainage of vitreous humor. Without this fluid, the eye fails to enlarge (hence *microphthalmia*, or "small eye"). This type of pleiotropy, in which several developing tissues are affected by the mutation even though they do not express the mutated gene, is called **relational pleiotropy** (Figure 21.5B).

Genetic heterogeneity

In pleiotropy, the same gene can produce different effects in different tissues. However, the opposite phenomenon is

FIGURE 21.5 Mosaic and relational pleiotropy. (A) In mosaic pleiotropy, a gene is independently expressed in several tissues. Each tissue needs the gene product and develops abnormally in its absence. (B) In relational pleiotropy, a gene product is needed by only one particular tissue. However, a second tissue needs a signal from the first tissue in order to develop properly. If the first tissue develops abnormally, the signal is not given, so the second tissue develops abnormally as well.

an equally important feature of syndromes: mutations in *different genes* can produce the *same phenotype*. If several genes are part of the same signal transduction pathway, a mutation in any of them have a similar phenotypic result. This production of similar phenotypes by mutations in different genes is called **genetic heterogeneity**. Cyclopia, for example, can be produced by mutations in the *sonic hedgehog* gene, or by mutations in the genes controlling cholesterol synthesis. Since cholesterol must bind to Sonic hedgehog protein and its receptor in order to use Shh as a paracrine factor (see Chapter 6), mutations that interfere with the synthesis of cholesterol also prevent the activation of Shh.

Similarly, we saw in Chapter 6 that mutations in the murine stem cell factor (*Steel*) gene produce a syndrome resembling that produced by mutations in the gene for its receptor, the Kit protein (*White*). Since mutations of either of these genes prevent *Mitf* from being activated, they produce a phenotype similar to that of the *Mitf*-deficient mouse (Figure 21.6; see also Figure 1.15). But there is a difference. Since the Kit and stem cell factor proteins are also used by migrating germ cells and blood cell precursors (which do not use Mitf), mice with mutations of the genes encoding Kit and stem cell factor also have fewer gametes and blood cells.

Phenotypic variability

Not only can different mutations produce the same phenotype, but the same mutation can produce a different phenotype in different individuals (Wolf 1995, 1997; Nijhout and Paulsen 1997). This phenomenon, called **phenotypic variability**, is discussed at length in Chapter 22. Phenotypic variability comes about because genes are not autonomous entities. Rather, genes interact with other genes and gene products, becoming integrated into complex pathways and networks.

Bellus and colleagues (1996) analyzed the phenotypes derived from the same mutation in the *FGFR3* gene in 10

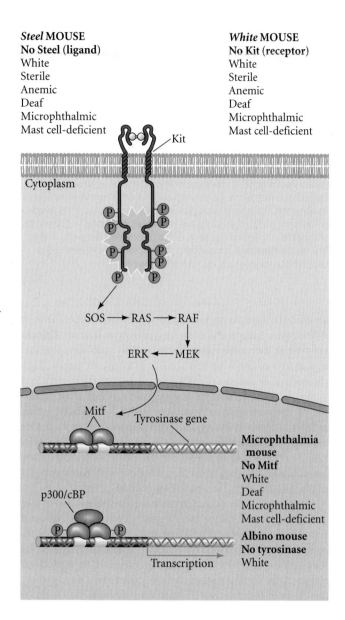

Steel MOUSE
No Steel (ligand)
White
Sterile
Anemic
Deaf
Microphthalmic
Mast cell-deficient

White MOUSE
No Kit (receptor)
White
Sterile
Anemic
Deaf
Microphthalmic
Mast cell-deficient

Kit

Cytoplasm

SOS → RAS → RAF

ERK ← MEK

Mitf   Tyrosinase gene

Microphthalmia mouse
No Mitf
White
Deaf
Microphthalmic
Mast cell-deficient

p300/cBP

Albino mouse
No tyrosinase
White

Transcription

FIGURE 21.6 Phenotypes of mice with mutations along the pigment synthesis pathway. The *Steel* and *White* mice have mutations in the genes encoding stem cell factor (the ligand) and Kit protein (the receptor), respectively. These proteins activate the *Mitf* transcription factor through the pathway shown in Figure 6.15. *Mitf* mutations give rise to the *microphthalmia* phenotype, which contains a subset of those anomalies seen upstream in the pathway. The albino mutation is farther down the pathway and contains a subset of those conditions found in the *microphthalmia* mutants.

mosomes containing the normal allele, and she had severe hemophilia (the inability to clot one's blood after injury). The other twin, who inactivated a lower percentage of her normal X chromosomes, had a milder case (Tiberio 1994; Valleix et al. 2002).

Mechanisms of dominance

Whether a syndrome is dominant or recessive can now be explained in many cases at a molecular level. First, it must be recognized that there are many syndromes that are referred to as "dominant" only because the homozygous condition is lethal to the embryo; therefore the homozygous mutant individual is never born and the homozygous condition is not seen. Second, there are at least three ways of achieving a "dominant" phenotype.

The first mechanism of dominance is **haploinsufficiency**. This term merely means that one wild-type copy of the gene (the heterozygous condition) is not enough to produce the quantity of gene product required for normal development. For example, individuals with Waardenburg syndrome type 2 have roughly half the wild-type amount of MITF. This is not enough for full pigment cell proliferation, mast cell differentiation, or inner ear development. Thus, an aberrant phenotype results when even a single copy of this gene is absent or nonfunctional.

The second mechanism of dominance is **gain-of-function mutations**. Thus, thanatophoric dysplasia (as well as milder forms of dwarfism such as achondroplasia) results from mutations in the *FGFR3* gene that causes the mutant FGF receptor 3 to be constitutively active (instead of being activated only by FGFs; see Chapter 6). This activity is enough to cause the premature differentiation of cartilage in the growth plates of the long bone, leading to the shortening of these bones.

The third mechanism of dominance is a **dominant negative allele**. When the active form of a protein is made up of multiple subunits, all of the subunits may have to be wild-type in order for the protein to function. In such cases, a mutation in just one allele—the dominant negative allele—can make the protein nonfunctional. A dominant negative allele is not just nonfunctional, it is deleterious. A dominant negative allele is the cause of **Marfan syndrome**, a disor-

different, unrelated human families. These phenotypes ranged from relatively mild anomalies to potentially lethal malformations. Similarly, Freire-Maia (1975) reported that within one family, the homozygous state of a mutant gene affecting limb development caused phenotypes ranging from severe phocomelia (lack of limb development; see Figure 1.16) to a mild abnormality of the thumb. The severity of a mutant gene's effect often depends on the *other* genes in the pathway.

Phenotypic variability in women can also be caused by statistical variation in X chromosome inactivation. One example is a case involving identical twins, each of whom carried one normal allele and one mutant allele for an X-linked clotting factor. One twin's random inactivation pattern resulted in the loss of a large percentage of the X chro-

der of the extracellular matrix. Marfan syndrome results in joint and connective tissue anomalies, not all of which are necessarily disadvantageous. Increased height, disproportionately long limbs and digits, and mild to moderate joint laxity are characteristic of this syndrome. However, patients with Marfan syndrome may also experience vertebral column deformities, myopia, loose or dislocated lenses, and (most importantly) aortic problems that can lead to aneurysm (tearing of the aorta) later in life.

The mutation responsible for Marfan syndrome is in the gene encoding fibrillin, a secreted glycoprotein that forms multimeric microfibrils in elastic connective tissue. The presence of even small amounts of mutant fibrillin inhibits the association of wild-type fibrillin into microfibrils. Eldadah and colleagues (1995) have shown that when a mutant human gene for fibrillin is transfected into fibroblast cells that already contain two wild-type genes, the incorporation of fibrillin into the extracellular matrix is inhibited.

We are beginning to understand the molecular bases for many of the inherited malformation syndromes in humans. This understanding constitutes a critically important synthesis of developmental biology, medical genetics,* and pediatric medicine.

Gene Expression and Human Disease

Candidate gene mapping and positional gene cloning techniques have brought together medical embryology and medical genetics. This fusion has enabled scientists and physicians to understand normal human development as well as the causes of many malformations. As would be expected, alterations in gene expression can occur at the levels of transcription, RNA processing, translation, and post-translational modification. The best-studied group of genes are those involved in regulating transcription (Table 21.1). These include genes encoding transcription factors, and the components of signal trans-

duction cascades. Several of these genes (e.g., *PAX6, MITF*) have already been discussed.

Inborn errors of nuclear RNA processing

MUTATIONS IN SPLICE SITES It is estimated that a majority of human genes produce RNAs that can be alternatively spliced (see Chapter 5). Therefore, even though the human genome may contain only 20,000–30,000 genes, its **proteome** (encompassing all the proteins encoded by the genome) is far more complex.

Mutations in the splice sites of genes can prevent certain isoforms from arising. It is estimated that 15 percent of all point mutations that result in human genetic disease create such splice site abnormalities (Krawczak et al. 1992; Cooper and Mattox 1997). One such disease is congenital adrenal hyperplasia (CAH), a syndrome usually caused by

TABLE 21.1 Some genes encoding human transcription factors and phenotypes resulting from their mutation

Gene	Mutation phenotype
Androgen receptor	Androgen insensitivity syndrome
AZF1	Azoospermia
CBFA1	Cleidocranial dysplasia
CSX	Heart defects
EMX2	Schizencephaly
Estrogen receptor	Growth regulation problems, sterility
Forkhead-like 15	Thyroid agenesis, cleft palate
GLI3	Grieg syndrome
HOXA13	Hand-foot-genital syndrome
HOXD13	Polysyndactyly
LMX1B	Nail-patella syndrome
MITF	Waardenburg syndrome type 2
PAX2	Renal-coloboma syndrome
PAX3	Waardenburg syndrome type 1
PAX6	Aniridia
PTX2	Reiger syndrome
PITX3	Congenital cataracts
POU3F4	Deafness and dystonia
SOX9	Campomelic dysplasia, male sex reversal
SRY	Male sex reversal
TBX3	Schinzel syndrome (ulna-mammary syndrome)
TBX5	Holt-Oram syndrome
TCOF	Treacher-Collins syndrome
TWIST	Seathre-Chotzen syndrome
WT1	Urogenital anomalies

*Although there is significant overlap between the disciplines, *human genetics* is the study of human variability, while *medical genetics* is the study of abnormal human variability. *Clinical genetics* is the medical specialty that cares for individuals with abnormal variability, and *genetic counseling* is that profession that supports in all respects the roles of the clinical geneticists and the needs of the patients and their families (J. M. Opitz, personal communication.)

FIGURE 21.7 Fragile X syndrome. (A) A 26-year-old man with fragile X syndrome. Affected individuals often have a large head with relatively large forehead, ears, and jaws. (B) The number of CGG repeats is critical in determining whether the *FMR1* gene is transcribed in the nervous system. This number can be determined by the polymerase chain reaction. PCR usually reveals 6–50 CGG repeats in the *FMR1* gene. However, in people with fragile X syndrome, over 200 repeats are seen. The product of this gene appears to be necessary for the translation of certain neuronal messages; without it, these messages are not brought to the ribosomes. (Photograph from De Boulle et al. 1993.)

(A)

(B)

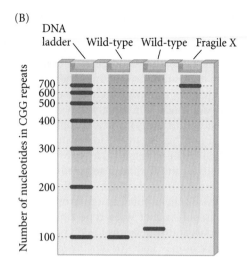

the absence of a functional gene for 21-steroid hydroxylase (*CYP21*). In the absence of this enzyme, steroids that would normally be used to make cortisol are shunted into making testosterone, and females with CAH are phenotypically masculinized. In many instances, people with this condition have a point mutation in the splice site for the second intron of the *CYP21* gene. This mutation prevents the intron from being skipped, and the resulting enzyme is ineffective. Mutations that create new splice sites or prevent the use of existing ones can both result in nonfunctional proteins (Hutchinson et al. 2001).

MUTATIONS IN SPLICING FACTORS There are some proteins and small nuclear RNAs that are used throughout the body to effect differential pre-mRNA splicing, and there are proteins that appear to regulate differential splicing in a manner characteristic of a particular cell type. If the genes encoding these cell type-specific splicing factors are mutated, one could expect several cell type-specific isoforms to be aberrant. This appears to be the case in one of the leading causes of hereditary infant mortality, spinal muscular atrophy. Here, mutations in the gene encoding the SMN (*s*urvival of *mo*tor *n*eurons) protein prevent the maintenance of motor neurons. SMN is involved in splicing nRNAs in motor neurons (Pellizzoni et al. 1999).

Inborn errors of translation

Many genetic diseases are due to mutations that create translation termination codons. For instance, androgen insensitivity syndrome (see Figure 17.10) can be caused by a guanine-to-adenine transition at nucleotide 2682 of the gene encoding the androgen receptor. This mutation changes the coding of codon 717 (in the middle of the message) from tryptophan to a translation termination codon (Sai et al. 1990). The result truncated receptor protein that lacks most of its androgen-binding domain.

Other mutations can alter the longevity of an mRNA, which can greatly affect the number of protein molecules synthesized from it. For example, *Hemoglobin Constant Spring* is a naturally occurring mutation wherein the translation termination codon of the α-globin gene has been mutated to an amino acid codon, and translation thus continues for an additional 31 codons (X. Wang et al. 1995). This read-through results in destabilization of the α-globin mRNA and a consequent reduction of greater than 95 percent in α-globin gene expression from the affected locus. The resultant clinical disease is one type of **α-thalassemia** (a condition characterized by anemia and undersized red blood cells).

In many instances, certain proteins are used to stabilize particular messenger RNAs or to bring them to ribosomes. The most prevalent form of inherited mental retardation, **fragile-X syndrome**, may result from a mutation in a gene whose product is critical for the translation of certain nerve-specific messages. Fragile-X syndrome is usually caused by the expansion and hypermethylation of CGG repeats in the 5′ untranslated region of the *FMR1* gene (Figure 21.7). Whereas most individuals have *FMR1* alleles with from 6 to 50 CGG repeats, people with fragile-X syndrome have alleles with over 200 such repeats. This CGG expansion blocks transcription of this gene.

The *FMR1* gene encodes an RNA-binding protein that appears to be critical for the *translation* of certain messages. Normally, nearly 85 percent of the FMR1 protein molecule is associated with translating polysomes. FMR1 proteins with mutations in the RNA-binding domains are not observed with cytoplasmic polysomes, and such mutations produce severe forms of fragile-X syndrome (Feng et al. 1997a,b). Recent studies have shown that a particular subset of mouse brain messenger RNA requires the FMR1 protein for proper translation. Most of these messages are involved with synapse function or neuronal development (Brown et al. 2001; Darnell et al. 2001). It is probable that FMR1 binds specific mRNAs, either regulating their translation or targeting them to the dendrite where they may await the signal for translation.

Recent studies have also revealed a condition brought about by mutations of translational initiation factor eIF2B. This factor is made of five proteins, and mutations in any of these proteins produce a devastating disease wherein the individual's brain loses its integrity after the first year of life (van der Knapp et al. 2002; Ainsworth 2005). Although this "vanishing white matter syndrome" is primarily characterized by the progressive demyelination of the brain, defects in several other tissues are also seen.

INFERTILITY

Infertility—the inability to achieve or sustain pregnancy—is not a disease in the usual sense of the word. It is not a symptom or condition that prevents the physical well-being of the infertile individual or couple. However, since the desire to have children can be exceptionally strong for biological and social reasons (indeed, empires have fallen due to a ruler's infertility), it is certainly an important condition in our society, and it is usually managed in the context of clinical medicine.

Diagnosing Infertility

There is no precise definition of infertility. Rather, infertility is the absence of pregnancy after an appropriate duration of attempting conception by regular intercourse. This duration differs between and within cultures. Population conception curves reveal that 80 percent of couples will have achieved a conception by 12 months and 90 percent by 18 months. The remaining 10 percent may be labeled as "infertile" or "subfertile."

Infertility can be caused by failure to ovulate a mature oocyte, by few or defective sperm, by physical blockage of the male or female ducts, or by incompatibilities between the sperm and the milieu of the egg or the reproductive tract (McVeigh and Barlow 2000). While there are numerous treatments for women that can lead to the maturation and ovulation of oocytes, there are relatively few treatments for men who are not making sufficient sperm. In women, exogenous gonadotropins or anti-estrogenic drugs (clomiphene or tamoxifen) can be used to stimulate the ovaries. In men, sperm may be concentrated and injected either into the oocyte or into the reproductive tract near the oocyte. There are also several **assisted reproductive techologies** (**ART**), which are medical techniques that enhance the probability of fertilization by manipulating the oocyte outside of the woman's body. The most widely practiced of these techniques is in vitro fertilization.

In Vitro Fertilization

In vitro fertilization (**IVF**) is an assisted reproductive technology in which oocytes and sperm retrieved from the male and female partners are placed together in a petri dish, where fertilization can take place. After the fertilized eggs have begun dividing, they are transferred into the female partner's uterus, where implantation and embryonic development can occur as in a typical pregnancy.

IVF was developed in the early 1970s to treat infertility caused by blocked or damaged fallopian tubes. The first IVF baby, Louise Brown, was born in England in 1978. Since then, the number of IVF procedures performed each year has increased and their success rate has improved. (IVF success rates compare favorably to natural pregnancy rates in any given month when the woman is under age 40 and there are no sperm problems; see Trounson and Gardner 2000.)

In part because it is so widely publicized (indeed, it is the subject of commercial advertising in some countries), many people mistakenly believe that IVF is the only treatment option for infertile couples. Actually, fewer than 5 percent of all patients who seek treatment for infertility receive IVF. Most infertile couples respond well to less complicated treatment options, such as hormonal therapies and artificial insemination. However, IVF remains the most commonly used of the ART procedures.

The IVF procedure

The IVF procedure has four basic steps (Figure 21.8):

- **Step 1: Ovarian stimulation and monitoring.** Having several mature oocytes* available for IVF increases the possibility that at least one will result in a pregnancy. Typically, women are injected with gonadotropins or anti-estrogens over a period of days or weeks in order to "hyperstimulate" the ovaries to produce mature oocytes.
- **Step 2: Egg retrieval.** Once the follicle has matured (but not yet ruptured), the physician retrieves as many oocytes as possible. This is done surgically, guiding an aspiration pipette to each mature follicle and sucking up the oocyte. Once recovered, those oocytes that are mature and healthy are transferred to a sterile container to await fertilization in the laboratory.
- **Step 3: Fertilization.** A semen sample is collected from the male partner approximately 2 hours before the female partner's oocytes are retrieved. These sperm are processed by a procedure called sperm washing. Sperm washing capacitates the sperm (see Chapter 7) and selects only the healthiest and most active sperm in the sample. The selected sperm are placed in a petri dish with the oocytes, and the gametes are incubated at body temperature. In general, each oocyte is incubated for 12–18 hours with 50,000–100,000 motile sperm. If fertilization is successful, the eggs will begin to divide. The success rate for achieving fertilization in this way is between 50 and 70 percent.
- **Step 4: Embryo transfer.** Embryo transfer is not complicated and can be performed without anesthesia or sur-

---

*Although they are usually called eggs, the female gametes about to be ovulated are actually metaphase II oocytes (see Chapter 19).

FIGURE 21.8   In vitro fertilization. (A) The IVF process can be divided into four basic steps: (1) ovarian stimulation, (2) egg retrieval; (3) fertilization and (4) transfer of the embryo into the uterus. (B) Assisted hatching, whereby a hole is poked in the zona pellucida, is a procedure to help ensure that the embryo implants in the uterus. (Photograph courtesy of The Institute for Reproductive Medicine and Science of St. Barnabas, Livingston, NJ.)

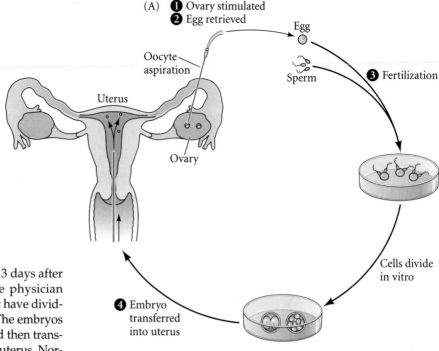

gery. The procedure is usually done 3 days after egg retrieval and fertilization. The physician looks for healthy embryos (those that have divided well and now contain 6–8 cells). The embryos are sucked into a tubular catheter and then transferred via the catheter directly to the uterus. Normal implantation and maturation of at least one embryo is required to achieve pregnancy.

In cases in which fertilization has been achieved in vitro, but after a number of cycles, implantation into the uterus fails, the physician may suggest "assisted hatching," in which a small hole is lysed in the zona pellucida prior to inserting the embryo into the uterus (Figure 21.8B; also see Chapter 7). This procedure ensures that the embryo will be able to hatch from the zona pellucida in time to adhere to the uterus.

 **WEBSITE**

21.2   Variations on IVF. In addition to the sperm-meets-egg-in-a-dish method of in vitro fertilization, other techniques have become available. This website describes intracytoplasmic sperm injection (ICSI), gamete intrafallopian transfer (GIFT), and zygote intrafallopian transfer (ZIFT, also known as tubal embryo transfer).

Success rates and complications of IVF

The rate of delivery of live babies per oocyte retrieval depends on the age of the female partner. Some recent statistics suggest that approximately 31 couples out of every 100 who try one retrieval with IVF are likely to achieve pregnancy and delivery. Compared with the one in four (25 percent) probability of achieving conception in a given cycle for normal healthy couples using unprotected intercourse, IVF offers improved chances of conception to some infertile couples. The success rate drops to 25.5 percent, however, for women 35–37 years of age, and to 17.1 percent for women 38–40. After 40 years of age, the success rate is less than 5 percent (CDC 2002a; Speroff and Fritz 2005). This decline may be due to the declining viability of eggs as women advance in age.

The IVF procedure has been very successful in achieving pregnancy, and the more embryos are transferred, the greater the chance of pregnancy. Thus physicians typically transfer several embryos—thus increasing the risk of multiple births. The rate of multiple births depends on the number of embryos transferred, and also on the woman's age. According to one set of statistics (Speroff and Fritz 2005), when three embryos were transferred, the multiple-birth rate was 46 percent for women aged 20–29; the rate was 39 percent for women aged 40–44 when seven or more embryos were transferred.

The risk of multiple births is a serious concern because multiple-birth infants are predisposed to many health problems, including malformations, infant death, premature delivery, and low birth weight (Lipshultz and Adamson 1999; Schieve et al. 1999; Bhattacharya and Templeton 2000; Gleicher et al. 2000). Babies born prematurely and at low birth weight are at risk for cerebral palsy and chronic respiratory problems. In addition, mothers who carry multiple infants are also at risk for many health conditions and complications (e.g., high blood pressure, diabetes), and the costs for multiple pregnancies are also greatly increased. (See Gilbert et al. 2005 for a discussion of this and other social considerations surrounding assisted reproductive technologies.)

WEBSITE **21.3** **Some ethical issues of IVF.** The widespread practice of IVF has given rise to numerous ethical and legal concerns over the safety of the techniques, the legal status of *ex utero* embryos, and the economic inequity that occurs when only a portion of the population has financial access to a medical technology.

SIDELIGHTS & SPECULATIONS

Prenatal Diagnosis and Preimplantation Genetics

One of the consequences of in vitro fertilization and the ability to detect genetic mutations early in development is a new area of medicine called **preimplantation genetics**. Preimplantation genetics seeks to test for genetic disease *before* the embryo enters the uterus. After that, many genetic diseases can still be diagnosed before a baby is born. This **prenatal diagnosis** can be done by chorionic villus sampling at 8–10 weeks of gestation, or by **amniocentesis** around the fourth or fifth month of pregnancy.

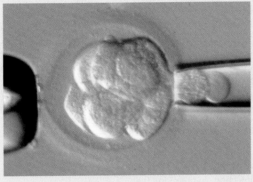

FIGURE 21.9 Preimplantation genetics is performed on one or two blastomeres (seen here in the pipette) taken from an early blastocyst. The polymerase chain reaction is then used to determine whether certain genes in these cells are present, absent, or mutant. (Photograph courtesy of The Institute for Reproductive Medicine and Science of St. Barnabas, Livingston, NJ.)

Chorionic villus sampling and amniocentesis

Chorionic villus sampling involves taking a sample of the placenta, whereas **amniocentesis** involves taking a sample of the amnionic fluid. In both cases, fetal cells from the sample are grown and then analyzed for the presence or absence of certain chromosomes, genes, or enzymes.

However useful these procedures have been in detecting genetic disease, they have brought with them a serious ethical concern: if a fetus is found to have a genetic disease, the only means of prevention presently available is to abort the pregnancy. The need to make such a choice can be overwhelming to prospective parents.* Indeed, the waiting time between knowledge of being pregnant and the results from amniocentesis or chorionic villus sampling has created a new phenomenon, the "tentative pregnancy." Many couples do not announce their pregnancy during this stressful period for fear that it might have to be terminated (Rothman et al. 1995).

By using IVF, one can consider implanting only those embryos that are most likely

to be healthy than aborting those fetuses that are most likely to produce malformed or nonviable children. This can be achieved by screening embryonic cells before the embryo is implanted in the womb. While the embryos are still in the petri dish (at the 6- to 8-cell stage), a small hole is made in the zona pellucida and two blastomeres are removed from the embryo (Figure 21.9). Since the mammalian egg undergoes regulative development (see Chapter 11), the removal of these blastomeres does not endanger the embryo, and the isolated blastomeres are tested immediately. The polymerase chain reaction technique can be used to determine the presence or absence of certain genes to be determined, and **fluorescent in situ hybridization** (**FISH**) can be used to determine whether the normal numbers and types of chromosomes are present (Kanavakis and Traeger-Synodinos 2002; Miny et al. 2002). Results are often available within 2 days. Presumptive wild-type embryos can be implanted into the uterus, while any presumptive embryos with deleterious mutations are discarded.

Sex selection and sperm selection

The same procedures that allow preimplantation genetics also enable the physician to know the sex of the embryo. Sometimes parents wish to have this information;

sometimes they do not. However, knowing the sex of an embryo prior to its implantation raises the possibility that parents could decide to have only embryos of the desired sex implanted. Sex selection using preimplantation genetics is seen by many as a beneficial way of preventing X-linked diseases, but in fact it is often used as a method of simply choosing your offspring's sex. Opponents of sex selection point to its possible use to prevent the birth of girls in cultures where women are not as highly valued as men (see Gilbert 2005). Different countries and even different hospitals have different policies permitting preimplantation genetic diagnosis solely for the purpose of sex determination.

Another way to accomplish sex selection is through sperm selection. The X chromosome is substantially larger than the Y chromosome; therefore, human sperm cells containing an X chromosome contain nearly 3 percent more total DNA than sperm cells containing a Y chromosome. This DNA difference can be measured, and the X- and Y-bearing sperm cells separated based on their size/mass ratio, using a flow cytometer. The separated sperm can then be used for artificial insemination or in vitro fertilization. Recent studies have shown that this technique is about 90 percent reliable for sorting X-bearing sperm, and about 78 percent reliable for sorting Y-bearing sperm (Stern et al. 2002).

---

*When does a human life begin? For a discussion of the differing views held by scientists, as well as an overview of the theological positions, see Gilbert et al. 2005 and Website 2.1.

TERATOGENESIS: ENVIRONMENTAL ASSAULTS ON HUMAN DEVELOPMENT

In addition to genetic mutations that can affect development, we now know that numerous environmental factors can disrupt development. The summer of 1962 brought two portentous discoveries. The first was the disclosure by Rachel Carson (1962) that the pesticide DDT was destroying bird eggs and was preventing reproduction in several species (see Chapter 22). The second (Lenz 1962) was the discovery that thalidomide, a sedative drug used to manage pregnancies, could cause limb and ear abnormalities in the fetus (see Chapter 1). These two discoveries showed that the embryo was vulnerable to environmental agents.* This was underscored in 1964, when an epidemic of rubella (German measles) spread across America. Adults showed relatively mild symptoms when infected by this virus, but over 20,000 fetuses infected by rubella became either blind, deaf, or both. Many of these infants were also born with heart defects or mental retardation (CDC 2002b).

*Indeed, Rachel Carson realized the connection, commenting that "It is all of a piece, thalidomide and pesticides. They represent our willingness to rush ahead and use something without knowing what the results will be" (Carson 1962).

Abnormalities caused by exogenous agents are called **disruptions**. The agents responsible for these disruptions are called **teratogens**.* Most teratogens produce their effects only during certain critical periods of development. Human development is usually divided into two periods, the **embryonic period** (to the end of week 8) and the **fetal period** (the remaining time *in utero*). It is during the embryonic period that most of the organ systems form; the fetal period is generally one of growth and modeling. Figure 21.10 indicates the times at which various organs are most susceptible to teratogens.

*In some cases, the same condition can be either a disruption (caused by an exogenous agent) or a malformation (caused by a genetic mutation). Chondrodysplasia punctata is a congenital defect of bone and cartilage, characterized by abnormal bone mineralization, underdevelopment of nasal cartilage, and shortened fingers. It is caused by a defective gene on the X chromosome. An identical phenotype is produced by exposure of fetuses to the anticoagulant (blood-thinning) compound warfarin. It appears that the defective gene is normally responsible for producing an arylsulfatase protein (CDPX2) necessary for cartilage growth. The warfarin compound inhibits this same enzyme (Franco et al. 1995).

FIGURE 21.10   Periods (weeks of gestation) and degrees of sensitivity of embryonic organs to teratogens. (After Moore and Persaud 1993.)

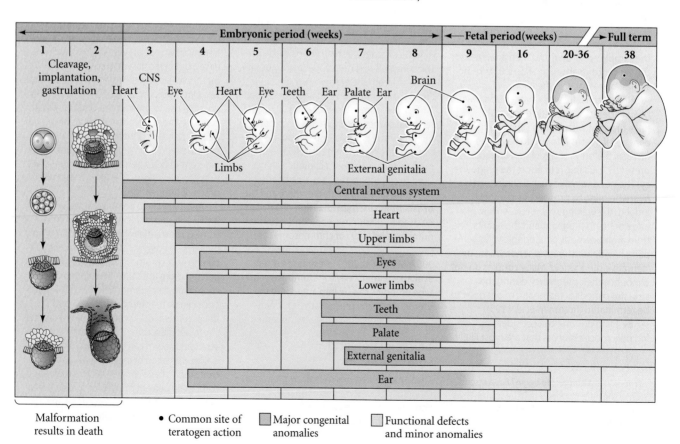

The period of maximum susceptibility to teratogens is between weeks 3 and 8, since that is when most organs are forming. The nervous system, however, is constantly forming and remains susceptible throughout development. Prior to week 3, exposure to teratogens does not usually produce congenital anomalies because a teratogen encountered at this time either damages most or all of the cells of an embryo, resulting in its death; or only a few cells, allowing the embryo to recover.

 **WEBSITE** **21.4 Thalidomide as a teratogen.** The drug thalidomide caused thousands of babies to be born with malformed arms and legs, and it provided the first major evidence that drugs could induce congenital anomalies. The mechanism of its action is still hotly debated.

Teratogenic Agents

Different agents are teratogenic in different organisms. The largest class of teratogens includes drugs and chemicals, but viruses, radiation, hyperthermia, and metabolic conditions in the mother can also act as teratogens. A partial list of agents that are teratogenic in humans is given in Table 21.2.

Some chemicals that are naturally found in the environment can cause birth defects. For example, jervine and cyclopamine are chemical products of the plant *Veratrum californicum* that can block cholesterol synthesis. As mentioned earlier in this chapter and in Chapter 6, blocking cholesterol synthesis can block Sonic hedgehog function and lead to cyclopia (see Figure 6.18).

Quinine and alcohol, two common substances derived from plants, also cause developmental disruptions. Quinine ingested by a pregnant mother can cause deafness, and alcohol can cause physical and mental retardation in the infant, as we will see in the next section. The widely used natural substances nicotine and caffeine have not been proved to cause congenital anomalies, but women who are heavy smokers (20 cigarettes a day or more) are likely to have infants that are smaller than those born to women who do not smoke. Smoking also significantly lowers the number, quality, and motility of sperm in the semen of males who smoke at least 4 cigarettes a day (Kulikauskas et al. 1985; Mak et al. 2000; Shi et al. 2001).

Alcohol as a teratogen

In terms of the frequency of its effects and its cost to society, the most devastating teratogen is undoubtedly ethanol. In 1968, Lemoine and colleagues noticed a syndrome of birth defects in the children of alcoholic mothers. This **fetal alcohol syndrome (FAS)** was confirmed by Jones and Smith (1973). Babies with FAS are characterized by small head size, an indistinct philtrum (the pair of ridges that runs between the nose and mouth above the center of the upper lip), a narrow upper lip, and a low nose bridge. The brain of such a child may be dramatically smaller than normal and often shows defects in neuronal and glial migration (Figure 21.11; Clarren 1986). There is also prominent abnormal cell death in the frontonasal process and the cranial nerve ganglia (Sulik et al. 1988). Fetal alcohol syndrome is the third most prevalent type of mental retardation (behind fragile-X syndrome and Down syndrome) and affects 1 out of every 500–750 children born in the United States (Abel and Sokol 1987).

Children with fetal alcohol syndrome are developmentally and mentally retarded, with a mean IQ of about 68 (Streissguth and LaDue 1987). FAS patients with a mean chronological age of 16.5 years were found to have the functional vocabulary of 6.5-year-olds and to have the mathematical abilities of fourth-graders. Most adults and adolescents with FAS cannot handle money or their own lives, and they have difficulty learning from past experi-

TABLE 21.2 Some agents thought to cause disruptions in human fetal development[a]

DRUGS AND CHEMICALS	IONIZING RADIATION (X-RAYS)
Alcohol	
Aminoglycosides (Gentamycin)	**HYPERTHERMIA (FEVER)**
Aminopterin	**INFECTIOUS MICROORGANISMS**
Antithyroid agents (PTU)	Coxsackie virus
Bromine	Cytomegalovirus
Cortisone	Herpes simplex
Diethylstilbesterol (DES)	Parvovirus
Diphenylhydantoin	Rubella (German measles)
Heroin	*Toxoplasma gondii* (toxoplasmosis)
Lead	*Treponema pallidum* (syphilis)
Methylmercury	
Penicillamine	**METABOLIC CONDITIONS IN THE MOTHER**
Retinoic acid (Isotretinoin, Accutane)	Autoimmune disease (including Rh incompatibility)
Streptomycin	Diabetes
Tetracycline	Dietary deficiencies, malnutrition
Thalidomide	Phenylketonuria
Trimethadione	
Valproic acid	
Warfarin	

Source: Adapted from Opitz 1991.
[a]This list includes known and possible teratogenic agents and is not exhaustive.

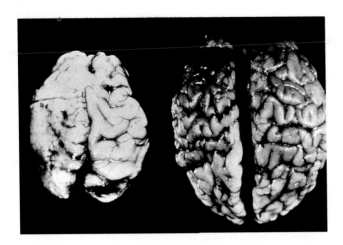

FIGURE 21.11 Comparison of a brain from an infant with fetal alcohol syndrome (FAS) with a brain from a normal infant of the same age. The brain from the infant with FAS (left) is smaller, and the pattern of convolutions is obscured by glial cells that have migrated over the top of the brain. (Photographs courtesy of S. Clarren.)

ences.* Moreover, in many cases of FAS, the behavioral abnormalities exist without any gross physical changes in head size or reductions in IQ (J. M. Opitz, personal communication).

There is great variabilty in the ability of both mothers and fetuses to metabolize ethanol, but it is believed that 30–40 percent of the children born to alcoholic mothers who drink during pregnancy will have FAS. Some alleles of alcohol-metabolizing enzymes (e.g., acohol dehydrogenase) appear to be better than others at detoxifying ethanol, and thus some fetuses are more susceptible to alcohol than others (Warren and Li 2005). It is also thought that lower amounts of ethanol ingestion by the mother can lead to fetal alcohol effect, a condition that is less severe than FAS, but which nevertheless lowers the functional and intellectual abilities of the affected person.

A mouse model system has been used to explain the effects of ethanol on the face and nervous system. When mice are exposed to ethanol at the time of gastrulation, it induces the same range of developmental defects as in humans. As early as 12 hours after the mother ingests alcohol, abnormalities of development are observed. The midline facial structures fail to form, allowing the abnormally close proximity of the medial processes of the face. Forebrain anomalies are also seen, and the most severely affected fetuses lack a forebrain entirely (Sulik et al. 1988). Studies on these mice suggest that ethanol may induce its teratogenic effects by more

*For remarkable accounts of raising children with fetal alcohol syndrome, read Michael Dorris's *The Broken Cord* (1989) and Liz and Jodee Kulp's *The Best I Can Be* (2000). For an excellent account of the debates within the medical profession about this syndrome, see Janet Golden's *Message in a Bottle* (2005).

than one mechanism. First, anatomical evidence suggests that neural crest cell migration is severely impaired. Instead of migrating and dividing, ethanol-treated neural crest cells prematurely initiate their differentiation into facial cartilage (Hoffman and Kulyk 1999).

Second, ethanol-induced apoptosis can delete millions of neurons from the developing forebrain, frontonasal (facial) process, and cranial nerve ganglia. One reason for this nerve cell death is that alcohol generates superoxide radicals that can oxidize cell membranes and lead to cytolysis (Figure 21.12A–C; Davis et al. 1990; Kotch et al. 1995; Sulik 2005). Those neurons that remain have impaired development of their mitochondria (Xu et al. 2005). In one study, ethanol injected into the amnion of a chick embryo resulted in the apoptosis of cranial neural crest cells and failure to form the frontonasal process (Ahlgren et al. 2002). This apoptosis was correlated with a loss of *sonic hedgehog* gene expression in the pharyngeal arches. Shh secreting cells placed in the head mesenchyme at this time prevented the ethanol-induced apoptosis of the cranial neural crest cells. Ethanol has recently been found to downregulate *sonic hedgehog* expression in the mouse embryo as well (Chrisman et al. 2004).

In addition, alcohol may directly interfere with the ability of the cell adhesion molecule L1 to hold cells together. Ramanathan and colleagues (1996) have shown that ethanol can block the adhesive function of the L1 protein in vitro at levels as low as 7 m*M*, a concentration of ethanol produced in the blood or brain with a single drink (Figure 21.12D). Moreover, mutations in the human *L1* gene cause a syndrome of mental retardation and malformations similar to that seen in severe cases of fetal alcohol syndrome. Taken together, these studies show that ethanol is exceptionally dangerous to the developing vertebrate because its effects are so numerous, because it can travel across the placenta, and because it can operate in several ways.

Retinoic acid as a teratogen

In some instances, even a compound involved in normal metabolism can have deleterious effects if it is present in large enough amounts and/or at particular times. Retinoic acid (RA) is important in forming the anterior-posterior axis of the mammalian embryo as well as in forming the jaws (see Chapter 11). In normal development, RA is secreted from discrete cells and works in a small area. However, if RA is present in large amounts, cells that normally would not receive high concentrations of this molecule are exposed to it and will respond to it.

13-*cis*-retinoic acid (also called isotretinoin and sold under the trade name Accutane®) has been useful in treating severe cystic acne and has been available for this purpose since 1982. Because the deleterious effects of administering large amounts of vitamin A or its analogues to pregnant animals have been known since the 1950s (Cohlan 1953; Giroud and Martinet 1959; Kochhar et al. 1984), the drug carries a label warning that it should not

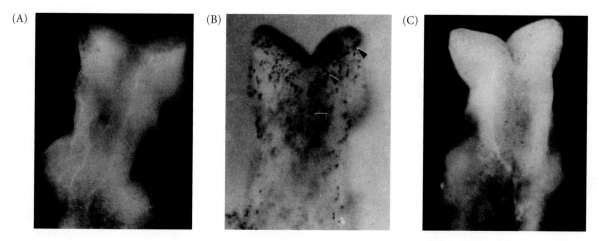

FIGURE 21.12 Possible mechanisms producing fetal alcohol syndrome (FAS). (A–C) Cell death caused by ethanol-induced superoxide radicals. Staining with Nile blue sulfate shows areas of cell death. (A) Control 9-day mouse embryo head region. (B) Head region of ethanol-treated embryo, showing areas of cell death. (C) Head region of embryo treated with both ethanol and superoxide dismutase, an inhibitor of superoxide radicals. The enzyme prevents the alcohol-induced cell death. (D) The inhibition of L1-mediated cell adhesion by ethanol. (A–C from Kotch et al. 1995, photographs courtesy of K. Sulik; D after Ramanathan et al. 1996.)

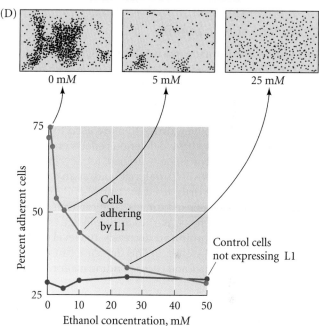

be used by pregnant women. However, about 160,000 women of childbearing age (15–45 years) have taken this drug since it was introduced, and some of them have used it during pregnancy. Lammer and his co-workers (1985) studied a group of women who inadvertently exposed themselves to retinoic acid and who elected to remain pregnant. Of their 59 fetuses, 26 were born without any noticeable anomalies, 12 aborted spontaneously, and 21 were born with obvious anomalies. The affected infants had a characteristic pattern of anomalies, including absent or defective ears, absent or small jaws, cleft palate, aortic arch abnormalities, thymic deficiencies, and abnormalities of the central nervous system.*

---

*Retinoic acid is a critical public health concern because there is significant overlap between the population using acne medicine and the population of women of childbearing age, and because it is estimated that half of the pregnancies in America are unplanned (Nulman et al. 1997). Vitamin A is itself teratogenic in megadose amounts. Rothman and colleagues (1995) found that pregnant women who took more than 10,000 international units of vitamin A per day (in the form of vitamin supplements) had a 2 percent chance of having a baby born with disruptions similar to those produced by retinoic acid. According to the new rules of the U. S. Food and Drug Administration, each patient using isotretinoin, each physician prescribing it, and each pharmacy selling it must sign a registry. Moreover, each woman using this drug will be expected to take a pregnancy test within seven days before filling their prescription and agree to use two methods of birth control and adhere to pregnancy testing on a monthly basis.

This pattern of multiple congenital anomalies is similar to that seen in rat and mouse embryos whose pregnant mothers were given these drugs. Goulding and Pratt (1986) placed 8-day mouse embryos in a solution containing 13-*cis*-retinoic acid at a very low concentration ($2 \times 10^{-6} M$). Even at this concentration, approximately one-third of the embryos developed a very specific pattern of anomalies, including dramatic reduction in the size of the first and second pharyngeal arches (see Figure 11.44). In normal mice, the first arch eventually forms the maxilla and mandible of the jaw and two ossicles of the middle ear, while the second arch forms the third ossicle of the middle ear as well as other facial bones.

The basis for this developmental disruption appears to reside in RA's ability to alter the expression of the Hox genes and thereby respecify portions of the anterior-posterior axis and inhibit neural crest cell migration from the cranial region of the neural tube (Moroni et al. 1994; Studer et al. 1994). Radioactively labeled retinoic acid binds to the cranial neural crest cells and arrests both their proliferation and their migration (Johnston et al. 1985; Goulding and Pratt 1986). This binding seems to be specific to the cranial neural crest cells, and the teratogenic effect of the drug is confined to a specific developmental period (days 8–10 in mice; days 20–35 in humans). Animal models of retinoic acid teratogenesis have been extremely successful in elucidating the mechanisms of teratogenesis at the cellular level (see Chapter 11).

 WEBSITE   21.5 Mechanisms of retinoic acid teratogenesis. Within the cell, numerous retinoid-binding proteins interact to influence the ability of retinoic acid to transcribe particular genes.

Other teratogenic agents

In addition to natural chemicals, hundreds of new artificial compounds come into general use each year in our industrial society. Pesticides and organic mercury compounds have caused neurological and behavioral abnormalities in infants whose mothers have ingested them during pregnancy. Moreover, drugs that are used to control diseases in adults may have deleterious effects on fetuses. Such drugs include cortisone, warfarin, tetracycline, and valproic acid (see Table 21.2). Valproic acid, as an example, is an anticonvulsant drug used to control epilepsy. It is known to be teratogenic in humans, as it can cause major and minor spinal defects. Barnes and colleagues (1996) have shown that valproic acid decreases the level of *Pax1* transcription in chick somites. This decrease causes malformation of the somites and corresponding malformations of the vertebrae and ribs.

Over 50,000 artificial chemicals are currently used in the United States, and between 200 and 500 new compounds are being made each year (Johnson 1980). Although teratogenic compounds have always been with us, the risks increase as more and more untested compounds enter our environment. Most industrial chemicals have not been screened for their teratogenic effects. Standard screening protocols are expensive, long, and subject to interspecies differences in metabolism. There is still no consensus on how to test a substance's teratogenicity for human embryos.

HEAVY METALS   Heavy metals such as zinc, lead, and mercury are powerful teratogens. Industrial pollution has resulted in high concentrations of heavy metals in the environment in many places. In the former Soviet Union, the unregulated "industrial production at all costs" approach left behind a legacy of soaring birth defect rates. In some regions of Kazakhstan, heavy metals are found in high concentrations in drinking water, vegetables, and the air. In such locations, nearly half the people tested have extensive chromosome breakage, and in some areas the incidence of birth defects has doubled since 1980 (Edwards 1994).

Lead and mercury can cause damage to the developing nervous system (Bellinger 2005). The polluting of Minamata Bay, Japan, with mercury in 1956 produced brain and eye deficiencies both by transmission of the mercury across the placenta and by its transmission through mother's milk. Mercury is selectively absorbed by regions of the developing cerebral cortex (Eto 2000; Kondo 2000; Eto et al. 2001), and when pregnant mice are given mercury on day 9 of gestation, nearly half of the pups are born with small brains or small eyes (O'Hara et al. 2002).

PATHOGENS   Another class of teratogens includes viruses and other pathogens. Gregg (1941) first documented the fact that women who contracted rubella (German measles) during the first trimester of their pregnancy had a 1 in 6 chance of giving birth to an infant with eye cataracts, heart malformations, or deafness. This study provided the first evidence that the mother could not fully protect the fetus from the outside environment. The rubella virus is able to enter many cell types, where it produces a protein that prevents mitosis by blocking kinases that allow the cell cycle to progress (Atreya et al. 2004). Thus, numerous organs are effected, and the earlier in pregnancy the rubella infection occurred, the greater the risk that the embryo would be malformed. The first 5 weeks of development appear to have been the most critical, because that is when the heart, eyes, and ears are formed (see Figure 21.10). The rubella epidemic of 1963–1965 probably resulted in over 10,000 fetal deaths and 20,000 infants with birth defects in the United States (CDC 2002b). Two other viruses, cytomegalovirus and the herpes simplex virus, are also teratogenic. Cytomegalovirus infection of early embryos is nearly always fatal, but infection of later embryos can lead to blindness, deafness, cerebral palsy, and mental retardation.

Bacteria and protists are rarely teratogenic, but three of them are known to damage human embryos. *Toxoplasma gondii*, a protist carried by rabbits and cats (and their feces), can cross the placenta and cause brain and eye defects in the fetus. *Treponema pallidum*, the bacterium that causes syphilis, can kill early fetuses and produce congenital deafness and facial damage in older ones.

Endocrine disruptors and human development

A newly appreciated set of teratogens are the **endocrine disruptors** (sometimes called hormone mimics or environmental signal modulators). These are exogenous chemicals that disrupt development by interfering with the normal function of hormones. Endocrine disruptors can disrupt hormonal function in several ways:

- They can mimic the effects of a natural hormone by binding to that hormone's receptors. Diethylstilbestrol, better known a DES, is one such example.
- They can block the binding of a hormone to its receptor, or they can block the synthesis of the hormone. Finasteride, a drug used to prevent male pattern baldness and

enlargement of the prostate gland, is an anti-androgen, since it blocks the synthesis of dihydrotestosterone. Dihydrotestosterone is necessary for the development of the male external genitalia, prostate differentiation, and the descent of the testes (see Figure 17.11); so women are warned not to handle this drug if they are pregnant.

- They can interfere with the transport of a hormone or its elimination from the body. Polychlorinated biphenyls (PCBs) act in this manner, as we will see in the next chapter.

This chapter will discuss the known and possible effects of endocrine disruptors on human development. The next chapter, on the environmental regulation of development, will provide evidence that environmental endocrine disruptors are causing disruptions in the development of wild animals. Some endocrine disruptors are the products of natural plant biochemistry. Others, such as stabilizers that are used in the manufacture of polystyrene and other plastics, are the products of human industry.

DIETHYLSTILBESTROL   One of the most potent environmental estrogens has been **diethylstilbestrol**, or **DES** (Figure 21.13A), a drug that was prescribed to approximately 5 million pregnant women from the 1940s through the 1960s to prevent miscarriage and premature labor (see Palmund 1996; Bell 1986). Although research from the mid-1950s showed that DES had no beneficial effects on pregnancy, it continued to be prescribed until the early 1970s, when it was shown that women exposed to this drug in utero had an increased chance of having cervical cancers and morphological abnormalities of the reproductive tract. Male offspring who had been exposed to DES in utero often had abnormal genitalia (see Robboy et al. 1982; Mittendorf 1995).

The female reproductive tract abnormalities resulting from prenatal DES exposure involve changes in the cell types along the Müllerian duct. In particular, there is a loss of the boundary between the oviduct and the uterus due to the lack of the uterotubal junction. Moreover, the distal Müllerian ducts often fail to come together to form a sin-

gle cervical canal. Symptoms similar to the human DES syndrome occur in mice exposed to DES in utero.

Ma and colleagues (1998) showed that the effects of DES on the female mouse reproductive tract could be explained as the result of altered *Hoxa10* expression in the Müllerian duct. They showed that estrogen and progesterone are able to regulate the 5′ genes of the HoxA cluster (*Hoxa9, Hoxa10, Hoxa11,* and *Hoxa13*). Normally, these genes are regulated in a nested fashion throughout the Müllerian duct. *Hoxa9* message is detected throughout the uterus and continues about halfway through the presumptive oviduct. *Hoxa10* expression exhibits a sharp anterior boundary at the junction between the future uterus and the future oviduct. *Hoxa11* has strong expression in the anterior regions where *Hoxa10* is expressed, but weakens in the posterior regions. *Hoxa13* expression is seen only in the cervix (Figure 21.13B).

To determine whether DES changed Hox gene expression patterns, DES was injected under the skin of pregnant mice, and the fetuses were allowed to develop almost to birth. When the fetuses from the DES-injected mothers were compared with fetuses from mothers that had not received DES, it was seen that DES almost completely repressed the expression of *Hoxa10* in the Müllerian duct (Figure 21.14). This repression was most pronounced in the stroma (mesenchyme) of the duct, the place where experimental embryologists had localized the effect of DES (Boutin and Cunha 1997). The case for DES acting through

(A)

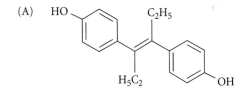

(B)

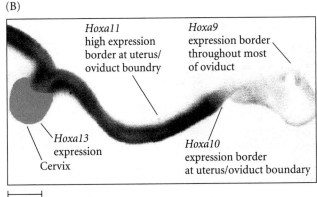

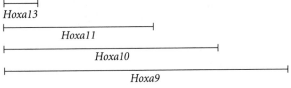

FIGURE 21.13   Effects of exposure to diethylstilbestrol on the female reproductive system. (A) The chemical structure of diethylstilbestrol (DES). (B) 5′ *HoxA* gene expression in the reproductive system of a normal 16.5-day embryonic female mouse. A whole-mount in situ hybridization of the *Hoxa13* probe is shown here (red) along with probe for *Hoxa10* (purple). *Hoxa9* expression extends from the cervix through the uterine anlage to about halfway up the Müllerian duct. *Hoxa10* expression has a sharp anterior border at the transition between the presumptive uterus and the oviduct. *Hoxa11* has the same anterior border as *Hoxa10*, but its expression diminishes closer to the cervix. The expression of *Hoxa13* is found only in the cervix and upper vagina. (After Ma et al. 1998.)

(A)   (B)

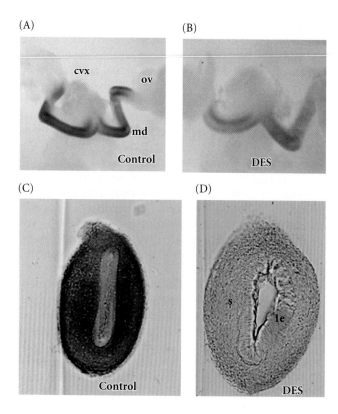

FIGURE 21.14   In situ hybridization of a *Hoxa10* probe shows that DES exposure represses *Hoxa10*. (A) Normal 16.5-day embryonic female mice show *Hoxa10* expression from the boundary of the cervix through the uterus primordium and through most of the oviduct. (B) In mice exposed prenatally to DES, this expression was severely repressed. (C) In control female mice at 5 days after birth (when the reproductive tissues are still forming), a section through the uterus shows abundant expression of the *Hoxa10* gene in the uterine mesenchyme. (D) In female mice that were given high doses of DES at 5 days after birth, *Hoxa10* gene expression in the mesenchyme is almost completely suppressed. cvx, cervix; md, Müllerian duct; ov, ovary; le, luminal epithelium; s, stroma. (After Ma et al. 1998.)

and the reproductive tracts of DES-exposed female mice also resemble those of *Wnt7a* knockout mice. Miller and colleagues (1998) have shown that the Hox genes and the Wnt genes communicate to keep each other activated. Moreover, the Hox and Wnt proteins are involved in the specification and morphogenesis of the reproductive tissues (Figure 21.15). However, DES, acting through the estrogen receptor, represses the *Wnt7a* gene. This repression prevents the maintenance of the Hox gene expression pattern, and also prevents the activation of another Wnt gene, *Wnt5a*, which encodes a protein necessary for cell proliferation.

The DES tragedy is a complex story of public policy, medicine, and developmental biology (Palmund 1996; Bell 1986). However, according to biologists Frederick vom Saal, Ana Soto, and others, the public and the pharmaceutical industry have learned little from the DES tragedy, and we may be experiencing the same endocrine disruption today, only on a much larger scale. These researchers claim that some of the major constituents of plastics are estrogenic compounds, that they are present in doses large enough to

repression of *Hoxa10* is strengthened by the phenotype of the *Hoxa10* knockout mouse (Benson et al. 1996; Ma et al. 1998), in which there is a transformation of the proximal quarter of the uterus into oviduct tissue, and there are abnormalities of the uterotubal junction.

The link between Hox gene expression and uterine morphology is the Wnt proteins. Wnt proteins are associated with cell proliferation and protection against apoptosis,

FIGURE 21.15   Misregulation of Müllerian duct morphogenesis by DES. (A) During normal morphogenesis, the *Hoxa10* and *Hoxa11* genes in the mesenchyme are activated and maintained by Wnt7a from the epithelium. Wnt7a also induces Wnt5a in the mesenchyme, and Wnt5a protein both maintains *Wnt7a* expression and causes mesenchymal cell proliferation. Together, these factors specify and order the morphogenesis of the uterus. (B) DES, acting through the estrogen receptor, blocks *Wnt7a* expression. The proper activation of the Hox genes and *Wnt5a* in the mesenchyme does not occur, leading to a radically altered morphology. (After Kitajewsky and Sassoon 2000.)

(A)   (B)

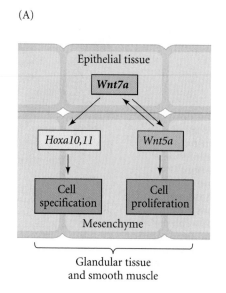
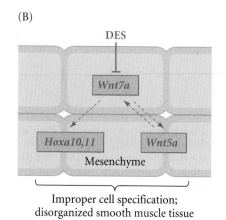

have profound effects on sexual development and behavior, and that the plastics industry is fighting against public awareness of this issue.

NONYLPHENOL Estrogenic compounds are in the food we eat and in the plastic wrapping that surrounds them. The discovery of the estrogenic effect of plastic stabilizers was made in a particularly alarming way. Investigators at Tufts University Medical School had been studying estrogen-responsive tumor cells, which require estrogen in order to proliferate. Their studies were going well until 1987, when the experiments suddenly went awry. Their control cells began to show high growth rates suggesting stimulation comparable to that of the estrogen-treated cells. Thus, it was as if someone had contaminated the medium by adding estrogen to it. What was the source of contamination? After spending 4 months testing all the components of their experimental system, the researchers discovered that the source of estrogen was the plastic containers that held their water and serum. The company that made the containers refused to describe its new process for stabilizing the polystyrene plastic, so the scientists had to discover it themselves.

The culprit turned out to be *p*-nonylphenol, a compound that is also used to harden the plastic of the pipes that bring us water and to stabilize the polystyrene plastics that hold water, milk, orange juice, and other common liquid food products (Soto et al. 1991; Colborn et al. 1996). This compound is also the degradation product of detergents, household cleaners, and contraceptive creams. Nonylphenol has been shown to alter reproductive physiology in female mice and to disrupt sperm function. It is also correlated with developmental anomalies in wildlife (Fairchild et al. 1999; Hill et al. 2002; Kim et al. 2002; Adeoya-Osiguwa et al. 2003).

BISPHENOL A Bisphenol A (BPA) was first synthesized as an estrogenic compound in the 1930s. In the early years of hormone research, the steroid hormones were very difficult to isolate, so chemists manufactured synthetic analogues that would accomplish the same tasks. Then polymer chemists discovered that bisphenol A could be used in plastic production, and today it is one of the top 50 chemicals produced worldwide. Four corporations in the United States make almost 2 billion pounds of it each year for use in (among many other things) the resin lining in most aluminum cans, the polycarbonate plastic in baby bottles and children's toys, and dental sealant. Bisphenol A is also used in making the brightly colored Nalgene® polycarbonate water bottles. Its modified form, tetrabromo-bisphenol A, is the major flame retardant used on the world's fabrics.

However, as several laboratories have shown, the BPA in plastic is not fixed forever (Krishnan et al. 1993; vom Saal 2000; Howdeshell et al. 2003). If one lets water sit in an old polycarbonate rat cage at room temperature for a week, you can measure up to 310 µg per liter of BPA in the water. That is a biologically active amount—a concentration that will reverse the sex of a frog. It also can cause chromosome anomalies. When a laboratory technician mistakenly rinsed some polycarbonate cages in an alkaline detergent, the female mice housed in these cages had meiotic abnormalities in 40 percent of their oocytes (the normal amount is about 1.5 percent). When bisphenol-A was administered to the pregnant mice under controlled circumstances, Hunt and her colleagues (2003) showed that a short, low-dose exposure to BPA was sufficient to cause meiotic defects in maturing mouse oocytes (Figure 21.16).

Bisphenol A at environmentally relevant concentrations can cause disruptions in the morphology of the fetal sex organs, low sperm counts, and behavioral changes when these fetuses become adults (vom Saal et al. 1998; Palanza et al. 2002; Kubo et al. 2003). Indeed, recent research has implicated bisphenol A (and related estrogenic compounds such as dioxins) in a suite of trends, including the lowering of human sperm counts, the increase in prostate gland size, and lowered age of female sexual maturation (see Sidelights & Speculations).

Testicular dysgenesis and declining sperm counts

Men born prior to 1959 have significantly higher sperm counts than men born after 1970, and the rate of testicular cancers has risen over 2.5 times from 1970 to 2000 (Carlsen et al. 1992; Aitken et al. 2004). Skakkebaek (2004) analyzed this data and has hypothesized that there exists a **testicular dysgenesis syndrome** characterized by disorganized testis development, testicular germ cell tumors, and low

(A) (B)

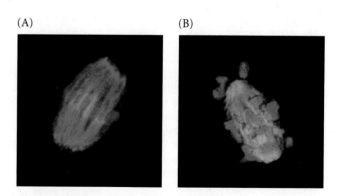

FIGURE 21.16 Bisphenol A causes meiotic defects in maturing oocytes. (A) Chromosomes (red) normally line up at the center of the spindle during first meiotic metaphase. (B) Short exposures to bisphenol-A cause chromosomes to randomly align on the spindle. Different numbers of chromosomes then enter the egg and polar body, resulting in chromosomal aneuploidy and infertility. (From Hunt et al. 2003, photographs courtesy of P. Hunt.)

SIDELIGHTS & SPECULATIONS

Bisphenol-A: Potencies and Politics

The age at which girls begin to express adult female sexual characteristics has declined over the past hundred years, probably due to better nutrition. However, new research suggests that today puberty in human females is starting extremely early (Herman-Giddens et al. 1997). Work in vom Saal's laboratory showed that when female mice are exposed in utero to low doses of bisphenol A, they undergo sexual maturation faster than unexposed mice (Howdeshell et al. 1999; vom Saal 2000; vom Saal and Hughes 2005). Moreover, female mice exposed to BPA as embryos showed altered mammary development at puberty. Such female mice had alterations in the organization of their breast tissue and ovaries and altered estrous cyclicity as adults (Markey et al. 2003). Each mammary gland produced more terminal buds and was more sensitive to estrogen, perhaps predisposing these mice to breast cancer as adults (Munoz-de-Toro et al. 2005). What is remarkable in these studies is that the exposure dosage of BPA was 2000 times lower than the dosage set as "safe" by the United States government.

Females aren't the only sex affected by bisphenol A. In 1992, Carlsen and colleagues reported a large average decline in sperm counts from studies around the world. This finding has been confirmed by Swan and colleagues (2000), who also came to the conclusion that, on average, human sperm density and quality have declined significantly over the past 5 decades. These studies have implicated bisphenol A (and other estrogenic compounds) in causing the increase in prostate enlargement seen in men. In the United States, 65 percent of men have enlarged prostate glands by age 65, and 40,000 men die of prostate cancer each year.

As an assay, researchers used the mouse prostate, a gland whose size is sensitive to estrogens. vom Saal and colleagues (1998) found that when they gave pregnant mice 2 parts per billion bisphenol A—that is, 2 nanograms per gram of body weight—for the 7 days at the end of the pregnancy (equivalent to the period when human

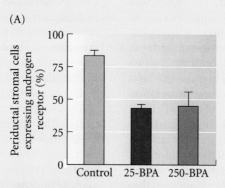 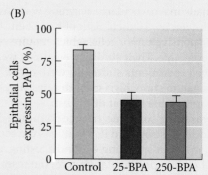

FIGURE 21.17 Bisphenol A reduces androgen receptor and prostatic acid phosphatase expression in the ventral prostate of rats exposed to it in utero. (A) Androgen receptor expression in periductal stromal cells of ventral rat prostates. Two bisphenol A-treated groups (whose mothers were given 25 μg/kg/day or 250 μg/kg/day bisphenol A from embryonic day 8 to birth) had smaller percentages of androgen receptor-containing stromal cells than did controls rats. (B) Prostatic acid phosphatase expression in epithelial cells in the ventral prostates of rats. In the control animals, many more glandular epithelial cells were positive for prostatic acid phosphatase than in the bisphenol A-treated groups. Means between controls and BPA-treated groups differ significantly ($p < 0.01$). (After Ramos et al. 2001.)

reproductive organs are developing), the male offspring of the treated animals showed an increase in prostate size of about 30 percent. Indeed, the whole area normally stimulated by dihydrotestosterone was enlarged, and the sperm count of the male mice was low. Adult male mice exposed to small amounts of bisphenol A have enlarged prostates, and bisphenol A increases the rate of mitosis in human prostate cells (Figure 21.17; Ramos et al. 2001; Wetherill 2002; Timms et al. 2005).

But is there any evidence that bisphenol A reaches the *human* fetus in concentrations that matter? Unfortunately, recent studies (Ikezuki et al. 2002; Schöenfelder et al. 2002) have shown that bisphenol A in the human placenta is neither eliminated nor metabolized into an inactive compound; rather, it

accumulates to concentrations that can alter development in laboratory animals.

The plastics industry counters that vom Saal's evidence is meager and that his experiments cannot be repeated (Cagen et al. 1999; see Lamb 2002). However, a review of that industry's own study (claiming that mice exposed in utero to bisphenol A do not have enlarged prostates or low sperm counts) points out that the positive control of the industry-sponsored research did not produce the expected effects. Reviewing the literature, vom Saal and Hughes (2005) conclude that BPA is one of the most dangerous chemicals known and that governments should consider banning the use of bisphenol A in products containing liquids that humans and animals might drink.

 WEBSITE 21.6 Our stolen future. This website monitors the environmental effects of endocrine disruptors. It is a political and consumer action site as well as a scientific clearinghouse for endocrine disruption. Run by the authors of the book *Our Stolen Future*, it also provides links to the websites of people who disagree with them.

sperm levels. This syndrome can be induced by administering phthalate derivatives to pregnant rats. Phthalates are ubiquitous in industrialized society, widely used in plastics and in cosmetics. Even that "new car smell" consists largely of volatilizing phthalates from the plastics in the car interior.

Evidence obtained from newborns indicates that male babies exposed in utero to relatively high phthalate levels had some morphological changes in their testes. DNA analysis has shown that men with infertility problems often have higher than expected levels of phthalates in their blood (Duty et al. 2003). Other endocrine disruptors that adversely affect sperm count include dioxins, nonylphenol, bisphenol A, acrylamide, and certain pesticides and herbicides* (see Aitken et al. 2004).

The link between pesticides and infertility has been known for a long time (Carson 1962; Colborn et al. 1996). One of the most important teratogenic pesticides is DDT, which is discussed at length in the following chapter. In humans, DDT has been linked to pre-term births and immature babies, and its use has been banned in the United States (Longnecker et al. 2001). Anway and colleagues (2005) published a paper citing evidence that two widely used chemicals—the fungicide vinclozolin (used throughout the wine industry on grapes) and the insecticide methoxychlor—not only impaired the fertility of males in the generation exposed in utero, but also impaired male fertility for at least three generations afterward. Both methoxychlor and vinclozolin are known endocrine disruptors. Both chemicals have been shown to cause sterility in male rats born to mothers who were injected with these pesticides late in their pregnancy. However, the new study showed that if the pregnant rats were injected at a slightly earlier time, their male offspring could still reproduce, but exhibited testis cells that underwent greater apoptosis than usual, sperm counts some 20 percent below normal, and remaining sperm that was significantly less motile. Moreover, when affected males were mated with normal females, the resultant male offspring also suffered from testicular dysgenesis syndrome; some of them were sterile and some had reduced fertility. The study ended after the fourth generation of males continued to show low sperm count, low sperm motility, and high testicular cell apoptosis (Figure 21.18).

One possible mechanism by which the effects of endocrine disruptors can be transmitted from one gener-

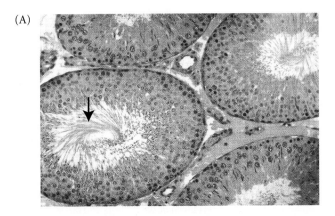

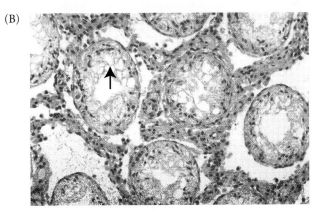

FIGURE 21.18 Cross-section of seminiferous tubules from the testes of control rat (A) and a rat (B) whose grandfather was born from a mother who had been injected with vinclozolin. This rat was infertile. The arrow in (A) shows the tails of the sperm. The arrow in (B) shows the lack of germ cells in the much smaller tubule. (From Anway et al. 2005; photographs courtesy of M. K. Skinner.)

ation to the next could involve the altered methylation of the male germline cells. Anway and colleagues found several genes whose methylation patterns were altered after exposure to the endocrine disruptors. The ability of endocrine disruptors to "re-program" germ cells has important implications for preventative medicine and public health.

DEVELOPMENTAL BIOLOGY AND THE FUTURE OF MEDICINE

Developmental Cancer Therapies

Many cancers result from an accumulation of mutations that puts certain cells back into the cell cycle and makes them unresponsive to their cellular environments (see Lengauer et al. 1998; Sandberg and Chen 2002). This

---

*Genistein, the estrogenic compound found in soy (and soy products such as tofu) is also being scrutinized (Newbold et al. 2001; Wiszniewski et al. 2005). For people eating an omnivorous diet with soy supplementation, genistein should not be a concern (indeed, this compound may protect against certain cancers). However, researchers at the National Institute of Environmental Health Science are worried that infants born to vegan or vegetarian mothers and who are fed solely soy-based formulae may develop abnormalities of the reproductive system and thyroid glands.

SIDELIGHTS & SPECULATIONS

The Embryonic Origins of Adult-Onset Illnesses

Teratogenesis is usually associated with *congenital* disease (i.e., a condition appearing at birth) and is also associated with disruptions of organogenesis during the embryonic period. However, D. J. P. Barker and colleagues (1994a,b) have offered evidence that certain *adult-onset* diseases may also result from conditions in the uterus prior to birth. Based on epidemiological evidence, they hypothesize that there are critical periods of development during which certain physiological insults or stimuli can cause specific changes in the body. The "Barker hypothesis" postulates that certain anatomical and physiological parameters get "programmed" during embryonic and fetal development, and that deficits in nutrition during this time can produce permanent changes in the pattern of metabolic activity—changes that can predispose the adult to particular diseases.

Specifically, Barker and colleagues showed that infants whose mother experienced protein deprivation (due to wars, famines, or migrations) during certain months of pregnancy were at high risk for having certain diseases as adults. Undernutrition during an individual's first trimester can lead to hypertension and strokes in adult life, while those fetuses experiencing undernutrition during the second trimester had a high risk of developing heart disease and diabetes as adults. Those fetuses experiencing undernutrition during the third trimester were prone to blood clotting defects as adults.

Recent studies have tried to determine if there are physiological or anatomical reasons for these correlations (Gluckman and Hanson 2004, 2005; Lau and Rogers 2005). Anatomically, undernutrition can change the number of cells produced during a critical time of organ formation. When pregnant rats are fed low-protein diets at certain times during their pregnancy, the resulting offspring are at high risk for hypertension as adult. The poor diet appears to cause low nephron numbers in the adult kidney (see Moritz et al. 2003). In humans, the number of nephrons present in the kidneys of men with hypertension was only about half the number found in men without hypertension (Figure 21.19A; Keller et al. 2003). In addition, the glomeruli (the blood-filtering unit of the nephron) of hypertensive men were larger than those in control subjects (Figure 21.19B).

Similar trends have been reported for non-insulin dependent (Type II) diabetes and glucose intolerance (Hales et al. 1991; Hales and Barker 1992). Here, poor nutrition reduces the number of β-cells in the pancreas and hence the ability to synthesize insulin. Moreover, the pancreas isn't the only organ involved. Undernutrition in rats changes the histological architecture in the liver as well. A low-protein diet during gestation appeared to *increase* the amount of periportal cells that produce the glucose-synthesizing enzyme phosphoenolpyruvate carboxykinase while *decreasing* the number of perivenous cells that synthesize the glucose-degrading enzyme glucokinase in the offspring, (Burns et al. 1997). These changes may be coordinated by glucocorticoid hormones that are stimulated by malnutrition and which act to conserve resources, even though such actions might make the person

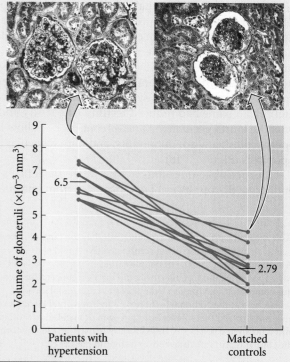

(B)

FIGURE 21.19 Anatomical changes associated with hypertension. (A) In age-matched individuals, the kidneys of men with hypertension had about half the number of nephrons as those with normal blood pressure. (B) The glomeruli of these nephrons in hypertensive kidneys were much larger than those glomeruli in control subjects. (After Keller 2003; photographs courtesy of G. Keller.)

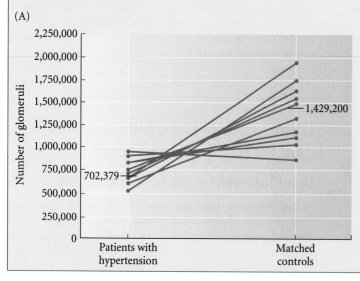

(A)

SIDELIGHTS & SPECULATIONS

FIGURE 21.20 Activity of the liver gene for peroxisomal proliferator-activated receptor (PPARα) is susceptible to dietary differences. (A) DNA methylation pattern of the PPARα promoter region, showing highly methylated control promoters compared with poorly methylated promoters from the livers of mice whose mothers had protein restricted diets ($p < 0.001$). Adding folate to the protein-restricted diet abolished this difference. (B) Levels of mRNA for the PPARα gene were much higher in the mice fed the protein-restricted diet ($p < 0.0001$). (After Lillycrop et al. 2005.)

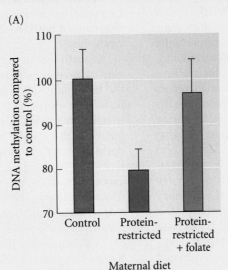

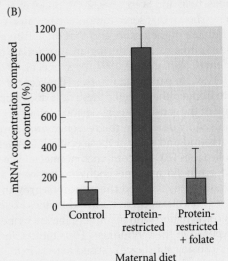

prone to hypertension later in life (see Fowden and Forhead 2004). (Since, historically, most humans died before age 50, this would not be a detrimental evolutionary trade-off.)

Hales and Barker (2001) have proposed a "thrifty phenotype" hypothesis wherein the malnourished fetus is "programmed" to expect an energy-deficient environment. The developing fetus sets its biochemical parameters to conserve energy and store fat.* Resulting adults who do indeed meet with the expected poor environment are ready for it and can survive better than individuals whose metabolisms were set to uti-

---

*In other words, the embryo has phenotypic plasticity—the ability to modulate its phenotype depending on the environment; this plasticity will be discussed further in Chapter 22.

lize energy and not store it as efficiently. However, if such a "deprivationally developed" person lives in an energy- and protein-rich environment, their cells store more fats and their heart and kidney have developed to survive more stringent conditions. Both these developments put the person at risk for several later-onset diseases.

How can conditions experienced in the uterus create anatomical and biochemical conditions that will be maintained throughout adulthood? One place to look is DNA methylation. Lillycrop and colleagues (2005) have shown that rats born to mothers having a low-protein diet had a different pattern of liver gene methylation than did the offspring of mothers fed a normal diet. These differences in methylation changed the metabolic profile of the rat's liver. For instance, the methylation of the promoter

region of the PPARα gene (which is critical in the regulation of carbohydrate and lipid metabolism) is 20 percent lower in the offspring of protein-restricted rats, and the gene's transcriptional activity becomes tenfold greater (Figure 21.20). Moreover, the difference between these methylation patterns could be abolished by including folic acid in the protein-restricted diet. Thus, the difference in methylation probably results from changes in folate metabolism caused by the limited amount of protein available to the fetus.

It does appear, then, that prenatal nutrition can induce long-lasting gene-specific alterations in transcriptional activity and metabolism. The prevention of adult disease through prenatal diet could thus become a public health issue in the coming decades.

---

genetic approach to malignancy has explained the formation of numerous tumor types, but it is not the whole story. As Folkman and colleagues (2000) recently noted, the genetic approach to cancer therapy must be complemented by a developmental approach that sees that "epigenetic, cell-cell, and extracellular interactions are also pivotal in tumor progression."

Cancer as a disease of altered development

There are many reasons to view cancer as a disease of development. First, many tumor cells have normal genomes, and whether or not they are malignant depends upon their environment. The most remarkable of these cases is the **teratocarcinoma**, which is a tumor of germ cells or stem cells (see

Chapter 19; Illmensee and Mintz 1976; Stewart and Mintz 1981). Teratocarcinomas are malignant growths of cells that resemble the inner cell mass of the mammalian blastocyst, and they can kill the organism. However, if a teratocarcinoma cell is placed on the inner cell mass of a mouse blastocyst, it will integrate into the blastocyst, lose its malignancy, and divide normally. Its cellular progeny can become part of numerous embryonic organs. Should its progeny form part of the germ line, sperm or egg cells can be formed from the tumor cell and transmit its genome to the next generation (see Figure 19.10).

Second, cancer is often caused by miscommunication between cells. The surroundings of the cell are critical in determining malignancy. Cell division is a "normal" function of cells, and in many cases, tissue interactions are required to prevent cells from dividing. Thus, there are tumors that arise through defects in tissue architecture (Sonnenschein and Soto 1999, 2000; Bissel et al. 2002). Studies have shown that some tumors can be caused by altering the structure of the tissue, and that these tumors can actually be suppressed by restoring an appropriate tissue environment (Coleman et al. 1997; Weaver et al. 1997; Sternlicht et al. 1999). In particular, whereas 80 percent of human tumors are from epithelial cells, these cells do not appear to be the site of the cancer-causing lesion. Rather, these cancers are caused by defects in the mesenchymal stromal cells that surround and sustain these epithelia. When Maffini and colleagues (2004) recombined normal and carcinogen-treated epithelia and mesenchyme between rat mammary glands, they showed the tumorous growth of mammary *epithelial* cells occurred only when these cells were placed in combination with mammary mesenchyme that had been exposed to the carcinogen. Thus the carcinogen (the cancer-causing substance) caused defects in the mesenchymal stroma of the mammary gland. These treated mammary cells could not hold back their epithelium from dividing.

Third, defects in the paracrine signaling pathways can cause cancers. Indeed, both carcinogenesis and teratogenesis are diseases of tissue organization and intercellular communication. They are opposite sides of the same coin. Some defects in signaling pathways can cause cancers, while other defects in the same pathways (even the same protein, but at different sites) can cause malformations (see Cohen 2003). By understanding the developmental events involved, one can design better therapies for these diseases.

For instance, medulloblastomas, the most common malignant brain tumors in children, appear to be due to abnormalities of the Hedgehog signaling pathway. In these brain tissues, the Smoothened protein is constitutively activated, due to either mutations in Patched (the normal negative regulator of Smoothened) or faulty methylation of the *Patched* gene. Moreover, the malignant growth of medulloblastomas in mice and humans (and digestive tract tumors in mice) can be halted by cyclopamine, a teratogen that functions by *blocking* the Hedgehog pathway (Berman et al. 2002, 2003; Thayer et al. 2003). Another set of tumors, those of the pancreas and esophagus, are generated when these cells make Sonic hedgehog protein in the adult (as they normally do in the embryo). It is possible that once the cells express these factors, an autocrine loop results whereby the cells respond to their own growth factors. Autocrine loops involving FGF and EGF signaling have also been suggested for asbestos-induced mesotheliomas (Stapelberg et al. 2005), and upregulation of epithelial growth factor receptors has been documented to occur in lung tumors.

One mutation that shows how closely related cancers are to developmental anomalies concerns the *AXIN2* gene in humans (Lammi et al. 2004). Axin2 is involved in Wnt signaling, and it has been shown that both inhibition and overstimulation of the Wnt pathway led to abnormalities of tooth formation. So it was not surprising that people who had gain-of-function mutations of *AXIN2* lacked several permanent teeth. However nearly all of these people also had rectal carcinoma as well. Thus, this single gene mutation caused both the developmental anomalies and carcinogenesis.

A fourth reason to consider cancers as diseases of disrupted development involves **metastasis**, the invasion of the malignant cell into other tissues. Tumor cells do not usually stay put. Rather, they migrate and form colonies in other organs. In Chapter 6, we discussed the roles of cadherins in the sorting out of cells to form tissues during development. Cells formed boundaries and segregated into tissues by altering the strengths of their attachments. In cancer metastasis, this property is lost; cadherin levels are downregulated, and the strength of attachment to the extracellular matrix and other types of cells becomes greater than the cohesive force binding the tissue together. As a result, the cells become able to spread into other tissues (Foty and Steinberg 1997, 2004).

Another phase of metastasis involves the digestion of extracellular matrices by **metalloproteinases**. These enzymes are used by migrating cells to digest a path to their destination. They are commonly seen to be secreted by trophoblast cells, axon growth cones, sperm cells, and somitic cells. However, metalloproteinases are reactivated in malignant cancer cells, allowing the cancer to invade other tissues. Thus the presence of these enzymes is a marker that the tumor is particularly dangerous (see Gu et al. 2005).

A fifth reason to think about cancers as diseases of development is that the properties of the cancer cells can often be explained by our knowledge of development. These suggest ways of treating the cancer. Melanomas for instance, are tumors of the pigment cells. Recent studies have demonstrated that MITF, in addition to activating tyrosinase and other melanin-forming genes, also activates the anti-apoptosis gene *BLC2*.* That is probably why melanomas are so resistant to treatment. Moreover, melanomas have amplified the MITF-containing portions of their chromosomes, so that they now have several copies of the MITF gene (McGill et al. 2002; Garraway et al. 2005). Because MITF is critical for the survival of these melanoma cells, one possible therapy against melanomas might be to block MITF activity. In one clinical trial, a malignant melanoma cell line was injected with an inhibitor of MITF (a retrovirus containing a dominant negative MITF mutant), and the melanoma cells became more sensitive to conventional chemotherapy (Garraway et al. 2005).

This brings us to a sixth, and very important, reason for thinking of cancer as a disease of development. MITF is important in the self-renewal of melanocyte stem cells and (through the inactivation of *BCL2*) in the survival of these cells (Nishimura et al. 2005; see Chapter 13). It is possible that the melanomas are not differentiated melanocytes that

---

*It is probable that many cancer cells survive only if their anti-apoptotic genes or anti-senescence genes are suppressed (see Sharpless and DePinho 2005).

have reverted to a primitive stage, but melanocyte stem cells that have amplified their MITF genes and are able to survive outside their niche. It is probable that most stem cells cannot survive outside their niche and will undergo apoptosis if they are withdrawn from the paracrine factors of their environment. But once it has extra copies of an anti-apoptosis gene, the stem cell can enter a new environment and not die. Rather, it acts as a normal stem cell, making more of itself and more differentiated progeny. In the absence of community controls, growth continues unchecked and eventually the cells become malignant. This thinking has led to the **cancer stem cell hypothesis** (Reya et al. 2001; Dean et

al. 2005) which postulates that certain cancers arise from adult stem cells. Human myeloid leukemias and astrocytic brain tumors (the most common type of brain tumor) have both been linked to stem cells that have managed to enter new domains and retain their stem cell properties (Lapidot et al. 1994; Bonnet and Dick 1997; Singh et al. 2004). Thus, adult stem cells, which are normally quiescent until "called to duty," are also cells that can become tumors when exposed to carcinogens. This hypothesis would explain how cancer cells could arise with these stem-cell-like abilities, and why tumors often contain a mixed population of stem cells and more differentiated cells.

SIDELIGHTS & SPECULATIONS

Differentiation Therapy

If cancer is a disease of development, then developmental biology should be able to explain some of the phenomena of tumors and also suggest ways of curing cancers. One way, as mentioned, is to use teratogens when appropriate. The same properties that make these chemicals teratogens can also make them useful for destroying tumor tissue (Blaheta et al. 2002). Thus, the teratogens cyclopamine and valproic acid have been shown to reduce several types of tumors.

Another area of cancer therapy in which developmental biology has played an important role has been **differentiation therapy**. If the cancerous cell is actually a stem cell, then it might be "domesticated" by the right mixture of paracrine factors. In 1978, Pierce and his colleagues noted that cancer cells were in many ways reversions to embryonic cells, and hypothesized that cancer cells should revert to normalcy if they were made to differentiate. That same year, Sachs (1978) discovered that certain leukemias could be controlled by making their cells differentiate rather than proliferate. One of these leukemias, acute promyelocytic leukemia (APL), is caused by a somatic recombination between chromosomes 15 and 17 in the precursor cells that give rise to neutrophils (see Figure 15.23). The fusion of these chromosomes creates a chimeric transcription factor, one of whose parts is retinoic acid receptor-α. The expression of this chimeric transcription factor causes the cell to become malignant (Miller et al. 1992; Grignani et al. 1998). Normally, retinoic acid receptor-α is involved in the differentiation

of myeloid precursor cells into neutrophils, but the chimeric transcription factor appears to prevent this from occurring. Treatment of APL patients with all-*trans* retinoic acid causes remission of APL in more than 90 percent of cases, since the additional retinoic acid is able to effect the differentiation of the leukemic cells into normal neutrophils (Hansen et al. 2000; Fontana and Rishi 2002).

The concept of differentiation therapy is being extended to other types of cancers (see Altucci and Gronemeyer 2001; Cao and Heng 2005). It appears that tumor cells can become "addicted" to the high expressions of transcriptional regulators that spur their rapid cell division. Jain and colleagues (2002)

have found that a brief inactivation of growth regulators such as Myc tells the tumor cells to either differentiate or die (Figure 21.21). An important question, then, is how to bring about such brief inactivation of growth regulators. The answer may be provided by RNA interference technology. Xia and colleagues (2002) demonstrated the efficacy of a viral-mediated delivery mechanism that caused the specific silencing of targeted genes through the expression of **small interfering RNA** (**siRNA**; see Chapter 4). Adenovirus vectors were used to transfect tumor cells with genes encoding a small antisense transcript targeted to a specific region of the gene of interest. As a result, the protein encoded by that gene was not translated.

(A)

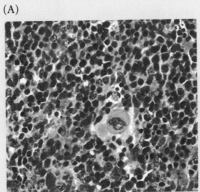

(B)

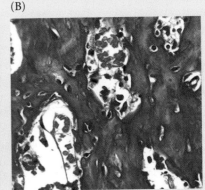

FIGURE 21.21 Differentiation of osteogenic sarcomas in mice due to inactivation of the Myc transcription factor. (A) Osteogenic sarcomas injected into mice remained undifferentiated and metastasized. (B) When transcription of the Myc transcription factor, which promotes cell division, was briefly blocked, the tumor cells differentiated into mature bone-like cells. (After Jain et al. 2002.)

Angiogenesis inhibition

Another area in which knowledge of development could contribute to cancer therapies is the inhibition of angiogenesis. Judah Folkman (1974) has estimated that as many as 350 billion mitoses occur in the human body every day. With each cell division comes the chance that the resulting cells will be malignant. Indeed, autopsies have shown that every person over 50 years old has microscopic tumors in their thyroid glands, although less than 1 in 1000 persons have thyroid cancer (Folkman and Kalluri 2004). Folkman suggested that cells capable of forming tumors develop at a certain frequency, but that most never form observable tumors. The reason is that a solid tumor, like any other rapidly dividing tissue, needs oxygen and nutrients to survive. Without a blood supply, potential tumors either die or remain dormant. Such "microtumors" remain as a stable cell population wherein dying cells are replaced by new cells.

The critical point at which a node of cancerous cells becomes a rapidly growing tumor occurs when it becomes vascularized. A microtumor can expand to 16,000 times its original volume in the 2 weeks after vascularization (Folkman 1974; Ausprunk and Folkman 1977). To accomplish this vascularization, the microtumor secretes substances called **tumor angiogenesis factors**. These often include the same factors that produce blood vessel growth in the embryo—VEGF, Fgf2, placenta-like growth factor, and others. Tumor angiogenesis factors stimulate mitosis in endothelial cells and direct the cell differentiation into blood vessels in the direction of the tumor.

Tumor angiogenesis can be demonstrated by implanting a piece of tumor tissue within the layers of a rabbit or mouse cornea. The cornea itself is not vascularized, but it is surrounded by a vascular border, or limbus. The tumor tissue induces blood vessels to form and grow toward the tumor (Figure 21.22; Muthukkaruppan and Auerbach 1979). Once the blood vessels enter the tumor, the tumor cells undergo explosive growth, eventually bursting the eye. Other adult solid tissues do not induce blood vessels to form. It might therefore be possible to block tumor development by blocking angiogenesis. Folkman and his colleagues have pioneered attempts to find and use angiogenesis inhibitors. There are now over 50 chemicals being tested as natural and artificial angiogenesis inhibitors. These compounds act to prevent the endothelial cells from responding to the angiogenetic signal of the tumor.* Interestingly, thalidomide, the teratogen responsible for birth defects in the 1960s, is on this list. Thalidomide has been found to be a potent anti-angiogenesis factor that can reduce the growth of cancers in rats and mice (D'Amato et al. 1994; Dredge et al. 2002). One of the advantages of these compounds is that the tumor cells are unlikely to evolve resistance to them, since the tumor cell itself is not the target of these agents (Boehm et al. 1997; Kerbel and Folkman 2002).

Hanahan and Folkman (1996) have hypothesized that tumor angiogenesis is mediated by a change in the balance between angiogenesis factors and angiogenesis inhibitors. Observations of human tumor progression and gene knockouts in mice suggest that tumor angiogenesis may be mediated either by a decrease in the production of angiogenesis inhibitors or an increase in the production of angiogenesis factors. The signal for new VEGF synthesis in tumor cells might even be the hypoxia the cells experience in the center of a premalignant cell mass (Shweiki et al. 1992).

---

*The consumption of green tea has been associated with lower incidences of human cancer and the inhibition of tumor cell growth in laboratory animals. Cao and Cao (1999) have shown that green tea, as well as one of its components, epigallocatechin-3-gallate (EGCG), prevents Fgf2- and VEGF-induced angiogenesis. EGCG may also inhibit the receptor tyrosine kinase pathway, repress the activity of certain transcription factors, and cause apoptosis in malignant cells (Nakazato et al. 2005; Shimizu and Weinstein 2005).

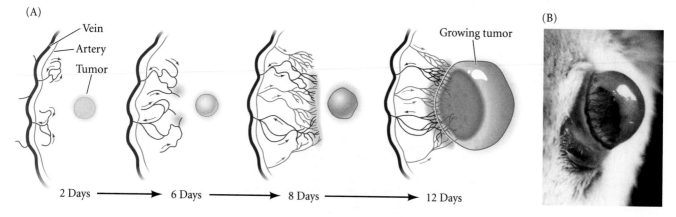

FIGURE 21.22 New blood vessel growth to the site of a mammary tumor transplanted into the cornea of an albino mouse. (A) Sequence of events leading to the vascularization of the tumor on days 2, 6, 8, and 12. The veins and arteries in the limbus surrounding the cornea both supply blood vessels to the tumor. (B) Photograph of living cornea of an albino mouse, with new blood vessels from the limbus approaching the tumor graft. (From Muthukkaruppan and Auerbach 1979; photograph courtesy of R. Auerbach.)

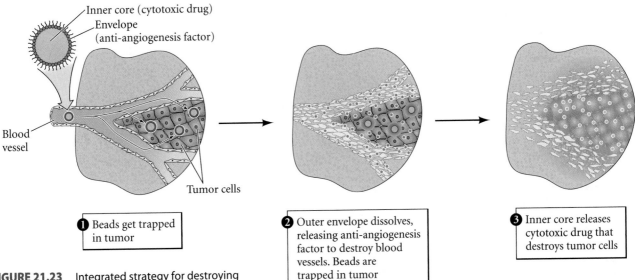

FIGURE 21.23 Integrated strategy for destroying tumors using anti-angiogenesis factors and cytotoxic drugs. Nanometer-sized beads comprise an inner core of cytotoxic drug and an outer envelope containing an anti-angiogenesis factor within a lipid envelope that preferentially adheres to tumor cells. The outer envelope degenerates first, releasing anti-angiogenesis factors. These factors destroy the vasculature and trap the bead's inner core inside the tumor. This cytotoxic core then kills the tumor cells. (After Mooney 2005.)

These ideas are being tested in clinical trials. In one set of trials, an antibody against VEGF was found to be successful against colon cancer, but not against mammary carcinoma. This is probably because colon cancer is more dependent on VEGF-induced angiogenesis than are mammary tumors (Whisenant and Bergsland 2005). But one of the problems with this type of cancer therapy is that the anti-angiogenesis factors, in starving the tumors for oxygen, induce hypoxia-inducible factor-1α (HIF1α)—which causes the tumors to produce more angiogenesis factors. To get around this, Sengupta and colleagues (2005) used a two-step method of cancer chemotherapy. They created a nanometer-sized bead that contained two drugs. The inner core of the bead contained a potent cytotoxic (cell-killing) drug. The outer envelope contained an anti-angiogenisis agent. The bead was encapsulated in a polymer whose size makes it preferentially stick to tumor cells. (The blood vessels of tumors are leaky and can take up larger particles than the blood vessels of normal tissues.) When the bead was absorbed onto the tumor surface, the outer envelope released the anti-angiogenesis factor. This not only stopped the flow of blood to the tumor, but it trapped the cytotoxic core of the bead inside the tumor. This inner core then released its drug, killing the tumor (Figure 21.23). This inventive "double-whammy" approach simultaneously prevents the remaining cells from producing more angiogenesis factors and permits the drug to destroy the tumor.

Cancer and congenital malformations are opposite sides of the same coin. Both involve disruptions of normal devel-

opment. Thus, as we have seen, agents that have been known to cause congenital malformations—thalidomide, retinoic acid, and cyclopamine—can be used as drugs to prevent cancers. Just as they disrupt normal development, these substances can disrupt the caricature of development that is caused by tumor cells. Moreover, by blocking angiogenesis, the tumor cells can be starved.

Gene Therapy

The technologies of gene therapy may give us the ability to genetically modify our bodies (and the bodies of our children) within the next decade. The ability to insert new genes into a fertilized egg and the ability to produce human embryonic stem cells, however, have raised concerns that the human genome could be manipulated. Some people look forward to the time when genetic diseases might be treated by gene therapy and, even further, eliminated from subsequent generations. Other people (or in some cases, the same people) worry that the ability to manipulate genes will result in misguided attempts to enhance human physical and mental abilities.

Somatic cell gene therapy raises the hope of curing a person's genetic disease by inserting a wild-type gene that would be activated at the appropriate times and places. This type of gene therapy is currently being tested in over 600 laboratories (Roberts 2002). In such experiments, a committed stem cell (such as a blood stem cell or liver stem cell) is cultured, given the new gene, and reinserted into the body (Anderson 1998; Gage 1998; Ye et al. 1999). The new gene is usually packaged in a viral or lipid vector that enables it to enter the cell and be transcribed (Figure 21.24). This technique was used in the small interfering RNA study mentioned in the Sidelights & Speculations. Plasmids containing small interfering RNAs have also been

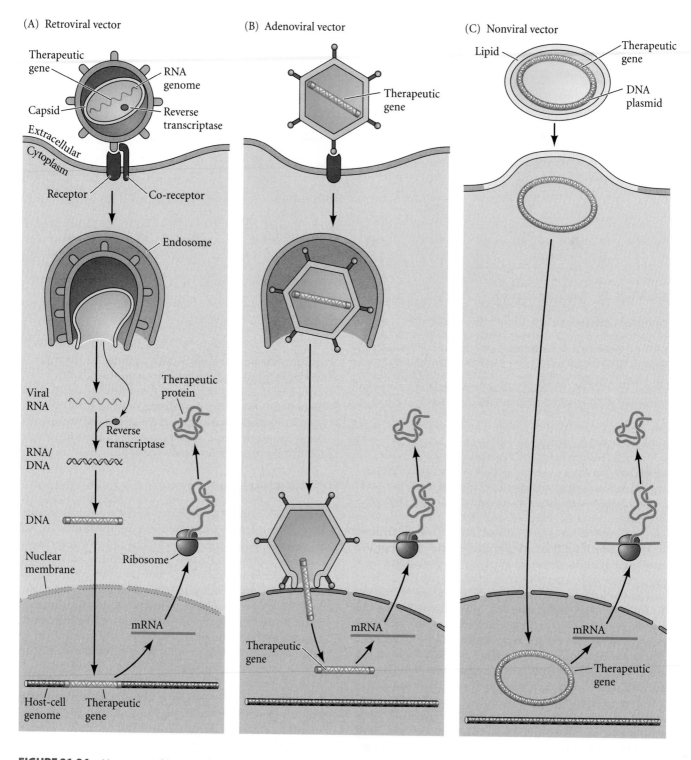

FIGURE 21.24 Vectors used in gene therapy. (A) When a gene (green) is placed in a retroviral vector, the virus is taken into the cell and releases its RNA into the cell. The reverse transcriptase of the virus copies a double-stranded DNA from the RNA, and this DNA is integrated into the host genome. The inserted gene can then be transcribed. (B) The adenovirus genome is already a double-stranded DNA. This vector uncoats itself on the host cell's nuclear envelope. The genes it carries can then be expressed without being incorporated into the host genome. (C) DNA inserted into a lipid membrane must be inserted in the host cell artificially. However, once in the host cell, the DNA can enter the host nucleus and be transcribed. (After Eck 1999.)

successful in curing the liver disease in mice infected with hepatitis B (McCaffrey et al. 2003). Here, the siRNAs were directed against mRNAs encoding essential parts of the hepatitis B virus replication complex.

There have been both stunning successes and prominent failures in the field of gene therapy. In 1990, W. French Anderson and his colleagues treated several patients suffering from immunodeficiency due to a mutation in their adenosine deaminase (ADA) gene. Their own T cells (which lack the ADA enzyme) were transfected with a wild-type ADA gene packaged in a retrovirus. These patients became immunocompetent, although they still have to return for treatment (see Blaese et al. 1995; Onodera et al. 1998). Somatic cell gene therapy has also been used successfully to treat genetic liver disease and hemophilia (Roth et al. 2001).

However, in 1999, an adult patient undergoing gene therapy for a kidney-specific enzyme defect died four days after starting the treatment. The cause of his death remains a mystery, but it is possible that the adenovirus vector stimulated an allergic reaction that shut down the patient's lungs. Similarly, a promising study targeting another inherited immune deficiency was halted when one of the patients developed leukemia-like symptoms (Check 2002). More recently, lentiviruses have been used for introducing genes into cells. One promising result was that mice with the rodent form of amyotrophic lateral sclerosis (ALS, or "Lou Gehrig disease," an incurable adult onset degeneration of motor neurons) were cured when various muscles were given a single injection of a lentivirus expressing the VEGF gene (Azzouz et al. 2004).

Somatic cell gene therapy is like other experimental medical treatments in that the ethical issues it raises are akin to those involved in human tests of any new drug or surgical procedure (see Gilbert et al. 2005). However, there is concern among some people that the same techniques that could be used to cure genetic diseases and cancers could also be used to alter the phenotypes of "normal" individuals. There is a very fine line between treatment and enhancement. Insertion of certain genes into muscles, for instance, could give recipients a stronger musculature or make them less likely to experience muscle fatigue. Are baldness, short stature, and mild obesity—all of which could conceivably be "treated" with gene therapy—diseases, or are they just normal conditions that our society has defined as suboptimal? If genes were identified that could increase longevity or protect memory function, should such genes be used to "enhance" those who could afford such treatments? The answers to such questions may mean a great deal to the commercial success of genetic engineering.

Germline Gene Therapy

In contrast to somatic cell gene therapy, which seeks to cure individuals, the goal of germline gene therapy is to eliminate "faulty" genes both from the individual and from that person's descendants. Germline gene therapy can be accomplished in two ways:

- A germ cell or fertilized egg is modified such that the new genome is present in every cell of the resultant individual's body, and is therefore transferred to next generation through the germ cells.
- Embryonic stem cells are modified so that the developing body contains a high percentage of cells derived from these genetically altered blastomeres.

Genes have been routinely added, subtracted, and replaced in mice for the past decade (see Chapter 5). There is little doubt that such procedures could be done in humans should we desire to do so. Already, transgenic rhesus monkeys and mice have been born who carry the gene for green fluorescent protein (GFP) in each cell of their bodies (Figure 21.25; Chan et al. 2001; Lois et al. 2002). A viral

(A)

(B)

FIGURE 21.25 Germline gene transmission. (A) ANDi (backwards for "inserted DNA"), a transgenic rhesus monkey. The gene for green fluorescent protein is expressed in each of his nuclei. (B) Fluorescent photograph of an 11.5-day mouse embryo whose egg was provided with a lentivirus carrying the GFP gene fused to a myogenin (muscle-specific) enhancer. The GFP gene is expressed in the somatic musculature as well as those muscles of the emerging eye, limb, and jaw. (A courtesy of G. Schatten; B from Lois et al. 2002, courtesy of D. Baltimore.)

vector carrying the gene for GFP was injected into the space between the oocyte and the zona pellucida. The virus inserted itself into the egg DNA, and the animal that developed from these infected eggs had GFP genes (from a jellyfish) in all of their cells.

Germline gene therapy raises many important social and ethical issues (see Gilbert et al. 2005). There are also scientific issues that might mediate against successful germline gene therapy. One of these concerns is the possibility that the introduction of a new gene might knock out a previously functioning gene. Viral insertion mechanisms may prefer to splice viral vectors into genes rather than into noncoding regions of DNA (Schröder et al. 2002). The *inversion of embryonic turning* gene (see Chapter 11) was discovered when Yokoyama and colleagues (1993) inadvertently knocked out this gene by inserting a transgene into a mouse embryo.

Stem Cells and Therapeutic Cloning

The cloning of mammals described in Chapter 4 has applications in agriculture, and the ability to clone transgenic animals producing large quantities of human proteins has important applications in medicine (see Di Berardino 2001). However, mammalian clones are not usually healthy animals. There are many reasons for the abnormal development of cloned embryos, including the faulty activation of imprinted genes, the failure of histone modifications, and methylation deficiencies (Wilmut et al. 2002). Microarray technology has documented that cloned mice have abnormal patterns of gene expression in many of their organs (Humphreys et al. 2002). One of the most important genes for establishing pluripotency, the *Oct4* gene, is aberrantly expressed in the blastocysts of most mouse clones made by somatic cell nuclear transfer, and the pattern of X chromosome inactivation is aberrant in cow clones (Boiani et al. 2002; Xue et al. 2002).

Such reproductive cloning does not appear to be a good technology to apply to humans. As embryologist Hans Schöler (quoted in Glausiusz 2002) has noted, "To obtain one normal organism, you're paving the way with a lot of dead or malformed fetuses." Ian Wilmut, the lead scientist of the research group that cloned Dolly the sheep, has similar negative views concerning human reproductive cloning (Wilmut et al. 2000). However, the technology may find applications in **therapeutic cloning**. In therapeutic cloning (sometimes called **somatic cell nuclear transplantation** to distinguish it from reproductive cloning), the result is the production of stem cells, not the production of a new child (Table 21.3).

Embryonic stem cells and therapeutic cloning

Embryonic stem cells (ES cells) are pluripotent, can be cultured indefinitely in an undifferentiated state, and retain their developmental potential after prolonged culture (Figure 21.26). Currently, human ES cells are obtained by two major techniques. First, they can be derived from the inner cell mass blastomeres of human blastocysts, such as those left over from in vitro fertilization (Thomson et al. 1998; see above). Second, they can be generated from germ cells derived from spontaneously aborted fetuses (Gearhart 1998). In both cases, the ES cells are pluripotent, since they are able to differentiate in culture to form more restricted stem cell types. Some experimental evidence (Strelchenko et al. 2004) suggests that it may also be possible to derive embryonic stem cells from late morulae before they go on to form blastocysts.

TABLE 21.3 Materials and techniques of stem cell research and cloning (SCNT)[a]

Technique	Purpose	Starting material	End product
Adult (or fetal) stem cell research	To obtain undifferentiated stem cells for research and therapy	Isolated stem cells from adult or fetal tissue	Cells produced in culture to repair diseased or injured tissue
Embryonic stem (ES) cell research	To obtain undifferentiated stem cells for research and therapy	Stem cells from a blastocyst-stage embryo	Cells produced in culture to repair diseased or injured tissue
Therapeutic cloning ("nuclear transplantation")	To obtain undifferentiated stem cells that are genetically matched to the recipient for therapy and tissue regeneration	Stem cells from a blastocyst-stage embryo produced from an enucleated egg supplied with nuclear material from the recipient's own somatic cells	Cells produced in culture to repair diseased or injured tissue
Reproductive cloning (embryos produced using SCNT; see Chapter 4)	To produce an embryo for implantation in womb, leading to birth of a child	Enucleated egg supplied with material from a donor somatic cell	Embryo (and eventual offspring) genetically identical to donor of nuclear material

Source: National Institutes of Health 2001: *Stem Cells: Scientific Progress and Future Research Directions*
[a] SCNT is the abbreviation for "somatic cell nuclear transfer," the technical term for the technique used to clone cells or organisms. The distinction between therapeutic and reproductive cloning is particularly important.

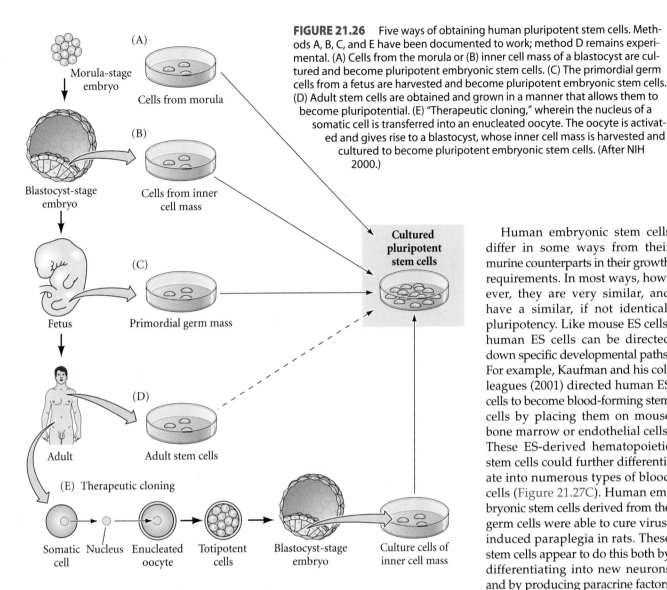

FIGURE 21.26 Five ways of obtaining human pluripotent stem cells. Methods A, B, C, and E have been documented to work; method D remains experimental. (A) Cells from the morula or (B) inner cell mass of a blastocyst are cultured and become pluripotent embryonic stem cells. (C) The primordial germ cells from a fetus are harvested and become pluripotent embryonic stem cells. (D) Adult stem cells are obtained and grown in a manner that allows them to become pluripotential. (E) "Therapeutic cloning," wherein the nucleus of a somatic cell is transferred into an enucleated oocyte. The oocyte is activated and gives rise to a blastocyst, whose inner cell mass is harvested and cultured to become pluripotent embryonic stem cells. (After NIH 2000.)

Morula-stage embryo

Cells from morula

Blastocyst-stage embryo

Cells from inner cell mass

Fetus

Primordial germ mass

Adult

Adult stem cells

(E) Therapeutic cloning

Somatic cell   Nucleus   Enucleated oocyte   Totipotent cells   Blastocyst-stage embryo   Culture cells of inner cell mass

Cultured pluripotent stem cells

Human embryonic stem cells differ in some ways from their murine counterparts in their growth requirements. In most ways, however, they are very similar, and have a similar, if not identical, pluripotency. Like mouse ES cells, human ES cells can be directed down specific developmental paths. For example, Kaufman and his colleagues (2001) directed human ES cells to become blood-forming stem cells by placing them on mouse bone marrow or endothelial cells. These ES-derived hematopoietic stem cells could further differentiate into numerous types of blood cells (Figure 21.27C). Human embryonic stem cells derived from the germ cells were able to cure virus-induced paraplegia in rats. These stem cells appear to do this both by differentiating into new neurons and by producing paracrine factors (BDNF and TGF-α) that prevent the death of the existing neurons (Kerr et al. 2003). Similarly, embryonic stem cells from monkey blastocysts have been able to cure a Parkinson-like disease in adult monkeys whose dopaminergic neurons had been destroyed (Takagi et al. 2005). Thus, embryonic stem cells may be able to provide a reusable and readily available source of cells to heal damaged tissue in adult men and women.

One big difference between human and mouse experimentation is that mice can be inbred and made genetically identical; humans, obviously, cannot. Moreover, as human ES cells differentiate, they express significant amounts of the major histocompatability proteins that can cause immune rejection (Drukker et al. 2002). To get around the problem of host rejection, human ES cells could be modified, or somatic cell nuclear transfer could be used to ensure that the stem cells are genetically identical to the person who would be receiving their progeny. This brings us to another potential way of obtaining embryonic stem cells:

The importance of pluripotent stem cells in medicine is potentially enormous. The hope is that human ES cells can be used to produce new neurons for people with degenerative brain disorders (such Alzheimer or Parkinson disease) or spinal cord injuries, to produce a new pancreas for people with diabetes, or to produce new blood cells for people with anemias. People with deteriorating hearts might be able to have damaged tissue replaced with new heart cells, and those suffering from immune deficiencies might be able to replenish their failing immune systems. Such therapies have already worked in mice. Murine ES cells have been cultured under conditions causing them to form insulin-secreting cells, muscle stem cells, glial stem cells, and neural stem cells (Figure 21.27A,B; Brüstle et al. 1999; McDonald et al. 1999). Dopaminergic neurons derived from ES cells have been shown to significantly reduce the symptoms of Parkinson disease in rodents (Bjorklund et al. 2002; Kim et al. 2002).

(A)

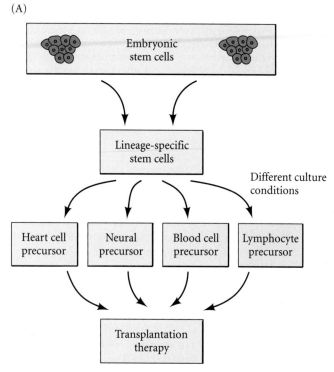

(B)

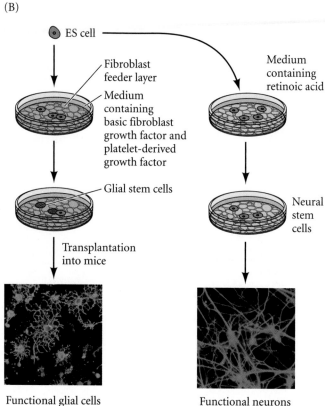

Functional glial cells

Functional neurons

FIGURE 21.27 Embryonic stem cell therapeutics. (A) Human embryonic stem cells (ES cells) can differentiate into lineage-specific stem cells, which can then be transplanted into a host. (B) The differentiation of mouse ES cells into lineage-restricted (neuronal and glial) stem cells can be accomplished by altering the media in which the ES cells grow. (C) Blood cells developing from human embryonic stem cells cultured on mouse bone marrow. (A after Gearhart 1998; B, photographs from Brüstle et al. 1999 and Wickelgren 1999, courtesy of O. Brüstle and J. W. McDonald; C, photograph courtesy of University of Wisconsin.)

(C)

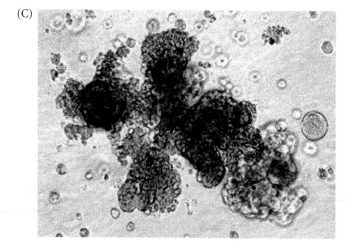

therapeutic cloning (see Figure 21.26E). In this technique, a nucleus from the patient is inserted into an enucleated oocyte (as in reproductive cloning). The resulting embryo is grown in vitro until it has developed an inner cell mass. Cells from the inner cell mass are then cultured to generate stem cells that are genetically identical to the patient.

Therapeutic cloning has been shown to work in mice to restore the dopaminergic neurons damaged by Parkinson disease. Barberi and his colleagues (2003) performed somatic cell nuclear transfer such that the nucleus of one type of mice was inside the oocyte of another strain of mice. This cell divided and became a blastocyst, and embryonic stem cells were derived from the blastocyst (Figure 21.28). Moreover, these embryonic stem cells were induced to become neural stem cells by growing them on mesoderm and providing Fgf2 in the medium. The neural stem cells were then induced to become *ventral* neural stem cells by adding another paracrine factor, Sonic hedgehog, to the medium. (As you will recall from Chapter 13, Sonic

hedgehog in the embryo is secreted by the notochord and induces ventral gene expression in the lower neural tube.) Further exposure of these cells to Fgf2 and Fgf8, followed by exposure to brain-derived neurotrophic factor (BDNF) produced cells that had the characteristics of dopaminergic neurons. Moreover, when these cells were injected into mice that had their dopaminergic neurons destroyed, the cloned ES-derived neurons were able to restore normal function to the mice.*

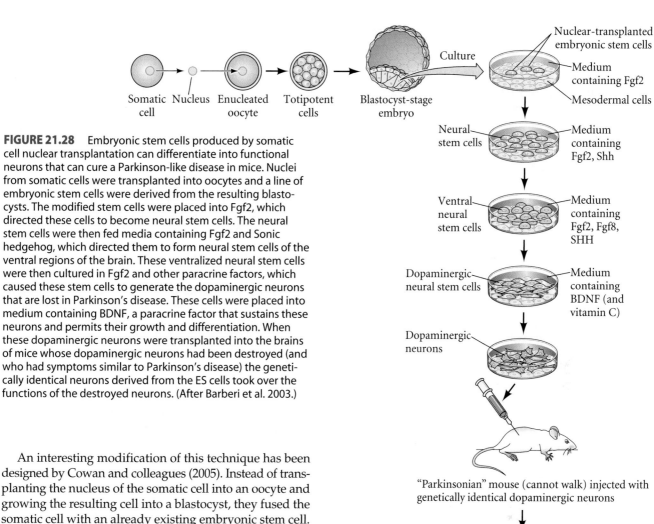

FIGURE 21.28 Embryonic stem cells produced by somatic cell nuclear transplantation can differentiate into functional neurons that can cure a Parkinson-like disease in mice. Nuclei from somatic cells were transplanted into oocytes and a line of embryonic stem cells were derived from the resulting blastocysts. The modified stem cells were placed into Fgf2, which directed these cells to become neural stem cells. The neural stem cells were then fed media containing Fgf2 and Sonic hedgehog, which directed them to form neural stem cells of the ventral regions of the brain. These ventralized neural stem cells were then cultured in Fgf2 and other paracrine factors, which caused these stem cells to generate the dopaminergic neurons that are lost in Parkinson's disease. These cells were placed into medium containing BDNF, a paracrine factor that sustains these neurons and permits their growth and differentiation. When these dopaminergic neurons were transplanted into the brains of mice whose dopaminergic neurons had been destroyed (and who had symptoms similar to Parkinson's disease) the genetically identical neurons derived from the ES cells took over the functions of the destroyed neurons. (After Barberi et al. 2003.)

An interesting modification of this technique has been designed by Cowan and colleagues (2005). Instead of transplanting the nucleus of the somatic cell into an oocyte and growing the resulting cell into a blastocyst, they fused the somatic cell with an already existing embryonic stem cell. In many instances, not only did the cells fuse, but so did their nuclei, to make a tetraploid nucleus. These hybrid cells kept the stem cell phenotype and could differentiate into the three major germ layers. This finding opens the possibility that transplantation of somatic nuclei into enucleated embryonic stem cells may allow patient-specific stem cells to be made without having to use early embryonic stages. This could circumvent many of the religious objections to using human embryonic stem cells.

Multipotent adult stem cells

Numerous organs contain committed multipotent stem cells, even in the adult. These multipotent stem cells can give rise to a limited set of adult tissue types. Earlier in this book we discussed blood stem cells, epidermal stem cells, neural stem cells, hair stem cells, melanocyte stem cells, muscle stem cells, gut stem cells, and germ stem cells. Such cells are not as easy to use as pluripotent embryonic stem cells; they are difficult to isolate, and are often fewer than one out of every thousand cells in an organ. In addition, they appear to have a relatively low rate of cell division and do not proliferate readily. However, neither of these facts precludes their usefulness. Bone marrow transplantation has been used to transfer hematopioetic stem cells from one person to another in over 40,000 operations each year. This pluripotential hematopoietic stem cell is rare (about one in every 15,000 bone marrow cells) but such transplantation works well for people who are suffering from red blood cell deficiencies or leukemias. In mice, very few (perhaps even one) blood stem cell will reconstitute the mouse's blood and immune systems (Osawa et al.

---

*Interestingly, the methylation problems that plague cloned animals do not appear to be a problem for embryonic stem cells derived by somatic cell nuclear transfer. It appears that the nuclei within the small population of ES cells that survive in culture have had their methylation patterns erased, thus enabling them to re-differentiate (Jaenisch 2004; Rugg-Gunn et al. 2005).

1996), and a single mammary stem cell will generate an entire mammary gland (including epithelium, muscles, and stroma) in mice (Shackleton et al. 2006).

When researchers have been able to isolate and culture such cells, they have proved to be very useful. Carvey and colleagues (2001) have shown that when neural stem cells from the midbrain of adult rats are cultured in a mixture of paracrine factors, they, too, will differentiate into dopaminergic neurons that can cure the rat version of Parkinson disease. Techniques to selectively allow the growth and isolation of multipotent stem cells may allow some organ deficiencies to be treated like blood cell deficiencies—by administering a source of committed stem cells.

Pluripotent adult stem cells

Whether multipotent stem cells can become pluripotent stem cells is controversial. It appears that most (if not all) adult multipotent stem cells are restricted to forming only a few cell types (Wagers et al. 2002). When hematopoietic stem cells were marked with green fluorescent protein (GFP) and placed in mice, their labeled descendants were found throughout the blood of these animals, but not in any other tissue. However, some scientists claim that under certain culture conditions, multipotent stem cells are able to generate progeny that can produce many different cell types.

MESENCHYMAL STEM CELLS Jiang and colleagues (2002a) isolated a population of stem cells from the bone marrow of mice that, when injected into adult mice, gave rise not only to blood cells (as expected), but also to neural cells and endodermal cells. LaBarge and Blau (2002) have shown that these cells can also become muscle cells. When mouse bone marrow was labeled with green fluorescent protein and transplanted into recipient mice, GFP was subsequently found in muscle stem cells, and later (after trauma to the muscles) in mature myocytes. These **mesenchymal stem cells** (sometimes called **bone marrow-derived stem cells**, or **BMDCs**) had other unexpected abilities: they were able to integrate into the inner cell mass of a blastocyst, and they expressed *Oct4*—just like ES cells.

This same type of stem cell has also been found among human bone marrow stem cells (Jiang et al. 2002b; Korbling et al. 2002). When biopsies were taken from individuals who had received hematopoietic stem cell transplants, hepatocytes and skin cells were found to be derived from the donor cells. These cells may represent an extremely small population of pluripotent stem cells that persist into adult life, or they might be normal multipotent organ-specific stem cells that regain pluripotency under the culture conditions.

When bone marrow-derived stem cells are incubated in a particular mixture of paracrine factors (including Fgf2 and PDGF), they can become muscle stem cells. These bone marrow derived muscle stem cells have been found to generate muscles and to repair muscle degeneration in rats. Moreover, human BMDCs, cultured to make muscle stem cells, can repair muscle lesions in adult rats and mice (Dezawa et al. 2005). In a different set of paracrine factors (including Sonic hedgehog and retinoic acid), the BMDCs from mouse tibias and femurs were able to differentiate into sensory neurons (Kondo et al. 2005). Such cells have also been isolated from human bone marrow and have been found to be pluripotent (Song and Tuan 2004; Baksh et al. 2004). Therefore, it is possible that our bone marrows might contain pluripotent stem cells that can be induced to produce different cell types, just as embryonic stem cells can.

UMBILICAL AND MATERNAL STEM CELLS In addition to the possibility of pluripotent stem cells in adult bone marrow, there are two other possible sources of such cells. The umbilical cord seems to harbor a population of pluripotential stem cells, and culturing the cord blood cells in different media promotes the differentiation of these stem cells into cartilage, liver cells, bone cells, astrocytes, or neurons (Kogler et al. 2004). Some physicians propose to prospective parents that they freeze and store their child's umbilical cord cells so that the cells will be available for transplantation later in life. Human umbilical cord cells were seen to differentiate into lymphocytes when transplanted into immune-deficient mice and to differentiate into heart cells when transplanted into rats induced to have damaged cardiac tissue (Traggiai et al. 2004; Hirata et al. 2004).

One of the many surprises of stem cell research (in addition to finding that there were functional committed stem cells in the adult brain, and that there were pluripotent stem cells in the bone marrow) was that the fetus may supply its mother with stem cells. Diana Bianchi and her colleagues found that fetal cells enter the maternal circulation, and that male cells of presumed fetal origin can be detected in about 50 percent of the healthy women who delivered sons (Khosrotchrani et al. 2004, 2005). Moreover, some of these cells appear to be stem cells that can be found in the mother's organs decades later! What is also amazing is that these pregnancy-associated stem cells appear to help the mother. Male cells (of presumed fetal origin) were found in the thyroid of a woman whose own thyroid tissue was diseased, and the pregnancy-associated progenitor cells appeared to cluster specifically at regions of diseased or damaged maternal tissue. It is not yet known whether these cells are mesenchymal stem cells, cord-like stem cells, or some new stem cell type. Pregnancy may therefore include the acquisition of cells that may help the mother for the rest of her life.

Transgenic stem cells

The combination of somatic cell nuclear transfer and gene therapy can be used to cure genetic diseases that affect specific organs or tissues. For example, mice deficient in the *Rag2* gene are unable to recombine their DNA to make antibodies. Rideout and his colleagues (2002) took tail tip cells from a *Rag2*-deficient mouse and implanted the nuclei from those cells into enucleated mouse oocytes (Figure 21.29).

The renucleated oocytes were activated and gave rise to blastocysts, from which *Rag2*-deficient ES cells were cultivated. One of the mutant *Rag2* genes in the ES cells was then converted into a wild-type gene by homologous recombination. The "repaired" ES cells were grown, labeled, and cultured to produce hematopoietic stem cells. These hematopoietic stem cells were injected back into the mouse from which the nuclei were originally taken, and they repopulated the mouse's immune system. Mature antibody-producing cells were detectable within a month of the transplantation. Transgenic human embryonic stem cells have recently been produced (Zwaka and Thompson 2003).

Regeneration therapy

Humans do not regenerate organs. Children can regenerate their fingertips, but even this ability is lost in adults. The ability to regenerate damaged human organs would constitute a medical revolution. Researchers are attempting to find ways of activating developmental programs that were used during organogenesis in the adult. One avenue of this research program is the above-mentioned search for adult stem cells that are still relatively undifferentiated, but which can form particular cell types if given the appropriate environment. The second avenue is a search for the environments that will allow such cells to initiate cell and organ formation. We will look at two of these medical efforts.

BONE REGENERATION While fractured bones can heal, bone cells in adults usually do not regrow to bridge wide gaps.

The finding that the same paracrine and endocrine factors involved in endochondral ossification are also involved in fracture repair (Vortkamp et al. 1998) raises the possibility that new bone could grow if the proper paracrine factors and extracellular environment were provided. Several methods are now being tried to develop new functional bone in patients with severely fractured or broken bones.

One solution to the problem of delivery was devised by Bonadio and his colleagues (1999), who developed a collagen gel containing plasmids carrying the human parathyroid hormone gene. The plasmid-impregnated gel was placed in the gap between the ends of a broken dog tibia or femur. As cells migrated into the collagen matrix, they incorporated the plasmid and made parathyroid hormone. A dose-dependent increase in new bone formation was seen in about a month (Figure 21.30). This type of treatment has the potential to help people with large bone fractures as well as those with osteoporosis.

Another approach is to find the right mixture of paracrine factors to recruit stem cells and produce normal bone. Peng and colleagues (2002) genetically modified muscle stem cells to secrete BMP4 or VEGF. These cells were placed in gel matrix discs, which were implanted in wounds made in mouse skulls. The researchers found that certain ratios of BMP4 and VEGF were able to heal the wounds by making new bone. Similarly, BMP2 has been used to heal large fractures of primate mandibles and rabbit femurs (Li et al. 2002; Marukawa et al. 2002).

A third approach is to make artificial scaffolds that resemble those of the bone and seed them with mesenchy-

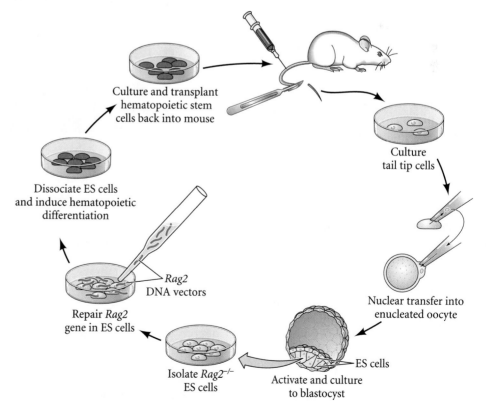

FIGURE 21.29 Repair of *Rag2* deficiency by therapeutic cloning. Tail tip cells from a *Rag2*-deficient (*Rag2*⁻/⁻) mouse are cultured, and the nuclei of these somatic cells are transplanted into activated enucleated oocytes. The oocytes develop to the blastocyst stage, and the *Rag2*-deficient inner cell mass cells are isolated and grown as stem cells. When these stem cells are incubated with wild-type *Rag2* DNA, homologous recombination allows some of these cells to replace the mutant allele with a wild-type allele. The repaired ES cells are aggregated, marked with a GFP gene, and cultured in a manner that directs them to differentiate into hematopoietic and lymphocytic stem cells. When these cells are injected into the mouse, they can restore its immune system by producing *Rag2*-positive cells that can make antibodies.

Culture and transplant hematopoietic stem cells back into mouse

Culture tail tip cells

Dissociate ES cells and induce hematopoietic differentiation

Rag2 DNA vectors

Repair *Rag2* gene in ES cells

Nuclear transfer into enucleated oocyte

Isolate *Rag2*⁻/⁻ ES cells

Activate and culture to blastocyst

ES cells

FIGURE 21.30 Bone formation from collagen matrix containing plasmids bearing the human parathyroid hormone. (A) A 1.6 cm gap was made in a dog femur and stabilized with screws. Plasmid-containing gel was placed on the edges of the break. Radiographs of the area at 2, 8, 12, 16, and 18 weeks after the surgery show the formation of bone bridging the gap at 18 weeks. (B) Control fracture (no plasmid in the gel) at 24 weeks. (C) Whole bone a year after surgery, showing repaired region. (After Bonadio et al. 1999; photographs courtesy of J. Bonadio.

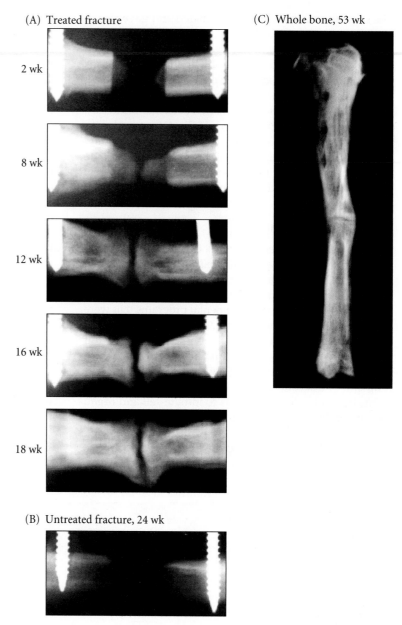

(A) Treated fracture

2 wk

8 wk

12 wk

16 wk

18 wk

(B) Untreated fracture, 24 wk

(C) Whole bone, 53 wk

mal stem cells. This approach, combining developmental biology and mechanical engineering is called **tissue engineering**. Li and colleagues (2005) made scaffolds of material that resembles normal extracellular matrix and which can be molded to form the shape of bone needed. The mesenchymal stem cells can be placed in these scaffolds and placed into the existing bone. These stem cells will be told how to differentiate by the local conditions, and histological and gene expression studies have shown that these cells formed bone with the appropriate amounts of osteocytes and chondrocytes.

A fourth approach has been to let the body do the work. While bones can't regenerate if wide gaps appear, many bones can undergo normal healing of small wounds. Here, cells in the periosteum (the sheath of cells that surrounds the bone) will differentiate into new cartilage, bone, and ligaments to fill the crack. Stevens and colleagues (2005) have used this normal healing process to make new bones. They injected saline solutions between the rabbit tibial bone and its periosteum, mimicking a fracture, and kept this space open by adding a gel containing calcium (to push the periosteal cells to differentiate into bone rather than cartilage). Within a few weeks, these cavities filled with new bone, which could be transplanted into sites where bone had been damaged. This technique might provide a relatively painless way to produce new bone for fusing vertebrae and other surgical techniques.

NEURONAL REGENERATION While the central nervous system is characterized by its ability to change and make new connections, it has very little regenerative capacity. The motor neurons of the peripheral nervous system, however, have significant regenerative powers, even in adult mammals. The regeneration of motor neurons involves the regrowth of a severed axon, not the replacement of a missing or diseased cell body. If the cell body of a motor neuron is destroyed, it cannot be replaced.

The myelin sheath that covers the axon of a motor neuron is necessary for its regeneration. This sheath is made by the Schwann cells, a type of glial cell in the peripheral nervous system (see Chapter 12). When an axon is severed, the Schwann cells divide to form a pathway along which the axon can regrow from the proximal stump. This proliferation of the Schwann cells is critical for directing the regenerating axon to the original Schwann cell basement membrane. If the regrowing axon can find that basement membrane, it can be guided to its target and restore the original connection. In turn, the regenerating neuron secretes mitogens that allow the Schwann cells to divide. Some of these mitogens are specific to the developing or regenerating nervous system (Livesey et al. 1997).

The neurons of the central nervous system cannot regenerate their axons under normal conditions. Thus, spinal cord injuries can cause permanent paralysis. As mentioned in Chapter 12, one strategy to get around this block is to find ways of enlarging the population of adult neural stem cells and to direct their development in ways that circumvent the lesions caused by disease or trauma. The neural stem cells found in adult mammals may be very similar to embryonic neural stem cells and may respond to the same growth factors (Johe et al. 1996; Johansson et al. 1999).

Another strategy for CNS neural regeneration is to create environments that encourage axonal growth. Unlike the Schwann cells of the peripheral nervous system, the myelinating glial cells of the central nervous system, the oligodendrocytes, produce substances that inhibit axon regeneration (Schwab and Caroni 1988). Schwann cells transplanted from the peripheral nervous system into a CNS lesion are able to encourage the growth of CNS axons to their targets (Keirstead et al. 1999; Weidner et al. 1999). Three substances that inhibit axonal outgrowth have been isolated from oligodendrocyte myelin: myelin-associated glycoprotein, Nogo-1, and oligodendrocyte-myelin glycoprotein (Mukhopadyay et al. 1994; Chen et al. 2000; Grand-Pré et al. 2000; K. Wang et al. 2002). Antibodies against

Nogo-1 and the blocking of the Nogo receptor allow partial axon regeneration after spinal cord injury (Schnell and Schwab 1990; GrandPré et al. 2002). Moreover, at sites of CNS injury, a glial scar forms, containing chondroitin sulfate proteoglycans in its extracellular matrix. These compounds inhibit axonal outgrowth as well. When Bradbury and colleagues (2002) delivered chondroitinase to lesioned spinal cords of adult rats, the axons were able to grow past the lesions.

As mentioned in the case with bone, there is a close relationship between wound healing and regeneration. Patients with multiple sclerosis suffer from the demyelination of these brain neurons. The transcription factor Olig1 is required to heal myelination defects in mammals. Thus, while multiple sclerosis may not be prevented, it is possible that remyelination might be reinitiated by activating Olig1 transcription in the oligodendrocyte precursors remaining in the adult central nervous system (Arnnett et al. 2004). Research into CNS axon regeneration may become one of the most important contributions of developmental biology to medicine.

 ## Snapshot Summary: Medical Implications of Developmental Biology

1. Positional gene cloning starts with pedigree analysis and makes a DNA map of the region differing between people having and lacking a particular phenotype.

2. Candidate gene mapping starts with a correlation between a genetic region and a gene whose mutation causes a similar phenotype in another species.

3. Pleiotropy occurs when several different effects are produced by a single gene. In mosaic pleiotropy, each effect is caused independently by the expression of the same gene in different tissues. In relational pleiotropy, abnormal gene expression in one tissue influences other tissues, even though those other tissues do not express that gene.

4. Genetic heterogeneity occurs when mutations in more than one gene can produce the same phenotype.

5. Phenotypic variability arises when the same gene can produce different defects (or differing severities of the same defect) in different individuals.

6. Dominant inheritance can be caused by haploinsufficiency, in which expression of a single copy of a gene is not sufficient to produce the wild-type phenotype; by gain-of-function mutations, in which a pathway is activated independently of its normal initiator; or by dominant negative alleles, mutant alleles encoding a subunit of a protein whose dysfunction makes the entire protein nonfunctional.

7. Inborn errors of gene expression can cause abnormalities at the levels of transcription, RNA processing, translation, and posttranslational modification.

8. Fragile X syndrome, the leading cause of mental retardation in humans, is caused by an overabundance of CGG repeats in the *FMR1* gene. These repeats prevent the *FMR1* gene from being transcribed. The protein encoded by *FMR1* appears to target certain neuronal genes to the ribosome for translation.

9. In vitro fertilization involves retrieving oocytes from the ovary and mixing them with sperm. The dividing eggs are then inserted into the uterus.

10. Preimplantation genetics involves testing for genetic abnormalities in early embryos in vitro, and implanting only those embryos that may develop normally. Selection for sex is also possible using preimplantation genetics.

11. Teratogenic agents include certain chemicals such as ethanol, retinoic acid, and valproic acid, as well as heavy metals, certain pathogens, and ionizing radiation.

12. Endocrine disruptors can bind to or block hormone receptors or block the synthesis, transport, or excretion of hormones. DES is a powerful endocrine disruptor. Presently, bisphenol A and other PCBs are being considered as possible agents of precocious puberty in women and low sperm counts in men.

13. Environmental estrogens can cause reproductive system anomalies by suppressing Hox gene expression and Wnt pathways.

14. Cancer can be seen as a disease of altered development. Some tumors revert to nonmalignancy when placed in environments that fail to support rapid cell division.

15. Differentiation therapy makes tumor cells nonmalignant by exposing them to factors that cause them to stop dividing and to differentiate.

16. Disrupting tumor-induced angiogenesis may become an important means of stopping tumor progression.

17. Somatic cell gene therapy aims to cure an individual's genetic disease by adding genes to some of that person's somatic cells.

18. Germ line gene therapy aims to cure the genetic disease of a person and that person's descendants by repairing the gene in the germ cells.

19. Therapeutic cloning involves inserting the nucleus of a somatic cell from an individual into an enucleated activated oocyte, growing the embryo to blastocyst stage, and making stem cells from the inner cell mass of that embryo. The pluripotent stem cells would not be rejected by that individual.

20. Adult stem cells may provide alternative ways of growing differentiated cells to repair damaged tissue. Mesenchymal stem cells from the bone marrow appear to be able to form many cell types. Most adult stem cells have a limited developmental repertoire and are difficult to culture.

21. By altering conditions to resemble those in the embryo and by providing surfaces on which adult stem cells might grow, some adult stem cells might be "tricked" into regenerating bones and neurons.

For Further Reading

Complete bibliographical citations for all literature cited in this chapter can be found on the VadeMecum[2] CD that accompanies this book and at the free access website www.devbio.com

Anway, M. D., A. S. Cupp, M. Uzumcu and M. K. Skipper. 2005. Epigenetic transgeneration effects of endocrine disruptors and male fertility. *Science* 308: 1466–1469.

Baksh, D., L. Song and R. S. Tuan. 2004. Adult mesenchymal stem cells: Characterization, differentiation, and application in cell and gene therapy. *J. Cell Mol. Med.* 8: 301–316.

Barberi, T. and 13 others. Neural subtype specification of fertilization and nuclear transfer embryonic stem cells and application in Parkinsonian mice. *Nature Biotechnol.* 21: 1200–1207.

Bissell, M. J., D. C. Radisky, A. Rizki, V. M. Weaver and O. W. Petersen. 2002. The organizing principle: Microenvironmental influences in the normal and malignant breast. *Differentiation* 70: 537–546.

Brüstle, O. and 7 others. 1999. Embryonic stem cell-derived glial precursors: A source of myelinating transplants. *Science* 285: 754–756.

Cao, T. and B. C. Heng. 2005. Differentiation therapy of cancer: Potential advantages over conventional thera-

peutic approaches targeting death of cancer/tumor. *Med. Hypotheses* 65: 1202–1203.

Dezawa, M. and 7 others. 2005. Bone marrow stromal cells generate muscle cells and repair muscle degeneration. *Science* 309: 314–317.

Foty, R. A. and M. S. Steinberg. 1997. Measurement of tumor cell cohesion and suppression of invasion by E- and P-cadherin. *Cancer Res.* 57: 5033–5036.

Gluckman, P. D. and M. A. Hanson. 2004. Living with the past: Evolution, development, and patterns of disease. *Science* 305: 1733–1739.

Howdeshell, K. L., A. K. Hotchkiss, K. A. Thayer, J. G. Vandenbergh and F. S. vom Saal. 1999. Plastic bisphenol A speeds growth and puberty. *Nature* 401: 762–764.

Keller, G., G. Zimmer, G. Mall, E. Ritz and K. Amann. 2003. Nephron number in patients with primary hypertension. *New Engl. J. Med.* 348: 101–108.

Lammer, E. J. and 11 others. 1985. Retinoic acid embryopathy. *N. Engl. J. Med.* 313: 837–841.

Lillycrop, K. A., E. S. Phillips, A. A. Jackson, M. A. Hanson and G. C. Burdge. 2005. Dietary protein restriction of pregnant rats induces and folic acid supplementation prevents epigenetic modification of hepatic gene expression in the offspring. *J. Nutrition* 135: 1382–1386.

Maffini, M. V., A. M. Soto, J. M. Calabro, A. A. Ucci and C. Sonnenschein. 2004. The stroma as a crucial target in mammary gland carcinogenesis. *J. Cell Sci.* 117: 1495–1502.

Nakayama, A., M. T. Nguyen, C. C. Chen, K. Opdecamp, C. A. Hodgkinson and H. Arnheiter. 1998. Mutations in *microphthalmia*, the mouse homolog of the human deafness gene *MITF*, affect neuroepithelial and neural crest-derived melanocytes differently. *Mech Dev.* 70: 155–166.

Reya, T., S. J. Morrison, M. F. Clarke and I. L. Weissman. 2001. Stem cells, cancer, and cancer stem cells. *Nature* 414: 105–111.

Sulik, K. K. 2005. Genesis of alcohol-induced craniofacial dysmorphism. *Exp. Biol. Med.* 230: 366–375.

Environmental Regulation of Animal Development

IT HAD LONG BEEN THOUGHT that the environment played only minor roles in development. Nearly all developmental phenomena were believed to be regulated by genes, and those organisms whose development *was* significantly controlled by the environment were considered interesting oddities. However, recent studies have shown that the environmental context plays significant roles in the development of almost all species, and that animal and plant genomes have evolved to respond to environmental conditions. Moreover, symbiotic associations, wherein the genes of one organism are regulated by the products of another organism, appear to be the rule rather than the exception.

One of the reasons developmental biologists have largely ignored the environment is that one criterion for selecting which animals to study has been their ability to develop regularly in the laboratory (Bolker 1995). Given adequate nutrition and temperature, all "model organisms"—*C. elegans*, *Drosophila*, sea urchins, *Xenopus*, chicks, and laboratory mice—develop independently of their particular environment, leaving us with the erroneous impression that everything needed to form the embryo is within the fertilized egg. Today, with new concerns about the loss of organismal diversity and the effects of environmental pollutants, there is renewed interest in the regulation of development by the environment (see van der Weele 1999; Gilbert 2001).

Phenotypic Plasticity

In most developmental interactions, the genome provides the specific instructions, while the environment is permissive. Dogs will generate dogs and cats will beget cats, even if they live in the same house. However, in most species, there are instances in development wherein the environment plays the instructive role and the genome is merely permissive. The ability of an individual to express one phenotype under one set of circumstances and a different phenotype under another set is called **phenotypic plasticity**.*

There are two main types of phenotypic plasticity. (Woltereck 1909; Schmalhausen 1949; Stearns et al. 1991). Under a **reaction norm**, the genome encodes a *continuous range* of potential phenotypes, and the environment the individual

> 66 We may now turn to consider adaptations towards the external environment; and firstly the direct adaptations ... in which an animal, during its development, becomes modified by external factors in such a way as to increase its efficiency in dealing with them. 99
> *C. H. WADDINGTON (1957)*

> 66 Honor thy symbionts. 99
> *JIAN XU AND JEFFREY I. GORDON (2003)*

---

*The ability of environmental cues to induce phenotypic change should be considered "tertiary induction." Primary induction involves the establishment of a single field within the embryo (such that one egg gives rise to just one embryo). Secondary induction refers to those cascades of inductive events within the embryo by which the organs are formed. Tertiary induction is the induction of developmental changes by factors in the environment.

(A)

Stationary morph

Migratory morph

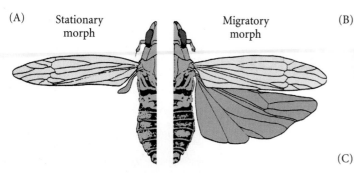

(B)

(C)

FIGURE 22.1 Density-induced polyphenism in planthoppers and grasshoppers. (A) Composite diagram showing the short-winged (left) and long-winged (right) forms of the planthopper *Prokelisia marginata*. The long-winged migratory morph is an excellent flier; the short-winged morph is flightless. (B,C) Density-induced changes in the "plague locust" *Schistocerca gregaria*. (B) Low-density morph, showing green pigmentation and miniature wings. (C) High-density morph, showing new pigmentation and wing and leg development. (A after Denno et al. 1985; B,C from Tawfik et al. 1999, photographs courtesy of S. Tanaka.)

encounters determines the phenotype (usually the most adaptive one). For instance, constant and intense labor can induce our muscles to grow larger; the muscular phenotype is thus determined by the amount of exercise the body is exposed to (even though there is a genetically defined limit to how much muscular hypertrophy is possible). Similarly, the microhabitat of a young salamander can cause its color to change (again, within genetically defined limits). The upper and lower limits of a reaction norm are also a property of the genome that can be selected. Different genotypes are expected to differ in the direction and amount of plasticity that they are able to express (Gotthard and Nylin 1995; Via et al. 1995).

The second type of phenotypic plasticity, **polyphenism**, refers to *discontinuous* ("either/or") phenotypes elicited by the environment. Migratory locusts, for instance, exist in two mutually exclusive forms: either a short-winged, uniformly colored, solitary insect; or a long-winged, brightly colored, gregarious morph (Figure 22.1; Pener 1991; Rogers et al. 2003, 2004). The phenotypic differences between these two forms of the desert locust *Schistocerca gregaria* are so striking that it wasn't until 1921 that Russian biologist Boris Usarov realized they were the same species! Cues in the environment (mainly population density) determine which morphology a young locust will develop. The major stimulus appears to be the rubbing together of their legs. When these locusts get crowded enough that a certain neuron in their hind femur is stimulated by other insects, the changes begin to take place.

There are numerous examples (and *Homo sapiens* provides some of the best) wherein the environment plays a critical role in determining the organism's phenotype (Table 22.1; Gilbert 2004). In Chapters 3 and 17, we encountered environmentally regulated sex determination and morphologies. In these and many other cases, the environment can elicit different phenotypes from the same nuclear genotype. The genetic ability to respond to such environmental factors has to be inherited, of course, but in these cases it is the environment that directs the formation of the particular phenotype. There are at least three pathways through which environmental agents can specify phenotype by inducing differential gene expression:

1. *By acting at the cell surface:* For instance, physical tension monitored by cell surface receptors of mesenchyme cells directly activates signal transduction cascades that change gene transcription. In other instances, products from bacteria can directly induce differential gene expression in intestinal cells.

2. *By modifying the genes, themselves:* Dietary supplementation can induce differential gene expression in the brain by changing the methylation patterns of genes there.

3. *Indirectly, by activating by the neuroendocrine system:* Here it is the organism's nervous system that monitors environmental agents and induces the secretion of hormones. These hormones alter gene expression. This mechanism is very often found in insects where the nervous system detects differences in photoperiod or temperature and, in response, induces ecdysone or juvenile hormone secretion. These hormones then regulate gene expression.

In this chapter, we will discuss how organisms use environmental cues in the course of their normal development, as well as how exogenous compounds found in the environment can divert development from its usual path and cause congenital abnormalities.

TABLE 22.1 Some topics covered by ecological developmental biology

CONTEXT-DEPENDENT NORMAL DEVELOPMENT

A. Morphological polyphenisms
1. Nutrition-dependent (*Nemoria*, hymenoptera castes, sea urchin larvae)
2. Temperature-dependent (*Arachnia, Bicyclus*)
3. Density-dependent (locusts)
4. Stress-dependent (*Scaphiopus*)

B. Sex determination polyphenisms
1. Location-dependent (*Bonellia, Crepidula*)
2. Temperature-dependent (*Menidia*, turtles)
3. Social-dependent (wrasses, gobys)

C. Predator-induced polyphenisms
1. Adaptive predator-avoidance morphologies (*Daphnia, Hyla*)
2. Adaptive immunological responses (*Gallus, Homo*)
3. Adaptive reproductive allocations (ant colonies)

D. Stress-induced bone formation
1. Prenatal (fibular crest in birds)
2. Postnatal (patella in mammals; lower jaw in humans?)

E. Environmentally responsive neural systems
1. Experience-mediated visual synapses (*Felix*, monkeys)
2. Cortical remodeling (phantom limbs; learning)

CONTEXT-DEPENDENT LIFE CYCLE PROGRESSION

A. Larval settlement
1. Substrate-induced metamorphosis (bivalves, gastropods)
2. Prey-induced metamorphosis (gastropods, chitons)
3. Temperature/photoperiod-dependent metamorphosis

B. Diapause
1. Overwintering in insects
2. Delayed implantation in mammals

C. Sexual/asexual progression
1. Temperature/photoperiod-induced (aphids, *Megoura*)
2. Temperature/colony-induced (*Volvox*)

D. Symbioses/parasitism
1. Blood meals (*Rhodnius, Aedes*)
2. Commensalism (*Euprymna/Vibrio*; eggs/algae; *Paleon/Alteromonas*; Mammalian gut microbiota)
3. Parasites (*Wollbachia* in wood lice)

E. Developmental plant-insect interactions

ADAPTATIONS OF EMBRYOS AND LARVAE TO ENVIRONMENTS

A. Egg protection
1. Sunscreens against radiation (*Rana*, sea urchins)
2. Plant-derived protection (*Utetheisa*)

B. Larval protection
1. Plant-derived protection (*Danaus*; tortoise beetles)

TERATOGENESIS

A. Chemical teratogens
1. Natural compounds (retinoids, alcohol, lead)
2. Synthetic compounds (thalidomine, warfarin)
3. Hormone mimics (diethylstilbesterol, PCBs)

B. Infectious agents
1. Viruses (*Coxsackie, Herpes, Rubella*)
2. Bacteria (*Toxoplasma, Treponema*)

C. Maternal conditions
1. Malnutrition
2. Diabetes
3. Autommunity

Source: After Gilbert 2001.

Note: This list should not be thought to be inclusive. For example, the list is limited to animals; plant developmental plasticity and many plant-animal interactions have not been included here.

The Environment as Part of Normal Development

If the environment contains predictable components (such as gravity) or predictable changes (such as seasons), these elements can become part of the development of an organism. Any developing animal can expect to encounter a 1-G gravitational field, bacteria, and fungi as part of its environment. Many species use these as agents in their normal development.

Gravity and pressure

In Chapters 10 and 11, we saw that gravity is critical for frog and chick axis formation. There are also several bones whose formation is dependent on pressure from the movement of the embryo. Such stresses are known to be responsible for the formation of the human patella (kneecap) after birth and have also been found to be critical for jaw growth.* Abnormal muscle and joint forces on bones have been seen as caus-

---

*Normal human jaw development may be predicated on expected tension due to grinding food. Mechanical tension appears to stimulate *indian hedgehog* expression in the mandibular cartilage and this paracrine factor stimulates cartilage growth (Tang et al. 2004). If infant monkeys are given soft food, their lower jaw becomes smaller than usual (Corrucini and Beacher 1982 1984). Those authors and Varella (1992) have shown that people in cultures where infants are fed hard food have jaws that "fit" better, and speculate that soft infant food explains why so many children in Western societies need braces on their teeth. The notion that mechanical tension can change jaw size and shape is the basis of the functional hypothesis of modern orthodontics (Moss 1962, 1997).

(A)

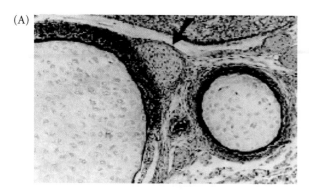

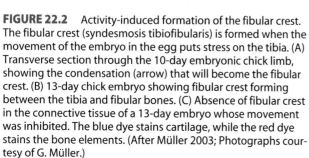

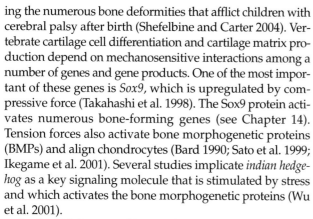

(B)

(C)

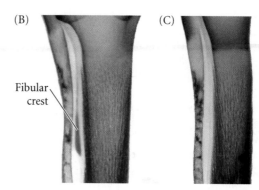

Fibular crest

FIGURE 22.2 Activity-induced formation of the fibular crest. The fibular crest (syndesmosis tibiofibularis) is formed when the movement of the embryo in the egg puts stress on the tibia. (A) Transverse section through the 10-day embryonic chick limb, showing the condensation (arrow) that will become the fibular crest. (B) 13-day chick embryo showing fibular crest forming between the tibia and fibular bones. (C) Absence of fibular crest in the connective tissue of a 13-day embryo whose movement was inhibited. The blue dye stains cartilage, while the red dye stains the bone elements. (After Müller 2003; Photographs courtesy of G. Müller.)

ing the numerous bone deformities that afflict children with cerebral palsy after birth (Shefelbine and Carter 2004). Vertebrate cartilage cell differentiation and cartilage matrix production depend on mechanosensitive interactions among a number of genes and gene products. One of the most important of these genes is *Sox9*, which is upregulated by compressive force (Takahashi et al. 1998). The Sox9 protein activates numerous bone-forming genes (see Chapter 14). Tension forces also activate bone morphogenetic proteins (BMPs) and align chondrocytes (Bard 1990; Sato et al. 1999; Ikegame et al. 2001). Several studies implicate *indian hedgehog* as a key signaling molecule that is stimulated by stress and which activates the bone morphogenetic proteins (Wu et al. 2001).

In the chick, several bones do not form if embryonic movement in the egg is suppressed. One of these bones is the fibular crest, which connects the tibia directly to the fibula. This direct connection is believed to be important in the evolution of birds, and the fibular crest is a universal feature of the bird hindlimb (Müller and Steicher 1989). When the chick is prevented from moving within its egg, the fibular crest fails to develop (Figure 22.2; K. C. Wu et al. 2001; Müller 2003).

Mechanical stress is also important for the differentiation of smooth muscle. Smooth muscle develops at sites of continuous mechanical tension, such as the vasculature and viscera. In the lungs, the pressure generated by breathing is necessary for bronchial smooth muscle development (Yang et al. 2000). These smooth muscles develop from bipotential mesenchymal cells that can become either adipocytes (fat cells) or myocytes (muscle cells). Physical pressure induces

the cells to become muscles by differentially expressing two proteins, TIP1 and TIP3. TIP1 is induced by mechanical stress, binds to and activates the promoter regions of myogenic genes, and causes the mesenchyme cells to follow the myogenic pathway. TIP3, an activator of fat-producing genes, is *repressed* by mechanical stress. In the absence of such stress, the cells to become adipocytes (McBeath et al. 2004; Jakkaraju et al. 2005). Therefore, even in the formation of important features such as bones and muscles, the environment can play a critical role.

Developmental symbiosis

In some cases, the development of one individual is brought about by the presence of organisms of a different species. In some organisms, this relationship has become symbiotic—the symbionts have become so tightly integrated into the host organism that the host cannot develop without them (Sapp 1994). Indeed, recent evidence indicates that developmental symbioses may be the rule rather than the exceptional case (McFall-Ngai 2002).

***EUPRYMNA-VIBRIO* SYMBIOSIS** One of the best-studied examples of developmental symbiosis is that between the squid *Euprymna scolopes* and the luminescent bacterium *Vibrio fischeri* (McFall-Ngai and Ruby 1991; Montgomery and McFall-Ngai 1995). The adult *Euprymna* is equipped with a light organ composed of sacs filled with these bacteria. The juvenile squid, however, does not contain these light-emitting symbionts, nor does it have the light organ to house them. The mature squid acquires *V. fischeri* from the seawater pumped through its mantle cavity. The bacteria bind to a ciliated epithelium that extends into this cavity; the epithelium only binds *V. fischeri*, allowing other bacteria to pass through. The bacteria then induce the apoptotic death of these epithelial cells, their replacement by a nonciliated epithelium, and the differentiation of the surrounding cells into storage sacs for the bacteria (Figure 22.3).

The substance *V. fischeri* secretes to effect these changes turns out to be fragments of the bacterial cell wall, and the active agents are tracheal cytotoxin and lipopolysaccharide (Koropatnick et al. 2004). This finding was surprising,

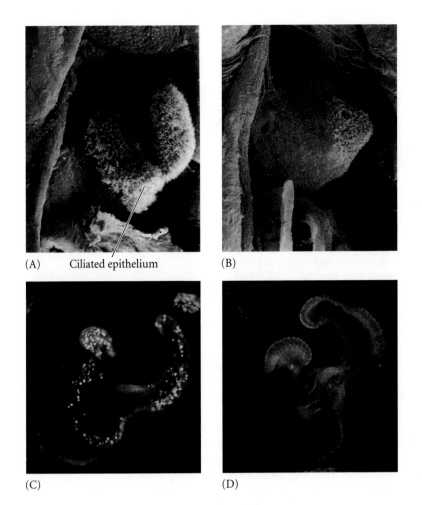

(A) Ciliated epithelium   (B)

(C)   (D)

FIGURE 22.3 Symbiosis in the squid *Euprymna*. (A,B) Scanning electron micrographs of a light organ primordium of a 3-day-old juvenile squid *E. scolopes*. (A) Light organ of an uninfected juvenile. (B) Light organ of a juvenile infected with the symbiotic *V. fischeri* bacteria. Regression of the epithelium is obvious. (C) Bacteria-induced apoptosis is shown by acridine orange staining at 12 hours after infection of the juvenile squid with the bacteria. The bright green areas indicate regions of cell death. (D) Light organ of a squid grown in the absence of *V. fischeri*. No areas of apoptosis are seen. (From Montgomery and McFall-Ngai 1995; photographs courtesy of M. McFall-Ngai.)

because these two agents have been long known to cause inflammation and disease. Indeed, tracheal cytotoxin is responsible for the tissue damage in both whooping cough and gonorrheal infections. The destruction and replacement of ciliated tissue in the respiratory tract and oviduct is due to these bacterial compounds. Although these compounds are deleterious to our bodies, the squid has been able to make use of them as morphogenetic signals.

EGG SYMBIOSES Symbioses between egg masses and photosynthetic algae are critical for the development of several species. Clutches of amphibian and snail eggs, for example, are packed together in tight masses. The supply of oxygen limits the rate of development, so embryos on the inside of the cluster develop more slowly than those near the surface. While there is a steep gradient of oxygen from the outside of the cluster to deep within it, the embryos seem to get around this problem by coating themselves with a thin film of photosynthetic algae. In clutches of amphibian and snail eggs, photosynthesis from this algal "fouling" enables net oxygen production in the light, while respiration exceeds photosynthesis in the dark (Bachmann et al. 1986; Pinder and Friet 1994; Cohen and Strathmann

1996). Thus the algal photosynthesis can "rescue" the eggs.

Symbioses between eggs and bacteria can also protect eggs from fungal pathogens. Lobster and shrimp eggs, for instance, are prone to fungal infection. (As anyone who owns an aquarium knows, uneaten fish food soon become surrounded by a halo of filamentous fungi.) The chorions of these crustacean eggs actually attract bacteria that produce fungicidal compounds (Gil-Turnes et al. 1989).

An even tighter link between morphogenesis and symbiosis is exemplified by the parasitic wasp *Asobara tabida* and the leafhopper *Euscelis incisus*. In these insects, symbiotic bacteria are found within the egg cytoplasm and are transferred through the generations, just like mitochondria. In the leafhopper, these bacteria have become so specialized that they can multiply only inside the leafhopper's cytoplasm, and the host has become so dependent on the bacteria that it cannot complete embryogenesis without them; the bacteria appear to be essential for the formation of the embryonic gut. If the bacterial symbionts are removed from the eggs surgically or by feeding antibiotics to larvae or adults, the symbiont-free oocytes develop into embryos that lack an abdomen (Figure 22.4; Sander 1968; Schwemmler 1974 1989). In *Asobara*, the bacteria enable the wasp to complete yolk production and egg maturation (Dedeinde et al. 2001). If the symbionts are removed, no eggs are produced.

INTESTINE-ANAEROBIC BACTERIA SYMBIOSIS Even mammals maintain developmental symbioses with bacteria. The polymerase chain reaction technique is able to identify bacterial species that cannot be cultured, and microarray analyses can show changes in the expression of a large population of genes. These techniques have revealed particular distributions of the bacterial symbionts within our bodies. The 500 to 1000 different bacterial species of the human colon are stratified into specific regions along the length and diameter of the gut tube, where they can attain

(A)

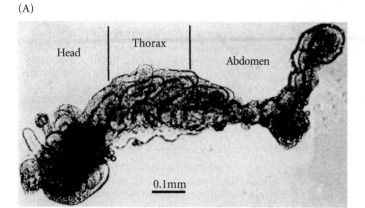

(B)

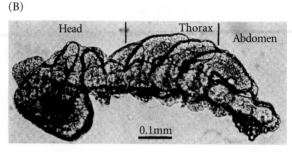

FIGURE 22.4   Microbial symbionts are necessary for gut formation in the leafhopper *Euscelis incisus*. (A) Control embryo, with symbionts, has normal gut formation. (B) Abnormal, gut-deficient embryo formed after antibiotics have eliminated most of the symbiotic bacteria from the egg. (From Schwemmler 1974; photographs courtesy of W. Schwemmler.)

densities of 10^{11} cells per milliliter (Hooper et al. 1998; Xu and Gordon 2003). We never lack these microbial components; we pick them up from the reproductive tract of our mother as soon as the amnion bursts. We have coevolved to share our spaces with them, and we have even co-developed such that our cells are primed to bind to them, and their cells are primed to induce gene expression in our nuclei (Bry et al. 1996).

Bacteria-induced expression of mammalian genes was first demonstrated in the mouse gut. Umesaki (1984) noticed that a particular fucosyl transferase enzyme characteristic of mouse intestinal villi was induced by bacteria, and more recent studies (Hooper et al. 1998) have shown that the intestines of germ-free mice can initiate, but not complete, their differentiation. For complete development, the microbial symbionts of the gut are needed. Normally occurring gut bacteria can upregulate the transcription of several mouse genes, including those encoding colipase, which is important in nutrient absorption; angiogenin-3, which helps form blood vessels; and sprr2a, a small, proline-rich protein that is thought to fortify matrices that line the intestine (Figure 22.5A; Hooper et al. 2001). Stappenbeck and colleagues (2002) have demonstrated that in the absence of particular intestinal microbes, the capillaries of the small intestinal villi fail to develop their complete vascular networks (Figure 22.5B,C).

Intestinal microbes also appear to be critical for the maturation of the mammalian gut-associated lymphoid tissue (GALT). GALT mediates mucosal immunity and oral immune tolerance, allowing us to eat food without making an immune response to it (see Rook and Stanford 1998; Cebra 1999; Steidler 2001). When introduced into germ-free rabbit appendices, neither *Bacillus fragilis* nor *B. subtilis*, alone, was capable of consistently inducing the proper formation of GALT. However, the combination of these two common mammalian gut bacteria consistently induced it (Rhee et al. 2004). The major inducer here appears to be bacterial polysaccharide-A (PSA), especially that of *B. fragilis*. The PSA-deficient mutant of *B. fragilis* is not able to restore normal immune function to germ-free mice (Mazmanian et al. 2005). Thus, a bacterial compound appears to be responsible for playing a major role in inducing the host's immune system.

In short, mammals have coevolved with bacteria to the point that our bodily phenotypes do not fully develop without them. The microbial community of our gut can be viewed as an "organ" that provides us with certain functions that we haven't evolved (such as the ability to process plant polysaccharides) and, like our developing organs, the microbes induce changes in neighboring tissues. As Mazmanian and colleagues have concluded, "The most impressive feature of this relationship may be that the host not only tolerates but has evolved to require colonization by commensal microorganisms for its own development and health."

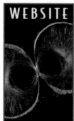

WEBSITE   **22.1 Developmental symbioses and parasitism.** Some embryos acquire protection and nutrients by forming symbiotic associations with other organisms. The mechanisms by which these associations form are now being elucidated. In other situations, one species uses material from another to support its development. Blood-sucking mosquitoes are examples of such parasites.

Larval settlement

Environmental cues are critical to metamorphosis in many species; some of the best-studied examples are the settlement cues used by marine larvae. A free-swimming marine

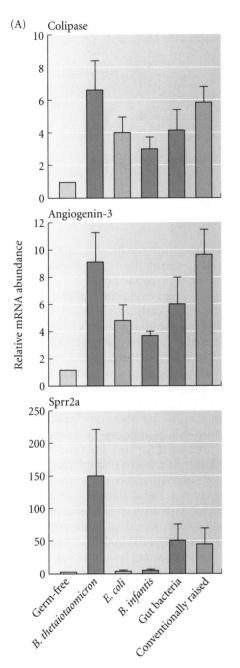

(A) Colipase

Angiogenin-3

Sprr2a

Relative mRNA abundance

Germ-free | *B. thetaiotaomicron* | *E. coli* | *B. infantis* | Gut bacteria | Conventionally raised

(B)

(C)

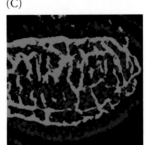

FIGURE 22.5 Specificity of host genome responses to different bacteria. (A) Mice raised in "germ-free" environments were either left alone or inoculated with one or more types of bacteria. After 10 days, their intestinal mRNAs were isolated and tested on microarrays. Mice grown in "germ-free" conditions had very little expression of the genes encoding colipase, angiogenin-3, or sprr2a. Several different bacteria—*Bacteroides thetaiotaomicron, E. coli, Bifidobacterium infantis*, and an assortment of gut bacteria harvested from conventionally raised mice—induced the genes for colipase and angionenin-3. *B. thetaiotaomicron* appeared to be totally responsible for the 205-fold increase in *sprr2a* expression over that of germ-free animals. (B) Confocal microscope section of an intestinal villus capillary bed in a mouse raised for 6 weeks in germ-free conditions. The capillaries are stained green. (C) Capillary network of an intestinal villus of a mouse raised for 6 weeks in germ-free conditions, then inoculated with conventional gut microbes 10 days before examination. The capillary network has fully developed. (A after Hooper et al. 2001; B,C after Stappenbeck et al. 2002; photograph courtesy of J. L. Gordon.)

ations of parthenogenetically (asexually) reproducing females. During the autumn, however, a particular type of female is produced whose eggs can give rise to both males and sexual females. These sexual forms mate, and their eggs are able to survive the winter. When the overwintering eggs hatch, each one gives rise to an asexual female.

Some of the mysteries of this type of development were solved in 1909 by Thomas Hunt Morgan (before he started working on fruit flies). Morgan analyzed the chromosomes of the hickory aphid through several generations (Figure 22.6). He found that the diploid number of female aphids is 12. In parthenogenetically reproducing females, only one polar body is extruded from the developing ovum during oogenesis, so the diploid number of 12 is retained in the egg. This type of egg develops without being fertilized. In the females that give rise to eggs that become male or female, a modification of oogenesis occurs. In the female-

larva often needs to settle near a source of food or on a firm substrate on which it can metamorphose. Among the molluscs, there are often very specific cues for settlement (Hadfield 1977). In some cases, the prey supply the cues, while in other cases the substrate itself gives off molecules used by the larvae to initiate settlement. These cues may not be constant, but they need to be part of the environment if further development is to occur* (Pechenik et al. 1998).

Sex in its season

Several species of aphids have a fascinating life cycle wherein an egg hatched in the spring gives rise to several gener-

*The importance of substrates for larval settlement and metamorphosis was first demonstrated in 1880, when William Keith Brooks, an embryologist at Johns Hopkins University, was asked to help the ailing oyster industry of Chesapeake Bay. For decades, oysters had been dredged from the bay, and there had always been a new crop to take their place. But recently, each year brought fewer oysters. What was responsible for the decline? Experimenting with larval oysters, Brooks discovered that the American oyster (unlike its better-studied European cousin) needs a hard substrate on which to metamorphose. For years, oystermen had thrown the shells back into the sea, but with the advent of suburban sidewalks, the oystermen were selling the shells to the cement factories. Brooks's solution: throw the shells back into the bay. The oyster population responded, and the Baltimore wharves still sell their descendants.

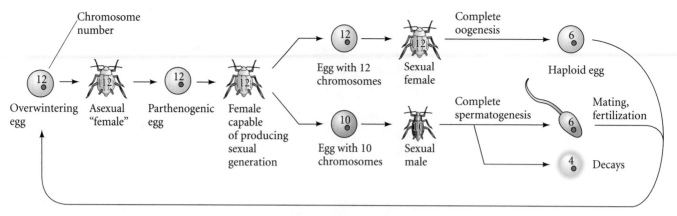

FIGURE 22.6 Chromosomal changes during the life cycle of the hickory aphid. Autumn weather induces the production of males and females, which mate to produce the overwintering eggs.

producing eggs, 6 chromosome pairs enter the sole polar body; the diploid number of 12 is thereby retained. In the male-producing eggs, however, an extra chromosome pair enters the polar body. The male diploid number is thus 10. The resulting males and females are sexual and produce gametes by complete meiotic divisions. The females produce oocytes with a haploid set of 6 chromosomes. The males, however, divide their 10 chromosomes to produce some sperm with a haploid number of 4 and other sperm with a haploid number of 6. The sperm with 4 chromosomes degenerate. The sperm with 6 chromosomes fertilize the eggs to restore the diploid chromosome number of 12. These eggs overwinter, and when they hatch in the spring, parthenogenetic females emerge.

Morgan solved one riddle, but the riddle of how the season determines whether the female reproduces sexually or parthenogenetically remains unsolved. Similarly, we do not know what regulates whether a diploid oocyte gives rise to male- or female-producing eggs. Moreover, the same environmental factors are used differently by other aphid species (Hardie 1981; Hardie and Lees 1985). We still do not know how the autumn weather (or perhaps declining hours of sunlight) causes the differential movement of chromosomes into the polar body.

Diapause: Suspended development

Many species of insects and mammals have evolved a developmental strategy called **diapause** to survive periodically harsh conditions. Diapause is a suspension of development that can occur at the embryonic, larval, pupal, or adult stage, depending on the species (see Chapter 18). Diapause is not a physiological response brought about by harsh conditions. Rather, it is induced by stimuli (such as changes in the duration of daylight) that *presage* a change in the environment—cues beginning before the severe conditions actually arise. Diapause is especially important for temperate-zone insects, enabling them to survive the win-

ter. The overwintering eggs of the hickory aphid (mentioned above) provide an example of this strategy. The development in the egg is suspended over the winter, so the larvae do not hatch when food is unavailable. In this case, diapause occurs during early development. The silkworm moth *Bombyx mori* similarly overwinters as an embryo, entering diapause just before segmentation. The gypsy moth *Lymantria dispar* initiates diapause as a larva, and needs an extended period of cold weather to end diapause, which is why this pest is not found in the southern regions of Europe or the United States.

Over 100 mammalian species undergo diapause. The two most common mammalian strategies are delayed fertilization (the sperm are stored for later use) and delayed implantation (the blastocyst remains unimplanted within the uterus, and the rate of cell divisions diminishes or vanishes). Some species have *seasonal* diapause, so embryos conceived in autumn will be born in spring rather than winter; in other species, diapause is induced by the presence of a newborn who is still getting milk. In the tammar kangaroo, *Macropus eugenii*, diapause can be a response to suckling-induced prolactin release, but it can also be induced by prolactin synthesized in response to changes in day length. In both cases, progesterone seems to be the signal that restores implantation and embryonic growth. Different groups of mammals use different hormones to induce or break diapause, but the result is the same: diapause lengthens the gestation period, allowing mating to occur and young to be born at times and seasons appropriate to the habitat of that species (Renfree and Shaw 2000).

 WEBSITE **22.2 Complex environmental effects on development.** The life cycles of certain insects are controlled by several environmental cues whose intersection provides a delicate timing mechanism.

 WEBSITE **22.3 Mechanisms of diapause.** Light and temperature are critical for the induction and maintenance of diapause. Different species use different signals for this event.

FIGURE 22.7 Polyphenic variation in *Pontia* (Pieridae) butterflies. The top row shows summer morphs: *P. protodice* female (left) and male (center); *P. occidentalis* male (right). The bottom row shows spring morphs: *P. protodice* female (left) and male (center); *P. occidentalis* male (right). (Photograph courtesy of T. Valente.)

Polyphenisms and Plasticity

Seasonal polyphenism in lepidopterans

The phenotype of the moth *Nemoria arizonaria* depends on its diet (see Figure 3.4). This type of polyphenism is not uncommon among insects. Throughout much of the Northern Hemisphere, one can see such a polyphenism in butterflies of the family Pieridae (the cabbage whites), with phenotypes that differ between individuals that eclose during the long days of summer and those that eclose at the beginning of the season, in the short, cooler days of spring. The hindwing pigments of the short-day forms are darker than those of the long-day butterflies. This pigmentation has a functional advantage during the cool months of spring: darker pigments absorb sunlight more efficiently than light ones, raising the body temperature more rapidly (Figure 22.7; Shapiro 1968; Watt 1968; see also Nijhout 1991).

In tropical parts of the world, there is often a hot wet season and a cooler dry season. In Africa, a polyphenism of the dimorphic Malawian butterfly *Bicyclus anynana* is adaptive to seasonal changes. The dry (cool) season morph is a mottled brown butterfly that survives by hiding in dead leaves on the forest floor. In contrast, the wet (hot) season morph, which routinely flies, has prominent ventral eyespots that deflect attacks from predatory birds and lizards (Figure 22.8).

The factor determining the seasonal pigmentation of *B. anynana* is not diet, but the temperature during pupation. Low temperatures produce the dry-season morph; higher temperatures produce the wet-season morph (Brakefield and Reitsma 1991). The mechanism by which temperature regulates the *Bicyclus* phenotype is becoming known. In the late larval stages, the transcription of the *distal-less* gene

FIGURE 22.8 Phenotypic plasticity in *Bicyclus anynana* is regulated by temperature. High temperature (either in the wild or in controlled laboratory conditions) allows the accumulation of 20-hydroxyecdysone (20E), a hormone that is able to sustain *Distal-less* expression in the pupal imaginal disc. The region of *Distal-less* expression becomes the focus of each eyespot. In cooler weather 20-hydroxyecdysone is not formed, *Distal-less* expression in the imaginal disc begins but is not sustained, and eyespots fail to form. (Photographs courtesy of S. B. Carroll and P. Brakefield.)

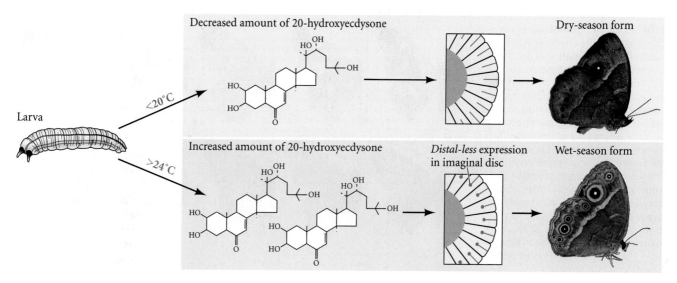

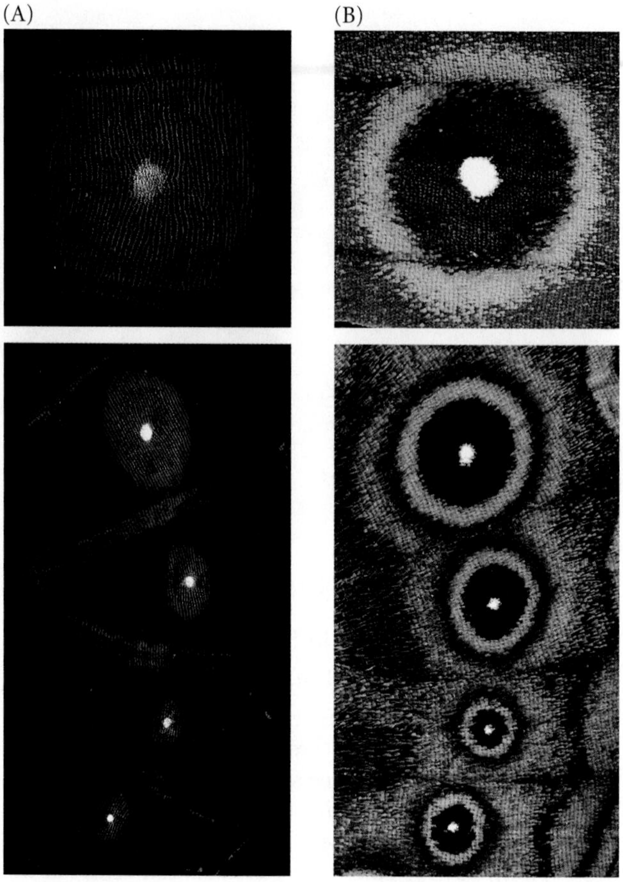

FIGURE 22.9   Distal-less and Spalt proteins define concentric areas corresponding to the rings of adult eyespots. (A) The transcription factors Distal-less (purple), Spalt (green), and their overlap (white) define the areas of the expanding pupal wing imaginal disc that correspond to the colored scales of the adult. (B) Portion of ventral hindwing of an adult *Bicyclus*, showing four of the seven eyespots and the large ventral eyespot of the forewing. (After Brunetti et al. 2001; photographs courtesy of S. Paddock and S. B. Carroll.)

in the wing imaginal discs is restricted to a set of cells that will become the signaling center of each eyespot. In the early pupa, higher temperatures elevate the formation of 20-hydroxyecdysone (20E; see Chapter 18). This hormone (in a manner not yet described) sustains and expands the expression of *distal-less* in those regions of the wing imaginal disc, resulting in prominent eyespots. The cooler dry season temperatures prevent the accumulation of 20E in the pupa, and the foci of Distal-less signaling are not sustained. In the absence of the Distal-less signal, the eyespots do not form (Brakefield et al. 1996; Koch et al. 1996). Distal-less protein is believed to be the activating signal that determines the size of the eyespot (Figure 22.9).

The importance of hormones such as 20E for mediating environmental signals controlling wing phenotypes has been documented in the *Araschnia* butterfly mentioned in Chapter 3 (see Figure 3.3). *Araschnia* develops alternative phenotypes depending on whether the fourth and fifth instars experience a photoperiod (hours of daylight) that is longer or shorter than a particular critical day length. Below this critical day length, ecdysone levels are low and the butterfly has the orange wings characteristic of spring butterflies. Above the critical point, ecdysone is made and the summer pigmentation forms. The summer form can be induced in diapause (spring) pupae by injecting 20E into the pupae. Moreover, by altering the timing of 20E injections, one can generate a series of intermediate forms not seen in the wild (Figure 22.10; Koch and Bückmann 1987; Nijhout 2003).

FIGURE 22.10   Hormonal regulation mediates the environmentally controlled pigmentation of *Araschnia*. In the wild, different generations experience significantly different photoperiods. In the short photoperiod (below the critical day length), there is no pulse of 20E during early pupation, and the spring form of the butterfly is generated. When these spring butterflies mate, the larvae experience a long photoperiod and generate the summer pigmentation. In the laboratory, injections of 20E at different times during pupation can induce both phenotypes, as well as intermediate phenotypes not seen in the wild. (From Nijhout 2003; photograph courtesy of H. F. Nijhout.)

Normal summer form

Normal spring form

FIGURE 22.11   Gyne and worker of the ant *Pheidologeton*. This picture shows the remarkable dimorphism between the large queen and the small worker (seen near the queen's antennae). The difference between these two sisters involves larval feeding and juvenile hormone synthesis. (Photograph © Mark Moffett/Minden Pictures.)

Nutritional polyphenism

HYMENOPTERAN QUEENS AND WORKERS   Not all polyphenisms are controlled by the seasons. In hymenopteran ants and bees, the determination of "castes" can be controlled by genes, by nutrition, by temperature, or even by volatile chemicals secreted by other individuals. In certain bees (such as honeybees), the size of the female larva at its metamorphic molt determines whether the individual is to be a worker or a queen. A larva fed nutrient-rich "royal jelly" retains the activity of her corpora allata during her last instar stage. The juvenile hormone secreted by these organs delays pupation, allowing the resulting bee to emerge larger and (in some species) more specialized in her anatomy (Brian 1974, 1980; Plowright and Pendrel 1977). The JH level in these queen larvae is 25 times greater than in larvae destined to become workers, and applying JH to worker larvae can transform them into queens (Wirtz 1973; Rachinsky and Hartfelder 1990).

Similarly, ant colonies are predominantly female, and the females can be extremely polymorphic. The much larger reproductive females ("gynes" or "queens") have functional ovaries and wings; the workers do not (Figure 22.11). These striking differences in anatomy and physiology are regulated through juvenile hormone (Wheeler 1991). The influence of the environment on hormone levels and gene expression in ants was analyzed by Abouheif and Wray (2002), who found that nutrition-induced JH levels regulated wing formation. In the queen, both the forewing and the hindwing disc undergo normal development, expressing the same genes as *Drosophila* wing discs (see Chapter 18). However, in the wing imaginal discs of workers, some of these genes remain unexpressed and the wings fail to form. Abouheif and Wray (2002) also found that although the end result—winged queens and wingless workers—is the same in numerous ant species, the actual genes downregulated in the workers differed from species to species.

WHEN DUNG REALLY MATTERS   For the male dung beetle (*Onthophagus*; see the photograph on p. 693), what really matters in life is the amount and quality of the dung he eats as a larva. The hornless female dung beetle digs tunnels, then gathers balls of dung and buries them in these tunnels. She then lays a single egg on each dung ball; when the larvae hatch, they eat the dung. Metamorphosis occurs when the dung ball is finished, and the phenotype and behavior of the male dung beetle is determined by the quality and quantity of this maternally provided food (Emlen 1997; Moczek and Emlen 2000). The amount of food determines the size of the larva at metamorphosis; the size of the larva at metamorphosis determines the titre of juvenile hormone during its last molt; and the titre of juvenile hormone effects the growth of the imaginal discs that make the horns (Figure 22.12A; Emlen and Nijhout 1999; Moczek 2005). If juvenile hormone is added to tiny *O. taurus* males during

(A)

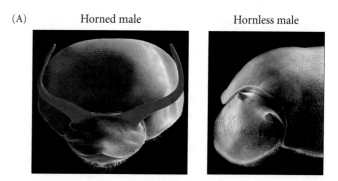

Horned male   Hornless male

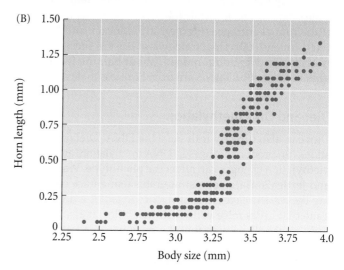

(B)

FIGURE 22.12   Horned and hornless male dung beetles. (A) The hornless and horned morphs of *Onthophagus taurus* are determined by the titre of juvenile hormone at the last molt. This hormone titre depends on the larva's body size at the time. (B) A study of 500 individuals of *O. acuminatus* reveals a sharp threshold of body size, before which horns fail to form and after which horn growth is linear with the size of the beetle. This sigmoidal distribution produces males with either no horns or with significant horns , and very few horns of intermediate size. (After Emlen 2000, photographs courtesy of D. Emlen.)

FIGURE 22.13 The presence or absence of horns determines the reproductive strategy of the male dung beetle (*Onthophagus*). Horned males mate repeatedly with the females and guard the entrances to brood tunnels dug by the females. They successfully prevent other males from entering the females' tunnels, and the males with the longest horns usually win these contests. Smaller, hornless males do not guard tunnels. Their equally successful strategy is to dig their own tunnels, connect their tunnel to that of a female, mate, and exit—all without a violent encounter. (After Emlen 2000.)

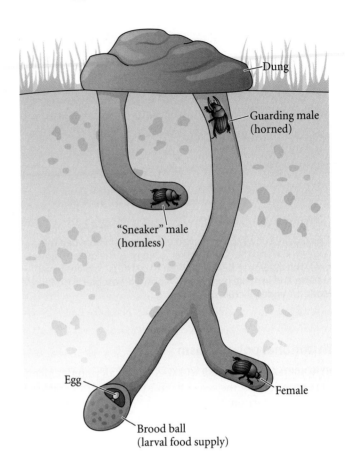

the sensitive period of their last molt, the cuticle in its head expands to produce a horn. Thus, whether a male is horned or hornless does not depend on the male's genes, but on the food his mother left for him.

The male horn does not grow until the beetle reaches a certain size. After this threshold body size, horn growth is very rapid.* Thus, although body size has a normal distribution, there is a bimodal distribution of horn sizes: about half the males have no horns, while the other half have horns of considerable length (Figure 22.12B).

The size of the horn determines a male's behavior and chances for reproductive success. Horned males guard the females' tunnels and use their horns to prevent other males from mating with the female; the male with the biggest horns wins such contests. But what about the males with no horns? Hornless males do not fight with the horned males for mates. Since they, like the females, lack horns, they are able to dig their own tunnels. These "sneaker males" dig tunnels that intersect those of the females and mate with the females while the horned male stands guard at the tunnel entrance (Figure 22.13; Moczek and Emlen 2000; Emlen 2000). This strategy is at least as successful as guarding and fighting. Indeed, about half the fertilized eggs in most populations are from hornless males.

Diet and DNA methylation

Diet can also directly influence the DNA. Dietary alterations can produce changes in DNA methylation, and these methylation changes can affect the phenotype. Waterland and Jertle (2003a) demonstrated this using mice containing the *viable-yellow* allele of *Agouti*. *Agouti* is a dominant gene that gives mice yellowish hair color; it also affects lipid metabolism such that the mice become fatter. The *viable-yellow* allele of *Agouti* has a transposable element inserted in the first exon. These transposon insertion sites are very interesting for regulation: whereas most regions of the adult genome have hardly any intraspecies variation in CpG methylation, there are large DNA methylation

*Interestingly, the threshold size at which the phenotype change from hornless to horned occurs *is* genetically transmitted and can change when conditions favor one morph over the other (Emlen 1996, 2000).

differences between individuals at the sites of transposon insertion. Such CpG methylation can block gene transcription. When the promoter of the *Agouti* gene is methylated, the gene is not transcribed. The mouse's fur remains black and lipid metabolism is not altered.

Waterland and Jirtle fed pregnant *viable-yellow Agouti* mice methyl donor supplements, including folate, choline, and betain. They found that the more methyl supplementation, the greater the methylation of the transposon insertion site, and the darker the pigmentation of the mouse's offspring. The mice in Figure 22.14 are genetically identical, but their mothers were fed different diets during pregnancy. The mouse whose mother did not receive methyl donor supplementation is fat and yellow—the *Agouti* gene promoter was unmethylated, so the gene was active. The mouse born to the mother who was given supplements is sleek and dark; the methylated *Agouti* gene was not transcribed.

As we saw in Chapter 21, such differential gene methylation has been linked to human health problems. Dietary restrictions during a woman's pregnancy may show up as heart or kidney problems in her adult children. Moreover, studies in rats showed that differences in protein and methyl donor concentration in the mother's prenatal diet affected metabolism in the pup's livers (Lillycrop et al. 2005).

FIGURE 22.14 Prenatal diet changes DNA methylation in the fetus and can alter adult phenotype. These genetically identical mice are the progeny of mothers who carried the *viable-yellow Agouti* gene. The mouse on the left was born to a mother who lacked methyl donor supplementation (e.g., folate) in her prenatal diet; the pup's *Agouti* gene promoter was unmethylated and the gene remained active, leading to the formation of yellow pigment and altered lipid metabolism. The mother of the sleek black mouse received dietery folate supplements during pregnancy, and methylation blocked transcription of the *Agouti* gene in her offspring. (Photograph courtesy of R. L. Jirtle.)

Environment-dependent sexual phenotype

As we saw in Chapter 17, there are many species in which the environment determines whether an individual is male or female. The temperature dependence of sex determination in fish and reptiles is the best studied case (see Figure 17.21). This type of environmental sex determination has advantages and disadvantages. One advantage is that it probably gives the species the benefits of sexual reproduction without tying the species to a 1:1 sex ratio. In crocodiles, in which temperature extremes produce females while moderate temperatures produce males, the sex ratio

---

SIDELIGHTS & SPECULATIONS

Fetal DNA Methylation and Adult Behavior

Environmentally derived methylation differences near birth may lead to important behavioral differences in adults. In rats, behavioral differences in the response to stressful situations have been correlated with the number of glucorticoid receptors in the brain's hippocampus. The more glucocorticoid receptors, the better the adult rat is able to downregulate adrenal hormones and deal with stress. The number of glucocorticoid receptors appears to depend on the quality of grooming and licking the rat pup experiences during the first week after birth.

How is the adult phenotype regulated by these perinatal (near the time of birth) experiences? Weaver and colleagues (2004) have shown that the difference involves the methylation of a particular site in the promoter region on the glucocorticoid receptor gene. Before birth, there is no methylation at this site; one day after birth, this site is methylated in all rat pups. However, in those pups that experience intensive grooming and licking during the first week after birth, this site *loses* its methylation, but methylation is retained in those rats who do not have such extensive care. Moreover, this methylation difference is not seen at other sites in or near the gene (Figure 22.15).

By switching pups and parents, they demonstrated that this methylation difference was dependent on the mother's care, and was not the result of differences in the pups themselves. When unmethylated, the target site binds the Egr1 transcription factor and is associated with "active" acetylated nucleosomes. The transcription factor does not bind to the methylated site, and the chromatin in such cases is not activated. These chromatin differences, established during the first week after birth, are retained throughout the life of the rat. Thus, adult rats that received extensive perinatal grooming have more glucocorticoid receptors and are able to deal with stress better than rats who received less care. (How grooming can alter DNA methylation patterns remains to be discovered.)

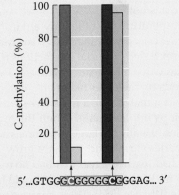

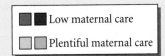

Low maternal care

Plentiful maternal care

5′...GTGGGCGGGGGCGGGAG... 3′

Figure 22.15 Differential DNA methylation due to behavioral differences in newborn care. A portion of an enhancer sequence of the rat glucocorticoid gene is shown along the bottom. The Egr1 binding site is boxed. Two cytidine residues within this site have the potential for methylation. The cytosine at the 5′ end of the box is completely methylated in the brains of pups that did not receive extensive maternal grooming (red bar). The Egr1 transcription factor did not bind to these sites and the gene was inactive. If the pups did receive sufficient maternal care, this same site was mostly unmethylated and the gene was transcribed in the brain (pink bar). The site at the 3′ end (blue bars) was always methylated and had no effect on Egr1 binding. (After

may be as great as 10 females to each male (Woodward and Murray 1993). The major disadvantage of temperature-dependent sex determination may be its narrowing of the temperature limits within which a species can exist. Thus thermal pollution (either locally or due to global warming) could conceivably eliminate a species in a given area (Janzen and Paukstis 1991). Researchers (Ferguson and Joanen 1982; Miller et al. 2004) have speculated that dinosaurs may have had temperature-dependent sex determination and that their sudden demise may have been caused by a slight change in temperature creating conditions wherein only males or only females hatched.

Charnov and Bull (1977) argued that environmental sex determination would be adaptive in those habitats characterized by patchiness—that is, a habitat having some regions where it is more advantageous to be male and other regions where it is more advantageous to be female. Conover and Heins (1987) provided evidence for this hypothesis. In certain fish species, females benefit from being larger, since larger size translates into higher fecundity. If you are a female Atlantic silverside (*Menidia menidia*), it is advantageous to be born early in the breeding season, because you have a longer feeding season and thus can grow larger. (Size is of no importance in males of this species.) In the southern range of *Menidia*, females are indeed born early in the breeding season, and temperature appears to play a major role in this pattern. However, in the northern reaches of its range, the species shows no environmental sex determination. Rather, a 1:1 sex ratio is generated at all temperatures (Figure 22.16). Conover and Heins speculated that the more northern populations have a very short feeding season, so there is no advantage for females in being born earlier. Thus, this fish has environmental sex determination in those regions where it is adaptive and genotypic sex determination in those regions where it is not.

Temperature isn't the only environmental factor that can affect sex determination in fish. Many fish can change their sex based on social interactions (Godwin et al. 2002). The sex of the blue-headed wrasse, a Panamanian reef fish, depends on the other fish it encounters. If a wrasse larva reaches a reef where a male lives with many females, it develops into a female. When the male dies, one of the females (usually the largest) becomes a male. Within a day, its ovaries shrink and its testes grow. If the same wrasse larva had reached a reef that had no males or that had territory undefended by a male, it would have developed into a male (Warner 1984). These changes appear to be mediated by glucocorticoid stress hormones that activate neuropeptides in the hypothalamus of these fish (Godwin et al. 2000, 2003; Perry and Grober 2003).

Marine gobys are among the few fish that can change their sex more than once—and in either direction. A female goby can become male if the male of the group dies. However, if a larger male enters the group, such males revert to being female (Black et al. 2005). Grober and Sunobe (1996) induced females to become males, males to become

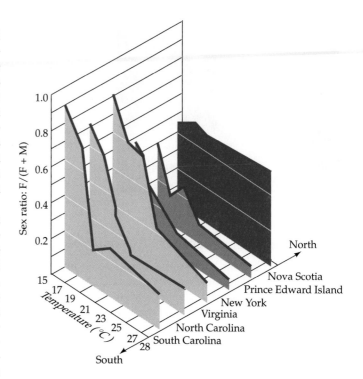

FIGURE 22.16 Relationship between temperature during the period of sex determination and sex ratio [F:(F + M)] in *Menidia menidia*. In fish collected from the northernmost portion of its range (Nova Scotia), temperature had little effect on sex determination. Among fish collected at more southerly locations (especially from Virginia through South Carolina), however, temperature had a large effect. (After Conover and Heins 1987.)

females, and females to become males and then females again, merely by changing their companions. A goby can change its sex in about 4 days, and these changes appear to be mediated by the same hypothalamic neuropeptides correlated with wrasse sex determination.

Polyphenisms for alternative environmental conditions

Most studies of adaptation concern the roles adult structures play in enabling the individual to survive in otherwise precarious or hostile environments. However, the developing animal, too, has to survive in its habitat, and its development must adapt to the conditions of its existence.

The spadefoot toad, *Scaphiopus couchii*, has a remarkable strategy for coping with a very harsh environment. These toads are called out from hibernation by the thunder that accompanies the first spring storm in the Sonoran desert. (Unfortunately, motorcycles produce the same sounds, causing the toads to come out of hibernation only to die in the scorching Arizona sun.) The toads breed in temporary ponds formed by the rain, and the embryos develop quickly into larvae. After the larvae metamorphose, the young toads return to the desert, burrowing into the sand until the next year's storms bring them out.

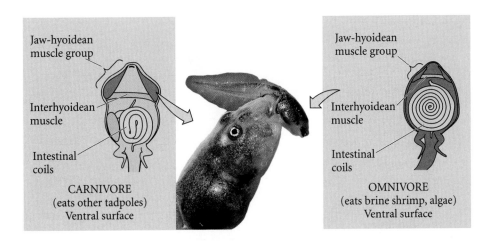

FIGURE 22.17 Polyphenism in the tadpoles of the spadefoot toad, *Scaphiopus couchii*. The typical morph is an omnivore, usually eating insects and algae. When ponds are drying out quickly, however, a carnivorous (cannibalistic) morph forms. It develops a wider mouth, larger jaw muscles, and an intestine modified for a carnivorous diet. The center photograph shows a cannibalistic tadpole eating a smaller pondmate. (Photograph © Thomas Wiewandt; drawings courtesy of R. Ruibel.)

Desert ponds are ephemeral pools that can either dry up quickly or persist, depending on the initial depth and the frequency of the rainfall. One might envision two alternative scenarios confronting a tadpole in such a pond: either (1) the pond persists until you have time to metamorphose and you live, or (2) the pond dries up before your metamorphosis is complete, and you die. *S. couchii* (and several other amphibians), however, have evolved a third alternative. The timing of their metamorphosis is controlled by the pond. If the pond persists at a viable level, development continues at its normal rate, and the algae-eating tadpoles develop into juvenile toads. However, if the pond is drying out and getting smaller, some of the tadpoles embark on an alternative developmental pathway. They develop a wider mouth and powerful jaw muscles, which enables them to eat (among other things) other *Scaphiopus* tadpoles (Figure 22.17). These carnivorous tadpoles metamorphose quickly, albeit into a smaller version of the juvenile spadefoot toad. But they survive while other *Scaphiopus* tadpoles perish from desiccation (Newman 1989, 1992).

The signal for accelerated metamorphosis appears to be the change in water volume. In the laboratory, *Scaphiopus* tadpoles are able to sense the removal of water from aquaria, and their acceleration of metamorphosis depends on the rate at which the water is removed. The stress-induced corticotropin-releasing hormone signaling system appears to modulate this effect (Denver et al. 1998). This increase in brain corticotropin-releasing hormone is thought to be responsible for the subsequent elevation of the thyroid hormones that initiate metamorphosis (Boorse and Denver 2003). As in many other cases of polyphenism, the developmental changes are mediated through the endocrine system. Sensory organs send a neural signal to regulate hormone release. The hormones then can alter gene expression in a coordinated and relatively rapid fashion.

Predator-induced polyphenisms

Imagine an animal who is frequently confronted by a particular predator. One could then imagine an individual who could recognize soluble molecules secreted by that predator and who could use those molecules to activate the development of structures that would make this individual less palatable to the predator. This ability to modulate development in the presence of predators is called predator-induced defense, or **predator-induced polyphenism**.

To demonstrate predator-induced polyphenism, one has to show that the phenotypic modification is caused by the presence of the predator, and that the modification increases the fitness of its bearers when the predator is present (Adler and Harvell 1990; Tollrian and Harvell 1999). Figure 22.18 shows both the typical and predator-induced morphs for several species. In each case, the induced morph is more successful at surviving the predator, and soluble filtrate from water surrounding the predator is able to induce the changes. Chemicals that are released by a predator and can induce defenses in its prey are called **kairomones**.

Several rotifer species will alter their morphology when they develop in pond water in which their predators were cultured (Dodson 1989; Adler and Harvell 1990). The predatory rotifer *Asplanchna* releases a soluble compound that induces the eggs of a prey rotifer species, *Keratella slacki*, to develop into individuals with slightly larger bodies and anterior spines 130 percent longer than they otherwise would be, making the prey more difficult to eat. The snail *Thais lamellosa* develops a thickened shell and a "tooth" in its aperture when exposed to the effluent of the crab species that preys on it. In a mixed snail population, crabs will not attack the thicker snails until more than half of the normal-morph snails are devoured (Palmer 1985).

The predator-induced polyphenism of the parthenogenetic water flea *Daphnia* is beneficial not only to itself, but also to its offspring. When *Daphnia cucullata* encounter the predatory larvae of the fly *Chaeoborus*, their "helmets" grow to twice the normal size (Figure 22.19). This increase lessens the chances that *Daphnia* will be eaten by the fly larvae. This same helmet induction occurs if the *Daphnia* are exposed to extracts of water in which the fly larvae had been swimming. Agrawal and colleagues (1999) have shown that the offspring of such an induced *Daphnia* are

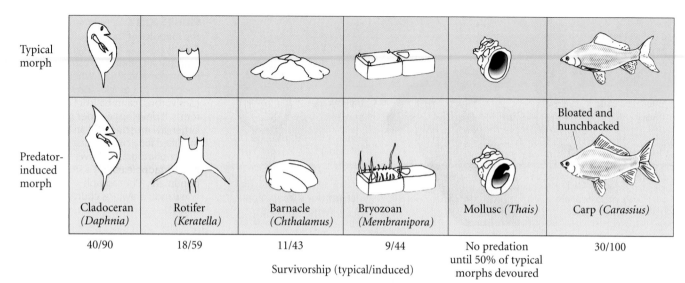

	Cladoceran (*Daphnia*)	Rotifer (*Keratella*)	Barnacle (*Chthalamus*)	Bryozoan (*Membranipora*)	Mollusc (*Thais*)	Carp (*Carassius*)
Typical morph						
Predator-induced morph						Bloated and hunchbacked
Survivorship (typical/induced)	40/90	18/59	11/43	9/44	No predation until 50% of typical morphs devoured	30/100

FIGURE 22.18 Predator-induced defenses. Typical (upper row) and predator-induced (lower row) morphs of various organisms are shown. The numbers beneath each column represent the percentage of organisms surviving predation when both induced and uninduced individuals were presented with predators (in various assays). (Data from Adler and Harvell 1990 and references cited therein.)

born with this same altered head morphology. It is possible that the *Chaeoborus* kairomone regulates gene expression both in the adult and in the developing embryo. We still do not know the identity of the kairomone, the identity of its receptor, or the mechanisms by which the binding of the kairomone to the receptor initiates the adaptive morphological changes.

Predator-induced polyphenism is not limited to invertebrates.* Indeed, predator-induced polyphenisms are abundant among amphibians. Tadpoles found in ponds or reared in the presence of other species may differ significantly from tadpoles reared by themselves in aquaria. For instance, newly hatched wood frog tadpoles (*Rana sylvetica*) reared in tanks

containing the predatory larval dragonfly *Anax* (confined in mesh cages so that they cannot eat the tadpoles) grow smaller than those reared in similar tanks without predators. Moreover, their tail musculature deepens, allowing faster turning and swimming speeds (van Buskirk and Relyea 1998). The addition of more predators to the tank causes a continuously deeper tail fin and tail musculature, and in fact what initially appeared to be a polyphenism may be a reaction norm that can assess the number (and type) of predators.

McCollum and Van Buskirk (1996) have shown that in the presence of its predators, the tail fin of the tadpole of the gray tree frog *Hyla chrysoscelis* grows larger and becomes bright red. This phenotype allows the tadpole to swim away faster and to deflect predator strikes toward the tail region. The trade-off is that noninduced tadpoles grow more slowly and survive better in predator-free envi-

---

*Indeed, when viewed biologically rather than medically, the vertebrate immune system is a wonderful example of predator-induced polyphenism. Here, our immune cells utilize chemicals from our predators (viruses and bacteria) to change our phenotype so that we can better resist them (see Frost 1999).

(A)

(B)   (C)

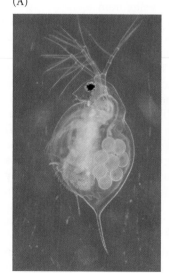
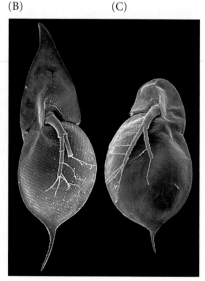

FIGURE 22.19 Predator-induced polyphenism in *Daphnia*. (A) *Daphnia* is an all-female species, producing eggs (visible within the adult organism) parthenogenetically. (B,C) Scanning electron micrographs showing predator-induced (B) and normal (C) morphs of the same clone. (A, photograph courtesy of R. Tollrian; B,C, photographs courtesy of A. A. Agrawal.)

ronments. In some species, phenotypic plasticity is reversible and removing the predators can restore the original phenotype (Relyea 2003a).

The metabolism of predator-induced morphs may differ significantly from that of the uninduced morphs, and this has important consequences. Relyea (2003b, 2004) has found that in the presence of the chemical cues emitted by predators, the toxicity of pesticides such as carbaryl (Sevin®) can become up to 46 times more lethal than it is without the predator cues. Bullfrog and green frog tadpoles were especially sensitive to carbaryl when exposed to predator chemicals (Figure 22.20). Relyea has related these findings to the global decline of amphibian populations, saying that governments should test the toxicity of the chemicals under more natural conditions, including that of predator stress. He concludes that "ignoring the relevant ecology can cause incorrect estimates of a pesticide's lethality in nature, yet it is the lethality of pesticides under natural conditions that is of utmost interest. The accumulated evidence strongly suggests that pesticides in nature could be playing a role in the decline of amphibians" (Relyea 2003b).

WEBSITE   **22.4** **Inducible caste determination in ant colonies.** In some species of ants, the loss of soldier ants creates conditions that induce more workers to become soldiers.

WEBSITE   **22.5** **Genetic assimilation.** One of the most exciting evolutionary speculations has been the idea that once an organism has the competence to respond to an environmental stimulus, it can transfer that competence to a stimulus within the embryo.

Learning: The Environmentally Adaptive Nervous System

We saw in Chapter 13 that neuronal activity can be a critical factor in deciding which synapses are retained by the adult organism. Here we extend that discussion to highlight those remarkable instances in which new experiences modify the original set of neuronal connections, causing the creation of new neurons, or the formation of new synapses between existing neurons.

The formation of new neurons

Since neurons, once formed, do not divide, the "birthday" of a neuron can be identified by treating the organism with radioactive thymidine. Normally, very little radioactive thymidine is taken up into the DNA of a neuron that has already been formed. However, if a new neuron differentiates by cell division during the treatment, it will incorporate radioactive thymidine into its DNA.

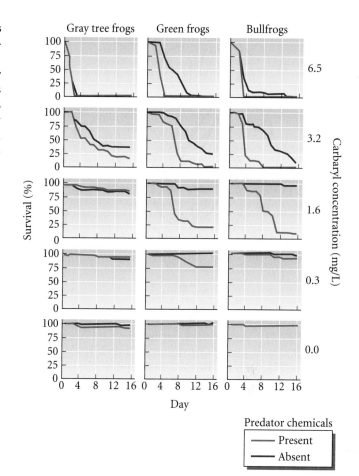

FIGURE 22.20 Survival of three tadpole species after being exposed to a combination of predator chemicals and four concentrations of the pesticide carbaryl. For gray tree frog tadpoles, predator stress plays little role in their response to carbaryl. At high concentrations, it is lethal, but there is little difference between the predator-induced morphs and the uninduced morphs. In green frog (*Rana clamitans*) and the bullfrog (*Rana catesbeiana*) tadpoles, carbaryl kills them at much lower doses if the tadpoles have responded to the predator cues (After Relyea 2003b).

Such new neurons are seen to be generated when male songbirds first learn their songs. Juvenile zebra finches memorize a model song and then learn the pattern of muscle contractions necessary to sing a particular phrase. In this learning and repetition process, new neurons are generated in the hyperstriatum of the finch's brain. Many of these new neurons send axons to the archistriatum, which is responsible for controlling the vocal musculature (Nordeen and Nordeen 1988). These changes are not seen in males who are too old to learn the song, nor are they seen in juvenile females (who do not sing these phrases). In white-crowned sparrows, where song is regulated by photoperiod and hormones, exposing adult males to long hours of light and to testosterone induces over 50,000 new neurons in their vocal centers (Tramontin et al. 2000). The

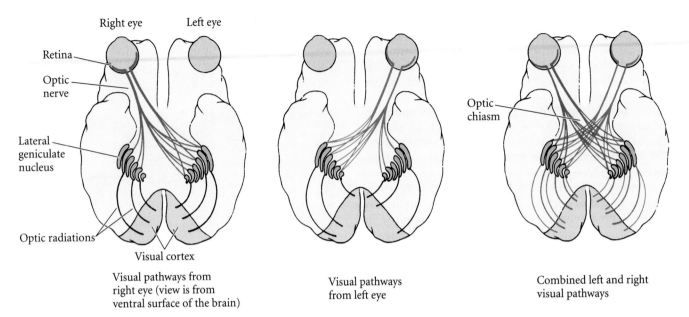

FIGURE 22.21 Major pathways of the mammalian visual system. In mammals, the optic nerve from each eye branches, sending nerve fibers to a lateral geniculate nucleus on each side of the brain. On the ipsilateral side, a particular part of the retina projects to a particular part of the lateral geniculate nucleus. On the contralateral side, the lateral geniculate nucleus receives input from all parts of the retina. Neurons from each lateral geniculate nucleus innervate the visual cortex on the same side.

neural circuitry of these birds' brains shows seasonal plasticity. Testosterone is believed to increase the level of brain-derived neurotropic factor (BDNF) in the song-producing vocal centers. If female birds are given BDNF, they also produce more neurons there (Rasika et al. 1999).

The cerebral cortices of young rats reared in stimulating environments are packed with more neurons, synapses, and dendrites than are found in rats reared in isolation (Turner and Greenough 1983). Even the adult brain continues to develop in response to new experiences. When adult canaries learn new songs, they generate new neurons whose axons project from one vocal region of the brain to another (Alvarez-Buylla et al. 1990). Studies on adult rats and mice indicate that environmental stimulation can increase the number of new neurons in the dentate gyrus of the hippocampus (Kempermann et al. 1997a,b; Gould et al. 1999; van Praag et al. 1999). Similarly, when adult rats learn to keep their balance on dowels, their cerebellar Purkinje neurons develop new synapses (Black et al. 1990).

In addition to inducing the formation of new neurons, learning and new experiences can also remodel old neurons into new patterns of connections. When students were taught the classic three-ball cascade juggling routine (which takes months to get right), the neurons in a specific area of the temporal lobe of the brain took on a new pattern—a pattern not seen in students who were not taught this skill

(Draganski et al. 2004). Similarly, mice reared in cages experienced changes in their neural circuitry when they were placed into more natural environments (Polley et al 2004). Thus, the pattern of neuronal connections is a product of inherited patterning and patterning produced by experiences. This interplay between innate and experiential development has been detailed most dramatically in studies on mammalian vision.

Experiential changes in mammalian visual pathways

Some of the most interesting research on mammalian neuronal patterning concerns the effects of sensory deprivation on the developing visual system in kittens and monkeys. The paths by which electric impulses pass from the retina to the brain in mammals are shown in Figure 22.21. Axons from the retinal ganglion cells form the two optic nerves, which meet at the optic chiasm. As in *Xenopus* tadpoles, some axons go to the opposite (contralateral) side of the brain, but, unlike most other vertebrates, mammalian retinal ganglion cells also send inputs into the same (ipsilateral) side of the brain (see Chapter 13). These axons end at the two lateral geniculate nuclei. Here the input from each eye is kept separate, with the uppermost and anterior layers receiving the axons from the contralateral eye, and the middle of the layers receiving input from the ipsilateral eye. The situation becomes even more complex as neurons from the lateral geniculate nuclei connect with the neurons of the visual cortex. Over 80 percent of the neural cells in the visual cortex receive input from both eyes. The result is binocular vision and depth perception.

A remarkable finding is that the retinocortical projection pattern is the same for both eyes. If a certain cortical neuron is stimulated by light flashing across a region of the left eye 5° above and 1° to the left of the fovea,* it will also be stimulated by a light flashing across a region of the *right* eye 5° above and 1° to the left of the fovea. Moreover, the response evoked in the cortical neuron when both eyes are stimulated is greater than the response when either retina is stimulated alone.

Hubel, Wiesel, and their co-workers demonstrated that the development of the nervous system depends to some degree on the experience of the individual during a critical period of development (see Hubel 1967). In other words, not all neuronal development is encoded in the genome; some is the result of learning. Experience appears to strengthen or stabilize some neuronal connections that are already present at birth and to weaken or eliminate others. These conclusions come from studies of partial sensory deprivation. Hubel and Wiesel (1962, 1963) sewed shut the right eyelids of newborn kittens and left them closed for 3 months. After this time, they unsewed the right eyelids. The cortical neurons of these kittens could not be stimulated by shining light into the right eye. Almost all the inputs into the visual cortex came from the left eye only. The behavior of the kittens revealed the inadequacy of their right eyes; when the left eyes of these kittens were covered, they became functionally blind. Because the lateral geniculate neurons appeared to be stimulated by input from

---

*The fovea is a depression in the center of the retina where only cones are present (rods and blood vessels are absent). In this instance, it serves as a convenient landmark.

both right and left eyes, the physiological defect appeared to be in the connections between the lateral geniculate nuclei and the visual cortex. Similar phenomena have been observed in rhesus monkeys, where the defect has been correlated with a lack of protein synthesis in the lateral geniculate neurons innervated by the covered eye (Kennedy et al. 1981).

Although it would be tempting to conclude that the blindness resulting from these experiments was the result of failure to form the proper visual connections, this is not the case. Rather, when a kitten or monkey is born, axons from lateral geniculate neurons receiving input from each eye overlap extensively in the visual cortex (Hubel and Wiesel 1963; Crair et al. 1998). However, when one eye is covered early in the animal's life, its connections in the visual cortex are taken over by those of the other eye (Figure 22.22). The axons compete for connections, and experience plays a role in strengthening and stabilizing the connections that are made. Thus, when *both* eyes of a kitten are sewn shut for 3 months, most cortical neurons can still be stimulated by appropriate illumination of one eye or the other. The critical time in kitten development for this validation of neuronal connections begins between 4 and 6 weeks after birth. Monocular deprivation up to the fourth week produces little or no physiological deficit, but through the sixth week, it produces all the characteristic neuronal changes. If a kitten has had normal visual experience for the first 3 months, any subsequent monocular deprivation (even for a year or more) has no effect. At that point, the synapses have been stabilized.

(A)

(B)

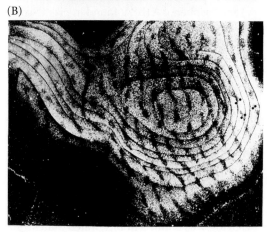

FIGURE 22.22   (A,B) Dark-field autoradiographs of monkey striate (visual) cortex 2 weeks after one eye was injected with [³H]proline in the vitreous humor. Each retinal neuron takes up the radioactive label and transfers it to the cells with which it forms synapses. (A) Normal labeling pattern. The white stripes indicate that roughly half the columns took up the label, while the other half did not—a pattern reflecting that half the cells were innervated by the labeled eye and half by the unlabeled eye. (B) Labeling pattern when the unlabeled eye was sutured shut for 18 months. Axonal projections from the normal (labeled) eye have taken over the regions that would normally have been innervated by the sutured eye. (C,D) Drawings of axons from the lateral geniculate nuclei of kittens in which one eye was occluded for 33 days. The terminal branching of axons receiving input from the occluded eye (C) was far less extensive than that of axons receiving input from the nonoccluded eye (D). (A,B from Wiesel 1982, photograph courtesy of T. Wiesel; C,D after Antonini and Stryker 1993.)

(C)   (D)

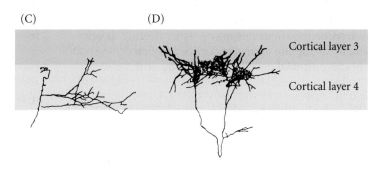

Two principles, then, can be seen in the patterning of the mammalian visual system. First, the neuronal connections involved in vision are present even before the animal sees. Second, experience plays an important role in determining whether or not certain connections persist.* Just as experience refines the original neuromuscular connections, experience plays a role in refining and improving the visual connections. It is possible, then, that adult functions such as learning and memory arise from the establishment and/or strengthening of different synapses by experience. As Purves and Lichtman (1985) remark:

> The interaction of individual animals and their world continues to shape the nervous system throughout life in ways that could never have been programmed. Modification of the nervous system by experience is thus the last and most subtle developmental strategy.

WEBSITE **22.6** **The phantom limb phenomenon.** Individuals who have a limb amputated sometimes feel pain in the absent appendage. This phenomenon appears to be caused by a reorganization of the human cerebral cortex following the amputation.

Endocrine Disruptors

If the developing organism is sensitive to environmental factors, it also becomes vulnerable to agents in the environment that can disrupt normal development. Compounds that can disrupt normal development are called teratogens. Some of these compounds were mentioned in Chapters 1 and 21. One class of teratogens is especially pertinent here: the set of compounds called endocrine disruptors, which are exogenous chemicals that interfere with the normal function of hormones.

Endocrine disruptors can alter hormonal function in many ways. As mentioned in regard to humans in Chapter 21, they can mimic the effects of natural hormones (as in DES); they can block the synthesis of a hormone, or block the binding of a hormone to its receptor (as in finasteride); or they can interfere with the transport or elimination of a hormone (as in PCBs).

Developmental toxicology and endocrine disruption are relatively new fields of research. While traditional toxicology has pursued the environmental causes of death, cancer, and genetic damage, endocrine disruptor research and the field of developmental toxicology focus on the roles environmental chemicals may play in altering development by disrupting normal endocrine function during both pre- and postnatal development (Bigsby et al. 1999).

---

*Studies have shown that differences in neurotransmitter release result in changes in synaptic adhesivity and cause the withdrawal of the axon providing the weaker stimulation (Colman et al. 1997). Studies in mice suggest that brain-derived neurotropic factor (BDNF) is crucial during the critical period (Huang et al. 1999; Katz 1999).

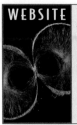

WEBSITE **22.7** **Environmental endocrine disruptors.** The Wingspread Consensus Statement of 1991 began a move by scientists to influence government policy concerning potential endocrine disruptors. Since then, there has been vehement disagreement between several scientists and representatives of the pesticide and herbicide industries.

Environmental estrogens

There are probably no greater controversies in the field of toxicology than the arguments over whether chemical pollutants are responsible for congenital malformations in wild animals, the decline of sperm counts in men, and the increase in breast cancer among women and men. One major focus of this research is whether pesticide pollutants can function as estrogens in adult and developing mammals (Figure 22.23). Estrogen is more than just a "sex hormone." In both sexes, estrogens are used to regulate muscle and skeletal growth, maintain bone density, develop the organs of the immune system, and help maintain the nervous system. Thus, any chemical that activates or suppresses estrogen receptors can potentially affect several reproductive and nonreproductive organs.

There are several sources of environmental estrogens. The first source comprises naturally occurring estrogenic compounds. Besides the estrogens that mammalian embryos acquire from the maternal blood, estrogenic compounds are also found in certain plants, such as soybeans. These plant-derived estrogens are referred to as **phytoestrogens**. The second source of estrogenic compounds is drugs that are specifically designed to work as estrogens, including the synthetic estrogens used in birth control pills as well as drugs such as diethylstilbesterol (DES; see Chapter 21), which caused birth defects in humans when administered to pregnant women (Hill 1997; Palanza et al. 2001). The third and fourth categories are pesticides and other industrial compounds. Americans use some 2 billion pounds of pesticides each year, and some pesticide residues stay in the food chain for decades. And industrial chemicals such as polychlorinated biphenyls (PCBs, discussed below) and bisphenol A (BPA, found in plastic-coated food cans and microwave pizza wrappings) are ubiquitous by-products of our technological culture.

In 1962, Rachel Carson, a fisheries biologist, published *Silent Spring*, one of the most influential books of the twentieth century. Carson warned that pesticides were destroying wildlife, that DDT in particular appeared to be destroying shorebird populations, and that pesticides were becoming a staple of the American diet. For this, she was reviled by the agricultural chemicals industry and called a fanatic, a Communist, and worse. But subsequent research bore out her claims, and when peregrine falcons and bald eagles were found to be endangered because of DDT-induced fragility of their eggshells (Cooke 1973), the use of this pesticide was banned in the United States.

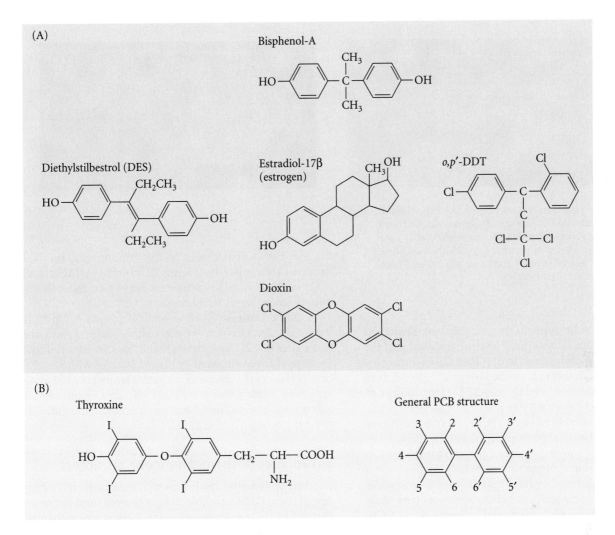

(A)

Bisphenol-A

Diethylstilbestrol (DES)

Estradiol-17β (estrogen)

o,p'-DDT

Dioxin

(B)

Thyroxine

General PCB structure

FIGURE 22.23 Structures of some endocrine disruptors. (A) Several anthropogenic estrogenic compounds present in the environment can disrupt the function of organismal estrogen. (B) The structure of some PCBs also resembles that of thyroid hormones such as thyroxine, and PCB exposure can alter thyroid hormone function.

DDT (dichloro-diphenyl-trichloroethane) and its chief metabolic by-product, DDE (which lacks one of the chlorine atoms) can act as estrogenic compounds, either by mimicking estrogen or by inhibiting androgen effectiveness (Davis et al. 1993; Kelce et al. 1995). Although banned in the United States in 1972, DDT has an environmental half-life of about 15 years. This means it can take 100 years or more for concentrations of DDT in the soil to get below active levels. DDE is a more potent estrogen than DDT and is able to inhibit androgen-responsive transcription at doses comparable to those found in contaminated soil in the United States and other countries. DDT and DDE have been linked to such environmental problems as the decrease in alligator populations in Florida, the feminization of fish in Lake Superior, the rise in breast cancers, and the worldwide decline in human sperm counts (Carlsen et al. 1992; Keid-

ing and Skakkebaek 1993; Stone 1994; Swan et al. 1997). Guillette and co-workers (1994; Matter et al. 1998) have linked a pollutant spill in Florida's Lake Apopka (a discharge including DDT, DDE, and numerous other polychlorinated biphenyls) to a 90 percent decline in the birth rate of alligators and reduced penis size in the young males.

Dioxin, a by-product of the chemical processes used to make pesticides and paper products, has been linked to reproductive anomalies in male rats. When female rats are exposed to this planar, lipophilic molecule while pregnant, their male offspring have reduced sperm counts, smaller testes, and fewer male-specific sexual behaviors. Fish

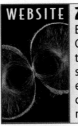

WEBSITE

22.8 Rachel Carson and the ban on DDT. Even before the age of molecular biology, Rachel Carson pointed out that DDT was having a disastrous effect on bird populations. DDT caused egg shells to thin, and birds would often crush their eggs when sitting on them. Her book, *Silent Spring*, caused the political movement leading to the banning of DDT in the United States.

(A)

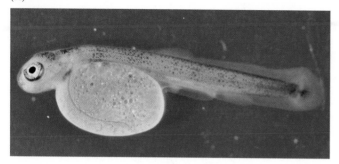

(B)

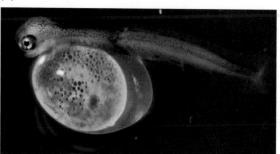

FIGURE 22.24 Lake trout 4 weeks after hatching. (A) Normal larva with its golden yellow yolk sac. (B) Dioxin-exposed larva exhibiting a blue yolk sac. The yolk sac has swelled with water and has numerous sites of hemorrhage. Such fish often have reduced growth, as well as heart and facial anomalies. (Photograph courtesy of R. E. Peterson.)

embryos seem to be particularly susceptible to dioxin and related compounds, and it has been speculated that the amount of these compounds in the Great Lakes during the 1940s was so high that none of the lake trout hatched there during that time survived (Figure 22.24; Hornung et al. 1996; Zabel and Peterson 1996; Johnson et al. 1998).

Polychlorinated biphenyls (PCBs) were widely used as refrigerants before they were banned in the 1970s, when they were shown to cause cancer in rats. However, PCBs remain circulating through the food chain (in both water and sediments), and they have been blamed for the widespread decline in the reproductive capacities of otters, seals, mink, and fish. An environmental estrogen, some PCBs resemble diethylstilbesterol in shape, and they may affect the estrogen receptor as DES does, perhaps by binding to another site on the estrogen receptor.

A similar molecule, methoxychlor, is found in many pesticides. Pickford and colleagues (1999) found that methoxychlor blocked progesterone-induced oocyte maturation in *Xenopus* at concentrations similar to those found in the environment. Such blockage would severely inhibit the fertility of the frogs, and it may be a component of the worldwide decline in amphibian populations.

Some scientists say that these claims are exaggerated. Tests on mice indicate that litter size, sperm concentration, and development were not affected by environmentally relevant concentrations of environmental estrogens. However, recent investigations by Spearow and colleagues (1999) have shown a remarkable genetic difference in sensitivity to estrogen among different strains of mice. The strain that was used for testing environmental estrogens, the CD-1 strain of laboratory mice, is at least 16 times more resistant to endocrine disruption than the most sensitive strains, such as B6. When estrogen-containing pellets were implanted beneath the skin of young male CD-1 mice, very little happened. However, when the same pellets were placed beneath the skin of B6 mice, their testes shrank and the number of sperm seen in the seminiferous tubules dropped dramatically (Figure 22.25). This widespread range of sensitivities has important consequences for determining safety limits for humans.

Another factor involved in establishing safety limits is the interaction of environmental estrogens. Silva and her colleagues (2002) have shown that if cells are exposed to a mixture of environmental estrogens, each of which is at a concentration that induces an estrogen-responsive gene only very weakly, the *mixture* can induce a response that is much more than the additive responses of the individual components (Figure 22.26).

Environmental thyroid hormone disruptors

In addition to being environmental estrogens, the structure of some PCBs resembles that of thyroid hormones (see Figure 22.23B), and exposure to them alters serum thyroid hormone levels in humans. Hydroxylated PCBs have high affinities for the thyroid hormone serum transport protein transthyretin and can block thyroxine from binding to this protein. This leads to the elevated excretion of thyroid hormones. Thyroid hormones are critical for the growth of the cochlea of the inner ear, and rats whose mothers were exposed to PCBs had poorly developed cochleas and hearing defects (Goldey and Crofton, cited in Stone 1995; Cheek et al. 1999).

Chains of causation

Whether in law or in science, establishing a **chain of causation** is a demanding and necessary task. In developmental toxicology, numerous endpoints must be checked, and many different levels of causation must be established (Crain and Guillette 1998; McNabb et al. 1999). For example, researchers might ask whether the pollutant spill in Lake Apopka was responsible for the decline in the alligator population there. To establish this, they first ask how the specific chemicals in the spill could contribute to physiological anomalies in alligators, and what would be the consequences of these anomalies. Table 22.2 shows the postulated chain of causation.

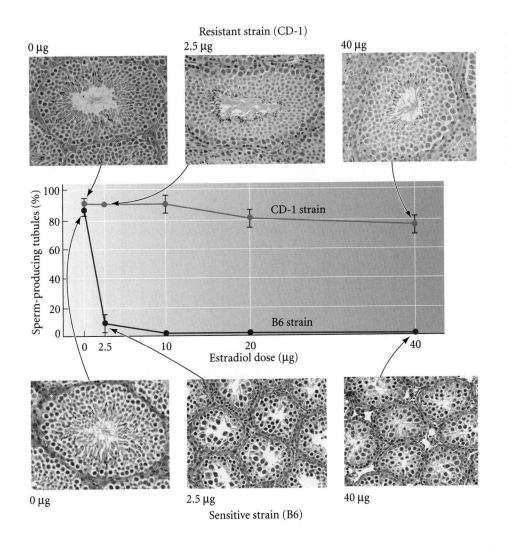

FIGURE 22.25 Effects of estrogen implants on different strains of mice. The graph shows the percentage of seminiferous tubules containing elongated spermatozoa. (The mean ± standard error is for an average of six individuals.) The photographs show cross sections of the testicles and are all at the same magnification. 40 μg of estradiol did not affect spermatogenesis in the CD1 strain, but as little as 2.5 μg almost completely abolished spermatogenesis in the B6 strain. (After Spearow et al. 1999; photographs courtesy of J. L. Spearow.)

After observing that the number of alligators in Lake Apopka had declined, *population level* observations revealed that there was a decrease in the number of alligators being born (the birth rate). At the *organism level*, researchers found unusually high levels of estrogens in the female alligators and unusually low levels of testosterone in the males. On the *tissue and organ level*, the researchers observed elevated production of estrogens from the juvenile testes, malformations of the testes and penis, and changes in enzyme activity in the female gonads. On the *cellular level*, ovarian abnormalities were correlated with unusually elevated estrogen levels. These cellular changes could be explained at the *molecular level* by the fact that many of the chemicals in the pollutant spill bind to alligator estrogen and progesterone receptors, and that these chemicals are able to circumvent the cell's usual defenses against the overproduction of steroid hormones (Crain et al. 1998). Thus, the pollutant chemicals can be linked to a reproductive anomaly that could explain the decreased birth rate among alligators in the lake.

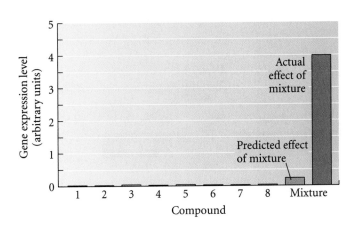

FIGURE 22.26 Ability of a mixture of estrogenic compounds at low concentrations to activate an estrogen-responsive gene. Plasmids containing an estrogen response element attached to a *lacZ* gene were added to cells, and each of eight environmental estrogens was added at concentrations that only weakly induced the expression of the *lacZ* gene. However, when a mixture of these eight compounds was added, the response was far greater than the additive values of the separate responses. (After Silva et al. 2002.)

TABLE 22.2 Chain of causation linking contaminant spill in Lake Apopka to endocrine disruption in juvenile alligators

Level	Evidence
Population	The juvenile alligator population in Lake Apopka has decreased.
Organism	Juvenile Apopka females have elevated circulating levels of estradiol-17β.
	Juvenile Apopka males have depressed circulating concentrations of testosterone.
Tissue/organ	Juvenile Apopka females have altered gonad aromatase activity.
	Juvenile Apopka males have poorly organized seminiferous tubules.
	Juvenile Apopka males have reduced penis size.
	Testes from juvenile Apopka males have elevated estradiol (estrogen) production.
Cellular	Juvenile Apopka females have polyovular follicles that are characteristic of estrogen excess.
Molecular	Many contaminants bind the alligator estrogen receptors and progesterone receptors.
	Many of these contaminants do not bind to the alligator cytosol proteins that blockade excess hormones.

Source: After Crain and Guillette 1998.

However, as Taylor (2005) has shown in his analysis of ecological problem-solving, a complete chain of causation must also take into account abiological agents such as geology, meteorology, economics, and social norms and practices. In Chapter 17, we discussed the ability of the weed-killer atrazine to induce ovaries in male frogs. Hayes (2005) has detailed a web of causation that relates the atrazine-induced decline of frog populations to the interactions of political, geological, and biological agents (Figure 22.27).

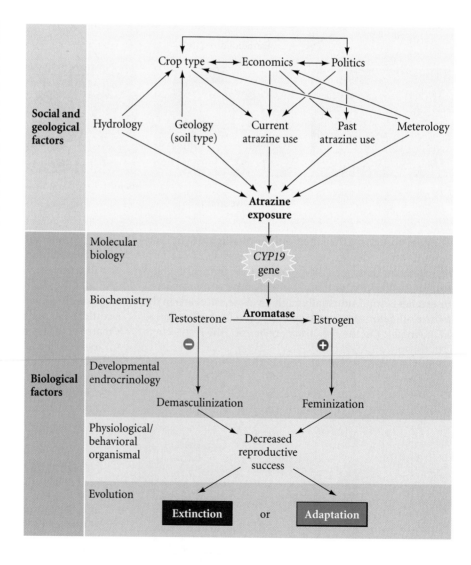

FIGURE 22.27 Possible chain of causation leading to the feminization of male frogs and the decline of frog populations in regions where atrazine has been used to control weed populations. Both biological and suprabiological agents (including human-driven social forces) are shown. The *Cyp19* gene encodes aromatase; it has also been shown that in humans (see Sanderson et al. 2000), transcription of the *CYP19* gene is induced by herbicides and other hormone disrupters. (After Hayes 2005.)

Deformed Frogs

Throughout the United States and southern Canada, dramatic numbers of deformed frogs and salamanders have been found in what seem to be pristine woodland ponds. In some local ponds, it is estimated that 60 percent of certain amphibian species populations have visible malformations (Ouellet et al. 1997; NARCAM 2002). These deformities include extra or missing limbs (Figure 22.28), missing or misplaced eyes, deformed jaws, and malformed hearts and guts. In recent years, three of the many hypotheses that have been put forward to explain the phenomenon appear to be gaining acceptance.

The first hypothesis proposes a combination of causes. The proximate cause of limb deformities may be the infection of larval limb buds by trematode parasites (Stopper et al. 2002; see Figure 16.4). Under most conditions, trematode larvae can be destroyed by the tadpole's immune system. In some ponds, however, it appears that tadpoles have acquired an immune deficiency syndrome. Frogs in habitats contaminated by certain pesticides appear less able to resist parasite infection. Kiesecker (2002) found that neither pesticides alone nor trematodes alone were sufficient to cause limb deformities in wild populations of frogs, but that the combination of the two factors resulted in significant proportions of the frog population showing limb deformities. In laboratory studies, Christin and colleagues (2004) showed that mixtures of pesticides that mimic the concentrations found in certain Canadian lakes can cause immune deficiencies in *Rana pipiens* (leopard frog) and *Xenopus* tadpoles. At least in some species, trematode infestation has to be coupled with pollutants (such as pesticides) in order for the limb anomalies to be observed. The pesticides may be suppressing the tadpole's immune resistance to the trematode larvae.

But some lakes with high proportions of malformed frogs do not appear to be infested with trematodes. A second hypothesis proposes that ultraviolet radiation can cause these malformations. Ankley and col-

Figure 22.28   A leopard frog (*Rana pipiens*) from a wild population, suffering from polymelia (an extra set of hindlimbs). (Photograph courtesy U.S. Geological Survey.)

leagues (2002) showed that, whereas UV irradiation can cause the death of amphibian embryos (see Chapter 3), those individuals that survive have high proportions of limb deformities. The spectrum of limb abnormalities seen in frogs that develop from UV-exposed embryos and tadpoles in the laboratory is somewhat different from that seen in natural populations (Meteyer et al. 2000), but irradiation may still account for some limb abnormalities.

There are other (nonlimb) malformations that do not seem to be explained by this route, however, and a third possibility is that endocrine-disrupting pesticides and herbicides are causing physiological disruptions. Certain herbicides, such as atrazine, are thought to be responsible for gonad malformations in many frog species (Hayes et al. 2002, 2003, 2004; Tavera-Mendoza et al. 2002; see Figure 17.22). Moreover, atrazine encountered *early* in development appears to cause precocious metamorphosis, leading to abnormalities of limb, gut, and head

development (Lenkowski and McLaughlin 2005). The spectrum of abnormalities seen in natural populations of deformed frogs resembles some of the malformations caused by exposing tadpoles to retinoic acid (Crawford and Vincenti 1998; Gardiner and Hoppe 1999). Ndayibagira and colleagues (2002) have shown that retinoic acid can in fact induce all the limb malformations that have recently been seen in wild populations of some frog species. Moreover, Grün and colleagues (2002), in the same laboratory, have purified an endocrine-disrupting retinoid from the water of a lake that had a high incidence of malformed amphibians.

If retinoids are in the water, how did they get there? One possibility for the activation of retinoic acid pathways involves the insecticidal compound methoprene. Methoprene is a juvenile hormone mimic that inhibits mosquito pupae from metamorphosing

(Continued on next page)

WEBSITE   **22.9 Deformed frogs and salamanders.** The original observations of malformed frogs were made by public school science classes in Minnesota. Since then, considerable efforts have been made to find the causes for both the recent decline of amphibian populations and the developmental anomalies being discovered in these animals. There are several websites for the recording and analysis of amphibian malformations.

into adults. Because vertebrates do not have JH, it was assumed that this pesticide would not harm fish, amphibians, or humans. Indeed, methoprene, itself, does not have teratogenic properties. However, on exposure to sunlight, methoprene breaks down into products that have significant teratogenic activity in frogs (Figure 22.29). These compounds have a structure similar to that of retinoic acid and will bind to the retinoid receptor (Harmon et al. 1995; La Claire et al. 1998). When *Xenopus* eggs are incubated in water containing these compounds, the tadpoles are often malformed, and show a spectrum of deformities similar to those seen in the wild (La Claire et al. 1998).

(A)

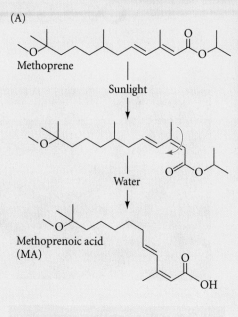

Methoprene

Sunlight

Water

Methoprenoic acid (MA)

Retinoic acid

(B)

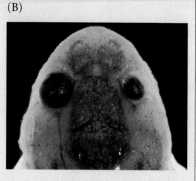

Figure 22.29 Retinoid-induced malformations. (A) One of several pathways by which methoprene can decay into teratogenic compounds such as methoprenic acid. The isomer of retinoic acid shows its structural similarities to methoprenic acid. (B) *Xenopus* tadpole with eye deformities caused by incubating newly fertilized eggs in water containing methoprenic acid. This type of deformity is often associated with RA exposure. (A after La Claire et al. 1998; B photograph courtesy of J. Bantle.)

Developmental Biology Meets the Real World

In a review of insect development, Fred Nijhout (1999) concluded:

> A single genotype can produce many phenotypes, depending on many contingencies encountered during development. That is, the phenotype is an outcome of a complex series of developmental processes that are influenced by environmental factors as well as by genes.

Development usually occurs in a rich environmental milieu, and most animals are sensitive to environmental cues. As we have seen in this chapter, the environment may induce remarkable structural and chemical adaptations according to the season, may induce specific morphological changes that allow an individual to escape predation, and can even determine sexual phenotype. The environment we experience can also alter the structure of our neurons and the specificity of our immunocompetent cells. Unfortunately, the environment can also be the source of chemicals that may disrupt normal developmental processes.

This concept that the environment is critical in phenotype production has many implications. First, the developmental plasticity of the nervous system assures that each person is an individual. Our brain adds experience to endowment. Fears that cloning could produce "thousands of Hitlers," for example, are unfounded. Not only have no genes been identified for bigotry, demagoguery, oratory skill, or political canniness, but one would have to reconstruct Hitler's personal, social, and political milieus to even come close to replicating the dictator's personality. Wolpe (1997) has pointed out that believing a genetically identical clone of Hitler would perforce become a bigoted dictator is buying into the same philosophy of genetic essentialism that Hitler himself espoused. Stephen Jay Gould (1997) noted that even Eng and Chang Bunker, the well-publicized, conjoined "Siamese twins" who most likely shared 100 percent of their genes and certainly shared the same environment, nevertheless became very different people. One was cheerful and abstained from drinking liquor. The other was a morose alcoholic (which was a problem, since they shared the same liver). The plasticity of the nervous system enables us to be individuals and "allows us to escape the tyranny of our genes" (Childs 1999).

As mentioned in Chapter 4, even sheep cloned from the same embryo (and therefore having the same genome) can be very different from one another. We can therefore give a definite answer to the question posed by Wolpert (p. 51) in 1994:

> Will the egg be computable? That is, given a total description of the fertilized egg—the total DNA sequence and the location of all proteins and RNA—could one predict how the embryo will develop?

The answer almost has to be "no." One cannot reduce phenotype completely to the inherited genes. Experience must be added to endowment.

The idea that one's phenotype is controlled in part by other species (as in developmental symbioses and predator-induced polyphenisms) certainly links us to our environment. Our "self" is partially constructed by "others," and co-development may be as important a concept as coevolution. We must learn to look for developmental signals coming not only from within but also from outside the organism. Our methods might have to expand. With the realization that animals in the wild develop differently than the same animals would in our laboratories, we may have to go outdoors to study developmental phenomena—certainly not the standard operating procedure of past research.

Indeed, recent events may necessitate the enlistment of developmental biology in the service of conservation biology, a field where it has not been very active. The pesticide studies of Hayes, Relyea and Mills, La Claire, and others indicate that the chemicals we use in agriculture may be depleting natural populations of many species by killing embryos or fetuses, or by rendering adults unable to mate. The studies of Morreale and colleagues (1982) show that the practices of conservation biology must incorporate an understanding of the way development works in the wild. Even the way we test compounds for their dangerous effects would have to change, since predation-induced polyphenism can alter responses significantly.

The study of environmental regulation of development opens up a whole new world to developmental biologists. Ecologists know numerous stories of developmental plasticity—animals obtaining different jaw morphologies with different diets (Corruccini and Beecher 1982; Stearns 1989; Hegrenes 2001), animals resorbing organs under certain conditions (Piersma and Gill 1998), and frogs that can change their development within minutes of a predator's presence (Warkentin 1995, 2000). The proximate causes of these incredible developmental changes have not been studied. The National Science Foundation has recently recommended an initiative, called "Integrative Developmental Biology," explicitly to study interactions between developing organisms and their environments.

By returning to the "real world," developmental biology is becoming the science that integrates ecology, evolution, genetics, cell biology, and physiology. In 1973, evolutionary biologist Leigh Van Valen claimed that evolution can be defined as "the control of development by ecology." We have reached a point where we can now begin to study the mechanisms by which this happens.

Snapshot Summary: Environmental Regulation of Development

1. Developmental plasticity makes it possible for environmental circumstances to elicit different phenotypes from the same genotype.

2. Development is sometimes cued to normal circumstances that the organism can expect to find in its environment. The larvae of many marine invertebrates do not begin metamorphosis until they find a suitable substrate. In other instances, symbiotic relationships between two or more species are necessary for the complete development of one or more of the species.

3. Some species exhibit polyphenisms, in which distinctly different phenotypes are evoked by different environmental cues. Many species have a broad reaction norm, wherein the genotype can respond in a graded way to environmental conditions.

4. Seasonal cues such as photoperiod, temperature, or type of food can alter development in ways that make the organism more fit under the conditions it encounters. Changes in temperature also are responsible for determining sex in several organisms, including many reptiles and fish.

5. Predator-induced polyphenisms have evolved such that prey species can respond morphologically to the presence of a specific predator. In some instances, this induced adaptation can be transmitted to the progeny of the prey.

6. There are several routes through which gene expression can be influenced by the environment: (a) environmental factors can methylate genes differentially; (b) environmental factors can induce gene expression in surrounding cells, and (c) environmental agents can be monitored by the nervous system, which then produces hormones that affect gene expression.

7. Numerous compounds present in the environment may act as hormone mimics or antagonists. These compounds may disrupt normal development by interfering with the endocrine system.

8. Genetic differences can predispose individuals to being affected differently by teratogens.

For Further Reading

Complete bibliographical citations for all literature cited in this chapter can be found on the Vade Mecum CD that accompanies the book and at the free access website www.devbio.com

Agrawal, A. A., C. Laforsch and R. Tollrian. 1999. Transgenerational induction of defenses in animals and plants. *Nature* 401: 60–63.

Brakefield, P. M. and N. Reitsma. 1991. Phenotypic plasticity, seasonal climate, and the population biology of *Bicyclus* butterflies (Satyridae) in Malawi. *Ecol. Entomol.* 16: 291–303.

Hayes, T. B. 2005. Welcome to the revolution. Integrative biology and assessing the impact of endocrine disruptors on environmental and public health. *Integr. Compar. Biol.* 45: 321–329.

Hooper, L. V., M. H. Wong, A. Thelin, L. Hansson, P. G. Falk and J. I. Gordon. 2001. Molecular analysis of commensal host-microbial relationships in the intestine. *Science* 291: 881–884.

McFall-Ngai, M. J. 2002. Unseen forces: The influence of bacteria on animal development. *Dev. Biol.* 242: 1–14.

Moczek, A. P. 2005. The evolution of development of novel traits, or how beetles got their horns. *BioScience* 55: 937–951.

Relyea, R. A. and N. Mills. 2001. Predator-induced stress makes the pesticide carbaryl more deadly to grey treefrog tadpoles (*Hyla versicolor*). *Proc. Natl. Acad. Sci. USA* 2491–2496.

Spearow, J. L., P. Doemeny, R. Sera, R. Leffler and M. Barkley. 1999. Genetic variation in susceptibility to endocrine disruption by estrogen in mice. *Science* 285: 1259–1261.

Stappenbeck, T. S., L. V. Hooper and J. I. Gordon. 2002. Developmental regulation of intestinal angiogenesis by indigenous microbes via Paneth cells. *Proc. Natl. Acad. Sci. USA* 99: 15451–15455.

Waterland, R. A. and R. L. Jirtle. 2003. Transposable elements: Targets for early nutritional effects of epigenetic gene regulation. *Mol. Cell. Biol.* 23: 5293–5300.

Developmental Mechanisms of Evolutionary Change

23

WHEN WILHELM ROUX ANNOUNCED the creation of experimental embryology in 1894, he broke many of the ties that linked embryology to evolutionary biology. However, he promised that embryology would someday return to evolutionary biology, bringing with it new knowledge of how animals were generated and how evolutionary changes might occur. He stated that "an ontogenetic and a phylogenetic developmental mechanics are to be perfected." Roux thought that research into the developmental mechanics of individual embryos (the ontogenetic branch) would proceed faster than the phylogenetic (evolutionary) branch, but he predicted that "in consequence of the intimate causal connections between the two, many of the conclusions drawn from the investigation of individual development [would] throw light on the phylogenetic processes." A little more than a century later, we are at the point of fulfilling Roux's prophecy. Developmental biology is returning to evolutionary biology, forging a new discipline, **evolutionary developmental biology** (sometimes called "evo-devo") and producing a new model of evolution that integrates developmental genetics and population genetics to explain and define the diversity of life on Earth.

The fundamental principle of this new evolutionary synthesis is that evolution is caused by heritable changes in the development of organisms. This view can be traced back to Darwin, and it is compatible with and complementary to the view of evolution based on population genetics—i.e., that evolution is caused by changes in gene frequency between generations. The merger is creating a more complete evolutionary biology that illuminates the origin of both species and higher taxa (Raff 1996; Hall 1999; Wilkins 2002; Carroll et al. 2005; Kirschner and Gerhart 2005).

"Unity of Type" and "Conditions of Existence"

Charles Darwin's synthesis

In the nineteenth century, debates over the origin of species pitted two ways of viewing nature against each other. One view, championed by Georges Cuvier and Charles Bell, focused on the *differences* among species that allowed each species to adapt to its environment. Thus, the hand of the human, the flipper of the seal, and the wings of birds and bats were seen as marvelous contrivances, each fashioned by the Creator, to allow these animals to adapt to their "conditions of existence." The other view, championed by Étienne Geoffroy Saint-Hilaire and Richard Owen, was that "unity of type" (the *similarities* among organisms, which Owen called "homologies") was critical. The human hand, the seal's flipper, and the wings of bats and birds were all modifications of the same basic plan

> " How does newness come into the world? How is it born? Of what fusions, translations, conjoinings is it made? How does it survive, extreme and dangerous as it is? What compromises, what deals, what betrayals of its secret nature must it make to stave off the wrecking crew, the exterminating angel, the guillotine? "
> *SALMAN RUSHDIE* (1988)

> " Biology points out the individuality of every being, and at the same time reminds us of the brotherhood of all. "
> *JEAN ROSTAND* (1962)

(see Figure 1.13). In discovering that plan, one could find the form upon which the Creator designed these animals. The adaptations were secondary.

Darwin acknowledged his debt to these earlier debates when he wrote in 1859, "It is generally acknowledged that all organic beings have been formed on two great laws—Unity of Type, and Conditions of Existence." Darwin went on to explain that his theory would explain unity of type by descent from a common ancestor. The changes creating the marvelous adaptations to the conditions of existence would be explained by natural selection. Darwin called this concept **descent with modification**. Darwin noted that the homologies between the embryonic and larval structures of different phyla provided excellent evidence for descent with modification. He also argued that adaptations that depart from the "type" and allow an organism to survive in the "conditions" of its particular environment develop late in the embryo. Thus, Darwin recognized two ways of looking at descent with modification. One could emphasize *common descent* in the embryonic homologies between two or more groups of animals, or one could emphasize the *modifications* by showing how development was altered to produce diverse adaptive structures (Gilbert 2003). Or, as Darwin's friend Thomas Huxley aptly remarked, "Evolution is not a speculation but a fact; and it takes place by epigenesis" (Huxley 1893, p. 202).

WEBSITE **23.1 Lillie and Wilson.** In the late 1800s, two eminent embryologists came forth with proposals relating embryology to molluscan evolution. Wilson stressed embryological homologies as showing common descent; Lillie stressed embryological adaptations as showing natural selection. Both approaches are still operating today.

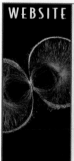

WEBSITE **23.2 Haeckel's biogenetic law.** In the early 1900s, a fusion of evolution and embryology was wrongly interpreted to support a "progressive" model of evolution. The interpretation of Ernst Haeckel was that every organism evolved by the terminal addition of a new stage to the end of the last "highest" organism. Thus, he saw the entire animal kingdom as representing truncated steps of human development.

"Life's splendid drama"

Until the present decade, "many invertebrate biologists saw the reconstruction of relationships among the phyla as an insoluble dilemma. ... Indeed, as late as 1990, a comprehensive summary concluded that the relationships between most of the higher animal groups were entirely unresolved" (Erwin et al. 1997). However, in the 1990s, a broad consensus on the general form of a phylogenetic tree of life began to emerge among paleontologists, molecular biologists, and developmental geneticists (see Winnepenninckx et al. 1998; Adoutte et al. 1999; Philippe et al. 2005). This consensus came about from (1) improved methods of using DNA for phylogenetic analysis, taking into account its variation within groups of animals; (2) new data on conserved regulatory gene sequences such as the Hox genes, which are usually stable within phyla but can diverge between phyla; (3) morphological evidence for the related nature of some structures that had once been thought to be distinct; and (4) computer programs that can sort out enormous amounts of data, without privileging any particular set of relationships over others. The results, one representation of which is shown in Figure 23.1, can be summarized as follows:

- The animal kingdom can be divided into Porifera (sponges); Cnidaria and Ctenophora (jellyfish and comb jellies); and the Bilateria (all other animals). The Porifera lack any coherent epithelium or symmetry. The Cnidaria and Ctenophora are diploblastic (with two epithelial layers, and with little mesoderm or none at all) and have radial symmetry. The Bilateria are triploblastic (with true endoderm, mesoderm, and ectoderm) and have bilateral symmetry.

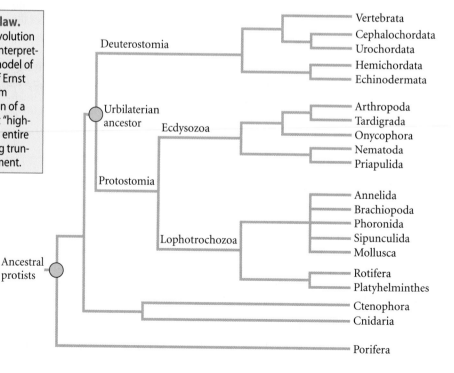

FIGURE 23.1 Relationships among phyla. One possible phylogeny of the animals, emphasizing deuterostomes, ecdysozoa, and lophotrochozoa. The relative positions of the urochordates and cephalochordates is in dispute due to the data concerning neural crest-like cells in tunicates. The positions of several invertebrate phyla are also controversial. (After Philippe et al. 2005.)

- The Bilateria can be divided into two groups, the **deuterostomes** and the **protostomes**. The protostomes form their mouth from the blastopore opening; the deuterostomes form the mouth secondarily, and the blastopore marks the site of the anus. The majority of invertebrate species are protostomes. Deuterostomes include the chordates and the echinoderms. (Although adult echinoderms display radial symmetry, echinoderm larvae are originally bilateral.)
- The protostomes can be divided into two groups, the **Ecdysozoa** (animals whose bodies are covered by an exoskeleton, which therefore molt as they grow) and the **Lophotrochozoa** (animals that have most or all of their soft tissues in contact with the environment and that generally use cilia in feeding or locomotion).

WEBSITE **23.3** **The emergence of embryos.** How did individual cells come to sacrifice their individual potentials and generate embryos? How did gastrulation evolve? How did the deuterostome gut emerge from the protostome gut? The answers may involve predation and the inability to divide and be ciliated at the same time.

WEBSITE **23.4** **How taxonomic groups are classified.** The advent of cladistics has put some order into the various ways of classifying animals. This does not mean, however, that there is unanimous agreement on the results.

Preconditions for Macroevolution through Developmental Change

Evolutionary biologists since the 1930s have usually divided evolution into two categories: microevolutionary changes within a species, and macroevolutionary changes between taxa. Evolutionary developmental biologists have shown that macroevolutionary changes are made possible by two sets of conditions: **modularity** and **molecular parsimony**.

Modularity: Divergence through dissociation

How can the development of an embryo change when development is so finely tuned and complex? How can such change occur without destroying the entire organism? It was once thought that the only way to promote evolution was to add a step to the end of embryonic development, but we now know that even early stages can be altered to produce evolutionary novelties. Such changes can occur because development occurs through a series of discrete and interacting **modules** (Riedl 1978; Bonner 1988; Bolker 2000). Examples of developmental modules include morphogenetic fields (for example, those described for the limb or eye), signal transduction pathways (described throughout this book), imaginal discs, cell lineages (such

as the inner cell mass or trophoblast), insect parasegments, and vertebrate organ rudiments (Gilbert et al. 1996; Raff 1996; Wagner 1996; Schlosser and Wagner 2004). The ability of one module to develop differently from other modules (a phenomenon sometimes called **dissociation**) was well known to early experimental embryologists. For instance, when Victor Twitty (Twitty and Schwind 1931; Twitty and Elliott 1934) grafted the limb bud from the early larva of a large salamander onto the embryonic trunk of a small salamander larva, the limb grew to its normal large dimensions, indicating that the limb field module was independent from the global growth patterning of the embryo. The same independence was seen for the eye field (Figure 23.2). Modular units allow certain parts of the body to change without interfering with the functions of other parts.

One of the most important aspects of evolutionary embryology is that not only are the *anatomical* units modular (such that one part of the body can develop differently than the others), but the DNA regions that form the *enhancers* of genes are modular. This was shown in Figures 5.7 and 5.12. The modularity of enhancer elements allows particular sets of genes to be activated together and permits a particular gene to become expressed in several discrete places. Thus, if a particular gene loses or gains a modular enhancer element, the organism containing that particular allele will express that gene in different places or at different times than those organisms retaining the original allele. This mutability can result in the development of different anatomical and physiological morphologies (Sucena and Stern 2000; Shapiro et al. 2004).

The importance of enhancer modularity in evolution has been dramatically demonstrated by David Kingsley's analysis of evolution in threespine stickleback fishes (*Gasterosteus aculeatus*). Freshwater sticklebacks evolved from marine sticklebacks about 12,000 years ago, when marine populations colonized the newly formed freshwater lakes at the end of the last ice age. Marine sticklebacks (Figure 23.3A) have a pelvic spine that serves as protection against predation, lacerating the mouths of those predatory fish who try to eat the stickleback. Freshwater sticklebacks, however, do not have pelvic spines (Figure 23.3B). This may be because freshwater species lack the piscine predators that the marine fish face, but must deal instead with invertebrate predators that can easily capture them by grasping onto such spines. Thus, a pelvis without lacerating spines evolved in the freshwater populations of this species.

To determine which genes might be involved in this difference, researchers mated individuals from marine (with spines) and freshwater (no spines) populations. The resulting offspring were bred to each other and produced numerous progeny, some of which had pelvic spines and some of which didn't. Using molecular markers to identify specific regions of the parental chromosomes, Shapiro and coworkers (2004) found that the major gene for pelvic spine development mapped to the distal end of chromo-

(A)

(B)

(C)

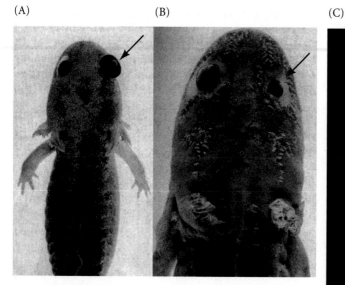

FIGURE 23.2 Modularity of development: morphogenetic fields. Transplantation of developmental modules between early larval stages *of Ambystoma tigrinum* (a large salamander) and *A. punctatum* (a smaller salamander) shows dissociation of the modules (here, the eye and limb fields) from the growth control of other tissues. (A) When the *A. tigrinum* optic field is grafted into an *A. punctatum* larva, the grafted field (right eye) develops at its intrinsic growth rate and becomes large. (B) The *A. punctatum* eye field grafted into an *A. tigrinum* larva keeps its intrinsic growth rate and is small, despite the larger size of the salamander. (C) When an *A. tigrinum* limb field is transplanted into an *A. punctatum* larval host, the leg grows to its normal large size (left). An *A. punctatum* limb field placed into an *A. tigrinum* host produces a small limb (right). (From Twitty and Ellliot 1934; Twitty and Schwind 1931.)

FIGURE 23.3 Modularity of development: enhancers. Loss of *Pitx1* gene expression in the pelvic region of freshwater threespine sticklebacks. (A) Bony plates and a pelvic spine characterize marine threespine sticklebacks. In freshwater sticklebacks (B), the pelvic spine is absent, as is much of the bony armor. In magnified ventral views of embryos (inset photos), in situ hybridization reveals *Pitx1* expression (purple) in the pelvic area (as well as in sensory neurons, thymic cells, and nasal regions) of the marine species. The staining in the pelvic region is absent in the freshwater sticklebacks, although it is still seen in the other areas (the arrow points to one of several expression points). (C) Model for the evolution of pelvic spine loss. Four enhancer regions are postulated to reside near the *Pitx1* coding region. These enhancers direct the expression of this gene in the thymus, pelvic spine, sensory neurons, and nose, respectively. In freshwater populations of threespine sticklebacks, the pelvic spine enhancer module has been mutated and the *Pitx1* gene fails to function there. (Photographs courtesy of D. M. Kingsley.)

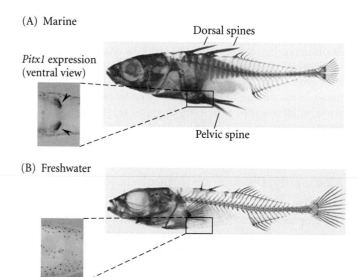

(A) Marine

Dorsal spines

Pitx1 expression
(ventral view)

Pelvic spine

(B) Freshwater

(C)

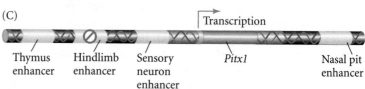

Transcription

Thymus enhancer

Hindlimb enhancer

Sensory neuron enhancer

Pitx1

Nasal pit enhancer

some 7. That is to say, nearly all the fish with pelvic spines inherited this "hindlimb-encoding" chromosomal region from the marine parent, while fish lacking pelvic spines obtained this region from the freshwater parent. They then tested numerous candidate genes (genes known to be present in the hindlimb structures of mice, for instance), and found that the gene encoding transcription factor *Pitx1* was located on this region of chromosome 7.

When they compared the amino acid sequences of the Pitx1 protein between marine and freshwater sticklebacks, there were no differences. However, there was a critically important difference when they compared the *expression patterns* of *Pitx1* between these species. In both species, *Pitx1* was seen to be expressed in the precursors of the thymus, nose, and sensory neurons. In the marine species, *Pitx1* was also expressed in the pelvic region. But in the freshwater populations, the pelvic expression of *Pitx1* was absent or severely reduced (Figure 23.3C). Since the coding region of *Pitx1* was not mutated (and since the gene involved in the pelvic spine differences maps to the site of the *Pitx1* gene, and the difference between the freshwater and marine species involves the expression of this gene at a particular site), it is reasonable to conclude that the *enhancer region* allowing expression of Pitx1 in the pelvic area no longer functions in the freshwater species. Thus, the modularity of the enhancer has enabled this particular expression domain to be lost, and with it the pelvic spine. No other function of Pitx1 had to be disturbed. Interestingly, the loss of the pelvic spines in other stickleback species appears to have been the result of independent losses of this Pitx1 expression domain. This finding suggests that if the loss of Pitx1 expression in the pelvis occurs, this trait can be readily selected (Colosimo et al. 2004). Here we see that combining population genetics approaches and developmental genetics approaches, one can determine the mechanisms by which evolution can occur.

WEBSITE **23.5** **Modularity as a principle of evolution.** Complex structures are created by the assortment of preexisting modules. It is foolish to consider a protein as a collection of atoms. It is an ordered assembly of amino acids that have already formed from atoms. Modularity allows evolution to occur by forming components that can be individually modified.

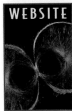

WEBSITE **23.6** **Correlated progression.** In many cases, modules must co-evolve. The upper and lower jaws, for instance, have to fit together properly. If one changes, so must the other. If the sperm-binding proteins on the egg change, then so must the egg-binding proteins on the sperm. This site looks at correlated changes during evolution.

WEBSITE **23.7** **Correlated progression in domestic animals.** Domestication appears to be selection for neotenic conditions. In selecting for behavioral plasticity, changes in skull shape and pigment patterns are also produced. This phenomenon can also be seen in current attempts to domesticate wild wolves and foxes.

Molecular parsimony: Gene duplication and diversion

The second precondition for macroevolution through developmental change is **molecular parsimony**, sometimes called the "small toolkit." In other words, although development differs enormously from lineage to lineage, development within all lineages uses the same types of molecules. The transcription factors, paracrine factors, adhesion molecules, and signal transduction cascades are remarkably similar from one phylum to another. As mentioned earlier, certain transcription factors such as Pax6 and Hox genes are found in all animal phyla, including cnidarians, insects, and primates. In fact, some "toolkit genes" appear to play the same roles in all animal lineages. Thus, Pax6 appears to be involved in specifying light-sensing organs, irrespective of whether the eye is that of a mollusc, an insect, or a primate (see Figure 23.4; Gehring 2005). Hox genes appear to specify the anterior-posterior body axis, even thought the way these genes are activated differs enormously between nematodes, flies, and birds. The Wnt and TGF-β paracrine factors and their signaling pathways can be found in the primitive cnidarians as well as in primitive bilaterians such as flatworms (Finnerty et al. 2004; Carroll et al. 2005). There is descent with modification for the signal transduction pathways and transcription factors, just as there is descent with modification of whole organisms.

One theme that echoes through this book is that Hox gene expression provides the basis for anterior-posterior axis specification throughout the animals. This means that the enormous variation of morphological form among animals is underlain by a common set of instructions. Indeed, Hox genes provide one of the most remarkable pieces of evidence for deep evolutionary homologies among all the animals of the world.

The similarity of all the Hox genes is best explained by descent from a common ancestor. First, the different Hox genes in the homeotic gene cluster each originated from duplications of an ancestral gene. This would mean that in *Drosophila*, the *Deformed*, *Ultrabithorax*, and *Antennnapedia* genes all emerged as duplications of an original gene. The sequence patterns of these three genes (especially in the homeodomain region) are extremely well conserved. Such tandem gene duplications are thought to be the result of errors in DNA replication, and they are very common. Many genomes have at least two genes (usually located close to each other) that resulted from a replication error.

Once replicated, the gene copies can diverge by random mutations in their coding sequences and enhancers, developing different expression patterns and new functions (Damen 2002; Lynch and Conery 2000; Locascio et al. 2002).

This **duplication-and-divergence** scenario is seen in the Hox genes, the globin genes, the collagen genes, the *distal-less* genes, and in many paracrine factor families (e.g., the Wnt genes). Each member of such a gene family is homologous to the others (that is, their sequence similarities are due to descent from a common ancestor and are not the result of convergence for a particular function), and are called **paralogues**. Thus, the *Antennapedia* gene is a paralogue of *Ultrabithorax*. Susumu Ohno (1970), one of the founders of the gene family concept, likened gene duplication to a method used by a sneaky criminal to circumvent surveillance. While the "police force" of natural selection makes certain that there is a "good" gene properly performing its function, that gene's duplicate, unencumbered by the constraints of selection, can mutate and reveal new functions.

SIDELIGHTS & SPECULATIONS

The Search for the Urbilaterian Ancestor

It is doubtful that we will find a fossilized representative of the ancestral lineage that gave rise to both the deuterostomes and the protostomes. This hypothetical animal is sometimes called the **Urbilaterian ancestor** or the **PDA** (protostome-deuterostome ancestor). Since it is doubtful that such an animal would have had either a bony endoskeleton (a deuterostome chordate trait) or a hard exoskeleton (characteristic of protostomate ecdysozoans), it would not have fossilized well. However, we can undertake what Sean Carroll has called "paleontology without fossils." The logic of this approach is to find homologous genes that are performing the same functions in both a deuterostome (usually a chick or a mouse) and a protostome (generally an arthropod such as *Drosophila*). Many such genes have been found (Table 23.1), and their similarities of structure and function in protostomes and deuterostomes make it likely that these genes emerged in an animal that is ancestral to both groups.

The Pax6 protein, for example, plays a role in forming eyes in both vertebrates and invertebrates (see Chapters 4 and 5). Ectopic expression of *Pax6* results in extra eyes in both *Drosophila* and *Xenopus*—representatives of the protostomes and deuterostomes, respectively (Chow et al. 1999). Moreover, the ectopic expression of a deuterostome (mouse) *Pax6* gene in a fly larva induces ectopic fly eyes (Figure 23.4), and the ectopic expression of the *Drosophila Pax6* gene in *Xenopus* ectoderm induces eye development in the frog tadpole (Halder et al. 1995; Onuma et al. 2002). Therefore, it is a safe assumption that the same *Pax6* gene is involved in eye production in both deuterostomes and protostomes. Moreover, at least three other genes—*sine oculis*, *eyes absent*, and *dachshund*—are also used to form eyes in both *Drosophila* and vertebrates (Jean et al. 1998; Relaix and Buckingham 1999). Since it is extremely unlikely that deuterostomes and protostomes would have evolved *Pax6* and its partners independently—and used them independently for the same function—it is likely that the PDA possessed a *Pax6* gene and used it for generating eyes.

Another gene shared by deuterostomes and protostomes is the homeobox-containing gene *tinman*. The Tinman protein is expressed in the *Drosophila* splanchnic mesoderm, eventually residing in the region of the cardiac mesoderm. Loss-of-function mutants of *tinman* lack a heart (hence its name). In mice, the homologous gene is *Nkx2-5*, and it, too, is originally expressed in the splanchnic mesoderm and continues to be expressed in those cells that form the heart tubes (see Chapter 15; Manak and Scott 1994). Thus, even though the hearts of vertebrates and the hearts of insects have little in common except their ability to pump fluids, they both appear to be predicated on the expression of the same gene, and it is therefore probable that the PDA had a circulatory system with a pump based on the expression of the ancestral *Nkx2-5/tinman* gene.

Another set of genes shared by deuterostomes and protostomes are those for the transcription factors involved in head formation (Finkelstein and Boncinelli 1994; Hirth and Reichert 1999). In *Drosophila*, the brain is composed of three segments, called **neuromeres**. These neu-

FIGURE 23.4   The *Pax6* gene for eye development is an example of a gene ancestral to both protostomes and deuterostomes. The micrograph shows ommatidia emerging in the leg of a fruit fly (a protostome) in which mouse (deuterostome) *Pax6* cDNA was expressed in the leg disc. (From Halder et al. 1995, photograph courtesy of W. J. Gehring and G. Halder.)

romeres are specified by three transcription factors. The genes encoding these factors are *tailless* (*tll*) and *orthodenticle* (*otd*), which are expressed predominantly in the anteriormost neuromere, and *empty spiracles* (*ems*), which is expressed in the posterior two neuromeres (Monaghan et al. 1995; Hirth et al. 1998). Loss-of-function mutations of *otd* eliminate the anteriormost neuromere of the developing *Drosophila* embryo, and loss-of-function mutations of *ems* eliminate the second and third neuromeres (Hirth et al. 1995). In frogs and mice, the homologues of these genes (*Otx1, Otx2, Emx1, Emx2*) are also expressed in the brain (Simeone et al. 1992), although the exact patterns of transcription are not identical (see Figure 11.42). The *Otx2* gene has been experimentally knocked out (Acampora et al. 1995; Matsuo et al. 1995; Ang et al. 1996), and the resulting mice have neural and mesodermal head deficiencies anterior to rhombomere 3. In humans, mutations of *EMX2* lead to a rare condition known as schizencephaly, in which there are clefts ripping through the entire cerebral cortex (Brunelli et al. 1996). Even though the *Drosophila otd* and *ems* genes are specified by the Bicoid and Hunchback gradients and the mammalian *Otx* and *Emx* genes are induced by the anterior dorsal

mesoderm and endoderm, it appears that the same genes are used for determining the anterior brain regions. It is therefore likely that the ancestor of all bilaterian organisms had sensory organs based on Pax6, a heart based on *tinman*, and a head based on *otd*, *ems*, and *tll*. It also had something else: an anterior-posterior polarity based on the expression of Hox genes.

Anatomical similarities: Larval forms

Those features held in common between protostomes and deuterostomes are thought to be derived from the PDA. In addition to the molecular similarities, there are also structural similarities between these two groups. The most basal forms of both

the deuterostomes and the protostomes arise from ciliated larvae. These larvae might form in different ways—the protostomes using the blastopore region as their foregut and mouths, and the deutersotomes using the blastopore region as their anal hindgut, but recent molecular evidence suggests that even this morphological distinction may have underlying similarities. Arendt and colleagues (2001) have shown that the *Brachyury* (*T*) gene is expressed in the ventral foreguts of pluteus and hemichordate larvae of basal deuterostomes and in the trochophore larvae of annelid worms (Figure 23.5). *Goosecoid* and *Otx* are also found in the foregut regions of deuterostome and protostome larvae. Convergent evolution of these three genes would be unlikely (espe-

(Continued on next page)

TABLE 23.1 Developmental regulatory genes conserved between protostomes and deuterostomes

Gene	Function	Distribution
achaete-scute group	Cell fate specification	Cnidarians, *Drosophila*, vertebrates
Bcl2/Drob-1/ced9	Programmed cell death	*Drosophila*, nematodes, vertebrates
Caudal	Posterior differentiation	*Drosophila*, vertebrates
delta/Xdelta-1	Primary neurogenesis	*Drosophila*, *Xenopus*
Distal-less/DLX	Appendage formation (proximal-distal axis)	Numerous phyla of protostomes and deuterostomes
Dorsal/NFkB	Immune response	*Drosophila*, vertebrates
forkhead/Fox	Terminal differentiation	*Drosophila*, vertebrates
Fringe/radical fringe	Formation of limb margin (apical ectodermal ridge in vertebrates)	*Drosophila*, chick
Hac-1/Apaf/ced 4	Programmed cell death	*Drosophila*, nematodes, vertebrates
Hox complex	Anterior-posterior patterning	Widespread among metazoans
lin-12/Notch	Cell fate specification	*C. elegans*, *Drosophila*, vertebrates
Otx-1, Otx-2/Otd, Emx-1, Emx-2/ems	Anterior patterning, cephalization	*Drosophila*, vertebrates
Pax6/eyeless; Eyes absent/eya	Anterior CNS/eye regulation	*Drosophila*, vertebrates
Polycomb group	Controls Hox expression/ cell differentiation	*Drosophila*, vertebrates
Netrins, Split proteins, and their receptors	Axon guidance	*Drosophila*, vertebrates
RAS	Signal transduction	*Drosophila*, vertebrates
sine occulus/Six3	Anterior CNS/eye pattern formation	*Drosophila*, vertebrates
sog/chordin, dpp/BMP4	Dorsal-ventral patterning, neurogenesis	*Drosophila*, *Xenopus*
tinman/Nkx 2-5	Heart/blood vascular system	*Drosophila*, mouse
vnd, msh	Neural tube patterning	*Drosophila*, vertebrates

Source: After Erwin 1999.

cially considering how specific their localizations are), suggesting that both protostomes and deuterostomes inherited a ciliated larval form from the PDA. Indeed, Arendt and colleagues (2001, 2004) have proposed that the Urbilaterian had a single blastoporal opening that extended along the surface of the embryo (like the blastopores of certain annelid embryos today), becoming a mouth in the protostomes and an anus in the deuterostomes.

Cnidarians and their larvae as PDA candidates

Current research has focused on cnidarians and their planula larvae as the possible bilaterian ancestor. Bilateral symmetry appears to have originated with the appearance of the third germ layer, the mesoderm; and while cnidarians are classified as diploblasts (having only ectoderm and endoderm) and have radial, not bilateral, symmetry, some cnidarians have been found to possess a rudimentary mesoderm as well as bilateral symmetry (Ball et al. 2004; Seipel and Schmid 2005). These "primitive" organisms have many of the transcription factors and signal transduction pathways seen in both protostomes and the deuterostomes. Thus, some cnidarians have the TGF-β and Wnt pathways, as well as transcription factors such as snail, Otx, and Hox. Moreover, the *Brachyury* (*T*) gene is expressed around the blastopore, just as it is in protostomes and deuterostomes (Technau 2001, Scholz and Technau 2003). Axis formation in the cnidarian *Hydra* is regulated both by the Wnt signaling pathway and by the chordin-BMP pathway, just as in bilaterians (Broun et al. 2005, Hobmayer, quoted in Meinhardt 2004). Indeed, the *Bmp4/dpp* homologue in anthozoans (the corals and sea anemones) is expressed on only one side of the blastopore, indicating both bilateral symmetry and the asymmetric expression of this gene—traits characteristic of bilaterian animals (Figure 23.6; Hayward et al. 2002; Finnerty et al. 2004).

One of the most far-reaching speculations concerning the cnidarian identity of the PDA is that the cnidarian body becomes the "brain" of the bilaterians. Meinhardt (2004) has noted that *Otx*, a gene seen in the forebrains of vertebrates and insects, is seen throughout the body of *Hydra*. Similarly, genes that are expressed at the posterior

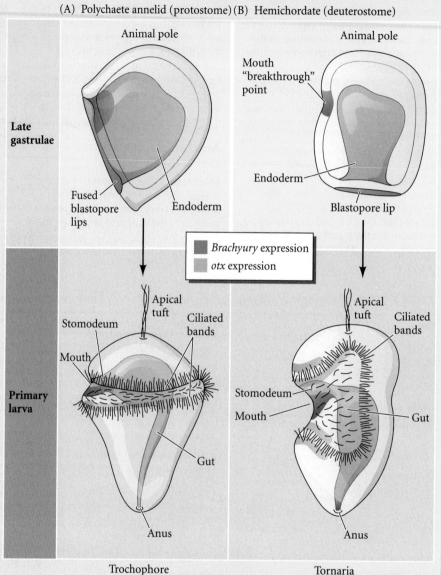

(A) Polychaete annelid (protostome) (B) Hemichordate (deuterostome)

FIGURE 23.5 Late gastrula embryos (top) develop into ciliated larvae (bottom). (A) In the polychaete annelid worms (Protostomia), the lateral blastopore lips fuse along the later ventral midline. The blastopore gives rise to mouth and anus at opposite ends. In the trochophore larva produced by these embryos, the *Brachyury* gene (green) is expressed in the ventral portion of the stomodeum (i.e., the mouth) and in the proctodeum (anus), while *otx* (gold) is expressed in two bands of cells along the ciliated bands. (B) In the hemichrodates (a deuterostome lineage that includes the acorn worms), the tip of the gastrulation cavity touches the lateral body wall on the future ventral side, where the mouth later breaks through. The blastopore gives rise to the anus only. In the early tornaria larva produced by these embryos, *Brachyury* is expressed in the ventral portion of the stomodeum and in the proctodeum, and *otx* is expressed in two upper bands parallel to the preoral ciliated band and in two lower bands parallel to the postoral ciliated band.

of the *Hydra* are expressed at the base of the brain in vertebrates and flies. Thus, the ancestor of the bilaterians may have been a radially symmetric cnidarian similar to

Hydra, but this cnidarian's body would have been similar to the bilaterian head! The trunk would be formed later, by convergent extension along the dorsal midline in deuterostomes and possibly by terminal addition of new segments in protostomes.

The combination of molecular and anatomical investigations offers us provocative clues as to what the proverbial Urbilatarian ancestor might have been.

(A) 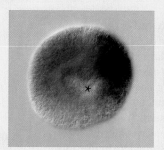 (B)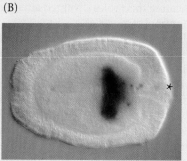

FIGURE 23.6 Evidence of the evolutionary conservation of regulatory genes. (A) The cnidarian homologue of the vertebrate *Bmp4* and *Drosophila Decapentaplegic* genes is expressed asymmetrically at the edge of the blastopore (*) in the embryo of the sea anemone *Nematostella*. This gene represents an ancestral form of the protostome and deuterostome forms of the gene. (B) The Hox gene *Anthox6*, a cnidarian member of the paralogue 1 group of Hox genes, is expressed at the blastopore side of the larval sea anemone. (From Finnerty et al. 2004, photographs courtesy of M. Martindale.)

Each Hox gene in *Drosophila* has a homologue in vertebrates. In some cases, the homologies go very deep and can also be seen in the gene's functions. Not only is the vertebrate *Hoxb4* gene similar in sequence to its *Drosophila* homologue, *Deformed*, but the human *HOXB4* gene can perform the functions of *Deformed* when introduced into *Dfd*-deficient *Drosophila* embryos (Malicki et al. 1992). As mentioned in Chapter 11, not only are the Hox genes of the different phyla homologous, but they are in the same order on their respective chromosomes. Their expression patterns are also remarkably similar: the genes at the 3' end are expressed anteriorly, while those at the 5' end are expressed posteriorly* (see Figure 11.42). Thus they are homologous genes between species (as opposed to members of a gene family being homologous within a species). Genes that are homologous between species are called **orthologues**.

All multicellular organisms—animals, plants, and fungi—have Hox-like genes, so it is most likely that an ancestral Hox gene existed that encoded a basic helix-loop-helix transcription factor in protozoans. In the earliest animal groups, this gene became replicated. One of the two Hox genes present in some extant cnidarians (e.g., jellyfish) corresponds to the anterior set of vertebrate Hox genes (and is expressed in the anterior portion of the cnidarian larva), while sequences in the other cnidarian gene is a posterior class Hox gene (see Figure 23.6; Yanze et al. 2001; Hill et al. 2003; Finnerty et al. 2004). Perhaps

even the ancient cnidarians used Hox genes to distinguish their anterior and posterior tissues. In bilateral phyla, the central Hox genes emerged as a duplication from one of the earlier genes (de Rosa et al. 1999).

In the deuterostome echinoderms, and in many protostome lineages (including the arthropods), a complex set of Hox genes became crucial in specifying the various parts of the organism. In the deuterostome chordate lineage, however, two large-scale duplications of the entire Hox cluster took place, such that vertebrates have four Hox clusters per haploid genome instead of just one. Thus, instead of having a single *Hox4* gene (orthologous to *Deformed* in *Drosophila*), vertebrates have *Hoxa4*, *Hoxb4*, *Hoxc4*, and *Hoxd4*. This constitutes the *Hox4* **paralogue group** in vertebrates. Such large-scale duplications have had several consequences. First, as seen in Chapters 14 and 16, these duplications create much redundancy. It is difficult to obtain a loss-of-function mutant phenotype, since to do so means all copies of these paralogue group genes must be deleted or made nonfunctional (Wellik and Capecchi 2003). However, in some instances, the genes *have* become specialized. *Hoxd11*, for instance, plays an important role in the mammalian limb bud, but not in the reproductive system. Mammalian *Hoxa11*, on the other hand, plays roles in both the limb (where it is critical in specifying the zeugopod) and in the female reproductive tract (where it helps construct the uterus; Wong et al. 2004).

The second duplication event of the Hox genes within the vertebrates took place during fish evolution. In the ancestral group that gave rise to the teleost fishes (such as zebrafish), the entire genome was duplicated. After this duplication, some clusters were lost. But most teleosts still have six or seven Hox clusters, whereas the other vertebrates have four. Again, there is much redundancy, causing a stabilization of phenotype. But the fish have also used some of these genes for different functions. For instance,

---

*The conservation of Hox genes and their colinearity demands an explanation. One recent proposal (Kmita et al. 2000, 2002) contends that the Hox genes "compete" for a remote enhancer that recognizes the Hox genes in a polar fashion. This enhancer most efficiently activates Hox genes at the 5' end. If the positions of the genes are changed by recombination or deletion, then different genes are activated in different regions of the body, and morphology changes.

the expression pattern of one zebrafish *Hoxc13* paralogue, *Hoxc13a*, is different from the second paralogue, *Hoxc13b*, indicating their roles diverged after the second duplication (Thummel et al. 2004).

In summary, then, gene duplication and divergence is an extremely important mechanism for evolution.* Duplication allows the formation of redundant structures, and divergence allows these structures to assume new roles. While one gene copy maintains its original role, the other copies are free to mutate and diverge functionally. Numerous transcription factors and paracrine factors are members of such paralogue families. *Hox* genes are used to pattern the body and limb axes, *Distal-less* genes are used to extend appendages and to pattern the vertebrate skull, and members of the *MyoD* family specify different stages of muscle development. These genes, each family derived from a single ancestral gene, are active in different tissues and provide instructions for the formation of different cell types. So perhaps "the most important difference between the genome of a fruit fly and that of a human is therefore not that the human has new genes but that where the fly only has one gene, our species has multigene families" (Morange 2001, p. 33).

Homologous Pathways of Development

One of the most exciting findings of the past decade has been the discovery not only of homologous regulatory genes, but also of homologous signal transduction pathways, many of which have been mentioned earlier in this book. In different organisms, these pathways are composed of homologous proteins arranged in a homologous manner (Zuckerkandl 1994; Gilbert 1996; Gilbert et al. 1996).

Homologous signal transduction pathways form the infrastructure of development. However, the targets of these pathways may differ among organisms. For example, the Dorsal-Cactus pathway used by *Drosophila* to specify dorsal-ventral polarity is also used by the mammalian immune system to activate inflammatory proteins. This does not mean that the *Drosophila* blastoderm is homologous to the human macrophage. It merely means that there is an ancient pathway that predates the deuterostome-protostome split, and that this pathway can be used in different systems. The pathways are homologous; the organs they form are not.

In some instances, homologous pathways made of homologous parts are used for the same function in both protostomes and deuterostomes. Conserved similarities in both the pathway and its function over millions of years of phylogenetic divergence are considered to be evidence of "deep homology" between these structures (Shubin et

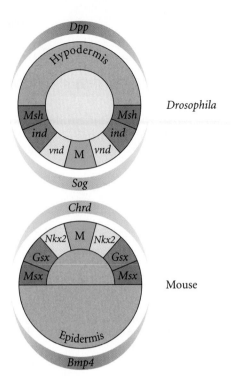

FIGURE 23.7 The same set of instructions forms the nervous systems of both protostomes and deuterostomes. In the fruit fly (protostome), the TGF-β family member *Dpp* (decapentaplegic) is expressed dorsally and is opposed by *Sog* ventrally. In the mouse (deuterostome), the TGF-β family member *Bmp4* is expressed ventrally and is countered dorsally by *chordin*. The highest concentration of chordin/Sog becomes the midline (M). The midline is dorsal in vertebrates and ventral in insects, and the concentration gradient of the TGF-β family protein (Bmp4 or Dpp) activates the genes specifying the regions of the nervous system in the same order in both groups: *vnd/Nkx2*, followed by *ind/Gsx*, and finally *Msx/Msh*. These genes have also been seen to be expressed in a similar fashion in cnidarians. (After Ball et al. 2004.)

al. 1997). One example of deep homology is the chordin/BMP4 pathway discussed in Chapter 10. In both vertebrates and invertebrates, chordin/Short-gastrulation (Sog) inhibits the lateralizing effects of Bmp4/Decapentaplegic (Dpp), thereby allowing the ectoderm protected by chordin/Sog to become the neurogenic ectoderm. These reactions are so similar that *Drosophila* Dpp protein can induce ventral fates in *Xenopus* and can substitute for Sog[†]

---

*The duplication-and-divergence scheme for generating paralogous and orthologous homologies was actually first described in the 1840s by Sir Richard Owen, a friend (and later rival) of Charles Darwin (see Gilbert 1980).

[†]In addition to this central inhibitory reaction, there are other reactions that add to the deep homology of the instructions for forming the protostome and deuterostome neural tube. For instance, the spread of Dpp in *Drosophila* is aided by Tolloid, a metalloprotease that degrades Sog. The gradient of Dpp concentration from dorsal to ventral is created by the opposing actions of Tolloid (increasing Dpp) and Sog (decreasing it) (Marqués et al. 1997). In *Xenopus* and zebrafish, the homologues of Tolloid (Xolloid and BMP1, respectively) have the same function: they degrade chordin. The gradient of BMP4 from ventral to dorsal is established by the antagonistic interactions of Xolloid or BMP1 (increasing BMP4) and chordin (decreasing BMP4) (Blader et al. 1997; Piccolo et al. 1997). Even the regulators of BMP stability are conserved between these two groups of animals and function in the same way (Larrain et al. 2001).

(Figure 23.7; Holley et al. 1995). Thus the protostome and deuterostome nervous systems, despite their obvious differences, seem to be formed by the same set of instructions. The plan for the animal nervous system may have been laid down only once.

Mechanisms of Macroevolutionary Change

In 1940, Richard Goldschmidt wrote that the accumulation of small genetic changes was not sufficient to generate evolutionarily novel structures such as the neural crest, teeth, turtle shells, feathers, or cnidocysts. He claimed that such evolution could occur only through inheritable changes in the genes that regulated development. For years, such genes remained in the realm of hypothesis. In 1977, the idea that change within regulatory genes was critical to evolution was extended by François Jacob, the Nobel laureate who helped establish the operon model of gene regulation. First, Jacob said, evolution works with what it has: it combines existing parts in new ways rather than creating new parts. Second, he predicted that such "tinkering" would be most likely to occur in those genes that construct the embryo, not in the genes that function in adults (Jacob 1977).

Wallace Arthur (2004) has catalogued four ways in which Jacob's "tinkering" can take place at the level of gene expression:

- Heterotopy (change in location)
- Heterochrony (change in time)
- Heterometry (change in amount)
- Heterotypy (change in kind)

These changes can only be accomplished if the gene expression patterns are modular, that is, if they are controlled by different enhancer elements. The modularity of development allows one part of the organism to change without necessarily affecting the other parts.

Heterotopy

WEBFOOTED DUCKS AND WINGED MAMMALS   In the past decade, many different scientists have confirmed that changes in development are responsible for the generation of novel structures. In Chapter 16, the difference between the webbed duck foot and the clawed chicken foot was ascribed to the expression of *gremlin* in the duck's interdigital space. The Gremlin protein is an inhibitor of BMPs, and BMP signaling is thought to be responsible for initiating apoptosis in the interdigital region (Figure 23.8; see also Figure 16.25; Merino et al. 1999). Thus by blocking BMPs in the interdigital mesenchyme, the duck hindlimb retains its webbing.

Bat forelimbs present a similar situation. Here the interdigital webbing that forms the wing has been maintained where it has disappeared in other mammals. Again, inhibition of BMP-induced apoptosis is critical. Instead of Gremlin, however, the agent of BMP inhibition in bats appears to be FGF signaling. Unlike other mammals, bats express Fgf8 in their interdigital webbing, and this protein is critical for maintaining the cells there. If FGF signaling is inhibited (by drugs such as SU5402), BMPs can induce the apoptosis of the forelimb webbing, just like in other mammals (Laufer et al. 1997; Weatherbee et al. 2005).

HOW SHRIMPS DIFFER FROM LOBSTERS   Next time you are fortunate enough to have a seafood platter in front of you,

FIGURE 23.8 Autopods of chicken feet (upper row) and duck feet (lower row) are shown at similar stages. Both show BMP4 expression (dark blue) in the interdigital webbing; BMP4 induces apoptosis. The duck foot (but not the chicken foot) expresses the BMP4-inhibitory protein Gremlin (dark brown; arrows) in the interdigital webbing. Thus the chicken foot undergoes interdigital apoptosis (as seen by neutral red dye accumulation in the dying cells) but the duck appendage does not. (Photographs courtesy of J. Hurle and E. Laufer.)

Chick hindlimb

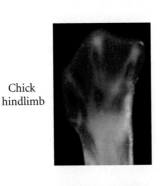

Duck hindlimb

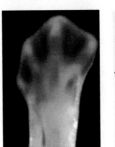

BMP   Gremlin   Apoptosis   Newborn

FIGURE 23.9 Hox gene expression and morphological change in crustaceans. At the bottom are the structures of a brine shrimp (crustacean) and a grasshopper (insect). The domains of Hox gene expression specifying the various structures are coded in color. Whereas the Hox gene expression domains segregate in the insect thorax and abdomen, they coincide in the crustacean thorax. Above them is a hypothetical model for the divergence of insect and crustacean body plans from a common ancestor. The *Antennapedia*, *Ultrabithorax*, and *abdominal A* genes are very similar and are thought to have emerged by gene duplication from a single gene in a distant ancestor of the arthropods. *Abdominal B* is expressed in the segments destined to become genitalia. Paleontological evidence suggests that the ancestral arthropod had identical thoracic segments, similar to those of extant crustaceans. In both cases, *Antennapedia*, *Ultrabithorax*, and *abdominal A* expression is seen throughout the thorax. The crustaceans later evolved tail segments. Insects, however, diversified their thoracic segments and used the Hox genes to specify the segments differently. The ancestors of the insects and crustaceans are hypothetical reconstructions based on fossils of the middle Cambrian. (After Averof and Akam 1995 and Manton 1977.)

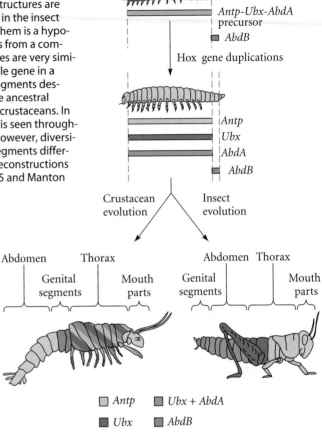

take a look at the specimen on your plate. Crustaceans are characterized by a pregnathal head (similar to the insect acron), gnathal (jawed) head segments, six thoracic segments, genital segments, abdominal segments, and a telson. The thoracic segments all look alike, and there is no specialization among them; each of them expresses *Antp*, *Ubx*, and *abdA*, and these genes appear to be interchangeable in the crustaceans. In the arthropod lineage that gave rise to the insects, however, each of these genes took on different (sometimes overlapping) functions (Figure 23.9; Averof and Akam 1995). But even within the crustacean lineages, there are interesting variations on this theme. Averof and Patel (1997) have shown that if a thoracic segment does not express both *Ubx* and *abdA*, the anterior locomotor limb becomes a feeding appendage called a **maxilliped**. Thus, brine shrimp such as *Artemia* have uniform expression of *Ubx* and *abdA* in their thoracic segments, and they lack maxillipeds. Lobsters lack *Ubx* and *abdA* expression in their first and second thoracic segments, and these segments have paired maxillipeds (Figure 23.10). The fossil record suggests that the earliest crustaceans lacked maxillipeds and had uniform thoracic segments. This would mean that the presence of maxillipeds is a derived characteristic that evolved in several crustacean lineages.*

HOW BIRDS GOT THEIR FEATHERS   Although it had long been thought that feathers emerged as modified reptilian scales (see Maderson 1972; Prum et al. 1999; Maderson and Alibardi 2000), the mechanism that produces feathers has remained elusive. Harris and his colleagues (2002) have provided a developmental mechanism for feather evolu-

tion, showing that the feather most likely evolved from the archosaurian (dinosaur/bird ancestor) scale through an alteration of the expression pattern of the Sonic hedgehog (Shh) and BMP2 proteins.

Scales and feathers start off the same way, with *Bmp2* and *Shh* expressed in separate domains. However, in the feather, both expression domains shift to the distal region of the appendage. This feather-specific pattern is repeated serially around the proximal-distal axis. The interaction between BMP2 and Shh proteins then causes each of these regions to form its own axis—the barbs of the feather (Figure 23.11). Moreover, when this serially repeated pattern was experimentally modified, the feather pattern was modified in a predictable manner (Harris et al. 2002; Yu et al. 2002).

HOW SNAKES LOST THEIR LIMBS   As described in Chapter 11, the expression pattern of Hox genes in vertebrates determines the type of vertebral structure formed. Thoracic vertebrae, for instance, have ribs, while cervical (neck) vertebrae and lumbar vertebrae do not. The type of vertebra produced is specified by the Hox genes expressed in the somite.

One of the most radical alterations of the vertebrate body plan is seen in snakes. Snakes evolved from lizards,

---

*This chapter concentrates on *transcriptional-level* changes that can generate new morphological forms. However, morphological changes can be instigated at the translational level as well. Abzhanov and Kaufman (1999), for instance, have shown that translational differences in the *Sex combs reduced* gene are critical in converting legs into maxillipeds in the crustacean pillbug, *Porcellio scaber*.

FIGURE 23.10 Schematic representation of the expression of *Ubx* and *abdA* (green) in the thoracic segments of different types of crustaceans. The generation of maxillipeds occurs in the thoracic segments that do not express either of these homeodomain proteins. (After Averof and Patel 1997.)

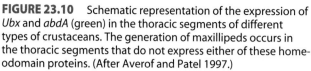

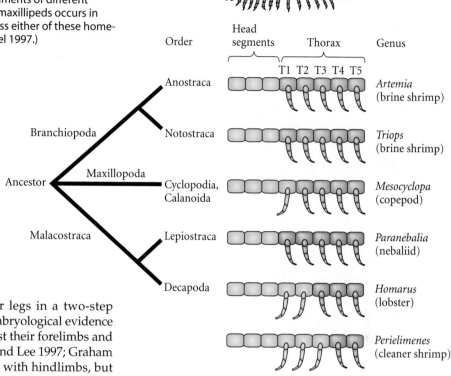

and they appear to have lost their legs in a two-step process. Both paleontological and embryological evidence support the view that snakes first lost their forelimbs and later lost their hindlimbs (Caldwell and Lee 1997; Graham and McGonnell 1999). Fossil snakes with hindlimbs, but no forelimbs, have been found. Moreover, while the most derived snakes (such as vipers) are completely limbless, the more primitive snakes (such as boas and pythons) have pelvic girdles and rudimentary femurs.

The missing forelimbs can be explained by the Hox expression pattern in the anterior portion of the snake. In most vertebrates, the forelimb forms just anterior to the most anterior expression domain of *Hoxc6* (Gaunt 1994; Burke et al. 1995; Gaunt 2000). Caudal to that point, *Hoxc6*, in combination with *Hoxc8*, helps specify vertebrae to be thoracic. During early python development, *Hoxc6* is expressed in the absence of *Hoxc8*, so the forelimbs do not form. Rather, the combination of *Hoxc6* and *Hoxc8* is expressed for most of the length of the organism, telling the vertebrae to form ribs throughout most of the body (Figure 23.12; Cohn and Tickle 1999).

The loss of hindlimbs apparently occurred by a different mechanism. Hindlimb buds do begin to form in some snakes, such as pythons, but do not produce anything more than

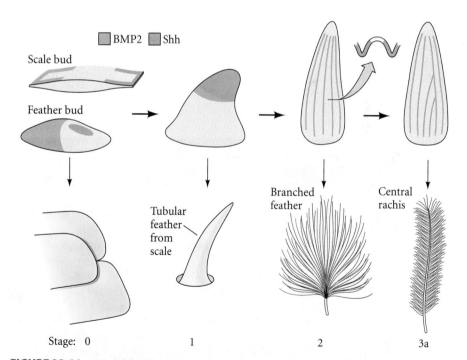

FIGURE 23.11 Model for the evolution of the feather by changes in the pattern of *Bmp2* and *Shh* expression. Stage 0 shows the *Shh* and *Bmp2* expression in the scale (above) and feather bud (below). Stage 1 represents a tubular feather as evolved from an archosaurian scale. The *Shh* and *Bmp2* expression patterns are postulated to be at the tip. Stage 2 represents the emergence of a branched feather evolved by further changing the expression patterns of *Bmp2* and *Shh* to form rows along the proximal-distal axis. In stage 3a, changes in feather morphology evolved by altering the pattern to produce a central rachis. (After Harris et al. 2002.)

(A)

(B)

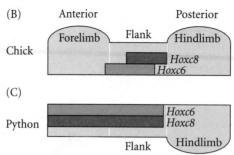

Chick

(C)

Python

FIGURE 23.12 Loss of limbs in snakes. (A) Skeleton of the garter snake, *Thamnophis*, stained with alcian blue. Ribbed vertebrae are seen from the head to the tail. (B,C) Hox expression patterns in chick (B) and python (C). (Photograph courtesy of A. C. Burke; B,C after Cohn and Tickle 1999.)

hindlimb resembles that of mouse embryos with loss-of-function mutations of *Sonic hedgehog* (Chiang et al. 1996).

HOW MAMMALS CHANGED THEIR TEETH Stephen J. Gould (1989) joked that paleontologists believe mammalian evolution occurs when two teeth mate to produce slightly altered descendant teeth. Since enamel is far more durable than ordinary bone, teeth often remain after all the other bones have decayed. Indeed, the study of tooth morphology has been critical to mammalian systematics and ecology. Changes in the cusp pattern of molars is regarded as especially important in allowing the radiation of mammals into new ecological niches. What mechanism allows molars to change their form so rapidly?

Jukka Jernvall and colleagues (Jernvall et al. 2000; Salazar-Ciudad and Jernvall 2002) pioneered a computer-based approach to phenotype production using Geographic Information Systems (GIS)—the same technology ecologists use to map the vegetation of hillsides. They have used this technology to map gene expression patterns in

a femur. This appears to be due to the lack of *Sonic hedgehog* expression by the limb bud mesenchyme. Sonic hedgehog is needed both for limb polarity and for the maintenance of the apical ectodermal ridge (AER). Python hindlimb buds lack the AER, and the phenotype of the python

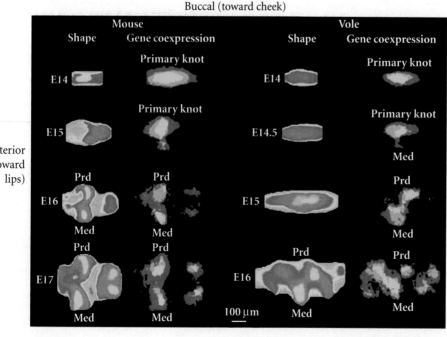

FIGURE 23.13 GIS analysis of gene activity in the formation of the first set of cusps in mouse and vole molars. (The first two cusps are the proconid, labeled Prd, and the metaconid, labeled Med). For both species, GIS mapping of molar shape is shown on the left and the expression of *fgf4* and *shh* (two genes expressed from the enamel knots) are shown at the right. The gene expression pattern on embryonic day 15 predicts the formation of the new cusp seen on day 16; the gene expression pattern on day 16 predicts the formation of cusps in those areas on day 17. Similarly, in the vole molar, whose cusps are diagonal to one another, gene expression predicts cusp formation. (After Jernvall et al. 2000; photograph courtesy of J. Jernvall.)

(A)

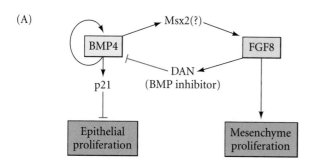

(B)

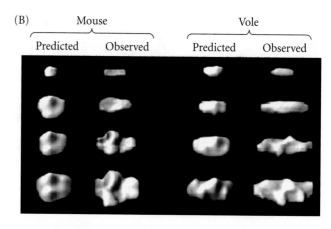

FIGURE 23.14 Basic model for cusp development in mice and voles. (A) Experimentally derived gene network wherein BMPs activate their own production as well as the production of their inhibitors, Shh and FGFs. The FGFs and Shh stimulate cell proliferation; the BMPs inhibit it. (B) Predicted and observed results from this model. The model can generate the final and intermediate forms of molar development in mice and voles, and the difference between mouse and vole molars can be reproduced by slight alterations in the rate of BMP diffusion and binding to inhibitors. (After Salazar-Ciudad and Jernvall 2002; photographs courtesy of J. Jernvall.)

incipient tooth buds (literally turning a mountain into a molar). Their studies have shown that gene expression patterns forecast the exact location of the tooth cusps. They have also shown that the differences between the molars of mice and voles are predicated on differences in gene expression patterns (Figure 23.13).

Salazar-Ciudad and Jernvall (2004) proposed a mathematical reaction-diffusion model (see Chapter 1) that would explain the differences in gene expression between mice and voles. Small changes in a gene network, again working through the interactions of the BMPs and Shh proteins, were crucial. Shh and FGFs (produced by the enamel knot signaling centers) inhibit BMP production, while BMPs stimulate both the production of more BMPs and the synthesis of its own inhibitors. This would create regions of activators (BMPs) that blocked epithelial proliferation, and regions of inhibitors (FGFs, Shh) that blocked BMP synthesis and independently stimulated mesenchymal proliferation. The result is a pattern of gene activity that changes as the shape of the tooth changes, and vice-versa (Figure 23.14).

Using this model, the large differences between mouse and vole molars can be generated by small changes in the binding constants and diffusion rates of the BMP and Shh proteins. A small increase in the diffusion rate of BMP4 and a stronger binding constant of its inhibitor is sufficient to change the vole pattern of tooth growth into that of the mouse. Thus, large morphological changes can result from very small changes in initial conditions. Another conclusion is that all the cells can start off with the same basic set of instructions; the specific instructions emerge as the cells interact. This mathematical model also predicts that some types of teeth are much more likely to evolve in certain ways than in others, and that certain shapes are most likely to evolve (see Figure 1.22). These predictions conform to what paleontologists have concluded about mammalian evolution.*

---

*One of the critical differences between the Jernvall and Salazar-Ciudad model and other reaction-diffusion models is that it allows the parameters to change while the tooth is developing. This assumption gives a great amount of flexibility to the developing system, allowing it to change at relatively rapid rates during evolution.

Heterochrony

Heterochrony is a shift in the relative timing of two developmental processes from one generation to the next. In other words, one module can change its time of expression relative to the other modules of the embryo. We have already come across this concept in our discussion of neoteny and progenesis in relation to metamorphosis. Heterochrony can also give larval characteristics to an adult organism, as in the fetal growth rate of human newborn brain tissue (see Chapter 12). This temporal modularity of development can be appreciated when we look at the development of eyes relative to other regions of the early amniote embryo (Figure 23.15). The eyes of birds and lizards initiate development earlier than the eyes of mammals, and are therefore proportionally much larger than mammalian eyes when they reach the pharyngula stage (Jeffery et al. 2002). Another example is found among marsupials, where the mouth and forelimb develop at a faster rate than the same organs do in placental mammals, allowing the marsupial embryo to climb into the maternal pouch and suckle (Smith 2003).

HETEROCHRONY ON THE MOLECULAR LEVEL In salamanders, heterochronic changes by which the larval stage is either retained or truncated are caused by genetic changes in the ability to induce or respond to the hormones initiating metamorphosis (see Chapter 18). Other phenotypes, however, can be the result of the heterochronic expression of certain genes. For instance, the direct development of some

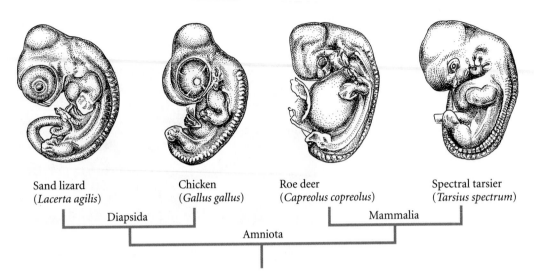

Sand lizard
(*Lacerta agilis*)

Chicken
(*Gallus gallus*)

Roe deer
(*Capreolus copreolus*)

Spectral tarsier
(*Tarsius spectrum*)

Diapsida

Mammalia

Amniota

FIGURE 23.15   Allometry and modularity in the vertebrate eye. In the diapsid amniotes (reptiles and birds), eye development starts relatively early, and by the pharyngula stage, the embryo is characterized by large eyes. In mammals, however, eye development is initiated later, and the eye is much smaller relative to the body. (After Jeffery et al. 2002; drawings by N. Haver.)

sea urchins involves the early activation of adult genes and the suppression of larval gene expression (Raff and Wray 1989; Ferkowicz and Raff 2001). In one species of direct developing urchin, *wnt5* is expressed in the same places as in a closely related indirect developer—but it is expressed much earlier. Whereas *wnt5* expression, which initiates production of juvenile structures, occurs in the larva of the indirect developer, it occurs in the embryo of the direct developer. Another example of molecular heterochrony occurs in the lizard genus *Hmiergis*, which includes species with 3, 4, or 5 digits on each limb. The number of digits is regulated by the length of time the *sonic hedgehog* gene is active in the limb bud's zone of polarizing activity. The shorter the duration of *shh* expression, the fewer the number of digits (Shapiro et al. 2003b).

THE ORIGIN OF JAWS   Heterochronic changes are thought to be responsible for the evolution of one of the most important adaptations in the history of vertebrates—the jaw. When the vertebrate face first evolved, it did not have a jaw. These earliest vertebrates were similar to today's **cyclostomes**—the lampreys and hagfish, which have a completely cartilagenous skeleton and a skull that does not differ greatly from the vertebrae. In gnathostomes (jawed vertebrates), cranial neural crest cells enter the pharyngeal arches, and those neural crest cells migrating into the first pharyngeal arch form the mandible (lower jaw) and the maxillary process (roof of the mouth). A migration of neural crest cells takes place in cyclostomes, but the first pharyngeal arch does not form Meckel's cartilage or the mandible derived from it. Instead, the neural crest cells

form a round, jawless mouth. It appears that two major events had to occur in order for this group of first pharyngeal arch neural crest cells to become a jaw: they had to have a permissive environment, and they had to receive a set of new instructions.

The *permissive environment* came via the removal of a barrier. The cyclostomes form a naso-hypophyseal plate from which the nasal epithelium and the hypophysis (pituitary) both develop. This epithelial plate forms a barrier to neural crest cell migration, and the only way for the neural crest cells to migrate forward is to travel beneath it. These neural crest cells then form the upper lip of the cyclostome mouth. In gnathostomes, the naso-hypophyseal plate has separated into the nasal placode and Rathke's pouch, leaving a space between these structures through which the neural crest cells can migrate. The cells that migrate into this region form the mandible (Figure 23.16; Kuratani et al. 2001). Thus, the difference allowing the formation of the jaw may be a difference in the timing of the separation between the rudiments of the nasal placodes and the rudiments of the pituitary. If the separation is early, jaws are possible. If the separation is late, a barrier exists that prevents neural crest cell migration into the region that would form the mandible. Such a shift in timing may be due to a slight change in the timing of a particular tissue interaction (Shigetani et al. 2002).

The *new set of instructions* come from a shift in Hox gene expression. In lampreys, there is Hox gene expression in the first pharyngeal arch (Cohn 2002); in gnathostomes, there is no such expression. Moreover, if Hox genes are ectopically expressed in the first pharyngeal arches of fish, frog, or chick embryos, jaw development is severely inhibited (indeed, the structures look reminiscent of the ancestral hyoid arches) (Alexandre et al. 1996; Pasqualetti et al. 2000; Grammatopoulos et al. 2000). Thus Hox genes appear to prevent the neural crest cells of the first pharyngeal arch from forming jaws. When Hox genes are inhibited, a new set of genetic instructions is made available.

(A) Lamprey
(jawless vertebrate)

(B) Gnathostomes
(jawed vertebrate)

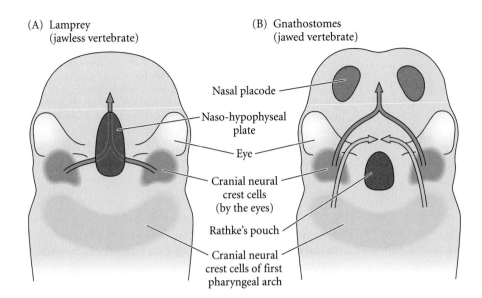

Nasal placode

Naso-hypophyseal plate

Eye

Cranial neural crest cells (by the eyes)

Rathke's pouch

Cranial neural crest cells of first pharyngeal arch

FIGURE 23.16 Neural crest cell migration patterns in lampreys and gnathostomes. (A) In the lamprey, the upper lip develops from cranial neural crest cells that migrate rostrally underneath the naso-hypophyseal plate (which will form both the nasal placode and the pituitary). (B) In gnathostomes, the naso-hyophyseal tissue has already split into Rathke's pouch (the pituitary rudiment) and the nasal (olfactory) placode. The cranial neural crest cells migrate rostrally between these two structures to form part of the lower jaw cartilage. The upper jaw forms from those cranial neural crest cells that migrate rostrally and fuse anterior to Rathke's pouch. (After Kuratani et al. 2001.)

Heterometry

DARWIN'S FINCHES **Heterometry** is the change in the amount of a gene product or structure. One of the best examples of heterometry involves Darwin's celebrated finches. Darwin's finches are a set of 14 closely related birds collected by Charles Darwin and his shipmates during his visit to the Galápagos and Cocos Islands in 1835. These birds helped him frame his evolutionary theory of descent with modification, and they still serve as one of the best examples of adaptive radiation and natural selection (see Weiner 1994; Grant 1999). Systematists have shown that these finch species evolved in a particular manner, with a major speciation event being the split between the cactus finches and the ground finches (Figure 23.17). The ground finches evolved deep, broad beaks that enable them to crack seeds open, whereas the cactus finches evolved narrow, pointed beaks that allow them to probe cactus flowers and fruits for insects and flower parts. Earlier research (Schneider and Helms 2003) had shown that species differences in the beak pattern were caused by changes in the growth of the neural crest-derived mesenchyme of the frontonasal process (i.e., those cells that form the facial bones). Abzhanov and his colleagues (2004) found a remarkable correlation between the beak shape of the finches and timing and amount of *Bmp4* expression. No other paracrine factor showed such differences. The expression of *Bmp4* in ground finches starts earlier and is much greater than *Bmp4* expression in cactus finches. In all cases, the *Bmp4* expression pattern correlated with the breadth and depth of the beak.

The importance of these expression differences was confirmed experimentally by changing the *Bmp4* expression pattern in chick embryos (Abzhanov et al. 2004; Wu et al. 2004). When *Bmp4* expression was enhanced in the frontonasal process mesenchyme, the chick developed a broad beak reminiscent of the beaks of the ground finches. Conversely, when BMP signaling was inhibited in this region (by Noggin, a BMP inhibitor), the beak became narrow and pointed, like those of cactus finches. Thus, enhancers controlling the amount of beak-specific BMP4 synthesis may be critically important in the evolution of Darwin's finches.

HUMAN INTERLEUKIN-4 Most human variation, whether pathological or nonpathological, does not come from changes in the structural genes. Rather, variation arises from mutations in the regulatory regions of these genes (Rockman and Wray 2002). Very often, these mutations involve changes in the rate of transcription (heterometry) of particular genes. One such example is the gene encoding **interleukin-4** (IL4).

IL4 is a paracrine factor released by cells of the immune system to promote the differentiation of IgE-secreting B cells.* Several alleles of the *IL4* gene are medically relevant. One of the most common alleles is *–524T* (that is, in this allele the base thymine—T—occupies the position 524 base pairs before the start of transcription). This allele is recent and is unique to human beings (Rockman et al. 2003). The wild-type allele, *–524C* (which has cytosine in the –524 position) is found in all primate populations—and indeed in every other mammal tested. The –524T allele, therefore, emerged in the lineage separating *Homo* from the other great apes. Individuals carrying the –524T allele are subject to several disease states, including severe allergies, asthma, contact dermatitis, and subacute sclerosing panencephalitis.

---

*IgEs are a type of antibody. The IgEs are responsible for allergic reactions, and also help protect us against parasitic worms.

Bmp4 expression

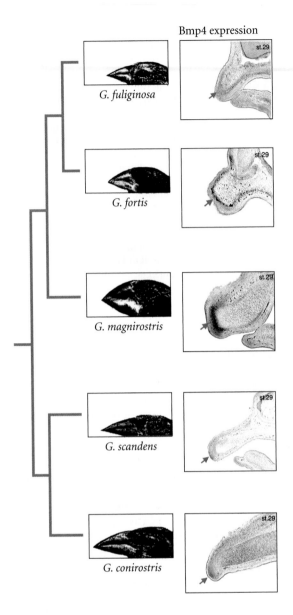

G. fuliginosa

G. fortis

G. magnirostris

G. scandens

G. conirostris

FIGURE 23.17 Correlation between beak shape and the expression of *Bmp4* in Darwin's finches. In the genus *Geospiza*, the ground finches (represented by *G. fuliginosa*, *G. fortis*, and *G. magnirostris*) diverged from the cactus finches (represented by *G. scandens* and *G. conirostris*). The differences in beak morphology correlate to heterochronic and heterometric changes in *Bmp4* expression in the beak. That is, *Bmp4* is expressed earlier and at higher levels in the seed-crushing ground finches. This gene expression difference provides one explanation for the role of natural selection on these birds. (After Abzhanov et al. 2004).

positively selected, and illustrates how medical genetics is providing evidence for the hypothesis that genetic changes in the enhancers of regulatory genes are forces for evolutionary change.

ALLOMETRY Another consequence of modularity associated with heterometry is **allometry**—changes that occur when different parts of an organism grow at different *rates* (Huxley and Teissier, 1936; Gayon 2000; see Chapter 1). Allometry can be very important in forming diverse body plans within a lineage. Such differential changes in growth rate can involve altering a target cell's sensitivity to growth factors, or altering the amounts of growth factors produced. Again, the vertebrate limb provides a useful illustration. Local differences among chondrocytes caused the central toes of ancestral, five-toed horses to grow at a rate 1.4 times that of the lateral toes (Wolpert 1983). As horses grew larger over evolutionary time, this regional difference in chondrocytes resulted in the one-toed state seen in modern horses.

A particularly dramatic example of allometry in evolution comes from skull development. In the very young (4- to 5-mm) whale embryo, the nose is in the usual mammalian position. However, the enormous growth of the maxilla and premaxilla (upper jaw) pushes over the frontal bone and forces the nose to the top of the skull (Figure 23.18). This new position turns the mammalian nose into the cetacean blowhole, allowing the whale to have a large and highly specialized jaw apparatus and (not incidentally) to breathe while swimming parallel to the water's surface (Slijper 1962).

Allometry can also generate evolutionary novelty by small, incremental changes that eventually cross some developmental threshold (sometimes called a **bifurcation point**). When such a threshold is crossed, a change in quantity eventually becomes a change in quality. It has been postulated that this type of mechanism produced the unique cheek pouches of pocket gophers and kangaroo rats that live in deserts. These external pouches differ from internal ones in that they are fur-lined and have no internal connection to the mouth. They allow these animals to store seeds outside their mouths, thus minimizing water

What does the *–524T* allele do differently than the *–524C* allele? The C → T mutation creates a new binding site for the transcription factor NFAT (*n*uclear *f*actor for *a*ctivated *T* cells). The wild-type promoters generally contain six NFAT binding sites that interact with AP1 to promote IL4 transcription. The presence of a new seventh site in *–524T* leads to more rapid production of IL4, yielding protein levels of at least threefold the norm. Moreover, population genetic statistics show that the new allele has been positively selected in particular populations but not in others. Having this allele appears to be advantageous in those populations exposed to intestinal helminth parasites. This example suggests that it may be relatively easy to change the regulation of a gene to produce a variant that can be

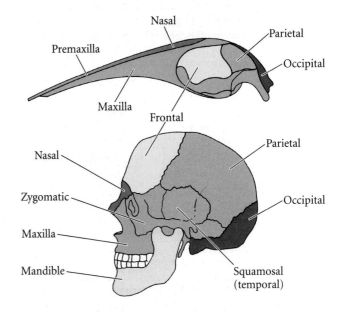

FIGURE 23.18 Allometric growth in the whale head. An adult human skull is shown for comparison. The whale's upper jaw (maxilla) has pushed forward, causing the nose to move to the top of the skull. The mandible is not shown. (The premaxilla is present in the early human fetus, but it fuses with the maxilla by the end of the third month of gestation. The human premaxilla was discovered by Goethe, among others, in 1786.) (After Slijper 1962.)

loss that would occur to seeds held inside the mouth. Brylski and Hall (1988) have dissected pocket gopher and kangaroo rat embryos to examine the way the external cheek pouch is constructed. They found that these external pouches and the internal cheek pouches of animals such as hamsters are actually formed in a very similar manner. Both pouch types are formed by outpocketings of the cheek (buccal) epithelium into the facial mesenchyme. In animals with internal cheek pouches, these evaginations stay within the cheek. However, in animals that form external pouches, the elongation of the snout draws the outpocketings up into the region of the lip. As the lip epithelium rolls out of the oral cavity, so do the outpockets. What had been internal becomes external. The fur lining is probably derived from the external pouches' coming into contact with dermal mesenchyme, which can induce hair to form in epithelia (see Chapter 12). The molecular bases for these growth rate changes has not yet been identified.

The transition from internal to external pouch is one of threshold. The placement of the evaginations—anteriorly or posteriorly—determines whether the pouch is internal or external. There is no "transitional stage" displaying two openings, one internal and one external.* One could envision this externalization occurring by a chance mutation or concatenation of alleles that shifted the outpocketings to a slightly more anterior location. Such a trait would be

selected for in desert environments, where dessication is a constant risk. As Van Valen reflected in 1976, evolution can be defined as "the control of development by ecology."

Heterotypy

In heterochrony, heterotopy, and heterometry, the mutations affect the regulatory regions of the gene. The gene's product—the protein—remains the same, although it may be synthesized in a new place, at a different time, or in different amounts. The changes of **heterotypy** affect the actual coding region of the gene, and thus can change the functional properties of the protein being synthesized.

WHY INSECTS HAVE ONLY SIX LEGS   Insects have six legs, whereas most other arthropod groups (think of spiders, millipedes, centipedes, and shrimp) have many more. How is it that the insects came to form legs only in their three thoracic segments? The answer seems to reside in the relationship, mentioned earlier, between Ultrabithorax protein and the *Distal-less* gene.

In most arthropod groups, Ubx protein does not inhibit the *Distal-less* gene. However, in the insect lineage, a mutation occurred in the *Ubx* gene wherein the original 3′ end of the protein-coding region was replaced by a group of nucleotides encoding a stretch of about 10 alanine residues (Figure 23.19; Galant and Carroll 2002; Ronshaugen et al. 2002). This polyalanine region represses *Distal-less* transcription. When a shrimp *Ubx* gene is experimentally modified to encode this polyalanine region, it, too, represses the *Distal-less* gene. The ability of insect Ubx to inhibit *Distal-less* thus appears to be the result of a gain-of-function mutation that characterizes the insect lineage.

---

*The lack of such transitional forms is often cited by creationists as evidence against evolution. For instance, in the transition from reptiles to mammals, three of the bones of the reptilian jaw became the incus and malleus, leaving only one bone (the dentary) in the lower jaw (see Chapter 1 and below). Gish (1973), a creationist, says that this is an impossible situation, since no fossil has been discovered showing two or three jaw bones and two or three ear ossicles. Such an animal, he claims, would have dragged its jaw on the ground. However, such a specific transitional form (and there are over a dozen documented transitional forms between reptilian and mammalian skulls) need never have existed. Hopson (1966) has shown on embryological grounds how the bones of the jaw could have divided and been used for different functions, and Romer (1970) has found reptilian fossils wherein the new jaw articulation was already functional while the older bones were becoming useless. Several species of therapsid reptiles had two jaw articulations, with the stapes brought into close proximity with the upper portion of the quadrate bone (which would become the incus).

FIGURE 23.19 Changes in Ubx protein associated with the insect clade in the evolution of arthropods. Of all arthropods, only the Ubx protein of insects is able to repress *Distal-less* gene expression, and thereby inhibit abdominal legs. This ability to repress *Distal-less* is due to a mutation that is only seen in the insect *Ubx* gene. (After Galant and Carroll 2002; Ronshaugen et al. 2002).

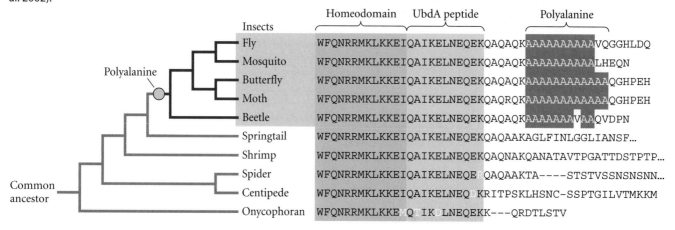

SIDELIGHTS & SPECULATIONS

Nature's Experiments: Coding Tandem Repeats

Heterotypy has been an important force in the generation of new regulatory proteins. Heterotypy can be generated by the normal mechanisms of mutation, but there are faster ways to generate new proteins. One such mechanism is **replication slippage**, whereby a codon becomes repeated numerous times. These "coding tandem repeats" can create long stretches of the same amino acid, substantially changing the function of a protein. One such repeat is the polyalanine sequence of the insect Ubx protein (see Figure 23.19), and the Hox family is the most polyalanine-rich group of proteins in the human genome (Lavoie et al. 2003). Glycine repeats have been observed to create variations in the rapidly evolving sea urchin bindins (Zigler and Lessios 2003), and amino acid repeats have also been seen in the gamete binding proteins of abalones (Galindo et al. 2003). The rate of change in the length of these repeats has been estimated

as being 100,000 times greater than the rate of new point mutations.

Repeats of serine, glycine, proline, and alanine have been found in numerous transcription factors (Caburet et al. 2005). Fondon and Garner (2004) have shown that repeat length variations in transcription factors correlate with specific morphologies in dogs. One allele of the *Alx4* gene with a particularly small number of repeats was found to be homozygous in only one breed of dogs, the Great Pyrenees. These dogs are characterized by polydactyly (an extra toe, or "dew claw"), and individuals of this breed without this characteristic trait were found to be homozygous for *Alx4* alleles that are found in all other dog breeds. Dogs with different tandem repeats of glutamine or alanine in the *Runx2* gene have different facial features. These phenotypic variations have been selected (in this case artificially, by breeders) and are responsible for the different facial angles and lengths of the many

dog breeds. The rapid length variations of these amino acid tracts could explain the ability to derive 200 different dog breeds in 135,000 years; this explanation is certainly more plausible than attributing canine diversity to point mutations (which have a rate of 10^{-7} to 10^{-9}).

Replication slippage, as mutation, has been referred to "nature's experimentation." Its results are not always beneficial. Huntington syndrome is a lethal hereditary condition characterized by the expansion of the CAG codon in the *Huntingtin* gene to produce a large tract of glutamine. This polyglutamic acid sequence interacts with the TBP (*TATA-binding protein*), a central component of the eukaryotic transcription complex, and prevents TBP from functioning (Schaffar et al. 2004). Other pathologies, such as fragile-X syndrome (the most common form of inherited mental retardation) are also thought to be caused by replication slippage (see Handa et al. 2005; Napierala et al. 2005).

Recruitment

No one structure is destined for a single purpose. A pencil can be used for writing, but it can also be used as a toothpick, a dagger, a hole-punch, or a drumstick. On the molecular level, the gene *engrailed* is first used for segmen-

tation in the *Drosophila* embryo, is used later to specify its neurons, and is used in the larval stages to provide an anterior-posterior axis to imaginal discs. Similarly, a protein that functions as an enolase or alcohol dehydrogenase enzyme in the liver can function as a structural crystallin

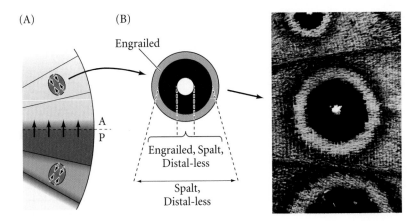

(A) (B)

Engrailed

Engrailed, Spalt, Distal-less

Spalt, Distal-less

A
P

FIGURE 23.20 Co-option of signal transduction pathways in the formation of butterfly eyespots. (A) The interaction between Hedgehog (green) and Engrailed (blue) is first seen in the insect blastoderm and is modified during the creation of a new structure, the wing. Hedgehog in the posterior compartment of the wing disc induces Engrailed expression in the cells immediately anterior to the anterior/posterior boundary. Within the wing, the same pathway is used to create the eyespot. In the posterior of the wing, Hedgehog is locally downregulated so that re-expression can induce Engrailed. (B) In *Bicyclus anynana* (see Figures 22.8 and 22.9), the signal to produce the eyespot activates several genes that were used in wing formation. The central portion of the eyespot expresses high levels of Engrailed, Spalt, and Distal-less; these activate genes that make the white pigment. The cells surrounding the center express Distal-less and Spalt, which activate melanin-synthesizing genes that make black pigment. The peripheral ring of cells that generates the orange pigment expresses only Engrailed. Other species use these same genes to produce very different color patterns (see Marcus 2005). (After Keys et al. 1999 and Brunetti et al. 2001; photograph courtesy of S. B. Carroll and S. Paddock.)

protein in the lens (Piatigorsky and Wistow 1991). In other words, preexisting units can be **recruited** (co-opted; redeployed; repurposed) for new functions. Sometimes, whole signaling pathways are co-opted. For instance, the same pathway by which Hedgehog protein induces Engrailed protein to pattern and extend the insect wing is later used within the wing blades to make the eyespots of butterflies and moths. Distal-less, another protein used to extend the wing imaginal disc, is later used to form the center of such eyespots (Figure 23.20). Recruitment can entail the creation of a new enhancer element such that a gene that usually functions in one area or time now functions in another area or at another time, *in addition to* its original temporal or spatial expression (Hinman et al. 2003).

Recruitment can also be accomplished by having a new region of the embryo creating conditions similar to those in which the gene originally functioned. For instance, when

a reporter *lacZ* gene is placed on the enhancer and promoter from the *Xenopus Distal-less2* gene and inserted into a mouse egg, that *Xenopus* gene directs transcription in places where it is normally seen in *Xenopus* (ventral brain and epidermis). However, in the mouse embryo, this same gene is also expressed in places that no frog gene has ever seen: mammary glands, whisker pads, and teeth (Figure 23.21;

(A)

(B)

(C)

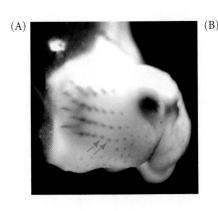
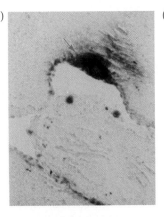
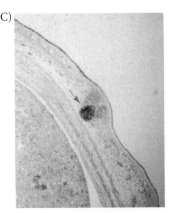

FIGURE 23.21 Expression of a frog *Distal-less* gene in the mouse. (A) Transgenic mice were made using a *Xenopus Distal-less2* promoter and enhancer fused to a *lacZ* reporter gene. The frog gene was expressed in the nasal, limb, and tail primordia, as in frog embryos. However, this gene also became expressed (arrows) in the mouse's whisker pads (A), teeth (B), and mammary glands (C)—anatomical structures that are not present in the frog. (From Morasso et al. 1995, photographs courtesy of M. Morasso.)

FIGURE 23.22 Evolution of the mammalian middle ear bones from the reptilian jaw. The quadrate and articular bones of reptiles were part of the lower jaw. Sound could be transmitted from these bones via the large stapes. When the dentary bone grew and took over the jaw functions of these two bones, the articular bone became the malleus and the quadrate bone became the incus. (After Romer 1949.)

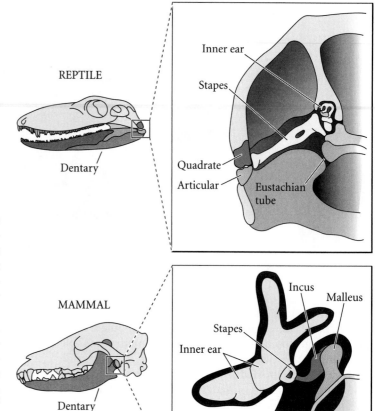

Morasso et al. 1995). These are places where the mouse orthologue of *Xenopus Distal-less2* is routinely expressed. The same signals that cause this gene's expression in frogs also induce its expression in mice; thus the amphibian gene responds to signals produced in mammalian structures such as mammary glands and whisker pads.

Co-option can also be seen on the morphological level. A famous case of co-option is the use of embryonic jaw parts in the creation of the mammalian middle ear, as described in Chapter 1 (see Figure 1.14; Gould 1990). First, the gill arches of jawless fishes became the jaws of their descendants. Millions of years later, the upper elements of the reptilian jawbone became the malleus and incus (hammer and stirrup) bones of the mammalian middle ear (Figure 23.22). Similarly, in arthropods, the rudiments of ancestral gills have become (in land arthropods) wings, spinnerets, and a variety of breathing organs (Damen et al. 2002).

SIDELIGHTS & SPECULATIONS

How the Chordates Got a Head

The vertebrate head is derived largely from the neural crest cells. Thus, Hall (2000) considers vertebrates to be not merely triploblastic animals, but *quadroblastic*, with the neural crest constituting a fourth germ layer. Holland and Chen (2001) have even proposed calling vertebrates and their fossilized precursors "cristozoa," the "crest-animals." But how did these unique neural crest cells arise?

This is a critically important question in evolutionary developmental biology because it goes to the heart of evolutionary novelty. A turtle can form bones that no other organism has ever made—but they are still bones. A crustacean can turn a leg into a mouthpart, but the mouthpart is made of the same materials as the leg. But

how does a completely new type of cell emerge? How did bone cells come into being for the first time? For this to happen, new combinations of transcription factors have to form a new stable network. Such networks were shown in Chapter 14, and scientists are figuring out how combinations of gene expression patterns can become stabilized into a new cell type.

Only vertebrates have a neural crest, and the cranial neural crest is responsible for forming the bones and cartilage of the face and much of the skull (see Chapter 13). Although we do not know how neural crest cells arose, there are cells at the neural plate/epidermal boundaries of cephalochordate and urochordate embryos that might be "latent homologues" of the neural

crest. Both the urochordate tunicates and the cephalochordate *Amphioxus* express in their dorsal midline ectoderm many of the same genes expressed in vertebrate neural crest cells; however, they haven't evolved the pathways that integrate these genes (Holland and Holland 2001; Wada 2001; Stone and Hall 2004).

Amphioxus is an invertebrate chordate that has a notochord, somites, and a hollow neural tube. It lacks a brain and facial structures, and, most importantly, it lacks neural crest cells. In *Amphioxus*, the neural plate/epidermal border contains cells expressing several of the same genes expressed in vertebrate neural crest cells—*Bmp2*, *Pax3/7*, *Msx*, *Dll*, and *Snail*. However, the protochordate cells expressing these

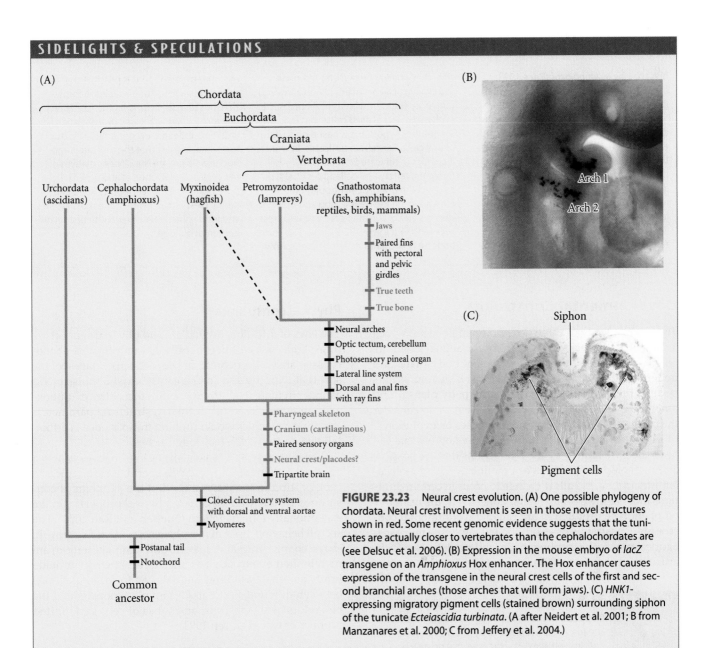

FIGURE 23.23 Neural crest evolution. (A) One possible phylogeny of chordata. Neural crest involvement is seen in those novel structures shown in red. Some recent genomic evidence suggests that the tunicates are actually closer to vertebrates than the cephalochordates are (see Delsuc et al. 2006). (B) Expression in the mouse embryo of *lacZ* transgene on an *Amphioxus* Hox enhancer. The Hox enhancer causes expression of the transgene in the neural crest cells of the first and second branchial arches (those arches that will form jaws). (C) *HNK1*-expressing migratory pigment cells (stained brown) surrounding siphon of the tunicate *Ecteiascidia turbinata*. (A after Neidert et al. 2001; B from Manzanares et al. 2000; C from Jeffery et al. 2004.)

genes do not migrate, nor do they differentiate into a wide range of tissues (Figure 23.23A; Holland and Holland 2001).

In addition, many of the *cis*-regulatory elements needed for the positioning of neural crest cells in the vertebrate body are also present in *Amphioxus*. Manzanares and colleagues (2000) added regulatory regions of *Amphioxus* Hox genes onto a reporter *lacZ* gene and observed the expression of these chimeric genes in mouse and chick embryos. Remarkably, some of the *Amphioxus* Hox regulatory sequences caused the expression of the reporter gene in the neural crest cells and neural placodes of mice and chicks (Figure 23.23B).

Two transcription factors, AP2 and Distal-less, may have been critical in the formation of neural crest cells. Vertebrate *AP2* is expressed in non-neural ectoderm, cranial neural crest, and neural tube cells, even in jawless fish. It appears to be essential for *Hoxa2* expression in the neural crest cells. However, *AP2* expression is confined to the non-neural ectoderm of *Amphioxus*. Meulemans and Bronner-Fraser (2002) speculate that altering the expression pattern of *AP2* was critical in establishing the neural crest cell lineage.

Holland and colleagues (1996) have suggested that the origin of the neural crest involves the duplication and divergence of the *Distal-less* genes, which are found throughout the animal kingdom, and are expressed in those tissues that stick out from the body axis, notably limbs and antennae (Panganiban et al. 1997). But in vertebrates, *Distal-less* has acquired new functions. Like *Drosophila*, *Amphioxus* has only one copy of the *Distal-less* gene per haploid genome and, also like *Drosophila*, this gene is expressed in the epidermis and central nervous system. Vertebrates, however, have five or six closely related copies of *Distal-less*, all of which probably originated from a single ancestral gene that resem-

(Continued on next page)

bles the one in *Amphioxus* (Price 1993; Boncinelli 1994; Neidert et al. 2001). These *Distal-less* homologues have found new functions. Some are expressed in the mesoderm, a place where *Distal-less* is not expressed in *Amphioxus*. At least three of the vertebrate *Distal-less* genes function in the patterning of neural crest cells, and deleting them results in the absence or malformation of the pharyngeal arches, face, jaws, teeth, and vestibular apparatus (Qiu et al. 1997; DePew et al. 1999). Although it remains to be proved, it is pos-

sible that a new type of *Distal-less* gene could have played a major role in allowing migratory ectodermal cells of *Amphioxus* to evolve into neural crest cells.

Another hypothesis is based on finding neural crest-like cells in tunicates. In working with the tunicate *Ecteinascidia turbinata*, Jeffery and colleagues (2004) discovered a migratory cell population that emerges from the neural tube and that synthesizes two of the markers for neural crest cells—HNK1 (Figure 23.23C) and Zic. These tunicate cells do not make skeletons, how-

ever, but pigment. So it is possible that the neural crest arose not for skeleton formation, but to make pigments that could protect an ascidian-like ancestor from solar radiation. Indeed, new genomic and structural data suggest that the tunicates, and not *Amphioxus*, are the closest relatives of the vertebrates (Delsuc et al. 2006). The development of the vertebrate head and neural crest remains one of the most active and interesting areas of evolutionary developmental biology.

Developmental Constraints

There are only about three dozen animal lineages, which represent all the major body plans of the animal kingdom. One can easily envision other types of body plans and imagine animals that do not exist (science fiction writers do it all the time). So why aren't more body plans found among the animals? To answer this, we have to consider the constraints that development imposes on evolution.

The number and forms of possible phenotypes that can be created are limited by the interactions that are possible among molecules and between modules.* These molecular interactions also allow change to occur in certain directions more easily than in others. Collectively, the restraints on phenotype production are called **developmental constraints.** Constraints on evolution fall into three major categories: physical, morphogenetic, and phyletic (see Richardson and Chipman 2003).

WEBSITE

23.8 Why are there no new animal phyla? It appears that all three dozen or so known animal lineages were in existence by 500 million years ago. Indeed, it may be the case that no new phylum has emerged since the late Cambrian. What is the evidence for the early formation of the phyla, and are any animal body plans left unused?

---

*G. W. Leibniz, who may have been the philosopher who most influenced Darwin, noted that existence must be limited not only to the possible but to the *compossible*—that is, whereas numerous things *can* come into existence, only those that are mutually compatible *will* actually exist (see Lovejoy 1964). So, even though many developmental changes are possible, only those that can integrate into the rest of the organism (or that can cause a compensatory change in the rest of the organism) will be seen. Developmental biologist see constraints as limiting the appearance of certain phenotypes, whereas population geneticists see constraints as limiting "ideal" adaptation (such as constraints on optimal foraging) (Amundson 1994, 2005).

Physical constraints

The laws of diffusion, hydraulics, and physical support are immutable and will permit only certain physical phenotypes to arise. A vertebrate on wheeled appendages (of the sort that Dorothy saw in Oz) cannot exist because blood cannot circulate to a rotating organ; this entire evolutionary avenue is closed off. Similarly, structural parameters and fluid dynamics would prohibit the existence of 6-foottall mosquitoes or 25-foot-long leeches.

The elasticity and tensile strength of tissues is also a physical constraint. In *Drosophila* sperm, for example, the type of tubulin that can be used in the axoneme is constrained by the need for certain physical properties in the exceptionally long flagellum (Nielsen and Raff 2002). The six cell behaviors used in morphogenesis (division, growth, shape change, migration, death, and matrix secretion) are each limited by physical parameters, and thereby provide limits on what structures animals can form. Interactions between different sets of tissues involve coordinating the behaviors of sheets, rods, and tubes of cells in a limited number of ways (Larsen 1992).

Morphogenetic constraints

Rules for morphogenetic construction also limit the phenotypes that are possible (Oster et al. 1988). Bateson (1894) and Alberch (1989) noted that when organisms depart from their normal development, they do so in only a limited number of ways. Some of the best examples of these types of constraints come from the analysis of limb formation in vertebrates. Although there have been many modifications of the vertebrate limb over 300 million years, some modifications (such as a middle digit shorter than its surrounding digits) are not found (Holder 1983). Moreover, analyses of natural populations suggest that there is a relatively small number of ways in which limb changes can occur (Wake and Larson 1987). If a longer limb is favorable in a given environment, the humerus may become elongated, but one never sees two smaller humeri joined together in tandem, although one could imagine the selective advantages that such an arrangement might have. This observation suggest a limb construction scheme that follows certain rules.

It may be that the rules of the reaction-diffusion model govern the architecture of the limb (Miura and Shiota 2000; Kiskowski et al. 2004; Newman and Müller 2005). Oster and colleagues (1988) found that this model can explain the known morphologies of the limb, and could explain why other morphologies are forbidden. The reaction-diffusion equations predicted the observed succession of bone development from stylopod (humerus/femur) to zeugopod (ulna-radius/tibia-fibula) to autopod (hand/foot). If limb morphology is indeed determined by the reaction-diffusion mechanism, then spatial features that cannot be generated by reaction-diffusion kinetics will not occur. Similarly, the reaction-diffusion mechanisms that produce the cusps and valleys of mammalian teeth (see Figures 23.14 and 1.22) predict that only certain types of teeth are possible (Salazar-Ciudad and Jernvall 2004).

Evidence supporting the reaction-diffusion model comes from experimental manipulations, comparative anatomy, and cell biology. When an axolotl limb bud is treated with the antimitotic drug colchicine, the number of cells in the limb bud is reduced. In these experimental limbs, there is not only a reduction in the number of digits, but a loss of certain digits in a certain order, as predicted by the mathematical model and by the "forbidden" morphologies. Moreover, the losses of specific digits produce limbs very similar to those of certain salamanders whose limbs develop from particularly small limb buds (Alberch and Gale 1983, 1985).

Phyletic constraints

Phyletic constraints on the evolution of new structures are historical restrictions based on the genetics of an organism's development (Gould and Lewontin 1979). Once a structure comes to be generated by inductive interactions, it is difficult to start over again. The notochord, for example, which is functional in adult protochordates such as *Amphioxus* (Berrill 1987), degenerates in adult vertebrates. Yet it is transiently necessary in vertebrate embryos, where it specifies the neural tube. Similarly, Waddington (1938) noted that, although the pronephric kidney of the chick embryo is considered vestigial (since it has no ability to concentrate urine), it is the source of the ureteric bud that induces the formation of a functional kidney during chick development (see Chapter 14).

One fascinating example of a phyletic constraint is the lack of variation among marsupial limbs. Although eutherian limbs show a dramatic range of diversity (claws, wings, paddles, flippers, hands), limbs among marsupial species are all pretty much the same. Sears (2004) has documented that the necessity for the marsupial fetus to crawl into its mother's pouch has constrained limb development such that the limb musculature and cartilage must develop very

early into a structure that can grasp and crawl. Any variation in this trait has effectively been eliminated.

As genes acquire new functions during the course of evolution, they may become involved in more than one module, making change difficult. Galis and colleagues provide evidence that the reason the segment polarity gene network is conserved in all types of insects is that these genes play roles in several different pathways (Galis et al. 2002). Such pleiotropy constrains the possibilities for alternative mechanisms, since it makes change difficult. **Pleiotropy**, the ability of a gene to play different roles in different cells, is the "opposite" of modularity, involving the connections between parts rather than their independence.

Pleiotropies may underlie the constraints seen in mammalian development. Galis speculates that mammals have only seven cervical vertebrae (while birds may have dozens) because the Hox genes that specify these vertebrae have become linked to cell proliferation in mammals (Galis 1999; Galis and Metz 2001). Thus, changes in Hox gene expression that might facilitate evolutionary changes in the skeleton might also misregulate cell proliferation and lead to cancers. She supports this speculation with epidemiological evidence showing that changes in skeletal morphology correlate with childhood cancer. The intraembryonic selection against having more or fewer than seven cervical ribs appears to be remarkably strong. At least 78 percent of human embryos with an extra anterior rib (i.e., six cervical vertebrae) die before birth and 83 percent have died by the end of the first year. These deaths appear to be caused by multiple congenital anomalies or cancers (Figure 23.24; Galis et al. 2006).

Until recently, it was thought that the earliest stages of development would be the hardest to change, because altering them would either destroy the embryo or generate a radically new phenotype. But recent work (and the reappraisal of older work; see Raff et al. 1991) has shown that certain alterations can be made to early cleavage without upsetting the organism's final form. Evolutionary modifications of cytoplasmic determinants in mollusc embryos can give rise to new types of larvae that still metamorphose into molluscs, and changes in sea urchin cytoplasmic determinants can generate sea urchins that develop directly (with no larval stage), but still become sea urchins.

The earliest stages of development, then, appear to be extremely plastic. The later stages are very different from species to species, as the different phenotypes of adult vertebrates amply demonstrate. There is something in the middle of development, however, that appears to be invariant. Sander (1983) and Raff (1994) have argued that the formation of new body plans (*Baupläne*) is inhibited by the global consequences of induction during the **phylotypic stage**—the stage that typifies a phylum. For instance, the late neurula, also known as the **pharyngula**, is the phylotypic stage that appears to be critical for all vertebrates (see Figure 1.5; Slack et al. 1993). And in fact, while all the vertebrates arrive at the pharyngula stage, they do so by very

(A)

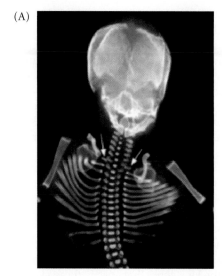

(B)

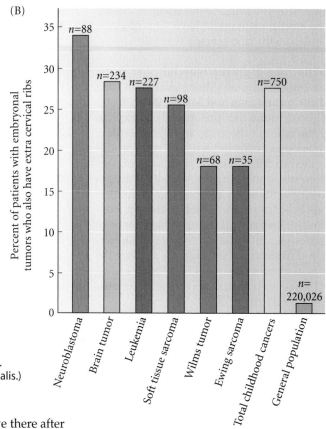

Percent of patients with embryonal tumors who also have extra cervical ribs

Neuroblastoma: n=88
Brain tumor: n=234
Leukemia: n=227
Soft tissue sarcoma: n=98
Wilms tumor: n=68
Ewing sarcoma: n=35
Total childhood cancers: n=750
General population: n=220,026

FIGURE 23.24   Extra cervical ribs are associated with childhood cancers. (A) Radiogram showing an extra cervical rib. (B) Nearly 80 percent of fetuses with extra cervical ribs die before birth. Those surviving often develop cancers very early in life. This indicates strong selection against changes in the number of mammalian cervical ribs. (After Galis et al. 2006, photograph courtesy of F. Galis.)

different means. Birds, reptiles, and fish arrive there after meroblastic cleavages of different sorts; amphibians get there by way of radial holoblastic cleavage; and mammals reach the same stage after constructing a blastocyst, chorion, and amnion (see Chapter 11).

Before the vertebrate pharyngula stage, there are few inductive events, and most of them are on global scales (involving axis specification). In these early stages of development, there is a great deal of regulation, so small changes in morphogen distributions or the position of cleavage planes can be accommodated (Figure 23.25; Henry et al. 1989). After the pharyngula stage, there are a great many inductive events, but almost all of them occur within discrete modules. The lens induces the cornea, but if it fails to do so, only the eye is affected. But during the phylotypic pharyngula stage, the modules interact. Failure to have

the heart in a certain place can affect the induction of eyes. Failure to induce the mesoderm in a certain region leads to malformations of the kidneys, limbs, and tail. By searching the literature on congenital anomalies, Galis and Metz (2001) have documented that the pharyngula is much more vulnerable than any other stage. Moreover, based on patterns of multiple organ anomalies within the same person, they concluded that the multiple malformations were due to the interactivity of the modules at this stage. Thus, this phylotypic stage that typifies the vertebrate phylum appears to constrain its evolution. Once an organism becomes a vertebrate, it is probably impossible for it to evolve into anything else.

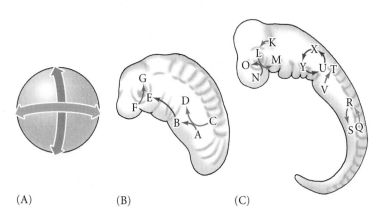

(A)   (B)   (C)

FIGURE 23.25   Mechanism for the bottleneck at the pharyngula stage of vertebrate development. (A) In the cleaving embryo, global interactions exist, but there are very few of them (mainly to specify the axes of the organism). (B) At the neurula to pharyngula stages, there are many global inductive interactions. (C) After the pharyngula stage, there are even more inductive interactions, but they are primarily local in effect, confined to their own modules. (After Raff 1994.)

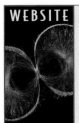

WEBSITE **23.9 Changing embryonic traits through natural selection.** Just as changes in embryos can produce new phenotypes, so natural selection on adults can favor certain types of embryos that produce favorable adult phenotypes. This selection for adult traits may explain the conservation of some patterns of development as well as deviations from the norm.

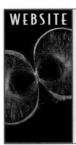

WEBSITE **23.10 Alternative mechanisms for evolutionary developmental biology.** Evolution is accomplished through heritable changes in development. In this textbook, these heritable changes are assumed to be those that alter gene expression patterns. However, other models have been proposed in which there is horizontal transmission of genetic information between phyla, or in which there is inheritance of cytoplasmic properties.

SIDELIGHTS & SPECULATIONS

Canalization and the Release of Developmental Constraints

Morphogenetic constraints are one way of preventing certain variations. They are not the only way, however. Surprisingly, many newly occurring mutations, even in vital genes, do not affect development. Rather, development appears to be buffered so that slight abnormalities of genotype or slight perturbations in the environment do not lead to the formation of abnormal phenotypes (Waddington 1942; Siegal and Bergman 2002). This phenomenon, called **canalization**, actually serves as an additional constraint on the evolution of new phenotypes.

It is difficult for a mutation to affect development (Nijhout and Paulsen 1997). It is the rare mutation that is 100 percent penetrant. Canalization allows mutations to accrue in the genotype without being expressed in the phenotype (and therefore without being immediately accessible to natural selection). Thus, in the short term, it limits the variability of the phenotype by promoting cryptic genetic variation. However, in the long term, canalization can act as a capacitor for phenotypic change, since it allows mutant alleles to accumulate in the genome without being expressed. Such cryptic genetic variability may be made manifest by a change in the environmental conditions, and can then be selected.

Canalization can be the result of genetic redundancy. One of the major discoveries of the last decade in developmental biology has been the stability of the phenotype even after the deletion of major developmentally important genes (Wilkins 1997; Morange 2001). In many instances, the loss of function of a particular gene is compensated for by the activation of another gene—sometimes not even in the same structural family as the deleted gene. In other instances, there is already a protein in the cell whose activities are partially redundant to those of the protein encoded by the deleted gene (Erickson 1993; Wilkins 1997). Nowak and colleagues (1997) have provided mathematical models to explain how such redundancy can be selected for, and how redundancy can be made evolutionarily stable.

Canalization can also result from the buffering capacity of heat shock proteins. It has long been known that stress, in the form of environmental factors such as temperature, can overpower the buffering systems of development and alter the phenotype. The altered phenotype then becomes subject to natural selection, and if selected, will eventually appear in the absence of the stress that originally induced it. Waddington called this phenomenon genetic assimilation (see Website 22.5). For instance, when Waddington subjected *Drosophila* larvae of a certain strain to high temperatures, they lost their wing crossveins. After a few generations of repeated heat shock, this "crossveinless" phenotype continued to be expressed even without the heat shock treatment. Although Waddington's results look like a case of "inheritance of acquired characteristics," there is no evidence for that view. Certainly, the crossveinless phenotype was not an adaptive response to heat, nor did heat shock cause the mutations that led to it. Rather, the heat shock overcame the buffering systems, allowing preexisting mutations to give rise to the new phenotype.

In 1998, Suzanne Rutherford and Susan Lindquist showed that a major agent responsible for this buffering was the "heat shock protein" Hsp90. Hsp90 binds to a set of signal transduction molecules that are inherently unstable. When it binds to them, it stabilizes their tertiary structure so that they can respond to upstream signaling molecules. Heat shock, however, causes other proteins in the cell to become unstable, and Hsp90 is diverted from its normal function (i.e., stabilizing signal transduction proteins) to the more general function of stabilizing any of the now partially denatured peptides in the cell (Jakob et al. 1995; Nathan et al. 1997). Since Hsp90 was known to be involved with inherently unstable proteins and could be diverted by stress, the researchers suspected that it might be involved in buffering developmental pathways against environmental contingencies.

Evidence for the role of Hsp90 as a developmental buffer first came from mutations of *Hsp83*, the gene for Hsp90. Homozygous mutations of *Hsp83* are lethal in *Drosophila*. Heterozygous mutations increase the proportion of developmental abnormalities; in *Drosophila* populations heterozygous for mutant *Hsp83*, deformed eyes, bristle duplications, and abnormalities of legs and wings appeared (Figure 23.26). When different mutant alleles of *Hsp83* were brought together in the same flies, both the incidence and severity of the abnormalities increased. Abnormalities were also seen when a specific inhibitor of Hsp90 (geldanamycin) was added to the food of wild-type flies, and the types of defects seen differed between different stocks of flies. The abnormalities observed did not show simple Mendelian inheritance, but were the outcome of the interactions of several gene

(Continued on next page)

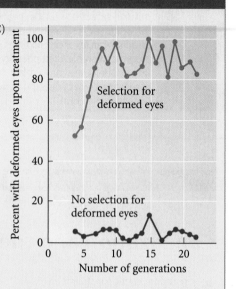

FIGURE 23.26 Hsp90 buffers development. (A,B) Developmental abnormalities in *Drosophila* associated with mutations in the *Hsp83* gene include (A) deformed eyes and (B) thickened wing veins. (C) The deformed eye trait seen in (A) was selected by breeding only those individuals expressing the trait. This abnormality was not observed in the original stock, but it can be seen in a high proportion of the descendants of individuals who are mated to heterozygous *Hsp83* flies. The strong response to selection showed that even though the population was small, it contained a large amount of hidden genetic variation. (After Rutherford and Lindquist 1998; photographs courtesy of the authors.)

products. Selective breeding of the flies with the abnormalities led over a few generations to populations in which 80 to 90 percent of the progeny had the mutant phenotype. But not all of the mutant progeny carried the *Hsp83* mutation. In other words, once a mutation in *Hsp83* had allowed cryptic mutations to be expressed, selective matings could retain the abnormal phenotype even in the absence of abnormal Hsp90.

Thus, Hsp90 is probably a major component of a buffering system that enables the canalization of development. It provides one way to resist phenotype fluctuations that would otherwise result from slight mutations or slight environmental changes. Hsp90 might also be responsible for allowing mutations to accumulate, but keeping them from being expressed until the environment changes. No individual mutation would change the phenotype, but mating would allow these mutations to be "collected" by members of the population. An environmental change (anything that might stress the cells) might thereby release the hidden phenotypic possibilities of the population. In other words, transient decreases in Hsp90 (resulting from its aiding stress-damaged proteins) would uncover preexisting genetic variations that would produce morphological variations. Most of these morphological variations would probably be deleterious, but some might be selected for in the new environment. Such release of hidden morphological variation may be responsible for the many examples of rapid speciation found in the fossil record.

A New Evolutionary Synthesis

In 1922, Walter Garstang declared that **ontogeny** (an individual's development) does not recapitulate **phylogeny** (evolutionary history); rather, it creates phylogeny. Evolution is generated by heritable changes in development. "The first bird," said Garstang, "was hatched from a reptile's egg." Thus, when we say that the contemporary one-toed horse evolved from a five-toed ancestor, we are saying that heritable changes occurred in the differentiation of the limb mesoderm into chondrocytes during embryogenesis in the horse lineage. Evolution, said Richard Goldschmidt, is the result of heritable changes in development, and this is as true for whether a fly has two or three bristles on its back as for whether an appendage is to become a fin or a limb (Goldschmidt 1940).

This view of evolution as the result of hereditary changes affecting development was largely lost during the 1940s, when the **Modern Synthesis** of population genet-

ics and evolutionary biology formed a new framework for research in evolutionary biology. The Modern Synthesis has been one of the great intellectual achievements of biology. By merging the traditions of Darwin and Mendel, evolution within a species could be explained: Diversity within a population arose from the random production of mutations, and the environment acted to select the most fit phenotypes. Those individuals most capable of reproducing would transmit the genes that gave them their advantage. Such genes included, for example, those encoding enzymes with higher rates of synthesis, or those encoding globins with greater oxygen-carrying capacity. It was assumed that the same kinds of changes (genetic or chromosomal mutations) that caused evolution within a species also caused the evolution of new species. There would need to be an accumulation of these mutations, and a mechanism of reproductive isolation to enable them to accumulate in new combinations, if a new phenotype was to be produced.

This population genetic model of natural selection is based on genetic differences in adult organisms of a given species competing for reproductive advantage. The developmental genetic model has been formulated to account for phylogeny—evolution above the species level. It is based on the similarities among regulatory genes that are active in embryos and larvae. We are still approaching evolution in the two ways that Darwin recognized. Both views involve descent with modification, and one can emphasize either the similarities or the differences between taxa. Thus, when confronted with the question of how the arthropod body plan arose, Hughes and Kaufman (2003) begin their study, "To answer this question by invoking natural selection is correct—but insufficient. The fangs of a centipede … and the claws of a lobster accord these organisms a fitness advantage. However, the crux of the mystery is this: From what developmental genetic changes did these novelties arise in the first place?"

There is emerging a rapprochement between the population genetic and the developmental genetic accounts of evolution. The population genetic approach has focused on variation within populations, while the developmental genetic approach has focused on variation between populations (Amundson 2001, 2005; Gilbert 2000). Similarly, population geneticists have been looking primarily at genes in adults competing for reproductive success, while developmental geneticists have been looking at genes involved in forming embryonic and larval organs. These differences are becoming blurred as both population geneticists and developmental geneticists begin looking at the regulatory genes that control development (see Arthur 1997; Macdonald and Goldstein 1999; Zeng et al. 1999). The two approaches complement each other: While the population genetic approach focuses on the survival of the fittest, the developmental genetic approach to evolution is more concerned with the *arrival* of the fittest (see Müller and Newman 2005).

WEBSITE

23.11 Population genetics versus developmental biology. The population genetic approach to evolution contains certain assumptions that had impeded the emergence of evolutionary developmental biology. Today, a rapprochement is being "negotiated" between these two approaches to the study of evolution.

Evolutionary developmental biology is already providing answers to classic evolutionary genetic questions. Evolutionary biologists have long studied problems such as mimicry and industrial melanism. Now, the genes involved in these processes are being identified so that the mechanisms of underlying phenomena can be explained (Koch et al. 1998; Brakefield 1998). To explain evolution, both the population genetic and the developmental genetic accounts are required (Figure 23.27).

Leaving developmental biology out of evolutionary biology has left evolutionary biology open to attacks from promoters of "intelligent design." According to one of

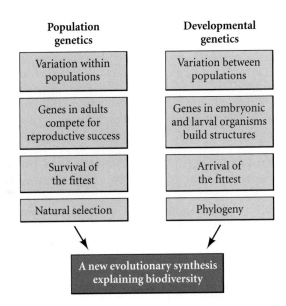

FIGURE 23.27   An emerging evolutionary synthesis. The classical approach to evolution has been that of population genetics, which emphasizes variations within a species that enable certain adult individuals to reproduce more frequently; thus, population genetics could explain natural selection. The developmental approach looks at variation between populations and emphasizes the regulatory genes responsible for organ formation. Developmental genetics is better able to explain evolutionary novelty and constraint. Taken together, the two approaches constitute a more complete genetic approach to evolution.

them, Michael Behe (1996), population genetics cannot explain the origin of structures such as the eye, so Darwinism must be false.* How could such a complicated structure have emerged by a collection of chance mutations? Mutations, claims Behe, would serve only to destroy complex organs, not create them.

But once development is added to the evolutionary synthesis, it is straightforward to see how the eye can develop through induction, and that the concepts of modularity and correlated progression can readily explain such a phenomenon (Waddington 1940; Gehring 1998). Moreover, when one sees that the formation of eyes in all known phylogenetic lineages is based on the same signal transduction pathway and uses the *Pax6* gene, it is not difficult to see descent with modification forming the various types of eyes. This "leap" was much more difficult before the similarity of eye induction mechanisms was understood. Indeed, one study based in population genetics claimed that photoreceptors of eyes arose independently more than

---

*Behe (1996) makes this point explicitly, using the example of the eye. Although he attempts to disprove the theory of evolution by using the eye as an example, he never mentions the myriad studies on *Pax6* or reciprocal induction. Rather, Behe mentions theories from the 1980s (based solely on population genetics) and puts them forth as contemporary science.

40 times during the course of animal evolution (Salvini-Plawen and Mayr 1977). By integrating population genetics with developmental genetics and embryology, we can now begin to explain the construction and evolution of different organs.

In his review of evolution in 1953, J. B. S. Haldane expressed his thoughts about evolution with the following developmental analogy: "The current instar of the evolutionary theory may be defined by such books as those of Huxley, Simpson, Dobzhansky, Mayr, and Stebbins [the founders of the Modern Synthesis]. We are certainly not ready for a new moult, but signs of new organs are perhaps visible." This recognition of developmental ideas "points forward to a broader synthesis in the future."

We have finally broken through the old pupal integument, and a new, broader, developmentally inclusive evolutionary synthesis is taking wing.

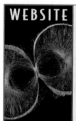

WEBSITE 23.12 "Scientific Creationism" and "Intelligent Design." Modern creationists often deny that genes play a role in development, because this would enable evolution to occur. These websites look at "scientific creationism" and "intelligent design" and include a lecture on how evolutionary developmental biology provides some of the best evidence against creationism.

Snapshot Summary **Evolutionary Developmental Biology**

1. Evolution is the result of inherited changes in development. Modifications of embryonic or larval development can create new phenotypes that can then be selected.

2. Darwin's concept of "descent with modification" explained both homologies and adaptations. The similarities of structure are due to common ancestry (homology), while the modifications were due to natural selection (adaptation to the environmental circumstances).

3. The Urbilaterian ancestor can be extrapolated by looking at the developmental genes that are common to protostomes and deuterostomes and that perform similar functions. These include the Hox genes that specify body segments; the *tinman* gene that regulates heart development; the *Pax6* gene that specifies those regions able to form eyes; and the genes that instruct head and tail formation.

4. The ways of effecting evolutionary change through development are: change in location (heterotopy), change in timing (heterochrony), change in amount (heterometry) and change in kind (heterotypy).

5. Changes in their sequence can give Hox genes new properties that may have significant developmental effects. The constraint on insect anatomy of having only six legs is one example.

6. Changes in the targets of Hox genes can alter the effects of those genes. In response to the Ubx protein, for example, the third thoracic segment specifies halteres in flies but hindwings in butterflies.

7. Changes of Hox gene expression within a region can alter the structures formed by that region. For instance, changes in the expression of *Ubx* and *abdA* in insects regulate the production of prolegs in the abdominal segments of the larvae.

8. Changes in Hox gene expression between body regions can alter the structures formed by each region. In crustaceans, different Hox expression patterns enable the body to have or to lack maxillipeds on its thoracic segments.

9. Changes in Hox gene expression are correlated with the limbless phenotypes of snakes.

10. Changes in Hox gene number may allow Hox genes to take on new functions. Large changes in the numbers of Hox genes correlate with major transitions in evolution.

11. Duplications of genes may enable these genes to become expressed in new places. The formation of new cell types may result from duplicated genes whose regulation has diverged.

12. In addition to structures and genes being homologous, signal transduction pathways can be homologous. In these cases, homologous proteins are organized in homologous ways. These pathways can be used for different developmental processes both in different organisms and within the same organism.

13. The modularity of development allows parts of the embryo to change without affecting other parts.

14. The dissociation of one module from another is shown by heterochrony (a shift in the timing of the development of one region with respect to another) and by allometry (a shift in the growth rates of different parts of the organism relative to one another).

15. Allometry can create new structures (such as the pocket gopher cheek pouch) by crossing a threshold.

16. Duplication and divergence are important mechanisms of evolution. On the genetic level, the Hox genes and many other gene families started as single genes that were duplicated. The divergent members of such a gene family can assume different functions.

17. Changes in gene expression appear to account for the evolution of the turtle shell, the loss of limbs in snakes, the emergence of feathers, and the evolution of differently shaped molars.

18. Co-option (recruitment) of existing genes and pathways for new functions is a fundamental mechanism for creating new phenotypes. One such case is the use of the limb development signaling pathway to form eyespots in butterfly wings.

19. Developmental constraints prevent certain phenotypes from occurring. Such constraints may be physical (no rotating limbs), morphogenetic (no middle digit smaller than its neighbors), or phyletic (no neural tube without a notochord).

20. The merging of the population genetic model with the developmental genetic model of evolution is creating a new evolutionary synthesis that can account for macroevolutionary as well as microevolutionary phenomena.

For Further Reading

Complete bibliographical citations for all literature cited in this chapter can be found on the Vade Mecum CD that accompanies the book and at the free access website www.devbio.com

Abzhanov, A., M. Protas, B. R. Grant, P. R. Grant and C. J. Tabin. 2004. *Bmp4* and morphological variation of beaks in Darwin's finches. *Science* 305: 1462–1465.

Cohn, M. J. and C. Tickle. 1999. Developmental basis of limblessness and axial patterning in snakes. *Nature* 399: 474–479.

Galant, R. and S. B. Carroll. 2002. Evolution of a transcriptional repression domain in an insect Hox protein. *Nature* 415: 910–913.

Gilbert, S. F., J. M. Opitz and R. A. Raff. 1996. Resynthesizing evolutionary and developmental biology. *Dev. Biol.* 173: 357–372.

Merino, R., J. Rodríguez-Leon, D. Macias, Y. Ganan, A. N. Economides and J. M. Hurle. 1999. The BMP antagonist Gremlin regulates outgrowth, chondrogenesis and programmed cell death in the developing limb. *Development* 126: 5515–5522.

Rockman, M. V., Hahn, M. W., Soranzo, N., Goldstein, D. B., and Wray, G. A. 2003. Positive selection on a human-specific transcription factor binding site regulating IL4 expression. *Curr. Biol.* 13: 2118–2123.

Ronshaugen, M., N. McGinnis and W. McGinnis. 2002. Hox protein mutation and macroevolution of the insect body plan. *Nature* 415: 914–917.

Shapiro, M. D. and 7 others. 2004. Genetic and developmental basis of evolutionary pelvic reduction in three-spine sticklebacks. *Nature* 428: 717–723.

Smith, K. 2003. Time's arrow: Heterochrony and the evolution of development. *Int. J. Dev. Biol.* 47: 613–621.

Illustration Sources

Preface

Lines from *Truckin'* © 1970 Ice Nine Publishing. All rights reserved. Music by Jerry Garcia, Phil Lesh, Bob Weir, lyrics by Robert Hunter.

Chapter 1

Chapter Opening Material

Virgil. 37 BCE. *Georgics* II: 490.

Oppenheimer, J. M. 1955. Analysis of development: Problems, concepts, and their history. In *Analysis of Development*, B. H. Willier, P. A. Weiss, and V. Hamburger (eds.). Saunders, Philadelpha, pp. 1–24.

Drawing by the artist/embryologist Emil Witschi (1890–1971) showing a human embryo at 4–5 weeks gestation. From E. Witschi, *Development of Vertebrates*, 1956. Frontispiece. Reprinted with permission of Elsevier.

Illustration Credits

Figure 1.2: (A) Malpighi, M. 1672. *De Formatione Pulli in Ovo* (London). Reprinted in H. B. Adelmannm, *Marcello Malpighi and the Evolution of Embryology*. Cornell University Press, Ithaca, NY, 1966. (B) Lillie, F. R. 1908. *The Embryology of the Chick*. Henry Holt, New York. (C) Pander, C. 1817. *Beiträge zur Entwickelungsgeschichte des Hünchens im Eye*. Brönner, Würzburg. (D) Carlson, B. M. 1981. *Patten's Foundations of Embryology*. McGraw-Hill, New York.

Figure 1.4A: Patten, B. M. 1951. *The Early Embryo of the Chick*. 4th Ed. McGraw-Hill, New York.

Figure 1.5: Richardson, M. K., J. Hanken, L. Selwood, G. M. Wright, R. J. Richards, C. Pieau

and A. Raynaud. 1998. Haeckel, embryos, and evolution. *Science* 280: 983–984.

Figure 1.7: (A) Nishida, H. 1987. Cell lineage analysis in ascidian embryos by intracellular injection of a tracer enzyme. III. Up to the tissue-restricted stage. *Dev. Biol.* 121: 526–541. (B) Conklin, E. G. 1905. The organization and cell lineage of the ascidian egg. *J. Acad. Nat. Sci. Phila.* 13: 1–119.

Figure 1.8: Vogt, W. 1929. Gestaltungsanalyse am Amphibienkeim mit örtlicher Vitalfärbung. II. Teil Gastrulation und Mesodermbildung bei Urodelen und Anuren. *Wilhelm Roux Arch. Entwicklungsmech. Org.* 120: 384–706.

Figure 1.9: (A,B) Kozlowski, D. J., T. Muramaki, R. K. Ho and E. S. Weinberg. 1998. Regional cell movement and tissue patterning in the zebrafish embryo revealed by fate mapping with caged fluorescein. *Biochem. Cell Biol.* 75: 551–562. (C) Woo, K. and S. E. Fraser. 1995. Order and coherence in the fate map of the zebrafish embryo. *Development* 121: 2595–2609.

Figure 1.10: Darnell, D. K. and G. C. Schoenwolf. 1997. Modern techniques for labeling in avian and murine embryos. *In* G. P. Daston (ed.), *Molecular and Cellular Methods in Developmental Toxicology*. CRC Press, Boca Raton, FL, pp. 231–272.

Figure 1.11B: Weston, J. 1963. A radiographic analysis of the migration and localization of trunk neural crest cells in the chick. *Dev. Biol.* 6: 274–310.

Figure 1.12: Müller, F. 1864. *Für Darwin*. Engelmann, Leipzig.

Figure 1.14A: Zangerl, R. and M. E. Williams. 1975. New evidence on the nature of the jaw suspen-

sion in Paleozoic anacanthus sharks. *Paleontology* 18: 333–341.

Figure 1.16B: Nowack, E. 1965. Die sensible Phase bei der Thalidomide-Embryopathie. *Humangenetik* 1: 516–536.

Figure 1.17B: Thompson, D. W. 1942. *On Growth and Form*. Cambridge University Press, Cambridge.

Figure 1.18: Moore, K. L. 1983. *The Developing Human*. 3rd Ed. Saunders, Philadelphia.

Figure 1.21: Meinhardt, M. 2003. *The Algorhythmic Beauty of Sea Shells*. 3rd Ed. Springer, Berlin.

Table 1.2: Thompson, D. W. 1942. *On Growth and Form*. Cambridge University Press, Cambridge.

Chapter 2

Chapter Opening Material

Bonner, J. T. 1965. *Size and Cycle: An Essay on the Structure of Biology*. Princeton University Press, Princeton, NJ, p. 3.

Lines from *The Circle of Life* © 1994 Wonderland Music Co. Inc. All rights reserved. Music by Elton John, lyrics by Tim Rice.

Photograph © Patrice Ceisel/ Visuals Unlimited. See Figure 2.4.

Illustration Credits

Figure 2.7A: Kirk, M. M. and D. L. 2004. Exploring germ-soma differentiation in *Volvox. J. Biosci.* 29: 143–152.

Figure 2.8: (A) Kirk, D. L. 1988. The ontogeny and phylogeny of cellular differentiation in *Volvox. Trends Genet.* 4: 32–36. (B) Cover photo by D. L. Kirk. November 1, 2001. *Dev. Biol.* 239(1).

Figure 2.11: (B) Tomchick, K. J. and P. N. Devreotes. 1981. Adenosine 3′,5′ monophosphate

waves in *Dictyostelium discoideum. Science* 212: 443–446. (C) Figure 2.11C: Siegert, F. and C. J. Weijer. 1989. Digital image processing of optical density wave propagation in *Dictyostelium discoideum* and analysis of the effects of caffeine and ammonia. *J. Cell Sci.* 93: 325–335. (D) Dallon, J. C. and H. G. Othmer. 1997. A discrete cell model with adaptive signalling for aggregation of *Dictyostelium discoideum. Phil. Trans. Roy. Soc. Lond.* [B] 352: 391–417.

Figure 2.15: Sultan, S. E. 2000. Phenotypic plasticity for plant development, function, and life history. *Trends Plant Sci.* 5: 537–542.

Figure 2.16: Glenner, H., A. J. Hansen, M. V. Sorensen, F. Ronquist, J. P. Huelsenbeck, and E. Willerslev. 2004. Bayesian inference of the metazoan phylogeny: A combined molecular and morphologicalapproach. *Curr. Biol.* 14: 1644 – 1649.

Figure 2.18: Newmark, P. A. and A. J. Alvarado. 2000. Bromodeoxyuridine specifically labels the regenerative stem cells of planarians. *Dev. Biol.* 220: 142 – 153.

Chapter 3

Chapter Opening Material

Bard, J. 1997. Explaining development. *BioEssays* 20: 598–599.

Morgan, T. H. 1898. "Some Problems of Regeneration." Biological Lectures Delivered at the Marine Biology Laboratory, Woods Hole, 1898, p. 207.

Photograph courtesy of M. S. Steinberg and R. A. Foty. See Figure 3.27.

Illustration Credits

Figure 3.2: Leutert, T. R. 1974. Zur Geschlechtsbestimmung und Gametogenese von *Bonellia viridis* Rolando. *J. Embryol. Exp. Morphol.* 32: 169–193.

Figure 3.5: Adams, N. L. and J. M. Shick. 2001. Mycosporine-like amino acids prevent UVB-induced abnormalities during early development of the green sea urchin *Strongylocentrotus droebachiensis. Mar. Biol.* 138: 267–280.

Figure 3.6: Blaustein, A. R., P. D. Hoffman, D. G. Hokit, J. M. Kiesecker, S. C. Walls and J. B. Hays. 1994. UV repair and resistance to solar UV-B in amphibian eggs: A link to population declines? *Proc. Natl. Acad. Sci. USA* 91: 1791–1795.

Figure 3.7: Wilson, E. B. 1904. Experimental studies on germinal location. *J. Exp. Zool.* 1: 1–72.

Figure 3.8: Reverberi, G. and A. Minganti. 1946. Fenomeni di evocazione nello sviluppo dell'uovo di Ascidie. Risultati dell'indagine spermentale sull'ouvo di *Ascidiella aspersa* e di *Ascidia malaca* allo stadio di 8 blastomeri. *Pubbl. Staz. Zool. Napoli* 20: 199–253. (Quoted in G. Reverberi,. *Experimental Embryology of Marine and Fresh-water Invertebrates.* North-Holland, Amsterdam 1971, p. 537.)

Figure 3.9: Whittaker, J. R. 1977. Segregation during cleavage of a factor determining endodermal alkaline phosphatase development in ascidian embryos. *J. Exp. Zool.* 202: 139–153.

Figure 3.10: Whittaker, J. R. 1982. Muscle cell lineage cytoplasm can change the developmental expression in epidermal lineage cells of ascidian embryos. *Dev. Biol.* 93: 463–470.

Figure 3.12: Wolpert, L. R. Beddington, T. Jessell, P. Lawrence, E. Meyerowitz and J. Smith. 2002. *Principles of Development*, 2nd Ed. Oxford University Press, New York.

Figure 3.15: Wilson, E. B. 1896. *The Cell in Development and Inheritance.* Macmillan, New York.

Figure 3.18: Huxley, J. and G. R. de Beer. 1934. *The Elements of Experimental Embryology.* Cambridge University Press, Cambridge.

Figure 3.19: Wolpert, L. 1978. Pattern formation in biological development. *Sci. Am.* 239(4): 154–164.

Figure 3.22: National Institutes of Health. 2001. *Stem Cells: Scientific Progress and Future Research Directions.* http://www.nih.gov/news/stemcell/fullrptstem.pdf

Figures 3.23 and 3.24: Townes, P. L. and J. Holtfreter. 1955. Directed movements and selective adhesion of embryonic amphibian cells. *J. Exp. Zool.* 128: 53–120.

Figure 3.25: Monroy, A. and A. A. Moscona. 1979. *Introductory Concepts in Developmental Biology.* University of Chicago Press, Chicago.

Figure 3.26: Armstrong, P. B. 1989. Cell sorting out: The self-assembly of tissues in vitro. *CRC Crit. Rev. Biochem. Mol. Biol.* 24: 119–149.

Figure 3.27: Foty, R. A., C. M. Pfleger, G. Forgacs and M. S. Steinberg. 1996. Surface tensions of embryonic cells predict their mutual envelopment behavior. *Development* 122: 1611–1620.

Figure 3.28: Takeichi, M. 1991. Cadherin cell adhesion receptors as a morphogenetic regulator. *Science* 251: 1451–1455.

Figure 3.29A: Heasman, J., D. Ginsberg, K. Goldstone, T. Pratt, C. Yoshidanaro and C. Wylie. 1994. A functional test for maternally inherited cadherin in *Xenopus* shows its importance in cell adhesion at the blastula stage. *Development* 120: 49–57.

Figure 3.29B: Kintner, C. 1993. Regulation of embryonic cell adhesion by the cadherin cytoplasm domain. *Cell* 69: 225–236.

Table 3.2: Davidson, E. H. 1991. Spatial mechanisms of gene regulation in metazoan embryos. *Development* 113: 1–26.

Chapter 4

Chapter Opening Material

Brenner, S. 1979. Quoted in H. F. Judson, *The Eighth Day of Creation.* Simon & Schuster, New York, p. 205.

Ozick, C. 1989. *Metaphor and Memory.* Alfred A. Knopf, New York, p. 111.

Photograph courtesy of J. Gurdon. See Figure 4.7.

Illustration Credits

Figure 4.2A: Baltzer, F. 1967. *Theodor Boveri: Life and Work of a Great Biologist.* (Trans. D. Rudnick.) University of California Press, Berkeley.

Figure 4.4B: Waddington, C. H. 1948. *The Scientific Attitude.* Pelican Books, New York.

Figure 4.5: King, T. J. 1966. Nuclear transplantation in amphibia. *Methods Cell Physiol.* 2: 1–36.

Figure 4.6: McKinnell, R. G. 1978. *Cloning: Nuclear Transplantation in Amphibia.* University of Minnesota Press, Minneapolis.

Figure 4.8B: Wilmut, I., K. Campbell and C. Tudge. 2000. *The Second Creation: Dolly and the Age of Biological Control.* Harvard University Press, Cambridge, MA.

Figure 4.11: (A) Burkholder, G. D. 1976. Whole mount electron microscopy of polytene chromosome from *Drosophila melanogaster. Can. J. Genet. Cytol.* 18: 67–77. (B) Barnett, T., C. Pachl, J. P. Gergen and P. C. Wensink. 1980. The isolation and characterization of *Drosophila* yolk protein genes. *Cell* 21: 729–738.

Figure 4.15: Grindley, J. C., D. R., Davidson and R. E. Hill. 1995. The role of *Pax-6* in eye and nasal development. *Development* 121: 1433–1442.

Figure 4.16: Li, H.-S., J.-M. Yang, R. D. Jacobson, D. Pasko and O. Sundin. 1994. *Pax-6* is first expressed in a region of ectoderm anterior to the early neural plate: Implications for stepwise determination of the lens. *Dev. Biol.* 162: 181–194.

Figure 4.17: Wagner, T. E., P. Hoppe, J. D. Jollick, D. R. Scholl, R. L. Hodinka and J. B. Gault. 1981. Microinjection of rabbit β-globin gene into zygotes and its subsequent expression in adult mice and offspring. *Proc. Natl. Acad. Sci. USA* 78: 6376–6380. |

Figure 4.20: Dudley, A. T., K. M. Lyons and E. J. Robertson. 1995. A requirement for bone morphogenetic protein 7 during development of the mammalian kidney and eye. *Genes Dev.* 9: 2795–2807.

Figure 4.21B: Rosenberg, U. B., A. Preiss, E. Seifert, H. Jäckle and D. C. Knipple. 1985. Production of phenocopies by *Krüppel* antisense RNA injection into *Drosophila* embryos. *Nature* 313: 703–706.

Figure 4.23: Wianny, F. and M. Zernicka-Goetz. 2000. Specific interference with gene function by double-stranded RNA in early mouse development. *Nature Cell Biol.* 2: 70–75.

Chapter 5

Chapter Opening Material

Claude, A. 1974. The coming of age of the cell. Nobel lecture, reprinted in *Science* 189: 433–435.

Waddington, C. H. 1956. *Principles of Embryology.* Macmillan, New York, p. 5.

Photograph courtesy of K. Luger et al., 1997. See Figure 5.1A.

Illustration Credits

Figure 5.1: (A) Luger , K., A. W. Mäder, R. K. Richmond, D. F. Sargent and T. J. Richmond. 1997. Crystal structure of the nucleosome core particle at 2.8 Å resolution. *Nature* 389: 251–260. (B,C) Wolfe, S. L. 1993. *Molecular and Cellular Biology.* Wadsworth, Belmont, CA.

Figure 5.2: Lawn, R. M., A. Efstratiadis, C. O'Connell and T. Maniatis. 1980. The nucleotide sequence of the human β-globin gene. *Cell* 21: 647–651.

Figure 5.5: Chen, J.-L., L. D. Attardi, C. P. Verrijzer, K. Yokomori and R. Tjian. 1995. Assembly of recombinant TFIID reveals differential cofactor requirements for distinct transcriptional activators. *Cell* 79: 93–105.

Figure 5.6B: Offield, M., F. N. Hirsch and R. M. Grainger. 2000. The use of *Xenopus tropicalis* transgenic lines for studying lens developmental timing in living embryos. *Development* 127: 1789–1797.

Figure 5.7: (A) Williams, S. C., C. R. Altmann, R. L. Chow, A. Hemmati-Brivanlou and R. A. Lang. 1998. A highly conserved lens transcriptional control element from the *Pax-6* gene. *Mech. Dev.* 73: 225–229. (B) Kammandel, B., K. Chowdhury, A. Stoykova, S. Aparicio, S. Brenner and P. Gruss. 1998. Distinct *cis*-essential modules direct the time-space pattern of *Pax6* gene activity. *Dev. Biol.* 205: 79–97.

Figure 5.8: Steingrímsson, E. and 10 others. 1994. Molecular basis of mouse *microphthalmia* (*mi*) mutations helps explain their developmental and phenotypic consequences. *Nature Genet.* 8: 256–263.

Figure 5.9: Nakayama, A., M.-T. Nguyen, C. C. Chen, K. Opdecamp, C. A. Hodgkinson and H. Arnheiter. 1998. Mutations in *microphthalmia*, the mouse homolog of the human deafness gene *MITF*, affect neuroepithelial and neural crest-derived melanocytes differently. *Mech. Dev.* 70: 155–166.

Figure 5.10: Pennisi, E. 2000. Matching the transcription machinery to the right DNA. *Science* 288: 1372–1373.

Figure 5.11: Xu, W., M. A. S. Rould, S. Jun, C. Desplan and S. O. Pääbo. 1995. Crystal structure of a paired domain-DNA complex at 2.5 Å resolution reveals structural basis for *Pax* developmental mutations. *Cell* 80: 639–650.

Figure 5.12: (A) Cvekl, A. and J. Piatigorsky. 1996. Lens development and crystallin gene expression: Many roles for *Pax-6. BioEssays* 18: 621–630. (B) Andersen, F. G., J. Jensen, R. S. Heller, H. V. Petersen, L.-I. Larsson, O. D. Madsen and P. Serup. 1999. *Pax6* and *Pdx1* form a functional complex on the rat somatostatin gene upstream enhancer. *FEBS Lett.* 445: 315–320.

Figure 5.13: Tapscott, S. J. 2005. The circuitry of a master switch: MyoD and the regulation of skeletal muscle gene transcription. *Development* 132: 2685–2695.

Figure 5.14: (A) Beimesche, S. and 8 others. 1999. Tissue-specific transcriptional activity of a pancreatic islet cell-specific enhancer sequence/Pax6 binding site determined in normal adult tissues in vivo using transgenic mice. *Mol. Endocrinol.* 13: 718–728. (B) Cvekl, A., C. M. Sax, X. Li, J. B. McDermott and J. Piatigorsky. 1995. Pax-6 and lens-specific transcription of the chicken δ1-crystallin gene. *Proc. Natl. Acad. Sci. USA* 92: 4681–4685.

Figure 5.18: Kallunki, P., G. M. Edelman and F. S. Jones. 1997. Tissue-specific expression of the L1 cell adhesion molecule is modulated by the neural restrictive silencer element. *J. Cell Biol.* 138: 1343–1355.

Figure 5.19: Mavilio, F. and 9 others. 1983. Molecular mechanisms for human hemoglobin switching: Selective undermethylation and expression of globin genes in embryonic, fetal, and adult erythroblasts. *Proc. Natl. Acad. Sci. USA* 80: 6907–6911.

Figure 5.20B: Walter, J. and M. Paulsen. 2003. Imprinting and disease. *Seminars Cell Dev. Biol.* 14: 101–110.

Figure 5.21C: Litt, M. D., M. Simpson, M. Gaszner, C. D. Allis and G. Felsenfeld. 2001. Correlation between histone lysine methylation and developmental changes at the chicken β-globin locus. *Science* 293: 2453–2455.

Figure 5.22B,C: Sugimoto, M., S.-S. Tan and N. Takagi. 2000. X chromosome inactivation revealed by the X-linked *lacZ* transgene activity in periimplantation mouse embryos. *Int. J. Dev. Biol.* 44: 177–182.

Figure 5.23B: Reik, W. and A. Lewis. 2005. Co-evolution of X-chromosome inactivation and imprinting in mammals. *Nature Rev. Genet.* 6: 403 –418.

Figure 5.24B–D: Sheardown, S. A. and 9 others. 1997. Stabilization of *Xist* RNA mediates initiation of X chromosome inactivation. *Cell* 91: 99–107.

Figure 5.25: Jeppesen, P. and B. M. Turner. 1993. The inactive X chromosome in female mammals is distinguished by a lack of histone H4 acetylation, a cytogenetic marker for gene expression. *Cell* 74: 281–289.

Figure 5.27: Gagnon, M. L., L. M. Angerer and R. C. Angerer. 1992. Posttranscriptional regulation of ectoderm-specific gene expres-sion in early sea urchin embryos. *Development* 114: 457–467.

Figure 5.28: McAlinden A., N. Havlioglu, and L. J. Sandell. Regulation of protein diversity by alternative pre-mRNA splicing with specific focus on chondrogenesis. *Birth Def. Res.* C 72: 51–68.

Figure 5.29: Breitbart, R. A., A. Andreadis and B. Nadal-Ginard. 1987. Alternative splicing: A ubiquitous mechanism for the generation of multiple protein isoforms from single genes. *Annu. Rev. Biochem.* 56: 481–495.

Figure 5.30: Yamakawa, K., Y. K. Huot , M. A. Haendelt, R. Hubert, X.-N. Chen, G. E. Lyons and J. R. Korenberg. 1998. *DSCAM*: A novel member of the immunoglobulin superfamily maps in a Down syndrome region and is involved in the development of the nervous system. *Hum. Mol. Genet.* 7: 227–237.

Figure 5.32: Guyette, W. A., R. J. Matusik and J. M. Rosen. 1979. Prolactin-mediated transcriptional and post-transcriptional control of casein gene expression. *Cell* 17: 1013–1023.

Figure 5.33: (A) Wells, S. E., P. E. Hilner, R. D. Vale, and A. B. Sachs. 1998. Circularization of mRNA by eukaryotic translation initiation factors. *Mol. Cell* 2: 135–40. (B,C) Mendez, R. and J. D. Richter. 2001. Translational control by CPEB: A means to the end. *Nature Rev. Mol. Cell Biol.* 2: 521–529.

Figure 5.34: Cho, P. F. and eight others. 2005. A New Paradigm for Translational Control: Inhibition via 5'-3' mRNA Tethering by Bicoid and the eIF4E Cognate 4EHP. *Cell* 121: 411–423.

Figure 5.35: Wang, H. and Tiedge, H. 2004. Translational control at the synapse. *Neuroscientist* 10:456–466.

Figure 5.36: Wickens, M. and K. Takayama. 1995. Deviants—or emissaries? *Nature* 367: 17–18.

Figure 5.37: He, L. and G. J. Hannon. 2004. MicroRNAs: small RNAs with a big role in gene regulation. *Nature Rev. Genet.* 5: 522–531.

Figure 5.39: Orphanides, G. and D. Reinberg. 2002. A unified theory of gene expression. *Cell* 108: 439–451.

Chapter 6

Chapter Opening Material

Butler, O. 1998. *Parable of the Talents*. Warner Books, New York, p. 3.

Jonas, H. 1966. *The Phenomenon of Life*. Dell, New York, p. x.

Photograph courtesy of W.-S. Kim and J. F. Fallon. See Figure 6.6B.

Illustration Credits

Figure 6.2: Fujiwara, M., T. Uchida, N. Osumi-Yamashita and K. Eto. 1994. Uchida rat (*rSey*): A new mutant rat with craniofacial abnormalities resembling those of the mouse *Sey* mutant. *Differentiation* 57: 31–38.

Figure 6.4 : (A) Jacobson, A. G. 1966. Inductive processes in embryonic development. *Science* 152: 25–34. (B) Grainger, R. M. 1992. Embryonic lens induction: Shedding light on vertebrate tissue determination. *Trends Genet.* 8: 349–356.

Figure 6.5A–E: Cvekl, A. and J. Piatigorsky. 1996. Lens development and crystallin gene expression: Many roles for Pax-6. *BioEssays* 18: 621–630.

Figure 6.7: Saunders, J. W., Jr. 1980. *Developmental Biology*. Macmillan, New York.

Figure 6.8: Hamburgh, M. 1970. *Theories of Differentiation*. Elsevier, New York.

Figure 6.9: (A) Muthukkarapan, V. R. 1965. Inductive tissue interaction in the development of the mouse lens in vitro. *J. Exp. Zool.* 159: 269–288. (B–D) Grobstein, C. 1956. Trans-filter induction of tubules in mouse metanephrogenic mesenchyme. *Exp. Cell Res.* 10: 424–440.

Figure 6.13: (A,B) Nakayama, A., M.-T. Nguyen, C. C. Chen, K. Opdecamp, C. A. Hodgkinson and H. Arnheiter. 1998. Mutations in *microphthalmia*, the mouse homolog of the human deafness gene *MITF*, affect neuroepithelial and neural crest-derived melanocytes differently. *Mech. Dev.* 70: 155–166. (C) Price, E. R. and 7 others. 1998. Lineage-specific signaling in melanocytes: c-Kit stimulation recruits p300/CBP to microphthalmia. *J. Biol. Chem.* 273: 17983–17986.

Figure 6.14: Groner, B. and F. Gouilleux. 1995. Prolactin-mediated gene activation in mammary epithelial cells. *Curr. Opin. Genet. Dev.* 5: 587–594.

Figure 6.15: Gilbert-Barness, E. and J. M. Opitz. 1996. Abnormal bone development: Histopathology and skeletal dysplasias. *In* M. E. Martini-Neri, G. Neri and J. M. Opitz (eds.), *Gene Regulation and Fetal Development*. March of Dimes Birth Defects Foundation Original Article Series 30: (1). Wiley-Liss, New York, pp. 103–156.

Figure 6.17: Johnson, R. L. and M. P. Scott. 1998. New players and puzzles in the hedgehog signal-ing pathway. *Curr. Opin. Genet. Dev.* 8: 450–456.

Figure 6.21: Hogan, B. L. M. 1996. Bone morphogenesis proteins: Multifunctional regulators of vertebrate development. *Genes Dev.* 10: 1580–1594.

Figure 6.22: Tulina, N. and E. Matunis. 2001. Control of stem cell self-renewal in *Drosophila* spermatogenesis by JAK-STAT signaling. *Science* 294: 2546–2549.

Figure 6.24: Adams, J. M. and S. Cory. 1998. The Bcl-2 protein family: Arbiters of cell survival. *Science* 281: 1322–1326.

Figure 6.25: Kuida, K. and 7 others. 1996. Decreased apoptosis in the brain and premature lethality in CPP32-deficient mice. *Nature* 384: 368–372.

Figure 6.27: Katz, W. S. and P. W. Sternberg. 1996. Intercellular signalling in *Caenorhabditis elegans* vulval pattern formation. *Semin. Cell Dev. Biol.* 7: 175–183.

Figure 6.28: Greenwald, I. and G. M. Rubin. 1992. Making a difference: The role of cell-cell interactions in establishing separate identities for equivalent cells. *Cell* 68: 271–281.

Figure 6.29: Greenwald, I. and G. M. Rubin. 1992. Making a difference: The role of cell-cell interactions in establishing separate identities for equivalent cells. *Cell* 68: 271–281.

Figure 6.30B: Dufour, S., J.-L. Duband, M. J. Humphries, M. Obara, K. M. Yamada and J. P. Thiery. 1988. Attachment, spreading and locomotion of avian neural crest cells are mediated by multiple adhesion sites on fibronectin molecules. *EMBO J.* 7: 2661–2671.

Figure 6.32: Luna, E. J. and A. L. Hitt. 1992. Cytoskeleton-plasma membrane interactions. *Science* 258: 955–964.

Figure 6.33: Hadley, M. A., S. W. Byers, C. A. Suárez-Quian, H. Kleinman and M. Dym. 1985. Extracellular matrix regulates Sertoli cell differentiation, testicular cord formation, and germ cell development in vitro. *J. Cell Biol.* 101: 1511–1512.

Figure 6.34: Bissell, M., I. S. Mian, D. Raditsky and E. Turley. 2003. Tissue-specificity: Structural cues allow diverse phenotypes form a constant genotype. *In* G. B. Müller and S. A. Newman (eds.), *Origin of Organismal Form*. MIT Press, Cambridge.

Figure 6.35A: Wary, K. K., A. Mariotti, C. Zurzolo and F. Giancotti. 1998. A requirement for caveolin-1 and associated kinase Fyn in integrin signaling and anchorage-dependent cell growth. *Cell* 94: 625–634.

Figure 6.36: Warner, A. E., S. C. Guthrie and N. B. Gilula. 1984. Antibodies to gap junctional protein selectively disrupt junctional communication in the early amphibian embryo. *Nature* 311: 127–131.

Chapter 7

Chapter Opening Material

Darwin, C. 1871. *The Descent of Man*. Murray, London, p. 893.

Whitman, W. 1855. "Song of Myself." In *Leaves of Grass and Selected Prose*. S. Bradley (ed.), 1949. Holt, Rinehart & Winston, New York, p. 25.

Photograph courtesy of S. Suarez. See Figure 7.29.

Illustration Credits

Figure 7.1: Hartsoeker, N. 1694. *Essai de dioptrique*. Paris.

Figure 7.2A: Clermont, Y. and C. P. Leblond. 1955. Spermiogenesis of man, monkey, and other animals as shown by the "periodic acid-Schiff" technique. *Am. J. Anat.* 96: 229–253.

Figure 7.2B: Sutovsky, P., C. S. Navara and G. Schatten. 1996. Fate of the sperm mitochondria and the incorporation, conversion, and disassembly of the sperm tail structures during bovine fertilization. *Biol. Reprod.* 55: 1195–1205.

Figure 7.3B: De Robertis, E. D. P., F. A. Saez and E. M. F. De Robertis. 1975. *Cell Biology*, 6th Ed. Saunders, Philadelphia; and Tilney, L. G., J. Bryan, D. J. Bush, K. Fujiwara, M. S. Mooseker, D. B. Murphy and D. H. Snyder. 1973. Microtubules: Evidence for 13 protofilaments. *J. Cell Biol.* 59: 267–275.

Figure 7.4: Epel, D. 1977. The program of fertilization. *Sci. Am.* 237(5): 128–138.

Figure 7.5: Austin, C. R. 1965. *Fertilization*. Prentice-Hall, Englewood Cliffs, NJ.

Figure 7.6: Schroeder, T. E. 1979. Surface area change at fertilization: Resorption of the mosaic membrane. *Dev. Biol.* 70: 306–327.

Figure 7.9: Ward, G. E., C. J. Brokaw, D. L. Garbers and V. D. Vacquier. 1985. Chemotaxis of *Arbacia punctulata* spermatozoa to resact, a peptide from the egg jelly layer. *J. Cell Biol.* 101: 2324–2329.

Figure 7.10: Kirkman-Brown, J. C., K. A. Sutton and H. M. Florman. 2003. How to attract a sperm. *Nature Cell Biol.* 5: 93–96.

Table 7.1: Whitaker, M. J. and R. Steinhardt. 1985. Ionic signalling in the sea urchin egg at fertilization. *In* C. B. Metz and A. Monroy (eds.), *Biology of Fertilization*, Vol. 3. Academic Press, Orlando, FL, pp 167–221.

Figure 7.11: Summers, R. G. and B. L. Hylander. 1974. An ultrastructural analysis of early fertilization in the sand dollar, *Echinarachnius parma*. *Cell Tissue Res.* 150: 343–368.

Figure 7.12: (A) Epel, D. 1977. The program of fertilization. *Sci. Am.* 237(5): 128–138. (B) Glabe, C. G. and V. D. Vacquier. 1978. Egg surface glycoprotein receptor for sea urchin sperm bindin. *Proc. Natl. Acad. Sci. USA* 75: 881–885.

Figure 7.13B,C: Moy, G. W. and V. D. Vacquier. 1979. Immunoperoxidase localization of bindin during the adhesion of sperm to sea urchin eggs. *Curr. Top. Dev. Biol.* 13: 31–44.

Figure 7.14: (B) Foltz, K. R., J. S. Partin and W. J. Lennarz. 1993. Sea urchin egg receptor for sperm: Sequence similarity of binding domain and hsp 70. *Science* 259: 1421–1425. (C) Kamei, N. and C. G. Glabe. 2003. The species-specific egg receptor for sea urchin sperm is ERB1, a novel ADAMTS protein. *Genes Dev.* 17: 2502–2507.

Figure 7.15A–C: Schatten, G. and D. Mazia. 1976. The penetration of the spermatozoon through the sea urchin egg surface at fertilization: Observations from the outside on whole eggs and from the inside on isolated surfaces. *Exp. Cell Res.* 98: 325–337.

Figure 7.16A–F: Boveri, T. 1907. Zellenstudien VI. Die Entwicklung dispermer Seeigeleier. Ein Beiträge zur Befruchtungslehre und zur Theorie des Kernes. *Jena Z. Naturwiss.* 43: 1–292.

Figure 7.16H: Simerly, C. and 7 others. 1999. Biparental inheritance of γ-tubulin during human fertilization: Molecular reconstitution of functional zygote centrosomes in inseminated human oocytes and in cell-free extracts nucleated by human sperm. *Mol. Cell Biol.* 10: 2955–2969.

Figure 7.17: Jaffe, L. A. 1980. Electrical polyspermy block in sea urchins: Nicotine and low sodium experiments. *Dev. Growth Diff.* 22: 503–507.

Figure 7.18: Vacquier, V. D. and J. E. Payne. 1973. Methods for quantitating sea urchin sperm in egg binding. *Exp. Cell Res.* 82: 227–235.

Figure 7.19: (A) Austin, C. R. 1965. *Fertilization*. Prentice-Hall, Englewood Cliffs, NJ. (B–E)

Chandler, D. E. and J. Heuser. 1979. Membrane fusion during secretion: Cortical granule exocytosis in sea urchin eggs as studied by quick-freezing and freeze fracture. *J. Cell Biol.* 83: 91–108.

Figure 7.21: (A) Luttmer, S. and F. J. Longo. 1985. Ultrastructural and morphometric observations of cortical endoplasmic reticulum in *Arbacia*, *Spisula*, and mouse eggs. *Dev. Growth Diff.* 27: 349–359. (B) McPherson, S. M., P. S. McPherson, L. Mathews, K. P. Campbell and F. J. Longo. 1992. Cortical localization of a calcium release channel in sea urchin eggs. *J. Cell Biol.* 116: 1111–1121.

Figure 7.22: Epel, D. 1980. Fertilization. *Endeavour* N.S. 4: 26–31.

Figure 7.23: (A) Gross, P. R., L. I. Malkin and W. Moyer. 1964. Templates for the first proteins of embryonic development. *Proc. Natl. Acad. Sci. USA* 51: 407–414. (B) Humphreys, T. 1971. Measurements of messenger RNA entering polysomes upon fertilization in sea urchins. *Dev. Biol.* 26: 201–208.

Figure 7.25: Voronina, E. and G. M. Wessel. 2003. βγ subunits of heterotrimeric G-proteins contribute to Ca²⁺ release at fertilization in sea urchins. *J. Cell Sci.* 117: 5995–6005.

Figure 7.27A: Hamaguchi, M. S. and Y. Hiramoto. 1980. Fertilization process in the heart-urchin, *Clypaester japonicus*, observed with a differential interference microscope. *Dev. Growth Diff.* 22: 517–530.

Figure 7.27B: Holy, J. and G. Schatten. 1991. Spindle pole centrosomes of sea urchin embryos are partially composed of material recruited from maternal stores. *Dev. Biol.* 147: 343–353.

Figure 7.28: Visconti, P. E. and G. S. Kopf. 1998. Regulation of protein phosphorylation during sperm capacitation. *Biol. Reprod.* 59: 1–6.

Figure 7.29: Lefebvre, R., P. J. Chenoweth, M. Drost, C. T. LeClear, M. MacCubbin, J. T. Dutton and S. S. Suarez. 1995. Characterization of the oviductal sperm receptor in cattle. *Biol. Reprod.* 53: 1066–1074.

Figure 7.31: (A) Bleil, J. D. and P. M. Wassarman. 1980. Mammalian sperm and egg interaction: Identification of a glycoprotein in mouse-egg zonae pellucidae possessing receptor activity for sperm. *Cell* 20: 873–882; and Florman, H. M. and P. M. Wassarman. 1985. O-linked oligosaccharides of mouse egg ZP account for its sperm receptor activity. *Cell* 41: 313–324. (B) Bleil, J. D. and P. M. Wassarman.

1986. Autoradiographic visualization of the mouse egg's sperm receptor bound to sperm. *J. Cell Biol.* 102: 1363–1371.

Figure 7.32: (A) Meizel, S. 1984. The importance of hydrolytic enzymes to an exocytotic event, the mammalian sperm acrosome reaction. *Biol. Rev.* 59: 125–157. (B) Yanagimachi, R. and Y. D. Noda. 1970. Electron microscope studies of sperm incorporation into the golden hamster egg. *Am. J. Anat.* 128: 429–462.

Figure 7.33A–C: Yanagimachi, R. and Y. D. Noda. 1970. Electron microscope studies of sperm incorporation into the golden hamster egg. *Am. J. Anat.* 128: 429–462; and Yanagimachi, R. 1994. Mammalian fertilization. *In* E. Knobil and J. D. Neill (eds.), *The Physiology of Reproduction*, 2nd Ed. Raven Press, New York.

Figure 7.34: Simerly, C. and 7 others. 1995. The paternal inheritance of the centrosome, the cell's microtubule-organizing center, in humans, and the implications for infertility. *Nature Med.* 1: 47–52.

Table 7.1: Whitaker, M. and R. Steinhardt. 1982. Ionic regulation of egg activation. *Q. Rev. Biophys.* 15: 593–667; and Mohri, T., P. I. Ivonnet and E. L. Chambers. 1995. Effect of sperm-induced activation current and increase of cytosolic Ca²⁺ by agents that modify the mobilization of [Ca²⁺]. I. Heparin and pentosan polysulfate. *Dev. Biol.* 172: 139–157.

Table 7.2: McGrath, J. and D. Solter. 1984. Completion of mouse embryogenesis requires both the maternal and paternal genome. *Cell* 37: 179–183.

Chapter 8

Chapter Opening Material

Just, E. E. 1939. *The Biology of the Cell Surface*. Blakiston, Philadelphia, p. 288.

Wolpert, L. 1986. *From Egg to Embryo: Determinative Events in Early Development*. Cambridge University Press, Cambridge, p. 1.

Photograph courtesy of G. von Dassow and the Center for Cell Dynamics. See Figure 8.33B.

Illustration Credits

Figure 8.1B: Nigg, E. A. 1995. Cyclin-dependent protein kinases: Key regulators of the eukaryotic cell cycle. *BioEssays* 17: 471–480.

Figure 8.8: Logan, C. Y. and D. R. McClay. 1999. Lineages that give rise to endoderm and mesoderm in the sea urchin embryo. In S. A. Moody (ed.), *Cell Lineage and Determination*. Academic Press, New

York, pp. 41–58; and Wray, G. A. 1999. Introduction to sea urchins. In S. A. Moody (ed.), *Cell Lineage and Determination*. Academic Press, New York, pp. 3–9.

Figure 8.9: Hörstadius, S. 1939. The mechanics of sea urchin development, studied by operative methods. *Biol. Rev.* 14: 132–179.

Figure 8.10: Ransick, A. and E. H. Davidson. 1993. A complete second gut induced by transplanted micromeres in the sea urchin embryo. *Science* 259: 1134–1138.

Figure 8.11: Logan, C. Y., J. R. Miller, M. J. Ferkowicz and D. R. McClay. 1998. Nuclear β-catenin? is required to specify vegetal cell fates in the sea urchin embryo. *Development* 126: 345–358.

Figure 8.12: Oliveri, P., D. M. Carrick and E. H. Davidson. 2002. A regulatory gene network that directs micromere specification in the sea urchin embryo. *Dev. Biol.* 246: 209–228.

Figure 8.13: (A) Davidson, E. H. and 24 others. 2002. A provisional regulatory gene network for specification of endomesoderm in the sea urchin embryo. *Dev. Biol.* 246: 162–190. (B) Yuh, C.-H, H. Bolouri and E. H. Davidson. 2001. *Cis*-regulatory logic in the *endo16* gene: Switching from a specification to a differentiation mode of control. *Development* 128: 617–629. (C) Oliveri, P., D. M. Carrick and E. H. Davidson. 2002. A regulatory gene network that directs micromere specification in the sea urchin embryo. *Dev. Biol.* 246: 209–228.

Figure 8.16B,C: Cherr, G. N., R. G. Summers, J. D. Baldwin and J. B. Morrill. 1992. Preservation and visualization of the sea urchin blastocoelic extracellular matrix. *Microsc. Res. Tech.* 22: 11–22.

Figure 8.17: (A) Ettensohn, C. A. 1990. The regulation of primary mesenchyme cell patterning. *Dev. Biol.* 140: 261–271. (B)Morrill, J. B. and L. L. Santos. 1985. A scanning electron micrographical overview of cellular and extracellular patterns during blastulation and gastrulation in the sea urchin, *Lytechinus variegatus*. In R. H. Sawyer and R. M. Showman (eds.), *The Cellular and Molecular Biology of Invertebrate Development*. University of South Carolina Press, Columbia, pp. 3–33.

Figure 8.18B: Miller, J. R., S. E. Fraser and D. R. McClay. 1995. Dynamics of thin filopodia during sea urchin gastrulation. *Development* 121: 2505–2511.

Figure 8.19: (A) Morrill, J. B. and L. L. Santos. 1985. A scanning electron micrographical overview of cellular and extracellular pat-

terns during blastulation and gastrulation in the sea urchin, *Lytechinus variegatus*. In R. H. Sawyer and R. M. Showman (eds.), *The Cellular and Molecular Biology of Invertebrate Development*. University of South Carolina Press, Columbia, pp. 3–33. (B) Logan, C. Y. and D. R. McClay. 1999. Lineages that give rise to endoderm and mesoderm in the sea urchin embryo. In S. A. Moody (ed.), *Cell Lineage and Determination*. Academic Press, New York, pp. 41–58.

Figure 8.20: Hardin, J. D. 1990. Context-dependent cell behaviors during gastrulation. *Semin. Dev. Biol.* 1: 335–345.

Figure 8.24B–E: Craig, M. M. and J. B. Morrill. 1986. Cellular arrangements and surface topography during early development in embryos of *Ilyanassa obsoleta*. *Int. J. Invert. Reprod. Dev.* 9: 209–228.

Figure 8.25: Morgan, T. H. 1927. *Experimental Embryology*. Columbia University Press, New York.

Figure 8.27: Raff, R. A. and T. C. Kaufman. 1983. *Embryos, Genes, and Evolution: The Developmental-Genetic Basis of Evolutionary Change*. Macmillan, New York.

Figure 8.29: Lambert, J. D. and L. M. Nagy. 2002. Asymmetric inheritance of centrosomally localized mRNAs during embryonic cleavage. *Nature* 420: 682–686.

Figure 8.30A: Wilson, E. B. 1904. Experimental studies on germinal localization. I. The germ regions of the egg of *Dentalium*. II. Experiments on the cleavage-mosaic in *Patella* and *Dentalium*. *J. Exp. Zool.* 1: 1–72.

Figure 8.31: Newrock, K. M. and R. A. Raff. 1975. Polar lobe specific regulation of translation in embryos of *Ilyanassa obsoleta*. *Dev. Biol.* 42: 242–261.

Figure 8.32: Lambert, J. D. L. and L. M. Nagy. 2001. MAPK signaling by the D quadrant embryonic organizer of the mollusk *Ilyanassa obsoleta*. *Development* 128: 45–56.

Figure 8.33A–C: Conklin, E. G. 1897. The embryology of *Crepidula. J. Morphol.* 13: 3–209.

Figure 8.34A: Balinsky, B. I. 1981. *Introduction to Embryology*, 5th Ed. Saunders, Philadelphia.

Figure 8.35: Conklin, E. G. 1905. The orientation and cell-lineage of the ascidian egg. *J. Acad. Nat. Sci. Phila.* 13: 5–119.

Figure 8.36: Swalla, B. J. 2004. Protochordate gastrulation: Lancelets and ascidians. In C. D. Stern (ed.), *Gastrulation: From Cells to Embryo*. Cold Spring Harbor Laboratory Press, Cold Spring Harbor, NY, pp. 139–149.

Figure 8.37: Nishida, H. and K. Sawada. 2001. *macho-1* encodes a localized mRNA in ascidian eggs that specifies muscle fate during embryogenesis. *Nature* 409: 724–729.

Figure 8.38: Imai, K. S., N. Takada, N. Satoh and Y. Satou. 2000. β-Catenin mediates the specification of endoderm cells in ascidian embryos. *Development* 127: 3009–3020.

Figure 8.39: Kobayashi, K., K. Sawada, H. Yamamoto, S. Wada, H. Saiga and H. Nishida. 2003. Maternal *Macho-1* is an intrinsic factor that makes cell response to the same FGF signal differ between mesenchyme and notochord induction in ascidian embryos. *Development* 130: 5179–5190.

Figure 8.40: Satoh, N. 1978. Cellular morphology and architecture during early morphogenesis of the ascidian egg: An SEM study. *Biol. Bull.* 155: 608–614; and Jeffery, W. R. and B. J. Swalla. 1997. Tunicates. In S. F. Gilbert and A. M. Raunio (eds.), *Embryology: Constructing the Organism*. Sinauer Associates, Sunderland, MA, pp. 331–364.

Figure 8.41: Deschet K,, Y. Nakatani and W. C. Smith. 2003. Generation of Ci-Brachyury-GFP stable transgenic lines in the ascidian *Ciona savignyi. Genesis* 35: 248–259.

Figure 8.42: Pines, M. (ed.). 1992. *From Egg to Adult*. Howard Hughes Medical Institute, Bethesda, MD, pp. 30–38; and Sulston, J. E. and H. R. Horvitz. 1977. Postembryonic cell lineages of the nematode *Caenorhabditis elegans. Dev. Biol.* 56: 110–156; and Sulston, J. E., J. Schierenberg, J. White and N. Thomson. 1983. The embryonic cell lineage of the nematode *Caenorhabditis elegans. Dev. Biol.* 100: 64–119.

Figure 8.43A–E: Nance, J. 2005. PAR proteins and the establishment of cell polarity during *C. elegans* development. *BioEssays* 27: 126–135.

Figure 8.45: Bowerman, B., B. A. Eaton and J. R. Priess. 1992a. *skn-1*, a maternally expressed gene required to specify the fate of ventral blastomeres in the early *C. elegans* embryo. *Cell* 68: 1061–1075.

Figure 8.46: Goldstein, B. 1992. Induction of gut in *Caenorhabditis elegans* embryos. *Nature* 357: 255–258.

Figure 8.47: Han, M. 1998. Gut reaction to Wnt signaling in worms. *Cell* 90: 581–584.

Figure 8.48: Schierenberg, E. 1997. Nematodes, the roundworms. In S. F. Gilbert and A. M. Raunio (eds.), *Embryology: Constructing the Organism*. Sinauer Associates, Sunderland, MA, pp. 131–148.

Table 8.2: Fink, R. D. and D. R. McClay. 1985. Three cell recognition changes accompany the ingression of sea urchin primary mesenchyme cells. *Dev. Biol.* 107: 66–74.

Chapter 9

Chapter Opening Material

Kohler, R. E. 1944. *Lords of the Fly:* Drosophila *Genetics and the Experimental Life*. University of Chicago Press, Chicago, p. 33.

Schultz, J. 1935. Aspects of the relation between genes and development in *Drosophila. American Naturalist* 69: 30–54.

Photograph courtesy of E. B. Lewis. See Figure 9.36.

Illustration Credits

Figure 9.2: Karr, T. L. and B. M. Alberts. 1986. Organization of the cytoskeleton in early *Drosophila* embryos. *J. Cell Biol.* 102: 1494–1509.

Figure 9.3: (A) Fullilove, S. L. and Jacobson, A. G. 1971. Nuclear elongation and cytokinesis in *Drosophila montana. Dev. Biol.* 26: 560–578. (B) Sullivan, W., P. Fogarty and W. E. Theurkauf. 1993. Mutations affecting the cytoskeletal organization of syncytial *Drosophila* embryos. *Development* 118: 1245–1254. (C) Zhang, C., M. P. Lee, A. D. Chen, S. D. Brown and T. Hsieh. 1996. Isolation and characterization of a *Drosophila* gene essential for embryonic development and cortical cleavage furrow formation. *J. Cell Biol.* 134: 923–934 (with photo as the journal cover).

Figure 9.4: Edgar, B. A. and P. H. O'Farrell. 1989. Genetic control of cell division patterns in the *Drosophila* embryo. *Cell* 57: 177–187.

Figure 9.5D and 9.6: Campos-Ortega, J. A. and V. Hartenstein. 1985. *The Embryonic Development of Drosophila melanogaster*. Springer-Verlag, New York.

Figure 9.8E: Stephanson, E. C., Y.-C. Chao and J. D. Frackenthal. 1988. Molecular analysis of the swallow gene of *Drosophila melanogaster. Genes Dev.* 2: 1655–1665.

Figure 9.9: (A) Ray, R. P. and T. Schüpbach. 1996. Intercellular signaling and the polarization of body axes during *Drosophila* oogenesis. *Genes Dev.* 10: 1711–1723. (B,C) Peri, F., C. Bökel and S. Roth. 1999. Local Gurken signaling and dynamic MAPK activation during *Drosophila* oogenesis. *Mech. Dev.* 81: 75–88.

Figure 9.11: Van Eeden, F. and D. St. Johnston. 1999. The polarisation of the anterior-posterior and dorsal-ventral axes during *Drosophila* oogenesis. *Curr. Opin. Genet. Devel.* 9: 396–404.

Figure 9.12: Anderson, K. V. and C. Nüsslein-Volhard. 1984. Information for the dorsal-ventral pattern of the *Drosophila* embryo is stored as maternal mRNA. *Nature* 311: 223–227.

Figure 9.13: (A) Rushlow, C. A., K. Han, J. L. Manley and M. Levine. 1989. The graded distribution of the dorsal morphogen is initiated by selective nuclear transport in *Drosophila. Cell* 59: 1165–1177. (B–D) Roth, S., D. Stein and C. Nüsslein-Volhard. 1989. A gradient of nuclear localization of the dorsal protein determines dorsoventral pattern in the *Drosophila* embryo. *Cell* 59: 1189–1202.

Figure 9.14: Leptin, M. 1991a. Mechanics and genetics of cell shape changes during *Drosophila* ventral furrow formation. In R. Keller et al. (eds.), *Gastrulation: Movements, Patterns, and Molecules.* Plenum, New York, pp. 199–212.

Figure 9.15: (A) Steward, R. and S. Govind. 1993. Dorsal-ventral polarity in the *Drosophila* embryo. *Curr. Opin. Genet. Dev.* 3: 556–561. (B) Furlong, E. E. M., E. C. Andersen, B. Null, K. P. White and M. P. Scott. 2001. Patterns of gene expression during *Drosophila* mesoderm development. *Science* 293: 1629–1633; Leptin, M. and M. Affolter. 2004. *Drosophila* gastrulation: Identification of a missing link. *Curr. Biol.* 14: R480–R482.

Figure 9.16: Kosman, D., C. M. Mizutani, D. Lemons, W. G. Cox, W. McGinnis and E. Bier. 2004. Multiplex detection of RNA expression in *Drosophila* embryos. *Science* 305: 846.

Figure 9.18: Kalthoff, K. 1969. Der Einfluss vershiedener Versuchparameter auf die Häufigkeit der Missbildung "Doppelabdomen" in UV-bestrahlten Eiern von *Smittia* sp. (Diptera, Chironomidae*). Zool. Anz. Suppl.* 33: 59–65.

Figure 9.19: St. Johnston, D. and C. Nüsslein-Volhard. 1992. The origin of pattern and polarity in the *Drosophila* embryo. *Cell* 68: 201–219.

Figure 9.20: Macdonald, P. M. and G. Struhl. 1986. A molecular gradient in early *Drosophila* embryos and its role in specifying body pattern. *Nature* 324: 537–545.

Figure 9.21: Wreden, C., A. C. Verrotti, J. A. Schisa, M. E. Lieberfarb and S. Strickland. 1997. Nanos and pumilio establish embryonic

polarity in *Drosophila* by promoting posterior deadenylation of *hunchback* mRNA. *Development* 124: 3015–3023.

Figure 9.22C: Macdonald, P. M. and C. A. Smibert. 1996. Translational regulation of maternal mRNAs. *Curr. Opin. Genet. Dev.* 6: 403–407.

Figure 9.23: (A) Kaufman, T. C., M. A. Seeger and G. Olsen. 1990. Molecular and genetic organization of the Antennapedia gene complex of *Drosophila melanogaster. Adv. Genet.* 27: 309–362. (B,C) Driever, W. and C. Nüsslein-Volhard. 1988b. A gradient of Bicoid protein in *Drosophila* embryos. *Cell* 54: 83–93. (D) Driever, W., V. Siegel and C. Nüsslein-Volhard. 1990. Autonomous determination of anterior structures in the early *Drosophila* embryo by the Bicoid morphogen. *Development* 109: 811–820.

Figure 9.24: Driever, W., V. Siegel and C. Nüsslein-Volhard. 1990. Autonomous determination of anterior structures in the early *Drosophila* embryo by the Bicoid morphogen. *Development* 109: 811–820.

Figure 9.25: (A) Gabay, L., R. Seger and B. Z. Shilo. 1997. MAP kinase in situ activation atlas during *Drosophila* embryogenesis. *Development* 124: 3535–3541. (B) Paroush, Z., S. M. Wainwright and D. Ish-Horowitz. 1997. Torso signaling mediates terminal patterning in *Drosophila* by antagonizing Groucho-mediated repression. *Development* 124: 3827–3834.

Figure 9.27: (A) Martinez-Arias, A. and P. A. Lawrence. 1985. Parasegments and compartments in the *Drosophila* embryo. *Nature* 313: 639–642. (B) Deutsch, J. 2004. Segments and parasegments in arthropods: A functional perspective. *BioEssays* 26: 1117–1125.

Figure 9.28: Monk, N. 2004. Development: Dissecting the dynamics of segment determination. *Curr. Biol.* R705–R707.

Figure 9.29: Fujioka, M., Y. Emi-Sarker, G. L. Yusibova, T. Goto and J. B. Jaynes. 1999. Analysis of an even-skipped rescue transgene reveals both composite and discrete neuronal and early blastoderm enhancers, and multi-stripe positioning by gap gene repressor gradients. *Development* 126: 2527–2538; Sackerson, C. M. Fujioka and T. Goto. 1999. The *even-skipped* locus is contained in a 16-kb chromatin domain. *Dev. Biol.* 210: 39–52.

Figure 9.30: (A) Small, S., A. Blair and M. Levine. 1992. Regulation of *even-skipped* stripe 2 in the

Drosophila embryo. *EMBO J.* 11: 4047–4057. (B) Clyde, D. E., M. S. G. Corado, X. Wu, A. Paré, D. Poaptsenko and S. Small. 2003. A self-organizing system of repressor gradients establishes segmental complexity in *Drosophila. Nature* 426: 849–853.

Figure 9.31: Kaufman, T. C., M. A. Seeger and G. Olsen. 1990. Molecular and genetic organization of the Antennapedia gene complex of *Drosophila melanogaster. Adv. Genet.* 27: 309–362.

Figure 9.32A–D: Karr, T. L. and T. B. Kornberg. 1989. Fushi tarazu protein expression in the cellular blastoderm of *Drosophila* detected using a novel imaging technique. *Development* 105: 95–103.

Figure 9.33C: Simmonds, A. J., G. dosSantos, I. Livne-Bar, and H. M. Krause. 2001. Apical localization of *wingless* transcripts is required for Wingless signaling. *Cell* 105: 197–207.

Figure 9.34: Heemskerk, J. and S. DiNardo. 1994. *Drosophila* hedgehog acts as a morphogen in cellular patterning. *Cell* 76: 449–460.

Figure 9.35A: Dessain, S., C. T. Gross, M. A. Kuziora and W. McGinnis. 1992. *Antp*-type homeodomains have distinct DNA-binding specificities that correlate with their different regulatory functions in embryos. *EMBO J.* 11: 991–1002; Kaufman, T. C., M. A. Seeger and G. Olsen. 1990. Molecular and genetic organization of the Antennapedia gene complex of *Drosophila melanogaster. Adv. Genet.* 27: 309–362.

Figure 9.37: Kaufman, T. C., M. A. Seeger and G. Olsen. 1990. Molecular and genetic organization of the Antennapedia gene complex of *Drosophila melanogaster. Adv. Genet.* 27: 309–362.

Figure 9.38: Weatherbee, S. D., G. Halder, J. Kim, A. Hudson and S. Carroll. 1998. Ultrabithorax regulates genes at several levels of the wing-patterning hierarchy to shape the development of the *Drosophila* haltere. *Genes Dev.* 12: 1474–1482.

Figure 9.39: Castelli-Gair, J. and M. Akam. 1995. How the Hox gene *Ultrabithorax* specifies two different segments: The significance of spatial and temporal regulation within metameres. *Development* 121: 2973–2982.

Figure 9.40: (A) Riddihough, G. 1992. Homing in on the homeobox. *Nature* 357: 643–644. (B) Hanes, S. D. and R. Brent. 1991. A genetic model for interaction of the homeodomain recognition helix with DNA. *Science* 251: 426–430.

Figure 9.41B: Panzer, S., D. Weigel and S. K. Beckendorf. 1992. Organogenesis in *Drosophila melanogaster*: Embryonic salivary gland determination is controlled by homeotic and dorsoventral patterning genes. *Development* 114: 49–57.

Table 9.1: Anderson, K. V. 1989. *Drosophila*: The maternal contributions. *In* D. M. Glover and B. D. Hames (eds.), *Genes and Embryos.* IRL, New York, pp. 1–37.

Chapter 10

Chapter Opening Material

Rostand, J. 1960. *Carnets d'un Biologiste.* Librairie Stock, Paris.

Spemann, H. 1943. *Forschung und Leben.* Quoted in T. J. Horder, J. A. Witkowski and C. C. Wylie, 1986, *A History of Embryology.* Cambridge University Press, Cambridge, p. 219.

Photograph courtesy of P. Hausen. See Figure 10.23A.

Illustration Credits

Figure 10.1: Cha, B. J. and D. L. Gard. 1999. XMAP230 is required for the organization of cortical microtubules and patterning of the dorsoventral axis in fertilized *Xenopus* eggs. *Dev. Biol.* 205: 275–286.

Figure 10.2: Gerhart, J. C., M. Danilchik, T. Doniach, S. Roberts, B. Rowning, and R.Stewart. 1989. Cortical rotation of the *Xenopus* egg: Consequences for the antero-posterior pattern of embryonic dorsal development. *Development* [Suppl.] 107: 37–51.

Figure 10.3: Carlson, B. M. 1981. *Patten's Foundations of Embryology.* McGraw-Hill, New York.

Figure 10.4A: Beams, H. W. and R. G. Kessel. 1976. Cytokinesis: A comparative study of cytoplasmic division in animal cells. *Am. Sci.* 64: 279–290.

Figure 10.5: Heasman, J., D. Ginsberg, K. Goldstone, T. Pratt, C. Yoshidanaro and C. Wylie. 1994b. A functional test for maternally inherited cadherin in *Xenopus* shows its importance in cell adhesion at the blastula stage. *Development* 120: 49–57.

Figure 10.6: Lane, M. C. and W. C. Smith. 1999. The origins of primitive blood in *Xenopus*: Implications for axial patterning. *Development* 126: 423–434; and Newman, C. S. and P. A. Krieg. 1999. Specification and differentiation of the heart in amphibia. *In* S. A. Moody, *Cell Lineage and Fate Determination.* Academic Press, New York, pp. 341–351.

Figure 10.7A–F: Keller, R. E. 1986. The cellular basis of amphibian gastrulation. In L. Browder (ed.), *Developmental Biology: A Comprehensive Synthesis,* Vol. 2. Plenum, New York, pp. 241–327.

Figure 10.9: Holtfreter, J. 1944. A study of the mechanics of gastrulation, Part II. *J. Exp. Zool.* 95: 171–212.

Figure 10.10: Winklbauer, R. and M. Schürfeld. 1999. Vegetal rotation, a new gastrulation movement involved in the internalization of the mesoderm and endoderm in *Xenopus. Development* 126: 3703–3713.

Figure 10.11A: Balinsky, B. I. 1975. *Introduction to Embryology,* 4th Ed. Saunders, Philadelphia.

Figure 10.12: Wilson, P. and R. Keller. 1991. Cell rearrangement during gastrulation of *Xenopus:* Direct observation of cultured explants. *Development* 112: 289–300; Winklbauer, R. and M. Schürfeld. 1999. Vegetal rotation, a new gastrulation movement involved in the internalization of the mesoderm and endoderm in *Xenopus. Development* 126: 3703–3713.

Figure 10.14: (A,B) Marsden, M. and D. W. DeSimeone. 2001. Regulation of cell polarity, radial intercalation, and epiboly in *Xenopus:* Novel roles for integrin and fibronectin. *Development* 128: 3635–3647. (C–F) Boucaut, J.-C., T. D'Arribère, T. J. Poole, H. Aoyama, K. M. Yamada and J.-P. Thiery. 1984. Biologically active synthetic peptides as probes of embryonic development: A competitive peptide inhibition of fibronectin function inhibits gastrulation in amphibian embryos and neural crest cell migration in avian embryos. *J. Cell Biol.* 99: 1822–1830.

Figure 10.15: (A,B) Saka, Y. and J. C. Smith. 2001. Spatial and temporal patterns of cell division during early *Xenopus* embryogenesis. *Dev. Biol.* 229: 307–318. (C) Keller, R. E. 1980. The cellular basis of epiboly: An SEM study of deep cell rearrangement during gastrulation of *Xenopus laevis. J. Embryol. Exp. Morphol.* 60: 201–243.

Figures 10.16 and 10.17: Spemann, H. 1938. *Embryonic Development and Induction.* Yale University Press, New Haven.

Figure 10.18: Saxén, L. and S. Toivonen. 1962. *Embryonic Induction.* Prentice-Hall, Englewood Cliffs, NJ.

Figure 10.19: Hamburger, V. 1988. *The Heritage of Experimental Embryology: Hans Spemann and the organizer.* Oxford University Press, Oxford.

Figure 10.20C: De Robertis, E. M., M. Blum, C. Niehrs and H. Steinbeisser. 1992. Goosecoid and the organizer. *Development* 1992 [Suppl.]: 167–171.

Figure 10.21: Gimlich, R. L. and J. C. Gerhart. 1984. Early cellular interactions promote embryonic axis formation in *Xenopus laevis. Dev. Biol.* 104: 117–130.

Figure 10.22: Dale, L. and J. M. W. Slack. 1987. Regional specificity within the mesoderm of early embryos of *Xenopus laevis. Development* 100: 279–295.

Figure 10.23: (A,D) Moon, R. T. and D. Kimelman. 1998. From cortical rotation to organizer gene expression: Toward a molecular explanation of axis specification in *Xenopus. BioEssays* 20: 536–545. (B,C) Schneider, S., H. Steinbeisser, R. M. Warga and P. Hausen. 1996. ?-catenin translocation into nuclei demarcates the dorsalizing centers in frog and fish embryos. *Mech. Dev.* 57: 191–198.

Figure 10.24: (A–E) Weaver, C. and D. Kimelman. 2004. Move it or lose it: Axis specification in *Xenopus. Development* 131: 3491–3499. (F) Pierce, S. B. and D. Kimelman. 1995. Regulation of Spemann organizer formation by the intracellular kinase Xgsk-3. *Development* 121: 755–765.

Figure 10.25: Moon, R. T. and D. Kimelman. 1998. From cortical rotation to organizer gene expression: Toward a molecular explanation of axis specification in *Xenopus. BioEssays* 20: 536–545.

Figure 10.26: Agius, E., M. Oelgeschläager, O. Wessely, C. Kemp and E. M. De Robertis. 2000. Endodermal Nodal-related signals and mesoderm induction in *Xenopus. Development* 127: 1151–1159.

Figure 10.27: Cho, K. W. Y., B. Blumberg, H. Steinbeisser and E. De Robertis. 1991a. Molecular nature of Spemann's organizer: The role of the *Xenopus* homeobox gene *goosecoid.* Cell 67: 1111–1120; Niehrs, C., R. Keller, K. W. Y. Cho and E. M. De Robertis. 1993. The homeobox gene *goosecoid* controls cell migration in *Xenopus* embryos. *Cell* 72: 491–503.

Figure 10.28: Toivonen, S. 1979. Transmission problem in primary induction. *Differentiation* 15: 177–181.

Figure 10.31: Sasai, Y., B. Lu, H. Steinbeisser, D. Geissert, L. K. Gont and E. M. De Robertis. 1994. *Xenopus* chordin: A novel dorsalizing factor activated by organizer-specific homeobox genes. *Cell* 79: 779–790.

Figure 10.32: (A) Dosch, R., V. Gawantka, H. Delius, C. Blumenstock and C. Niehrs. 1997. BMP-4 acts as a morphogen in dorsolateral mesoderm patterning in *Xenopus. Development* 124: 2325–2334; De Robertis, E. M., J. Larraín, M. Oelgeschländer and O. Wessley. 2000. The establishment of Spemann's organizer and patterning of the vertebrate embryo. *Nature Rev. Genet.* 1: 171–181. (B) Kurata, T., J. Nakabayashi, T. S. Yamamoto, M. Mochii and N. Ueno. 2000. Visualization of endogenous BMP signaling during *Xenopus* development. *Differentiation* 67: 33–40.

Figure 10.33: (A,B) Khokha, M. K., J. Yeh, T. C. Grammer and R. M. Harland. 2005. Depletion of three BMP antagonists from Spemann's organizer leads to catastrophic loss of dorsal structures. *Dev. Cell* 8: 401–411. (C,D) Reversade, B. and E. M. De Robertis. 2006. Reciprocal regulation of Admp and Bmp2/4/7 at opposite embryonic poles generates a self-regulating morphogenetic field. *Cell,* in press.

Figure 10.34: Mangold, O. 1933. Über die Induktionsfähigkeit der verschiedenen Bezirke der Neurula von Urodelen. *Naturwissenschaften* 21: 761–766.

Figure 10.35: Saxén, L. and S. Toivonen. 1962. *Embryonic Induction.* Prentice-Hall, Englewood Cliffs, NJ.

Figure 10.37: Bouwmeester, T., S.-H. Kim, Y. Sasai, B. Lu and E. M. De Robertis. 1996. Cerberus is a head-inducing secreted factor expressed in the anterior endoderm of Spemann's organizer. *Nature* 382: 595–601.

Figure 10.38: Leyns, L., T. Bouwmeester, S.-H. Kim, S. Piccolo and E. M. De Robertis. 1997. Frzb-1 is a secreted antagonist of Wnt signaling expressed in the Spemann organizer. *Cell* 88: 747–756.

Figure 10.39: Pera, E. M., O. Wessely, S.-S. Li and E. M. De Robertis. 2001. Neural and head induction by insulin-like growth factor signals. *Dev. Cell* 1: 655–665.

Figure 10.40: Kiecker, C. and C. Niehrs. 2001. A morphogen gradient of Wnt/?-catenin signalling regulates anteroposterior neural patterning in *Xenopus. Development* 128: 4189–4201; Niehrs, C. 2004. Regionally specific induction by the Spemann-Mangold organizer. *Nature Rev. Genet.* 5: 425–434.

Figure 10.42: Ryan, A. and 14 others. 1998. Pitx2 determines left-right asymmetries in vertebrates. *Nature* 394: 54–55.

Chapter 11

Chapter Opening Material

Doyle, A. C. 1891. "A Case of Identity." In *The Adventures of Sherlock Holmes.* Reprinted in *The Complete Sherlock Holmes Treasury,* 1976. Crown, New York, p. 31.

Holub, M. 1990. "From the Intimate Life of Nude Mice." In *The Dimension of the Present Moment.* Trans. D. Habova and D. Young. Faber and Faber, London, p. 38.

Photograph courtesy of J. Rossant. See Figure 11.30B.

Illustration Credits

Figure 11.1: Langeland, J. and C. B. Kimmel. 1997. The embryology of fish. In S. F. Gilbert and A. M. Raunio (eds.), *Embryology: Constructing the Organism.* Sinauer Associates, Sunderland, MA, pp. 383–407.

Figure 11.2: Haffter, P. and 16 others. 1996. The identification of genes with unique and essential functions in the development of the zebrafish, *Danio rerio. Development* 123: 1–36.

Figure 11.4: Beams, H. W. and R. G. Kessel. 1976. Cytokines: A comparative study of cytoplasmic division in animal cells. *Am. Sci.* 64: 279–290.

Figure 11.5: (A,C) Langeland, J. and C. B. Kimmel. 1997. The embryology of fish. In S. F. Gilbert and A. M. Raunio (eds.), *Embryology: Constructing the Organism.* Sinauer Associates, Sunderland, MA, pp. 383–407. (B) Trinkaus, J. P. 1993. The yolk syncitial layer of *Fundulus:* Its origin and history and its significance for early embryogenesis. *J. Exp. Zool.* 265: 258–284.

Figure 11.6: Driever, W. 1995. Axis formation in zebrafish. *Curr. Opin. Genet. Dev.* 5: 610–618; Langeland, J. and C. B. Kimmel. 1997. The embryology of fish. In S. F. Gilbert and A. M. Raunio (eds.), *Embryology: Constructing the Organism.* Sinauer Associates, Sunderland, MA, pp. 383–407.

Figure 11.7: Langeland, J. and C. B. Kimmel. 1997. The embryology of fish. In S. F. Gilbert and A. M. Raunio (eds.), *Embryology: Constructing the Organism.* Sinauer Associates, Sunderland, MA, pp. 383–407.

Figure 11.8A: Shinya, M., M. Furutani-Seiki, A. Kuroiwa and H. Takeda. 1999. Mosaic analysis with *oep* mutant reveals a repressive interaction between floorplate and non-floor-plate mutant cells in the zebrafish neural tube. *Dev. Growth Diff.* 41: 135–142.

Figure 11.9: Schier, A. F. and W. S. Talbot. 1998. The zebrafish organizer. *Curr. Opin. Genet. Dev.* 8: 464–471.

Figure 11.10B: Schier, A. F. and W. S. Talbot. 2001. Nodal signaling and the zebrafish organizer. *Int. J. Dev. Biol.* 45: 289–297.

Figure 11.11: Driever, W. 2005. A message to the back side. *Nature* 438: 926–927.

Figure 11.12: Kudoh, T., S. W. Wilson and I. B. Dawid. 2002. Distinct roles for FGF, Wnt, and retinoic acid in posteriorizing the neural ectoderm. *Development* 129: 4335–4346.

Figure 11.13: (A) Halpern, M. E., J. O. Liang and J. T. Gamse. 2003. Leaning to the left: Laterality in the zebrafish forebrain. *Trends Neurosci.* 26: 308–313. (B) Gamse , J. T., Y. S. Kuan, M. Macurak, C. Brosamle, B. Thisse, C. Thisse and M. E. Halpern. 2005. Directional asymmetry of the zebrafish epithalamus guides dorsoventral innervation of the midbrain target. *Development* 132: 4869–4881.

Figure 11.14: Bellairs, R., F. W. Lorenz and T. Dunlap. 1978. Cleavage in the chick embryo. *J. Embryol. Exp. Morphol.* 43: 55–69.

Figure 11.15: Stern, C. D. 2004. Gastrulation in the chick. In *Gastrulation: From Cells to Embryo.* (C. D. Stern, editor). Cold Spring Harbor Laboratiory Press, Cold Spring Harbor, NY. Pp. 219–232.

Figure 11.16: Spratt, N. T., Jr. 1946. Formation of the primitive streak in the explanted chick blastoderm marked with carbon particles. *J. Exp. Zool.* 103: 259–304; Smith, J. L. and G. C. Schoenwolf. 1998. Getting organized: New insights into the organizer of higher vertebrates. *Curr. Top. Dev. Biol.* 40: 79–110; Stern, C. D. 2005a. The chick: A great model system becomes even greater. *Dev. Cell* 8: 9–17; Stern, C. D. 2005b. Neural induction: Old problems, new findings, yet more questions. *Development* 132: 2007–2021.

Figure 11.17: (A) Balinsky, B. I. 1975. *Introduction to Embryology*, 4th Ed. Saunders, Philadelphia. (B) Solursh, M. and J. P. Revel. 1978. A scanning electron microscope study of cell shape and cell appendages in the primitive streak region of the rat and chick embryo. *Differentiation* 11: 185–190.

Figure 11.18E: Spratt, N. T., Jr. 1947. Regression and shortening of the primitive streak in the explanted chick blastoderm. *J. Exp. Zool.* 104: 69–100.

Figure 11.19: Wolpert, L., R. Beddington, J. Brockes, T. Jessell, P. Lawrence and E. Meyerowitz.

1998. *Principles of Development.* Current Biology Ltd., London.

Figure 11.20: Bachvarova, R. F., I. Skromne and C. D. Stern. 1998. Induction of primitive streak and Hensen's node by the posterior marginal zone in the early chick embryo. *Development* 125: 3521–3534.

Figure 11.21: (A,B) Waddington, C. H. 1933. Induction by the primitive streak and its derivatives in the chick. *J. Exp. Zool.* 10: 38–46. (C) Boettger, T., H. Knoetgen, L. Wittler and M. Kessel. 2001. The avian organizer. *Int. J. Dev. Biol.* 45: 281–287.

Figure 11.22: (A) Boettger, T., H. Knoetgen, L. Wittler and M. Kessel. 2001. The avian organizer. *Int. J. Dev. Biol.* 45: 281–287. (B,C) Lawson, A., J.-F. Colas and G. C. Schoenwolf. 2001. Classification scheme for genes expressed during formation and progression of the anterior primitive streak. *Anat. Rec.* 262: 221–226.

Figure 11.23: Faure, S., P. de Santa Barbara, D. J. Roberts and M. Whitman. 2002. Endogenous patterns of BMP signaling during early chick development. *Dev. Biol.* 244: 44–65.

Figure 11.24: Sheng, G., M. dos Reis and C. D. Stern. 2003. Churchill, a zinc finger transcriptional activator, regulates the transition between gastrulation and neurulation. *Cell* 115: 603–613.

Figure 11.25: Blentic, A. E. Gale and M. Maden. 2003. Retinoic acid signaling centres in the avian embryo identified by sites of expression of synthesising and catabolising enzymes. *Dev. Dyn.* 227: 114–127.

Figure 11.26: (A) Raya, A. and J. C. Izpisua-Belmonte. 2004. Unveiling the establishment of left-right asymmetry in the chick embryo. *Mech. Dev.* 121: 1043–1054. (B) Rodriguez-Esteban, C., J. Capdevilla, A. N. Economides, J. Pascual, Á. Ortiz and J. C. Izpisúa-Belmonte. 1999. The novel Cer-like protein Caronte mediates the establishment of embryonic left-right asymmetry. *Nature* 401: 243–251. (D) Logan, M., S. M. Pagán-Westphal, D. M. Smith, L. Paganessi and C. J. Tabin. 1998. The transcription factor Pitx2 mediates situs-specific morphogenesis in response to left-right asymmetric signals. *Cell* 94: 307–317.

Figure 11.27: Tuchmann-Duplessis, H., G. David and P. Haegel. 1972. *Illustrated Human Embryology*, Vol. 1. Springer-Verlag, New York.

Figure 11.28: Gulyas, B. J. 1975. A reexamination of the cleavage

patterns in eutherian mammalian eggs: Rotation of the blastomere pairs during second cleavage in the rabbit. *J. Exp. Zool.* 193: 235–248.

Figure 11.29: (A–F) Mulnard, J. G. 1967. Analyse microcinematographique du developpement de l'oeuf de souris du stade II au blastocyste. *Arch. Biol.* (Liege) 78: 107–138. (G) Ducibella, T., D. F. Albertini, E. Anderson and J. D. Biggers. 1975. The preimplantation mammalian embryo: Characterization of intercellular junctions and their appearance during development. *Dev. Biol.* 45: 231–250.

Figure 11.30: Mitsui, K. and 8 others. 2003. The homeoprotein Nanog is required for maintenance of pluripotency in mouse epiblast and ES cells. *Cell* 113: 631–642; Strumpf, D. C.-A. Mao, Y. Yamanaka, A. Ralston, K. Chawengsaksophak, F. Beck and J. Rossant. 2005. Cdx2 is required for correct cell fate specification and differentiation of trophectoderm in the mouse blastocyst. *Development* 132: 2093–2102.

Figure 11.31: (A) Mark, W. H., K. Signorelli and E. Lacy. 1985. An inserted mutation in a transgenic mouse line results in developmental arrest at day 5 of gestation. *Cold Spring Harb. Symp. Quant. Biol.* 50: 453–463. (B) Rugh, R. 1967. *The Mouse.* Burgess, Minneapolis.

Figure 11.32: Luckett, W. P. 1978. Origin and differentiation of the yolk sac and extraembryonic mesoderm in presomite human and rhesus monkey embryos. *Am. J. Anat.* 152: 59–98; Bianchi, D. W., L. E. Wilkins-Haug, A. C. Enders and E. D. Hay. 1993. Origin of extraembryonic mesoderm in experimental animals: Relevance to chorionic mosaicism in humans. *Am. J. Med. Genet.* 46: 542–550.

Figure 11.34: Larsen, W. J. 1993. *Human Embryology.* Churchill Livingston, New York.

Figure 11.37: Langman, J. 1981. *Medical Embryology*, 4th Ed. Williams & Wilkins, Baltimore.

Figure 11.38B: Markert, C. L. and R. M. Petters. 1978. Manufactured hexaparental mice show that adults are derived from three embryonic cells. *Science* 202: 56–58.

Figure 11.40: Bachiller, D. and 10 others. 2000. The organizer factors Chordin and Noggin are required for mouse forebrain development. *Nature* 403: 658–661.

Figure 11.41: (A) Robb, L. and P. P. L. Tam. 2004. Gastrula organiser and embryonic patterning in the mouse. *Sem. Cell Dev. Biol.* 15: 543–554. (B) Dubrulle, J. and O.

Pourquié. 2004. Fgf8 mRNA decay establishes a gradient that couples axial elongation to patterning in the vertebrate embryo. *Nature* 427: 419–422. (C) Lohnes, D. 2003. The Cdx1 homeodomain protein: An integrator of posterior signaling in the mouse. *Bioessays* 25: 971–980.

Figure 11.42: Carroll, S. B. 1995. Homeotic genes and the evolution of arthropods and chordates. *Nature* 376: 479–485.

Figure 11.43: Wellik, D. M. and M. R. Capecchi. 2003. *Hox10* and *Hox11* genes are required to globally pattern the mammalian skeleton. *Science* 301: 363–367.

Figure 11.44: Houle, M., J. R. Sylvestre and D. Lohnes. 2003. RA is a critical regulator of *Cdx1* in vivo. *Development* 130: 6555–6567.

Figure 11.45: (A) Kmita, M. and D. Duboule 2003. Organizing axes in time and speace: 25 years of collinear tinkering. *Science* 301: 331–333. (B) Burke, A. C., A. C. Nelson, B. A. Morgan and C. Tabin. 1995. Hox genes and the evolution of vertebrate axial morphology. *Development* 121: 333–346.

Figure 11.46: (A,B) Hiiragi, T. and D. Solter. 2004. First cleavage plane of the mouse egg not predetermined but defined by the topology of the two apposing nuclei. *Nature* 430: 360–364. (C) Plusa, B. and 7 others. 2005. The first cleavage of the mouse zygote predicts the blastocyst axis. *Nature* 434: 391–395.

Figure 11.47: Kosaki, K. and B. Casey. 1998. Genetics of human left-right axis malformations. *Semin. Cell Dev.* 9: 89–99.

Figure 11.48B: Tanaka, Y., Y. Okada and N. Hirokawa. 2005. FGF-induced vesicular release of sonic hedgehog and retinoic acid in leftward nodal flow is critical for left-right determination. *Nature* 435: 172–177.

Figure 11.49: Solnicka-Krezel, L. 2005. Conserved patterns of cell movements during vertebrate gastrulation. *Curr. Biol.* 15: R213–R228.

Chapter 12

Chapter Opening Material

Thomas, L. 1979. "On Embryology." In *The Medusa and the Snail*, Viking Press, New York, p. 157. Photograph courtesy of T. Jessell. See Figure 12.13F.

Illustration Credits

Figure 12.2F: Patten, B. M. 1971. *Early Embryology of the Chick*, 5th Ed. McGraw-Hill, New York.

Figure 12.3: Smith, J. L. and G. C. Schoenwolf. 1997. Neurulation: Coming to closure. *Trends Neurosci.* 11: 510–517.

Figure 12.4: Balinsky, B. I. 1975. *Introduction to Embryology*, 4th Ed. Saunders, Philadelphia.

Figure 12.5B: Nakatsu, T., C. Uwabe and K. Shiota. 2000. Neural tube closure in humans initiates at multiple sites: Evidence from human embryos and implications for the pathogenesis of neural tube defects. *Anat. Embryol.* 201: 455–466.

Figure 12.7: Saitsu, H., M. Ishibashi, H. Nakano and K. Shiota. 2003. Spatial and temporal expression of folate-binding protein 1 (Fbp1) is closely associated with anterior neural tube closure in mice. *Dev. Dynam.* 226: 112–117.

Figure 12.8: Catala, M., M.-A. Teillet and N. M. Le Douarin. 1995. Organization and development of the tail bud analyzed with the quail-chick chimaera system. *Mech. Dev.* 51: 51–65.

Figure 12.9: Moore, K. L. and T. V. N. Persaud. 1993. *Before We Are Born: Essentials of Embryology and Birth Defects.* W. B. Saunders, Philadelphia.

Figure 12.10: Lumsden, A. 2004. Segmentation and compartition in the early avian hindbrain. *Mech. Dev.* 121: 1081- 1088.

Figure 12.11: Lowery, L. A. and H. Sive. 2005. Initial formation of zebrafish brain ventricles occurs independently of circulation and requires the *nagie oko* and *snakehead/atp1a1a.1* gene products. *Development* 132: 2057–2067.

Figure 12.13F: Jessell, T. M. 2000. Neuronal specification in the spinal cord: Inductive signals and transcriptional codes. *Nature Rev. Genet.* 1: 20–29.

Figure 12.14: (A) Briscoe, J., L. Sussel, D. Hartigan-O'Connor, T. M. Jessell, J. L. R. Rubenstein and J. Ericson. 1999. Homeobox gene *Nkx2.2* and specification of neuronal identity by graded Sonic hedgehog signaling. *Nature* 398: 622–627. (B,C) Placzek, M., M. Tessier-Lavigne, T. Yamada, T. Jessell and J. Dodd. 1990. Mesodermal control of neural cell identity: Floor plate induction by the notochord. *Science* 250: 985–988.

Figure 12.15B: Sauer, F. C. 1935. Mitosis in the neural tube. *J. Comp. Neurol.* 62: 377–405.

Figure 12.16: Jacobson, M. 1991. *Developmental Neurobiology*, 2nd Ed. Plenum, New York.

Figure 12.17: Larsen, W. J. 1993. *Human Embryology.* Churchill Livingstone, New York.

Figure 12.19: (A) Rakic, P. 1975. Cell migration and neuronal ectopias in the brain. *In* D. Bergsma (ed.), *Morphogenesis and Malformations of Face and Brain.* Birth Defects Original Article Series, vol. 11, no. 7. Alan R. Liss, New York, pp. 95–129. (B) Hatten, M. E. 1990. Riding the glial monorail: A common mechanism for glial-guided neuronal migration in different regions of the mammalian brain. *Trends Neurosci.* 13: 179–184.

Figure 12.20: McConnell, S. K. and C. E. Kaznowski. 1991. Cell cycle dependence of laminar determination in developing cerebral cortex. *Science* 254: 282–285.

Figure 12.21: Bogin, B. 1997. Evolutionary hypotheses for human childhood. *Yrbk. Phys. Anthropol.* 40: 63–89.

Figure 12.23: (A) van Praag, H., A. F. Schinder, B. R. Christie, N. Toni, T. D. Palmer and F. H. Gage. 2002. Functional neurogenesis in the adult hippocampus. *Nature* 415: 1030–1034. (B) Erikksson, P. S., E. Perfiliea, T. Björn-Erikksson, A.-M. Alborn, C. Nordberg, D. A. Peterson and F. H. Gage. 1998. Neurogenesis in the adult human hippocampus. *Nature Med.* 4: 1313–1317.

Figure 12.25B: Letourneau, P. C. 1979. Cell substratum adhesion of neurite growth cones and its role in neurite elongation. *Exp. Cell Res.* 124: 127–138.

Figure 12.27A–C: Hilfer, S. R. and J.-J. W. Yang. 1980. Accumulation of CPC-precipitable material at apical cell surfaces during formation of the optic cup. *Anat. Rec.* 197: 423–433.

Figure 12.28A–C: Cooper, M. K., J. A. Porter, K. E. Young and P. A. Beachy. 1998. Teratogen-mediated inhibition of target tissue response to Shh signaling. *Science* 280: 1603–1607.

Figure 12.29: Yamamoto, Y., D. Stock and W. R. Jeffery. 2004. Hedgehog signaling controls eye degeneration in blind cavefish. *Nature* 431: 844–8477.

Figure 12.30: Bailey, T. J., H. El-Hodiri, L. Zhang, R. Shah, P. H. Mathers and M. Jamrich. 2004. Regulation of vertebrate eye development by Rx genes. *Int. J. Dev. Biol.* 48: 761–770.

Figure 12.32: (A–D) Cvekl, A. and E. R. Tamm. 2004. Anterior eye development and ocular mesenchyme: New insights from mouse models and human diseases. *BioEssays* 26: 374–386. (E) Kondoh, H., M. Uchikawa and Y. Kamachi. 2004. Interplay of Pax6 and Sox2 in lens development as a paradigm of genetic switch

mechanisms for cell differentiation. *Int. J. Dev. Biol.* 48: 819–827.

Figure 12.33: Montagna, W. and P. F. Parakkal. 1974. The piliary apparatus. *In* W. Montagna (ed.), *The Structure and Formation of Skin.* Academic Press, New York, pp. 172–258.

Figure 12.34: (A–F) Philpott, M. and R. Paus. 1998. Principles of hair follicle morphogenesis. *In* C.-M. Chuong (ed.), *Molecular Basis of Epithelial Appendage Morphogenesis.* R. G. Landes, Austin, TX, pp. 75–110. (G) Alonso, L. and E. Fuchs. 2003. Stem cells of the skin epithelium. *Proc. Natl. Acad. Sci. USA* 100: 11830–11835.

Chapter 13

Chapter Opening Material

Ramon y Cajal, S. 1937. *Recollections of My Life.* Trans. E. H. Craigie and J. Cano. MIT Press, Cambridge, MA, pp. 36–37.

Whitehead, A. N. 1934. *Nature and Life.* Cambridge University Press, Cambridge, p. 41.

Photograph courtesy of R. B. Levin and R. Luedemanan. See Figure 12.25A.

Illustration Credits

Figure 13.1: Trainor, P. and R. Krumlauf. 2000. Plasticity in mouse neural crest cells reveals a new patterning role for cranial mesoderm. *Nature Cell Biol.* 2: 96–102.

Figure 13.2: Le Douarin, N. M. 1982. *The Neural Crest.* Cambridge University Press, New York.

Figure 13.3: (B) Wang, H. U. and D. J. Anderson. 1997. Eph family transmembrane ligands can mediate repulsive guidance of trunk neural crest migration and motor axon outgrowth. *Neuron* 18: 383–396. (C,D) Bronner-Fraser, M. 1986. Analysis of the early stages of trunk neural crest migration in avian embryos using monoclonal antibody HNK-1. *Dev. Biol.* 115: 44–55.

Figure 13.4: Liu, J.-P. and T. M. Jessell. 1998. A role for rhoB in the delamination of neural crest cells from the dorsal neural tube. *Development* 125: 5055–5067.

Figure 13.5: (A,B) Krull, C. and and 7 others. 1997. Interactions between Eph-related receptors and ligands confer rostrocaudal pattern to trunk neural crest migration. *Curr. Biol.* 7: 571–580. (C) O'Leary, D. D. M. and D. G. Wilkinson. 1999. Eph receptors and ephrins in neural development. *Curr. Opin. Neurobiol.* 9: 65–73.

Figure 13.6: Lang, D. and 9 others. 2005. Pax3 functions at a nodal point in melanocyte stem cell differentiation. *Nature* 433: 884–887; Nishimura, E. K. , S. R. Granter, and D. E. Fisher. 2005. Mechanisms of hair graying: Incomplete melanocyte stem cell maintenance in the niche. *Science* 307: 720–724.

Figure 13.7: (A) Lumsden, A. G. S. 1988a. Multipotent cells in the avian neural crest. *Trends Neurosci.* 12: 81–83. (B) Le Douarin, N. M. 2004. The avian embryo as a model to study the development of the neural crest: A long and still ongoing study. *Mech. Dev.* 121: 1089–1102.

Figure 13.8: Bronner-Fraser, M. 2004. Making sense of the sensory lineage. *Science* 303: 966–968.

Figure 13.9: (A) Le Douarin, N. M. 2004. The avian embryo as a model to study the development of the neural crest: A long and still ongoing study. *Mech. Dev.* 121: 1089–1102. (B) Helms, J. A., D. Cordero, and M. D. Tapadia. 2005. New insights into craniofacial morphogenesis. *Development* 132: 851–861. (C) Carlson, B. M. 1999. *Human Embryology and Developmental Biology.*, 2nd Ed. Mosby. St. Louis.

Figure 13.10: Trainor, P. and R. Krumlauf. 2001. Hox genes, neural crest cells, and branchial arch patterning. *Curr. Opin. Cell Biol.* 13: 698–705.

Figure 13.11: Santagati, F. and F. M. Rijli. 2003. Cranial neural crest and the building of the vertebrate head. *Nature Rev. Neurosci.* 4: 806–818.

Figure 13.12: Otto, F. and 11 others. 1997. Cba1, a candidate gene for cleidocranial dysplasia syndrome, is essential for osteoblast differentiation and bone development. Cell 89: 765–771.

Figure 13.13: Jiang, X., S. Iseki, R. E. Maxson, H. M. Sucov and G. Morriss-Kay. 2002. Tissue origins and interactions in the mammalian skull vault. *Dev. Biol.* 241: 106–116.

Figure 13.14: (A) Kirby, M. L. and K. L. Waldo. 1990. Role of neural crest in congenital heart disease. *Circulation* 82: 332–340. (B) Waldo, K., S. Miyagawa-Tomita, D. Kumiski and M. L. Kirby. 1998. Cardiac neural crest cells provide new insight into septation of the cardiac outflow tract: Aortic sac to ventricular septal closure. *Dev. Biol.* 196: 129–144.

Figure 13.17: Tsushida, T., M. Ensini, S. B. Morton, M. Baldassare, T. Edlund, T. M. Jessell and S. L. Pfaff. 1994. Topographic organization of embryonic motor neurons defined by expression of LIM homeobox genes. *Cell* 79: 957–970; Tosney, K. W., K. B. Hotary and C. Lance-Jones. 1995. Specificity of motoneurons. *BioEssays* 17: 379–382.

Figure 13.18: Lance-Jones, C. and L. Landmesser. 1980. Motor neuron projection patterns in chick hindlimb following partial reversals of the spinal cord. *J. Physiol.* 302: 581–602.

Figure 13.19: Gundersen, R. W. 1987. Response of sensory neurites and growth cones to patterned substrata of laminin and fibronectin in vitro. *Dev. Biol.* 121: 423–431.

Figure 13.20: Wang, H. U. and D. J. Anderson. 1997. Eph family transmembrane ligands can mediate repulsive guidance of trunk neural crest migration and motor axon outgrowth. *Neuron* 18: 383–396.

Figure 13.21: Kolodkin, A. L., D. J. Matthes and C. S. Goodman. 1993. The semaphorin genes encode a family of transmembrane and secreted growth cone guidance molecules. *Cell* 75: 1389–1399.

Figure 13.22: (A) Marx, J. 1995. Helping neurons find their way. *Science* 268: 971–973. (B,C) Messersmith, E. K., E. D. Leonardo, C. J. Shatz, M. Tessier-Lavigne, C. S. Goodman and A. Kolodkin. 1995. Semaphorin III can function as a selective chemorepellent to pattern sensory projections in the spinal cord. *Neuron* 14: 949–959.

Figure 13.23B: Kennedy, T. E., T. Serafini, J. R. de la Torre and M. Tessier-Lavigne. 1994. Netrins are diffusible chemotropic factors for commissural axons in the embryonic spinal cord. *Cell* 78: 425–435.

Figure 13.24: Kennedy, T. E., T. Serafini, J. R. de la Torre and M. Tessier-Lavigne. 1994. Netrins are diffusible chemotropic factors for commissural axons in the embryonic spinal cord. *Cell* 78: 425–435.

Figure 13.25: Goodman, C. S. 1994. The likeness of being: Phylogenetically conserved molecular mechanisms of growth cone guidance. *Cell* 78: 353–356.

Figure 13.26: Colamarino, S. A. and M. Tessier-Lavigne. 1995. The axonal chemoattractant netrin-1 is also a chemorepellent for trochlear motor axons. *Cell* 81: 621–629.

Figure 13.27: (A–D) Kidd, T., K. S. Bland and C. S. Goodman. 1999. Slit is the midline repellent for the Robo receptor in *Drosophila*. *Cell* 96: 785–794. (E) Woods, C. G. 2004. Neuroscience. Crossing the midline. *Science* 304: 1455–1456.

Figure 13.28: Paves, H. and M. Saarma. 1997. Neurotrophins as in vitro growth cone guidance molecules for embryonic sensory neurons. *Cell Tissue Res.* 290: 285–297.

Figure 13.29: Hall, Z. W. and J. R. Sanes. 1993. Synaptic structure and development: The neuromuscular junction. *Neuron* 10 [Suppl.]: 99–121; Purves, D. 1994. *Neural Activity and the Growth of the Brain*. Cambridge University Press, New York; Hall, Z. W. 1995. Laminin α2 (S-laminin): A new player at the synapse. *Science* 269: 362–363.

Figure 13.30: Ibáñez, C. F., T. Ebendal and H. Persson. 1991. Chimeric molecules with multiple neurotrophic activities reveal structural elements determining the specificities of NGF and BDNF. *EMBO J.* 10: 2105–2110.

Figure 13.31: Schmidt, M. and S. B. Kater. 1993. Fibroblast growth factors, depolarization, and substrate interact in a combinatorial way to promote neuronal survival. *Dev. Biol.* 158: 228–237.

Figure 13.32: Hynes, R. O. and A. D. Lander. 1992. Contact and adhesive specificities in the associations, migrations, and targeting of cells and axons. *Cell* 68: 303–322.

Figure 13.33: Holt, C. 2002. Retinal-tectal projection. *Macmillan Encyclopedia of Life Sciences*. Macmillan, London.

Figure 13.34: Walter, J., S. Henke-Fahle and F. Bonhoeffer. 1987. Avoidance of posterior tectal membranes by temporal retinal axons. *Development* 101: 909–913.

Figure 13.35: Barinaga, M. 1995. Receptors find work as guides. *Science* 269: 1668–1670; Hansen, M. J., G. E. Dallal and J. G. Flanagan. 2004. Retinal axon response to ephrin-A shows a graded, concentration-dependent transition from growth promotion to inhibition. *Neuron* 42: 717–730.

Table 13.1: Jacobson, M. 1991. *Developmental Neurobiology*, 2nd Ed. Plenum, New York.

Table 13.2: Larsen, W. J. 1993. *Human Embryology*. Churchill Livingstone, New York.

Chapter 14
Chapter Opening Material

Angier, N. *New York Times* November 1, 1994.

Whitman, W. 1867. "Inscriptions." In *Leaves of Grass and Selected Prose*. S. Bradley (ed.), 1949. Holt, Rinehart & Winston, New York, p. 1.

Photograph courtesy of T. J. Mauch. See Figure 14.1B.

Illustration Credits

Figure 14.1B: Denkers, N., P. Garcia-Villalba, C. K. Rodesch, K. R. Nielson, and T. J. Mauch. 2004. FISHing for chick genes: Triple-label whole mount fluorescence in situ hybridization detects simultaneous and overlapping gene expression in avian embryos. *Dev. Dyn.* 229: 651–657.

Figure 14.3: Tonegawa, A. and Y. Takahashi. 1998. Somitogenesis controlled by Noggin. *Dev. Biol.* 202: 172–182.

Figure 14.5: (A–C) Sato, Y., T. Watanabe, J. Kouyama, H. Okano, and Y. Takahashi. 2005. Activation of Notch in the posterior compartment of a somite determines the differentiation, directed migration, and patterning of the dorsal aorta precursors. (Abstract) *Dev. Biol.* 283: 695. (D,E) Dunwoodie, S. L., M. Clements, D. B. Sparrow, X. Sa, R. A. Conlon and R. S. P. Beddington. 2002. Axial skeletal defects caused by mutation in the spondylocostal dysplasia/pudgy gene *Dll3* are associated with disruption of the segmentation clock within the presomitic mesoderm. *Development* 129: 1795–1806.

Figure 14.8A: Durbin, L. and 8 others. 1998. Eph signaling is required for segmentation and differentiation of the somites. *Genes Dev.* 12: 3096–3109.

Figure 14.9: Nakaya, Y., S. Kuroda, Y. T. Katagiri, K. Kaibuchi, and Y. Takahashi. 2004. Mesenchymal-epithelial transition during somitic segmentation is regulated by differential roles of Cdc42 and Rac1. *Dev. Cell* 7: 425–438.

Figure 14.10: Kieny, M., A. Mauger and P. Segel. 1972. Early regionalization of somitic mesoderm as studied by the development of the axial skeleton of the chick embryo. *Dev. Biol.* 28: 142–161.

Figure 14.11 : (A,B) Langman, J. 1981. *Medical Embryology*, 4th Ed. Williams & Wilkins, Baltimore. (C,D) Ordahl, C. P. 1993. Myogenic lineages within the developing somite. *In* M. Bernfield (ed.), *Molecular Basis of Morphogenesis*. Wiley-Liss, New York, pp. 165–170.

Figure 14.12: Cossu, G., S. Tajbakhsh and M. Buckingham. 1996b. How is myogenesis initiated in the embryo? *Trends Genet.* 12: 218–223.

Figure 14.13: (A–E) Wolpert, L. 1998. *Principles of Development*. Current Biology Press, London. (F) Nameroff, M. and E. Munar. 1976. Inhibition of cellular differentiation by phospholipase C. II. Separation of fusion and recognition among myogenic cells. *Dev. Biol.* 49: 288–293.

Figure 14.14: Horton, W. A. 1990. The biology of bone growth. *Growth Genet. Horm.* 6(2): 1–3.

Figure 14.16: Tuan, R. S. and M. H. Lynch. 1983. Effect of experimentally induced calcium deficiency on the developmental expression of collagen types in chick embryonic skeleton. *Dev. Biol.* 100: 374–386.

Figure 14.17: Larson, W. J. 1998. *Essentials of Human Embryology*. Churchill-Livingstone, New York.

Figure 14.18A: Schweitzer, R. and 7 others. 2001. Analysis of the tendon cell fate using Scleraxis, a specific marker for tendons and ligaments. *Development* 128: 3855–3866.

Figure 14.19A,C: Brent, A. E., R. Schweitzer, and C. J. Tabin. 2003. A somitic compartment of tendon precursors. *Cell* 113: 235–248.

Figure 14.20A,B: Mauch, T. J., G. Yang, M. Wright, D. Smith and G. C. Schoenwolf. 2000. Signals from trunk paraxial mesoderm induce pronephros formation in chick intermediate mesoderm. *Dev. Biol.* 220: 62–75.

Figure 14.21A–C: Saxén, L. 1987. *Organogenesis of the Kidney*. Cambridge University Press, Cambridge.

Figure 14.22: Saxén, L. 1987. *Organogenesis of the Kidney*. Cambridge University Press, Cambridge; Sariola, H. 2002. Nephron induction revisited: From caps to condensates. *Curr. Opin. Nephrol. Hypertens.* 11: 17–21.

Figure 14.23: Srinivas, S., M. R. Goldberg, T. Watanabe, V. D'Agati, Q. al-Awqati, and F. Costantini. 1999. Expression of green fluorescent protein in the ureteric bud of transgenic mice: a new tool for the analysis of ureteric bud morphogenesis. *Dev. Genet.* 24: 241–251.

Figure 14.24: (A) Schuchardt, A., V. D-Agati, V. Pachnis and F. Constantini. 1996. Renal agenesis and hypodysplasia in *ret-k–* mutant mice result from defects in ureteric bud development. *Development* 122: 1919–1929. (B–D) Pichel, J. G. and 11 others. 1996. Defects in enteric innervation and kidney development in mice lacking GDNF. *Nature* 382: 73–76.

Figure 14.25: Carroll, T. J., J. S. Park, S. Hayashi, A. Majumdar, and A. P. McMahon. 2005. Wnt9b plays a central role in the regulation of mesenchymal to epithelial transitions underlying the organogenesis of the mammalian urogenital system. *Dev. Cell* 9: 283–292.

Figure 14.26: Karavanov, A. A., I. Karavanova, A., Perantoni and I. B. Dawid. 1998. Expression pattern of the rat Lim-1 homeobox gene suggests a dual role during kidney development. *Int. J. Dev. Biol.* 42: 61–66.

Figure 14.27: Sainio, K. and 10 others. 1997. Glial cell derived neurotrophic factor is required for bud initiation from ureteric epithelium. *Development* 124: 4077–4087.

Figure 14.28: (A–C) Ritvos, O., T. Tuuri, M. Erämaa, K. Sainio, K. Hilden, L. Saxén and S. F. Gilbert. 1995. Activin disrupts epithelial branching morphogenesis in developing murine kidney, pancreas, and salivary gland. *Mech. Dev.* 50: 229–245. (D) Lin, Y. and 8 others. 2001. Induced repatterning of type XVIII collagen expression in ureter bud from kidney to lung: association with sonic hedgehog and ectopic surfactant protein C. *Development* 128: 1573–1585.

Figure 14.29: (A,B) Cochard, L. R. 2002. *Netter's Atlas of Human Embryology*. MediMedia USA, Peterboro, NJ. (C–F) Batourina, E. and 8 others. 2002. Distal ureter morphogenesis depends on epithelial cell remodeling mediated by vitamin A and Ret. *Nature Genet.* 32: 109–115.

Chapter 15

Chapter Opening Material

Goethe, J. W. von. 1805. *Faust*, Part I. Trans. R. Jarrell, 1976.

Harvey, W. 1628. *Exercitio Anatomica de Motu Cordis et Sanguinis Animalibus*. Reprinted in 1928, C. C. Thomas, Baltimore, p. A2.

Photograph courtesy of N. Ferrara and K. Alitalo. See Figure 15.15.

Illustration Credits

Figure 15.1: (A) Rugh, R. 1951. *The Frog: Its Reproduction and Development*. Blakiston, Philadelphia. (B,C) Patten, B. M. 1951. *Early Embryology of the Chick*, 4th Ed. McGraw-Hill, New York.

Figure 15.2: (A) Redkar, A., M. Montgomery and J. Litvin. 2001. Fate map of the early avian cardiac progenitor cells. *Development* 128: 2269–2279. (C) Mikawa, T. 1999. Determination of heart cell lineages. *In* S. A. Moody (ed.), *Cell Lineage and Fate Determination*. Academic Press, San Diego. Pp. 451–462.

Figure 15.3: Carlson, B. M. 1981. *Patten's Foundations of Embryology*. McGraw-Hill, New York.

Figure 15.4: (B,C) Kupperman, E., S. An, N. Osborne, S. Waldron and D. Y. Stainier. 2000. A sphingosine-1-phosphate receptor regulates cell migration during vertebrate heart development. *Nature* 406: 192–195. (D,E) Li, S., D. Zhou, M. M. Lu, and E. E. Morrisey. 2004. Advanced cardiac morphogenesis does not require heart tube fusion. *Science* 305: 1619–1622.

Figure 15.5: Simões-Costa, M. S. and 8 others. 2005. The evolutionary origin of heart chambers. *Dev. Biol.* 277: 1–15.

Figure 15.7: (A) Srivastava, D. and E. N. Olson. 2000. A genetic blueprint for cardiac development. *Nature* 407: 221–232. (B,C) Wang, D.-Z. and 8 others. 1999. Requirement of a novel gene, *Xin*, in cardiac morphogenesis. *Development* 126: 1281–1294.

Figure 15.7D,E: Xavier-Neto, J. and 7 others. 1999. A retinoic acid-inducible transgenic marker of sino-atrial development in the mouse heart. *Development* 126: 2667–2687.

Figure 15.8: Larsen, W. J. 1993. *Human Embryology*. Churchill-Livingstone, New York.

Figure 15.9: Carlson, B. M. 1981. *Patten's Foundations of Embryology*. McGraw-Hill, New York.

Figure 15.12: Langman, J. 1981. *Medical Embryology*, 4th Ed. Williams & Wilkins, Baltimore.

Figure 15.13: Hanahan, D. 1997. Signaling vascular morphogenesis and maintenance. *Science* 277: 48–50 and Risau, W. 1997. Mechanisms of angiogenesis. *Nature* 386: 671–674.

Figure 15.14: (A) Langman, J. 1981. *Medical Embryology*, 4th Ed. Williams & Wilkins, Baltimore. (B) Katayama, I. and H. Kayano. 1999. Yolk sac with blood island. *New Engl. J. Med.* 340: 617.

Figure 15.15: Ferrara, N. and K. Alitalo. 1999. Clinical applications of angiogenic growth factors and their inhibitors. *Nature Med.* 5: 1359–1364.

Figure 15.16: Yancopopoulos, G. D., M. Klagsbrun and J. Folkman. 1998. Vasculogenesis, angiogenesis, and growth factors: Ephrins enter the fray at the border. *Cell* 93: 661–664.

Figure 15.18: (A) Weinstein, B. M. and N. D. Lawson. 2003. Arteries, veins, Notch, and VEGF. *Cold Spring Harb. Symp. Quant. Biol.* 67: 155–162. (B–D) Popoff, D. 1894. *Dottersack-gafässe des Huhnes*. Kreidl's Verlag, Wiesbaden.

Figure 15.19: Mukouyama, Y. S., D. Shin, S. Britsch, M. Taniguchi and D. J. Anderson. 2002. Sensory nerves determine the pattern of arterial differentiation and blood vessel branching in the skin. *Cell* 109: 693–705.

Figure 15.20: Karkkainenen, M. J. and 11 others. 2004. Vascular endothelial growth factor C is required for sprouting of the first lymphatic vessels from embryonic veins. *Nature Immunol.* 5: 74–80.

Figure 15.22: (A) Dieterlen-Lièvre, F. and C. Martin. 1981. Diffuse intraembryonic hemopoiesis in normal and chimeric avian development. *Dev. Biol.* 88: 180–191. (B) Ottersbach, K. and E. Dzierak. 2005. The murine placenta contains hematopoietic stem cells within the vascular labyrinth region. *Dev. Cell* 8: 377–387.

Figure 15.23: Kluger, Y. Z. Lian, X. Zang, P. E. Newburger, and S. M. Weissman. 2004. A panorama of lineage-specific transcription in hematopoiesis. *BioEssays* 26: 1276–1287.

Figure 15.24: Crelin, E. S. 1961. Development of the gastrointestinal tract. *Clin. Symp.* 13: 68–82.

Figure 15.25: Carlson, B. M. 1981. *Patten's Foundations of Embryology*. McGraw-Hill, New York.

Figure 15.26A: Grapin-Botton, A., A. R. Majithia and D. A. Melton. 2001. Key events of pancreas formation are triggered in gut endoderm by ectopic expression of pancreatic regulatory genes. *Genes Dev.* 15: 444–454.

Figure 15.27: Langman, J. 1981. *Medical Embryology*, 4th Ed. Williams & Wilkins, Baltimore.

Figure 15.28: Gualdi, R., P. Bossard, M. Zheng, Y. Hamada, J. R. Coleman and K. S. Zaret. 1996. Hepatic specification of the gut endoderm in vitro: Cell signaling and transcriptional control. *Genes Dev.* 10: 1670–1682.

Figure 15.29: (A–C) Lammert, E., O. Cleaver and D. A. Melton. 2001. Induction of pancreatic differentiation by signals from blood vessels. *Science* 294: 564–567. (D) Grapin-Botton, A., A. R. Majithia and D. A. Melton. 2001. Key events of pancreas formation are triggered in gut endoderm by ectopic expression of pancreatic regulatory genes. *Genes Dev.* 15: 444–454.

Figure 15.30: Langman, J. 1981. *Medical Embryology*, 4th Ed. Williams & Wilkins, Baltimore.

Figure 15.31: Wessells, N. K. 1970. Mammalian lung development: Interactions in formulation and morphogenesis of tracheal buds. *J. Exp. Zool.* 175: 455–466.

Figure 15.33: Carlson, B. M. 1981. *Patten's Foundations of Embryology*. McGraw-Hill, New York.

Chapter 16

Chapter Opening Material

Darwin, C. 1859. *On the Origin of Species*. Reprinted by New American Library, New York, p. 403

Niehoff, D. 2005. *The Language of Life: How Cells Communicate in Health and Disease*. Joseph Henry Press, Chevy Chase, MD.

Photograph courtesy of S. Sessions. See Figure 16.4.

Illustration Credits

Figure 16.2: Stocum, D. L. and J. F. Fallon. 1982. Control of pattern formation in urodele limb ontogeny: A review and hypothesis. *J. Embryol. Exp. Morphol.* 69: 7–36.

Figure 16.5: Ohuchi, H. and 11 others. 1997. The mesenchymal factor, Fgf10, initiates and maintains the outgrowth of the chick limb bud through interaction with Fgf8, and apical ectodermal factor. *Development* 124: 2235–2244.

Figure 16.6: Kawakami, Y., J. Capdevila, D. Büscher, T. Itoh, C. Rodriguez Esteban and J. C. Izpisúa Belmonte. 2001. WNT signals control FGF-dependent limb initiation and AER induction in the chick embryo. *Cell* 104: 891–900.

Figure 16.7A,B: Ohuchi, H. and 7 others. 1998. Correlation of wing-leg identity in ectopic FGF-induced chimeric limbs with the differential expression of chick *Tbx5* and *Tbx4*. *Development* 125: 51–60; Ohuchi, H. and S. Noji. 1999. Fibroblast growth factor-induced additional limbs in the study of initiation of limb formation, limb identity, myogenesis, and innervation. *Cell Tissue Res.* 296: 45–56.

Figure 16.9: Wessells, N. K. 1977. *Tissue Interaction and Development*. Benjamin Cummings, Menlo Park, CA.

Figure 16.11: Iten, L. E. 1982. Pattern specification and pattern regulation in the embryonic chick limb bud. *Am. Zool.* 22: 117–129.

Figure 16.12: Summerbell, D. and J. H. Lewis. 1975. Time, place, and positional value in the chick limb bud. *J. Embryol. Exp. Morphol.* 33: 621–643.

Figure 16.14: (A,B) Wellik, D. M. and M. R. Capecchi. 2003. *Hox10* and *Hox11* genes are required to globally pattern the mammalian skeleton. *Science* 301: 363–367. (C) Davis, A. P., D. P. Witte, H. M. Hsieh-Li, S. Potter and M. R. Capecchi. 1995. Absence of radius and ulna in mice lacking *hoxa-11* and *hoxd-11*. *Nature* 375: 791–795. (D) Muragaki, Y., S. Mundlos, J. Upton and B. Olsen. 1996. Altered growth and branching patterns in synpolydactyly caused by mutations in *HOXD13*. *Science* 272: 548–551.

Figure 16.15: Honig, L. S. and D. Summerbell. 1985. Maps of strength of positional signaling activity in the developing chick wing bud. *J. Embryol. Exp. Morphol.* 87: 163–174.

Figure 16.16B: Riddle, R. D., R. L. Johnson, E. Laufer and C. Tabin. 1993. Sonic hedgehog mediates the polarizing activity of the ZPA. *Cell* 75: 1401–1416.

Figure 16.17: Maas, S. A. and J. F. Fallon. 2005. Single base pair change in the long-range Sonic hedgehog limb-specific enhancer is a genetic basis for preaxial polydactyly. *Dev. Dyn.* 232: 345–348.

Figure 16.18: Deschamps, J. 2004. Hox genes in the limb: a play in two acts. *Science* 304: 1610–1611.

Figure 16.20: Harfe, B. D., P. J. Scherz, S. Nissim, H. Tian, A. P. McMahon, and C. J. tabin. 2004. Evidence for an expansion-based tempral Shh gradient in specifying vertebrate digit identities. *Cell* 118: 517–528.

Figure 16.21: Dahn, R. D. and J. F. Fallon. 2000. Interdigital regulation of digit identity and homeotic transformation by modulating BMP signaling. *Science* 289: 438–441; Litingtung, Y., R. D. Dahn, Y. Li, J. F. Fallon and C. Chiang. 2002. Shh and Gli3 are dispensable for limb skeleton formation but regulate digit number and identity. *Nature* 418: 979–983.

Figure 16.22: Parr, B. A. and A. P. McMahon. 1995. Dorsalizing signal *wnt-7a* required for normal polarity of D-V and A-P axes of the mouse limb. *Nature* 374: 350–353.

Figure 16.23: te Welscher, P., A. Zuniga, S. Kuijper, T. Drenth, H. J. Goedemans, F. Meijlink and R. Zeller. 2002. Progression of vertebrate limb development through SHH-mediated counteraction of GLI3. *Science* 298: 827–830.

Figure 16.24: Saunders, J. W., Jr. and J. F. Fallon. 1966. Cell death in morphogenesis. *In* M. Locke (ed.), *Major Problems of Developmental Biology*. Academic Press, New York, pp. 289–314.

Figure 16.25: Merino, R, J. Rodriguez-Leon, D. Macias, Y. Gañan, A. N. Economides and J. M. Hurle. 1999. The BMP antagonist Gremlin regulates outgrowth, chondrogenesis and programmed cell death in the developing limb. *Development* 126: 5515–5522.

Figure 16.26: (A) Grotewold, L. and U. Rüther. 2002. The Wnt antagonist Dicckopf-1 is regulated by BMP signaling and c-Jun and modulates programmed cell death. *EMBO J.* 21: 966–975. (B,C) Brunet, L. J., J. A. McMahon, A. P. McMahon and R. M. Harland. 1998. Noggin, cartilage morphogenesis and joint formation in the mammalian skeleton. *Science* 280: 1455–1457.

Chapter 17

Chapter Opening Material

Darwin, E. 1791. Quoted in M. T. Ghiselin, 1974, *The Economy of Nature and the Evolution of Sex*, University of California Press, Berkeley, p. 49.

Thomson, J. A. 1926. *Heredity.* Putnam, New York, p. 477.

Photograph courtesy of J. Adams. See Figure 17.14B.

Illustration Credits

Figure 17.1B: Marx, J. 1995. Mammalian sex determination: Snaring the genes that divide sexes for mammals. *Science* 269: 1824–1825.

Figure 17.2: Langman, J. 1981. *Medical Embryology*, 4th Ed. Williams & Wilkins, Baltimore.

Figure 17.4: Koopman, P., J. Gubbay, N. Vivian, P. Goodfellow and R. Lovell-Badge. 1991. Male development of chromosomally female mice transgenic for *Sry*. *Nature* 351: 117–121.

Figure 17.5: Vidal, V. P. I., M.-C. Chaboissier, D. G. de Rooij and A. Schedl. 2001. Sox9 induces testis development in XX transgenic mice. *Nature Genet.* 28: 216–217.

Figure 17.6: (A, B) Capel, B., K. H. Albrecht, L. L. Washburn and E. M. Eicher. 1999. Migration of mesonephric cells into the mammalian gonad depends on Sry. *Mech. Dev.* 84: 127–131. (C) Sariola, H. and M. Saarma. 1999. GDNF and its receptors in the regulation of ureter branching. *Int. J. Dev. Biol.* 43: 413–418.

Figure 17.7: Genetics Review Group. 1995. One for a boy, two for a girl? *Curr. Biol.* 5: 37–39.

Figure 17.8: Swain, A., V. Narvaez, P. Burgoyne, G. Camerino and R. Lovell-Badge. 1998. Dax1 antagonizes Sry action in mammalian sex determination. *Nature* 391: 761–767.

Figure 17.11: Imperato-McGinley, J., L. Guerrero, T. Gautier and R. E. Peterson. 1974. Steroid 5α-reductase deficiency in man: An inherited form of male pseudohermaphroditism. *Science* 186: 1213–1215.

Figure 17.12: Agate, R. J. and 7 others. 2002. Neural, not gonadal, origin of brain sex differences in a gynandromorphic finch. *Proc. Natl. Acad. Sci. USA* 100: 4873–4878.

Figure 17.13: Amateau, S. K. and M. M. McCarthy. Induction of PGE$_2$ by estradiol mediates developmental masculinization of sex behavior. *Nature Neurosci.* 7: 643–650.

Figure 17.14: Morgan, T. H. and C. B. Bridges. 1919. The origin of gynandromorphs. *In Contributions to the Study of* Drosophila. Publication no. 278. Carnegie Institution of Washington, Washington, DC, pp. 1–122.

Figure 17.15: Baker, B. S., R. N. Nagoshi and K. C. Burtis. 1987. Molecular genetic aspects of sex determination in *Drosophila*. *BioEssays* 6: 66–70.

Figure 17.16: Keyes, L. N., T. W. Cline and P. Schedl. 1992. The primary sex determination signal of *Drosophila* acts at the level of transcription. *Cell* 68: 933–943.

Figure 17.17 Baker, B. S. 1989. Sex in flies: The splice of life. *Nature* 340: 521–524; and Baker, B. S., B. J. Taylor and J. C. Hall. 2001. Are complex behaviors specified by directed regulatory genes? Reasoning from fruit flies. *Cell* 105: 13–24.

Figure 17.18: Handa, N. and 7 others. 1999. Structural basis for recognition of the *tra* mRNA precursor by the Sex-lethal protein. *Nature* 398: 579–585.

Figure 17.19: Christiansen, A. E., E. L. Keisman, S. M. Ahmad and B. S. Baker. 2002. Sex comes in from the cold: The integration of sex and pattern. *Trends Genet.* 18: 510–516.

Figure 17.20: Stockinger, P. D. Kvitsiani, S. Rotkopf, L. Tirián and B. J. Dickson. 2005. Neural circuitry that governs *Drosophila* male courtship behavior. *Cell* 121: 795–807.

Figure 17.21: Crain, D. A. and L. J. Guillette Jr. 1998. Reptiles as models of contaminant-induced endocrine disruption. *Anim. Reprod. Sci.* 53: 77–86.

Figure 17.22: (A, B) Hayes, T., K. Haston, M. Tsui, A. Hoang, C. Haeffele and A. Vonk. 2003. Atrazine-induced hermaphroditism at 0.1 ppb in American leopard frogs (*Rana pipiens*): laboratory and field evidence. *Envir. Health Perspec.* 111: 568–575. (C) Hayes, T. B., A. Collins, M. Lee, M. Mendoza, N. Noriega, A. Stuart and A. Vonk. 2002a. Hermaphroditic, demasculinized frogs after exposure to the herbicide atrazine at low ecologically relevant doses. *Proc. Natl. Acad. Sci. USA* 99: 5476–5480.

Chapter 18

Chapter Opening Material

Schotte, O. 1950. Quoted in R. J. Goss, The natural history (and mystery) of regeneration. In C. E. Dinsmore (ed.), *A History of Regeneration Research*, 1991. Cambridge University Press, Cambridge, p. 12.

Williams, C. M. 1959. Hormonal regulation of insect metamorphosis. In *The Chemical Basis of Development*, W. D. McElroy and B. Glass (eds.). Johns Hopkins University Press, Baltimore, p. 794.

Photograph courtesy of K. Crawford. See Figure 18.6B.

Illustration Credits

Figure 18.1: (A,B) Hoskins, S. G. and P. Grobstein. 1984. Thyroxine induces the ipsilateral retinothalamic projection in *Xenopus laevis*. *Nature* 307: 730–733. (C,D) Nakagawa, S., C. Brennan, K. G. Johnson, D. Shewan, W. A. Harris and C. E. Holt. 2000. Ephrin-B regulates the ipsilateral routing of retinal axons at the optic chiasm. *Neuron* 25: 599–610.

Figure 18.2: Berry, D. L., C. S. Rose, B. F. Remo and D. D. Brown. 1998. The expression pattern of thyroid hormone response genes in remodeling tadpole tissues defines distinct growth and resorption gene expression patterns. *Dev. Biol.* 203: 24–35.

Figure 18.3: Cohen, P. P. 1970. Biochemical differentiation during amphibian metamorphosis. *Science* 168: 533–543.

Figure 18.5: Schwind, J. L. 1933. Tissue specificity at the time of metamorphosis in frog larvae. *J. Exp. Zool.* 66: 1–14.

Figure 18.10: Kalm, L. von, D. Fristrom and J. Fristrom. 1995. The making of a fly leg: A model for epithelial morphogenesis. *BioEssays* 17: 693–702.

Figure 18.11: Fristrom, J. W., D. Fristrom, E. Fekete and A. H. Kuniyuki. 1977. The mechanism of evagination of imaginal discs of *Drosophila melanogaster*. *Am. Zool.* 17: 671–684.

Figure 18.12: Fristrom, D. and J. W. Fristrom. 1975. The mechanisms of evagination of imaginal disks of *Drosophila melanogaster*. I. General considerations. *Dev. Biol.* 43: 1–23; Kalm, L. von, D. Fristrom and J. Fristrom. 1995. The making of a fly leg: A model for epithelial morphogenesis. *BioEssays* 17: 693–702.

Figure 18.13: Abu-Shaar, M. and R. S. Mann. 1998. Generation of multiple antagonistic domains along the proximodistal axis during *Drosophila* leg development. *Development* 125: 3821–3830.

Figure 18.14: Bangi, E. and K. Wharton. 2006. Two BMPs exhibit different effective ranges in the establishment of the BMP activity gradient critical for *Drosophila* wing patterning. *Development*, in press.

Figure 18.16: Gilbert, L. I. and W. Goodman. 1981. Chemistry, metabolism, and transport of hormones controlling insect metamorphosis. *In* L. I. Gilbert and E. Frieden (eds.), *Metamorphosis: A Problem in Developmental Biology*. Plenum, New York, pp. 139–176.

Figure 18.18: (A) King-Jones, K., J.-P. Charles, G. lam, and C. S. Thummel. 2005. The ecdysone-induced DHR4 orphan nuclear receptor coordinates growth and maturation in *Drosophila*. *Cell* 121: 773–784. (B) Buszczak, M. and W. A. Segraves. 2000. Insect metamorphosis: Out with the old, in with the new. *Curr. Biol.* 10: R830–R833.

Figure 18.19: Goss, R. J. 1969. *Principles of Regeneration*. Academic Press, New York.

Figure 18.20: Stocum, D. L. 1979. Stages of forelimb regeneration in *Ambystoma maculatum*. *J. Exp. Zool.* 209: 395–416.

Figure 18.21: Yokoyama, H., H. Ide and K. Tamura. 2001. Fgf-10 stimulates limb regeneration ability in *Xenopus laevis*. *Dev. Biol.* 233: 72–79.

Figure 18.22: Endo, T., S. V. Bryant, and D. M. Gardiner. 2004. A stepwise model system for limb regeneration. *Dev. Biol.* 270: 135–145.

Figure 18.23: Maden, M. 1982. Vitamin A and pattern formation of the regenerating limb. *Nature* 295: 672–675.

Figure 18.24: Crawford, K. and D. L. Stocum. 1988a. Retinoic acid coordinates proximalizes regenerate pattern and blastema differential affinity in axolotl limbs. *Development* 102: 687–698; Crawford, K. and D. L. Stocum. 1988b. Retinoic acid proximalizes level-specific properties responsible for intercalary regeneration in axolotl limbs. *Development* 104: 703–712.

Figure 18.25B: Steele, R. 2002. Developmental signaling in *Hydra*: What does it take to build a "simple" animal? *Dev. Biol.* 248: 199–219.

Figure 18.26: Newman, S. A. 1974. The interaction of the organizing regions of hydra and its possible relation to the role of the cut end of regeneration. *J. Embryol. Exp. Morphol.* 31: 541–555.

Figure 18.27: Newman, S. A. 1974. The interaction of the organizing regions of hydra and its possible relation to the role of the cut end of regeneration. *J. Embryol. Exp. Morphol.* 31: 541–555.

Figure 18.28: Broun, M. and H. R. Bode. 2002. Characterization of the head organizer in hydra. *Development* 129: 875–884.

Figure 18.29: Hobmayer, B., F. Rentzsch, K. Kuhn, C. M. Happel, C. C. von Laue, P. Snyder, U. Rothbäcker and T. W. Holstein. 2000. WNT signalling molecules act in axis formation in the diploblastic metazoan *Hydra*. *Nature* 407: 186–189.

Figure 18.30: (A) Bode, P. M. and H. R. Bode. 1984. Patterning in *Hydra*. *In* G. M. Malacinski and S. V. Bryant (eds.), *Pattern Formation*. Macmillan, New York, pp. 213–241. (B) Bridge, D. M., N. A. Stover and R. E. Steele. 2000. Expression of a novel receptor tyrosine kinase gene and a *paired*-like homeobox gene provides evidence of differences in patterning at the oral and aboral ends of hydra. *Dev. Biol.* 220: 253–262.

Figure 18.31: Taub, R. 2004. Liver regeneration: from myth to mechanism. *Nature Rev. Mol. Cell Biol.* 5: 836–847.

Figure 18.32: Arking, R. 1998. *The Biology of Aging*, 2nd Ed. Sinauer Associates, Sunderland, MA.

Figure 18.33: Hart, R. and R. B. Setlow. 1974. Correlation between deoxyribonucleic acid excision repair and life-span in a number of mammalian species. *Proc. Natl. Acad. Sci. USA* 71: 2169–2173.

Figure 18.35: Trifunovic, A. and 12 others. 2004. Premature ageing in mice expressing defective mitochondrial DNA polymerase. *Nature* 429: 417–423.

Table 18.1: Turner, C. D. and J. T. Bagnara. 1976. *General Endocrinology*, 6th Ed. Saunders, Philadelphia.

Chapter 19

Chapter Opening Material

Elliot, T. S. 1942. "Little Gidding." In *Four Quartets*. Harcourt, Brace and Company, New York. 1943. p. 39. Copyright © T. S. Elliot. All rights reserved.

Hadorn, E. 1955. *Developmental Genetics and Lethal Factors*. London 1961, p. 105.

Photograph courtesy of B. M. Mechler. See Figure 19.4D.

Illustration Credits

Figure 19.1: Waddington, C. H. 1966. *Principles of Development and Differentiation*. Macmillan, New York.

Figure 19.2: Seydoux, G. and A. Fire. 1994. Soma-germline asymmetry in the distributions of embryonic RNAs in *Caenorhabditis elegans*. *Development* 120: 2823–2834.

Figure 19.4: (A) Ruohola, H., K. A. Bremer, D. Baker, J. R. Swedlow, L. Y. Jan and Y. N. Jan. 1991. Role of neurogenic genes in establishment of follicle cell fate and oocyte polarity during oogenesis in *Drosophila*. *Cell* 66: 433–449. (B) Lin, H. and A. C. Spradling. 1995. Fusome asymmetry and oocyte determination in *Drosophila*. *Dev. Genet.* 16: 6–12.

Figure 19.5: Jongens, T. A., B. Hay, L. Y. Jan and Y. N. Jan. 1992. The *germ cell-less* gene product: A posteriorly localized component necessary for germ cell development in *Drosophila*. *Cell* 70: 569–584.

Figure 19.6: Kloc, M., C. Larabell, A. P. Y. Chan and L. D. Etkin. 1998. Contributions of METRO pathway localized molecules to the organization of the germ cell lineage. *Mech. Dev.* 75: 81–93.

Figure 19.7: Hogan, B. 2002. Decisions, decisions! *Nature* 418: 282–283.

Figure 19.8: (A–C) Yeom, Y. I. and 7 others. 1996. Germline regulatory element of Oct-4 specific for the totipotent cycle of embryonal cells. *Development* 122: 881–894. (D,E) Pesce, M., X. Wang, D. J. Wolgemuth and H. Schöler. 1998. Differential expression of the Oct-4 transcription factor during mouse germ cell differentiation. *Mech. Dev.* 71: 89–98.

Figure 19.9: Gardner, R. L. 1982. Manipulation of development. *In* C. R. Austin and R. V. Short (eds.), *Embryonic and Fetal Development*. Cambridge University Press, Cambridge, pp. 159–180.

Figure 19.10: Stewart, T. A. and B. Mintz. 1981. Successful generations of mice produced from an established culture line of euploid teratocarcinoma cells. *Proc. Natl. Acad. Sci. USA* 78: 6314–6318.

Figure 19.11 diagrams: Howard, K. 1998. Organogenesis: *Drosophila* goes gonadal. *Curr. Biol.* 8: R415–R417.

Figure 19.12: (A,B) Yoon, C., K. Kamami and N. Hopkins. 1997. Zebrafish *vasa* homologue RNA is localized to the cleavage planes of 2- and 4-cell-stage embryos and is expressed in the primordial germ cells. *Development* 124: 3157–3165. (C) Weidinger, G., U. Wolke, M. Koprunner, M. Klinger and E. Raz. 1999. Identification of tissues and patterning events required for distinct steps in early migration of zebrafish primordial germ cells. *Development* 126: 5295–5307.

Figure 19.13 : (A–C) Bounoure, L. 1934. Recherches sur lignée germinale chez la grenouille rousse aux premiers stades au développement. *Ann. Sci. Zool.* Ser. 17, 10: 67–248. (E) Heasman, J., T. Mohun and C. C. Wylie. 1977. Studies on the locomotion of primordial germ cells from *Xenopus laevis* in vitro. *J. Embryol. Exp. Morphol.* 42: 149–162.

Figure 19.14: (A) Heath, J. K. 1978. Mammalian primordial germ cells. *Dev. Mammals* 3: 272–298. (C) Mintz, B. 1957. Embryological development of primordial germ cells in the mouse: Influence of a new mutation. *J. Embryol. Exp. Morphol.* 5: 396–403.

Figure 19.15: Tsunekawa, N., M. Naito, T. Nidhida and T. Noce. 2000. Isolation of chicken *vasa* homolog gene and tracing the origin of primordial germ cells. *Development* 127: 2741–2750.

Figure 19.16: (A) Kuwana, T. 1993. Migration of avian primordial germ cells toward the gonadal anlage. *Dev. Growth Diff.* 35: 237–243. (B) Romanoff, A. L. 1960. *The Avian Embryo*. Macmillan, New York. (C) Tsunekawa, N., M. Naito, T. Nidhida and T. Noce. 2000. Isolation of chicken *vasa* homolog gene and tracing the origin of primordial germ cells. *Development* 127: 2741–2750.

Figure 19.17: (A) Schmekel, K. and B. Daneholt. 1995. The central region of the synaptonemal complex revealed in three dimensions. *Trends Cell Biol.* 5: 239–242. (B) von Wettstein, D. 1971. The synaptonemal complex and four-strand crossing over. *Proc. Natl. Acad. Sci. USA* 68: 851–855.

Figure 19.18: Tay, J. and J. D. Richter. 2001. Germ cell differentiation and synaptonemal complex formation are disrupted in CPEB knockout mice. *Dev. Cell* 1: 201–213.

Figure 19.19C: Eckmann, C. R., S. L. Crittenden, N. Suh and J. Kimble. 2004. GLD-3 and control of the mitosis/meiosis decision in the germline of *Caenorhabditis elegans*. *Genetics* 168: 147–160.

Figure 19.20: Eckmann, C. R., S. L. Crittenden, N. Suh and J. Kimble. 2004. GLD-3 and control of the mitosis/meiosis decision in the germline of *Caenorhabditis elegans*. *Genetics* 168: 147–160.

Figure 19.21: Dym, M. 1977. The male reproductive system. *In* L. Weiss and R. O. Greep (eds.), *Histology*, 4th Ed. McGraw-Hill, New York, pp. 979–1038.

Figure 19.22: Bloom, W. and D. W. Fawcett. 1975. *Textbook of Histology*, 10th Ed. Saunders, Philadelphia.

Figure 19.23: Baker, T. G. 1970. Primordial germ cells. *In* C. R. Austin and R. V. Short (eds.), *Reproduction in Mammals, Vol. 1: Germ Cells and Fertilization*. Cambridge University Press, Cambridge, pp. 1–13.

Figure 19.24: De Vos, L. 2002. "Méiose." http://www.ulb. ac.be/sciences/biodic/ImCellulles3.html.

Figure 19.25: Grant, P. 1953. Phosphate metabolism during oogenesis in *Rana temporaria. J. Exp. Zool.* 124: 513–543.

Figure 19.27A: Old, R. W., H. G. Callan and K. W. Gross. 1977. Localization of histone gene transcripts in newt lampbrush chromosomes by in situ hybridization. *J. Cell Sci.* 27: 57–80.

Figure 19.28A: Gurdon, J. B. 1976. *The Control of Gene Expression in Animal Development.* Harvard University Press, Cambridge, MA.

Figure 19.29A: Carlson, B. M. 1981. *Patten's Foundations of Embryology.* McGraw-Hill, New York.

Figure 19.30: (A) Roller, R. J., R. A. Kinloch, B. Y. Hiraoka, S. S.-L. Li and P. M. Wassarman. 1989. Gene expression during mammalian oogenesis and early embryogenesis: Quantification of three messenger RNAs abundant in full grown mouse oocytes. *Development* 106: 251–261. (B,C) Lira, A. A., R. A. Kinloch, S. Mortillo and P. A. Wassarman. 1990. An upstream region of the mouse *ZP3* gene directs expression of firefly luciferinase specifically to growing oocytes in transgenic mice. *Proc. Natl. Acad. Sci. USA* 87: 7215–7219.

Table 19.1: Handel, M. A. and J. J. Eppig. 1998. Sexual dimorphism in the regulation of meiosis. *In* M. A. Handel (ed.), *Meiosis and Gametogenesis.* Academic Press, San Diego. pp. 333-358.

Table 19.2: Laskey, R. A. 1979. Biochemical processes in early development. *In* A. T. Bull et al. (eds.), *Companion to Biochemistry,* Vol. 2. Longman, London, pp. 137–160.

Chapter 20

Chapter Opening Material

Thompson, D. W. 1942. *On Growth and Form.* Cambridge University Press, Cambridge.

Photograph courtesy of P. Benfey. See Figure 20.23C.

Illustration Credits

Figure 20.7: Franklin-Tong, V. E., T. L. Holdaway-Clarke, K. R. Straatman, J. G. Kunkel and P. K. Hepler. 2002. Involvement of extracellular calcium influx in the self-incompatibility response of *Papaver rhoeas. Plant J.* 29: 333–345.

Figure 20.8: Nasrallah, J. B. 2002. Recognition and rejection of self in plant reproduction. *Science* 296: 305–308.

Figure 20.23A: Scheres, G. and R. Heidstra. 1999. Digging out roots:

Pattern formation, cell division, and morphogenesis in plants. *Curr. Top. Dev. Biol.* 45: 207–247.

Figure 20.27: Bharathan, G., T. E. Goliber, C. Moore, S. Kessler, T. Pham and N. R. Sinha. 2002. Homologies in leaf form inferred from *KNOX1* gene expression during development. *Science* 296: 1858–1860.

Figure 20.29: McDaniel, C. N. and R. S. Poethig. 1988. Cell-lineage patterns in the shoot apical meristem of the germinating maize embryo. *Planta* 175: 13–22.

Figure 20.33: Coen, E. S. and E. M. Meyerowitz. 1991. The war of the whorls: Genetic interactions controlling flower development. *Nature* 353: 31–37.

Chapter 21

Chapter Opening Material

Gibson, W. 1999. Quoted in M. Payser, "The Home of the Gay," *Newsweek* March 1, 1999, p. 50.

van Heyningen, V. 2000. Gene games of the future. *Nature* 408: 769–771.

Computer-generated image of a 7-week human embryo, reconstructed from Micro-MRI. The image shows the left side of the embryo, whose actual size is about 13 mm. Copyright © Anatomical Travelogue/Photo Researchers, Inc.

Illustration Credits

Figure 21.1: Volpe, E. P. 1987. Developmental biology and human concerns. *Am. Zool.* 27: 697–714.

Figure 21.3: Ton, C. T. and 14 others. 1991. Positional cloning and characterization of a paired-box and homeobox-containing gene from the aniridia region. *Cell* 67: 1059–1074.

Figure 21.7: De Boulle, K. and 9 others. 1993. A point mutation in the *FMR-1* gene associated with fragile X mental retardation. *Nat. Genet.* 3: 31–35.

Figure 21.10: Moore, K. L. and T. N. N. Persaud. 1993. *Before We Are Born: Essentials of Embryology and Birth Defects.* W. B. Saunders, Philadelphia.

Figure 21.12A–C: Kotch, L. E., S.-Y. Chen and K. K. Sulik. 1995. Ethanol-induced teratogenesis: Free radical damage as a possible mechanism. *Teratology* 52: 128–136.

Figure 21.13: Ma, L., G. V. Benson, H. Lim, S. K. Dey and R. Maas. 1998. *Abdominal B* (*AbdB*) Hoxa genes: Regulation in adult uterus by estrogen and progesterone and repression in Müllerian duct by the synthetic estrogen diethyl-

stilbestrol (DES). *Dev. Biol.* 197: 141–154.

Figure 21.14: Ma, L., G. V. Benson, H. Lim, S. K. Dey and R. Maas. 1998. *Abdominal B* (*AbdB*) Hoxa genes: Regulation in adult uterus by estrogen and progesterone and repression in Müllerian duct by the synthetic estrogen diethylstilbestrol (DES). *Dev. Biol.* 197: 141–154.

Figure 21.15: Kitajewsky, J. and D. Sassoon. 2000. The emergence of molecular gynecology: Homeobox and Wnt genes in the female reproductive tract. *BioEssays* 22: 902–910.

Figure 21.16: Hunt, P. A. and 8 others. 2003. Bisphenol A exposure causes meiotic aneuploidy in the female mouse. *Curr. Biol.* 13: 546–553.

Figure 21.17: Ramos, J. G., J. Varayoud, C. Sonnenschein, A. M. Soto, M. Munoz De Toro and E. H. Luque. 2001. Prenatal exposure to low doses of bisphenol A alters the periductal stroma and glandular cell function in the rat ventral prostate. *Biol. Reprod.* 65: 1271–1277.

Figure 21.18: Anway, M. D., A. S. Cupp, M. Uzumcu, and M. K. Skipper. 2005. Epigenetic transgeneration effects of endocrine disruptors and male fertility. *Science* 308: 1466–1469.

Figure 21.19: Keller, G., G. Zimmer, G. Mall, E. Ritz, and K. Amann. 2003. Nephron number in patients with primary hypertension. *New Engl. J. Med.* 348: 101–108.

Figure 21.20: Lillycrop, K.A., E. S.Phillips, A. A. Jackson, M. A. Hanson, and G. C. Burdge. 2005. Dietary protein restriction of pregnant rats induces and folic acid supplementation prevents epigenetic modification of hepatic gene expression in the offspring. *J. Nutrition* 135: 1382–1386.

Figure 21.21: Jain, M. and 8 others. 2002. Sustained loss of neoplastic phenotype by brief inactivation of MYC. *Science* 297: 102–104.

Figure 21.22: Muthukkaruppan, V. R. and R. Auerbach. 1979. Angiogenesis in the mouse cornea. *Science* 205: 1416–1418.

Figure 21.23: Mooney, D. 2005. One step at a time. *Nature* 436: 468–469.

Figure 21.24: Eck, S. L. 1999. The prospects for gene therapy. *Hosp. Pract.* 34: 67–75.

Figure 21.25B: Lois, S., E. J. Hong, S. Pease, E. J. Brown, and D. Baltimore. 2002. Germline transmission and tissue-specific expression of transgenes delivered by

lentiviral vectors. *Science* 295: 868–872.

Figure 21.26: NIH (National Institutes of Health). 2000. *Stem Cells: A Primer.* http://www.nih.gov/news/stemcells/primer.htm.

Figure 21.27: (A) Gearhart, J. 1998. New potential for human embryonic stem cells. *Science* 282: 1061–1062. (B) Brüstle, O. and 7 others. 1999. Embryonic stem cell-derived glial precursors: A source of myelinating transplants. *Science* 285: 754–756.

Figure 21.28: Barberi, T. and 13 others. Neural subtype specification of fertilization and nuclear transfer embryonic stem cells and application in Parkinsonian mice. *Nature Biotechnol.* 21: 1200–1207.

Figure 21.30: Bonadio, J., E. Smiley, P. Patil and S. Goldstein. 1999. Localized, direct plasmid gene delivery in vivo: Prolonged therapy results in reproducible tissue regeneration. *Nature Med.* 5: 753–759.

Chapter 22

Chapter Opening Material

Waddington, C. H. 1957. *The Strategy of the Genes.* Allen & Unwin, London, pp. 154–155.

Xu, J. and J. I. Gordon. 2003. Honor thy symbionts. *Proc. Natl. Acad. Sci. USA* 100: 10452–10459.

Photograph of the dung beetle *Onthophagus mouhoti* courtesy of A. Moczek. From Moczek, A. P. 2005. The evolution of development of novel traits, or how beetles got their horns. *BioScience* 55: 937–951 (cover photograph).

Illustration Credits

Figure 22.1: (A) Denno, R. F., L. W. Douglass and D. Jacobs. 1985. Crowding and host plant nutrition: Environmental determinants of wing form in *Prokelisia marginata. Ecology* 66: 1588–1596. (B,C) Tawfik, A. I. and 9 others. 1999. Identification of the gregarization-associated dark-pigmentotropin in locusts through an albino mutant. *Proc. Natl. Acad. Sci USA* 96: 7083–7087.

Figure 22.2: Müller, G. B. 2003. Embryonic motility: Environmental influences on evolutionary innovation. *Evo. Dev.* 5: 56–60.

Figure 22.3: Montgomery, M. K. and M. J. McFall-Ngai. 1995. The inductive role of bacterial symbionts in the morphogenesis of a squid light organ. *Am. Zool.* 35: 372–380.

Figure 22.4: Schwemmler, W. 1974. Endosymbionts: Factors of egg patterning. *J. Insect Physiol.* 20: 1467–1474.

Figure 22.5: (A) Hooper, L. V., M. H. Wong, A. Thelin, L. Hansson, P. G. Falk and J. I. Gordon. 2001. Molecular analysis of commensal host-microbial relationships in the intestine. *Science* 291: 881–884. (B,C) Stappenbeck, T. S., L. V. Hooper and J. I. Gordon. 2002. Developmental regulation of intestinal angiogenesis by indigenous microbes via Paneth cells. *Proc. Natl. Acad. Sci. USA* 99: 15451–15455.

Figure 22.9: Brunetti, C. R., J. E. Selegue, A. Monteiro, V. French, P. M. Brakefield and S. B. Carroll. 2001. The generation and diversification of butterfly eyespot color patterns. *Curr. Biol.* 11: 1578–1585.

Figure 22.10: Nijhout, H. F. 2003. Development and evolution of adaptive polyphenisms. *Evo. Dev.* 5: 9–18.

Figure 22.12: Emlen, D. J. 2000. Integrating development with evolution: A case study with beetle horns. *Bioscience* 50: 403–418.

Figure 22.15: Weaver, I.C. G. and 8 others. 2004. Epigenetic programming by maternal behavior. *Nature Neuosci.* 7: 847–854.

Figure 22.16: Conover, D. O. and S. W. Heins. 1987. Adaptive variation in environmental and genetic sex determination in a fish. *Nature* 326: 496–498.

Figure 22.18: Adler, F. R. and C. D. Harvell. 1990. Inducible defenses, phenotypic variability, and biotic environments. *Trends Ecol. Evol.* 5: 407–410.

Figure 22.20: Relyea, R. A. 2003b. Predator cues and pesticides: A double dose of danger for amphibians. *Ecolog. Applic.* 13: 1515–1521.

Figure 22.22: (A,B) Wiesel, T. N.1982. Postnatal development of the visual cortex and the influence of environment. *Nature* 299: 583–591. (C,D) Antonini, A. and M. P. Stryker. 1993. Rapid remodeling of axonal arbors in the visual cortex. *Science* 260: 1818–1821.

Figure 22.25: Spearow, J. L., P. Doemeny, R. Sera, R. Leffler and M. Barkley. 1999. Genetic variation in susceptibility to endocrine disruption by estrogen in mice. *Science* 285: 1259–1261.

Figure 22.26: Silva, E., N. Rajapakse and A. Kortenkamp. 2002. Something from "nothing": Eight weak estrogenic chemicals combined at concentration below NOECs produce significant mixture effects. *Environ. Sci. Technol.* 36: 1751–1756.

Figure 22.27: Hayes, T. B. 2005. Welcome to the revolution. Integrative biology and assessing the impact of endocrine disruptors on environmental and public health. *Integr. Compar. Biol.* 45: 321–329.

Figure 22.29A: La Claire, J. J., J. A. Bantle and J. Dumont. 1998. Photoproducts and metabolites of a common insect growth regulator produce developmental deformities in *Xenopus*. *Environ. Sci. Technol.* 32: 1453–1461.

Chapter 23
Chapter Opening Material

Rostand, J. 1962. *The Substance of Man*. Doubleday, Garden City, New York, p. 12.

Rushdie, S. 1989. *The Satanic Verses*. Viking, New York, p. 8.

Photograph courtesy of A. C. Burke. See Figure 23.12A.

Illustration Credits

Figure 23.1: Philippe, H., N. Lartillot and H. Brinkmann. 2005. Multigene analyses of bilaterian animals correlate the monophyly of ecdysozoa, lophotrochozoa, and protostomia. *Mol. Biol. Evol.* 22: 1246–1253.

Figure 23.2: Twitty, V. C. and H. A. Elliott. 1934. The relative growth of the amphibian eye, studied by means of transplantation. *J. Exp. Zool.* 68: 247–291; Twitty, V. C. and J. L. Schwind. 1931. The growth of eyes and limbs transplanted heteroplastically between two species of *Amblystoma*. *J. Exp. Zool.* 59: 61–86.

Figure 23.4: Halder, G., P. Callaerts and W. J. Gehring. 1995. Induction of ectopic eyes by targeted expression of the *eyeless* gene in *Drosophila*. *Science* 267: 1788–1792.

Table 23.1: Erwin, D. H. 1999. The origin of bodyplans. *Am. Zool.* 39: 617–629.

Figure 23.6: Finnerty. J. R., K. Pang, P. Burton, D. Paulson and M. Q. Martindale. 2004. Origins of bilateral symmetry: *Hox* and *Dpp* expression in a sea anemone. *Science* 304: 1335–1337.

Figure 23.7: Ball, E. E., D. C. Hayward, R. Saint, and D. J. Miller.

2004. A simple plan: Cnidarians and the origins of developmental mechanisms. *Nature Rev. Genet.* 5: 567–577.

Figure 23.9: Averof, M. and M. Akam. 1995. Hox genes and the diversification of insect and crustacean body plans. *Nature* 376: 420–423; Manton, S. M. 1977. *The Arthropoda: Habits, Functional Morphology, and Evolution*. Clarendon Press, Oxford.

Figure 23.10: Averof, M. and N. H. Patel. 1997. Crustacean appendage evolution associated with changes in Hox gene expression. *Nature* 388: 682–686.

Figure 23.11: Harris, M. P., J. F. Fallon and R. O. Prum. 2002. Shh-Bmp2 signaling module and the evolutionary origin and diversification of feathers. *J. Exp. Zool.* 294: 160–176.

Figure 23.12B,C: Cohn, M. J. and C. Tickle. 1999. Developmental basis of limblessness and axial patterning in snakes. *Nature* 399: 474–479.

Figure 23.13: Jernvall, J., S. V. Keranen and I. Thesleff. 2000. Evolutionary modification of development in mammalian teeth: Quantifying gene expression patterns and topography. *Proc. Natl. Acad. Sci. USA* 97: 14444–14448.

Figure 23.14: Salazar-Ciudad, I. and J. Jernvall. 2002. A gene network model accounting for development and evolution of mammalian teeth. *Proc. Natl. Acad. Sci. USA* 99: 8116–8120.

Figure 23.15: Jeffery, J. E., O. R. P. Beninda-Emonds, M. I. Coates and M. K. Richardson. 2002. Analyzing evolutionary patterns in amniote embryonic development. *Evol. Dev.* 4: 292–302.

Figure 23.16: Kuratani, S., Y. Nobusada, N. Horigome and Y. Shigetani. 2001. Embryology of the lamprey and evolution of the vertebrate jaw: Insights from molecular and developmental perspectives. *Philos. Trans. Roy. Soc. Lond. B* 356: 1615–1632.

Figure 23.17: Abzhanov, A., M. Protas, B. R. Grant, P. R. Grant and C. J. Tabin. 2004. *Bmp4* and morphological variation of beaks in Darwin's finches. *Science* 305: 1462–1465.

Figure 23.18: Slijper, E. J. 1962. *Whales*. A. J. Pomerans (transl.). Basic Books, New York.

Figure 23.19: Galant, R. and S. B. Carroll. 2002. Evolution of a transcriptional repression domain in an insect Hox protein. *Nature* 415: 910–913; Ronshaugen, M., N. McGinnis and W. McGinnis. 2002. Hox protein mutation and macroevolution of the insect body plan. *Nature* 415: 914–917.

Figure 23.20: Keys, D. N. and 8 others. 1999. Recruitment of a *hedgehog* regulatory circuit in butterfly eyespot evolution. *Science* 283: 532–534; Brunetti, C. R., J. E. Selegue, A. Monteiro, V. French, P. M. Brakefield and S. B. Carroll. 2001. The generation and diversification of butterfly eyespot color pattern. *Curr. Biol.* 11: 1578–1585.

Figure 23.21: Morasso, M. I., K. A. Mahon, and T. D. Sargent. 1995. A *Xenopus Distal-less* gene in transgenic mice: Conserved regulation in distal limb epidermis and other sites of epithelial-mesenchymal interaction. *Proc. Natl. Acad. Sci. USA* 92: 3968–972.

Figure 23.22: Romer, A. S. 1949. *The Vertebrate Body*. Saunders, Philadelphia.

Figure 23.23: (A) Neidert, A. H., V. Virupannavar, G. W. Hooker and J. A. Langeland. 2001. Lamprey *Dlx* genes and early vertebrate evolution. *Proc. Natl. Acad. Sci. USA* 98: 1665–1670. (B) Manzanares, M., H. Wada, N. Itasaki, P. A. Trainor, R. Krumlauf and P. W. H. Holland. 2000. Conservation and elaboration of Hox gene regulation during evolution of the vertebrate head. *Nature* 408: 854–857. (C) Jeffery, W. R., A. G. Stickler and Y. Yamamoto. 2004. Migratory neural crest-like cells form body pigmentation in a urochordate embryo. *Nature* 431: 696–699.

Figure 23.24: Galis, F. and 7 others. 2006. Extreme selection against homeotic transformations of cervical vertebrae in humans. In press.

Figure 23.25: Raff, R. A. 1994. Developmental mechanisms in the evolution of animal form: Origins and evolvability of body plans. *In* S. Bengston (ed.), *Early Life on Earth*. Columbia University Press, New York, pp. 489–500.

Figure 23.26: Rutherford, S. L. and S. Lindquist. 1998. Hsp90 as a capacitor for morphological evolution. *Nature* 396: 336–342.

Author Index

Author Index

Subject Index

In-text definitions of terms are indexed in **boldface** type.

Italic type indicates the information will be found in an illustration.

The designation "*n*" indicates the information is found in a footnote.

A

A23187, 191, 454
Abaxial muscles, 451
Abaxial myoblasts, 452
Abaxial surface, 644
Abaxial/adaxial asymmetry, 652
ABC model, 651–652
Abdominal A (AbdA) gene, 361
abdominal A (abdA) gene, 284, 285, 732, 733
Abdominal A (AbdA) protein, 286
Abdominal B (AbdB) gene, 284, 285, 361
Abortion, spontaneous, 655
Abscisic acid, 640
Abystoma, 724
acacia (tl) mutant, 647
Accutane®, *667*, 668
Acetylcholine, 396, 414
Acetylcholinesterase, 56
N-Acetylglucosamine, 205
achaete gene, 727
Achondroplasic dwarfism, 151, 526, 660
Acidic FGF, 147. *See also* Fibroblast growth factor 1
Acoustic ganglia, 397
Acron, 274
Acrosomal membrane, 201
Acrosomal process, 176, 183, 184, 185
Acrosomal vesicle, 176, 183
Acrosome, 176
Acrosome reaction, 183–184, 200, 201, 203–204
Acrylamide, 675
Actin. *See also* Microfilaments
    acrosomal process and, 184
    fertilization cone, 186, *187*
    in meroistic oogenesis, 620
    in mitosis, 213

mRNAs stored in oocyte cytoplasm, *131*
α-Actinin, 165, *166*
Activin, 90
    distance at which effective, 146n
    functions of, 156
    left-right axis formation, 347
    mesoderm induction, 311
    neural tube dorsal-ventral axis formation, *384*, 385
Acute promyelocytic leukemia (APL), 679
ADA gene, 683
Adaptation, of embryonic cleavage, 233
Adaxial surface, 644
Adenosine deaminase (ADA) gene, 683
Adenovirus vectors, *682*, 683
Adenyl cyclase, 620
Adenylate cyclase, *200*
Adepithelial cells, 565
Adhesion proteins. *See* Cadherins
Adipocytes, 696
ADMP. *See* Anti-dorsalizing morphogenetic protein
Adrenal gland, 412, 415
Adrenergic neurons, 414
Adult stem cells, 687–688
Adult-onset illnesses, 676–677
Aequorin, 190
AER. *See* Apical ectodermal ridge
afilia (afl) mutant, 647
After-ripening, 640
AGAMOUS (AG) gene, 651
Aggregation, *Dictyostelium*, 37–38
Aging, 585–591
AGM. *See* Aorta-gonad-mesonephros region
Agouti gene, 704, *705*
Albumen, 336
Alcohol, 667–668
Algae, 697
Alisphenoid bone, 421
Allantois, 45, 502–503
Alligators, 50, *550*
Allometric growth, 20–21
Allometry, 738–739

ALS. *See* Amyotrophic lateral sclerosis
Alternation of generations, 40, 627
Alternative nRNA splicing, 126–128, 129
Alternative splicing factor, 126
Alx4 gene, 740
Alzheimer disease, 658
Amacrine neurons, 399
Amborella, 633, 646
    A. trichopoda, 634
Ambystoma. See also Salamanders
    A. maculatum, 506–507, *576*
    A. mexicanum, 8, *562*
    A. tigrinum, 562
American alligator, *550*
American oyster, 699n
Ametabolous development, 564, *565*
AMH. *See* Anti-Müllerian hormone
γ-Aminobutyric acid (GABA), 396, 633
Aminoglycosides, 667
Aminopterin, 667
Ammonium ions, 195
Ammonotelic animals, 558
Amnion, 45, 356, 357, **501**
Amnionic cavity, 352
Amnionic fluid, 352, 501
Amnionic sac, 356–357
Amnioserosa, *257*
Amniote egg, 45, 501
Amniotes, 501
Amoebas, 33
Amphibian axis determination
    anterior-posterior axis, 316–322
    dorsal-ventral axis, 306–309
    formation of the organizer, 306–311
    left-right axis, 322, *323*
    primary embryonic induction, 305
    regulative development, 302–304
Amphibians. *See also* Amphibian axis determination; Salamanders; *Xenopus*
    cell reaggregation experiments, 68–69
    cloning, 81, 82–83
    egg symbioses, 697
    environmental estrogens and, 714, *715*
    fate map, 295, 302